HAESE MATHEMATICS

# Mathematics
## for the international student
## Mathematics SL

third edition

Robert Haese
Sandra Haese
Michael Haese
Marjut Mäenpää
Mark Humphries

for use with

IB Diploma Programme

# WORKED SOLUTIONS

James Foley
Michael Haese
Robert Haese
Sandra Haese
Mark Humphries

Haese Mathematics

# MATHEMATICS FOR THE INTERNATIONAL STUDENT
## Mathematics SL third edition – WORKED SOLUTIONS
## IB Diploma Programme

James Foley B.Ma.Comp.Sc.(Hons.)
Michael Haese B.Sc.(Hons.), Ph.D.
Robert Haese B.Sc.
Sandra Haese B.Sc.
Mark Humphries B.Sc.(Hons.)

Haese Mathematics
3 Frank Collopy Court, Adelaide Airport, SA 5950, AUSTRALIA
Telephone: +61 8 8355 9444, Fax: +61 8 8355 9471
Email: info@haesemathematics.com.au
Web: www.haesemathematics.com.au

National Library of Australia Card Number & ISBN 978-1-921972-09-6



Published by Haese Mathematics
3 Frank Collopy Court, Adelaide Airport, SA 5950, AUSTRALIA

| | |
|---|---|
| First Edition | 2005 |
| Second Edition | 2009, 2011 |
| Third Edition | 2012 |

Artwork and cover design by Piotr Poturaj.

Typeset in Australia by Deanne Gallasch and Charlotte Frost
Typeset in Times Roman $8\frac{1}{2}/10$

Printed in Malaysia through Bookpac Production Services, Singapore.





**Acknowledgements**: While every attempt has been made to trace and acknowledge copyright, the authors and publishers apologise for any accidental infringement where copyright has proved untraceable. They would be pleased to come to a suitable agreement with the rightful owner.

## FOREWORD

This book gives you fully worked solutions for every question in each chapter of our textbook *Mathematics SL third edition* which is one of the textbooks in our series **Mathematics for the International Student** intended for use with IB Diploma and Middle Years courses.

Correct answers can sometimes be obtained by different methods. In this book, where applicable, each worked solution is modelled on the worked example in the textbook.

Be aware of the limitations of calculators and computer modelling packages. Understand that when your calculator gives an answer that is different from the answer you find in the book, you have not necessarily made a mistake, but the book may not be wrong either.

We have a list of errata for our books on our website. Please contact us if you notice any errors in this book.

*JMF PMH*
*RCH SHH MAH*

e-mail: info@haesemathematics.com.au
web: www.haesemathematics.com.au

## TABLE OF CONTENTS

# Chapter 1
## QUADRATICS

### EXERCISE 1A.1

**1** **a** $4x^2 + 7x = 0$
$\therefore\ x(4x+7) = 0$
$\therefore\ x = 0$ or $4x + 7 = 0$ {Null Factor law}
$\therefore\ x = 0$ or $-\frac{7}{4}$

**b** $6x^2 + 2x = 0$
$\therefore\ 2x(3x+1) = 0$
$\therefore\ x = 0$ or $3x + 1 = 0$ {Null Factor law}
$\therefore\ x = 0$ or $-\frac{1}{3}$

**c** $3x^2 - 7x = 0$
$\therefore\ x(3x-7) = 0$
$\therefore\ x = 0$ or $3x - 7 = 0$ {Null Factor law}
$\therefore\ x = 0$ or $\frac{7}{3}$

**d** $2x^2 - 11x = 0$
$\therefore\ x(2x-11) = 0$
$\therefore\ x = 0$ or $2x - 11 = 0$ {Null Factor law}
$\therefore\ x = 0$ or $\frac{11}{2}$

**e** $3x^2 = 8x$
$\therefore\ 3x^2 - 8x = 0$
$\therefore\ x(3x-8) = 0$
$\therefore\ x = 0$ or $3x - 8 = 0$ {Null Factor law}
$\therefore\ x = 0$ or $\frac{8}{3}$

**f** $9x = 6x^2$
$\therefore\ 6x^2 - 9x = 0$
$\therefore\ 3x(2x-3) = 0$
$\therefore\ x = 0$ or $2x - 3 = 0$ {Null Factor law}
$\therefore\ x = 0$ or $\frac{3}{2}$

**g** $x^2 - 5x + 6 = 0$
$\therefore\ (x-2)(x-3) = 0$
$\therefore\ x - 2 = 0$ or $x - 3 = 0$ {Null Factor law}
$\therefore\ x = 2$ or $3$

**h** $x^2 = 2x + 8$
$\therefore\ x^2 - 2x - 8 = 0$
$\therefore\ (x-4)(x+2) = 0$
$\therefore\ x - 4 = 0$ or $x + 2 = 0$ {Null Factor law}
$\therefore\ x = -2$ or $4$

**i** $x^2 + 21 = 10x$
$\therefore\ x^2 - 10x + 21 = 0$
$\therefore\ (x-3)(x-7) = 0$
$\therefore\ x - 3 = 0$ or $x - 7 = 0$ {Null Factor law}
$\therefore\ x = 3$ or $7$

**j** $9 + x^2 = 6x$
$\therefore\ x^2 - 6x + 9 = 0$
$\therefore\ (x-3)^2 = 0$
$\therefore\ x - 3 = 0$
$\therefore\ x = 3$

**k** $x^2 + x = 12$
$\therefore\ x^2 + x - 12 = 0$
$\therefore\ (x+4)(x-3) = 0$
$\therefore\ x + 4 = 0$ or $x - 3 = 0$ {Null Factor law}
$\therefore\ x = -4$ or $3$

**l** $x^2 + 8x = 33$
$\therefore\ x^2 + 8x - 33 = 0$
$\therefore\ (x+11)(x-3) = 0$
$\therefore\ x + 11 = 0$ or $x - 3 = 0$ {Null Factor law}
$\therefore\ x = -11$ or $3$

**2** **a** $9x^2 - 12x + 4 = 0$
$\therefore\ (3x-2)^2 = 0$
$\therefore\ x = \frac{2}{3}$

**b** $2x^2 - 13x - 7 = 0$
$\therefore\ (2x+1)(x-7) = 0$
$\therefore\ x = -\frac{1}{2}$ or $7$

**c** $3x^2 = 16x + 12$
$\therefore\ 3x^2 - 16x - 12 = 0$
$\therefore\ (3x+2)(x-6) = 0$
$\therefore\ x = -\frac{2}{3}$ or $6$

**d** $3x^2 + 5x = 2$
$\therefore\ 3x^2 + 5x - 2 = 0$
$\therefore\ (3x-1)(x+2) = 0$
$\therefore\ x = \frac{1}{3}$ or $-2$

**e** $2x^2 + 3 = 5x$
$\therefore\ 2x^2 - 5x + 3 = 0$
$\therefore\ (2x-3)(x-1) = 0$
$\therefore\ x = \frac{3}{2}$ or $1$

**f** $3x^2 + 8x + 4 = 0$
$\therefore\ (3x+2)(x+2) = 0$
$\therefore\ x = -\frac{2}{3}$ or $-2$

**g** $3x^2 = 10x + 8$
$\therefore\ 3x^2 - 10x - 8 = 0$
$\therefore\ (3x+2)(x-4) = 0$
$\therefore\ x = -\frac{2}{3}$ or $4$

**h** $4x^2 + 4x = 3$
$\therefore\ 4x^2 + 4x - 3 = 0$
$\therefore\ (2x+3)(2x-1) = 0$
$\therefore\ x = -\frac{3}{2}$ or $\frac{1}{2}$

**i** $4x^2 = 11x + 3$
$\therefore\ 4x^2 - 11x - 3 = 0$
$\therefore\ (4x+1)(x-3) = 0$
$\therefore\ x = -\frac{1}{4}$ or $3$

**j** $12x^2 = 11x + 15$
$\therefore\ 12x^2 - 11x - 15 = 0$
$\therefore\ (4x+3)(3x-5) = 0$
$\therefore\ x = -\frac{3}{4}$ or $\frac{5}{3}$

**k** $7x^2 + 6x = 1$
$\therefore\ 7x^2 + 6x - 1 = 0$
$\therefore\ (7x-1)(x+1) = 0$
$\therefore\ x = \frac{1}{7}$ or $-1$

**l** $15x^2 + 2x = 56$
$\therefore\ 15x^2 + 2x - 56 = 0$
$\therefore\ (15x-28)(x+2) = 0$
$\therefore\ x = \frac{28}{15}$ or $-2$

**3** **a**
$$\begin{aligned} (x+1)^2 &= 2x^2-5x+11 \\ \therefore\ x^2+2x+1 &= 2x^2-5x+11 \\ \therefore\ x^2-7x+10 &= 0 \\ \therefore\ (x-2)(x-5) &= 0 \\ \therefore\ x &= 2 \text{ or } 5 \end{aligned}$$

**b**
$$\begin{aligned} (x+2)(1-x) &= -4 \\ \therefore\ x-x^2+2-2x &= -4 \\ \therefore\ x^2+x-6 &= 0 \\ \therefore\ (x+3)(x-2) &= 0 \\ \therefore\ x &= -3 \text{ or } 2 \end{aligned}$$

**c**
$$\begin{aligned} 5-4x^2 &= 3(2x+1)+2 \\ \therefore\ 5-4x^2 &= 6x+3+2 \\ \therefore\ 4x^2+6x &= 0 \\ \therefore\ 2x(2x+3) &= 0 \\ \therefore\ x &= 0 \text{ or } -\tfrac{3}{2} \end{aligned}$$

**d**
$$\begin{aligned} x+\frac{2}{x} &= 3 \\ \therefore\ x^2+2 &= 3x \\ \therefore\ x^2-3x+2 &= 0 \\ \therefore\ (x-1)(x-2) &= 0 \\ \therefore\ x &= 1 \text{ or } 2 \end{aligned}$$

**e**
$$\begin{aligned} 2x-\frac{1}{x} &= -1 \\ \therefore\ 2x^2-1 &= -x \\ \therefore\ 2x^2+x-1 &= 0 \\ \therefore\ (2x-1)(x+1) &= 0 \\ \therefore\ x &= \tfrac{1}{2} \text{ or } -1 \end{aligned}$$

**f**
$$\begin{aligned} \frac{x+3}{1-x} &= -\frac{9}{x} \\ \therefore\ x(x+3) &= -9(1-x) \\ \therefore\ x^2+3x &= -9+9x \\ \therefore\ x^2-6x+9 &= 0 \\ \therefore\ (x-3)^2 &= 0 \\ \therefore\ x &= 3 \end{aligned}$$

## EXERCISE 1A.2

**1** **a**
$$\begin{aligned} (x+5)^2 &= 2 \\ \therefore\ x+5 &= \pm\sqrt{2} \\ \therefore\ x &= -5\pm\sqrt{2} \end{aligned}$$

**b** $(x+6)^2 = -11$
has no real solutions as
$(x+6)^2$ cannot be negative.

**c**
$$\begin{aligned} (x-4)^2 &= 8 \\ \therefore\ x-4 &= \pm\sqrt{8} \\ \therefore\ x &= 4\pm2\sqrt{2} \end{aligned}$$

**d**
$$\begin{aligned} (x-8)^2 &= 7 \\ \therefore\ x-8 &= \pm\sqrt{7} \\ \therefore\ x &= 8\pm\sqrt{7} \end{aligned}$$

**e**
$$\begin{aligned} 2(x+3)^2 &= 10 \\ \therefore\ (x+3)^2 &= 5 \\ \therefore\ x+3 &= \pm\sqrt{5} \\ \therefore\ x &= -3\pm\sqrt{5} \end{aligned}$$

**f**
$$\begin{aligned} 3(x-2)^2 &= 18 \\ \therefore\ (x-2)^2 &= 6 \\ \therefore\ x-2 &= \pm\sqrt{6} \\ \therefore\ x &= 2\pm\sqrt{6} \end{aligned}$$

**g**
$$\begin{aligned} (x+1)^2+1 &= 11 \\ \therefore\ (x+1)^2 &= 10 \\ \therefore\ x+1 &= \pm\sqrt{10} \\ \therefore\ x &= -1\pm\sqrt{10} \end{aligned}$$

**h**
$$\begin{aligned} (2x+1)^2 &= 3 \\ \therefore\ 2x+1 &= \pm\sqrt{3} \\ \therefore\ 2x &= -1\pm\sqrt{3} \\ \therefore\ x &= -\tfrac{1}{2}\pm\tfrac{1}{2}\sqrt{3} \end{aligned}$$

**i**
$$\begin{aligned} (1-3x)^2-7 &= 0 \\ \therefore\ (1-3x)^2 &= 7 \\ \therefore\ 1-3x &= \pm\sqrt{7} \\ \therefore\ 3x &= 1\pm\sqrt{7} \\ \therefore\ x &= \tfrac{1}{3}\pm\tfrac{\sqrt{7}}{3} \end{aligned}$$

**2** **a**
$$\begin{aligned} x^2-4x+1 &= 0 \\ \therefore\ x^2-4x &= -1 \\ \therefore\ x^2-4x+(-2)^2 &= -1+(-2)^2 \\ \therefore\ (x-2)^2 &= 3 \\ \therefore\ x-2 &= \pm\sqrt{3} \\ \therefore\ x &= 2\pm\sqrt{3} \end{aligned}$$

**b**
$$\begin{aligned} x^2+6x+2 &= 0 \\ \therefore\ x^2+6x &= -2 \\ \therefore\ x^2+6x+3^2 &= -2+3^2 \\ \therefore\ (x+3)^2 &= 7 \\ \therefore\ x+3 &= \pm\sqrt{7} \\ \therefore\ x &= -3\pm\sqrt{7} \end{aligned}$$

**c**
$$\begin{aligned} x^2-14x+46 &= 0 \\ \therefore\ x^2-14x &= -46 \\ \therefore\ x^2-14x+(-7)^2 &= -46+(-7)^2 \\ \therefore\ (x-7)^2 &= 3 \\ \therefore\ x-7 &= \pm\sqrt{3} \\ \therefore\ x &= 7\pm\sqrt{3} \end{aligned}$$

**d**
$$\begin{aligned} x^2 &= 4x+3 \\ \therefore\ x^2-4x &= 3 \\ \therefore\ x^2-4x+(-2)^2 &= 3+(-2)^2 \\ \therefore\ (x-2)^2 &= 7 \\ \therefore\ x-2 &= \pm\sqrt{7} \\ \therefore\ x &= 2\pm\sqrt{7} \end{aligned}$$

**e**
$$\begin{aligned} x^2+6x+7&=0\\ \therefore\ x^2+6x&=-7\\ \therefore\ x^2+6x+3^2&=-7+3^2\\ \therefore\ (x+3)^2&=2\\ \therefore\ x+3&=\pm\sqrt{2}\\ \therefore\ x&=-3\pm\sqrt{2} \end{aligned}$$

**f**
$$\begin{aligned} x^2&=2x+6\\ \therefore\ x^2-2x&=6\\ \therefore\ x^2-2x+(-1)^2&=6+(-1)^2\\ \therefore\ (x-1)^2&=7\\ \therefore\ x-1&=\pm\sqrt{7}\\ \therefore\ x&=1\pm\sqrt{7} \end{aligned}$$

**g**
$$\begin{aligned} x^2+6x&=2\\ \therefore\ x^2+6x+3^2&=2+3^2\\ \therefore\ (x+3)^2&=11\\ \therefore\ x+3&=\pm\sqrt{11}\\ \therefore\ x&=-3\pm\sqrt{11} \end{aligned}$$

**h**
$$\begin{aligned} x^2+10&=8x\\ \therefore\ x^2-8x&=-10\\ \therefore\ x^2-8x+(-4)^2&=-10+(-4)^2\\ \therefore\ (x-4)^2&=6\\ \therefore\ x-4&=\pm\sqrt{6}\\ \therefore\ x&=4\pm\sqrt{6} \end{aligned}$$

**i**
$$\begin{aligned} x^2+6x&=-11\\ \therefore\ x^2+6x+3^2&=-11+3^2\\ \therefore\ (x+3)^2&=-2 \end{aligned}$$

$\therefore$ $x$ has no real solutions, since the perfect square cannot be negative.

**3** **a**
$$\begin{aligned} 2x^2+4x+1&=0\\ \therefore\ x^2+2x+\tfrac{1}{2}&=0\\ \therefore\ x^2+2x&=-\tfrac{1}{2}\\ \therefore\ x^2+2x+1^2&=-\tfrac{1}{2}+1^2\\ \therefore\ (x+1)^2&=\tfrac{1}{2}\\ \therefore\ x+1&=\pm\tfrac{1}{\sqrt{2}}\\ \therefore\ x&=-1\pm\tfrac{1}{\sqrt{2}} \end{aligned}$$

**b**
$$\begin{aligned} 2x^2-10x+3&=0\\ \therefore\ x^2-5x+\tfrac{3}{2}&=0\\ \therefore\ x^2-5x&=-\tfrac{3}{2}\\ \therefore\ x^2-5x+(-\tfrac{5}{2})^2&=-\tfrac{3}{2}+(-\tfrac{5}{2})^2\\ \therefore\ (x-\tfrac{5}{2})^2&=-\tfrac{3}{2}+\tfrac{25}{4}\\ \therefore\ (x-\tfrac{5}{2})^2&=\tfrac{19}{4}\\ \therefore\ x-\tfrac{5}{2}&=\pm\tfrac{\sqrt{19}}{2}\\ \therefore\ x&=\tfrac{5}{2}\pm\tfrac{\sqrt{19}}{2} \end{aligned}$$

**c**
$$\begin{aligned} 3x^2+12x+5&=0\\ \therefore\ x^2+4x+\tfrac{5}{3}&=0\\ \therefore\ x^2+4x&=-\tfrac{5}{3}\\ \therefore\ x^2+4x+2^2&=-\tfrac{5}{3}+2^2\\ \therefore\ (x+2)^2&=\tfrac{7}{3}\\ \therefore\ x+2&=\pm\sqrt{\tfrac{7}{3}}\\ \therefore\ x&=-2\pm\sqrt{\tfrac{7}{3}} \end{aligned}$$

**d**
$$\begin{aligned} 3x^2&=6x+4\\ \therefore\ x^2&=2x+\tfrac{4}{3}\\ \therefore\ x^2-2x&=\tfrac{4}{3}\\ \therefore\ x^2-2x+(-1)^2&=\tfrac{4}{3}+(-1)^2\\ \therefore\ (x-1)^2&=\tfrac{7}{3}\\ \therefore\ x-1&=\pm\sqrt{\tfrac{7}{3}}\\ \therefore\ x&=1\pm\sqrt{\tfrac{7}{3}} \end{aligned}$$

**e**
$$\begin{aligned} 5x^2-15x+2&=0\\ \therefore\ x^2-3x+\tfrac{2}{5}&=0\\ \therefore\ x^2-3x&=-\tfrac{2}{5}\\ \therefore\ x^2-3x+(-\tfrac{3}{2})^2&=-\tfrac{2}{5}+(-\tfrac{3}{2})^2\\ \therefore\ (x-\tfrac{3}{2})^2&=-\tfrac{2}{5}+\tfrac{9}{4}=\tfrac{37}{20}\\ \therefore\ x-\tfrac{3}{2}&=\pm\sqrt{\tfrac{37}{20}}\\ \therefore\ x&=\tfrac{3}{2}\pm\sqrt{\tfrac{37}{20}} \end{aligned}$$

**f**
$$\begin{aligned} 4x^2+4x&=5\\ \therefore\ x^2+x&=\tfrac{5}{4}\\ \therefore\ x^2+x+(\tfrac{1}{2})^2&=\tfrac{5}{4}+(\tfrac{1}{2})^2\\ \therefore\ (x+\tfrac{1}{2})^2&=\tfrac{6}{4}\\ \therefore\ x+\tfrac{1}{2}&=\pm\tfrac{\sqrt{6}}{2}\\ \therefore\ x&=-\tfrac{1}{2}\pm\tfrac{\sqrt{6}}{2} \end{aligned}$$

## EXERCISE 1A.3

**1** **a** $x^2-4x-3=0$ has $a=1,\quad b=-4,\quad c=-3$

$$\therefore\quad x=\frac{-(-4)\pm\sqrt{(-4)^2-4(1)(-3)}}{2(1)}=\frac{4\pm\sqrt{28}}{2}=\frac{4\pm2\sqrt{7}}{2}=2\pm\sqrt{7}$$

**b** $x^2+6x+7=0$ has $a=1,\quad b=6,\quad c=7$

$$\therefore\quad x=\frac{-6\pm\sqrt{6^2-4(1)(7)}}{2(1)}=\frac{-6\pm\sqrt{8}}{2}=\frac{-6\pm2\sqrt{2}}{2}=-3\pm\sqrt{2}$$

**c** $x^2+1=4x$

$\therefore\quad x^2-4x+1=0$

which has $a=1,\quad b=-4,\quad c=1$

$$\therefore\quad x=\frac{-(-4)\pm\sqrt{(-4)^2-4(1)(1)}}{2(1)}=\frac{4\pm\sqrt{12}}{2}=\frac{4\pm2\sqrt{3}}{2}=2\pm\sqrt{3}$$

**d** $x^2+4x=1$

$\therefore\quad x^2+4x-1=0$

which has $a=1,\quad b=4,\quad c=-1$

$$\therefore\quad x=\frac{-4\pm\sqrt{4^2-4(1)(-1)}}{2(1)}=\frac{-4\pm\sqrt{20}}{2}=\frac{-4\pm2\sqrt{5}}{2}=-2\pm\sqrt{5}$$

**e** $x^2-4x+2=0$ has $a=1,\quad b=-4,\quad c=2$

$$\therefore\quad x=\frac{-(-4)\pm\sqrt{(-4)^2-4(1)(2)}}{2(1)}=\frac{4\pm\sqrt{8}}{2}=\frac{4\pm2\sqrt{2}}{2}=2\pm\sqrt{2}$$

**f** $2x^2-2x-3=0$ has $a=2,\quad b=-2,\quad c=-3$

$$\therefore\quad x=\frac{-(-2)\pm\sqrt{(-2)^2-4(2)(-3)}}{2(2)}=\frac{2\pm\sqrt{28}}{4}=\frac{2\pm2\sqrt{7}}{4}=\tfrac{1}{2}\pm\tfrac{\sqrt{7}}{2}$$

**g** $(3x+1)^2=-2x$

$\therefore\quad 9x^2+6x+1=-2x$

$\therefore\quad 9x^2+8x+1=0$

which has $a=9,\quad b=8,\quad c=1$

$$\therefore\quad x=\frac{-8\pm\sqrt{8^2-4(9)(1)}}{2(9)}=\frac{-8\pm\sqrt{28}}{18}=\frac{-8\pm2\sqrt{7}}{18}\quad\text{or}\quad-\tfrac{4}{9}\pm\tfrac{\sqrt{7}}{9}$$

**h** $(x+3)(2x+1)=9$

$\therefore\quad 2x^2+x+6x+3=9$

$\therefore\quad 2x^2+7x-6=0$

which has $a=2,\quad b=7,\quad c=-6$

$$\therefore\quad x=\frac{-7\pm\sqrt{7^2-4(2)(-6)}}{2(2)}=\frac{-7\pm\sqrt{49+48}}{4}=-\tfrac{7}{4}\pm\tfrac{\sqrt{97}}{4}$$

**2** **a**
$$(x+2)(x-1) = 2-3x$$
$$\therefore \quad x^2 - x + 2x - 2 = 2 - 3x$$
$$\therefore \quad x^2 + 4x - 4 = 0$$
which has $a = 1$, $b = 4$, $c = -4$
$$\therefore \quad x = \frac{-4 \pm \sqrt{4^2 - 4(1)(-4)}}{2(1)} = \frac{-4 \pm \sqrt{32}}{2} = \frac{-4 \pm 4\sqrt{2}}{2} = -2 \pm 2\sqrt{2}$$

**b**
$$(2x+1)^2 = 3 - x$$
$$\therefore \quad 4x^2 + 4x + 1 = 3 - x$$
$$\therefore \quad 4x^2 + 5x - 2 = 0$$
which has $a = 4$, $b = 5$, $c = -2$
$$\therefore \quad x = \frac{-5 \pm \sqrt{5^2 - 4(4)(-2)}}{2(4)} = \frac{-5 \pm \sqrt{25+32}}{8} = -\tfrac{5}{8} \pm \tfrac{\sqrt{57}}{8}$$

**c**
$$(x-2)^2 = 1 + x$$
$$\therefore \quad x^2 - 4x + 4 = 1 + x$$
$$\therefore \quad x^2 - 5x + 3 = 0$$
which has $a = 1$, $b = -5$, $c = 3$
$$\therefore \quad x = \frac{-(-5) \pm \sqrt{(-5)^2 - 4(1)(3)}}{2(1)} = \frac{5 \pm \sqrt{25-12}}{2} = \tfrac{5}{2} \pm \tfrac{\sqrt{13}}{2}$$

**d**
$$\frac{x-1}{2-x} = 2x + 1$$
$$\therefore \quad x - 1 = (2x+1)(2-x)$$
$$\therefore \quad x - 1 = 4x - 2x^2 + 2 - x$$
$$\therefore \quad 2x^2 - 2x - 3 = 0$$
which has $a = 2$, $b = -2$, $c = -3$
$$\therefore \quad x = \frac{-(-2) \pm \sqrt{(-2)^2 - 4(2)(-3)}}{2(2)} = \frac{2 \pm \sqrt{28}}{4} = \frac{2 \pm 2\sqrt{7}}{4} \quad \text{or} \quad \tfrac{1}{2} \pm \tfrac{\sqrt{7}}{2}$$

**e**
$$x - \frac{1}{x} = 1$$
$$\therefore \quad x^2 - 1 = x$$
$$\therefore \quad x^2 - x - 1 = 0$$
which has $a = 1$, $b = -1$, $c = -1$
$$\therefore \quad x = \frac{-(-1) \pm \sqrt{(-1)^2 - 4(1)(-1)}}{2(1)} = \frac{1 \pm \sqrt{1+4}}{2} = \tfrac{1}{2} \pm \tfrac{\sqrt{5}}{2}$$

**f**
$$2x - \frac{1}{x} = 3$$
$$\therefore \quad 2x^2 - 1 = 3x$$
$$\therefore \quad 2x^2 - 3x - 1 = 0$$
which has $a = 2$, $b = -3$, $c = -1$
$$\therefore \quad x = \frac{-(-3) \pm \sqrt{(-3)^2 - 4(2)(-1)}}{2(2)} = \frac{3 \pm \sqrt{9+8}}{4} = \tfrac{3}{4} \pm \tfrac{\sqrt{17}}{4}$$

## EXERCISE 1B

**1** **a** $x^2 + 7x - 3 = 0$
has $a = 1$, $b = 7$, $c = -3$
$$\therefore \quad \Delta = b^2 - 4ac = 7^2 - 4(1)(-3) = 61$$
Since $\Delta > 0$, there are two distinct real solutions.

**b** $x^2 - 3x + 2 = 0$
has $a = 1$, $b = -3$, $c = 2$
$$\therefore \quad \Delta = b^2 - 4ac = (-3)^2 - 4(1)(2) = 1$$
Since $\Delta > 0$, there are two distinct real solutions.

**c** $3x^2 + 2x - 1 = 0$
has $a = 3, \quad b = 2, \quad c = -1$
$\therefore \quad \Delta = b^2 - 4ac$
$= 2^2 - 4(3)(-1)$
$= 16$
Since $\Delta > 0$, there are two distinct real solutions.

**d** $5x^2 + 4x - 3 = 0$
has $a = 5, \quad b = 4, \quad c = -3$
$\therefore \quad \Delta = b^2 - 4ac$
$= 4^2 - 4(5)(-3)$
$= 76$
Since $\Delta > 0$, there are two distinct real solutions.

**e** $x^2 + x + 5 = 0$
has $a = 1, \quad b = 1, \quad c = 5$
$\therefore \quad \Delta = b^2 - 4ac$
$= 1^2 - 4(1)(5)$
$= -19$
Since $\Delta < 0$, there are no real roots.

**f** $16x^2 - 8x + 1 = 0$
has $a = 16, \quad b = -8, \quad c = 1$
$\therefore \quad \Delta = b^2 - 4ac$
$= (-8)^2 - 4(16)(1)$
$= 0$
$\therefore$ there is one repeated real root.

**2** **a** $6x^2 - 5x - 6 = 0$
has $a = 6, \quad b = -5, \quad c = -6$
$\therefore \quad \Delta = b^2 - 4ac$
$= (-5)^2 - 4(6)(-6)$
$= 169$
$\therefore \quad \sqrt{\Delta} = 13$, so the equation has rational roots.

**b** $2x^2 - 7x - 5 = 0$
has $a = 2, \quad b = -7, \quad c = -5$
$\therefore \quad \Delta = b^2 - 4ac$
$= (-7)^2 - 4(2)(-5)$
$= 89$
$\therefore \quad \sqrt{\Delta} = \sqrt{89}$, so the equation does not have rational roots.

**c** $3x^2 + 4x + 1 = 0$
has $a = 3, \quad b = 4, \quad c = 1$
$\therefore \quad \Delta = b^2 - 4ac$
$= 4^2 - 4(3)(1)$
$= 4$
$\therefore \quad \sqrt{\Delta} = 2$, so the equation has rational roots.

**d** $6x^2 - 47x - 8 = 0$
has $a = 6, \quad b = -47, \quad c = -8$
$\therefore \quad \Delta = b^2 - 4ac$
$= (-47)^2 - 4(6)(-8)$
$= 2401$
$\therefore \quad \sqrt{\Delta} = 49$,
so the equation has rational roots.

**e** $4x^2 - 3x + 2 = 0$
has $a = 4, \quad b = -3, \quad c = 2$
$\therefore \quad \Delta = b^2 - 4ac$
$= (-3)^2 - 4(4)2$
$= -23$
Since $\Delta < 0$, the equation does not have rational roots.

**f** $8x^2 + 2x - 3 = 0$
has $a = 8, \quad b = 2, \quad c = -3$
$\therefore \quad \Delta = b^2 - 4ac$
$= 2^2 - 4(8)(-3)$
$= 100$
$\therefore \quad \sqrt{\Delta} = 10$, so the equation has rational roots.

**3** **a** For $x^2 + 4x + m = 0$,
$a = 1, \quad b = 4, \quad c = m$
So, $\Delta = b^2 - 4ac$
$= 4^2 - 4(1)(m)$
$= 16 - 4m$
which has sign diagram

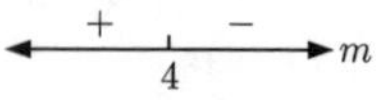


**i** For a repeated root, $\Delta = 0$
$\therefore \quad m = 4$
**ii** For two distinct real roots, $\Delta > 0$
$\therefore \quad m < 4$
**iii** For no real roots, $\Delta < 0$
$\therefore \quad m > 4$

**b** For $mx^2 + 3x + 2 = 0$,
$a = m, \quad b = 3, \quad c = 2$
So, $\Delta = b^2 - 4ac$
$= 3^2 - 4(m)(2)$
$= 9 - 8m$
which has sign diagram

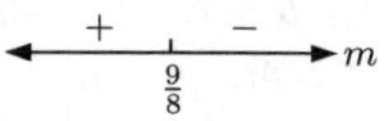


**i** For a repeated root, $\Delta = 0$
$\therefore \quad m = \frac{9}{8}$
**ii** For two distinct real roots, $\Delta > 0$
$\therefore \quad m < \frac{9}{8}$
**iii** For no real roots, $\Delta < 0$
$\therefore \quad m > \frac{9}{8}$

**c** For $mx^2 - 3x + 1 = 0$,
$a = m, \quad b = -3, \quad c = 1$

So, $\Delta = b^2 - 4ac$
$= (-3)^2 - 4(m)(1)$
$= 9 - 4m$

which has sign diagram

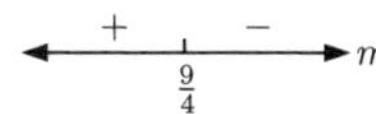

**i** For a repeated root, $\Delta = 0 \quad \therefore \quad m = \frac{9}{4}$

**ii** For two distinct real roots, $\Delta > 0 \quad \therefore \quad m < \frac{9}{4}$

**iii** For no real roots, $\Delta < 0 \quad \therefore \quad m > \frac{9}{4}$

**4** **a** For $2x^2 + kx - k = 0$,
$a = 2, \quad b = k, \quad c = -k$

So, $\Delta = b^2 - 4ac$
$= k^2 - 4(2)(-k)$
$= k^2 + 8k$
$= k(k + 8)$

which has sign diagram

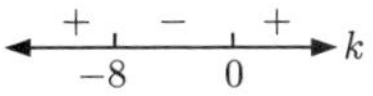

+ | − | + on $k$, critical values −8, 0

**i** For two distinct real roots, $\Delta > 0$
$\therefore \quad k < -8 \text{ or } k > 0$

**ii** For two real roots, $\Delta \geqslant 0$
$\therefore \quad k \leqslant -8 \text{ or } k \geqslant 0$

**iii** For a repeated root, $\Delta = 0$
$\therefore \quad k = -8 \text{ or } 0$

**iv** For no real roots, $\Delta < 0$
$\therefore \quad -8 < k < 0$

**b** For $kx^2 - 2x + k = 0$,
$a = k, \quad b = -2, \quad c = k$

So, $\Delta = b^2 - 4ac$
$= (-2)^2 - 4(k)(k)$
$= 4 - 4k^2$
$= 4(1 + k)(1 - k)$

which has sign diagram

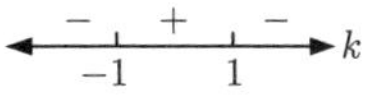

− | + | − on $k$, critical values −1, 1

**i** For two distinct real roots, $\Delta > 0$
$\therefore \quad -1 < k < 1$

**ii** For two real roots, $\Delta \geqslant 0$
$\therefore \quad -1 \leqslant k \leqslant 1$

**iii** For a repeated root, $\Delta = 0$
$\therefore \quad k = -1 \text{ or } 1$

**iv** For no real roots, $\Delta < 0$
$\therefore \quad k < -1 \text{ or } k > 1$

**c** For $x^2 + (k + 2)x + 4 = 0$,
$a = 1, \quad b = k + 2, \quad c = 4$

So, $\Delta = b^2 - 4ac$
$= (k + 2)^2 - 4(1)(4)$
$= k^2 + 4k + 4 - 16$
$= k^2 + 4k - 12$
$= (k + 6)(k - 2)$

which has sign diagram

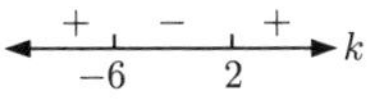

+ | − | + on $k$, critical values −6, 2

**i** For two distinct real roots, $\Delta > 0$
$\therefore \quad k < -6 \text{ or } k > 2$

**ii** For two real roots, $\Delta \geqslant 0$
$\therefore \quad k \leqslant -6 \text{ or } k \geqslant 2$

**iii** For a repeated root, $\Delta = 0$
$\therefore \quad k = -6 \text{ or } 2$

**iv** For no real roots, $\Delta < 0$
$\therefore \quad -6 < k < 2$

**d** For $2x^2 + (k - 2)x + 2 = 0$,
$a = 2, \quad b = k - 2, \quad c = 2$

So, $\Delta = b^2 - 4ac$
$= (k - 2)^2 - 4(2)(2)$
$= k^2 - 4k + 4 - 16$
$= k^2 - 4k - 12$
$= (k - 6)(k + 2)$

which has sign diagram

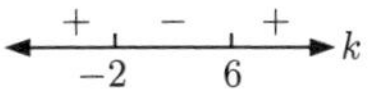

+ | − | + on $k$, critical values −2, 6

**i** For two distinct real roots, $\Delta > 0$
$\therefore \quad k < -2 \text{ or } k > 6$

**ii** For two real roots, $\Delta \geqslant 0$
$\therefore \quad k \leqslant -2 \text{ or } k \geqslant 6$

**iii** For a repeated root, $\Delta = 0$
$\therefore \quad k = -2 \text{ or } 6$

**iv** For no real roots, $\Delta < 0$
$\therefore \quad -2 < k < 6$

**e** For $x^2 + (3k - 1)x + (2k + 10) = 0$,
$a = 1, \quad b = 3k - 1, \quad c = 2k + 10$

So, $\Delta = b^2 - 4ac$
$= (3k - 1)^2 - 4(1)(2k + 10)$
$= 9k^2 - 6k + 1 - 8k - 40$
$= 9k^2 - 14k - 39$
$= (9k + 13)(k - 3)$

which has sign diagram

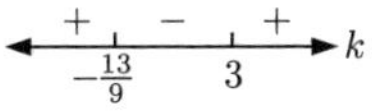

+ | − | + on $k$, critical values $-\frac{13}{9}$, 3

**i** For two distinct real roots, $\Delta > 0$
$\therefore \quad k < -\frac{13}{9} \text{ or } k > 3$

**ii** For two real roots, $\Delta \geqslant 0$
$\therefore \quad k \leqslant -\frac{13}{9} \text{ or } k \geqslant 3$

**iii** For a repeated root, $\Delta = 0$
$\therefore \quad k = -\frac{13}{9} \text{ or } 3$

**iv** For no real roots, $\Delta < 0$
$\therefore \quad -\frac{13}{9} < k < 3$

**f** For $(k+1)x^2 + kx + k = 0$,
$a = k+1, \quad b = k, \quad c = k$

So, $\Delta = b^2 - 4ac$
$= k^2 - 4(k+1)(k)$
$= k^2 - 4k^2 - 4k$
$= -3k^2 - 4k$
$= -k(3k+4)$

which has sign diagram

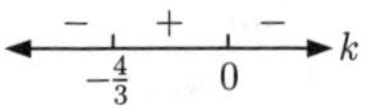

− | + | − on the $k$ axis, with critical values $-\frac{4}{3}$ and $0$

**i** For two distinct real roots, $\Delta > 0$
$\therefore \quad -\frac{4}{3} < k < 0$

**ii** For two real roots, $\Delta \geqslant 0$
$\therefore \quad -\frac{4}{3} \leqslant k \leqslant 0$

**iii** For a repeated root, $\Delta = 0$
$\therefore \quad k = -\frac{4}{3}$ or $0$

**iv** For no real roots, $\Delta < 0$
$\therefore \quad k < -\frac{4}{3}$ or $k > 0$

## EXERCISE 1C.1

**1** **a** $y = (x-4)(x+2)$ has $x$-intercepts $-2$ and $4$ and $y$-intercept $-8$

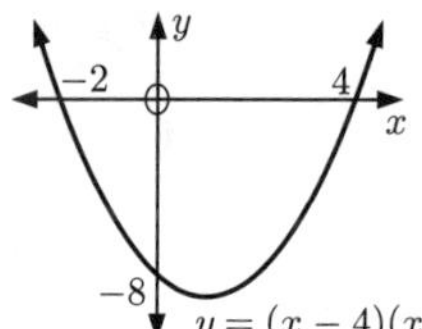


**b** $y = -(x-4)(x+2)$ has $x$-intercepts $-2$ and $4$ and $y$-intercept $8$

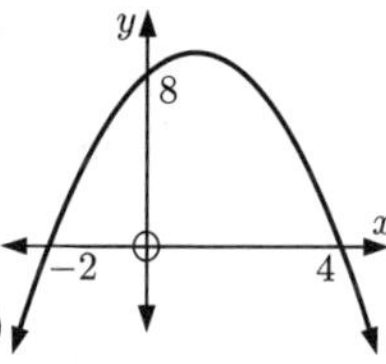


**c** $y = 2(x+3)(x+5)$ has $x$-intercepts $-5$ and $-3$ and $y$-intercept $30$

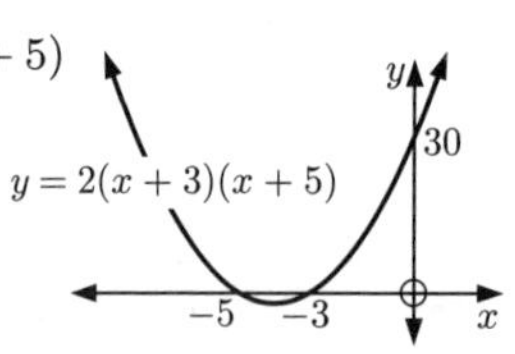


**d** $y = -3x(x+4)$ has $x$-intercepts $0$ and $-4$ and $y$-intercept $0$

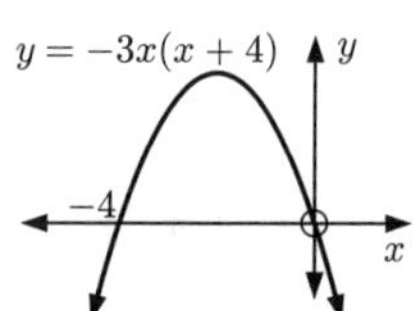


**e** $y = 2(x+3)^2$ has $x$-intercept $-3$ and $y$-intercept $18$

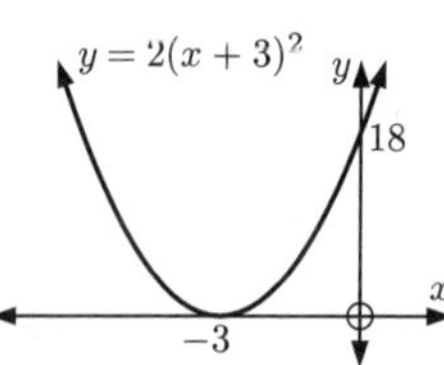


**f** $y = -\frac{1}{4}(x+2)^2$ has $x$-intercept $-2$ and $y$-intercept $-1$

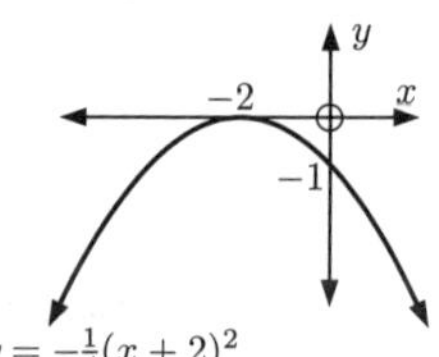


**2** **a** The average of the $x$-intercepts is $1$, so the axis of symmetry is $x = 1$.

**b** The average of the $x$-intercepts is $1$, so the axis of symmetry is $x = 1$.

**c** The average of the $x$-intercepts is $-4$, so the axis of symmetry is $x = -4$.

**d** The average of the $x$-intercepts is $-2$, so the axis of symmetry is $x = -2$.

**e** The only $x$-intercept is $-3$, so the axis of symmetry is $x = -3$.

**f** The only $x$-intercept is $-2$, so the axis of symmetry is $x = -2$.

**3** **a** **C** **b** **E** **c** **B** **d** **F** **e** **G** **f** **H** **g** **A** **h** **D**

**4** **a** The vertex is $(1, 3)$. The axis of symmetry is $x = 1$. The $y$-intercept is $4$.

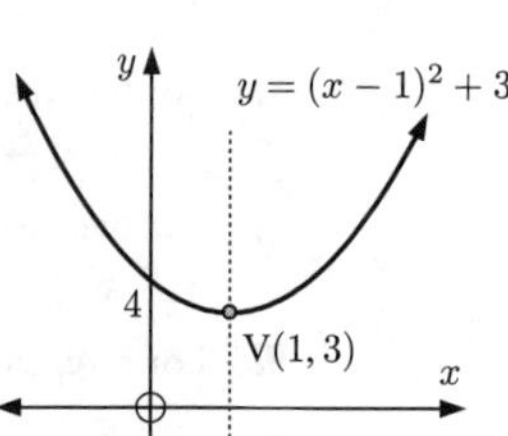


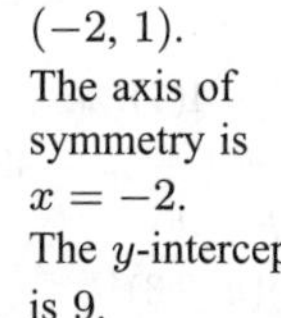

**b** The vertex is $(-2, 1)$. The axis of symmetry is $x = -2$. The $y$-intercept is $9$.

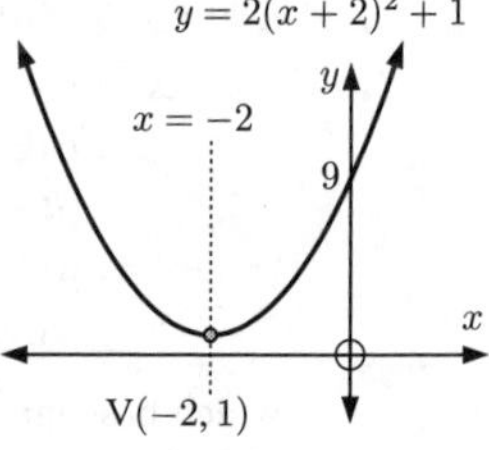

**c** The vertex is $(1,\ -3)$.
The axis of symmetry is $x = 1$.
The $y$-intercept is $-5$.

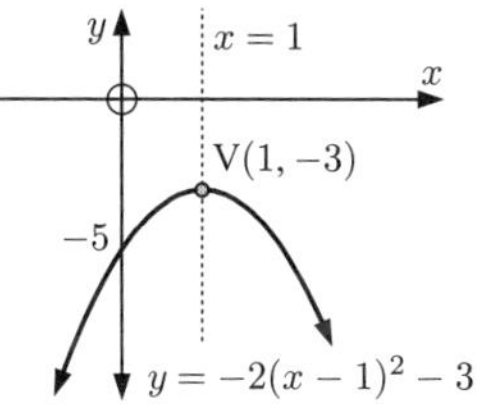


**d** The vertex is $(3,\ 2)$.
The axis of symmetry is $x = 3$.
The $y$-intercept is $\frac{13}{2}$.

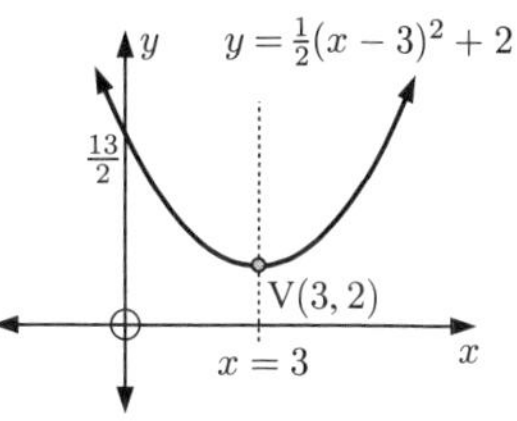


**e** The vertex is $(1,\ 4)$.
The axis of symmetry is $x = 1$.
The $y$-intercept is $3\frac{2}{3}$.

$3\frac{2}{3}$ V(1, 4) $x = 1$ $y = -\frac{1}{3}(x-1)^2 + 4$

**f** The vertex is $(-2,\ -3)$.
The axis of symmetry is $x = -2$.
The $y$-intercept is $-3\frac{2}{5}$.

V(−2, −3) $x = -2$ $-3\frac{2}{5}$ $y = -\frac{1}{10}(x+2)^2 - 3$

**5** **a** G **b** A **c** E **d** B **e** I **f** C **g** D **h** F **i** H

**6** **a** $y = x^2 - 4x + 2$
has $a = 1,\ \ b = -4,\ \ c = 2$
$\therefore\ \ -\dfrac{b}{2a} = -\dfrac{(-4)}{2(1)} = 2$
$\therefore$ the axis of symmetry is $x = 2$.
When $x = 2$,
$y = 2^2 - 4 \times 2 + 2 = -2$
$\therefore$ the vertex is at $(2,\ -2)$.

**b** $y = x^2 + 2x - 3$
has $a = 1,\ \ b = 2,\ \ c = -3$
$\therefore\ \ -\dfrac{b}{2a} = -\dfrac{2}{2(1)} = -1$
$\therefore$ the axis of symmetry is $x = -1$.
When $x = -1$,
$y = (-1)^2 + 2(-1) - 3 = -4$
$\therefore$ the vertex is at $(-1,\ -4)$.

**c** $y = 2x^2 + 4$
has $a = 2,\ \ b = 0,\ \ c = 4$
$\therefore\ \ -\dfrac{b}{2a} = -\dfrac{0}{2(2)} = 0$
$\therefore$ the axis of symmetry is $x = 0$.
When $x = 0,\ \ y = 4$
$\therefore$ the vertex is at $(0,\ 4)$.

**d** $y = -3x^2 + 1$
has $a = -3,\ \ b = 0,\ \ c = 1$
$\therefore\ \ -\dfrac{b}{2a} = -\dfrac{0}{2(-3)} = 0$
$\therefore$ the axis of symmetry is $x = 0$.
When $x = 0,\ \ y = 1$
$\therefore$ the vertex is at $(0,\ 1)$.

**e** $y = 2x^2 + 8x - 7$
has $a = 2,\ \ b = 8,\ \ c = -7$
$\therefore\ \ -\dfrac{b}{2a} = -\dfrac{8}{2(2)} = -2$
$\therefore$ the axis of symmetry is $x = -2$.
When $x = -2$,
$y = 2(-2)^2 + 8(-2) - 7 = -15$
$\therefore$ the vertex is at $(-2,\ -15)$.

**f** $y = -x^2 - 4x - 9$
has $a = -1,\ \ b = -4,\ \ c = -9$
$\therefore\ \ -\dfrac{b}{2a} = -\dfrac{(-4)}{2(-1)} = -2$
$\therefore$ the axis of symmetry is $x = -2$.
When $x = -2,\ \ y = -(-2)^2 - 4(-2) - 9$
$= -4 + 8 - 9$
$= -5$
$\therefore$ the vertex is at $(-2,\ -5)$.

**g** $y = 2x^2 + 6x - 1$
has $a = 2,\ \ b = 6,\ \ c = -1$
$\therefore\ \ -\dfrac{b}{2a} = -\dfrac{6}{2(2)} = -\dfrac{3}{2}$
$\therefore$ the axis of symmetry is $x = -\frac{3}{2}$.
When $x = -\frac{3}{2},\ \ y = 2(-\frac{3}{2})^2 + 6(-\frac{3}{2}) - 1$
$= \frac{9}{2} - 9 - 1$
$= -\frac{11}{2}$
$\therefore$ the vertex is at $(-\frac{3}{2},\ -\frac{11}{2})$.

**h** $y = 2x^2 - 10x + 3$
has $a = 2,\ \ b = -10,\ \ c = 3$
$\therefore\ \ -\dfrac{b}{2a} = -\dfrac{(-10)}{2(2)} = \dfrac{5}{2}$
$\therefore$ the axis of symmetry is $x = \frac{5}{2}$.
When $x = \frac{5}{2},\ \ y = 2(\frac{5}{2})^2 - 10(\frac{5}{2}) + 3$
$= \frac{25}{2} - \frac{50}{2} + 3$
$= -\frac{19}{2}$
$\therefore$ the vertex is at $(\frac{5}{2},\ -\frac{19}{2})$.

**i** $y = -\frac{1}{2}x^2 + x - 5$

has $a = -\frac{1}{2}$, $b = 1$, $c = -5$

$\therefore \; -\frac{b}{2a} = -\frac{1}{2(-\frac{1}{2})} = 1$

$\therefore$ the axis of symmetry is $x = 1$.

When $x = 1$, $y = -\frac{1}{2}(1)^2 + 1 - 5 = -\frac{9}{2}$

$\therefore$ the vertex is at $(1, -\frac{9}{2})$.

**7** **a** When $y = 0$, $x^2 - 9 = 0$

$\therefore \; (x+3)(x-3) = 0$

$\therefore \; x = \pm 3$

$\therefore$ the $x$-intercepts are $\pm 3$

**b** When $y = 0$, $2x^2 - 6 = 0$

$\therefore \; x^2 - 3 = 0$

$\therefore \; (x + \sqrt{3})(x - \sqrt{3}) = 0$

$\therefore \; x = \pm\sqrt{3}$

$\therefore$ the $x$-intercepts are $\pm\sqrt{3}$

**c** When $y = 0$, $x^2 + 7x + 10 = 0$

$\therefore \; (x+5)(x+2) = 0$

$\therefore \; x = -5$ or $-2$

$\therefore$ the $x$-intercepts are $-5$ and $-2$

**d** When $y = 0$, $x^2 + x - 12 = 0$

$\therefore \; (x+4)(x-3) = 0$

$\therefore \; x = -4$ or $3$

$\therefore$ the $x$-intercepts are $-4$ and $3$

**e** When $y = 0$, $4x - x^2 = 0$

$\therefore \; x(4-x) = 0$

$\therefore \; x = 0$ or $4$

$\therefore$ the $x$-intercepts are $0$ and $4$

**f** When $y = 0$, $-x^2 - 6x - 8 = 0$

$\therefore \; x^2 + 6x + 8 = 0$

$\therefore \; (x+4)(x+2) = 0$

$\therefore \; x = -4$ or $-2$

$\therefore$ the $x$-intercepts are $-4$ and $-2$

**g** When $y = 0$, $-2x^2 - 4x - 2 = 0$

$\therefore \; x^2 + 2x + 1 = 0$

$\therefore \; (x+1)^2 = 0$

$\therefore \; x = -1$

$\therefore$ the $x$-intercept is $-1$ (touching)

**h** When $y = 0$, $4x^2 - 24x + 36 = 0$

$\therefore \; x^2 - 6x + 9 = 0$

$\therefore \; (x-3)^2 = 0$

$\therefore \; x = 3$

$\therefore$ the $x$-intercept is $3$ (touching)

**i** When $y = 0$, $x^2 - 4x + 1 = 0$

$a = 1$, $b = -4$, and $c = 1$

$$\therefore \; x = \frac{-(-4) \pm \sqrt{(-4)^2 - 4(1)(1)}}{2(1)} = \frac{4 \pm \sqrt{12}}{2} = \frac{4 \pm 2\sqrt{3}}{2} = 2 \pm \sqrt{3}$$

$\therefore$ the $x$-intercepts are $2 \pm \sqrt{3}$

**j** When $y = 0$, $x^2 + 4x - 3 = 0$

$a = 1$, $b = 4$, and $c = -3$

$$\therefore \; x = \frac{-4 \pm \sqrt{4^2 - 4(1)(-3)}}{2(1)} = \frac{-4 \pm \sqrt{28}}{2} = \frac{-4 \pm 2\sqrt{7}}{2} = -2 \pm \sqrt{7}$$

$\therefore$ the $x$-intercepts are $-2 \pm \sqrt{7}$

**k** When $y = 0$, $x^2 - 6x - 2 = 0$

$a = 1$, $b = -6$, and $c = -2$

$$\therefore \; x = \frac{-(-6) \pm \sqrt{(-6)^2 - 4(1)(-2)}}{2(1)} = \frac{6 \pm \sqrt{44}}{2} = \frac{6 \pm 2\sqrt{11}}{2} = 3 \pm \sqrt{11}$$

$\therefore$ the $x$-intercepts are $3 \pm \sqrt{11}$

**l** When $y = 0$, $x^2 + 8x + 11 = 0$

$a = 1$, $b = 8$, and $c = 11$

$$\therefore \; x = \frac{-8 \pm \sqrt{8^2 - 4(1)(11)}}{2(1)} = \frac{-8 \pm \sqrt{20}}{2} = \frac{-8 \pm 2\sqrt{5}}{2} = -4 \pm \sqrt{5}$$

$\therefore$ the $x$-intercepts are $-4 \pm \sqrt{5}$

**8** **a** **i** $y = x^2 - 2x + 5$ has $a = 1, \quad b = -2, \quad c = 5$

$\therefore \quad -\dfrac{b}{2a} = -\dfrac{(-2)}{2(1)} = 1$

$\therefore$ the axis of symmetry is $x = 1$

**ii** When $x = 1$,

$$\begin{aligned} y &= 1^2 - 2(1) + 5 \\ &= 1 - 2 + 5 \\ &= 4 \end{aligned}$$

$\therefore$ the vertex is at $(1, 4)$

**iii** When $x = 0, \quad y = 5$, so the $y$-intercept is 5

When $y = 0, \quad x^2 - 2x + 5 = 0$

$$\therefore \quad x = \frac{-(-2) \pm \sqrt{(-2)^2 - 4(1)(5)}}{2(1)} = \frac{2 \pm \sqrt{4 - 20}}{2}$$

This has no real solutions, so there are no $x$-intercepts.

**iv**

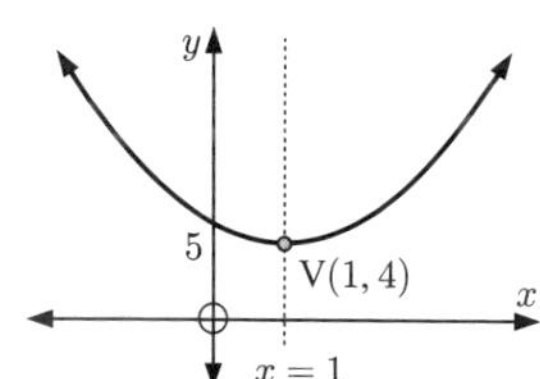


**b** **i** $y = x^2 + 4x - 1$ has $a = 1, \quad b = 4, \quad c = -1$

$\therefore \quad -\dfrac{b}{2a} = -\dfrac{4}{2(1)} = -2$

$\therefore$ the axis of symmetry is $x = -2$

**ii** When $x = -2$,

$$\begin{aligned} y &= (-2)^2 + 4(-2) - 1 \\ &= 4 - 8 - 1 \\ &= -5 \end{aligned}$$

$\therefore$ the vertex is at $(-2, -5)$

**iii** When $x = 0, \quad y = -1$, so the $y$-intercept is $-1$.

When $y = 0, \quad x^2 + 4x - 1 = 0$

$$\begin{aligned} \therefore \quad x &= \frac{-4 \pm \sqrt{4^2 - 4(1)(-1)}}{2(1)} \\ &= \frac{-4 \pm \sqrt{20}}{2} \\ &= \frac{-4 \pm 2\sqrt{5}}{2} \\ &= -2 \pm \sqrt{5} \end{aligned}$$

$\therefore$ the $x$-intercepts are $-2 \pm \sqrt{5}$

**iv**

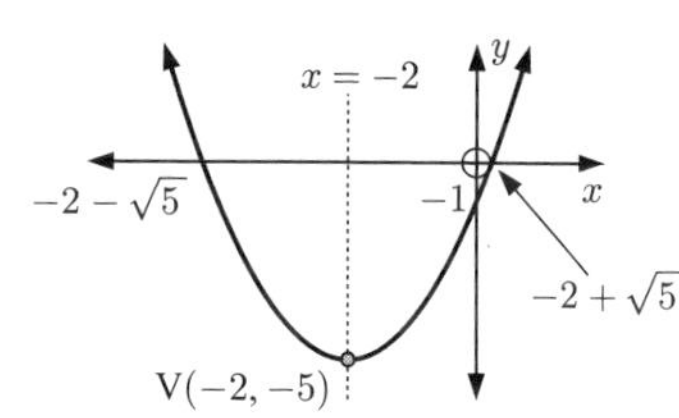


**c** **i** $y = 2x^2 - 5x + 2$ has $a = 2, \quad b = -5, \quad c = 2$

$\therefore \quad -\dfrac{b}{2a} = -\dfrac{(-5)}{2(2)} = \dfrac{5}{4}$

$\therefore$ the axis of symmetry is $x = \frac{5}{4}$

**ii** When $x = \frac{5}{4}$,

$$\begin{aligned} y &= 2(\tfrac{5}{4})^2 - 5(\tfrac{5}{4}) + 2 \\ &= \tfrac{25}{8} - \tfrac{25}{4} + 2 \\ &= -\tfrac{9}{8} \end{aligned}$$

$\therefore$ the vertex is at $(\frac{5}{4}, -\frac{9}{8})$

**iii** When $x = 0, \ y = 2$, so the $y$-intercept is 2.

When $y = 0, \quad 2x^2 - 5x + 2 = 0$

$\therefore \quad (2x - 1)(x - 2) = 0$

$\therefore \quad x = \frac{1}{2}$ or 2

$\therefore$ the $x$-intercepts are $\frac{1}{2}$ and 2

**iv**

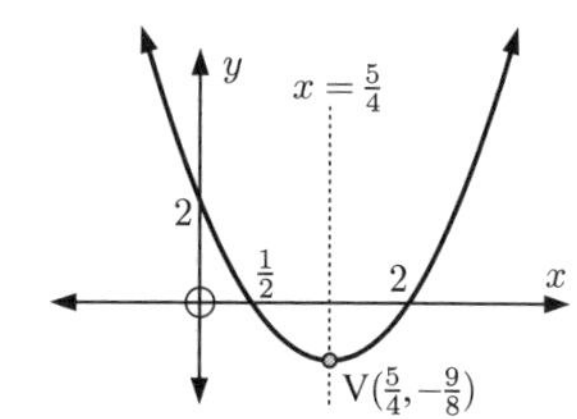

**d** **i** $y = -x^2 + 3x - 2$ has $a = -1$, $b = 3$, $c = -2$

$\therefore \quad -\dfrac{b}{2a} = -\dfrac{3}{2(-1)} = \dfrac{3}{2}$

$\therefore$ the axis of symmetry is $x = \frac{3}{2}$

**ii** When $x = \frac{3}{2}$, $y = -(\frac{3}{2})^2 + 3(\frac{3}{2}) - 2$
$= -\frac{9}{4} + \frac{9}{2} - 2$
$= \frac{1}{4}$

$\therefore$ the vertex is at $(\frac{3}{2}, \frac{1}{4})$

**iii** When $x = 0$, $y = -2$,
so the $y$-intercept is $-2$.
When $y = 0$, $-x^2 + 3x - 2 = 0$
$\therefore \quad x^2 - 3x + 2 = 0$
$\therefore \quad (x-1)(x-2) = 0$
$\therefore \quad x = 1$ or $2$

$\therefore$ the $x$-intercepts are 1 and 2

**iv**

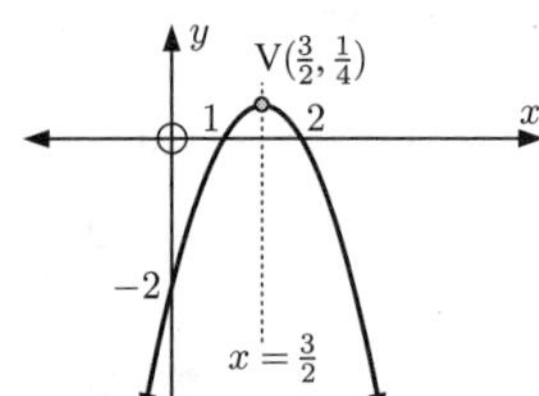


**e** **i** $y = -3x^2 + 4x - 1$ has $a = -3$, $b = 4$, $c = -1$

$\therefore \quad -\dfrac{b}{2a} = -\dfrac{4}{2(-3)} = \dfrac{2}{3}$

$\therefore$ the axis of symmetry is $x = \frac{2}{3}$

**ii** When $x = \frac{2}{3}$, $y = -3(\frac{2}{3})^2 + 4(\frac{2}{3}) - 1$
$= -\frac{4}{3} + \frac{8}{3} - 1$
$= \frac{1}{3}$

$\therefore$ the vertex is at $(\frac{2}{3}, \frac{1}{3})$

**iii** When $x = 0$, $y = -1$,
so the $y$-intercept is $-1$.
When $y = 0$, $-3x^2 + 4x - 1 = 0$
$\therefore \quad 3x^2 - 4x + 1 = 0$
$\therefore \quad (3x-1)(x-1) = 0$
$\therefore \quad x = \frac{1}{3}$ or $1$

$\therefore$ the $x$-intercepts are $\frac{1}{3}$ and 1

**iv**

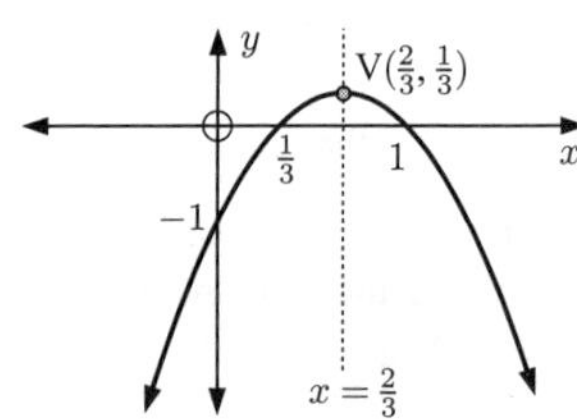


**f** **i** $y = -2x^2 + x + 1$ has $a = -2$, $b = 1$, $c = 1$

$\therefore \quad -\dfrac{b}{2a} = -\dfrac{1}{2(-2)} = \dfrac{1}{4}$

$\therefore$ the axis of symmetry is $x = \frac{1}{4}$

**ii** When $x = \frac{1}{4}$, $y = -2(\frac{1}{4})^2 + \frac{1}{4} + 1$
$= -\frac{1}{8} + \frac{1}{4} + 1$
$= \frac{9}{8}$

$\therefore$ the vertex is at $(\frac{1}{4}, \frac{9}{8})$

**iii** When $x = 0$, $y = 1$,
so the $y$-intercept is 1.
When $y = 0$, $-2x^2 + x + 1 = 0$
$\therefore \quad 2x^2 - x - 1 = 0$
$\therefore \quad (2x+1)(x-1) = 0$
$\therefore \quad x = -\frac{1}{2}$ or $1$

$\therefore$ the $x$-intercepts are $-\frac{1}{2}$ and 1

**iv**

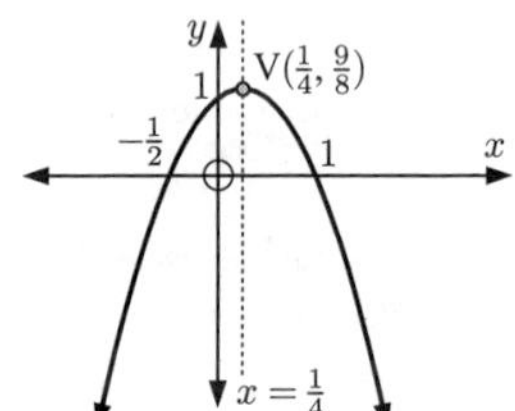


**g** **i** $y = 6x - x^2$ has $a = -1$, $b = 6$, $c = 0$

$\therefore \quad -\dfrac{b}{2a} = -\dfrac{6}{2(-1)} = 3$

$\therefore$ the axis of symmetry is $x = 3$

**ii** When $x = 3$, $y = 6 \times 3 - 3^2$
$= 9$

$\therefore$ the vertex is at $(3, 9)$

**iii** When $x = 0$, $y = 0$,
so the $y$-intercept is 0.
When $y = 0$, $6x - x^2 = 0$
$\therefore \quad x(6 - x) = 0$
$\therefore \quad x = 0$ or $6$

$\therefore$ the $x$-intercepts are 0 and 6

**iv**

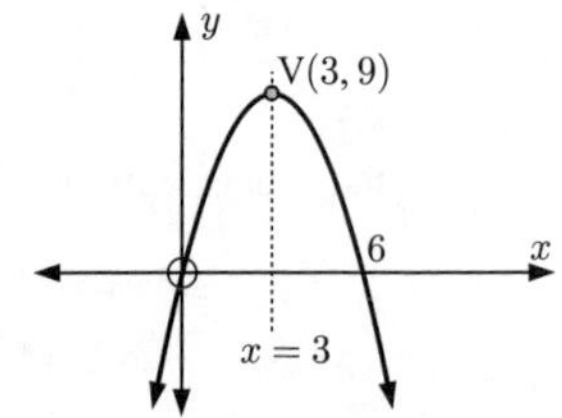

**h** **i** $y = -x^2 - 6x - 8$
has $a = -1$, $b = -6$, $c = -8$

$\therefore \; -\dfrac{b}{2a} = -\dfrac{(-6)}{2(-1)} = -3$

$\therefore$ the axis of symmetry is $x = -3$

**ii** When $x = -3$,
$y = -(-3)^2 - 6(-3) - 8$
$= -9 + 18 - 8$
$= 1$

$\therefore$ the vertex is at $(-3, 1)$

**iii** When $x = 0$, $y = -8$,
so the $y$-intercept is $-8$.

When $y = 0$, $-x^2 - 6x - 8 = 0$
$\therefore \; x^2 + 6x + 8 = 0$
$\therefore \; (x + 4)(x + 2) = 0$
$\therefore \; x = -4$ or $-2$

$\therefore$ the $x$-intercepts are $-4$ and $-2$

**iv**

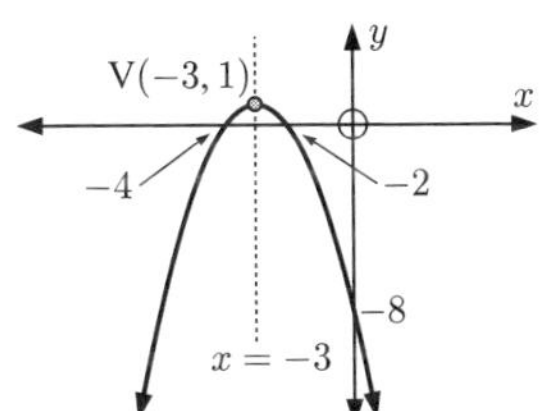


**i** **i** $y = -\frac{1}{4}x^2 + 2x + 1$
has $a = -\frac{1}{4}$, $b = 2$, $c = 1$

$\therefore \; -\dfrac{b}{2a} = -\dfrac{2}{2(-\frac{1}{4})} = 4$

$\therefore$ the axis of symmetry is $x = 4$

**ii** When $x = 4$, $y = -\frac{1}{4}(4)^2 + 2(4) + 1$
$= -4 + 8 + 1$
$= 5$

$\therefore$ the vertex is at $(4, 5)$

**iii** When $x = 0$, $y = 1$,
so the $y$-intercept is 1.

When $y = 0$, $-\frac{1}{4}x^2 + 2x + 1 = 0$
$\therefore \; x^2 - 8x - 4 = 0$

$$\therefore \; x = \frac{-(-8) \pm \sqrt{(-8)^2 - 4(1)(-4)}}{2(1)}$$
$$= \frac{8 \pm \sqrt{80}}{2}$$
$$= \frac{8 \pm 4\sqrt{5}}{2}$$
$$= 4 \pm 2\sqrt{5}$$

$\therefore$ the $x$-intercepts are $4 \pm 2\sqrt{5}$.

**iv**

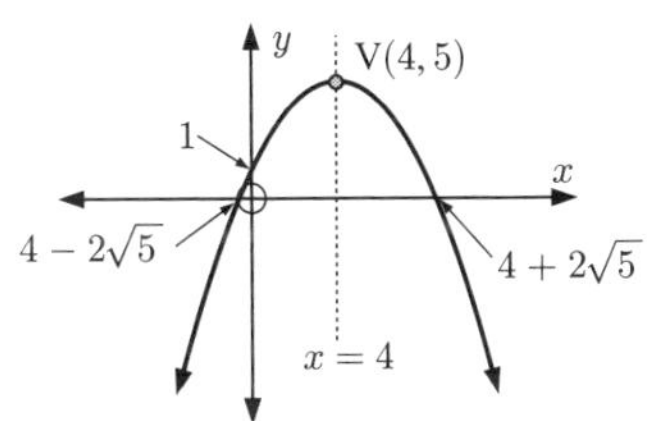


## EXERCISE 1C.2

**1** **a** $y = x^2 - 2x + 3$
$\therefore \; y = x^2 - 2x + 1^2 + 3 - 1^2$
$\therefore \; y = (x - 1)^2 + 2$
$\therefore$ vertex is $(1, 2)$, $y$-intercept is 3

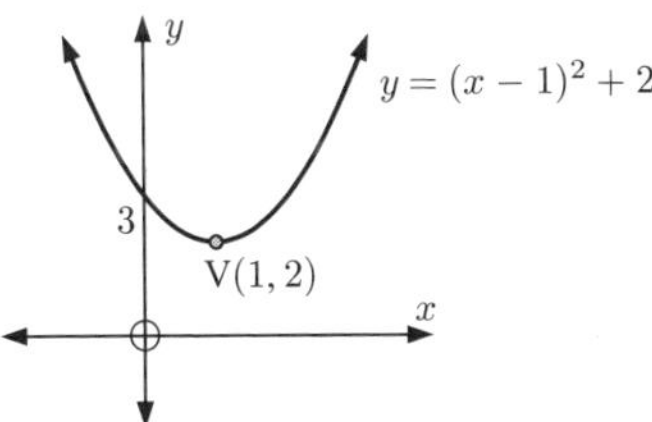


**b** $y = x^2 + 4x - 2$
$\therefore \; y = x^2 + 4x + 2^2 - 2 - 2^2$
$\therefore \; y = (x + 2)^2 - 6$
$\therefore$ vertex is $(-2, -6)$, $y$-intercept is $-2$

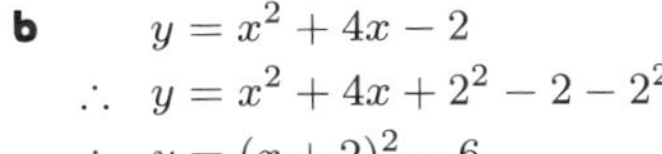

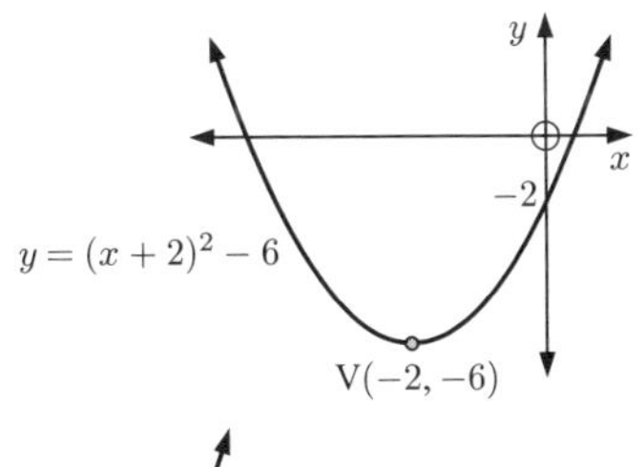


**c** $y = x^2 - 4x$
$\therefore \; y = x^2 - 4x + 2^2 - 2^2$
$\therefore \; y = (x - 2)^2 - 4$
$\therefore$ vertex is $(2, -4)$, $y$-intercept is 0

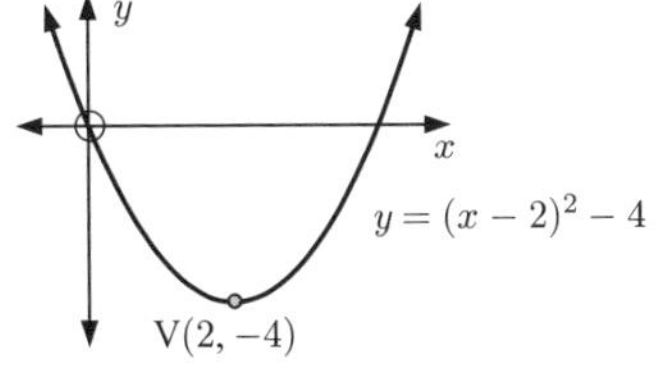

**d** $y = x^2 + 3x$

$\therefore \; y = x^2 + 3x + (\frac{3}{2})^2 - (\frac{3}{2})^2$

$\therefore \; y = (x + \frac{3}{2})^2 - \frac{9}{4}$

$\therefore$ vertex is $(-\frac{3}{2}, -\frac{9}{4})$, $y$-intercept is 0

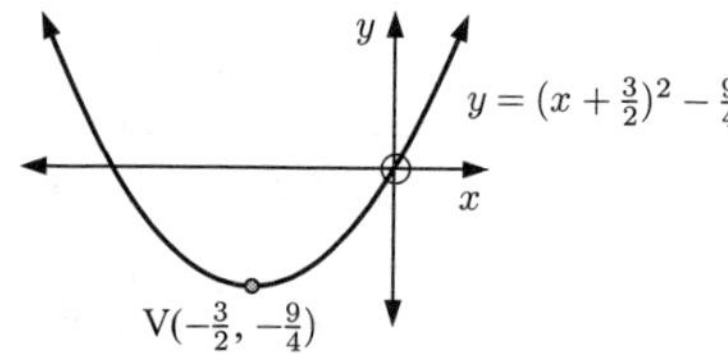


**e** $y = x^2 + 5x - 2$

$\therefore \; y = x^2 + 5x + (\frac{5}{2})^2 - 2 - (\frac{5}{2})^2$

$\therefore \; y = (x + \frac{5}{2})^2 - \frac{33}{4}$

$\therefore$ vertex is $(-\frac{5}{2}, -\frac{33}{4})$, $y$-intercept is $-2$

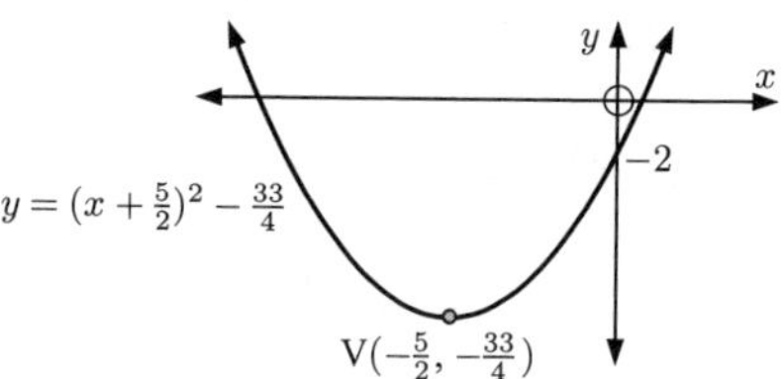


**f** $y = x^2 - 3x + 2$

$\therefore \; y = x^2 - 3x + (\frac{3}{2})^2 + 2 - (\frac{3}{2})^2$

$\therefore \; y = (x - \frac{3}{2})^2 - \frac{1}{4}$

$\therefore$ vertex is $(\frac{3}{2}, -\frac{1}{4})$, $y$-intercept is 2

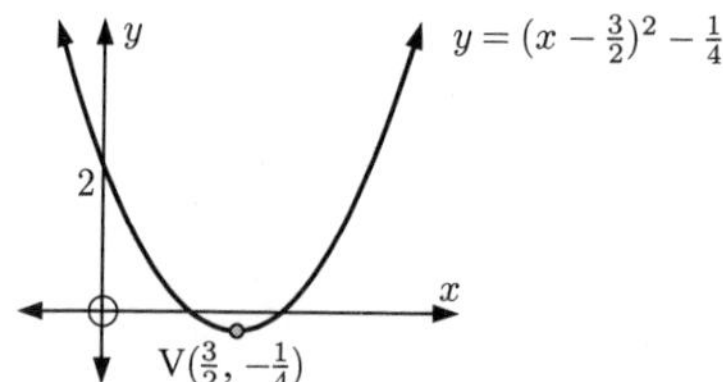


**g** $y = x^2 - 6x + 5$

$\therefore \; y = x^2 - 6x + 3^2 + 5 - 3^2$

$\therefore \; y = (x - 3)^2 - 4$

$\therefore$ vertex is $(3, -4)$, $y$-intercept is 5

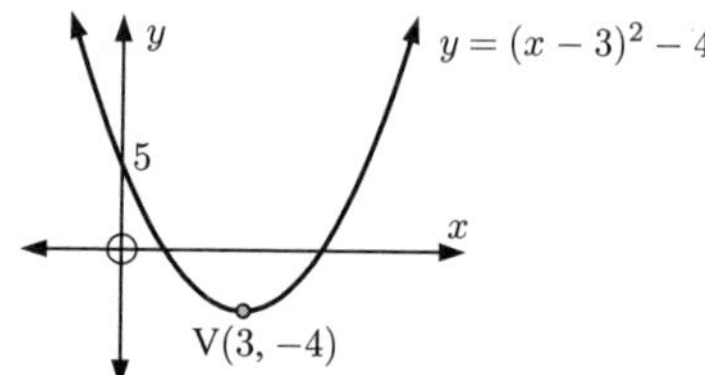


**h** $y = x^2 + 8x - 2$

$\therefore \; y = x^2 + 8x + 4^2 - 2 - 4^2$

$\therefore \; y = (x + 4)^2 - 18$

$\therefore$ vertex is $(-4, -18)$, $y$-intercept is $-2$

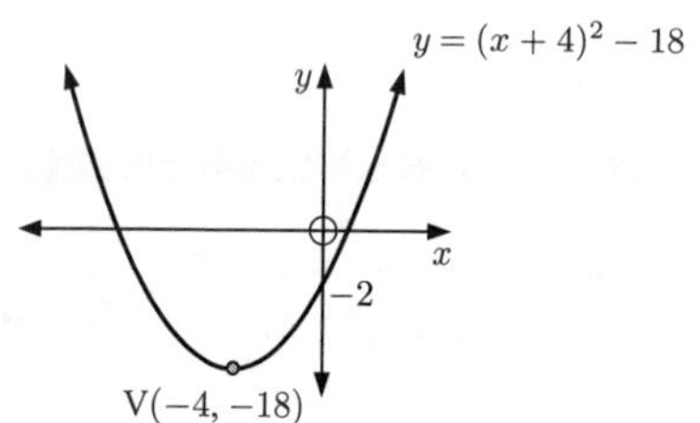


**i** $y = x^2 - 5x + 1$

$\therefore \; y = x^2 - 5x + (\frac{5}{2})^2 + 1 - (\frac{5}{2})^2$

$\therefore \; y = (x - \frac{5}{2})^2 - \frac{21}{4}$

$\therefore$ vertex is $(\frac{5}{2}, -5\frac{1}{4})$, $y$-intercept is 1

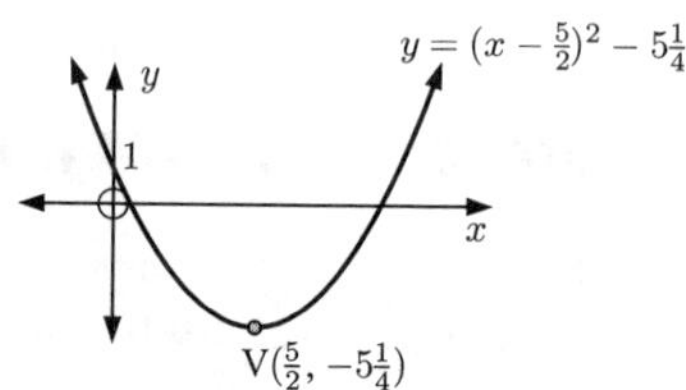


**2** **a** **i** $y = 2x^2 + 4x + 5$

$= 2[x^2 + 2x + \frac{5}{2}]$

$= 2[x^2 + 2x + 1^2 - 1^2 + \frac{5}{2}]$

$= 2[(x+1)^2 + \frac{3}{2}]$

$= 2(x+1)^2 + 3$

**ii** The vertex is $(-1, 3)$.

**iii** When $x = 0$, $y = 5$

$\therefore$ the $y$-intercept is 5

**iv**

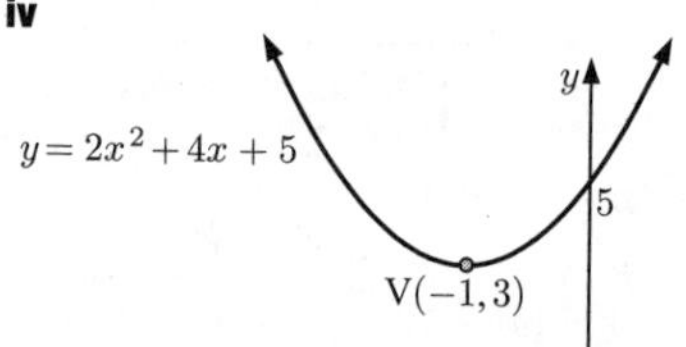

**b** **i** $y = 2x^2 - 8x + 3$
$= 2[x^2 - 4x + \frac{3}{2}]$
$= 2[x^2 - 4x + 2^2 - 2^2 + \frac{3}{2}]$
$= 2[(x-2)^2 - \frac{5}{2}]$
$= 2(x-2)^2 - 5$

**ii** The vertex is $(2, -5)$.

**iii** When $x = 0$, $y = 3$
$\therefore$ the $y$-intercept is 3

**iv**

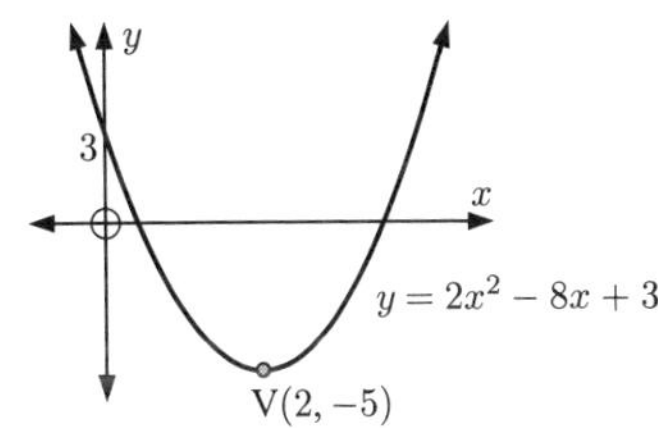


**c** **i** $y = 2x^2 - 6x + 1$
$= 2[x^2 - 3x + \frac{1}{2}]$
$= 2[x^2 - 3x + (\frac{3}{2})^2 - (\frac{3}{2})^2 + \frac{1}{2}]$
$= 2[(x - \frac{3}{2})^2 - \frac{7}{4}]$
$= 2(x - \frac{3}{2})^2 - \frac{7}{2}$

**ii** The vertex is $(\frac{3}{2}, -\frac{7}{2})$.

**iii** When $x = 0$, $y = 1$
$\therefore$ the $y$-intercept is 1

**iv**

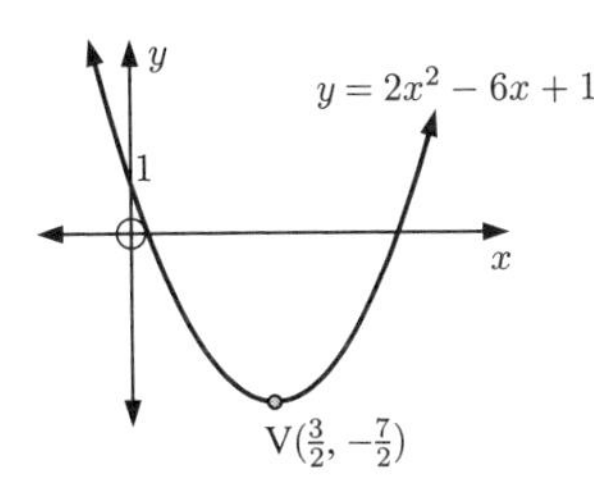


**d** **i** $y = 3x^2 - 6x + 5$
$= 3[x^2 - 2x + \frac{5}{3}]$
$= 3[x^2 - 2x + 1^2 - 1^2 + \frac{5}{3}]$
$= 3[(x-1)^2 + \frac{2}{3}]$
$= 3(x-1)^2 + 2$

**ii** The vertex is $(1, 2)$.

**iii** When $x = 0$, $y = 5$
$\therefore$ the $y$-intercept is 5

**iv**

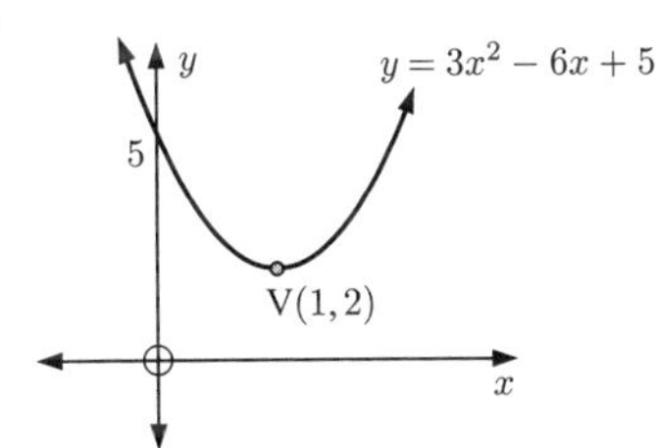


**e** **i** $y = -x^2 + 4x + 2$
$= -[x^2 - 4x - 2]$
$= -[x^2 - 4x + 2^2 - 2^2 - 2]$
$= -[(x-2)^2 - 6]$
$= -(x-2)^2 + 6$

**ii** The vertex is $(2, 6)$.

**iii** When $x = 0$, $y = 2$
$\therefore$ the $y$-intercept is 2

**iv**

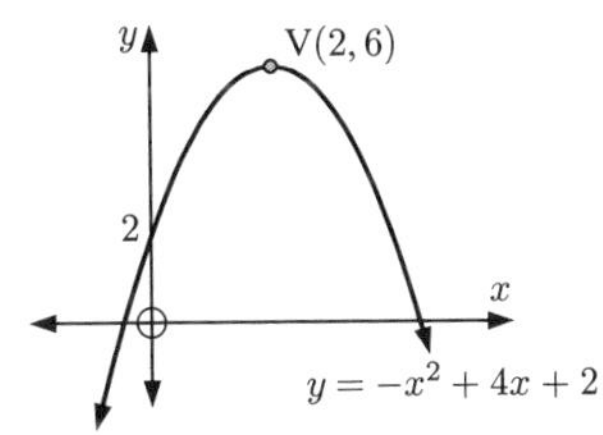


**f** **i** $y = -2x^2 - 5x + 3$
$= -2[x^2 + \frac{5}{2}x - \frac{3}{2}]$
$= -2[x^2 + \frac{5}{2}x + (\frac{5}{4})^2 - (\frac{5}{4})^2 - \frac{3}{2}]$
$= -2[(x + \frac{5}{4})^2 - \frac{25}{16} - \frac{24}{16}]$
$= -2[(x + \frac{5}{4})^2 - \frac{49}{16}]$
$= -2(x + \frac{5}{4})^2 + \frac{49}{8}$

**ii** The vertex is $(-\frac{5}{4}, \frac{49}{8})$.

**iii** When $x = 0$, $y = 3$
$\therefore$ the $y$-intercept is 3

**iv**

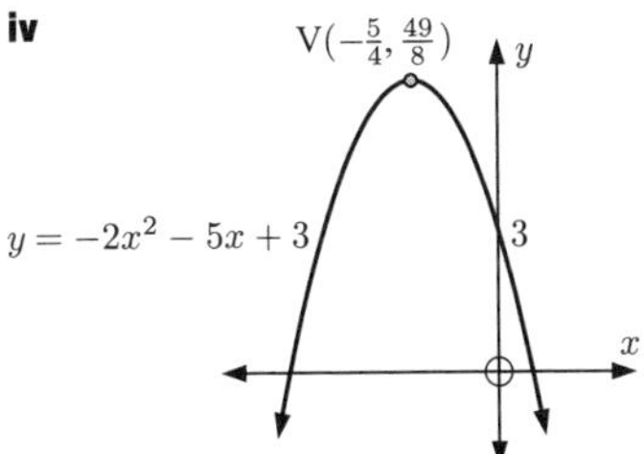

**3** **a** Using technology, the graph is

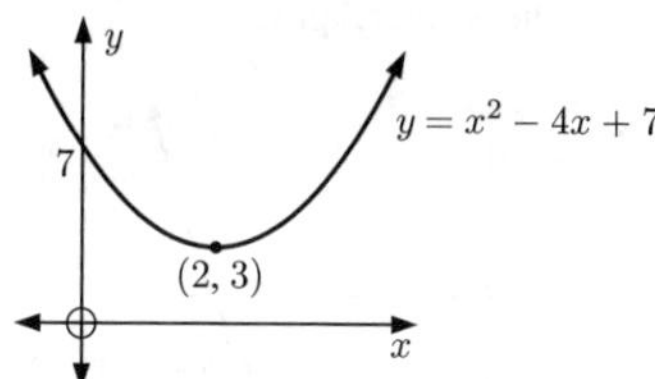


Since the vertex is at $(2,\ 3)$,
the function must be of the form
$y = a(x-2)^2 + 3$ for some $a$.
$\therefore$ when $x = 0$,
$y = a(-2)^2 + 3 = 4a + 3$
but the $y$-intercept is 7
$\therefore\ 4a + 3 = 7$
$\therefore\ a = 1$
$\therefore$ the equation is $y = (x-2)^2 + 3$

**b** Using technology, the graph is

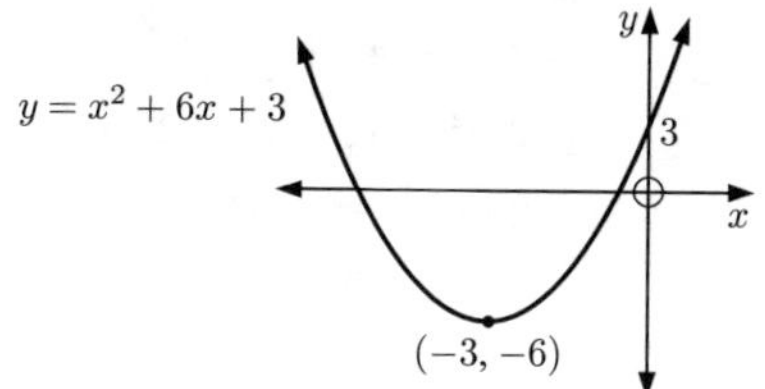


Since the vertex is at $(-3,\ -6)$,
the function must be of the form
$y = a(x+3)^2 - 6$ for some $a$.
$\therefore$ when $x = 0$,
$y = a \times 3^2 - 6 = 9a - 6$
but the $y$-intercept is 3
$\therefore\ 9a - 6 = 3$
$\therefore\ a = 1$
$\therefore$ the equation is $y = (x+3)^2 - 6$

**c** Using technology, the graph is

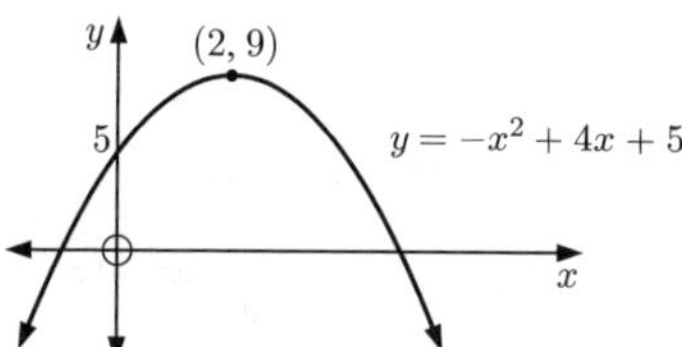


Since the vertex is at $(2,\ 9)$,
the function must be of the form
$y = a(x-2)^2 + 9$ for some $a$.
$\therefore$ when $x = 0$,
$y = a(-2)^2 + 9 = 4a + 9$
but the $y$-intercept is 5
$\therefore\ 4a + 9 = 5$
$\therefore\ a = -1$
$\therefore$ the equation is $y = -(x-2)^2 + 9$

**d** Using technology, the graph is

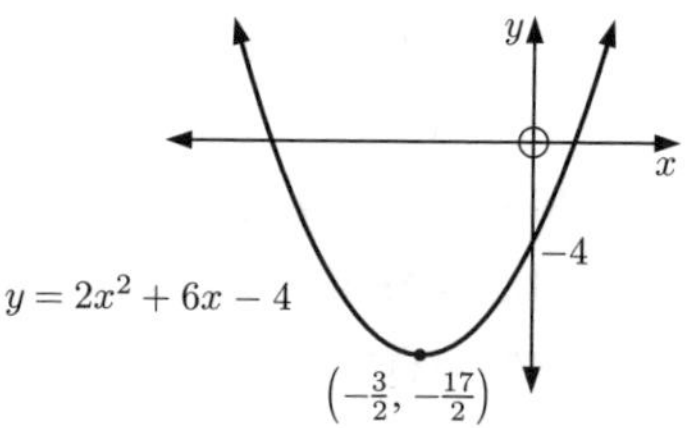


Since the vertex is at $(-\frac{3}{2},\ -\frac{17}{2})$,
the function must be of the form
$y = a(x+\frac{3}{2})^2 - \frac{17}{2}$ for some $a$.
$\therefore$ when $x = 0$,
$y = a(\frac{3}{2})^2 - \frac{17}{2} = \frac{9}{4}a - \frac{17}{2}$
but the $y$-intercept is $-4$
$\therefore\ \frac{9}{4}a - \frac{17}{2} = -4$
$\therefore\ \frac{9}{4}a = \frac{9}{2}$
$\therefore\ a = 2$
$\therefore$ the equation is $y = 2(x+\frac{3}{2})^2 - \frac{17}{2}$

**e** Using technology, the graph is

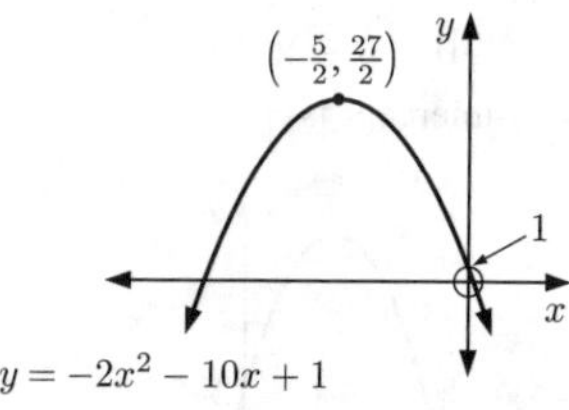


Since the vertex is at $(-\frac{5}{2},\ \frac{27}{2})$,
the function must be of the form
$y = a(x+\frac{5}{2})^2 + \frac{27}{2}$ for some $a$.
$\therefore$ when $x = 0$,
$y = a(\frac{5}{2})^2 + \frac{27}{2} = \frac{25}{4}a + \frac{27}{2}$
but the $y$-intercept is 1
$\therefore\ \frac{25}{4}a + \frac{27}{2} = 1$
$\therefore\ \frac{25}{4}a = -\frac{25}{2}$
$\therefore\ a = -2$
$\therefore$ the equation is $y = -2(x+\frac{5}{2})^2 + \frac{27}{2}$

**f** Using technology, the graph is

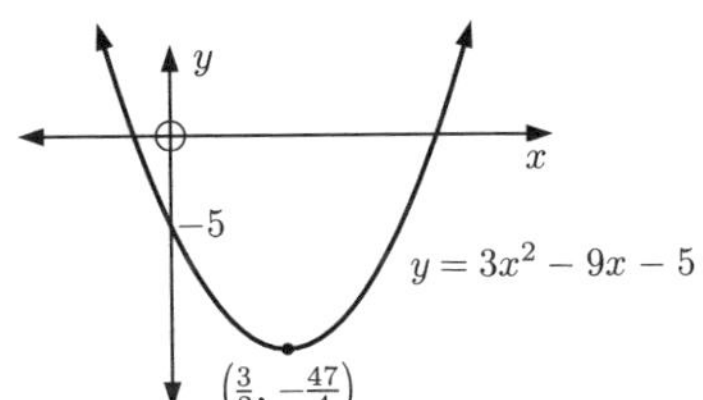


Since the vertex is at $(\frac{3}{2}, -\frac{47}{4})$,
the function must be of the form

$y = a(x - \frac{3}{2})^2 - \frac{47}{4}$ for some $a$.

$\therefore$ when $x = 0$, $y = a(-\frac{3}{2})^2 - \frac{47}{4} = \frac{9}{4}a - \frac{47}{4}$

but the $y$-intercept is $-5$

$\therefore \frac{9}{4}a - \frac{47}{4} = -5$

$\therefore \frac{9}{4}a = \frac{27}{4}$

$\therefore a = 3$

$\therefore$ the equation is $y = 3(x - \frac{3}{2})^2 - \frac{47}{4}$

## EXERCISE 1C.3

**1** **a** $y = x^2 + x - 2$
has $a = 1$, $b = 1$, $c = -2$

$\therefore \Delta = b^2 - 4ac$
$= 1^2 - 4(1)(-2)$
$= 9 > 0$

$\therefore$ the graph cuts the $x$-axis twice, and since $a > 0$, the graph is concave up.

**b** $y = x^2 - 4x + 1$
has $a = 1$, $b = -4$, $c = 1$

$\therefore \Delta = b^2 - 4ac$
$= (-4)^2 - 4(1)(1)$
$= 12 > 0$

$\therefore$ the graph cuts the $x$-axis twice, and since $a > 0$, the graph is concave up.

**c** $y = -x^2 - 3$
has $a = -1$, $b = 0$, $c = -3$

$\therefore \Delta = b^2 - 4ac$
$= 0^2 - 4(-1)(-3)$
$= -12 < 0$

$\therefore$ the graph does not cut the $x$-axis, and since $a < 0$, the graph is concave down.
$\therefore$ it is entirely below the $x$-axis.
$\therefore$ the graph is negative definite.

**d** $y = x^2 + 7x - 2$
has $a = 1$, $b = 7$, $c = -2$

$\therefore \Delta = b^2 - 4ac$
$= 7^2 - 4(1)(-2)$
$= 57 > 0$

$\therefore$ the graph cuts the $x$-axis twice, and since $a > 0$, the graph is concave up.

**e** $y = x^2 + 8x + 16$
has $a = 1$, $b = 8$, $c = 16$

$\therefore \Delta = b^2 - 4ac$
$= 8^2 - 4(1)(16)$
$= 0$

$\therefore$ the graph touches the $x$-axis, and since $a > 0$, the graph is concave up.

**f** $y = -2x^2 + 3x + 1$
has $a = -2$, $b = 3$, $c = 1$

$\therefore \Delta = b^2 - 4ac$
$= 3^2 - 4(-2)(1)$
$= 17 > 0$

$\therefore$ the graph cuts the $x$-axis twice, and since $a < 0$, the graph is concave down.

**g** $y = 6x^2 + 5x - 4$
has $a = 6$, $b = 5$, $c = -4$

$\therefore \Delta = b^2 - 4ac$
$= 5^2 - 4(6)(-4)$
$= 121 > 0$

$\therefore$ the graph cuts the $x$-axis twice, and since $a > 0$, the graph is concave up.

**h** $y = -x^2 + x + 6$
has $a = -1$, $b = 1$, $c = 6$

$\therefore \Delta = b^2 - 4ac$
$= 1^2 - 4(-1)(6)$
$= 25 > 0$

$\therefore$ the graph cuts the $x$-axis twice, and since $a < 0$, the graph is concave down.

**i** $y = 9x^2 + 6x + 1$
has $a = 9$, $b = 6$, $c = 1$

$\therefore \Delta = b^2 - 4ac$
$= 6^2 - 4(9)(1)$
$= 0$

$\therefore$ the graph touches the $x$-axis, and since $a > 0$, the graph is concave up.

**2** **a** $x^2 - 3x + 6$
has $a = 1$, $b = -3$, $c = 6$

$\therefore \Delta = b^2 - 4ac$
$= (-3)^2 - 4(1)(6)$
$= -15$

Since $a > 0$ and $\Delta < 0$,
$x^2 - 3x + 6$ is positive definite.
$\therefore$ $x^2 - 3x + 6 > 0$ for all $x$.

**b** $4x - x^2 - 6$
has $a = -1$, $b = 4$, $c = -6$

$\therefore \Delta = b^2 - 4ac$
$= 4^2 - 4(-1)(-6)$
$= -8$

Since $a < 0$ and $\Delta < 0$,
$4x - x^2 - 6$ is negative definite.
$\therefore$ $4x - x^2 - 6 < 0$ for all $x$.

**c** $2x^2 - 4x + 7$
has $a = 2$, $b = -4$, $c = 7$

$\therefore \Delta = b^2 - 4ac$
$= (-4)^2 - 4(2)(7)$
$= -40$

Since $a > 0$ and $\Delta < 0$,
$2x^2 - 4x + 7$ is positive definite.

**d** $-2x^2 + 3x - 4$
has $a = -2$, $b = 3$, $c = -4$

$\therefore \Delta = b^2 - 4ac$
$= 3^2 - 4(-2)(-4)$
$= -23$

Since $a < 0$ and $\Delta < 0$,
$-2x^2 + 3x - 4$ is negative definite.

**3** $3x^2 + kx - 1$
has $a = 3$, $b = k$, $c = -1$

$\therefore \Delta = b^2 - 4ac$
$= k^2 - 4(3)(-1)$
$= k^2 + 12$

$\therefore$ $\Delta > 0$ for all $k$
{as $k^2 \geqslant 0$ for all $k$}

$\therefore$ $3x^2 + kx - 1$ has two real distinct roots for all $k$.
$\therefore$ it can never be positive definite.

**4** $2x^2 + kx + 2$
has $a = 2$, $b = k$, $c = 2$

$\therefore \Delta = b^2 - 4ac$
$= k^2 - 4(2)(2)$
$= k^2 - 16$

Now $2x^2 + kx + 2$ has $a > 0$.
$\therefore$ it is positive definite provided $k^2 - 16 < 0$
$\therefore$ $k^2 < 16$
$\therefore$ $-4 < k < 4$

## EXERCISE 1D

**1** **a** The $x$-intercepts are 1 and 2.
$\therefore$ $y = a(x-1)(x-2)$
for some $a \neq 0$.
But the $y$-intercept is 4.
$\therefore$ $a(-1)(-2) = 4$
$\therefore$ $2a = 4$
$\therefore$ $a = 2$
$\therefore$ $y = 2(x-1)(x-2)$

**b** The graph touches the $x$-axis when $x = 2$.
$\therefore$ $y = a(x-2)^2$
for some $a \neq 0$.
But the $y$-intercept is 8.
$\therefore$ $a(-2)^2 = 8$
$\therefore$ $4a = 8$
$\therefore$ $a = 2$
$\therefore$ $y = 2(x-2)^2$

**c** The $x$-intercepts are 1 and 3.
$\therefore$ $y = a(x-1)(x-3)$
for some $a \neq 0$.
But the $y$-intercept is 3.
$\therefore$ $a(-1)(-3) = 3$
$\therefore$ $3a = 3$
$\therefore$ $a = 1$
$\therefore$ $y = (x-1)(x-3)$

**d** The $x$-intercepts are $-1$ and 3.
$\therefore$ $y = a(x+1)(x-3)$
for some $a \neq 0$.
But the $y$-intercept is 3.
$\therefore$ $a(1)(-3) = 3$
$\therefore$ $-3a = 3$
$\therefore$ $a = -1$
$\therefore$ $y = -(x+1)(x-3)$

**e** The graph touches the $x$-axis when $x = 1$.
$\therefore$ $y = a(x-1)^2$
for some $a \neq 0$.
But the $y$-intercept is $-3$.
$\therefore$ $a(-1)^2 = -3$
$\therefore$ $a = -3$
$\therefore$ $y = -3(x-1)^2$

**f** The $x$-intercepts are $-2$ and 3.
$\therefore$ $y = a(x+2)(x-3)$
for some $a \neq 0$.
But the $y$-intercept is 12.
$\therefore$ $a(2)(-3) = 12$
$\therefore$ $-6a = 12$
$\therefore$ $a = -2$
$\therefore$ $y = -2(x+2)(x-3)$

**2** **a** As the axis of symmetry is $x = 3$, the other $x$-intercept is 4.

$\therefore\ y = a(x-2)(x-4)$ for some $a \neq 0$.

But the $y$-intercept $= 12$

$\therefore\ a(-2)(-4) = 12$

$\therefore\ 8a = 12$

$\therefore\ a = \frac{12}{8} = \frac{3}{2}$

$\therefore\ y = \frac{3}{2}(x-2)(x-4)$

**b** As the axis of symmetry is $x = -1$, the other $x$-intercept is 2.

$\therefore\ y = a(x+4)(x-2)$ for some $a \neq 0$.

But the $y$-intercept $= 4$

$\therefore\ a(4)(-2) = 4$

$\therefore\ -8a = 4$

$\therefore\ a = -\frac{1}{2}$

$\therefore\ y = -\frac{1}{2}(x+4)(x-2)$

**c** The graph touches the $x$-axis at $x = -3$,

$\therefore\ y = a(x+3)^2$ for some $a \neq 0$.

But the $y$-intercept is $-12$, so $a(3)^2 = -12$

$\therefore\ 9a = -12$

$\therefore\ a = -\frac{12}{9} = -\frac{4}{3}$

$\therefore\ y = -\frac{4}{3}(x+3)^2$

**3** **a** Since the $x$-intercepts are 5 and 1, the equation is $y = a(x-5)(x-1)$ for some $a \neq 0$.

But when $x = 2$, $y = -9$

$\therefore\ -9 = a(2-5)(2-1)$

$\therefore\ -9 = a(-3)(1)$

$\therefore\ -3a = -9$

$\therefore\ a = 3$

$\therefore$ the equation is $y = 3(x-5)(x-1)$

$\therefore\ y = 3(x^2 - 6x + 5)$

$\therefore\ y = 3x^2 - 18x + 15$

**b** Since the $x$-intercepts are 2 and $-\frac{1}{2}$, the equation is $y = a(x-2)(x+\frac{1}{2})$ for some $a \neq 0$.

But when $x = 3$, $y = -14$

$\therefore\ -14 = a(3-2)(3+\frac{1}{2})$

$\therefore\ -14 = a(1)(\frac{7}{2})$

$\therefore\ \frac{7}{2}a = -14$

$\therefore\ a = -4$

$\therefore$ the equation is $y = -4(x-2)(x+\frac{1}{2})$

$\therefore\ y = -4(x^2 - \frac{3}{2}x - 1)$

$\therefore\ y = -4x^2 + 6x + 4$

**c** Since the graph touches the $x$-axis at 3, its equation is $y = a(x-3)^2$, for some $a \neq 0$.

But when $x = -2$, $y = -25$

$\therefore\ -25 = a(-2-3)^2$

$\therefore\ -25 = 25a$

$\therefore\ a = -1$

$\therefore$ the equation is $y = -(x-3)^2$

$\therefore\ y = -(x^2 - 6x + 9)$

$\therefore\ y = -x^2 + 6x - 9$

**d** Since the graph touches the $x$-axis at $-2$, its equation is $y = a(x+2)^2$, for some $a \neq 0$.

But when $x = -1$, $y = 4$

$\therefore\ 4 = a(-1+2)^2$

$\therefore\ 4 = a$

$\therefore$ the equation is $y = 4(x+2)^2$

$\therefore\ y = 4(x^2 + 4x + 4)$

$\therefore\ y = 4x^2 + 16x + 16$

**e** Since the graph cuts the $x$-axis at 3 and has axis of symmetry $x = 2$, it must also cut the $x$-axis at 1.

$\therefore$ the $x$-intercepts are 3 and 1, and the equation is $y = a(x-3)(x-1)$ for some $a \neq 0$.

But when $x = 5$, $y = 12$

$\therefore\ 12 = a(5-3)(5-1)$

$\therefore\ 12 = a(2)(4)$

$\therefore\ 8a = 12$

$\therefore\ a = \frac{12}{8} = \frac{3}{2}$

$\therefore$ the equation is $y = \frac{3}{2}(x-3)(x-1)$

$\therefore\ y = \frac{3}{2}(x^2 - 4x + 3)$

$\therefore\ y = \frac{3}{2}x^2 - 6x + \frac{9}{2}$

**f** Since the graph cuts the $x$-axis at 5 and has axis of symmetry $x = 1$, it must also cut the $x$-axis at $-3$.
$\therefore$ the $x$-intercepts are 5 and $-3$, and the equation is $y = a(x-5)(x+3)$ for some $a \neq 0$.
But when $x = 2$, $y = 5$
$\therefore\ 5 = a(2-5)(2+3)$
$\therefore\ 5 = a(-3)(5)$
$\therefore\ -15a = 5$
$\therefore\ a = -\frac{5}{15} = -\frac{1}{3}$

$\therefore$ the equation is $y = -\frac{1}{3}(x-5)(x+3)$
$\therefore\ y = -\frac{1}{3}(x^2 - 2x - 15)$
$\therefore\ y = -\frac{1}{3}x^2 + \frac{2}{3}x + 5$

**4** **a** The vertex is $(2, 4)$, so the quadratic has equation
$y = a(x-2)^2 + 4$ for some $a \neq 0$.
But the graph passes through the origin
$\therefore\ 0 = a(0-2)^2 + 4$
$\therefore\ 4a + 4 = 0$
$\therefore\ a = -1$
$\therefore$ the equation is $y = -(x-2)^2 + 4$

**b** The vertex is $(2, -1)$, so the quadratic has equation
$y = a(x-2)^2 - 1$ for some $a \neq 0$.
But the graph passes through $(0, 7)$
$\therefore\ 7 = a(0-2)^2 - 1$
$\therefore\ 7 = 4a - 1$
$\therefore\ a = 2$
$\therefore$ the equation is $y = 2(x-2)^2 - 1$

**c** The vertex is $(3, 8)$, so the quadratic has equation
$y = a(x-3)^2 + 8$ for some $a \neq 0$.
But the graph passes through $(1, 0)$
$\therefore\ 0 = a(1-3)^2 + 8$
$\therefore\ 0 = 4a + 8$
$\therefore\ a = -2$
$\therefore$ the equation is $y = -2(x-3)^2 + 8$

**d** The vertex is $(4, -6)$, so the quadratic has equation
$y = a(x-4)^2 - 6$ for some $a \neq 0$.
But the graph passes through $(7, 0)$
$\therefore\ 0 = a(7-4)^2 - 6$
$\therefore\ 9a - 6 = 0$
$\therefore\ a = \frac{2}{3}$
$\therefore$ the equation is $y = \frac{2}{3}(x-4)^2 - 6$

**e** The vertex is $(2, 3)$, so the quadratic has equation
$y = a(x-2)^2 + 3$ for some $a \neq 0$.
But the graph passes through $(3, 1)$
$\therefore\ 1 = a(3-2)^2 + 3$
$\therefore\ 1 = a + 3$
$\therefore\ a = -2$
$\therefore$ the equation is $y = -2(x-2)^2 + 3$

**f** The vertex is $(\frac{1}{2}, -\frac{3}{2})$, so the quadratic has equation
$y = a(x - \frac{1}{2})^2 - \frac{3}{2}$ for some $a \neq 0$.
But the graph passes through $(\frac{3}{2}, \frac{1}{2})$
$\therefore\ \frac{1}{2} = a(\frac{3}{2} - \frac{1}{2})^2 - \frac{3}{2}$
$\therefore\ \frac{1}{2} = a - \frac{3}{2}$
$\therefore\ a = 2$
$\therefore$ the equation is $y = 2(x - \frac{1}{2})^2 - \frac{3}{2}$

## EXERCISE 1E

**1** **a** $y = x^2 - 2x + 8$ meets $y = x + 6$
when $x^2 - 2x + 8 = x + 6$
$\therefore\ x^2 - 3x + 2 = 0$
$\therefore\ (x-1)(x-2) = 0$
$\therefore\ x = 1$ or $2$
Substituting into $y = x + 6$,
when $x = 1$, $y = 7$
and when $x = 2$, $y = 8$
$\therefore$ the graphs meet at $(1, 7)$ and $(2, 8)$.

**b** $y = -x^2 + 3x + 9$ meets $y = 2x - 3$
when $-x^2 + 3x + 9 = 2x - 3$
$\therefore\ x^2 - x - 12 = 0$
$\therefore\ (x-4)(x+3) = 0$
$\therefore\ x = 4$ or $-3$
Substituting into $y = 2x - 3$,
when $x = -3$, $y = 2(-3) - 3 = -9$
and when $x = 4$, $y = 2(4) - 3 = 5$
$\therefore$ the graphs meet at $(-3, -9)$ and $(4, 5)$.

**c** $y = x^2 - 4x + 3$ meets $y = 2x - 6$

when $x^2 - 4x + 3 = 2x - 6$

$\therefore\ x^2 - 6x + 9 = 0$

$\therefore\ (x-3)^2 = 0$

$\therefore\ x = 3$

Substituting into $y = 2x - 6$,

when $x = 3$, $y = 0$

$\therefore$ the graphs touch at $(3, 0)$.

**d** $y = -x^2 + 4x - 7$ meets $y = 5x - 4$

when $-x^2 + 4x - 7 = 5x - 4$

$\therefore\ x^2 + x + 3 = 0$

which has $a = 1$, $b = 1$, $c = 3$

$$\therefore\ x = \frac{-1 \pm \sqrt{1^2 - 4(1)(3)}}{2(1)}$$

$$\therefore\ x = \frac{-1 \pm \sqrt{-11}}{2}$$

$\therefore$ there are no real solutions

$\therefore$ the graphs do not meet.

**2** **a** $(0.59, 5.59)$ and $(3.41, 8.41)$ **b** $(3, -4)$ touching

**c** graphs do not meet **d** $(-2.56, -18.81)$ and $(1.56, 1.81)$

**3** **a** **i**

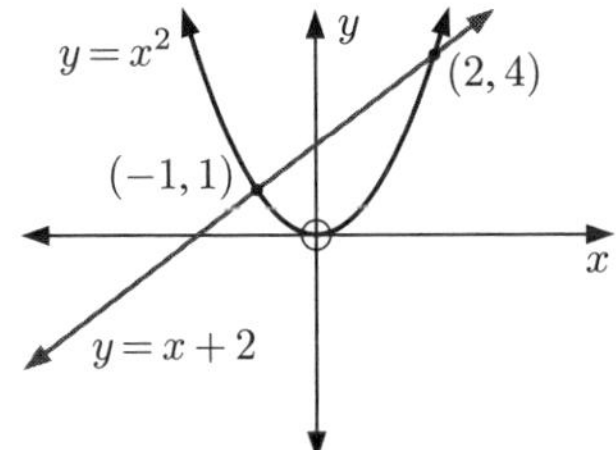


$\therefore\ y = x^2$ meets $y = x + 2$ at the points $(-1, 1)$ and $(2, 4)$.

**ii** Using the graph in **a i**, $x^2 > x + 2$ when $x < -1$ or $x > 2$.

**b** **i**

$y=x^2+2x-3$ $y$

$y=x-1$

$(1,0)$ $x$

$(-2,-3)$

$\therefore\ y = x^2 + 2x - 3$ meets $y = x - 1$ at the points $(-2, -3)$ and $(1, 0)$.

**ii** Using the graph in **b i**, $x^2 + 2x - 3 > x - 1$ when $x < -2$ or $x > 1$.

**c** **i**

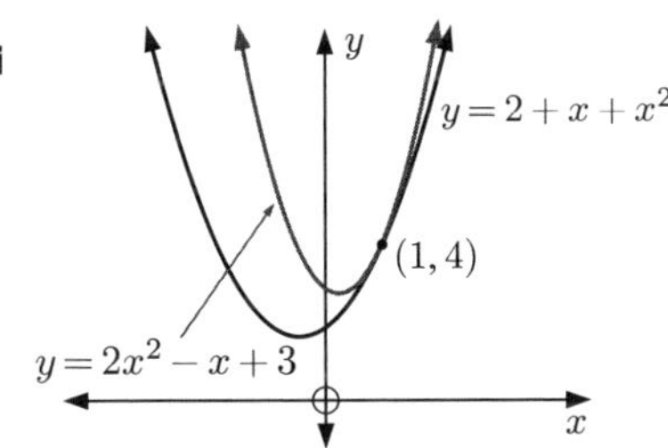


$\therefore\ y = 2x^2 - x + 3$ meets $y = 2 + x + x^2$ at the point $(1, 4)$.

**ii** Using the graph in **c i**, $2x^2 - x + 3 > 2 + x + x^2$ when $x \neq 1$.

**d** **i**

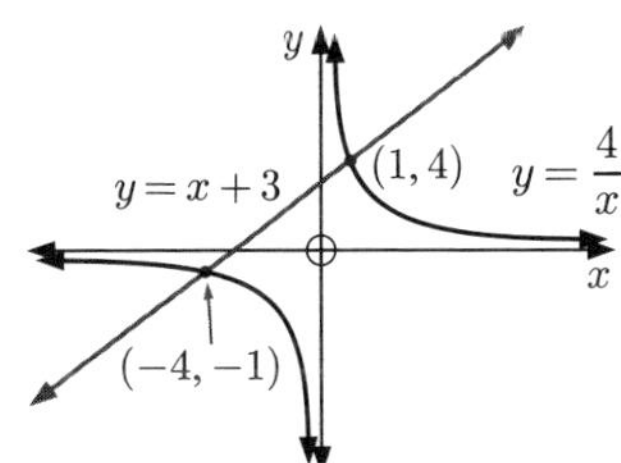


$\therefore\ y = \dfrac{4}{x}$ meets $y = x + 3$ at the points $(-4, -1)$ and $(1, 4)$.

**ii** Using the graph in **d i**, $\dfrac{4}{x} > x + 3$ when $x < -4$ or $0 < x < 1$.

**4** $y = 3x + c$ is a tangent to $y = x^2 - 5x + 7$ if they meet at exactly one point (touch).

$y = x^2 - 5x + 7$ meets $y = 3x + c$ when $x^2 - 5x + 7 = 3x + c$

$\therefore\ x^2 - 8x + 7 - c = 0$

The graphs meet exactly once when this equation has a repeated root $\therefore\ \Delta = 0$

$\therefore\ (-8)^2 - 4(1)(7 - c) = 0$

$\therefore\ 64 - 28 + 4c = 0$

$\therefore\ 4c = -36$

$\therefore\ c = -9$

**5** $y = mx - 2$ is a tangent to $y = x^2 - 4x + 2$ if they meet at exactly one point (touch).

$y = x^2 - 4x + 2$ meets $y = mx - 2$ when $x^2 - 4x + 2 = mx - 2$

$\therefore \quad x^2 - (m+4)x + 4 = 0$

The graphs meet exactly once when this equation has a repeated root $\therefore \quad \Delta = 0$

$\therefore \quad (-(m+4))^2 - 4(1)(4) = 0$

$\therefore \quad m^2 + 8m + 16 - 16 = 0$

$\therefore \quad m(m+8) = 0$

$\therefore \quad m = 0$ or $-8$

**6** Lines with $y$-intercept 1 have the form $y = mx + 1$.

$y = mx + 1$ is a tangent to $y = 3x^2 + 5x + 4$ if they meet at exactly one point (touch).

$y = 3x^2 + 5x + 4$ meets $y = mx + 1$ when $3x^2 + 5x + 4 = mx + 1$

$\therefore \quad 3x^2 + (5-m)x + 3 = 0$

The graphs meet exactly once when this equation has a repeated root $\therefore \quad \Delta = 0$

$\therefore \quad (5-m)^2 - 4(3)(3) = 0$

$\therefore \quad 25 - 10m + m^2 - 36 = 0$

$\therefore \quad m^2 - 10m - 11 = 0$

$\therefore \quad (m+1)(m-11) = 0$

$\therefore \quad m = -1$ or $11$

$\therefore$ the required lines have gradient $-1$ or $11$.

**7** **a** $y = x + c$ meets $y = 2x^2 - 3x - 7$

when $2x^2 - 3x - 7 = x + c$

$\therefore \quad 2x^2 - 4x - 7 - c = 0$

The graphs will never meet if this equation has no real roots $\therefore \quad \Delta < 0$

$\therefore \quad (-4)^2 - 4(2)(-7-c) < 0$

$\therefore \quad 16 + 56 + 8c < 0$

$\therefore \quad 8c < -72$

$\therefore \quad c < -9$

**b** Choose $c$ such that $c < -9$, for example $c = -10$:

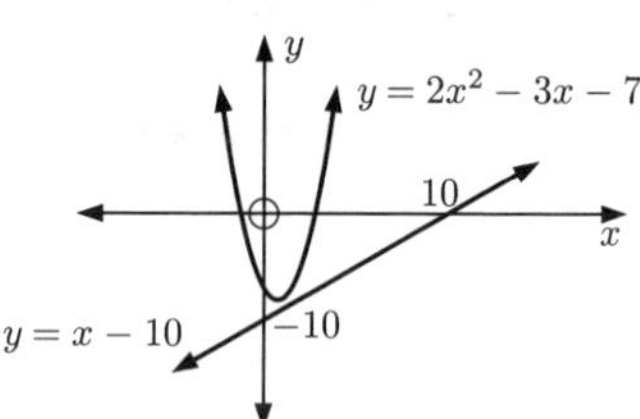


## EXERCISE 1F

**1** Let the smaller of the integers be $x$.

The other integer is $(x + 12)$.

$\therefore$ the sum of their squares is

$x^2 + (x+12)^2 = 74$

$\therefore \quad x^2 + x^2 + 24x + 144 = 74$

$\therefore \quad 2x^2 + 24x + 70 = 0$

$\therefore \quad x^2 + 12x + 35 = 0$

$\therefore \quad (x+7)(x+5) = 0$

$\therefore \quad x = -7$ or $-5$

So, the integers are $-7$ and $5$, or $-5$ and $7$.

**2** Let the number be $x$, so its reciprocal is $\dfrac{1}{x}$.

They have sum $x + \dfrac{1}{x} = 5\frac{1}{5}$

$\therefore \quad x^2 + 1 = \frac{26}{5}x$

$\therefore \quad x^2 - \frac{26}{5}x + 1 = 0$

$\therefore \quad 5x^2 - 26x + 5 = 0$

$\therefore \quad (5x-1)(x-5) = 0$

$\therefore \quad x = \frac{1}{5}$ or $5$

So, the number is either $\frac{1}{5}$ or $5$.

**3** Let the number be $x$ so its square is $x^2$.

$\therefore$ the sum is $x + x^2 = 210$

$\therefore \quad x^2 + x - 210 = 0$

$\therefore \quad (x+15)(x-14) = 0$

$\therefore \quad x = -15$ or $14$

But $x$ is a natural number, so $x > 0$,

$\therefore$ the number is 14.

**4** Suppose the numbers are $x$ and $(x+2)$.

Then $x(x+2) = 360$

$\therefore \quad x^2 + 2x - 360 = 0$

$\therefore \quad (x+20)(x-18) = 0$

$\therefore \quad x = -20$ or $18$

$\therefore$ the numbers are $-20$ and $-18$, or $18$ and $20$.

**5** Suppose the numbers are $x$ and $(x+2)$.

$$\text{Then}\quad x(x+2) = 255$$
$$\therefore\quad x^2 + 2x - 255 = 0$$
$$\therefore\quad (x+17)(x-15) = 0$$
$$\therefore\quad x = -17 \text{ or } 15$$

$\therefore$ the numbers are $-17$ and $-15$,
or $15$ and $17$.

**6** If the polygon has $n$ sides, then

$$\frac{n}{2}(n-3) = 90$$
$$\therefore\quad \tfrac{1}{2}n^2 - \tfrac{3}{2}n = 90$$
$$\therefore\quad n^2 - 3n - 180 = 0$$
$$\therefore\quad (n-15)(n+12) = 0$$
$$\therefore\quad n = -12 \text{ or } 15$$

$\therefore$ the polygon has 15 sides. $\{\text{as } n > 0\}$

**7** If the width of the rectangle is $w$ cm,
then its length is $(w+4)$ cm.

$\therefore$ the area is $w(w+4) = 26$

$$\therefore\quad w^2 + 4w - 26 = 0$$

which has $a = 1$, $b = 4$, $c = -26$

$$\therefore\quad w = \frac{-4 \pm \sqrt{4^2 - 4(1)(-26)}}{2(1)}$$
$$\therefore\quad w = \frac{-4 \pm \sqrt{120}}{2} = -2 \pm \sqrt{30}$$

But $w > 0$, so $w = -2 + \sqrt{30}$
$\approx 3.477$

So, the width is approximately 3.48 cm.

**8** **a** The base has sides of length $x$ cm, so the areas of the top and bottom surfaces are both $x^2$ cm$^2$.

The box has height $(x+1)$ cm,
so the area of each of the side faces
is $x(x+1)$ cm$^2$.

$\therefore$ the total surface area is

$$A = 2x^2 + 4x(x+1)$$
$$= 2x^2 + 4x^2 + 4x$$
$$= 6x^2 + 4x \text{ cm}^2$$

**b**
$$6x^2 + 4x = 240$$
$$\therefore\quad 3x^2 + 2x - 120 = 0$$
$$\therefore\quad (3x+20)(x-6) = 0$$
$$\therefore\quad x = -\tfrac{20}{3} \text{ or } 6$$

but $x > 0$, so $x = 6$

$\therefore$ the box is $6 \text{ cm} \times 6 \text{ cm} \times 7 \text{ cm}$.

**9** Suppose the tinplate was $x \text{ cm} \times x \text{ cm}$.
When $3 \text{ cm} \times 3 \text{ cm}$ squares are cut from the corners, the base of the open box formed is $(x-6) \text{ cm} \times (x-6) \text{ cm}$.
The open box has height 3 cm, so its volume is $3 \times (x-6) \times (x-6) = 80$

$$\therefore\quad 3(x^2 - 12x + 36) = 80$$
$$3x^2 - 36x + 108 = 80$$
$$\therefore\quad 3x^2 - 36x + 28 = 0$$

which has $a = 3$, $b = -36$, $c = 28$

$$\therefore\quad x = \frac{-(-36) \pm \sqrt{(-36)^2 - 4(3)(28)}}{2(3)}$$
$$= \frac{36 \pm \sqrt{960}}{6} \quad \text{and since } x > 6,$$
$$x = 6 + \frac{\sqrt{960}}{6} \approx 11.16$$

$\therefore$ the original piece of tinplate was about 11.2 cm square.

**10**

y cm

x cm

Suppose one side of the rectangle has length $x$ cm and the other has length $y$ cm.
The perimeter is $(2x+2y)$ cm,

$$\text{so}\quad 2x + 2y = 20$$
$$\therefore\quad 2y = 20 - 2x$$
$$\therefore\quad y = 10 - x$$

The area of the rectangle is therefore
$x(10-x)$ cm$^2$.

If the area is 30 cm$^2$, then

$$x(10-x) = 30$$
$$\therefore\quad 10x - x^2 = 30$$
$$\therefore\quad x^2 - 10x + 30 = 0$$

which has $a = 1$, $b = -10$, $c = 30$

$$\therefore\quad x = \frac{-(-10) \pm \sqrt{(-10)^2 - 4(1)(30)}}{2(1)}$$
$$= \frac{10 \pm \sqrt{100 - 120}}{2}$$
$$= \frac{10 \pm \sqrt{-20}}{2}$$

$\therefore$ $x$ has no real solutions, so it is not possible.

**11** The smaller rectangle is similar to the original rectangle.

$$\therefore \quad \frac{AB}{AD} = \frac{BC}{BY}$$

Suppose AB $= x$ units, and AD $=$ BC $= 1$ unit

$$\therefore \quad \frac{x}{1} = \frac{1}{x-1}$$

$\therefore \quad x(x-1) = 1$

$\therefore \quad x^2 - x - 1 = 0$

which has $a = 1, \quad b = -1, \quad c = -1$

$$\therefore \quad x = \frac{-(-1) \pm \sqrt{(-1)^2 - 4(1)(-1)}}{2(1)} = \frac{1 \pm \sqrt{1+4}}{2} = \frac{1 \pm \sqrt{5}}{2}$$

$\therefore \quad x = \dfrac{1+\sqrt{5}}{2}$, since $x > 0$

But $x = \dfrac{AB}{AD}$, which is the golden ratio

$\therefore$ the golden ratio is $\dfrac{1+\sqrt{5}}{2}$.

**12** Suppose the express train travels at $x$ km h$^{-1}$. We know speed $= \dfrac{\text{distance}}{\text{time}}$, so time $= \dfrac{\text{distance}}{\text{speed}}$.

$\therefore$ it takes the express train $\dfrac{160}{x}$ hours and the normal train $\dfrac{160}{x-10}$ hours.

$$\therefore \quad \frac{160}{x} + \tfrac{1}{2} = \frac{160}{x-10}$$

$\therefore \quad 160(x-10) + \frac{1}{2}x(x-10) = 160x$

$\therefore \quad 160x - 1600 + \frac{1}{2}x^2 - 5x = 160x$

$\therefore \quad x^2 - 10x - 3200 = 0$ which has $a = 1, \quad b = -10, \quad c = -3200$

$$\therefore \quad x = \frac{-(-10) \pm \sqrt{(-10)^2 - 4(1)(-3200)}}{2(1)} = \frac{10 \pm \sqrt{12\,900}}{2}$$

But $x > 0$, so $x = \dfrac{10 + \sqrt{12\,900}}{2} \approx 61.8$ km h$^{-1}$

$\therefore$ the express train travels on average at about 61.8 km h$^{-1}$.

**13** Suppose $n$ elderly citizens ended up going on the trip, so the cost per person was $\$\dfrac{160}{n}$.

If the original number of elderly citizens had gone, there would have been $(n+8)$, and the cost per person would have been $\$\dfrac{160}{n+8}$.

$$\text{Hence} \quad \frac{160}{n} = \frac{160}{n+8} + 1$$

$\therefore \quad 160(n+8) = 160n + n(n+8)$

$\therefore \quad 160n + 1280 = 160n + n^2 + 8n$

$\therefore \quad n^2 + 8n - 1280 = 0$

$\therefore \quad (n-32)(n+40) = 0$

$\therefore$ since $n > 0$, $n = 32$. So, 32 elderly citizens went on the trip.

**14** **a**

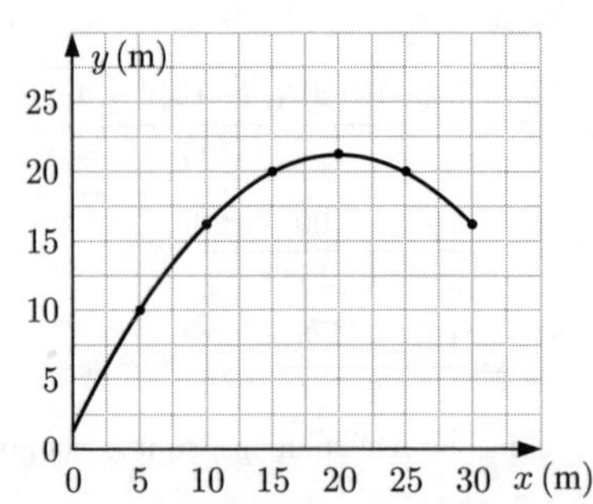


**b** The graph is a parabola.

**c** The maximum height reached by the ball is 21.25m.

**d** Let $f(x) = ax^2+bx+c$ be the form of the formula for height given horizontal distance $x$.

$$f(0) = 1.25$$
$$\therefore \; a(0)^2 + b(0) + c = 1.25$$
$$\therefore \; c = 1.25$$

Also $f(5) = 10$

$$\therefore \; a(5)^2 + b(5) + 1.25 = 10$$
$$\therefore \; 25a + 5b = 8.75 \quad (1)$$

And $f(15) = 20$

$$\therefore \; a(15)^2 + b(15) + 1.25 = 20$$
$$\therefore \; 225a + 15b = 18.75$$
$$\therefore \; 75a + 5b = 6.25 \quad \{\div 3\} \quad (2)$$

Solving simultaneously (or using technology):

$$\begin{aligned} 75a + 5b &= 6.25 \quad (2) \\ -25a - 5b &= -8.75 \quad -(1) \\ \hline 50a \quad &= -2.5 \\ \therefore \; a &= -0.05 \end{aligned}$$

$$\therefore \; -1.25 + 5b = 8.75 \quad \{\text{substitute in (1)}\}$$
$$\therefore \; 5b = 10$$
$$\therefore \; b = 2$$

So $f(x) = -0.05x^2 + 2x + 1.25$

**e** The ball bounces when it hits the ground

$$\therefore \; f(x) = 0$$
$$\therefore \; -0.05x^2 + 2x + 1.25 = 0$$

Using technology,

$$x = -0.616 \quad \text{or} \quad x = 40.6$$

But $x \geqslant 0$ $\therefore$ ball bounces at 40.6 m.

But Badrani is 40 m from Abiola, so the ball will reach Badrani before it bounces.

**15** **a** The parabola has vertex (0, 8), so it has equation

$$y = a(x-0)^2 + 8$$
$$\therefore \; y = ax^2 + 8$$

When $x = 3$, $y = 0$, so

$$0 = a(3^2) + 8$$
$$\therefore \; 9a = -8$$
$$\therefore \; a = -\tfrac{8}{9}$$

$\therefore$ the equation of the parabola is $y = -\frac{8}{9}x^2 + 8$.

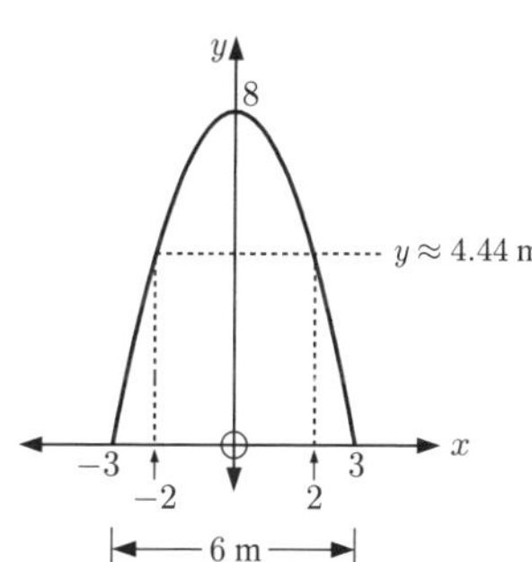


**b** The truck is 4 m wide, so we use the equation in **a** to find the height of the tunnel when it is 4 m wide.

When $x = \pm 2$, $y = -\frac{8}{9}(2)^2 + 8$
$= -\frac{32}{9} + 8$
$= \frac{40}{9} \approx 4.44$ m

For heights greater than 4.44 m, the tunnel is less than 4 m wide. But the truck is 5 m high.

$\therefore$ the truck will not fit through the tunnel.

## EXERCISE 1G

**1** **a** For $y = x^2 - 2x$,
$a = 1$, $b = -2$, $c = 0$.

As $a > 0$, the shape is $\cup$

$\therefore$ the minimum value occurs when

$$x = \frac{-b}{2a} = \frac{2}{2} = 1$$

and $y = 1^2 - 2(1) = -1$

$\therefore$ the minimum value of $y = x^2 - 2x$ is $-1$, occurring when $x = 1$.

**b** For $y = 7 - 2x - x^2$,
$a = -1$, $b = -2$, $c = 7$.

As $a < 0$, the shape is $\cap$

$\therefore$ the maximum value occurs when

$$x = \frac{-b}{2a} = -\frac{-2}{2(-1)} = -1$$

and $y = 7 - 2(-1) - (-1)^2 = 8$

$\therefore$ the maximum value of $y = 7 - 2x - x^2$ is 8, occuring when $x = -1$.

**c** For $y = 8 + 2x - 3x^2$,
$a = -3, \quad b = 2, \quad c = 8.$

As $a < 0$, the shape is $\frown$

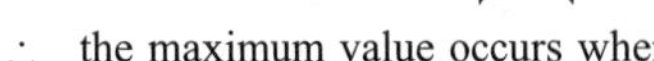

$\therefore$ the maximum value occurs when

$$x = \frac{-b}{2a} = \frac{-2}{-6} = \tfrac{1}{3}$$

and $y = 8 + 2(\tfrac{1}{3}) - 3(\tfrac{1}{3})^2 = 8\tfrac{1}{3}$

$\therefore$ the maximum value of $y = 8 + 2x - 3x^2$ is $8\tfrac{1}{3}$, occuring when $x = \tfrac{1}{3}$.

**d** For $y = 2x^2 + x - 1$,
$a = 2, \quad b = 1, \quad c = -1.$

As $a > 0$, the shape is 

$\therefore$ the minimum value occurs when

$$x = \frac{-b}{2a} = -\tfrac{1}{4}$$

and $y = 2(-\tfrac{1}{4})^2 + (-\tfrac{1}{4}) - 1$
$= \tfrac{1}{8} - \tfrac{1}{4} - 1 = -1\tfrac{1}{8}$

$\therefore$ the minimum value of $y = 2x^2 + x - 1$ is $-1\tfrac{1}{8}$, occuring when $x = -\tfrac{1}{4}$.

**e** For $y = 4x^2 - x + 5$,
$a = 4, \quad b = -1, \quad c = 5.$

As $a > 0$, the shape is $\smile$

$\therefore$ the minimum value occurs when

$$x = \frac{-b}{2a} = \tfrac{1}{8}$$

and $y = 4(\tfrac{1}{8})^2 - \tfrac{1}{8} + 5$
$= \tfrac{1}{16} - \tfrac{1}{8} + 5$
$= 4\tfrac{15}{16}$

$\therefore$ the minimum value of $y = 4x^2 - x + 5$ is $4\tfrac{15}{16}$, occuring when $x = \tfrac{1}{8}$.

**f** For $y = 7x - 2x^2$,
$a = -2, \quad b = 7, \quad c = 0.$

As $a < 0$, the shape is 

$\therefore$ the maximum value occurs when

$$x = \frac{-b}{2a} = \frac{-7}{-4} = \tfrac{7}{4}$$

and $y = 7(\tfrac{7}{4}) - 2(\tfrac{7}{4})^2$
$= \tfrac{49}{4} - \tfrac{49}{8}$
$= \tfrac{49}{8}$ or $6\tfrac{1}{8}$

$\therefore$ the maximum value of $y = 7x - 2x^2$ is $6\tfrac{1}{8}$, occuring when $x = \tfrac{7}{4}$.

**2** **a** For $P = -3x^2 + 240x - 800$,
$a = -3, \quad b = 240, \quad c = -800.$

As $a < 0$, the shape is $\frown$

$\therefore$ the maximum profit occurs when

$$x = \frac{-b}{2a} = \frac{-240}{-6} = 40$$

$\therefore$ 40 refrigerators should be made each day to maximise the total profit.

**b** $P = -3(40)^2 + 240(40) - 800$
$= 4000$

$\therefore$ the maximum profit is \$4000.

**3** **a** Let the other side be $y$ m long.
The perimeter is 200 m.

$\therefore \quad 2x + 2y = 200$
$\therefore \quad x + y = 100$
$\therefore \quad y = 100 - x$

$\therefore$ the area $A = xy$
$\therefore \quad A = x(100 - x)$
$\therefore \quad A = 100x - x^2$

**b** $A = 100x - x^2$ is a quadratic function with
$a = -1, \quad b = 100, \quad c = 0.$

As $a < 0$, the shape is 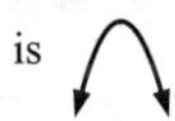

$\therefore$ the area is maximised when

$$x = \frac{-b}{2a} = \frac{-100}{-2} = 50$$

and $y = 100 - 50 = 50$

$\therefore$ the area of the rectangle is maximised when $x = y = 50$, which is when the rectangle is a square.

**4** Let the dimensions of the paddock be $x$ m $\times$ $y$ m.
If 1000 m of fence is available, then

$2x + y = 1000$ {perimeter}

$\therefore \quad y = 1000 - 2x \quad .... (1)$

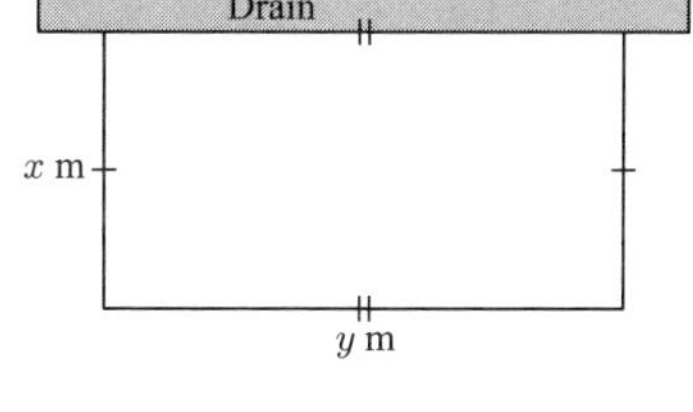


The area of the enclosure $A = xy$

Since $y = 1000 - 2x$, $A = x(1000 - 2x)$

$= 1000x - 2x^2$

$\therefore \quad A = -2x^2 + 1000x$

$A$ is a quadratic and $a < 0$, so its shape is

So, area is maximised when $x = \dfrac{-b}{2a} = \dfrac{-1000}{2 \times (-2)} = 250$

and when $x = 250$, $y = 1000 - 2(250) = 500$

$\therefore$ the paddock has a maximum area when the dimensions are 250 m $\times$ 500 m.

**5** **a**

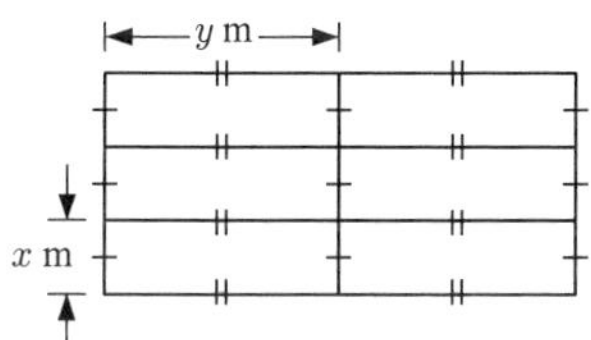


The length of fence required for this enclosure is $9x + 8y$. If 1800 m is available for this enclosure, then $9x + 8y = 1800$.

**b** If $9x + 8y = 1800$, then $y = \dfrac{1800 - 9x}{8}$.

The area of each pen is $A = xy$.

Substituting $y = \dfrac{1800 - 9x}{8}$ into $A$ we get

$$A = x\left(\frac{1800 - 9x}{8}\right)$$

$$\therefore \quad A = \frac{1800x}{8} - \frac{9x^2}{8}$$

$$\therefore \quad A = -\tfrac{9}{8}x^2 + 225x \text{ m}^2$$

**c** The area is a quadratic function with $a < 0$, so its shape is

So, at $x = \dfrac{-b}{2a}$ we have a maximum

$\therefore \quad x = \dfrac{-225}{2 \times (-\frac{9}{8})} = 100$, and when $x = 100$, $y = \dfrac{1800 - 9(100)}{8} = 112.5$

Hence, the area is maximised when the dimensions are 100 m $\times$ 112.5 m.

**6** **a** Let $x$ m $\times$ $y$ m be the dimensions of a single pen as shown below.

Hence, the total length of fence required is $6x + 6y$.

If there is 500 m of fence available, then $6x + 6y = 500$

$\therefore \quad x + y = 83\frac{1}{3}$

$\therefore \quad y = 83\frac{1}{3} - x \quad .... (1)$

The area of each pen will be $A = xy$ and substituting equation (1), we have

$$A = x(83\tfrac{1}{3} - x)$$

$$\therefore \quad A = -x^2 + 83\tfrac{1}{3}x$$

which is a quadratic with $a < 0$ $\therefore$ its shape is

Hence, at $x = \dfrac{-b}{2a}$ we have a maximum value of $A$.

$\therefore \quad x = \dfrac{-83\frac{1}{3}}{2(-1)} = 41\frac{2}{3}$ and so $y = 83\frac{1}{3} - x = 41\frac{2}{3}$

$\therefore$ the dimensions that maximise the area are $41\frac{2}{3}$ m $\times$ $41\frac{2}{3}$ m.

**b** Let $x$ m $\times$ $y$ m be the dimensions of a single pen as shown below.

Hence, the total length of fence required is $5x + 8y$.

If there is 500 m of fence available, then

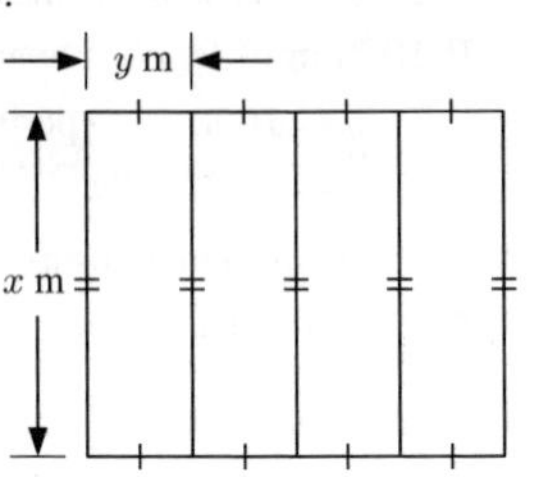


$5x + 8y = 500$

$\therefore \quad 8y = 500 - 5x$

$\therefore \quad y = \dfrac{500 - 5x}{8}$

$\therefore \quad y = 62\frac{1}{2} - \frac{5}{8}x$ .... (1)

The area of each pen will be $A = xy$ and substituting equation (1), we have

$A = x(62\frac{1}{2} - \frac{5}{8}x)$

$\therefore \quad A = -\frac{5}{8}x^2 + 62\frac{1}{2}x$ which is a quadratic with $a < 0$, $\therefore$ its shape is 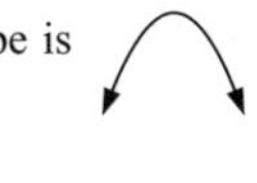

Hence, when $x = \dfrac{-b}{2a}$ we have a maximum value of $A$.

$\therefore \quad x = \dfrac{-62\frac{1}{2}}{2 \times (-\frac{5}{8})} = 50$ and substituting $x = 50$ into $y = 62\frac{1}{2} - \frac{5}{8}x$, we have $y = 31\frac{1}{4}$

$\therefore$ the dimensions that maximise the area are $50$ m $\times$ $31\frac{1}{4}$ m.

**7** **a** The graphs of $y = x^2 - 3x$ and $y = 2x - x^2$

meet where $x^2 - 3x = 2x - x^2$

$\therefore \quad 2x^2 - 5x = 0$

$\therefore \quad x(2x - 5) = 0$

$\therefore \quad x = 0$ or $2\frac{1}{2}$

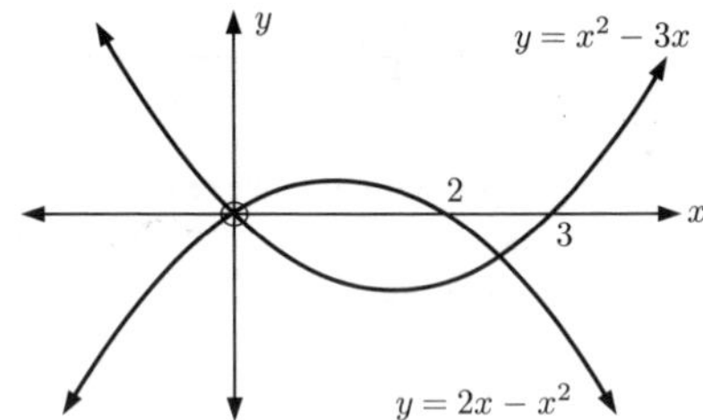


**b** The vertical separation between the curves is given by

$S = (2x - x^2) - (x^2 - 3x)$ $\{y = 2x - x^2$ is above $y = x^2 - 3x$ for $0 \leqslant x \leqslant 2\frac{1}{2}\}$

$\therefore \quad S = 2x - x^2 - x^2 + 3x$

$\therefore \quad S = -2x^2 + 5x$

Thus $S$ is a quadratic function with $a < 0$ so the shape is 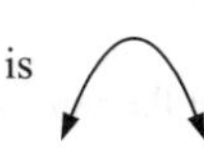

$\therefore$ the maximum separation occurs when $x = \dfrac{-b}{2a} = \dfrac{-5}{-4} = \frac{5}{4}$

and $S = -2(\frac{5}{4})^2 + 5(\frac{5}{4})$

$= -\frac{25}{8} + \frac{25}{4}$

$= \frac{25}{8}$ or $3\frac{1}{8}$

$\therefore$ the maximum vertical separation between the curves for $0 \leqslant x \leqslant 2\frac{1}{2}$ is $3\frac{1}{8}$ units.

**8** **a**

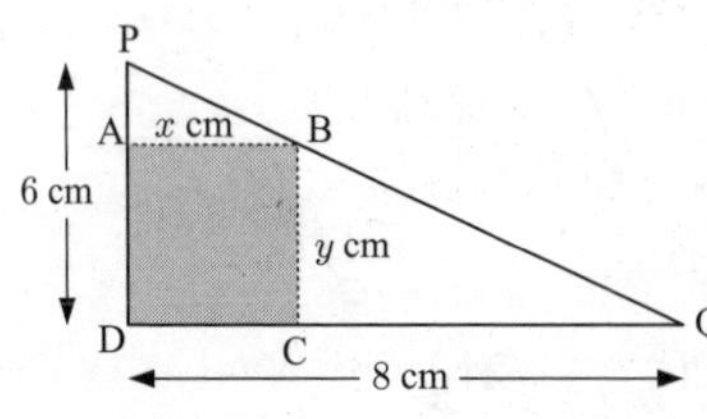


$\triangle$s PAB and PDQ are similar

$\{\widehat{APB}$ is common, $\widehat{ABP} = \widehat{DQP}$ as $[AB] \parallel [DQ]\}$

$\therefore \quad \dfrac{PA}{PD} = \dfrac{AB}{DQ}$

$\therefore \quad \dfrac{6 - y}{6} = \dfrac{x}{8}$

$\therefore \quad 6 - y = \frac{3}{4}x$

$\therefore \quad y = 6 - \frac{3}{4}x$

**b** Rectangle ABCD has area
$$\begin{aligned} A &= xy \\ &= x(6 - \tfrac{3}{4}x) \\ &= -\tfrac{3}{4}x^2 + 6x \end{aligned}$$
which is a quadratic with $a < 0$ $\therefore$ the shape is

$\therefore$ the area is maximised when $x = \dfrac{-b}{2a} = \dfrac{-6}{-\frac{3}{2}} = 4$

and when $x = 4$, $y = 6 - \frac{3}{4}(4) = 3$

$\therefore$ the dimensions of rectangle ABCD of maximum area are $4\text{ cm} \times 3\text{ cm}$.

## REVIEW SET 1A

**1** **a** The $x$-intercepts are $-2$ and $1$.

**b** The axis of symmetry lies midway between the $x$-intercepts, so its equation is $x = -\frac{1}{2}$.

**c** When $x = 0$, $y = -2(2)(-1) = 4$

$\therefore$ the $y$-intercept is $4$

**d** When $x = -\frac{1}{2}$, $y = -2(-\frac{1}{2} + 2)(-\frac{1}{2} - 1)$
$= -2(\frac{3}{2})(-\frac{3}{2}) = \frac{9}{2}$

$\therefore$ the vertex is $(-\frac{1}{2}, \frac{9}{2})$.

**e**

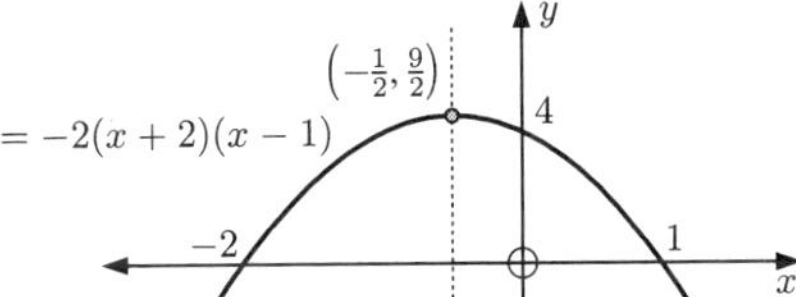


**2** **a**
$$\begin{aligned} 3x^2 - 12x &= 0 \\ \therefore\ 3x(x-4) &= 0 \\ \therefore\ x &= 0 \text{ or } 4 \end{aligned}$$

**b**
$$\begin{aligned} 3x^2 - x - 10 &= 0 \\ \therefore\ (3x+5)(x-2) &= 0 \\ \therefore\ x &= -\tfrac{5}{3} \text{ or } 2 \end{aligned}$$

**c**
$$\begin{aligned} x^2 - 11x &= 60 \\ \therefore\ x^2 - 11x - 60 &= 0 \\ \therefore\ (x+4)(x-15) &= 0 \\ \therefore\ x &= -4 \text{ or } 15 \end{aligned}$$

**3** **a** $x^2 + 5x + 3 = 0$
has $a = 1$, $b = 5$, $c = 3$

$\therefore$ $x = \dfrac{-5 \pm \sqrt{5^2 - 4(1)(3)}}{2(1)}$

$\therefore$ $x = -\frac{5}{2} \pm \frac{\sqrt{13}}{2}$

**b** $3x^2 + 11x - 2 = 0$
has $a = 3$, $b = 11$, $c = -2$

$\therefore$ $x = \dfrac{-11 \pm \sqrt{11^2 - 4(3)(-2)}}{2(3)}$

$\therefore$ $x = -\frac{11}{6} \pm \frac{\sqrt{145}}{6}$

**4**
$$x^2 + 7x - 4 = 0$$

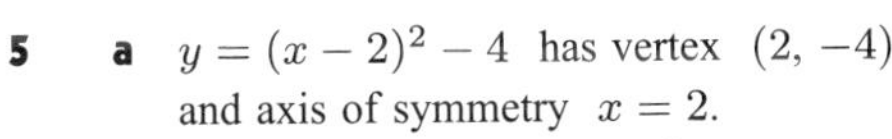

$$\begin{aligned} \therefore\ x^2 + 7x + (\tfrac{7}{2})^2 - (\tfrac{7}{2})^2 - 4 &= 0 \\ \therefore\ (x + \tfrac{7}{2})^2 - \tfrac{49}{4} - 4 &= 0 \end{aligned}$$

$$\begin{aligned} \therefore\ (x + \tfrac{7}{2})^2 &= \tfrac{65}{4} \\ \therefore\ x + \tfrac{7}{2} &= \pm\tfrac{\sqrt{65}}{2} \\ \therefore\ x &= -\tfrac{7}{2} \pm \tfrac{\sqrt{65}}{2} \end{aligned}$$

**5** **a** $y = (x-2)^2 - 4$ has vertex $(2, -4)$ and axis of symmetry $x = 2$.
When $x = 0$, $y = (-2)^2 - 4 = 0$
so the $y$-intercept is $0$.

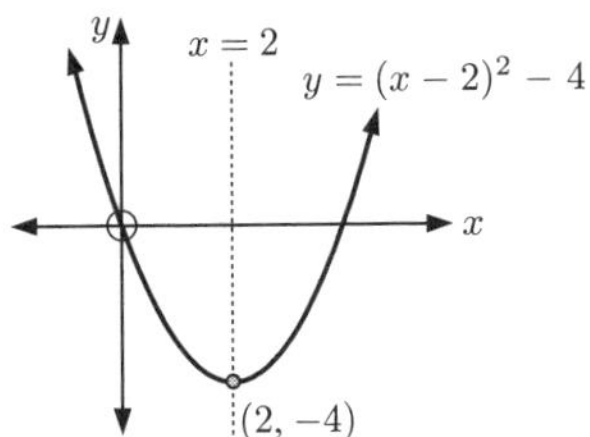


**b** $y = -\frac{1}{2}(x+4)^2 + 6$ has vertex $(-4, 6)$ and axis of symmetry $x = -4$.
When $x = 0$, $y = -\frac{1}{2}(4)^2 + 6 = -2$
so the $y$-intercept is $-2$.

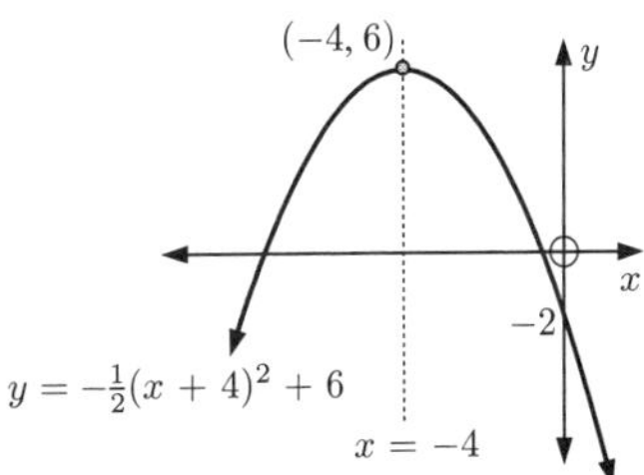

**6** **a** The graph touches the $x$-axis at 4, so its vertex is $(4, 0)$.

$\therefore$ its equation is $y = a(x-4)^2$ for some $a \neq 0$.

The graph also passes through $(2, 12)$ $\therefore$ $a(2-4)^2 = 12$

$\therefore$ $4a = 12$

$\therefore$ $a = 3$

$\therefore$ the equation is $y = 3(x-4)^2$ which is $y = 3(x^2 - 8x + 16)$

or $y = 3x^2 - 24x + 48$

**b** The quadratic has vertex $(-4, 1)$, so its equation is $y = a(x+4)^2 + 1$ for some $a \neq 0$.

The graph also passes through $(1, 11)$ $\therefore$ $11 = a(1+4)^2 + 1$

$\therefore$ $25a = 10$

$\therefore$ $a = \frac{2}{5}$

$\therefore$ the equation is $y = \frac{2}{5}(x+4)^2 + 1$ which is $y = \frac{2}{5}(x^2 + 8x + 16) + 1$

or $y = \frac{2}{5}x^2 + \frac{16}{5}x + \frac{37}{5}$

**7** $y = -2x^2 + 4x + 3$ has $a = -2$, $b = 4$, $c = 3$

Since $a < 0$, the graph has shape $\frown$ and will have a maximum.

The axis of symmetry is $x = -\dfrac{b}{2a} = -\dfrac{4}{2(-2)} = 1$

When $x = 1$, $y = -2(1)^2 + 4(1) + 3$

$= 5$

$\therefore$ the maximum is 5, and this occurs when $x = 1$.

**8** $y = x^2 - 3x$ meets $y = 3x^2 - 5x - 24$

when $x^2 - 3x = 3x^2 - 5x - 24$

$\therefore$ $2x^2 - 2x - 24 = 0$

$\therefore$ $x^2 - x - 12 = 0$

$\therefore$ $(x-4)(x+3) = 0$

$\therefore$ $x = 4$ or $-3$

Substituting into $y = x^2 - 3x$,

when $x = 4$, $y = 4^2 - 3 \times 4 = 4$

and when $x = -3$, $y = (-3)^2 - 3(-3)$

$= 9 + 9 = 18$

$\therefore$ the graphs meet at $(4, 4)$ and $(-3, 18)$.

**9** $y = -2x^2 + 5x + k$

has $a = -2$, $b = 5$, $c = k$.

$\therefore$ $\Delta = b^2 - 4ac$

$= 5^2 - 4(-2)k$

$= 25 + 8k$

The graph does not cut the $x$-axis if $\Delta < 0$

$\therefore$ $25 + 8k < 0$

$\therefore$ $8k < -25$

$\therefore$ $k < -\frac{25}{8}$

So, $k < -3\frac{1}{8}$

**10** $2x^2 - 3x + m = 0$

has $a = 2$, $b = -3$, $c = m$

$\therefore$ $\Delta = b^2 - 4ac$

$= (-3)^2 - 4(2)m$

$= 9 - 8m$

**a** There is a repeated root if $\Delta = 0$

$\therefore$ $9 - 8m = 0$

$\therefore$ $m = \frac{9}{8}$

**b** There are two distinct real roots if $\Delta > 0$

$\therefore$ $9 - 8m > 0$

$\therefore$ $8m < 9$

$\therefore$ $m < \frac{9}{8}$

**c** There are no real roots if $\Delta < 0$

$\therefore$ $9 - 8m < 0$

$\therefore$ $8m > 9$

$\therefore$ $m > \frac{9}{8}$

**11** Let the number be $x$, so its reciprocal is $\frac{1}{x}$.

$$\therefore \quad x + \frac{1}{x} = 2\tfrac{1}{30} = \tfrac{61}{30}$$
$$\therefore \quad x^2 + 1 = \tfrac{61}{30}x$$
$$\therefore \quad 30x^2 + 30 = 61x$$
$$\therefore \quad 30x^2 - 61x + 30 = 0$$
$$\therefore \quad (6x - 5)(5x - 6) = 0$$
$$\therefore \quad x = \tfrac{5}{6} \text{ or } \tfrac{6}{5} \qquad \therefore \text{ the number is } \tfrac{5}{6} \text{ or } \tfrac{6}{5}$$

**12** Let the line with $y$-intercept $(0, 10)$ have equation $y = mx + 10$.

$y = 3x^2 + 7x - 2$ meets this line when $3x^2 + 7x - 2 = mx + 10$

$$\therefore \quad 3x^2 + (7 - m)x - 12 = 0$$

For $y = mx + 10$ to be tangential to $y = 3x^2 + 7x - 2$, this equation must have exactly one solution, so there is a repeated root.

$$\therefore \quad \Delta = 0$$
$$\therefore \quad (7 - m)^2 - 4(3)(-12) = 0$$
$$\therefore \quad 49 - 14m + m^2 + 144 = 0$$
$$\therefore \quad m^2 - 14m + 193 = 0$$
$$\therefore \quad m = \frac{14 \pm \sqrt{(-14)^2 - 4(1)(193)}}{2}$$
$$\therefore \quad m = \frac{14 \pm \sqrt{-576}}{2} \quad \text{which has no real solutions.}$$

$\therefore$ no line with $y$-intercept $(0, 10)$ can be tangential to $y = 3x^2 + 7x - 2$.

**13** **a** The axis of symmetry is $x = 1$.

$$\therefore \quad x = -\frac{b}{2a} = -\frac{m}{2 \times 1} = 1 \quad \text{and so } m = -2.$$

Now when $x = 1$, $y = 3$ $\therefore$ $1^2 - 2(1) + n = 3$

$$\therefore \quad n = 4$$

So, $m = -2$ and $n = 4$.

**b** $y = x^2 - 2x + 4$

$\therefore$ when $x = 3$, $y = 3^2 - 2(3) + 4$

$= 7$

$\therefore$ $k = 7$

## REVIEW SET 1B

**1** **a**
$$\begin{aligned} y &= 2x^2 + 6x - 3 \\ &= 2[x^2 + 3x - \tfrac{3}{2}] \\ &= 2[x^2 + 3x + (\tfrac{3}{2})^2 - (\tfrac{3}{2})^2 - \tfrac{3}{2}] \\ &= 2[(x + \tfrac{3}{2})^2 - \tfrac{9}{4} - \tfrac{3}{2}] \\ &= 2[(x + \tfrac{3}{2})^2 - \tfrac{15}{4}] \\ &= 2(x + \tfrac{3}{2})^2 - \tfrac{15}{2} \end{aligned}$$

**b** The vertex is $(-\tfrac{3}{2}, -\tfrac{15}{2})$.

**c** When $x = 0$, $y = -3$

$\therefore$ the $y$-intercept is $-3$.

**d**

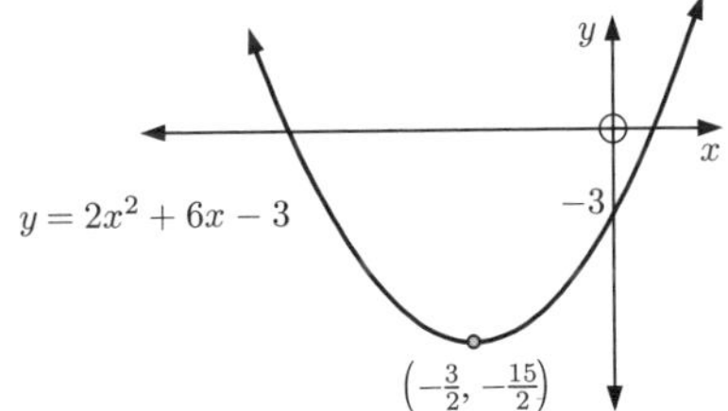


**2** **a** Using technology, $x \approx 0.586$ or $3.41$

**b** Using technology, $x \approx -0.186$ or $2.69$

**3** $y = -x^2 + 2x = x(2 - x)$

$\therefore$ the graph has $x$-intercepts 0 and 2, and $y$-intercept 0

Its axis of symmetry is midway between the $x$-intercepts, at $x = 1$, and when $x = 1$, $y = -1^2 + 2 = 1$

$\therefore$ the vertex is $(1, 1)$.

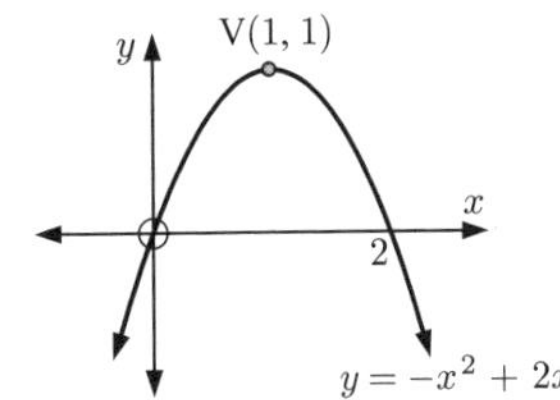

**4** $y=-3x^2+8x+7$ has $a=-3$, $b=8$, $c=7$

The axis of symmetry is $x=-\dfrac{b}{2a}=-\dfrac{8}{2(-3)}=\frac{4}{3}$

When $x=\frac{4}{3}$, $y=-3(\frac{4}{3})^2+8(\frac{4}{3})+7$

$=-\frac{16}{3}+\frac{32}{3}+7=\frac{37}{3}$

$\therefore$ the axis of symmetry is $x=\frac{4}{3}$ and the vertex is $(\frac{4}{3}, \frac{37}{3})$ or $(\frac{4}{3}, 12\frac{1}{3})$.

**5** **a** $2x^2-5x-7=0$

has $a=2$, $b=-5$, $c=-7$

$\therefore$ $\Delta=b^2-4ac$

$=(-5)^2-4(2)(-7)$

$=25+56=81$

$\therefore$ $\Delta>0$ and $\sqrt{\Delta}=9$

$\therefore$ there are two distinct real rational roots

**b** $3x^2-24x+48=0$

has $a=3$, $b=-24$, $c=48$

$\therefore$ $\Delta=b^2-4ac$

$=(-24)^2-4(3)(48)$

$=576-576$

$=0$

$\therefore$ there is a repeated real root

**6** **a** $y=3x+c$ intersects the parabola $y=x^2+x-5$ when $x^2+x-5=3x+c$

$\therefore$ $x^2-2x-5-c=0$

The graphs meet in two distinct points when this equation has two distinct real roots.

$\therefore$ $\Delta>0$

$\therefore$ $(-2)^2-4(1)(-5-c)>0$

$\therefore$ $4+20+4c>0$

$\therefore$ $4c>-24$

$\therefore$ $c>-6$

**b** Choose $c$ such that $c>-6$, for example, $c=-2$.

The graphs meet where $x^2+x-5=3x-2$

$\therefore$ $x^2-2x-3=0$

$\therefore$ $(x+1)(x-3)=0$

$\therefore$ $x=-1$ or $3$

Using the line $y=3x-2$, when $x=-1$, $y=3(-1)-2=-5$

and when $x=3$, $y=3(3)-2=7$

$\therefore$ the points of intersection are $(-1, -5)$ and $(3, 7)$.

**7** **a** $y=2x^2+4x-1$

has $a=2$, $b=4$, $c=-1$

The axis of symmetry is $x=-\dfrac{b}{2a}$

$\therefore$ $x=-\dfrac{4}{2\times 2}$

$\therefore$ $x=-1$

**b** When $x=-1$, $y=2(-1)^2+4(-1)-1$

$=2-4-1$

$=-3$

$\therefore$ the vertex is $(-1, -3)$

**c** When $x=0$, $y=-1$,

so the $y$-intercept is $-1$.

When $y=0$, $2x^2+4x-1=0$

$\therefore$ $x=\dfrac{-4\pm\sqrt{4^2-4(2)(-1)}}{2(2)}$

$=\dfrac{-4\pm\sqrt{24}}{4}$

$=\dfrac{-4\pm 2\sqrt{6}}{4}=-1\pm\frac{1}{2}\sqrt{6}$

$\therefore$ the $x$-intercepts are $-1\pm\frac{1}{2}\sqrt{6}$.

**d**

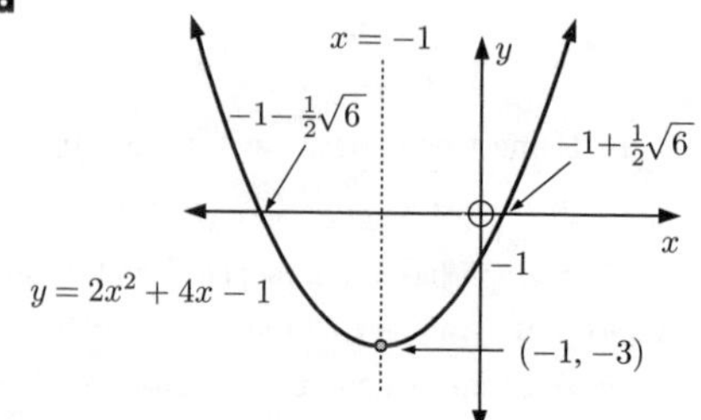

**8** Since the container has a square base, the original tinplate must have been square.

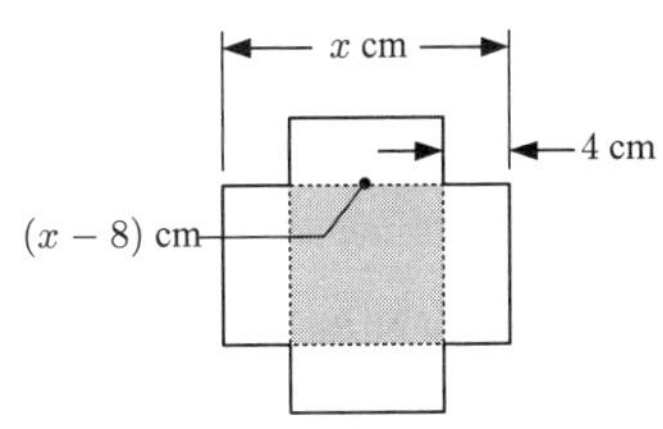


Suppose its side was $x$ cm long, so the base of the container is $(x-8)$ cm by $(x-8)$ cm.

The height of the container is 4 cm, so its capacity is $4(x-8)(x-8)$ cm$^3$.

$\therefore \quad 4(x-8)^2 = 120$

$\therefore \quad (x-8)^2 = 30$

$\therefore \quad x-8 = \pm\sqrt{30}$

$\therefore \quad x = 8 \pm \sqrt{30}$

Clearly, $x > 8$, so $x = 8 + \sqrt{30} \approx 13.48$

$\therefore$ the tinplate was about 13.5 cm by 13.5 cm.

**9** **a** $-x^2 - 5x + 3 = x^2 + 3x + 11$

$\therefore \quad 2x^2 + 8x + 8 = 0$

$\therefore \quad x^2 + 4x + 4 = 0$

$\therefore \quad (x+2)^2 = 0$

$\therefore \quad x = -2$

**b** From **a**, $x^2 + 3x + 11 = -x^2 - 5x + 3$ has only one solution.

$\therefore$ the lines touch but do not cross.

So $y = x^2 + 3x + 11$ is either above or below $y = -x^2 - 5x + 3$ for all $x \neq -2$.

If $x = 1$, $(1)^2 + 3(1) + 11 = 15$ and $-(1)^2 - 5(1) + 3 = -3$

$\therefore \quad x^2 + 3x + 11 > -x^2 - 5x + 3$ for $x \neq -2$.

**10** **a** $y = 3x^2 + 4x + 7$

has $a = 3$, $b = 4$, and $c = 7$

Since $a > 0$, the graph has shape ∪

and so has a minimum.

This occurs on the axis of symmetry

$$x = -\frac{b}{2a}$$

$$\therefore \quad x = -\frac{4}{2(3)} = -\tfrac{2}{3}$$

When $x = -\tfrac{2}{3}$,

$$y = 3(-\tfrac{2}{3})^2 + 4(-\tfrac{2}{3}) + 7$$
$$= \tfrac{4}{3} - \tfrac{8}{3} + 7$$
$$= \tfrac{17}{3} = 5\tfrac{2}{3}$$

$\therefore$ the minimum is $5\tfrac{2}{3}$ when $x = -\tfrac{2}{3}$

**b** $y = -2x^2 - 5x + 2$

has $a = -2$, $b = -5$, and $c = 2$

Since $a < 0$, the graph has shape ∩

and so has a maximum.

This occurs on the axis of symmetry

$$x = -\frac{b}{2a}$$

$$\therefore \quad x = -\frac{-5}{2(-2)} = -\tfrac{5}{4}$$

When $x = -\tfrac{5}{4}$,

$$y = -2(-\tfrac{5}{4})^2 - 5(-\tfrac{5}{4}) + 2$$
$$= -\tfrac{25}{8} + \tfrac{25}{4} + 2$$
$$= \frac{-25 + 50 + 16}{8}$$
$$= \tfrac{41}{8} = 5\tfrac{1}{8}$$

$\therefore$ the maximum is $5\tfrac{1}{8}$ when $x = -\tfrac{5}{4}$

**11** **a** The total length of fencing is $(8x+9y)$ m

$\therefore\ 8x+9y=600$

$\therefore\ 9y=600-8x$

$\therefore\ y=\dfrac{600-8x}{9}$

The area of each pen is

$A=xy$

$=x\left(\dfrac{600-8x}{9}\right)$ m$^2$

**b** $A=x\left(\dfrac{600-8x}{9}\right)$

$=\frac{600}{9}x-\frac{8}{9}x^2$

which has $a=-\frac{8}{9},\quad b=\frac{600}{9}$

Since $a<0$, $A$ is maximised at the axis of symmetry, which is $x=-\dfrac{b}{2a}$

$\therefore\ x=-\dfrac{\frac{600}{9}}{2(-\frac{8}{9})}=\dfrac{600}{16}$

$\therefore\ x=\frac{75}{2}$

When $x=\frac{75}{2},\quad y=\dfrac{600-8(\frac{75}{2})}{9}=33\frac{1}{3}$

$\therefore$ for maximum area, each pen should be $37\frac{1}{2}$ m $\times\ 33\frac{1}{3}$ m.

**c** The maximum area of each pen is

$37\frac{1}{2}\times 33\frac{1}{3}$

$=\frac{75}{2}\times\frac{100}{3}$

$=1250$ m$^2$

**12** **a** $9x^2-kx+4$ touches the $x$-axis if

$\Delta=0$

$\therefore\ (-k)^2-4(9)(4)=0$

$\therefore\ k^2-144=0$

$\therefore\ k=\pm 12$

**b** The functions intersect when

$9x^2-12x+4=9x^2+12x+4$

$\therefore\ 24x=0$

$\therefore\ x=0$

$f(0)=9(0)^2-12(0)+4=4$

The two functions intersect at $(0,\ 4)$.

## REVIEW SET 1C

**1** **a** The axis of symmetry is $x=2$.

**b** When $x=2,\quad y=\frac{1}{2}(2-2)^2-4$

$=-4$

$\therefore$ the vertex is $(2,\ -4)$

**c** When $x=0,\quad y=\frac{1}{2}(-2)^2-4$

$=-2$

$\therefore$ the $y$-intercept is $-2$

**d**

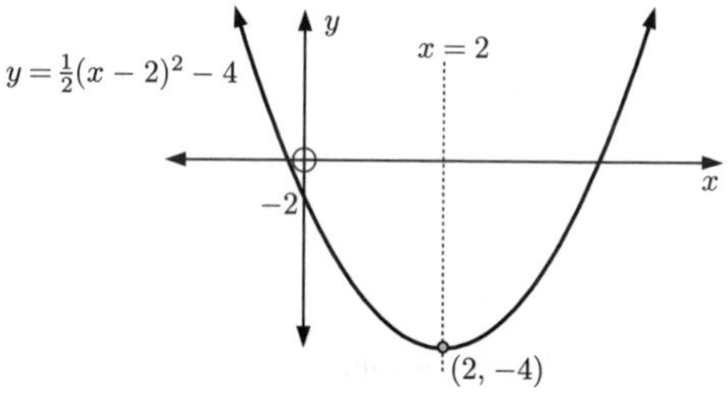

**2** **a** $x^2-5x-3=0$

has $a=1,\quad b=-5,\quad c=-3$

$\therefore\ x=\dfrac{-(-5)\pm\sqrt{(-5)^2-4(1)(-3)}}{2(1)}$

$=\frac{5}{2}\pm\frac{\sqrt{37}}{2}$

**b** $2x^2-7x-3=0$

has $a=2,\quad b=-7,\quad c=-3$

$\therefore\ x=\dfrac{-(-7)\pm\sqrt{(-7)^2-4(2)(-3)}}{2(2)}$

$=\frac{7}{4}\pm\frac{\sqrt{73}}{4}$

**3** **a** $x^2-7x+3=0$

has $a=1,\quad b=-7,\quad c=3$

$\therefore\ x=\dfrac{-(-7)\pm\sqrt{(-7)^2-4(1)(3)}}{2(1)}$

$=\dfrac{7\pm\sqrt{49-12}}{2}$

$=\dfrac{7\pm\sqrt{37}}{2}$

$\therefore\ x=\frac{7}{2}\pm\frac{\sqrt{37}}{2}$

**b** $2x^2-5x+4=0$

has $a=2,\quad b=-5,\quad c=4$

$\therefore\ x=\dfrac{-(-5)\pm\sqrt{(-5)^2-4(2)(4)}}{2(2)}$

$=\dfrac{5\pm\sqrt{25-32}}{4}$

$\therefore\ x=\dfrac{5\pm\sqrt{-7}}{4}$

$\therefore$ $x$ has no real solutions.

**4** **a** The graph has vertex $(2, -20)$, so its equation is
$y = a(x-2)^2 - 20$ for some $a \neq 0$.
Now an $x$-intercept is 5
$\therefore \quad a(5-2)^2 - 20 = 0$
$\therefore \quad 9a = 20$ and so $a = \frac{20}{9}$
So the equation is $y = \frac{20}{9}(x-2)^2 - 20$.

**c** The graph has vertex $(-3, 0)$, so its equation is $y = a(x+3)^2$ for some $a \neq 0$.
The $y$-intercept is 2
$\therefore \quad a(3)^2 = 2$
$\therefore \quad 9a = 2$ and so $a = \frac{2}{9}$
So the equation is $y = \frac{2}{9}(x+3)^2$.

**b** Since one $x$-intercept is 7 and the axis of symmetry is $x = 4$, the other $x$-intercept is $x = 1$.
$\therefore$ the graph has equation
$y = a(x-7)(x-1)$ for some $a \neq 0$.
The $y$-intercept is $-2$
$\therefore \quad a(-7)(-1) = -2$
$\therefore \quad a = -\frac{2}{7}$
$\therefore$ the equation is $y = -\frac{2}{7}(x-7)(x-1)$.

**5** **a** $y = 2x^2 + 3x - 7$
has $a = 2, \quad b = 3, \quad c = -7$
$\therefore \quad \Delta = b^2 - 4ac$
$= 3^2 - 4(2)(-7)$
$= 65$
Since $\Delta > 0$, the graph cuts the $x$-axis twice.
Note that since $a > 0$, the graph is

**b** $y = -3x^2 - 7x + 4$
has $a = -3, \quad b = -7, \quad c = 4$
$\therefore \quad \Delta = b^2 - 4ac$
$= (-7)^2 - 4(-3)4$
$= 97$
Since $\Delta > 0$, the graph cuts the $x$-axis twice.
Note that since $a < 0$, the graph is

**6** **a** $y = -2x^2 + 3x + 2$
has $a = -2, \quad b = 3, \quad c = 2$
$\therefore \quad \Delta = b^2 - 4ac$
$= 3^2 - 4(-2)(2)$
$= 25$
Since $\Delta > 0$, the function is neither positive definite nor negative definite.

**b** $y = 3x^2 + x + 11$
has $a = 3, \quad b = 1, \quad c = 11$
$\therefore \quad \Delta = b^2 - 4ac$
$= 1^2 - 4(3)(11)$
$= -131$
$\therefore \quad \Delta < 0$, and since $a > 0$, the function is positive definite.

**7** The quadratic has vertex $(2, 25)$.
$\therefore$ its equation is $y = a(x-2)^2 + 25$
The $y$-intercept is 1, so $a(-2)^2 + 25 = 1$
$\therefore \quad 4a = -24$
$\therefore \quad a = -6$
$\therefore$ the equation is $y = -6(x-2)^2 + 25$

**8** Let the line with gradient $-3$ and $y$-intercept $c$ have equation $y = -3x + c$.
$y = -3x + c$ is tangential to $y = 2x^2 - 5x + 1$ if they meet at exactly one point.
$y = 2x^2 - 5x + 1$ meets $y = -3x + c$ when $2x^2 - 5x + 1 = -3x + c$
$\therefore \quad 2x^2 - 2x + 1 - c = 0$
The graphs meet exactly once when this equation has a repeated root $\therefore \quad \Delta = 0$
$\therefore \quad (-2)^2 - 4(2)(1-c) = 0$
$\therefore \quad 4 - 8 + 8c = 0$
$\therefore \quad 8c = 4$
$\therefore \quad c = \frac{1}{2}$
$\therefore$ the $y$-intercept of the line is $\frac{1}{2}$.

**9** $y = x^2 - 2x + k$
has $a = 1$, $b = -2$, $c = k$

$\therefore \Delta = b^2 - 4ac$
$= (-2)^2 - 4(1)k$
$= 4 - 4k$

The graph cuts the $x$-axis twice if $\Delta > 0$
$\therefore 4 - 4k > 0$
$\therefore 4k < 4$
$\therefore k < 1$

**10** The $x$-intercepts are 3 and $-2$, so the equation is $y = a(x-3)(x+2)$ for some $a \neq 0$.
But the $y$-intercept is 24 $\therefore a(-3)(2) = 24$
$\therefore -6a = 24$
$\therefore a = -4$

$\therefore$ the equation is $y = -4(x-3)(x+2)$
$\therefore y = -4(x^2 - x - 6)$
$\therefore y = -4x^2 + 4x + 24$

**11** $y = mx - 10$ is a tangent to $y = 3x^2 + 7x + 2$ if they meet at exactly one point (touch).
$y = 3x^2 + 7x + 2$ meets $y = mx - 10$ when $3x^2 + 7x + 2 = mx - 10$
$\therefore 3x^2 + (7-m)x + 12 = 0$

The graphs meet exactly once when this equation has a repeated root $\therefore \Delta = 0$
$\therefore (7-m)^2 - 4(3)(12) = 0$
$\therefore 49 - 14m + m^2 - 144 = 0$
$\therefore m^2 - 14m - 95 = 0$
$\therefore (m+5)(m-19) = 0$
$\therefore m = -5$ or $19$

**12** **a** **i** The $x$-intercepts are $-m$ and $-n$. Since $m > n$, $-m < -n$.
$\therefore$ A is $(-m, 0)$, B is $(-n, 0)$.

**ii** The axis of symmetry is $x = \dfrac{-m-n}{2}$.

**b** **i** The graph cuts the $x$-axis twice, so $\Delta$ has positive sign.

**ii** The graph has shape $\cap$, so $a$ has negative sign.

**13** **a** **i** The quadratic has $x$-intercepts $\pm 3$, so its equation is
$y = a(x+3)(x-3)$ for some $a \neq 0$.
Its $y$-intercept is $-27$, so
$a(3)(-3) = -27$
$\therefore -9a = -27$
$\therefore a = 3$
$\therefore$ the equation is $y = 3(x+3)(x-3)$
or $y = 3x^2 - 27$

**ii** The straight line has gradient $\dfrac{0-(-27)}{3-0} = 9$
and its $y$-intercept is $-27$,
so its equations is $y = 9x - 27$.

**b** From the graph, the straight line is above the graph when $0 < x < 3$.

# Chapter 2
## FUNCTIONS

### EXERCISE 2A

**1** **a** $\{(1,\ 3),\ (2,\ 4),\ (3,\ 5),\ (4,\ 6)\}$ is a function since no two ordered pairs have the same $x$-coordinate.

**b** $\{(1,\ 3),\ (3,\ 2),\ (1,\ 7),\ (-1,\ 4)\}$ is not a function since two of the ordered pairs, $(1,\ 3)$ and $(1,\ 7)$, have the same $x$-coordinate 1.

**c** $\{(2,-1),(2,0),(2,3),(2,11)\}$ is not a function since each ordered pair has the same $x$-coordinate 2.

**d** $\{(7,\ 6),\ (5,\ 6),\ (3,\ 6),\ (-4,\ 6)\}$ is a function since no two ordered pairs have the same $x$-coordinate.

**e** $\{(0,\ 0),\ (1,\ 0),\ (3,\ 0),\ (5,\ 0)\}$ is a function since no two ordered pairs have the same $x$-coordinate.

**f** $\{(0,0),\ (0,-2),\ (0,2),\ (0,4)\}$ is not a function since each ordered pair has the same $x$-coordinate 0.

**2** **a**

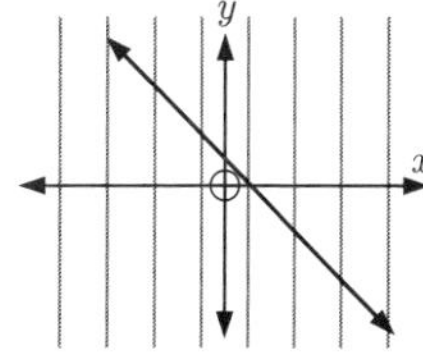

Each line cuts the graph no more than once, so it is a function.

**b**

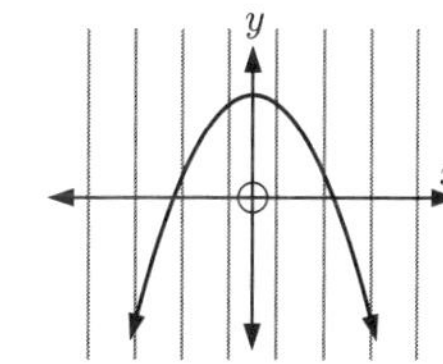

Each line cuts the graph no more than once, so it is a function.

**c**

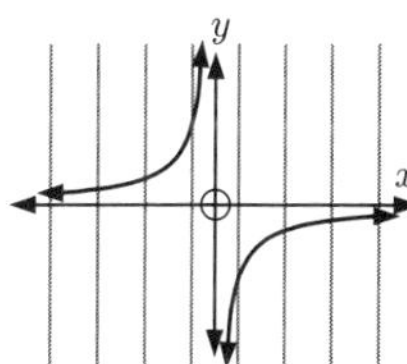

Each line cuts the graph no more than once, so it is a function.

**d**

Some lines cut the graph more than once, so it is not a function.

**e**

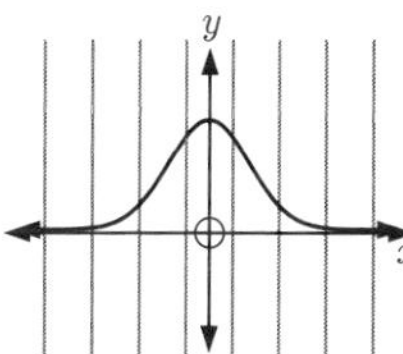

Each line cuts the graph no more than once, so it is a function.

**f**

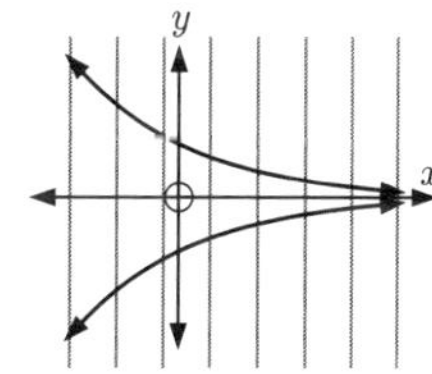

The lines cut the graph more than once, so it is not a function.

**g**

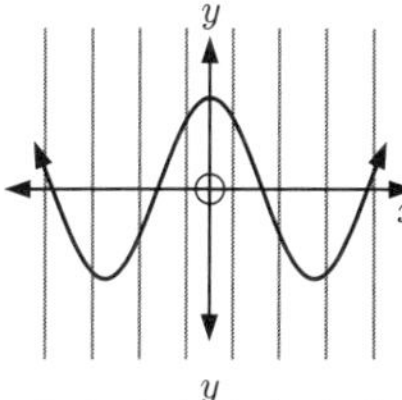

Each line cuts the graph no more than once, so it is a function.

**h**

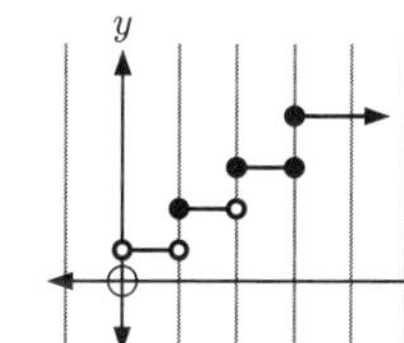

One line cuts the graph more than once, so it is not a function.

**i**

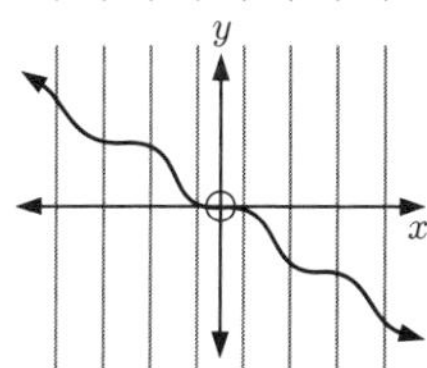

Each line cuts the graph no more than once, so it is a function.

**3** The graph of a straight line is not a function if the graph is a vertical line. So, it is not a function if it has the form $x = a$ for some constant $a$.

The vertical line through $x = a$ cuts the graph at every point, so it is not a function.

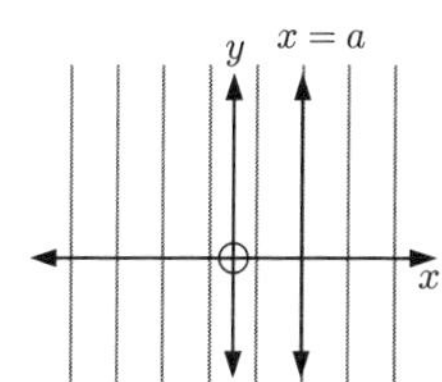

**4** $x^2 + y^2 = 9$ is the equation of a circle, centre (0, 0) and radius 3.

Now $x^2 + y^2 = 9$

$\therefore \ y^2 = 9 - x^2$

$\therefore \ y = \pm\sqrt{9 - x^2}$

For any value of $x$ where $-3 < x < 3$, $y$ has two real values. Hence $x^2 + y^2 = 9$ is not a function.

## EXERCISE 2B

**1** **a** $f(0) = 3(0) + 2 = 2$ **b** $f(2) = 3(2) + 2 = 8$ **c** $f(-1) = 3(-1) + 2 = -1$

**d** $f(-5) = 3(-5) + 2 = -13$ **e** $f(-\frac{1}{3}) = 3(-\frac{1}{3}) + 2 = 1$

**2** **a** $f(0) = 3(0) - 0^2 + 2$
$= 2$

**b** $f(3) = 3(3) - 3^2 + 2$
$= 9 - 9 + 2$
$= 2$

**c** $f(-3) = 3(-3) - (-3)^2 + 2$
$= -9 - 9 + 2$
$= -16$

**d** $f(-7) = 3(-7) - (-7)^2 + 2$
$= -21 - 49 + 2$
$= -68$

**e** $f(\frac{3}{2}) = 3(\frac{3}{2}) - (\frac{3}{2})^2 + 2$
$= \frac{9}{2} - \frac{9}{4} + 2$
$= \frac{17}{4}$

**3** **a** $g(1) = 1 - \frac{4}{1} = -3$ **b** $g(4) = 4 - \frac{4}{4} = 3$ **c** $g(-1) = -1 - \frac{4}{(-1)} = 3$

**d** $g(-4) = -4 - \frac{4}{(-4)} = -3$ **e** $g(-\frac{1}{2}) = -\frac{1}{2} - \dfrac{4}{(-\frac{1}{2})} = -\frac{1}{2} + 8 = \frac{15}{2}$

**4** **a** $f(a) = 7 - 3a$

**b** $f(-a) = 7 - 3(-a)$
$= 7 + 3a$

**c** $f(a+3) = 7 - 3(a+3)$
$= 7 - 3a - 9$
$= -3a - 2$

**d** $f(b-1) = 7 - 3(b-1)$
$= 7 - 3b + 3$
$= 10 - 3b$

**e** $f(x+2) = 7 - 3(x+2)$
$= 7 - 3x - 6$
$= 1 - 3x$

**f** $f(x+h) = 7 - 3(x+h)$
$= 7 - 3x - 3h$

**5** **a** $F(x+4)$
$= 2(x+4)^2 + 3(x+4) - 1$
$= 2(x^2 + 8x + 16) + 3x + 12 - 1$
$= 2x^2 + 16x + 32 + 3x + 11$
$= 2x^2 + 19x + 43$

**b** $F(2-x)$
$= 2(2-x)^2 + 3(2-x) - 1$
$= 2(4 - 4x + x^2) + 6 - 3x - 1$
$= 8 - 8x + 2x^2 + 5 - 3x$
$= 2x^2 - 11x + 13$

**c** $F(-x)$
$= 2(-x)^2 + 3(-x) - 1$
$= 2x^2 - 3x - 1$

**d** $F(x^2)$
$= 2(x^2)^2 + 3(x^2) - 1$
$= 2x^4 + 3x^2 - 1$

**e** $F(x^2 - 1)$
$= 2(x^2-1)^2 + 3(x^2-1) - 1$
$= 2(x^4 - 2x^2 + 1) + 3x^2 - 3 - 1$
$= 2x^4 - 4x^2 + 2 + 3x^2 - 4$
$= 2x^4 - x^2 - 2$

**f** $F(x+h)$
$= 2(x+h)^2 + 3(x+h) - 1$
$= 2(x^2 + 2xh + h^2) + 3x + 3h - 1$
$= 2x^2 + 4xh + 2h^2 + 3x + 3h - 1$

**6** **a** **i** $G(2) = \dfrac{2(2)+3}{2-4} = \dfrac{7}{-2} = -\dfrac{7}{2}$

**ii** $G(0) = \dfrac{2(0)+3}{0-4} = \dfrac{3}{-4} = -\dfrac{3}{4}$

**iii** $G(-\frac{1}{2}) = \dfrac{2(-\frac{1}{2})+3}{-\frac{1}{2}-4} = \dfrac{-1+3}{(-\frac{9}{2})} = \dfrac{2}{(-\frac{9}{2})} = -\dfrac{4}{9}$

**b** $G(x) = \dfrac{2x+3}{x-4}$ is undefined when $x - 4 = 0$

$\therefore \quad x = 4$

So, when $x = 4$, $G(x)$ does not exist.

**c** $G(x+2) = \dfrac{2(x+2)+3}{(x+2)-4} = \dfrac{2x+4+3}{x+2-4} = \dfrac{2x+7}{x-2}$

**d** $G(x) = -3$, so $\dfrac{2x+3}{x-4} = -3$ $\quad\therefore\quad 2x+3 = -3(x-4)$

$\therefore \quad 2x + 3 = -3x + 12$

$\therefore \quad 5x = 9$ and so $x = \frac{9}{5}$

**7** $f$ is the function which converts $x$ into $f(x)$ whereas $f(x)$ is the value of the function at any value of $x$.

**8** **a** $V(4) = 9650 - 860(4)$
$= 9650 - 3440$
$= 6210$

The value of the photocopier 4 years after purchase is 6210 euros.

**b** If $V(t) = 5780$,
then $9650 - 860t = 5780$
$\therefore \quad 860t = 3870$
$\therefore \quad t = 4.5$

The value of the photocopier is 5780 euros after $4\frac{1}{2}$ years.

**c** Original purchase price is when $t = 0$,
$V(0) = 9650 - 860(0)$
$= 9650$

The original purchase price was 9650 euros.

**9**

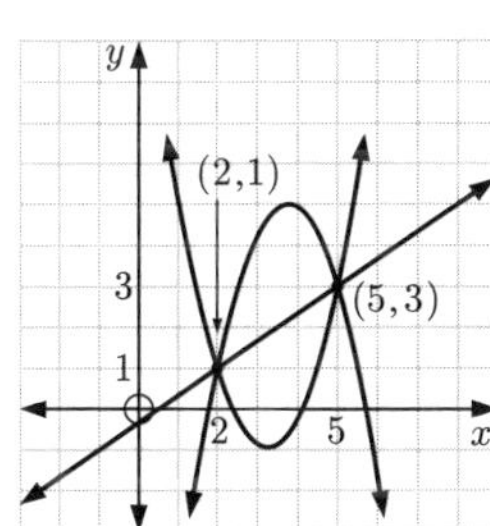


First sketch the linear function which passes through the two points (2, 1) and (5, 3).

Then sketch two quadratic functions which also pass through the two points.

**10** $f(x) = ax + b$ where $f(2) = 1$ and $f(-3) = 11$

So, $a(2) + b = 1$ and $a(-3) + b = 11$

$\therefore \quad 2a + b = 1$ $\qquad \therefore \quad -3a + b = 11$

$\therefore \quad b = 1 - 2a$ .... (1) $\qquad \therefore \quad b = 11 + 3a$ .... (2)

Solving (1) and (2) simultaneously, $1 - 2a = 11 + 3a$

$\therefore \quad 5a = -10$

$\therefore \quad a = -2$

Substituting $a = -2$ into (1) gives $b = 1 - 2(-2) = 5$. So, $a = -2$, $b = 5$

Hence $f(x) = -2x + 5$

**11** $f(x) = ax + \frac{b}{x}$ where $f(1) = 1$ and $f(2) = 5$

So, $a(1) + \frac{b}{1} = 1$ and $a(2) + \frac{b}{2} = 5$

$\therefore\ a + b = 1$

$\therefore\ a = 1 - b$ .... (1)

$\therefore\ 2a + \frac{b}{2} = 5$ .... (2)

Substituting (1) into (2), $2(1-b) + \frac{b}{2} = 5$

$\therefore\ 2 - 2b + \frac{b}{2} = 5$

$\therefore\ -\frac{3b}{2} = 3$

$\therefore\ b = -2$

Substituting $b = -2$ into (1) gives $a = 1 - (-2) = 3$.

So, $a = 3,\ b = -2$.

**12** $T(x) = ax^2 + bx + c$ where $T(0) = -4$, $T(1) = -2$, and $T(2) = 6$

So, $a(0)^2 + b(0) + c = -4$

$\therefore\ c = -4$

Also, $a(1)^2 + b(1) + c = -2$ and $a(2)^2 + b(2) + c = 6$

$\therefore\ a + b + c = -2$ and $\therefore\ 4a + 2b + c = 6$

Substituting $c = -4$ into both equations gives

$a + b + (-4) = -2$ and $4a + 2b + (-4) = 6$

$\therefore\ a + b = 2$ $\qquad$ $\therefore\ 4a + 2b = 10$ .... (2)

$\therefore\ a = 2 - b$ .... (1)

Substituting (1) into (2) gives $4(2-b) + 2b = 10$ $\quad\therefore\ 8 - 4b + 2b = 10$

$\therefore\ -2b = 2$

$\therefore\ b = -1$

Substituting $b = -1$ into (1) gives $a = 2 - (-1) = 3$.

$\therefore\ a = 3,\ b = -1,$ and $c = -4$. So, $T(x) = 3x^2 - x - 4$.

## EXERCISE 2C

**1** **a** Domain is $\{x \mid x \geqslant -1\}$
Range is $\{y \mid y \leqslant 3\}$

**b** Domain is $\{x \mid -1 < x \leqslant 5\}$
Range is $\{y \mid 1 < y \leqslant 3\}$

**c** Domain is $\{x \mid x \neq 2\}$
Range is $\{y \mid y \neq -1\}$

**d** Domain is $\{x \mid x \in \mathbb{R}\}$
Range is $\{y \mid 0 < y \leqslant 2\}$

**e** Domain is $\{x \mid x \in \mathbb{R}\}$
Range is $\{y \mid y \geqslant -1\}$

**f** Domain is $\{x \mid x \in \mathbb{R}\}$
Range is $\{y \mid y \leqslant 6\frac{1}{4}\}$ *or* $\{y \mid y \leqslant \frac{25}{4}\}$

**g** Domain is $\{x \mid x \geqslant -4\}$
Range is $\{y \mid y \geqslant -3\}$

**h** Domain is $\{x \mid x \in \mathbb{R}\}$
Range is $\{y \mid y > -2\}$

**i** Domain is $\{x \mid x \neq \pm 2\}$
Range is $\{y \mid y \leqslant -1$ or $y > 0\}$

**2** **a** $f(x)$ is defined when $x + 6 \geqslant 0$
$\therefore\ f(x)$ is defined for $x \geqslant -6$
$\therefore$ the domain is $\{x \mid x \geqslant -6\}$.

**b** $f(x)$ is defined when $x^2 \neq 0$
$\therefore\ f(x)$ is defined for $x \neq 0$
$\therefore$ the domain is $\{x \mid x \neq 0\}$.

**c** $f(x)$ is defined when $3 - 2x > 0$
$\therefore\ f(x)$ is defined for $x < \frac{3}{2}$
$\therefore$ the domain is $\{x \mid x < \frac{3}{2}\}$.

**3** **a** $y = 2x - 1$ can take any $x$-value and any $y$-value.

$\therefore$ the domain is $\{x \mid x \in \mathbb{R}\}$ and the range is $\{y \mid y \in \mathbb{R}\}$.

**b** $y = 3$ can take any value of $x$, but the only permissible value for $y$ is 3.

$\therefore$ the domain is $\{x \mid x \in \mathbb{R}\}$ and the range is $\{3\}$.

**c** $y = \sqrt{x}$ is defined when $x \geqslant 0$, and a square root cannot be negative.

$\therefore$ the domain is $\{x \mid x \geqslant 0\}$ and the range is $\{y \mid y \geqslant 0\}$.

**d** $y = \dfrac{1}{x+1}$ is defined when $x + 1 \neq 0$, or when $x \neq -1$.

$y = \dfrac{1}{x+1}$ cannot be 0 for any value of $x$.

$\therefore$ the domain is $\{x \mid x \neq -1\}$ and the range is $\{y \mid y \neq 0\}$.

**e** $y = -\dfrac{1}{\sqrt{x}}$ is defined when $x > 0$.

If $x$ is always positive, then $y = -\dfrac{1}{\sqrt{x}}$ is always negative.

$\therefore$ the domain is $\{x \mid x > 0\}$ and the range is $\{y \mid y < 0\}$.

**f** $y = \dfrac{1}{3-x}$ is defined when $3 - x \neq 0$, or when $x \neq 3$.

$y = \dfrac{1}{3-x}$ cannot be 0 for any value of $x$.

$\therefore$ the domain is $\{x \mid x \neq 3\}$ and the range is $\{y \mid y \neq 0\}$.

**4** **a**

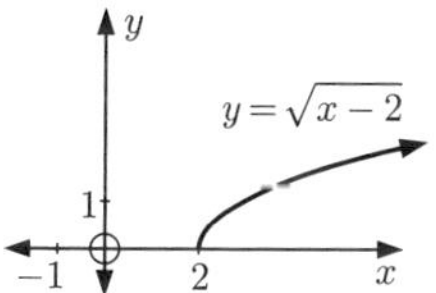


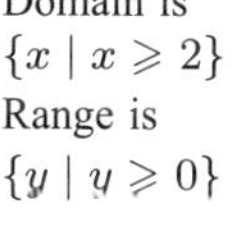

Domain is $\{x \mid x \geqslant 2\}$
Range is $\{y \mid y \geqslant 0\}$

**b**

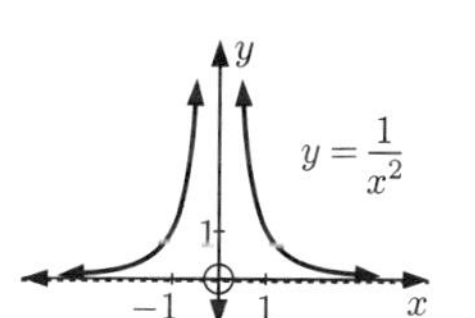


Domain is $\{x \mid x \neq 0\}$
Range is $\{y \mid y > 0\}$

**c**

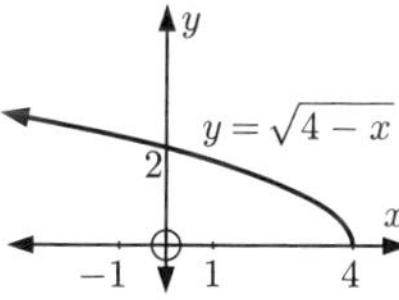


Domain is $\{x \mid x \leqslant 4\}$
Range is $\{y \mid y \geqslant 0\}$

**d**

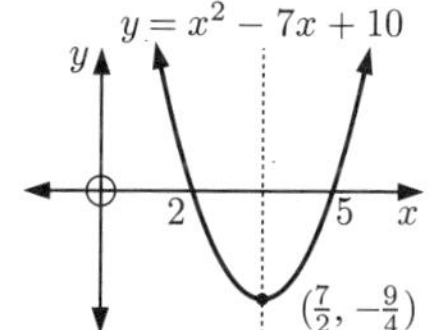


Domain is $\{x \mid x \in \mathbb{R}\}$
Range is $\{y \mid y \geqslant -2\frac{1}{4}\}$

**e**

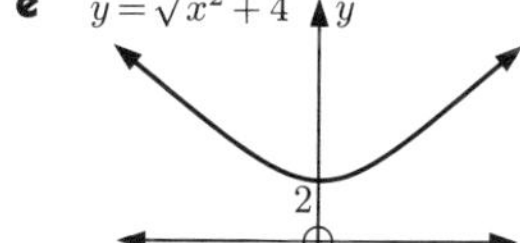


Domain is $\{x \mid x \in \mathbb{R}\}$
Range is $\{y \mid y \geqslant 2\}$

**f**

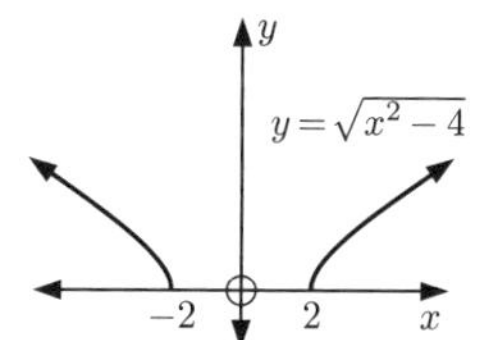


Domain is $\{x \mid x \leqslant -2$ or $x \geqslant 2\}$
Range is $\{y \mid y \geqslant 0\}$

**g**

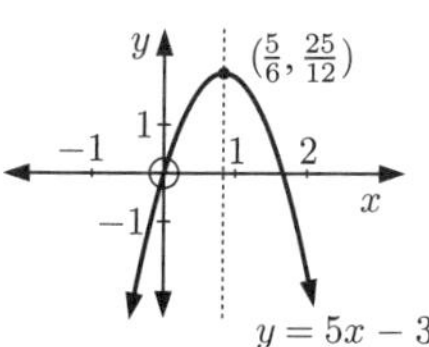


Domain is $\{x \mid x \in \mathbb{R}\}$
Range is $\{y \mid y \leqslant \frac{25}{12}\}$

**h**

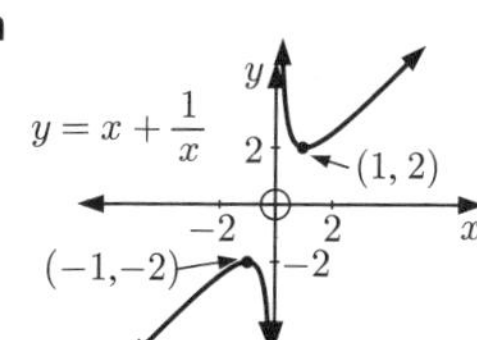


Domain is $\{x \mid x \neq 0\}$
Range is $\{y \mid y \leqslant -2$ or $y \geqslant 2\}$

**i**

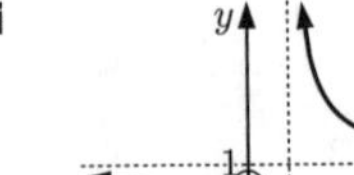

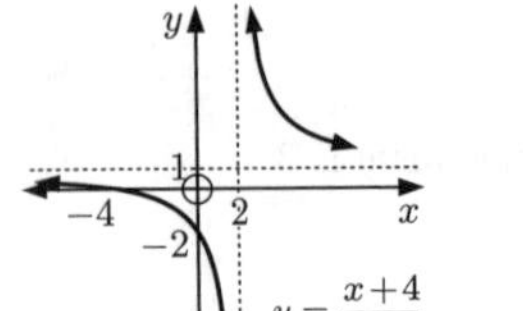


Domain is $\{x \mid x \neq 2\}$
Range is $\{y \mid y \neq 1\}$

**j**

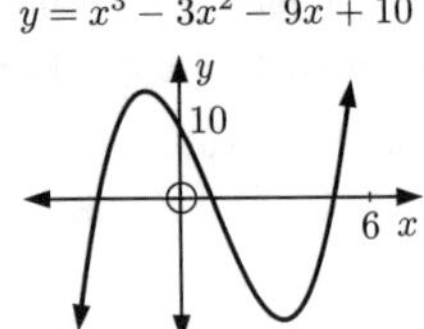


Domain is $\{x \mid x \in \mathbb{R}\}$
Range is $\{y \mid y \in \mathbb{R}\}$

**k**

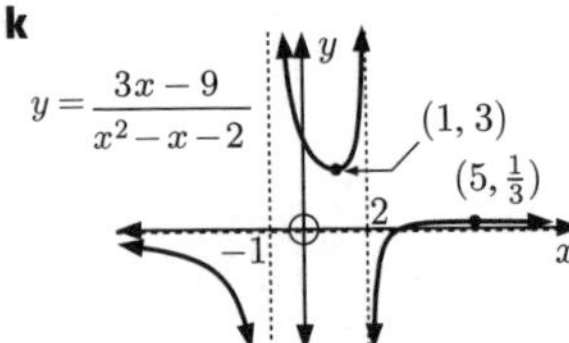


Domain is $\{x \mid x \neq -1$ and $x \neq 2\}$
Range is $\{y \mid y \leqslant \frac{1}{3}$ or $y \geqslant 3\}$

**l**

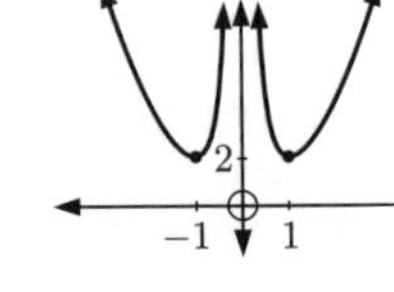


Domain is $\{x \mid x \neq 0\}$
Range is $\{y \mid y \geqslant 2\}$

**m**

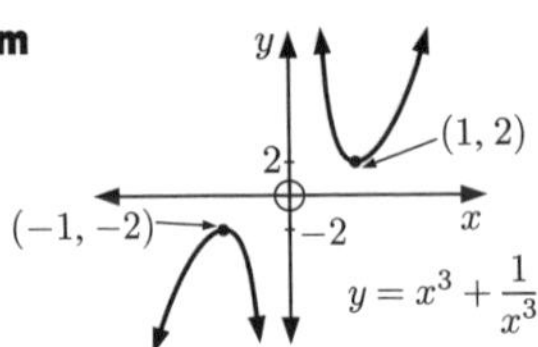

Domain is $\{x \mid x \neq 0\}$
Range is $\{y \mid y \leqslant -2$ or $y \geqslant 2\}$

**n**

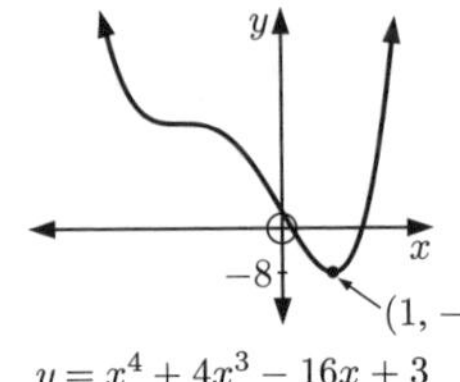


Domain is $\{x \mid x \in \mathbb{R}\}$
Range is $\{y \mid y \geqslant -8\}$

## EXERCISE 2D

**1** **a**
$$\begin{aligned}&(f \circ g)(x)\\&= f(g(x))\\&= f(1-x)\\&= 2(1-x)+3\\&= 2-2x+3\\&= 5-2x\end{aligned}$$

**b**
$$\begin{aligned}&(g \circ f)(x)\\&= g(f(x))\\&= g(2x+3)\\&= 1-(2x+3)\\&= 1-2x-3\\&= -2x-2\end{aligned}$$

**c**
$$\begin{aligned}&(f \circ g)(-3)\\&= 5-2(-3) \quad \{\text{from } \mathbf{a}\}\\&= 11\end{aligned}$$

**2** **a**
$$\begin{aligned}&(f \circ f)(x)\\&= f(f(x))\\&= f(2+x)\\&= 2+(2+x)\\&= 4+x\end{aligned}$$

**b**
$$\begin{aligned}&(f \circ g)(x)\\&= f(g(x))\\&= f(3-x)\\&= 2+(3-x)\\&= 5-x\end{aligned}$$

**c**
$$\begin{aligned}&(g \circ f)(x)\\&= g(f(x))\\&= g(2+x)\\&= 3-(2+x)\\&= 3-2-x\\&= 1-x\end{aligned}$$

**3** **a**
$$\begin{aligned}&(g \circ g)(x)\\&= g(g(x))\\&= g(5x-7)\\&= 5(5x-7)-7\\&= 25x-35-7\\&= 25x-42\end{aligned}$$

**b**
$$\begin{aligned}(f \circ g)(1) &= f(g(1))\\ \text{Now } g(1) &= 5(1)-7\\ &= -2\\ \therefore (f \circ g)(1) &= f(-2)\\ &= \sqrt{6-(-2)}\\ &= \sqrt{8}\end{aligned}$$

**c**
$$\begin{aligned}(g \circ f)(6) &= g(f(6))\\ \text{Now } f(6) &= \sqrt{6-6}\\ &= 0\\ \therefore (g \circ f)(6) &= g(0)\\ &= 5(0)-7\\ &= -7\end{aligned}$$

**4**
$$\begin{aligned}(f \circ g)(x) &= f(g(x))\\ &= f(2-x)\\ &= (2-x)^2\end{aligned}$$

Domain is $\{x \mid x \in \mathbb{R}\}$
Range is $\{y \mid y \geqslant 0\}$

$$\begin{aligned}(g \circ f)(x) &= g(f(x))\\ &= g(x^2)\\ &= 2-x^2\end{aligned}$$

Domain is $\{x \mid x \in \mathbb{R}\}$
Range is $\{y \mid y \leqslant 2\}$

**5** **a** **i** $(f \circ g)(x)$
$= f(g(x))$
$= f(3-x)$
$= (3-x)^2 + 1$
$= 9 - 6x + x^2 + 1$
$= x^2 - 6x + 10$

**ii** $(g \circ f)(x)$
$= g(f(x))$
$= g(x^2+1)$
$= 3 - (x^2+1)$
$= 3 - x^2 - 1$
$= 2 - x^2$

**b** $(g \circ f)(x) = f(x)$
$\therefore \; 2 - x^2 = f(x)$ {from **a ii**}
$\therefore \; 2 - x^2 = x^2 + 1$
$\therefore \; 2x^2 = 1$
$\therefore \; x^2 = \frac{1}{2}$
$\therefore \; x = \pm\frac{1}{\sqrt{2}}$

**6** **a** $ax + b = cx + d$ is true for all $x$ {given}
When $x = 0$, $a(0) + b = c(0) + d$
$\therefore \; b = d$ .... $(*)$

When $x = 1$, $a(1) + b = c(1) + d$
$\therefore \; a + b = c + d$
But from $(*)$, $b = d$, so $a + d = c + d$
$\therefore \; a = c$

**b** $(f \circ g)(x) = x$ for all $x$ {given}
$\therefore \; f(g(x)) = x$
$\therefore \; f(ax+b) = x$
$\therefore \; 2(ax+b) + 3 = x$
$\therefore \; 2ax + 2b + 3 = x$ for all $x$
$\therefore \; 2a = 1$ and $2b + 3 = 0$ {using **a**}
$\therefore \; a = \frac{1}{2}$ and $2b = -3$
So, $a = \frac{1}{2}$ and $b = -\frac{3}{2}$ as required.

**c** If $(g \circ f)(x) = x$
then $g(f(x)) = x$
$\therefore \; g(2x+3) = x$
$\therefore \; a(2x+3) + b = x$
$\therefore \; 2ax + 3a + b = x$
$\therefore \; 2a = 1$ and $3a + b = 0$ {using **a**}
$\therefore \; a = \frac{1}{2}$ and $b = -3a$
So, $a = \frac{1}{2}$ and $b = -\frac{3}{2}$
$\therefore$ the result in **b** is also true if $(g \circ f)(x) = x$ for all $x$.

**7** **a** $(f \circ g)(x) = f(g(x))$
$= f(x^2)$
$= \sqrt{1-x^2}$

**b** $(f \circ g)(x) = \sqrt{1-x^2}$ is defined when $1 - x^2 \geqslant 0$
$\therefore \; x^2 \leqslant 1$
$\therefore \; -1 \leqslant x \leqslant 1$
$(f \circ g)(x) = \sqrt{1-x^2}$ is always positive, and always $\leqslant 1$.
$\therefore$ the domain is $\{x \mid -1 \leqslant x \leqslant 1\}$
and the range is $\{y \mid 0 \leqslant y \leqslant 1\}$

## EXERCISE 2E

**1** **a**
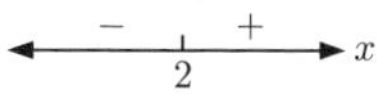


**b**
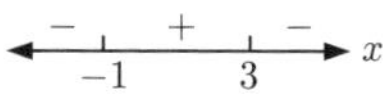


**c**
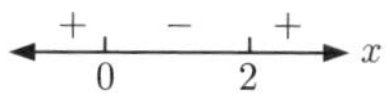


**d**
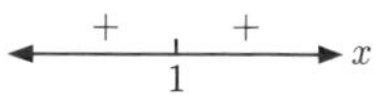


**e**
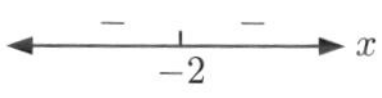


**f**
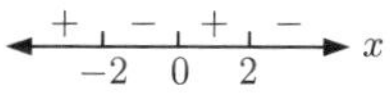


**g**
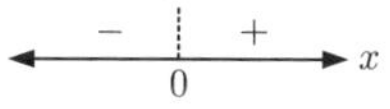


**h**
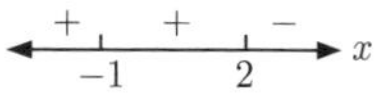


**i**
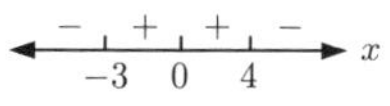


**j**
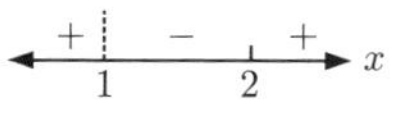


**k**
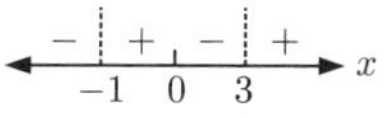


**l**
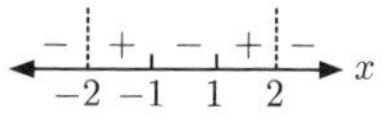


**2** **a** $y = (x+4)(x-2)$ is zero when $x = -4$ or 2.
When $x = 0$, $y = (4)(-2) = -8 < 0$.
The factors are single, so the signs alternate.
$\therefore$ sign diagram is:
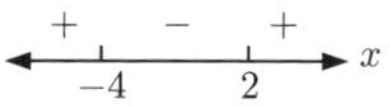


**b** $y = x(x-3)$ is zero when $x = 0$ or 3.
When $x = 10$, $y = 10(7) = 70 > 0$.
The factors are single, so the signs alternate.
$\therefore$ sign diagram is:
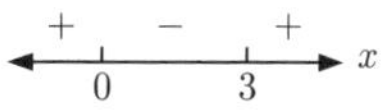

**c** $y = x(x+2)$ is zero when $x = -2$ or $0$.
When $x = 10$, $y = 10(12) = 120 > 0$.
The factors are single, so the signs alternate.

$\therefore$ sign diagram is: $+$ $-2$ $-$ $0$ $+$ $x$

**d** $y = -(x+1)(x-3)$ is zero when $x = -1$ or $3$.
When $x = 0$, $y = -(1)(-3) = 3 > 0$.
The factors are single, so the signs alternate.

$\therefore$ sign diagram is: $-$ $-1$ $+$ $3$ $-$ $x$

**e** $y = (2x-1)(3-x)$ is zero when $x = \frac{1}{2}$ or $3$.
When $x = 0$, $y = (-1)(3) = -3 < 0$.
The factors are single, so the signs alternate.

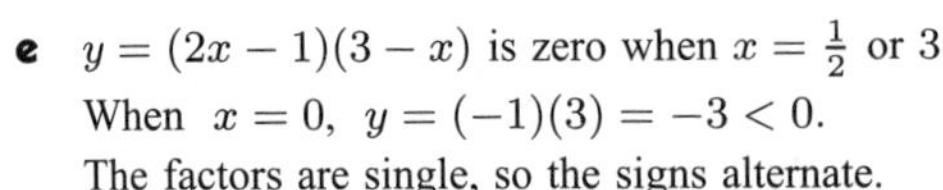

$\therefore$ sign diagram is: $-$ $\frac{1}{2}$ $+$ $3$ $-$ $x$

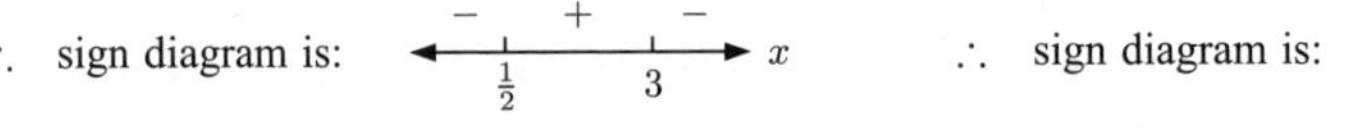

**f** $y = (5-x)(1-2x)$ is zero when $x = \frac{1}{2}$ or $5$.
When $x = 0$, $y = (5)(1) = 5 > 0$.
The factors are single, so the signs alternate.

$\therefore$ sign diagram is: $+$ $\frac{1}{2}$ $-$ $5$ $+$ $x$

**g** $y = x^2 - 9 = (x+3)(x-3)$ is zero when $x = -3$ or $3$.
When $x = 0$, $y = (3)(-3) = -9 < 0$.
The factors are single, so the signs alternate.

$\therefore$ sign diagram is: $+$ $-3$ $-$ $3$ $+$ $x$

**h** $y = 4 - x^2 = (2+x)(2-x)$ is zero when $x = -2$ or $2$.
When $x = 0$, $y = (2)(2) = 4 > 0$.
The factors are single, so the signs alternate.

$\therefore$ sign diagram is: $-$ $-2$ $+$ $2$ $-$ $x$

**i** $y = 5x - x^2 = x(5-x)$ is zero when $x = 0$ or $5$.
When $x = 10$, $y = 10(-5) = -50 < 0$.
The factors are single, so the signs alternate.

$\therefore$ sign diagram is: $-$ $0$ $+$ $5$ $-$ $x$

**j** $y = x^2 - 3x + 2 = (x-1)(x-2)$ is zero when $x = 1$ or $2$.
When $x = 0$, $y = (-1)(-2) = 2 > 0$.
The factors are single, so the signs alternate.

$\therefore$ sign diagram is: $+$ $1$ $-$ $2$ $+$ $x$

**k** $y = 2 - 8x^2 = 2(1+2x)(1-2x)$ is zero when $x = -\frac{1}{2}$ or $\frac{1}{2}$.
When $x = 0$, $y = 2(1)(1) = 2 > 0$.
The factors are single, so the signs alternate.

$\therefore$ sign diagram is:

$-$ $-\frac{1}{2}$ $+$ $\frac{1}{2}$ $-$ $x$

**l** $y = 6x^2 + x - 2 = (3x+2)(2x-1)$ is zero when $x = -\frac{2}{3}$ or $\frac{1}{2}$.
When $x = 0$, $y = (2)(-1) = -2 < 0$.
The factors are single, so the signs alternate.

$\therefore$ sign diagram is: $+$ $-\frac{2}{3}$ $-$ $\frac{1}{2}$ $+$ $x$

**m** $y = 6 - 16x - 6x^2 = 2(3+x)(1-3x)$ is zero when $x = -3$ or $\frac{1}{3}$.
When $x = 0$, $y = 2(3)(1) = 6 > 0$.
The factors are single, so the signs alternate.

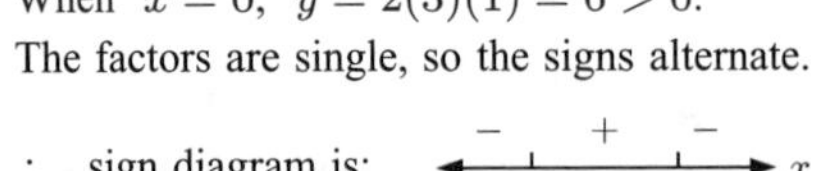

$\therefore$ sign diagram is: $-$ $-3$ $+$ $\frac{1}{3}$ $-$ $x$

**n** $y = -2x^2 + 9x + 5 = (2x+1)(5-x)$ is zero when $x = -\frac{1}{2}$ or $5$.
When $x = 0$, $y = (1)(5) = 5 > 0$.
The factors are single, so the signs alternate.

$\therefore$ sign diagram is:

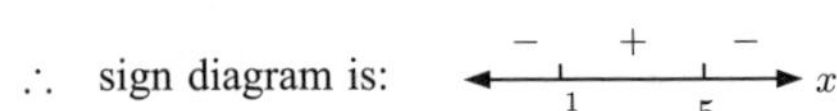

$-$ $-\frac{1}{2}$ $+$ $5$ $-$ $x$

**o** $y = -15x^2 - x + 2 = (5x+2)(1-3x)$ is zero when $x = -\frac{2}{5}$ or $\frac{1}{3}$.
When $x = 0$, $y = (2)(1) = 2 > 0$.
The factors are single, so the signs alternate.

$\therefore$ sign diagram is:

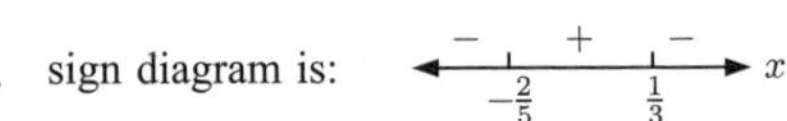

$-$ $-\frac{2}{5}$ $+$ $\frac{1}{3}$ $-$ $x$

**3** **a** $y = (x+2)^2$ is zero when $x = -2$.
When $x = 0$, $y = 2^2 = 4 > 0$.
The factor is squared, so the sign does not change.
$\therefore$ sign diagram is: $+$ $-2$ $+$ $x$

**b** $y = (x-3)^2$ is zero when $x = 3$.
When $x = 0$, $y = (-3)^2 = 9 > 0$.
The factor is squared, so the sign does not change.
$\therefore$ sign diagram is: $+$ $3$ $+$ $x$

**c** $y = -(x+2)^2$ is zero when $x = -2$.
When $x = 0$, $y = -(2^2) = -4 < 0$.
The factor is squared, so the sign does not change.
$\therefore$ sign diagram is: (− | − at −2)

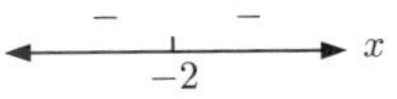

**d** $y = -(x-4)^2$ is zero when $x = 4$.
When $x = 0$, $y = -(-4)^2 = -16 < 0$.
The factor is squared, so the sign does not change.
$\therefore$ sign diagram is: (− | − at 4)

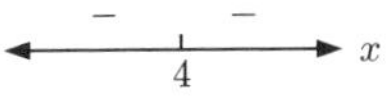

**e** $y = x^2 - 2x + 1 = (x-1)^2$ is zero when $x = 1$.
When $x = 0$, $y = (-1)^2 = 1 > 0$.
The factor is squared, so the sign does not change.
$\therefore$ sign diagram is: (+ | + at 1)

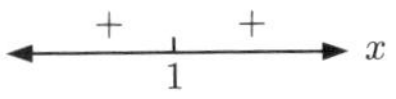

**f** $y = -x^2 + 4x - 4 = -(x-2)^2$ is zero when $x = 2$.
When $x = 0$, $y = -(-2)^2 = -4 < 0$.
The factor is squared, so the sign does not change.
$\therefore$ sign diagram is: (− | − at 2)

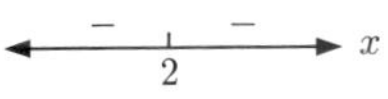

**g** $y = 4x^2 - 4x + 1 = (2x-1)^2$ is zero when $x = \frac{1}{2}$.
When $x = 0$, $y = (-1)^2 = 1 > 0$.
The factor is squared, so the sign does not change.
$\therefore$ sign diagram is: (+ | + at $\frac{1}{2}$)

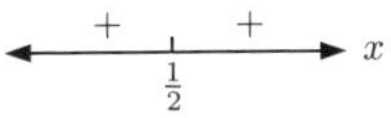

**h** $y = -x^2 - 6x - 9 = -(x+3)^2$ is zero when $x = -3$.
When $x = 0$, $y = -(3^2) = -9 < 0$.
The factor is squared, so the sign does not change.
$\therefore$ sign diagram is: (− | − at −3)

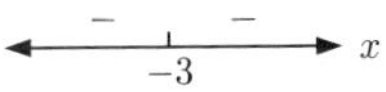

**i** $y = -4x^2 + 12x - 9 = -(2x-3)^2$ is zero when $x = \frac{3}{2}$.
When $x = 0$, $y = -(-3)^2 = -9 < 0$.
The factor is squared, so the sign does not change.
$\therefore$ sign diagram is: (− | − at $\frac{3}{2}$)

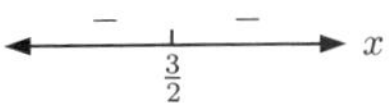

**4** **a** $y = \dfrac{x+2}{x-1}$ is zero when $x = -2$ and undefined when $x = 1$.
When $x = 0$, $y = \frac{2}{-1} = -2 < 0$.
Since the factors are single, the signs alternate.
$\therefore$ sign diagram is: (+ | − | + ; zero at −2, undefined at 1)

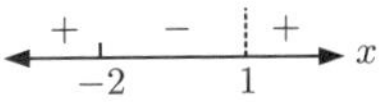

**b** $y = \dfrac{x}{x+3}$ is zero when $x = 0$ and undefined when $x = -3$.
When $x = 10$, $y = \frac{10}{13} > 0$.
Since the factors are single, the signs alternate.
$\therefore$ sign diagram is: (+ | − | + ; undefined at −3, zero at 0)

**c** $y = \dfrac{2x+3}{4-x}$ is zero when $x = -\frac{3}{2}$ and undefined when $x = 4$.
When $x = 0$, $y = \frac{3}{4} > 0$.
Since the factors are single, the signs alternate.
$\therefore$ sign diagram is: (− | + | − ; zero at $-\frac{3}{2}$, undefined at 4)

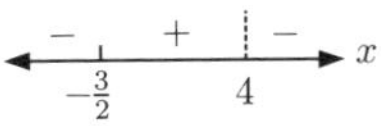

**d** $y = \dfrac{4x-1}{2-x}$ is zero when $x = \frac{1}{4}$ and undefined when $x = 2$.
When $x = 0$, $y = \frac{-1}{2} = -\frac{1}{2} < 0$.
Since the factors are single, the signs alternate.
$\therefore$ sign diagram is: (− | + | − ; zero at $\frac{1}{4}$, undefined at 2)

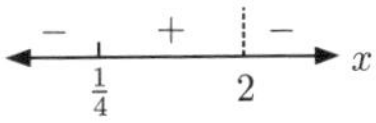

**e** $y = \dfrac{3x}{x-2}$ is zero when $x = 0$ and undefined when $x = 2$.
When $x = 5$, $y = \frac{15}{3} = 5 > 0$.
Since the factors are single, the signs alternate.
$\therefore$ sign diagram is: (+ | − | + ; zero at 0, undefined at 2)

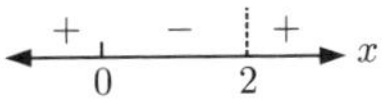

**f** $y = \dfrac{-8x}{3-x}$ is zero when $x = 0$ and undefined when $x = 3$.
When $x = 5$, $y = \frac{-40}{-2} = 20 > 0$.
Since the factors are single, the signs alternate.
$\therefore$ sign diagram is: (+ | − | + ; zero at 0, undefined at 3)

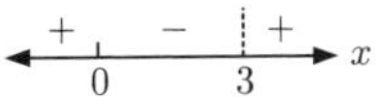

**g** $y = \dfrac{(x-1)^2}{x}$ is zero when $x = 1$ and undefined when $x = 0$.

When $x = 2$, $y = \frac{1^2}{2} = \frac{1}{2} > 0$.

Since the $(x-1)$ factor is squared, the sign does not change at $x = 1$.

$\therefore$ sign diagram is: $-$ | 0 | $+$ | 1 | $+$ ($x$)

**h** $y = \dfrac{4x}{(x+1)^2}$ is zero when $x = 0$ and undefined when $x = -1$.

When $x = 1$, $y = \frac{4}{2^2} = 1 > 0$.

Since the $(x+1)$ factor is squared, the sign does not change at $x = -1$.

$\therefore$ sign diagram is: $-$ | $-1$ | $-$ | 0 | $+$ ($x$)

**i** $y = \dfrac{(x+2)(x-1)}{3-x}$ is zero when $x = -2$ or 1 and undefined when $x = 3$.

When $x = 0$, $y = \frac{(2)(-1)}{3} = -\frac{2}{3} < 0$.

Since the factors are single, the signs alternate.

$\therefore$ sign diagram is: $+$ | $-2$ | $-$ | 1 | $+$ | 3 | $-$ ($x$)

**j** $y = \dfrac{x(x-1)}{2-x}$ is zero when $x = 0$ or 1 and undefined when $x = 2$.

When $x = 3$, $y = \frac{3(2)}{-1} = -6 < 0$.

Since the factors are single, the signs alternate.

$\therefore$ sign diagram is: $+$ | 0 | $-$ | 1 | $+$ | 2 | $-$ ($x$)

**k** $y = \dfrac{x^2-4}{-x} = \dfrac{(x-2)(x+2)}{-x}$ is zero when $x = \pm 2$ and undefined when $x = 0$.

When $x = 1$, $y = \frac{(-1)(3)}{-1} = 3 > 0$.

Since the factors are single, the signs alternate.

$\therefore$ sign diagram is: $+$ | $-2$ | $-$ | 0 | $+$ | 2 | $-$ ($x$)

**l** $y = \dfrac{3-x}{2x^2-x-6} = \dfrac{3-x}{(2x+3)(x-2)}$ is zero when $x = 3$ and undefined when $x = -\frac{3}{2}$ or 2.

When $x = 0$, $y = \frac{3}{-6} = -\frac{1}{2} < 0$.

Since the factors are single, the signs alternate.

$\therefore$ sign diagram is: $+$ | $-\frac{3}{2}$ | $-$ | 2 | $+$ | 3 | $-$ ($x$)

## EXERCISE 2F

**1** **a** **i** $f : x \mapsto \dfrac{3}{x-2}$ is undefined when $x = 2$, so $x = 2$ is a vertical asymptote.

As $|x| \to \infty$, $f(x) \to 0$, so $y = 0$ is a horizontal asymptote.

**ii** Domain is $\{x \mid x \neq 2\}$, Range is $\{y \mid y \neq 0\}$

**iii** $f(0) = \dfrac{3}{0-2} = -\frac{3}{2}$

So, the $y$-intercept is $-\frac{3}{2}$.

$f(x) = 0$ when $\dfrac{3}{x-2} = 0$,

which has no solutions.

$\therefore$ there is no $x$-intercept.

**iv** As $x \to 2^-$, $y \to -\infty$.

As $x \to 2^+$, $y \to \infty$.

As $x \to \infty$, $y \to 0^+$.

As $x \to -\infty$, $y \to 0^-$.

**v**

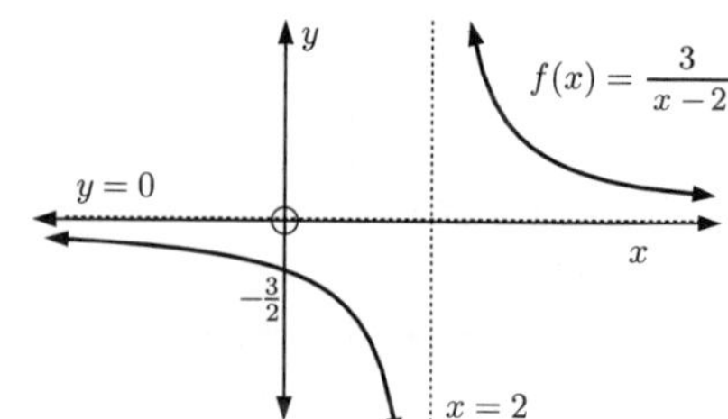


**b** **i** $f(x) = 2 - \dfrac{3}{x+1}$ is undefined when $x = -1$, so $x = -1$ is a vertical asymptote.

As $|x| \to \infty$, $\dfrac{3}{x+1} \to 0$, so $f(x) \to 2$ $\therefore$ $y = 2$ is a horizontal asymptote.

**ii** Domain is $\{x \mid x \neq -1\}$, Range is $\{y \mid y \neq 2\}$

**iii** $f(0) = 2 - \dfrac{3}{0+1} = -1$

So, the $y$-intercept is $-1$.

$f(x) = 0$ when $2 - \dfrac{3}{x+1} = 0$

$\therefore \quad \dfrac{3}{x+1} = 2$

$\therefore \quad x + 1 = \frac{3}{2}$

$\therefore \quad x = \frac{1}{2}$

So, the $x$-intercept is $\frac{1}{2}$.

**iv** As $x \to -1^-$, $y \to \infty$.
As $x \to -1^+$, $y \to -\infty$.
As $x \to \infty$, $y \to 2^-$.
As $x \to -\infty$, $y \to 2^+$.

**v**

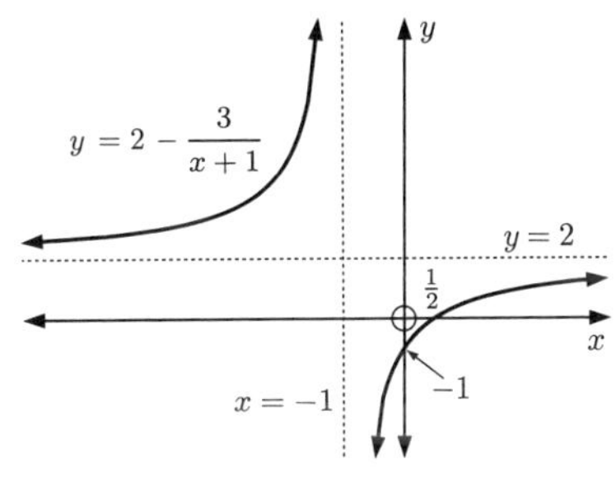


**c** **i** $f : x \mapsto \dfrac{x+3}{x-2}$ is undefined when $x = 2$, so $x = 2$ is a vertical asymptote.

Now $f(x) = \dfrac{x+3}{x-2} = \dfrac{1 + \frac{3}{x}}{1 - \frac{2}{x}}$

$\therefore$ as $|x| \to \infty$, $f(x) \to \frac{1}{1} = 1$, and so $y = 1$ is a horizontal asymptote.

**ii** Domain is $\{x \mid x \neq 2\}$, Range is $\{y \mid y \neq 1\}$

**iii** $f(0) = \dfrac{0+3}{0-2} = -\frac{3}{2}$

So, the $y$-intercept is $-\frac{3}{2}$.

$f(x) = 0$ when $\dfrac{x+3}{x-2} = 0$

$\therefore \quad x + 3 = 0$

$\therefore \quad x = -3$

So, the $x$-intercept is $-3$.

**iv** As $x \to 2^-$, $y \to -\infty$.
As $x \to 2^+$, $y \to \infty$.
As $x \to \infty$, $y \to 1^+$.
As $x \to -\infty$, $y \to 1^-$.

**v**

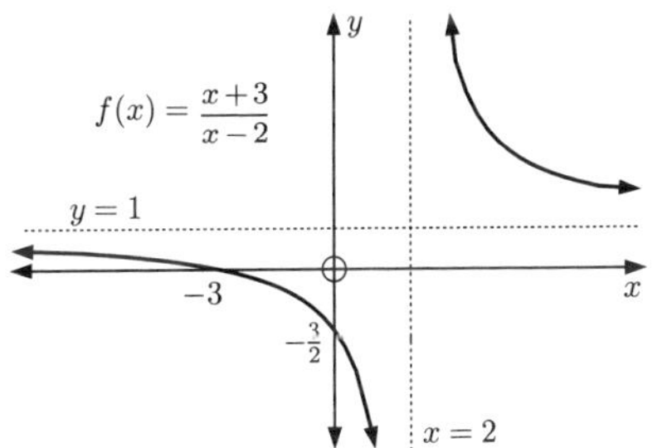


**d** **i** $f(x) = \dfrac{3x-1}{x+2}$ is undefined when $x = -2$, so $x = -2$ is a vertical asymptote.

$f(x) = \dfrac{3x-1}{x+2} = \dfrac{3 - \frac{1}{x}}{1 + \frac{2}{x}}$

As $|x| \to \infty$, $f(x) \to \frac{3}{1} = 3$ and so $y = 3$ is a horizontal asymptote.

**ii** Domain is $\{x \mid x \neq -2\}$, Range is $\{y \mid y \neq 3\}$

**iii** $f(0) = \dfrac{3(0)-1}{0+2} = -\frac{1}{2}$

So, the $y$-intercept is $-\frac{1}{2}$.

$f(x) = 0$ when $\dfrac{3x-1}{x+2} = 0$

$\therefore \quad 3x - 1 = 0$

$\therefore \quad x = \frac{1}{3}$

So, the $x$-intercept is $\frac{1}{3}$.

**iv** As $x \to -2^-$, $y \to \infty$.
As $x \to -2^+$, $y \to -\infty$.
As $x \to \infty$, $y \to 3^-$.
As $x \to -\infty$, $y \to 3^+$.

**v**

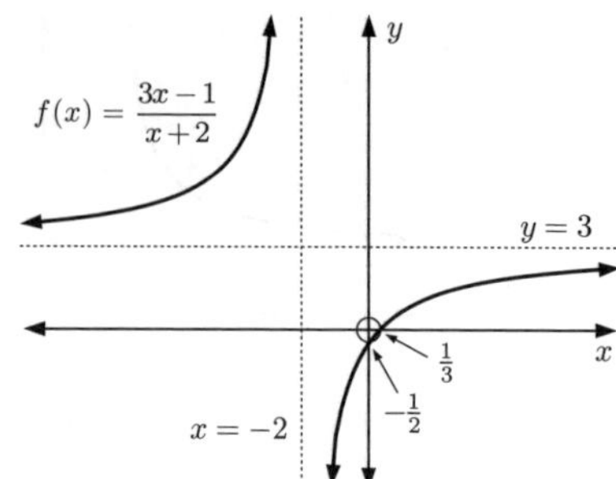


**2** **a** The function is defined when $cx+d\neq 0$, or when $x\neq -\frac{d}{c}$.

So, the domain is $\{x \mid x\neq -\frac{d}{c}\}$.

**b** The equation of the vertical asymptote is $x=-\frac{d}{c}$.

**c** To find the horizontal asymptote of $y=\frac{ax+b}{cx+d}$, we consider the function's behavior as $|x|\to\infty$.

Now $y=\frac{ax+b}{cx+d}=\frac{a+\frac{b}{x}}{c+\frac{d}{x}}$

$\therefore$ as $|x|\to\infty$, $y\to\frac{a}{c}$, and so $y=\frac{a}{c}$ is a horizontal asymptote.

## EXERCISE 2G

**1** **a** **i**

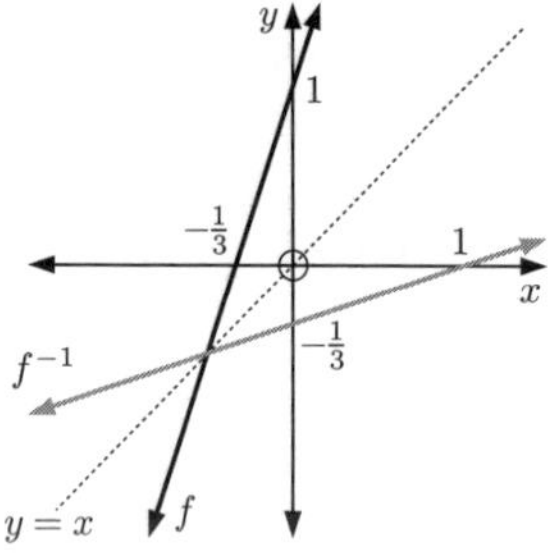


**ii** $f(x)$ passes through $(0, 1)$ and $(-\frac{1}{3}, 0)$

$\therefore$ $f^{-1}(x)$ passes through $(1, 0)$ and $(0, -\frac{1}{3})$

$f^{-1}(x)$ has gradient $\frac{-\frac{1}{3}-0}{0-1}=\frac{-\frac{1}{3}}{-1}=\frac{1}{3}$

So, its equation is $\frac{y-0}{x-1}=\frac{1}{3}$

which is $y=\frac{x-1}{3}$.

So, $f^{-1}(x)=\frac{x-1}{3}$

**iii** $f$ is $y=3x+1$

so $f^{-1}$ is $x=3y+1$

$\therefore$ $x-1=3y$

$\therefore$ $y=\frac{x-1}{3}$. So, $f^{-1}(x)=\frac{x-1}{3}$

**b** **i**

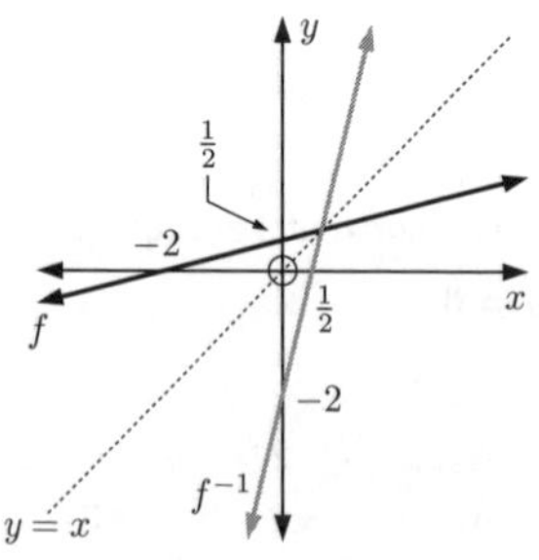


**ii** $f(x)$ passes through $(0, \frac{1}{2})$ and $(-2, 0)$

$\therefore$ $f^{-1}(x)$ passes through $(\frac{1}{2}, 0)$ and $(0, -2)$

$f^{-1}(x)$ has gradient $\frac{-2-0}{0-\frac{1}{2}}=\frac{-2}{-\frac{1}{2}}=4$

So, its equation is $\frac{y-0}{x-\frac{1}{2}}=4$

which is $y=4x-2$.

So, $f^{-1}(x)=4x-2$

**iii** $f$ is $y=\frac{x+2}{4}$

so $f^{-1}$ is $x=\frac{y+2}{4}$

$\therefore$ $4x=y+2$

$\therefore$ $y=4x-2$. So, $f^{-1}(x)=4x-2$

**2** **a** **i** $f$ is $y = 2x + 5$

so $f^{-1}$ is $x = 2y + 5$

$\therefore\ x - 5 = 2y$

$\therefore\ y = \dfrac{x-5}{2}$

So, $f^{-1}(x) = \dfrac{x-5}{2}$

**ii**

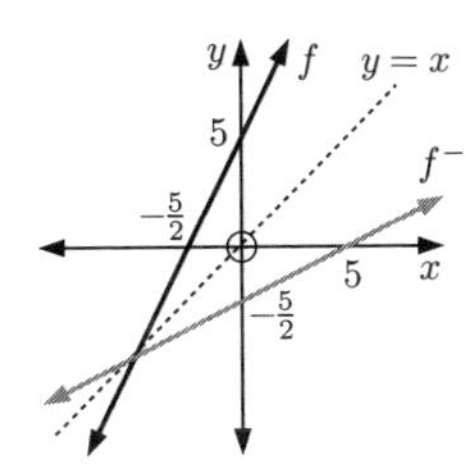


$f(x)$ passes through $(0, 5)$ and $(-\frac{5}{2}, 0)$

$\therefore\ f^{-1}(x)$ passes through $(5, 0)$ and $(0, -\frac{5}{2})$.

**iii** $(f^{-1} \circ f)(x)$ and $(f \circ f^{-1})(x)$

$$(f^{-1} \circ f)(x) = f^{-1}(2x+5) = \frac{2x+5-5}{2} = \frac{2x}{2} = x$$

$$(f \circ f^{-1})(x) = f(f^{-1}(x)) = f\left(\frac{x-5}{2}\right) = 2\left(\frac{x-5}{2}\right) + 5 = x - 5 + 5 = x$$

**b** **i** $f$ is $y = \dfrac{3-2x}{4}$

so $f^{-1}$ is $x = \dfrac{3-2y}{4}$

$\therefore\ 4x = 3 - 2y$

$\therefore\ 4x - 3 = -2y$

$\therefore\ y = -2x + \frac{3}{2}$

So, $f^{-1}(x) = -2x + \frac{3}{2}$

**ii**

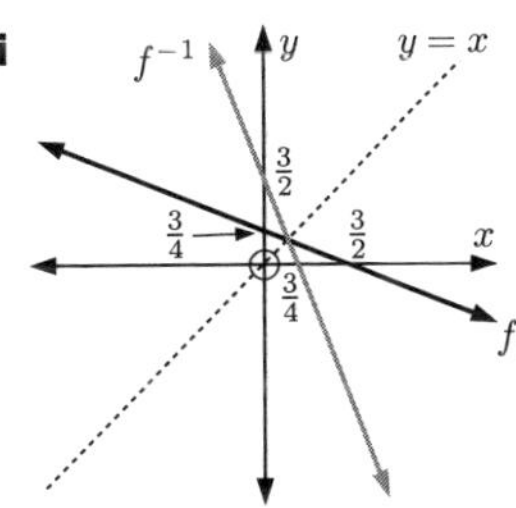


$f(x)$ passes through $(0, \frac{3}{4})$ and $(\frac{3}{2}, 0)$

$\therefore\ f^{-1}(x)$ passes through $(\frac{3}{4}, 0)$ and $(0, \frac{3}{2})$.

**iii** $(f^{-1} \circ f)(x)$ and $(f \circ f^{-1})(x)$

$$(f^{-1} \circ f)(x) = f^{-1}(f(x)) = f^{-1}\left(\frac{3-2x}{4}\right) = -2\left(\frac{3-2x}{4}\right) + \tfrac{3}{2} = \frac{3-2x}{-2} + \tfrac{3}{2} = -\tfrac{3}{2} + x + \tfrac{3}{2} = x$$

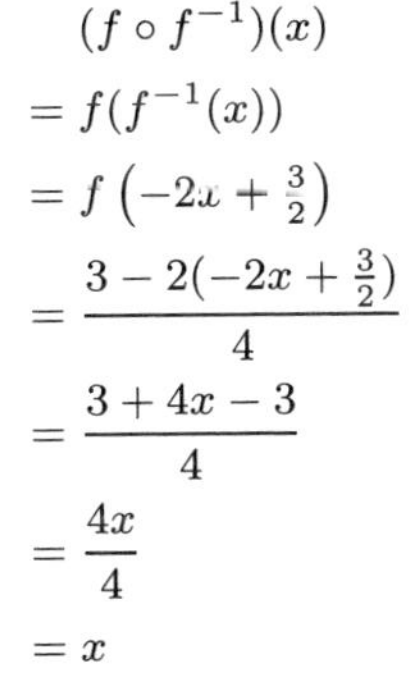

$$(f \circ f^{-1})(x) = f(f^{-1}(x)) = f\left(-2x + \tfrac{3}{2}\right) = \frac{3 - 2(-2x + \frac{3}{2})}{4} = \frac{3 + 4x - 3}{4} = \frac{4x}{4} = x$$

**c** **i**

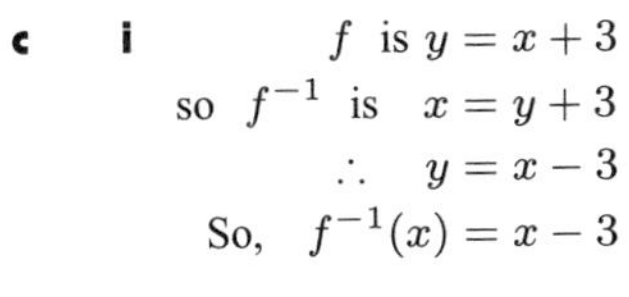

$f$ is $y = x + 3$

so $f^{-1}$ is $x = y + 3$

$\therefore\ y = x - 3$

So, $f^{-1}(x) = x - 3$

**ii**

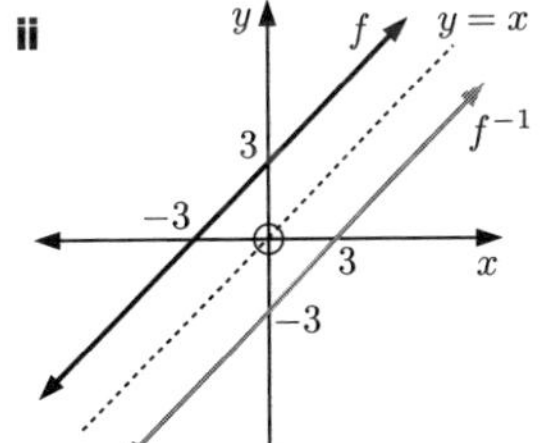


$f(x)$ passes through $(0, 3)$ and $(-3, 0)$

$\therefore\ f^{-1}(x)$ passes through $(3, 0)$ and $(0, -3)$.

**iii** $(f^{-1} \circ f)(x) = f^{-1}(f(x)) = f^{-1}(x+3) = (x+3) - 3 = x$

and $(f \circ f^{-1})(x) = f(f^{-1}(x)) = f(x-3) = (x-3) + 3 = x$

**3** **a**

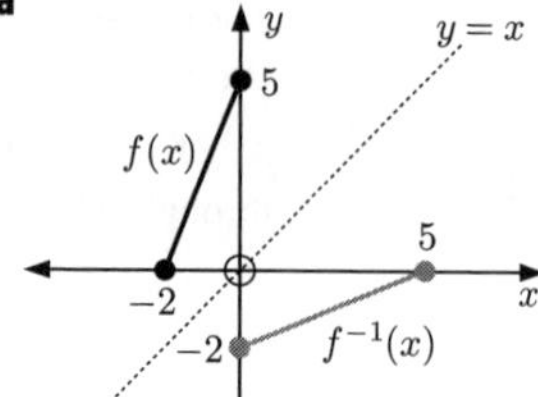


**b**

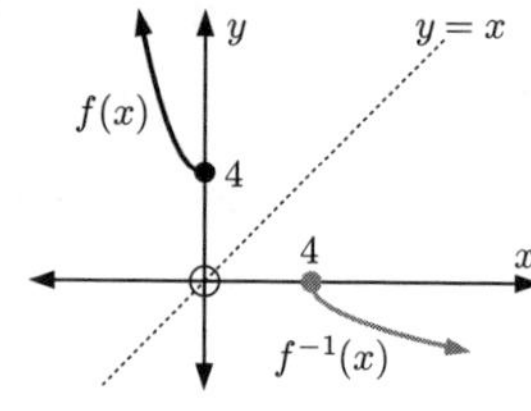


**c**

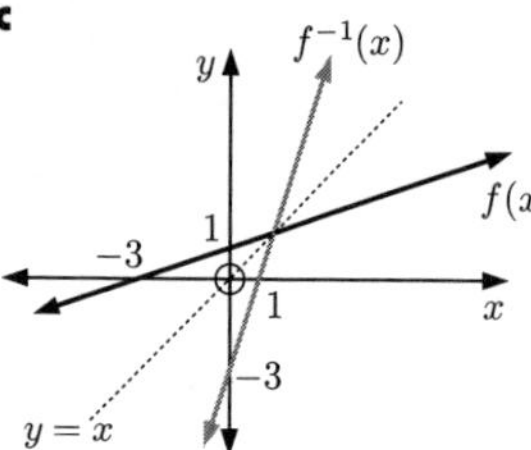


**d**

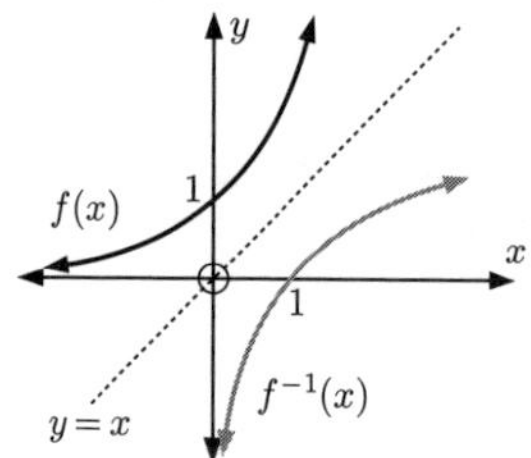


**e**

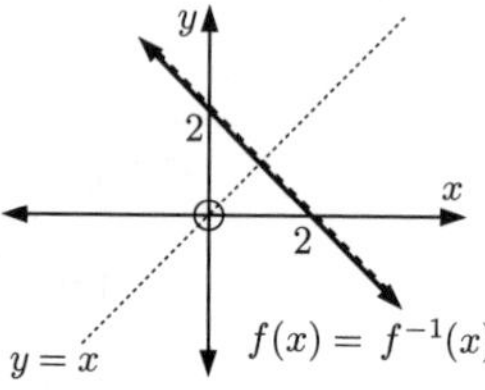


**f**

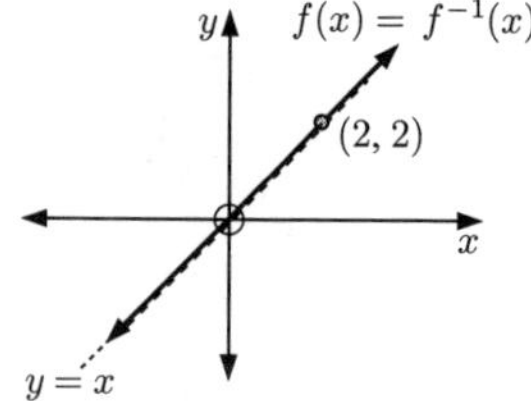


**4** **a** Domain of $f(x)$ is $\{x \mid -2 \leqslant x \leqslant 0\}$

**b** Range of $f(x)$ is $\{y \mid 0 \leqslant y \leqslant 5\}$

**c** Domain of $f^{-1}(x)$ is $\{x \mid 0 \leqslant x \leqslant 5\}$

**d** Range of $f^{-1}(x)$ is $\{y \mid -2 \leqslant y \leqslant 0\}$

**5** **a** The functions in **3 e** and **3 f** are self-inverse functions.

**b** Any linear function of the form $y = a - x$ will be a self-inverse function, for example $y = -x$ (where $a = 0$):

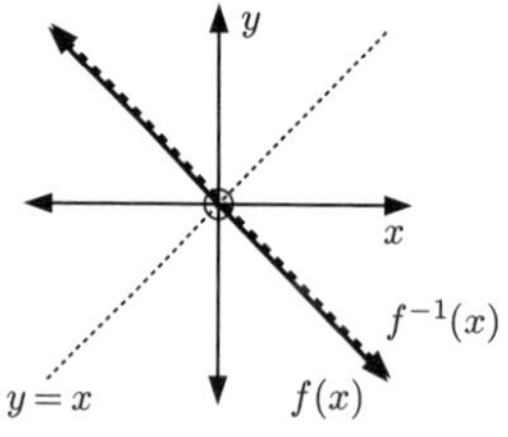


**c** Any rational function of the form $y = \dfrac{a}{x}$ will be a self-inverse function, for example $y = \dfrac{2}{x}$ (where $a = 2$):

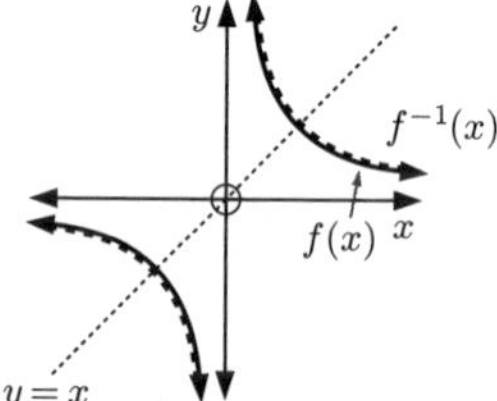


**6** Range of $H^{-1}(x)$ is $\{y \mid -2 \leqslant y < 3\}$

**7** $f$ is $y = 2x - 5$

$\therefore$ the inverse function is $x = 2y - 5$

$\therefore \quad 2y = x + 5$

$\therefore \quad y = \dfrac{x+5}{2}$

$\therefore \quad f^{-1}(x) = \dfrac{x+5}{2}$.

To find $f^{-1}(f^{-1}(x))$, we need to find the inverse function for $y = \dfrac{x+5}{2}$

This is $x = \dfrac{y+5}{2}$

$\therefore \quad 2x = y + 5$

$\therefore \quad y = 2x - 5$

This is the original function $f(x)$.

So, $f^{-1}(f^{-1}(x)) = f(x)$.

**8**

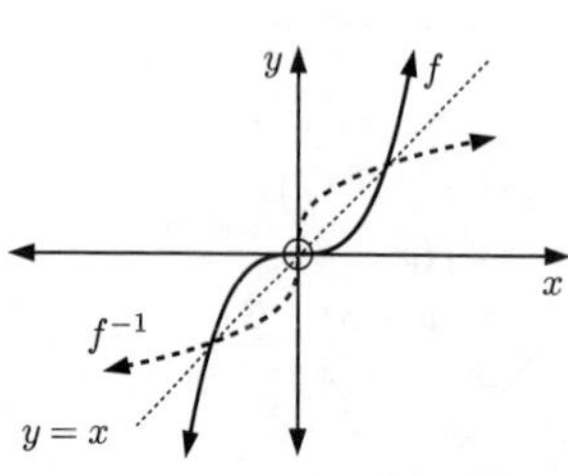

**9** $f(x) = \dfrac{1}{x}$ has inverse function $x = \dfrac{1}{y}$ or $y = \dfrac{1}{x}$

So, $f^{-1}(x) = \dfrac{1}{x}$, which means $f(x)$ is a self-inverse function.

**10** **a** $f(x) = \dfrac{3x-8}{x-3}$ has graph

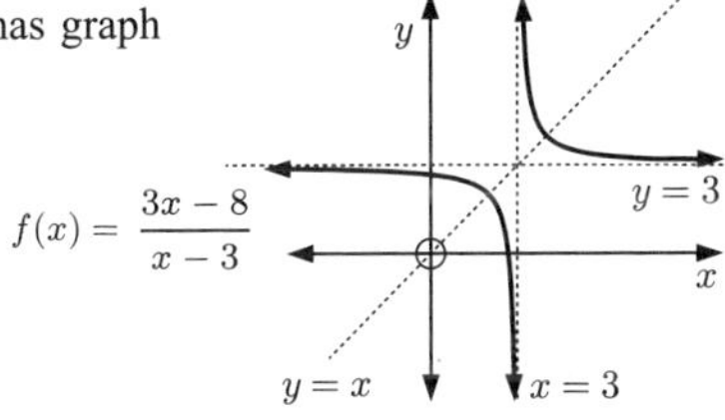


The vertical line test shows it to be a function.
Symmetry about $y = x$ shows it is a self-inverse function.

**b** $f(x) = \dfrac{3x-8}{x-3}$ has inverse function $x = \dfrac{3y-8}{y-3}$

$$\begin{aligned}
\therefore\ x(y-3) &= 3y-8 \\
\therefore\ xy - 3x &= 3y - 8 \\
\therefore\ xy - 3y &= 3x - 8 \\
\therefore\ y(x-3) &= 3x-8 \\
\therefore\ y &= \frac{3x-8}{x-3} \\
\therefore\ f^{-1}(x) &= \frac{3x-8}{x-3}
\end{aligned}$$

So, $f(x) = f^{-1}(x)$, which means $f(x)$ is a self-inverse function.

**11** **a** $f(x) = \frac{1}{2}x - 1$ has inverse function

$$\begin{aligned}
x &= \tfrac{1}{2}y - 1 \\
\therefore\ x + 1 &= \tfrac{1}{2}y \\
\therefore\ y &= 2x + 2
\end{aligned}$$

So, $f^{-1}(x) = 2x + 2$

**b** **i**

$$\begin{aligned}
&(f \circ f^{-1})(x) \\
&= f(f^{-1}(x)) \\
&= f(2x+2) \quad \{\text{using } \mathbf{a}\} \\
&= \tfrac{1}{2}(2x+2) - 1 \\
&= x + 1 - 1 \\
&= x
\end{aligned}$$

**ii**

$$\begin{aligned}
&(f^{-1} \circ f)(x) \\
&= f^{-1}(f(x)) \\
&= f^{-1}(\tfrac{1}{2}x - 1) \\
&= 2(\tfrac{1}{2}x - 1) + 2 \quad \{\text{using } \mathbf{a}\} \\
&= x - 2 + 2 \\
&= x
\end{aligned}$$

**12** **a**

$g$ is $y = \dfrac{8-x}{2}$

so $g^{-1}$ is $x = \dfrac{8-y}{2}$

$$\begin{aligned}
\therefore\ 2x &= 8 - y \\
\therefore\ y &= 8 - 2x
\end{aligned}$$

So, $g^{-1}(x) = 8 - 2x$

$\therefore\ g^{-1}(-1) = 8 - 2(-1) = 10$

**b**

$f$ is $y = 2x + 5$

so $f^{-1}$ is $x = 2y + 5$

$$\begin{aligned}
\therefore\ 2y &= x - 5 \\
\therefore\ y &= \frac{x-5}{2}
\end{aligned}$$

So, $f^{-1}(x) = \dfrac{x-5}{2}$

$$\begin{aligned}
\therefore\ f^{-1}(-3) &= \frac{-3-5}{2} \\
&= \frac{-8}{2} \\
&= -4
\end{aligned}$$

and

$$\begin{aligned}
g^{-1}(6) &= 8 - 2 \times 6 \\
&= 8 - 12 \\
&= -4
\end{aligned}$$

$$\begin{aligned}
\therefore\ f^{-1}(-3) - g^{-1}(6) &= -4 - (-4) \\
&= -4 + 4 \\
&= 0 \quad \text{as required.}
\end{aligned}$$

**c**

$$\begin{aligned}
(f \circ g^{-1})(x) &= 9 \\
\therefore\ f(g^{-1}(x)) &= 9 \\
\therefore\ f(8 - 2x) &= 9 \\
\therefore\ 2(8-2x) + 5 &= 9 \\
\therefore\ 16 - 4x + 5 &= 9 \\
\therefore\ -4x &= -12 \\
\therefore\ x &= 3
\end{aligned}$$

**13** **a** **i** $f$ is $y = 5^x$
so $f(2) = 5^2$
$= 25$

**ii** $g$ is $y = \sqrt{x}$ where $y \geqslant 0$
so $g^{-1}$ is $x = \sqrt{y}$ where $x \geqslant 0$
$\therefore\ y = x^2$
$\therefore\ g^{-1}(x) = x^2,\ x \geqslant 0$
$\therefore\ g^{-1}(4) = 4^2$
$= 16$

**b** $(g^{-1} \circ f)(x) = 25$
$\therefore\ g^{-1}(f(x)) = 25$
$\therefore\ g^{-1}(5^x) = 25$
$\therefore\ (5^x)^2 = 25$ $\{$as $g^{-1}(x) = x^2,\ x \geqslant 0\}$
$\therefore\ 5^{2x} = 5^2$
$\therefore\ 2x = 2$
$\therefore\ x = 1$

**14** $f$ is $y = 2x$
so $f^{-1}$ is $x = 2y$
$\therefore\ y = \frac{x}{2}$
$\therefore\ f^{-1}(x) = \frac{x}{2}$

$g$ is $y = 4x - 3$
so $g^{-1}$ is $x = 4y - 3$
$\therefore\ 4y = x + 3$
$\therefore\ y = \frac{x+3}{4}$
$\therefore\ g^{-1}(x) = \frac{x+3}{4}$

$(g \circ f)(x) = g(f(x))$
$= g(2x)$
$= 4(2x) - 3$
$\therefore\ (g \circ f)(x) = 8x - 3$
$\therefore\ g \circ f$ is $y = 8x - 3$
so $(g \circ f)^{-1}$ is $x = 8y - 3$
$\therefore\ y = \frac{x+3}{8}$
So, $(g \circ f)^{-1}(x) = \frac{x+3}{8}$

Now $(f^{-1} \circ g^{-1})(x) = f^{-1}(g^{-1}(x))$
$= f^{-1}\left(\frac{x+3}{4}\right)$
$= \frac{\left(\frac{x+3}{4}\right)}{2}$
$\therefore\ (f^{-1} \circ g^{-1})(x) = \frac{x+3}{8} = (g \circ f)^{-1}(x)$ as required

**15** **a** $f$ is $y = 2x$
so $f^{-1}$ is $x = 2y$
$\therefore\ y = \frac{x}{2}$
so $f^{-1}(x) = \frac{x}{2} \neq 2x$
So, $f^{-1}(x) \neq f(x)$

**b** $f$ is $y = x$
so $f^{-1}$ is $x = y$
$\therefore\ y = x$
so $f^{-1}(x) = x$
So, $f^{-1}(x) = f(x)$

**c** $f$ is $y = -x$
so $f^{-1}$ is $x = -y$
$\therefore\ y = -x$
so $f^{-1}(x) = -x$
So, $f^{-1}(x) = f(x)$

**d** $f$ is $y = \frac{2}{x}$
so $f^{-1}$ is $x = \frac{2}{y}$
$\therefore\ y = \frac{2}{x}$
so $f^{-1}(x) = \frac{2}{x}$
So, $f^{-1}(x) = f(x)$

**e** $f$ is $y = -\frac{6}{x}$
so $f^{-1}$ is $x = -\frac{6}{y}$
$\therefore\ y = -\frac{6}{x}$
so $f^{-1}(x) = -\frac{6}{x}$
So, $f^{-1}(x) = f(x)$

So, $f^{-1}(x) = f(x)$ is true for parts **b**, **c**, **d**, and **e**.

**16** **a** If $y = f(x)$ has an inverse function, then the inverse function must also be a function. It must satisfy the 'vertical line test', so no vertical line can cut it more than once. This condition for the inverse function cannot be satisfied if the original function does not satisfy the 'horizontal line test'. Thus, the 'horizontal line test' is a valid test for the existence of an inverse function.

**b** **i** This graph satisfies the 'horizontal line test' and therefore has an inverse function.

**ii, iii** These graphs both fail the 'horizontal line test' so neither of these have inverse functions.

**i**

**ii**

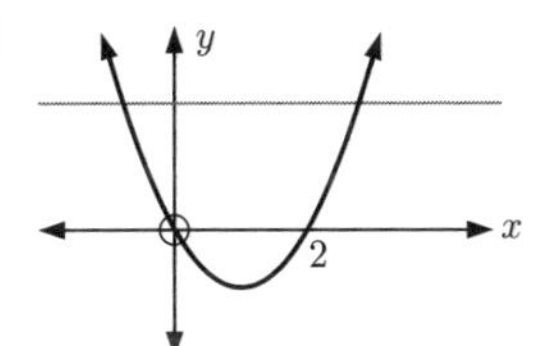


**iii**

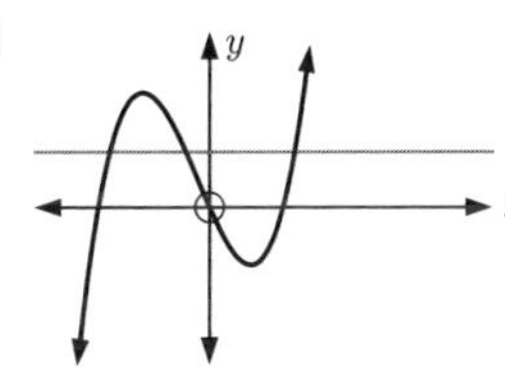


## REVIEW SET 2A

**1** **a** **i** Domain is $\{x \mid x \in \mathbb{R}\}$.

**ii** Range is $\{y \mid y > -4\}$.

**iii**

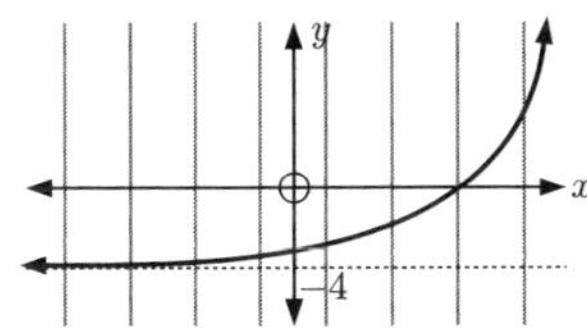


Each line cuts the graph no more than once, so the graph shows a function.

**b** **i** Domain is $\{x \mid x \in \mathbb{R}\}$.

**ii** Range is $\{2\}$.

**iii**

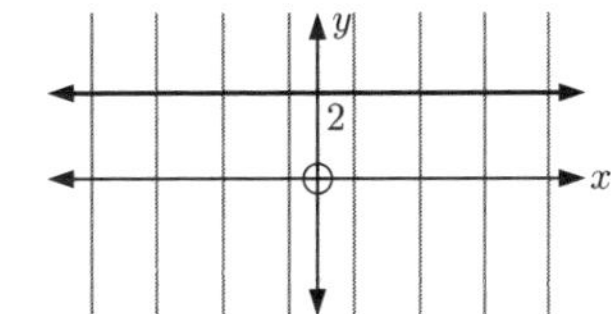


Each line cuts the graph no more than once, so the graph shows a function.

**c** **i** Domain is $\{x \mid x \in \mathbb{R}\}$.

**ii** Range is $\{y \mid y \leqslant -1 \text{ or } y \geqslant 1\}$.

**iii**

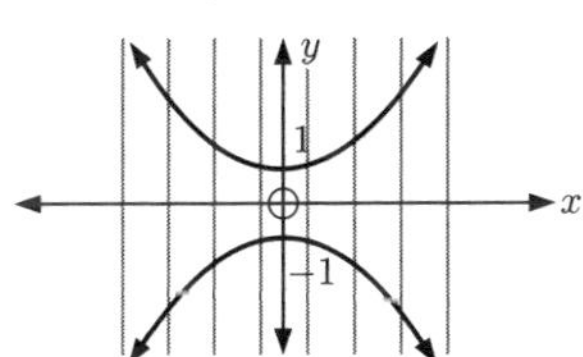


The lines cut the graph more than once, so the graph does not show a function.

**d** **i** Domain is $\{x \mid x \in \mathbb{R}\}$.

**ii** Range is $\{y \mid -5 \leqslant y \leqslant 5\}$.

**iii**

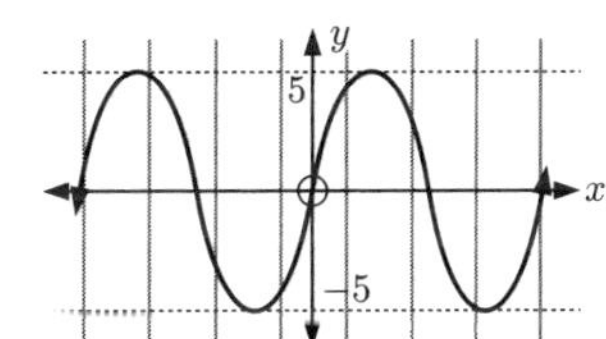


Each line cuts the graph no more than once, so the graph shows a function.

**2** **a** $f(x) = 2x - x^2$
$f(2) = 2(2) - 2^2$
$= 0$

**b** $f(-3) = 2(-3) - (-3)^2$
$= -6 - 9$
$= -15$

**c** $f(-\frac{1}{2}) = 2(-\frac{1}{2}) - (-\frac{1}{2})^2$
$= -1 - \frac{1}{4}$
$= -\frac{5}{4}$

**3** $f(x) = ax + b$, where $f(1) = 7$ and $f(3) = -5$

When $f(1) = 7$,
$7 = a(1) + b$
$\therefore\ 7 = a + b$
$\therefore\ a = 7 - b$ .... (1)

When $f(3) = -5$,
$-5 = a(3) + b$
$\therefore\ -5 = 3a + b$
$\therefore\ -5 = 3(7 - b) + b$ {using (1)}
$\therefore\ -5 = 21 - 3b + b$
$\therefore\ 2b = 26$ and so $b = 13$

Substituting $b = 13$ into (1), $a = 7 - 13 = -6$

$\therefore\ a = -6$ and $b = 13$

**4** $g(x) = x^2 - 3x$

**a** $g(x+1) = (x+1)^2 - 3(x+1)$
$= x^2 + 2x + 1 - 3x - 3$
$= x^2 - x - 2$

**b** $g(x^2 - 2) = (x^2 - 2)^2 - 3(x^2 - 2)$
$= x^4 - 4x^2 + 4 - 3x^2 + 6$
$= x^4 - 7x^2 + 10$

**5** **a** **i** Domain is $\{x \mid x \in \mathbb{R}\}$. Range is $\{y \mid y \geqslant -5\}$.

**ii** $x$-intercepts are $-1$ and $5$, $y$-intercept is $-\frac{25}{9}$

**iii** The graph passes the 'vertical line test' so is therefore a function.

**b** **i** Domain is $\{x \mid x \in \mathbb{R}\}$. Range is $\{y \mid y = 1 \text{ or } -3\}$.

**ii** There are no $x$-intercepts, $y$-intercept is 1.

**iii** The graph passes the 'vertical line test' so is therefore a function.

**6** **a** $y = (3x+2)(4-x)$ is zero when $x = -\frac{2}{3}$ or 4.

When $x = 0$, $y = (2)(4) = 8 > 0$.

Since the factors are single, the signs alternate.

$\therefore$ sign diagram is: $-$ | $-\frac{2}{3}$ | $+$ | $4$ | $-$ ($x$)

**b** $y = \dfrac{x-3}{x^2+4x+4} = \dfrac{x-3}{(x+2)^2}$ is zero when $x = 3$ and undefined when $x = -2$.

When $x = 0$, $y = \dfrac{-3}{2^2} = -\frac{3}{4} < 0$.

Since the $(x+2)$ factor is squared, the sign does not change at $x = -2$

$\therefore$ sign diagram is: $-$ | $-2$ | $-$ | $3$ | $+$ ($x$)

**7** $f(x) = ax + b$

Now $f(2) = 1$, so $a(2) + b = 1$

$\therefore\ b = 1 - 2a \quad .... (*)$

Now $f^{-1}(3) = 4$, so $f(4) = 3$

$\therefore\ a(4) + b = 3$

$\therefore\ 4a + (1 - 2a) = 3 \quad \{\text{from } (*)\}$

$\therefore\ 2a = 2$

$\therefore\ a = 1$

Substituting $a = 1$ into $(*)$, $b = 1 - 2(1) = -1$

So, $a = 1$ and $b = -1$.

**8** **a**

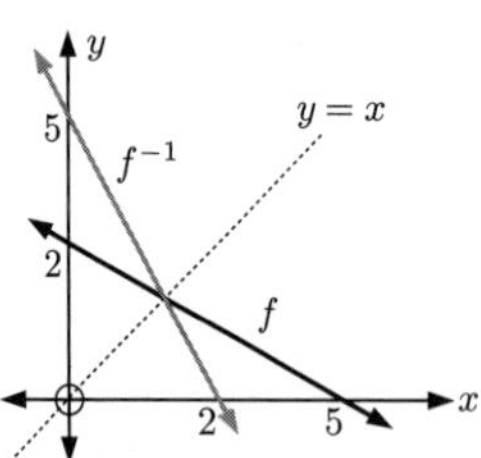


**b**

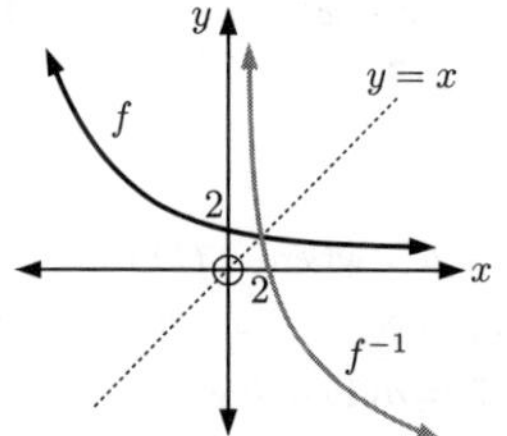


**9** **a** $f$ is $y = 4x + 2$

$\therefore\ f^{-1}(x)$ is $x = 4y + 2$

$\therefore\ y = \dfrac{x-2}{4}$

$\therefore\ f^{-1}(x) = \dfrac{x-2}{4}$

**b** $f$ is $y = \dfrac{3-5x}{4}$

so $f^{-1}(x)$ is $x = \dfrac{3-5y}{4}$

$\therefore\ 4x = 3 - 5y$

$\therefore\ y = \dfrac{3-4x}{5}$

$\therefore\ f^{-1}(x) = \dfrac{3-4x}{5}$

**10** **a** $f(x) = x^2$

$\therefore\ f(-3) = (-3)^2 = 9$

$g(x) = 1 - 6x$

$\therefore\ g(-\frac{4}{3}) = 1 - 6(-\frac{4}{3})$

$= 1 + 8 = 9$

$\therefore\ f(-3) = g(-\frac{4}{3})$

**b** $(f \circ g)(-2) = f(g(-2))$

Now, $g(-2) = 1 - 6(-2)$

$= 13$

$\therefore\ (f \circ g)(-2) = f(13)$

$= 13^2$

$= 169$

**c** $f(5) = 5^2 = 25$

So, we need to find $x$ such that $1 - 6x = 25$

$\therefore\ -6x = 24$

$\therefore\ x = -4$

**11** $f$ is $y = 3x + 6$

so $f^{-1}(x)$ is $x = 3y + 6$

$\therefore\ y = \dfrac{x-6}{3}$

$\therefore\ f^{-1}(x) = \dfrac{x-6}{3}$

$h$ is $y = \dfrac{x}{3}$

so $h^{-1}(x)$ is $x = \dfrac{y}{3}$

$\therefore\ y = 3x$

$\therefore\ h^{-1}(x) = 3x$

Now $(f^{-1} \circ h^{-1})(x) = f^{-1}(h^{-1}(x))$

$= f^{-1}(3x)$

$= \dfrac{3x-6}{3}$

$= x - 2$

$(h \circ f)(x) = h(f(x))$

$= h(3x+6)$

$= \dfrac{3x+6}{3}$

$\therefore\ h \circ f$ is $y = x + 2$

so $(h \circ f)^{-1}(x)$ is $x = y + 2$

$\therefore\ y = x - 2$

$\therefore\ (h \circ f)^{-1}(x) = x - 2$

$\therefore\ (f^{-1} \circ h^{-1})(x) = (h \circ f)^{-1}(x)$ as required.

## REVIEW SET 2B

**1** **a** $y = (x-1)(x-5)$

$\therefore$ the $x$-intercepts are $x = 1$ and $5$

The vertex is at $x = 3$, with $y = (3-1)(3-5) = 2 \times (-2) = -4$

$\therefore$ the vertex is at $(3, -4)$

The domain is $\{x \mid x \in \mathbb{R}\}$. The range is $\{y \mid y \geqslant -4\}$.

**b** From the graph, the domain is $\{x \mid x \neq 0,\ x \neq 2\}$ and the range is $\{y \mid y \leqslant -1$ or $y > 0\}$.

**2** **a** $(f \circ g)(x) = f(g(x))$

$= f(x^2 + 2)$

$= 2(x^2 + 2) - 3$

$= 2x^2 + 4 - 3$

$= 2x^2 + 1$

**b** $(g \circ f)(x) = g(f(x))$

$= g(2x - 3)$

$= (2x-3)^2 + 2$

$= 4x^2 - 12x + 9 + 2$

$= 4x^2 - 12x + 11$

**3** **a** $y = \dfrac{x^2 - 6x - 16}{x-3} = \dfrac{(x+2)(x-8)}{x-3}$ is zero when $x = -2$ or $8$ and undefined when $x = 3$.

When $x = 0$, $y = \frac{-16}{-3} > 0$.

Since the factors are single, the signs alternate. So, the sign diagram is:

$-$ | $-2$ | $+$ | $3$ | $-$ | $8$ | $+$ $x$

**b** $y = \dfrac{x+9}{x+5} + x\left(\dfrac{x+5}{x+5}\right) = \dfrac{x^2+6x+9}{x+5} = \dfrac{(x+3)^2}{x+5}$ is zero when $x = -3$

and undefined when $x = -5$.

When $x = 0$, $y = \dfrac{3^2}{5} > 0$. The $(x+3)$ factor is squared, so the sign does not change at $x = -3$.

So, the sign diagram is:

$-$ | $-5$ | $+$ | $-3$ | $+$ $x$

**4** **a** $f(x) = \dfrac{1}{x^2}$ is meaningless when $x = 0$.

**b**

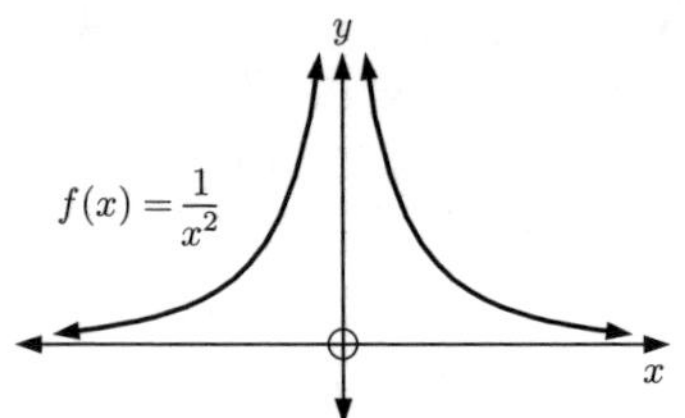


**c** Domain of $f(x)$ is $\{x \mid x \neq 0\}$.
Range of $f(x)$ is $\{y \mid y > 0\}$.

**5** **a** $f(x) = \dfrac{ax+3}{x-b}$ has asymptotes $x = -1$, $y = 2$.

$f(x)$ is undefined when $x - b = 0$

$\therefore$ $x = b$ is the vertical asymptote.

But $x = -1$ is the vertical asymptote, so $b = -1$.

So, $f(x) = \dfrac{ax+3}{x-(-1)} = \dfrac{ax+3}{x+1} = \dfrac{a+\frac{3}{x}}{1+\frac{1}{x}}$

As $|x| \to \infty$, $f(x) \to \dfrac{a}{1} = a$ so the horizontal asymptote is $y = a$.

But $y = 2$ is the horizontal asymptote, so $a = 2$.

$\therefore$ $a = 2$ and $b = -1$.

**b** Domain of $f$ is $\{x \mid x \neq -1\}$ and range of $f$ is $\{y \mid y \neq 2\}$.

$\therefore$ domain of $f^{-1}$ is $\{x \mid x \neq 2\}$ and range of $f^{-1}$ is $\{y \mid y \neq -1\}$.

**6** **a** $f : x \mapsto \dfrac{4x+1}{2-x}$ is undefined when $x = 2$

$\therefore$ $x = 2$ is a vertical asymptote.

Now $f(x) = \dfrac{4x+1}{2-x} = \dfrac{4+\frac{1}{x}}{-1+\frac{2}{x}}$

$\therefore$ as $|x| \to \infty$, $f(x) \to \frac{4}{-1} = -4$,

and so $y = -4$ is a horizontal asymptote.

**b** The domain is $\{x \mid x \neq 2\}$.
The range is $\{y \mid y \neq -4\}$.

**c** As $x \to 2^-$, $y \to \infty$.
As $x \to 2^+$, $y \to -\infty$.
As $x \to \infty$, $y \to -4^-$.
As $x \to -\infty$, $y \to -4^+$.

**d** $f(0) = \dfrac{4(0)+1}{2-0} = \frac{1}{2}$

So, the $y$-intercept is $\frac{1}{2}$.

$f(x) = 0$ when $\dfrac{4x+1}{2-x} = 0$

$\therefore$ $4x + 1 = 0$

$\therefore$ $x = -\frac{1}{4}$

So, the $x$-intercept is $-\frac{1}{4}$.

**e**

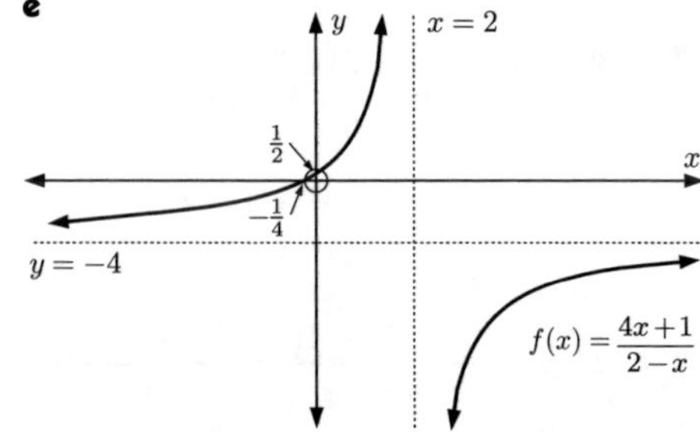


**7** **a** $(g \circ f)(x) = g(f(x))$

$= g(3x+1)$

$= \dfrac{2}{3x+1}$

**b** $(g \circ f)(x) = -4$

$\therefore$ $\dfrac{2}{3x+1} = -4$

$\therefore$ $-4(3x+1) = 2$

$\therefore$ $-12x - 4 = 2$

$\therefore$ $12x = -6$

$\therefore$ $x = -\frac{1}{2}$

**c** **i** $h(x) = \dfrac{2}{3x+1}$ is undefined when
$3x + 1 = 0$ or $x = -\frac{1}{3}$.
So, $x = -\frac{1}{3}$ is a vertical asymptote.
As $|x| \to \infty$, $h(x) \to 0$
$\therefore$ $y = 0$ is a horizontal asymptote.

**ii**

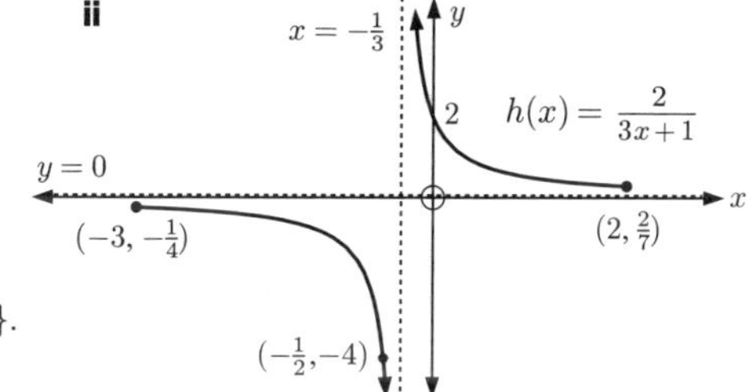


**iii** Range of $h$ is $\{y \mid y \leqslant -\frac{1}{4}$ or $y \geqslant \frac{2}{7}\}$.

**8** **a**

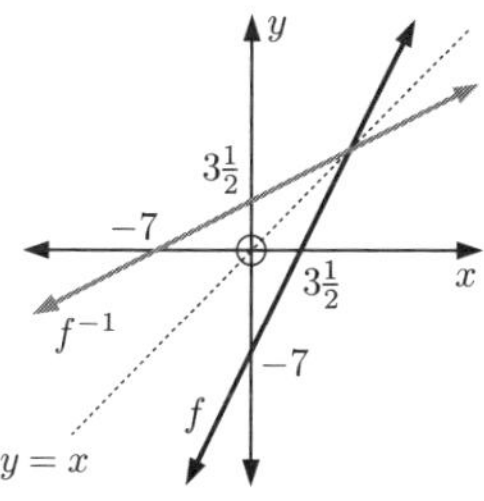


**b** The function $f$ is $y = 2x - 7$
so $f^{-1}$ is $x = 2y - 7$
$\therefore$ $y = \dfrac{x+7}{2}$
So, $f^{-1}(x) = \dfrac{x+7}{2}$

**c**

| $f \circ f^{-1}$ | and | $f^{-1} \circ f$ |
|---|---|---|
| $= f\left(f^{-1}(x)\right)$ | | $= f^{-1}(f(x))$ |
| $= f\left(\dfrac{x+7}{2}\right)$ | | $= f^{-1}(2x-7)$ |
| $= 2\left(\dfrac{x+7}{2}\right) - 7$ | | $= \dfrac{2x-7+7}{2}$ |
| $= x + 7 - 7$ | | $= \dfrac{2x}{2}$ |
| $= x$ | | $= x$ |

So, $f \circ f^{-1} = f^{-1} \circ f = x$.

**9** **a**

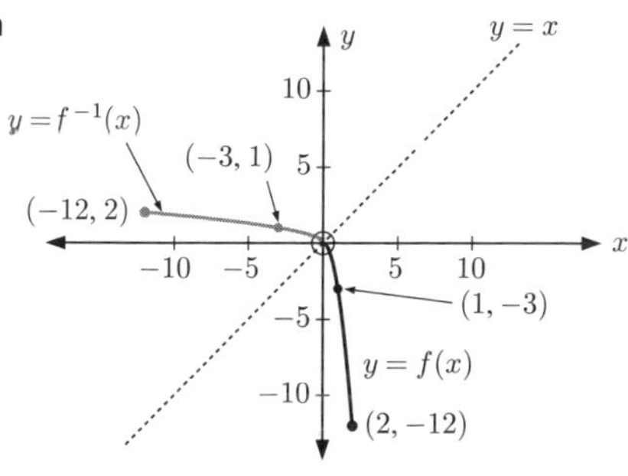


**b** Range of $f^{-1}$ is $\{y \mid 0 \leqslant y \leqslant 2\}$.

**c** **i** $f(x) = -10$
$-3x^2 = -10$, $0 \leqslant x \leqslant 2$
$\therefore$ $x^2 = \frac{10}{3}$
$\therefore$ $x = \sqrt{\frac{10}{3}}$ $\{0 \leqslant x \leqslant 2\}$
$\therefore$ $x \approx 1.83$

**ii** $f^{-1}(x) = 1$ so $f(1) = x$
The graph shows that $f(1) = -3$
$\therefore$ $x = -3$

## REVIEW SET 2C

**1** **a** Domain is $\{x \mid x \geqslant -2\}$. Range is $\{y \mid 1 \leqslant y < 3\}$.
**b** Domain is $\{x \mid x \in \mathbb{R}\}$. Range is $\{y \mid y = -1, 1, \text{ or } 2\}$.

**2** **a** $f(x) = x^2 + 3$
$\therefore$ $f(-3) = (-3)^2 + 3$
$= 9 + 3$
$= 12$

**b** $x^2 + 3 = 4$
$\therefore$ $x^2 = 1$
$\therefore$ $x = \pm 1$

**3** **a** $f(x) = 10 + \dfrac{3}{2x-1}$
is undefined when $2x - 1 = 0$
$\therefore$ $x = \frac{1}{2}$

**b** $f(x) = \sqrt{x+7}$
is undefined when $x + 7 < 0$
$\therefore$ $x < -7$

**4** **a** $f(x) = x(x+4)(3x+1)$ is zero when $x = 0,\ -4,$ and $-\frac{1}{3}$.

When $x = 10$, $y = 10(14)(31) = 4340 > 0$.

The factors are single so the signs alternate.

$\therefore$ sign diagram is: $-$ | $+$ | $-$ | $+$ with critical values $-4$, $-\frac{1}{3}$, $0$ on the $x$-axis

**b** $f(x) = \dfrac{-11}{(x+1)(x+8)}$ is undefined when $x = -1$ or $-8$.

When $x = 0$, $y = \frac{-11}{(1)(8)} = -\frac{11}{8} < 0$.

The factors are single, so the signs alternate.

$\therefore$ sign diagram is: $-$ ⁞ $+$ ⁞ $-$ with critical values $-8$, $-1$ on the $x$-axis

**5** **a**
$$\begin{aligned} h(x) &= 7 - 3x \\ h(2x-1) &= 7 - 3(2x-1) \\ &= 7 - 6x + 3 \\ &= 10 - 6x \end{aligned}$$

**b**
$$\begin{aligned} h(2x-1) &= -2 \\ \therefore\ 10 - 6x &= -2 \quad \{\text{using } \mathbf{a}\} \\ \therefore\ -6x &= -12 \\ \therefore\ x &= 2 \end{aligned}$$

**6** **a**
$$\begin{aligned} (f \circ g)(x) &= f(g(x)) \\ &= f(\sqrt{x}) \\ &= 1 - 2\sqrt{x} \end{aligned}$$

**b**
$$\begin{aligned} (g \circ f)(x) &= g(f(x)) \\ &= g(1-2x) \\ &= \sqrt{1-2x} \end{aligned}$$

**7** $f(x) = ax^2 + bx + c$, where $f(0) = 5$, $f(-2) = 21$, and $f(3) = -4$

When $f(0) = 5$,
$$\begin{aligned} 5 &= a(0)^2 + b(0) + c \\ \therefore\ 5 &= c \\ \therefore\ c &= 5 \quad .... (1) \end{aligned}$$

When $f(-2) = 21$,
$$\begin{aligned} 21 &= a(-2)^2 + b(-2) + c \\ &= 4a - 2b + c \\ &= 4a - 2b + 5 \quad \{\text{using (1)}\} \\ \therefore\ 4a - 2b &= 16 \\ \therefore\ 2a - b &= 8 \quad \text{and so} \quad b = 2a - 8 \quad .... (2) \end{aligned}$$

When $f(3) = -4$,
$$\begin{aligned} -4 &= a(3)^2 + b(3) + c \\ \therefore\ -4 &= 9a + 3b + c \\ \therefore\ -4 &= 9a + 3b + 5 \quad \{\text{using (1)}\} \\ \therefore\ -4 &= 9a + 3(2a-8) + 5 \quad \{\text{using (2)}\} \\ \therefore\ -4 &= 9a + 6a - 24 + 5 \\ \therefore\ 15 &= 15a \quad \text{and so} \quad a = 1 \end{aligned}$$

Now, substituting $a = 1$ into (2) gives $b = 2(1) - 8 = -6$

So, $a = 1$, $b = -6$, $c = 5$.

**8** **a**

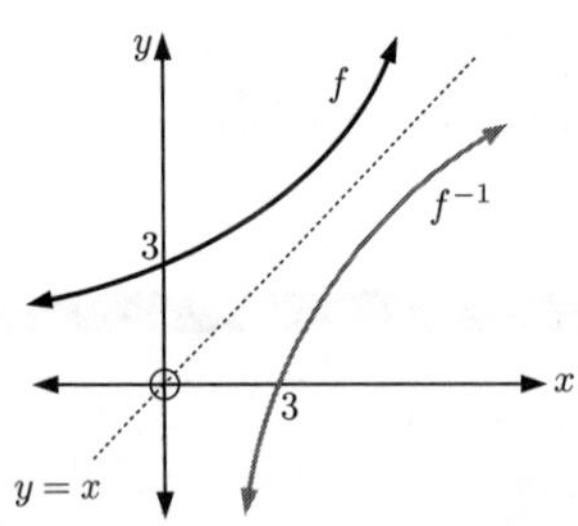


**b**

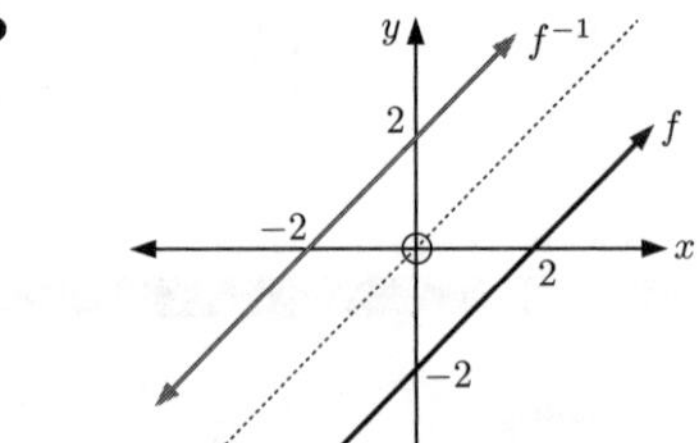


**9** **a**
$$\begin{aligned} f \text{ is } y &= 7 - 4x \\ \therefore\ f^{-1} \text{ is } x &= 7 - 4y \\ \therefore\ y &= \frac{7-x}{4} \\ \text{So, } f^{-1}(x) &= \frac{7-x}{4} \end{aligned}$$

**b**
$$\begin{aligned} f \text{ is } y &= \frac{3+2x}{5} \\ \therefore\ f^{-1} \text{ is } x &= \frac{3+2y}{5} \\ \therefore\ 5x &= 3 + 2y \\ \therefore\ y &= \frac{5x-3}{2} \\ \text{So, } f^{-1}(x) &= \frac{5x-3}{2} \end{aligned}$$

**10** $f$ is $y = 5x - 2$

$\therefore\ f^{-1}$ is $x = 5y - 2$

$\therefore\ y = \dfrac{x+2}{5}$

$\therefore\ f^{-1}(x) = \dfrac{x+2}{5}$

$h$ is $y = \dfrac{3x}{4}$

$\therefore\ h^{-1}$ is $x = \dfrac{3y}{4}$

$\therefore\ y = \dfrac{4x}{3}$

$\therefore\ h^{-1}(x) = \dfrac{4x}{3}$

Now $(f^{-1} \circ h^{-1})(x) = f^{-1}(h^{-1}(x))$

$$= f^{-1}\left(\frac{4x}{3}\right) = \frac{\frac{4x}{3}+2}{5} = \frac{4x+6}{15}$$

and $(h \circ f)(x) = h(f(x))$

$$= h(5x-2) = \frac{3(5x-2)}{4}$$

So, $y = \dfrac{15x-6}{4}$

$\therefore\ (h \circ f)^{-1}(x)$ is $x = \dfrac{15y-6}{4}$

$\therefore\ 4x = 15y - 6$

$\therefore\ y = \dfrac{4x+6}{15}$

$\therefore\ (h \circ f)^{-1}(x) = \dfrac{4x+6}{15}$

Hence, $(f^{-1} \circ h^{-1})(x) = (h \circ f)^{-1}(x)$ as required.

**11** $f$ is $y = 2x + 11$

so $f^{-1}(x)$ is $x = 2y + 11$

$\therefore\ y = \dfrac{x-11}{2}$

$\therefore\ f^{-1}(x) = \dfrac{x-11}{2}$

$g(x) = x^2$

$$(g \circ f^{-1})(x) = g(f^{-1}(x)) = g\left(\frac{x-11}{2}\right) = \left(\frac{x-11}{2}\right)^2$$

$$\therefore\ (g \circ f^{-1})(3) = \left(\frac{3-11}{2}\right)^2 = (-4)^2 = 16$$

**12** The domain is $\{x \mid x \neq 4\}$, so $x = 4$ is a vertical asymptote.
The range is $\{y \mid y \neq -1\}$, so $y = -1$ is a horizontal asymptote.
We now consider the behaviour of the function near the asymptotes, using the sign diagram

to help us.

As $x \to \infty$, $y \to -1$
As $x \to -\infty$, $y \to -1$

Note that we cannot tell whether the function tends to $-1$ from above or below.

As $x \to 4^-$, $y \to \infty$
As $x \to 4^+$, $y \to -\infty$

So, the function could be:
(**Note:** There may be other answers.)

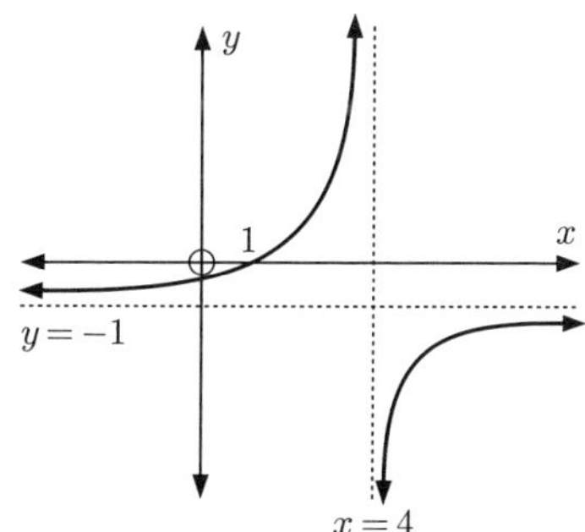

# Chapter 3
## EXPONENTIALS

### EXERCISE 3A

**1** **a** $2^1 = 2,\ 2^2 = 4,\ 2^3 = 8,\ 2^4 = 16,\ 2^5 = 32,\ 2^6 = 64$
**b** $3^1 = 3,\ 3^2 = 9,\ 3^3 = 27,\ 3^4 = 81,\ 3^5 = 243,\ 3^6 = 729$
**c** $4^1 = 4,\ 4^2 = 16,\ 4^3 = 64,\ 4^4 = 256,\ 4^5 = 1024,\ 4^6 = 4096$

**2** **a** $5^1 = 5,\ 5^2 = 25,\ 5^3 = 125,\ 5^4 = 625$
**b** $6^1 = 6,\ 6^2 = 36,\ 6^3 = 216,\ 6^4 = 1296$
**c** $7^1 = 7,\ 7^2 = 49,\ 7^3 = 343,\ 7^4 = 2401$

**3** **a** $(-1)^5$
$= (-1) \times (-1) \times (-1) \times (-1) \times (-1)$
$= 1 \times 1 \times (-1)$
$= -1$

**b** $(-1)^6$
$= (-1)^5 \times (-1)$
$= (-1) \times (-1)$
$= 1$

**c** $(-1)^{14}$
$= 1$

**d** $(-1)^{19}$
$= -1$

**e** $(-1)^8$
$= 1$

**f** $-1^8$
$= -(1^8)$
$= -1$

**g** $-(-1)^8$
$= -(1)$
$= -1$

**h** $(-2)^5$
$= (-2) \times (-2) \times (-2) \times (-2) \times (-2)$
$= 4 \times 4 \times (-2)$
$= -32$

**i** $-2^5$
$= -(2^5)$
$= -32$

**j** $-(-2)^6$
$= -(-2)^5 \times (-2)$
$= 32 \times (-2)$
$= -64$

**k** $(-5)^4$
$= (-5) \times (-5) \times (-5) \times (-5)$
$= 25 \times 25$
$= 625$

**l** $-(-5)^4$
$= -(-5) \times (-5) \times (-5) \times (-5)$
$= -25 \times 25$
$= -625$

**4** **a** $4^7 = 16\,384$ **b** $7^4 = 2401$ **c** $-5^5 = -3125$ **d** $(-5)^5 = -3125$
**e** $8^6 = 262\,144$ **f** $(-8)^6 = 262\,144$ **g** $-8^6 = -262\,144$
**h** $2.13^9 \approx 902.436\,039\,6$ **i** $-2.13^9 \approx -902.436\,039\,6$ **j** $(-2.13)^9 \approx -902.436\,039\,6$

**5** **a** $9^{-1} = 0.\overline{1}$ **b** $\dfrac{1}{9^1} = 0.\overline{1}$ **c** $6^{-2} = 0.02\overline{7}$ **d** $\dfrac{1}{6^2} = 0.02\overline{7}$
**e** $3^{-4} \approx 0.012\,345\,679$ **f** $\dfrac{1}{3^4} \approx 0.012\,345\,679$ **g** $17^0 = 1$ **h** $(0.366)^0 = 1$

We notice that $a^{-n} = \dfrac{1}{a^n}$ and $a^0 = 1$ for $a \neq 0$.

**6** $3^{101} = \underbrace{3^4 \times 3^4 \times 3^4 \times .... \times 3^4}_{25 \text{ of these}} \times 3^1$

But $3^4 = 81$ which ends in a 1
$\therefore\ \underbrace{3^4 \times 3^4 \times 3^4 \times .... \times 3^4}_{25 \text{ of these}}$ ends in a 1
$\therefore\ 3^{101}$ ends in a 3

**7** $7^1 = 7,\ \ 7^2 = 49,\ \ 7^3 = 343,\ \ 7^4 = 2401,\ \ 7^5 = 16\,807$
Now $7^{217} = \underbrace{7^4 \times 7^4 \times 7^4 \times .... \times 7^4}_{54 \text{ of these so this part ends in a 1}} \times 7^1$

$\therefore\ 7^{217}$ ends in $1 \times 7 = 7$.

## EXERCISE 3B

**1** **a** $5^4 \times 5^7 = 5^{4+7} = 5^{11}$

**b** $d^2 \times d^6 = d^{2+6} = d^8$

**c** $\dfrac{k^8}{k^3} = k^{8-3} = k^5$

**d** $\dfrac{7^5}{7^6} = 7^{5-6} = 7^{-1} = \frac{1}{7}$

**e** $(x^2)^5 = x^{2\times 5} = x^{10}$

**f** $(3^4)^4 = 3^{4\times 4} = 3^{16}$

**g** $\dfrac{p^3}{p^7} = p^{3-7} = p^{-4}$ or $\dfrac{1}{p^4}$

**h** $n^3 \times n^9 = n^{3+9} = n^{12}$

**i** $(5^t)^3 = 5^{t\times 3} = 5^{3t}$

**j** $7^x \times 7^2 = 7^{x+2}$

**k** $\dfrac{10^3}{10^q} = 10^{3-q}$

**l** $(c^4)^m = c^{4\times m} = c^{4m}$

**2** **a** $4 = 2\times 2 = 2^2$

**b** $\dfrac{1}{4} = \dfrac{1}{2^2} = 2^{-2}$

**c** $8 = 2\times 2\times 2 = 2^3$

**d** $\dfrac{1}{8} = \dfrac{1}{2^3} = 2^{-3}$

**e** $32 = 2\times 2\times 2\times 2\times 2 = 2^5$

**f** $\dfrac{1}{32} = \dfrac{1}{2^5} = 2^{-5}$

**g** $2 = 2^1$

**h** $\dfrac{1}{2} = \dfrac{1}{2^1} = 2^{-1}$

**i** $64 = 32\times 2 = 2^5\times 2^1 = 2^6$

**j** $\dfrac{1}{64} = \dfrac{1}{2^6} = 2^{-6}$

**k** $128 = 64\times 2 = 2^6\times 2^1 = 2^7$

**l** $\dfrac{1}{128} = \dfrac{1}{2^7} = 2^{-7}$

**3** **a** $9 = 3\times 3 = 3^2$

**b** $\dfrac{1}{9} = \dfrac{1}{3^2} = 3^{-2}$

**c** $27 = 3\times 3\times 3 = 3^3$

**d** $\dfrac{1}{27} = \dfrac{1}{3^3} = 3^{-3}$

**e** $3 = 3^1$

**f** $\dfrac{1}{3} = \dfrac{1}{3^1} = 3^{-1}$

**g** $81 = 3\times 3\times 3\times 3 = 3^4$

**h** $\dfrac{1}{81} = \dfrac{1}{3^4} = 3^{-4}$

**i** $1 = 3^0$

**j** $243 = 81\times 3 = 3^4\times 3^1 = 3^5$

**k** $\dfrac{1}{243} = \dfrac{1}{3^5} = 3^{-5}$

**4** **a** $2\times 2^a = 2^1\times 2^a = 2^{a+1}$

**b** $4\times 2^b = 2^2\times 2^b = 2^{b+2}$

**c** $8\times 2^t = 2^3\times 2^t = 2^{t+3}$

**d** $(2^{x+1})^2 = 2^{2(x+1)} = 2^{2x+2}$

**e** $(2^{1-n})^{-1} = 2^{-(1-n)} = 2^{n-1}$

**f** $\dfrac{2^c}{4} = \dfrac{2^c}{2^2} = 2^{c-2}$

**g** $\dfrac{2^m}{2^{-m}} = 2^{m-(-m)} = 2^{2m}$

**h** $\dfrac{4}{2^{1-n}} = \dfrac{2^2}{2^{1-n}} = 2^{2-(1-n)} = 2^{n+1}$

**i** $\dfrac{2^{x+1}}{2^x} = 2^{x+1-x} = 2^1$

**j** $\dfrac{4^x}{2^{1-x}} = \dfrac{(2^2)^x}{2^{1-x}} = 2^{2x-(1-x)} = 2^{3x-1}$

**5** **a** $9 \times 3^p = 3^2 \times 3^p = 3^{p+2}$

**b** $27^a = (3^3)^a = 3^{3a}$

**c** $3 \times 9^n = 3^1 \times (3^2)^n = 3^{2n+1}$

**d** $27 \times 3^d = 3^3 \times 3^d = 3^{d+3}$

**e** $9 \times 27^t = 3^2 \times (3^3)^t = 3^{3t+2}$

**f** $\dfrac{3^y}{3} = \dfrac{3^y}{3^1} = 3^{y-1}$

**g** $\dfrac{3}{3^y} = \dfrac{3^1}{3^y} = 3^{1-y}$

**h** $\dfrac{9}{27^t} = \dfrac{3^2}{(3^3)^t} = 3^{2-3t}$

**i** $\dfrac{9^a}{3^{1-a}} = \dfrac{(3^2)^a}{3^{1-a}} = 3^{2a-(1-a)} = 3^{3a-1}$

**j** $\dfrac{9^{n+1}}{3^{2n-1}} = \dfrac{(3^2)^{n+1}}{3^{2n-1}} = 3^{2n+2-(2n-1)} = 3^3$

**6** **a** $(2a)^2 = 2^2 \times a^2 = 4a^2$

**b** $(3b)^3 = 3^3 \times b^3 = 27b^3$

**c** $(ab)^4 = a^4 \times b^4 = a^4b^4$

**d** $(pq)^3 = p^3 \times q^3 = p^3q^3$

**e** $\left(\dfrac{m}{n}\right)^2 = \dfrac{m^2}{n^2}$

**f** $\left(\dfrac{a}{3}\right)^3 = \dfrac{a^3}{3^3} = \dfrac{a^3}{27}$

**g** $\left(\dfrac{b}{c}\right)^4 = \dfrac{b^4}{c^4}$

**h** $\left(\dfrac{2a}{b}\right)^0 = 1, \ b \neq 0$

**i** $\left(\dfrac{m}{3n}\right)^4 = \dfrac{m^4}{3^4 \times n^4} = \dfrac{m^4}{81n^4}$

**j** $\left(\dfrac{xy}{2}\right)^3 = \dfrac{x^3y^3}{2^3} = \dfrac{x^3y^3}{8}$

**7** **a** $(-2a)^2 = (-2)^2a^2 = 4a^2$

**b** $(-6b^2)^2 = (-6)^2b^4 = 36b^4$

**c** $(-2a)^3 = (-2)^3a^3 = -8a^3$

**d** $(-3m^2n^2)^3 = (-3)^3m^6n^6 = -27m^6n^6$

**e** $(-2ab^4)^4 = (-2)^4a^4b^{16} = 16a^4b^{16}$

**f** $\left(\dfrac{-2a^2}{b^2}\right)^3 = \dfrac{(-2)^3a^6}{b^6} = -\dfrac{8a^6}{b^6}$

**g** $\left(\dfrac{-4a^3}{b}\right)^2 = \dfrac{(-4)^2a^6}{b^2} = \dfrac{16a^6}{b^2}$

**h** $\left(\dfrac{-3p^2}{q^3}\right)^2 = \dfrac{(-3)^2p^4}{q^6} = \dfrac{9p^4}{q^6}$

**8** **a** $ab^{-2} = \dfrac{a}{b^2}$

**b** $(ab)^{-2} = \dfrac{1}{(ab)^2} = \dfrac{1}{a^2b^2}$

**c** $(2ab^{-1})^2 = 2^2a^2b^{-2} = \dfrac{4a^2}{b^2}$

**d** $(3a^{-2}b)^2 = 3^2a^{-4}b^2 = \dfrac{9b^2}{a^4}$

**e** $\dfrac{a^2b^{-1}}{c^2} = \dfrac{a^2}{bc^2}$

**f** $\dfrac{a^2b^{-1}}{c^{-2}} = \dfrac{a^2c^2}{b}$

**g** $\dfrac{1}{a^{-3}} = a^3$

**h** $\dfrac{a^{-2}}{b^{-3}} = \dfrac{b^3}{a^2}$

**i** $\dfrac{2a^{-1}}{d^2} = \dfrac{2}{ad^2}$

**j** $\dfrac{12a}{m^{-3}} = 12am^3$

**9** **a** $\dfrac{1}{a^n} = a^{-n}$

**b** $\dfrac{1}{b^{-n}} = b^n$

**c** $\dfrac{1}{3^{2-n}} = 3^{n-2}$

**d** $\dfrac{a^n}{b^{-m}} = a^nb^m$

**e** $\dfrac{a^{-n}}{a^{2+n}} = a^{-n-(2+n)} = a^{-2n-2}$

**10** **a** $(\frac{5}{3})^0 = 1$ **b** $(\frac{7}{4})^{-1} = \frac{4}{7}$ **c** $(\frac{1}{6})^{-1} = \frac{6}{1} = 6$ **d** $\frac{3^3}{3^0} = \frac{27}{1} = 27$

**e** $(\frac{4}{3})^{-2} = \frac{3^2}{4^2} = \frac{9}{16}$ **f** $2^1 + 2^{-1} = 2 + \frac{1}{2} = \frac{5}{2}$ **g** $(1\frac{2}{3})^{-3} = (\frac{5}{3})^{-3} = \frac{3^3}{5^3} = \frac{27}{125}$ **h** $5^2 + 5^1 + 5^{-1} = 25 + 5 + \frac{1}{5} = \frac{151}{5}$

**11** **a** $\frac{1}{9} = \frac{1}{3^2} = 3^{-2}$ **b** $\frac{1}{16} = \frac{1}{2^4} = 2^{-4}$ **c** $\frac{1}{125} = \frac{1}{5^3} = 5^{-3}$ **d** $\frac{3}{5} = 3 \times \frac{1}{5} = 3^1 \times 5^{-1}$ **e** $\frac{4}{27} = \frac{2^2}{3^3} = 2^2 \times 3^{-3}$

**f** $\frac{2^c}{8 \times 9} = \frac{2^c}{2^3 \times 3^2} = 2^{c-3} \times 3^{-2}$ **g** $\frac{9^k}{10} = \frac{(3^2)^k}{2 \times 5} = 3^{2k} \times 2^{-1} \times 5^{-1}$ **h** $\frac{6^p}{75} = \frac{(2 \times 3)^p}{3 \times 5^2} = \frac{2^p \times 3^p}{3 \times 5^2} = 2^p \times 3^{p-1} \times 5^{-2}$

**12** **a** $5^3 = 21 + 23 + 25 + 27 + 29$ **b** $7^3 = 43 + 45 + 47 + 49 + 51 + 53 + 55$

**c** $12^3 = 133 + 135 + 137 + 139 + 141 + 143 + 145 + 147 + 149 + 151 + 153 + 155$

## EXERCISE 3C

**1** **a** $\sqrt[5]{2} = 2^{\frac{1}{5}}$ **b** $\frac{1}{\sqrt[5]{2}} = \frac{1}{2^{\frac{1}{5}}} = 2^{-\frac{1}{5}}$ **c** $2\sqrt{2} = 2^1 \times 2^{\frac{1}{2}} = 2^{\frac{3}{2}}$ **d** $4\sqrt{2} = 2^2 \times 2^{\frac{1}{2}} = 2^{\frac{5}{2}}$

**e** $\frac{1}{\sqrt[3]{2}} = \frac{1}{2^{\frac{1}{3}}} = 2^{-\frac{1}{3}}$ **f** $2 \times \sqrt[3]{2} = 2^1 \times 2^{\frac{1}{3}} = 2^{\frac{4}{3}}$ **g** $\frac{4}{\sqrt{2}} = \frac{2^2}{2^{\frac{1}{2}}} = 2^{\frac{3}{2}}$

**h** $(\sqrt{2})^3 = (2^{\frac{1}{2}})^3 = 2^{\frac{3}{2}}$ **i** $\frac{1}{\sqrt[3]{16}} = \frac{1}{\sqrt[3]{2^4}} = \frac{1}{2^{\frac{4}{3}}} = 2^{-\frac{4}{3}}$ **j** $\frac{1}{\sqrt{8}} = \frac{1}{\sqrt{2^3}} = \frac{1}{2^{\frac{3}{2}}} = 2^{-\frac{3}{2}}$

**2** **a** $\sqrt[3]{3} = 3^{\frac{1}{3}}$ **b** $\frac{1}{\sqrt[3]{3}} = \frac{1}{3^{\frac{1}{3}}} = 3^{-\frac{1}{3}}$ **c** $\sqrt[4]{3} = 3^{\frac{1}{4}}$ **d** $3\sqrt{3} = 3^1 \times 3^{\frac{1}{2}} = 3^{\frac{3}{2}}$

**e** $\frac{1}{9\sqrt{3}} = \frac{1}{3^2 \times 3^{\frac{1}{2}}} = \frac{1}{3^{\frac{5}{2}}} = 3^{-\frac{5}{2}}$

**3** **a** $\sqrt[3]{7} = 7^{\frac{1}{3}}$ **b** $\sqrt[4]{27} = \sqrt[4]{3^3} = 3^{\frac{3}{4}}$ **c** $\sqrt[5]{16} = \sqrt[5]{2^4} = 2^{\frac{4}{5}}$ **d** $\sqrt[3]{32} = \sqrt[3]{2^5} = 2^{\frac{5}{3}}$

**e** $\sqrt[7]{49} = \sqrt[7]{7^2} = 7^{\frac{2}{7}}$ **f** $\frac{1}{\sqrt[3]{7}} = \frac{1}{7^{\frac{1}{3}}} = 7^{-\frac{1}{3}}$ **g** $\frac{1}{\sqrt[4]{27}} = \frac{1}{3^{\frac{3}{4}}} = 3^{-\frac{3}{4}}$ **h** $\frac{1}{\sqrt[5]{16}} = \frac{1}{2^{\frac{4}{5}}} = 2^{-\frac{4}{5}}$

**i** $\frac{1}{\sqrt[3]{32}} = \frac{1}{2^{\frac{5}{3}}} = 2^{-\frac{5}{3}}$ **j** $\frac{1}{\sqrt[7]{49}} = \frac{1}{7^{\frac{2}{7}}} = 7^{-\frac{2}{7}}$

**4** **a** $3^{\frac{3}{4}} \approx 2.28$ **b** $2^{\frac{7}{8}} \approx 1.83$ **c** $2^{-\frac{1}{3}} \approx 0.794$ **d** $4^{-\frac{3}{5}} \approx 0.435$

**e** $\sqrt[4]{8} \approx 1.68$ **f** $\sqrt[5]{27} \approx 1.93$ **g** $\dfrac{1}{\sqrt[3]{7}} \approx 0.523$

**5** **a** $4^{\frac{3}{2}} = (2^2)^{\frac{3}{2}} = 2^3 = 8$

**b** $8^{\frac{5}{3}} = (2^3)^{\frac{5}{3}} = 2^5 = 32$

**c** $16^{\frac{3}{4}} = (2^4)^{\frac{3}{4}} = 2^3 = 8$

**d** $25^{\frac{3}{2}} = (5^2)^{\frac{3}{2}} = 5^3 = 125$

**e** $32^{\frac{2}{5}} = (2^5)^{\frac{2}{5}} = 2^2 = 4$

**f** $4^{-\frac{1}{2}} = (2^2)^{-\frac{1}{2}} = 2^{-1} = \frac{1}{2}$

**g** $9^{-\frac{3}{2}} = (3^2)^{-\frac{3}{2}} = 3^{-3} = \frac{1}{27}$

**h** $8^{-\frac{4}{3}} = (2^3)^{-\frac{4}{3}} = 2^{-4} = \frac{1}{16}$

**i** $27^{-\frac{4}{3}} = (3^3)^{-\frac{4}{3}} = 3^{-4} = \frac{1}{81}$

**j** $125^{-\frac{2}{3}} = (5^3)^{-\frac{2}{3}} = 5^{-2} = \frac{1}{25}$

## EXERCISE 3D.1

**1** **a**
$$\begin{aligned} &x^2(x^3+2x^2+1) \\ &= x^2 \times x^3 + x^2 \times 2x^2 + x^2 \times 1 \\ &= x^5 + 2x^4 + x^2 \end{aligned}$$

**b**
$$\begin{aligned} &2^x(2^x+1) \\ &= 2^x \times 2^x + 2^x \times 1 \\ &= 2^{2x} + 2^x \end{aligned}$$

**c**
$$\begin{aligned} &x^{\frac{1}{2}}(x^{\frac{1}{2}} + x^{-\frac{1}{2}}) \\ &= x^{\frac{1}{2}} \times x^{\frac{1}{2}} + x^{\frac{1}{2}} \times x^{-\frac{1}{2}} \\ &= x^1 + x^0 \\ &= x + 1 \end{aligned}$$

**d**
$$\begin{aligned} &7^x(7^x+2) \\ &= 7^x \times 7^x + 7^x \times 2 \\ &= 7^{2x} + 2(7^x) \\ \text{or } &49^x + 2(7^x) \end{aligned}$$

**e**
$$\begin{aligned} &3^x(2-3^{-x}) \\ &= 3^x \times 2 - 3^x \times 3^{-x} \\ &= 2(3^x) - 3^0 \\ &= 2(3^x) - 1 \end{aligned}$$

**f**
$$\begin{aligned} &x^{\frac{1}{2}}(x^{\frac{3}{2}} + 2x^{\frac{1}{2}} + 3x^{-\frac{1}{2}}) \\ &= x^{\frac{1}{2}} \times x^{\frac{3}{2}} + x^{\frac{1}{2}} \times 2x^{\frac{1}{2}} + x^{\frac{1}{2}} \times 3x^{-\frac{1}{2}} \\ &= x^2 + 2x^1 + 3x^0 \\ &= x^2 + 2x + 3 \end{aligned}$$

**g**
$$\begin{aligned} &2^{-x}(2^x+5) \\ &= 2^{-x} \times 2^x + 2^{-x} \times 5 \\ &= 2^0 + 5(2^{-x}) \\ &= 1 + 5(2^{-x}) \end{aligned}$$

**h**
$$\begin{aligned} &5^{-x}(5^{2x}+5^x) \\ &= 5^{-x} \times 5^{2x} + 5^{-x} \times 5^x \\ &= 5^x + 5^0 \\ &= 5^x + 1 \end{aligned}$$

**i**
$$\begin{aligned} &x^{-\frac{1}{2}}(x^2 + x + x^{\frac{1}{2}}) \\ &= x^{-\frac{1}{2}} \times x^2 + x^{-\frac{1}{2}} \times x^1 + x^{-\frac{1}{2}} \times x^{\frac{1}{2}} \\ &= x^{\frac{3}{2}} + x^{\frac{1}{2}} + x^0 \\ &= x^{\frac{3}{2}} + x^{\frac{1}{2}} + 1 \end{aligned}$$

**2** **a**
$$\begin{aligned} &(2^x-1)(2^x+3) \\ &= 2^x \times 2^x + 2^x \times 3 - 1 \times 2^x - 3 \\ &= 2^{2x} + 2(2^x) - 3 \\ &= 4^x + 2^{x+1} - 3 \end{aligned}$$

**b**
$$\begin{aligned} &(3^x+2)(3^x+5) \\ &= 3^x \times 3^x + 3^x \times 5 + 2 \times 3^x + 10 \\ &= 3^{2x} + 7(3^x) + 10 \\ &= 9^x + 7(3^x) + 10 \end{aligned}$$

**c**
$$\begin{aligned} &(5^x-2)(5^x-4) \\ &= 5^x \times 5^x - 5^x \times 4 - 2 \times 5^x + 8 \\ &= 5^{2x} - 6(5^x) + 8 \\ &= 25^x - 6(5^x) + 8 \end{aligned}$$

**d**
$$\begin{aligned} &(2^x+3)^2 \\ &= (2^x)^2 + 2 \times 2^x \times 3 + 3^2 \\ &= 2^{2x} + 6(2^x) + 9 \\ &= 4^x + 6(2^x) + 9 \end{aligned}$$

**e**
$$\begin{aligned} &(3^x-1)^2 \\ &= (3^x)^2 - 2 \times 3^x \times 1 + 1^2 \\ &= 3^{2x} - 2(3^x) + 1 \\ &= 9^x - 2(3^x) + 1 \end{aligned}$$

**f**
$$\begin{aligned} &(4^x+7)^2 \\ &= (4^x)^2 + 2 \times 4^x \times 7 + 7^2 \\ &= 4^{2x} + 14(4^x) + 49 \\ &= 16^x + 14(4^x) + 49 \end{aligned}$$

**g** $(x^{\frac{1}{2}}+2)(x^{\frac{1}{2}}-2)$
$=(x^{\frac{1}{2}})^2-2^2$
$=x-4$

**h** $(2^x+3)(2^x-3)$
$=(2^x)^2-3^2$
$=2^{2x}-9$
$=4^x-9$

**i** $(x^{\frac{1}{2}}+x^{-\frac{1}{2}})(x^{\frac{1}{2}}-x^{-\frac{1}{2}})$
$=(x^{\frac{1}{2}})^2-(x^{-\frac{1}{2}})^2$
$=x^1-x^{-1}$
$=x-x^{-1}$

**j** $\left(x+\frac{2}{x}\right)^2$
$=x^2+2\times x\times\left(\frac{2}{x}\right)+\left(\frac{2}{x}\right)^2$
$=x^2+4+\frac{4}{x^2}$

**k** $(7^x-7^{-x})^2$
$=(7^x)^2-2\times 7^x\times 7^{-x}+(7^{-x})^2$
$=7^{2x}-2\times 7^0+7^{-2x}$
$=7^{2x}-2+7^{-2x}$

**l** $(5-2^{-x})^2$
$=5^2-2\times 5\times 2^{-x}+(2^{-x})^2$
$=25-10(2^{-x})+2^{-2x}$
$=25-10(2^{-x})+4^{-x}$

## EXERCISE 3D.2

**1** **a** $5^{2x}+5^x$
$=5^x\times 5^x+5^x$
$=5^x(5^x+1)$

**b** $3^{n+2}+3^n$
$=3^n\times 3^2+3^n$
$=3^n(3^2+1)$
$=10(3^n)$

**c** $7^n+7^{3n}$
$=7^n+7^n\times 7^{2n}$
$=7^n(1+7^{2n})$

**d** $5^{n+1}-5$
$=5\times 5^n-5$
$=5(5^n-1)$

**e** $6^{n+2}-6$
$=6\times 6^{n+1}-6$
$=6(6^{n+1}-1)$

**f** $4^{n+2}-16$
$=4^2\times 4^n-16$
$=16\times 4^n-16$
$=16(4^n-1)$

**2** **a** $9^x-4$
$=(3^x)^2-2^2$
$=(3^x+2)(3^x-2)$

**b** $4^x-25$
$=(2^x)^2-5^2$
$=(2^x+5)(2^x-5)$

**c** $16-9^x$
$=4^2-(3^x)^2$
$=(4+3^x)(4-3^x)$

**d** $25-4^x$
$=5^2-(2^x)^2$
$=(5+2^x)(5-2^x)$

**e** $9^x-4^x$
$=(3^x)^2-(2^x)^2$
$=(3^x+2^x)(3^x-2^x)$

**f** $4^x+6(2^x)+9$
$=(2^x)^2+6(2^x)+9$
$=(2^x+3)^2$
$\{a^2+6a+9=(a+3)^2\}$

**g** $9^x+10(3^x)+25$
$=(3^x)^2+10(3^x)+25$
$=(3^x+5)^2$
$\{a^2+10a+25=(a+5)^2\}$

**h** $4^x-14(2^x)+49$
$=(2^x)^2-14(2^x)+49$
$=(2^x-7)^2$
$\{a^2-14a+49=(a-7)^2\}$

**i** $25^x-4(5^x)+4$
$=(5^x)^2-4(5^x)+4$
$=(5^x-2)^2$
$\{a^2-4a+4=(a-2)^2\}$

**3** **a** $4^x+9(2^x)+18$
$=(2^x)^2+9(2^x)+18$
$=(2^x+3)(2^x+6)$
$\{a^2+9a+18=(a+3)(a+6)\}$

**b** $4^x-2^x-20$
$=(2^x)^2-2^x-20$
$=(2^x+4)(2^x-5)$
$\{a^2-a-20=(a+4)(a-5)\}$

**c** $9^x+9(3^x)+14$
$=(3^x)^2+9(3^x)+14$
$=(3^x+2)(3^x+7)$
$\{a^2+9a+14=(a+2)(a+7)\}$

**d** $9^x+4(3^x)-5$
$=(3^x)^2+4(3^x)-5$
$=(3^x+5)(3^x-1)$
$\{a^2+4a-5=(a+5)(a-1)\}$

**e** $25^x + 5^x - 2$
$= (5^x)^2 + 5^x - 2$
$= (5^x + 2)(5^x - 1)$
$\{a^2 + a - 2 = (a+2)(a-1)\}$

**f** $49^x - 7^{x+1} + 12$
$= (7^x)^2 - 7(7^x) + 12$
$= (7^x - 4)(7^x - 3)$
$\{a^2 - 7a + 12 = (a-4)(a-3)\}$

**4** **a** $\frac{12^n}{6^n} = \left(\frac{12}{6}\right)^n$
$= 2^n$

**b** $\frac{20^a}{2^a} = \left(\frac{20}{2}\right)^a$
$= 10^a$

**c** $\frac{6^b}{2^b} = \left(\frac{6}{2}\right)^b$
$= 3^b$

**d** $\frac{4^n}{20^n} = \left(\frac{4}{20}\right)^n$
$= \left(\frac{1}{5}\right)^n$
$= \frac{1}{5^n}$

**e** $\frac{35^x}{7^x} = \left(\frac{35}{7}\right)^x$
$= 5^x$

**f** $\frac{6^a}{8^a} = \left(\frac{6}{8}\right)^a$
$= \left(\frac{3}{4}\right)^a$

**g** $\frac{5^{n+1}}{5^n} = \frac{5 \times \cancel{5^n}}{\cancel{5^n}_1}$
$= 5$

**h** $\frac{5^{n+1}}{5} = \frac{\cancel{5} \times 5^n}{\cancel{5}_1}$
$= 5^n$

**5** **a** $\frac{6^m + 2^m}{2^m}$
$= \frac{2^m 3^m + 2^m}{2^m}$
$= \frac{\cancel{2^m}(3^m + 1)}{\cancel{2^m}_1}$
$= 3^m + 1$

**b** $\frac{2^n + 12^n}{2^n}$
$= \frac{2^n + 2^n 6^n}{2^n}$
$= \frac{\cancel{2^n}(1 + 6^n)}{\cancel{2^n}_1}$
$= 1 + 6^n$

**c** $\frac{8^n + 4^n}{2^n}$
$= \frac{2^n 4^n + 2^n 2^n}{2^n}$
$= \frac{\cancel{2^n}(4^n + 2^n)}{\cancel{2^n}_1}$
$= 4^n + 2^n$

**d** $\frac{12^x - 3^x}{3^x}$
$= \frac{3^x 4^x - 3^x}{3^x}$
$= \frac{\cancel{3^x}(4^x - 1)}{\cancel{3^x}_1}$
$= 4^x - 1$

**e** $\frac{6^n + 12^n}{1 + 2^n}$
$= \frac{6^n + 6^n 2^n}{1 + 2^n}$
$= \frac{6^n \cancel{(1 + 2^n)}}{\cancel{1 + 2^n}_1}$
$= 6^n$

**f** $\frac{5^{n+1} - 5^n}{4}$
$= \frac{5^n \times 5 - 5^n}{4}$
$= \frac{5^n \cancel{(5 - 1)}}{\cancel{4}_1}$
$= 5^n$

**g** $\frac{5^{n+1} - 5^n}{5^n}$
$= \frac{5^n \times 5 - 5^n}{5^n}$
$= \frac{\cancel{5^n}(5 - 1)}{\cancel{5^n}_1}$
$= 4$

**h** $\frac{4^n - 2^n}{2^n}$
$= \frac{2^n 2^n - 2^n}{2^n}$
$= \frac{\cancel{2^n}(2^n - 1)}{\cancel{2^n}_1}$
$= 2^n - 1$

**i** $\frac{2^n - 2^{n-1}}{2^n}$
$= \frac{2^{n-1} \times 2 - 2^{n-1}}{2^{n-1} \times 2}$
$= \frac{\cancel{2^{n-1}}(2 - 1)}{\cancel{2^{n-1}} \times 2}$
$= \frac{1}{2}$

**6** **a** $2^n(n+1) + 2^n(n-1)$
$= 2^n(n + 1 + n - 1)$
$= 2^n(2n)$
$= n2^{n+1}$

**b** $3^n\left(\frac{n-1}{6}\right) - 3^n\left(\frac{n+1}{6}\right)$
$= 3^n\left(\frac{n-1}{6} - \frac{n+1}{6}\right)$
$= 3^n(-\frac{1}{3})$
$= 3^n \times -3^{-1}$
$= -3^{n-1}$

## EXERCISE 3E

**1** **a** $2^x = 8$ $\therefore\ 2^x = 2^3$ $\therefore\ x = 3$

**b** $5^x = 25$ $\therefore\ 5^x = 5^2$ $\therefore\ x = 2$

**c** $3^x = 81$ $\therefore\ 3^x = 3^4$ $\therefore\ x = 4$

**d** $7^x = 1$ $\therefore\ 7^x = 7^0$ $\therefore\ x = 0$

**e** $3^x = \frac{1}{3}$ $\therefore\ 3^x = 3^{-1}$ $\therefore\ x = -1$

**f** $2^x = \sqrt{2}$ $\therefore\ 2^x = 2^{\frac{1}{2}}$ $\therefore\ x = \frac{1}{2}$

**g** $5^x = \frac{1}{125}$ $\therefore\ 5^x = 5^{-3}$ $\therefore\ x = -3$

**h** $4^{x+1} = 64$ $\therefore\ 4^{x+1} = 4^3$ $\therefore\ x+1 = 3$ $\therefore\ x = 2$

**i** $2^{x-2} = \frac{1}{32}$ $\therefore\ 2^{x-2} = 2^{-5}$ $\therefore\ x-2 = -5$ $\therefore\ x = -3$

**j** $3^{x+1} = \frac{1}{27}$ $\therefore\ 3^{x+1} = 3^{-3}$ $\therefore\ x+1 = -3$ $\therefore\ x = -4$

**k** $7^{x+1} = 343$ $\therefore\ 7^{x+1} = 7^3$ $\therefore\ x+1 = 3$ $\therefore\ x = 2$

**l** $5^{1-2x} = \frac{1}{5}$ $\therefore\ 5^{1-2x} = 5^{-1}$ $\therefore\ 1-2x = -1$ $\therefore\ -2x = -2$ $\therefore\ x = 1$

**2** **a** $8^x = 32$ $\therefore\ 2^{3x} = 2^5$ $\therefore\ 3x = 5$ $\therefore\ x = \frac{5}{3}$

**b** $4^x = \frac{1}{8}$ $\therefore\ 2^{2x} = 2^{-3}$ $\therefore\ 2x = -3$ $\therefore\ x = -\frac{3}{2}$

**c** $9^x = \frac{1}{27}$ $\therefore\ 3^{2x} = 3^{-3}$ $\therefore\ 2x = -3$ $\therefore\ x = -\frac{3}{2}$

**d** $25^x = \frac{1}{5}$ $\therefore\ 5^{2x} = 5^{-1}$ $\therefore\ 2x = -1$ $\therefore\ x = -\frac{1}{2}$

**e** $27^x = \frac{1}{9}$ $\therefore\ 3^{3x} = 3^{-2}$ $\therefore\ 3x = -2$ $\therefore\ x = -\frac{2}{3}$

**f** $16^x = \frac{1}{32}$ $\therefore\ 2^{4x} = 2^{-5}$ $\therefore\ 4x = -5$ $\therefore\ x = -\frac{5}{4}$

**g** $4^{x+2} = 128$ $\therefore\ 2^{2(x+2)} = 2^7$ $\therefore\ 2x+4 = 7$ $\therefore\ 2x = 3$ $\therefore\ x = \frac{3}{2}$

**h** $25^{1-x} = \frac{1}{125}$ $\therefore\ 5^{2(1-x)} = 5^{-3}$ $\therefore\ 2-2x = -3$ $\therefore\ -2x = -5$ $\therefore\ x = \frac{5}{2}$

**i** $4^{4x-1} = \frac{1}{2}$ $\therefore\ 2^{2(4x-1)} = 2^{-1}$ $\therefore\ 8x-2 = -1$ $\therefore\ 8x = 1$ $\therefore\ x = \frac{1}{8}$

**j** $9^{x-3} = 27$ $\therefore\ 3^{2(x-3)} = 3^3$ $\therefore\ 2x-6 = 3$ $\therefore\ 2x = 9$ $\therefore\ x = \frac{9}{2}$

**k** $\left(\frac{1}{2}\right)^{x+1} = 8$ $\therefore\ \left(2^{-1}\right)^{x+1} = 2^3$ $\therefore\ -x-1 = 3$ $\therefore\ -x = 4$ $\therefore\ x = -4$

**l** $\left(\frac{1}{3}\right)^{x+2} = 9$ $\therefore\ \left(3^{-1}\right)^{x+2} = 3^2$ $\therefore\ -x-2 = 2$ $\therefore\ -x = 4$ $\therefore\ x = -4$

**m** $81^x = 27^{-x}$ $\therefore\ 3^{4x} = 3^{-3x}$ $\therefore\ 4x = -3x$ $\therefore\ 7x = 0$ $\therefore\ x = 0$

**n** $\left(\frac{1}{4}\right)^{1-x} = 32$ $\therefore\ \left(2^{-2}\right)^{1-x} = 2^5$ $\therefore\ -2+2x = 5$ $\therefore\ 2x = 7$ $\therefore\ x = \frac{7}{2}$

**o** $\left(\frac{1}{7}\right)^x = 49$ $\therefore\ 7^{-x} = 7^2$ $\therefore\ -x = 2$ $\therefore\ x = -2$

**p** $\left(\frac{1}{3}\right)^{x+1} = 243$ $\therefore\ \left(3^{-1}\right)^{x+1} = 3^5$ $\therefore\ -x-1 = 5$ $\therefore\ -x = 6$ $\therefore\ x = -6$

**3** **a**
$$4^{2x+1} = 8^{1-x}$$
$$\therefore \left(2^2\right)^{2x+1} = \left(2^3\right)^{1-x}$$
$$\therefore 4x+2 = 3-3x$$
$$\therefore 7x = 1$$
$$\therefore x = \tfrac{1}{7}$$

**b**
$$9^{2-x} = \left(\tfrac{1}{3}\right)^{2x+1}$$
$$\therefore \left(3^2\right)^{2-x} = \left(3^{-1}\right)^{2x+1}$$
$$\therefore 4-2x = -2x-1$$
$$\therefore 4 = -1$$
This is clearly false, so no solutions exist.

**c**
$$2^x \times 8^{1-x} = \tfrac{1}{4}$$
$$\therefore 2^x \times \left(2^3\right)^{1-x} = 2^{-2}$$
$$\therefore x+3-3x = -2$$
$$\therefore -2x = -5$$
$$\therefore x = \tfrac{5}{2}$$
(or $2\frac{1}{2}$)

**4** **a**
$$3 \times 2^x = 24$$
$$\therefore 2^x = 8$$
$$\therefore 2^x = 2^3$$
$$\therefore x = 3$$

**b**
$$7 \times 2^x = 56$$
$$\therefore 2^x = 8$$
$$\therefore 2^x = 2^3$$
$$\therefore x = 3$$

**c**
$$3 \times 2^{x+1} = 24$$
$$\therefore 2^{x+1} = 8$$
$$\therefore 2^{x+1} = 2^3$$
$$\therefore x+1 = 3$$
$$\therefore x = 2$$

**d**
$$12 \times 3^{-x} = \tfrac{4}{3}$$
$$\therefore 3^{-x} = \tfrac{4}{3} \div 12$$
$$\therefore 3^{-x} = \tfrac{4}{3} \times \tfrac{1}{12}$$
$$\therefore 3^{-x} = \tfrac{1}{9}$$
$$\therefore 3^{-x} = 3^{-2}$$
$$\therefore x = 2$$

**e**
$$4 \times \left(\tfrac{1}{3}\right)^x = 36$$
$$\therefore \left(\tfrac{1}{3}\right)^x = 9$$
$$\therefore \left(3^{-1}\right)^x = 3^2$$
$$\therefore 3^{-x} = 3^2$$
$$\therefore -x = 2$$
$$\therefore x = -2$$

**f**
$$5 \times \left(\tfrac{1}{2}\right)^x = 20$$
$$\therefore \left(\tfrac{1}{2}\right)^x = 4$$
$$\therefore \left(2^{-1}\right)^x = 2^2$$
$$\therefore -x = 2$$
$$\therefore x = -2$$

**5** **a**
$$4^x - 6(2^x) + 8 = 0$$
$$\therefore (2^x)^2 - 6(2^x) + 8 = 0$$
$$\therefore (2^x-2)(2^x-4) = 0 \qquad \{a^2-6a+8 = (a-2)(a-4)\}$$
$$\therefore 2^x = 2 \text{ or } 4$$
$$\therefore 2^x = 2^1 \text{ or } 2^2$$
$$\therefore x = 1 \text{ or } 2$$

**b**
$$4^x - 2^x - 2 = 0$$
$$\therefore (2^x)^2 - 2^x - 2 = 0$$
$$\therefore (2^x-2)(2^x+1) = 0 \qquad \{a^2-a-2 = (a-2)(a+1)\}$$
$$\therefore 2^x = 2 \text{ or } -1$$
$$\therefore 2^x = 2^1 \qquad \{\text{since } 2^x \text{ cannot be negative}\}$$
$$\therefore x = 1$$

**c**
$$9^x - 12(3^x) + 27 = 0$$
$$\therefore (3^x)^2 - 12(3^x) + 27 = 0$$
$$\therefore (3^x-3)(3^x-9) = 0 \qquad \{a^2-12a+27 = (a-3)(a-9)\}$$
$$\therefore 3^x = 3 \text{ or } 9$$
$$\therefore 3^x = 3^1 \text{ or } 3^2$$
$$\therefore x = 1 \text{ or } 2$$

**d**
$$9^x = 3^x + 6$$
$$\therefore (3^x)^2 - 3^x - 6 = 0$$
$$\therefore (3^x-3)(3^x+2) = 0 \qquad \{a^2-a-6 = (a-3)(a+2)\}$$
$$\therefore 3^x = 3 \text{ or } -2$$
$$\therefore 3^x = 3^1 \qquad \{\text{since } 3^x \text{ cannot be negative}\}$$
$$\therefore x = 1$$

**e** $25^x - 23(5^x) - 50 = 0$

$\therefore\ (5^x)^2 - 23(5^x) - 50 = 0$

$\therefore\ (5^x - 25)(5^x + 2) = 0$ $\{a^2 - 23a - 50 = (a-25)(a+2)\}$

$\therefore\ 5^x = 25$ or $-2$

$\therefore\ 5^x = 5^2$ {since $5^x$ cannot be negative}

$\therefore\ x = 2$

**f** $49^x + 1 = 2(7^x)$

$\therefore\ (7^x)^2 - 2(7^x) + 1 = 0$

$\therefore\ (7^x - 1)^2 = 0$ $\{a^2 - 2a + 1 = (a-1)^2\}$

$\therefore\ 7^x = 1$

$\therefore\ 7^x = 7^0$

$\therefore\ x = 0$

## EXERCISE 3F

**1** **a** When $x = \frac{1}{2}$, $y = 2^{\frac{1}{2}}$

From point A, $y \approx 1.4$

$\therefore\ 2^{\frac{1}{2}} \approx 1.4$

**b** When $x = 0.8$, $y = 2^{0.8}$

From point B, $y \approx 1.7$

$\therefore\ 2^{0.8} \approx 1.7$

**c** When $x = 1.5$, $y = 2^{1.5}$

From point C, $y \approx 2.8$

$\therefore\ 2^{1.5} \approx 2.8$

**d** When $x = -\sqrt{2}$, $y = 2^{-\sqrt{2}}$

Using **a** we know $x \approx -1.4$

From point D, $y \approx 0.4$

$\therefore\ 2^{-\sqrt{2}} \approx 0.4$

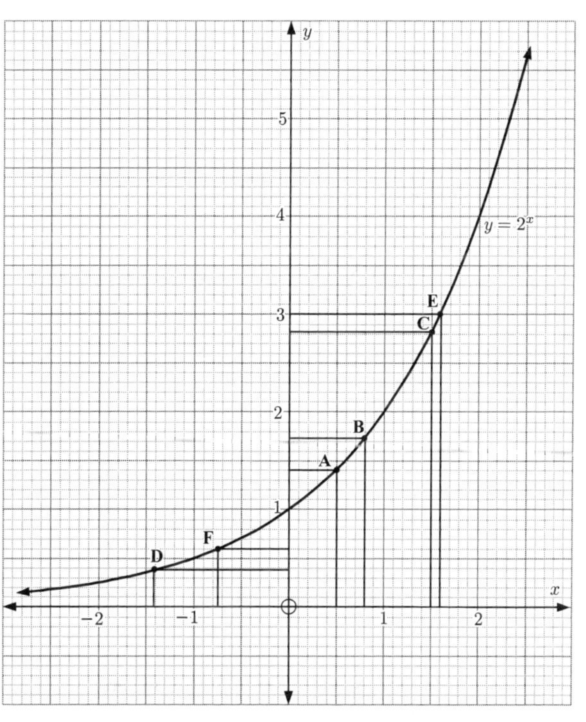


**2** **a** When $2^x = 3$, $x \approx 1.6$ from point E. **b** When $2^x = 0.6$, $x \approx -0.7$ from point F.

**3** The graph of $y = 2^x$ has a horizontal asymptote of $y = 0$.

$\therefore$ there is no value of $x$ such that $2^x = 0$.

$\therefore\ 2^x = 0$ has no solutions.

**4** **a**

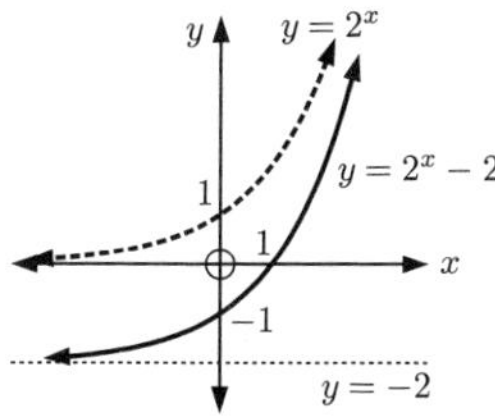


a vertical translation of 2 units downwards

$y = -2$ is the H.A.

**b**

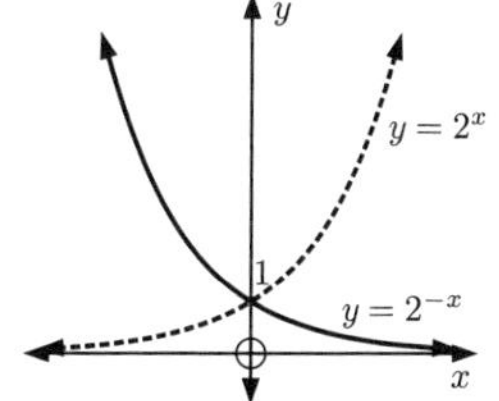


a reflection in the $y$-axis

**c**

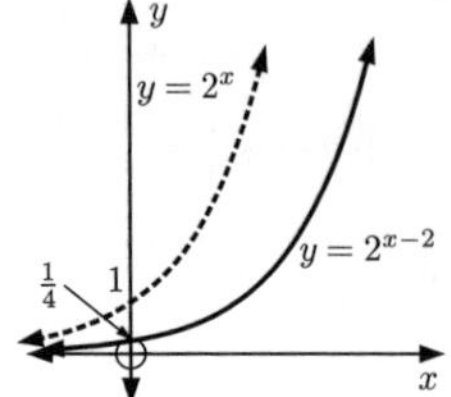


a horizontal translation of 2 units right

**d**

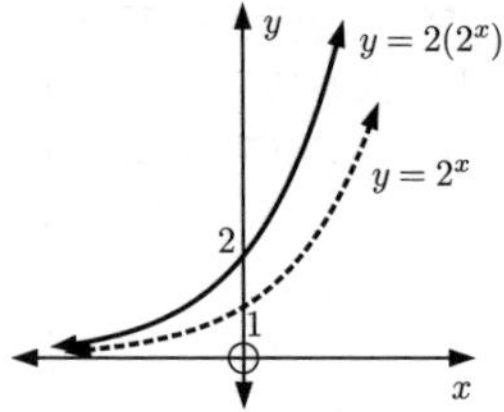


a vertical stretch of factor 2

**5** **a**

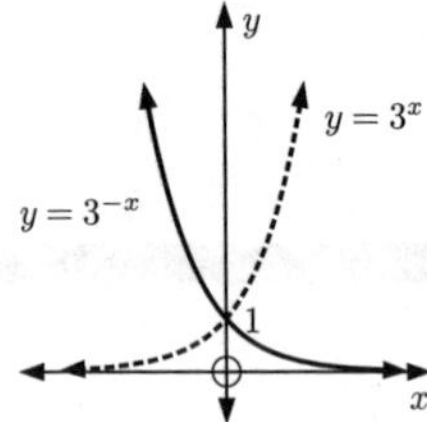


a reflection in the $y$-axis

**b**

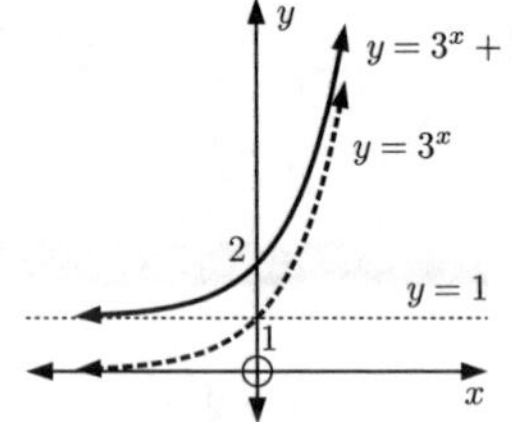


a vertical translation of 1 unit upwards

$y = 1$ is the H.A.

**c**

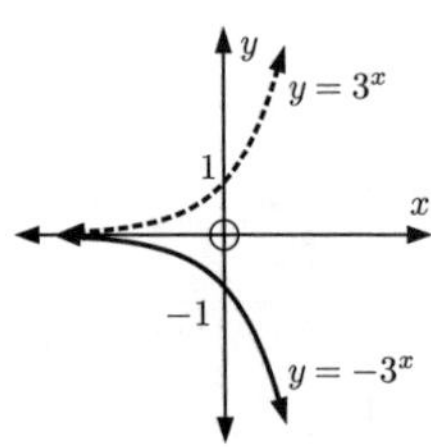


a reflection in the $x$-axis

**d**

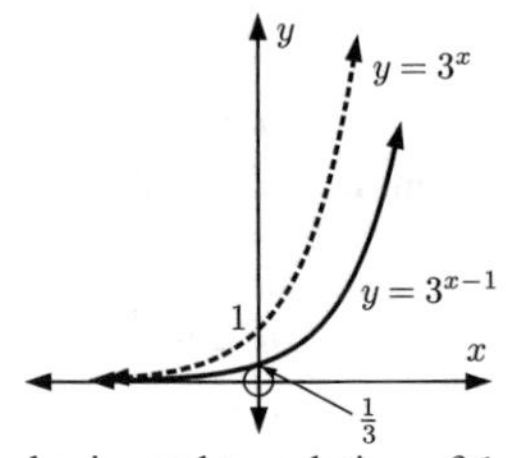


a horizontal translation of 1 unit right

**6** **a** **i**

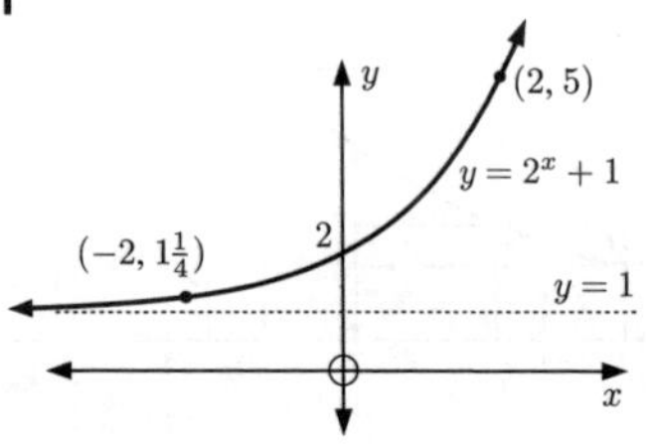


a vertical translation of 1 unit upwards

When $x = 2$, $y = 4 + 1 = 5$

When $x = -2$, $y = \frac{1}{4} + 1 = 1\frac{1}{4}$

**ii** Domain $= \{x \mid x \in \mathbb{R}\}$,

Range $= \{y \mid y > 1\}$

**iii** Using technology, when

$x = \sqrt{2}$, $y \approx 3.67$

**iv** As $x \to \infty$, $y \to \infty$

As $x \to -\infty$, $y \to 1^+$

**v** The horizontal asymptote is $y = 1$.

**b** **i**

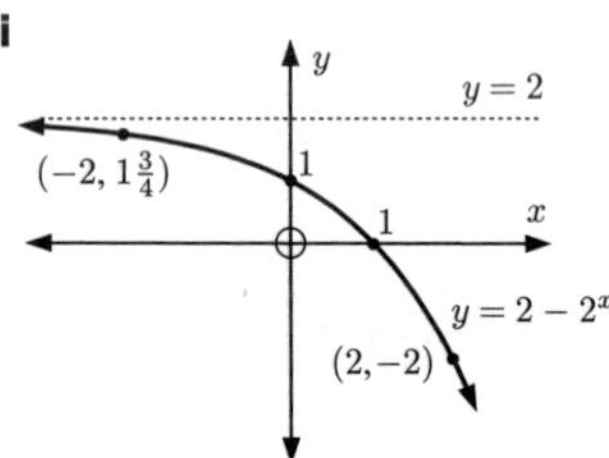


When $x = 0$, $y = 2 - 2^0 = 2 - 1 = 1$

$\therefore$ the $y$-intercept is 1

When $x = 1$, $y = 2 - 2 = 0$

When $x = 2$, $y = 2 - 4 = -2$

When $x = -2$, $y = 2 - \frac{1}{4} = 1\frac{3}{4}$

**ii** Domain $= \{x \mid x \in \mathbb{R}\}$,

Range $= \{y \mid y < 2\}$

**iii** Using technology, when

$x = \sqrt{2}$, $y \approx -0.665$

**iv** As $x \to \infty$, $y \to -\infty$

As $x \to -\infty$, $y \to 2^-$

**v** The horizontal asymptote is $y = 2$.

**c** **i**

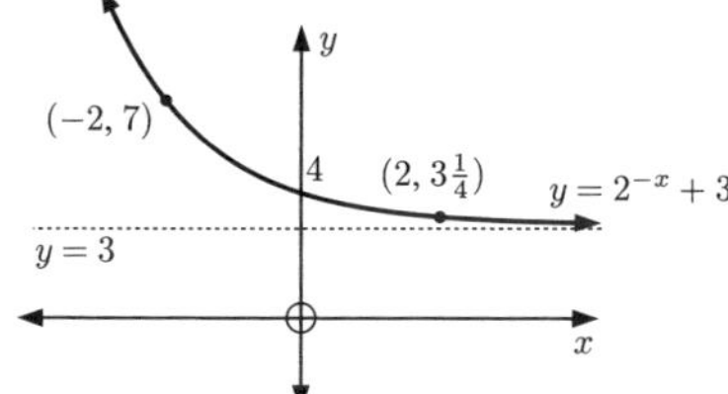


When $x = 0$, $y = 1 + 3 = 4$
When $x = 2$, $y = \frac{1}{4} + 3 = 3\frac{1}{4}$
When $x = -2$, $y = 2^2 + 3 = 7$

**ii** Domain $= \{x \mid x \in \mathbb{R}\}$,
Range $= \{y \mid y > 3\}$

**iii** Using technology, when
$x = \sqrt{2}$, $y \approx 3.38$

**iv** As $x \to \infty$, $y \to 3^+$
As $x \to -\infty$, $y \to \infty$

**v** The horizontal asymptote is $y = 3$.

**d** **i**

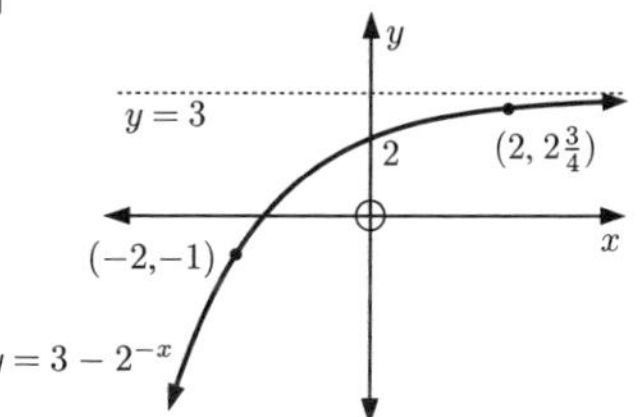


When $x = 0$, $y = 3 - 1 = 2$
When $x = 2$, $y = 3 - \frac{1}{4} = 2\frac{3}{4}$
When $x = -2$, $y = 3 - 4 = -1$

**ii** Domain $= \{x \mid x \in \mathbb{R}\}$,
Range $= \{y \mid y < 3\}$

**iii** Using technology, when
$x = \sqrt{2}$, $y \approx 2.62$

**iv** As $x \to \infty$, $y \to 3^-$
As $x \to -\infty$, $y \to -\infty$

**v** The horizontal asymptote is $y = 3$.

## EXERCISE 3G.1

**1** **a** When $t = 0$, $W_0 = 100$ grams = the initial weight

**b** **i** When $t = 4$,
$W_4 = 100 \times 2^{0.1\times 4}$
$= 100 \times 2^{0.4}$
$\approx 132$ grams

**ii** When $t = 10$,
$W_{10} = 100 \times 2^1$
$= 200$ grams

**iii** When $t = 24$, $W_{24} = 100 \times 2^{0.1\times 24}$
$= 100 \times 2^{2.4}$
$\approx 528$ grams

**c**

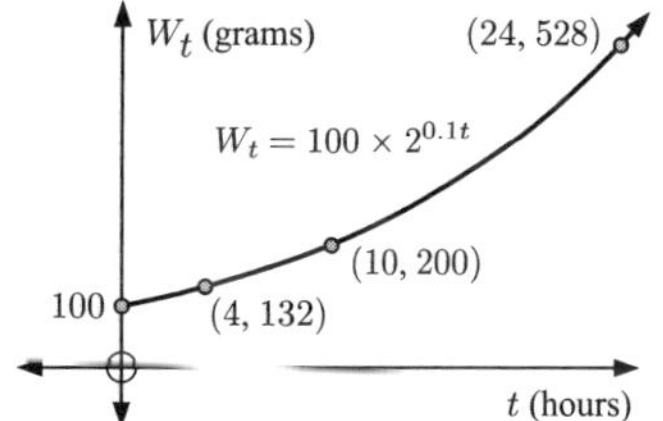


**2** **a** $P_0 = 50$ (the initial population)

**b** **i** When $n = 2$,
$P_2 = 50 \times 2^{0.3\times 2}$
$= 50 \times 2^{0.6}$
$\approx 75.785$
So, the expected population is 76 possums.

**ii** When $n = 5$,
$P_5 = 50 \times 2^{0.3\times 5}$
$= 50 \times 2^{1.5}$
$\approx 141.421$
So, the expected population is 141 possums.

**iii** When $n = 10$, $P_{10} = 50 \times 2^{0.3\times 10}$
$= 50 \times 2^3 = 400$ So, the expected population is 400 possums.

**c**

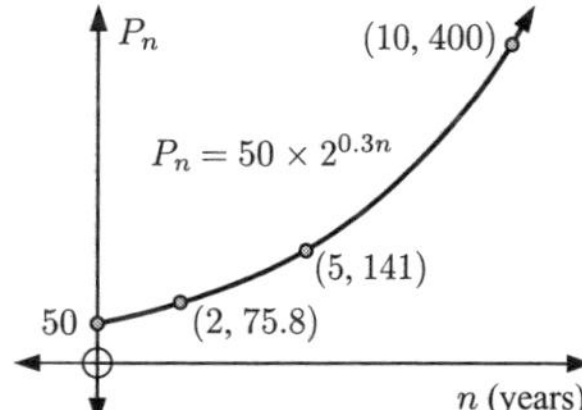


**3** **a** $B_0 = 6$ pairs $= 12$ bears

**b** In 2018, $t = 20$
$\therefore$ $B_{20} = 12 \times 2^{0.18\times 20}$
$= 12 \times 2^{3.6}$
$\approx 145.509$
$\approx 146$ bears

**c** In 2008, $t = 10$

$$\therefore \quad \% \text{ increase} = \left(\frac{B_{20} - B_{10}}{B_{10}}\right) \times 100\%$$
$$= \left(\frac{12 \times 2^{3.6} - 12 \times 2^{1.8}}{12 \times 2^{1.8}}\right) \times 100\%$$
$$= \left(\frac{2^{3.6} - 2^{1.8}}{2^{1.8}}\right) \times 100\%$$
$$\approx 248\%$$

**4** **a** **i** When $t = 0$,
$V_0 = V_0 \times 2^0$
$= V_0$
So, the speed is $V_0$.

**ii** When $t = 20$,
$V_{20} = V_0 \times 2^{0.05\times 20}$
$= V_0 \times 2^1$
$= 2V_0$
So, the speed is $2V_0$.

**b** $V_0$ becomes $2V_0$
So, there was a 100% increase in speed.

**c** $\left(\dfrac{V_{50} - V_{20}}{V_{20}}\right) \times 100\%$

$= \left(\dfrac{V_0 \times 2^{2.5} - V_0 \times 2^1}{V_0 \times 2^1}\right) \times 100\%$

$= \left(\dfrac{2^{2.5} - 2^1}{2^1}\right) \times 100\%$

$\approx 183\%$

This expression is the percentage increase in speed from the speed at 20°C to the speed at 50°C.
($V_{50} - V_{20}$ is the increase in speed.)

## EXERCISE 3G.2

**1** $W(t) = 250 \times (0.998)^t$ grams

**a** $W(0) = 250 \times (0.998)^0$
$= 250 \times 1$
$= 250$ grams $\quad\therefore$ 250 g of radioactive substance was put aside.

**b** **i** When $t = 400$
$W(400)$
$= 250 \times (0.998)^{400}$
$\approx 112$ grams

**ii** When $t = 800$
$W(800)$
$= 250 \times (0.998)^{800}$
$\approx 50.4$ grams

**iii** When $t = 1200$
$W(1200)$
$= 250 \times (0.998)^{1200}$
$\approx 22.6$ grams

**c**

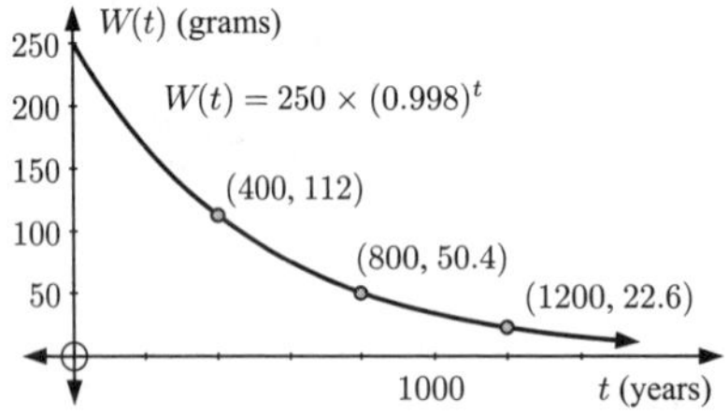

**d** When $W(t) = 125$
$250 \times (0.998)^t = 125$
$\therefore\ (0.998)^t = 0.5$
$\therefore\ t \approx 346.2$ {using technology}
It takes approximately 346 years.

**2** $T(t) = 100 \times 2^{-0.02t}$

**a** $T(0) = 100 \times 2^0$
$= 100 \times 1$
$= 100°C$

**b** **i** $T(15) = 100 \times 2^{-0.02\times 15}$
$= 100 \times 2^{-0.3}$
$\approx 81.2°C$

**ii** $T(20) = 100 \times 2^{-0.02\times 20}$
$= 100 \times 2^{-0.4}$
$\approx 75.8°C$

**iii** $T(78) = 100 \times 2^{-0.02\times 78}$
$= 100 \times 2^{-1.56}$
$\approx 33.9°C$

**c**

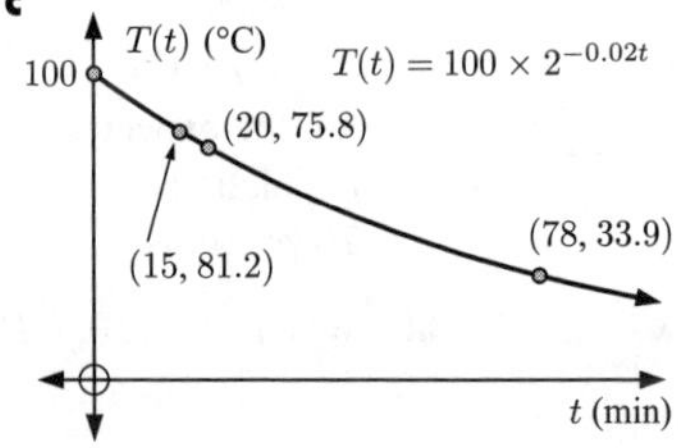

**3** **a** **i** $T(0) = -10 + 32 \times 2^0$
$= -10 + 32 \times 1$
$= 22°C$

**ii** $T(5) = -10 + 32 \times 2^{-0.2\times 5}$
$= -10 + 32 \times 2^{-1}$
$= 6°C$

**iii** $T(10) = -10 + 32 \times 2^{-0.2\times 10}$
$= -10 + 32 \times 2^{-2}$
$= -2°C$

**iv** $T(15) = -10 + 32 \times 2^{-0.2\times 15}$
$= -10 + 32 \times 2^{-3}$
$= -6°C$

**b**

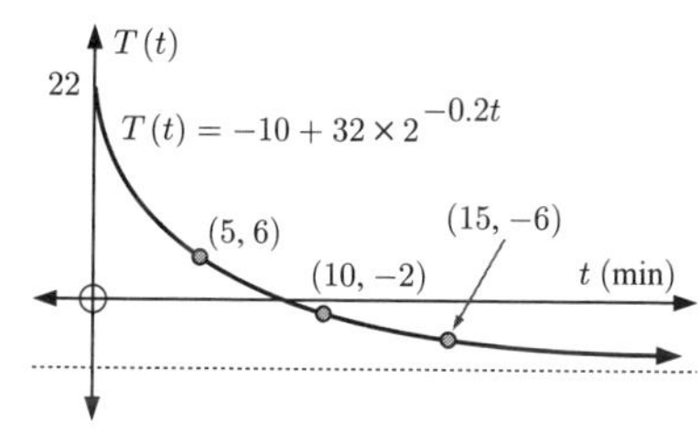


**c** $32 \times 2^{-0.2t}$ is always $> 0$ since $2^t$ is always $> 0$

$\therefore$ $-10 + 32 \times 2^{-0.2t}$ is always $> -10$

$\therefore$ the temperature of the packet of peas will never reach $-10°C$.

**4** $W_t = 1000 \times 2^{-0.03t}$

**a** $W_0 = 1000 \times 2^0$
$= 1000 \times 1$
$= 1000$ g

**b** **i** $W_{10}$
$= 1000 \times 2^{-0.3}$
$\approx 812$ g

**ii** $W_{100}$
$= 1000 \times 2^{-3}$
$= 125$ g

**iii** $W_{1000}$
$= 1000 \times 2^{-30}$
$\approx 9.31 \times 10^{-7}$ g

**c**

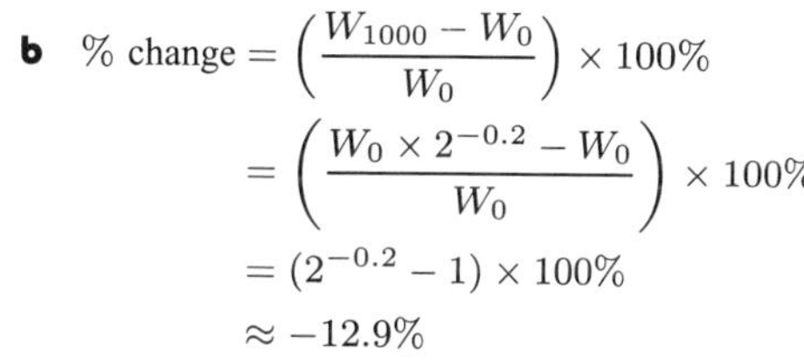


**d** When $W_t = 10$,
$1000 \times 2^{-0.03t} = 10$
$\therefore$ $(2^{-0.03})^t = 0.01$
$\therefore$ $t \approx 221.46$ {using technology}

There is 10 g of the substance remaining after approximately 221 years.

**e** Initial weight $= W_0 = 1000$ g

Amount remaining after $t$ years $= W_t = 1000 \times 2^{-0.03t}$

Amount that has decayed after $t$ years $= W_0 - W_t$
$= 1000 - 1000 \times 2^{-0.03t}$
$= 1000(1 - 2^{-0.03t})$ g

**5** **a** When $t = 0$, $W_0 = W_0 2^0$
$= W_0$ grams

$\therefore$ the original weight was $W_0$ grams.

**b** $\%$ change $= \left(\dfrac{W_{1000} - W_0}{W_0}\right) \times 100\%$
$= \left(\dfrac{W_0 \times 2^{-0.2} - W_0}{W_0}\right) \times 100\%$
$= (2^{-0.2} - 1) \times 100\%$
$\approx -12.9\%$

The weight loss was about 12.9%.

**c** $W_0 \times 2^{-0.0002t} = \frac{1}{512} W_0$
$\therefore$ $(2^{-0.0002})^t = \frac{1}{512}$
$\therefore$ $t = 45\,000$ {using technology}

It would take 45 000 years.

## EXERCISE 3H

**1**

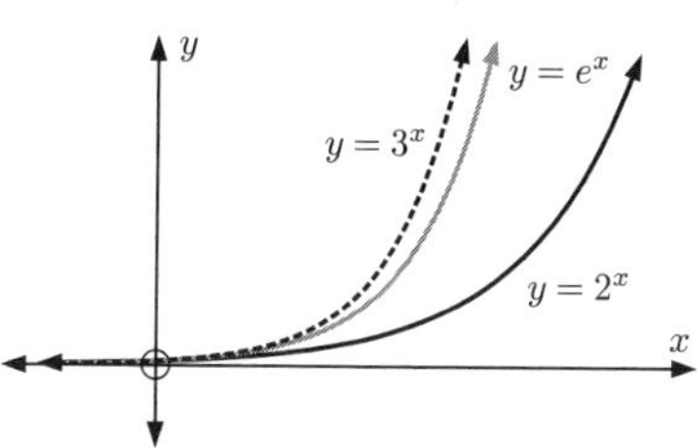


The graph of $y = e^x$ lies between $y = 2^x$ and $y = 3^x$.

**2**

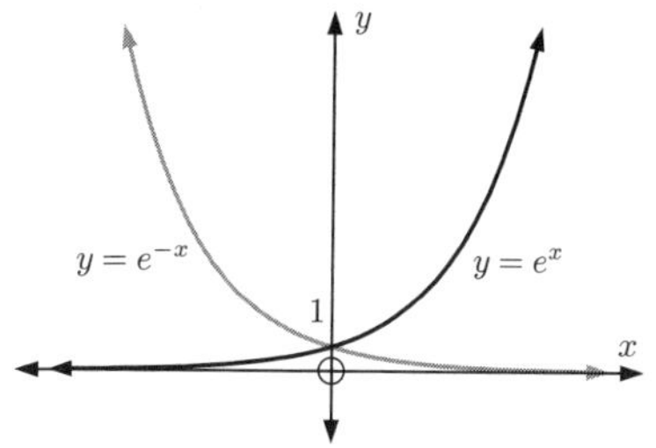


One is the other reflected in the $y$-axis.

**3** When $x=0$, $y=ae^0=a\times 1=a$ $\quad\therefore$ the $y$-intercept is $a$.

**4** **a** The graph of $y=e^x$ is entirely above the $x$-axis.

$y>0$ for all $x$

$\therefore\ e^x>0$ for all $x$

$\therefore\ 2e^x>0$ for all $x$

$\therefore\ y=2e^x$ cannot be negative.

**b** **i** When $x=-20$, $y=2e^{-20}\approx 4.12\times 10^{-9}$
$\approx 0.000\,000\,004\,12$

**ii** When $x=20$, $y=2e^{20}\approx 9.70\times 10^8$
$\approx 970\,000\,000$

**5** **a** $e^2\approx 7.39$ **b** $e^3\approx 20.1$ **c** $e^{0.7}\approx 2.01$ **d** $\sqrt{e}\approx 1.65$ **e** $e^{-1}\approx 0.368$

**6** **a** $\sqrt{e}=e^{\frac{1}{2}}$

**b** $\dfrac{1}{\sqrt{e}}$
$=\dfrac{1}{e^{\frac{1}{2}}}$
$=e^{-\frac{1}{2}}$

**c** $\dfrac{1}{e^2}=e^{-2}$

**d** $e\sqrt{e}$
$=e^1e^{\frac{1}{2}}$
$=e^{\frac{3}{2}}$

**7** **a** $\left(e^{0.36}\right)^{\frac{t}{2}}$
$=e^{0.36\times\frac{t}{2}}$
$=e^{0.18t}$

**b** $\left(e^{0.064}\right)^{\frac{t}{16}}$
$=e^{0.064\times\frac{t}{16}}$
$=e^{0.004t}$

**c** $\left(e^{-0.04}\right)^{\frac{t}{8}}$
$=e^{-0.04\times\frac{t}{8}}$
$=e^{-0.005t}$

**d** $\left(e^{-0.836}\right)^{\frac{t}{5}}$
$=e^{-0.836\times\frac{t}{5}}$
$\approx e^{-0.167t}$

**8** **a** $\approx 10.074$ **b** $\approx 0.099\,261$ **c** $\approx 125.09$ **d** $\approx 0.007\,994\,5$
**e** $\approx 41.914$ **f** $\approx 42.429$ **g** $\approx 3540.3$ **h** $\approx 0.006\,342\,4$

**9**

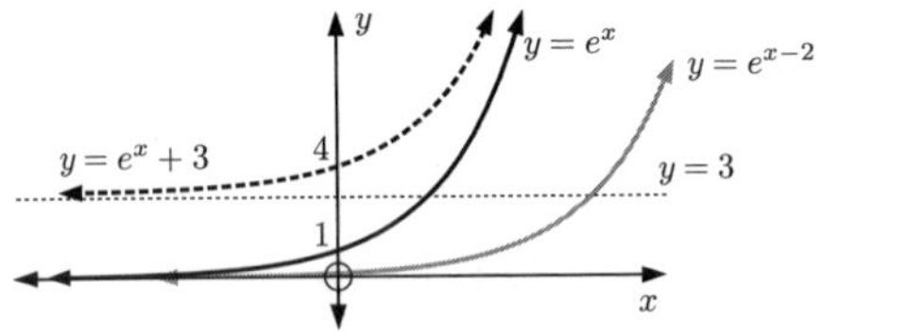

Domain of $f$, $g$, and $h$ is $\{x \mid x\in\mathbb{R}\}$
Range of $f$ is $\{y \mid y>0\}$
Range of $g$ is $\{y \mid y>0\}$
Range of $h$ is $\{y \mid y>3\}$

**10**

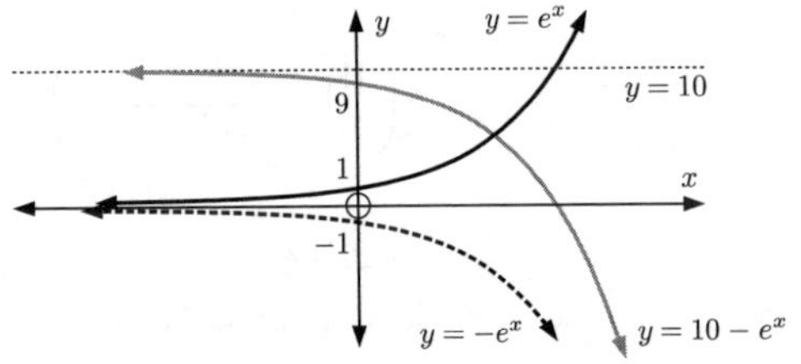

Domain of $f$, $g$, and $h$ is $\{x \mid x\in\mathbb{R}\}$
Range of $f$ is $\{y \mid y>0\}$
Range of $g$ is $\{y \mid y<0\}$
Range of $h$ is $\{y \mid y<10\}$

**11** **a** $(e^x+1)^2$
$=(e^x)^2+2\times e^x\times 1+1^2$
$=e^{2x}+2e^x+1$

**b** $(1+e^x)(1-e^x)$
$=1^2-(e^x)^2$
$=1-e^{2x}$

**c** $e^x(e^{-x}-3)$
$=e^x\times e^{-x}-e^x\times 3$
$=e^0-3e^x$
$=1-3e^x$

**12** $W(t)=2e^{\frac{t}{2}}$ grams

**a** **i** $W(0)=2e^0$
$=2\times 1$
$=2$ g

**ii** $W(\frac{1}{2})=2e^{\frac{1}{4}}$
$\approx 2.57$ g

**iii** $W(1\frac{1}{2})=2e^{\frac{3}{4}}$
$\approx 4.23$ g

**iv** $W(6)=2e^3$
$\approx 40.2$ g

**b**

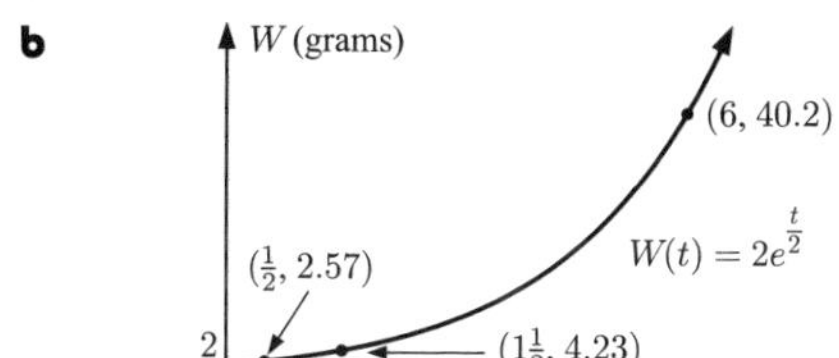


**13** **a**

$$e^x = \sqrt{e}$$
$$\therefore\ e^x = e^{\frac{1}{2}}$$
$$\therefore\ x = \tfrac{1}{2}$$

**b**

$$e^{\frac{1}{2}x} = \frac{1}{e^2}$$
$$\therefore\ e^{\frac{1}{2}x} = e^{-2}$$
$$\therefore\ \tfrac{1}{2}x = -2$$
$$\therefore\ x = -4$$

**14** $I(t) = 75e^{-0.15t}$

**a** **i** $I(1) = 75e^{-0.15}$
$\approx 64.6$ amps

**ii** $I(10) = 75e^{-1.5}$
$\approx 16.7$ amps

**b**

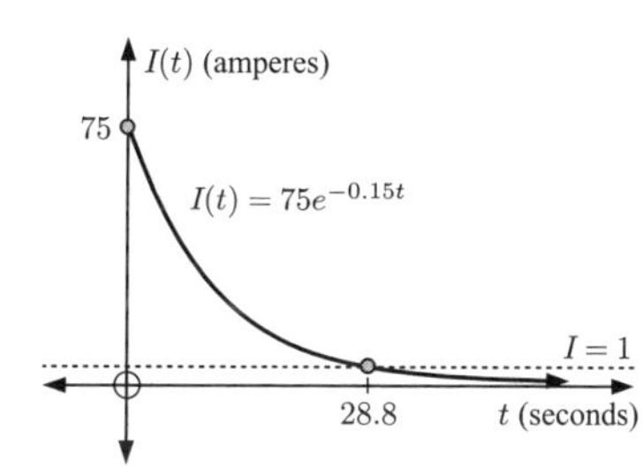


**c** We need to solve $75e^{-0.15t} = 1$.
Using technology, $t \approx 28.8$ s

**15** $f(x) = e^x$

**a**

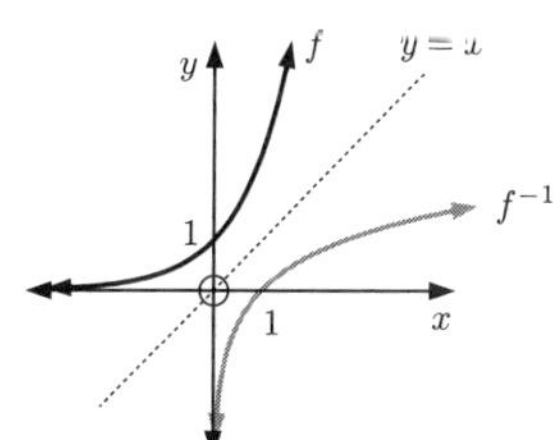


**b** Domain of $f^{-1}$ is $\{x \mid x > 0\}$
Range of $f^{-1}$ is $\{y \mid y \in \mathbb{R}\}$

## REVIEW SET 3A

**1** **a**

$$-(-1)^{10}$$
$$= -1$$

**b**

$$-(-3)^3$$
$$= -(-27)$$
$$= 27$$

**c**

$$3^0 - 3^{-1}$$
$$= 1 - \tfrac{1}{3}$$
$$= \tfrac{2}{3}$$

**2** **a**

$$a^4b^5 \times a^2b^2$$
$$= a^{4+2} \times b^{5+2}$$
$$= a^6b^7$$

**b**

$$6xy^5 \div 9x^2y^5$$
$$= \tfrac{6}{9}x^{1-2}y^{5-5}$$
$$= \tfrac{2}{3}x^{-1}y^0$$
$$= \frac{2}{3x}$$

**c**

$$\frac{5(x^2y)^2}{(5x^2)^2}$$
$$= \frac{5 \times x^4y^2}{25x^4}$$
$$= \tfrac{1}{5}x^0y^2$$
$$= \frac{y^2}{5}$$

**3** **a** **i** $f(4) = 3^4 = 81$

**ii** $f(-1) = 3^{-1} = \frac{1}{3}$

**b** $f(x+2) = kf(x)$
$\therefore\ 3^{x+2} = k \times 3^x$
$\therefore\ 3^2 \times 3^x = k \times 3^x$
$\therefore\ 3^2 = k$
$\therefore\ k = 9$

**4** **a** $x^{-2} \times x^{-3} = x^{-2+(-3)} = x^{-5} = \dfrac{1}{x^5}$

**b** $2(ab)^{-2} = 2 \times \dfrac{1}{(ab)^2} = \dfrac{2}{a^2b^2}$

**c** $2ab^{-2} = 2a \times \left(\dfrac{1}{b^2}\right) = \dfrac{2a}{b^2}$

**5** **a** $\dfrac{27}{9^a} = \dfrac{3^3}{(3^2)^a} = 3^{3-2a}$

**b** $\left(\sqrt{3}\right)^{1-x} \times 9^{1-2x} = (3^{\frac{1}{2}})^{1-x} \times (3^2)^{1-2x} = 3^{\frac{1}{2}-\frac{1}{2}x+2-4x} = 3^{\frac{5}{2}-\frac{9}{2}x}$

**6** **a** $8^{\frac{2}{3}} = \left(2^3\right)^{\frac{2}{3}} = 2^2 = 4$

**b** $27^{-\frac{2}{3}} = \left(3^3\right)^{-\frac{2}{3}} = 3^{-2} = \dfrac{1}{3^2} = \frac{1}{9}$

**7** **a** $mn^{-2} = m \times \dfrac{1}{n^2} = \dfrac{m}{n^2}$

**b** $(mn)^{-3} = \dfrac{1}{(mn)^3} = \dfrac{1}{m^3n^3}$

**c** $\dfrac{m^2n^{-1}}{p^{-2}} = m^2\left(\dfrac{1}{n}\right)p^2 = \dfrac{m^2p^2}{n}$

**d** $(4m^{-1}n)^2 = 4^2m^{-2}n^2 = \dfrac{16n^2}{m^2}$

**8** **a** $(3-e^x)^2 = 3^2 - 2 \times 3 \times e^x + (e^x)^2 = 9 - 6e^x + e^{2x}$

**b** $(\sqrt{x}+2)(\sqrt{x}-2) = (\sqrt{x})^2 - 2^2 = x - 4$

**c** $2^{-x}(2^{2x} + 2^x) = 2^{-x+2x} + 2^{-x+x} = 2^x + 2^0 = 2^x + 1$

**9** **a** $2^{x-3} = \frac{1}{32}$
$\therefore\ 2^{x-3} = 2^{-5}$
$\therefore\ x - 3 = -5$
$\therefore\ x = -2$

**b** $9^x = 27^{2-2x}$
$\therefore\ (3^2)^x = (3^3)^{2-2x}$
$\therefore\ 2x = 6 - 6x$
$\therefore\ 8x = 6$
$\therefore\ x = \frac{6}{8} = \frac{3}{4}$

**c** $e^{2x} = \dfrac{1}{\sqrt{e}}$
$\therefore\ e^{2x} = e^{-\frac{1}{2}}$
$\therefore\ 2x = -\frac{1}{2}$
$\therefore\ x = -\frac{1}{4}$

**10** Use the general exponential function $y = a \times b^{x-c} + d$.

**a** $y = -e^x$

$a = -1 \quad \therefore\ a < 0$, $b = e \quad \therefore\ b > 1$ } function is decreasing

When $x = 0$, $y = -e^0 = -1$
$\therefore$ $y$-intercept is $y = -1$.
$\therefore$ the graph is **C**.

**b** $y = 3 \times 2^x$

$a = 3 \quad \therefore\ a > 0$, $b = 2 \quad \therefore\ b > 1$ } function is increasing

When $x = 0$, $y = 3 \times 2^0 = 3$
$\therefore$ $y$-intercept is $y = 3$.
$\therefore$ the graph is **E**.

**c** $y = e^x + 1$

$a = 1 \quad \therefore \quad a > 0$, $b = e \quad \therefore \quad b > 1$ } function is increasing

When $x = 0$, $y = e^0 + 1 = 2$

$\therefore$ $y$-intercept is $y = 2$.

$d = 1 \quad \therefore \quad y = 1$ is a horizontal asymptote.

$\therefore$ the graph is **A**.

**d** $y = 3^{-x} = \dfrac{1}{3^x} = \left(\frac{1}{3}\right)^x$

$a = 1 \quad \therefore \quad a > 0$, $b = \frac{1}{3} \quad \therefore \quad 0 < b < 1$ } function is decreasing

When $x = 0$, $y = 3^0 = 1$

$\therefore$ $y$-intercept is $y = 1$.

$\therefore$ the graph is **B**.

**e** $y = -e^{-x} = -\dfrac{1}{e^x} = -\left(\dfrac{1}{e}\right)^x$

$a = -1 \quad \therefore \quad a < 0$, $b = \dfrac{1}{e} \quad \therefore \quad 0 < b < 1$ } function is increasing

When $x = 0$, $y = -e^0 = -1$

$\therefore$ $y$-intercept is $y = -1$.

$\therefore$ the graph is **D**.

**11** $y = a^x$

**a** $a^{2x} = (a^x)^2 = y^2$

**b** $a^{-x} = (a^x)^{-1} = y^{-1}$

**c** $\dfrac{1}{\sqrt{a^x}} = \dfrac{1}{\sqrt{y}} = y^{-\frac{1}{2}}$

## REVIEW SET 3B

**1** **a** $4 \times 2^n = 2^2 \times 2^n = 2^{n+2}$

**b** $7^{-1} - 7^0 = \frac{1}{7} - 1 = -\frac{6}{7}$

**c** $\left(\frac{2}{3}\right)^{-3} = \left(\frac{3}{2}\right)^3 = \frac{27}{8} = 3\frac{3}{8}$

**d** $\left(\dfrac{2a^{-1}}{b^2}\right)^2 = \dfrac{2^2a^{-2}}{b^4} = \dfrac{4}{a^2b^4}$

**2** **a** $3^{\frac{3}{4}} \approx 2.28$ **b** $27^{-\frac{1}{5}} \approx 0.517$ **c** $\sqrt[4]{100} \approx 3.16$

**3** $f(x) = 3 \times 2^x$

**a** $f(0) = 3 \times 2^0 = 3 \times 1 = 3$

**b** $f(3) = 3 \times 2^3 = 3 \times 8 = 24$

**c** $f(-2) = 3 \times 2^{-2} = 3 \times \dfrac{1}{2^2} = \frac{3}{4}$

**4** $f(x) = 2^{-x} + 1$

**a** $f(\frac{1}{2}) = 2^{-\frac{1}{2}} + 1 = \dfrac{1}{\sqrt{2}} + 1 \approx 1.71$

**b** $f(a) = 3$

$\therefore \quad 2^{-a} + 1 = 3$

$\therefore \quad 2^{-a} = 2$

$\therefore \quad 2^{-a} = 2^1$

$\therefore \quad -a = 1$

$\therefore \quad a = -1$

**5**

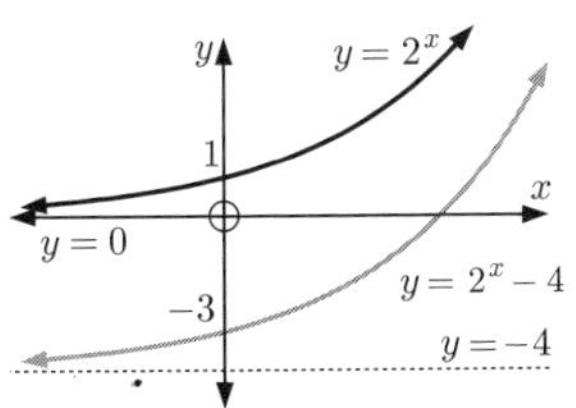


$y = 2^x$ has $y$-intercept 1 and horizontal asymptote $y = 0$

$y = 2^x - 4$ has $y$-intercept $-3$ and horizontal asymptote $y = -4$

**6** $T = 80 \times (0.913)^t$ °C

**a** When $t = 0$, $T = 80 \times (0.913)^0$
$= 80 \times 1$
$= 80$ $\therefore$ the initial temperature was 80°C.

**b** **i** When $t = 12$,
$T = 80 \times (0.913)^{12}$
$\approx 26.8$°C

**ii** When $t = 24$,
$T = 80 \times (0.913)^{24}$
$\approx 9.00$°C

**iii** When $t = 36$,
$T = 80 \times (0.913)^{36}$
$\approx 3.02$°C

**c**

T (°C)
80
60
40
20
$T = 80 \times (0.913)^t$
(12, 26.8)
(24, 9.00)
(36, 3.02)
10 20 30 40
t (minutes)

**d** When $T = 25$
$80 \times (0.913)^t = 25$
$\therefore\ 0.913^t = 0.3125$
$\therefore\ t \approx 12.8$ min {using technology}

**7** **a** When $x = 0$, $y = 3^0 - 5 = 1 - 5 = -4$
When $x = 1$, $y = 3^1 - 5 = 3 - 5 = -2$
When $x = 2$, $y = 3^2 - 5 = 9 - 5 = 4$
When $x = -1$, $y = 3^{-1} - 5 = \frac{1}{3} - 5 = -4\frac{2}{3}$
When $x = -2$, $y = 3^{-2} - 5 = \frac{1}{9} - 5 = -4\frac{8}{9}$

**b** As $x \to \infty$, $3^x \to \infty$
and so $y \to \infty$
As $x \to -\infty$, $3^x \to 0$
and so $y \to -5^+$

**c**

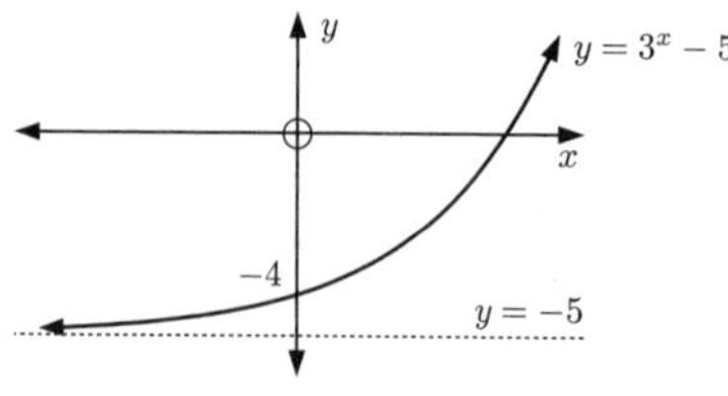


**d** $y = -5$ is the horizontal asymptote.

**8** **a**

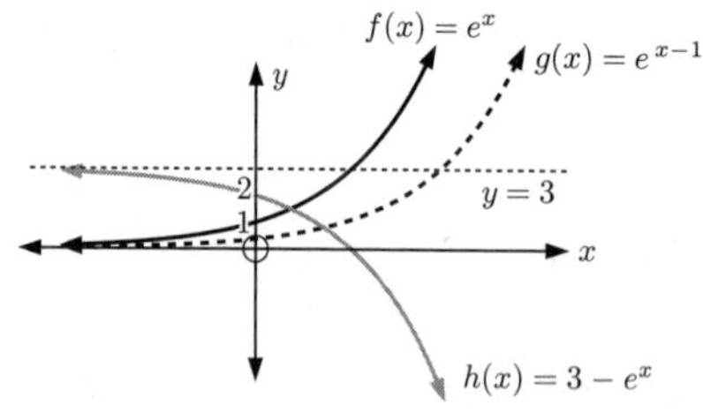


**b** Domain of $f$, $g$, and $h$ is $\{x \mid x \in \mathbb{R}\}$
Range of $f$ is $\{y \mid y > 0\}$
Range of $g$ is $\{y \mid y > 0\}$
Range of $h$ is $\{y \mid y < 3\}$

**9** **a** When $x = 0$, $y = 3 - 2^0 = 3 - 1 = 2$
When $x = 1$, $y = 3 - 2^{-1} = 3 - \frac{1}{2} = 2\frac{1}{2}$
When $x = 2$, $y = 3 - 2^{-2} = 3 - \frac{1}{4} = 2\frac{3}{4}$
When $x = -1$, $y = 3 - 2^1 = 3 - 2 = 1$
When $x = -2$, $y = 3 - 2^2 = 3 - 4 = -1$

**b** As $x \to \infty$, $2^{-x} \to 0$,
$\therefore\ y \to 3^-$
As $x \to -\infty$, $2^{-x} \to \infty$,
$\therefore\ y \to -\infty$

**c**

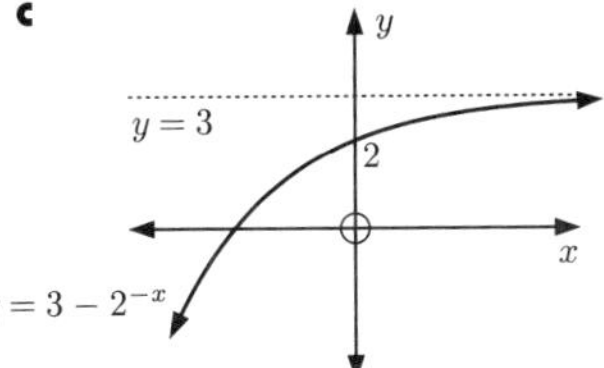


**d** horizontal asymptote is $y = 3$

**10** $W = 1500 \times (0.993)^t$ grams

**a** When $t = 0$,
$W = 1500 \times (0.993)^0$
$= 1500 \times 1$
$= 1500$ grams

**b** **i** When $t = 400$,
$W = 1500 \times (0.993)^{400}$
$\approx 90.3$ grams

**ii** When $t = 800$,
$W = 1500 \times (0.993)^{800}$
$\approx 5.44$ grams

**c**

W (grams)
1500
$W = 1500 \times (0.993)^t$
t (years)
400
800

**d** When $W = 100$,
$1500 \times (0.993)^t = 100$
$\therefore\ (0.993)^t \approx 0.0667$
$\therefore\ t \approx 385.5$ {using technology}
So, it will take about 386 years.

## REVIEW SET 3C

**1** **a** When $y = 3^x = 5$,
$x \approx 1.5$ from point A.

**b** When $y = 3^x = \frac{1}{2}$,
$x \approx -0.6$ from point B.

**c** $6 \times 3^x = 20$
$\therefore\ 3^x = \frac{20}{6} = 3\frac{1}{3}$
When $y = 3^x = 3\frac{1}{3}$,
$x \approx 1.1$ from point C.

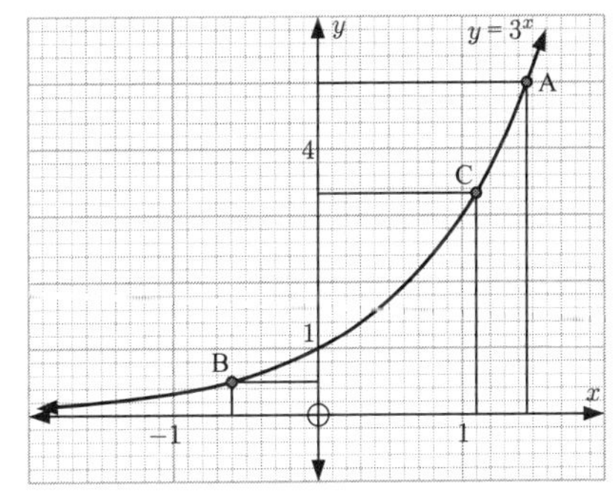


**2** **a** $(a^7)^3$
$= a^{7\times 3}$
$= a^{21}$

**b** $pq^2 \times p^3q^4$
$= p^{1+3}q^{2+4}$
$= p^4q^6$

**c** $\dfrac{8ab^5}{2a^4b^4}$
$= \frac{8}{2}a^{1-4}b^{5-4}$
$= 4a^{-3}b^1$
$= \dfrac{4b}{a^3}$

**3** **a** $2 \times 2^{-4}$
$= 2^1 \times 2^{-4}$
$= 2^{1+(-4)}$
$= 2^{-3}$

**b** $16 \div 2^{-3}$
$= 2^4 \div 2^{-3}$
$= 2^{4-(-3)}$
$= 2^7$

**c** $8^4$
$= (2^3)^4$
$= 2^{12}$

**4** **a** $b^{-3} = \dfrac{1}{b^3}$

**b** $(ab)^{-1}$
$= a^{-1}b^{-1}$
$= \dfrac{1}{ab}$

**c** $ab^{-1}$
$= a \times \dfrac{1}{b}$
$= \dfrac{a}{b}$

**5** $\dfrac{2^{x+1}}{2^{1-x}} = 2^{x+1-(1-x)}$
$= 2^{x+1-1+x}$
$= 2^{2x}$

**6** **a** $1 = 5^0$

**b** $5\sqrt{5}$
$= 5^1 \times 5^{\frac{1}{2}}$
$= 5^{\frac{3}{2}}$

**c** $\dfrac{1}{\sqrt[4]{5}}$
$= \dfrac{1}{5^{\frac{1}{4}}}$
$= 5^{-\frac{1}{4}}$

**d** $25^{a+3}$
$= (5^2)^{a+3}$
$= 5^{2a+6}$

**7** **a** $e^x(e^{-x} + e^x)$
$= e^0 + e^{2x}$
$= 1 + e^{2x}$

**b** $(2^x + 5)^2$
$= (2^x)^2 + 2 \times 2^x \times 5 + 5^2$
$= 2^{2x} + 5 \times 2^{x+1} + 25$
$= 4^x + 5 \times 2^{x+1} + 25$
$\{or \;\; 2^{2x} + 10(2^x) + 25\}$

**c** $(x^{\frac{1}{2}} - 7)(x^{\frac{1}{2}} + 7)$
$= (x^{\frac{1}{2}})^2 - 7^2$
$= x^1 - 49$
$= x - 49$

**8** **a** $6 \times 2^x = 192$
$\therefore \; 2^x = 32$
$\therefore \; 2^x = 2^5$
$\therefore \; x = 5$

**b** $4 \times (\frac{1}{3})^x = 324$
$\therefore \; (\frac{1}{3})^x = 81$
$\therefore \; (3^{-1})^x = 3^4$
$\therefore \; 3^{-x} = 3^4$
$\therefore \; x = -4$

**9** The point $(1, \sqrt{8})$ lies on the graph of $y = 2^{kx}$.
$\therefore \; 2^{k \times 1} = \sqrt{8}$
$\therefore \; 2^k = \sqrt{2^3}$
$\therefore \; 2^k = 2^{\frac{3}{2}}$
$\therefore \; k = \frac{3}{2}$

**10** **a** $2^{x+1} = 32$
$\therefore \; 2^{x+1} = 2^5$
$\therefore \; x + 1 = 5$
$\therefore \; x = 4$

**b** $4^{x+1} = \left(\frac{1}{8}\right)^x$
$\therefore \; (2^2)^{x+1} = (2^{-3})^x$
$\therefore \; 2x + 2 = -3x$
$\therefore \; 5x = -2$
$\therefore \; x = -\frac{2}{5}$

**11** **a** When $x = 0$, $\quad y = 2e^{-0} + 1 = 3$
When $x = 1$, $\quad y = 2e^{-1} + 1 \approx 1.74$
When $x = 2$, $\quad y = 2e^{-2} + 1 \approx 1.27$
When $x = -1$, $\quad y = 2e^{1} + 1 \approx 6.44$
When $x = -2$, $\quad y = 2e^{2} + 1 \approx 15.8$

**b** 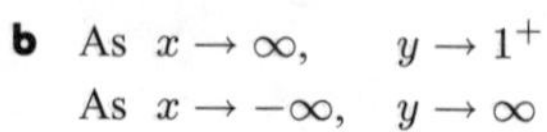
As $x \to \infty$, $\quad y \to 1^+$
As $x \to -\infty$, $\quad y \to \infty$

**c** 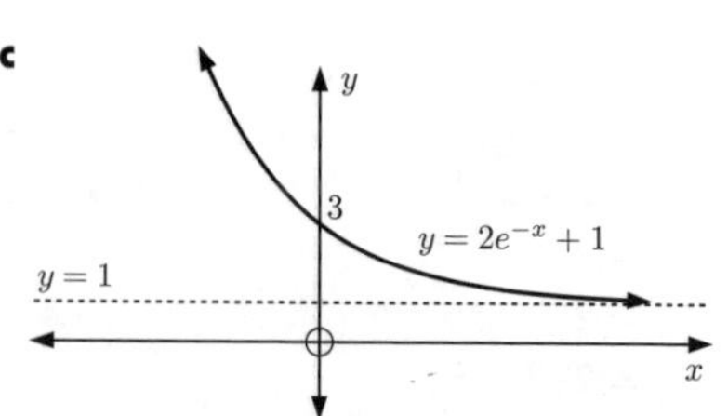


**d** $y = 1$ is a horizontal asymptote.

# Chapter 4
## LOGARITHMS

### EXERCISE 4A

**1** **a** $\log 10\,000 = \log_{10} 10^4 = 4$

**b** $\log 0.001 = \log_{10} 10^{-3} = -3$

**c** $\log 10 = \log_{10} 10^1 = 1$

**d** $\log 1 = \log_{10} 10^0 = 0$

**e** $\log \sqrt{10} = \log_{10} 10^{\frac{1}{2}} = \frac{1}{2}$

**f** $\log \sqrt[3]{10} = \log_{10} 10^{\frac{1}{3}} = \frac{1}{3}$

**g** $\log\left(\frac{1}{\sqrt[4]{10}}\right) = \log_{10} 10^{-\frac{1}{4}} = -\frac{1}{4}$

**h** $\log 10\sqrt{10} = \log_{10} 10^{\frac{3}{2}} = \frac{3}{2}$

**i** $\log \sqrt[3]{100} = \log_{10}(10^2)^{\frac{1}{3}} = \log_{10} 10^{\frac{2}{3}} = \frac{2}{3}$

**j** $\log\left(\frac{100}{\sqrt{10}}\right) = \log_{10}\left(\frac{10^2}{10^{\frac{1}{2}}}\right) = \log_{10} 10^{\frac{3}{2}} = \frac{3}{2}$

**k** $\log\left(10 \times \sqrt[3]{10}\right) = \log_{10}(10^1 \times 10^{\frac{1}{3}}) = \log_{10} 10^{\frac{4}{3}} = \frac{4}{3}$

**l** $\log 1000\sqrt{10} = \log_{10}(10^3 \times 10^{\frac{1}{2}}) = \log_{10} 10^{\frac{7}{2}} = \frac{7}{2}$

**2** **a** $\log 10^n = \log_{10} 10^n = n$

**b** $\log(10^a \times 100) = \log_{10}\left(10^a \times 10^2\right) = \log_{10}\left(10^{a+2}\right) = a + 2$

**c** $\log\left(\frac{10}{10^m}\right) = \log_{10}(10^{1-m}) = 1 - m$

**d** $\log\left(\frac{10^a}{10^b}\right) = \log_{10}(10^{a-b}) = a - b$

**3** **a** $6 = 10^{\log 6} \approx 10^{0.7782}$

**b** $60 = 10^{\log 60} \approx 10^{1.7782}$

**c** $6000 = 10^{\log 6000} \approx 10^{3.7782}$

**d** $0.6 = 10^{\log(0.6)} \approx 10^{-0.2218}$

**e** $0.006 = 10^{\log(0.006)} \approx 10^{-2.2218}$

**f** $15 = 10^{\log 15} \approx 10^{1.1761}$

**g** $1500 = 10^{\log 1500} \approx 10^{3.1761}$

**h** $1.5 = 10^{\log 1.5} \approx 10^{0.1761}$

**i** $0.15 = 10^{\log(0.15)} \approx 10^{-0.8239}$

**j** $0.000\,15 = 10^{\log(0.000\,15)} \approx 10^{-3.8239}$

**4** **a** **i** $\log 3 \approx 0.477$ **ii** $\log 300 \approx 2.477$

**b** $300 = 3 \times 10^2 = 10^{\log 3} \times 10^2 = 10^{\log 3 + 2}$

$\therefore \log 300 = \log 3 + 2$

**5** **a** **i** $\log 5 \approx 0.699$ **ii** $\log 0.05 \approx -1.301$

**b** $0.05 = 5 \times 10^{-2} = 10^{\log 5} \times 10^{-2} = 10^{\log 5 - 2}$

$\therefore \log 0.05 = \log 5 - 2$

**6** **a** $\log x = 2$ $\therefore x = 10^2$ $\therefore x = 100$

**b** $\log x = 1$ $\therefore x = 10^1$ $\therefore x = 10$

**c** $\log x = 0$ $\therefore x = 10^0$ $\therefore x = 1$

**d** $\log x = -1$ $\therefore x = 10^{-1}$ $\therefore x = \frac{1}{10}$

**e** $\log x = \frac{1}{2}$
$\therefore\ x = 10^{\frac{1}{2}}$
$\therefore\ x = \sqrt{10}$

**f** $\log x = -\frac{1}{2}$
$\therefore\ x = 10^{-\frac{1}{2}}$
$\therefore\ x = \dfrac{1}{10^{\frac{1}{2}}}$
$\therefore\ x = \dfrac{1}{\sqrt{10}}$

**g** $\log x = 4$
$\therefore\ x = 10^4$
$\therefore\ x = 10\,000$

**h** $\log x = -5$
$\therefore\ x = 10^{-5}$
$\therefore\ x = 0.000\,01$

**i** $\log x \approx 0.8351$
$\therefore\ x \approx 10^{0.8351}$
$\therefore\ x \approx 6.84$

**j** $\log x \approx 2.1457$
$\therefore\ x \approx 10^{2.1457}$
$\therefore\ x \approx 140$

**k** $\log x \approx -1.378$
$\therefore\ x \approx 10^{-1.378}$
$\therefore\ x \approx 0.0419$

**l** $\log x \approx -3.1997$
$\therefore\ x \approx 10^{-3.1997}$
$\therefore\ x \approx 0.000\,631$

## EXERCISE 4B

**1** **a** $10^2 = 100$ **b** $10^4 = 10\,000$ **c** $10^{-1} = 0.1$
**d** $10^{\frac{1}{2}} = \sqrt{10}$ **e** $2^3 = 8$ **f** $3^2 = 9$
**g** $2^{-2} = \frac{1}{4}$ **h** $3^{1.5} = \sqrt{27}$ **i** $5^{-\frac{1}{2}} = \frac{1}{\sqrt{5}}$

**2** **a** $\log_2 4 = 2$ **b** $\log_4 64 = 3$ **c** $\log_5 25 = 2$
**d** $\log_7 49 = 2$ **e** $\log_2 64 = 6$ **f** $\log_2\left(\frac{1}{8}\right) = -3$
**g** $\log_{10}(0.01) = -2$ **h** $\log_2\left(\frac{1}{2}\right) = -1$ **i** $\log_3\left(\frac{1}{27}\right) = -3$

**3** **a** $\log_{10} 100\,000$
$= \log_{10} 10^5$
$= 5$

**b** $\log_{10}(0.01)$
$= \log_{10} 10^{-2}$
$= -2$

**c** $\log_3 \sqrt{3}$
$= \log_3 3^{\frac{1}{2}}$
$= \frac{1}{2}$

**d** $\log_2 8$
$= \log_2 2^3$
$= 3$

**e** $\log_2 64$
$= \log_2 2^6$
$= 6$

**f** $\log_2 128$
$= \log_2 2^7$
$= 7$

**g** $\log_5 25$
$= \log_5 5^2$
$= 2$

**h** $\log_5 125$
$= \log_5 5^3$
$= 3$

**i** $\log_2(0.125)$
$= \log_2\left(\frac{1}{8}\right)$
$= \log_2\left(2^{-3}\right)$
$= -3$

**j** $\log_9 3$
$= \log_9 9^{\frac{1}{2}}$
$= \frac{1}{2}$

**k** $\log_4 16$
$= \log_4 4^2$
$= 2$

**l** $\log_{36} 6$
$= \log_{36} 36^{\frac{1}{2}}$
$= \frac{1}{2}$

**m** $\log_3 243$
$= \log_3 3^5$
$= 5$

**n** $\log_2 \sqrt[3]{2}$
$= \log_2 2^{\frac{1}{3}}$
$= \frac{1}{3}$

**o** $\log_a a^n$
$= n,\ a > 0$

**p** $\log_8 2$
$= \log_8 8^{\frac{1}{3}}$
$= \frac{1}{3}$

**q** $\log_t\left(\frac{1}{t}\right)$
$= \log_t t^{-1}$
$= -1,\ t > 0$

**r** $\log_6 6\sqrt{6}$
$= \log_6(6^1 \times 6^{\frac{1}{2}})$
$= \log_6 6^{\frac{3}{2}}$
$= \frac{3}{2}$

**s** $\log_4 1$
$= \log_4 4^0$
$= 0$

**t** $\log_9 9$
$= \log_9 9^1$
$= 1$

**4** **a** $\log_{10} 152 \approx 2.18$ **b** $\log_{10} 25 \approx 1.40$ **c** $\log_{10} 74 \approx 1.87$ **d** $\log_{10} 0.8 \approx -0.0969$

**5** **a** $\log_2 x = 3$
$\therefore\ x = 2^3$
$\therefore\ x = 8$

**b** $\log_4 x = \frac{1}{2}$
$\therefore\ x = 4^{\frac{1}{2}}$
$\therefore\ x = 2$

**c** $\log_x 81 = 4$
$\therefore\ 81 = x^4$
$\therefore\ x = \pm\sqrt[4]{81}$
$\therefore\ x = \pm 3$
$\therefore\ x = 3\ \{\text{as } x > 0\}$

**d** $\log_2(x-6) = 3$
$\therefore\ x - 6 = 2^3$
$\therefore\ x - 6 = 8$
$\therefore\ x = 14$

**6** **a** $\log_4 16$
$= \log_4 4^2$
$= 2$

**b** $\log_2 4$
$= \log_2 2^2$
$= 2$

**c** $\log_3(\frac{1}{3})$
$= \log_3 3^{-1}$
$= -1$

**d** $\log_{10} \sqrt[4]{1000}$
$= \log_{10}(10^3)^{\frac{1}{4}}$
$= \log_{10} 10^{\frac{3}{4}}$
$= \frac{3}{4}$

**e** $\log_7\left(\frac{1}{\sqrt{7}}\right)$
$= \log_7 7^{-\frac{1}{2}}$
$= -\frac{1}{2}$

**f** $\log_5(25\sqrt{5})$
$= \log_5(5^2 5^{\frac{1}{2}})$
$= \log_5 5^{\frac{5}{2}}$
$= \frac{5}{2}$

**g** $\log_3\left(\frac{1}{\sqrt{27}}\right)$
$= \log_3\left(\frac{1}{(3^3)^{\frac{1}{2}}}\right)$
$= \log_3 3^{-\frac{3}{2}}$
$= -\frac{3}{2}$

**h** $\log_4\left(\frac{1}{2\sqrt{2}}\right)$
$= \log_4\left(2^{-\frac{3}{2}}\right)$
$= \log_4\left((2^2)^{-\frac{3}{4}}\right)$
$= \log_4 4^{-\frac{3}{4}}$
$= -\frac{3}{4}$

**i** $\log_x x^2$
$= 2,\ x > 0$

**j** $\log_x \sqrt{x}$
$= \log_x x^{\frac{1}{2}}$
$= \frac{1}{2},\ x > 0$

**k** $\log_m m^3$
$= 3,\ m > 0$

**l** $\log_x\left(x\sqrt{x}\right)$
$= \log_x(x^1 \times x^{\frac{1}{2}})$
$= \log_x x^{\frac{3}{2}}$
$= \frac{3}{2},\ x > 0$

**m** $\log_n\left(\frac{1}{n}\right)$
$= \log_n n^{-1}$
$= -1,\ n > 0$

**n** $\log_a\left(\frac{1}{a^2}\right)$
$= \log_a a^{-2}$
$= -2,\ a > 0$

**o** $\log_a\left(\frac{1}{\sqrt{a}}\right)$
$= \log_a a^{-\frac{1}{2}}$
$= -\frac{1}{2},\ a > 0$

**p** $\log_m \sqrt{m^5}$
$= \log_m\left(m^5\right)^{\frac{1}{2}}$
$= \log_m m^{\frac{5}{2}}$
$= \frac{5}{2},\ m > 0$

## EXERCISE 4C.1

**1** **a** $\log 8 + \log 2$
$= \log(8 \times 2)$
$= \log 16$

**b** $\log 4 + \log 5$
$= \log(4 \times 5)$
$= \log 20$

**c** $\log 40 - \log 5$
$= \log\left(\frac{40}{5}\right)$
$= \log 8$

**d** $\log p - \log m$
$= \log\left(\frac{p}{m}\right)$

**e** $\log_4 8 - \log_4 2$
$= \log_4\left(\frac{8}{2}\right)$
$= \log_4 4$
$= 1$

**f** $\log 5 + \log(0.4)$
$= \log(5 \times 0.4)$
$= \log 2$

**g** $\log 2 + \log 3 + \log 4$
$= \log(2 \times 3 \times 4)$
$= \log 24$

**h** $1 + \log_2 3$
$= \log_2 2^1 + \log_2 3$
$= \log_2(2 \times 3)$
$= \log_2 6$

**i** $\log 4 - 1$
$= \log 4 - \log 10^1$
$= \log\left(\frac{4}{10}\right)$
$= \log 0.4$

**j** $\log 5 + \log 4 - \log 2$
$= \log\left(\frac{5\times 4}{2}\right)$
$= \log 10$
$= 1$

**k** $2 + \log 2$
$= \log 10^2 + \log 2$
$= \log(100 \times 2)$
$= \log 200$

**l** $t + \log w$
$= \log 10^t + \log w$
$= \log\left(10^t \times w\right)$

**m** $\log_m 40 - 2$
$= \log_m 40 - \log_m m^2$
$= \log_m\left(\frac{40}{m^2}\right)$

**n** $\log_3 6 - \log_3 2 - \log_3 3$
$= \log_3(6 \div 2 \div 3)$
$= \log_3 1$
$(= 0)$

**o** $\log 50 - 4$
$= \log 50 - \log 10^4$
$= \log\left(\frac{50}{10^4}\right)$
$= \log 0.005$

**p** $3 - \log_5 50$
$= \log_5 5^3 - \log_5 50$
$= \log_5\left(\frac{125}{50}\right)$
$= \log_5\left(\frac{5}{2}\right)$

**q** $\log_5 100 - \log_5 4$
$= \log_5\left(\frac{100}{4}\right)$
$= \log_5 25$
$= \log_5 5^2$
$= 2$

**r** $\log\left(\frac{4}{3}\right) + \log 3 + \log 7$
$= \log\left(\frac{4}{3} \times 3 \times 7\right)$
$= \log 28$

**2** **a** $5\log 2 + \log 3$
$= \log 2^5 + \log 3$
$= \log(2^5 \times 3)$
$= \log 96$

**b** $2\log 3 + 3\log 2$
$= \log 3^2 + \log 2^3$
$= \log(9 \times 8)$
$= \log 72$

**c** $3\log 4 - \log 8$
$= \log 4^3 - \log 8$
$= \log\left(\frac{64}{8}\right)$
$= \log 8$

**d** $2\log_3 5 - 3\log_3 2$
$= \log_3 5^2 - \log_3 2^3$
$= \log_3\left(\frac{25}{8}\right)$

**e** $\frac{1}{2}\log_6 4 + \log_6 3$
$= \log_6 4^{\frac{1}{2}} + \log_6 3$
$= \log_6(2 \times 3)$
$= \log_6 6$
$= 1$

**f** $\frac{1}{3}\log\left(\frac{1}{8}\right)$
$= \log\left(\frac{1}{8}\right)^{\frac{1}{3}}$
$= \log\left(2^{-3}\right)^{\frac{1}{3}}$
$= \log 2^{-1}$
$= \log\left(\frac{1}{2}\right)$ or $-\log 2$

**g** $3 - \log 2 - 2\log 5$
$= \log 10^3 - \log 2 - \log 5^2$
$= \log(1000 \div 2 \div 25)$
$= \log 20$

**h** $1 - 3\log 2 + \log 20$
$= \log 10^1 - \log 2^3 + \log 20$
$= \log(10 \div 8 \times 20)$
$= \log 25$

**i** $2 - \frac{1}{2}\log_n 4 - \log_n 5$
$= \log_n n^2 - \log_n 4^{\frac{1}{2}} - \log_n 5$
$= \log_n(n^2 \div 2 \div 5)$
$= \log_n\left(\frac{n^2}{10}\right)$

**3** **a** $\frac{\log 4}{\log 2}$
$= \frac{\log 2^2}{\log 2}$
$= \frac{2\log 2}{\log 2}$
$= 2$

**b** $\frac{\log_5 27}{\log_5 9}$
$= \frac{\log_5 3^3}{\log_5 3^2}$
$= \frac{3\log_5 3}{2\log_5 3}$
$= \frac{3}{2}$

**c** $\frac{\log 8}{\log 2}$
$= \frac{\log 2^3}{\log 2}$
$= \frac{3\log 2}{\log 2}$
$= 3$

**d** $\frac{\log 3}{\log 9}$
$= \frac{\log 3}{\log 3^2}$
$= \frac{\log 3}{2\log 3}$
$= \frac{1}{2}$

**e** $\frac{\log_3 25}{\log_3(0.2)}$
$= \frac{\log_3 5^2}{\log_3 5^{-1}}$
$= \frac{2\log_3 5}{-1\log_3 5}$
$= -2$

**f** $\frac{\log_4 8}{\log_4(0.25)}$
$= \frac{\log_4 2^3}{\log_4 2^{-2}}$ $\{0.25 = \frac{1}{4} = \frac{1}{2^2}\}$
$= \frac{3\log_4 2}{-2\log_4 2}$
$= -\frac{3}{2}$

**4** **a** $\log 9 = \log 3^2$
$= 2\log 3$

**b** $\log\sqrt{2} = \log 2^{\frac{1}{2}}$
$= \frac{1}{2}\log 2$

**c** $\log\left(\frac{1}{8}\right) = \log\left(\frac{1}{2^3}\right)$
$= \log 2^{-3}$
$= -3\log 2$

**d** $\log\left(\frac{1}{5}\right) = \log 5^{-1}$
$= -1\log 5$
$= -\log 5$

**e** $\log 5 = \log\left(\frac{10}{2}\right)$
$= \log 10^1 - \log 2$
$= 1 - \log 2$

**f** $\log 5000 = \log\left(\frac{10\,000}{2}\right)$
$= \log 10^4 - \log 2$
$= 4 - \log 2$

**5** **a** $\log_b 6$
$= \log_b(2\times 3)$
$= \log_b 2 + \log_b 3$
$= p + q$

**b** $\log_b 45$
$= \log_b(3^2 5)$
$= 2\log_b 3 + \log_b 5$
$= 2q + r$

**c** $\log_b 108$
$= \log_b(2^2 3^3)$
$= 2\log_b 2 + 3\log_b 3$
$= 2p + 3q$

**d** $\log_b\left(\frac{5\sqrt{3}}{2}\right)$
$= \log_b(5\times 3^{\frac{1}{2}}) - \log_b 2$
$= \log_b 5 + \frac{1}{2}\log_b 3 - \log_b 2$
$= r + \frac{1}{2}q - p$

**e** $\log_b\left(\frac{5}{32}\right)$
$= \log_b 5 - \log_b 2^5$
$= \log_b 5 - 5\log_b 2$
$= r - 5p$

**f** $\log_b(0.\overline{2})$
$= \log_b(\frac{2}{9})$
$= \log_b 2 - \log_b 3^2$
$= p - 2q$

**6** **a** $\log_2(PR)$
$= \log_2 P + \log_2 R$
$= x + z$

**b** $\log_2(RQ^2)$
$= \log_2 R + \log_2 Q^2$
$= \log_2 R + 2\log_2 Q$
$= z + 2y$

**c** $\log_2\left(\frac{PR}{Q}\right)$
$= \log_2(PR) - \log_2 Q$
$= \log_2 P + \log_2 R - \log_2 Q$
$= x + z - y$

**d** $\log_2\left(P^2\sqrt{Q}\right)$
$= \log_2 P^2 + \log_2 Q^{\frac{1}{2}}$
$= 2\log_2 P + \frac{1}{2}\log_2 Q$
$= 2x + \frac{1}{2}y$

**e** $\log_2\left(\frac{Q^3}{\sqrt{R}}\right)$
$= \log_2 Q^3 - \log_2 R^{\frac{1}{2}}$
$= 3\log_2 Q - \frac{1}{2}\log_2 R$
$= 3y - \frac{1}{2}z$

**f** $\log_2\left(\frac{R^2\sqrt{Q}}{P^3}\right)$
$= \log_2 R^2 + \log_2 Q^{\frac{1}{2}} - \log_2 P^3$
$= 2\log_2 R + \frac{1}{2}\log_2 Q - 3\log_2 P$
$= 2z + \frac{1}{2}y - 3x$

**7** **a** $\log_t N^2 = 1.72$
$\therefore\ 2\log_t N = 1.72$
$\therefore\ \log_t N = 1.72 \div 2$
$= 0.86$

**b** $\log_t(MN)$
$= \log_t M + \log_t N$
$= 1.29 + 0.86$
$= 2.15$

**c** $\log_t\left(\frac{N^2}{\sqrt{M}}\right)$
$= \log_t N^2 - \log_t M^{\frac{1}{2}}$
$= 1.72 - \frac{1}{2}\log_t M$
$= 1.72 - \frac{1}{2}(1.29)$
$= 1.075$

## EXERCISE 4C.2

**1** **a** $y = 2^x$
$\therefore\ \log y = \log 2^x$
$\therefore\ \log y = x\log 2$

**b** $y = 20b^3$
$\therefore\ \log y = \log(20b^3)$
$\therefore\ \log y = \log 20 + \log b^3$
$\therefore\ \log y \approx 1.30 + 3\log b$

**c** $M = ad^4$
$\therefore\ \log M = \log(ad^4)$
$\therefore\ \log M = \log a + \log d^4$
$\therefore\ \log M = \log a + 4\log d$

**d** $T = 5\sqrt{d} = 5d^{\frac{1}{2}}$
$\therefore\ \log T = \log(5d^{\frac{1}{2}})$
$\therefore\ \log T = \log 5 + \log d^{\frac{1}{2}}$
$\therefore\ \log T \approx 0.699 + \frac{1}{2}\log d$

**e** $R = b\sqrt{l} = bl^{\frac{1}{2}}$
$\therefore\ \log R = \log(bl^{\frac{1}{2}})$
$\therefore\ \log R = \log b + \log l^{\frac{1}{2}}$
$\therefore\ \log R = \log b + \frac{1}{2}\log l$

**f** $Q = \dfrac{a}{b^n}$
$\therefore\ \log Q = \log\left(\frac{a}{b^n}\right)$
$\therefore\ \log Q = \log a - \log b^n$
$\therefore\ \log Q = \log a - n\log b$

**g** $y = ab^x$
$\therefore \log y = \log(ab^x)$
$\therefore \log y = \log a + \log b^x$
$\therefore \log y = \log a + x \log b$

**h** $F = \dfrac{20}{\sqrt{n}} = \dfrac{20}{n^{\frac{1}{2}}}$
$\therefore \log F = \log\left(\dfrac{20}{n^{\frac{1}{2}}}\right)$
$\therefore \log F = \log 20 - \log n^{\frac{1}{2}}$
$\therefore \log F \approx 1.30 - \frac{1}{2}\log n$

**i** $L = \dfrac{ab}{c}$
$\therefore \log L = \log\left(\dfrac{ab}{c}\right)$
$\therefore \log L = \log ab - \log c$
$\therefore \log L = \log a + \log b - \log c$

**j** $N = \sqrt{\dfrac{a}{b}}$
$\therefore N = \left(\dfrac{a}{b}\right)^{\frac{1}{2}}$
$\therefore \log N = \log\left(\dfrac{a}{b}\right)^{\frac{1}{2}}$
$\therefore \log N = \frac{1}{2}\log\left(\dfrac{a}{b}\right)$
$\therefore \log N = \frac{1}{2}\log a - \frac{1}{2}\log b$

**k** $S = 200 \times 2^t$
$\therefore \log S = \log(200 \times 2^t)$
$\therefore \log S = \log 200 + \log 2^t$
$\therefore \log S = \log 200 + t \log 2$
$\therefore \log S \approx 2.30 + t \log 2$

**l** $y = \dfrac{a^m}{b^n}$
$\therefore \log y = \log\left(\dfrac{a^m}{b^n}\right)$
$\therefore \log y = \log a^m - \log b^n$
$\therefore \log y = m \log a - n \log b$

**2** **a** $\log D = \log e + \log 2$
$= \log(e \times 2)$
$\therefore D = 2e$

**b** $\log_a F = \log_a 5 - \log_a t$
$= \log_a\left(\dfrac{5}{t}\right)$
$\therefore F = \dfrac{5}{t}$

**c** $\log P = \frac{1}{2}\log x$
$= \log x^{\frac{1}{2}}$
$\therefore P = \sqrt{x}$

**d** $\log_n M = 2\log_n b + \log_n c$
$= \log_n b^2 + \log_n c$
$= \log_n (b^2 c)$
$\therefore M = b^2 c$

**e** $\log B = 3\log m - 2\log n$
$= \log m^3 - \log n^2$
$= \log\left(\dfrac{m^3}{n^2}\right)$
$\therefore B = \dfrac{m^3}{n^2}$

**f** $\log N = -\frac{1}{3}\log p$
$= \log p^{-\frac{1}{3}}$
$= \log\left(\dfrac{1}{\sqrt[3]{p}}\right)$
$\therefore N = \dfrac{1}{\sqrt[3]{p}}$

**g** $\log P = 3\log x + 1$
$= \log x^3 + \log 10^1$
$= \log(10x^3)$
$\therefore P = 10x^3$

**h** $\log_a Q = 2 - \log_a x$
$= \log_a a^2 - \log_a x$
$= \log_a\left(\dfrac{a^2}{x}\right)$
$\therefore Q = \dfrac{a^2}{x}$

**3** **a** $y = 3 \times 2^x$
$\therefore \log_2 y = \log_2(3 \times 2^x)$
$\therefore \log_2 y = \log_2 3 + \log_2 2^x$
$\therefore \log_2 y = \log_2 3 + x$

**b** $\log_2 y = \log_2 3 + x$ {from **a**}
$\therefore x = \log_2 y - \log_2 3$
$\therefore x = \log_2\left(\dfrac{y}{3}\right)$

**c** **i** When $y = 3$,
$x = \log_2\left(\frac{3}{3}\right)$
$\therefore x = \log_2 1$
$\therefore x = 0$

**ii** When $y = 12$,
$x = \log_2\left(\frac{12}{3}\right)$
$\therefore x = \log_2 4$
$\therefore x = \log_2 2^2$
$\therefore x = 2$

**iii** When $y = 30$,
$x = \log_2\left(\frac{30}{3}\right)$
$\therefore x = \log_2 10$
$\therefore x \approx 3.32$

**4** **a** $\log_3 27 + \log_3\left(\frac{1}{3}\right) = \log_3 x$
$\therefore\ \log_3\left(27 \times \frac{1}{3}\right) = \log_3 x$
$\therefore\ \log_3 9 = \log_3 x$
$\therefore\ x = 9$

**b** $\log_5 x = \log_5 8 - \log_5(6-x)$
$\therefore\ \log_5 x = \log_5\left(\dfrac{8}{6-x}\right)$
$\therefore\ x = \dfrac{8}{6-x}$
$\therefore\ 6x - x^2 = 8$
$\therefore\ x^2 - 6x + 8 = 0$
$\therefore\ (x-2)(x-4) = 0$
$\therefore\ x = 2$ or $4$

**Note:** $x > 0$ and $6 - x > 0$ so $0 < x < 6$

**c** $\log_5 125 - \log_5 \sqrt{5} = \log_5 x$
$\therefore\ \log_5\left(\frac{125}{\sqrt{5}}\right) = \log_5 x$
$\therefore\ x = \frac{125}{\sqrt{5}}\ (= 25\sqrt{5})$

**d** $\log_{20} x = 1 + \log_{20} 10$
$\therefore\ \log_{20} x = \log_{20} 20^1 + \log_{20} 10$
$= \log_{20} 200$
$\therefore\ x = 200$

**e** $\log x + \log(x+1) = \log 30$
$\therefore\ \log[x(x+1)] = \log 30$
$\therefore\ x^2 + x = 30$
$\therefore\ x^2 + x - 30 = 0$
$\therefore\ (x+6)(x-5) = 0$
$\therefore\ x = -6$ or $5$
but $x > 0$ for $\log x$ to exist
$\therefore\ x = 5$

**f** $\log(x+2) - \log(x-2) = \log 5$
$\therefore\ \log\left(\dfrac{x+2}{x-2}\right) = \log 5$
$\therefore\ \dfrac{x+2}{x-2} = 5$
$\therefore\ x + 2 = 5x - 10$
$\therefore\ -4x = -12$
$\therefore\ x = 3$

**Note:** $x + 2 > 0$ and $x - 2 > 0$
$\therefore\ x > 2$ ✓

## EXERCISE 4D.1

**1** **a** $\ln e^2$
$= 2\ \ \{\ln e^x = x\}$

**b** $\ln e^3$
$= 3$

**c** $\ln \sqrt{e}$
$= \ln e^{\frac{1}{2}}$
$= \frac{1}{2}$

**d** $\ln 1$
$= \ln e^0$
$= 0$

**e** $\ln\left(\dfrac{1}{e}\right)$
$= \ln e^{-1}$
$= -1$

**f** $\ln \sqrt[3]{e}$
$= \ln e^{\frac{1}{3}}$
$= \frac{1}{3}$

**g** $\ln\left(\dfrac{1}{e^2}\right)$
$= \ln e^{-2}$
$= -2$

**h** $\ln\left(\dfrac{1}{\sqrt{e}}\right)$
$= \ln e^{-\frac{1}{2}}$
$= -\frac{1}{2}$

**2** **a** $e^{\ln 3}$
$= 3$
$\{$using $e^{\ln x} = x\}$

**b** $e^{2\ln 3}$
$= 3^2$
$\{$using $e^{x \ln a} = a^x\}$
$= 9$

**c** $e^{-\ln 5}$
$= e^{-1\ln 5}$
$= 5^{-1}$
$= \frac{1}{5}$

**d** $e^{-2\ln 2}$
$= 2^{-2}$
$= \dfrac{1}{2^2}$
$= \frac{1}{4}$

**3** $\ln x$ exists only when $x > 0$.
$\therefore\ \ln(-2)$ and $\ln(0)$ do not exist.

**Note:** If $\ln(-2) = a$ then $-2 = e^a$
and $e^a = -2$ has no solutions as $e^a > 0$ for all $a$.

**4** **a** $\ln e^a$
$= a$

**b** $\ln(e \times e^a)$
$= \ln e^{1+a}$
$= a + 1$

**c** $\ln(e^a \times e^b)$
$= \ln(e^{a+b})$
$= a + b$

**d** $\ln(e^a)^b$
$= \ln e^{ab}$
$= ab$

**e** $\ln\left(\dfrac{e^a}{e^b}\right)$
$= \ln(e^{a-b})$
$= a - b$

**5** **a** $6 = e^{1.7918}$ **b** $60 = e^{4.0943}$ **c** $6000 = e^{8.6995}$ **d** $0.6 = e^{-0.5108}$
**e** $0.006 = e^{-5.1160}$ **f** $15 = e^{2.7081}$ **g** $1500 = e^{7.3132}$ **h** $1.5 = e^{0.4055}$
**i** $0.15 = e^{-1.8971}$ **j** $0.000\,15 = e^{-8.8049}$

**6** **a** $\ln x = 3$
$\therefore\ x = e^3$
$\therefore\ x \approx 20.1$

**b** $\ln x = 1$
$\therefore\ x = e^1$
$\therefore\ x = e \approx 2.72$

**c** $\ln x = 0$
$\therefore\ x = e^0$
$\therefore\ x = 1$

**d** $\ln x = -1$
$\therefore\ x = e^{-1}$
$\therefore\ x \approx 0.368$

**e** $\ln x = -5$
$\therefore\ x = e^{-5}$
$\therefore\ x \approx 0.006\,74$

**f** $\ln x \approx 0.835$
$\therefore\ x \approx e^{0.835}$
$\therefore\ x \approx 2.30$

**g** $\ln x \approx 2.145$
$\therefore\ x \approx e^{2.145}$
$\therefore\ x \approx 8.54$

**h** $\ln x \approx -3.2971$
$\therefore\ x \approx e^{-3.2971}$
$\therefore\ x \approx 0.0370$

## EXERCISE 4D.2

**1** **a** $\ln 15 + \ln 3$
$= \ln(15 \times 3)$
$= \ln 45$

**b** $\ln 15 - \ln 3$
$= \ln(\frac{15}{3})$
$= \ln 5$

**c** $\ln 20 - \ln 5$
$= \ln(\frac{20}{5})$
$= \ln 4$

**d** $\ln 4 + \ln 6$
$= \ln(4 \times 6)$
$= \ln 24$

**e** $\ln 5 + \ln(0.2)$
$= \ln(5 \times 0.2)$
$= \ln 1$
$= 0$

**f** $\ln 2 + \ln 3 + \ln 5$
$= \ln(2 \times 3 \times 5)$
$= \ln 30$

**g** $1 + \ln 4$
$= \ln e^1 + \ln 4$
$= \ln(e \times 4)$
$= \ln(4e)$

**h** $\ln 6 - 1$
$= \ln 6 - \ln e^1$
$= \ln\left(\dfrac{6}{e}\right)$

**i** $\ln 5 + \ln 8 - \ln 2$
$= \ln(5 \times 8 \div 2)$
$= \ln 20$

**j** $2 + \ln 4$
$= \ln e^2 + \ln 4$
$= \ln(e^2 \times 4)$
$= \ln(4e^2)$

**k** $\ln 20 - 2$
$= \ln 20 - \ln e^2$
$= \ln\left(\dfrac{20}{e^2}\right)$

**l** $\ln 12 - \ln 4 - \ln 3$
$= \ln(12 \div 4 \div 3)$
$= \ln 1$
$= 0$

**2** **a** $5\ln 3 + \ln 4$
$= \ln(3^5) + \ln 4$
$= \ln(243 \times 4)$
$= \ln 972$

**b** $3\ln 2 + 2\ln 5$
$= \ln(2^3) + \ln(5^2)$
$= \ln(8 \times 25)$
$= \ln 200$

**c** $3\ln 2 - \ln 8$
$= \ln(2^3) - \ln 8$
$= \ln(\frac{8}{8})$
$= \ln 1 = 0$

**d** $3\ln 4 - 2\ln 2$
$= \ln(4^3) - \ln(2^2)$
$= \ln(\frac{64}{4})$
$= \ln 16$

**e** $\frac{1}{3}\ln 8 + \ln 3$
$= \ln(8^{\frac{1}{3}}) + \ln 3$
$= \ln(2 \times 3)$
$= \ln 6$

**f** $\frac{1}{3}\ln(\frac{1}{27})$
$= \ln\left((\frac{1}{27})^{\frac{1}{3}}\right)$
$= \ln\left(\dfrac{1}{27^{\frac{1}{3}}}\right)$
$= \ln(\frac{1}{3})$

**g** $-\ln 2$
$= \ln(2^{-1})$
$= \ln(\frac{1}{2})$

**h** $-\ln(\frac{1}{2})$
$= \ln((\frac{1}{2})^{-1})$
$= \ln 2$

**i** $-2\ln(\frac{1}{4})$
$= \ln((\frac{1}{4})^{-2})$
$= \ln(4^2)$
$= \ln 16$

**3** **a** $\ln 27 = \ln 3^3 = 3\ln 3$

**b** $\ln\sqrt{3} = \ln 3^{\frac{1}{2}} = \frac{1}{2}\ln 3$

**c** $\ln(\frac{1}{16}) = \ln\left(\frac{1}{2^4}\right) = \ln(2^{-4}) = -4\ln 2$

**d** $\ln(\frac{1}{6}) = \ln 6^{-1} = -1\ln 6 = -\ln 6$

**e** $\ln\left(\frac{1}{\sqrt{2}}\right) = \ln 2^{-\frac{1}{2}} = -\frac{1}{2}\ln 2$

**f** $\ln\left(\frac{e}{5}\right) = \ln e^1 - \ln 5 = 1 - \ln 5$

**4** **a** $\ln\sqrt[3]{5} = \ln 5^{\frac{1}{3}} = \frac{1}{3}\ln 5$

**b** $\ln(\frac{1}{32}) = \ln 2^{-5} = -5\ln 2$

**c** $\ln\left(\frac{1}{\sqrt[5]{2}}\right) = \ln\left(\frac{1}{2^{\frac{1}{5}}}\right) = \ln 2^{-\frac{1}{5}} = -\frac{1}{5}\ln 2$

**d** $\ln\left(\frac{e^2}{8}\right) = \ln e^2 - \ln 8 = 2 - \ln 2^3 = 2 - 3\ln 2$

**5** **a**
$$\ln D = \ln x + 1$$
$$\therefore \ln D - \ln x = 1$$
$$\therefore \ln\left(\frac{D}{x}\right) = 1$$
$$\therefore \frac{D}{x} = e^1$$
$$\therefore D = ex$$

**b**
$$\ln F = -\ln p + 2$$
$$\therefore \ln F + \ln p = 2$$
$$\therefore \ln(Fp) = 2$$
$$\therefore Fp = e^2$$
$$\therefore F = \frac{e^2}{p}$$

**c**
$$\ln P = \frac{1}{2}\ln x$$
$$\therefore \ln P = \ln x^{\frac{1}{2}}$$
$$\therefore P = \sqrt{x}$$

**d**
$$\ln M = 2\ln y + 3$$
$$\therefore \ln M - 2\ln y = 3$$
$$\therefore \ln\left(\frac{M}{y^2}\right) = 3$$
$$\therefore \frac{M}{y^2} = e^3$$
$$\therefore M = e^3y^2$$

**e**
$$\ln B = 3\ln t - 1$$
$$\therefore \ln B - \ln t^3 = -1$$
$$\therefore \ln\left(\frac{B}{t^3}\right) = -1$$
$$\therefore \frac{B}{t^3} = e^{-1}$$
$$\therefore B = \frac{t^3}{e}$$

**f**
$$\ln N = -\frac{1}{3}\ln g$$
$$\therefore \ln N = \ln g^{-\frac{1}{3}}$$
$$\therefore N = g^{-\frac{1}{3}}$$
$$\therefore N = \frac{1}{\sqrt[3]{g}}$$

**g**
$$\ln Q \approx 3\ln x + 2.159$$
$$\therefore \ln Q - 3\ln x \approx 2.159$$
$$\therefore \ln\left(\frac{Q}{x^3}\right) \approx 2.159$$
$$\therefore \frac{Q}{x^3} \approx e^{2.159}$$
$$\therefore \frac{Q}{x^3} \approx 8.66$$
$$\therefore Q \approx 8.66x^3$$

**h**
$$\ln D \approx 0.4\ln n - 0.6582$$
$$\therefore \ln D - \ln n^{0.4} \approx -0.6582$$
$$\therefore \ln\left(\frac{D}{n^{0.4}}\right) \approx -0.6582$$
$$\therefore \frac{D}{n^{0.4}} \approx e^{-0.6582}$$
$$\therefore \frac{D}{n^{0.4}} \approx 0.518$$
$$\therefore D \approx 0.518n^{0.4}$$

## EXERCISE 4E

**1** **a**
$$2^x = 10$$
$$\therefore \log 2^x = \log 10$$
$$\therefore x\log 2 = \log 10^1$$
$$\therefore x = \frac{1}{\log 2}$$

**b**
$$3^x = 20$$
$$\therefore \log 3^x = \log 20$$
$$\therefore x\log 3 = \log 20$$
$$\therefore x = \frac{\log 20}{\log 3}$$

**c**
$$4^x = 100$$
$$\therefore \log 4^x = \log 100$$
$$\therefore x\log 4 = \log 10^2$$
$$\therefore x = \frac{2}{\log 4}$$

**d**
$$\begin{aligned} \left(\tfrac{1}{2}\right)^x &= 0.0625 \\ \therefore\ \log(\tfrac{1}{2})^x &= \log(\tfrac{1}{16}) \\ \therefore\ x\log(2^{-1}) &= \log(2^{-4}) \\ \therefore\ x &= \frac{-4\log 2}{-\log 2} \\ \therefore\ x &= 4 \end{aligned}$$

**e**
$$\begin{aligned} \left(\tfrac{3}{4}\right)^x &= 0.1 \\ \therefore\ \log(\tfrac{3}{4})^x &= \log 10^{-1} \\ \therefore\ x\log(\tfrac{3}{4}) &= -1 \\ \therefore\ x &= -\frac{1}{\log(\tfrac{3}{4})} \end{aligned}$$

**f**
$$\begin{aligned} 10^x &= 0.000\,01 \\ \therefore\ \log 10^x &= \log 0.000\,01 \\ \therefore\ \log 10^x &= \log 10^{-5} \\ \therefore\ x\log 10 &= -5\log 10 \\ \therefore\ x &= -5 \end{aligned}$$

**2** **a**
$$\begin{aligned} e^x &= 10 \\ \therefore\ x &= \ln 10 \end{aligned}$$

**b**
$$\begin{aligned} e^x &= 1000 \\ \therefore\ x &= \ln 1000 \end{aligned}$$

**c**
$$\begin{aligned} 2e^x &= 0.3 \\ \therefore\ e^x &= 0.15 \\ \therefore\ x &= \ln 0.15 \end{aligned}$$

**d**
$$\begin{aligned} e^{\frac{x}{2}} &= 5 \\ \therefore\ \frac{x}{2} &= \ln 5 \\ \therefore\ x &= 2\ln 5 \end{aligned}$$

**e**
$$\begin{aligned} e^{2x} &= 18 \\ \therefore\ 2x &= \ln 18 \\ \therefore\ x &= \tfrac{1}{2}\ln 18 \end{aligned}$$

**f**
$$\begin{aligned} e^{-\frac{x}{2}} &= 1 \\ \therefore\ -\frac{x}{2} &= \ln 1 \\ \therefore\ -\frac{x}{2} &= 0 \\ \therefore\ x &= 0 \end{aligned}$$

**3** **a**
$$\begin{aligned} R &= 200 \times 2^{0.25t} \\ \therefore\ 2^{0.25t} &= \frac{R}{200} \\ \therefore\ \log 2^{0.25t} &= \log\left(\frac{R}{200}\right) \\ \therefore\ 0.25t\log 2 &= \log R - \log 200 \\ \therefore\ t &= \frac{\log R - \log 200}{0.25\log 2} \end{aligned}$$

**b** **i** When $R = 600$,
$$\begin{aligned} t &= \frac{\log 600 - \log 200}{0.25\log 2} \\ \therefore\ t &\approx 6.34 \end{aligned}$$

**ii** When $R = 1425$,
$$\begin{aligned} t &= \frac{\log 1425 - \log 200}{0.25\log 2} \\ \therefore\ t &\approx 11.3 \end{aligned}$$

**4** **a**
$$\begin{aligned} M &= 20 \times 5^{-0.02x} \\ \therefore\ 5^{-0.02x} &= \frac{M}{20} \\ \therefore\ \log 5^{-0.02x} &= \log\left(\frac{M}{20}\right) \\ \therefore\ -0.02x\log 5 &= \log M - \log 20 \\ \therefore\ x &= \frac{\log M - \log 20}{-0.02\log 5} \end{aligned}$$

**b** **i** When $M = 100$,
$$\begin{aligned} x &= \frac{\log 100 - \log 20}{-0.02\log 5} \\ \therefore\ x &= -50 \end{aligned}$$

**ii** When $M = 232$,
$$\begin{aligned} x &= \frac{\log 232 - \log 20}{-0.02\log 5} \\ \therefore\ x &\approx -76.1 \end{aligned}$$

**5** **a**
$$\begin{aligned} 4 \times 2^{-x} &= 0.12 \\ \therefore\ 2^{-x} &= 0.03 \\ \therefore\ \log 2^{-x} &= \log(0.03) \\ \therefore\ -x\log 2 &= \log(0.03) \\ \therefore\ x &= -\frac{\log(0.03)}{\log 2} \end{aligned}$$

**b**
$$\begin{aligned} 300 \times 5^{0.1x} &= 1000 \\ \therefore\ 5^{0.1x} &= \tfrac{10}{3} \\ \therefore\ \log 5^{0.1x} &= \log\left(\tfrac{10}{3}\right) \\ \therefore\ 0.1x\log 5 &= \log\left(\tfrac{10}{3}\right) \\ \therefore\ x\log 5 &= 10\log\left(\tfrac{10}{3}\right) \\ \therefore\ x &= \frac{10\log\left(\tfrac{10}{3}\right)}{\log 5} \end{aligned}$$

**c**
$$32 \times 3^{-0.25x} = 4$$
$$\therefore\ 3^{-0.25x} = \tfrac{1}{8}$$
$$\therefore\ \log 3^{-0.25x} = \log\left(\tfrac{1}{8}\right)$$
$$\therefore\ -0.25x \log 3 = \log\left(\tfrac{1}{8}\right)$$
$$\therefore\ x \log 3 = -4 \log\left(\tfrac{1}{8}\right)$$
$$\therefore\ x = \frac{-4\log\left(\frac{1}{8}\right)}{\log 3}$$

**d**
$$20 \times e^{2x} = 840$$
$$\therefore\ e^{2x} = 42$$
$$\therefore\ \ln e^{2x} = \ln 42$$
$$\therefore\ 2x = \ln 42$$
$$\therefore\ x = \tfrac{1}{2}\ln 42$$

**e**
$$50 \times e^{-0.03x} = 0.05$$
$$\therefore\ e^{-0.03x} = 0.001$$
$$\therefore\ \ln e^{-0.03x} = \ln(0.001)$$
$$\therefore\ -0.03x = \ln(0.001)$$
$$\therefore\ -\tfrac{3}{100}x = \ln(0.001)$$
$$\therefore\ x = -\tfrac{100}{3}\ln(0.001)$$

**f**
$$41e^{0.3x} - 27 = 0$$
$$\therefore\ 41e^{0.3x} = 27$$
$$\therefore\ e^{0.3x} = \tfrac{27}{41}$$
$$\therefore\ \ln e^{0.3x} = \ln\left(\tfrac{27}{41}\right)$$
$$\therefore\ 0.3x = \ln\left(\tfrac{27}{41}\right)$$
$$\therefore\ \tfrac{3}{10}x = \ln\left(\tfrac{27}{41}\right)$$
$$\therefore\ x = \tfrac{10}{3}\ln\left(\tfrac{27}{41}\right)$$

**6** **a**
$$e^{2x} = 2e^x$$
$$\therefore\ e^x(e^x - 2) = 0$$
$$\therefore\ e^x = 2 \quad \{\text{as } e^x > 0\}$$
$$\therefore\ x = \ln 2$$

**b**
$$e^x = e^{-x}$$
$$\therefore\ x = -x$$
$$\therefore\ 2x = 0$$
$$\therefore\ x = 0$$

**c**
$$e^{2x} - 5e^x + 6 = 0$$
$$\therefore\ (e^x - 3)(e^x - 2) = 0$$
$$\therefore\ e^x = 3 \text{ or } 2$$
$$\therefore\ x = \ln 3 \text{ or } \ln 2$$

**d**
$$e^x + 2 = 3e^{-x}$$
$$\therefore\ e^{2x} + 2e^x = 3 \quad \{\times e^x\}$$
$$\therefore\ e^{2x} + 2e^x - 3 = 0$$
$$\therefore\ (e^x + 3)(e^x - 1) = 0$$
$$\therefore\ e^x = -3 \text{ or } 1$$
$$\therefore\ e^x = 1 \quad \{\text{as } e^x > 0\}$$
$$\therefore\ x = \ln 1$$
$$\therefore\ x = 0$$

**e**
$$1 + 12e^{-x} = e^x$$
$$\therefore\ e^x + 12 = e^{2x} \quad \{\times e^x\}$$
$$\therefore\ e^{2x} - e^x - 12 = 0$$
$$\therefore\ (e^x - 4)(e^x + 3) = 0$$
$$\therefore\ e^x = 4 \text{ or } -3$$
$$\therefore\ e^x = 4 \quad \{\text{as } e^x > 0\}$$
$$\therefore\ x = \ln 4$$

**f**
$$e^x + e^{-x} = 3$$
$$\therefore\ e^{2x} + 1 = 3e^x \quad \{\times e^x\}$$
$$\therefore\ e^{2x} - 3e^x + 1 = 0$$
$$\therefore\ e^x = \frac{3 \pm \sqrt{9-4}}{2}$$
$$\therefore\ e^x = \frac{3 \pm \sqrt{5}}{2}$$
$$\therefore\ x = \ln\left(\tfrac{3+\sqrt{5}}{2}\right) \text{ or } \ln\left(\tfrac{3-\sqrt{5}}{2}\right)$$
$$\approx 0.962 \text{ or } -0.962$$

**7** **a**
$y = e^x$ and $y = e^{2x} - 6$
meet when $e^x = e^{2x} - 6$
$$\therefore\ e^{2x} - e^x - 6 = 0$$
$$\therefore\ (e^x - 3)(e^x + 2) = 0$$
$$\therefore\ e^x = 3 \text{ or } -2$$
$$\therefore\ e^x = 3 \quad \{\text{as } e^x > 0\}$$
$\therefore$ $x = \ln 3$ and $y = e^x = 3$
$\therefore$ they meet at $(\ln 3, 3)$.

**b**
$y = 2e^x + 1$ and $y = 7 - e^x$
meet when $2e^x + 1 = 7 - e^x$
$$\therefore\ 3e^x = 6$$
$$\therefore\ e^x = 2$$
$\therefore$ $x = \ln 2$ and $y = 7 - e^x = 5$
$\therefore$ they meet at $(\ln 2, 5)$.

**c** $y = 3 - e^x$ and $y = 5e^{-x} - 3$

meet when $3 - e^x = 5e^{-x} - 3$

$\therefore\ 3e^x - e^{2x} = 5 - 3e^x \quad \{\times e^x\}$

$\therefore\ e^{2x} - 6e^x + 5 = 0$

$\therefore\ (e^x - 5)(e^x - 1) = 0$

$\therefore\ e^x = 1$ or $5$

$\therefore\ x = 0$ or $\ln 5$

When $x = 0$, $y = 3 - e^0 = 3 - 1 = 2$

When $x = \ln 5$, $y = 3 - e^{\ln 5} = 3 - 5 = -2$

$\therefore$ they meet at $(0, 2)$ and $(\ln 5, -2)$.

## EXERCISE 4F

**1** **a** $\log_3 12 = \dfrac{\log_{10} 12}{\log_{10} 3} \approx 2.26$

**b** $\log_{\frac{1}{2}} 1250 = \dfrac{\log_{10} 1250}{\log_{10}(0.5)} \approx -10.3$

**c** $\log_3(0.067) = \dfrac{\log_{10}(0.067)}{\log_{10} 3} \approx -2.46$

**d** $\log_{0.4}(0.006\,984) = \dfrac{\log_{10}(0.006\,984)}{\log_{10}(0.4)} \approx 5.42$

**2** **a** $2^x = 0.051$

$\therefore\ x = \log_2(0.051)$

$\therefore\ x = \dfrac{\ln(0.051)}{\ln 2}$

$\therefore\ x \approx -4.29$

**b** $4^x = 213.8$

$\therefore\ x = \log_4 213.8$

$\therefore\ x = \dfrac{\ln(213.8)}{\ln 4}$

$\therefore\ x \approx 3.87$

**c** $3^{2x+1} = 4.069$

$\therefore\ 2x + 1 = \log_3(4.069)$

$\therefore\ 2x + 1 = \dfrac{\ln(4.069)}{\ln 3}$

$\therefore\ 2x + 1 \approx 1.2774$

$\therefore\ 2x \approx 0.2774$

$\therefore\ x \approx 0.139$

**3** **a** $25^x - 3(5^x) = 0$

$\therefore\ 5^{2x} - 3(5^x) = 0$

$\therefore\ 5^x(5^x - 3) = 0$

$\therefore\ 5^x = 3$

$\{$as $5^x > 0$ for all $x\}$

$\therefore\ x = \log_5 3$

$\therefore\ x = \dfrac{\log 3}{\log 5}$

**b** $8(9^x) - 3^x = 0$

$\therefore\ 8 \times 3^{2x} - 3^x = 0$

$\therefore\ 3^x(8 \times 3^x - 1) = 0$

$\therefore\ 8 \times 3^x - 1 = 0$

$\{$as $3^x > 0$ for all $x\}$

$\therefore\ 3^x = \frac{1}{8}$

$\therefore\ x = \log_3(\frac{1}{8})$

$\therefore\ x = \dfrac{\log(\frac{1}{8})}{\log 3}$

**c** $2^x - 2(4^x) = 0$

$\therefore\ 2^x - 2 \times 2^{2x} = 0$

$\therefore\ 2^x(1 - 2 \times 2^x) = 0$

$\therefore\ 1 - 2 \times 2^x = 0$

$\{$as $2^x > 0$ for all $x\}$

$\therefore\ 2^x = \frac{1}{2}$

$\therefore\ x = \log_2(\frac{1}{2})$

$\therefore\ x = \dfrac{\log(\frac{1}{2})}{\log 2}$

$\therefore\ x = -1$

**4** $\log_4 x^3 + \log_2 \sqrt{x} = 8$

$\therefore\ \dfrac{\log x^3}{\log 4} + \dfrac{\log x^{\frac{1}{2}}}{\log 2} = 8$

$\therefore\ \dfrac{3 \log x}{2 \log 2} + \dfrac{\frac{1}{2} \log x}{\log 2} = 8$

$\therefore\ \dfrac{3 \log x}{2 \log 2} + \dfrac{\log x}{2 \log 2} = 8$

$\therefore\ \dfrac{4 \log x}{2 \log 2} = 8$

$\therefore\ \log x = 4 \log 2$

$\therefore\ \log x = \log 2^4$

$\therefore\ x = 16$

## EXERCISE 4G

**1** **a** $f(x) = \log_3(x+1), \quad x > -1$

**i** We require $x + 1 > 0 \quad \therefore \quad x > -1$
So, the domain is $\{x \mid x > -1\}$ and the range is $\{y \mid y \in \mathbb{R}\}$.

**ii** As $x \to -1^+, \quad y \to -\infty$,
so $x = -1$ is a vertical asymptote.
As $x \to \infty, \quad y \to \infty$.
When $x = 0, \quad y = \log_3 1 = 0$
So, the $y$-intercept is 0.
When $y = 0, \quad \log_3(x+1) = 0$
$\therefore \quad x + 1 = 3^0$
$\therefore \quad x + 1 = 1$
$\therefore \quad x = 0$
So, the $x$-intercept is 0.

**iii** We graph, using $y = \dfrac{\log(x+1)}{\log 3}$

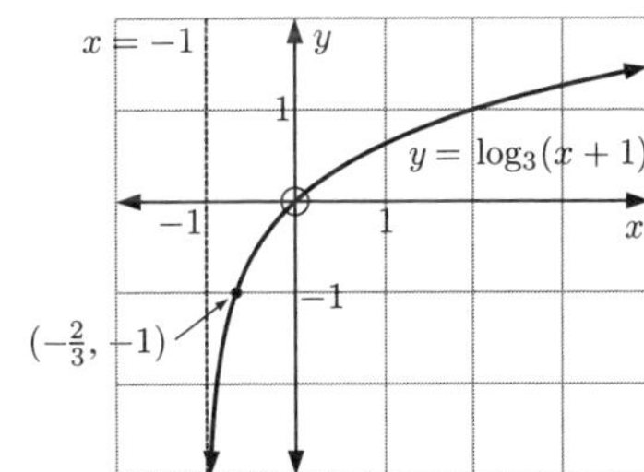


**iv** If $f(x) = -1$
then $\log_3(x+1) = -1$
$\therefore \quad x + 1 = 3^{-1}$
$\therefore \quad x = \frac{1}{3} - 1$
$\therefore \quad x = -\frac{2}{3}$
which checks with the graph

**v** $f$ is defined by $y = \log_3(x+1)$
$\therefore \quad f^{-1}$ is defined by $x = \log_3(y+1)$
$\therefore \quad y + 1 = 3^x$
$\therefore \quad y = 3^x - 1$
$\therefore \quad f^{-1}(x) = 3^x - 1$
Horizontal asymptote is $y = -1$.
Domain is $x \in \mathbb{R}$.
Range is $\{y \mid y > -1\}$.

**b** $f(x) = 1 - \log_3(x+1), \quad x > -1$

**i** We require $x + 1 > 0 \quad \therefore \quad x > -1$
So, the domain is $\{x \mid x > -1\}$ and the range is $\{y \mid y \in \mathbb{R}\}$.

**ii** As $x \to -1^+, \quad y \to \infty$,
so $x = -1$ is a vertical asymptote.
As $x \to \infty, \quad y \to -\infty$.
When $x = 0, \quad y = 1 - \log_3 1$
$= 1 - 0 = 1$
So, the $y$-intercept is 1.
When $y = 0, \quad 1 - \log_3(x+1) = 0$
$\therefore \quad \log_3(x+1) = 1$
$\therefore \quad x + 1 = 3^1$
$= 3$
$\therefore \quad x = 2$
So, the $x$-intercept is 2.

**iii** We graph using $y = 1 - \dfrac{\log(x+1)}{\log 3}$

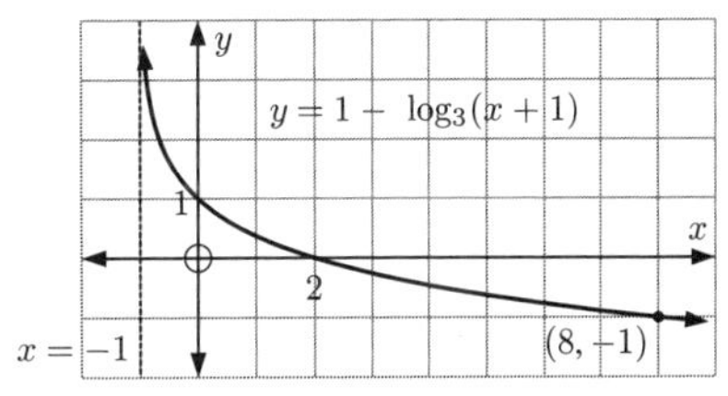


**iv** If $f(x) = -1$
then $1 - \log_3(x+1) = -1$
$\therefore \quad \log_3(x+1) = 2$
$\therefore \quad x + 1 = 3^2$
$\therefore \quad x = 8$
which checks with the graph

**v** $f$ is defined by $y = 1 - \log_3(x+1)$
$\therefore \quad f^{-1}$ is defined by $x = 1 - \log_3(y+1)$
$\therefore \quad \log_3(y+1) = 1 - x$
$\therefore \quad y + 1 = 3^{1-x}$
$\therefore \quad y = 3^{1-x} - 1$
$\therefore \quad f^{-1}(x) = 3^{1-x} - 1$
Horizontal asymptote is $y = -1$.
Domain is $x \in \mathbb{R}$.
Range is $\{y \mid y > -1\}$.

**c** $f(x) = \log_5(x-2) - 2, \quad x > 0$

**i** We require $x - 2 > 0 \quad \therefore \quad x > 2.$
So, the domain is $\{x \mid x > 2\}$ and the range is $\{y \mid y \in \mathbb{R}\}$.

**ii** As $x \to 2^+, \quad y \to -\infty,$
so $x = 2$ is a vertical asymptote.
As $x \to \infty, \quad y \to \infty.$
When $x = 0$, $y$ is undefined.
$\therefore$ there is no $y$-intercept.
When $y = 0$, $\log_5(x-2) = 2$

$$\therefore \quad x - 2 = 5^2$$
$$= 25$$
$$\therefore \quad x = 27$$

So, the $x$-intercept is 27.

**iii** We graph using $y = \dfrac{\log(x-2)}{\log 5} - 2$

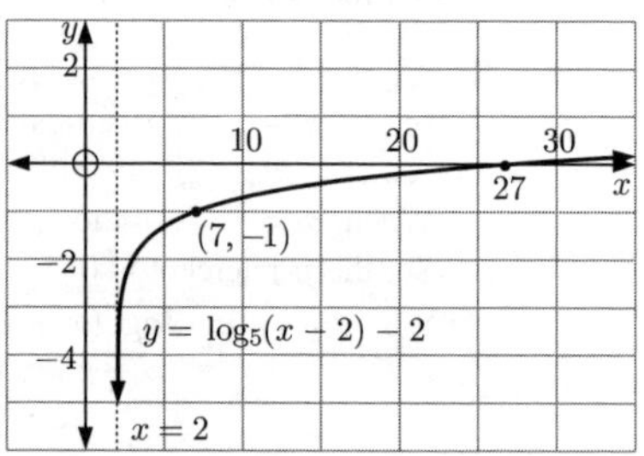


**iv** If $f(x) = -1$

then $\log_5(x-2) - 2 = -1$

$$\therefore \quad \log_5(x-2) = 1$$
$$\therefore \quad x - 2 = 5^1$$
$$\therefore \quad x = 5 + 2$$
$$\therefore \quad x = 7$$

which checks with the graph

**v** $f$ is defined by $y = \log_5(x-2) - 2$
$\therefore$ $f^{-1}$ is defined by $x = \log_5(y-2) - 2$

$$\therefore \quad x + 2 = \log_5(y-2)$$
$$\therefore \quad y - 2 = 5^{x+2}$$
$$\therefore \quad y = 5^{x+2} + 2$$
$$\therefore \quad f^{-1}(x) = 5^{x+2} + 2$$

Horizontal asymptote is $y = 2$.
Domain is $x \in \mathbb{R}$.
Range is $\{y \mid y > 2\}$.

**d** $f(x) = 1 - \log_5(x-2), \quad x > 2$

**i** We require $x - 2 > 0 \quad \therefore \quad x > 2.$
So, the domain is $\{x \mid x > 2\}$ and the range is $\{y \mid y \in \mathbb{R}\}$.

**ii** As $x \to 2^+, \quad y \to \infty,$
so $x = 2$ is a vertical asymptote.
As $x \to \infty, \quad y \to -\infty.$
When $x = 0$, $y$ is undefined.
$\therefore$ there is no $y$-intercept.
When $y = 0$, $1 - \log_5(x-2) = 0$

$$\therefore \quad \log_5(x-2) = 1$$
$$\therefore \quad x - 2 = 5^1$$
$$\therefore \quad x = 7$$

So, $x$-intercept is 7.

**iii** We graph using $y = 1 - \dfrac{\log(x-2)}{\log 5}$

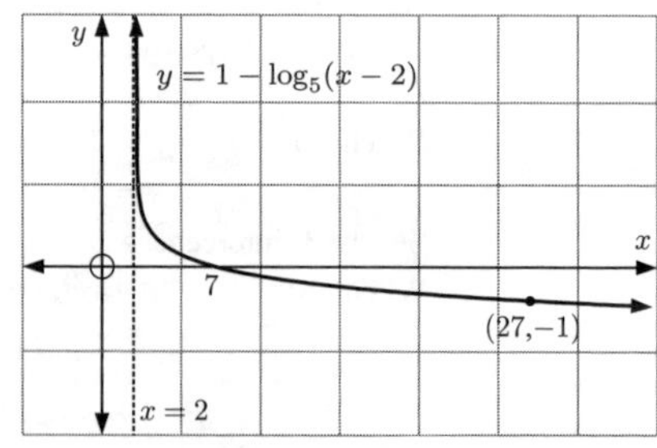


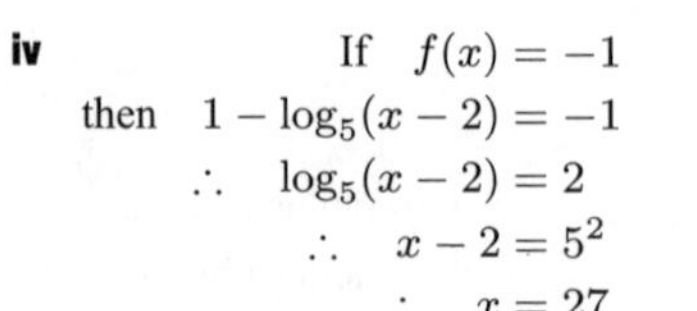

**iv** If $f(x) = -1$

then $1 - \log_5(x-2) = -1$

$$\therefore \quad \log_5(x-2) = 2$$
$$\therefore \quad x - 2 = 5^2$$
$$\therefore \quad x = 27$$

which checks with the graph

**v** $f$ is defined by $y = 1 - \log_5(x-2)$
$\therefore$ $f^{-1}$ is defined by $x = 1 - \log_5(y-2)$

$$\therefore \quad \log_5(y-2) = 1 - x$$
$$\therefore \quad y - 2 = 5^{1-x}$$
$$\therefore \quad y = 5^{1-x} + 2$$
$$\therefore \quad f^{-1}(x) = 5^{1-x} + 2$$

Horizontal asymptote is $y = 2$.
Domain is $x \in \mathbb{R}$.
Range is $\{y \mid y > 2\}$.

**e** $f(x) = 1 - 2\log_2 x, \quad x > 0$

**i** We require $x > 0$.
So, the domain is $\{x \mid x > 0\}$ and the range is $\{y \mid y \in \mathbb{R}\}$.

**ii** As $x \to 0^+$, $y \to \infty$,
so $x = 0$ is a vertical asymptote.
As $x \to \infty$, $y \to -\infty$.
When $x = 0$, $y$ is undefined.
$\therefore$ there is no $y$-intercept.
When $y = 0$, $\log_2 x = \frac{1}{2}$

$\therefore \quad x = 2^{\frac{1}{2}}$

$\therefore \quad x = \sqrt{2}$

$\therefore$ $x$-intercept is $\sqrt{2} \approx 1.41$.

**iii** We graph using $y = 1 - \dfrac{2\log x}{\log 2}$

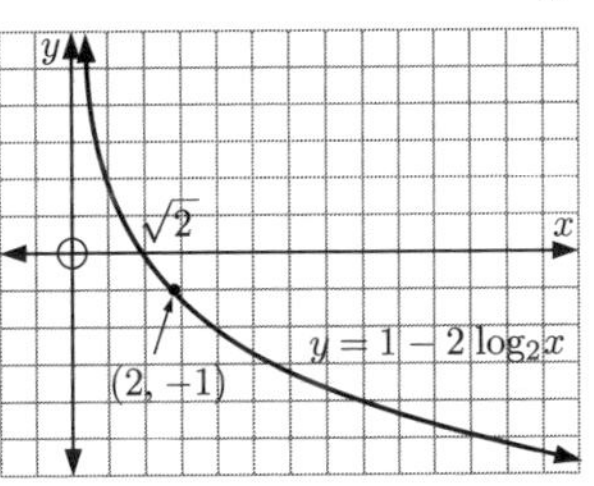


**iv** If $f(x) = -1$
then $1 - 2\log_2 x = -1$
$\therefore \quad -2\log_2 x = -2$
$\therefore \quad \log_2 x = 1$
$\therefore \quad x = 2^1$
$\therefore \quad x = 2$
which checks with the graph

**v** $f$ is defined by $y = 1 - 2\log_2 x$
$\therefore$ $f^{-1}$ is defined by $x = 1 - 2\log_2 y$
$\therefore \quad 2\log_2 y = 1 - x$
$\therefore \quad \log_2 y = \dfrac{1-x}{2}$
$\therefore \quad y = 2^{\frac{1-x}{2}}$
$\therefore \quad f^{-1}(x) = 2^{\frac{1-x}{2}}$
Horizontal asymptote is $y = 0$.
Domain is $x \in \mathbb{R}$. Range is $\{y \mid y > 0\}$.

**2 a i** $f(x) = e^x + 5$
or $y = e^x + 5$
has inverse function
$x = e^y + 5$
$\therefore \quad x - 5 = e^y$
$\therefore \quad y = \ln(x - 5)$
$\therefore \quad f^{-1}(x) = \ln(x - 5)$

**ii**

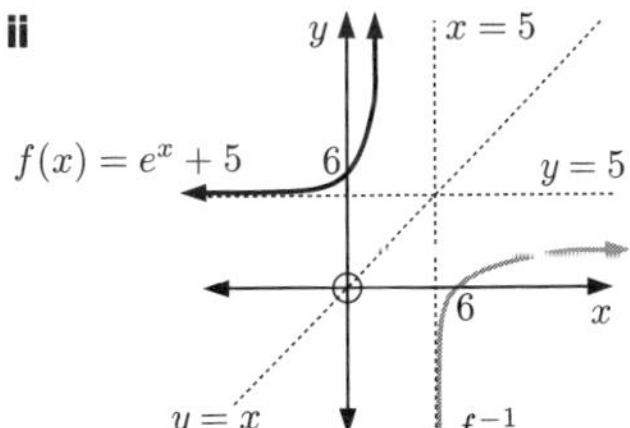


**iii** Domain of $f$ is $\{x \mid x \in \mathbb{R}\}$, range is $\{y \mid y > 5\}$.
Domain of $f^{-1}$ is $\{x \mid x > 5\}$, range is $\{y \mid y \in \mathbb{R}\}$.

**iv** $f$ has a H.A. $y = 5$. $f^{-1}$ has a V.A. $x = 5$.
When $x = 0$, $y = e^0 + 5$
$\therefore \quad y = 6$
$\therefore$ $y$-intercept of $f$ is 6.
When $y = 0$, $e^x + 5$ is undefined
$\therefore$ $f$ has no $x$-intercept.
$\therefore$ the $x$-intercept of $f^{-1}$ is 6, and $f^{-1}$ has no $y$-intercept.

**b i** $f(x) = e^{x+1} - 3$
or $y = e^{x+1} - 3$
has inverse function
$x = e^{y+1} - 3$
$\therefore \quad x + 3 = e^{y+1}$
$\therefore \quad y + 1 = \ln(x + 3)$
$\therefore \quad f^{-1}(x) = \ln(x + 3) - 1$

**ii**

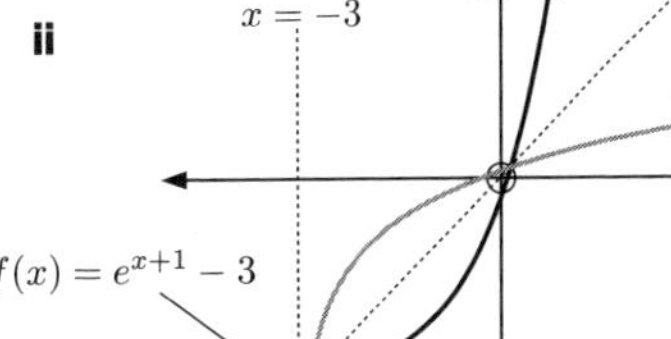

**iii** Domain of $f$ is $\{x \mid x \in \mathbb{R}\}$, range is $\{y \mid y > -3\}$.
Domain of $f^{-1}$ is $\{x \mid x > -3\}$, range is $\{y \mid y \in \mathbb{R}\}$.

**iv** $f$ has a H.A. $y = -3$. $f^{-1}$ has a V.A. $x = -3$.
When $x = 0$, $y = e^{0+1} - 3 = e - 3$
$\therefore$ the $y$-intercept of $f$ is $e - 3 \approx -0.282$
When $y = 0$, $e^{x+1} - 3 = 0$

$$\therefore\ e^{x+1} = 3$$
$$\therefore\ x + 1 = \ln 3$$
$$\therefore\ x = \ln 3 - 1$$

$\therefore$ the $x$-intercept of $f$ is $\ln 3 - 1 \approx 0.0986$
$\therefore$ the $x$-intercept of $f^{-1}$ is $e - 3 \approx -0.282$
and the $y$-intercept of $f^{-1}$ is $\ln 3 - 1 \approx 0.0986$

**c** **i**
$$f(x) = \ln x - 4,\ x > 0$$
$$\therefore\ y = \ln x - 4$$
and has inverse function
$$x = \ln y - 4$$
$$\therefore\ x + 4 = \ln y$$
$$\therefore\ y = e^{x+4}$$
$$\therefore\ f^{-1}(x) = e^{x+4}$$

**ii**

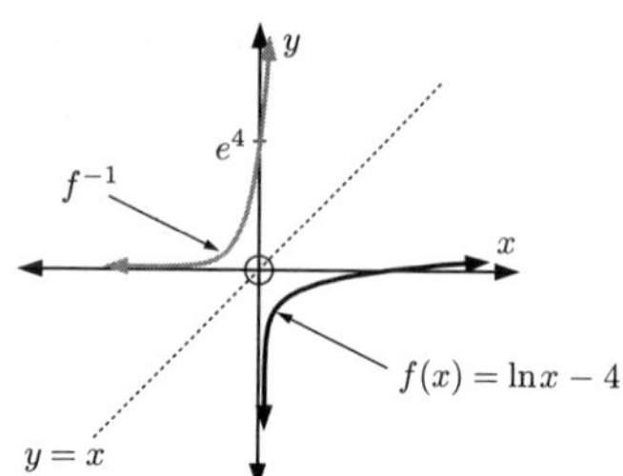


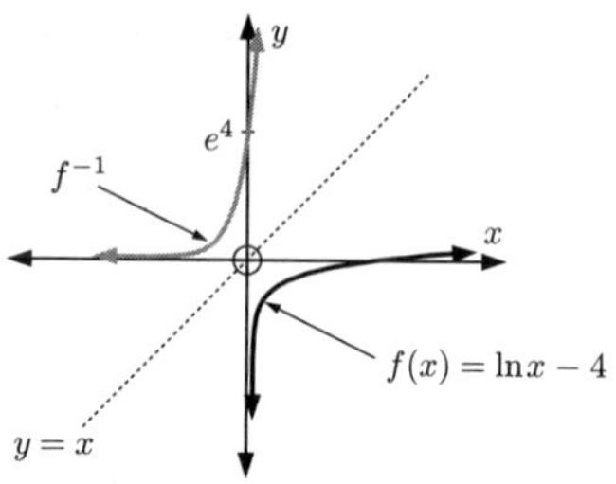

**iii** Domain of $f$ is $\{x \mid x > 0\}$, range is $\{y \mid y \in \mathbb{R}\}$.
Domain of $f^{-1}$ is $\{x \mid x \in \mathbb{R}\}$, range is $\{y \mid y > 0\}$.

**iv** $f$ has a V.A. $x = 0$. $f^{-1}$ has a H.A. $y = 0$.
When $x = 0$, $\ln x - 4$ is undefined.
$\therefore$ $f$ has no $y$-intercept.
When $y = 0$, $\ln x - 4 = 0$

$$\therefore\ \ln x = 4$$
$$\therefore\ x = e^4 \approx 54.6$$

$\therefore$ the $x$-intercept of $f$ is $e^4$.
$\therefore$ $f^{-1}$ has no $x$-intercept and the $y$-intercept of $f^{-1}$ is $e^4$.

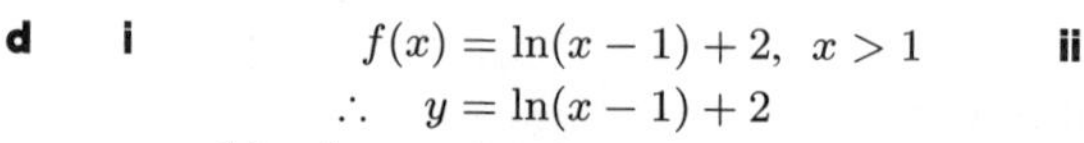

**d** **i**
$$f(x) = \ln(x-1) + 2,\ x > 1$$
$$\therefore\ y = \ln(x-1) + 2$$
and has inverse function
$$x = \ln(y-1) + 2$$
$$\therefore\ \ln(y-1) = x - 2$$
$$\therefore\ y - 1 = e^{x-2}$$
$$\therefore\ y = e^{x-2} + 1$$
$$\therefore\ f^{-1}(x) = e^{x-2} + 1$$

**ii**

**iii** Domain of $f$ is $\{x \mid x > 1\}$, range is $\{y \mid y \in \mathbb{R}\}$.
Domain of $f^{-1}$ is $\{x \mid x \in \mathbb{R}\}$, range is $\{y \mid y > 1\}$.

**iv** $f$ has a V.A. $x = 1$. $f^{-1}$ has a H.A. $y = 1$.
When $x = 0$, $\ln(x-1) + 2$ is undefined.
$\therefore$ $f$ has no $y$-intercept.
When $y = 0$, $\ln(x-1) + 2 = 0$

$$\therefore\ \ln(x-1) = -2$$
$$\therefore\ x - 1 = e^{-2}$$
$$\therefore\ x = 1 + e^{-2}$$

$\therefore$ the $x$-intercept of $f$ is $1 + e^{-2}$.
$\therefore$ $f^{-1}$ has no $x$-intercept and the $y$-intercept of $f^{-1}$ is $1 + e^{-2}$.

**3** **a** $f$ is $y = e^{2x}$
so the inverse function $f^{-1}$ is

$$x = e^{2y}$$
$$\therefore \quad 2y = \ln x$$
$$\therefore \quad y = \tfrac{1}{2}\ln x$$
$$\therefore \quad f^{-1}(x) = \tfrac{1}{2}\ln x$$
$$\therefore \quad (f^{-1} \circ g)(x) = f^{-1}(g(x)) = f^{-1}(2x-1) = \tfrac{1}{2}\ln(2x-1)$$

**b** $(g \circ f)(x) = g(f(x)) = g(e^{2x}) = 2(e^{2x}) - 1$

So, $y = 2e^{2x} - 1$ which has inverse

$$x = 2e^{2y} - 1$$
$$\therefore \quad x + 1 = 2e^{2y}$$
$$\therefore \quad \tfrac{1}{2}(x+1) = e^{2y}$$
$$\therefore \quad 2y = \ln\left(\frac{x+1}{2}\right)$$
$$\therefore \quad y = \tfrac{1}{2}\ln\left(\frac{x+1}{2}\right)$$
$$\therefore \quad (g \circ f)^{-1}(x) = \tfrac{1}{2}\ln\left(\frac{x+1}{2}\right)$$

**4** **a** $y = \ln x$ cuts the $x$-axis when $y = 0$

$$\therefore \quad \ln x = 0$$
$$\therefore \quad x = e^0 = 1$$

So, graph A is that of $y = \ln x$.

**Note:** $x$-intercept of $y = \ln(x-2)$ is when $x - 2 = e^0 = 1$

$$\therefore \quad x = 3$$

**b** The $x$-intercept of $y = \ln(x+2)$ occurs when $x + 2 = e^0 = 1$

$$\therefore \quad x = -1$$



**c** $y = \ln x$ has a V.A. of $x = 0$.
$y = \ln(x-2)$ has a V.A. of $x = 2$.
$y = \ln(x+2)$ has a V.A. of $x = -2$.

**5** Since $y = \ln(x^2)$, $y = 2\ln x$ {log law}
$\therefore$ the new $y$-values are twice the old $y$-values.
$\therefore$ Kelly is correct.
(Note that $y = \ln x^2$ is also defined for $x < 0$. However, we are only concerned with $y = \ln x^2$ for $x > 0$.)

**6** **a**

$$f(x) = e^{x+3} + 2$$

or $y = e^{x+3} + 2$ has inverse function

$$x = e^{y+3} + 2$$
$$\therefore \quad x - 2 = e^{y+3}$$
$$\therefore \quad \ln(x-2) = y + 3$$
$$\therefore \quad y = \ln(x-2) - 3$$

So, $f^{-1}(x) = \ln(x-2) - 3$

**b** **i** $f(x) < 2.1$ when $e^{x+3} + 2 < 2.1$

$$\therefore \quad e^{x+3} < 0.1$$
$$\therefore \quad x + 3 < \ln(0.1)$$
$$\therefore \quad x < \ln(0.1) - 3$$
$$\therefore \quad x < -5.30$$

**ii** Similarly, $f(x) < 2.01$ when

$$x < \ln(0.01) - 3$$
$$\therefore \quad x < -7.61$$

**iii** $f(x) < 2.001$ when

$$x < \ln(0.001) - 3$$
$$\therefore \quad x < -9.91$$

**iv** $f(x) < 2.0001$ when

$$x < \ln(0.0001) - 3$$
$$\therefore \quad x < -12.2$$

We conjecture that the H.A. is $y = 2$.

**c** As $x \to \infty$, $y \to \infty$
As $x \to -\infty$, $e^{x+3} \to 0$ $\therefore$ $y \to 2$
$\therefore$ H.A. is $y = 2$.

**d** $f$ has a H.A. $y = 2$ and range $\{y \mid y > 2\}$
$\therefore$ $f^{-1}$ has a V.A. $x = 2$ and domain $\{x \mid x > 2\}$

## EXERCISE 4H

**1** $W_t = 20 \times 2^{0.15t}$ grams

**a** When $W_t = 30$,
$$20 \times 2^{0.15t} = 30$$
$$\therefore\ 2^{0.15t} = 1.5$$
$$\therefore\ \log 2^{0.15t} = \log(1.5)$$
$$\therefore\ 0.15t \log 2 = \log(1.5)$$
$$\therefore\ t = \frac{\log(1.5)}{0.15 \times \log 2}$$
$$\therefore\ t \approx 3.90 \text{ hours}$$
$\therefore$ it takes about 3.90 hours to reach 30 g.

**b** When $W_t = 100$,
$$20 \times 2^{0.15t} = 100$$
$$\therefore\ 2^{0.15t} = 5$$
$$\therefore\ \log 2^{0.15t} = \log 5$$
$$\therefore\ 0.15t \log 2 = \log 5$$
$$\therefore\ t = \frac{\log 5}{0.15 \times \log 2}$$
$$\therefore\ t \approx 15.5 \text{ hours}$$
$\therefore$ it takes about 15.5 hours to reach 100 g.

**2** When $M_t = 50$, $25 \times e^{0.1t} = 50$
$$\therefore\ e^{0.1t} = 2$$
$$\therefore\ \ln e^{0.1t} = \ln 2$$
$$\therefore\ 0.1t = \ln 2$$
$$\therefore\ t = 10 \ln 2$$
$\therefore$ it takes $10 \ln 2$ hours to reach 50 g.

**3** **a**

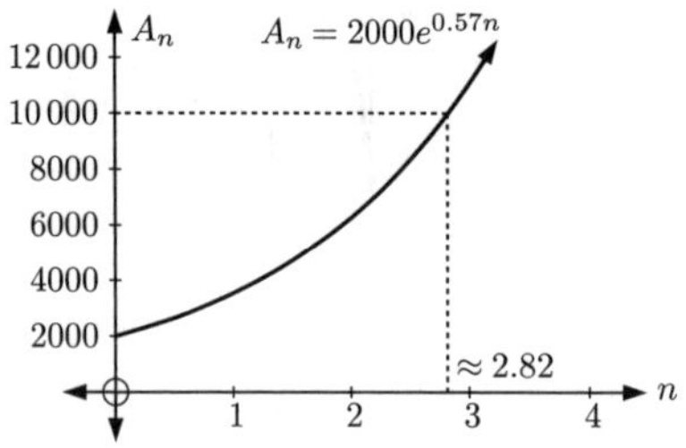

**b** When $A_n = 10\,000$, $t \approx 2.82$

$\therefore$ we estimate that it will take 2.82 weeks for the infested area to reach 10 000 ha.

**c** **i** When $A_n = 10\,000$, $2000 \times e^{0.57n} = 10\,000$
$$\therefore\ e^{0.57n} = 5$$
$$\therefore\ \ln e^{0.57n} = \ln 5$$
$$\therefore\ 0.57n = \ln 5$$
$$\therefore\ n = \frac{\ln 5}{0.57}$$
$$\therefore\ n \approx 2.82$$
$\therefore$ it takes about 2.82 weeks for the infested area to reach 10 000 hectares.

**4** $r = 107.5\% = 1.075$, $u_1 = 160\,000$, $u_{n+1} = 250\,000$
$$u_{n+1} = u_1 \times r^n$$
$$\therefore\ 250\,000 = 160\,000 \times (1.075)^n$$
$$\therefore\ (1.075)^n = \tfrac{25}{16}$$
$$\therefore\ \log(1.075)^n = \log(\tfrac{25}{16})$$
$$\therefore\ n\log(1.075) = \log(\tfrac{25}{16})$$
$$\therefore\ n = \frac{\log(\frac{25}{16})}{\log(1.075)} \approx 6.1709$$
$\therefore$ it would take 6.17 years or 6 years 62 days.

**5** $u_1 = 10\,000$, $u_{n+1} = 15\,000$, $r = 104.8\% = 1.048$
$$u_{n+1} = u_1 \times r^n$$
$$\therefore\ 15\,000 = 10\,000 \times (1.048)^n$$
$$\therefore\ (1.048)^n = 1.5$$
$$\therefore\ \log(1.048)^n = \log(1.5)$$
$$\therefore\ n\log(1.048) = \log(1.5)$$
$$\therefore\ n = \frac{\log(1.5)}{\log(1.048)}$$
$$\therefore\ n \approx 8.648$$
$\therefore$ it would take 9 years.
{interest compounded annually}

**6** **a** 8.4% p.a. compounded monthly

is $\dfrac{8.4\%}{12} = 0.7\%$ a month

$= 0.007$

So $r = 1 + 0.007$

$\therefore\ r = 1.007$

**b** $u_1 = 15\,000$ and $u_{n+1} = 25\,000$

$$u_{n+1} = u_1 \times r^n$$
$$\therefore\ 25\,000 = 15\,000 \times (1.007)^n$$
$$\therefore\ (1.007)^n = \tfrac{25}{15} = \tfrac{5}{3}$$
$$\therefore\ \log(1.007)^n = \log(\tfrac{5}{3})$$
$$\therefore\ n\log(1.007) = \log(\tfrac{5}{3})$$
$$\therefore\ n = \frac{\log(\frac{5}{3})}{\log(1.007)} \approx 73.23$$

$\therefore$ he will withdraw the money after 74 months.

**7** $M_t = 1000e^{-0.04t}$ $\therefore\ M_0 = 1000e^0 = 1000$ g

**a** For $M_t$ to halve, $M_t = 500$

$$\therefore\ 1000e^{-0.04t} = 500$$
$$\therefore\ e^{-0.04t} = 0.5$$
$$\therefore\ -0.04t = \ln(0.5)$$
$$\therefore\ t = \frac{\ln(0.5)}{-0.04}$$
$$\therefore\ t \approx 17.3 \text{ years}$$

**b** For $M_t = 25$ g,

$$\therefore\ 1000e^{-0.04t} = 25$$
$$\therefore\ e^{-0.04t} = 0.025$$
$$\therefore\ -0.04t = \ln(0.025)$$
$$\therefore\ t = \frac{\ln(0.025)}{-0.04}$$
$$\therefore\ t \approx 92.2 \text{ years}$$

**c** For $M_t = 1\%$ of $M_0$

$$\therefore\ 1000e^{-0.04t} = 0.01 \times 1000$$
$$\therefore\ e^{-0.04t} = 0.01$$
$$\therefore\ -0.04t = \ln(0.01)$$
$$\therefore\ t = \frac{\ln(0.01)}{-0.04}$$
$$\therefore\ t \approx 115 \text{ years}$$

**8** $V = 50(1 - e^{-0.2t})$ m s$^{-1}$

So, when $V = 40$, $50(1 - e^{-0.2t}) = 40$

$$\therefore\ 1 - e^{-0.2t} = 0.8$$
$$\therefore\ e^{-0.2t} = 0.2$$
$$\therefore\ -0.2t = \ln(0.2)$$
$$\therefore\ -\tfrac{1}{5}t = \ln(\tfrac{1}{5})$$
$$\therefore\ -\tfrac{1}{5}t = \ln(5^{-1})$$
$$\therefore\ -\tfrac{1}{5}t = -\ln 5$$
$$\therefore\ t = 5\ln 5$$

$\therefore$ it will take $5\ln 5$ seconds for the man's speed to reach 40 m s$^{-1}$.

**9** **a**

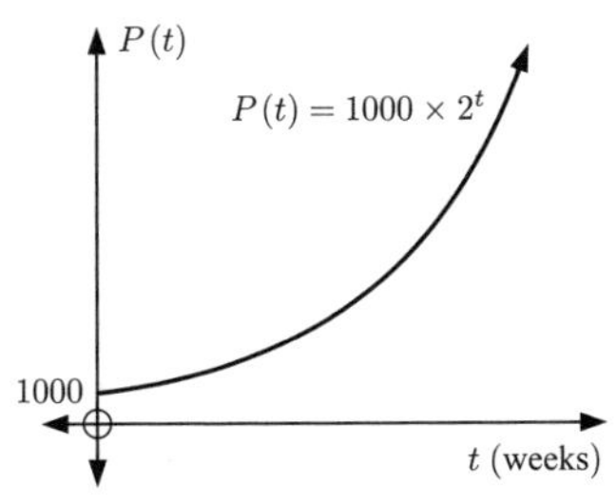


**b** When $P(t) = 20\,000$,

$$1000 \times 2^t = 20\,000$$
$$\therefore\ 2^t = 20$$
$$\therefore\ \log 2^t = \log 20$$
$$\therefore\ t\log 2 = \log 20$$
$$\therefore\ t = \frac{\log 20}{\log 2} \approx 4.32$$

$\therefore$ it will take 4.32 weeks for the population to reach 20 000 mice.

**c**
$$P = 1000 \times 2^t$$
$$\therefore \quad 2^t = \frac{P}{1000}$$
$$\therefore \quad \log 2^t = \log\left(\frac{P}{1000}\right)$$
$$\therefore \quad t \log 2 = \log P - \log 1000$$
$$\therefore \quad t = \frac{\log P - 3}{\log 2}$$

**d**

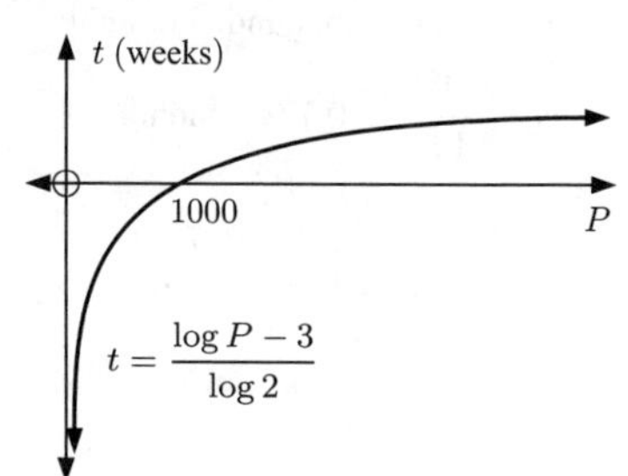


**10** $T = 4 + 96 \times e^{-0.03t}$ °C

**a** When $T = 25$,
$$4 + 96 \times e^{-0.03t} = 25$$
$$\therefore \quad 96 \times e^{-0.03t} = 21$$
$$\therefore \quad e^{-0.03t} = \tfrac{21}{96}$$
$$\therefore \quad -0.03t = \ln(\tfrac{21}{96})$$
$$\therefore \quad t = \frac{\ln(\frac{21}{96})}{-0.03}$$
$$\therefore \quad t \approx 50.7 \text{ minutes}$$

**b** When $T = 5$,
$$4 + 96 \times e^{-0.03t} = 5$$
$$\therefore \quad 96 \times e^{-0.03t} = 1$$
$$\therefore \quad e^{-0.03t} = \tfrac{1}{96}$$
$$\therefore \quad -0.03t = \ln(\tfrac{1}{96})$$
$$\therefore \quad t = \frac{\ln(\frac{1}{96})}{-0.03}$$
$$\therefore \quad t \approx 152 \text{ minutes}$$

**11** **a**



**b**
$$W = 1000 \times 2^{-0.04t}$$
$$\therefore \quad 2^{-0.04t} = \frac{W}{1000}$$
$$\therefore \quad \log 2^{-0.04t} = \log\left(\frac{W}{1000}\right)$$
$$\therefore \quad -0.04t \log 2 = \log W - \log 1000$$
$$\therefore \quad 0.04t \log 2 = 3 - \log W$$
$$\therefore \quad t = \frac{3 - \log W}{0.04 \log 2}$$

**c** **i** When $W = 20$,
$$t = \frac{3 - \log 20}{0.04 \log 2} \approx 141$$
$\therefore$ it will take about 141 years for the weight to reach 20 grams.

**ii** When $W = 0.001$,
$$t = \frac{3 - \log(0.001)}{0.04 \log 2} \approx 498$$
$\therefore$ it will take about 498 years for the weight to reach 0.001 grams.

**12** $W = W_0 \times 2^{-0.0002t}$ grams

**a** When $W$ is 25% of original,
$$W = \tfrac{1}{4} \text{ of } W_0$$
$$\therefore \quad W_0 \times 2^{-0.0002t} = \tfrac{1}{4} \times W_0$$
$$\therefore \quad 2^{-0.0002t} = 2^{-2}$$
$$\therefore \quad 0.0002t = 2$$
$$\therefore \quad t = \frac{2}{0.0002}$$
$$\therefore \quad t = 10\,000$$
$\therefore$ it would take 10 000 years.

**b** When $W$ is 0.1% of original,
$$W = \tfrac{0.1}{100} \text{ of } W_0$$
$$\therefore \quad W_0 \times 2^{-0.0002t} = \tfrac{1}{1000} \times W_0$$
$$\therefore \quad \log 2^{-0.0002t} = \log(0.001)$$
$$\therefore \quad -0.0002t \log 2 = \log(0.001)$$
$$\therefore \quad t = \frac{\log(0.001)}{-0.0002 \times \log 2}$$
$$\therefore \quad t \approx 49\,829$$
$\therefore$ it would take about 49 800 years.

**13** $I = I_0 \times 2^{-0.02t}$ amps

When $I$ is 10% of its original value,

$$\begin{aligned} I &= 10\% \text{ of } I_0 \\ \therefore\ I_0 \times 2^{-0.02t} &= 0.1 \times I_0 \\ \therefore\ 2^{-0.02t} &= 0.1 \\ \therefore\ \log 2^{-0.02t} &= \log(0.1) \\ \therefore\ -0.02t \log 2 &= \log(0.1) \\ \therefore\ -\tfrac{1}{50} t \log 2 &= -1 \\ \therefore\ t &= \frac{50}{\log 2} \text{ seconds} \end{aligned}$$

**14** $V = 60(1 - 2^{-0.2t})$ $\text{m s}^{-1}$

When $V = 50$, $60(1 - 2^{-0.2t}) = 50$

$$\begin{aligned} \therefore\ 1 - 2^{-0.2t} &= 0.8\overline{3} \\ \therefore\ 2^{-0.2t} &= 0.1\overline{6} \\ \therefore\ \log 2^{-0.2t} &= \log 0.1\overline{6} \\ \therefore\ -0.2t \log 2 &= \log 0.1\overline{6} \\ \therefore\ t &= \frac{\log 0.1\overline{6}}{-0.2 \times \log 2} \\ \therefore\ t &\approx 12.9 \text{ seconds} \end{aligned}$$

## REVIEW SET 4A

**1** **a** $\log_4 64 = \log_4 4^3 = 3$

**b** $\log_2 256 = \log_2 2^8 = 8$

**c** $\log_2(0.25) = \log_2(\tfrac{1}{4}) = \log_2 2^{-2} = -2$

**d** $\log_{25} 5 = \log_{25} 25^{\frac{1}{2}} = \tfrac{1}{2}$

**e** $\log_8 1 = \log_8 8^0 = 0$

**f** $\log_{81} 3 = \log_{81} 81^{\frac{1}{4}} = \tfrac{1}{4}$

**g** $\log_9(0.\overline{1}) = \log_9(\tfrac{1}{9}) = \log_9 9^{-1} = -1$

**h** $\log_k \sqrt{k} = \log_k k^{\frac{1}{2}} = \tfrac{1}{2}$ provided $k > 0$, $k \neq 1$

**2** **a** $\log \sqrt{10} = \log 10^{\frac{1}{2}} = \tfrac{1}{2}$

**b** $\log\left(\frac{1}{\sqrt[3]{10}}\right) = \log 10^{-\frac{1}{3}} = -\tfrac{1}{3}$

**c** $\log(10^a \times 10^{b+1}) = \log 10^{a+b+1} = a + b + 1$

**3** **a** $4\ln 2 + 2\ln 3 = \ln 2^4 + \ln 3^2 = \ln(16 \times 9) = \ln 144$

**b** $\tfrac{1}{2}\ln 9 - \ln 2 = \ln 9^{\frac{1}{2}} - \ln 2 = \ln 3 - \ln 2 = \ln(\tfrac{3}{2})$

**c** $2\ln 5 - 1 = \ln 5^2 - \ln e^1 = \ln\left(\frac{25}{e}\right)$

**d** $\tfrac{1}{4}\ln 81 = \ln\left(3^4\right)^{\frac{1}{4}} = \ln 3^1 = \ln 3$

**4** **a** $\ln\left(e\sqrt{e}\right) = \ln(e^1 e^{\frac{1}{2}}) = \ln e^{\frac{3}{2}} = \tfrac{3}{2}$

**b** $\ln\left(\frac{1}{e^3}\right) = \ln e^{-3} = -3$

**c** $\ln(e^{2x}) = 2x$

**d** $\ln\left(\frac{e}{e^x}\right) = \ln(e^{1-x}) = 1 - x$

**5** **a** $\log 16 + 2\log 3 = \log 16 + \log 3^2 = \log(16 \times 9) = \log 144$

**b** $\log_2 16 - 2\log_2 3 = \log_2 16 - \log_2 3^2 = \log_2(\tfrac{16}{9})$

**c** $2 + \log_4 5 = \log_4 4^2 + \log_4 5 = \log_4(16 \times 5) = \log_4 80$

**6** **a**
$$P = 3 \times b^x$$
$$\therefore \log P = \log(3 \times b^x)$$
$$\therefore \log P = \log 3 + \log b^x$$
$$\therefore \log P = \log 3 + x \log b$$

**b**
$$m = \frac{n^3}{p^2}$$
$$\therefore \log m = \log\left(\frac{n^3}{p^2}\right)$$
$$\therefore \log m = \log n^3 - \log p^2$$
$$\therefore \log m = 3\log n - 2\log p$$

**7**
$$\log_3 7 \times 2\log_7 x$$
$$= \overset{1}{\cancel{\log_3 7}} \times 2 \times \frac{\log_3 x}{\cancel{\log_3 7}_1}$$
$$= 2\log_3 x$$

**8** **a**
$$\log T = 2\log x - \log y$$
$$\therefore \log T = \log x^2 - \log y$$
$$\therefore \log T = \log\left(\frac{x^2}{y}\right)$$
$$\therefore T = \frac{x^2}{y}$$

**b**
$$\log_2 K = \log_2 n + \tfrac{1}{2}\log_2 t$$
$$\therefore \log_2 K = \log_2 n + \log_2 t^{\frac{1}{2}}$$
$$\therefore \log_2 K = \log_2(n \times \sqrt{t})$$
$$\therefore K = n\sqrt{t}$$

**9** **a** $\ln 32 = \ln 2^5$
$= 5\ln 2$

**b** $\ln 125 = \ln 5^3$
$= 3\ln 5$

**c** $\ln 729 = \ln 3^6$
$= 6\ln 3$

**10** $\log_2 x$ is defined for all $x > 0$
$\therefore$ the domain is $\{x \mid x > 0\}$
and the range is $y \in \mathbb{R}$.
$\ln(x+5)$ is defined for all $x > -5$
$\therefore$ the domain is $\{x \mid x > -5\}$
and the range is $y \in \mathbb{R}$.

So, the completed table is:

| *Function* | $y = \log_2 x$ | $y = \ln(x+5)$ |
|---|---|---|
| *Domain* | $x > 0$ | $x > -5$ |
| *Range* | $y \in \mathbb{R}$ | $y \in \mathbb{R}$ |

**11** **a**
$$\log_5 36$$
$$= \log_5(2^2 \times 3^2)$$
$$= \log_5 2^2 + \log_5 3^2$$
$$= 2\log_5 2 + 2\log_5 3$$
$$= 2A + 2B$$

**b**
$$\log_5 54$$
$$= \log_5(2 \times 3^3)$$
$$= \log_5 2 + \log_5 3^3$$
$$= \log_5 2 + 3\log_5 3$$
$$= A + 3B$$

**c**
$$\log_5(8\sqrt{3})$$
$$= \log_5(2^3 \times 3^{\frac{1}{2}})$$
$$= \log_5 2^3 + \log_5 3^{\frac{1}{2}}$$
$$= 3\log_5 2 + \tfrac{1}{2}\log_5 3$$
$$= 3A + \tfrac{1}{2}B$$

**d**
$$\log_5(20.25)$$
$$= \log_5(\tfrac{81}{4})$$
$$= \log_5\left(\frac{3^4}{2^2}\right)$$
$$= \log_5 3^4 - \log_5 2^2$$
$$= 4\log_5 3 - 2\log_5 2$$
$$= 4B - 2A$$

**e**
$$\log_5(0.\overline{8})$$
$$= \log_5(\tfrac{8}{9})$$
$$= \log_5\left(\frac{2^3}{3^2}\right)$$
$$= \log_5 2^3 - \log_5 3^2$$
$$= 3\log_5 2 - 2\log_5 3$$
$$= 3A - 2B$$

**12** **a**
$$3e^x - 5 = -2e^{-x}$$
$$\therefore\ 3e^{2x} - 5e^x = -2$$
{multiplying both sides by $e^x$}
$$\therefore\ 3e^{2x} - 5e^x + 2 = 0$$
$$\therefore\ (3e^x - 2)(e^x - 1) = 0$$
$$\therefore\ 3e^x - 2 = 0 \quad \text{or} \quad e^x - 1 = 0$$
$$\therefore\ e^x = \tfrac{2}{3} \quad \text{or} \quad e^x = 1$$
$$\therefore\ x = \ln(\tfrac{2}{3}) \quad \text{or} \quad x = 0$$

**b**
$$2\ln x - 3\ln\left(\frac{1}{x}\right) = 10$$
$$\therefore\ \ln x^2 - \ln\left(\frac{1}{x^3}\right) = 10$$
$$\therefore\ \ln x^2 + \ln\left(\frac{1}{x^3}\right)^{-1} = 10$$
$$\therefore\ \ln(x^2 \times x^3) = 10$$
$$\therefore\ \ln x^5 = 10$$
$$\therefore\ x^5 = e^{10}$$
$$\therefore\ x = \sqrt[5]{e^{10}}$$
$$\therefore\ x = e^2$$

## REVIEW SET 4B

**1** **a** $32 = 10^{\log 32}$
$\approx 10^{1.5051}$

**b** $0.0013$
$= 10^{\log(0.0013)}$
$\approx 10^{-2.8861}$

**c** $8.963 \times 10^{-5}$
$= 10^{\log(8.963)} \times 10^{-5}$
$\approx 10^{0.952} \times 10^{-5}$
$\approx 10^{-4.0475}$

**2** **a** $\log_2 x = -3$
$\therefore\ x = 2^{-3}$
$\therefore\ x = \frac{1}{8}$

**b** $\log_5 x \approx 2.743$
$\therefore\ x \approx 5^{2.743}$
$\therefore\ x \approx 82.7$

**c** $\log_3 x \approx -3.145$
$\therefore\ x \approx 3^{-3.145}$
$\therefore\ x \approx 0.0316$

**3** **a** $\log_2 k \approx 1.699 + x$
$\therefore\ k \approx 2^{1.699+x}$
$\therefore\ k \approx 2^{1.699} \times 2^x$
$\therefore\ k \approx 3.25 \times 2^x$

**b** $\log_a Q = 3\log_a P + \log_a R$
$= \log_a P^3 + \log_a R$
$= \log_a(P^3 \times R)$
$\therefore\ Q = P^3 R$

**c** $\log A \approx 5\log B - 2.602$
$\therefore\ \log A - \log B^5 \approx -2.602$
$$\therefore\ \log\left(\frac{A}{B^5}\right) \approx -2.602$$
$$\therefore\ \frac{A}{B^5} \approx 10^{-2.602} \approx 0.0025$$
$$\therefore\ A \approx \frac{B^5}{400}$$

**4** **a**
$$5^x = 7$$
$$\therefore\ \log 5^x = \log 7$$
$$\therefore\ x\log 5 = \log 7$$
$$\therefore\ x = \frac{\log 7}{\log 5}$$

**b**
$$20 \times 2^{2x+1} = 640$$
$$\therefore\ 2^{2x+1} = 32$$
$$\therefore\ \log 2^{2x+1} = \log 32$$
$$\therefore\ (2x+1)\log 2 = \log 32$$
$$\therefore\ 2x + 1 = \frac{\log 32}{\log 2} = 5$$
$$\therefore\ 2x = 4$$
$$\therefore\ x = 2$$

**5** $W_t = 2500 \times 3^{-\frac{t}{3000}}$ grams

**a** $W_0 = 2500 \times 3^0$
$= 2500 \times 1$
$= 2500$ grams

**b** We need $t$ when $W_t = 30\%$ of 2500 g
$$\therefore\ 2500 \times 3^{-\frac{t}{3000}} = 0.3 \times 2500$$
$$\therefore\ \log 3^{-\frac{t}{3000}} = \log(0.3)$$
$$\therefore\ -\frac{t}{3000} \times \log 3 = \log(0.3)$$
$$\therefore\ t = \frac{-\log(0.3) \times 3000}{\log 3}$$
$$\therefore\ t \approx 3287.7$$
$\therefore$ about 3290 years

**c** % change

$= \left(\dfrac{W_{1500} - W_0}{W_0}\right) \times 100\%$

$= \left(\dfrac{2500 \times 3^{-\frac{1500}{3000}} - 2500}{2500}\right) \times 100\%$

$= (3^{-\frac{1}{2}} - 1) \times 100\%$

$\approx -42.3\%$

So, a loss of 42.3%.

**d**

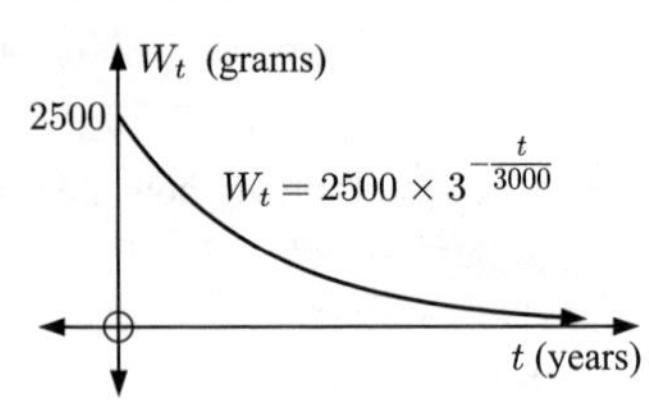


**6** $16^x - 5 \times 8^x = 0$

$\therefore\ 2^x \times 8^x - 5 \times 8^x = 0$

$\therefore\ 8^x(2^x - 5) = 0$

$\therefore\ 2^x = 5$ as $8^x > 0$ for all $x$

$\therefore\ x = \log_2 5$ as required

**7** **a** $\ln x = 5$

$\therefore\ x = e^5$

**b** $3\ln x + 2 = 0$

$\therefore\ 3\ln x = -2$

$\therefore\ \ln x = -\frac{2}{3}$

$\therefore\ x = e^{-\frac{2}{3}}$

**c** $e^x = 400$

$\therefore\ x = \ln 400$

**d** $e^{2x+1} = 11$

$\therefore\ 2x + 1 = \ln 11$

$\therefore\ 2x = \ln 11 - 1$

$\therefore\ x = \dfrac{\ln 11 - 1}{2}$

**e** $25e^{\frac{x}{2}} = 750$

$\therefore\ e^{\frac{x}{2}} = 30$

$\therefore\ \dfrac{x}{2} = \ln 30$

$\therefore\ x = 2\ln 30$

**8** $P_t = P_0 \times 2^{\frac{t}{3}}, \quad t \geqslant 0$

When $t = 0$, $P_0 = P_0 \times 2^0 = P_0$. So the initial population was $P_0$.

**a** If $P_t$ doubles, $P_t = 2P_0$

$\therefore\ P_0 2^{\frac{t}{3}} = 2P_0$

$\therefore\ 2^{\frac{t}{3}} = 2^1$

$\therefore\ \dfrac{t}{3} = 1$

$\therefore\ t = 3$ So, it would take 3 years.

**b** % increase $= \left(\dfrac{P_4 - P_0}{P_0}\right) \times 100\%$

$= \left(\dfrac{P_0 \times 2^{\frac{4}{3}} - P_0}{P_0}\right) \times 100\%$

$= (2^{\frac{4}{3}} - 1) \times 100\%$

$\approx 151.98\%$

So, the % increase is 152%.

*or* $P_4 = P_0 \times 2^{\frac{4}{3}}$

$\approx P_0 \times 2.52$

$\approx 252\%$ of $P_0$

So, an increase of 152%.

**9** **a** $g(x) = 2e^x - 5$ has inverse function $x = 2e^y - 5$

$\therefore \quad 2e^y = x + 5$

$\therefore \quad e^y = \dfrac{x+5}{2}$

$\therefore \quad y = \ln\left(\dfrac{x+5}{2}\right)$

$\therefore \quad g^{-1}(x) = \ln\left(\dfrac{x+5}{2}\right)$

**b**

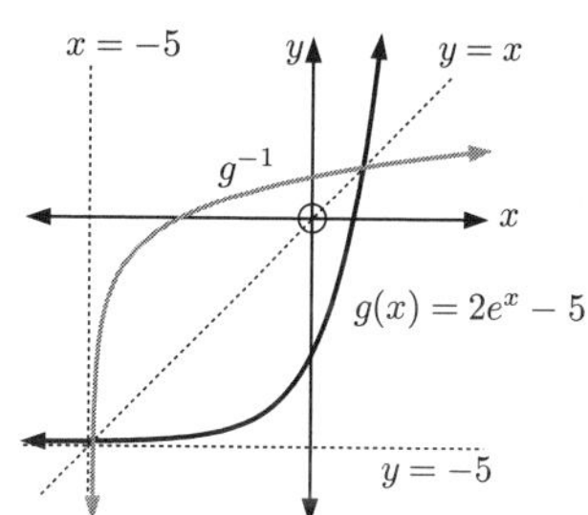


**c** domain of $g$ is $\{x \mid x \in \mathbb{R}\}$, range is $\{y \mid y > -5\}$
domain of $g^{-1}$ is $\{x \mid x > -5\}$, range is $\{y \mid y \in \mathbb{R}\}$

**d** When $x = 0$, $y = 2e^0 - 5 = -3$

$\therefore$ the $y$-intercept of $g$ is $-3$.

When $y = 0$, $2e^x - 5 = 0$

$\therefore \quad e^x = \frac{5}{2}$

$\therefore \quad x = \ln\left(\frac{5}{2}\right)$

$\therefore$ the $x$-intercept of $g$ is $\ln\left(\frac{5}{2}\right) \approx 0.916$.

$\therefore$ $g^{-1}$ has $x$-intercept $-3$ and $y$-intercept $\ln\left(\frac{5}{2}\right) \approx 0.916$.

As $x \to -\infty$, $y \to -5^+$

$\therefore$ the H.A. of $g$ is $y = -5$.

$\therefore$ $g^{-1}$ has V.A. $x = -5$.

**10** **a**
$$\begin{aligned} g(5) &= \ln(5+4) \\ &= \ln 9 \\ \therefore \quad (f \circ g)(5) &= f(g(5)) \\ &= e^{\ln 9} \\ &= 9 \end{aligned}$$

**b**
$$\begin{aligned} f(0) &= e^0 \\ &= 1 \\ \therefore \quad (g \circ f)(0) &= g(f(0)) \\ &= \ln(1+4) \\ &= \ln 5 \end{aligned}$$

## REVIEW SET 4C

**1** **a**
$$\begin{aligned} &\log\sqrt{1000} \\ &= \log\left(10^3\right)^{\frac{1}{2}} \\ &= \log 10^{\frac{3}{2}} \\ &= \tfrac{3}{2} \end{aligned}$$

**b**
$$\begin{aligned} &\log\left(\frac{10}{\sqrt[3]{10}}\right) \\ &= \log\left(\frac{10^1}{10^{\frac{1}{3}}}\right) \\ &= \log 10^{\frac{2}{3}} = \tfrac{2}{3} \end{aligned}$$

**c**
$$\begin{aligned} &\log\left(\frac{10^a}{10^{-b}}\right) \\ &= \log\left(10^{a-(-b)}\right) \\ &= \log 10^{a+b} \\ &= a + b \end{aligned}$$

**2** **a**
$$\begin{aligned} &e^{4\ln x} \\ &= \left(e^{\ln x}\right)^4 \\ &= x^4 \end{aligned}$$

**b** $\ln(e^5) = 5$
$\{$as $\ln e^a = a\}$

**c** $\ln\left(\sqrt{e}\right) = \ln e^{\frac{1}{2}}$
$= \frac{1}{2}$

**d**
$$\begin{aligned} &10^{\log x + \log 3} \\ &= 10^{\log x} \times 10^{\log 3} \\ &= x \times 3 \\ &= 3x \end{aligned}$$

**e** $\ln\left(\frac{1}{e^x}\right) = \ln e^{-x}$
$= -x$

**f**
$$\begin{aligned} &\frac{\log x^2}{\log_3 9} \\ &= \frac{\log x^2}{2} \\ &= \tfrac{1}{2}\log x^2 \\ &= \log\left(x^2\right)^{\frac{1}{2}} \\ &= \log x \end{aligned}$$

**3** **a** $20 = e^{\ln 20}$
$\approx e^{2.9957}$

**b** $3000 = e^{\ln 3000}$
$\approx e^{8.0064}$

**c** $0.075 = e^{\ln(0.075)}$
$\approx e^{-2.5903}$

**4** **a** $\log x = 3$
$\therefore\ x = 10^3$
$\therefore\ x = 1000$

**b** $\log_3(x+2) = 1.732$
$\therefore\ x + 2 = 3^{1.732}$
$\therefore\ x + 2 \approx 6.7046$
$\therefore\ x \approx 4.7046$
$\therefore\ x \approx 4.70$

**c** $\log_2\left(\frac{x}{10}\right) = -0.671$
$\therefore\ \frac{x}{10} = 2^{-0.671}$
$\therefore\ \frac{x}{10} \approx 0.628\,07$
$\therefore\ x \approx 6.28$

**5** **a** $\ln 60 - \ln 20$
$= \ln(\frac{60}{20})$
$= \ln 3$

**b** $\ln 4 + \ln 1$
$= \ln 4 + 0$
$= \ln 4$

**c** $\ln 200 - \ln 8 + \ln 5$
$= \ln(\frac{200}{8}) + \ln 5$
$= \ln\left(\frac{200}{8} \times 5\right)$
$= \ln 125$

**6** **a** $M = ab^n$
$\therefore\ \log M = \log(ab^n)$
$\therefore\ \log M = \log a + \log b^n$
$\therefore\ \log M = \log a + n \log b$

**b** $T = \frac{5}{\sqrt{l}}$
$\therefore\ \log T = \log\left(\frac{5}{l^{\frac{1}{2}}}\right)$
$\therefore\ \log T = \log 5 - \log l^{\frac{1}{2}}$
$\therefore\ \log T = \log 5 - \frac{1}{2}\log l$

**c** $G = \frac{a^2 b}{c}$
$\therefore\ \log G = \log\left(\frac{a^2 b}{c}\right)$
$\therefore\ \log G = \log(a^2 b) - \log c$
$\therefore\ \log G = \log a^2 + \log b - \log c$
$\therefore\ \log G = 2\log a + \log b - \log c$

**7** **a** $3^x = 300$
$\therefore\ \log 3^x = \log 300$
$\therefore\ x \log 3 = \log 300$
$\therefore\ x = \frac{\log 300}{\log 3}$
$\therefore\ x \approx 5.19$

**b** $30 \times 5^{1-x} = 0.15$
$\therefore\ 5^{1-x} = 0.005$
$\therefore\ \log 5^{1-x} = \log(0.005)$
$\therefore\ (1-x)\log 5 = \log(0.005)$
$\therefore\ 1 - x = \frac{\log(0.005)}{\log 5}$
$\therefore\ 1 - x \approx -3.292$
$\therefore\ x \approx 4.29$

**c** $3^{x+2} = 2^{1-x}$
$\therefore\ \log 3^{x+2} = \log 2^{1-x}$
$\therefore\ (x+2)\log 3 = (1-x)\log 2$
$\therefore\ x\log 3 + 2\log 3 = \log 2 - x\log 2$
$\therefore\ x(\log 3 + \log 2) = \log 2 - 2\log 3$
$\therefore\ x\log 6 = \log(\frac{2}{9})$
$\therefore\ x = \frac{\log(\frac{2}{9})}{\log 6}$
$\therefore\ x \approx -0.839$

**8** **a** $e^{2x} = 3e^x$

$\therefore\ e^{2x} - 3e^x = 0$

$\therefore\ e^x(e^x - 3) = 0$

$\therefore\ e^x - 3 = 0 \quad \{e^x > 0 \text{ for all } x\}$

$\therefore\ e^x = 3$

$\therefore\ x = \ln 3$

**b** $e^{2x} - 7e^x + 12 = 0$

$\therefore\ (e^x - 3)(e^x - 4) = 0$

$\therefore\ e^x - 3 = 0$ or $e^x - 4 = 0$

$\therefore\ e^x = 3$ or $e^x = 4$

$\therefore\ x = \ln 3$ or $\ln 4$

**9** **a** $\ln P = 1.5 \ln Q + \ln T$

$\therefore\ \ln P = \ln Q^{1.5} + \ln T$

$= \ln(TQ^{1.5})$

$\therefore\ P = TQ^{1.5}$

**b** $\ln M = 1.2 - 0.5 \ln N$

$\therefore\ \ln M + \ln N^{\frac{1}{2}} = 1.2$

$\therefore\ \ln\left(M\sqrt{N}\right) = 1.2$

$\therefore\ M\sqrt{N} = e^{1.2}$

$\therefore\ M = \dfrac{e^{1.2}}{\sqrt{N}}$

**10** $g(x) = \log_3(x+2) - 2$

**a** We require $x + 2 > 0$, so $x > -2$

$\therefore$ the domain is $\{x \mid x > -2\}$ and the range is $y \in \mathbb{R}$.

**b** If $x \to -2^+$, $y \to -\infty$ $\therefore$ V.A. is $x = -2$.

As $x \to \infty$, $y \to \infty$.

When $x = 0$, $g(0) = \log_3 2 - 2 \approx -1.37$ So, the $y$-intercept $\approx -1.37$

When $y = 0$, $\log_3(x+2) = 2$ $\therefore\ x + 2 = 3^2$

$\therefore\ x = 7$ So, the $x$-intercept is 7.

**c** $g^{-1}$ is defined by $x = \log_3(y+2) - 2$

$\therefore\ \log_3(y+2) = x + 2$

$\therefore\ y + 2 = 3^{x+2}$

$\therefore\ y = 3^{x+2} - 2$

$\therefore\ g^{-1}(x) = 3^{x+2} - 2$

Horizontal asymptote is $y = -2$.

Domain is $x \in \mathbb{R}$.

Range is $\{y \mid y > -2\}$.

**d**

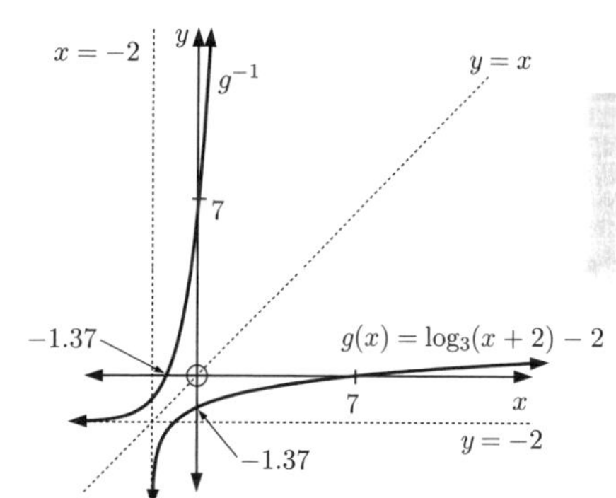


**11** $W_t = 8000 \times e^{-\frac{t}{20}}$ grams

$W_0 = 8000e^0$

$= 8000 \times 1$

$= 8000$ grams

**a** When $W_t = \frac{1}{2} \times 8000$ grams, $8000e^{-\frac{t}{20}} = 4000$

$\therefore\ e^{-\frac{t}{20}} = 0.5$

$\therefore\ -\dfrac{t}{20} = \ln(0.5)$

$\therefore\ t = -20\ln(0.5) \approx 13.9$ weeks

**b** When $W_t = 1000$ g,

$8000e^{-\frac{t}{20}} = 1000$

$\therefore\ e^{-\frac{t}{20}} = \frac{1}{8}$

$\therefore\ -\dfrac{t}{20} = \ln(\frac{1}{8})$

$\therefore\ t = -20\ln(\frac{1}{8})$

$\therefore\ t \approx 41.6$ weeks

**c** When $W_t = 0.1\%$ of $W_0$

$= \frac{1}{1000} \times 8000 = 8$ g,

$8000e^{-\frac{t}{20}} = 8$

$\therefore\ e^{-\frac{t}{20}} = 0.001$

$\therefore\ -\dfrac{t}{20} = \ln(0.001)$

$\therefore\ t = -20\ln(0.001) \approx 138$ weeks

# Chapter 5

## TRANSFORMING FUNCTIONS

### EXERCISE 5A

**1** **a** **i** $x$-intercepts are $-3$, 0, and 4, $y$-intercept is 0 {using technology}

**ii** $(-1.69, 6.30)$ is a local maximum, $(2.36, -10.4)$ is a local minimum {using technology}

**iii** $y = \frac{1}{2}x(x-4)(x+3)$ is defined for all $x \in \mathbb{R}$.

$\therefore$ there are no vertical asymptotes.

As $x \to \infty$, $y \to \infty$

As $x \to -\infty$, $y \to -\infty$

$\therefore$ there are no horizontal asymptotes.

**iv** Domain is $\{x \mid x \in \mathbb{R}\}$.

Range is $\{y \mid y \in \mathbb{R}\}$.

**v**

$y = \frac{1}{2}x(x-4)(x+3)$

$(-1.69, 6.30)$

$-3$   4

$(2.36, -10.4)$

**b** **i** $x$-intercepts are $\approx -4.97$ and $\approx -1.55$, $y$-intercept is 2 {using technology}

**ii** $(-3.88, -33.5)$ and $(0, 2)$ are local minima,

$(-0.805, 2.97)$ is a local maximum {using technology}

**iii** $y = \frac{4}{5}x^4 + 5x^3 + 5x^2 + 2$ is defined for $-5 \leqslant x \leqslant 1$.

$\therefore$ there are no vertical asymptotes.

The function has no horizontal asymptotes.

**iv** We need to check the endpoints of the function.

When $x = -5$, $y = 2$.

When $x = 1$, $y = 12.8$.

So, the domain is $\{x \mid -5 \leqslant x \leqslant 1\}$,

and the range is $\{y \mid -33.5 \leqslant y \leqslant 12.8\}$

**v**

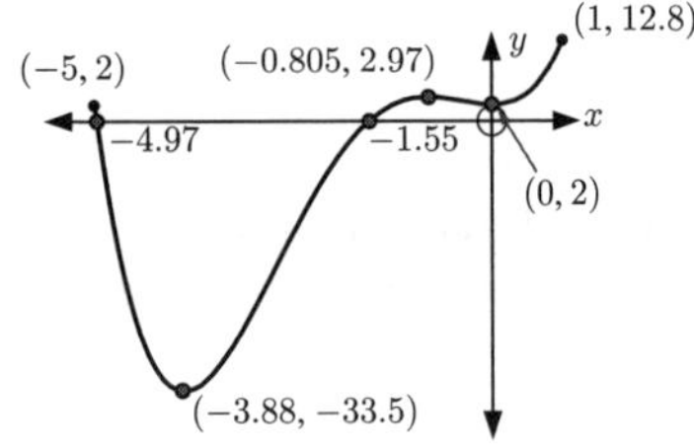


$y = \frac{4}{5}x^4 + 5x^3 + 5x^2 + 2$
$-5 \leqslant x \leqslant 1$

**c** **i** $x$-intercept is 0, $y$-intercept is 0 {using technology}

**ii** $(1.44, 0.531)$ is a local maximum {using technology}

**iii** $y = x2^{-x}$ is defined for all $x$.

$\therefore$ there are no vertical asymptotes.

As $x \to \infty$, $y \to 0^+$

As $x \to -\infty$, $y \to -\infty$

$\therefore$ the horizontal asymptote is $y = 0$.

**iv** Domain is $\{x \mid x \in \mathbb{R}\}$

Range is $\{y \mid y \leqslant 0.531\}$

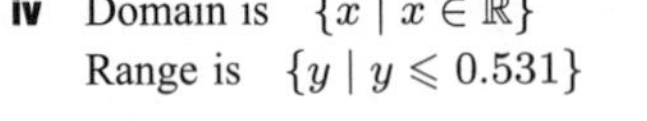

**v**

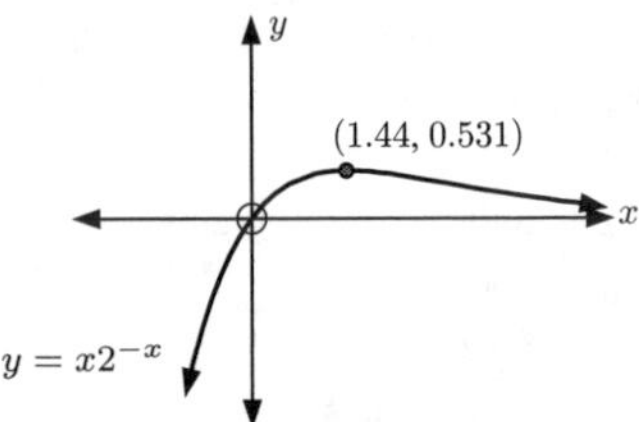


**d** **i** $x$-intercepts are $-1$ and 1, $y$-intercept is 0.25 {using technology}

**ii** $(-0.5, 0.333)$ is a local maximum {using technology}

**iii** As $x \to -2^-$, $y \to -\infty$

As $x \to -2^+$, $y \to -\infty$

$\therefore$ the vertical asymptote is $x = -2$.

As $x \to \infty$, $y \to -1^+$

As $x \to -\infty$, $y \to -1^-$

$\therefore$ the horizontal asymptote is $y = -1$.

**iv** Domain is $\{x \mid -5 \leqslant x \leqslant 5, \ x \neq 2\}$

Range is $\{y \mid y < -1, \ -1 < y \leqslant 0.333\}$

**v**

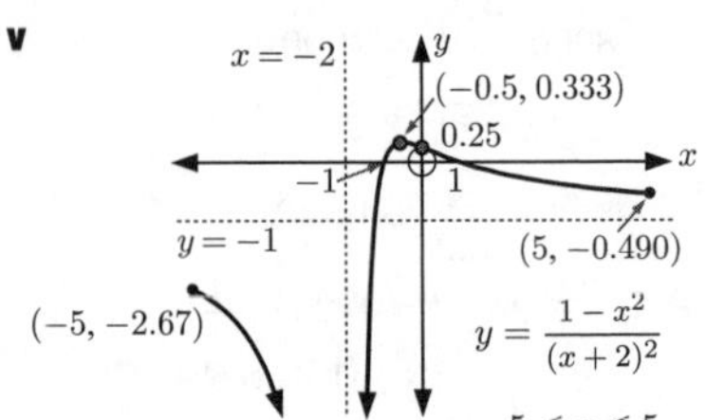

**e** **i** $y = \dfrac{1}{1+2^{-x}} > 0$ for all $x$

$\therefore$ there are no $x$-intercepts.

The $y$-intercept is $0.5$. {using technology}

**ii** The function has no turning points. {using technology}

**iii** $y = \dfrac{1}{1+2^{-x}}$ is defined for all $x \in \mathbb{R}$.

$\therefore$ there are no vertical asymptotes.

As $x \to \infty$, $y \to 1^-$

As $x \to -\infty$, $y \to 0^+$

$\therefore$ the horizontal asymptotes are $y = 0$ and $y = 1$.

**iv** Domain is $\{x \mid x \in \mathbb{R}\}$

Range is $\{y \mid 0 < y < 1\}$

**v**

**f** **i** $x$-intercepts are $-0.767$, 2, and 4, $y$-intercept is $-1$ {using technology}

**ii** $(0, -1)$ is a local minimum, $(2.89, 0.127)$ is a local maximum {using technology}

**iii** $y = \dfrac{x^2}{2^x} - 1$ is defined for all $x \in \mathbb{R}$.

$\therefore$ there are no vertical asymptotes.

As $x \to \infty$, $y \to -1^+$

As $x \to -\infty$, $y \to \infty$

$\therefore$ the horizontal asymptote is $y = -1$.

**iv** Domain is $\{x \mid x \in \mathbb{R}\}$

Range is $\{y \mid y \geqslant -1\}$

**v**

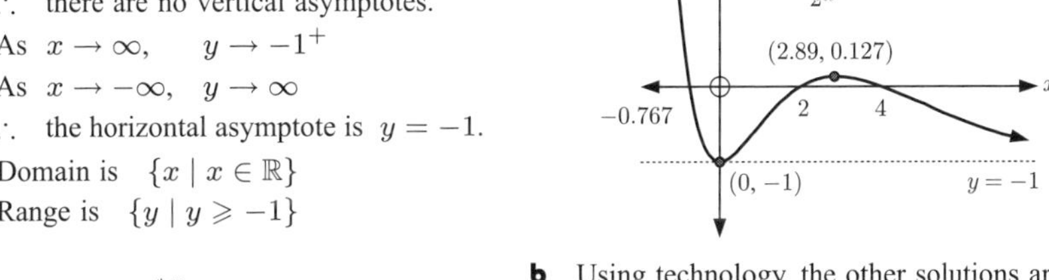


**2** **a**

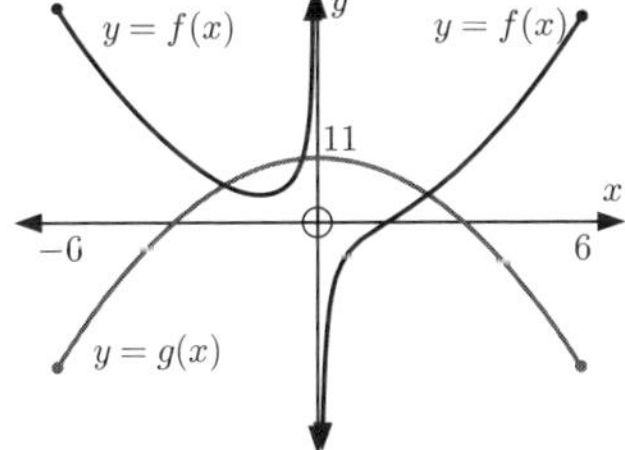


**b** Using technology, the other solutions are $x \approx -2.14$ and $x \approx -0.373$.

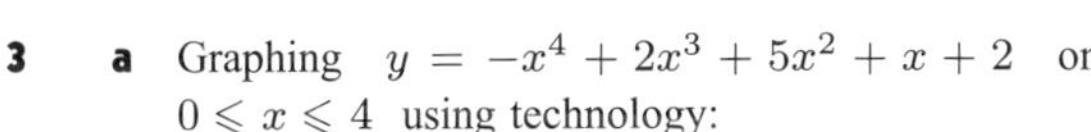

**3** **a** Graphing $y = -x^4 + 2x^3 + 5x^2 + x + 2$ on $0 \leqslant x \leqslant 4$ using technology:

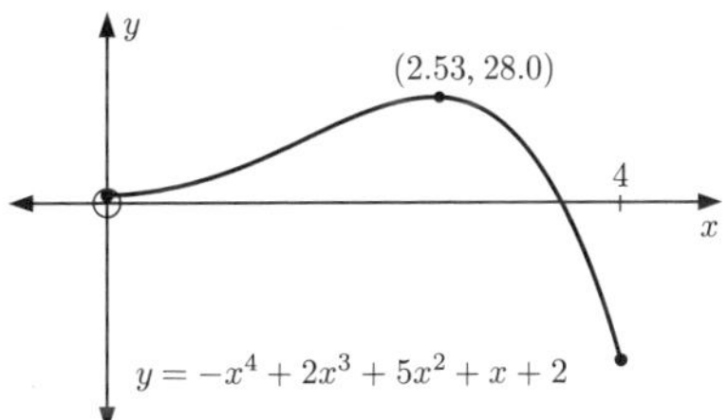


Clearly the maximum value occurs at the local maximum, which is $(2.53, 28.0)$, so the maximum value is $28.0$.

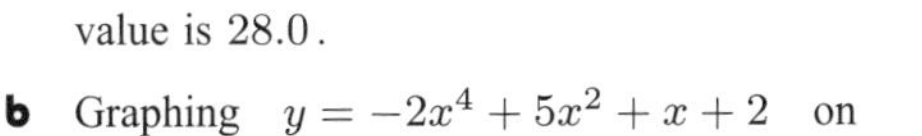

**b** Graphing $y = -2x^4 + 5x^2 + x + 2$ on $-2 \leqslant x \leqslant 2$ using technology:

**i** The higher of the two maxima $(1.17, 6.27)$ is the global maximum on $-2 \leqslant x \leqslant 2$, so the maximum value is $6.27$.

**ii** The global maximum on $-2 \leqslant x \leqslant 0$ occurs at the lower maximum $(-1.06, 4.03)$, so the maximum value on this interval is $4.03$.

**iii** The higher maximum $(1.17, 6.27)$ is in $0 \leqslant x \leqslant 2$, so the maximum value on this interval is $6.27$.

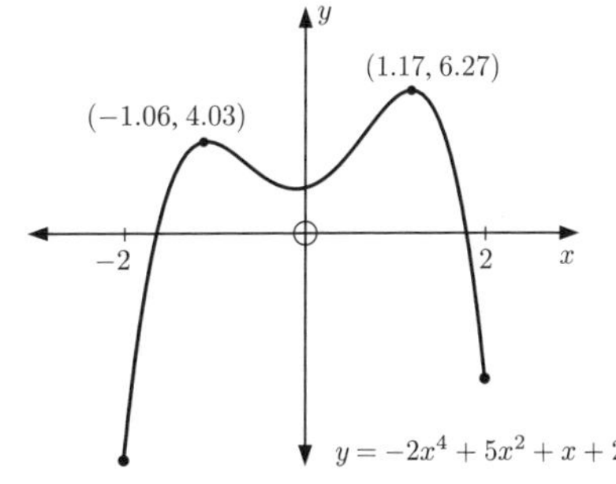

**4**

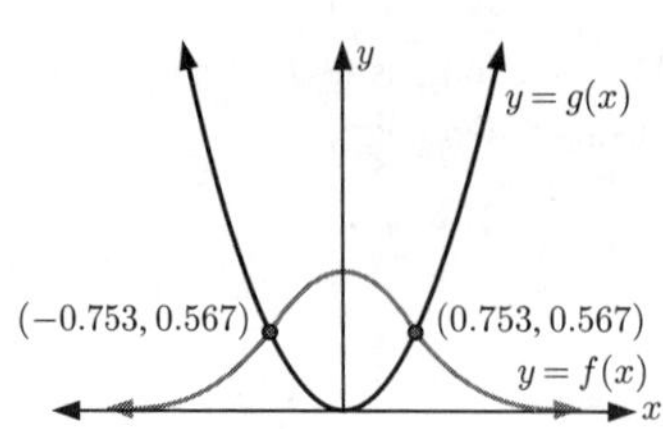


$e^{-x^2} = x^2$ when $x \approx -0.753$ or $0.753$

**5** **a**

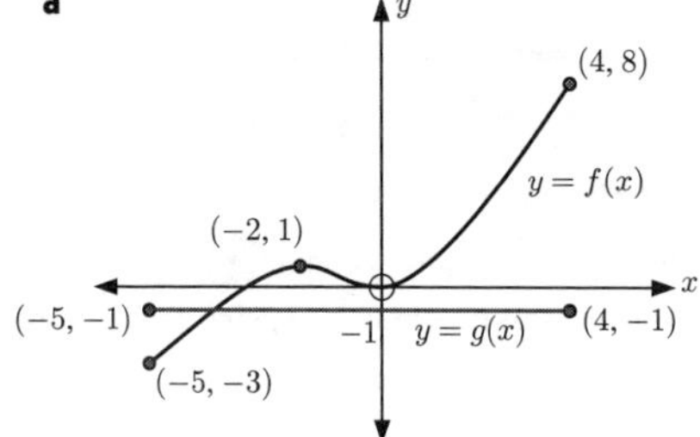


**b** $f(x)$ and $g(x)$ intersect at one point, so $f(x) = g(x)$ has one solution on the domain $-5 \leqslant x \leqslant 4$.

**c** **i** $f(x) = h(x)$ on the domain $-5 \leqslant x \leqslant 4$ has three solutions when $0 < k < 1$.

**ii** $f(x) = h(x)$ on the domain $-5 \leqslant x \leqslant 4$ has two solutions when $k = 0$ or $k = 1$.

**iii** $f(x) = h(x)$ on the domain $-5 \leqslant x \leqslant 4$ has one solution when $-3 \leqslant k < 0$ or $1 < k \leqslant 8$.

**iv** $f(x) = h(x)$ on the domain $-5 \leqslant x \leqslant 4$ has no solutions when $k < -3$ or $k > 8$.

**6** **a**

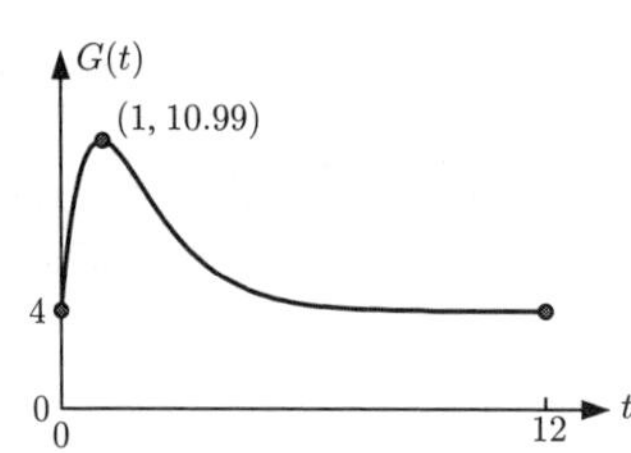


**b** The maximum value occurs at the local maximum, which is (1, 10.99), using technology. So the highest blood glucose level reached in the time period $0 \leqslant t \leqslant 12$ is $\approx 11.0$ mmol/L. This maximum occurs when $t = 1$, that is, 1 hour after the evening meal.

**c** Using technology, $G(t) = 8$ when $t \approx 0.278$ and $t \approx 2.46$. From the graph, $G(t)$ decreases after $t \approx 2.46$.
So the patient should wait 2.46 hours following their evening meal before sleep.

## EXERCISE 5B

**1** $f(x) = x$

**a** $f(2x) = 2x$

**b** $f(x) + 2$
$= x + 2$

**c** $\frac{1}{2} f(x) = \dfrac{x}{2}$

**d** $2\,f(x) + 3$
$= 2x + 3$

**2** $f(x) = x^2$

**a** $f(3x) = (3x)^2$
$= 9x^2$

**b** $f\left(\dfrac{x}{2}\right) = \left(\dfrac{x}{2}\right)^2$
$= \dfrac{x^2}{4}$

**c** $3\,f(x) = 3x^2$

**d** $2\,f(x-1) + 5$
$= 2(x-1)^2 + 5$
$= 2(x^2 - 2x + 1) + 5$
$= 2x^2 - 4x + 7$

**3** $f(x) = x^3$

**a** $f(4x)$
$= (4x)^3$
$= 64x^3$

**b** $\frac{1}{2} f(2x)$
$= \frac{1}{2}(2x)^3$
$= \frac{1}{2} \times 8x^3$
$= 4x^3$

**c** $f(x+1)$
$= (x+1)^3$
$= x^3 + 3x^2 + 3x + 1$

**d** $2\,f(x+1) - 3$
$= 2(x+1)^3 - 3$
$= 2(x^3 + 3x^2 + 3x + 1) - 3$
$= 2x^3 + 6x^2 + 6x - 1$

**4** $f(x) = 2^x$

**a** $f(2x) = 2^{2x}$
$= 4^x$

**b** $f(-x) + 1$
$= 2^{-x} + 1$

**c** $f(x-2) + 3$
$= 2^{x-2} + 3$

**d** $2\,f(x) + 3$
$= 2 \times 2^x + 3$
$= 2^{x+1} + 3$

**5** $f(x) = \dfrac{1}{x}$

**a** $f(-x)$
$= \dfrac{1}{(-x)}$
$= -\dfrac{1}{x}$

**b** $f(\frac{1}{2}x)$
$= \dfrac{1}{\frac{1}{2}x}$
$= \dfrac{2}{x}$

**c** $2\,f(x) + 3$
$= 2\left(\dfrac{1}{x}\right) + 3$
$= \dfrac{2}{x} + 3$
$= \dfrac{2+3x}{x}$

**d** $3\,f(x-1) + 2$
$= 3\left(\dfrac{1}{x-1}\right) + 2$
$= \dfrac{3 + 2(x-1)}{x-1}$
$= \dfrac{2x+1}{x-1}$

## EXERCISE 5C

**1** **a, b**

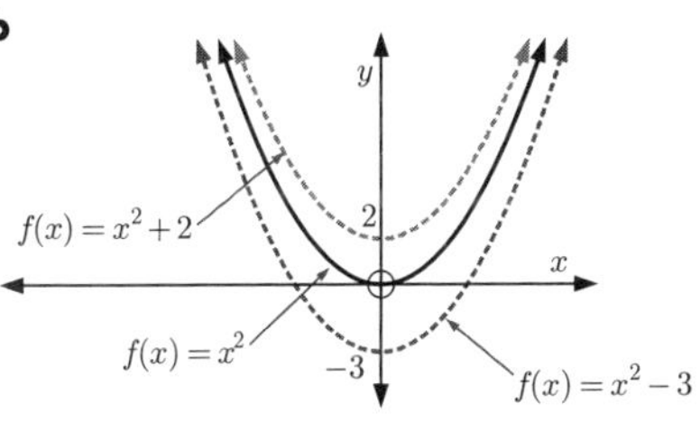


**c** **i** If $b > 0$, the function is translated vertically upwards through $b$ units.

**ii** If $b < 0$, the function is translated vertically downwards $|b|$ units.

**2** **a**

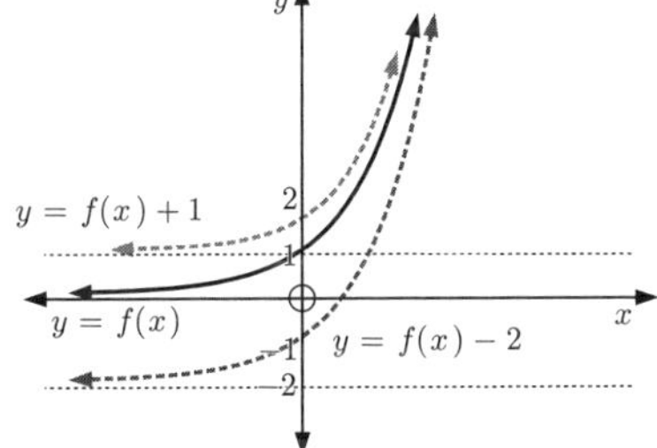


**b**

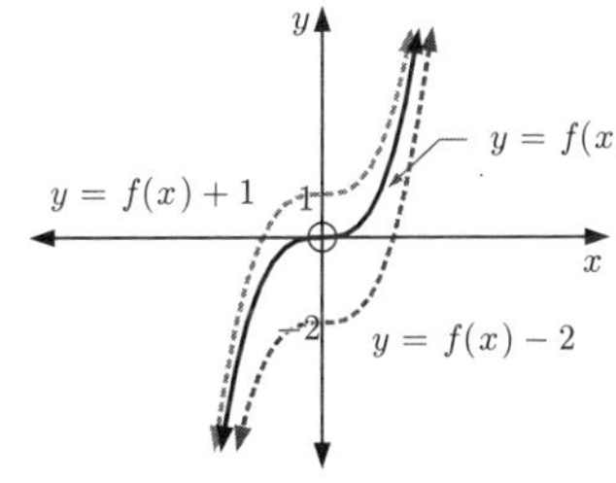


**c**

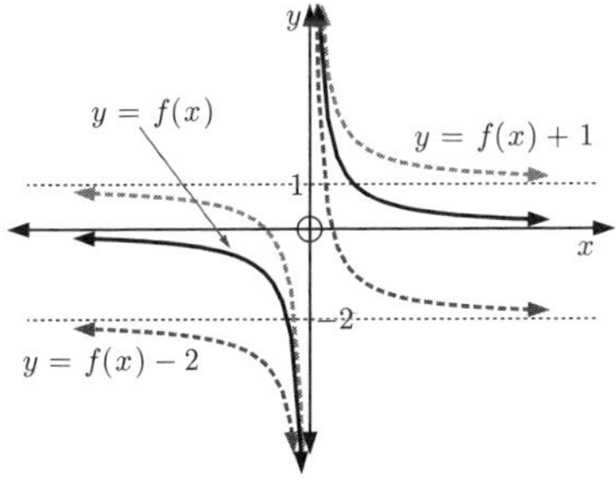


**d**

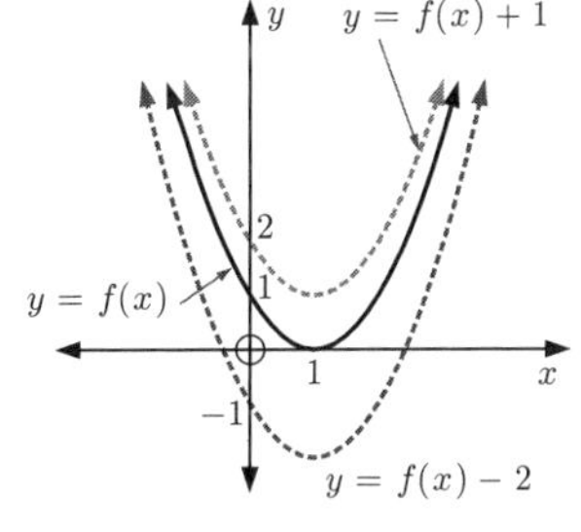


**Summary:** For $y = f(x) + b$, $y = f(x)$ is translated vertically through $b$ units.
If $b > 0$ movement is vertically upwards $b$ units.
If $b < 0$ movement is vertically downwards $|b|$ units.

**3** **a**

**b** **i** If $a > 0$, the graph is translated $a$ units right.

**ii** If $a < 0$, the graph is translated $|a|$ units left.

**4** **a**

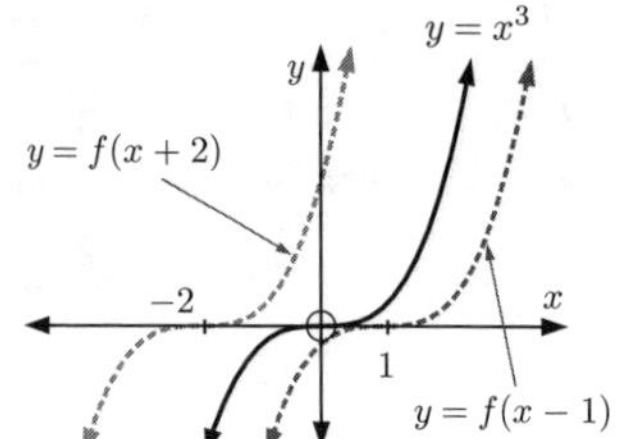


**b**

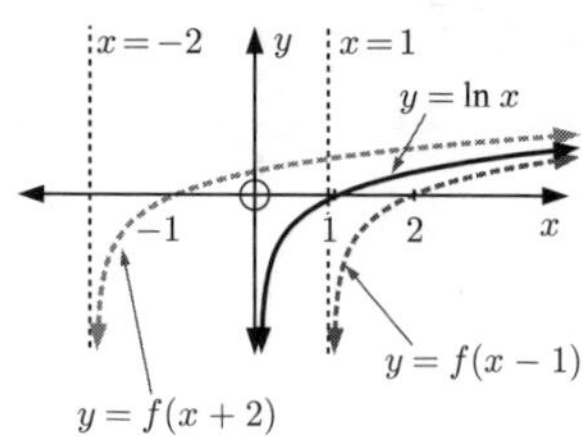


**c**

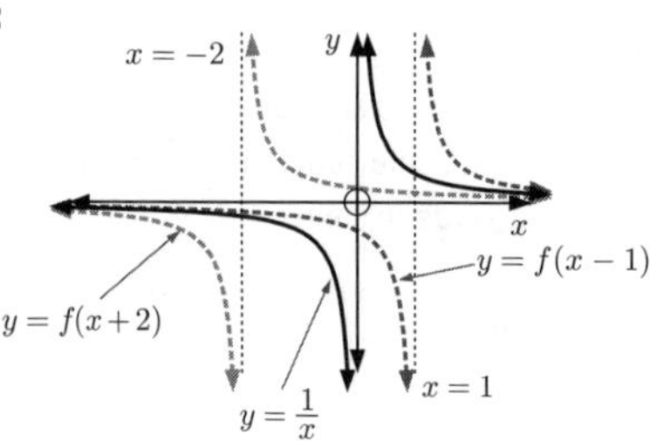


**d**

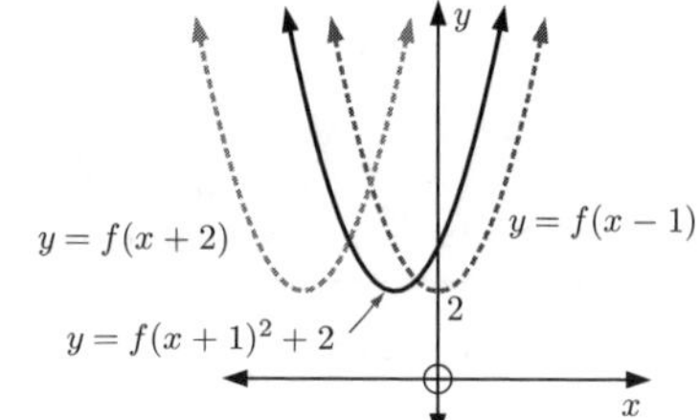


**Summary:** For $y = f(x - a)$, $y = f(x)$ is translated horizontally through $a$ units.
If $a > 0$ movement is to the right. If $a < 0$ movement is to the left.

**5** **a**

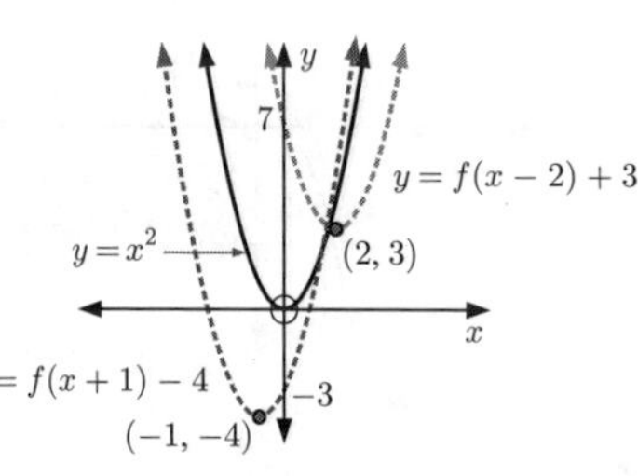


**b**

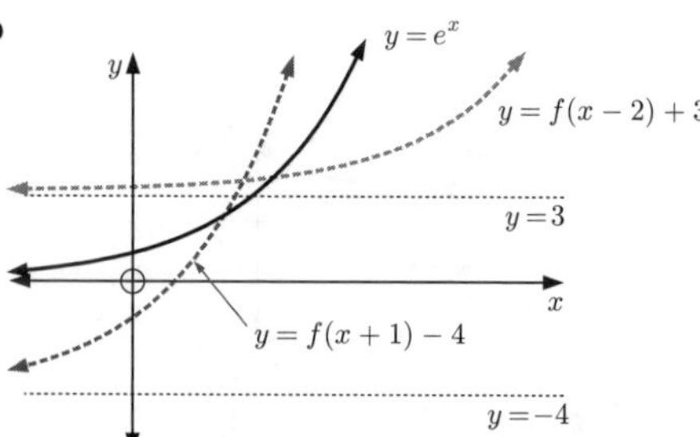


**c**

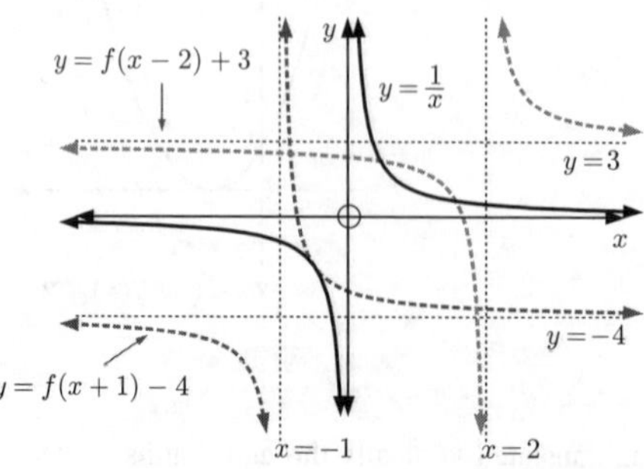

**6** A translation of 2 units right and 3 units down.

**a**

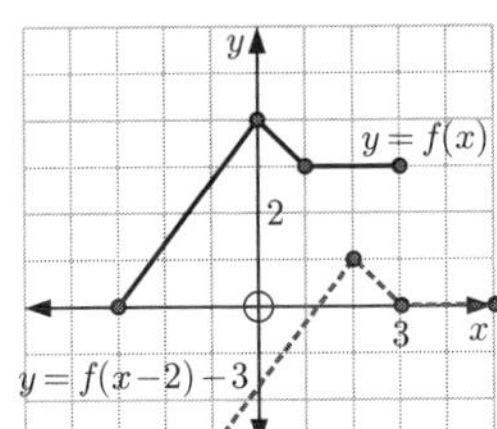


**b**

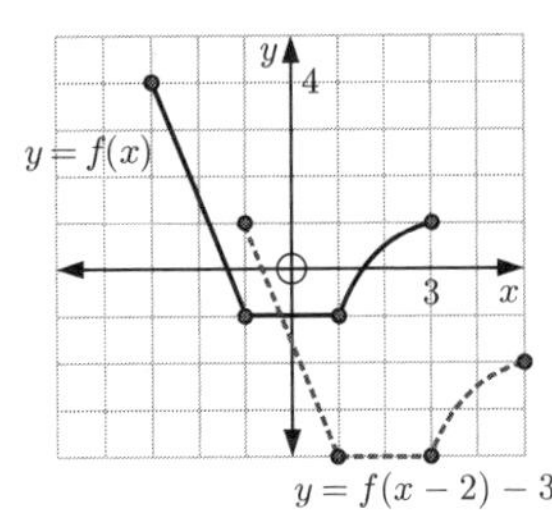


**7** To translate $f(x)$ 3 units right, we need to find $f(x-3)$.

$\therefore\ g(x) = f(x-3)$
$= (x-3)^2 - 2(x-3) + 2$
$= x^2 - 6x + 9 - 2x + 6 + 2$
$\therefore\ g(x) = x^2 - 8x + 17$

**8** **a** The transformation from $f(x) = x^2$ to $g(x) = (x-3)^2 + 2$ is a translation of 3 units right and 2 units up.

**i** (0, 0) is translated to (3, 2).

**ii** (−3, 9) is translated to (0, 11).

**iii** $f(2) = 2^2 = 4$
$\therefore$ (2, 4) is translated to (5, 6).

**b** The transformation from $g(x)$ back to $f(x)$ is a translation of $\begin{pmatrix} -3 \\ -2 \end{pmatrix}$.

**i** (1, 6) is translated to (−2, 4).

**ii** (−2, 27) is translated to (−5, 25).

**iii** $(1\frac{1}{2}, 4\frac{1}{4})$ is translated to $(-1\frac{1}{2}, 2\frac{1}{4})$.

## EXERCISE 5D

**1** **a**

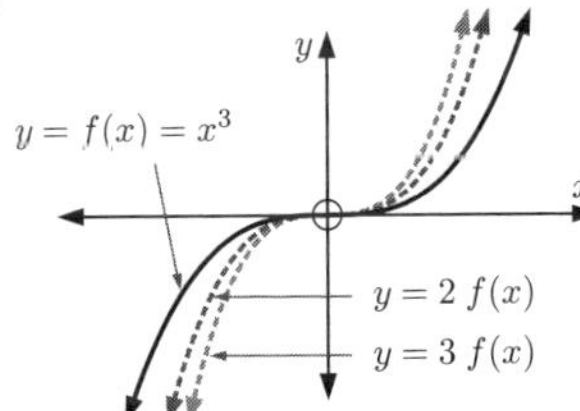


**b**

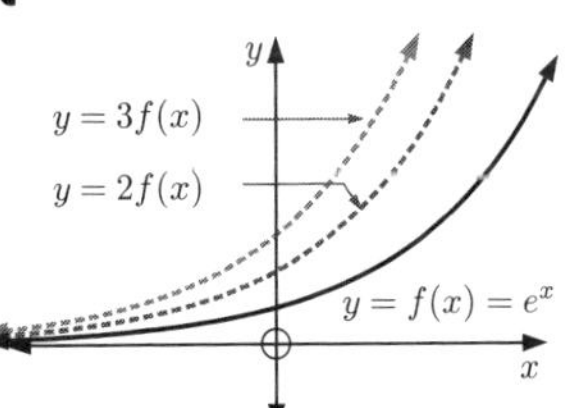


**c**

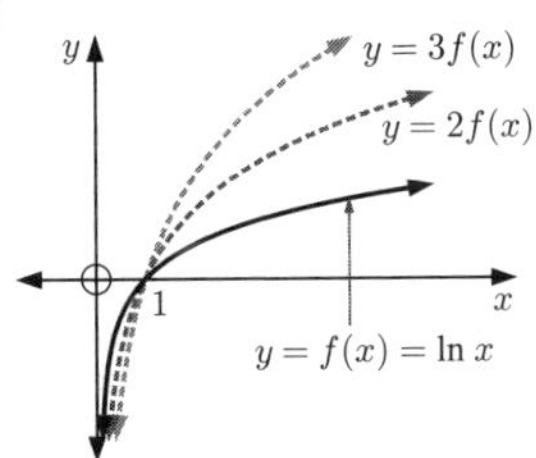


**d**

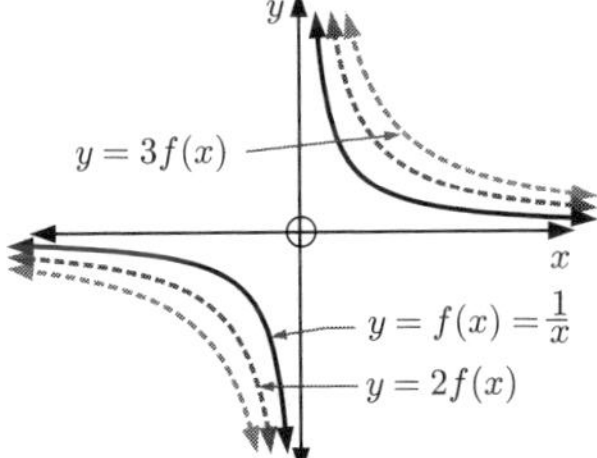


**e**

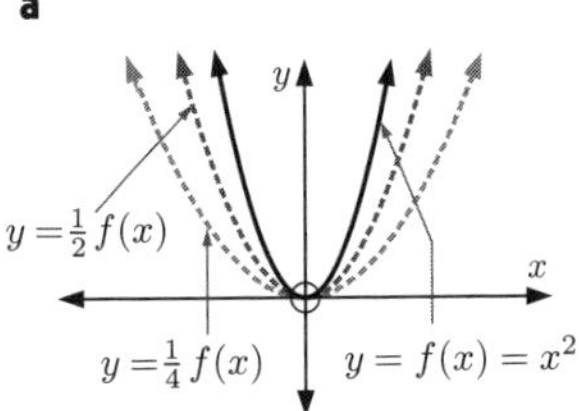


**2** **a**

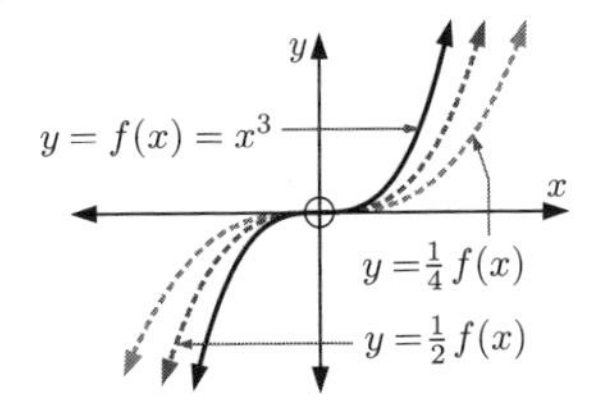


**b**

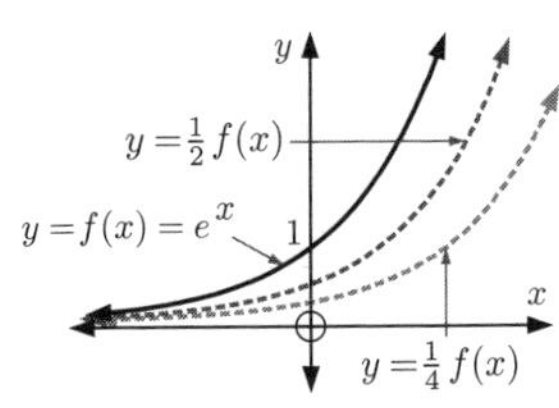


**c**

**3** **a**

**b**

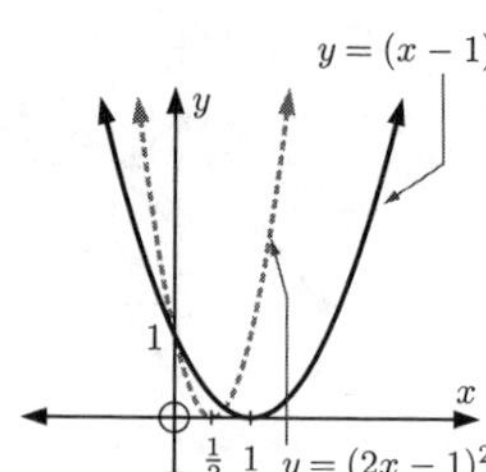


**c**

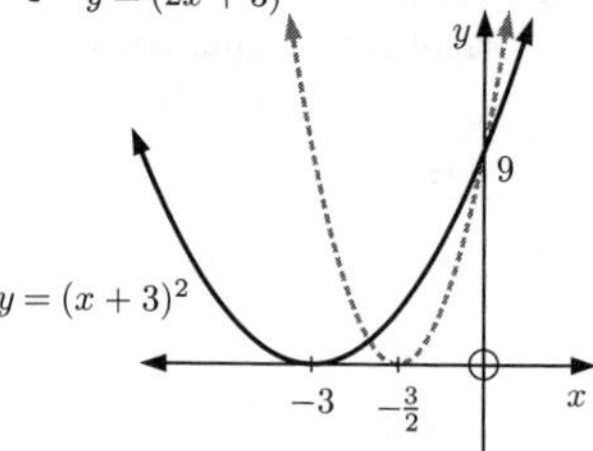


**4** **a**

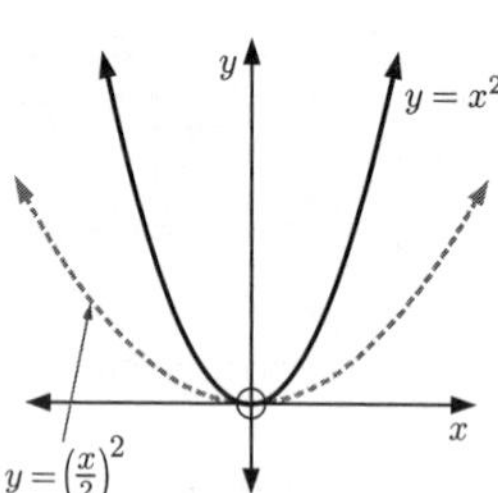


**b**

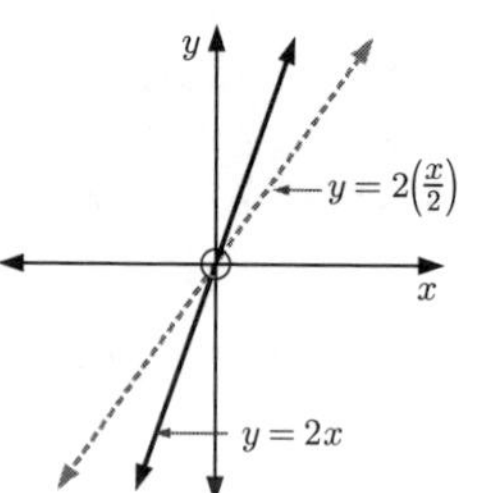


**c**

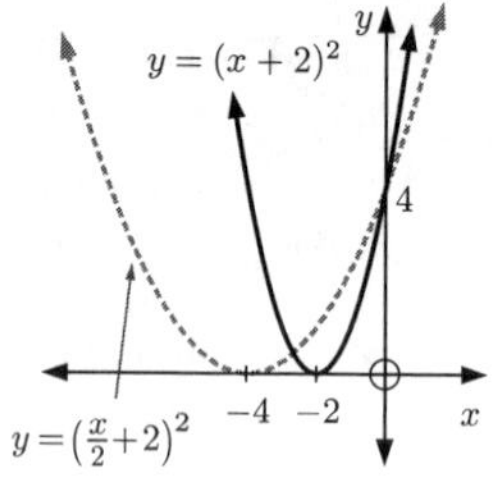


**5** **a**

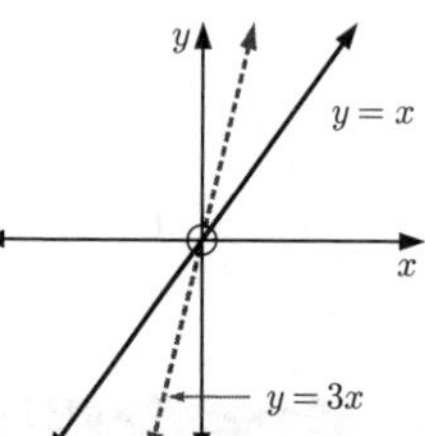


**b**

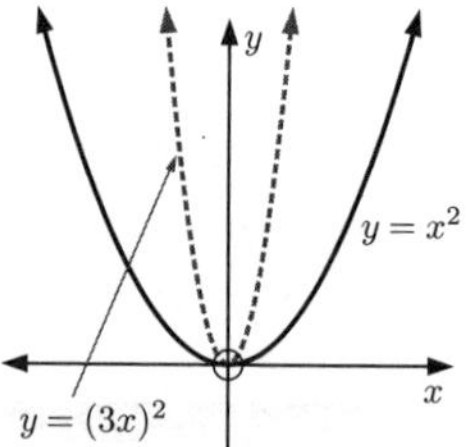


**c**

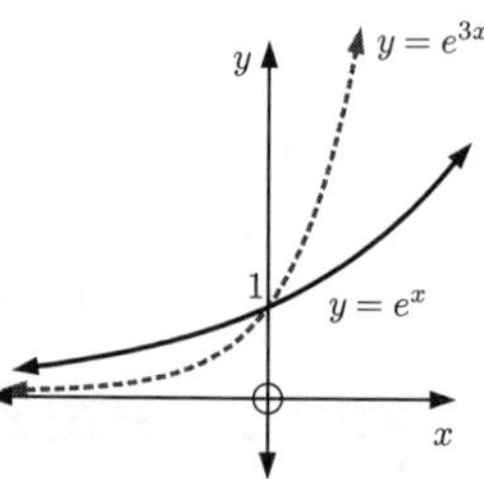


**6** **a**

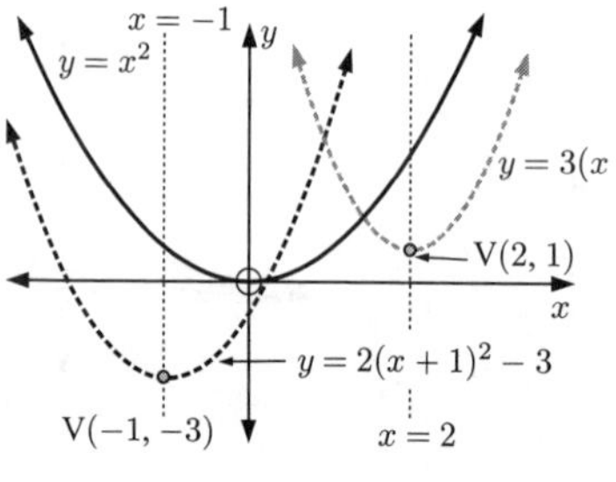


**b**

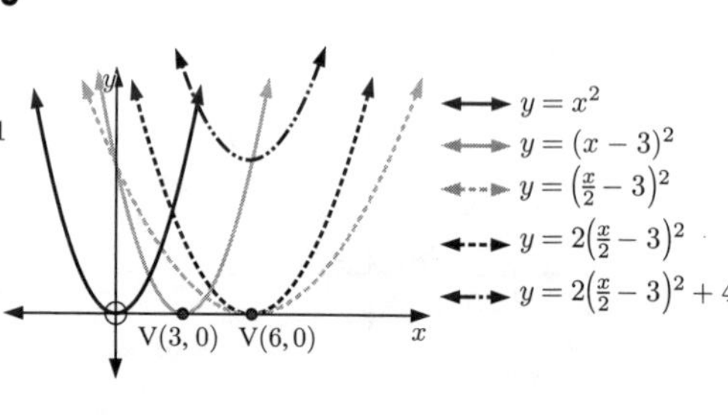


**c**

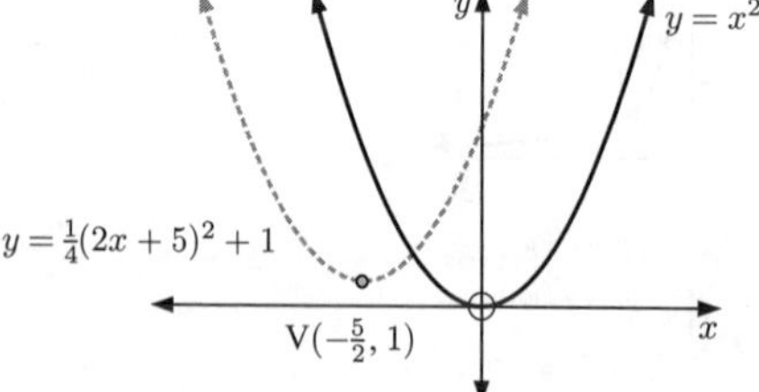


**7** **a** The transformation from $y = f(x)$ to $y = 3f(2x)$ is a horizontal stretch of factor $\frac{1}{2}$ followed by a vertical stretch of factor 3.

**i** $(3, -5) \to (\frac{3}{2}, -5) \to (\frac{3}{2}, -15)$ $\quad\therefore$ $(3, -5)$ is transformed to $(\frac{3}{2}, -15)$

**ii** $(1, 2) \to (\frac{1}{2}, 2) \to (\frac{1}{2}, 6)$ $\quad\therefore$ $(1, 2)$ is transformed to $(\frac{1}{2}, 6)$

**iii** $(-2, 1) \to (-1, 1) \to (-1, 3)$ $\quad\therefore$ $(-2, 1)$ is transformed to $(-1, 3)$

**b** The transformation from $y = 3\,f(2x)$ back to $y = f(x)$ is a vertical stretch of factor $\frac{1}{3}$ followed by a horizontal stretch of factor 2.

**i** $(2, 1) \to (2, \frac{1}{3}) \to (4, \frac{1}{3})$ $\quad\therefore$ $(4, \frac{1}{3})$ is the point on $y = f(x)$

**ii** $(-3, 2) \to (-3, \frac{2}{3}) \to (-6, \frac{2}{3})$ $\quad\therefore$ $(-6, \frac{2}{3})$ is the point on $y = f(x)$

**iii** $(-7, 3) \to (-7, 1) \to (-14, 1)$ $\quad\therefore$ $(-14, 1)$ is the point on $y = f(x)$

## EXERCISE 5E

**1** **a** If $f(x) = 3x$
then $-f(x) = -(3x)$
$= -3x$

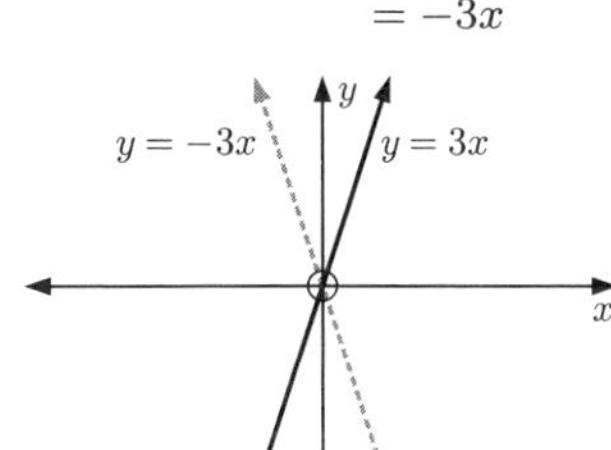


**b** If $f(x) = e^x$
then $-f(x) = -(e^x)$
$= -e^x$

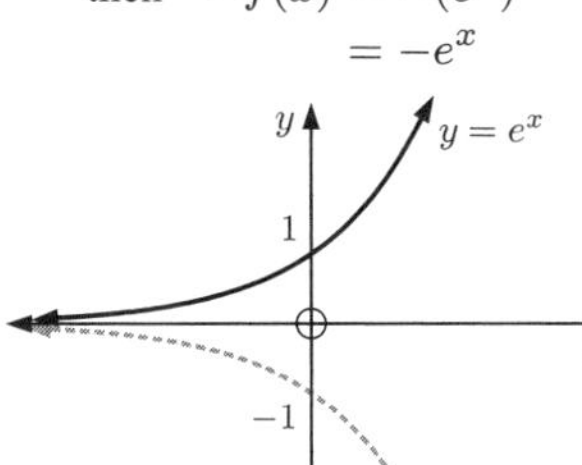


**c** If $f(x) = x^2$
then $-f(x) = -(x^2)$
$= -x^2$

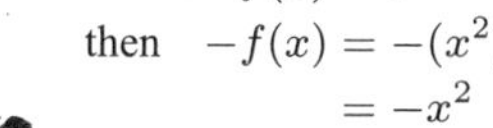

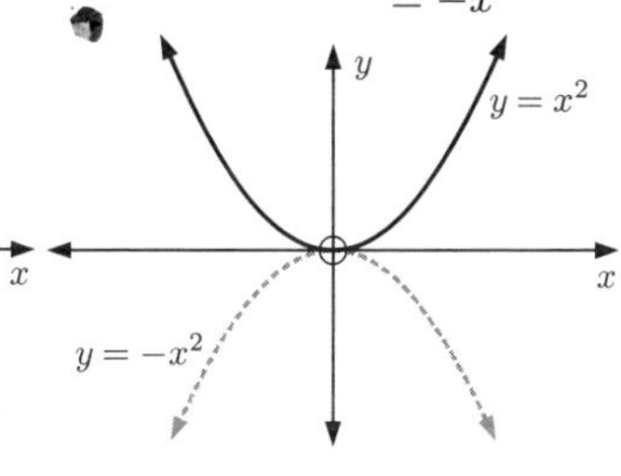


**d** If $f(x) = \ln x$
then $-f(x) = -(\ln x)$
$= -\ln x$

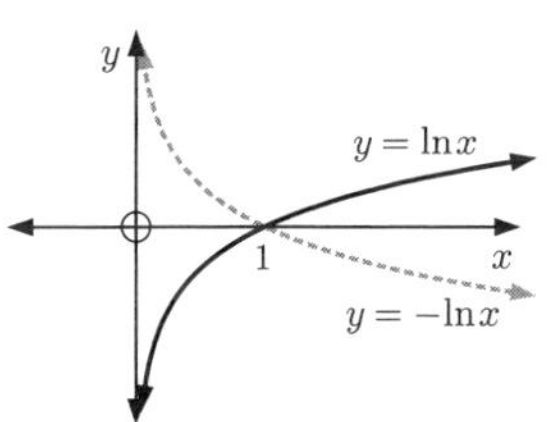


**e** If $f(x) = x^3 - 2$
then $-f(x) = -(x^3 - 2)$
$= -x^3 + 2$

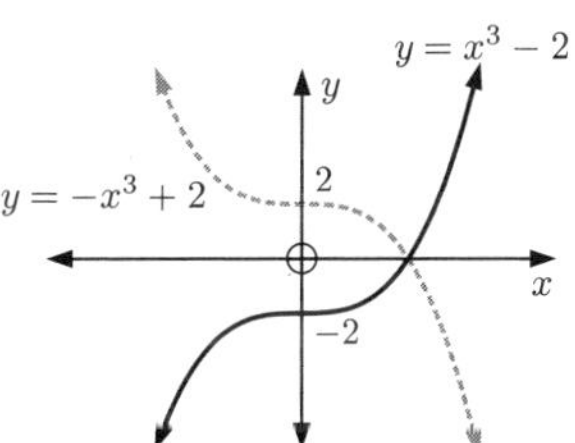


**f** If $f(x) = 2(x+1)^2$
then $-f(x) = -(2(x+1)^2)$
$= -2(x+1)^2$

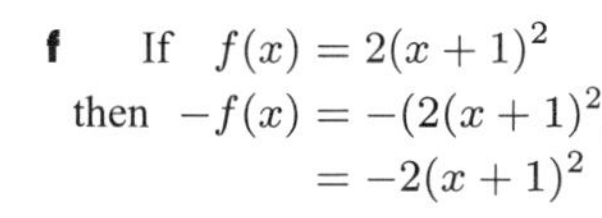

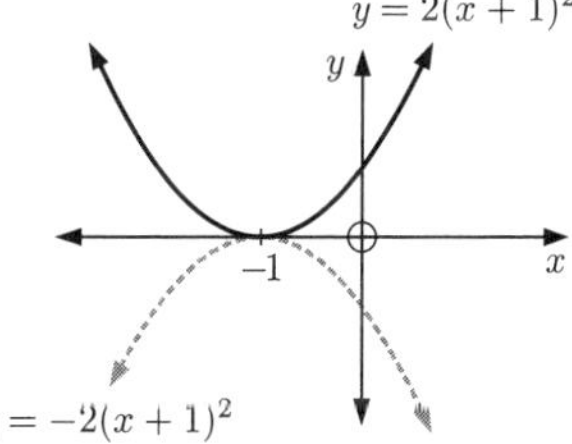


**2** **a** **i** $f(x) = 2x + 1$
$\therefore\ f(-x) = 2(-x) + 1$
$= -2x + 1$

**ii** $f(x) = x^2 + 2x + 1$
$\therefore\ f(-x) = (-x)^2 + 2(-x) + 1$
$= x^2 - 2x + 1$

**iii** $f(x) = x^3$
$\therefore\ f(-x) = (-x)^3$
$= -x^3$

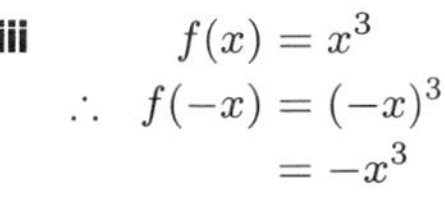

**b** **i** $y = 2x + 1$, $y = -2x + 1$ (graph with intercepts $1$, $-\frac{1}{2}$, $\frac{1}{2}$)

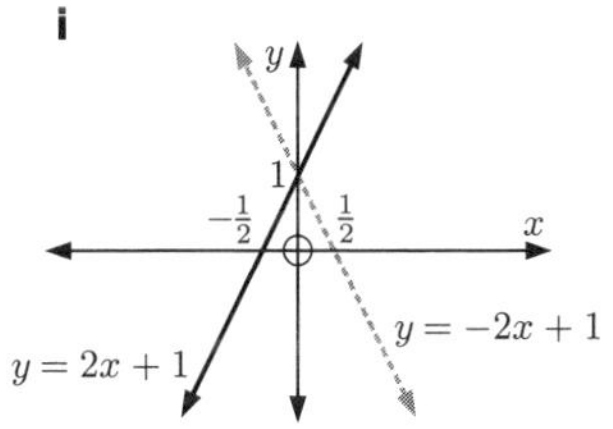

**ii** $y = x^2 + 2x + 1$, $y = x^2 - 2x + 1$ (graph with $1$, $-1$, $1$ marked)

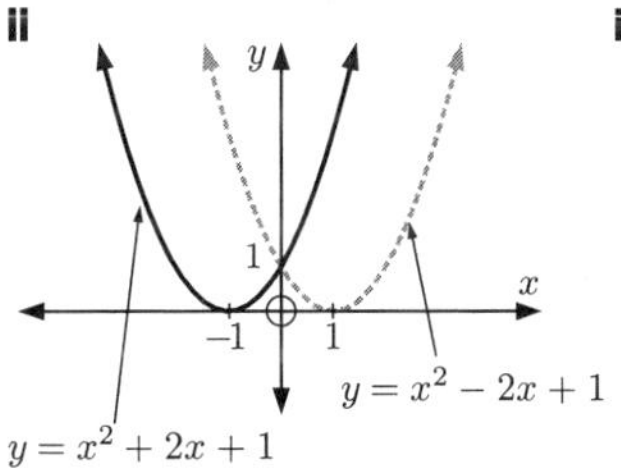

**iii**

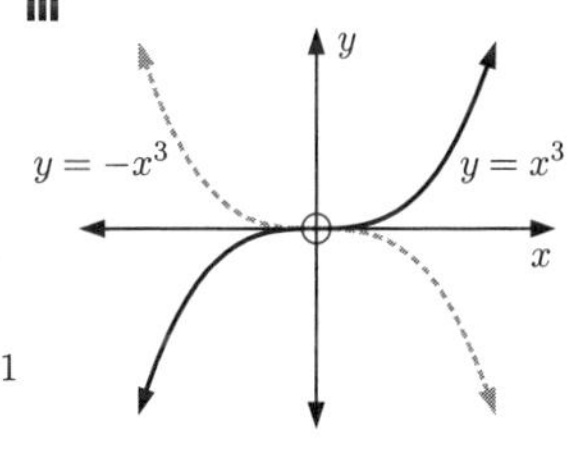


**3** If $g(x)$ is the reflection of $f(x)$ in the $x$-axis, then $g(x) = -f(x)$
$\therefore\ g(x) = -(x^3 - \ln x)$
$= -x^3 + \ln x$

**4** If $g(x)$ is the reflection of $f(x)$ in the $y$-axis,
then $g(x) = f(-x)$
$\therefore\ g(x) = (-x)^4 - 2(-x)^3 - 3(-x)^2 + 5(-x) - 7$
$= x^4 + 2x^3 - 3x^2 - 5x - 7$

**5** **a** To transform $y = f(x)$ to $g(x) = -f(x)$, we reflect $y = f(x)$ in the $x$-axis. To do this we keep the $x$-coordinates the same and take the negative of the $y$-coordinates.

**i** $(3, 0)$ is transformed to $(3, 0)$ **ii** $(2, -1)$ is transformed to $(2, 1)$

**iii** $(-3, 2)$ is transformed to $(-3, -2)$

**b** To find the points on $f(x)$ corresponding to $g(x)$, we again take the negative of the $y$-coordinates.

**i** The point transformed to $(7, -1)$ is $(7, 1)$.

**ii** The point transformed to $(-5, 0)$ is $(-5, 0)$.

**iii** The point transformed to $(-3, -2)$ is $(-3, 2)$.

**6** **a** To transform $y = f(x)$ to $h(x) = f(-x)$, we reflect $y = f(x)$ in the $y$-axis.
To do this we keep the $y$-coordinates the same and take the negative of the $x$-coordinates.

**i** $(2, -1)$ is transformed to $(-2, -1)$. **ii** $(0, 3)$ is transformed to $(0, 3)$.

**iii** $(-1, 2)$ is transformed to $(1, 2)$.

**b** To find the points on $f(x)$ corresponding to $h(x)$, we again take the negative of the $x$-coordinates.

**i** The point transformed to $(5, -4)$ is $(-5, -4)$.

**ii** The point transformed to $(0, 3)$ is $(0, 3)$.

**iii** The point transformed to $(2, 3)$ is $(-2, 3)$.

**7** **a**

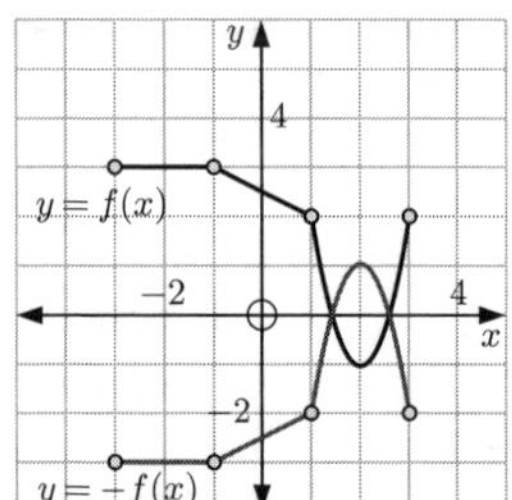


**b**

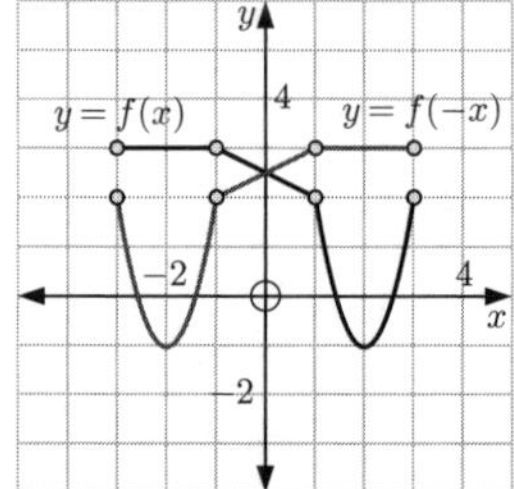


**8** **a** $f(x)$ is reflected in the $y$-axis to give $y = f(-x)$, then reflected in the $x$-axis to give $y = -f(-x)$. This has the effect of rotating the point about the origin through $180°$.

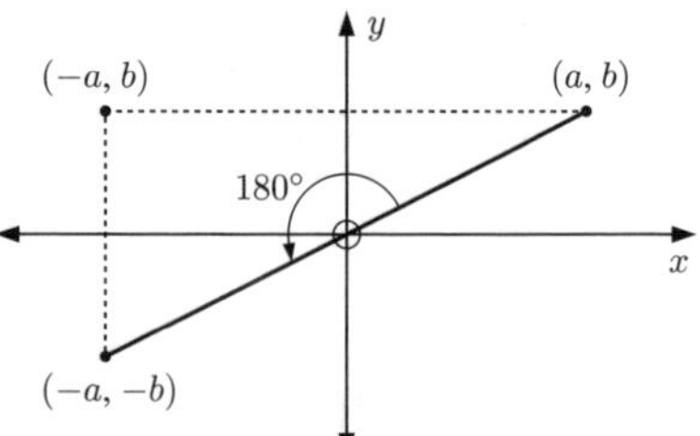


**b** The point $(a, b)$ is transformed to the point $(-a, -b)$.

$\therefore$ $(3, -7)$ is transformed to $(-3, 7)$

**c** The point that transforms to $(-5, -1)$ is $(5, 1)$.

## EXERCISE 5F

**1** **a**

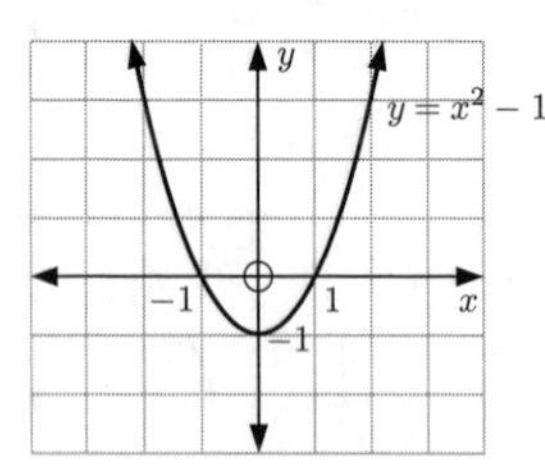


$y = x^2 - 1$ has $x$-intercepts $-1$ and $1$, and $y$-intercept $-1$.

**b** **i, ii**

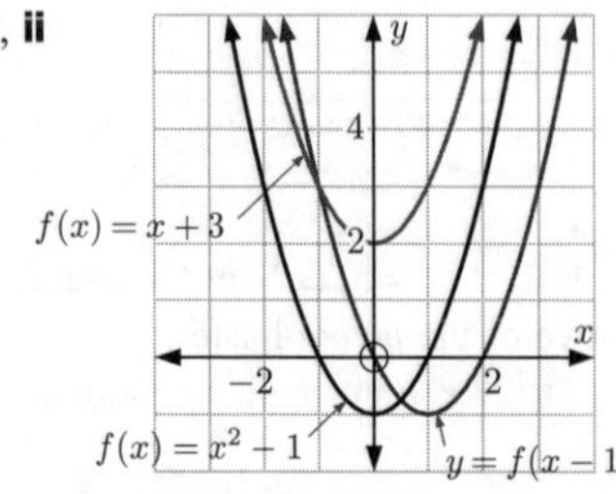


**iii, iv**

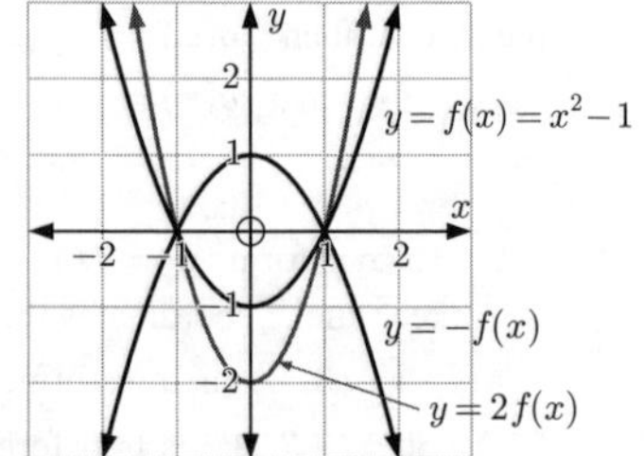

**c** **i** a vertical translation of $\binom{0}{3}$ **ii** a horizontal translation of $\binom{1}{0}$

**iii** a vertical stretch with scale factor 2 **iv** a reflection in the $x$-axis

**d**

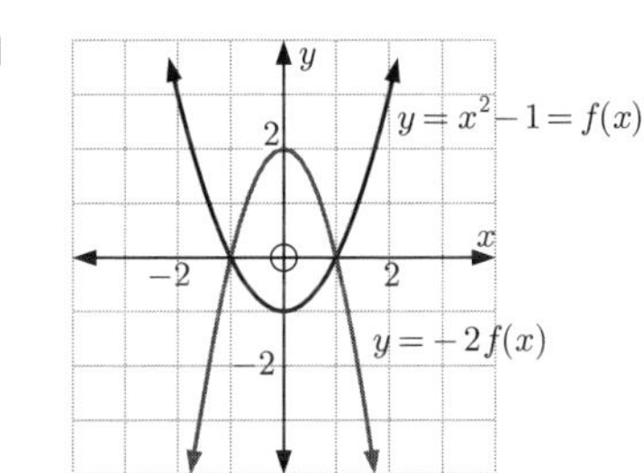


A reflection in the $x$-axis, followed by a vertical stretch with scale factor 2.

**e** $(-1,\ 0)$ and $(1,\ 0)$

**2** **a** **i** A vertical stretch with scale factor 3. **ii** $g(x) = 3f(x)$

**b** **i** A translation of 2 units downwards. **ii** $g(x) = f(x) - 2$

**c** **i** A vertical stretch with scale factor $\frac{1}{2}$. **ii** $g(x) = \frac{1}{2}f(x)$

**3** $y = -f(x)$ is obtained from $y = f(x)$ by reflecting it in the $x$-axis.

**a**

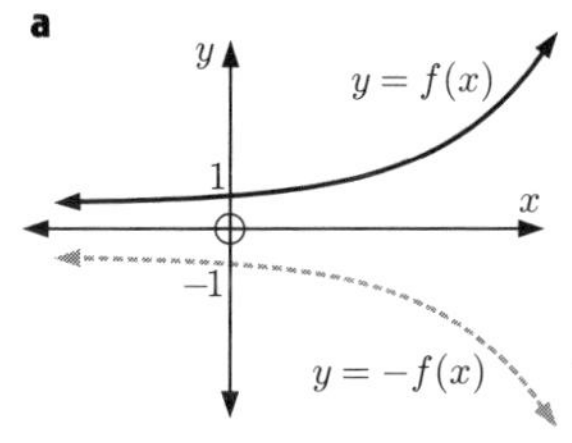


**b**

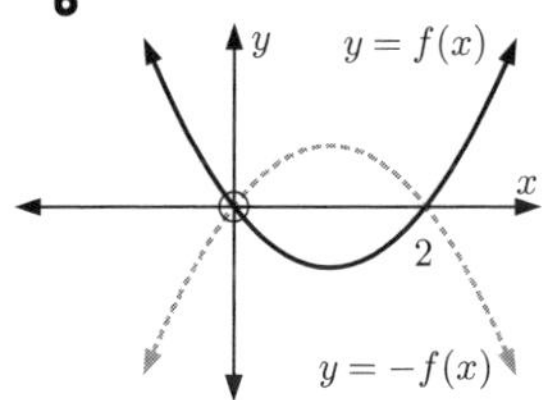


**c**

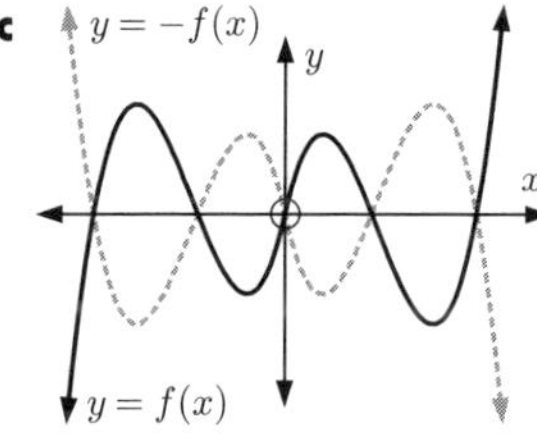


**4** $y = f(-x)$ is obtained from $y = f(x)$ by reflecting it in the $y$-axis.

**a**

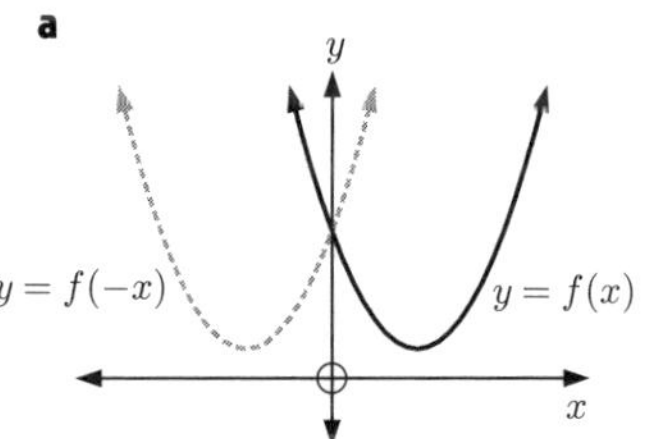


**b**

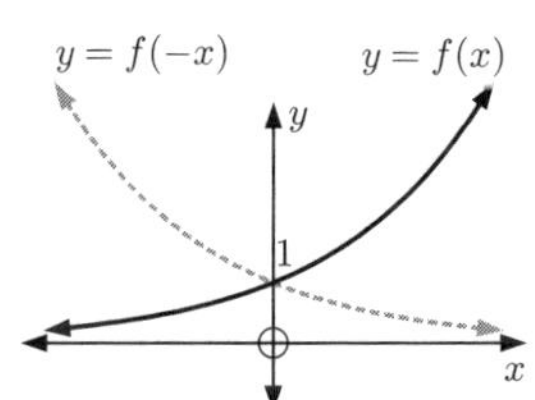


**c**

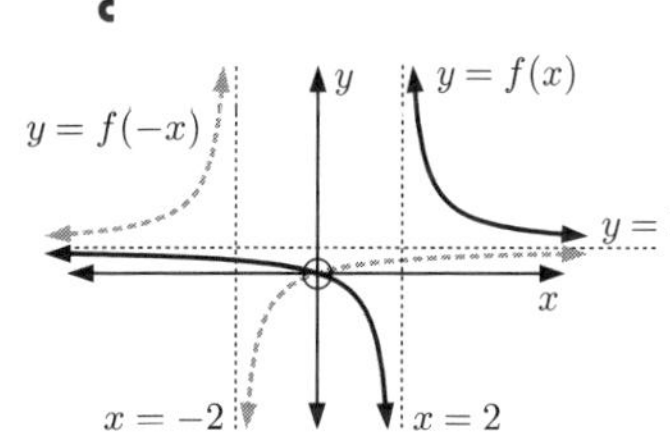


**5** $y = 2x^4$ and $y = 6x^4$ are 'thinner' than $y = x^4$ and $y = \frac{1}{2}x^4$ is 'fatter'.

$\therefore$ **a** is **A**, **b** is **B**, **c** is **D**, and **d** is **C**.

**6** **a**

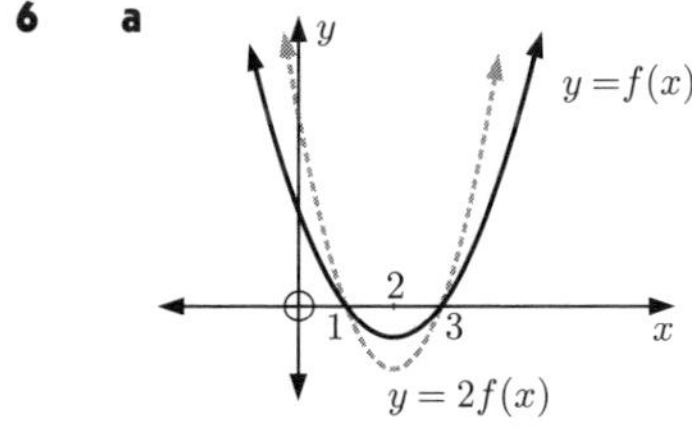


**b**

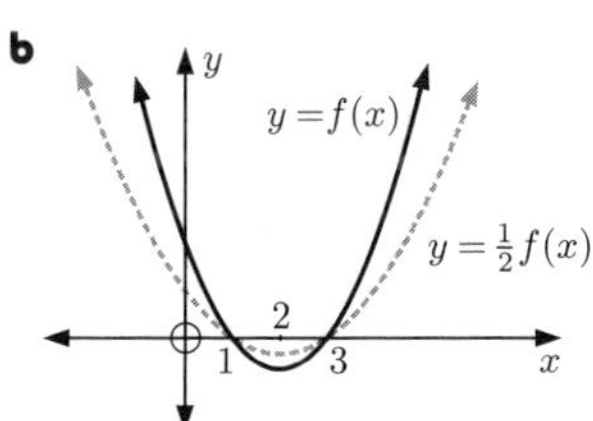


**c**

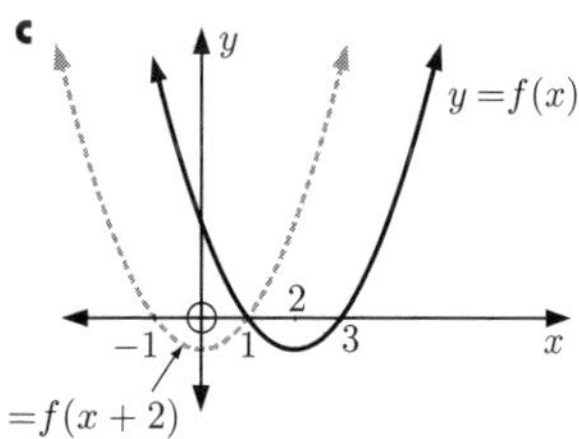


**d**

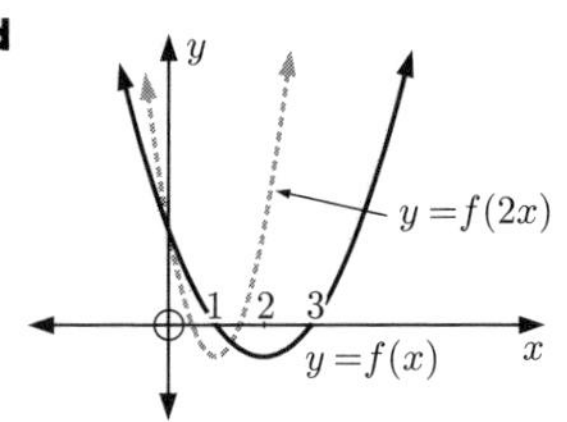


**e**

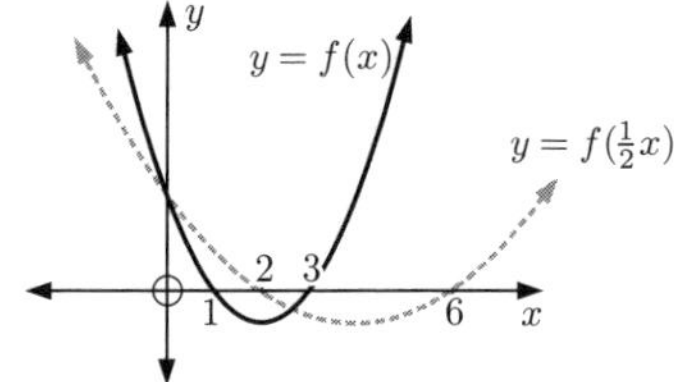

**7**

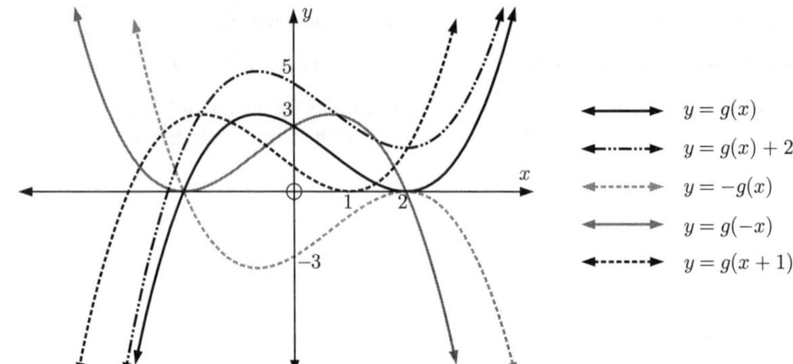


**8**

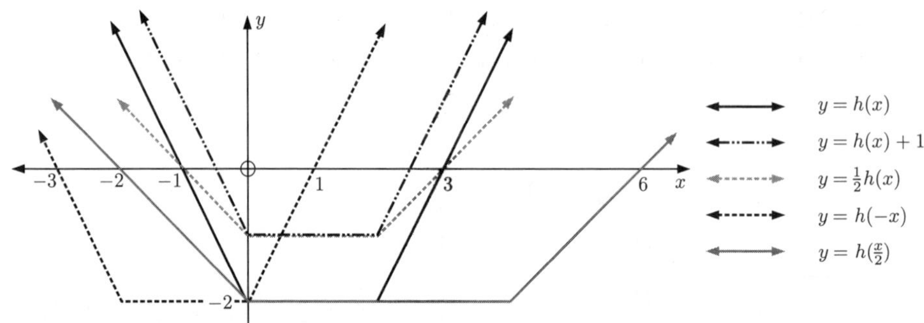


## REVIEW SET 5A

**1** $f(x) = x^2 - 2x$

**a** $f(3)$
$= 3^2 - 2(3)$
$= 9 - 6$
$= 3$

**b** $f(2x)$
$= (2x)^2 - 2(2x)$
$= 4x^2 - 4x$

**c** $f(-x)$
$= (-x)^2 - 2(-x)$
$= x^2 + 2x$

**d** $3\,f(x) - 2$
$= 3(x^2 - 2x) - 2$
$= 3x^2 - 6x - 2$

**2** $f(x) = 5 - x - x^2$

**a** $f(-1)$
$= 5 - (-1) - (-1)^2$
$= 5 + 1 - 1$
$= 5$

**b** $f(x-1)$
$= 5 - (x-1) - (x-1)^2$
$= 5 - x + 1 - [x^2 - 2x + 1]$
$= 6 - x - x^2 + 2x - 1$
$= -x^2 + x + 5$

**c** $f\left(\frac{x}{2}\right)$
$= 5 - \left(\frac{x}{2}\right) - \left(\frac{x}{2}\right)^2$
$= 5 - \frac{1}{2}x - \frac{1}{4}x^2$

**d** $2\,f(x) - f(-x)$
$= 2(5 - x - x^2) - \left[5 - (-x) - (-x)^2\right]$
$= 10 - 2x - 2x^2 - [5 + x - x^2]$
$= 10 - 2x - 2x^2 - 5 - x + x^2$
$= -x^2 - 3x + 5$

**3** $f(x) = 3x^3 - 2x^2 + x + 2$

If $g(x)$ is $f(x)$ translated $\begin{pmatrix} 1 \\ -2 \end{pmatrix}$, then $g(x) = f(x-1) - 2$
$= 3(x-1)^3 - 2(x-1)^2 + (x-1) + 2 - 2$
$= 3(x^3 - 3x^2 + 3x - 1) - 2(x^2 - 2x + 1) + x - 1$
$= 3x^3 - 9x^2 + 9x - 3 - 2x^2 + 4x - 2 + x - 1$
$= 3x^3 - 11x^2 + 14x - 6$

**4**

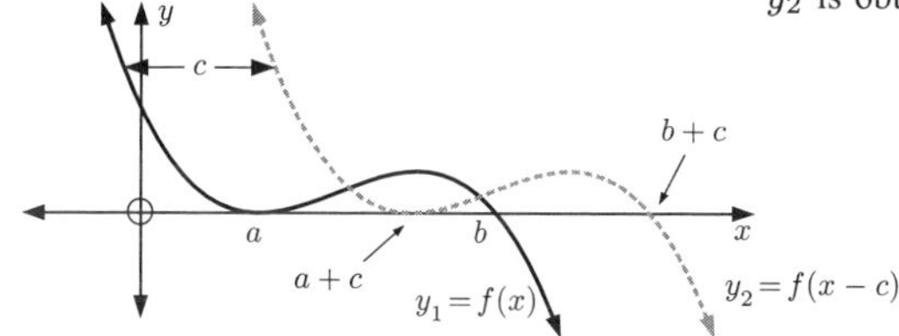


$y_2$ is obtained by translating $y_1$ $c$ units to the right.

**5** **a, b**

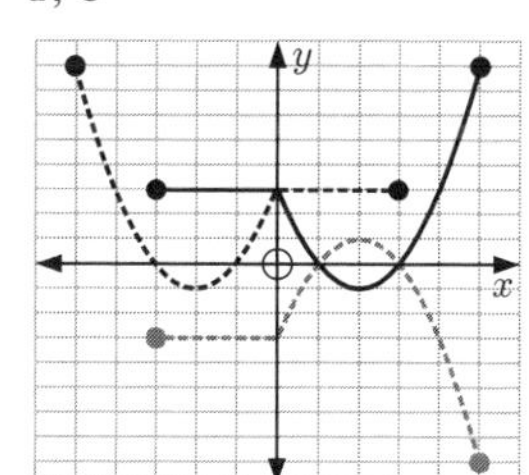

$y = f(x)$
$y = f(-x)$
$y = -f(x)$

(drawn on two graphs)

**c, d**

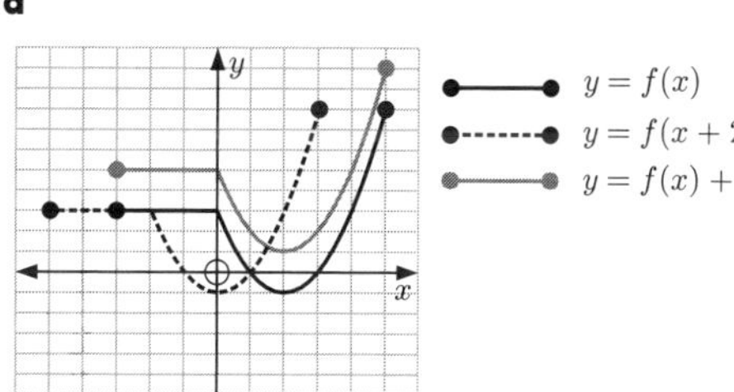


**6**

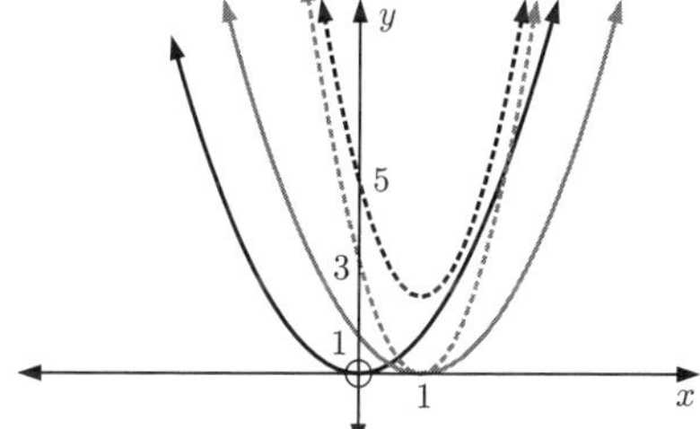


$y = f(x) = x^2$
$y = f(x - 1)$
$y = 3f(x - 1)$
$y = 3f(x - 1) + 2$

**7** **a**

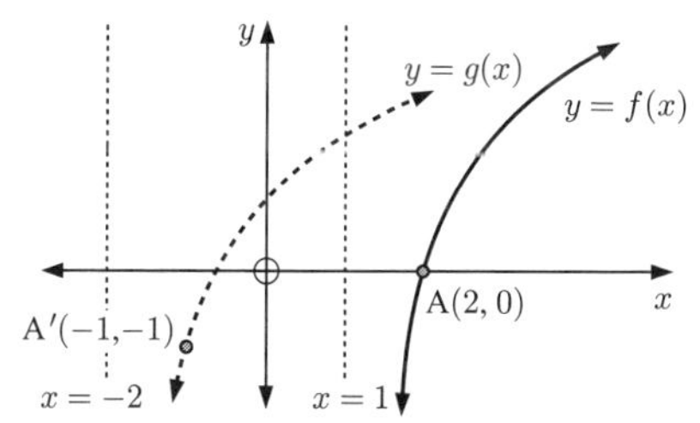


**b** $f(x + 3) - 1$ is a translation of $f(x)$ by $\binom{-3}{-1}$.

$\therefore$ vertical asymptote is at $x = 1 - 3 = -2$.

**c** A(2, 0) translated by $\binom{-3}{-1}$ gives $(2 - 3, 0 - 1)$ which is A′(−1, −1).

## REVIEW SET 5B

**1** When $y = 0$, $(x + 1)^2 - 4 = 0$

$\therefore (x + 1)^2 = 4$

$\therefore x + 1 = \pm 2$

$\therefore x = 2 - 1$ or $-2 - 1$

$\therefore x = 1$ or $-3$

$\therefore$ $x$-intercepts are 1, −3

When $x = 0$, $y = 1^2 - 4$
$= -3$

$\therefore$ $y$-intercept is −3

$y = (x + 1)^2 - 4$ is obtained from $y = x^2$ under a translation of $\binom{-1}{-4}$.

$y = x^2$ has its vertex at (0, 0), so the vertex of $y = (x + 1)^2 - 4$ must be (−1, −4).

So, the graph of $y = f(x)$ is:

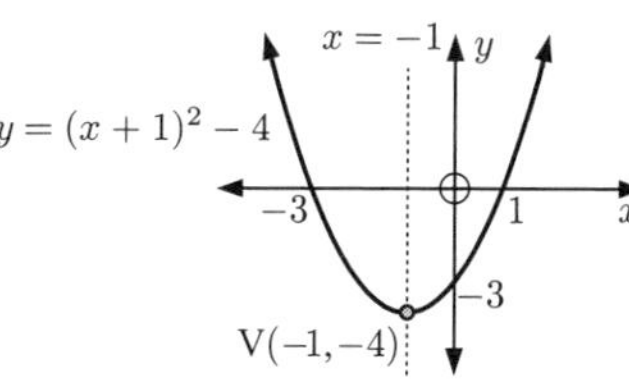

**2**

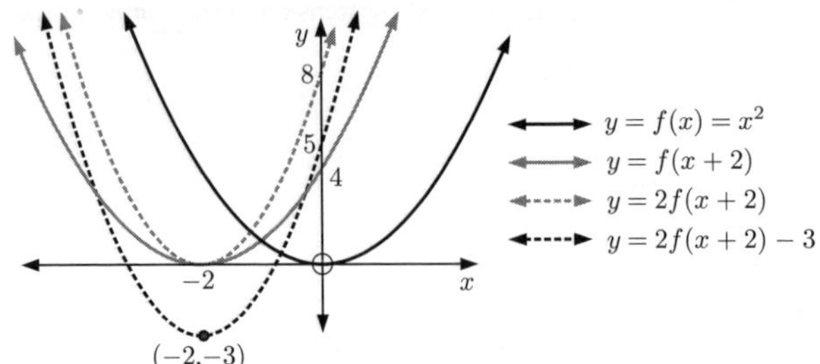


**3** **a** The function does not have any axes intercepts.

**b** As $x \to 0^-$, $y \to -\infty$

As $x \to 0^+$, $y \to \infty$

$\therefore$ the vertical asymptote is $x = 0$.

As $x \to \infty$, $y \to \infty$

As $x \to -\infty$, $y \to 0^-$

$\therefore$ the horizontal asymptote is $y = 0$.

**c** There is a local minimum at $(1.44, 1.88)$.

**d**

(4, 4)

(1.44, 1.88)

(−4, −0.0156)

$y = \dfrac{2^x}{x}$

**4** **a**

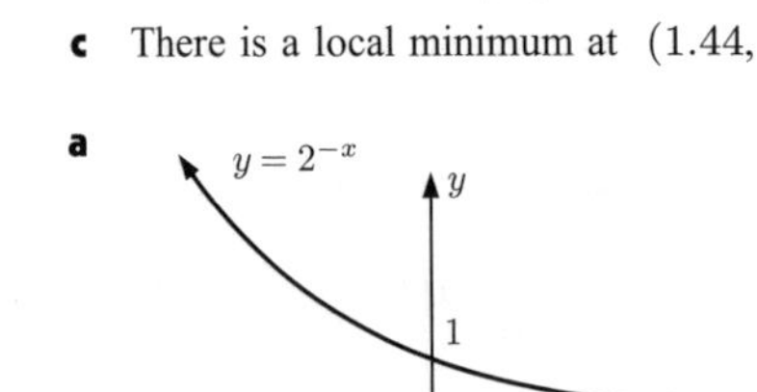


**b** **i** $x \to \infty$ means $x$ is very large and positive. We see the graph approaching the $x$-axis.

$\therefore$ $y \to 0$ $\therefore$ **true.**

**ii** $x \to -\infty$ means $x$ is very large and negative. We see the graph heading for $\infty$. $\therefore$ **false.**

**iii** When $x = 0$, $y = 2^0 = 1 \neq \frac{1}{2}$ $\therefore$ **false.**

**iv** The graph is above the $x$-axis for all $x$.

$\therefore$ $2^{-x} > 0$ for all $x$ $\therefore$ **true.**

**5** **a** $f(x) = (x+1)^2 + 4$ is translated by $\binom{2}{4}$ to get $g(x)$.

$$\begin{aligned}\therefore\ g(x) &= f(x-2) + 4 \\ &= \left[((x-2)+1)^2 + 4\right] + 4 \\ &= (x-1)^2 + 8\end{aligned}$$

**b** 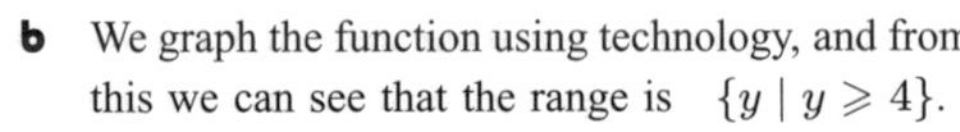
We graph the function using technology, and from this we can see that the range is $\{y \mid y \geqslant 4\}$.

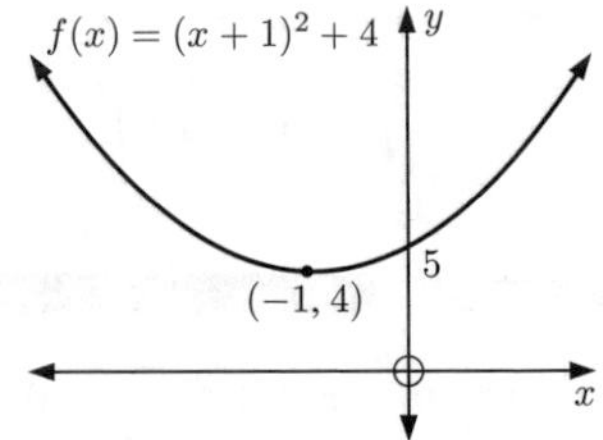


**c** $g(x)$ is $f(x)$ translated by $\binom{2}{4}$, so the minimum value of $g(x)$ is $4 + 4 = 8$.

$\therefore$ the range of $g(x)$ is $\{y \mid y \geqslant 8\}$.

**6** **a** **i** Under translation $\binom{1}{-2}$, $y = \dfrac{1}{x}$ becomes $y = \dfrac{1}{x-1} - 2$

**ii**

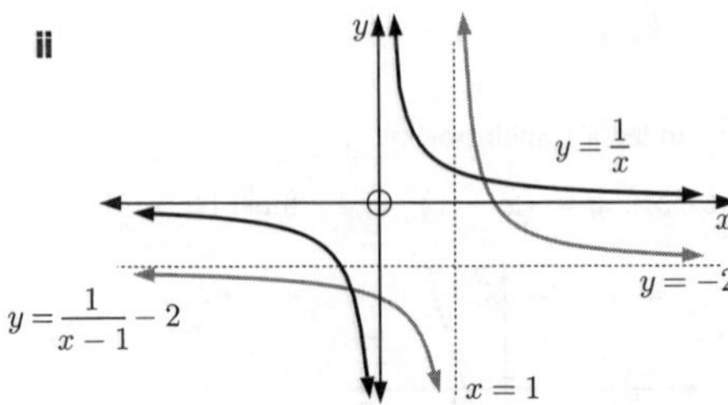


For $y = \dfrac{1}{x}$, V.A. is $x = 0$, H.A. is $y = 0$.

For $y = \dfrac{1}{x-1} - 2$, V.A. is $x = 1$, H.A. is $y = -2$.

**iii** For $y = \dfrac{1}{x}$, domain is $\{x \mid x \neq 0\}$, range is $\{y \mid y \neq 0\}$.

For $y = \dfrac{1}{x-1} - 2$, domain is $\{x \mid x \neq 1\}$, range is $\{y \mid y \neq -2\}$.

**b** **i** Under translation $\binom{1}{-2}$, $y = 2^x$ becomes $y = 2^{x-1} - 2$

**ii**

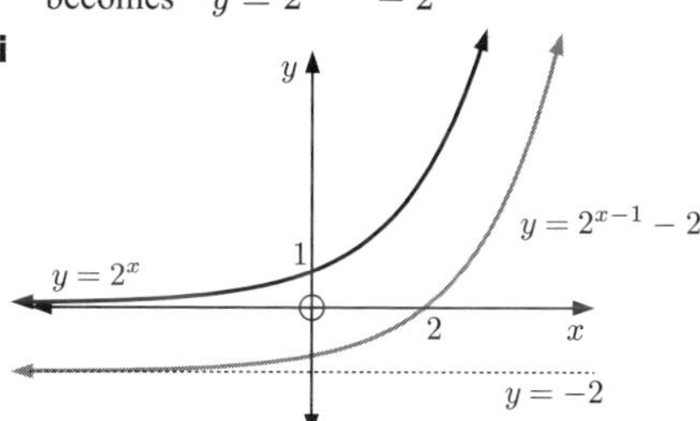


For $y = 2^x$, H.A. is $y = 0$, no V.A.
For $y = 2^{x-1} - 2$, H.A. is $y = -2$, no V.A.

**iii** For $y = 2^x$,
domain is $\{x \mid x \in \mathbb{R}\}$,
range is $\{y \mid y > 0\}$.

For $y = 2^{x-1} - 2$,
domain is $\{x \mid x \in \mathbb{R}\}$,
range is $\{y \mid y > -2\}$.

**c** **i** Under translation $\binom{1}{-2}$, $y = \log_4 x$ becomes $y = \log_4(x-1) - 2$

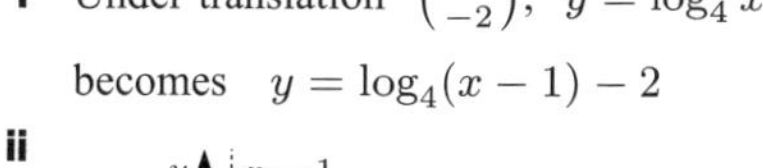

**ii**

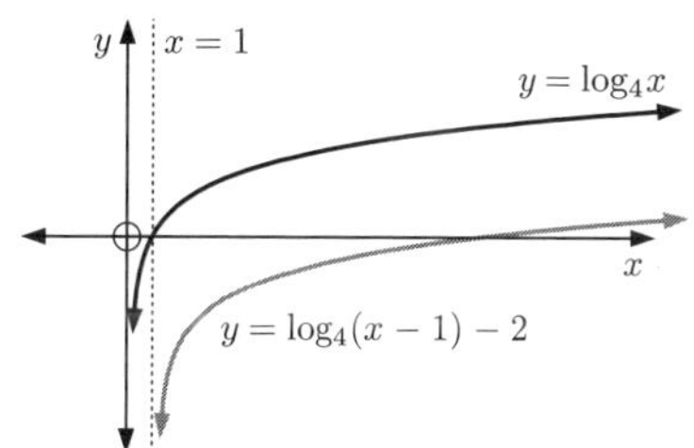


For $y = \log_4 x$, V.A. is $x = 0$, no H.A.
For $y = \log_4(x-1) - 2$, V.A. is $x = 1$, no H.A.

**iii** For $y = \log_4 x$,
domain is $\{x \mid x > 0\}$,
range is $\{y \mid y \in \mathbb{R}\}$.

For $y = \log_4(x-1) - 2$,
domain is $\{x \mid x > 1\}$,
range is $\{y \mid y \in \mathbb{R}\}$.

**7**

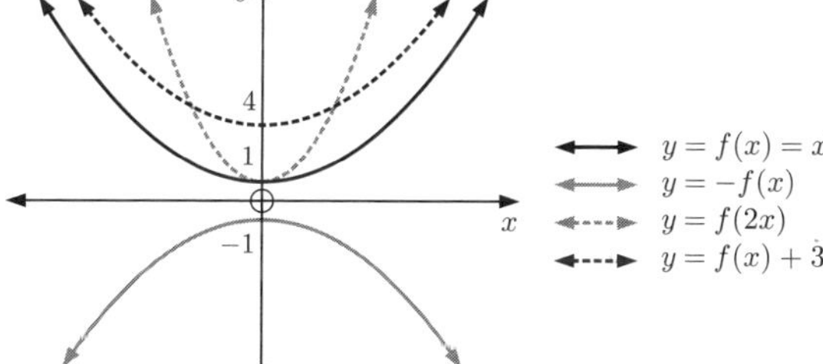


## REVIEW SET 5C

**1** **a**

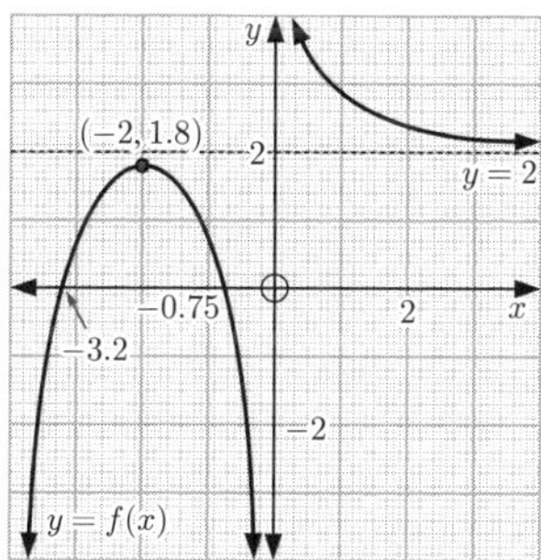


**i** The coordinates of the turning point are $(-2, 1.8)$.

**ii** The equation of the vertical asymptote is $x = 0$.

**iii** The equation of the horizontal asymptote is $y = 2$.

**iv** The $x$-intercepts are $-3.2$ and $-0.75$.

**b**

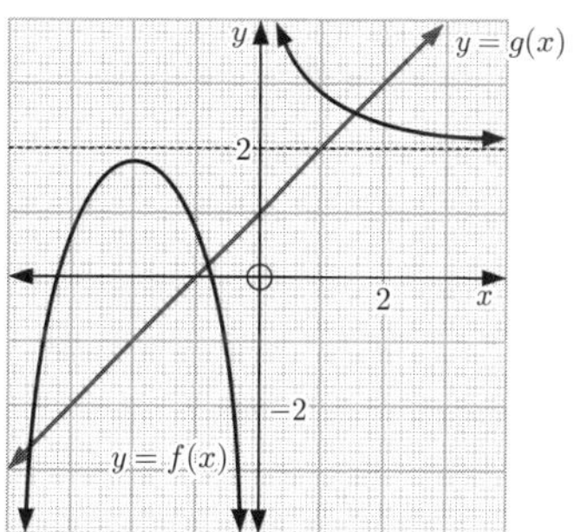


**c** The coordinates of the points of intersection are $(-3.65, -2.65)$, $(-0.8, 0.2)$, and $(1.55, 2.55)$.

**2** So that you can see the answers more easily, they have been drawn on two graphs.

**a, b**

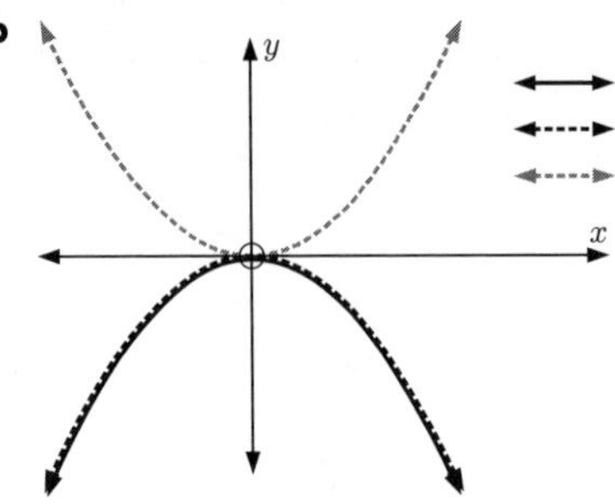


**c, d**

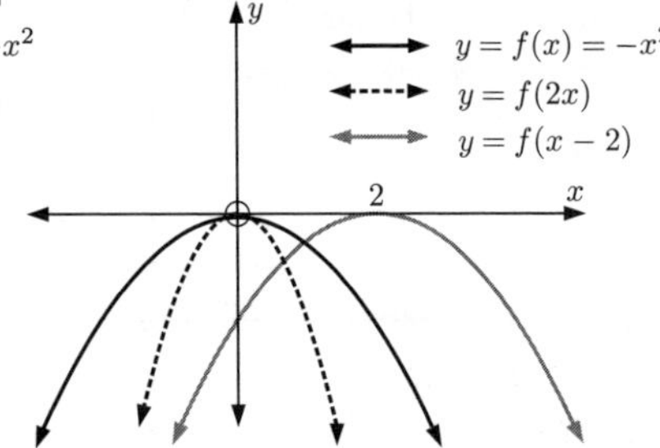


**3** **a**

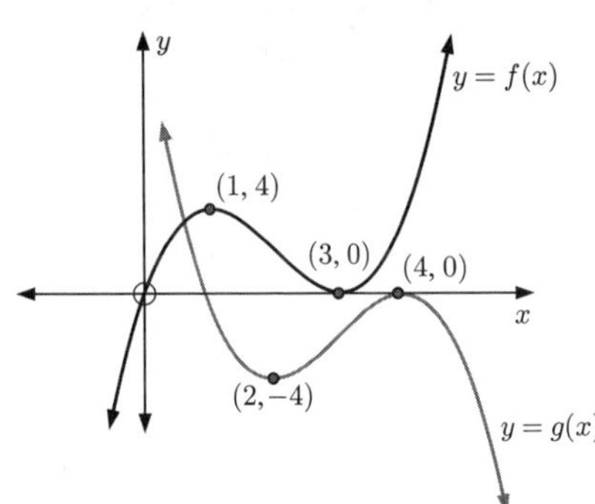


**b** $g(x)$ is obtained from $f(x)$ by a translation of $\binom{1}{0}$ and then a reflection in the $x$-axis. So, to get the turning point coordinates we add 1 to the $x$-coordinate and find the negative of the $y$-coordinate.

$(1, 4) \mapsto (2, -4)$ and $(3, 0) \mapsto (4, 0)$.

So, the turning points of $g(x)$ are $(2, -4)$ and $(4, 0)$.

**4** $f(x) = x^2$ is first reflected in the $x$-axis to become $-f(x) = -x^2$

The function is then translated by $\binom{-3}{2}$ to become

$$\begin{aligned} -f(x+3)+2 &= -(x+3)^2+2 \\ &= -(x^2+6x+9)+2 \\ \therefore \quad g(x) &= -x^2-6x-7 \end{aligned}$$

**5**

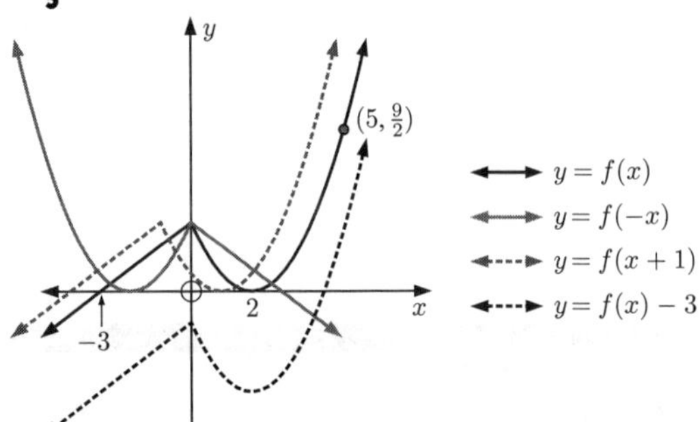


**6** $f(x) = x^3 + 3x^2 - x + 4$

$$\begin{aligned} g(x) &= f(x+1)+3 \\ &= \left[(x+1)^3+3(x+1)^2-(x+1)+4\right]+3 \\ &= x^3+3x^2+3x+1+3(x^2+2x+1)-x-1+4+3 \\ &= x^3+3x^2+3x+1+3x^2+6x+3-x-1+4+3 \\ &= x^3+6x^2+8x+10 \end{aligned}$$

**7** **a** $f(x) = 3x + 2$

**i** A translation of 2 units to the left gives

$$\begin{aligned} y &= f(x+2) \\ &= 3(x+2)+2 \\ &= 3x+8 \end{aligned}$$

**ii** A translation of 6 units upwards gives

$$\begin{aligned} y &= f(x)+6 \\ &= 3x+2+6 \\ &= 3x+8 \end{aligned}$$

**b** $f(x) = ax + b$ translated $k$ units to the left gives

$$\begin{aligned} y &= f(x+k) \\ &= a(x+k)+b \\ &= ax+ak+b \\ &= (ax+b)+ka \\ &= f(x)+ka \end{aligned}$$

which is $f(x)$ translated $ka$ units upwards.

# Chapter 6

## SEQUENCES AND SERIES

### EXERCISE 6A

1 a 4, 13, 22, 31 b 45, 39, 33, 27 c 2, 6, 18, 54 d 96, 48, 24, 12

2 a The sequence starts at 8 and each term is 8 more than the previous term. The next two terms are 40 and 48.

b The sequence starts at 2 and each term is 3 more than the previous term. The next two terms are 14 and 17.

c The sequence starts at 36 and each term is 5 less than the previous term. The next two terms are 16 and 11.

d The sequence starts at 96 and each term is 7 less than the previous term. The next two terms are 68 and 61.

e The sequence starts at 1 and each term is 4 times the previous term. The next two terms are 256 and 1024.

f The sequence starts at 2 and each term is 3 times the previous term. The next two terms are 162 and 486.

g The sequence starts at 480 and each term is half the previous term. The next two terms are 30 and 15.

h The sequence starts at 243 and each term is one third of the previous term. The next two terms are 3 and 1.

i The sequence starts at 50 000 and each term is one fifth of the previous term. The next two terms are 80 and 16.

3 a Each term is the square of the number of the term. The next three terms are 25, 36, and 49.

b Each term is the cube of the number of the term. The next three terms are 125, 216, and 343.

c Each term is $n \times (n+1)$ where $n$ is the number of the term. The next three terms are 30, 42, and 56.

4 a 79, 75 (subtracting 4 each time) b 1280, 5120 (multiplying by 4 each time)

c 625, 1296 ($1^4, 2^4, 3^4, 4^4, ....$) d 13, 17 (prime numbers)

e 16, 22 (the difference between terms increases by 1) f 6, 12 $(-1, +2, -3, +4, ....)$

### EXERCISE 6B

1 a $u_1 = 3(1) - 2$
$= 3 - 2$
$= 1$

b $u_5 = 3(5) - 2$
$= 15 - 2$
$= 13$

c $u_{27} = 3(27) - 2$
$= 81 - 2$
$= 79$

2 a $u_1 = 2(1) + 5$
$= 2 + 5$
$= 7$

$u_2 = 2(2) + 5$
$= 4 + 5$
$= 9$

$u_3 = 2(3) + 5$
$= 6 + 5$
$= 11$

$u_4 = 2(4) + 5$
$= 8 + 5$
$= 13$

b 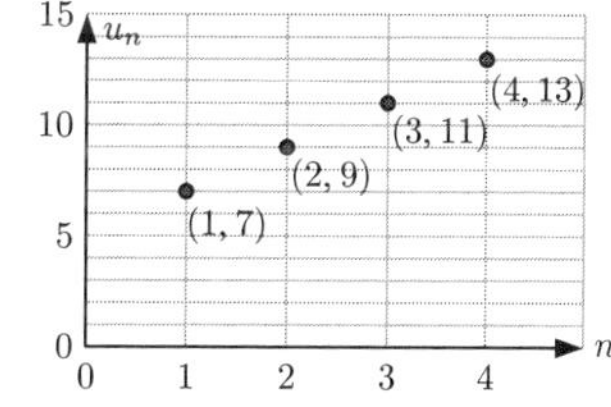


3 a The sequence $\{2n\}$ begins 2, 4, 6, 8, 10 (letting $n = 1, 2, 3, 4, 5, ....$).

b The sequence $\{2n+2\}$ begins 4, 6, 8, 10, 12 (letting $n = 1, 2, 3, 4, 5, ....$).

c The sequence $\{2n-1)$ begins 1, 3, 5, 7, 9 (letting $n = 1, 2, 3, 4, 5, ....$).

d The sequence $\{2n-3\}$ begins $-1$, 1, 3, 5, 7 (letting $n = 1, 2, 3, 4, 5, ....$).

e The sequence $\{2n+3\}$ begins 5, 7, 9, 11, 13 (letting $n = 1, 2, 3, 4, 5, ....$).

f The sequence $\{2n+11\}$ begins 13, 15, 17, 19, 21 (letting $n = 1, 2, 3, 4, 5, ....$).

**g** The sequence $\{3n+1\}$ begins 4, 7, 10, 13, 16 (letting $n = 1, 2, 3, 4, 5, ....$).

**h** The sequence $\{4n-3\}$ begins 1, 5, 9, 13, 17 (letting $n = 1, 2, 3, 4, 5, ....$).

**4** **a** The sequence $\{2^n\}$ begins 2, 4, 8, 16, 32 (letting $n = 1, 2, 3, 4, 5, ....$).

**b** The sequence $\{3 \times 2^n\}$ begins 6, 12, 24, 48, 96 (letting $n = 1, 2, 3, 4, 5, ....$).

**c** The sequence $\{6 \times (\frac{1}{2})^n\}$ begins $3, \frac{3}{2}, \frac{3}{4}, \frac{3}{8}, \frac{3}{16}$ (letting $n = 1, 2, 3, 4, 5, ....$).

**d** The sequence $\{(-2)^n\}$ begins $-2, 4, -8, 16, -32$ (letting $n = 1, 2, 3, 4, 5, ....$).

**5** $\{15-(-2)^n\}$ generates the sequence with first five terms:

$t_1 = 15-(-2)^1 = 17,\quad t_2 = 15-(-2)^2 = 11,\quad t_3 = 15-(-2)^3 = 23,$

$t_4 = 15-(-2)^4 = -1,\quad t_5 = 15-(-2)^5 = 47$

## EXERCISE 6C.1

**1** **a** The sequence begins with 19 and the common difference is $25-19=6$.

$$\begin{aligned} u_n &= u_1+(n-1)d \\ \therefore\ u_n &= 19+6(n-1) \\ \text{So,}\ u_{10} &= 19+6(10-1) \\ &= 19+6\times 9 \\ &= 73 \end{aligned}$$

**b** The sequence begins with 101 and the common difference is $97-101=-4$.

$$\begin{aligned} u_n &= u_1+(n-1)d \\ \therefore\ u_n &= 101+(-4)(n-1) \\ \text{So,}\ u_{10} &= 101+(-4)(10-1) \\ &= 101-4\times 9 \\ &= 65 \end{aligned}$$

**c** The sequence begins with 8 and the common difference is $9\frac{1}{2}-8=\frac{3}{2}$.

$$\begin{aligned} u_n &= u_1+(n-1)d \\ \therefore\ u_n &= 8+\tfrac{3}{2}(n-1) \\ \text{So,}\ u_{10} &= 8+\tfrac{3}{2}(10-1) \\ &= 8+\tfrac{3}{2}\times 9 \\ &= 21\tfrac{1}{2} \end{aligned}$$

**2** **a** The first term of the arithmetic sequence is 31 and the common difference is $36-31=5$.

$$\begin{aligned} u_n &= u_1+(n-1)d \\ \therefore\ u_n &= 31+5(n-1) \\ \text{So,}\ u_{15} &= 31+5(15-1) \\ &= 31+5\times 14 \\ &= 101 \end{aligned}$$

**b** The first term of the arithmetic sequence is 5 and the common difference is $-3-5=-8$.

$$\begin{aligned} u_n &= u_1+(n-1)d \\ \therefore\ u_n &= 5+(-8)(n-1) \\ \text{So,}\ u_{15} &= 5+(-8)(15-1) \\ &= 5-8\times 14 \\ &= -107 \end{aligned}$$

**c** The first term of the arithmetic sequence is $a$ and the common difference is $a+d-a=d$.

$$\begin{aligned} u_n &= u_1+(n-1)d \\ \therefore\ u_n &= a+(n-1)d \\ \text{So,}\ u_{15} &= a+d(15-1) \\ &= a+14d \end{aligned}$$

**3** **a** $17-6=11$

$28-17=11$

$39-28=11$

$50-39=11$

Assuming that the pattern continues, consecutive terms differ by 11.

$\therefore$ the sequence is arithmetic with $u_1=6,\ d=11$.

**b**
$$\begin{aligned} u_n &= u_1+(n-1)d \\ &= 6+(n-1)11 \\ &= 11n-5 \end{aligned}$$

**c**
$$\begin{aligned} u_{50} &= 11(50)-5 \\ &= 545 \end{aligned}$$

**d** Let $u_n = 325 = 11n-5$

$\therefore\ 330 = 11n$

$\therefore\ n = 30$

So, 325 is the 30th member.

**e** Let $u_n = 761 = 11n-5$

$\therefore\ 766 = 11n$

$\therefore\ n = 69\frac{7}{11}$, but $n$ must be an integer, so 761 is not a member of the sequence.

**4** **a** $83 - 87 = -4$
$79 - 83 = -4$
$75 - 79 = -4$

Assuming that the pattern continues, consecutive terms differ by $-4$.
$\therefore$ the sequence is arithmetic with $u_1 = 87$, $d = -4$.

**b** $u_n = u_1 + (n-1)d$
$= 87 + (n-1)(-4)$
$= 87 - 4n + 4$
$= 91 - 4n$

**c** $u_{40} = 91 - 4(40)$
$= 91 - 160$
$= -69$

**d** Let $u_n = -297 = 91 - 4n$
$\therefore$ $4n = 388$
$\therefore$ $n = 97$
So, $-297$ is the 97th term of the sequence.

**5** **a** $u_n = 3n - 2$, $u_{n+1} = 3(n+1) - 2 = 3n + 1$
$u_{n+1} - u_n = (3n+1) - (3n-2)$
$= 3$, a constant

Consecutive terms differ by 3.
$\therefore$ the sequence is arithmetic.

**b** $u_1 = 3(1) - 2 = 1$, $d = 3$

**c** $u_{57} = 3(57) - 2 = 169$

**d** Let $u_n = 450 = 3n - 2$, so $3n = 452$ and hence $n = 150\frac{2}{3}$.
We try the two values on either side of $n = 150\frac{2}{3}$, which are $n = 150$ and $n = 151$:
$u_{150} = 3(150) - 2 = 448$ and $u_{151} = 3(151) - 2 = 451$
So, $u_{150} = 448$ is the largest term which is smaller than 450.

**6** **a** $u_n = \dfrac{71 - 7n}{2} = 35\frac{1}{2} - \frac{7}{2}n$
$u_{n+1} = \dfrac{71 - 7(n+1)}{2} = \dfrac{71 - 7n - 7}{2} = \dfrac{64 - 7n}{2} = 32 - \frac{7}{2}n$
$u_{n+1} - u_n = (32 - \frac{7}{2}n) - (35\frac{1}{2} - \frac{7}{2}n) = -\frac{7}{2}$, a constant
So, consecutive terms differ by $-\frac{7}{2}$. $\therefore$ the sequence is arithmetic.

**b** $u_1 = \dfrac{71 - 7(1)}{2} = 32$, $d = -\frac{7}{2}$

**c** $u_{75} = \dfrac{71 - 7(75)}{2} = -227$

**d** Let $u_n = -200 = \dfrac{71 - 7n}{2}$ so $-400 = 71 - 7n$ $\therefore$ $7n = 471$
$\therefore$ $n = 67\frac{2}{7}$
We try the two values on either side of $n = 67\frac{2}{7}$, which are $n = 67$ and $n = 68$:
$u_{67} = \dfrac{71 - 7(67)}{2} = -199$ and $u_{68} = \dfrac{71 - 7(68)}{2} = -202\frac{1}{2}$
So, the terms of the sequence are less than $-200$ for $n \geqslant 68$.

**7** **a** The terms are consecutive, so we equate common differences:
$k - 32 = 3 - k$
$\therefore$ $2k = 35$
$\therefore$ $k = 17\frac{1}{2}$

**b** The terms are consecutive, so we equate common differences:
$7 - k = 10 - 7$
$\therefore$ $7 - k = 3$
$\therefore$ $k = 4$

**c** The terms are consecutive, so we equate common differences:
$(2k+1) - (k+1) = 13 - (2k+1)$
$\therefore$ $k = 12 - 2k$
$\therefore$ $3k = 12$
$\therefore$ $k = 4$

**d** The terms are consecutive, so we equate common differences:
$(2k+3) - (k-1) = (7-k) - (2k+3)$
$\therefore$ $k + 4 = 4 - 3k$
$\therefore$ $4k = 0$
$\therefore$ $k = 0$

**e** The terms are consecutive, so we equate common differences:
$k^2 - k = (k^2 + 6) - k^2$
$\therefore$ $k^2 - k - 6 = 0$
$\therefore$ $(k+2)(k-3) = 0$
$\therefore$ $k = -2$ or $3$

**f** The terms are consecutive, so we equate common differences:
$k - 5 = k^2 - 8 - k$
$\therefore$ $k^2 - 2k - 3 = 0$
$\therefore$ $(k-3)(k+1) = 0$
$\therefore$ $k = -1$ or $3$

**8** **a** $u_7 = 41 \quad \therefore u_1 + 6d = 41$ .... (1)
$u_{13} = 77 \quad \therefore u_1 + 12d = 77$ .... (2)
Solving simultaneously,
$-u_1 - 6d = -41$
$u_1 + 12d = 77$
$\therefore 6d = 36$ {adding the equations}
$\therefore d = 6$
So in (1), $u_1 + 6(6) = 41$
$\therefore u_1 + 36 = 41$
$\therefore u_1 = 5$
Now $u_n = u_1 + (n-1)d$
$\therefore u_n = 5 + (n-1)6$
$\therefore u_n = 6n - 1$

**b** $u_5 = -2 \quad \therefore u_1 + 4d = -2$ .... (1)
$u_{12} = -12\frac{1}{2} \quad \therefore u_1 + 11d = -12\frac{1}{2}$ .... (2)
Solving simultaneously,
$-u_1 - 4d = 2$
$u_1 + 11d = -12\frac{1}{2}$
$\therefore 7d = -10\frac{1}{2}$ {adding the equations}
$\therefore d = -\frac{3}{2}$
So in (1), $u_1 + 4(-\frac{3}{2}) = -2$
$\therefore u_1 = 4$
Now $u_n = u_1 + (n-1)d$
$\therefore u_n = 4 + (n-1)(-\frac{3}{2})$
$\therefore u_n = -\frac{3}{2}n + \frac{11}{2}$

**c** $u_7 = 1 \quad \therefore u_1 + 6d = 1$ .... (1)
$u_{15} = -39 \quad \therefore u_1 + 14d = -39$ .... (2)
Solving simultaneously,
$-u_1 - 6d = -1$
$u_1 + 14d = -39$
$\therefore 8d = -40$ {adding the equations}
$\therefore d = -5$
So in (1), $u_1 + 6(-5) = 1$
$\therefore u_1 - 30 = 1$
$\therefore u_1 = 31$
Now $u_n = u_1 + (n-1)d$
$\therefore u_n = 31 + (n-1)(-5)$
$\therefore u_n = 31 - 5n + 5$
$\therefore u_n = -5n + 36$

**d** $u_{11} = -16 \quad \therefore u_1 + 10d = -16$ .... (1)
$u_8 = -11\frac{1}{2} \quad \therefore u_1 + 7d = -11\frac{1}{2}$ .... (2)
Solving simultaneously,
$-u_1 - 10d = 16$
$u_1 + 7d = -11\frac{1}{2}$
$\therefore -3d = 4\frac{1}{2}$ {adding the equations}
$\therefore d = -\frac{3}{2}$
So in (1), $u_1 + 10(-\frac{3}{2}) = -16$
$\therefore u_1 - 15 = -16$
$\therefore u_1 = -1$
Now $u_n = u_1 + (n-1)d$
$\therefore u_n = -1 + (n-1)(-\frac{3}{2})$
$\therefore u_n = -\frac{3}{2}n + \frac{1}{2}$

**9** **a** Let the numbers be
$5,\ 5+d,\ 5+2d,\ 5+3d,\ 10.$
Then $5 + 4d = 10$
$\therefore 4d = 5$
$\therefore d = \frac{5}{4} = 1\frac{1}{4}$
So, the numbers are
$5, 6\frac{1}{4}, 7\frac{1}{2}, 8\frac{3}{4}, 10.$

**b** Let the numbers be $-1,\ -1+d,\ -1+2d,$ $-1+3d,\ -1+4d,\ -1+5d,\ -1+6d,\ 32.$
Then $-1 + 7d = 32$
$\therefore 7d = 33$
$\therefore d = \frac{33}{7} = 4\frac{5}{7}$
So, the numbers are
$-1,\ 3\frac{5}{7},\ 8\frac{3}{7},\ 13\frac{1}{7},\ 17\frac{6}{7},\ 22\frac{4}{7},\ 27\frac{2}{7},\ 32.$

**10** **a** $u_1 = 36,\ 35\frac{1}{3} - 36 = -\frac{2}{3},$
$34\frac{2}{3} - 35\frac{1}{3} = -\frac{2}{3}$
So, $d = -\frac{2}{3}.$

**b** $u_n = u_1 + (n-1)d$
$\therefore -30 = 36 + (n-1)(-\frac{2}{3})$ {letting $u_n = -30$}
$\therefore -66 = -\frac{2}{3}n + \frac{2}{3}$
$\therefore \frac{2}{3}n = 66\frac{2}{3}$
$\therefore n = 100$ So, $-30$ is the 100th term of the sequence.

**11** $u_1 = 23,\ 36 - 23 = 13$
$49 - 36 = 13$
$62 - 49 = 13$
so $d = 13$

$u_n = u_1 + (n-1)d$
$\therefore u_n = 23 + (n-1)13$
$= 23 + 13n - 13$
$\therefore u_n = 13n + 10$

Let $u_n = 100\,000$
$= 13n + 10$
$\therefore 99\,990 = 13n$
$\therefore n = 7691\frac{7}{13}$

We try the two values on either side of $n = 7691\frac{7}{13}$, which are $n = 7691$ and $n = 7692$:
$u_{7691} = 13(7691) + 10 = 99\,993$ and $u_{7692} = 13(7692) + 10 = 100\,006$
So, the first term to exceed 100 000 is $u_{7692} = 100\,006$.

## EXERCISE 6C.2

**1** **a** *Month 1:* 5 cars  *Month 2:* $5+13=18$ cars  *Month 3:* $18+13=31$ cars
*Month 4:* $31+13=44$ cars  *Month 5:* $44+13=57$ cars  *Month 6:* $57+13=70$ cars

**b** Every month after the first, the factory assembles 13 cars, so the difference between successive months is always 13. Thus we have an arithmetic sequence with $u_1=5$ and $d=13$.

**c** $u_n=u_1+(n-1)d$
$=5+(n-1)\times 13$
$=13n-8$
$\therefore\ u_{12}=13\times 12-8$ {12 months = 1 year}
$=148$
So, 148 cars are made in the first year.

**d** $u_n=250=13n-8$
$\therefore\ 258=13n$
$\therefore\ n=\dfrac{258}{13}\approx 19.85$
So, the 250th car is made in the 20th month.

**2** **a** $41-34=48-41=55-48=7$
Assuming the pattern continues, there is a common difference of 7, and so the number of Valéria's friends forms an arithmetic sequence with $u_1=34$ and $d=7$.

**b** $u_n=u_1+(n-1)d$
$=34+(n-1)\times 7$
$=27+7n$
$\therefore\ u_{12}=27+7\times 12=111$
So, after 12 weeks Valéria will have 111 online friends.

**c** $u_n=150=27+7n$
$\therefore\ 123=7n$
$\therefore\ n=\frac{123}{7}\approx 17.57$
So, Valéria will have 150 online friends after the 18th week.

**3** **a** *July 1st:* $100-2.7\times 1=97.3$ tonnes of hay
*July 2nd:* $100-2.7\times 2=94.6$ tonnes of hay
*July 3rd:* $100-2.7\times 3=91.9$ tonnes of hay

**b** $94.6-97.3=91.9-94.6=-2.7$
So $d=-2.7$, which means the cows eat 2.7 tonnes of hay per day.

**c** $u_{25}=100-2.7\times 25=32.5$ tonnes
So, at the end of July 25th there are 32.5 tonnes of hay remaining in the barn.

**d** July has 31 days, so the end of day 31 is the start of August.
$u_{31}=100-2.7\times 31=16.3$ tonnes.
Hence there are 16.3 tonnes of hay in the barn at the beginning of August.

## EXERCISE 6D.1

**1** **a** $\frac{6}{2}=3\ \therefore\ r=3,\ u_1=2\quad\therefore\ b=6\times 3=18$ and $c=18\times 3=54$

**b** $\frac{5}{10}=\frac{1}{2}\ \therefore\ r=\frac{1}{2},\ u_1=10\quad\therefore\ b=5\times\frac{1}{2}=2\frac{1}{2}$ and $c=2\frac{1}{2}\times\frac{1}{2}=1\frac{1}{4}$

**c** $\frac{-6}{12}=-\frac{1}{2}\ \therefore\ r=-\frac{1}{2},\ u_1=12\quad\therefore\ b=-6\times -\frac{1}{2}=3$ and $c=3\times -\frac{1}{2}=-1\frac{1}{2}$

**2** **a** $\frac{6}{3}=2\ \therefore\ r=2,\ u_1=3\ \therefore\ u_6=3\times 2^{6-1}=96$

**b** $\frac{10}{2}=5\ \therefore\ r=5,\ u_1=2\ \therefore\ u_6=2\times 5^{6-1}=6250$

**c** $\frac{256}{512}=\frac{1}{2}\ \therefore\ r=\frac{1}{2},\ u_1=512\ \therefore\ u_6=512\times(\frac{1}{2})^{6-1}=16$

**3** **a** $\frac{3}{1}=3\ \therefore\ r=3,\ u_1=1\ \therefore\ u_9=1\times 3^{9-1}=6561$

**b** $\frac{18}{12}=\frac{3}{2}\ \therefore\ r=\frac{3}{2},\ u_1=12\ \therefore\ u_9=12\times(\frac{3}{2})^{9-1}=307\frac{35}{64}$ or $\frac{19\,683}{64}$

**c** $\dfrac{-\frac{1}{8}}{\frac{1}{16}}=-2\ \therefore\ r=-2,\ u_1=\frac{1}{16}\ \therefore\ u_9=\frac{1}{16}\times(-2)^{9-1}=16$

**d** $\dfrac{ar}{a}=r\ \therefore\ r=r,\ u_1=a\ \therefore\ u_9=a\times r^{9-1}=ar^8$

**4** **a** $\frac{10}{5} = \frac{20}{10} = \frac{40}{20} = 2$

Assuming the pattern continues, consecutive terms have a common ratio of 2.

$\therefore$ the sequence is geometric with $u_1 = 5$ and $r = 2$.

**b** $u_n = u_1 r^{n-1}$

$\therefore$ $u_n = 5 \times 2^{n-1}$

so $u_{15} = 5 \times 2^{14} = 81\,920$

**5** **a** $\frac{-6}{12} = -\frac{1}{2}$

$\frac{3}{-6} = -\frac{1}{2}$

$\dfrac{(-\frac{3}{2})}{3} = -\frac{1}{2}$

Assuming the pattern continues, consecutive terms have a common ratio of $-\frac{1}{2}$.

$\therefore$ the sequence is geometric with $u_1 = 12$ and $r = -\frac{1}{2}$.

**b** $u_n = u_1 r^{n-1}$

$\therefore$ $u_n = 12 \times \left(-\frac{1}{2}\right)^{n-1}$

so $u_{13} = 12 \times (-\frac{1}{2})^{13-1}$

$= 12 \times (-\frac{1}{2})^{12}$

$= 12 \times \frac{1}{4096}$

$= 3 \times \frac{1}{1024} = \frac{3}{1024}$

**6** $\frac{-6}{8} = -\frac{3}{4}$

$\dfrac{4.5}{-6} = -\dfrac{(\frac{9}{2})}{6} = -\frac{3}{4}$

$\dfrac{-3.375}{4.5} = \dfrac{(-\frac{27}{8})}{(\frac{9}{2})} = -\frac{3}{4}$

Assuming the pattern continues, consecutive terms have a common ratio of $-\frac{3}{4}$.

$\therefore$ the sequence is geometric with $u_1 = 8$ and $r = -\frac{3}{4}$.

$u_n = u_1 r^{n-1} = 8 \times (-\frac{3}{4})^{n-1}$

So, $u_{10} = 8 \times (-\frac{3}{4})^9 = -0.600\,677\,490\,2$

$\approx -0.601$

**7** $\dfrac{4\sqrt{2}}{8} = \dfrac{\sqrt{2}}{2} = \frac{1}{\sqrt{2}}$

$\dfrac{4}{4\sqrt{2}} = \frac{1}{\sqrt{2}}$

$\dfrac{2\sqrt{2}}{4} = \dfrac{\sqrt{2}}{2} = \frac{1}{\sqrt{2}}$

Assuming the pattern continues, consecutive terms have a common ratio of $\frac{1}{\sqrt{2}}$.

$\therefore$ the sequence is geometric with $u_1 = 8$ and $r = \frac{1}{\sqrt{2}}$.

$$u_n = u_1 r^{n-1} = 8\left(\frac{1}{\sqrt{2}}\right)^{n-1} = 2^3 \times \left(2^{-\frac{1}{2}}\right)^{n-1} = 2^3 \times 2^{-\frac{1}{2}n+\frac{1}{2}}$$

So, $u_n = 2^{\frac{7}{2}-\frac{1}{2}n}$

**8** **a** Since the terms are geometric,

$\dfrac{k}{7} = \dfrac{28}{k}$ $\therefore$ $k^2 = 196$

$\therefore$ $k = \pm 14$

**b** Since the terms are geometric,

$\dfrac{3k}{k} = \dfrac{20-k}{3k} = 3$

$\therefore$ $20 - k = 9k$

$\therefore$ $20 = 10k$

$\therefore$ $k = 2$

**c** Since the terms are geometric,

$\dfrac{k+8}{k} = \dfrac{9k}{k+8}$

$\therefore$ $(k+8)^2 = 9k^2$

$\therefore$ $k^2 + 16k + 64 = 9k^2$

$\therefore$ $8k^2 - 16k - 64 = 0$

$\therefore$ $8(k^2 - 2k - 8) = 0$

$\therefore$ $8(k+2)(k-4) = 0$ and so $k = -2$ or $4$

**9** **a** $u_4 = 24 \quad \therefore \quad u_1 \times r^3 = 24 \quad .... (1)$

$u_7 = 192 \quad \therefore \quad u_1 \times r^6 = 192 \quad .... (2)$

So, $\dfrac{u_1 r^6}{u_1 r^3} = \dfrac{192}{24} \qquad \{(2) \div (1)\}$

$\therefore \quad r^3 = 8 \qquad \therefore \quad r = 2$

So in (1), $u_1 \times 2^3 = 24$

$\therefore \quad u_1 \times 8 = 24$

$\therefore \quad u_1 = 3$

$\therefore \quad u_n = 3 \times 2^{n-1}$

**b** $u_3 = 8 \quad \therefore \quad u_1 \times r^2 = 8 \quad .... (1)$

$u_6 = -1 \quad \therefore \quad u_1 \times r^5 = -1 \quad .... (2)$

So, $\dfrac{u_1 r^5}{u_1 r^2} = -\dfrac{1}{8} \qquad \{(2) \div (1)\}$

$\therefore \quad r^3 = -\frac{1}{8} \qquad \therefore \quad r = -\frac{1}{2}$

So in (1), $u_1 \times (-\frac{1}{2})^2 = 8$

$\therefore \quad u_1 \times \frac{1}{4} = 8$

$\therefore \quad u_1 = 32$

$\therefore \quad u_n = 32 \times (-\frac{1}{2})^{n-1}$

**c** $u_7 = 24 \quad \therefore \quad u_1 \times r^6 = 24 \quad .... (1)$

$u_{15} = 384 \quad \therefore \quad u_1 \times r^{14} = 384 \quad .... (2)$

So, $\dfrac{u_1 r^{14}}{u_1 r^6} = \dfrac{384}{24} \qquad \{(2) \div (1)\}$

$\therefore \quad r^8 = 16 \qquad \therefore \quad r = \pm\sqrt{2}$

So in (1), $u_1 \times (\pm\sqrt{2})^6 = 24$

$\therefore \quad u_1 \times 8 = 24$

$\therefore \quad u_1 = 3$

Now $u_n = u_1 r^{n-1}$

$\therefore \quad u_n = 3 \times (\sqrt{2})^{n-1}$

or $u_n = 3 \times (-\sqrt{2})^{n-1}$

**d** $u_3 = 5 \quad \therefore \quad u_1 \times r^2 = 5 \quad .... (1)$

$u_7 = \frac{5}{4} \quad \therefore \quad u_1 \times r^6 = \frac{5}{4} \quad .... (2)$

So, $\dfrac{u_1 r^6}{u_1 r^2} = \dfrac{(\frac{5}{4})}{5} \qquad \{(2) \div (1)\}$

$\therefore \quad r^4 = \frac{1}{4} \qquad \therefore \quad r = \pm\frac{1}{\sqrt{2}}$

So in (1), $u_1 \times (\pm\frac{1}{\sqrt{2}})^2 = 5$

$\therefore \quad u_1 \times \frac{1}{2} = 5$

$\therefore \quad u_1 = 10$

Now $u_n = u_1 r^{n-1}$

$\therefore \quad u_n = 10 \times (\frac{1}{\sqrt{2}})^{n-1}$

$= 10 \times (\sqrt{2})^{1-n}$

or $u_n = 10 \times (-\frac{1}{\sqrt{2}})^{n-1}$

$= 10 \times (-\sqrt{2})^{1-n}$

**10** **a** 2, 6, 18, 54, .... has $u_1 = 2$ and $r = 3$

$u_n = u_1 r^{n-1} \quad \therefore \quad u_n = 2 \times 3^{n-1}$

Let $u_n = 10\,000 = 2 \times 3^{n-1}$, so $5000 = 3^{n-1}$

$\therefore \quad n \approx 8.7527$ {using technology}

We try the two values on either side of $n = 8.7527$, which are $n = 8$ and $n = 9$:

$u_8 = 2 \times 3^7 = 4374$ and $u_9 = 2 \times 3^8 = 13\,122$

So, the first term to exceed 10 000 is $u_9 = 13\,122$.

**b** $4, 4\sqrt{3}, 12, 12\sqrt{3}, ....$ has $u_1 = 4$ and $r = \sqrt{3}$

$u_n = u_1 r^{n-1} \quad \therefore \quad u_n = 4 \times \left(\sqrt{3}\right)^{n-1}$

Let $u_n = 4800 = 4 \times \left(\sqrt{3}\right)^{n-1}$, so $1200 = \left(\sqrt{3}\right)^{n-1}$

$\therefore \quad n \approx 13.91$ {using technology}

We try the two values on either side of $n \approx 13.91$, which are $n = 13$ and $n = 14$:

$u_{13} = 4 \times \left(\sqrt{3}\right)^{12} = 2916$ and $u_{14} = 4 \times \left(\sqrt{3}\right)^{13} = 2916\sqrt{3} \approx 5050.66$

So, the first term to exceed 4800 is $u_{14} = 2916\sqrt{3} \approx 5050.66$.

**c** 12, 6, 3, 1.5, .... has $u_1 = 12$ and $r = \frac{1}{2}$

$u_n = u_1 r^{n-1} \quad \therefore \quad u_n = 12 \times (\frac{1}{2})^{n-1}$

Let $0.0001 = u_n = 12 \times \left(\frac{1}{2}\right)^{n-1}$

$\therefore \quad 0.000\,008\,\overline{3} = \left(\frac{1}{2}\right)^{n-1}$

$\therefore \quad n \approx 17.87$ {using technology}

We try the two values on either side of $n \approx 17.87$, which are $n = 17$ and $n = 18$:

$u_{17} = 12 \times \left(\frac{1}{2}\right)^{16} \approx 0.000\,183\,1$ and $u_{18} = 12 \times \left(\frac{1}{2}\right)^{17} \approx 0.000\,091\,55$

So, the first term of the sequence which is less than 0.0001 is $u_{18} \approx 0.000\,091\,55$.

## EXERCISE 6D.2

**1** There is a fixed percentage increase each week, so the population forms a geometric sequence.
$u_{n+1} = u_1 \times r^n$, where $u_1 = 500$, $r = 1.12$

**a** **i** $u_{11} = 500 \times (1.12)^{10}$
$\approx 1552.92$
There will be approximately 1550 ants.

**ii** $u_{21} = 500 \times (1.12)^{20}$
$\approx 4823.15$
There will be approximately 4820 ants.

**b** For the population to reach 2000, $u_{n+1} = 500 \times (1.12)^n = 2000$
$\therefore \quad (1.12)^n = 4$
$\therefore \quad n \approx 12.23$ {using technology}

It will take approximately 12.2 weeks.

**2** $u_{n+1} = u_1 \times r^n$, where $u_1 = 555$, $r = 0.955$

**a** $u_{16} = 555 \times (0.955)^{15}$
$\approx 278.19$ The population is approximately 278 animals in the year 2010.

**b** For the population to have declined to 50,
$u_{n+1} = 555 \times (0.955)^n = 50$
$\therefore \quad (0.955)^n = 0.0\overline{900}$
$\therefore \quad n \approx 52.3$ {using technology}
So, in the 53rd year the population is 50. This is the year 2047.

**3** $u_{n+1} = u_1 \times r^n$, where $u_1 = 32$, $r = 1.18$

**a** **i** $u_6 = 32 \times (1.18)^5$
$\approx 73.21$
There will be approximately 73 deer.

**ii** $u_{11} = 32 \times (1.18)^{10}$
$\approx 167.48$
There will be approximately 167 deer.

**b** For the population to reach 5000, $u_{n+1} = 32 \times (1.18)^n = 5000$
$\therefore \quad n \approx 30.52$ {using technology}

So, it will take approximately 30.5 years.

**4** $u_{n+1} = u_1 \times r^n$, where $u_1 = 178$, $r = 1.32$

**a** **i** $u_{11} = 178 \times (1.32)^{10}$
$\approx 2858.6$
There will be approximately 2860 marsupials.

**ii** $u_{26} = 178 \times (1.32)^{25}$
$\approx 183\,979.0$
There will be approximately 184 000 marsupials.

**b** For the population to reach 10 000, $u_{n+1} = 178 \times (1.32)^n = 10\,000$
$\therefore \quad n \approx 14.5$ {using technology}

So, it will take approximately 14.5 years.

## EXERCISE 6D.3

**1** **a** $u_{n+1} = u_1 \times r^n$
where $u_1 = 3000$, $r = 1.1$, $n = 3$
$\therefore \quad u_4 = 3000 \times (1.1)^3$
$= 3993$
The investment will amount to \$3993.

**b** Interest = amount after 3 years – initial amount
= \$3993 – \$3000
= \$993

**2** $u_{n+1} = u_1 \times r^n$ where $u_1 = 20\,000$, $r = 1.12$, $n = 4$
$\therefore \quad u_5 = 20\,000 \times (1.12)^4$
$\approx 31\,470.39$
Interest = €31 470.39 – €20 000
= €11 470.39

**3** **a** $u_{n+1} = u_1 \times r^n$
where $u_1 = 30\,000$, $r = 1.1$, $n = 4$
$\therefore$ $u_5 = 30\,000 \times (1.1)^4$
$= 43\,923$

The investment amounts to ¥43 923.

**b** Interest = amount after 4 years − initial amount
= ¥43 923 − ¥30 000
= ¥13 923

**4** $u_{n+1} = u_1 \times r^n$ where $u_1 = 80\,000$, $r = 1.09$, $n = 3$
$\therefore$ $u_4 = 80\,000 \times (1.09)^3$
$= 103\,602.32$
Interest = amount after 3 years − initial amount
= \$103 602.32 − \$80 000
= \$23 602.32

**5** $u_{n+1} = u_1 \times r^n$ where $u_1 = 100\,000$, $r = 1 + \dfrac{0.08}{2} = 1.04$, $n = 10$
$\therefore$ $u_{11} = 100\,000 \times (1.04)^{10}$
$\approx 148\,024.43$ It amounts to ¥148 024.43.

**6** $u_{n+1} = u_1 \times r^n$ where $u_1 = 45\,000$, $r = 1 + \dfrac{0.075}{4} = 1.018\,75$, $n = 7$ {21 months = 7 'quarters'}
$\therefore$ $u_{10} = 45\,000 \times (1.018\,75)^7$
$\approx 51\,249.06$ It amounts to £51 249.06.

**7** The initial investment $u_1$ is unknown.
The interest rate is compounded annually, so the multiplier $r = 1 + 0.075 = 1.075$.
There are 4 compounding periods, so $n = 4$
$\therefore$ $u_{n+1} = u_5$
Now $u_5 = u_1 \times r^4$ {using $u_{n+1} = u_1 \times r^n$}
$\therefore$ $20\,000 = u_1 \times (1.075)^4$
$\therefore$ $u_1 = \dfrac{20\,000}{(1.075)^4}$
$\therefore$ $u_1 \approx 14\,976.01$ So, \$14 976.01 should be invested now.

**8** The initial investment is unknown.
The interest rate is compounded annually, so the multiplier $r = 1 + 0.055 = 1.055$.
There are $\frac{60}{12} = 5$ compounding periods, so $n = 5$
$\therefore$ $u_{n+1} = u_6$
Now $u_6 = u_1 \times r^5$ {using $u_{n+1} = u_1 \times r^n$}
$\therefore$ $15\,000 = u_1 \times (1.055)^5$
$\therefore$ $u_1 = \dfrac{15\,000}{(1.055)^5}$
$\therefore$ $u_1 \approx 11\,477.02$ The initial investment required is £11 477.02.

**9** The initial investment is unknown.
The interest rate is compounded quarterly, so the multiplier $r = 1 + \frac{0.08}{4} = 1.02$.
There are $3 \times 4 = 12$ compounding periods, so $n = 12$
$\therefore$ $u_{n+1} = u_{13}$
Now $u_{13} = u_1 \times r^{12}$ {using $u_{n+1} = u_1 \times r^n$}
$\therefore$ $25\,000 = u_1 \times (1.02)^{12}$
$\therefore$ $u_1 = \dfrac{25\,000}{(1.02)^{12}}$
$\therefore$ $u_1 \approx 19\,712.33$ I should invest €19 712.33 now.

**10** The initial investment is unknown.

The interest rate is compounded monthly, so the multiplier $r = 1 + \frac{0.09}{12} = 1.0075$.

There are $8 \times 12 = 96$ compounding periods, so $n = 96$

$\therefore\ u_{n+1} = u_{97}$

Now $u_{97} = u_1 \times r^{96}$ {using $u_{n+1} = u_1 \times r^n$}

$\therefore\ 40\,000 = u_1 \times (1.0075)^{96}$

$\therefore\ u_1 = \dfrac{40\,000}{(1.0075)^{96}}$

$\therefore\ u_1 \approx 19\,522.47$ The initial investment should be ¥19 522.47.

## EXERCISE 6E

**1** **a** **i** 3, 11, 19, 27, .... is arithmetic with $u_1 = 3$, $d = 8$, so $u_n = 3 + (n-1)8 = 8n - 5$

$S_n = 3 + 11 + 19 + 27 + .... + (8n - 5)$

**ii** $S_5 = 3 + 11 + 19 + 27 + 35 = 95$

**b** **i** 42, 37, 32, 27, .... is arithmetic with $u_1 = 42$, $d = -5$,

so $u_n = 42 + (n-1)(-5) = 47 - 5n$

$\therefore\ S_n = 42 + 37 + 32 + 27 + .... + (47 - 5n)$

**ii** $S_5 = 42 + 37 + 32 + 27 + 22 = 160$

**c** **i** 12, 6, 3, $1\frac{1}{2}$, .... is geometric with $u_1 = 12$, $r = \frac{1}{2}$, so $u_n = 12 \times (\frac{1}{2})^{n-1}$

$S_n = 12 + 6 + 3 + 1\frac{1}{2} + .... + 12(\frac{1}{2})^{n-1}$

**ii** $S_5 = 12 + 6 + 3 + 1\frac{1}{2} + \frac{3}{4} = 23\frac{1}{4}$

**d** **i** 2, 3, $4\frac{1}{2}$, $6\frac{3}{4}$, .... is geometric with $u_1 = 2$, $r = \frac{3}{2}$, so $u_n = 2 \times (\frac{3}{2})^{n-1}$

$S_n = 2 + 3 + 4\frac{1}{2} + 6\frac{3}{4} + .... + 2(\frac{3}{2})^{n-1}$

**ii** $S_5 = 2 + 3 + 4\frac{1}{2} + 6\frac{3}{4} + 10\frac{1}{8} = 26\frac{3}{8}$

**e** **i** 1, $\frac{1}{2}$, $\frac{1}{4}$, $\frac{1}{8}$, .... is geometric with $u_1 = 1$, $r = \frac{1}{2}$, so $u_n = 1 \times (\frac{1}{2})^{n-1} = \dfrac{1}{2^{n-1}}$

$S_n = 1 + \frac{1}{2} + \frac{1}{4} + \frac{1}{8} + .... + \dfrac{1}{2^{n-1}}$

**ii** $S_5 = 1 + \frac{1}{2} + \frac{1}{4} + \frac{1}{8} + \frac{1}{16} = 1\frac{15}{16}$

**f** **i** 1, 8, 27, 64, ....

$S_n = 1 + 8 + 27 + 64 + .... + n^3$ {since $1 = 1^3$, $8 = 2^3$, $27 = 3^3$, $64 = 4^3$}

**ii** $S_5 = 1 + 8 + 27 + 64 + 125 = 225$

**2** **a** $\sum_{k=1}^{3} 4k = 4 + 8 + 12 = 24$

**b** $\sum_{k=1}^{6} (k+1) = 2 + 3 + 4 + 5 + 6 + 7 = 27$

**c** $\sum_{k=1}^{4} (3k - 5) = -2 + 1 + 4 + 7 = 10$

**d** $\sum_{k=1}^{5} (11 - 2k) = 9 + 7 + 5 + 3 + 1 = 25$

**e** $\sum_{k=1}^{7} k(k+1) = 2 + 6 + 12 + 20 + 30 + 42 + 56 = 168$

**f** $\sum_{k=1}^{5} 10 \times 2^{k-1} = 10 + 20 + 40 + 80 + 160 = 310$

**3** $u_n = 3n - 1$

$\therefore \quad u_1 + u_2 + u_3 + .... + u_{20} = \sum_{n=1}^{20} (3n-1)$

$= 2 + 5 + 8 + 11 + 14 + 17 + 20 + 23 + 26 + 29 + 32 + 35 + 38 + 41 + 44 + 47 + 50 + 53 + 56 + 59$

$= 610$

**4** **a** $\sum_{k=1}^{n} c = \underbrace{c + c + c + .... + c}_{n \text{ times}} = cn$

**b** $\sum_{k=1}^{n} ca_k = ca_1 + ca_2 + .... + ca_n$

$= c(a_1 + a_2 + .... + a_n)$

$= c\sum_{k=1}^{n} a_k$

**c** $\sum_{k=1}^{n} (a_k + b_k) = (a_1 + b_1) + (a_2 + b_2) + .... + (a_n + b_n)$

$= (a_1 + a_2 + .... + a_n) + (b_1 + b_2 + .... + b_n)$

$= \sum_{k=1}^{n} a_k + \sum_{k=1}^{n} b_k$

**5** **a** $\sum_{k=1}^{n} k = 1 + 2 + 3 + .... + (n-2) + (n-1) + n$

**b**

$$\begin{array}{ccccccccccccc} 1 & + & 2 & + & 3 & + & .... & + & (n-2) & + & (n-1) & + & n \\ n & + & (n-1) & + & (n-2) & + & .... & + & 3 & + & 2 & + & 1 \\ \hline \end{array}$$

$(n+1) + (n+1) + (n+1) + \; .... \; + (n+1) + (n+1) + (n+1) = n(n+1)$

**c** $S_n = \sum_{k=1}^{n} k = 1 + 2 + 3 + .... + (n-2) + (n-1) + n$ {from **a**}

and $2[1 + 2 + 3 + .... + (n-2) + (n-1) + n] = n(n+1)$ {from **b**}

$\therefore \quad 2S_n = n(n+1)$

$\therefore \quad S_n = \dfrac{n(n+1)}{2}$

**d** $\sum_{k=1}^{n} (ak + b) = \sum_{k=1}^{n} ak + \sum_{k=1}^{n} b$

$= a\sum_{k=1}^{n} k + nb$

$= \dfrac{an(n+1)}{2} + nb$ {using **c**}

But $\sum_{k=1}^{n} (ak + b) = 8n^2 + 11n$

$\therefore \quad \dfrac{an(n+1)}{2} + nb = 8n^2 + 11n$

$\therefore \quad \dfrac{an^2 + an}{2} + nb = 8n^2 + 11n$

$\therefore \quad \dfrac{a}{2}n^2 + \dfrac{a}{2}n + nb = 8n^2 + 11n$

$\therefore \quad \dfrac{a}{2}n^2 + \left(\dfrac{a}{2} + b\right) n = 8n^2 + 11n$

Comparing coefficients, we get $\dfrac{a}{2} = 8$ and $\dfrac{a}{2} + b = 11$

$\therefore \quad a = 16 \qquad \therefore \quad 8 + b = 11$

$\therefore \quad b = 3$

## EXERCISE 6F

**1** **a** The series is arithmetic with $u_1 = 3, \quad d = 4, \quad n = 20$

$$S_n = \frac{n}{2}(2u_1 + (n-1)d)$$

So, $S_{20} = \frac{20}{2}(2 \times 3 + 19 \times 4)$
$= 10(6 + 76)$
$= 820$

**b** The series is arithmetic with $u_1 = \frac{1}{2}, \quad d = \frac{5}{2}, \quad n = 50$

$$S_n = \frac{n}{2}(2u_1 + (n-1)d)$$

So, $S_{50} = \frac{50}{2}\left(2 \times \frac{1}{2} + 49 \times \frac{5}{2}\right)$
$= 25(1 + 122\frac{1}{2})$
$= 3087\frac{1}{2}$

**c** The series is arithmetic with $u_1 = 100, \quad d = -7, \quad n = 40$

$$S_n = \frac{n}{2}(2u_1 + (n-1)d)$$

So, $S_{40} = \frac{40}{2}(2 \times 100 + 39 \times (-7))$
$= 20(200 - 273)$
$= -1460$

**d** The series is arithmetic with $u_1 = 50, \quad d = -\frac{3}{2}, \quad n = 80$

$$S_n = \frac{n}{2}(2u_1 + (n-1)d)$$

So, $S_{80} = \frac{80}{2}\left(2 \times 50 + 79 \times (-\frac{3}{2})\right)$
$= 40(100 - \frac{237}{2})$
$= -740$

**2** **a** The series is arithmetic with $u_1 = 5, \quad d = 3, \quad u_n = 101$

Since $u_n = 101$,
then $u_1 + (n-1)d = 101$
$\therefore \quad 5 + 3(n-1) = 101$
$\therefore \quad 5 + 3n - 3 = 101$
$\therefore \quad 3n = 99$
$\therefore \quad n = 33$

So, $S_n = \frac{n}{2}(u_1 + u_n)$
$= \frac{33}{2}(5 + 101)$
$= 1749$

**b** The series is arithmetic with $u_1 = 50, \quad d = -\frac{1}{2}, \quad u_n = -20$

Since $u_n = -20$,
then $u_1 + (n-1)d = -20$
$\therefore \quad 50 + (-\frac{1}{2})(n-1) = -20$
$\therefore \quad -\frac{1}{2}n + \frac{1}{2} = -70$
$\therefore \quad -\frac{1}{2}n = -\frac{141}{2}$
$\therefore \quad n = 141$

So, $S_n = \frac{n}{2}(u_1 + u_n)$
$= \frac{141}{2}(50 + (-20))$
$= 2115$

**c** The series is arithmetic with $u_1 = 8, \quad d = \frac{5}{2}, \quad u_n = 83$

Since $u_n = 83$,
then $u_1 + (n-1)d = 83$
$\therefore \quad 8 + \frac{5}{2}(n-1) = 83$
$\therefore \quad \frac{5}{2}n - \frac{5}{2} = 75$
$\therefore \quad \frac{5}{2}n = \frac{155}{2}$
$\therefore \quad n = 31$

So, $S_n = \frac{n}{2}(u_1 + u_n)$
$= \frac{31}{2}(8 + 83)$
$= 1410\frac{1}{2}$

**3** **a** $\sum_{k=1}^{10}(2k + 5) = 7 + 9 + 11 + .... + 25$

This series is arithmetic with $u_1 = 7, \quad d = 2,$ and $n = 10$.

$\therefore \quad \text{sum} = \frac{n}{2}[2u_1 + (n-1)d] = \frac{10}{2}[14 + 9 \times 2] = 160$

**b** $\sum_{k=1}^{15}(k - 50) = (-49) + (-48) + (-47) + .... + (-35)$

This series is arithmetic with $u_1 = -49, \quad d = 1,$ and $n = 15$.

$\therefore \quad \text{sum} = \frac{n}{2}[2u_1 + (n-1)d] = \frac{15}{2}[-98 + 14 \times 1] = -630$

**c** $\sum_{k=1}^{20} \left(\frac{k+3}{2}\right) = 2 + \frac{5}{2} + 3 + .... + \frac{23}{2}$

This series is arithmetic with $u_1 = 2$, $r = \frac{1}{2}$, and $n = 20$.

$\therefore$ sum $= \frac{n}{2}[2u_1 + (n-1)d] = \frac{20}{2}[4 + 19 \times \frac{1}{2}] = 135$

**4** $u_1 = 5$, $n = 7$, $u_n = 53$

$S_n = \frac{n}{2}(u_1 + u_n)$
$= \frac{7}{2}(5 + 53)$
$= 203$

**5** $u_1 = 6$, $n = 11$, $u_n = -27$

$S_n = \frac{n}{2}(u_1 + u_n)$
$= \frac{11}{2}(6 + (-27))$
$= -115\frac{1}{2}$

**6** The total number of bricks can be expressed as an arithmetic series: $1 + 2 + 3 + 4 + .... + n$
We know that the total number of bricks is 171, so $S_n = 171$.
Also, $u_1 = 1$, $d = 1$ and we need to find $n$, the number of members (layers) of the series.

$S_n = \frac{n}{2}(2u_1 + (n-1)d) = 171$

$\therefore \frac{n}{2}(2 \times 1 + (n-1) \times 1) = 171$

$\therefore n(2 + n - 1) = 342$

$\therefore n(n+1) = 342$

$\therefore n^2 + n - 342 = 0$

$\therefore (n-18)(n+19) = 0$

$\therefore n = -19$ or $18$

But $n > 0$, so $n = 18$. So, there are 18 layers placed.

**7** The total number of seats in $n$ rows can be expressed as an arithmetic series:
$22 + 23 + 24 + .... + u_n$
Row 1 has 22 seats, so $u_1 = 22$. Row 2 has 23 seats, so $d = 1$.

$S_n = \frac{n}{2}(2u_1 + (n-1)d)$
$= \frac{n}{2}(2 \times 22 + 1(n-1))$
$= \frac{n}{2}(44 + n - 1)$

$\therefore S_n = \frac{n}{2}(n + 43)$ which is the total number of seats in $n$ rows.

**a** Number of seats in row 44 = total no. of seats in every row – no. of seats in the first 43 rows
$= S_{44} - S_{43}$
$= \frac{44}{2}(44 + 43) - \frac{43}{2}(43 + 43)$
$= 1914 - 1849$
$= 65$

**b** Number of seats in a section $= S_{44} = 1914$ (from **a**)

**c** Number of seats in 25 sections $= S_{44} \times 25 = 1914 \times 25 = 47\,850$

**8** **a** The first 50 multiples of 11 can be expressed as an arithmetic series:
$11 + 22 + 33 + .... + u_{50}$ where $u_1 = 11$, $d = 11$, $n = 50$

$S_n = \frac{n}{2}(2u_1 + (n-1)d)$ $\therefore S_{50} = \frac{50}{2}(2 \times 11 + 11(50-1))$
$= 25(22 + 539)$
$= 14\,025$

**b** The multiples of 7 between 0 and 1000 can be expressed as an arithmetic series:
$7 + 14 + 21 + 28 + .... + u_n$ where $u_1 = 7$, $d = 7$
To find $u_n$, we need to find the largest multiple of 7 less than 1000.
Now $\frac{1000}{7} \approx 142.9$, so $u_n = 142 \times 7 = 994$

But $u_n = u_1 + (n-1)d$
$\therefore\ 994 = 7 + 7(n-1)$
$\therefore\ 987 = 7n - 7$
$\therefore\ 7n = 994$
$\therefore\ n = 142$

So, $S_{142} = \frac{142}{2}(7 + 994) = 71\,071$

**c** The integers between 1 and 100 which are not divisible by 3 can be expressed as:
1, 2, 4, 5, 7, 8, ...., 100 where $u_1 = 1$, $u_n = 100$.
Alternatively, these integers can be expressed as two separate arithmetic series:
$S_A = 1 + 4 + 7 + .... + 97 + 100$ where $u_1 = 1$, $d = 3$, $u_n = 100$
and $S_B = 2 + 5 + 8 + .... + 95 + 98$ where $u_1 = 2$, $d = 3$, $u_n = 98$

Now for $S_A$, $u_n = u_1 + (n-1)d$
$\therefore\ 100 = 1 + 3(n-1)$
$\therefore\ 99 = 3n - 3$
$\therefore\ 3n = 102$
$\therefore\ n = 34$

and for $S_B$, $u_n = u_1 + (n-1)d$
$\therefore\ 98 = 2 + 3(n-1)$
$\therefore\ 96 = 3n - 3$
$\therefore\ 3n = 99$
$\therefore\ n = 33$

So, $S_A = \frac{34}{2}(1 + 100) = 1717$ and $S_B = \frac{33}{2}(2 + 98) = 1650$
The total sum $= S_A + S_B = 1717 + 1650 = 3367$

**9** The series of odd numbers can be expressed as an arithmetic series:
$1 + 3 + 5 + 7 + ....$ where $u_1 = 1$, $d = 2$

**a** Now $u_n = u_1 + (n-1)d = 1 + 2(n-1)$
$\therefore\ u_n = 2n - 1$

**b** We need to show that $S_n$ is $n^2$.
The sum of the first $n$ odd numbers can be expressed as an arithmetic series:
$1 + 3 + 5 + 7 + .... + (2n-1)$ {using $u_n = 2n - 1$ from **a**}

So, $S_n = \frac{n}{2}(u_1 + u_n)$
$= \frac{n}{2}(1 + (2n-1))$
$= \frac{n}{2}(2n)$ Hence $S_n = n^2$ as required.

**c** $S_1 = 1 = 1 = 1^2 = n^2$ for $n = 1$ ✓
$S_2 = 1 + 3 = 4 = 2^2 = n^2$ for $n = 2$ ✓
$S_3 = 1 + 3 + 5 = 9 = 3^2 = n^2$ for $n = 3$ ✓
$S_4 = 1 + 3 + 5 + 7 = 16 = 4^2 = n^2$ for $n = 4$ ✓

**10** $u_6 = 21$, $S_{17} = 0$. We need to find $u_1$ and $u_2$.

$S_n = \frac{n}{2}(2u_1 + (n-1)d)$
$\therefore\ S_{17} = \frac{17}{2}(2u_1 + 16d) = 0$
$\therefore\ u_1 + 8d = 0$
$\therefore\ u_1 = -8d$ .... (1)

Also, $u_n = u_1 + (n-1)d$
$\therefore\ u_6 = u_1 + 5d$
$\therefore\ 21 = -8d + 5d$ {using (1)}
$\therefore\ 21 = -3d$
$\therefore\ d = -7$

So, $u_1 = -8(-7) = 56$ and $u_2 = 56 - 7 = 49$
The first two terms are 56 and 49.

**11** Let the three consecutive terms be $x-d,\ x,$ and $x+d$.

Now, sum of terms $= 12$

$\therefore\ (x-d)+x+(x+d)=12$

$\therefore\ 3x=12$

$\therefore\ x=4$

So, the terms are $4-d,\ 4,\ 4+d$

Also, product of terms $= -80$

$\therefore\ (4-d)4(4+d)=-80$

$\therefore\ 4(4^2-d^2)=-80$

$\therefore\ 16-d^2=-20$

$\therefore\ d^2=36 \quad \therefore\ d=\pm 6$

So, the three terms could be $4-6,\ 4,\ 4+6,$ which are $-2,\ 4,\ 10$

*or* $4-(-6),\ 4,\ 4+(-6),$ which are $10,\ 4,\ -2$.

**12** Let the five consecutive terms be $n-2d,\ n-d,\ n,\ n+d,\ n+2d$.

Now, sum of terms $= 40$ $\quad\therefore\ (n-2d)+(n-d)+n+(n+d)+(n+2d)=40$

$\therefore\ 5n=40$

$\therefore\ n=8$

So the terms are $8-2d,\ 8-d,\ 8,\ 8+d,\ 8+2d$

Also, the product of the first, middle and last terms $=(8-2d)\times 8\times(8+2d)=224$

$\therefore\ 8(8^2-4d^2)=224$

$\therefore\ 64-4d^2=28$

$\therefore\ 4d^2=36$

$\therefore\ d^2=9$

$\therefore\ d=\pm 3$

So, the five terms could be $8-2(3),\ 8-3,\ 8,\ 8+3,\ 8+2(3),$ which are $2, 5, 8, 11, 14$

*or* $8-2(-3),\ 8-(-3),\ 8,\ 8+(-3),\ 8+2(-3),$ which are $14, 11, 8, 5, 2$.

## EXERCISE 6G.1

**1** **a** The series is geometric with $u_1=12,\quad r=\frac{1}{2},\quad n=10$

Now $S_n=\dfrac{u_1(1-r^n)}{1-r}$

$\therefore\ S_{10}=\dfrac{12\left(1-(\frac{1}{2})^{10}\right)}{1-\frac{1}{2}}$

$\approx 23.9766 \approx 24.0$

**b** The series is geometric with $u_1=\sqrt{7},\quad r=\sqrt{7},\quad n=12$

Now $S_n=\dfrac{u_1(r^n-1)}{r-1}$

$\therefore\ S_{12}=\dfrac{\sqrt{7}\left((\sqrt{7})^{12}-1\right)}{\sqrt{7}-1}$

$\approx 189\,134$

**c** The series is geometric with $u_1=6,\quad r=-\frac{1}{2},\quad n=15$

Now $S_n=\dfrac{u_1(1-r^n)}{1-r}$

$\therefore\ S_{15}=\dfrac{6\left(1-(-\frac{1}{2})^{15}\right)}{1-(-\frac{1}{2})}\approx 4.000$

**d** The series is geometric with $u_1=1,\quad r=-\frac{1}{\sqrt{2}},\quad n=20$

Now $S_n=\dfrac{u_1(1-r^n)}{1-r}$

$\therefore\ S_{20}=\dfrac{1\left(1-(-\frac{1}{\sqrt{2}})^{20}\right)}{1-(-\frac{1}{\sqrt{2}})}\approx 0.5852$

**2** **a** The series is geometric with $u_1=\sqrt{3},\ r=\sqrt{3}$

$S_n=\dfrac{u_1(r^n-1)}{r-1}$

$=\dfrac{\sqrt{3}\left((\sqrt{3})^n-1\right)}{\sqrt{3}-1}\times\left(\dfrac{\sqrt{3}+1}{\sqrt{3}+1}\right)$

$=\dfrac{\left(3+\sqrt{3}\right)\left((\sqrt{3})^n-1\right)}{3-1}$

$=\dfrac{3+\sqrt{3}}{2}\left((\sqrt{3})^n-1\right)$

**b** The series is geometric with $u_1=12,\ r=\frac{1}{2}$

$S_n=\dfrac{u_1(1-r^n)}{1-r}$

$=\dfrac{12\left(1-(\frac{1}{2})^n\right)}{1-\frac{1}{2}}$

$=24\left(1-(\frac{1}{2})^n\right)$

**c** The series is geometric with
$u_1 = 0.9, \quad r = 0.1$

$$S_n = \frac{u_1(1-r^n)}{1-r} = \frac{0.9\,(1-(0.1)^n)}{1-0.1} = 1-(0.1)^n$$

**d** The series is geometric with
$u_1 = 20, \quad r = -\frac{1}{2}$

$$S_n = \frac{u_1(1-r^n)}{1-r} = \frac{20\left(1-(-\frac{1}{2})^n\right)}{1-(-\frac{1}{2})} = \frac{20\left(1-(-\frac{1}{2})^n\right)}{(\frac{3}{2})} = \frac{40}{3}\left(1-(-\frac{1}{2})^n\right)$$

**3** **a** $S_1 = u_1 \quad \therefore \quad u_1 = 3$

**b** $u_2 = S_2 - S_1 = 4 - 3 = 1$
So, $r = \frac{1}{3}$

**c** $u_1 = 3, \quad r = \frac{1}{3}$
so $u_n = 3 \times (\frac{1}{3})^{n-1}$
$\therefore \quad u_5 = 3 \times (\frac{1}{3})^4 = \frac{1}{27}$

**4** **a** $\sum_{k=1}^{10} 3 \times 2^{k-1} = 3 + 6 + 12 + .... + 384 + 768 + 1536$

This series is geometric with $u_1 = 3, \quad r = 2,$ and $n = 10.$

$\therefore \quad \text{sum} = \frac{u_1(r^n-1)}{r-1} = \frac{3(2^{10}-1)}{1} = 3069$

**b** $\sum_{k=1}^{12} \left(\frac{1}{2}\right)^{k-2} = 2 + 1 + \frac{1}{2} + .... + \frac{1}{256} + \frac{1}{512} + \frac{1}{1024}$

This series is geometric with $u_1 = 2, \quad r = \frac{1}{2},$ and $n = 12.$

$\therefore \quad \text{sum} = \frac{u_1(1-r^n)}{1-r} = \frac{2\left(1-(\frac{1}{2})^{12}\right)}{\frac{1}{2}} = 4\left(1-\left(\frac{1}{2}\right)^{12}\right) = \frac{2^{12}-1}{2^{10}}$

$\therefore \quad \text{sum} = \frac{4095}{1024} \approx 4.00$

**c** $\sum_{k=1}^{25} 6 \times (-2)^k = -12 + 24 + (-48) + .... + 100\,663\,296 + (-201\,326\,592)$

This series is geometric with $u_1 = -12, \quad r = -2,$ and $n = 25.$

$\therefore \quad \text{sum} = \frac{u_1(1-r^n)}{1-r} = \frac{-12\left(1-(-2)^{25}\right)}{1+2} = -4\left(1-(-2)^{25}\right)$

$\therefore \quad \text{sum} = -134\,217\,732$

**5** **a**
$$\begin{aligned} A_2 &= A_1 \times 1.06 + 2000 \\ &= (A_0 \times 1.06 + 2000) \times 1.06 + 2000 \\ &= (2000 \times 1.06 + 2000) \times 1.06 + 2000 \end{aligned}$$
$\therefore \quad A_2 = 2000 + 2000 \times 1.06 + 2000 \times (1.06)^2$ as required

**b**
$$\begin{aligned} A_3 &= A_2 \times 1.06 + 2000 \\ &= \left[2000 + 2000 \times 1.06 + 2000 \times (1.06)^2\right] \times 1.06 + 2000 \quad \{\text{from } \mathbf{a}\} \end{aligned}$$
$\therefore \quad A_3 = 2000\left[1 + 1.06 + (1.06)^2 + (1.06)^3\right]$ as required

**c**
$$A_9 = 2000[1 + 1.06 + (1.06)^2 + (1.06)^3 + (1.06)^4 + (1.06)^5 + (1.06)^6 + (1.06)^7 + (1.06)^8 + (1.06)^9]$$
$\therefore \quad A_9 \approx 26\,361.59$

$\therefore$ the total bank balance after 10 years is $26 361.59

**6** **a** $S_1 = \frac{1}{2}, \quad S_2 = \frac{1}{2} + \frac{1}{4} = \frac{3}{4}, \quad S_3 = \frac{1}{2} + \frac{1}{4} + \frac{1}{8} = \frac{7}{8}, \quad S_4 = \frac{1}{2} + \frac{1}{4} + \frac{1}{8} + \frac{1}{16} = \frac{15}{16},$

$S_5 = \frac{1}{2} + \frac{1}{4} + \frac{1}{8} + \frac{1}{16} + \frac{1}{32} = \frac{31}{32}$

**b** $S_n = \dfrac{2^n - 1}{2^n}$

**c** $S_n = \dfrac{u_1(1-r^n)}{1-r}$, where $u_1 = \frac{1}{2}$, $r = \frac{1}{2}$

$= \dfrac{\frac{1}{2}\left(1 - (\frac{1}{2})^n\right)}{1 - \frac{1}{2}}$

$\therefore \quad S_n = 1 - (\frac{1}{2})^n = 1 - \dfrac{1}{2^n} = \dfrac{2^n - 1}{2^n}$

**d** As $n \to \infty$, $\left(\frac{1}{2}\right)^n \to 0$, and so $S_n \to 1$ (from below)

**e** The diagram represents one whole unit divided into smaller and smaller fractions.
As $n \to \infty$, the area which the fraction represents becomes smaller and smaller, and the total area approaches the area of a $1 \times 1$ unit square.

## EXERCISE 6G.2

**1** **a** **i** $u_1 = \frac{3}{10}$

**ii** $r = \dfrac{(\frac{3}{100})}{(\frac{3}{10})} = \frac{1}{10}$
$= 0.1$

**b** We need to show that $0.\overline{3} = \frac{1}{3}$.

Now $0.\overline{3} = \frac{3}{10} + \frac{3}{100} + \frac{3}{1000} + ....$

So, let $S_n = \frac{3}{10} + \frac{3}{100} + \frac{3}{1000} + ....$

Since $n \to \infty$, then $S = \dfrac{u_1}{1-r} = \dfrac{\frac{3}{10}}{1 - (\frac{1}{10})} = \frac{1}{3}$

So, $0.\overline{3} = \frac{1}{3}$ as required.

**2** **a** $0.\overline{4} = 0.444444....$
$= \frac{4}{10} + \frac{4}{100} + \frac{4}{1000} + ....$
which is a geometric series with
$u_1 = 0.4$, $r = 0.1$

$\therefore \quad S = \dfrac{u_1}{1-r} = \dfrac{0.4}{1-0.1} = \dfrac{0.4}{0.9}$
$= \frac{4}{9}$

So, $0.\overline{4} = \frac{4}{9}$

**b** $0.\overline{16} = 0.161616....$
$= \frac{16}{10^2} + \frac{16}{10^4} + \frac{16}{10^6} + ....$
which is a geometric series with
$u_1 = 0.16$, $r = 0.01$

$\therefore \quad S = \dfrac{u_1}{1-r} = \dfrac{0.16}{0.99} = \dfrac{16}{99}$

So, $0.\overline{16} = \frac{16}{99}$

**c** $0.\overline{312} = 0.312\,312\,312\,....$
$= \frac{312}{10^3} + \frac{312}{10^6} + \frac{312}{10^9} + ....$
which is a geometric series with $u_1 = 0.312$, $r = 0.001$

$\therefore \quad S = \dfrac{u_1}{1-r} = \dfrac{0.312}{0.999} = \dfrac{312}{999} = \dfrac{104}{333}$ So, $0.\overline{312} = \frac{104}{333}$

**3** Checking **Exercise 6G.1 6d**: $S = \dfrac{u_1}{1-r} = \dfrac{\frac{1}{2}}{1 - \frac{1}{2}} = 1$ ✓

**4** **a** $18 + 12 + 8 + ....$ is an infinite geometric series with $u_1 = 18$, $r = \frac{2}{3}$.

$\therefore \quad S = \dfrac{u_1}{1-r} = \dfrac{18}{\frac{1}{3}} = 54$

**b** $18.9 - 6.3 + 2.1 - ....$ is an infinite geometric series with $u_1 = 18.9$, $r = -\frac{1}{3}$.

$\therefore \quad S = \dfrac{u_1}{1-r} = \dfrac{18.9}{\frac{4}{3}} = 14.175$

**5** **a** $\displaystyle\sum_{k=1}^{\infty} \frac{3}{4^k} = \frac{3}{4} + \frac{3}{16} + \frac{3}{64} + ....$
is an infinite geometric series with
$u_1 = \frac{3}{4}$, $r = \frac{1}{4}$.

$\therefore \quad S = \dfrac{u_1}{1-r} = \dfrac{\frac{3}{4}}{\frac{3}{4}} = 1$

**b** $\displaystyle\sum_{k=0}^{\infty} 6(-\tfrac{2}{5})^k = 6 - 6 \times (\tfrac{2}{5}) + 6 \times (\tfrac{2}{5})^2 - ....$
is an infinite geometric series with
$u_1 = 6$, $r = -\frac{2}{5}$.

$\therefore \quad S = \dfrac{u_1}{1-r} = \dfrac{6}{\frac{7}{5}} = \frac{30}{7} \quad (\approx 4\frac{2}{7})$

**6** Let the terms of the geometric series be $u_1,\ u_1r,\ u_1r^2,\ ....$

Then $u_1 + u_1r + u_1r^2 = 19$

$\therefore\ u_1(1+r+r^2) = 19$

$\therefore\ u_1 = \dfrac{19}{1+r+r^2}$ .... (1)

and $\dfrac{u_1}{1-r} = 27$

$\therefore\ u_1 = 27(1-r)$ .... (2)

Equating (1) and (2), $\dfrac{19}{1+r+r^2} = 27(1-r)$

$\therefore\ \frac{19}{27} = (1-r)(1+r+r^2)$

$\therefore\ \frac{19}{27} = 1+r+r^2-r-r^2-r^3$

$\therefore\ \frac{19}{27} = 1-r^3$

$\therefore\ r^3 = \frac{8}{27}$

$\therefore\ r = \frac{2}{3}$

Substituting $r = \frac{2}{3}$ into (2) gives $u_1 = 27(1-\frac{2}{3}) = 9$

$\therefore$ the first term is 9 and the common ratio is $\frac{2}{3}$.

**7** Let the terms of the geometric series be $u_1,\ u_1r,\ u_1r^2,\ ....$

Then $u_1r = \frac{8}{5}$

$\therefore\ u_1 = \dfrac{8}{5r}$ .... (1)

and $\dfrac{u_1}{1-r} = 10$

$\therefore\ u_1 = 10 - 10r$ .... (2)

Equating (1) and (2), $\dfrac{8}{5r} = 10 - 10r$

$\therefore\ 8 = 50r - 50r^2$

$\therefore\ 50r^2 - 50r + 8 = 0$

$\therefore\ 2(25r^2 - 25r + 4) = 0$

$\therefore\ 2(5r-1)(5r-4) = 0$

$\therefore\ r = \frac{1}{5}$ or $\frac{4}{5}$

Using (2), if $r = \frac{1}{5}$, $u_1 = 10 - 10(\frac{1}{5}) = 8$

if $r = \frac{4}{5}$, $u_1 = 10 - 10(\frac{4}{5}) = 2$

$\therefore$ either $u_1 = 8,\ r = \frac{1}{5}$ or $u_1 = 2,\ r = \frac{4}{5}$

**8** **a** Total time of motion $= 1 + (90\% \times 1) + (90\% \times 1) + (90\% \times 90\% \times 1)$
$+ (90\% \times 90\% \times 1) + (90\% \times 90\% \times 90\% \times 1) + ....$
$= 1 + 0.9 + 0.9 + (0.9)^2 + (0.9)^2 + (0.9)^3 + ....$
$= 1 + 2(0.9) + 2(0.9)^2 + 2(0.9)^3 + ....$ as required

**b** The total time of motion can be written as $[2 + 2(0.9) + 2(0.9)^2 + 2(0.9)^3 + ....] - 1$

So, $S_n = \dfrac{u_1(1-r^n)}{1-r} - 1$, where $u_1 = 2,\ r = 0.9$

$\therefore\ S_n = \dfrac{2(1-0.9^n)}{1-0.9} - 1$

$\therefore\ S_n = \dfrac{2(1-0.9^n)}{0.1} - 1$

$\therefore\ S_n = 20(1-0.9^n) - 1$

$\therefore\ S_n = 20 - 20 \times 0.9^n - 1$

$\therefore\ S_n = 19 - 20 \times 0.9^n$

**c** For the ball to come to rest, $n$ must approach infinity.

As $n \to \infty$, $0.9^n \to 0$ and so $20 \times 0.9^n \to 0$ also.

$\therefore\ S_n \to 19^-$.

So, it takes 19 seconds for the ball to come to rest.

## REVIEW SET 6A

**1** **a** arithmetic, $d=-8$ **b** geometric, $r=1$ or arithmetic, $d=0$

**c** geometric, $r=-\frac{1}{2}$ **d** neither **e** 4, 8, 12, 16, .... arithmetic, $d=4$

**2** Since the terms are consecutive,

$$(k-2)-3k = k+7-(k-2) \quad \{\text{equating common differences}\}$$
$$\therefore\ k-2-3k = k+7-k+2$$
$$\therefore\ -2-2k = 9$$
$$\therefore\ 2k = -11$$
$$\therefore\ k = -\tfrac{11}{2}$$

**3** 28, 23, 18, 13, ....

$23-28=-5$
$18-23=-5$
$13-18=-5$

Assuming that the pattern continues, consecutive terms differ by $-5$.
$\therefore$ the sequence is arithmetic with $u_1=28$, $d=-5$.

$$\begin{aligned} u_n &= u_1+(n-1)d \\ &= 28+(n-1)(-5) \\ &= 28-5n+5 \\ &= 33-5n \end{aligned}$$

$$\begin{aligned} S_n &= \frac{n}{2}(2u_1+(n-1)d) \\ &= \frac{n}{2}(2\times 28+(n-1)(-5)) \\ &= \frac{n}{2}(56-5n+5) \\ &= \frac{n}{2}(61-5n) \end{aligned}$$

**4** The terms are geometric, so $\dfrac{k}{4}=\dfrac{k^2-1}{k}$

$$\therefore\ k^2=4(k^2-1)$$
$$\therefore\ 3k^2=4$$
$$\therefore\ k^2=\tfrac{4}{3}$$
$$\therefore\ k=\pm\tfrac{2}{\sqrt{3}}=\pm\tfrac{2\sqrt{3}}{3}$$

**5** $u_6=\frac{16}{3}$ $\quad\therefore\ u_1\times r^5=\frac{16}{3}$ .... (1)

$u_{10}=\frac{256}{3}$ $\quad\therefore\ u_1\times r^9=\frac{256}{3}$ .... (2)

So, $\dfrac{u_1r^9}{u_1r^5}=\dfrac{(\frac{256}{3})}{(\frac{16}{3})}$ $\quad\{(2)\div(1)\}$

$$\therefore\ r^4=16$$
$$\therefore\ r=\pm 2$$

Substituting $r=2$ into (1) gives

$$u_1\times 2^5=\tfrac{16}{3}$$
$$\therefore\ u_1\times 32=\tfrac{16}{3}$$
$$\therefore\ u_1=\tfrac{1}{6}$$

Substituting $r=-2$ into (1) gives

$$u_1\times(-2)^5=\tfrac{16}{3}$$
$$\therefore\ u_1\times(-32)=\tfrac{16}{3}$$
$$\therefore\ u_1=-\tfrac{1}{6}$$

Now $u_n=u_1r^{n-1}$ $\quad\therefore\ u_n=\frac{1}{6}\times 2^{n-1}$ or $-\frac{1}{6}\times(-2)^{n-1}$

**6** Let the numbers be $23$, $23+d$, $23+2d$, $23+3d$, $23+4d$, $23+5d$, $23+6d$, $9$

Then $23+7d=9$

$$\therefore\ 7d=-14$$
$$\therefore\ d=-2$$

So, the numbers are 23, 21, 19, 17, 15, 13, 11, 9.

**7** **a** The sequence 86, 83, 80, 77, .... is arithmetic with $u_1=86$, $d=-3$.

$$u_n=u_1+(n-1)d$$
$$\therefore\ u_n=86+(n-1)(-3)=86-3n+3$$
$$\therefore\ u_n=89-3n$$

**b** $\frac{3}{4}, 1, \frac{7}{6}, \frac{9}{7}, ....$ can also be written as $\frac{3}{4}, \frac{5}{5}, \frac{7}{6}, \frac{9}{7}, ....$

So, the numerator starts at 3 and increases by 2 each time, whilst the denominator starts at 4 and increases by 1 each time.

The $n$th term is $\dfrac{2n+1}{n+3}$, and so $u_n = \dfrac{2n+1}{n+3}$

**c** The sequence 100, 90, 81, 72.9, .... is geometric with $u_1 = 100$, $r = \frac{90}{100} = 0.9$

$u_n = u_1 r^{n-1}$

$\therefore \quad u_n = 100(0.9)^{n-1}$

**8** **a** $\displaystyle\sum_{k=1}^{7} k^2 = 1^2 + 2^2 + 3^2 + 4^2 + 5^2 + 6^2 + 7^2$
$= 1 + 4 + 9 + 16 + 25 + 36 + 49$
$= 140$

**b** $\displaystyle\sum_{k=1}^{8} \frac{k+3}{k+2} = \frac{4}{3} + \frac{5}{4} + \frac{6}{5} + \frac{7}{6} + \frac{8}{7} + \frac{9}{8} + \frac{10}{9} + \frac{11}{10}$
$= \dfrac{23\,761}{2520}$
$\approx 9.43$

**9** **a** $18 - 12 + 8 - ....$

The series is geometric with

$u_1 = 18$, $r = -\frac{2}{3}$

$\therefore \quad S = \dfrac{u_1}{1-r}$
$= \dfrac{18}{\frac{5}{3}}$
$= \frac{54}{5}$ or $10\frac{4}{5}$

**b** $8 + 4\sqrt{2} + 4 + ....$

The series is geometric with

$u_1 = 8$, $r = \frac{1}{\sqrt{2}}$

$\therefore \quad S = \dfrac{u_1}{1-r} = \dfrac{8}{(1-\frac{1}{\sqrt{2}})} \times \dfrac{(1+\frac{1}{\sqrt{2}})}{(1+\frac{1}{\sqrt{2}})}$
$= \dfrac{8 + \frac{8}{\sqrt{2}}}{1 - \frac{1}{2}}$
$= \dfrac{8 + 4\sqrt{2}}{\frac{1}{2}}$
$= 16 + 8\sqrt{2}$

**10** 3 m

Total distance travelled

$= 3 + 3 \times 0.8 \times 2 + 3 \times (0.8)^2 \times 2 + 3 \times (0.8)^3 \times 2 + ....$
$= 3 + 3 \times 0.8 \times 2\left[1 + 0.8 + (0.8)^2 + (0.8)^3 + ....\right]$
$= 3 + 4.8 \times \dfrac{1}{1 - 0.8}$ $\left\{\text{as } r = 0.8, \ |r| < 1 \text{ so converges to } \dfrac{u_1}{1-r}\right\}$
$= 3 + \frac{4.8}{0.2}$
$= 3 + 24 = 27$ metres

**11** **a** $S_n = \dfrac{3n^2 + 5n}{2}$

$\therefore \quad u_n = S_n - S_{n-1}$
$= \dfrac{3n^2+5n}{2} - \dfrac{3(n-1)^2 + 5(n-1)}{2}$
$= \dfrac{3n^2 + 5n - 3(n^2 - 2n + 1) - 5(n-1)}{2}$
$= \dfrac{3n^2 + 5n - 3n^2 + 6n - 3 - 5n + 5}{2}$
$= \dfrac{6n+2}{2}$

$\therefore \quad u_n = 3n + 1$

**b** Using part **a**,

$u_n - u_{n-1} = [3n+1] - [3(n-1)+1]$
$= 3n + 1 - 3n + 3 - 1$
$= 3$

The difference between consecutive terms is constant for all $n$, so the sequence is arithmetic.

## REVIEW SET 6B

**1** $u_n = 6\left(\frac{1}{2}\right)^{n-1}$

**a** $\dfrac{u_{n+1}}{u_n} = \dfrac{6\left(\frac{1}{2}\right)^{n+1-1}}{6\left(\frac{1}{2}\right)^{n-1}} = \frac{1}{2}$ for all $n$

$\therefore$ $\{u_n\}$ is a geometric sequence.

**b** $u_1 = 6,$
$r = \frac{1}{2}$

**c** $u_{16} = 6\left(\frac{1}{2}\right)^{15}$
$\approx 0.000\,183$

**2** **a** Given $24, 23\frac{1}{4}, 22\frac{1}{2}, ....., -36$ we have $u_1 = 24,$ $u_n = -36,$ and we need to find $n$.

The sequence is arithmetic with $d = -\frac{3}{4}$.

Now $u_n = u_1 + (n-1)d$

$\therefore$ $-36 = 24 + (n-1)(-\frac{3}{4})$

$\therefore$ $-60 = -\frac{3}{4}n + \frac{3}{4}$

$\therefore$ $\frac{3}{4}n = \frac{243}{4}$

$\therefore$ $n = 81$ So, $-36$ is the 81st term in the sequence.

**b** $u_{35} = 24 + (35-1)(-\frac{3}{4})$
$= 24 - \frac{102}{4}$
$= -\frac{3}{2}$ $(= -1\frac{1}{2})$

**c** $S_n = \dfrac{n}{2}(2u_1 + (n-1)d)$

$\therefore$ $S_{40} = \frac{40}{2}(2 \times 24 + (40-1)(-\frac{3}{4}))$
$= 20(48 - \frac{117}{4})$
$= 375$

**3** **a** $3 + 9 + 15 + 21 + ....$
The series is arithmetic with
$u_1 = 3,$ $d = 6,$ $n = 23$

Now $S_n = \dfrac{n}{2}(2u_1 + (n-1)d)$

$\therefore$ $S_{23} = \frac{23}{2}(2 \times 3 + 6(23-1))$

$\therefore$ $S_{23} = \frac{23}{2}(6 + 132)$
$= 1587$

**b** $24 + 12 + 6 + 3 + ....$
The series is geometric with
$u_1 = 24,$ $r = \frac{1}{2},$ $n = 12$

$S_n = \dfrac{u_1(1 - r^n)}{1 - r}$

$\therefore$ $S_{12} = \dfrac{24\left(1 - (\frac{1}{2})^{12}\right)}{1 - \frac{1}{2}}$
$= 48\left(1 - (\frac{1}{2})^{12}\right)$
$= 47\frac{253}{256} \approx 48.0$

**4** $5, 10, 20, 40, ....$ The sequence is geometric with $u_1 = 5,$ $r = 2$

$u_n = u_1 r^{n-1} = 5 \times 2^{n-1}$

Let $u_n = 10\,000 = 5 \times 2^{n-1}$

$\therefore$ $2000 = 2^{n-1}$

$\therefore$ $n \approx 11.97$ {using technology}

We try the two values on either side of $n \approx 11.97$, which are $n = 11$ and $n = 12$:

$u_{11} = 5 \times 2^{10}$ and $u_{12} = 5 \times 2^{11}$
$= 5120$ $= 10\,240$

So, the first term to exceed 10 000 is $u_{12} = 10\,240$.

**5** **a** $u_6 = u_1 \times r^5$ is the amount after 5 years, where $u_1 = 6000,$ $r = 1.07$
$= 6000 \times (1.07)^5$
$\approx 8415.31$ So, the value of the investment will be €8415.31.

**b** If interest is compounded quarterly, then $r = 1 + \dfrac{0.07}{4} = 1.0175$

and $n = 5 \times 4 = 20$

$$\begin{aligned} \therefore \quad u_{21} &= u_1 \times r^{20} \\ &= 6000 \times (1.0175)^{20} \\ &\approx 8488.67 \end{aligned}$$

So, the value of the investment will be €8488.67.

**c** If interest is compounded monthly, then $r = 1 + \dfrac{0.07}{12} = 1.005\,8\bar{3}$

and $n = 5 \times 12 = 60$

$$\begin{aligned} \therefore \quad u_{61} &= u_1 \times r^{60} \\ &= 6000 \times (1.005\,8\bar{3})^{60} \\ &\approx 8505.75 \end{aligned}$$

So, the value of the investment will be €8505.75

**6** **a** $u_n = 5n - 8$

$\therefore \; u_{10} = 5 \times 10 - 8 = 42$

**b** $$\begin{aligned} u_{n+1} - u_n &= (5(n+1) - 8) - (5n - 8) \\ &= 5n + 5 - 8 - 5n + 8 \\ &= 5 \end{aligned}$$

**c** The difference between consecutive terms $u_n$ and $u_{n+1}$ is constant for all $n$, so the sequence is arithmetic.

**d** $S_n = \dfrac{n}{2}\left(2u_1 + (n-1)d\right)$ where $d = 5$ and $u_1 = 5 \times 1 - 8 = -3$

Now, $$\begin{aligned} u_{15} + u_{16} + u_{17} + .... + u_{30} &= S_{30} - S_{14} \\ &= \tfrac{30}{2}\left(2(-3) + (30-1) \times 5\right) - \tfrac{14}{2}\left(2(-3) + (14-1) \times 5\right) \\ &= 1672 \end{aligned}$$

**7** $u_6 = 24 \qquad \therefore \; u_1 \times r^5 = 24 \quad .... (1)$

$u_{11} = 768 \qquad \therefore \; u_1 \times r^{10} = 768 \quad .... (2)$

So $\dfrac{u_1 r^{10}}{u_1 r^5} = \dfrac{768}{24}$ $\{(2) \div (1)\}$

$\therefore \; r^5 = 32$

$\therefore \; r = 2$

Substituting $r = 2$ into (1) gives $u_1 \times 2^5 = 24$

$\therefore \; u_1 = \frac{24}{32} = \frac{3}{4}$

$u_n = u_1 r^{n-1} = \left(\frac{3}{4}\right) 2^{n-1}$

**a** $$\begin{aligned} u_{17} &= \left(\tfrac{3}{4}\right) 2^{17-1} \\ &= 49\,152 \end{aligned}$$

**b** $$\begin{aligned} S_n &= \frac{u_1(r^n - 1)}{r - 1} = \frac{\frac{3}{4}(2^n - 1)}{2 - 1} \\ &= \tfrac{3}{4}(2^n - 1) \end{aligned}$$

$\therefore \; S_{15} = \frac{3}{4}(2^{15} - 1) = 24\,575.25$

**8** 24, 8, $\frac{8}{3}$, $\frac{8}{9}$, .... is geometric with $u_1 = 24$, $r = \frac{1}{3}$

$u_n = u_1 r^{n-1} = 24\left(\frac{1}{3}\right)^{n-1}$

Given $u_n = 0.001$, we need to find $n$, so $u_n = 24\left(\frac{1}{3}\right)^{n-1} = 0.001$

$\therefore \; \left(\frac{1}{3}\right)^{n-1} = \dfrac{0.001}{24}$

$\therefore \; n \approx 10.18$ {using technology}

We try the two values on either side of $n \approx 10.18$, which are $n = 10$ and $n = 11$:

$$u_{10} = 24\left(\tfrac{1}{3}\right)^9 = \frac{8}{6561} \approx 0.001\,22 \qquad \text{and} \qquad u_{11} = 24\left(\tfrac{1}{3}\right)^{10} = \frac{8}{19\,683} \approx 0.000\,406$$

$\therefore \; u_{11} \approx 0.000\,406$ is the first term of the sequence which is less than $0.001$.

**9** **a** 128, 64, 32, 16, ...., $\frac{1}{512}$ is geometric with:
$u_1 = 128$, $r = \frac{1}{2}$, $u_n = \frac{1}{512}$

$$u_n = u_1 r^{n-1} = 128\left(\tfrac{1}{2}\right)^{n-1} = 2^7 \times 2^{1-n}$$

$\therefore \frac{1}{512} = 2^7 \times 2^{1-n}$

$\therefore 2^{-9} = 2^{8-n}$

$\therefore -9 = 8 - n$

$\therefore n = 17$ So, there are 17 terms in the sequence.

**b** $S_n = \dfrac{u_1(1-r^n)}{1-r}$

$\therefore S_{17} = \dfrac{128\left(1-(\frac{1}{2})^{17}\right)}{1-\frac{1}{2}} = 255\frac{511}{512} \approx 255.998 \approx 256$

**10** **a** $1.21 - 1.1 + 1 - ....$ is an infinite geometric series with $u_1 = 1.21$, $r = -\frac{10}{11}$.

$\therefore S = \dfrac{u_1}{1-r} = \dfrac{1.21}{\frac{21}{11}} = \frac{1331}{2100}$

$\therefore S \approx 0.634$

**b** $\frac{14}{3} + \frac{4}{3} + \frac{8}{21} + ....$ is an infinite geometric series with $u_1 = \frac{14}{3}$, $r = \frac{2}{7}$.

$\therefore S = \dfrac{u_1}{1-r} = \dfrac{\frac{14}{3}}{\frac{5}{7}}$

$\therefore S = \frac{98}{15}$ or $6\frac{8}{15}$

**11** $u_{n+1} = u_1 \times r^n$ where $u_{n+1} = 20\,000$, $r = 1 + \dfrac{0.09}{12} = 1.0075$, $n = 4 \times 12 = 48$

$\therefore 20\,000 = u_1 \times (1.0075)^{48}$

$\therefore u_1 = \dfrac{20\,000}{(1.0075)^{48}}$

$\therefore u_1 \approx 13\,972.28$ So, \$13 972.28 should be invested.

**12** **a** $u_{n+1} = u_1 \times r^n$ where $u_1 = 3000$, $r = 1.05$, $n = 3$

$\therefore u_{n+1} = 3000 \times (1.05)^3 = 3472.875$ There were approximately 3470 iguanas.

**b** $u_{n+1} = u_1 \times r^n$ where $u_1 = 3000$, $u_{n+1} = 5000$, $r = 1.05$

$\therefore 10\,000 = 3000 \times (1.05)^n$

$\therefore n \approx 24.68$ {using technology}

After 24.68 years the population will exceed 10 000. This is during the year 2029.

## REVIEW SET 6C

**1** $u_n = 68 - 5n$

**a** $u_{n+1} - u_n = [68 - 5(n+1)] - [68 - 5n] = 68 - 5n - 5 - 68 + 5n = -5$ for all $n$

$\therefore$ the sequence is arithmetic with common difference $d = -5$.

**b** $u_1 = 68 - 5(1) = 63$, $d = -5$

**c** $u_{37} = 68 - 5(37) = -117$

**d** Let $u_n = -200$, and we need to find $n$.

$u_n = 68 - 5n = -200$

$\therefore 5n = 268$

$\therefore n = 53\frac{3}{5}$

We try the two values on either side of $n = 53\frac{3}{5}$, which are $n = 53$ and $n = 54$:

$u_{53} = 68 - 5(53) = -197$ and $u_{54} = 68 - 5(54) = -202$

So, the first term of the sequence less than $-200$ is $u_{54} = -202$.

**2** **a** 3, 12, 48, 192, ....

$\frac{12}{3} = 4 \quad \frac{48}{12} = 4 \quad \frac{192}{48} = 4$

Assuming the pattern continues, consecutive terms have a common ratio of 4.

$\therefore$ the sequence is geometric with $u_1 = 3$ and $r = 4$.

**b** $u_n = u_1 r^{n-1}$

$\therefore \quad u_n = 3 \times 4^{n-1}$

$\therefore \quad u_9 = 3 \times 4^8 = 196\,608$

**3** $u_7 = 31 \quad \therefore \quad u_1 + 6d = 31 \quad .... (1)$

$u_{15} = -17 \quad \therefore \quad u_1 + 14d = -17 \quad .... (2)$

So, $(u_1 + 14d) - (u_1 + 6d) = -17 - 31 \quad \{(2) - (1)\}$

$\therefore \quad 8d = -48$

$\therefore \quad d = -6$

So in (1), $u_1 + 6(-6) = 31$

$\therefore \quad u_1 - 36 = 31$

$\therefore \quad u_1 = 67$

Now $u_n = u_1 + (n-1)d$

$\therefore \quad u_n = 67 + (n-1)(-6)$

$\therefore \quad u_n = 67 - 6n + 6$

$\therefore \quad u_n = 73 - 6n$

So, $u_{34} = 73 - 6(34) = -131$

**4** **a** $4 + 11 + 18 + 25 + ....$

The series is arithmetic with $u_1 = 4$, $d = 7$, $u_k = u_1 + (k-1)d$

$= 4 + 7(k-1)$

$= 7k - 3$

So, the series is $\sum_{k=1}^{n} (7k - 3)$.

**b** $\frac{1}{4} + \frac{1}{8} + \frac{1}{16} + \frac{1}{32} + ....$

The series is geometric with $u_1 = \frac{1}{4}$, $r = \frac{1}{2}$,

$$u_k = u_1 r^{k-1} = \tfrac{1}{4} \times \left(\tfrac{1}{2}\right)^{k-1} = \left(\tfrac{1}{2}\right)^2 \left(\tfrac{1}{2}\right)^{k-1} = \left(\tfrac{1}{2}\right)^{k+1}$$

So, the series is $\sum_{k=1}^{n} (\frac{1}{2})^{k+1}$.

**5** **a** $\sum_{k=1}^{8} \left(\frac{31 - 3k}{2}\right) = 14 + 12\frac{1}{2} + 11 + 9\frac{1}{2} + 8 + 6\frac{1}{2} + 5 + 3\frac{1}{2}$

This series is arithmetic with $u_1 = 14$, $n = 8$, and $u_n = 3\frac{1}{2}$.

$\therefore$ the sum is $\frac{8}{2}(14 + 3\frac{1}{2}) = 70$

**b** $\sum_{k=1}^{15} 50(0.8)^{k-1} \approx 50 + 40 + 32 + .... + 3.436 + 2.749 + 2.199$

This series is geometric with $u_1 = 50$, $r = 0.8$, and $n = 15$.

$\therefore$ the sum is $\dfrac{50\left[1 - (0.8)^{15}\right]}{1 - 0.8} \approx 241$

**c** $\sum_{k=7}^{\infty} 5(\frac{2}{5})^{k-1} = 5(\frac{2}{5})^6 + 5(\frac{2}{5})^7 + 5(\frac{2}{5})^8 + ....$

The series is an infinite geometric series with $u_1 = 5(\frac{2}{5})^6$, $r = \frac{2}{5}$

$$\therefore \quad S = \frac{u_1}{1-r} = \frac{5(\frac{2}{5})^6}{\frac{3}{5}} = \frac{2^6}{3 \times 5^4} = \frac{64}{1875}$$

**6** $11+16+21+26+....$ is arithmetic with $u_1=11$, $d=5$

$\therefore\ S_n=\frac{n}{2}(2u_1+(n-1)d)$
$=\frac{n}{2}(2\times 11+5(n-1))$
$=\frac{n}{2}(22+5n-5)$
$=\frac{n}{2}(5n+17)$

Given $S_n=450$, we need to find $n$,
so $S_n=\frac{n}{2}(5n+17)=450$
$\therefore\ \frac{5}{2}n^2+\frac{17}{2}n-450=0$
$\therefore\ 5n^2+17n-900=0$
$\therefore\ n\approx -15.2,\ 11.8$ {using technology}
But $n>0$, so $n\approx 11.8$

We try the two values on either side of $n\approx 11.8$, which are $n=11$ and $n=12$:

$S_{11}=\frac{11}{2}(5(11)+17)=396$ and $S_{12}=\frac{12}{2}(5(12)+17)=462$

$\therefore$ 12 terms of the series are required to exceed a sum of 450.

**7** **a** $u_{n+1}=u_1\times r^n$ where $u_1=12\,500$, $r=1+\frac{0.0825}{2}=1.041\,25$, $n=5\times 2=10$

So, $u_{n+1}=12\,500\times(1.041\,25)^{10}$
$\approx 18\,726.65$ The value of the invcstment is £18 726.65.

**b** $u_{n+1}=u_1\times r^n$ where $u_1=12\,500$, $r=1+\frac{0.0825}{12}=1.006\,875$, $n=5\times 12=60$

So, $u_{n+1}=12\,500\times(1.006\,875)^{60}$
$\approx 18\,855.74$ The value of the investment is £18 855.74.

**8** **a** Let the terms of the geometric series be $u_1,\ u_1r,\ u_1r^2,\ ....$

Then $u_1+u_1r=90$ and $u_1r^2=24$

$\therefore\ u_1(1+r)=90$
$\therefore\ u_1=\frac{90}{1+r}$ .... (1)

$\therefore\ u_1=\frac{24}{r^2}$ .... (2)

Equating (1) and (2) gives $\frac{90}{1+r}=\frac{24}{r^2}$

$\therefore\ 90r^2=24r+24$
$\therefore\ 90r^2-24r-24=0$
$\therefore\ 6(15r^2-4r-4)=0$
$\therefore\ 6(5r+2)(3r-2)=0$
$\therefore\ r=-\frac{2}{5}$ or $\frac{2}{3}$

Using (2), if $r=-\frac{2}{5}$ then $u_1=\frac{24}{(-\frac{2}{5})^2}=\frac{24}{\frac{4}{25}}=150$

if $r=\frac{2}{3}$ then $u_1=\frac{24}{(\frac{2}{3})^2}=\frac{24}{\frac{4}{9}}=54$

$\therefore$ either $u_1=150$, $r=-\frac{2}{5}$ or $u_1=54$, $r=\frac{2}{3}$

**b** Since $|r|<1$ in each case, both series converge.

When $u_1=150$, $r=-\frac{2}{5}$
$\therefore\ S=\frac{u_1}{1-r}$
$=\frac{150}{\frac{7}{5}}$
$=\frac{750}{7}$ or $107\frac{1}{7}$

When $u_1=54$, $r=\frac{2}{3}$
$\therefore\ S=\frac{u_1}{1-r}$
$=\frac{54}{\frac{1}{3}}$
$=162$

**9** Since Seve walks an additional 500 m $=$ 0.5 km each week, we have an arithmetic sequence with $u_1 = 10$ and constant difference $d = 0.5$.

$$u_n = u_1 + (n-1)d$$
$$\therefore\ u_n = 10 + (n-1)0.5$$

**a** $u_{52} = 10 + (52-1)0.5$ {52 weeks in a year}
$= 35.5$

$\therefore$ Seve walks 35.5 km in the last week.

**b** In total, Seve walks $10 + 10.5 + 11 + .... + 35.5$, which is an arithmetic series.

$$S_n = \frac{n}{2}(u_1 + u_n)$$
$$\therefore\ S_{52} = \tfrac{52}{2}(10 + 35.5)$$
$$= 1183$$

$\therefore$ Seve walks 1183 km in total.

**10** **a** $\sum\limits_{k=1}^{\infty} 50\,(2x-1)^{k-1}$ is a geometric series with $r = 2x - 1$ and converges if $-1 < r < 1$

$\therefore\ -1 < 2x - 1 < 1$
$\therefore\ 0 < 2x < 2$
$\therefore\ 0 < x < 1$

**b** When $x = 0.3$, $2x - 1 = 0.6 - 1 = -0.4$

and $\sum\limits_{k=1}^{\infty} 50\,(2x-1)^{k-1} = 50(-0.4)^0 + 50(-0.4)^1 + 50(-0.4)^2 + ....$

which is geometric with $u_1 = 50$, $r = -0.4$

Now as $0 < 0.3 < 1$, the series converges and $S = \dfrac{u_1}{1-r} = \dfrac{50}{1+0.4} = \dfrac{50}{\frac{7}{5}} = 35\frac{5}{7}$

# Chapter 7

## THE BINOMIAL EXPANSION

### EXERCISE 7A

**1** **a** $(p+q)^3$
$= p^3 + 3p^2q + 3pq^2 + q^3$

**b** $(x+1)^3$
$= x^3 + 3x^2(1)^1 + 3x(1)^2 + (1)^3$
$= x^3 + 3x^2 + 3x + 1$

**c** $(x-3)^3$
$= x^3 + 3x^2(-3) + 3x(-3)^2 + (-3)^3$
$= x^3 - 9x^2 + 27x - 27$

**d** $(2+x)^3$
$= 2^3 + 3(2)^2x + 3(2)x^2 + x^3$
$= 8 + 12x + 6x^2 + x^3$

**e** $(3x-1)^3$
$= (3x)^3 + 3(3x)^2(-1) + 3(3x)(-1)^2 + (-1)^3$
$= 27x^3 - 27x^2 + 9x - 1$

**f** $(2x+5)^3$
$= (2x)^3 + 3(2x)^2(5) + 3(2x)(5)^2 + (5)^3$
$= 8x^3 + 60x^2 + 150x + 125$

**g** $(2a-b)^3$
$= (2a)^3 + 3(2a)^2(-b) + 3(2a)(-b)^2 + (-b)^3$
$= 8a^3 - 12a^2b + 6ab^2 - b^3$

**h** $(3x - \frac{1}{3})^3 = (3x)^3 + 3(3x)^2(-\frac{1}{3}) + 3(3x)(-\frac{1}{3})^2 + (-\frac{1}{3})^3$
$= 27x^3 - 9x^2 + x - \frac{1}{27}$

**i** $\left(2x + \frac{1}{x}\right)^3 = (2x)^3 + 3(2x)^2\left(\frac{1}{x}\right) + 3(2x)\left(\frac{1}{x}\right)^2 + \left(\frac{1}{x}\right)^3$
$= 8x^3 + 12x + \frac{6}{x} + \frac{1}{x^3}$

**2** **a** $(1+x)^4 = 1^4 + 4(1)^3x + 6(1)^2x^2 + 4(1)x^3 + x^4$
$= 1 + 4x + 6x^2 + 4x^3 + x^4$

**b** $(p-q)^4 = p^4 + 4p^3(-q) + 6p^2(-q)^2 + 4p(-q)^3 + (-q)^4$
$= p^4 - 4p^3q + 6p^2q^2 - 4pq^3 + q^4$

**c** $(x-2)^4 = x^4 + 4x^3(-2)^1 + 6x^2(-2)^2 + 4x(-2)^3 + (-2)^4$
$= x^4 - 8x^3 + 24x^2 - 32x + 16$

**d** $(3-x)^4 = 3^4 + 4(3)^3(-x) + 6(3)^2(-x)^2 + 4(3)(-x)^3 + (-x)^4$
$= 81 - 108x + 54x^2 - 12x^3 + x^4$

**e** $(1+2x)^4 = 1^4 + 4(1)^3(2x) + 6(1)^2(2x)^2 + 4(1)(2x)^3 + (2x)^4$
$= 1 + 8x + 24x^2 + 32x^3 + 16x^4$

**f** $(2x-3)^4 = (2x)^4 + 4(2x)^3(-3)^1 + 6(2x)^2(-3)^2 + 4(2x)(-3)^3 + (-3)^4$
$= 16x^4 - 12 \times 8x^3 + 54 \times 4x^2 - 108 \times 2x + 81$
$= 16x^4 - 96x^3 + 216x^2 - 216x + 81$

**g** $(2x+b)^4 = (2x)^4 + 4(2x)^3b + 6(2x)^2b^2 + 4(2x)b^3 + b^4$
$= 16x^4 + 32x^3b + 24x^2b^2 + 8xb^3 + b^4$

**h** $\left(x + \frac{1}{x}\right)^4 = x^4 + 4x^3\left(\frac{1}{x}\right) + 6x^2\left(\frac{1}{x}\right)^2 + 4x\left(\frac{1}{x}\right)^3 + \left(\frac{1}{x}\right)^4$
$= x^4 + 4x^2 + 6 + \frac{4}{x^2} + \frac{1}{x^4}$

**i** $\left(2x-\frac{1}{x}\right)^4 = (2x)^4 + 4(2x)^3\left(-\frac{1}{x}\right) + 6(2x)^2\left(-\frac{1}{x}\right)^2 + 4(2x)\left(-\frac{1}{x}\right)^3 + \left(-\frac{1}{x}\right)^4$
$= 16x^4 - 32x^2 + 24 - \frac{8}{x^2} + \frac{1}{x^4}$

**3** **a** $(x+2)^5 = x^5 + 5x^4(2) + 10x^3(2)^2 + 10x^2(2)^3 + 5x(2)^4 + 2^5$
$= x^5 + 10x^4 + 40x^3 + 80x^2 + 80x + 32$

**b** $(x-2y)^5 = x^5 + 5x^4(-2y) + 10x^3(-2y)^2 + 10x^2(-2y)^3 + 5x(-2y)^4 + (-2y)^5$
$= x^5 - 10x^4y + 40x^3y^2 - 80x^2y^3 + 80xy^4 - 32y^5$

**c** $(1+2x)^5 = 1^5 + 5(1)^4(2x) + 10(1)^3(2x)^2 + 10(1)^2(2x)^3 + 5(1)(2x)^4 + (2x)^5$
$= 1 + 10x + 40x^2 + 80x^3 + 80x^4 + 32x^5$

**d** $\left(x-\frac{1}{x}\right)^5 = x^5 + 5x^4\left(-\frac{1}{x}\right) + 10x^3\left(-\frac{1}{x}\right)^2 + 10x^2\left(-\frac{1}{x}\right)^3 + 5x\left(-\frac{1}{x}\right)^4 + \left(-\frac{1}{x}\right)^5$
$= x^5 - 5x^3 + 10x - \frac{10}{x} + \frac{5}{x^3} - \frac{1}{x^5}$

**4** **a** 1 5 10 10 5 1 ⟵ the 5th row
1 6 15 20 15 6 1 ⟵ the 6th row

**b** **i** $(x+2)^6 = x^6 + 6x^5(2) + 15x^4(2)^2 + 20x^3(2)^3 + 15x^2(2)^4 + 6x(2)^5 + (2)^6$
$= x^6 + 12x^5 + 60x^4 + 160x^3 + 240x^2 + 192x + 64$

**ii** $(2x-1)^6 = (2x)^6 + 6(2x)^5(-1) + 15(2x)^4(-1)^2 + 20(2x)^3(-1)^3 + 15(2x)^2(-1)^4$
$+ 6(2x)(-1)^5 + (-1)^6$
$= 64x^6 - 6\times 32x^5 + 15\times 16x^4 - 20\times 8x^3 + 15\times 4x^2 - 6\times 2x + 1$
$= 64x^6 - 192x^5 + 240x^4 - 160x^3 + 60x^2 - 12x + 1$

**iii** $\left(x+\frac{1}{x}\right)^6 = x^6 + 6x^5\left(\frac{1}{x}\right) + 15x^4\left(\frac{1}{x}\right)^2 + 20x^3\left(\frac{1}{x}\right)^3 + 15x^2\left(\frac{1}{x}\right)^4 + 6x\left(\frac{1}{x}\right)^5 + \left(\frac{1}{x}\right)^6$
$= x^6 + 6x^4 + 15x^2 + 20 + \frac{15}{x^2} + \frac{6}{x^4} + \frac{1}{x^6}$

**5** **a** $(a+b)^3 = 8 + 12e^x + ....$
and $(a+b)^3 = a^3 + 3a^2b + 3ab^2 + b^3$
$\therefore\ a^3 = 8$
$\therefore\ a = 2$
and $3a^2b = 12e^x$
$\therefore\ 3(2)^2b = 12e^x$
$\therefore\ 12b = e^x$
$\therefore\ b = e^x$
So, $a = 2,\ b = e^x$.

**b** $(a+b)^3 = (2+e^x)^3$
$= 2^3 + 3(2)^2e^x + 3(2)(e^x)^2 + (e^x)^3$
$= 8 + 12e^x + 6e^{2x} + e^{3x}$
So, the remaining two terms are $6e^{2x}$ and $e^{3x}$.

**6** **a** $\left(1+\sqrt{2}\right)^3 = (1)^3 + 3(1)^2\left(\sqrt{2}\right) + 3(1)\left(\sqrt{2}\right)^2 + \left(\sqrt{2}\right)^3$
$= 1 + 3\sqrt{2} + 3\times 2 + 2\times\sqrt{2}$
$= 1 + 3\sqrt{2} + 6 + 2\sqrt{2}$
$= 7 + 5\sqrt{2}$

**b** $\left(\sqrt{5}+2\right)^4 = \left(\sqrt{5}\right)^4 + 4\left(\sqrt{5}\right)^3(2) + 6\left(\sqrt{5}\right)^2(2)^2 + 4\left(\sqrt{5}\right)(2)^3 + 2^4$
$= 25 + 8\times 5\sqrt{5} + 24\times 5 + 32\sqrt{5} + 16$
$= 25 + 40\sqrt{5} + 120 + 32\sqrt{5} + 16$
$= 161 + 72\sqrt{5}$

**c** $\left(2-\sqrt{2}\right)^5$
$= (2)^5 + 5(2)^4\left(-\sqrt{2}\right) + 10(2)^3\left(-\sqrt{2}\right)^2 + 10(2)^2\left(-\sqrt{2}\right)^3 + 5(2)^1\left(-\sqrt{2}\right)^4 + \left(-\sqrt{2}\right)^5$
$= 32 - 80\sqrt{2} + 160 - 80\sqrt{2} + 40 - 4\sqrt{2}$
$= 232 - 164\sqrt{2}$

**7** **a** $(2+x)^6 = (2)^6 + 6(2)^5x + 15(2)^4x^2 + 20(2)^3x^3 + 15(2)^2x^4 + 6(2)x^5 + x^6$
$= 64 + 192x + 240x^2 + 160x^3 + 60x^4 + 12x^5 + x^6$

**b** $(2.01)^6$ is obtained by letting $x = 0.01$
$\therefore\ (2.01)^6 = 64 + 192\times(0.01) + 240\times(0.01)^2 + 160\times(0.01)^3$
$+ 60\times(0.01)^4 + 12\times(0.01)^5 + (0.01)^6$
$= 65.944\,160\,601\,201$

$$\begin{array}{rr} & 64 \\ & 1.92 \\ & 0.024 \\ & 0.000\,16 \\ & 0.000\,000\,6 \\ & 0.000\,000\,001\,2 \\ + & 0.000\,000\,000\,001 \\ \hline & 65.944\,160\,601\,201 \end{array}$$

**8** $(2x+3)(x+1)^4$
$= (2x+3)(x^4 + 4x^3 + 6x^2 + 4x + 1)$
$= 2x^5 + 8x^4 + 12x^3 + 8x^2 + 2x + 3x^4 + 12x^3 + 18x^2 + 12x + 3$
$= 2x^5 + 11x^4 + 24x^3 + 26x^2 + 14x + 3$

**9** **a** $(3a+b)^5 = (3a)^5 + 5(3a)^4b + 10(3a)^3b^2 + \ldots.$
$\therefore$ the coefficient of $a^3b^2$ is $10\times 3^3 = 270$

**b** $(2a+3b)^6 = (2a)^6 + 6(2a)^5(3b) + 15(2a)^4(3b)^2 + 20(2a)^3(3b)^3 + \ldots.$
$\therefore$ the coefficient of $a^3b^3$ is $20\times 2^3\times 3^3 = 4320$

## EXERCISE 7B

**1** $0! = 1$
$1! = 1$
$2! = 2\times 1 = 2$
$3! = 3\times 2\times 1 = 6$
$4! = 4\times 3\times 2\times 1 = 24$
$5! = 5\times 4\times 3\times 2\times 1 = 120$
$6! = 6\times 5! = 6\times 120 = 720$
$7! = 7\times 6! = 7\times 720 = 5040$
$8! = 8\times 7! = 8\times 5040 = 40\,320$
$9! = 9\times 8! = 9\times 40\,320 = 362\,880$
$10! = 10\times 9! = 10\times 362\,880 = 3\,628\,800$

**2** **a** $\dfrac{6!}{4!} = \dfrac{6\times 5\times 4!}{4!} = 30$

**b** $\dfrac{100!}{99!} = \dfrac{100\times 99!}{99!} = 100$

**c** $\dfrac{7!}{5!\times 2!} = \dfrac{7\times 6\times 5!}{5!\times 2}$
$= 21$

**3** **a** $\dfrac{n!}{(n-1)!} = \dfrac{n\times \cancel{(n-1)!}}{\cancel{(n-1)!}_1}$
$= n$

**b** $\dfrac{(n+2)!}{n!} = \dfrac{(n+2)(n+1)\cancel{n!}}{\cancel{n!}_1}$
$= (n+2)(n+1)$

**4** **a** $\binom{3}{1} = \dfrac{3!}{1!(3-1)!}$
$= \dfrac{3!}{1!\times 2!}$
$= \dfrac{3\times \cancel{2\times 1}}{1\times \cancel{2\times 1}}$
$= 3$

**b** $\binom{4}{2} = \dfrac{4!}{2!(4-2)!}$
$= \dfrac{4!}{2!\times 2!}$
$= \dfrac{4\times 3\times \cancel{2\times 1}}{2\times 1\times \cancel{2\times 1}}$
$= \dfrac{12}{2}$
$= 6$

**c** $\binom{7}{3} = \dfrac{7!}{3!(7-3)!}$
$= \dfrac{7!}{3! \times 4!}$
$= \dfrac{7 \times 6 \times 5 \times \cancel{4 \times 3 \times 2 \times 1}}{3 \times 2 \times 1 \times \cancel{4 \times 3 \times 2 \times 1}}$
$= \dfrac{210}{6}$
$= 35$

**d** $\binom{10}{4} = \dfrac{10!}{4!(10-4)!}$
$= \dfrac{10!}{4! \times 6!}$
$= \dfrac{10 \times 9 \times 8 \times 7 \times \cancel{6 \times 5 \times 4 \times 3 \times 2 \times 1}}{4 \times 3 \times 2 \times 1 \times \cancel{6 \times 5 \times 4 \times 3 \times 2 \times 1}}$
$= \dfrac{5040}{24}$
$= 210$

**5** **a** **i** $\binom{8}{2} = \dfrac{8!}{2!(8-2)!}$
$= \dfrac{8!}{2! \times 6!}$
$= \dfrac{8 \times 7 \times \cancel{6 \times 5 \times 4 \times 3 \times 2 \times 1}}{2 \times 1 \times \cancel{6 \times 5 \times 4 \times 3 \times 2 \times 1}}$
$= \dfrac{56}{2}$
$= 28$

**ii** $\binom{8}{6} = \dfrac{8!}{6!(8-6)!}$
$= \dfrac{8!}{6! \times 2!}$
$= \dfrac{8 \times 7 \times \cancel{6 \times 5 \times 4 \times 3 \times 2 \times 1}}{\cancel{6 \times 5 \times 4 \times 3 \times 2 \times 1} \times 2 \times 1}$
$= \dfrac{56}{2}$
$= 28$

**b**

|  | |
|---|---|
| Row 0 | 1 |
| Row 1 | 1 1 |
| Row 2 | 1 2 1 |
| Row 3 | 1 3 3 1 |
| Row 4 | 1 4 6 4 1 |
| Row 5 | 1 5 10 10 5 1 |
| Row 6 | 1 6 15 20 15 6 1 |
| Row 7 | 1 7 21 35 35 21 7 1 |
| Row 8 | 1 8 [28] 56 70 56 [28] 8 1 |

$\binom{8}{2}$ and $\binom{8}{6}$ correspond to terms 3 and 7 of row 8 in Pascal's triangle. These are equal because of the triangle's symmetry.

## EXERCISE 7C

**1** **a** $(1+2x)^{11} = 1^{11} + \binom{11}{1}1^{10}(2x)^1 + \binom{11}{2}1^9(2x)^2 + .... + \binom{11}{10}1^1(2x)^{10} + (2x)^{11}$

**b** $\left(3x + \dfrac{2}{x}\right)^{15}$
$= (3x)^{15} + \binom{15}{1}(3x)^{14}\left(\dfrac{2}{x}\right)^1 + \binom{15}{2}(3x)^{13}\left(\dfrac{2}{x}\right)^2 + .... + \binom{15}{14}(3x)^1\left(\dfrac{2}{x}\right)^{14} + \left(\dfrac{2}{x}\right)^{15}$

**c** $\left(2x - \dfrac{3}{x}\right)^{20}$
$= (2x)^{20} + \binom{20}{1}(2x)^{19}\left(-\dfrac{3}{x}\right)^1 + \binom{20}{2}(2x)^{18}\left(-\dfrac{3}{x}\right)^2 + .... + \binom{20}{19}(2x)^1\left(-\dfrac{3}{x}\right)^{19} + \left(-\dfrac{3}{x}\right)^{20}$

**2** **a** For $(2x+5)^{15}$, $a = (2x)$, $b = 5$, and $n = 15$
Now $T_{r+1} = \binom{n}{r}a^{n-r}b^r$ and letting $r = 5$ gives $T_6 = \binom{15}{5}(2x)^{10}5^5$.

**b** For $\left(x^2 + y\right)^9$, $a = (x^2)$, $b = y$, and $n = 9$
Now $T_{r+1} = \binom{n}{r}a^{n-r}b^r$ and letting $r = 3$ gives $T_4 = \binom{9}{3}(x^2)^6(y)^3$.

**c** For $\left(x - \dfrac{2}{x}\right)^{17}$, $a = x$, $b = \left(-\dfrac{2}{x}\right)$, and $n = 17$
Now $T_{r+1} = \binom{n}{r}a^{n-r}b^r$ and letting $r = 9$ gives $T_{10} = \binom{17}{9}x^8\left(-\dfrac{2}{x}\right)^9$.

**d** For $\left(2x^2 - \frac{1}{x}\right)^{21}$, $a = (2x^2)$, $b = \left(-\frac{1}{x}\right)$, and $n = 21$

Now $T_{r+1} = \binom{n}{r} a^{n-r} b^r$ and letting $r = 8$ gives $T_9 = \binom{21}{8}(2x^2)^{13}\left(-\frac{1}{x}\right)^8$.

**3** **a** For $(x+b)^7$, $a = x$, $b = b$, and $n = 7$

$\therefore$ the general term $T_{r+1} = \binom{7}{r} x^{7-r} b^r$

**b** If $x^{7-r} = x^4$

then $7 - r = 4$

$\therefore\ r = 3$

Now $T_4 = \binom{7}{3} x^4 b^3$

$\therefore$ the coefficient of $x^4$ is $\binom{7}{3} b^3 = 35b^3$

But the coefficient of $x^4$ is $-280$

So, $35b^3 = -280$

$\therefore\ b^3 = -8$

$\therefore\ b = \sqrt[3]{-8}$

$\therefore\ b = -2$

**4** **a** In $\left(3 + 2x^2\right)^{10}$, $a = 3$, $b = (2x^2)$, and $n = 10$

Now $T_{r+1} = \binom{n}{r} a^{n-r} b^r$

$= \binom{10}{r} 3^{10-r} (2x^2)^r$

$= \binom{10}{r} 3^{10-r} 2^r x^{2r}$

We now let $2r = 10$

$\therefore\ r = 5$

So, $T_6 = \underbrace{\binom{10}{5} 3^5 2^5}\, x^{10}$

$\therefore$ the coefficient is $\binom{10}{5} 3^5 2^5$.

**b** In $\left(2x^2 - \frac{3}{x}\right)^6$, $a = (2x^2)$, $b = \left(-\frac{3}{x}\right)$, and $n = 6$

Now $T_{r+1} = \binom{n}{r} a^{n-r} b^r$

$= \binom{6}{r} \left(2x^2\right)^{6-r} \left(-\frac{3}{x}\right)^r$

$= \binom{6}{r} 2^{6-r} x^{12-2r} \frac{(-3)^r}{x^r}$

$= \binom{6}{r} 2^{6-r} (-3)^r x^{12-3r}$

We now let $12 - 3r = 3$

$\therefore\ 3r = 9$

$\therefore\ r = 3$

So, $T_4 = \underbrace{\binom{6}{3} 2^3 (-3)^3}\, x^3$

$\therefore$ the coefficient is $\binom{6}{3} 2^3 (-3)^3$.

**c** In $\left(2x^2 - 3y\right)^6$, $a = (2x^2)$, $b = (-3y)$, and $n = 6$

Now $T_{r+1} = \binom{n}{r} a^{n-r} b^r$

$= \binom{6}{r} \left(2x^2\right)^{6-r} (-3y)^r$

$= \binom{6}{r} 2^{6-r} x^{12-2r} (-3)^r y^r$

$= \binom{6}{r} 2^{6-r} (-3)^r x^{12-2r} y^r$

We find $r$ such that

$12 - 2r = 6$ and $r = 3$

$\therefore\ r = 3$ is the solution

So, $T_4 = \underbrace{\binom{6}{3} 2^3 (-3)^3}\, x^6 y^3$

$\therefore$ the coefficient is $\binom{6}{3} 2^3 (-3)^3$.

**d** In $\left(2x^2 - \frac{1}{x}\right)^{12}$, $a = (2x^2)$, $b = \left(-\frac{1}{x}\right)$, and $n = 12$

Now $T_{r+1} = \binom{n}{r} a^{n-r} b^r$

$= \binom{12}{r} \left(2x^2\right)^{12-r} \left(-\frac{1}{x}\right)^r$

$= \binom{12}{r} 2^{12-r} x^{24-2r} \frac{(-1)^r}{x^r}$

$= \binom{12}{r} 2^{12-r} (-1)^r x^{24-3r}$

We now let $24 - 3r = 12$

$\therefore\ 3r = 12$

$\therefore\ r = 4$

So, $T_5 = \underbrace{\binom{12}{4} 2^8 (-1)^4}\, x^{12}$

$\therefore$ the coefficient is $\binom{12}{4} 2^8 (-1)^4$.

**5** **a** For $\left(x+\frac{2}{x^2}\right)^{15}$, $a=x$, $b=\frac{2}{x^2}$, and $n=15$

Now $T_{r+1} = \binom{n}{r}a^{n-r}b^r$
$= \binom{15}{r}x^{15-r}\left(\frac{2}{x^2}\right)^r$
$= \binom{15}{r}x^{15-r}\frac{2^r}{x^{2r}}$
$= \binom{15}{r}2^r x^{15-3r}$

The constant term does not contain $x$.
$\therefore\ 15-3r=0$
$\therefore\ r=5$
so $T_6 = \binom{15}{5}2^5x^0$
$\therefore$ the constant term is $\binom{15}{5}2^5$.

**b** For $\left(x-\frac{3}{x^2}\right)^9$, $a=x$, $b=\left(-\frac{3}{x^2}\right)$, and $n=9$

Now $T_{r+1} = \binom{n}{r}a^{n-r}b^r$
$= \binom{9}{r}x^{9-r}\left(-\frac{3}{x^2}\right)^r$
$= \binom{9}{r}x^{9-r}\frac{(-3)^r}{x^{2r}}$
$= \binom{9}{r}(-3)^r x^{9-3r}$

The constant term does not contain $x$.
$\therefore\ 9-3r=0$
$\therefore\ r=3$
so $T_4 = \binom{9}{3}(-3)^3x^0$
$\therefore$ the constant term is $\binom{9}{3}(-3)^3$.

**6** **a**

| | Row | | **b** | sum | | |
|---|---|---|---|---|---|---|
| Row 1 | 1 1 | ← | **i** | sum $=1+1$ | $=2$ | $=2^1$ |
| Row 2 | 1 2 1 | ← | **ii** | sum $=1+2+1$ | $=4$ | $=2^2$ |
| Row 3 | 1 3 3 1 | ← | **iii** | sum $=1+3+3+1$ | $=8$ | $=2^3$ |
| Row 4 | 1 4 6 4 1 | ← | **iv** | sum $=1+4+6+4+1$ | $=16$ | $=2^4$ |
| Row 5 | 1 5 10 10 5 1 | ← | **v** | sum $=1+5+10+10+5+1$ | $=32$ | $=2^5$ |
| Row 6 | 1 6 15 20 15 6 1 | | | | | |

**c** The sum of the numbers in row $n$ of Pascal's triangle is $2^n$.

**d** $(1+x)^n$
$= \binom{n}{0}1^n + \binom{n}{1}1^{n-1}x + \binom{n}{2}1^{n-2}x^2 + \binom{n}{3}1^{n-3}x^3 + \dots + \binom{n}{n-1}1^1x^{n-1} + \binom{n}{n}x^n$
$= \binom{n}{0} + \binom{n}{1}x + \binom{n}{2}x^2 + \binom{n}{3}x^3 + \dots + \binom{n}{n-1}x^{n-1} + \binom{n}{n}x^n$ {as all powers of 1 are 1}

Now letting $x=1$ gives LHS $=(1+1)^n = 2^n$
and RHS $= \binom{n}{0} + \binom{n}{1} + \binom{n}{2} + \binom{n}{3} + \dots + \binom{n}{n-1} + \binom{n}{n}$

$\therefore\ \binom{n}{0} + \binom{n}{1} + \binom{n}{2} + \dots + \binom{n}{n-1} + \binom{n}{n} = 2^n$

**7** **a** $(x+2)(x^2+1)^8$
$= (x+2)\left[(x^2)^8 + \binom{8}{1}(x^2)^7 1 + \binom{8}{2}(x^2)^6 1^2 + \dots + \binom{8}{6}(x^2)^2 1^6 + \binom{8}{7}(x^2)^1 1^7 + \binom{8}{8}1^8\right]$

only terms which when multiplied give an $x^5$

$\therefore$ coefficient of $x^5$ is $1\times\binom{8}{6} = \binom{8}{6} = 28$.

**b** $(2-x)(3x+1)^9$
$= (2-x)\left[(3x)^9 + \binom{9}{1}(3x)^8 + \binom{9}{2}(3x)^7 + \binom{9}{3}(3x)^6 + \binom{9}{4}(3x)^5 + \dots\right]$

$\therefore$ term containing $x^6$ is $2\times\binom{9}{3}3^6x^6 + (-x)\times\binom{9}{4}3^5x^5 = 2\binom{9}{3}3^6x^6 - \binom{9}{4}3^5x^6 = 91\,854x^6$

**8** In $(x^2y-2y^2)^6$, $a=(x^2y)$, $b=(-2y^2)$, and $n=6$.

Now $T_{r+1} = \binom{n}{r}a^{n-r}b^r$
$= \binom{6}{r}(x^2y)^{6-r}(-2y^2)^r$
$= \binom{6}{r}x^{12-2r}y^{6-r}(-2)^r y^{2r}$
$= \binom{6}{r}(-2)^r x^{12-2r}y^{6+r}$

Since $x$ and $y$ are raised to the same power,
$12-2r = 6+r$
$\therefore\ 3r=6$
$\therefore\ r=2$
$T_3 = \binom{6}{2}(-2)^2x^8y^8 = 60x^8y^8$

**9** **a** $(1+x)^n$ has $T_3 = \binom{n}{2}1^{n-2}x^2 = \binom{n}{2}x^2$ and $n \geqslant 2$

But this term is $36x^2$ $\quad \therefore \binom{n}{2} = 36$

$$\therefore \frac{n(n-1)}{2} = 36$$
$$\therefore n(n-1) = 72$$
$$\therefore n^2 - n - 72 = 0$$
$$\therefore (n-9)(n+8) = 0$$
$$\therefore n = 9 \text{ or } -8$$

But $n \geqslant 2$, so $n = 9$

and $T_4 = \binom{n}{3}1^{n-3}x^3$

$\therefore T_4 = \binom{9}{3}x^3$

$= 84x^3$

**b** $(1+kx)^n = 1^n + \binom{n}{1}1^{n-1}(kx)^1 + \binom{n}{2}1^{n-2}(kx)^2 + ....$

$= 1 + \binom{n}{1}kx + \binom{n}{2}k^2x^2 + ....$

$\therefore \binom{n}{1}k = -12$ and $\binom{n}{2}k^2 = 60$

$\therefore nk = -12$ and $\dfrac{n(n-1)}{2}k^2 = 60$

$\therefore n(n-1)k^2 = 120$

But $k = -\dfrac{12}{n}$ $\quad \therefore n(n-1)\dfrac{144}{n^2} = 120$

$\therefore 144(n-1) = 120n \quad \{n \geqslant 2\}$

$\therefore 144n - 120n = 144$

$\therefore 24n = 144$

$\therefore n = 6$ and so $k = -2$

**10** $T_{r+1} = \binom{n}{r}a^{n-r}b^r$ where $n = 10$, $a = (x^2)$, $b = \left(\dfrac{1}{ax}\right)$

$= \binom{10}{r}(x^2)^{10-r}\left(\dfrac{1}{ax}\right)^r$

$= \binom{10}{r}x^{20-2r} \times \dfrac{1}{a^r x^r}$

$= \binom{10}{r}x^{20-3r} \times \dfrac{1}{a^r}$

We let $20 - 3r = 11$

$\therefore 3r = 9$

$\therefore r = 3$

and $T_4 = \binom{10}{3}x^{11} \times \dfrac{1}{a^3}$

$= \dfrac{\binom{10}{3}}{a^3}x^{11}$

So, $\dfrac{\binom{10}{3}}{a^3} = 15$

$\therefore \dfrac{120}{a^3} = 15$

$\therefore a^3 = 8$

$\therefore a = 2$

## REVIEW SET 7A

**1** $\binom{6}{4} = \dfrac{6!}{4!(6-4)!}$

$= \dfrac{6!}{4! \times 2!}$

$= \dfrac{6 \times 5 \times 4 \times 3 \times 2 \times 1}{4 \times 3 \times 2 \times 1 \times 2 \times 1}$ (with $4 \times 3 \times 2 \times 1$ cancelled)

$= \dfrac{30}{2} = 15$

**2** **a** $(x-2y)^3$

$= x^3 + 3x^2(-2y) + 3x(-2y)^2 + (-2y)^3$

$= x^3 - 6x^2y + 12xy^2 - 8y^3$

**b** $(3x+2)^4$

$= (3x)^4 + 4(3x)^3(2) + 6(3x)^2(2)^2 + 4(3x)(2)^3 + (2)^4$

$= 81x^4 + 216x^3 + 216x^2 + 96x + 16$

**3** In the expansion of $\left(2x^2-\frac{1}{x}\right)^6$, $a=(2x^2)$, $b=\left(-\frac{1}{x}\right)$, $n=6$

$$\begin{aligned}T_{r+1}&=\binom{n}{r}a^{n-r}b^r\\&=\binom{6}{r}(2x^2)^{6-r}\left(-\frac{1}{x}\right)^r\\&=\binom{6}{r}2^{6-r}x^{12-2r}(-1)^rx^{-r}\\&=\binom{6}{r}2^{6-r}(-1)^rx^{12-3r}\end{aligned}$$

For the constant term we let $12-3r=0$

$\therefore\ r=4$

and $T_5=\underbrace{\binom{6}{4}2^2(-1)^4}x^0$

$\therefore$ the constant term is $\binom{6}{4}2^2(-1)^4=60$.

**4** The sixth row of Pascal's triangle is 1 6 15 20 15 6 1

$\therefore\ (a+b)^6=a^6+6a^5b+15a^4b^2+20a^3b^3+15a^2b^4+6ab^5+b^6$

**a** $$\begin{aligned}(x-3)^6&=x^6+6x^5(-3)+15x^4(-3)^2+20x^3(-3)^3+15x^2(-3)^4+6x(-3)^5+(-3)^6\\&=x^6-18x^5+135x^4-540x^3+1215x^2-1458x+729\end{aligned}$$

**b** $$\begin{aligned}&\left(1+\frac{1}{x}\right)^6\\&=(1)^6+6(1)^5\left(\frac{1}{x}\right)+15(1)^4\left(\frac{1}{x}\right)^2+20(1)^3\left(\frac{1}{x}\right)^3+15(1)^2\left(\frac{1}{x}\right)^4+6(1)\left(\frac{1}{x}\right)^5+\left(\frac{1}{x}\right)^6\\&=1+\frac{6}{x}+\frac{15}{x^2}+\frac{20}{x^3}+\frac{15}{x^4}+\frac{6}{x^5}+\frac{1}{x^6}\end{aligned}$$

**5** $$\begin{aligned}\left(\sqrt{3}+2\right)^5&=\left(\sqrt{3}\right)^5+5\left(\sqrt{3}\right)^4(2)+10\left(\sqrt{3}\right)^3(2)^2+10\left(\sqrt{3}\right)^2(2)^3+5\left(\sqrt{3}\right)^1(2)^4+2^5\\&=9\sqrt{3}+90+120\sqrt{3}+240+80\sqrt{3}+32\\&=362+209\sqrt{3}\end{aligned}$$

**6** **a** $n=6$ so there are 7 terms.

**b** In $\left(3x^2+\frac{1}{x}\right)^6$, $a=(3x^2)$, $b=\left(\frac{1}{x}\right)$, and $n=6$.

Now $$\begin{aligned}T_{r+1}&=\binom{n}{r}a^{n-r}b^r\\&=\binom{6}{r}(3x^2)^{6-r}\left(\frac{1}{x}\right)^r\\&=\binom{6}{r}3^{6-r}x^{12-2r}\frac{1}{x^r}\\&=\binom{6}{r}3^{6-r}x^{12-3r}\end{aligned}$$

The constant term does not contain $x$

$\therefore\ 12-3r=0$

$\therefore\ r=4$

So, $T_5=\binom{6}{4}3^2x^0=135$

**c** To find a term involving $x^5$, we solve

$12-3r=5$

$\therefore\ 3r=7$

$\therefore\ r=\frac{7}{3}$

But $r$ must be an integer.

$\therefore$ no term in the expansion involves $x^5$.

**7** **a** $(a+b)^4=e^{4x}-4e^{2x}+\ldots.$

and $(a+b)^4=a^4+4a^3b+6a^2b^2+4ab^3+b^4$

$\therefore\ a^4=e^{4x}$ and $4a^3b=-4e^{2x}$

$\therefore\ a^4=(e^x)^4$ $\quad\therefore\ 4e^{3x}b=-4e^{2x}$

$\therefore\ a=e^x$ $\quad\therefore\ b=-e^{-x}$

**b** $$\begin{aligned}(a+b)^4&=e^{4x}-4e^{2x}+6(e^x)^2(-e^{-x})^2+4e^x(-e^{-x})^3+(-e^{-x})^4\\&=e^{4x}-4e^{2x}+6e^{2x}e^{-2x}-4e^xe^{-3x}+e^{-4x}\\&=e^{4x}-4e^{2x}+6-4e^{-2x}+e^{-4x}\end{aligned}$$

## REVIEW SET 7B

**1** In the expansion of $(x+5)^6$, $a = x$, $b = 5$, $n = 6$

$T_{r+1} = \binom{n}{r}a^{n-r}b^r$
$= \binom{6}{r}x^{6-r}5^r$

For the coefficient of $x^3$ we let $6 - r = 3$
$\therefore \quad r = 3$

and $T_4 = \underbrace{\binom{6}{3}5^3}\, x^3$

$\therefore$ the coefficient is $\binom{6}{3}5^3 = 2500$.

**2** $(4+x)^3 = 4^3 + 3(4)^2x^1 + 3(4)^1x^2 + x^3$
$= 64 + 48x + 12x^2 + x^3$

Letting $x = 0.02$ gives $(4.02)^3 = 64 + 48(0.02) + 12(0.02)^2 + (0.02)^3$
$= 64 + 0.96 + 0.0048 + 0.000\,008$
$= 64.964\,808$

**3** In $\left(2x - \dfrac{3}{x^2}\right)^{12}$, $a = (2x)$, $b = \left(-\dfrac{3}{x^2}\right)$, $n = 12$

$T_{r+1} = \binom{n}{r}a^{n-r}b^r$
$= \binom{12}{r}(2x)^{12-r}\left(-\dfrac{3}{x^2}\right)^r$
$= \binom{12}{r}2^{12-r}x^{12-r}\dfrac{(-3)^r}{x^{2r}}$
$= \binom{12}{r}2^{12-r}(-3)^r\,x^{12-3r}$

For the coefficient of $x^{-6}$ we let $12 - 3r = -6$
$\therefore \quad 3r = 18$
$\therefore \quad r = 6$

So, $T_7 = \underbrace{\binom{12}{6}2^6(-3)^6}\, x^{-6}$

$\therefore$ the coefficient is $\binom{12}{6}2^6(-3)^6 = 43\,110\,144$.

**4** $(2x+3)(x-2)^6$
$= (2x+3)\left[x^6 + \binom{6}{1}x^5(-2) + \binom{6}{2}x^4(-2)^2 + .....\right]$

$\therefore$ the term containing $x^5$ is $2x \times \binom{6}{2}(-2)^2x^4 + 3 \times \binom{6}{1}(-2)x^5 = \left[8\binom{6}{2} - 6\binom{6}{1}\right]x^5 = 84x^5$

**5** $(m-2n)^{10} = m^{10} + \binom{10}{1}m^9(-2n) + \binom{10}{2}m^8(-2n)^2 + .... + (-2n)^{10}$
$= m^{10} - 20m^9n + 45m^8(4n^2) + .... + 1024n^{10}$
$= m^{10} - 20m^9n + 180m^8n^2 + .... + 1024n^{10}$
$\therefore \quad k = 180$

**6** $(1+cx)(1+x)^4 = (1+cx)(1^4 + \binom{4}{1}1^3x + \binom{4}{2}1^2x^2 + \binom{4}{3}1x^3 + x^4)$

$\therefore$ coefficient of $x^3$ is $1 \times \binom{4}{3} \times 1 + c \times \binom{4}{2} \times 1^2 = 4 + 6c$

But the coefficient of $x^3$ is 22, so $4 + 6c = 22$
$\therefore \quad 6c = 18$
$\therefore \quad c = 3$

**7** **a** $(2+x)^n = 2^n + \binom{n}{1}2^{n-1}x^1 + \binom{n}{2}2^{n-2}x^2 + \binom{n}{3}2^{n-3}x^3 + .... + \binom{n}{n-1}2^1x^{n-1} + x^n$

**b** $2^n + \binom{n}{1}2^{n-1} + \binom{n}{2}2^{n-2} + \binom{n}{3}2^{n-3} + .... + 2n + 1$
$= 2^n + \binom{n}{1}2^{n-1}x^1 + \binom{n}{2}2^{n-2}x^2 + \binom{n}{3}2^{n-3}x^3 + .... + \binom{n}{n-1}2^1x^{n-1} + x^n$, where $x = 1$
$= (2+1)^n$
$= 3^n$

# Chapter 8
## THE UNIT CIRCLE AND RADIAN MEASURE

### EXERCISE 8A

**1** **a** $180° = \pi$ radians
$\therefore\ 90° = \frac{\pi}{2}$ radians

**b** $180° = \pi$ radians
$\therefore\ 60° = \frac{\pi}{3}$ radians

**c** $180° = \pi$ radians
$\therefore\ 30° = \frac{\pi}{6}$ radians

**d** $180° = \pi$ radians
$\therefore\ 18° = \frac{\pi}{10}$ radians

**e** $180° = \pi$ radians
$\therefore\ 9° = \frac{\pi}{20}$ radians

**f** $180° = \pi$ radians
$\therefore\ 45° = \frac{\pi}{4}$ radians
$\therefore\ 135° = \frac{3\pi}{4}$ radians

**g** $180° = \pi$ radians
$\therefore\ 45° = \frac{\pi}{4}$ radians
$\therefore\ 225° = \frac{5\pi}{4}$ radians

**h** $180° = \pi$ radians
$\therefore\ 90° = \frac{\pi}{2}$ radians
$\therefore\ 270° = \frac{3\pi}{2}$ radians

**i** $360° = 2 \times 180°$
$= 2\pi$ radians

**j** $720° = 4 \times 180°$
$= 4\pi$ radians

**k** $180° = \pi$ radians
$\therefore\ 45° = \frac{\pi}{4}$ radians
$\therefore\ 315° = \frac{7\pi}{4}$ radians

**l** $180° = \pi$ radians
$\therefore\ 540° = 3\pi$ radians

**m** $180° = \pi$ radians
$\therefore\ 36° = \frac{\pi}{5}$ radians

**n** $180° = \pi$ radians
$\therefore\ 10° = \frac{\pi}{18}$ radians
$\therefore\ 80° = \frac{8\pi}{18}$ radians
$= \frac{4\pi}{9}$ radians

**o** $180° = \pi$ radians
$\therefore\ 10° = \frac{\pi}{18}$ radians
$\therefore\ 230° = \frac{23\pi}{18}$ radians

**2** **a** $36.7°$
$= 36.7 \times \frac{\pi}{180}$ radians
$\approx 0.641$ radians

**b** $137.2°$
$= 137.2 \times \frac{\pi}{180}$ radians
$\approx 2.39$ radians

**c** $317.9°$
$= 317.9 \times \frac{\pi}{180}$ radians
$\approx 5.55$ radians

**d** $219.6°$
$= 219.6 \times \frac{\pi}{180}$ radians
$\approx 3.83$ radians

**e** $396.7°$
$= 396.7 \times \frac{\pi}{180}$ radians
$\approx 6.92$ radians

**3** **a** $\frac{\pi}{5} = \frac{180°}{5} = 36°$

**b** $\frac{3\pi}{5} = \frac{3\times180°}{5} = 108°$

**c** $\frac{3\pi}{4} = \frac{3\times180°}{4} = 135°$

**d** $\frac{\pi}{18} = \frac{180°}{18} = 10°$

**e** $\frac{\pi}{9} = \frac{180°}{9} = 20°$

**f** $\frac{7\pi}{9} = \frac{7\times180°}{9} = 140°$

**g** $\frac{\pi}{10} = \frac{180°}{10} = 18°$

**h** $\frac{3\pi}{20} = \frac{3\times180°}{20} = 27°$

**i** $\frac{7\pi}{6} = \frac{7\times180°}{6} = 210°$

**j** $\frac{\pi}{8} = \frac{180°}{8} = 22.5°$

**4** **a** $2^c$
$= 2 \times \frac{180}{\pi}$ degrees
$\approx 114.59°$

**b** $1.53^c$
$= 1.53 \times \frac{180}{\pi}$ degrees
$\approx 87.66°$

**c** $0.867^c$
$= 0.867 \times \frac{180}{\pi}$ degrees
$\approx 49.68°$

**d** $3.179^c$
$= 3.179 \times \frac{180}{\pi}$ degrees
$\approx 182.14°$

**e** $5.267^c$
$= 5.267 \times \frac{180}{\pi}$ degrees
$\approx 301.78°$

**5** **a**

| *Degrees* | 0 | 45 | 90 | 135 | 180 | 225 | 270 | 315 | 360 |
|---|---|---|---|---|---|---|---|---|---|
| *Radians* | 0 | $\frac{\pi}{4}$ | $\frac{\pi}{2}$ | $\frac{3\pi}{4}$ | $\pi$ | $\frac{5\pi}{4}$ | $\frac{3\pi}{2}$ | $\frac{7\pi}{4}$ | $2\pi$ |

**b**

| *Degrees* | 0 | 30 | 60 | 90 | 120 | 150 | 180 | 210 | 240 | 270 | 300 | 330 | 360 |
|---|---|---|---|---|---|---|---|---|---|---|---|---|---|
| *Radians* | 0 | $\frac{\pi}{6}$ | $\frac{\pi}{3}$ | $\frac{\pi}{2}$ | $\frac{2\pi}{3}$ | $\frac{5\pi}{6}$ | $\pi$ | $\frac{7\pi}{6}$ | $\frac{4\pi}{3}$ | $\frac{3\pi}{2}$ | $\frac{5\pi}{3}$ | $\frac{11\pi}{6}$ | $2\pi$ |

## EXERCISE 8B

**1** **a** arc length $= \frac{7\pi}{4} \times 9$
$\approx 49.5$ cm

area $= \frac{1}{2} \times \frac{7\pi}{4} \times 9^2$
$\approx 223$ cm$^2$

**b** arc length $= 4.67 \times 4.93$
$\approx 23.0$ cm

area $= \frac{1}{2}(4.67) \times 4.93^2$
$\approx 56.8$ cm$^2$

**2** **a** $\theta = 107.9°, \quad l = 5.92$

$\therefore \left(\frac{107.9}{360}\right) \times 2\pi \times r = 5.92$

$\therefore r = \dfrac{5.92 \times 360}{107.9 \times 2 \times \pi}$

$\therefore r \approx 3.14$ m

**b** area $= \left(\frac{107.9}{360}\right) \times \pi \times (3.1436)^2$
$\approx 9.30$ m$^2$

**3** **a** area $= \frac{1}{2}\theta r^2$

$\therefore 20.8 = \frac{1}{2}(1.19) \times r^2$

$\therefore \dfrac{20.8 \times 2}{1.19} = r^2$

$\therefore r = \sqrt{\dfrac{20.8 \times 2}{1.19}}$

$\therefore r \approx 5.91$ cm

**b** perimeter
$= l + 2r$
$\approx 1.19 \times 5.912 + 2 \times 5.912$
$\approx 18.9$ cm

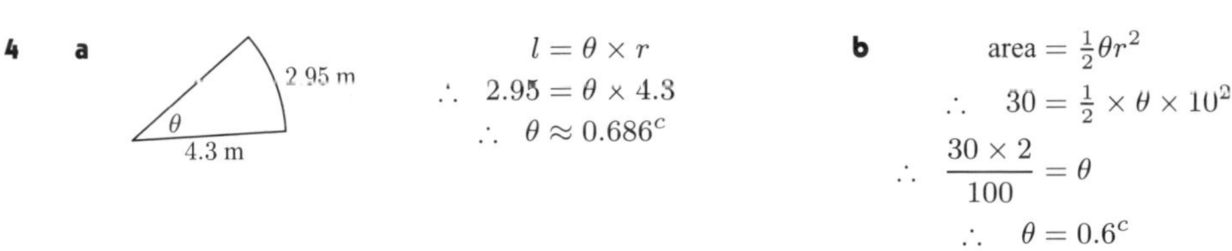


**4** **a** $l = \theta \times r$

$\therefore 2.95 = \theta \times 4.3$

$\therefore \theta \approx 0.686^c$

**b** area $= \frac{1}{2}\theta r^2$

$\therefore 30 = \frac{1}{2} \times \theta \times 10^2$

$\therefore \dfrac{30 \times 2}{100} = \theta$

$\therefore \theta = 0.6^c$

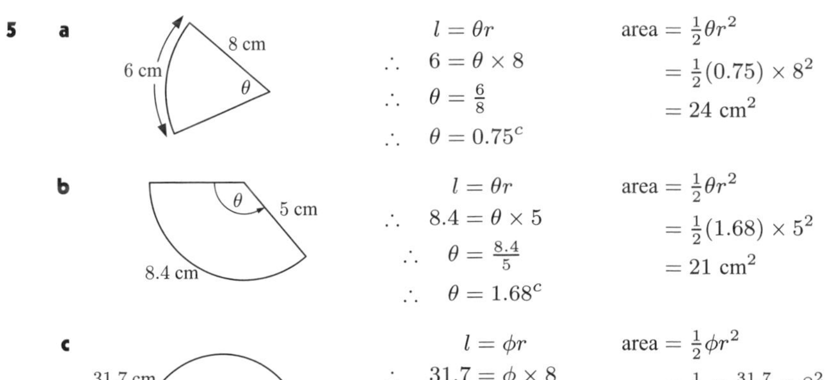


**5** **a** $l = \theta r$

$\therefore 6 = \theta \times 8$

$\therefore \theta = \frac{6}{8}$

$\therefore \theta = 0.75^c$

area $= \frac{1}{2}\theta r^2$
$= \frac{1}{2}(0.75) \times 8^2$
$= 24$ cm$^2$

**b** $l = \theta r$

$\therefore 8.4 = \theta \times 5$

$\therefore \theta = \frac{8.4}{5}$

$\therefore \theta = 1.68^c$

area $= \frac{1}{2}\theta r^2$
$= \frac{1}{2}(1.68) \times 5^2$
$= 21$ cm$^2$

**c** $l = \phi r$

$\therefore 31.7 = \phi \times 8$

$\therefore \phi = \frac{31.7}{8}$

$\therefore \phi \approx 3.96^c$

But $\theta = 2\pi - \phi$

$\therefore \theta \approx 2.32^c$

area $= \frac{1}{2}\phi r^2$
$= \frac{1}{2} \times \frac{31.7}{8} \times 8^2$
$= 126.8$ cm$^2$

**6**

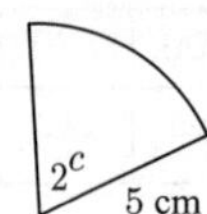


$$\text{arc length} = \theta r = 2 \times 5 = 10 \text{ cm}$$

$$\text{area} = \tfrac{1}{2}\theta r^2 = \tfrac{1}{2} \times 2 \times 5^2 = 25 \text{ cm}^2$$

**7**

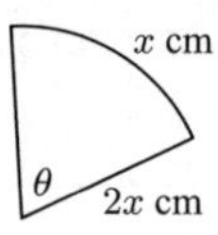


$$\text{arc length} = \theta r$$

$$\therefore \quad x = \theta(2x)$$

$$\therefore \quad \theta = \tfrac{1}{2}$$

$$\text{area} = \tfrac{1}{2}\theta r^2 = \tfrac{1}{2} \times (\tfrac{1}{2}) \times (2x)^2 = x^2 \text{ cm}^2$$

**8** **a**

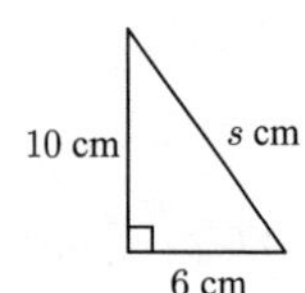


$$s^2 = 6^2 + 10^2 \quad \{\text{Pythagoras}\}$$

$$\therefore \quad s = \sqrt{6^2 + 10^2}$$

$$\therefore \quad s \approx 11.6619$$

$$\therefore \quad s \approx 11.7$$

$\therefore$ slant length is 11.7 cm.

**b** $r = s \approx 11.7$

**c** arclength = circumference of cone base

$$= 2\pi \times 6 \approx 37.6991 \approx 37.7 \text{ cm}$$

**d** arc length $= \theta r$

$$\therefore \quad 37.6991 \approx \theta \times 11.6619$$

$$\therefore \quad \theta \approx \frac{37.6991}{11.6619}$$

$$\therefore \quad \theta \approx 3.23 \text{ radians}$$

**9**

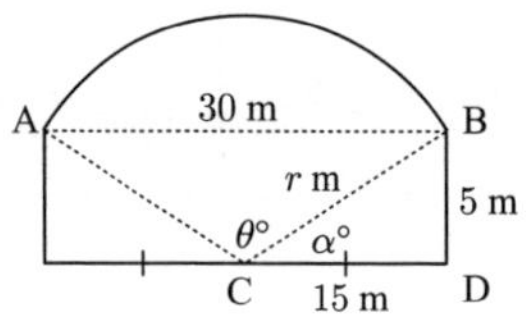


**a** $\tan\alpha = \frac{5}{15}$

$$\therefore \quad \alpha = \tan^{-1}\left(\tfrac{1}{3}\right)$$

$$\therefore \quad \alpha \approx 18.43$$

**b** $\theta + 2\alpha = 180 \quad \{\text{angles on a line}\}$

$$\therefore \quad \theta \approx 180 - 2 \times 18.43$$

$$\therefore \quad \theta \approx 143.1$$

**c** area $= 2 \times$ area of $\triangle$CDB + area of sector

$$= 2 \times \tfrac{1}{2} \times \text{CD} \times \text{BD} + \left(\tfrac{\theta}{360}\right) \times \pi \times r^2$$

Now $r^2 = 5^2 + 15^2 = 250$

$$\therefore \quad \text{area} \approx 2 \times \tfrac{1}{2} \times 15 \times 5 + \left(\tfrac{143.1}{360}\right) \times \pi \times 250 \approx 387 \text{ m}^2$$

**10** Since [AT] is a tangent, $\text{O}\widehat{\text{T}}\text{A}$ is a right angle.

$$\therefore \quad \cos\theta = \tfrac{5}{13}$$

$$\therefore \quad \theta \approx 67.38°$$

$$\text{arc length BT} = \left(\tfrac{\theta}{360}\right) \times 2\pi r \approx \tfrac{67.38}{360} \times 2 \times \pi \times 5 \approx 5.88 \text{ cm}$$

$$\text{AT}^2 + \text{OT}^2 = \text{OA}^2 \quad \{\text{Pythagoras}\}$$

$$\therefore \quad \text{AT}^2 = 13^2 - 5^2$$

$$\therefore \quad \text{AT} = 12 \text{ cm}$$

$$\therefore \quad \text{perimeter} = \text{AT} + \text{arc length BT} + \text{AB} \approx 12 + 5.88 + (13 - 5) \approx 25.9 \text{ cm}$$

**11** **a** $l = \left(\tfrac{\theta}{360}\right) \times 2\pi r$

$$= \frac{\frac{1}{60}}{360} \times 2 \times \pi \times 6370 \text{ km} \approx 1.853 \text{ km}$$

**b** $\text{speed} = \dfrac{\text{distance}}{\text{time}}$

$$\therefore \quad \text{time} = \frac{\text{distance}}{\text{speed}} = \frac{2130 \text{ km}}{480 \text{ n miles h}^{-1}} = \frac{2130 \text{ km}}{480 \times 1.853 \text{ km h}^{-1}} \approx 2.395 \text{ hours} \approx 2 \text{ hours } 24 \text{ min}$$

**12** 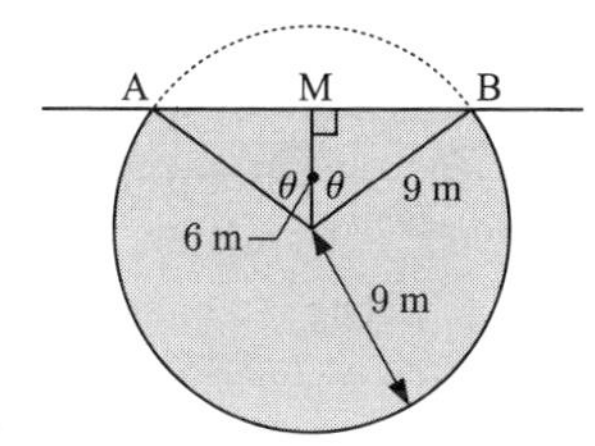


$\cos\theta = \frac{6}{9} = \frac{2}{3}$

$\therefore \quad \theta = \cos^{-1}\left(\frac{2}{3}\right)$

$\therefore \quad \theta \approx 48.19°$

So, $360 - 2\theta \approx 263.62°$

Now $\text{MB} = \sqrt{9^2 - 6^2}$
$= \sqrt{45}$

$\therefore$ available feeding area
$=$ area of $\triangle$ + area of sector
$\approx \frac{1}{2} \times 2 \times \sqrt{45} \times 6$
$+ \left(\frac{263.62}{360}\right) \times \pi \times 9^2$
$\approx 227\ \text{m}^2$

## EXERCISE 8C

**1** **a** **i** A$(\cos 26°, \sin 26°)$, B$(\cos 146°, \sin 146°)$, C$(\cos 199°, \sin 199°)$
**ii** A$(0.899, 0.438)$, B$(-0.829, 0.559)$, C$(-0.946, -0.326)$

**b** **i** A$(\cos 123°, \sin 123°)$, B$(\cos 251°, \sin 251°)$, C$(\cos(-35°), \sin(-35°))$
**ii** A$(-0.545, 0.839)$, B$(-0.326, -0.946)$, C$(0.819, -0.574)$

**2**

| $\theta$ (degrees) | 0° | 90° | 180° | 270° | 360° | 450° |
|---|---|---|---|---|---|---|
| $\theta$ (radians) | 0 | $\frac{\pi}{2}$ | $\pi$ | $\frac{3\pi}{2}$ | $2\pi$ | $\frac{5\pi}{2}$ |
| sine | 0 | 1 | 0 | $-1$ | 0 | 1 |
| cosine | 1 | 0 | $-1$ | 0 | 1 | 0 |
| tangent | 0 | undef. | 0 | undef. | 0 | undef. |

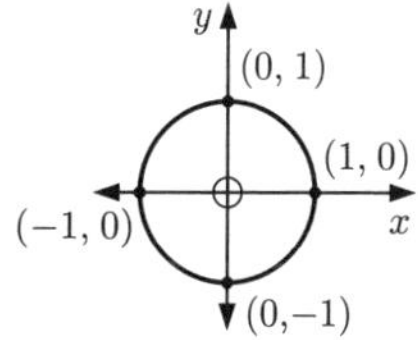


**3** **a** **i** $\frac{1}{\sqrt{2}} \approx 0.707$
**ii** $\frac{\sqrt{3}}{2} \approx 0.866$

**b**

| $\theta$ (degrees) | 30° | 45° | 60° | 135° | 150° | 240° | 315° |
|---|---|---|---|---|---|---|---|
| $\theta$ (radians) | $\frac{\pi}{6}$ | $\frac{\pi}{4}$ | $\frac{\pi}{3}$ | $\frac{3\pi}{4}$ | $\frac{5\pi}{6}$ | $\frac{4\pi}{3}$ | $\frac{7\pi}{4}$ |
| sine | $\frac{1}{2}$ | $\frac{1}{\sqrt{2}}$ | $\frac{\sqrt{3}}{2}$ | $\frac{1}{\sqrt{2}}$ | $\frac{1}{2}$ | $-\frac{\sqrt{3}}{2}$ | $-\frac{1}{\sqrt{2}}$ |
| cosine | $\frac{\sqrt{3}}{2}$ | $\frac{1}{\sqrt{2}}$ | $\frac{1}{2}$ | $-\frac{1}{\sqrt{2}}$ | $-\frac{\sqrt{3}}{2}$ | $-\frac{1}{2}$ | $\frac{1}{\sqrt{2}}$ |
| tangent | $\frac{1}{\sqrt{3}}$ | 1 | $\sqrt{3}$ | $-1$ | $-\frac{1}{\sqrt{3}}$ | $\sqrt{3}$ | $-1$ |

**4** **a** **i** 0.985 **ii** 0.985 **iii** 0.866 **iv** 0.866 **v** 0.5 **vi** 0.5
**vii** 0.707 **viii** 0.707

**b** $\sin(180° - \theta) = \sin\theta$ as the points have the same $y$-coordinate.

**c** 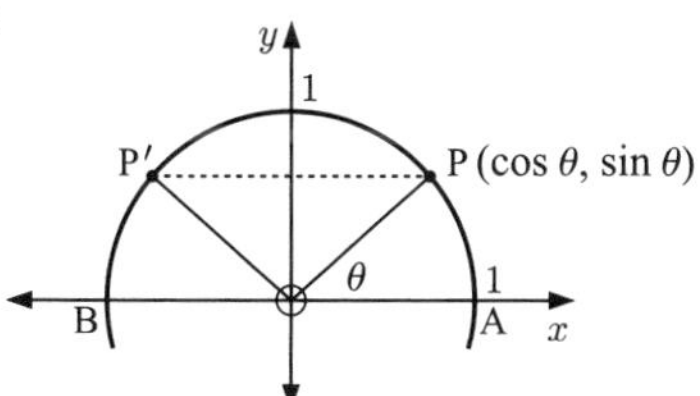


The diagram shows P reflected in the $y$-axis to P′, so $\text{P}'\widehat{\text{O}}\text{B} = \text{P}\widehat{\text{O}}\text{A} = \theta$, and P′ has coordinates $(-\cos\theta, \sin\theta)$.
But $\text{A}\widehat{\text{O}}\text{P}' = 180° - \theta$ $\{\text{A}\widehat{\text{O}}\text{P}' + \text{P}'\widehat{\text{O}}\text{B} = 180°\}$,
so P′ has coordinates $(\cos(180° - \theta), \sin(180° - \theta))$.
$\therefore \sin(180° - \theta) = \sin\theta$ {equating $y$-coordinates of P′}

**d** **i** $180° - 45° = 135°$ **ii** $180° - 51° = 129°$ **iii** $\pi - \frac{\pi}{3} = \frac{2\pi}{3}$
**iv** $\pi - \frac{\pi}{6} = \frac{5\pi}{6}$ {using $\sin(180° - \theta) = \theta$}

**5** **a** **i** 0.342 **ii** $-0.342$ **iii** 0.5 **iv** $-0.5$ **v** 0.906 **vi** $-0.906$
**vii** 0.174 **viii** $-0.174$

**b** $\cos(180° - \theta) = -\cos\theta$

**c** 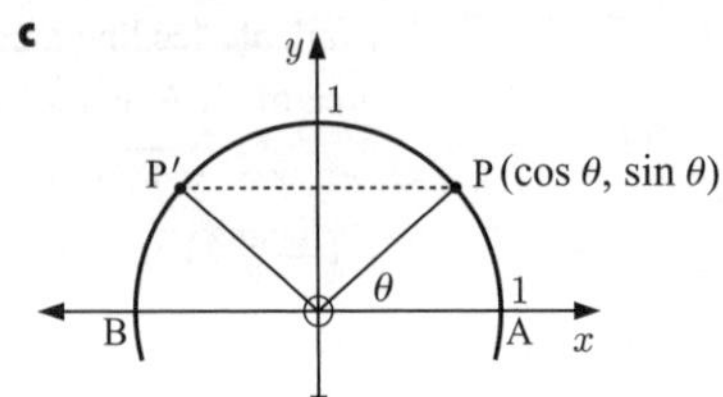


The diagram shows P reflected in the $y$-axis to P′, so $P'\widehat{O}B = P\widehat{O}A = \theta$, and P′ has coordinates $(-\cos\theta, \sin\theta)$.
But $A\widehat{O}P' = 180° - \theta$ $\{A\widehat{O}P' + P'\widehat{O}B = 180°\}$,
so P′ has coordinates $(\cos(180° - \theta), \sin(180° - \theta))$.
$\therefore$ $\cos(180° - \theta) = -\cos\theta$ {equating $x$-coordinates of P′}

**d** **i** $180° - 40° = 140°$ **ii** $180° - 19° = 161°$ **iii** $\pi - \frac{\pi}{5} = \frac{4\pi}{5}$
**iv** $\pi - \frac{2\pi}{5} = \frac{3\pi}{5}$ {using $\cos(180° - \theta) = -\cos\theta$}

**6** **a** $\sin 137°$
$= \sin(180 - 137)°$
$= \sin 43°$
$\approx 0.6820$

**b** $\sin 59°$
$= \sin(180 - 59)°$
$= \sin 121°$
$\approx 0.8572$

**c** $\cos 143°$
$= -\cos(180 - 143)°$
$= -\cos 37°$
$\approx -0.7986$

**d** $\cos 24°$
$= -\cos(180 - 24)°$
$= -\cos 156°$
$\approx 0.9135$

**e** $\sin 115°$
$= \sin(180 - 115)°$
$= \sin 65°$
$\approx 0.9063$

**f** $\cos 132°$
$= -\cos(180 - 132)°$
$= -\cos 48°$
$\approx -0.6691$

**7** **a**

| *Quadrant* | *Degree measure* | *Radian measure* | $\cos\theta$ | $\sin\theta$ | $\tan\theta$ |
|---|---|---|---|---|---|
| 1 | $0° < \theta < 90°$ | $0 < \theta < \frac{\pi}{2}$ | +ve | +ve | +ve |
| 2 | $90° < \theta < 180°$ | $\frac{\pi}{2} < \theta < \pi$ | −ve | +ve | −ve |
| 3 | $180° < \theta < 270°$ | $\pi < \theta < \frac{3\pi}{2}$ | −ve | −ve | +ve |
| 4 | $270° < \theta < 360°$ | $\frac{3\pi}{2} < \theta < 2\pi$ | +ve | −ve | −ve |

**b** **i** 1 and 4
**ii** 2 and 3
**iii** 3
**iv** 2

**8** **a** $A\widehat{O}Q = 180° - \theta$ or $\pi - \theta$ radians
**b** [OQ] is a reflection of [OP] in the $y$-axis and so Q has coordinates $(-\cos\theta, \sin\theta)$.
**c** $\cos(180° - \theta) = -\cos\theta$, $\sin(180° - \theta) = \sin\theta$

**9** **a**

| $\theta^c$ | $\sin\theta$ | $\sin(-\theta)$ | $\cos\theta$ | $\cos(-\theta)$ |
|---|---|---|---|---|
| 0.75 | 0.682 | −0.682 | 0.732 | 0.732 |
| 1.772 | 0.980 | −0.980 | −0.200 | −0.200 |
| 3.414 | −0.269 | 0.269 | −0.963 | −0.963 |
| 6.25 | −0.0332 | 0.0332 | 0.999 | 0.999 |
| −1.17 | −0.921 | 0.921 | 0.390 | 0.390 |

**b** Suspect that $\sin(-\theta) = -\sin\theta$ and $\cos(-\theta) = \cos\theta$.

**c** **i** P is reflected in the $x$-axis to Q, so Q has coordinates $(\cos\theta, -\sin\theta)$.
But Q has coordinates $(\cos(-\theta), \sin(-\theta))$.
$\therefore$ $Q(\cos(-\theta), \sin(-\theta)) = Q(\cos\theta, -\sin\theta)$.
So the suspicion is correct.

**ii** The point Q on the unit circle corresponds to the angle $(2\pi - \theta)$ and the angle $(-\theta)$.
$\therefore$ $\cos(2\pi - \theta) = \cos(-\theta)$
But $\cos(-\theta) = \cos\theta$ {from **c i**}
$\therefore$ $\cos(2\pi - \theta) = \cos\theta$

## EXERCISE 8D.1

**1** **a** $\cos^2\theta+\sin^2\theta=1$
$\therefore\ \cos^2\theta+(\frac{1}{2})^2=1$
$\therefore\ \cos^2\theta=\frac{3}{4}$
$\therefore\ \cos\theta=\pm\frac{\sqrt{3}}{2}$

**b** $\cos^2\theta+\sin^2\theta=1$
$\therefore\ \cos^2\theta+(-\frac{1}{3})^2=1$
$\therefore\ \cos^2\theta=\frac{8}{9}$
$\therefore\ \cos\theta=\pm\frac{\sqrt{8}}{3}$
$\therefore\ \cos\theta=\pm\frac{2\sqrt{2}}{3}$

**c** $\cos^2\theta+\sin^2\theta=1$
$\therefore\ \cos^2\theta+0^2=1$
$\therefore\ \cos\theta=\pm1$

**d** $\cos^2\theta+\sin^2\theta=1$
$\therefore\ \cos^2\theta+(-1)^2=1$
$\therefore\ \cos\theta=0$

**2** **a** $\cos^2\theta+\sin^2\theta=1$
$\therefore\ (\frac{4}{5})^2+\sin^2\theta=1$
$\therefore\ \sin^2\theta=\frac{9}{25}$
$\therefore\ \sin\theta=\pm\frac{3}{5}$

**b** $\cos^2\theta+\sin^2\theta=1$
$\therefore\ (-\frac{3}{4})^2+\sin^2\theta=1$
$\therefore\ \sin^2\theta=\frac{7}{16}$
$\therefore\ \sin\theta=\pm\frac{\sqrt{7}}{4}$

**c** $\cos^2\theta+\sin^2\theta=1$
$\therefore\ 1^2+\sin^2\theta=1$
$\therefore\ \sin^2\theta=0$
$\therefore\ \sin\theta=0$

**d** $\cos^2\theta+\sin^2\theta=1$
$\therefore\ 0^2+\sin^2\theta=1$
$\therefore\ \sin\theta=\pm1$

**3** **a** $\cos^2\theta+\sin^2\theta=1$
$\therefore\ \frac{4}{9}+\sin^2\theta=1$
$\therefore\ \sin^2\theta=\frac{5}{9}$
$\therefore\ \sin\theta=\pm\frac{\sqrt{5}}{3}$
But $\theta$ is in quadrant 1
where $\sin\theta>0$
$\therefore\ \sin\theta=\frac{\sqrt{5}}{3}$

**b** $\cos^2\theta+\sin^2\theta=1$
$\therefore\ \cos^2\theta+\frac{4}{25}=1$
$\therefore\ \cos^2\theta=\frac{21}{25}$
$\therefore\ \cos\theta=\pm\frac{\sqrt{21}}{5}$
But $\theta$ is in quadrant 2
where $\cos\theta<0$
$\therefore\ \cos\theta=-\frac{\sqrt{21}}{5}$

**c** $\cos^2\theta+\sin^2\theta=1$
$\therefore\ \cos^2\theta+\frac{9}{25}=1$
$\therefore\ \cos^2\theta=\frac{16}{25}$
$\therefore\ \cos\theta=\pm\frac{4}{5}$
But $\theta$ is in quadrant 4
where $\cos\theta>0$
$\therefore\ \cos\theta=\frac{4}{5}$

**d** $\cos^2\theta+\sin^2\theta=1$
$\therefore\ \frac{25}{169}+\sin^2\theta=1$
$\therefore\ \sin^2\theta=\frac{144}{169}$
$\therefore\ \sin\theta=\pm\frac{12}{13}$
But $\theta$ is in quadrant 3
where $\sin\theta<0$
$\therefore\ \sin\theta=-\frac{12}{13}$

**4** **a** $\cos^2x+\sin^2x=1$
$\therefore\ \cos^2x+\frac{1}{9}=1$
$\therefore\ \cos^2x=\frac{8}{9}$
$\therefore\ \cos x=\pm\frac{2\sqrt{2}}{3}$
But $x$ is in quadrant 2
where $\cos x<0$
$\therefore\ \cos x=-\frac{2\sqrt{2}}{3}$
and so $\tan x=\dfrac{\sin x}{\cos x}=\dfrac{\frac{1}{3}}{-\frac{2\sqrt{2}}{3}}=-\frac{1}{2\sqrt{2}}$

**b** $\cos^2x+\sin^2x=1$
$\therefore\ \frac{1}{25}+\sin^2x=1$
$\therefore\ \sin^2x=\frac{24}{25}$
$\therefore\ \sin x=\pm\frac{2\sqrt{6}}{5}$
But $x$ is in quadrant 4
where $\sin x<0$
$\therefore\ \sin x=-\frac{2\sqrt{6}}{5}$
and so $\tan x=\dfrac{\sin x}{\cos x}=\dfrac{-\frac{2\sqrt{6}}{5}}{\frac{1}{5}}=-2\sqrt{6}$

**c** $\cos^2 x + \sin^2 x = 1$

$\therefore \cos^2 x + \frac{1}{3} = 1$

$\therefore \cos^2 x = \frac{2}{3}$

$\therefore \cos x = \pm\frac{\sqrt{2}}{\sqrt{3}}$

But $x$ is in quadrant 3 where $\cos x < 0$

$\therefore \cos x = -\frac{\sqrt{2}}{\sqrt{3}}$

and so $\tan x = \dfrac{\sin x}{\cos x} = \dfrac{-\frac{1}{\sqrt{3}}}{-\frac{\sqrt{2}}{\sqrt{3}}} = \frac{1}{\sqrt{2}}$

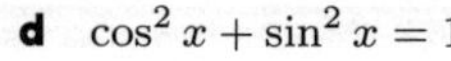

**d** $\cos^2 x + \sin^2 x = 1$

$\therefore \frac{9}{16} + \sin^2 x = 1$

$\therefore \sin^2 x = \frac{7}{16}$

$\therefore \sin x = \pm\frac{\sqrt{7}}{4}$

But $x$ is in quadrant 2 where $\sin x > 0$

$\therefore \sin x = \frac{\sqrt{7}}{4}$

and so $\tan x = \dfrac{\sin x}{\cos x} = \dfrac{\frac{\sqrt{7}}{4}}{-\frac{3}{4}} = -\frac{\sqrt{7}}{3}$

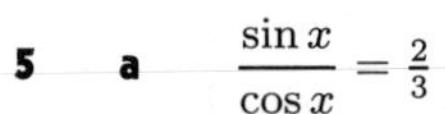

**5** **a** $\dfrac{\sin x}{\cos x} = \frac{2}{3}$

$\therefore \sin x = \frac{2}{3}\cos x$

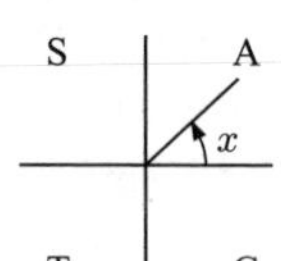


Now $\cos^2 x + \sin^2 x = 1$

$\therefore \cos^2 x + \frac{4}{9}\cos^2 x = 1$

$\therefore \frac{13}{9}\cos^2 x = 1$

$\therefore \cos x = \pm\frac{3}{\sqrt{13}}$

But $x$ is in quadrant 1

$\therefore \cos x$ and $\sin x$ are positive.

$\therefore \cos x = \frac{3}{\sqrt{13}}, \quad \sin x = \frac{2}{\sqrt{13}}$

**b** $\dfrac{\sin x}{\cos x} = -\frac{4}{3}$

$\therefore \sin x = -\frac{4}{3}\cos x$

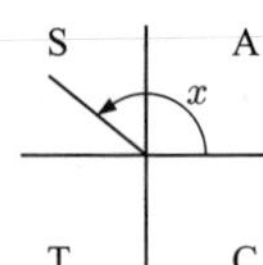


Now $\cos^2 x + \sin^2 x = 1$

$\therefore \cos^2 x + \frac{16}{9}\cos^2 x = 1$

$\therefore \frac{25}{9}\cos^2 x = 1$

$\therefore \cos x = \pm\frac{3}{5}$

But $x$ is in quadrant 2

$\therefore \cos x$ is negative and $\sin x$ is positive.

$\therefore \cos x = -\frac{3}{5}, \quad \sin x = \frac{4}{5}$

**c** $\dfrac{\sin x}{\cos x} = \frac{\sqrt{5}}{3}$

$\therefore \sin x = \frac{\sqrt{5}}{3}\cos x$

S A x T C

Now $\cos^2 x + \sin^2 x = 1$

$\therefore \cos^2 x + \frac{5}{9}\cos^2 x = 1$

$\therefore \frac{14}{9}\cos^2 x = 1$

$\therefore \cos x = \pm\frac{3}{\sqrt{14}}$

But $x$ is in quadrant 3

$\therefore \cos x$ and $\sin x$ are both negative.

$\therefore \cos x = -\frac{3}{\sqrt{14}}, \quad \sin x = -\frac{\sqrt{5}}{\sqrt{14}}$

**d** $\dfrac{\sin x}{\cos x} = -\frac{12}{5}$

$\therefore \sin x = -\frac{12}{5}\cos x$

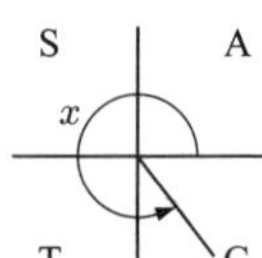


Now $\cos^2 x + \sin^2 x = 1$

$\therefore \cos^2 x + \frac{144}{25}\cos^2 x = 1$

$\therefore \frac{169}{25}\cos^2 x = 1$

$\therefore \cos x = \pm\frac{5}{13}$

But $x$ is in quadrant 4

$\therefore \cos x$ is positive and $\sin x$ is negative.

$\therefore \cos x = \frac{5}{13}, \quad \sin x = -\frac{12}{13}$

**6** $\dfrac{\sin x}{\cos x} = k$

$\therefore \sin x = k\cos x$

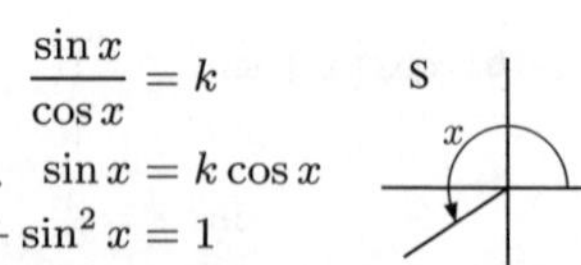


Now $\cos^2 x + \sin^2 x = 1$

$\therefore \cos^2 x + k^2\cos^2 x = 1$

$\therefore (k^2+1)\cos^2 x = 1$

$\therefore \cos x = \dfrac{\pm 1}{\sqrt{k^2+1}}$

But $x$ is in quadrant 3, $\therefore$ $\cos x$ and $\sin x$ are both negative.

$\therefore \cos x = \dfrac{-1}{\sqrt{k^2+1}}, \quad \sin x = \dfrac{-k}{\sqrt{k^2+1}}$

## EXERCISE 8D.2

**1** **a** $\tan\theta = 4$

Using technology,

$\tan^{-1}(4) \approx 1.33$

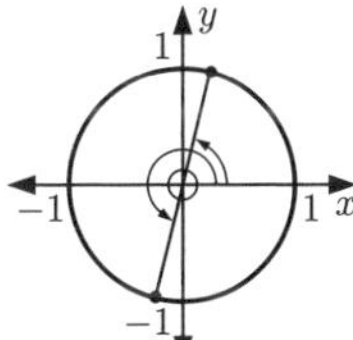


$\therefore\ \theta \approx 1.33$ or $\pi + 1.33$

$\therefore\ \theta \approx 1.33$ or $4.47$

**b** $\cos\theta = 0.83$

Using technology,

$\cos^{-1}(0.83) \approx 0.592$

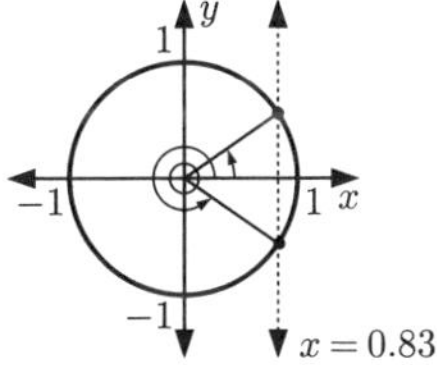


$\therefore\ \theta \approx 0.592$ or $2\pi - 0.592$

$\therefore\ \theta \approx 0.592$ or $5.69$

**c** $\sin\theta = \frac{3}{5}$

Using technology,

$\sin^{-1}(\frac{3}{5}) \approx 0.644$

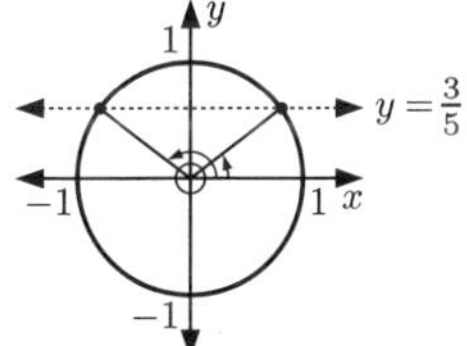


$\therefore\ \theta \approx 0.644$ or $\pi - 0.644$

$\therefore\ \theta \approx 0.644$ or $2.50$

**d** $\cos\theta = 0$

$\therefore\ \cos^{-1}(0) = \frac{\pi}{2}$

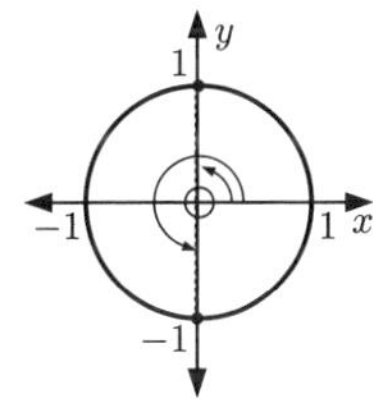


$\therefore\ \theta = \frac{\pi}{2}$ or $2\pi - \frac{\pi}{2}$

$\therefore\ \theta = \frac{\pi}{2}$ or $\frac{3\pi}{2}$

**e** $\tan\theta = 1.2$

Using technology,

$\tan^{-1}(1.2) \approx 0.876$

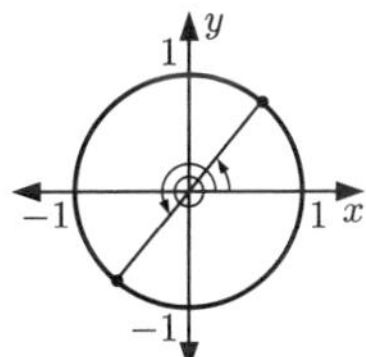


$\therefore\ \theta \approx 0.876$ or $\pi + 0.876$

$\therefore\ \theta \approx 0.876$ or $4.02$

**f** $\cos\theta = 0.7816$

Using technology,

$\cos^{-1}(0.7816) \approx 0.674$

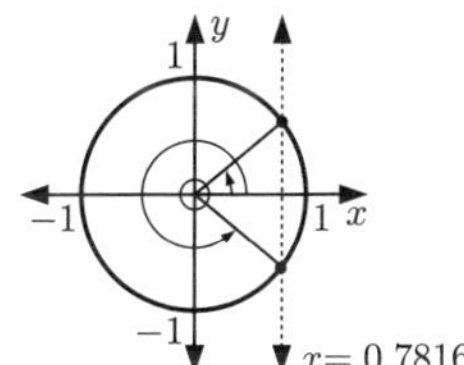


$\therefore\ \theta \approx 0.674$ or $2\pi - 0.674$

$\therefore\ \theta \approx 0.674$ or $5.61$

**g** $\sin\theta = \frac{1}{11}$

Using technology,

$\sin^{-1}(\frac{1}{11}) \approx 0.0910$

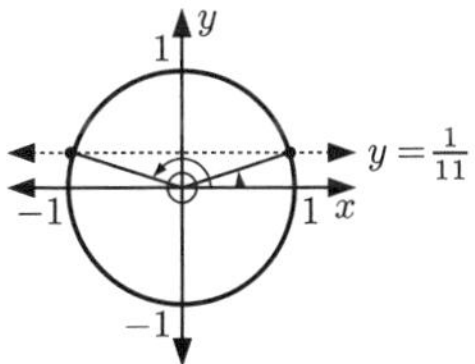


$\therefore\ \theta \approx 0.0910$ or $\pi - 0.0910$

$\therefore\ \theta \approx 0.0910$ or $3.05$

**h** $\tan\theta = 20.2$

Using technology,

$\tan^{-1}(20.2) \approx 1.52$

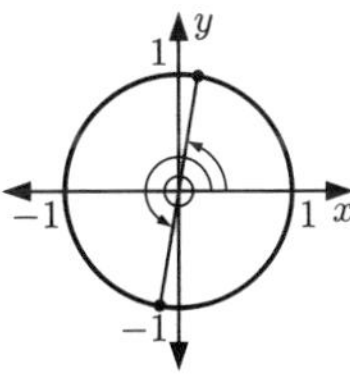


$\therefore\ \theta \approx 1.52$ or $\pi + 1.52$

$\therefore\ \theta \approx 1.52$ or $4.66$

**i** $\sin\theta = \frac{39}{40}$

Using technology,

$\sin^{-1}(\frac{39}{40}) \approx 1.35$

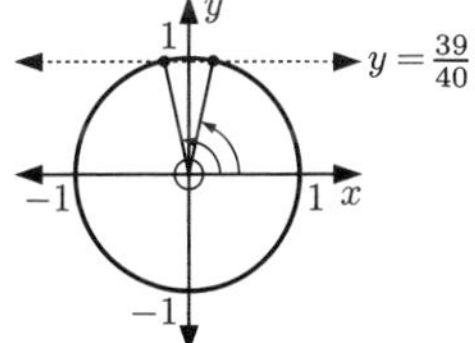


$\therefore\ \theta \approx 1.35$ or $\pi - 1.35$

$\therefore\ \theta \approx 1.35$ or $1.79$

**2** **a** $\cos\theta = -\frac{1}{4}$

Using technology,

$\cos^{-1}(-\frac{1}{4}) \approx 1.82$

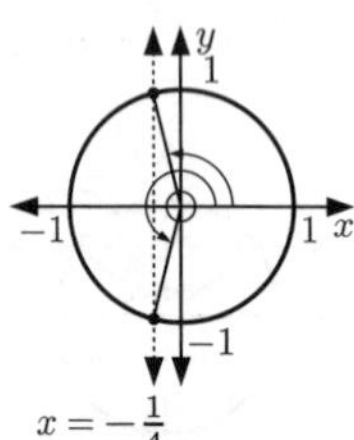


$\therefore \quad \theta \approx 1.82$ or $2\pi - 1.82$

$\therefore \quad \theta \approx 1.82$ or $4.46$

**b** $\sin\theta = 0$

$\therefore \quad \sin^{-1}(0) = 0$

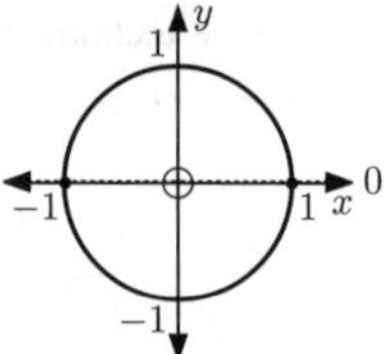


$\therefore \quad \theta = 0$ or $\pi - 0$ or $2\pi$

$\therefore \quad \theta = 0, \pi,$ or $2\pi$

**c** $\tan\theta = -3.1$

Using technology,

$\tan^{-1}(-3.1) \approx -1.26$

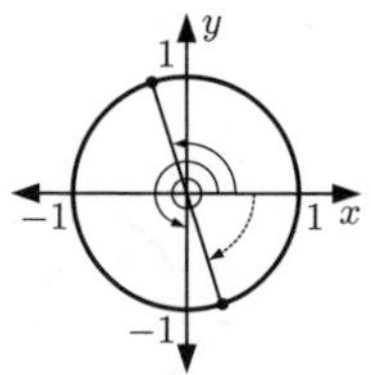


But $0 \leqslant \theta \leqslant 2\pi$

$\therefore \quad \theta \approx \pi - 1.26$ or $2\pi - 1.26$

$\therefore \quad \theta \approx 1.88$ or $5.02$

**d** $\sin\theta = -0.421$

Using technology,

$\sin^{-1}(-0.421) \approx -0.435$

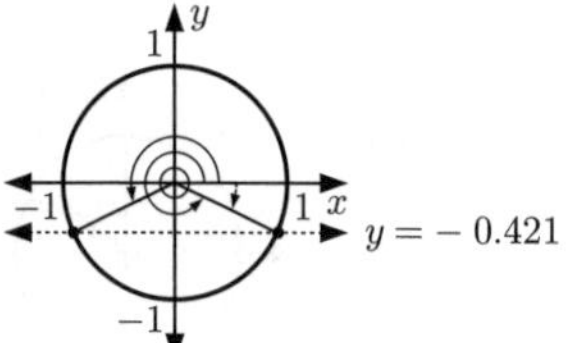


But $0 \leqslant \theta \leqslant 2\pi$

$\therefore \quad \theta \approx \pi + 0.435$ or $2\pi - 0.435$

$\therefore \quad \theta \approx 3.58$ or $5.85$

**e** $\tan\theta = -6.67$

Using technology,

$\tan^{-1}(-6.67) \approx -1.42$

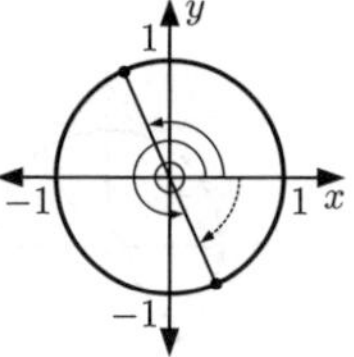


But $0 \leqslant \theta \leqslant 2\pi$

$\therefore \quad \theta \approx \pi - 1.42$ or $2\pi - 1.42$

$\therefore \quad \theta \approx 1.72$ or $4.86$

**f** $\cos\theta = -\frac{2}{17}$

Using technology,

$\cos^{-1}(-\frac{2}{17}) \approx 1.69$

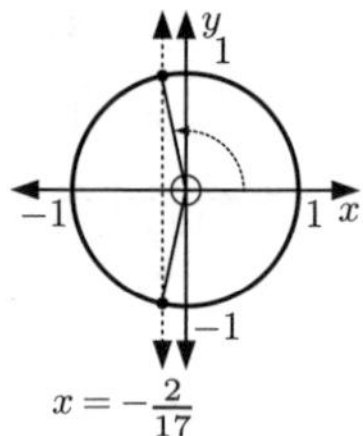


$\therefore \quad \theta \approx 1.69$ or $2\pi - 1.69$

$\therefore \quad \theta \approx 1.69$ or $4.59$

**g** $\tan\theta = -\sqrt{5}$

Using technology,

$\tan^{-1}(-\sqrt{5}) \approx -1.15$

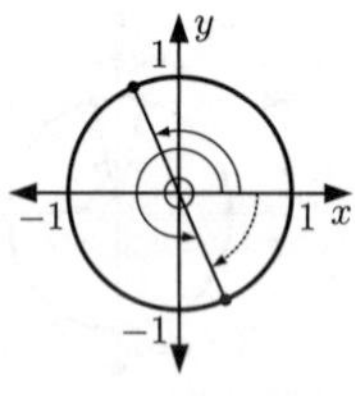


But $0 \leqslant \theta \leqslant 2\pi$

$\therefore \quad \theta \approx \pi - 1.15$ or $2\pi - 1.15$

$\therefore \quad \theta \approx 1.99$ or $5.13$

**h** $\cos\theta = -\frac{1}{\sqrt{3}}$

Using technology,

$\cos^{-1}(-\frac{1}{\sqrt{3}}) \approx 2.19$

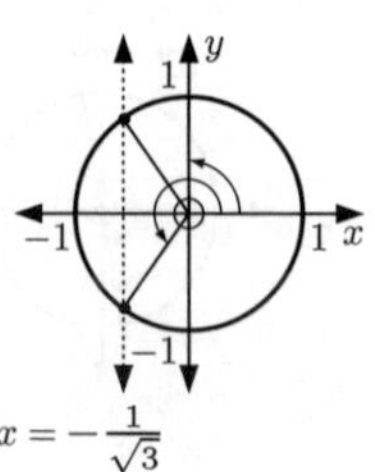


$\therefore \quad \theta \approx 2.19$ or $2\pi - 2.19$

$\therefore \quad \theta \approx 2.19$ or $4.10$

**i** $\sin\theta = -\frac{\sqrt{2}}{\sqrt{5}}$

Using technology,

$\sin^{-1}(-\frac{\sqrt{2}}{\sqrt{5}}) \approx -0.685$

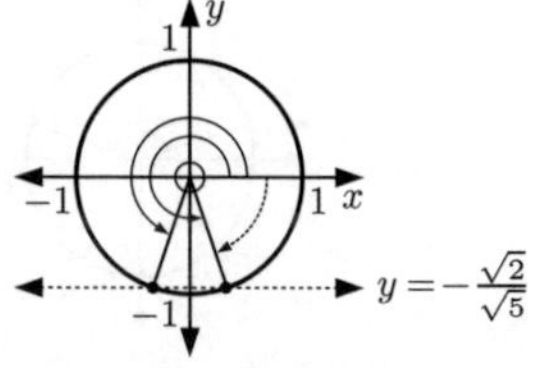


But $0 \leqslant \theta \leqslant 2\pi$

$\therefore \quad \theta \approx \pi + 0.685$ or $2\pi - 0.685$

$\therefore \quad \theta \approx 3.83$ or $5.60$

## EXERCISE 8E

**1**

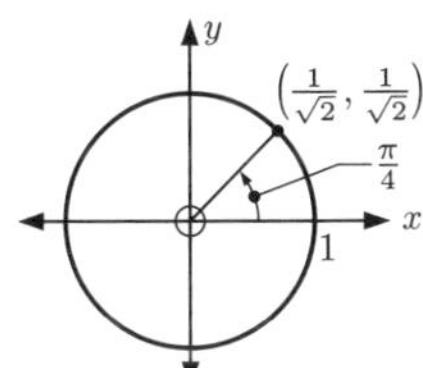


So, $\cos(\frac{\pi}{4}) = \frac{1}{\sqrt{2}}$

$\sin(\frac{\pi}{4}) = \frac{1}{\sqrt{2}}$

$\tan(\frac{\pi}{4}) = \dfrac{\frac{1}{\sqrt{2}}}{\frac{1}{\sqrt{2}}} = 1$

You should draw separate unit circle diagrams for each case.

| | **a** | **b** | **c** | **d** | **e** |
|---|---|---|---|---|---|
| $\sin\theta$ | $\frac{1}{\sqrt{2}}$ | $\frac{1}{\sqrt{2}}$ | $-\frac{1}{\sqrt{2}}$ | $0$ | $-\frac{1}{\sqrt{2}}$ |
| $\cos\theta$ | $\frac{1}{\sqrt{2}}$ | $-\frac{1}{\sqrt{2}}$ | $\frac{1}{\sqrt{2}}$ | $-1$ | $-\frac{1}{\sqrt{2}}$ |
| $\tan\theta$ | $1$ | $-1$ | $-1$ | $0$ | $1$ |

**2**

So, $\cos(\frac{\pi}{6}) = \frac{\sqrt{3}}{2}$

$\sin(\frac{\pi}{6}) = \frac{1}{2}$

$\tan(\frac{\pi}{6}) = \dfrac{\frac{1}{2}}{\frac{\sqrt{3}}{2}} = \frac{1}{\sqrt{3}}$

You should draw separate unit circle diagrams for each case.

| | **a** | **b** | **c** | **d** | **e** |
|---|---|---|---|---|---|
| $\sin\beta$ | $\frac{1}{2}$ | $\frac{\sqrt{3}}{2}$ | $-\frac{1}{2}$ | $-\frac{\sqrt{3}}{2}$ | $-\frac{1}{2}$ |
| $\cos\beta$ | $\frac{\sqrt{3}}{2}$ | $-\frac{1}{2}$ | $-\frac{\sqrt{3}}{2}$ | $\frac{1}{2}$ | $\frac{\sqrt{3}}{2}$ |
| $\tan\beta$ | $\frac{1}{\sqrt{3}}$ | $-\sqrt{3}$ | $\frac{1}{\sqrt{3}}$ | $-\sqrt{3}$ | $-\frac{1}{\sqrt{3}}$ |

**3** **a** $120° = \frac{2\pi}{3}$ which is a multiple of $\frac{\pi}{6}$

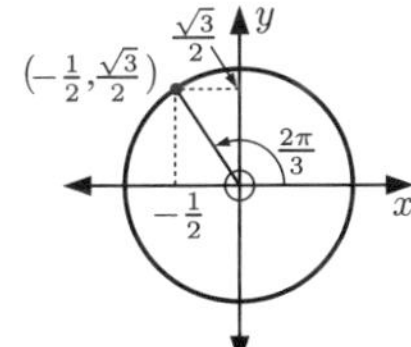


So, $\cos 120° = -\frac{1}{2}$

$\sin 120° = \frac{\sqrt{3}}{2}$

$\tan 120° = -\sqrt{3}$

**b** $-45° = -\frac{\pi}{4}$ which is a multiple of $\frac{\pi}{4}$

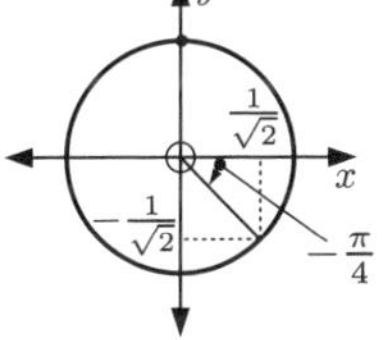


So, $\cos(-45°) = \frac{1}{\sqrt{2}}$

$\sin(-45°) = -\frac{1}{\sqrt{2}}$

$\tan(-45°) = -1$

**4** **a** $90° = \frac{\pi}{2}$

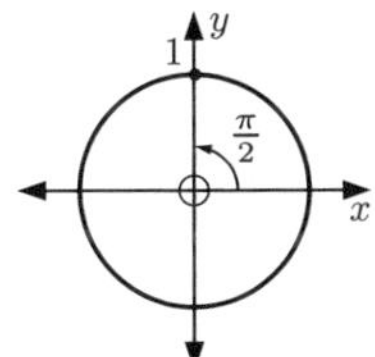


$\cos 90° = 0, \quad \sin 90° = 1$

**b** $\tan 90° = \dfrac{\sin 90°}{\cos 90°} = \dfrac{1}{0}$

$\tan 90°$ is undefined.

**5** **a**

$\sin^2 60°$
$= \sin 60° \times \sin 60°$
$= \frac{\sqrt{3}}{2} \times \frac{\sqrt{3}}{2}$
$= \frac{3}{4}$

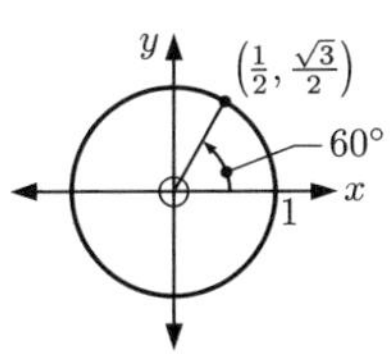


**b**

$\sin 30° \cos 60°$
$= \frac{1}{2} \times \frac{1}{2}$
$= \frac{1}{4}$

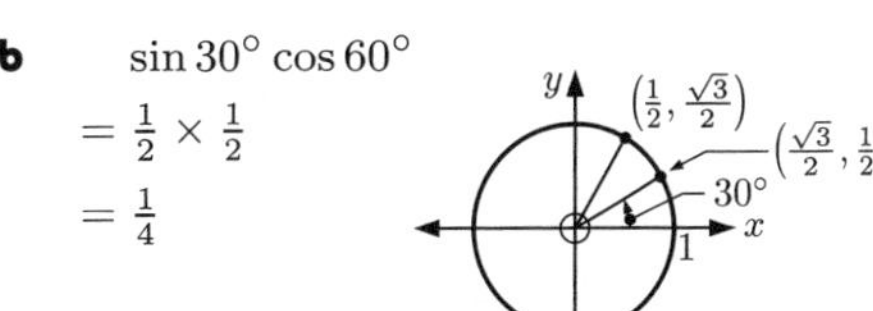

**c** $4\sin 60° \cos 30°$

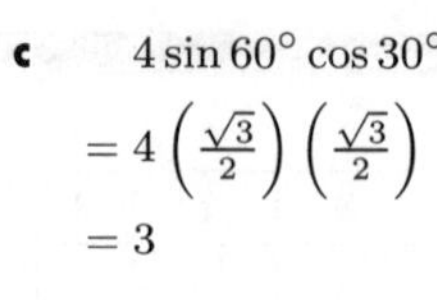

$= 4\left(\frac{\sqrt{3}}{2}\right)\left(\frac{\sqrt{3}}{2}\right)$

$= 3$

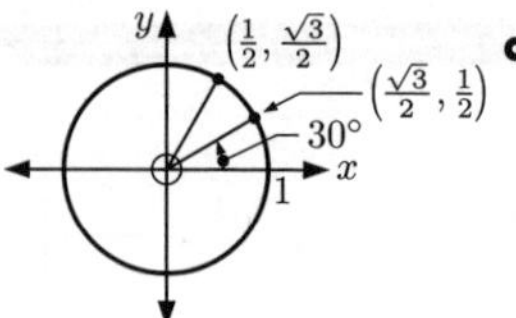


**d** $1 - \cos^2(\frac{\pi}{6})$

$= 1 - \left(\frac{\sqrt{3}}{2}\right)^2$

$= 1 - \frac{3}{4}$

$= \frac{1}{4}$

**e** $\sin^2(\frac{2\pi}{3}) - 1$

$= \left(\frac{\sqrt{3}}{2}\right)^2 - 1$

$= \frac{3}{4} - 1$

$= -\frac{1}{4}$

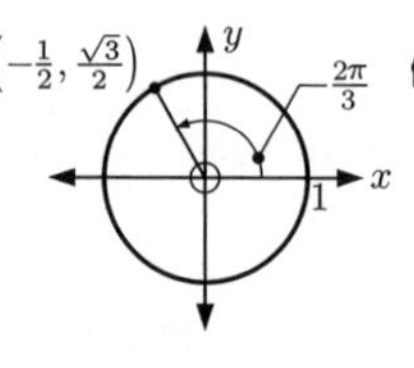


**f** $\cos^2(\frac{\pi}{4}) - \sin(\frac{7\pi}{6})$

$= \left(\frac{1}{\sqrt{2}}\right)^2 - (-\frac{1}{2})$

$= \frac{1}{2} + \frac{1}{2}$

$= 1$

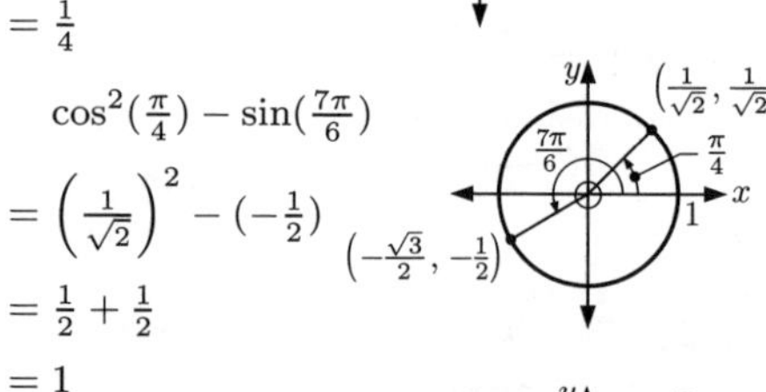


**g** $\sin(\frac{3\pi}{4}) - \cos(\frac{5\pi}{4})$

$= \frac{1}{\sqrt{2}} - \left(-\frac{1}{\sqrt{2}}\right)$

$= \frac{1}{\sqrt{2}} + \frac{1}{\sqrt{2}}$

$= \frac{2}{\sqrt{2}}$ or $\sqrt{2}$

**h** $1 - 2\sin^2(\frac{7\pi}{6})$

$= 1 - 2(-\frac{1}{2})^2$

$= 1 - 2 \times \frac{1}{4}$

$= \frac{1}{2}$

**i** $\cos^2(\frac{5\pi}{6}) - \sin^2(\frac{5\pi}{6})$

$= \left(-\frac{\sqrt{3}}{2}\right)^2 - (\frac{1}{2})^2$

$= \frac{3}{4} - \frac{1}{4}$

$= \frac{1}{2}$

**j** $\tan^2(\frac{\pi}{3}) - 2\sin^2(\frac{\pi}{4})$

$= \left(\sqrt{3}\right)^2 - 2\left(\frac{1}{\sqrt{2}}\right)^2$

$= 3 - 2(\frac{1}{2})$

$= 2$

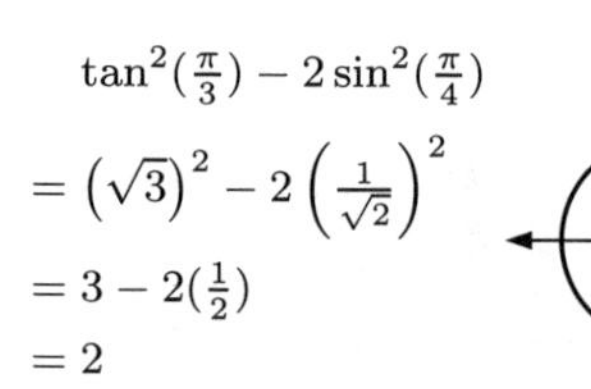

**k** $2\tan(-\frac{5\pi}{4}) - \sin(\frac{3\pi}{2})$

$= 2(-1) - (-1)$

$= -1$

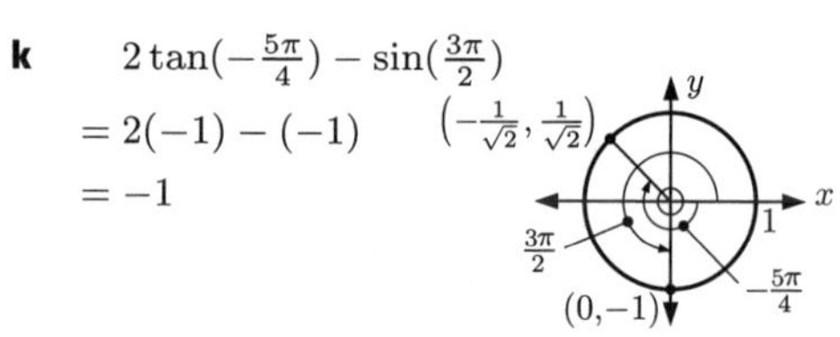


**l** $\dfrac{2\tan 150°}{1 - \tan^2 150°}$

$= \dfrac{2(-\frac{1}{\sqrt{3}})}{1 - (-\frac{1}{\sqrt{3}})^2}$

$= \dfrac{-\frac{2}{\sqrt{3}}}{1 - \frac{1}{3}}$

$= \dfrac{-\frac{2}{\sqrt{3}}}{\frac{2}{3}} = -\frac{3}{\sqrt{3}} = -\sqrt{3}$

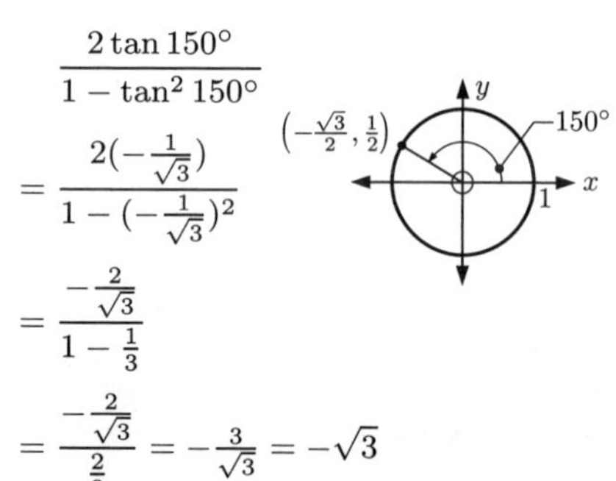


**6** **a**

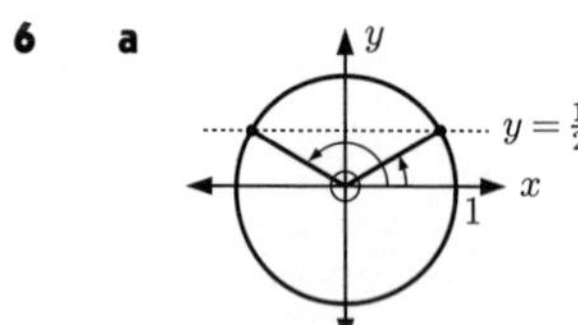


$\theta = 30°, 150°$

**b**

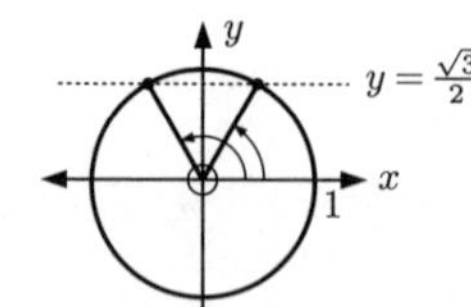


$\theta = 60°, \ 120°$

**c**

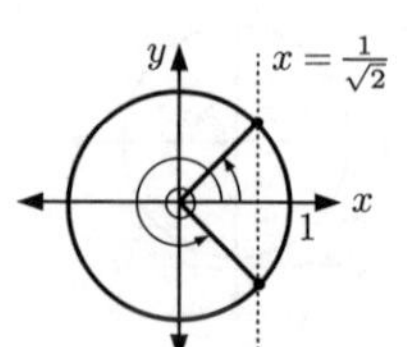


$\theta = 45°, 315°$

**d**

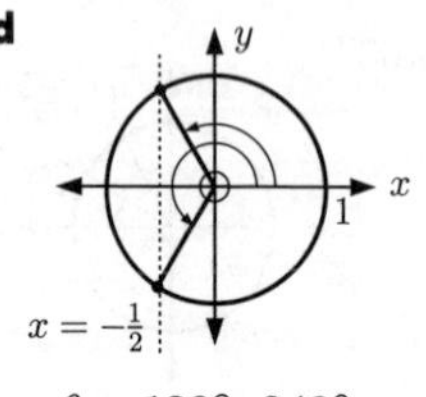


$\theta = 120°, 240°$

**e**

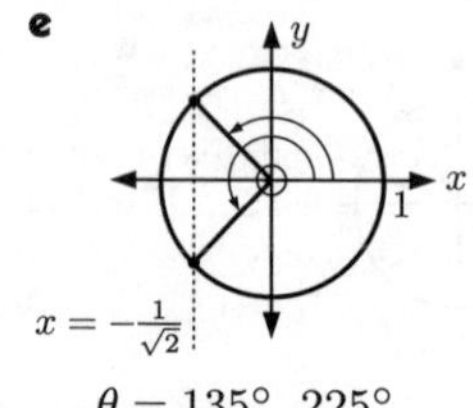


$\theta = 135°, 225°$

**f**

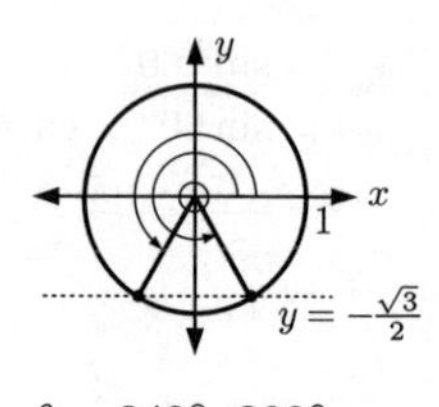


$\theta = 240°, 300°$

**7** **a**

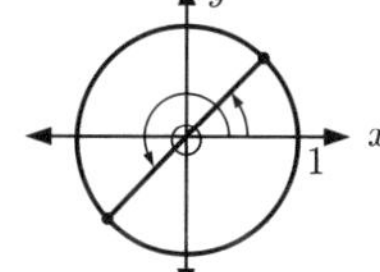


$\theta = \frac{\pi}{4}, \frac{5\pi}{4}$

**b**

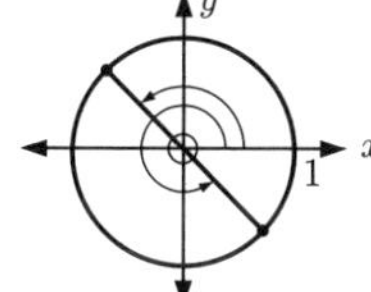


$\theta = \frac{3\pi}{4}, \frac{7\pi}{4}$

**c**

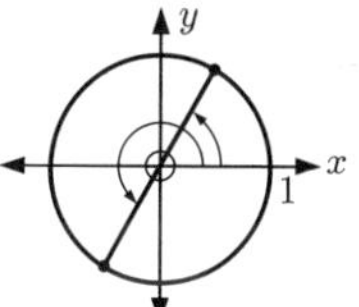


$\theta = \frac{\pi}{3}, \frac{4\pi}{3}$

**d**

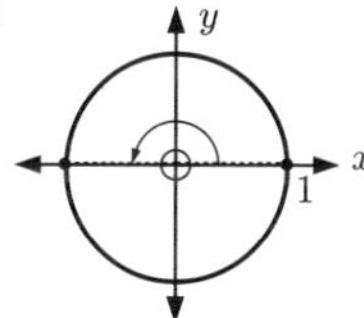


$\theta = 0, \pi, 2\pi$

**e**

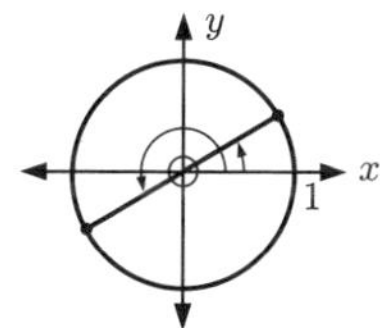


$\theta = \frac{\pi}{6}, \frac{7\pi}{6}$

**f**

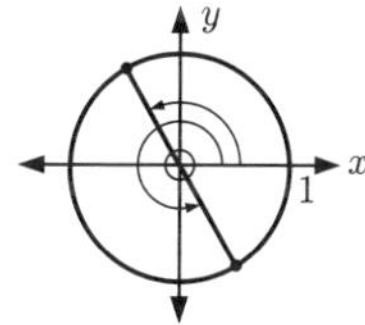


$\theta = \frac{2\pi}{3}, \frac{5\pi}{3}$

**8** **a**

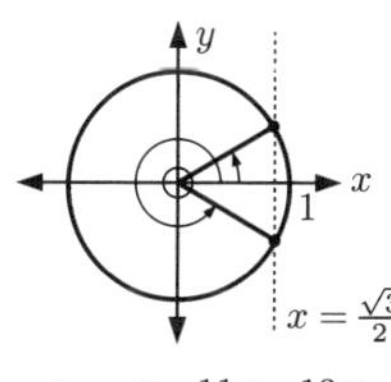


$\theta = \frac{\pi}{6}, \frac{11\pi}{6}, \frac{13\pi}{6}, \frac{23\pi}{6}$

**b**

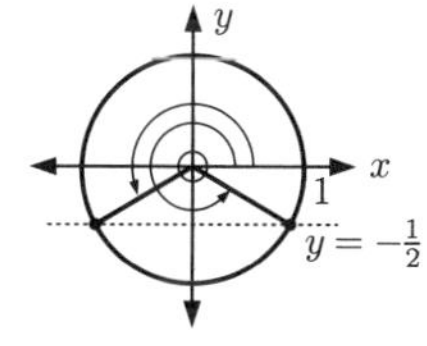


$\theta = \frac{7\pi}{6}, \frac{11\pi}{6}, \frac{19\pi}{6}, \frac{23\pi}{6}$

**c**

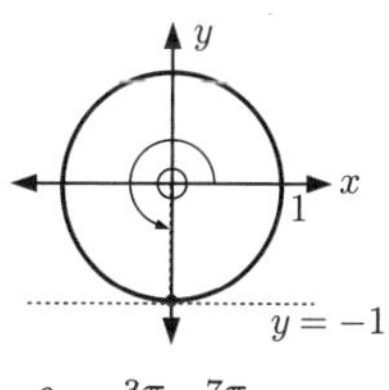


$\theta = \frac{3\pi}{2}, \frac{7\pi}{2}$

**9** **a** $\cos\theta = \frac{1}{2}$

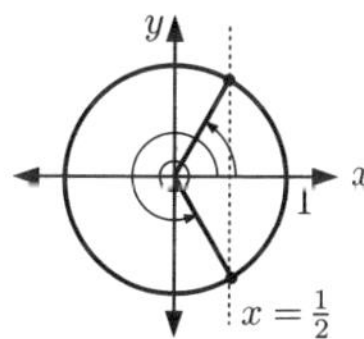


$\therefore \quad \theta = \frac{\pi}{3}, \frac{5\pi}{3}$

**b** $\sin\theta = \frac{\sqrt{3}}{2}$

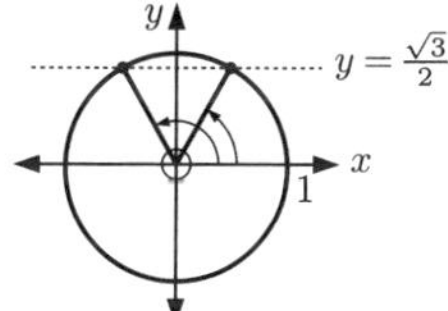


$\therefore \quad \theta = \frac{\pi}{3}, \frac{2\pi}{3}$

**c** $\cos\theta = -1$

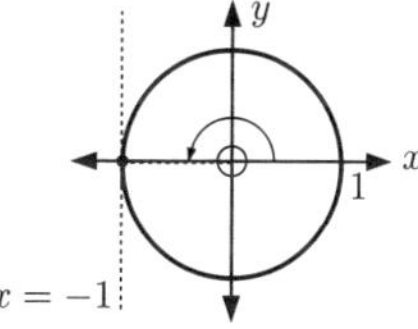


$\therefore \quad \theta = \pi$

**d** $\sin\theta = 1$

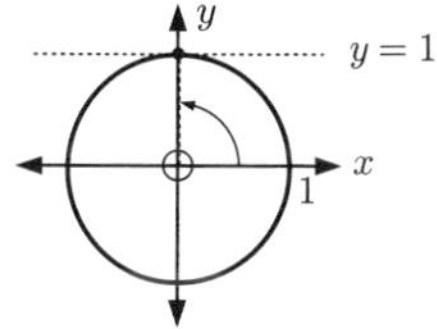


$\therefore \quad \theta = \frac{\pi}{2}$

**e** $\cos\theta = -\frac{1}{\sqrt{2}}$

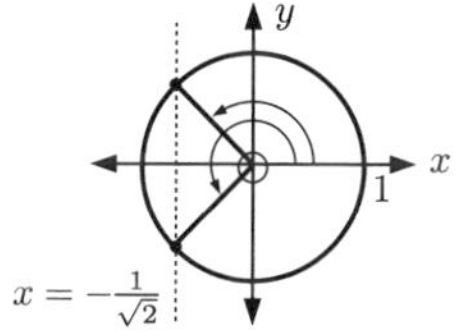


$\therefore \quad \theta = \frac{3\pi}{4}, \frac{5\pi}{4}$

**f** $\sin^2\theta = 1$

$\therefore \quad \sin\theta = \pm 1$

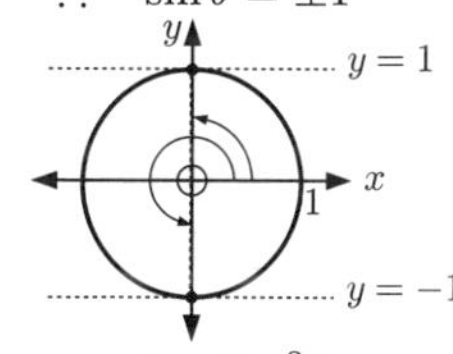


$\therefore \quad \theta = \frac{\pi}{2}, \frac{3\pi}{2}$

**g** $\cos^2\theta = 1$

$\therefore \quad \cos\theta = \pm 1$

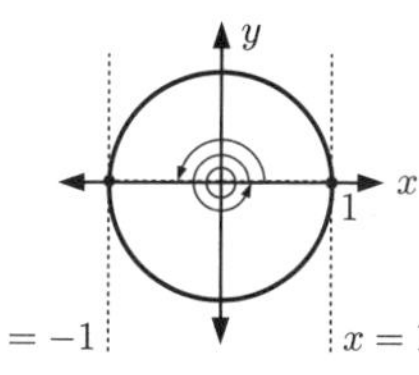


$\therefore \quad \theta = 0, \pi, 2\pi$

**h** $\cos^2\theta = \frac{1}{2}$

$\therefore \quad \cos\theta = \pm\frac{1}{\sqrt{2}}$

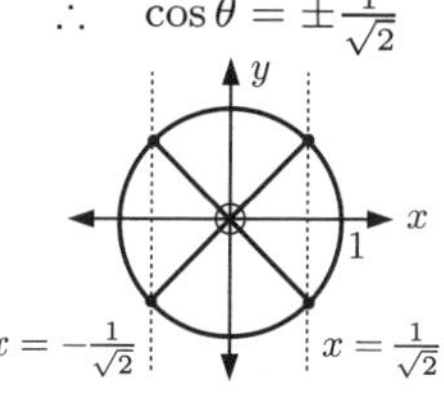


$\therefore \quad \theta = \frac{\pi}{4}, \frac{3\pi}{4}, \frac{5\pi}{4}, \frac{7\pi}{4}$

**i** $\tan\theta = -\frac{1}{\sqrt{3}}$

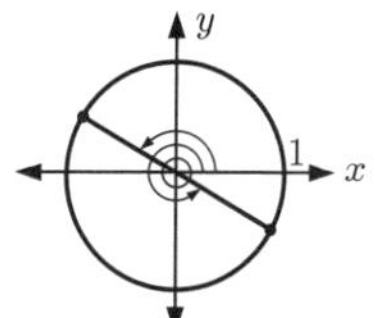


$\therefore \quad \theta = \frac{5\pi}{6}, \frac{11\pi}{6}$

**j** 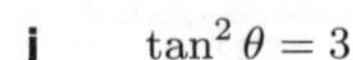
$\tan^2\theta = 3$

$\therefore \quad \tan\theta = \pm\sqrt{3}$

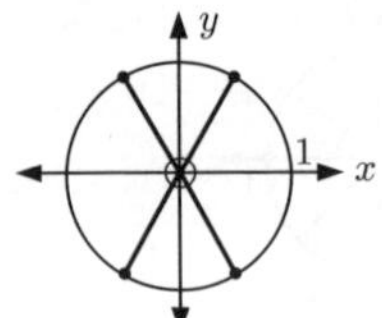


$\therefore \quad \theta = \frac{\pi}{3}, \frac{2\pi}{3}, \frac{4\pi}{3}, \frac{5\pi}{3}$

**10** **a** $\tan\theta$ is zero when $\dfrac{\sin\theta}{\cos\theta} = \dfrac{0}{\cos\theta}$

$\therefore$ when $\sin\theta = 0$

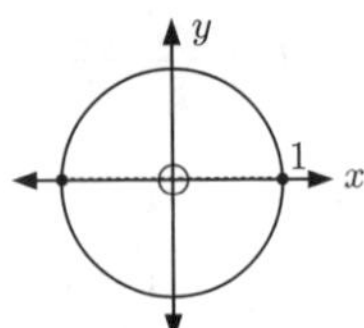


$\therefore \quad \theta = ...., -\pi, 0, \pi, 2\pi, ....$

$\therefore \quad \theta = k\pi, \text{ for } k \in \mathbb{Z}$

**b** $\tan\theta$ is undefined when $\dfrac{\sin\theta}{\cos\theta} = \dfrac{\sin\theta}{0}$

$\therefore$ when $\cos\theta = 0$

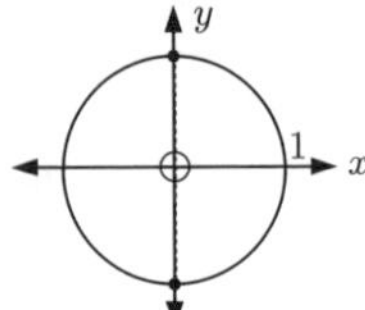


$\therefore \quad \theta = ...., -\frac{\pi}{2}, \frac{\pi}{2}, \frac{3\pi}{2}, ....$

$\therefore \quad \theta = \frac{\pi}{2} + k\pi, \text{ for } k \in \mathbb{Z}$

## EXERCISE 8F

**1** **a** The line has gradient $m = \tan 60° = \sqrt{3}$ and $y$-intercept 0.

$\therefore$ the line has equation $y = \sqrt{3}x$.

**b** The line makes an angle of $\frac{\pi}{2} - \frac{\pi}{4} = \frac{\pi}{4}$ with the positive $x$-axis.

$\therefore$ the line has gradient $m = \tan\frac{\pi}{4} = 1$ and $y$-intercept 0.

$\therefore$ the line has equation $y = x$.

**c** The line makes an angle of $\pi - \frac{\pi}{6} = \frac{5\pi}{6}$ with the positive $x$-axis.

$\therefore$ the line has gradient $m = \tan\frac{5\pi}{6} = -\frac{1}{\sqrt{3}}$ and $y$-intercept 0.

$\therefore$ the line has equation $y = -\frac{1}{\sqrt{3}}x$.

**2** **a** The line makes an angle of $\frac{\pi}{2} - \frac{\pi}{6} = \frac{\pi}{3}$ with the positive $x$-axis.

$\therefore$ the line has gradient $m = \tan\frac{\pi}{3} = \sqrt{3}$ and $y$-intercept 2.

$\therefore$ the line has equation $y = \sqrt{3}x + 2$.

**b** The line makes an angle of $\pi - \frac{\pi}{3} = \frac{2\pi}{3}$ with the positive $x$-axis.

$\therefore$ the line has gradient $m = \tan\frac{2\pi}{3} = -\sqrt{3}$ and $y$-intercept 0.

$\therefore$ the line has equation $y = -\sqrt{3}x$.

**c** The line makes an angle of $\frac{\pi}{2} - \frac{\pi}{3} = \frac{\pi}{6}$ with the positive $x$-axis.

$\therefore$ the line has gradient $m = \tan\frac{\pi}{6} = \frac{1}{\sqrt{3}}$.

$\therefore \quad y = \frac{1}{\sqrt{3}}x + c$, where $c$ is a constant.

But when $x = 2\sqrt{3}$, $y = 0$, so $0 = \frac{2\sqrt{3}}{\sqrt{3}} + c$

$\therefore \quad c = -2$

So, the line has equation $y = \frac{1}{\sqrt{3}}x - 2$.

## REVIEW SET 8A

**1** **a** $120°$
$= \left(120 \times \frac{\pi}{180}\right)^c$
$= \frac{2\pi}{3}^c$

**b** $225°$
$= 5 \times 45°$
$= 5 \times \frac{\pi}{4}^c$
$= \frac{5\pi}{4}^c$

**c** $150°$
$= 5 \times 30°$
$= 5 \times \frac{\pi}{6}^c$
$= \frac{5\pi}{6}^c$

**d** $540°$
$= 3 \times 180°$
$= 3\pi^c$

**2** **a** $\sin \frac{2\pi}{3} = \sin(\pi - \frac{2\pi}{3}) = \sin \frac{\pi}{3}$
$\therefore \quad \theta = \frac{\pi}{3}$

**b** $\sin 165° = \sin(180 - 165)° = \sin 15°$
$\therefore \quad \theta = 15°$

**c** $\cos 276° = \cos(360 - 276)° = \cos 84°$
$\therefore \quad \theta = 84°$

**3** **a** $\sin 159°$
$= \sin(180 - 159)°$
$= \sin 21°$
$\approx 0.358$

**b** $\cos 92°$
$= -\cos(180 - 92)°$
$= -\cos 88°$
$\approx -0.035$

**c** $\cos 75°$
$= -\cos(180 - 75)°$
$= -\cos 105°$
$\approx 0.259$

**d** $\sin(-133°) = \sin(-47)°$
$= -\sin 47°$
$\approx -0.731$

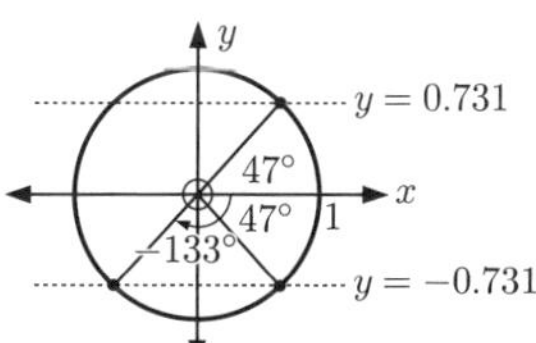


**4**

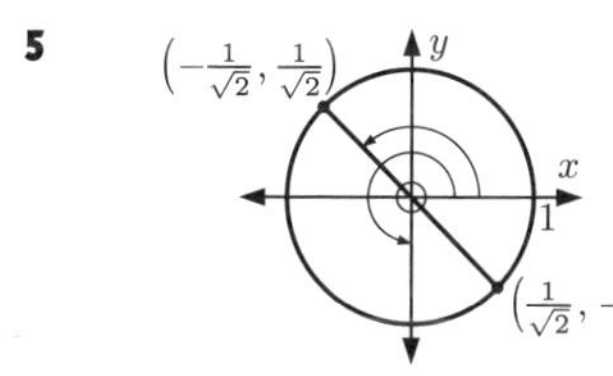


**a** $\cos 360° = 1, \quad \sin 360° = 0$

**b** $\cos(-\pi) = -1, \quad \sin(-\pi) = 0$

**5** When $\cos\theta = -\sin\theta,$

$\frac{\sin\theta}{\cos\theta} = -1$

$\therefore \quad \tan\theta = -1$

and this only occurs at the two points shown.

So, $\theta = \frac{3\pi}{4}, \frac{7\pi}{4}$

**6** **a**

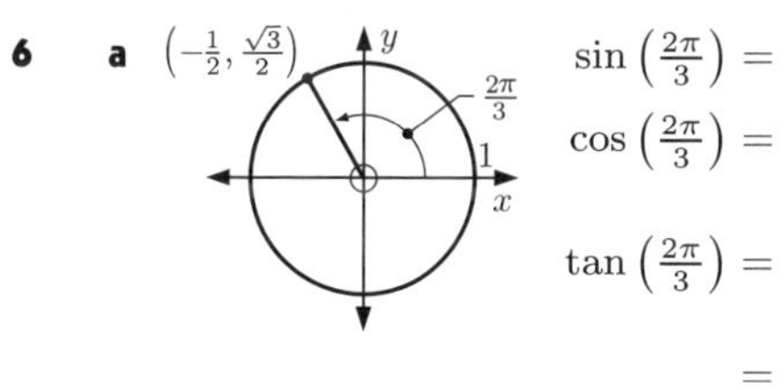

$\sin\left(\frac{2\pi}{3}\right) = \frac{\sqrt{3}}{2}$

$\cos\left(\frac{2\pi}{3}\right) = -\frac{1}{2}$

$\tan\left(\frac{2\pi}{3}\right) = \frac{\frac{\sqrt{3}}{2}}{-\frac{1}{2}}$

$= -\sqrt{3}$

**b**

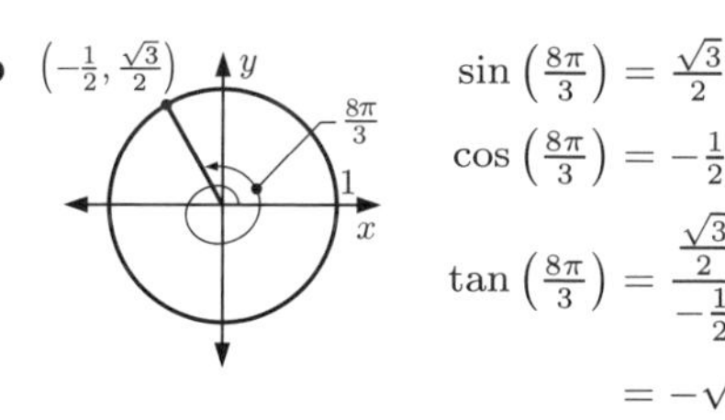

$\sin\left(\frac{8\pi}{3}\right) = \frac{\sqrt{3}}{2}$

$\cos\left(\frac{8\pi}{3}\right) = -\frac{1}{2}$

$\tan\left(\frac{8\pi}{3}\right) = \frac{\frac{\sqrt{3}}{2}}{-\frac{1}{2}}$

$= -\sqrt{3}$

**7** $\cos^2 x + \sin^2 x = 1$
$\therefore \cos^2 x + \frac{1}{16} = 1$
$\therefore \cos^2 x = \frac{15}{16}$
$\therefore \cos x = \pm\frac{\sqrt{15}}{4}$

But $x$ is in quadrant 3 where $\cos x < 0$

$\therefore \cos x = -\frac{\sqrt{15}}{4}$

and so $\tan x = \frac{\sin x}{\cos x} = \frac{-\frac{1}{4}}{-\frac{\sqrt{15}}{4}} = \frac{1}{\sqrt{15}}$

**8** $\cos^2\theta + \sin^2\theta = 1$
$\therefore \quad \frac{9}{16} + \sin^2\theta = 1$
$\therefore \quad \sin^2\theta = \frac{7}{16}$
$\therefore \quad \sin\theta = \pm\frac{\sqrt{7}}{4}$

**9** **a**

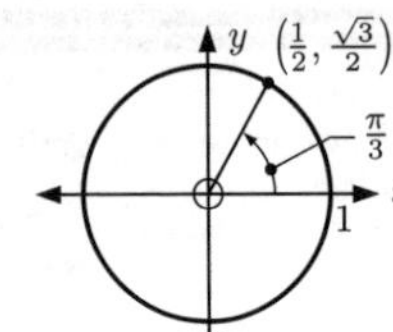


$2\sin(\frac{\pi}{3})\cos(\frac{\pi}{3})$

$= 2(\frac{\sqrt{3}}{2})(\frac{1}{2})$

$= \frac{\sqrt{3}}{2}$

**b**

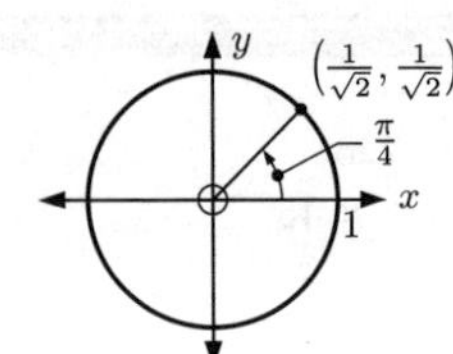


$\tan^2(\frac{\pi}{4}) - 1$

$= 1^2 - 1$

$= 0$

**c**

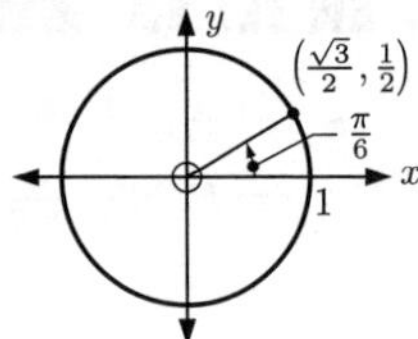


$\cos^2(\frac{\pi}{6}) - \sin^2(\frac{\pi}{6})$

$= \left(\frac{\sqrt{3}}{2}\right)^2 - \left(\frac{1}{2}\right)^2$

$= \frac{3}{4} - \frac{1}{4} = \frac{1}{2}$

**10** $\dfrac{\sin x}{\cos x} = -\frac{3}{2}$

$\therefore \sin x = -\frac{3}{2}\cos x$

Now $\cos^2 x + \sin^2 x = 1$

$\therefore \cos^2 x + \frac{9}{4}\cos^2 x = 1$

$\therefore \frac{13}{4}\cos^2 x = 1$

$\therefore \cos x = \pm\frac{2}{\sqrt{13}}$

But $x$ is in quadrant 4, so $\cos x$ is positive and $\sin x$ is negative.

$\therefore \cos x = \frac{2}{\sqrt{13}}, \quad \sin x = -\frac{3}{\sqrt{13}}$

So, **a** $\sin x = -\frac{3}{\sqrt{13}}$ **b** $\cos x = \frac{2}{\sqrt{13}}$

**11** arc length $= \theta r$

$= 1 \times 4$

$= 4$ units

$\therefore$ perimeter $= 2 \times 4 + 4$

$= 12$ units

area $= \frac{1}{2}\theta r^2$

$= \frac{1}{2} \times 1 \times 4^2$

$= 8$ units$^2$

**12** $\cos^2\theta + \sin^2\theta = 1$

$\therefore \left(\frac{\sqrt{11}}{\sqrt{17}}\right)^2 + \sin^2\theta = 1$

$\therefore \sin^2\theta = \frac{6}{17}$

$\therefore \sin\theta = \pm\frac{\sqrt{6}}{\sqrt{17}}$

But $\theta$ is acute, $\therefore \sin\theta = \frac{\sqrt{6}}{\sqrt{17}}$

$\tan\theta = \dfrac{\frac{\sqrt{6}}{\sqrt{17}}}{\frac{\sqrt{11}}{\sqrt{17}}} = \frac{\sqrt{6}}{\sqrt{11}}$

## REVIEW SET 8B

**1** **a** The point is $(\cos 320°, \sin 320°) \approx (0.766, -0.643)$.

**b** The point is $(\cos 163°, \sin 163°) \approx (-0.956, 0.292)$.

**2** **a** $71°$ $= \left(71 \times \frac{\pi}{180}\right)^c$ $\approx 1.239^c$

**b** $124.6°$ $= \left(124.6 \times \frac{\pi}{180}\right)^c$ $\approx 2.175^c$

**c** $-142°$ $= \left(-142 \times \frac{\pi}{180}\right)^c$ $\approx -2.478^c$

**3** **a** $3^c$ $= \left(3 \times \frac{180}{\pi}\right)^\circ$ $\approx 171.89°$

**b** $1.46^c$ $= \left(1.46 \times \frac{180}{\pi}\right)^\circ$ $\approx 83.65°$

**c** $0.435^c$ $= \left(0.435 \times \frac{180}{\pi}\right)^\circ$ $\approx 24.92°$

**d** $-5.271^c$ $= \left(-5.271 \times \frac{180}{\pi}\right)^\circ$ $\approx -302.01°$

**4** area $= \frac{1}{2} \times \frac{5\pi}{12} \times 13^2$

$\approx 111$ cm$^2$

**5** $M(\cos 73°, \sin 73°) \approx (0.292, 0.956)$,

$N(\cos 190°, \sin 190°) \approx (-0.985, -0.174)$,

$P(\cos(-53°), \sin(-53°)) = P(\cos 307°, \sin 307°) \approx (0.602, -0.799)$

**6** The $x$-coordinate of A $= -0.222$

$\therefore \quad \cos\theta = -0.222$

$\therefore \quad \theta = \cos^{-1}(-0.222)$

$\therefore \quad \theta \approx 102.8°, 257.2°$

$\therefore \quad \theta \approx 103°$ {taking angle to positive $x$-axis}

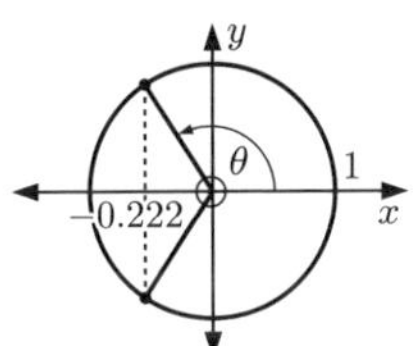


**7** **a**

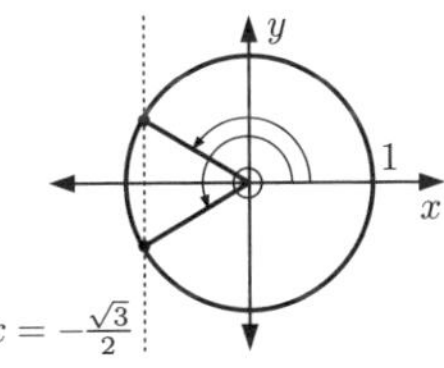


$\therefore \quad \theta = 150°$ or $210°$

**b**

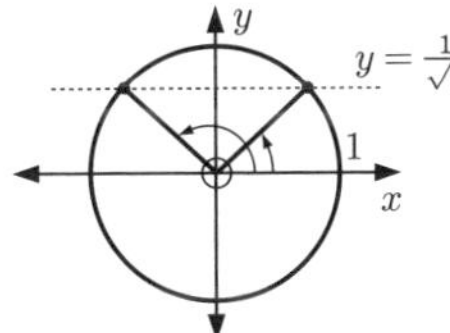


$\therefore \quad \theta = 45°$ or $135°$

**c**

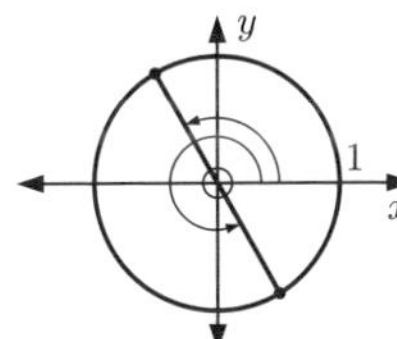


$\therefore \quad \theta = 120°$ or $300°$

**8** **a**

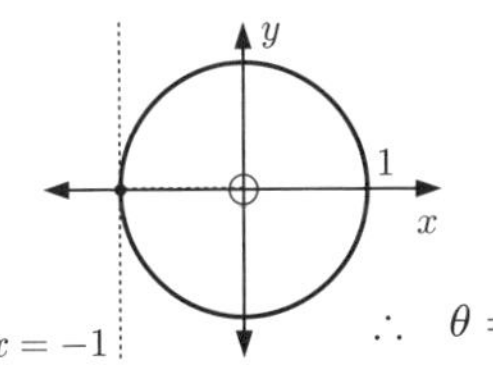


$\therefore \quad \theta = \pi$

**b** $\sin^2\theta = \frac{3}{4}$

$\therefore \quad \sin\theta = \pm\frac{\sqrt{3}}{2}$

$\therefore \quad \theta = \frac{\pi}{3}, \frac{2\pi}{3}, \frac{4\pi}{3}, \frac{5\pi}{3}$

**9** **a** $\sin 47° = \sin(180 - 47)°$

$= \sin 133°$

$\therefore \quad \theta = 133°$

**b** $\sin(\frac{\pi}{15}) = \sin(\pi - \frac{\pi}{15})$

$= \sin(\frac{14\pi}{15})$

$\therefore \quad \theta = \frac{14\pi}{15}$

**c** $\cos 186° = \cos(360 - 186)°$

$= \cos 174°$

$\therefore \quad \theta = 174°$

**10**

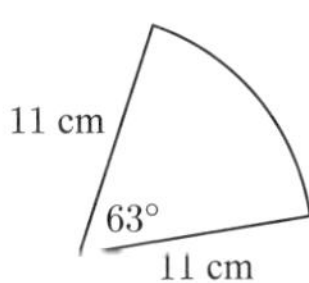


perimeter $= 2 \times 11 + \left(\frac{63}{360}\right) \times 2\pi \times 11$

$\approx 34.1$ cm

area $= \left(\frac{63}{360}\right) \times \pi \times 11^2$

$\approx 66.5\text{ cm}^2$

**11**

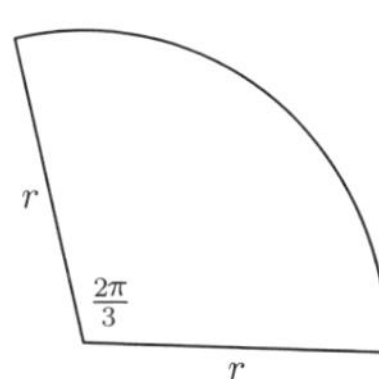


perimeter $= 2r + (\frac{2\pi}{3})r$

$\therefore \quad 36 = r\left(2 + \frac{2\pi}{3}\right)$

$\therefore \quad r = \dfrac{36}{2 + \frac{2\pi}{3}}$ cm

$\therefore \quad r \approx 8.79$ cm

area $\approx \frac{1}{2}(\frac{2\pi}{3}) \times (8.7925)^2$

$\approx 81.0\text{ cm}^2$

**12** **a** $\cos\theta = \frac{2}{3}$

Using technology,

$\cos^{-1}(\frac{2}{3}) \approx 0.841$

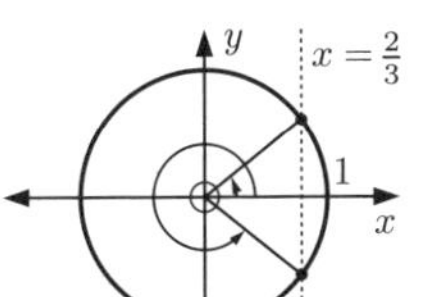


$\therefore \quad \theta \approx 0.841$ or $2\pi - 0.841$

$\therefore \quad \theta \approx 0.841$ or $5.44$

**b** $\sin\theta = -\frac{1}{4}$

Using technology,

$\sin^{-1}(-\frac{1}{4}) \approx -0.253$

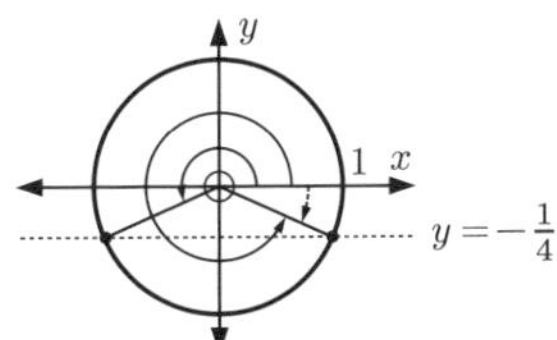


But $0 \leqslant \theta \leqslant 2\pi$

$\therefore \quad \theta \approx \pi + 0.253$ or $2\pi + (-0.253)$

$\therefore \quad \theta \approx 3.39$ or $6.03$

**c** $\tan\theta = 3$

Using technology,

$\tan^{-1}(3) \approx 1.25$

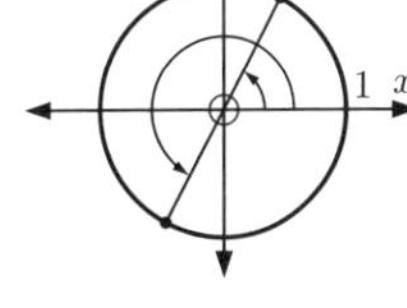


$\therefore \quad \theta \approx 1.25$ or $\pi + 1.25$

$\therefore \quad \theta \approx 1.25$ or $4.39$

## REVIEW SET 8C

**1** **a** $\frac{2\pi}{5} = \frac{2\times 180^\circ}{5}$
$= 72^\circ$

**b** $\frac{5\pi}{4} = \frac{5\times 180^\circ}{4}$
$= 225^\circ$

**c** $\frac{7\pi}{9} = \frac{7\times 180^\circ}{9}$
$= 140^\circ$

**d** $\frac{11\pi}{6} = \frac{11\times 180^\circ}{6}$
$= 330^\circ$

**2**

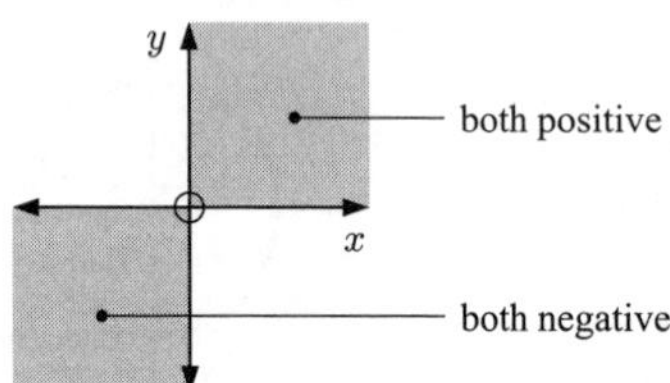


**3** **a**

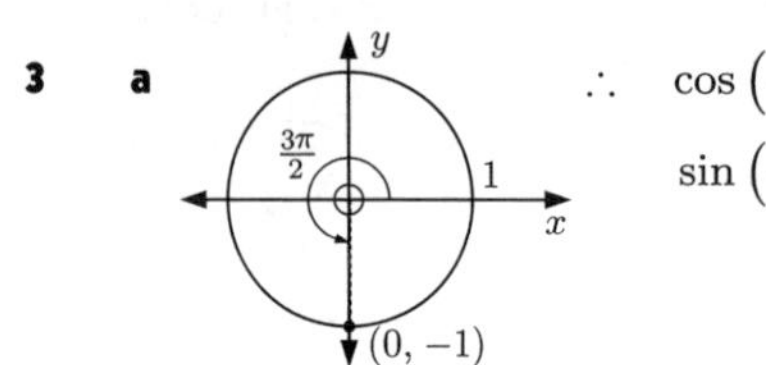


$\therefore \quad \cos\left(\frac{3\pi}{2}\right) = 0$

$\sin\left(\frac{3\pi}{2}\right) = -1$

**b** $\therefore \quad \cos\left(-\frac{\pi}{2}\right) = 0$

$\sin\left(-\frac{\pi}{2}\right) = -1$

**4** **a** $\sin(\pi - \theta) = \sin\theta$
$\therefore \sin(\pi - p) = \sin p$
$= m$

**b** $\sin(\theta + 2\pi) = \sin\theta$
$\therefore \sin(p + 2\pi) = \sin p$
$= m$

**c** $\cos^2 p + \sin^2 p = 1$
$\therefore \cos^2 p + m^2 = 1$
$\therefore \cos^2 p = 1 - m^2$
$\therefore \cos p = \pm\sqrt{1 - m^2}$
But $p$ is acute, $\therefore \cos p = \sqrt{1 - m^2}$

**d** $\tan p = \dfrac{\sin p}{\cos p}$

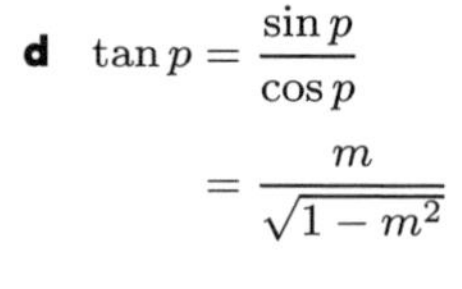

$= \dfrac{m}{\sqrt{1 - m^2}}$

**5** **a** **i** $\theta = 60^\circ$ {equilateral triangle} **ii** $\theta = \frac{\pi}{3}$ radians

**b** arc length $= \theta r = \frac{\pi}{3}$ units **c** sector area $= \frac{1}{2}\theta r^2 = \frac{\pi}{6}$ units$^2$

**6**

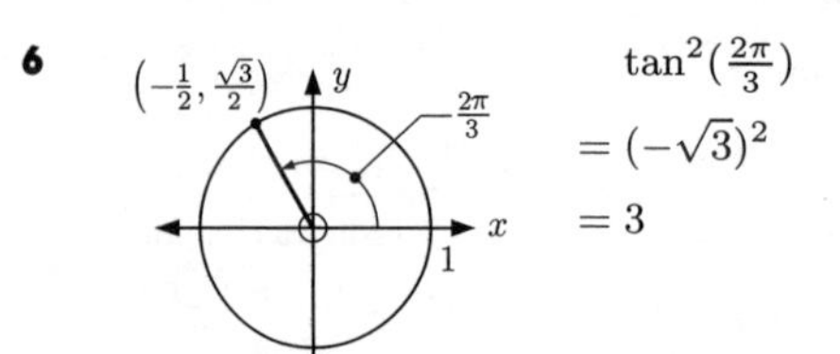


$\tan^2\left(\frac{2\pi}{3}\right)$
$= (-\sqrt{3})^2$
$= 3$

**7**

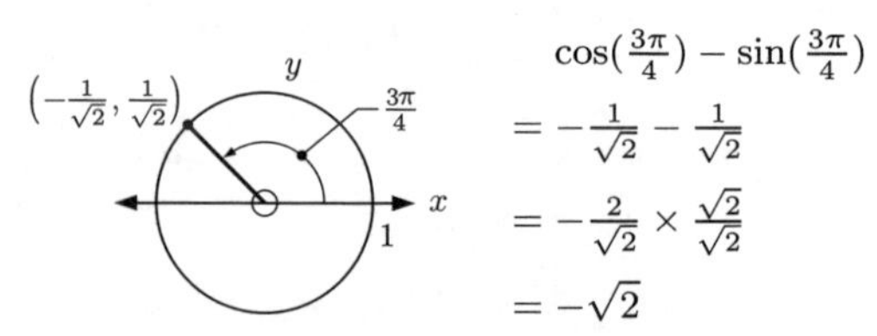


$\cos\left(\frac{3\pi}{4}\right) - \sin\left(\frac{3\pi}{4}\right)$
$= -\frac{1}{\sqrt{2}} - \frac{1}{\sqrt{2}}$
$= -\frac{2}{\sqrt{2}} \times \frac{\sqrt{2}}{\sqrt{2}}$
$= -\sqrt{2}$

**8** **a** $\cos^2\theta + \sin^2\theta = 1$
$\therefore \frac{9}{16} + \sin^2\theta = 1$
$\therefore \sin^2\theta = \frac{7}{16}$
$\therefore \sin\theta = \pm\frac{\sqrt{7}}{4}$
But $\theta$ is in quadrant 2 where $\sin\theta > 0$
$\therefore \sin\theta = \frac{\sqrt{7}}{4}$

**b** $\tan\theta = \dfrac{\sin\theta}{\cos\theta} = \dfrac{\frac{\sqrt{7}}{4}}{-\frac{3}{4}} = -\frac{\sqrt{7}}{3}$

**c** $\sin(\theta + \pi)$
$= -\sin\theta$
$= -\frac{\sqrt{7}}{4}$

**9** **a**

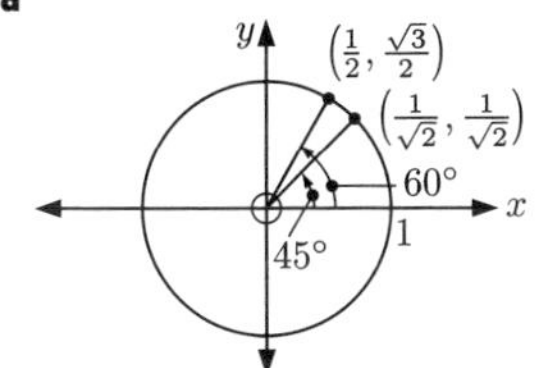


**b**

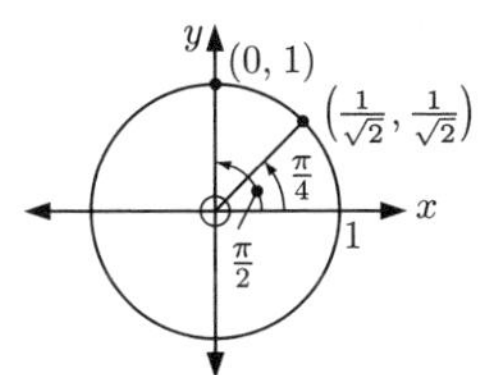


**c**

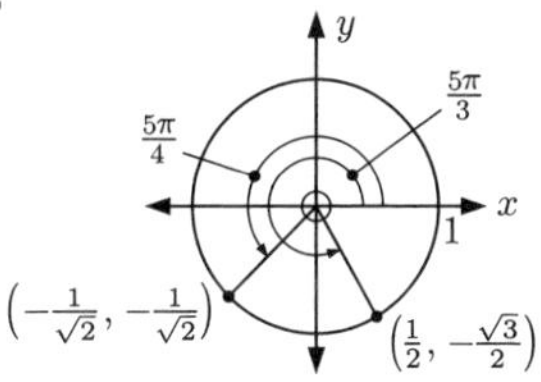


**a**

$\tan^2 60° - \sin^2 45°$

$= (\sqrt{3})^2 - \left(\frac{1}{\sqrt{2}}\right)^2$

$= 3 - \frac{1}{2}$

$= 2\frac{1}{2}$

**b**

$\cos^2(\frac{\pi}{4}) + \sin(\frac{\pi}{2})$

$= \left(\frac{1}{\sqrt{2}}\right)^2 + 1$

$= \frac{1}{2} + 1$

$= 1\frac{1}{2}$

**c**

$\cos(\frac{5\pi}{3}) - \tan(\frac{5\pi}{4})$

$= \frac{1}{2} - 1$

$= -\frac{1}{2}$

**10** **a** $\sin(\pi - \theta) - \sin\theta = \sin\theta - \sin\theta$
$= 0$

**b** $\cos\theta\tan\theta = \cos\theta\left(\dfrac{\sin\theta}{\cos\theta}\right)$
$= \sin\theta$

**11** **a** The line has gradient $m = \tan(-30°) = -\frac{1}{\sqrt{3}}$ and $y$-intercept 0.

$\therefore$ the line has equation $y = -\frac{1}{\sqrt{3}}x$.

**b** When $x = k$, $y = 2$ $\quad\therefore\ 2 = -\frac{1}{\sqrt{3}}k$

$\therefore\ k = -2\sqrt{3}$

**12** [AB], [AC], and [BC] are all radii,
so AB = AC = BC = $r$.
Hence $\triangle$ABC is equilateral
and so $\widehat{CAB} = 60°$.

$\therefore\ \sin 60° = \dfrac{CD}{AC}$

$\therefore\ CD = \sin 60° \times AC = \frac{\sqrt{3}}{2}r$

$\therefore$ area of $\triangle = \frac{1}{2}(r)(\frac{\sqrt{3}}{2}r)$

$= \frac{\sqrt{3}}{4}r^2$

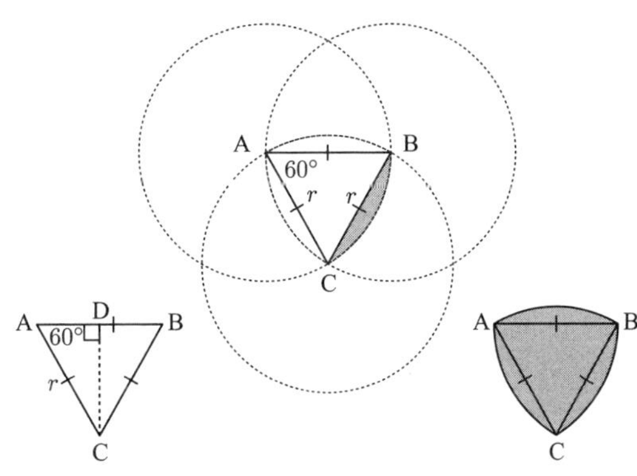


shaded area of sector

$= \text{area of sector} - \text{area of }\triangle$

$= \frac{60}{360}\pi r^2 - \frac{\sqrt{3}}{4}r^2$

$= \frac{\pi}{6}r^2 - \frac{\sqrt{3}}{4}r^2$

$\therefore$ shaded area of figure

$= 3\left[\frac{\pi}{6}r^2 - \frac{\sqrt{3}}{4}r^2\right] + \frac{\sqrt{3}}{4}r^2$

$= \frac{\pi}{2}r^2 - \frac{3\sqrt{3}}{4}r^2 + \frac{\sqrt{3}}{4}r^2$

$= \frac{\pi}{2}r^2 - \frac{2}{4}\sqrt{3}r^2$

$= \dfrac{r^2}{2}\left(\pi - \sqrt{3}\right)$

# Chapter 9

## NON-RIGHT ANGLED TRIANGLE TRIGONOMETRY

### EXERCISE 9A

**1** **a** area
$= \frac{1}{2} \times 9 \times 10 \times \sin 40°$
$\approx 28.9\ \text{cm}^2$

**b** area
$= \frac{1}{2} \times 25 \times 31 \times \sin 82°$
$\approx 384\ \text{km}^2$

**c** area
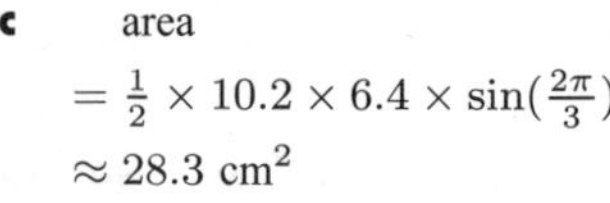
$= \frac{1}{2} \times 10.2 \times 6.4 \times \sin(\frac{2\pi}{3})$
$\approx 28.3\ \text{cm}^2$

**2** area $= 150\ \text{cm}^2$
$\therefore\ \frac{1}{2} \times 17 \times x \times \sin 68° = 150$
$\therefore\ x = \dfrac{2 \times 150}{17 \times \sin 68°}$
$\therefore\ x \approx 19.0$

**3**

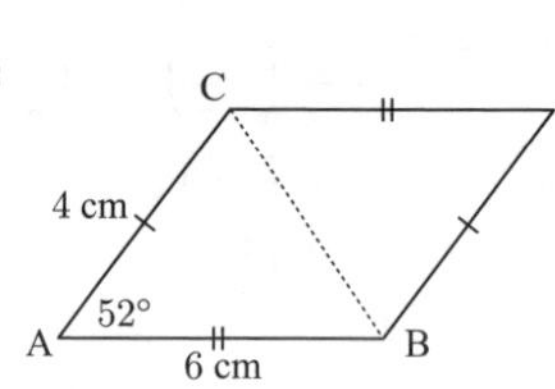


area
$= 2 \times \text{area } \triangle\text{ABC}$
$= 2 \times \frac{1}{2} \times 4 \times 6 \times \sin 52°$
$\approx 18.9\ \text{cm}^2$

**4**

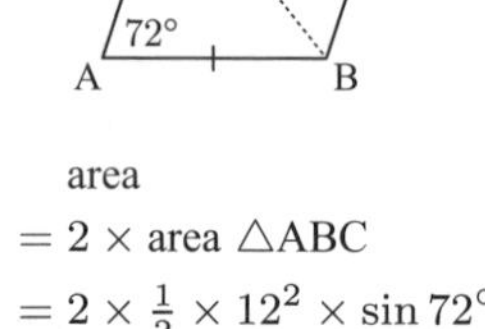

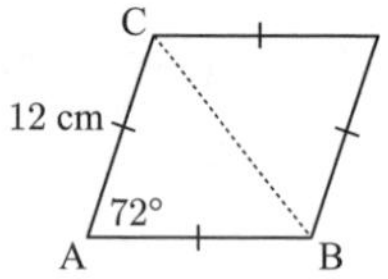

area
$= 2 \times \text{area } \triangle\text{ABC}$
$= 2 \times \frac{1}{2} \times 12^2 \times \sin 72°$
$\approx 137\ \text{cm}^2$

**5**

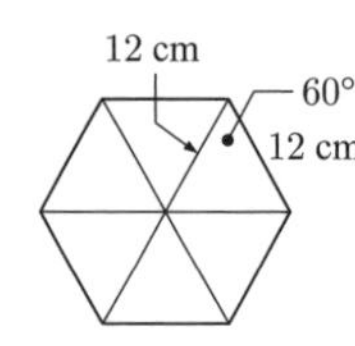


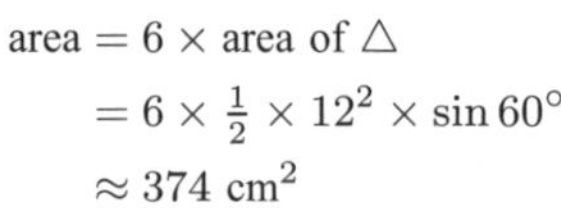

area $= 6 \times$ area of $\triangle$
$= 6 \times \frac{1}{2} \times 12^2 \times \sin 60°$
$\approx 374\ \text{cm}^2$

**6**

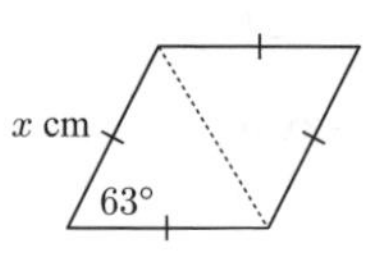


area $= 2 \times \frac{1}{2}x^2 \sin 63°$
$\therefore\ x^2 \sin 63° = 50$
$\therefore\ x^2 = \dfrac{50}{\sin 63°}$
$\therefore\ x = \sqrt{\dfrac{50}{\sin 63°}}$
$\therefore\ x \approx 7.49 \quad \{x > 0\}$
So, sides are 7.49 cm long.

**7**

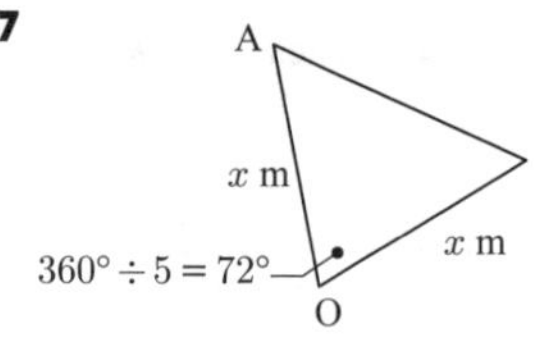


area of $\triangle = \frac{338}{5}$
$\therefore\ \frac{1}{2}x^2 \sin 72° = \frac{338}{5}$
$\therefore\ x^2 = \dfrac{2 \times 338}{5 \times \sin 72°}$
$\therefore\ x = \sqrt{\dfrac{2 \times 338}{5 \times \sin 72°}}$
$\therefore\ x \approx 11.9 \quad \{x > 0\}$
So, $\text{OA} \approx 11.9$ m

**8** **a** If the included angle is $\theta$
then $\frac{1}{2} \times 5 \times 8 \times \sin\theta = 15$
$\therefore\ 20 \sin\theta = 15$
$\therefore\ \sin\theta = \frac{3}{4}$
Now $\arcsin(\frac{3}{4}) \approx 48.6°$
$\therefore\ \theta \approx 48.6°$ or $(180 - 48.6)°$
$\therefore\ \theta \approx 48.6°$ or $131.4°$

**b** Likewise,
$\frac{1}{2} \times 45 \times 53 \times \sin\theta = 800$
$\therefore\ \sin\theta = \frac{800 \times 2}{45 \times 53}$
Now $\arcsin\left(\frac{800 \times 2}{45 \times 53}\right) \approx 42.1°$
$\therefore\ \theta \approx 42.1°$ or $(180 - 42.1)°$
$\therefore\ \theta \approx 42.1°$ or $137.9°$

**9** (diagram: triangle with sides $r$, $r$; $360° \div 12 = 30°$)

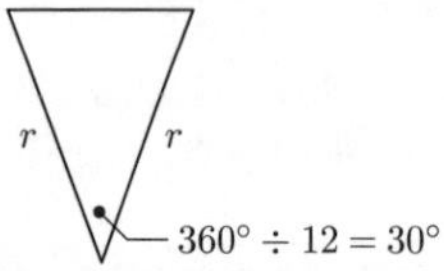

total area of 8 coins
$= 8 \times 12 \times \frac{1}{2}r^2 \sin 30°$
$= 48r^2(\frac{1}{2})$
$= 24r^2$

area of \$10 note
$= 8r \times 4r$
$= 32r^2$

fraction covered
$= \dfrac{24r^2}{32r^2}$
$= \frac{3}{4}\ \therefore\ \frac{1}{4}$ is not covered

**10** **a** shaded area
$= \text{area of sector} - \text{area of triangle}$
$= \frac{1}{2} \times 1.5 \times 12^2 - \frac{1}{2} \times 12^2 \times \sin(1.5)$
$\approx 36.2 \text{ cm}^2$

**b** shaded area
$= \text{area of triangle} - \text{area of sector}$
$= \frac{1}{2} \times 12 \times 30 \times \sin(0.66) - \frac{1}{2} \times 0.66 \times 12^2$
$\approx 62.8 \text{ cm}^2$

**c** shaded area
$= \text{area of sector} - \text{area of triangle}$
$= \left(\frac{135}{360}\right) \times \pi \times 7^2 - \frac{1}{2} \times 7^2 \times \sin 135°$
$\approx 40.4 \text{ mm}^2$

**11** area segment AXBD
$= \text{area sector ACBD} - \text{area } \triangle\text{ACB}$
$= \left(\frac{100}{360}\right) \times \pi \times 7.3^2 - \frac{1}{2} \times 7.3 \times 7.3 \times \sin 100°$
$\approx 20.264 \text{ cm}^2$

area segment AXBE
$= \text{area sector AFBE} - \text{area } \triangle\text{AFB}$
$= \left(\frac{80}{360}\right) \times \pi \times 8.7^2 - \frac{1}{2} \times 8.7 \times 8.7 \times \sin 80°$
$\approx 15.572 \text{ cm}^2$

$\therefore$ shaded area = area segment AXBD − area segment AXBE
$\approx 20.264 - 15.572 \approx 4.69 \text{ cm}^2$

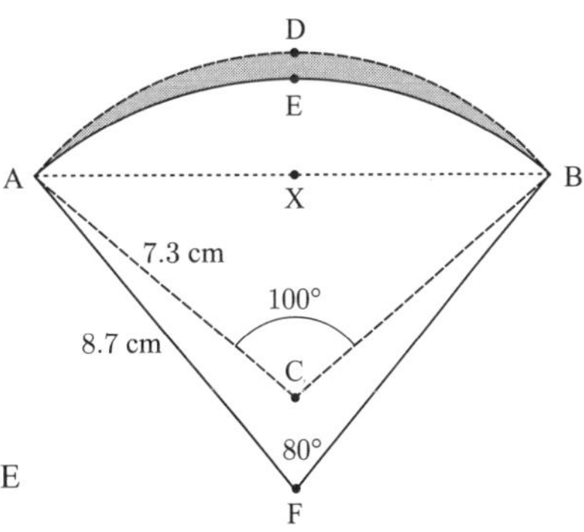


## EXERCISE 9B

**1** **a** $BC^2 = 21^2 + 15^2 - 2 \times 21 \times 15 \times \cos 105°$
$\therefore \quad BC = \sqrt{21^2 + 15^2 - 2 \times 21 \times 15 \times \cos 105°} \approx 28.8 \text{ cm}$

**b** $PQ^2 = 6.3^2 + 4.8^2 - 2 \times 6.3 \times 4.8 \times \cos 32°$
$\therefore \quad PQ = \sqrt{6.3^2 + 4.8^2 - 2 \times 6.3 \times 4.8 \times \cos 32°} \approx 3.38 \text{ km}$

**c** $KM^2 = 6.2^2 + 14.8^2 - 2 \times 6.2 \times 14.8 \times \cos 72°$
$\therefore \quad KM = \sqrt{6.2^2 + 14.8^2 - 2 \times 6.2 \times 14.8 \times \cos 72°} \approx 14.2 \text{ m}$

**2** $\cos B\widehat{A}C = \dfrac{12^2 + 13^2 - 11^2}{2 \times 12 \times 13}$
$\therefore \quad B\widehat{A}C = \cos^{-1}\left(\frac{192}{312}\right)$
$\therefore \quad B\widehat{A}C \approx 52.0°$

$\cos A\widehat{B}C = \dfrac{13^2 + 11^2 - 12^2}{2 \times 13 \times 11}$
$\therefore \quad A\widehat{B}C = \cos^{-1}\left(\frac{146}{286}\right)$
$\therefore \quad A\widehat{B}C \approx 59.3°$

$A\widehat{C}B = 180° - B\widehat{A}C - A\widehat{B}C$
$\approx 68.7°$

**3** **a** $\cos P\widehat{Q}R = \dfrac{5^2 + 7^2 - 10^2}{2 \times 5 \times 7}$
$\therefore \quad P\widehat{Q}R = \cos^{-1}\left(\frac{-26}{70}\right) \approx 112°$

**b** $\text{area} \approx \frac{1}{2} \times 5 \times 7 \times \sin 112°$
$\approx 16.2 \text{ cm}^2$

**4** **a**

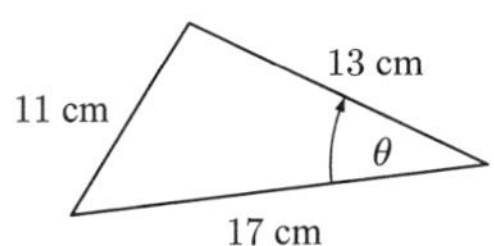


The smallest angle is opposite the shortest side.
$\cos\theta = \dfrac{13^2 + 17^2 - 11^2}{2 \times 13 \times 17}$
$\therefore \quad \theta = \cos^{-1}\left(\frac{337}{442}\right) \approx 40.3°$
So, the smallest angle measures 40.3°.

**b**

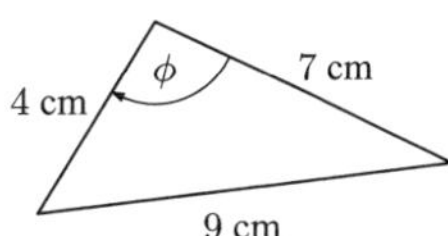


The largest angle is opposite the longest side.
$\cos\phi = \dfrac{4^2 + 7^2 - 9^2}{2 \times 4 \times 7}$
$\therefore \quad \phi = \cos^{-1}\left(-\frac{16}{56}\right) \approx 106.60°$
So, the largest angle measures about 107°.

**5** **a** $\cos\theta = \dfrac{2^2+5^2-4^2}{2\times2\times5}$
$= \frac{13}{20}$
$= 0.65$

**b** $x^2 = 5^2+3^2-2\times5\times3\times\cos\theta$
$\therefore\ x = \sqrt{5^2+3^2-2\times5\times3\times0.65}$
$\therefore\ x \approx 3.81$

**6** **a** $7^2 = x^2+6^2-2\times x\times6\times\cos60°$
$\therefore\ 49 = x^2+36-12x\times(\frac{1}{2})$
$\therefore\ x^2-6x-13=0$
$\therefore\ x = \dfrac{6\pm\sqrt{36-4(1)(-13)}}{2}$
$= \dfrac{6\pm\sqrt{88}}{2}$
$= 3\pm\sqrt{22}$
But $x>0$, so $x = 3+\sqrt{22}$

**b** $5^2 = x^2+3^2-2\times x\times3\times\cos120°$
$\therefore\ 25 = x^2+9-6x\times(-\frac{1}{2})$
$\therefore\ x^2+3x-16=0$
$\therefore\ x = \dfrac{-3\pm\sqrt{9-4(1)(-16)}}{2}$
$= \dfrac{-3\pm\sqrt{73}}{2}$
But $x>0$, so $x = \dfrac{-3+\sqrt{73}}{2}$

**c** $5^2 = (2x)^2+x^2-2\times(2x)\times x\times\cos60°$
$\therefore\ 25 = 4x^2+x^2-4x^2(\frac{1}{2})$
$\therefore\ 3x^2 = 25$
$\therefore\ x^2 = \frac{25}{3}$
$\therefore\ x = \pm\frac{5}{\sqrt{3}}$
But $x>0$, so $x = \frac{5}{\sqrt{3}}$

**7** **a** Area $= 11.6\ \text{m}^2$
$\therefore\ 11.6 = \frac{1}{2}\times6\times4\times\sin\theta$
$\therefore\ \sin\theta = \frac{29}{30}$
$\therefore\ \theta = \sin^{-1}\left(\frac{29}{30}\right)$
$\therefore\ \theta \approx 75.2°$

**b** Let the third side have length $x$ m.
By the cosine rule,
$x^2 = 6^2+4^2-2\times6\times4\times\cos75.2°$
$\therefore\ x = \sqrt{6^2+4^2-2\times6\times4\times\cos75.2°}$
$\therefore\ x\approx 6.30$
The third side has length 6.30 m.

**8** **a** $11^2 = x^2+8^2-2\times x\times8\times\cos70°$
$\therefore\ 121 = x^2+64-16x\cos70°$
$\therefore\ x^2-(16\cos70°)x-57=0$
Using the quadratic formula or technology,
$x\approx -5.29$ or $10.8$.
But $x>0$, so $x\approx 10.8$.

**b** $13^2 = x^2+5^2-2\times x\times5\times\cos130°$
$\therefore\ 169 = x^2+25-10x\cos130°$
$\therefore\ x^2-(10\cos130°)x-144=0$
Using the quadratic formula or technology,
$x\approx -15.6$ or $9.21$.
But $x>0$, so $x\approx 9.21$.

**9** $5^2 = x^2+6^2-2\times x\times6\times\cos40°$
$\therefore\ 25 = x^2+36-12x\cos40°$
$\therefore\ x^2-(12\cos40°)x+11=0$
Using the quadratic formula or technology, $x\approx 1.41$ or $7.78$.

## EXERCISE 9C.1

**1** **a** By the sine rule,
$\dfrac{x}{\sin48°} = \dfrac{23}{\sin37°}$
$\therefore\ x = \dfrac{23\times\sin48°}{\sin37°}$
$\therefore\ x\approx 28.4$

**b** By the sine rule,
$\dfrac{x}{\sin115°} = \dfrac{11}{\sin48°}$
$\therefore\ x = \dfrac{11\times\sin115°}{\sin48°}$
$\therefore\ x\approx 13.4$

**c** By the sine rule,
$\dfrac{x}{\sin51°} = \dfrac{4.8}{\sin80°}$
$\therefore\ x = \dfrac{4.8\times\sin51°}{\sin80°}$
$\therefore\ x\approx 3.79$

**2** **a**

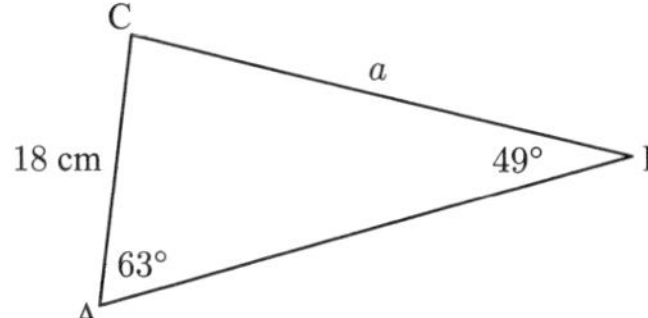


By the sine rule, $\dfrac{a}{\sin 63°} = \dfrac{18}{\sin 49°}$

$\therefore\ a = \dfrac{18 \times \sin 63°}{\sin 49°}$

$\therefore\ a \approx 21.3$ cm

**b**

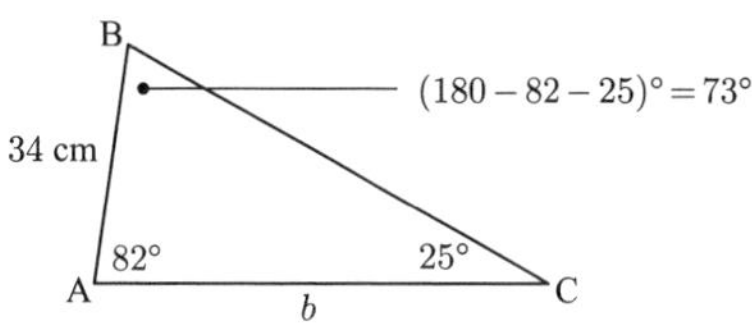


By the sine rule, $\dfrac{b}{\sin 73°} = \dfrac{34}{\sin 25°}$

$\therefore\ b = \dfrac{34 \times \sin 73°}{\sin 25°}$

$\therefore\ b \approx 76.9$ cm

**c**

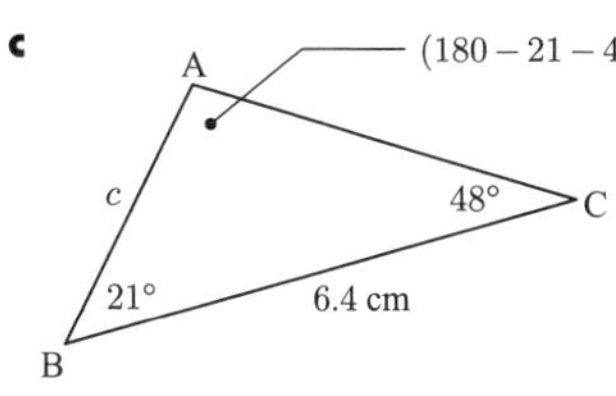


By the sine rule, $\dfrac{c}{\sin 48°} = \dfrac{6.4}{\sin 111°}$

$\therefore\ c = \dfrac{6.4 \times \sin 48°}{\sin 111°}$

$\therefore\ c \approx 5.09$ cm

## EXERCISE 9C.2

**1** By the sine rule, $\dfrac{\sin C}{11} = \dfrac{\sin 40°}{8}$

$\therefore\ \sin C = \dfrac{11 \times \sin 40°}{8}$

$\therefore\ C = \sin^{-1}\left(\dfrac{11 \times \sin 40°}{8}\right)$ or its supplement

$\therefore\ C \approx 62.1°$ or $(180 - 62.1)°$

$\therefore\ C \approx 62.1°$ or $117.9°$

**2** **a** $\dfrac{\sin B\widehat{A}C}{a} = \dfrac{\sin A\widehat{B}C}{b}$

$\therefore\ \sin B\widehat{A}C = \dfrac{14.6 \times \sin 65°}{17.4}$

$\therefore\ B\widehat{A}C = \sin^{-1}\left(\dfrac{14.6 \times \sin 65°}{17.4}\right)$

or its supplement

$\therefore\ B\widehat{A}C \approx 49.5°$ or $180° - 49.5°$

$\therefore\ B\widehat{A}C \approx 49.5°$ or $130.5°$

*Check*: $B\widehat{A}C = 130.5°$ is impossible as $B\widehat{A}C + A\widehat{B}C = 130.5° + 65°$ is already over $180°$. $\therefore\ B\widehat{A}C \approx 49.5°$

**b** $\dfrac{\sin A\widehat{D}C}{43.8} = \dfrac{\sin 43°}{31.4}$

$\therefore\ \sin A\widehat{B}C = \dfrac{43.8 \times \sin 43°}{31.4}$

$\therefore\ A\widehat{B}C = \sin^{-1}\left(\dfrac{43.8 \times \sin 43°}{31.4}\right)$

or its supplement

$\therefore\ A\widehat{B}C \approx 72.0°$ or $108°$

both of which are possible as $108 + 43 = 151$ which is $< 180$.

**c** $\dfrac{\sin A\widehat{C}B}{4.8} = \dfrac{\sin 71°}{6.5}$

$\therefore\ \sin A\widehat{C}B = \dfrac{4.8 \times \sin 71°}{6.5}$

$\therefore\ A\widehat{C}B = \sin^{-1}\left(\dfrac{4.8 \times \sin 71°}{6.5}\right)$ or its supplement

$\therefore\ A\widehat{C}B \approx 44.3°$ or $135.7°$

But $135.7 + 71 > 180$, so this case is impossible. $\therefore\ A\widehat{C}B \approx 44.3°$

**3** The third angle is $180° - 85° - 68° = 27°$

Now $\dfrac{\sin 85°}{11.4} \approx 0.087\,38$ and $\dfrac{\sin 27°}{9.8} \approx 0.046\,32$

This is not possible since $\dfrac{\sin 85°}{11.4} \neq \dfrac{\sin 27°}{9.8}$ violates the sine rule.

**4**

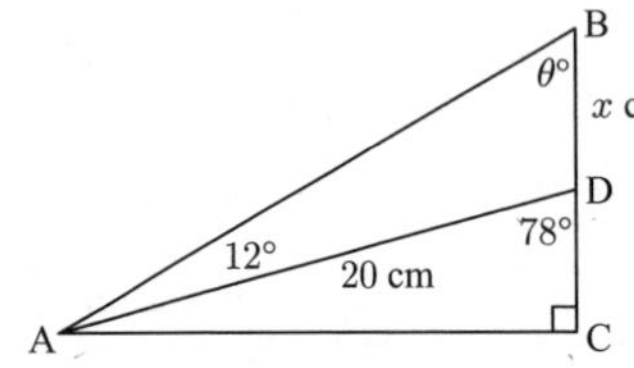


In △ABD,

$\theta = 78 - 12$

$\therefore\ \widehat{ABC} = 66°$

Now $\dfrac{x}{\sin 12°} = \dfrac{20}{\sin 66°}$

$\therefore\ x = \dfrac{20 \times \sin 12°}{\sin 66°}$

$\therefore\ x \approx 4.55$

$\therefore$ BD $\approx 4.55$ cm

**5** First we find the length of the diagonal, $d$ m.

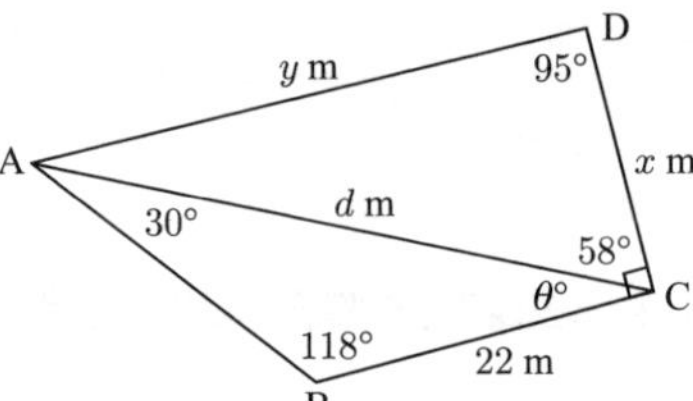


$$\frac{d}{\sin 118°} = \frac{22}{\sin 30°}$$

$$\therefore\ d = \frac{22 \times \sin 118°}{\sin 30°}$$

$$\therefore\ d \approx 38.85$$

Now $\theta = 180 - 30 - 118 = 32$

$\therefore\ \widehat{ACD} = 90° - 32° = 58°$

Using the sine rule, $\dfrac{y}{\sin 58°} = \dfrac{38.85}{\sin 95°}$ and $\dfrac{x}{\sin(180 - 95 - 58)°} \approx \dfrac{38.85}{\sin 95°}$

$\therefore\ y \approx \dfrac{38.85 \times \sin 58°}{\sin 95°}$

$\therefore\ y \approx 33.1$

$\therefore\ x \approx \dfrac{38.85 \times \sin 27°}{\sin 95°}$

$\therefore\ x \approx 17.7$

**6**

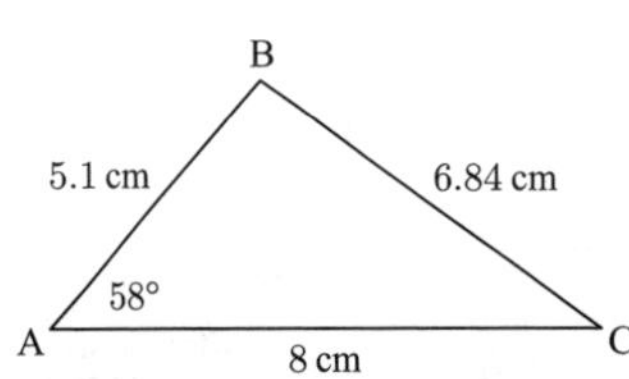


**a** $\dfrac{\sin \widehat{B}}{8} = \dfrac{\sin 58°}{6.84}$

$\therefore\ \sin \widehat{B} = \dfrac{8 \sin 58°}{6.84}$

$\therefore\ \widehat{B} = \sin^{-1}\left(\dfrac{8 \sin 58°}{6.84}\right)$ or its supplement

$\therefore\ \widehat{B} \approx 83°$ or $(180 - 83)°$

$\therefore\ \widehat{B} \approx 83°$ or $97°$

**b** $\cos \widehat{B} = \dfrac{5.1^2 + 6.84^2 - 8^2}{2 \times 5.1 \times 6.84}$

$\therefore\ \widehat{B} = \cos^{-1}\left(\dfrac{5.1^2 + 6.84^2 - 8^2}{2 \times 5.1 \times 6.84}\right)$

$\therefore\ \widehat{B} \approx 83°$

**c** When faced with using either the sine rule or the cosine rule, it is better to use the cosine rule as it avoids the ambiguous case.

**7**

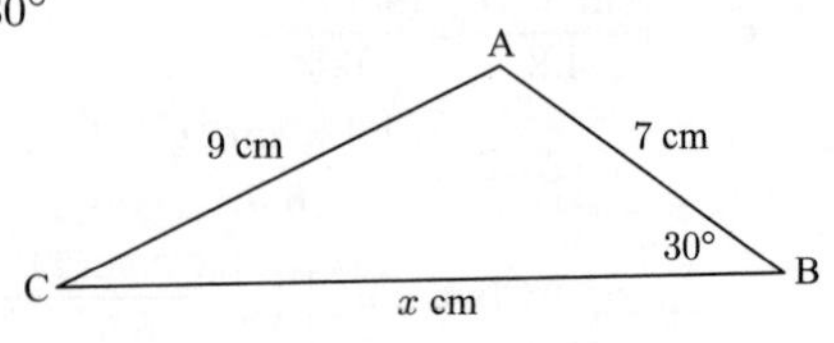


$9^2 = x^2 + 7^2 - 2 \times x \times 7 \times \cos 30°$

$\therefore\ 81 = x^2 + 49 - 14x(\frac{\sqrt{3}}{2})$

$\therefore\ x^2 - \frac{14\sqrt{3}}{2}x - 32 = 0$

Using the quadratic formula or technology,

$x \approx -2.23$ or $14.35$

but $x > 0$, so $x \approx 14.35$

$\therefore$ area of triangle $\approx \frac{1}{2} \times 7 \times 14.35 \times \sin 30° \approx 25.1$ cm$^2$

**8**
$$\frac{2x-5}{\sin 45^\circ} = \frac{x+3}{\sin 30^\circ}$$
$$\therefore \quad (2x-5)\sin 30^\circ = (x+3)\sin 45^\circ$$
$$\therefore \quad \frac{2x-5}{2} = \frac{x+3}{\sqrt{2}}$$
$$\therefore \quad 2\sqrt{2}x - 5\sqrt{2} = 2x+6$$
$$\therefore \quad -6-5\sqrt{2} = x(2-2\sqrt{2})$$
$$\therefore \quad x = \left(\frac{-6-5\sqrt{2}}{2-2\sqrt{2}}\right)\left(\frac{2+2\sqrt{2}}{2+2\sqrt{2}}\right)$$
$$= \frac{-12-12\sqrt{2}-10\sqrt{2}-10(2)}{4-4(2)}$$
$$= \frac{-32-22\sqrt{2}}{-4}$$
$$= 8 + \tfrac{11}{2}\sqrt{2}$$

## EXERCISE 9D

**1**

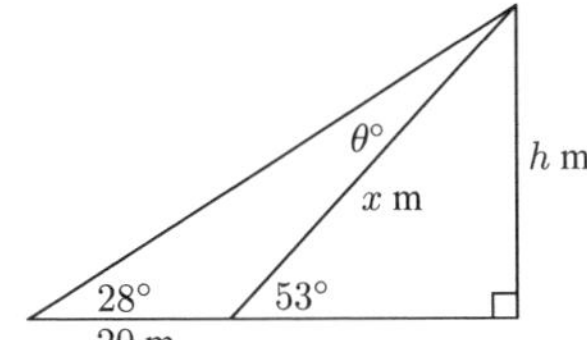


By the sine rule,
$$\frac{x}{\sin 28^\circ} = \frac{20}{\sin 25^\circ}$$
$$\therefore \quad x \approx \frac{20 \times \sin 28^\circ}{\sin 25^\circ}$$
$$\therefore \quad x \approx 22.22$$

and $\sin 53^\circ = \dfrac{h}{x}$
$$\therefore \quad h = x\sin 53^\circ$$
$$\approx 22.22 \times \sin 53^\circ$$
$$\approx 17.7 \text{ m}$$
$\therefore$ the pole is 17.7 m high.

$$\theta^\circ + 28^\circ = 53^\circ$$
{exterior angle of a $\triangle$ theorem}
$$\therefore \quad \theta = 25$$

**2**
$$\text{PR}^2 = 63^2 + 175^2 - 2\times 63\times 175\times\cos 112^\circ$$
$$\therefore \quad \text{PR} = \sqrt{63^2 + 175^2 - 2\times 63\times 175\times\cos 112^\circ}$$
$$\therefore \quad \text{PR} \approx 207 \text{ m}$$

**3**
$$\cos T = \frac{220^2 + 340^2 - 165^2}{2\times 220\times 340}$$
$$\therefore \quad T = \cos^{-1}\left(\frac{136\,775}{149\,600}\right)$$
$$\therefore \quad T \approx 23.9$$
$\therefore$ the tee shot was 23.9° off line.

**4**

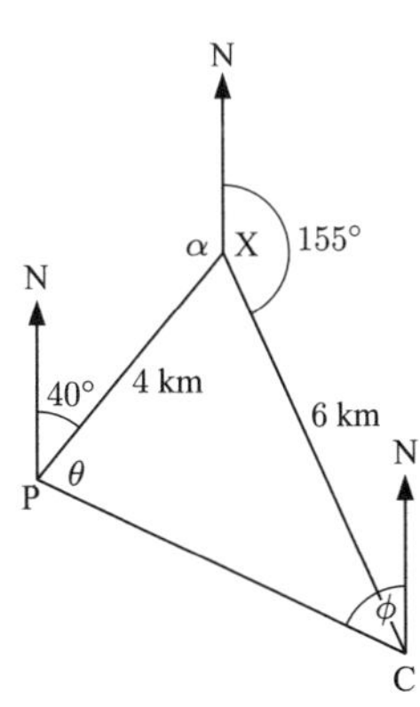


In $\triangle$ABD,
$$\cos(23.6+15.9)^\circ = \frac{200}{x}$$
$$\therefore \quad x = \frac{200}{\cos 39.5^\circ}$$
$$\therefore \quad x \approx 259.2$$

In $\triangle$ACD,
$$\frac{h}{\sin 15.9^\circ} = \frac{x}{\sin 113.6^\circ}$$
$$\therefore \quad h \approx \frac{259.2\times\sin 15.9^\circ}{\sin 113.6^\circ}$$
$$\therefore \quad h \approx 77.5$$
$\therefore$ the tower is 77.5 m high.

**5**

N
N
N
α X 155°
40° 4 km
6 km
P θ
ϕ
C

**a** $\alpha = 140^\circ$ {co-interior angles}
$$\therefore \quad \text{P}\widehat{\text{X}}\text{C} = 360^\circ - 140^\circ - 155^\circ \quad \text{\{angles at a point\}}$$
$$= 65^\circ$$
So, $\text{PC}^2 = 4^2 + 6^2 - 2\times 4\times 6\cos 65^\circ$
$$\therefore \quad \text{PC} = \sqrt{16+36-48\cos 65^\circ}$$
$$\approx 5.6315 \text{ km}$$
$\therefore$ Esko hikes 5.63 km.

**b**
$$\cos\theta \approx \frac{4^2 + 5.6315^2 - 6^2}{2\times 4\times 5.6315}$$
$$\therefore \quad \theta \approx 74.9^\circ$$
$$\therefore \quad \text{bearing} = 40^\circ + \theta$$
$$\approx 114.9^\circ$$
$\therefore$ Esko hikes on a bearing of 115°.

**c** **i** $\text{speed} = \frac{\text{distance}}{\text{time}} \Rightarrow \text{time} = \frac{\text{distance}}{\text{speed}}$

$\therefore \text{time}_{\text{Ritva}} = \frac{4+6}{10} = 1 \text{ hour}$ and $\text{time}_{\text{Esko}} \approx \frac{5.6315}{6} \approx 0.9386 \text{ hours} \approx 56.32 \text{ min}$

So Esko arrives at the campsite first.

**ii** $60 - 56.32 = 3.68$

Esko needs to wait about 3.68 minutes before Ritva arrives.

**d** $\phi \approx 180° - 114.9° \approx 65.1°$ {co-interior angles}

$\therefore 360° - \phi \approx 295°$

The return bearing is 295°.

**6**

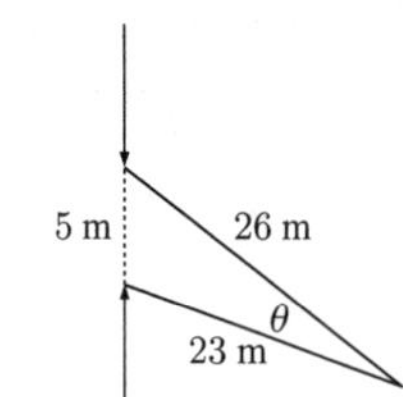


$$\cos\theta = \frac{23^2 + 26^2 - 5^2}{2 \times 23 \times 26}$$

$\therefore \theta = \cos^{-1}\left(\frac{1180}{1196}\right)$

$\therefore \theta \approx 9.38°$

$\therefore$ the angle of view is 9.38°.

**7**

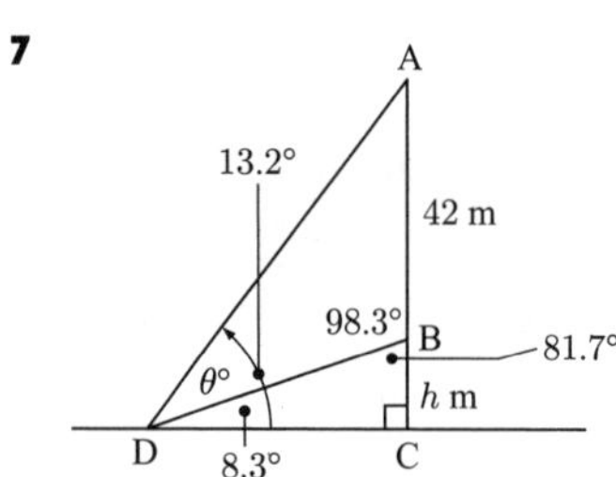


$\theta = 13.2° - 8.3° = 4.9°$

In △ABD,

$$\frac{AD}{\sin 98.3°} = \frac{42}{\sin 4.9°}$$

$\therefore AD = \frac{42 \times \sin 98.3°}{\sin 4.9°}$

$\therefore AD \approx 486.56$ m

In △ADC,

$$\sin 13.2° = \frac{h+42}{AD}$$

$\therefore h + 42 \approx 486.56 \times \sin 13.2°$

$\therefore h + 42 \approx 111.1$

$\therefore h \approx 69.1$

$\therefore$ the hill is 69.1 m high.

**8**

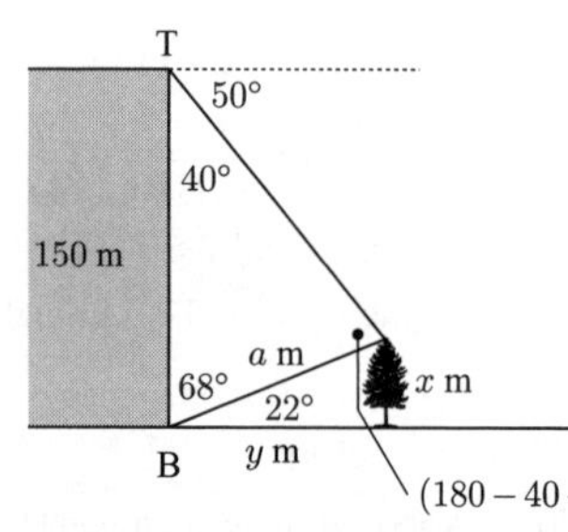


By the sine rule, $\frac{a}{\sin 40°} = \frac{150}{\sin 72°}$

$\therefore a = \frac{150 \times \sin 40°}{\sin 72°}$

$\therefore a \approx 101.38$

**a** $\sin 22° \approx \frac{x}{101.38}$

$\therefore x \approx 101.38 \times \sin 22°$

$\therefore x \approx 38.0$

$\therefore$ the tree is 38.0 m high.

**b** $\cos 22° \approx \frac{y}{101.38}$

$\therefore y \approx 101.38 \times \cos 22°$

$\therefore y \approx 94.0$

$\therefore$ the tree is 94.0 m from the building.

**9** Using Pythagoras' theorem

$RQ = \sqrt{4^2 + 7^2} = \sqrt{65}$ m

$PQ = \sqrt{8^2 + 7^2} = \sqrt{113}$ cm

$PR = \sqrt{8^2 + 4^2} = \sqrt{80}$ cm

Now $\cos Q = \frac{(\sqrt{113})^2 + (\sqrt{65})^2 - (\sqrt{80})^2}{2 \times \sqrt{113} \times \sqrt{65}}$

$\therefore \cos Q \approx \left(\frac{98}{171.4}\right)$

$\therefore Q \approx \cos^{-1}\left(\frac{98}{171.4}\right)$

$\therefore Q \approx 55.1$ So, $P\widehat{Q}R$ measures 55.1°.

**10**

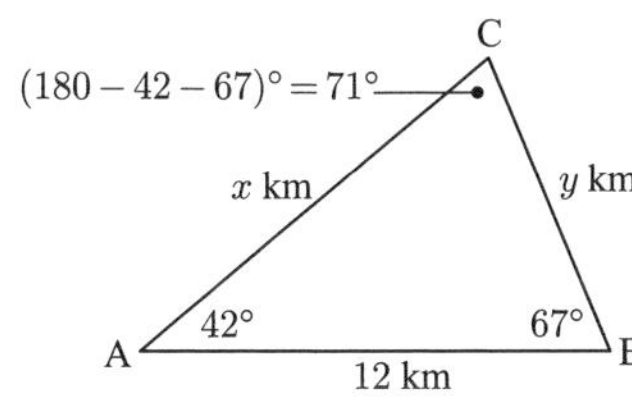


$$\frac{x}{\sin 67°} = \frac{12}{\sin 71°} = \frac{y}{\sin 42°}$$

$$\therefore \quad x = \frac{12 \times \sin 67°}{\sin 71°} \quad \text{and} \quad y = \frac{12 \times \sin 42°}{\sin 71°}$$

$$\therefore \quad x \approx 11.7 \qquad \therefore \quad y \approx 8.49$$

So, C is 11.7 km from A and 8.49 km from B.

**11** **a** $\text{QS} = \sqrt{8^2 + 12^2 - 2 \times 8 \times 12 \times \cos 70°}$

$\approx 11.93$

$\therefore$ area $\approx \frac{1}{2} \times 8 \times 12 \times \sin 70° + \frac{1}{2} \times 10 \times 11.93 \times \sin 30°$

$\approx 74.9 \text{ km}^2$

**b** 1 ha is 100 m × 100 m

$= 0.1 \text{ km} \times 0.1 \text{ km}$

$= 0.01 \text{ km}^2$

$\therefore$ 1 km$^2$ = 100 ha

$\therefore$ area $\approx 7490$ ha

**12**

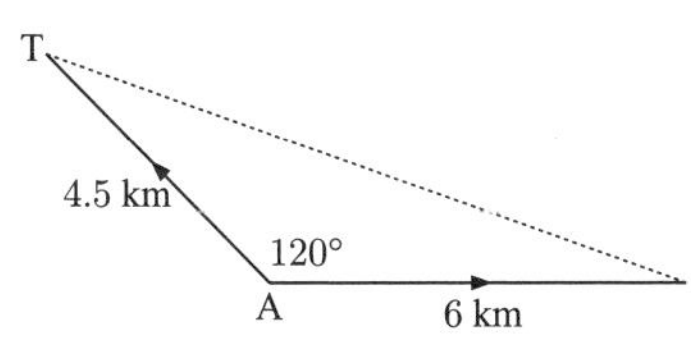


Distance = speed × time

So, after 45 min = 0.75 h,

AT = 6 × 0.75 = 4.5 km

AP = 8 × 0.75 = 6 km

Now $\text{PT} = \sqrt{4.5^2 + 6^2 - 2 \times 4.5 \times 6 \times \cos 120°}$

$\therefore$ PT $\approx 9.12$

So, they are 9.12 km apart.

**13**

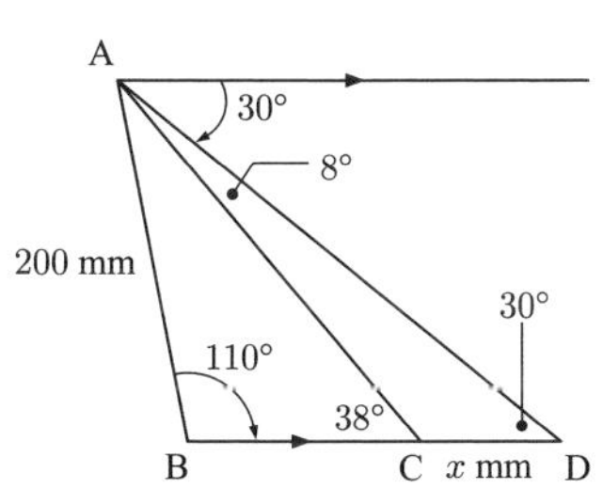


In $\triangle$ABC, $\dfrac{\text{AC}}{\sin 110°} = \dfrac{200}{\sin 38°}$

$\therefore$ $\text{AC} = \dfrac{200 \times \sin 110°}{\sin 38°} \approx 305.26$

and in $\triangle$ACD, $\dfrac{x}{\sin 8°} \approx \dfrac{305.26}{\sin 30°}$

$\therefore$ $x \approx \dfrac{305.26 \times \sin 8°}{\sin 30°} \approx 84.968$

$\therefore$ the metal strip is 85.0 mm wide.

**14**

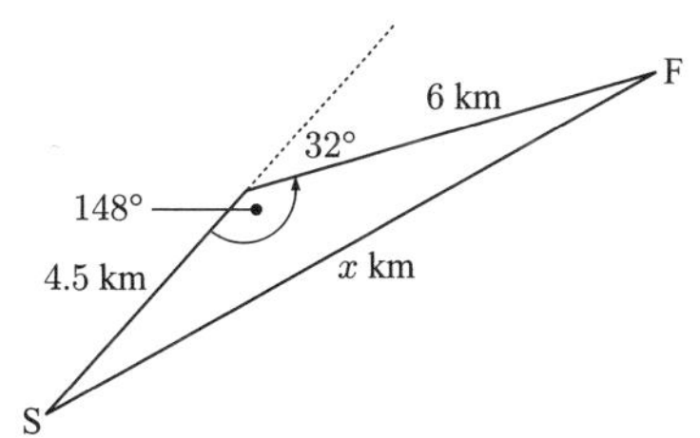


$x = \sqrt{6^2 + (4.5)^2 - 2 \times 6 \times 4.5 \times \cos 148°}$

$\therefore$ $x \approx 10.1$

$\therefore$ the orienteer is 10.1 km from the start.

**15**

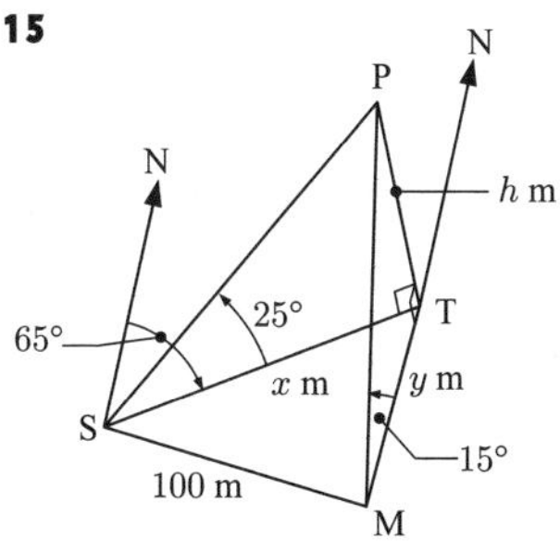


In $\triangle$PST, $\tan 25° = \dfrac{h}{x}$

$\therefore$ $x = \dfrac{h}{\tan 25°} \approx 2.145h$

In $\triangle$PMT, $\tan 15° = \dfrac{h}{y}$

$\therefore$ $y = \dfrac{h}{\tan 15°} \approx 3.732h$

But $\widehat{\text{STM}} = 65°$ {equal alternate angles}

and $100^2 = x^2 + y^2 - 2xy \cos 65°$

$\therefore$ $10\,000 \approx (2.145h)^2 + (3.732h)^2 - 2 \times (2.145)(3.732)h^2 \cos 65°$

$\therefore$ $10\,000 \approx 11.762\,h^2$

$\therefore$ $h^2 \approx 850.17$

$\therefore$ $h \approx 29.2$ So, the tree is 29.2 m high.

**16**

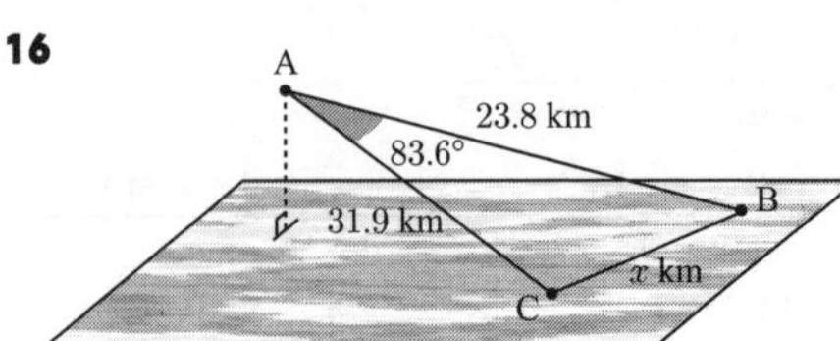


By the cosine rule

$x^2 = 23.8^2 + 31.9^2 - 2 \times 23.8 \times 31.9 \times \cos 83.6°$

$\therefore\ x = \sqrt{23.8^2 + 31.9^2 - 2 \times 23.8 \times 31.9 \times \cos 83.6°}$

$\therefore\ x \approx 37.6$

$\therefore$ B and C are 37.6 km apart.

## REVIEW SET 9A

**1** area $= \frac{1}{2} \times 7 \times 8 \times \sin 30°$

$= 28 \times \frac{1}{2}$

$= 14$ km$^2$

**2** If the unknown is an angle, use the cosine rule to avoid the ambiguous case.

**3** **a** By the cosine rule, $7^2 = 8^2 + x^2 - 2 \times 8 \times x \times \cos 60°$

$\therefore\ 49 = 64 + x^2 - 16x\left(\frac{1}{2}\right)$

$\therefore\ 49 = 64 + x^2 - 8x$

$\therefore\ x^2 - 8x + 15 = 0$

$\therefore\ (x-3)(x-5) = 0$

$\therefore\ x = 3$ or $5$

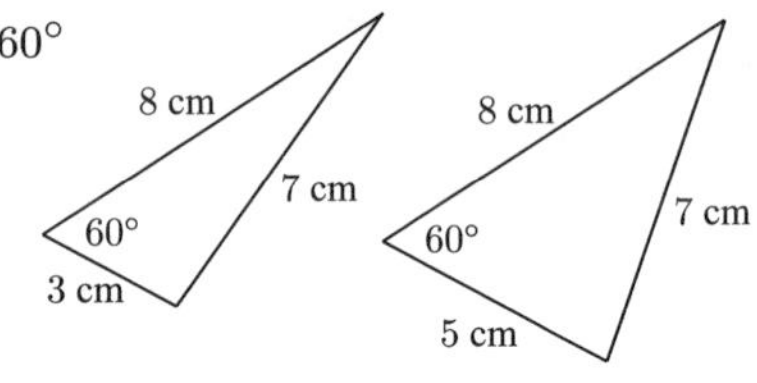


**b** Kady's response should be "Please supply me with additional information as there are two possibilities. Which one do you want?"

**4** area $= 42$ cm$^2$

$\therefore\ \frac{1}{2} \times 7 \times 13 \times \sin x° = 42$

$\therefore\ \sin x° = \dfrac{42 \times 2}{7 \times 13}$

$= \frac{12}{13}$

7 cm

$x°$

13 cm

**5** Total distance travelled $= x + 10$ km

$\therefore$ AB $= (x + 10) - 4 = x + 6$ km

Now $(x+6)^2 = x^2 + 10^2 - 2 \times x \times 10 \times \cos 120°$

$\therefore\ \cancel{x^2} + 12x + 36 = \cancel{x^2} + 100 - 20x(-\frac{1}{2})$

$\therefore\ 12x + 36 = 100 + 10x$

$\therefore\ 2x = 64$

$\therefore\ x = 32$

$\therefore$ the boat travelled $x + 10 = 42$ km.

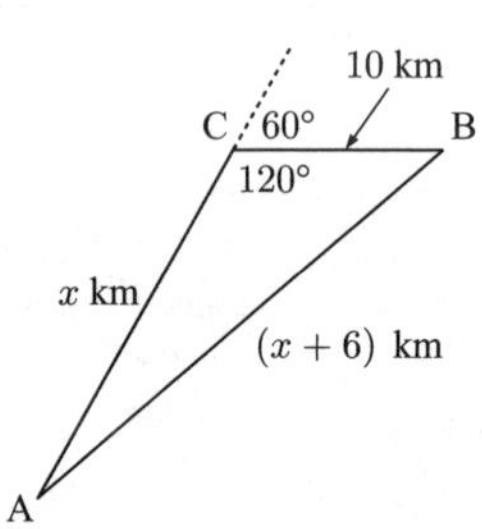


**6** shaded area = area of sector − area of $\triangle$

$= \frac{1}{2} \times \frac{13\pi}{18} \times 7^2 - \frac{1}{2} \times 7 \times 7 \times \sin\left(\frac{13\pi}{18}\right)$

$= \frac{49}{2}\left(\frac{13\pi}{18} - \sin\left(\frac{13\pi}{18}\right)\right)$

## REVIEW SET 9B

**1** **a** $\cos x° = \dfrac{13^2 + 19^2 - 11^2}{2 \times 13 \times 19}$

$\therefore\ \cos x° = \frac{409}{494}$

$\therefore\ x° = \cos^{-1}\left(\frac{409}{494}\right)$

$\therefore\ x \approx 34.1$

**b** $x^2 = 15^2 + 17^2 - 2 \times 15 \times 17 \times \cos 72°$

$\therefore\ x = \sqrt{15^2 + 17^2 - 2 \times 15 \times 17 \times \cos 72°}$

$\therefore\ x \approx 18.9$

**2** $\quad \text{AC}^2 = 11^2 + 9.8^2 - 2 \times 11 \times 9.8 \times \cos 74°$

$\therefore \quad \text{AC} = \sqrt{11^2 + 9.8^2 - 2 \times 11 \times 9.8 \times \cos 74°}$

$\therefore \quad \text{AC} \approx 12.554$ cm

$\therefore \quad \text{AC} \approx 12.6$ cm

Now $\dfrac{\sin C}{11} = \dfrac{\sin 74°}{\text{AC}}$

$\therefore \quad \sin C \approx \dfrac{11 \times \sin 74°}{12.554}$

$\therefore \quad C \approx \sin^{-1}\left(\dfrac{11 \times \sin 74°}{12.554}\right)$ or its supplement

$\therefore \quad C \approx 57.4°$ or $122.6°$

(impossible as $122.6 + 74 > 180$)

$\therefore \quad C$ measures $57.4°$

$\therefore \quad A$ measures $48.6°$.

**3** $\quad \text{DB}^2 = 7^2 + 11^2 - 2 \times 7 \times 11 \times \cos 110°$

$\therefore \quad \text{DB} = \sqrt{7^2 + 11^2 - 2 \times 7 \times 11 \times \cos 110°} \approx 14.922$ cm

$\therefore$ total area = area $\triangle$ABD + area $\triangle$BCD

$\approx \frac{1}{2} \times 7 \times 11 \times \sin 110° + \frac{1}{2} \times 16 \times 14.922 \times \sin 40°$

$\approx 113 \text{ cm}^2$

**4**

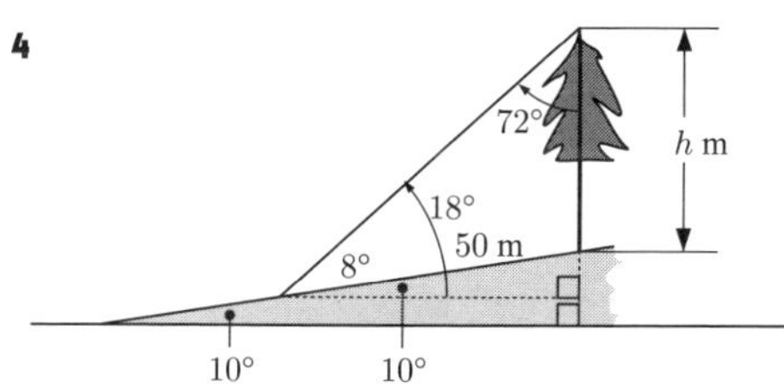


$\dfrac{h}{\sin 8°} = \dfrac{50}{\sin 72°}$

$\therefore \quad h = \dfrac{50 \times \sin 8°}{\sin 72°}$

$\therefore \quad h \approx 7.32$

So, the tree is 7.32 m high.

**5**

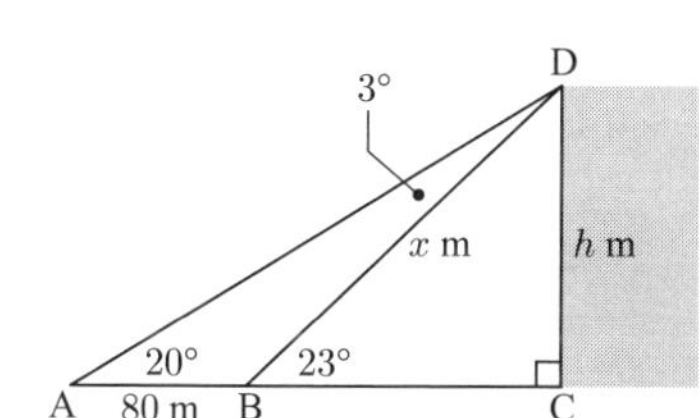


In $\triangle$ABD, $\dfrac{x}{\sin 20°} = \dfrac{80}{\sin 3°}$

$\therefore \quad x = \dfrac{80 \times \sin 20°}{\sin 3°} \approx 522.8$

Now $\sin 23° = \dfrac{h}{x}$

$\therefore \quad h \approx 522.8 \times \sin 23°$

$\therefore \quad h \approx 204$

So the building is 204 m tall.

**6**

$\text{A}\widehat{\text{S}}\text{P} = 210° - 113° = 97°$

$\therefore \quad x^2 = 310^2 + 430^2 - 2 \times 310 \times 430 \times \cos 97°$

$\therefore \quad x = \sqrt{310^2 + 430^2 - 2 \times 310 \times 430 \times \cos 97°}$

$\therefore \quad x \approx 559.9$

$\therefore$ Peter and Alix are 560 m apart.

and $\quad \cos\theta \approx \dfrac{310^2 + 559.9^2 - 430^2}{2 \times 310 \times 559.9}$

$\therefore \quad \theta \approx 49.7$

and $\quad 30 + \theta \approx 79.7$

$\therefore$ the bearing of Peter from Alix is $079.7°$.

## REVIEW SET 9C

**1** **a** $\cos x^\circ = \dfrac{11^2 + 19^2 - 13^2}{2 \times 11 \times 19}$

$\therefore \cos x^\circ = \frac{313}{418}$

$\therefore x^\circ = \cos^{-1}\left(\frac{313}{418}\right)$

$\therefore x \approx 41.5$

**b** $x^2 = 14^2 + 21^2 - 2 \times 14 \times 21 \times \cos 47^\circ$

$\therefore x = \sqrt{14^2 + 21^2 - 2 \times 14 \times 21 \times \cos 47^\circ}$

$\therefore x \approx 15.4$

**2** area $= 80$ cm$^2$

$\therefore \frac{1}{2} \times 11.3 \times 19.2 \sin x^\circ = 80$

$\therefore \sin x^\circ = \dfrac{160}{11.3 \times 19.2}$

Now $\arcsin\left(\dfrac{160}{11.3 \times 19.2}\right) \approx 47.5^\circ$

$\therefore x \approx 47.5$ or $180 - 47.5$

$\therefore x \approx 47.5$ or $132.5$

**3** Using Pythagoras,

$\text{ED} = \sqrt{6^2 + 3^2} = \sqrt{45}$ m

$\text{DG} = \sqrt{4^2 + 3^2} = \sqrt{25} = 5$ m

$\text{EG} = \sqrt{6^2 + 4^2} = \sqrt{52}$ m

Using the cosine rule, $\cos\theta = \dfrac{(\sqrt{45})^2 + 5^2 - (\sqrt{52})^2}{2 \times \sqrt{45} \times 5}$

$\therefore \theta = \cos^{-1}\left(\frac{18}{10\sqrt{45}}\right)$

$\therefore \theta \approx 74.4^\circ$ Thus $\widehat{\text{EDG}}$ measures $74.4^\circ$.

**4** **a**

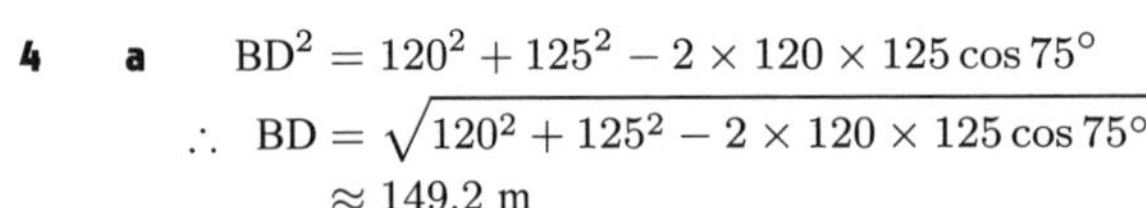

$\text{BD}^2 = 120^2 + 125^2 - 2 \times 120 \times 125 \cos 75^\circ$

$\therefore \text{BD} = \sqrt{120^2 + 125^2 - 2 \times 120 \times 125 \cos 75^\circ}$

$\approx 149.2$ m

The area of the block = area of $\triangle$ABD + area of $\triangle$BCD

$\approx \frac{1}{2} \times 120 \times 125 \times \sin 75^\circ + \frac{1}{2} \times 149.2 \times 90 \times \sin 30^\circ$

$\approx 10\,600$ m$^2$

**b** $\approx 1.06$ ha $\{10\,000 \text{ m}^2 = 1 \text{ ha}\}$

**5**

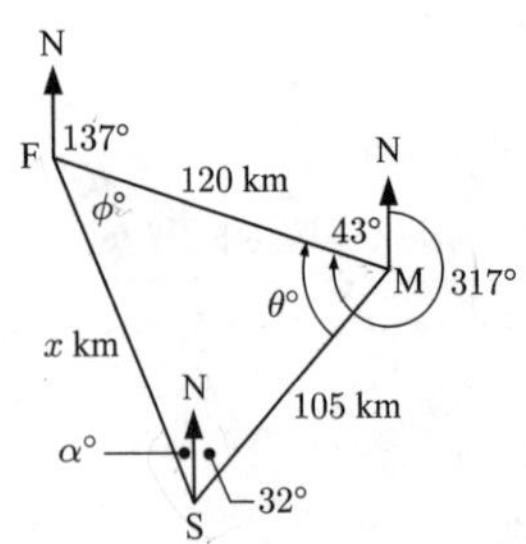


In 45 minutes, $140 \times \frac{3}{4} = 105$ km is travelled.

In 40 minutes, $180 \times \frac{2}{3} = 120$ km is travelled.

We notice that $\theta + 43 + 32 = 180$ {co-interior angles add to $180^\circ$}

$\therefore \theta = 105$

Using the cosine rule, $x^2 = 120^2 + 105^2 - 2 \times 120 \times 105 \times \cos 105^\circ$

$\therefore x = \sqrt{120^2 + 105^2 - 2 \times 120 \times 105 \times \cos 105^\circ}$

$\therefore x \approx 178.74$

So, the car is 179 km from the start.

Now $\dfrac{\sin\phi^\circ}{105} \approx \dfrac{\sin 105^\circ}{178.74}$

$\therefore \sin\phi^\circ \approx \dfrac{105 \times \sin 105^\circ}{178.74}$

$\therefore \phi \approx 34.6$

$\therefore \alpha \approx 180 - 105 - 34.6 - 32 \approx 8.4 \approx 8$

So, the bearing from its starting point is $352^\circ$.

**6** **a**

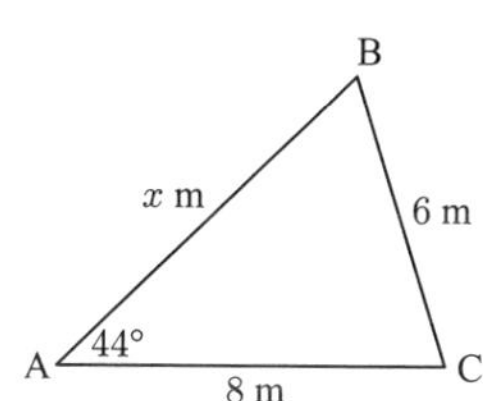


By the cosine rule, $6^2 = x^2 + 8^2 - 2 \times x \times 8 \times \cos 44°$

$\therefore\ 36 = x^2 + 64 - 16x \times \cos 44°$

$\therefore\ x^2 - 11.51x + 28 \approx 0$

$$\therefore\ x \approx \frac{11.51 \pm \sqrt{11.51^2 - 4(1)(28)}}{2}$$

$$\therefore\ x \approx \frac{11.51 \pm 4.524}{2}$$

$\therefore\ x \approx 8.02$ or $3.49$

Frank needs additional information as there are two possible cases:

(1) when AB $\approx 8.02$ m and

(2) when AB $\approx 3.49$ m

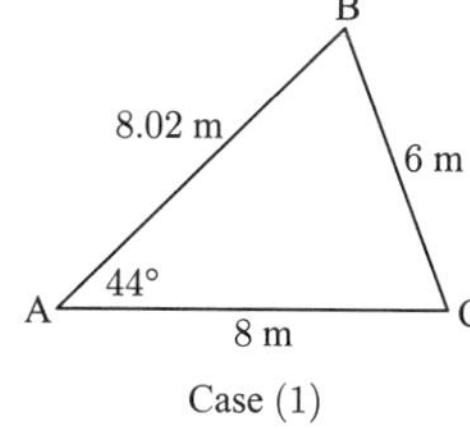


Case (1)

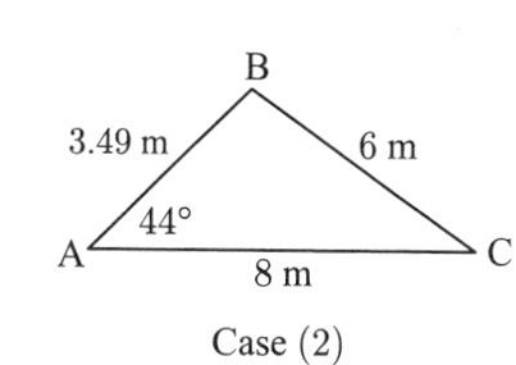


Case (2)

**b** Volume = area × depth

$= \frac{1}{2} \times 8 \times x \times \sin 44° \times 0.1$ and is a maximum when $x \approx 8.02$ m

$\approx 4 \times 8.02 \times \sin 44° \times 0.1$

$\approx 2.23\ \text{m}^3$

# Chapter 10
## TRIGONOMETRIC FUNCTIONS

### EXERCISE 10A

**1** **a** periodic **b** periodic **c** periodic **d** not periodic **e** periodic
**f** periodic **g** not periodic **h** not periodic

**2** **a**

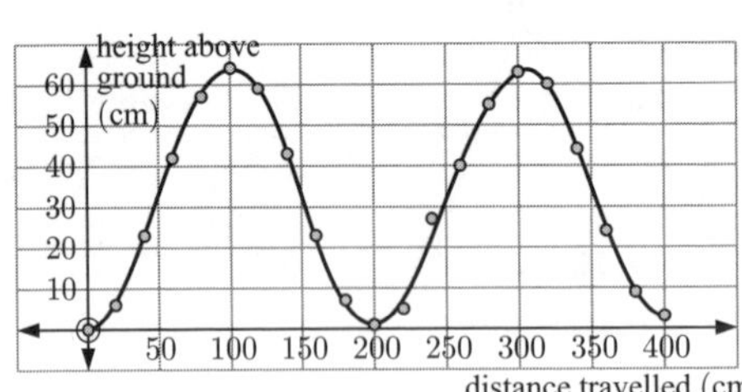


**b** A curve can be fitted to the data as the distance travelled is continuous.

**c** The data is periodic.

**i** The minimum value from the table is 0 and the maximum value is 64.

So, the principal axis is $y \approx \frac{0+64}{2}$

$\therefore \; y \approx 32$

**ii** The maximum value is $\approx 64$ cm.

**iii** The period is $\approx 200$ cm.

**iv** The amplitude is $\approx 32$ cm.

**3** **a**

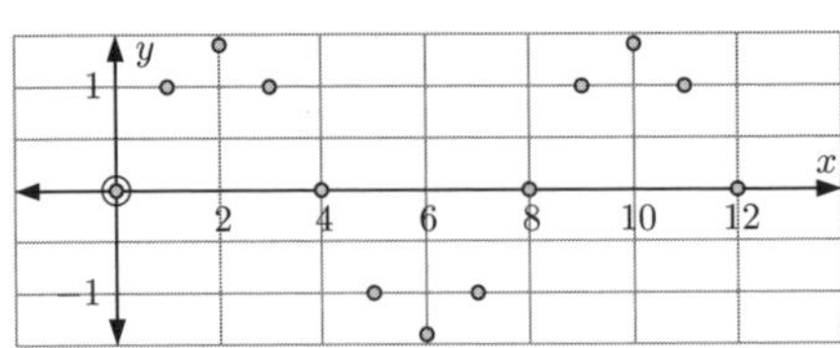


Data exhibits periodic behaviour.

**b**

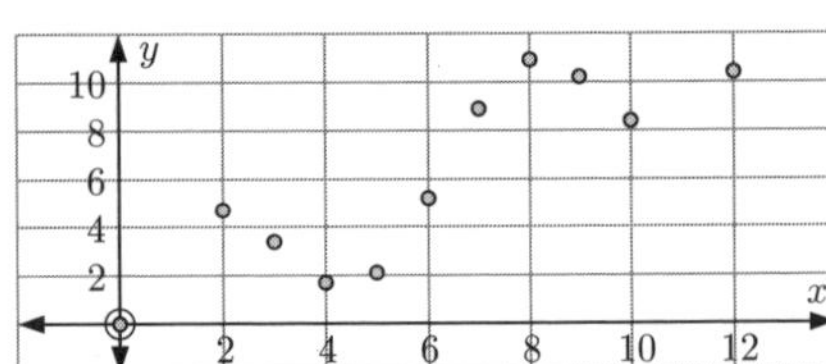


Not enough information to say data is periodic.

### EXERCISE 10B.1

**1** **a** $y = 3\sin x$
has amplitude 3 and period $\frac{2\pi}{1} = 2\pi$
When $x = 0$, $y = 0$.

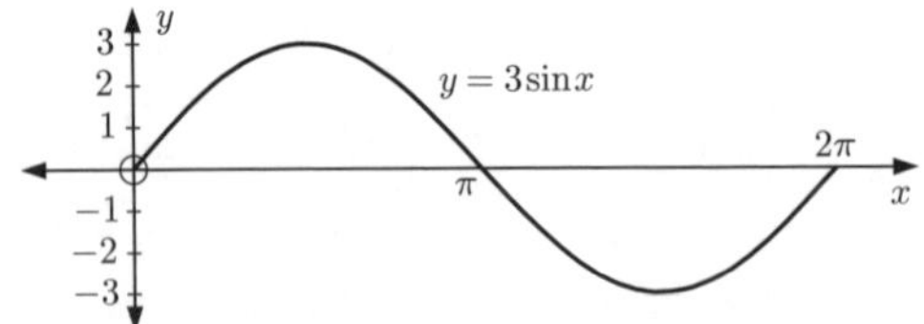


**b** $y = -3\sin x$
has amplitude $|-3| = 3$
and period $\frac{2\pi}{1} = 2\pi$.
When $x = 0$, $y = 0$.

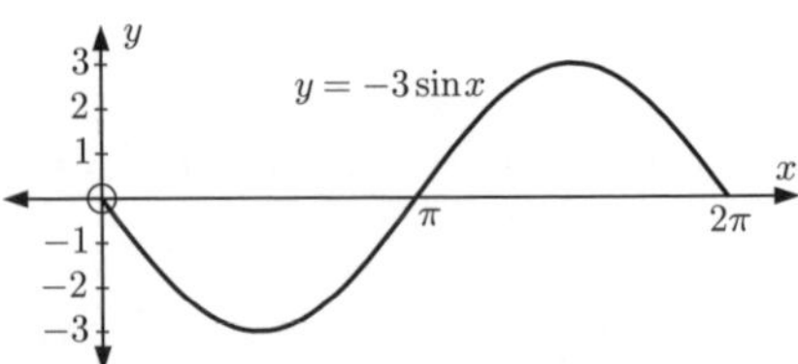


It is the reflection of $y = 3\sin x$ in the $x$-axis.

**c** $y = \frac{3}{2}\sin x$
has amplitude $\frac{3}{2}$ and period $\frac{2\pi}{1} = 2\pi$.
When $x = 0$, $y = 0$.

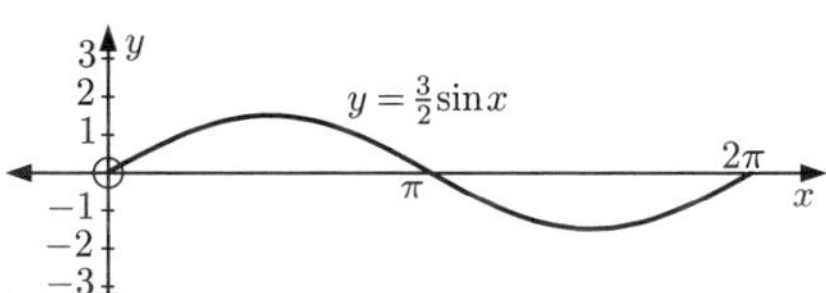


**d** $y = -\frac{3}{2}\sin x$
has amplitude $\left|-\frac{3}{2}\right| = \frac{3}{2}$
and period $\frac{2\pi}{1} = 2\pi$.

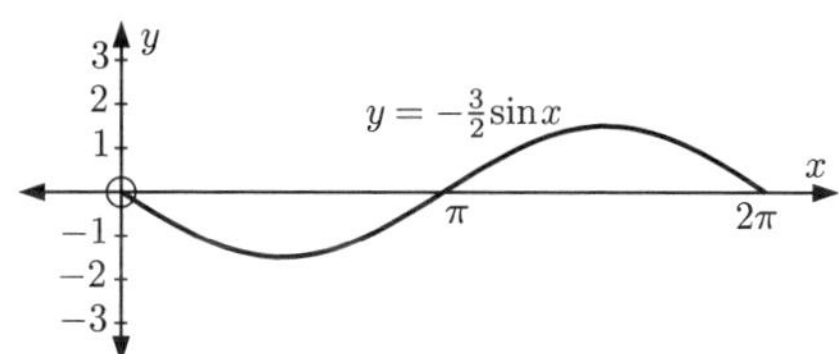


It is the reflection of $y = \frac{3}{2}\sin x$ in the $x$-axis.

**2** **a** $y = \sin 3x$
has amplitude 1 and period $\frac{2\pi}{3}$.
When $x = 0$, $y = 0$.

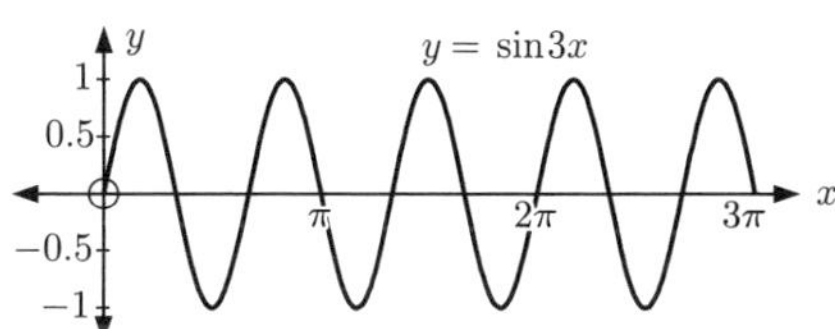


**b** $y = \sin\left(\frac{x}{2}\right)$
has amplitude 1 and period $\frac{2\pi}{\frac{1}{2}} = 4\pi$.
When $x = 0$, $y = 0$.

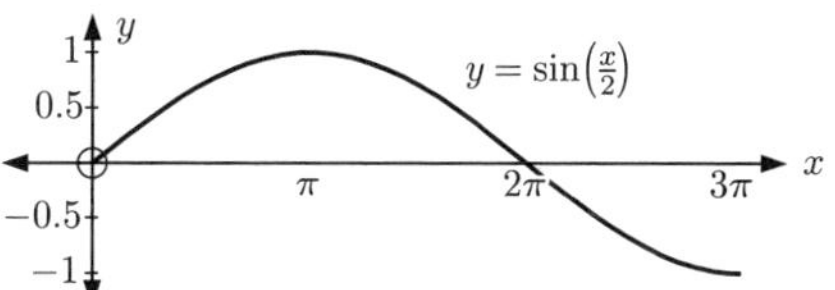


**c** $y = \sin(-2x)$
has amplitude 1 and period $\frac{2\pi}{|-2|} = \pi$.
When $x = 0$, $y = 0$.

It is the reflection of $y = \sin 2x$ in the $y$-axis.

**3** **a** period $= \frac{2\pi}{4} = \frac{\pi}{2}$ **b** period $= \frac{2\pi}{|-4|} = \frac{\pi}{2}$ **c** period $= \frac{2\pi}{\left(\frac{1}{3}\right)} = 6\pi$ **d** period $= \frac{2\pi}{0.6} = \frac{20\pi}{6} = \frac{10\pi}{3}$

**4** **a** $\frac{2\pi}{b} = 5\pi$ $\therefore b = \frac{2}{5}$ **b** $\frac{2\pi}{b} = \frac{2\pi}{3}$ $\therefore b = 3$ **c** $\frac{2\pi}{b} = 12\pi$ $\therefore b = \frac{1}{6}$ **d** $\frac{2\pi}{b} = 4$ $\therefore b = \frac{\pi}{2}$ **e** $\frac{2\pi}{b} = 100$ $\therefore b = \frac{2\pi}{100} = \frac{\pi}{50}$

## EXERCISE 10B.2

**1** **a**

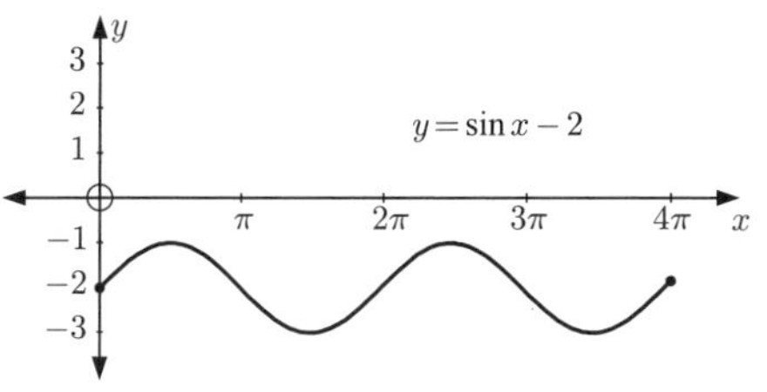


This is the graph of $y = \sin x$ translated by $\begin{pmatrix}0\\-2\end{pmatrix}$.

**b**

This is the graph of $y = \sin x$ translated by $\begin{pmatrix}2\\0\end{pmatrix}$.

**c**

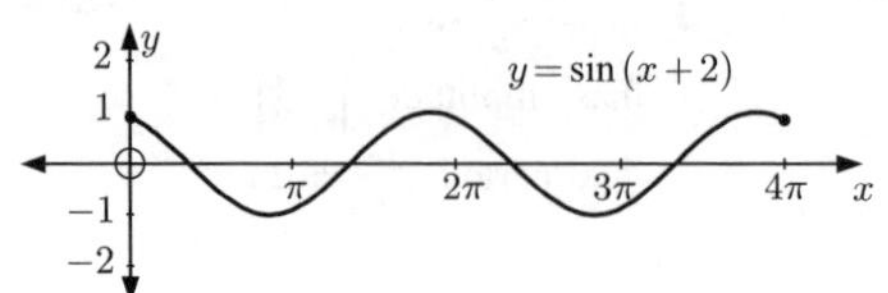


This is the graph of $y = \sin x$ translated by $\binom{-2}{0}$.

**d**

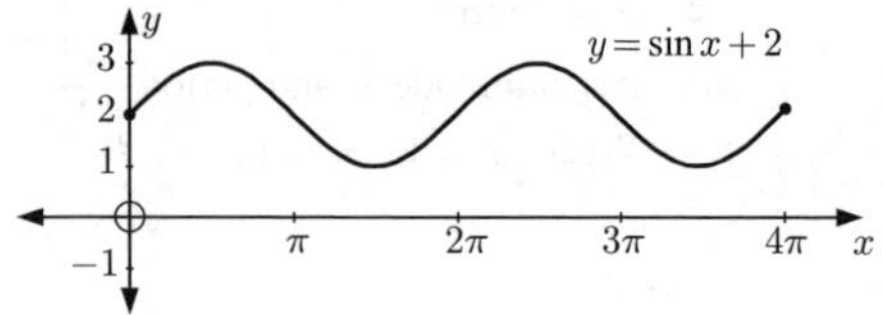


This is the graph of $y = \sin x$ translated by $\binom{0}{2}$.

**e**

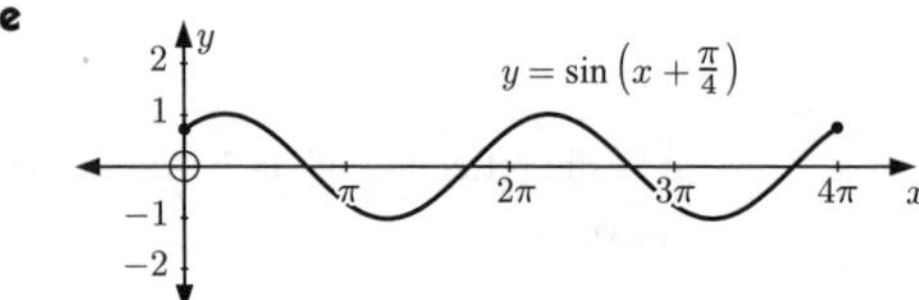


This is the graph of $y = \sin x$ translated by $\begin{pmatrix} -\frac{\pi}{4} \\ 0 \end{pmatrix}$.

**f**

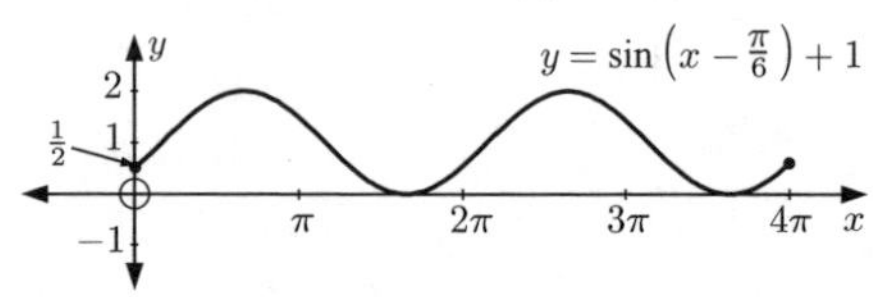


This is the graph of $y = \sin x$ translated by $\begin{pmatrix} \frac{\pi}{6} \\ 1 \end{pmatrix}$.

**2** **a** period $= \dfrac{2\pi}{5} = \frac{2\pi}{5}$ **b** period $= \dfrac{2\pi}{(\frac{1}{4})} = 8\pi$ **c** period $= \dfrac{2\pi}{|-2|} = \pi$

**3** **a** $\dfrac{2\pi}{b} = 3\pi$ $\therefore\ b = \frac{2}{3}$

**b** $\dfrac{2\pi}{b} = \dfrac{\pi}{10}$ $\therefore\ b = 20$

**c** $\dfrac{2\pi}{b} = 100\pi$ $\therefore\ b = \frac{2}{100} = \frac{1}{50}$

**d** $\dfrac{2\pi}{b} = 50$ $\therefore\ b = \frac{2\pi}{50} = \frac{\pi}{25}$

**4** **a** A translation of $\binom{0}{-1}$, or vertically down 1 unit.

**b** A translation of $\begin{pmatrix} \frac{\pi}{4} \\ 0 \end{pmatrix}$, or horizontally $\frac{\pi}{4}$ units right.

**c** A vertical stretch of factor 2.

**d** A horizontal stretch of factor $\frac{1}{4}$.

**e** A vertical stretch of factor $\frac{1}{2}$.

**f** A horizontal stretch of factor 4.

**g** A reflection in the $x$-axis.

**h** A translation of $\binom{-2}{-3}$.

**i** A vertical stretch of factor 2 followed by a horizontal stretch of factor $\frac{1}{3}$.

**j** A translation of $\begin{pmatrix} \frac{\pi}{3} \\ 2 \end{pmatrix}$.

## EXERCISE 10C

**1** **a**

| *Month, t* | 1 | 2 | 3 | 4 | 5 | 6 | 7 | 8 | 9 | 10 | 11 | 12 |
|---|---|---|---|---|---|---|---|---|---|---|---|---|
| *Temp, T* | 15 | 14 | 15 | 18 | 21 | 25 | 27 | 26 | 24 | 20 | 18 | 16 |

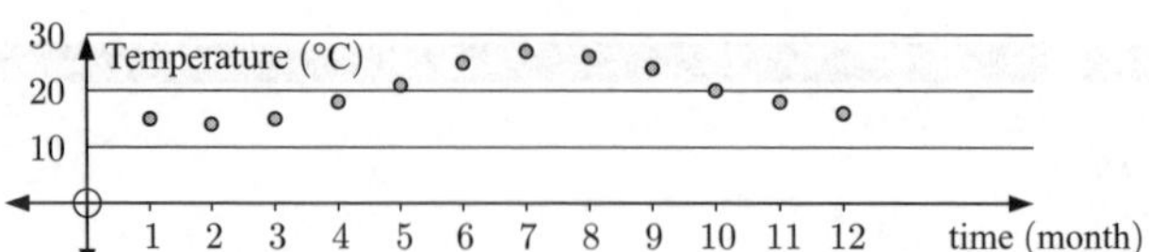


The period is 12 months so $\dfrac{2\pi}{b} = 12$

$\therefore\ b = \frac{\pi}{6}$ {assuming $b > 0$}.

Amplitude, $a \approx \dfrac{\text{max.} - \text{min.}}{2} \approx \dfrac{27 - 14}{2} \approx 6.5$

As the principal axis is midway between min. and max., then $d \approx \dfrac{27 + 14}{2} \approx 20.5$

When $T$ is 20.5 (midway between min. and max.), $c \approx \dfrac{2 + 7}{2} \approx 4.5$ {average of $t$ values}

$\therefore\ T \approx 6.5 \sin(\frac{\pi}{6}(t - 4.5)) + 20.5$ where $\frac{\pi}{6} \approx 0.524$

**b** Using technology, $T \approx 6.14\sin(0.575t - 2.70) + 20.4$

$\therefore \quad T \approx 6.14\sin(0.575(t - 4.70)) + 20.4$

The model fits reasonably well.

**2** **a**

| *Month, t* | 1 | 2 | 3 | 4 | 5 | 6 | 7 | 8 | 9 | 10 | 11 | 12 |
|---|---|---|---|---|---|---|---|---|---|---|---|---|
| *Temp, T* | 15 | 16 | $14\frac{1}{2}$ | 12 | 10 | $7\frac{1}{2}$ | 7 | $7\frac{1}{2}$ | $8\frac{1}{2}$ | $10\frac{1}{2}$ | $12\frac{1}{2}$ | 14 |

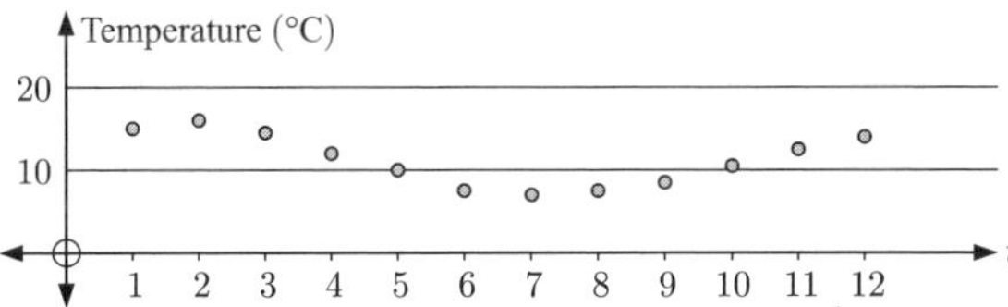


The period is $\dfrac{2\pi}{b} = 12 \quad \therefore \quad b = \frac{\pi}{6} \quad \{b > 0\}$

Amplitude, $a \approx \dfrac{\text{max.} - \text{min.}}{2} \approx \dfrac{16 - 7}{2} \approx 4.5$

As the principal axis is midway between min. and max. then $d \approx \dfrac{16 + 7}{2} \approx 11.5$

At min., $t = 7$ and at max., $t = 2 + 12 = 14 \quad \therefore \quad c \approx \dfrac{7 + 14}{2} \approx 10.5$

So, $T \approx 4.5\sin(\frac{\pi}{6}(t - 10.5)) + 11.5$

**b** Using technology, $T \approx 4.29\sin(0.533t + 0.769) + 11.2$

$\therefore \quad T \approx 4.29\sin(0.533(t + 1.44)) + 11.2$

**Note:** (1) $\frac{\pi}{6} \approx 0.524$ ✓

(2) $1.44 - (-10.5)$
$= 11.94 \approx 12$

**3** **a**

| *Month, t* | 1 | 2 | 3 | 4 | 5 | 6 | 7 | 8 | 9 | 10 | 11 | 12 |
|---|---|---|---|---|---|---|---|---|---|---|---|---|
| *Temp, T* | 0 | 0 | $-4$ | $-9$ | $-14$ | $-17$ | $-18$ | $-19$ | $-17$ | $-13$ | $-6$ | $-2$ |

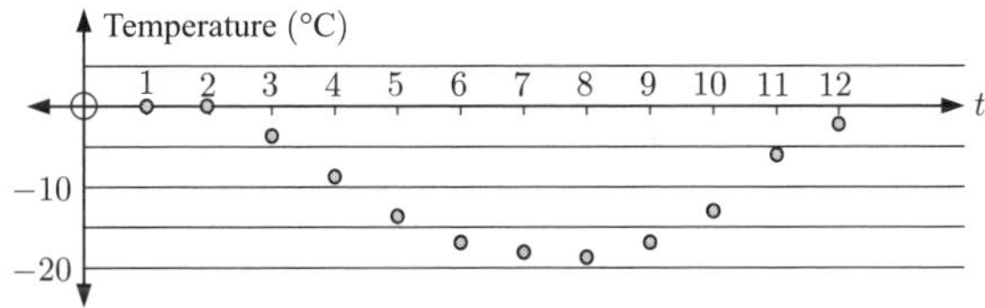


The period is $\dfrac{2\pi}{b} = 12 \quad \therefore \quad b = \frac{\pi}{6} \quad \{b > 0\}$

Amplitude, $a \approx \dfrac{\text{max.} - \text{min.}}{2} \approx \dfrac{0 - (-19)}{2} \approx 9.5$

$d \approx \dfrac{\text{max.} + \text{min.}}{2} \approx \dfrac{0 + (-19)}{2} \approx -9.5$

At min., $t = 8$ and at max., $t = 1 + 12 = 13 \quad \therefore \quad c \approx \dfrac{8 + 13}{2} \approx 10.5$

So, $T \approx 9.5\sin(\frac{\pi}{6}(t - 10.5)) - 9.5$

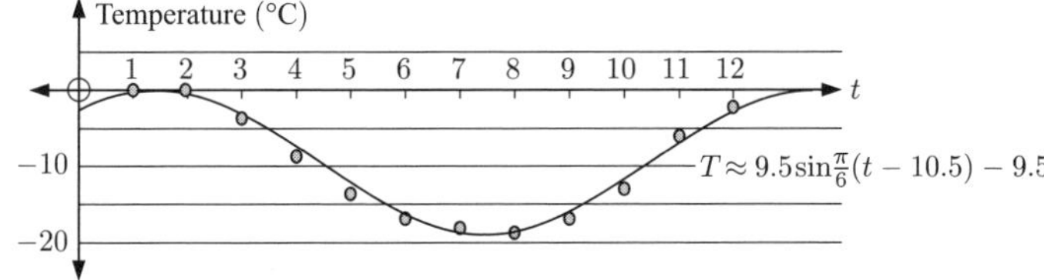


**b** The model is reasonably appropriate.

**4** **a** For the model $H = a\sin(b(t-c)) + d$

$$\text{period} = \frac{2\pi}{b} = 12.4 \text{ hours} \quad \therefore \quad b = \frac{2\pi}{12.4} \approx 0.507$$

We let the principal axis be 0, so $d = 0$

$\therefore$ the amplitude $a = 7$, so the min. is $-7$, and the max. is $+7$

Let $t = 0$ correspond to 'low tide' $\therefore$ $t = 6.2$ corresponds to 'high tide'

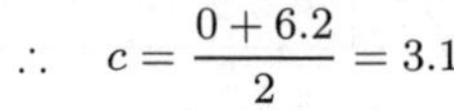

$$\therefore \quad c = \frac{0 + 6.2}{2} = 3.1$$

So, $H \approx 7\sin(0.507(t-3.1)) + 0$

$\therefore$ $H \approx 7\sin(0.507(t-3.1))$

**b**

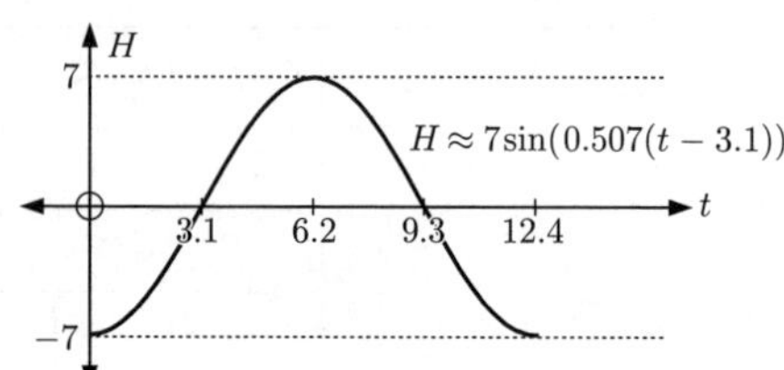


**5** Let the model be $H = a\sin(b(t-c)) + d$ metres

When $t = 0$, $H = 2$ and when $t = 50$, $H = 22$
($H = 2$ is the min.; $H = 22$ is the max.)

$$\text{period} = \frac{2\pi}{b} = 100 \quad \therefore \quad b = \frac{2\pi}{100} = \frac{\pi}{50}$$

$a = 10$ {from the diagram}, $d = \dfrac{\text{max.} + \text{min.}}{2} = \dfrac{22+2}{2} = 12$

$c = \dfrac{0+50}{2} = 25$ {values of $t$ at min. and max.} $\therefore$ $H = 10\sin(\frac{\pi}{50}(t-25)) + 12$

## EXERCISE 10D

**1** **a** $y = \cos x + 2$

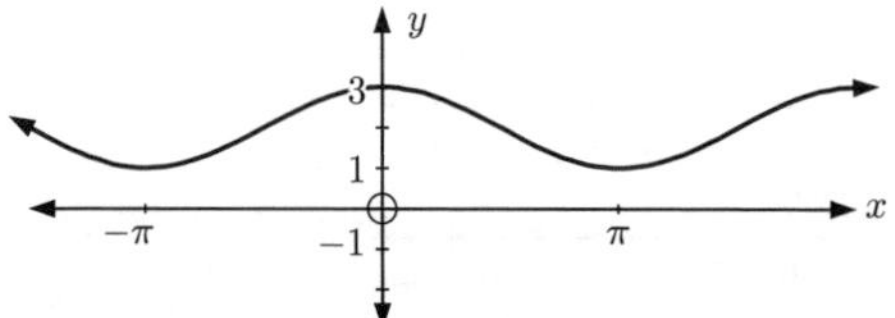


This is a vertical translation of $y = \cos x$ through $\binom{0}{2}$.

**b** $y = \cos x - 1$

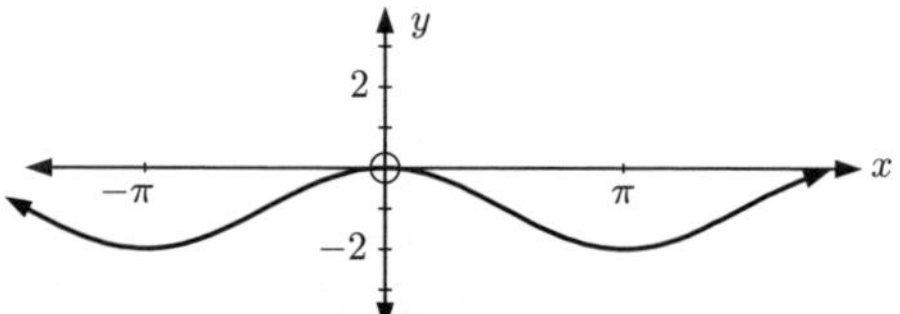


This is a vertical translation of $y = \cos x$ through $\binom{0}{-1}$.

**c** $y = \cos(x - \frac{\pi}{4})$

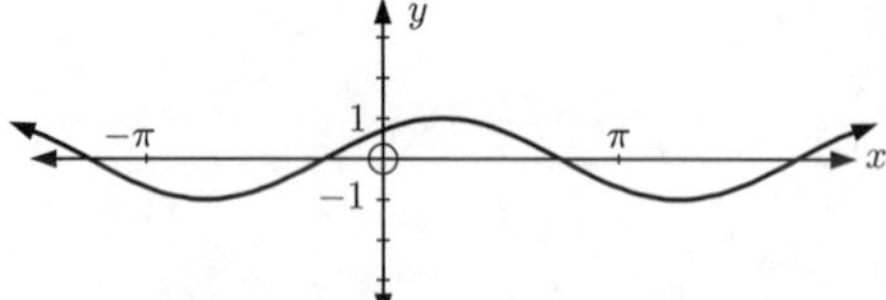


This is a horizontal translation of $y = \cos x$ through $\begin{pmatrix} \frac{\pi}{4} \\ 0 \end{pmatrix}$.

**d** $y = \cos(x + \frac{\pi}{6})$

This is a horizontal translation of $y = \cos x$ through $\begin{pmatrix} -\frac{\pi}{6} \\ 0 \end{pmatrix}$.

**e** $y = \frac{2}{3}\cos x$

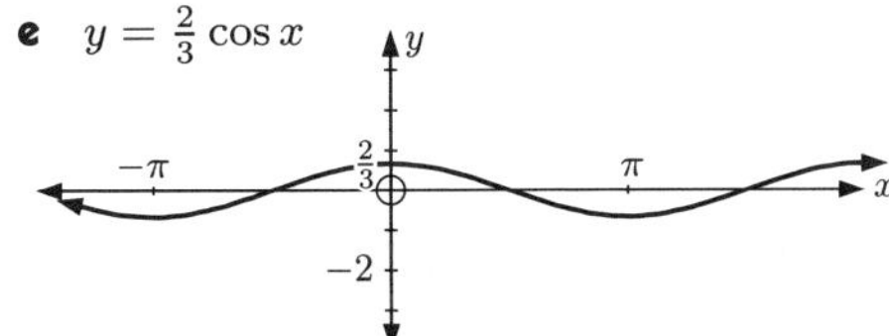


This is a vertical stretch of $y = \cos x$ with factor $\frac{2}{3}$.

**f** $y = \frac{3}{2}\cos x$

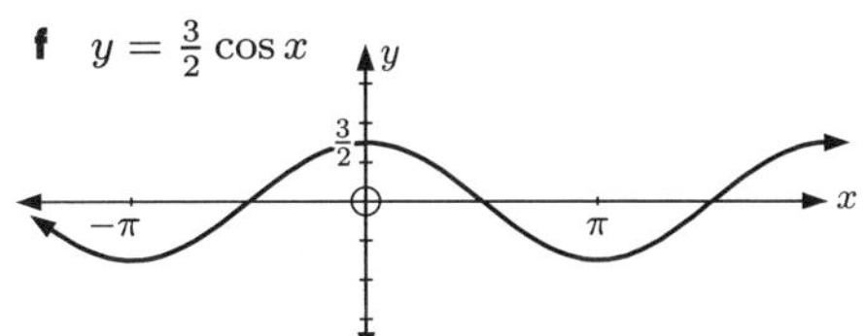


This is a vertical stretch of $y = \cos x$ with factor $\frac{3}{2}$.

**g** $y = -\cos x$

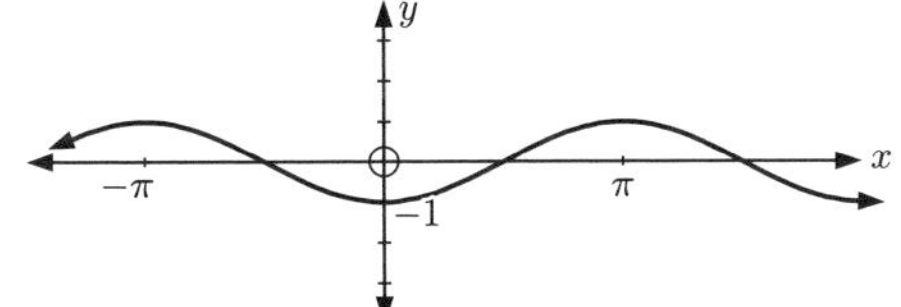


This is a reflection of $y = \cos x$ in the $x$-axis.

**h** $y = \cos(x - \frac{\pi}{6}) + 1$

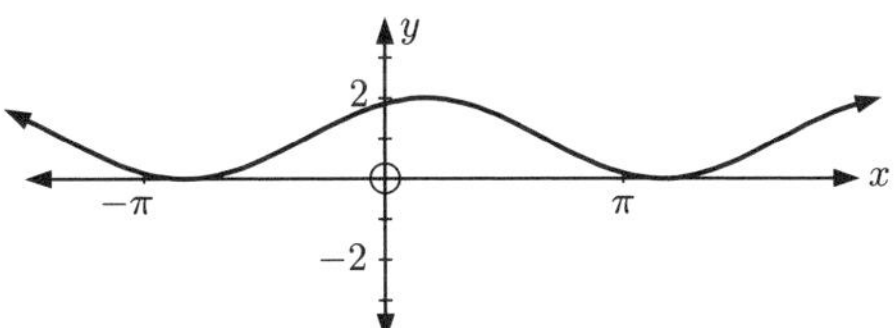


This is a translation of $\begin{pmatrix} \frac{\pi}{6} \\ 1 \end{pmatrix}$.

**i** $y = \cos(x + \frac{\pi}{4}) - 1$

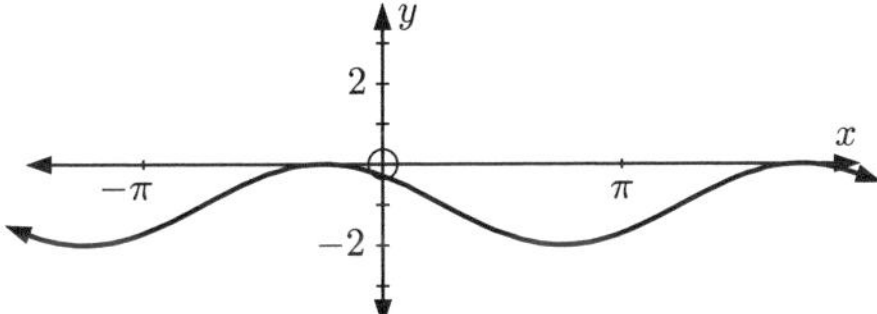


This is a translation of $\begin{pmatrix} -\frac{\pi}{4} \\ -1 \end{pmatrix}$.

**j** $y = \cos 2x$

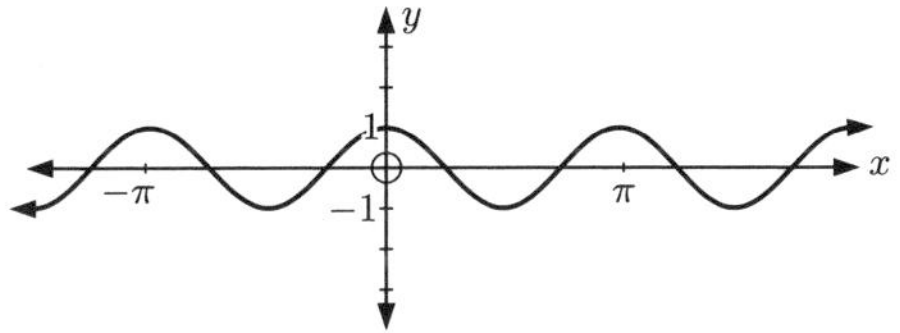


This is a horizontal stretch of factor $\frac{1}{2}$.

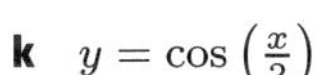
**k** $y = \cos\left(\frac{x}{2}\right)$

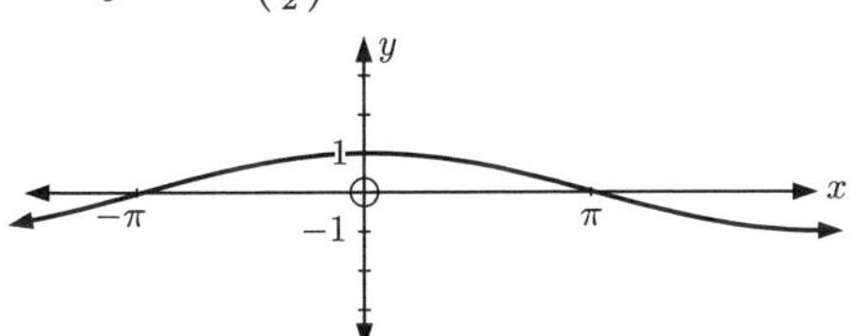


This is a horizontal stretch of factor 2.

**l** $y = 3\cos 2x$

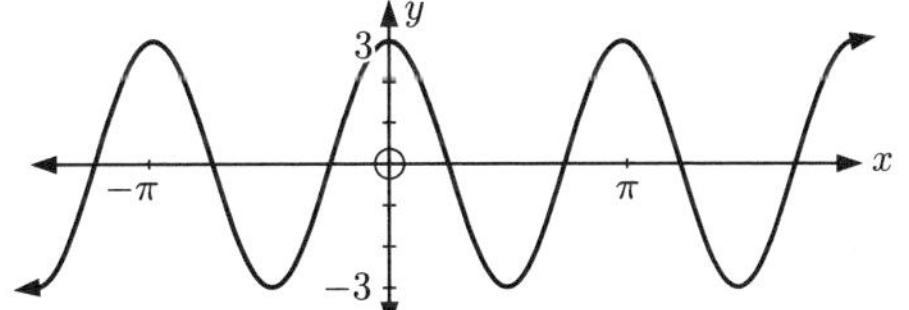


This is a horizontal stretch of factor $\frac{1}{2}$ followed by a vertical stretch of factor 3.

**2** **a** period $= \frac{2\pi}{3}$ **b** period $= \dfrac{2\pi}{\frac{1}{3}} = 6\pi$ **c** period $= \dfrac{2\pi}{\frac{\pi}{50}} = 100$

**3** $a$ controls the amplitude {amplitude $= |a|$}. $b$ controls the period {period $= \dfrac{2\pi}{|b|}$}.

$c$ controls the horizontal translation. $d$ controls the vertical translation.

**4** **a** If $y = a\cos(b(x - c)) + d$, then $a = 2$, $\pi = \dfrac{2\pi}{b}$ $\therefore$ $b = 2$

$c$ and $d$ are 0 as there is no horizontal or vertical shift. $\therefore$ $y = 2\cos(2x)$

**b** If $y = a\cos(b(x - c)) + d$, then $a = 1$, $4\pi = \dfrac{2\pi}{b}$ $\therefore$ $b = \frac{1}{2}$

A vertical shift of 2 units, no horizontal shift $\therefore$ $d = 2$, $c = 0$.

So, $y = \cos(\frac{1}{2}x) + 2$ or $y = \cos\left(\frac{x}{2}\right) + 2$.

**c** If $y = a\cos(b(x-c)) + d$, then $a = -5$, $6 = \dfrac{2\pi}{b}$ $\therefore$ $b = \frac{\pi}{3}$

$c = d = 0$ {as there is no translation} $\therefore$ $y = -5\cos\left(\frac{\pi}{3}x\right)$

## EXERCISE 10E

**1** **a** **i** $y = \tan(x - \frac{\pi}{2})$ is $y = \tan x$ translated 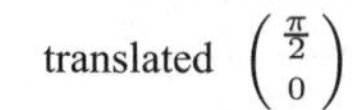.

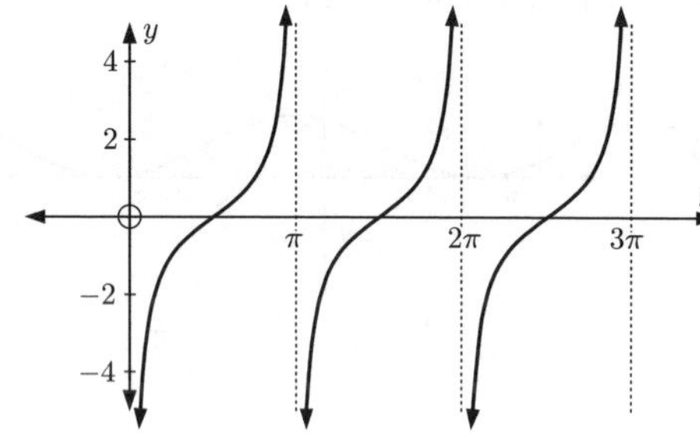


**ii** $y = -\tan x$ is $y = \tan x$ reflected in the $x$-axis.

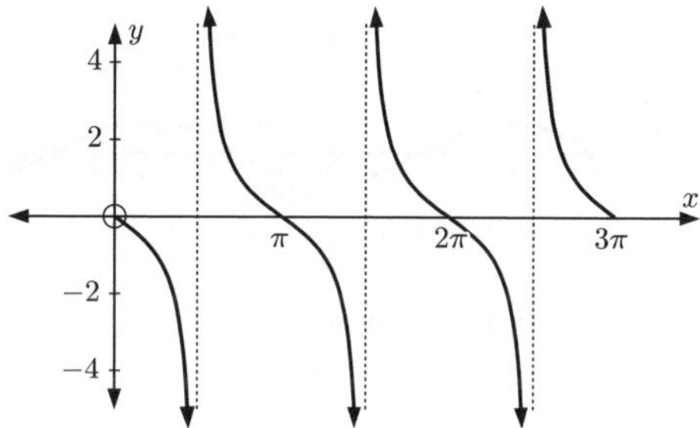


**iii** $y = \tan 3x$ comes from $y = \tan x$ under a horizontal stretch of factor $\frac{1}{3}$.

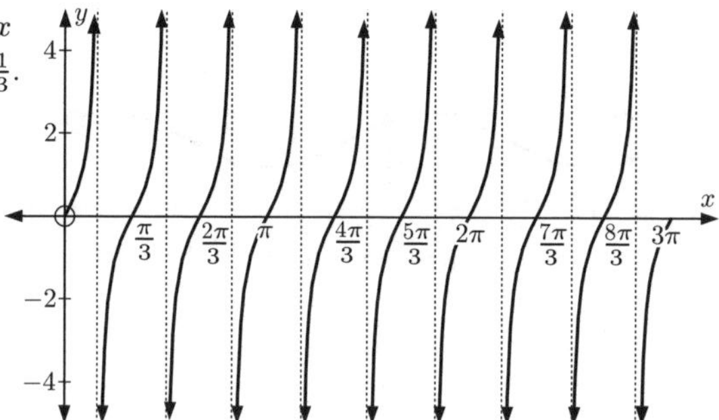


**2** **a** translation through $\binom{1}{2}$ **b** reflection in $x$-axis

**c** horizontal stretch, factor 2; vertical stretch, factor 2

**3** **a** period $= \frac{\pi}{1} = \pi$ **b** period $= \frac{\pi}{3}$ **c** period $= \frac{\pi}{n}$

## EXERCISE 10F

**1** **a** amplitude $= |1| = 1$ **b** amplitude undefined **c** amplitude $= |-1| = 1$

**2** **a** period $= \dfrac{\pi}{1} = \pi$ **b** period $= \dfrac{2\pi}{\frac{1}{3}} = 6\pi$ **c** period $= \dfrac{2\pi}{2} = \pi$

**3** **a** $\dfrac{2\pi}{b} = 2\pi$ $\therefore$ $b = 1$ **b** $\dfrac{2\pi}{b} = \dfrac{2\pi}{3}$ $\therefore$ $b = 3$ **c** $\dfrac{\pi}{b} = \dfrac{\pi}{2}$ $\therefore$ $b = 2$ **d** $\dfrac{2\pi}{b} = 4$ $\therefore$ $b = \frac{\pi}{2}$

**4** **a**

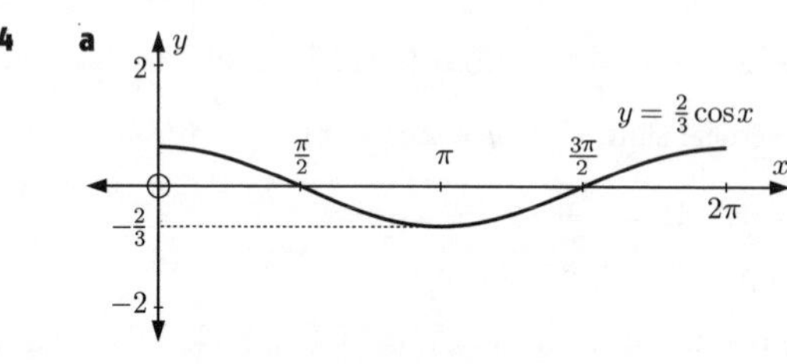


**b**

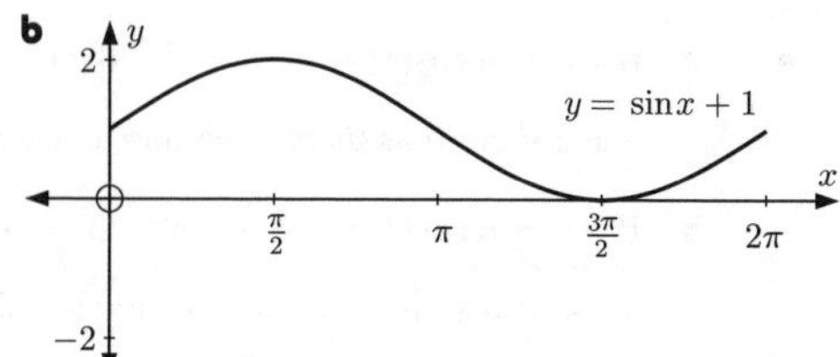

**c** $y = \tan\left(x + \frac{\pi}{2}\right)$

**d** $y = 3\cos 2x$

**e** $y = \sin\left(x + \frac{\pi}{4}\right) - 1$

**f**

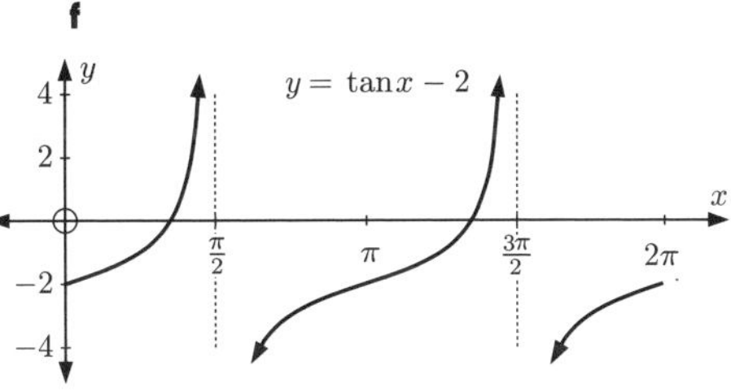


**5** **a** $y = -\sin 5x$ has maximum value $-(-1) = 1$ $\{$when $\sin 5x = -1\}$
and minimum value $-(1) = -1$ $\{$when $\sin 5x = 1\}$

**b** $y = 3\cos x$ has maximum value $3(1) = 3$ $\{$when $\cos x = 1\}$
and minimum value $3(-1) = -3$ $\{$when $\cos x = -1\}$

**c** $y = 2\tan x$ has no maximum or minimum values.

**d** $y = -\cos 2x + 3$ has maximum value $-(-1) + 3 = 4$ $\{$when $\cos 2x = -1\}$
and minimum value $-(1) + 3 = 2$ $\{$when $\cos 2x = 1\}$

**e** $y = 1 + 2\sin x$ has maximum value $1 + 2(1) = 3$ $\{$when $\sin x = 1\}$
and minimum value $1 + 2(-1) = -1$ $\{$when $\sin x = -1\}$

**f** $y = \sin\left(x - \frac{\pi}{2}\right) - 3$ has maximum value $1 - 3 = -2$ $\{$when $\sin\left(x - \frac{\pi}{2}\right) = 1\}$
and minimum value $-1 - 3 = -4$ $\{$when $\sin\left(x - \frac{\pi}{2}\right) = -1\}$

**6** **a** vertical stretch, factor $\frac{1}{2}$ **b** horizontal stretch, factor 4
**c** reflection in the $x$-axis **d** vertical translation down 2 units
**e** horizontal translation $\frac{\pi}{4}$ units to the left **f** reflection in the $y$-axis

**7** The amplitude is 2, so $m = 2$.
The principal axis is $y = -3$, so $n = -3$.

**8** The period is $2\pi$, so $\frac{\pi}{p} = 2\pi$

$\therefore\ p = \frac{1}{2}$

The graph has undergone a vertical translation of 1 unit, so $q = 1$.

## REVIEW SET 10A

**1** **a** not periodic **b** periodic

**2** $y = 4\sin x$ has amplitude 4.

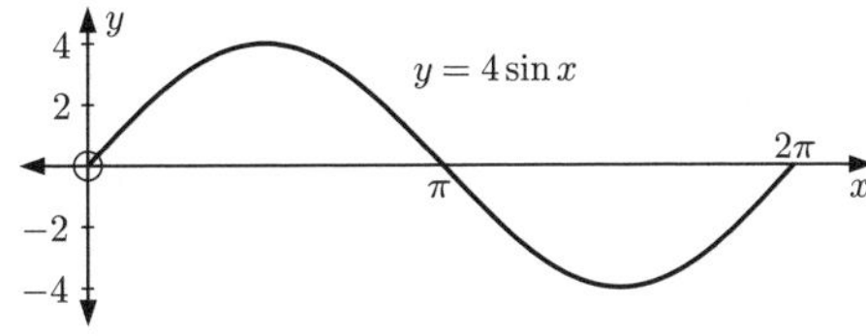

**3** **a** $-1 \leqslant \sin x \leqslant 1$

$\therefore$ $1 + \sin x$ has minimum $1 + (-1) = 0$ and maximum $1 + 1 = 2$.

**b** $-1 \leqslant \cos 3x \leqslant 1$

$\therefore$ $-2\cos 3x$ has minimum $-2(1) = -2$ and maximum $-2(-1) = 2$.

**4** **a** period $= \dfrac{2\pi}{\frac{1}{5}} = 10\pi$

**b** period $= \dfrac{2\pi}{4} = \frac{\pi}{2}$

**c** period $= \dfrac{2\pi}{\frac{1}{2}} = 4\pi$

**d** period $= \frac{\pi}{3}$

**5**

| *Function* | *Period* | *Amplitude* | *Domain* | *Range* |
|---|---|---|---|---|
| $y = -3\sin(\frac{x}{4}) + 1$ | $8\pi$ | 3 | $x \in \mathbb{R}$ | $-2 \leqslant y \leqslant 4$ |
| $y = \tan 2x$ | $\frac{\pi}{2}$ | undefined | $x \neq \pm\frac{\pi}{4}, \pm\frac{3\pi}{4}, ....$ | $y \in \mathbb{R}$ |
| $y = 3\cos \pi x$ | 2 | 3 | $x \in \mathbb{R}$ | $-3 \leqslant y \leqslant 3$ |

**6** **a** If $y = a\cos(b(t - c)) + d$

then $a = -4$, $\dfrac{2\pi}{b} = \pi$

$\therefore$ $b = 2$

$c = d = 0$

$\therefore$ $y = -4\cos 2x$

**b** If $y = a\cos(b(x - c)) + d$

then $a = 1$, $\dfrac{2\pi}{b} = 8$ $\therefore$ $b = \frac{\pi}{4}$

$d = \dfrac{\text{max.} + \text{min.}}{2} = \dfrac{3+1}{2} = 2$

$c = 0$

So, $y = \cos\left(\frac{\pi}{4}x\right) + 2$

## REVIEW SET 10B

**1** **a**

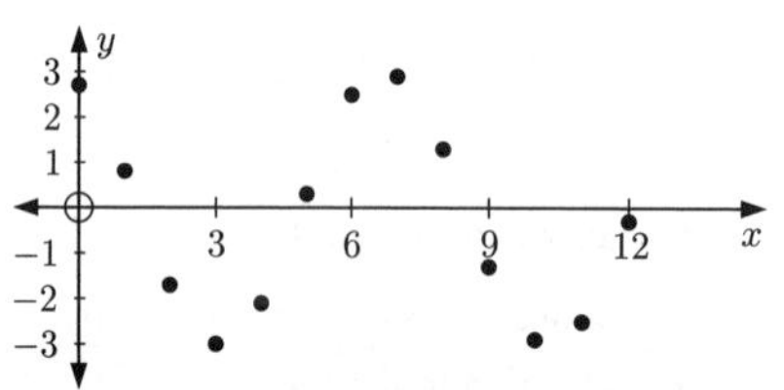


approximately periodic

**b**

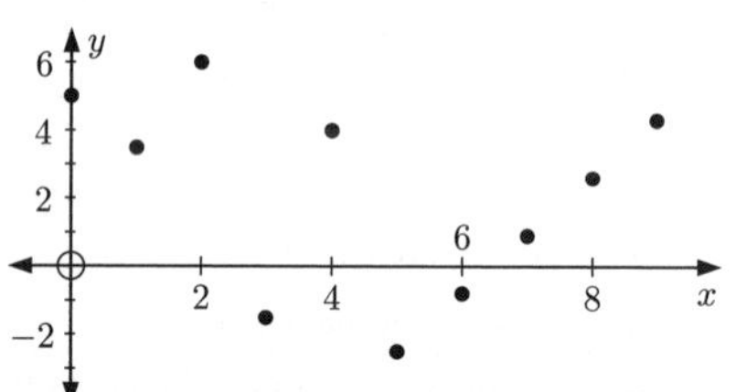


not periodic

**2** $y = \sin 3x$ has period $\frac{2\pi}{3}$.

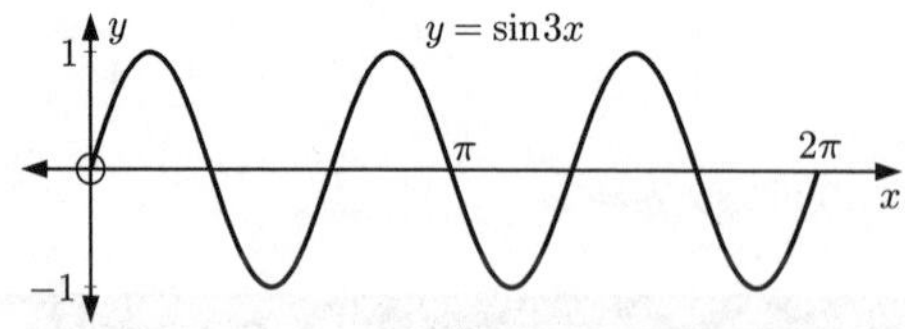


**3** **a** period $= \dfrac{2\pi}{\frac{1}{3}} = 6\pi$

**b** period $= \frac{\pi}{4}$

**4** $y = 0.6\cos(2.3x)$ has period $\frac{2\pi}{2.3} \approx 2.73$

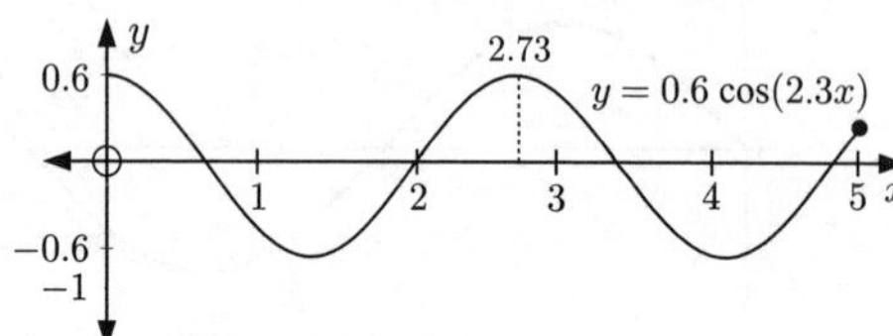

**5** **a** maximum $= -5°\text{C}$, minimum $= -79°\text{C}$

**b** amplitude $= \dfrac{-5 - -79}{2} = 37°\text{C}$, so $a = 37$

principal axis is $y = \dfrac{-5 + -79}{2} = -42$, so $c = -42$

Now, we see that the temperature is $-68°\text{C}$ and rising on days 600 and 1300, so we estimate the period to be 700 days.

$\therefore\ b \approx \frac{2\pi}{700} \approx 0.008\,98$

So, $T \approx 37\sin(0.008\,98n) - 42$ °C

**c** A Mars year is equivalent to one period of the temperature pattern, so 1 Mars year ≈ 700 Mars days.

**6** Minimum = mean value − amplitude $= c - |a|$, maximum = mean value + amplitude $= c + |a|$.

**a** $y = 5\sin x - 3$ has $a = 5$, $c = -3$
so $\min = -3 - 5 = -8$
and $\max = -3 + 5 = 2$

**b** $y = \frac{1}{3}\cos x + 1$ has $a = \frac{1}{3}$, $c = 1$
so $\min = 1 - \frac{1}{3} = \frac{2}{3}$
and $\max = 1 + \frac{1}{3} = 1\frac{1}{3}$

## REVIEW SET 10C

**1** **a** The graph is periodic because it repeats itself over and over in a horizontal direction in intervals of the same length.

**b** **i** period = 8 **ii** maximum value = 5 **iii** minimum value = −1

**2** **a** period $= \dfrac{2\pi}{b} = 6\pi$ $\therefore\ b = \frac{1}{3}$

**b** period $= \dfrac{2\pi}{b} = \frac{\pi}{12}$ $\therefore\ b = 24$

**c** period $= \dfrac{2\pi}{b} = 9$ $\therefore\ b = \frac{2\pi}{9}$

**3** **a**

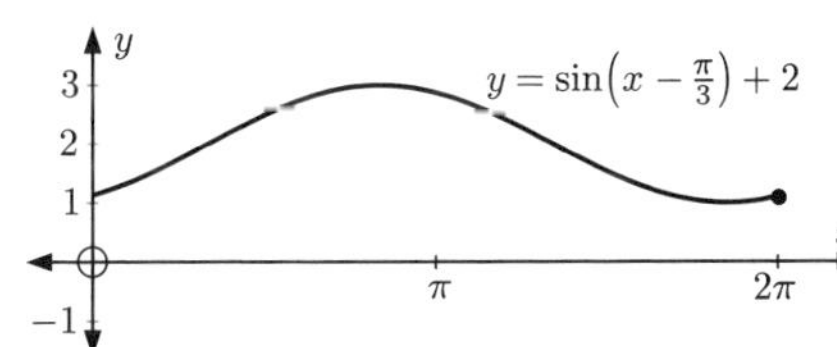


**b** $f(x)$ has minimum value $-1 + 2 = 1$
and maximum value $1 + 2 = 3$
$\therefore\ f(x) = k$ will have solutions for $1 \leqslant k \leqslant 3$

**4** **a**

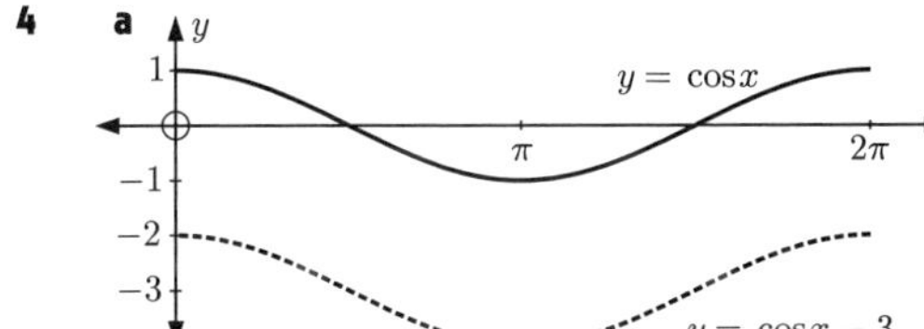


**b**

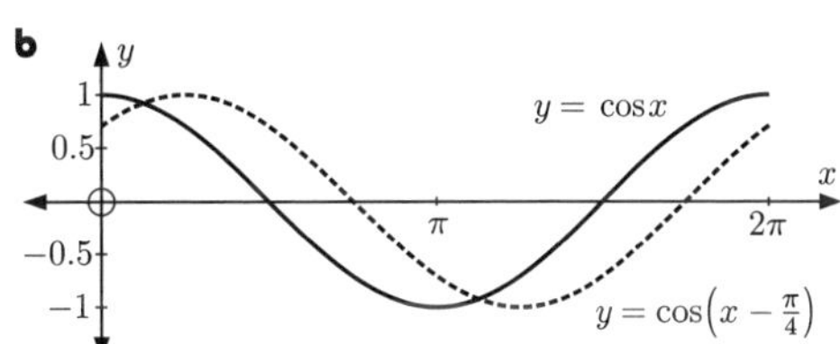


**c**

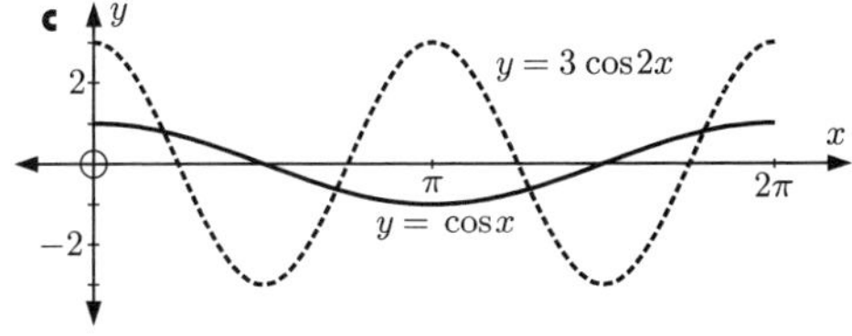


**d**

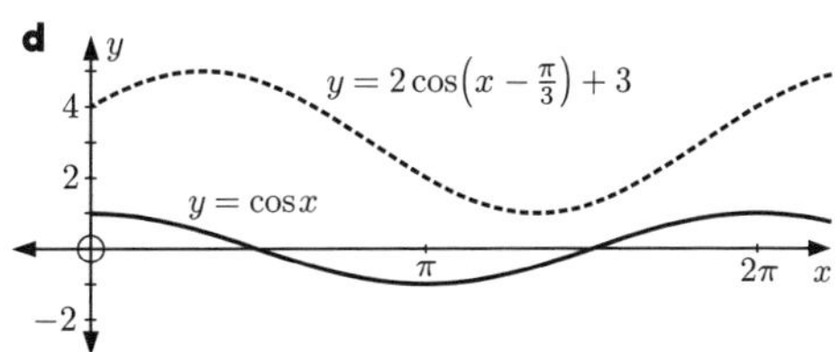

**5**

| *Month* | 1 | 2 | 3 | 4 | 5 | 6 | 7 | 8 | 9 | 10 | 11 | 12 |
|---|---|---|---|---|---|---|---|---|---|---|---|---|
| *Temp* | 31.5 | 31.8 | 29.5 | 25.4 | 21.5 | 18.8 | 17.7 | 18.3 | 20.1 | 22.4 | 25.5 | 28.8 |

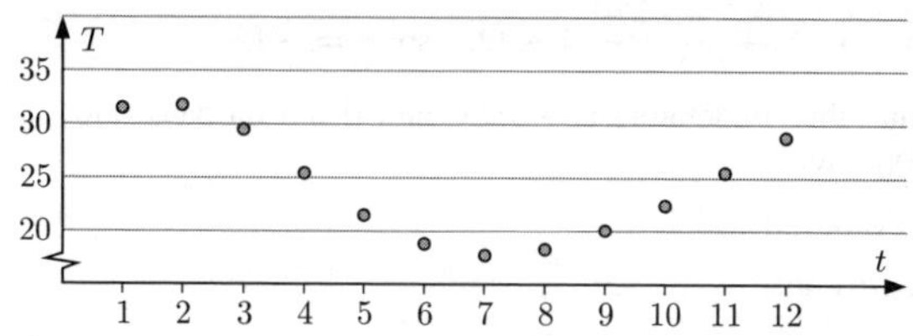


**a** $T = a\sin b(t-c)+d$ period $= \dfrac{2\pi}{b} = 12$, $\therefore$ $b = \dfrac{2\pi}{12} = \dfrac{\pi}{6}$

max. $= 31.8$ $\therefore$ $a = \dfrac{\text{max.} - \text{min.}}{2} \approx \dfrac{31.8-17.7}{2} \approx 7.05$

min. $= 17.7$

$d = \dfrac{\text{max.} + \text{min.}}{2} \approx \dfrac{31.8+17.7}{2} \approx 24.75$

$c = \dfrac{7+14}{2} = 10.5$ {values of $t$ at min. and max.}

So, $T \approx 7.05\sin(\frac{\pi}{6}(t-10.5))+24.75$

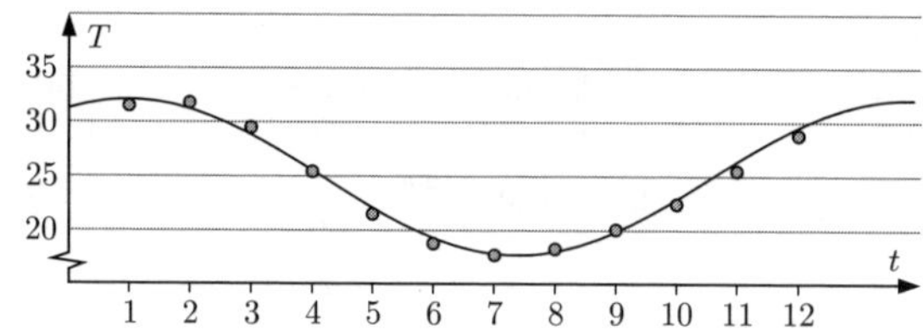


**b** From technology, $T \approx 7.21\sin(0.488t+1.082)+24.75$

$\approx 7.21\sin(0.488(t+2.22))+24.75$

The model fits reasonably well.

**6** **a** translation through $\begin{pmatrix} \frac{\pi}{3} \\ 1 \end{pmatrix}$

**b** vertical stretch with scale factor 2, followed by a reflection in the $x$-axis

**c** horizontal stretch with scale factor $\frac{1}{3}$

# Chapter 11

## TRIGONOMETRIC EQUATIONS AND IDENTITIES

### EXERCISE 11A.1

**1**

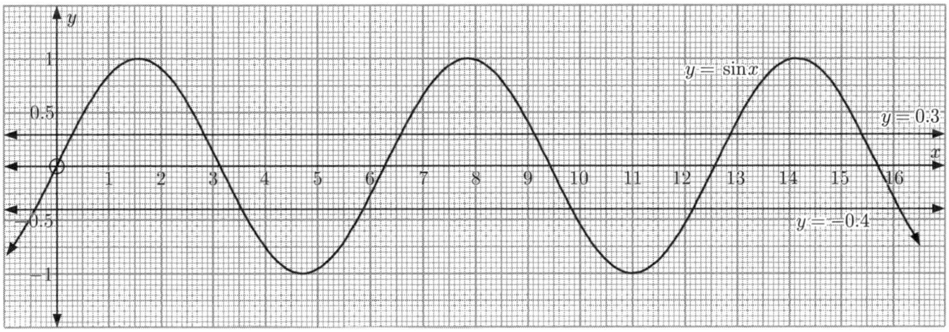

**a** When $\sin x = 0.3$, $x \approx 0.3, 2.8, 6.6, 9.1, 12.9$ **b** When $\sin x = -0.4$, $x \approx 5.9, 9.8, 12.2$

**2**

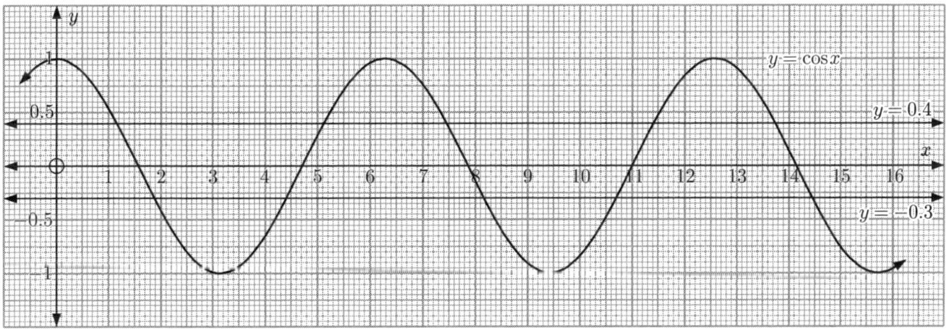

**a** When $\cos x = 0.4$, $x \approx 1.2, 5.1, 7.4$ **b** When $\cos x = -0.3$, $x \approx 4.4, 8.2, 10.7$

**3**

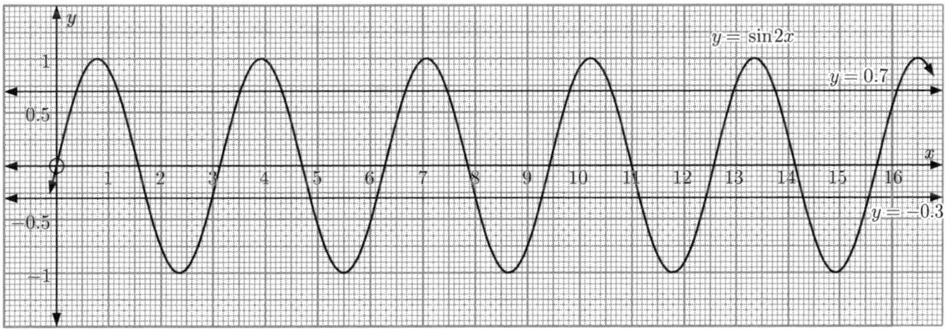

**a** When $\sin 2x = 0.7$, $x \approx 0.4, 1.2, 3.5, 4.3, 6.7, 7.5, 9.8, 10.6, 13.0, 13.7$

**b** When $\sin 2x = -0.3$, $x \approx 1.7, 3.0, 4.9, 6.1, 8.0, 9.3, 11.1, 12.4, 14.3, 15.6$

**4**

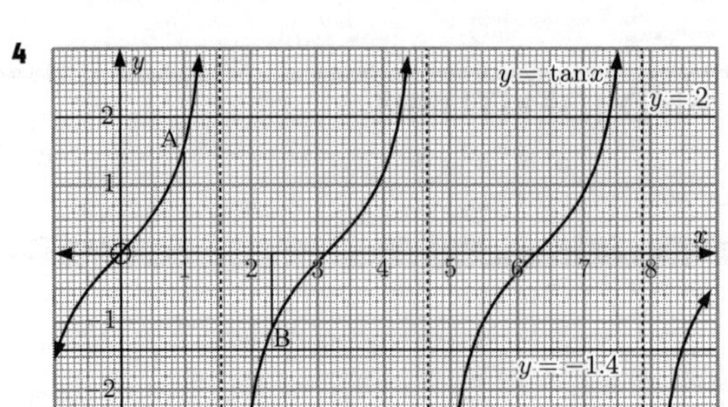


**a** **i** $\tan 1 \approx 1.6$ {point A}

**ii** $\tan 2.3 \approx -1.1$ {point B}

Using technology, we see that
$\tan 1 \approx 1.557$ and $\tan 2.3 \approx -1.119$.

**b** **i** When $\tan x = 2$, $x \approx 1.1, 4.2, 7.4$

**ii** When $\tan x = -1.4$, $x \approx 2.2, 5.3$

## EXERCISE 11A.2

**1** Using technology:

**a** $\sin x = 0.431$ when $x \approx 0.446, 2.70, 6.73, 8.98$

**b** $\cos x = -0.814$ when $x \approx 2.52, 3.76, 8.80, 10.0$

**c** $3\tan x - 2 = 0$ when $x \approx 0.588, 3.73, 6.87, 10.0$

**2** Using technology:

**a** $5\cos x - 4 = 0$ when $x \approx -0.644, 0.644$

**b** $2\tan x + 13 = 0$ when $x \approx -4.56, -1.42, 1.72, 4.87$

**c** $8\sin x + 3 = 0$ when $x \approx -2.76, -0.384, 3.53$

**3** Using technology:

**a** $\sin(x+2) = 0.0652$ when $x \approx 1.08, 4.35$

**b** $\sin^2 x + \sin x - 1 = 0$ when $x \approx 0.666, 2.48$

**c** $x\tan(\frac{x^2}{10}) = x^2 - 6x + 1$ when $x \approx 0.171, 4.92$

**d** $2\sin(2x)\cos x = \ln x$ when $x \approx 1.31, 2.03, 2.85$

**4** $\cos(x-1) + \sin(x+1) = 6x + 5x^2 - x^3$ when $x \approx -0.951, 0.234, 5.98$

## EXERCISE 11A.3

**1** **a** $2\cos x - 1 = 0$

$\therefore \cos x = \frac{1}{2}$

There are two points on the unit circle with cosine $\frac{1}{2}$.

They correspond to angles $\frac{\pi}{3}$ and $\frac{5\pi}{3}$.

For the domain $0 \leqslant x \leqslant 4\pi$ we have 4 solutions: $x = \frac{\pi}{3}, \frac{5\pi}{3}, \frac{7\pi}{3}$, or $\frac{11\pi}{3}$.

**b** $\sqrt{2}\sin x = 1$

$\therefore \sin x = \frac{1}{\sqrt{2}}$

There are two points on the unit circle with sine $\frac{1}{\sqrt{2}}$.

They correspond to angles $\frac{\pi}{4}$ and $\frac{3\pi}{4}$.

For the domain $0 \leqslant x \leqslant 4\pi$ we have 4 solutions: $x = \frac{\pi}{4}, \frac{3\pi}{4}, \frac{9\pi}{4}$, or $\frac{11\pi}{4}$.

**c** $\tan x = 1$

There are two points on the unit circle with tangent 1.

They correspond to angles $\frac{\pi}{4}$ and $\frac{5\pi}{4}$.

For the domain $0 \leqslant x \leqslant 4\pi$ we have 4 solutions: $x = \frac{\pi}{4}, \frac{5\pi}{4}, \frac{9\pi}{4}$, or $\frac{13\pi}{4}$.

**2** **a** $2\sin x - \sqrt{3} = 0$

$\therefore \quad \sin x = \frac{\sqrt{3}}{2}$

There are two points on the unit circle with sine $\frac{\sqrt{3}}{2}$.

They correspond to angles $\frac{\pi}{3}$ and $\frac{2\pi}{3}$.

For the domain $-2\pi \leqslant x \leqslant 2\pi$ we have 4 solutions: $x = -\frac{5\pi}{3}, -\frac{4\pi}{3}, \frac{\pi}{3}$, or $\frac{2\pi}{3}$.

**b** $\sqrt{2}\cos x + 1 = 0$

$\therefore \quad \cos x = -\frac{1}{\sqrt{2}}$

There are two points on the unit circle with cosine $-\frac{1}{\sqrt{2}}$.

They correspond to angles $\frac{3\pi}{4}$ and $\frac{5\pi}{4}$.

For the domain $-2\pi \leqslant x \leqslant 2\pi$ we have 4 solutions: $x = -\frac{5\pi}{4}, -\frac{3\pi}{4}, \frac{3\pi}{4}$, or $\frac{5\pi}{4}$.

**c** $\tan x = -1$

There are two points on the unit circle with tangent $-1$.

They correspond to angles $\frac{3\pi}{4}$ and $\frac{7\pi}{4}$.

For the domain $-2\pi \leqslant x \leqslant 2\pi$ we have 4 solutions: $x = -\frac{5\pi}{4}, -\frac{\pi}{4}, \frac{3\pi}{4}$, or $\frac{7\pi}{4}$.

**3** **a** If $0 \leqslant x \leqslant 2\pi$ then $0 \leqslant 2x \leqslant 4\pi$

**b** If $0 \leqslant x \leqslant 2\pi$ then $0 \leqslant \frac{x}{3} \leqslant \frac{2\pi}{3}$

**c** If $0 \leqslant x \leqslant 2\pi$ then $\frac{\pi}{2} \leqslant x + \frac{\pi}{2} \leqslant \frac{5\pi}{2}$

**d** If $0 \leqslant x \leqslant 2\pi$ then $-\frac{\pi}{6} \leqslant x - \frac{\pi}{6} \leqslant \frac{11\pi}{6}$

**e** If $0 \leqslant x \leqslant 2\pi$ then $-\frac{\pi}{4} \leqslant (x - \frac{\pi}{4}) \leqslant \frac{7\pi}{4}$ and so $-\frac{\pi}{2} \leqslant 2(x - \frac{\pi}{4}) \leqslant \frac{7\pi}{2}$

**f** If $0 \leqslant x \leqslant 2\pi$ then $-2\pi \leqslant -x \leqslant 0$

**4** **a** If $-\pi \leqslant x \leqslant \pi$ then $-3\pi \leqslant 3x \leqslant 3\pi$

**b** If $-\pi \leqslant x \leqslant \pi$ then $-\frac{\pi}{4} \leqslant \frac{x}{4} \leqslant \frac{\pi}{4}$

**c** If $-\pi \leqslant x \leqslant \pi$ then $-\frac{3\pi}{2} \leqslant x - \frac{\pi}{2} \leqslant \frac{\pi}{2}$

**d** If $-\pi \leqslant x \leqslant \pi$ then $-2\pi \leqslant 2x \leqslant 2\pi$ and so $-\frac{3\pi}{2} \leqslant 2x + \frac{\pi}{2} \leqslant \frac{5\pi}{2}$

**e** If $-\pi \leqslant x \leqslant \pi$ then $2\pi \geqslant -2x \geqslant -2\pi$ and so $-2\pi \leqslant -2x \leqslant 2\pi$

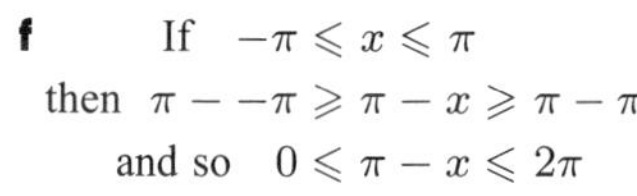

**f** If $-\pi \leqslant x \leqslant \pi$ then $\pi - -\pi \geqslant \pi - x \geqslant \pi - \pi$ and so $0 \leqslant \pi - x \leqslant 2\pi$

**5** The three equations all have the form $\cos\theta = \frac{1}{2}$.

There are two points on the unit circle with cosine $\frac{1}{2}$.

They correspond to angles $\frac{\pi}{3}$ and $\frac{5\pi}{3}$.

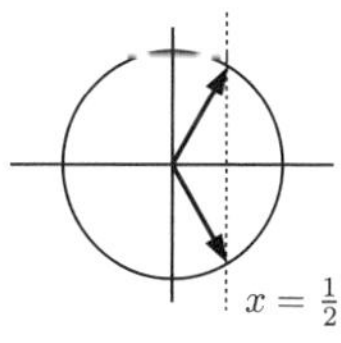


**a** In this case $\theta$ is simply $x$, so we have the domain $0 \leqslant x \leqslant 3\pi$.

The solutions for this domain are $x = \frac{\pi}{3}, \frac{5\pi}{3}$, or $\frac{7\pi}{3}$.

**b** In this case $\theta$ is $2x$.

If $0 \leqslant x \leqslant 3\pi$ then $0 \leqslant 2x \leqslant 6\pi$.

$\therefore \quad 2x = \frac{\pi}{3}, \frac{5\pi}{3}, \frac{7\pi}{3}, \frac{11\pi}{3}, \frac{13\pi}{3}$, or $\frac{17\pi}{3}$

$\therefore \quad x = \frac{\pi}{6}, \frac{5\pi}{6}, \frac{7\pi}{6}, \frac{11\pi}{6}, \frac{13\pi}{6}$, or $\frac{17\pi}{6}$

**c** In this case $\theta$ is $x + \frac{\pi}{3}$.

If $0 \leqslant x \leqslant 3\pi$ then $\frac{\pi}{3} \leqslant x + \frac{\pi}{3} \leqslant \frac{10\pi}{3}$

$\therefore \quad x + \frac{\pi}{3} = \frac{\pi}{3}, \frac{5\pi}{3}$, or $\frac{7\pi}{3}$

$\therefore \quad x = 0, \frac{4\pi}{3}$, or $2\pi$

**6** **a** $\cos x = -\frac{1}{2}, \quad 0 \leqslant x \leqslant 5\pi$

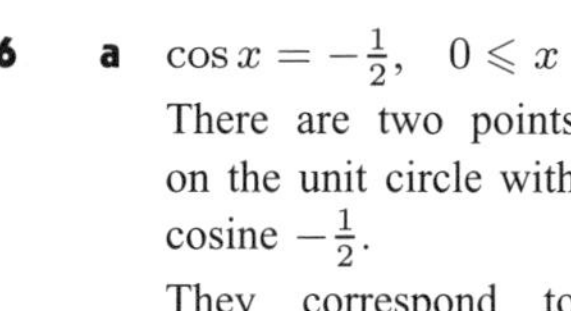

There are two points on the unit circle with cosine $-\frac{1}{2}$.

They correspond to angles $\frac{2\pi}{3}$ and $\frac{4\pi}{3}$.

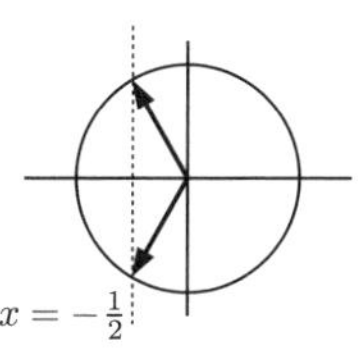


For the domain $0 \leqslant x \leqslant 5\pi$:

$x = \frac{2\pi}{3}, \frac{4\pi}{3}, \frac{8\pi}{3}, \frac{10\pi}{3}, \frac{14\pi}{3}$

**b** $2\sin x - 1 = 0, \quad -360° \leqslant x \leqslant 360°$

$\therefore \quad \sin x = \frac{1}{2}$

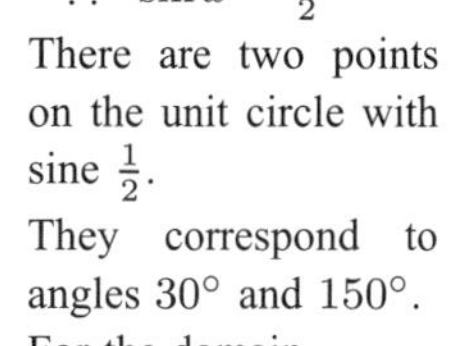

There are two points on the unit circle with sine $\frac{1}{2}$.

They correspond to angles $30°$ and $150°$.

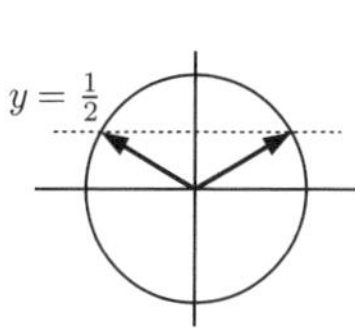


For the domain $-360° \leqslant x \leqslant 360°$:

$x = -330°, -210°, 30°, 150°$

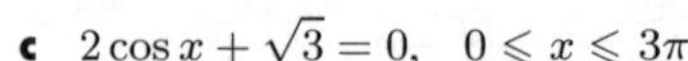

**c** $2\cos x + \sqrt{3} = 0, \quad 0 \leqslant x \leqslant 3\pi$

$\therefore \quad \cos x = -\frac{\sqrt{3}}{2}$

There are two points on the unit circle with cosine $-\frac{\sqrt{3}}{2}$.

They correspond to angles $\frac{5\pi}{6}$ and $\frac{7\pi}{6}$.

For the domain $0 \leqslant x \leqslant 3\pi$:

$x = \frac{5\pi}{6}, \frac{7\pi}{6}, \frac{17\pi}{6}$

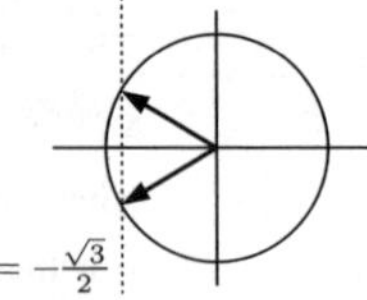


**d** $\cos(x - \frac{2\pi}{3}) = \frac{1}{2}, \quad -2\pi \leqslant x \leqslant 2\pi$

There are two points on the unit circle with cosine $\frac{1}{2}$.

They correspond to angles $\frac{\pi}{3}$ and $\frac{5\pi}{3}$.

$-2\pi \leqslant x \leqslant 2\pi$

$\therefore \quad -\frac{8\pi}{3} \leqslant x - \frac{2\pi}{3} \leqslant \frac{4\pi}{3}$

So, $x - \frac{2\pi}{3} = -\frac{7\pi}{3}, -\frac{5\pi}{3}, -\frac{\pi}{3}, \frac{\pi}{3}$

$\therefore \quad x = -\frac{5\pi}{3}, -\pi, \frac{\pi}{3}, \pi$

$x = \frac{1}{2}$

**e** $2\sin(x + \frac{\pi}{3}) = 1, \quad -3\pi \leqslant x \leqslant 3\pi$

$\therefore \quad \sin(x + \frac{\pi}{3}) = \frac{1}{2}$

There are two points on the unit circle with sine $\frac{1}{2}$.

They correspond to angles $\frac{\pi}{6}$ and $\frac{5\pi}{6}$.

$-3\pi \leqslant x \leqslant 3\pi$

$\therefore \quad -\frac{8\pi}{3} \leqslant x + \frac{\pi}{3} \leqslant \frac{10\pi}{3}$

So, $x + \frac{\pi}{3} = -\frac{11\pi}{6}, -\frac{7\pi}{6}, \frac{\pi}{6}, \frac{5\pi}{6}, \frac{13\pi}{6}, \frac{17\pi}{6}$

$\therefore \quad x = -\frac{13\pi}{6}, -\frac{3\pi}{2}, -\frac{\pi}{6}, \frac{\pi}{2}, \frac{11\pi}{6}, \frac{5\pi}{2}$

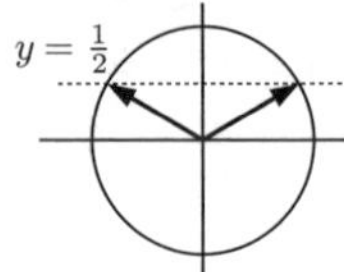


**f** $\sqrt{2}\sin(x - \frac{\pi}{4}) + 1 = 0, \quad 0 \leqslant x \leqslant 3\pi$

$\therefore \quad \sin(x - \frac{\pi}{4}) = -\frac{1}{\sqrt{2}}$

There are two points on the unit circle with sine $-\frac{1}{\sqrt{2}}$.

They correspond to angles $\frac{5\pi}{4}$ and $\frac{7\pi}{4}$.

$0 \leqslant x \leqslant 3\pi$

$\therefore \quad -\frac{\pi}{4} \leqslant x - \frac{\pi}{4} \leqslant \frac{11\pi}{4}$

So, $x - \frac{\pi}{4} = -\frac{\pi}{4}, \frac{5\pi}{4}, \frac{7\pi}{4}$

$\therefore \quad x = 0, \frac{3\pi}{2}, 2\pi$

$y = -\frac{1}{\sqrt{2}}$

**g** $3\cos 2x + 3 = 0, \quad 0 \leqslant x \leqslant 3\pi$

$\therefore \quad \cos 2x = -1$

The one point on the unit circle with cosine $-1$ corresponds to angle $\pi$.

$0 \leqslant x \leqslant 3\pi$

$\therefore \quad 0 \leqslant 2x \leqslant 6\pi$

So, $2x = \pi, 3\pi, 5\pi$

$\therefore \quad x = \frac{\pi}{2}, \frac{3\pi}{2}, \frac{5\pi}{2}$

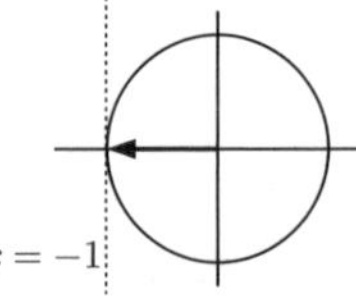


**h** $4\cos 3x + 2 = 0, \quad -\pi \leqslant x \leqslant \pi$

$\therefore \quad \cos 3x = -\frac{1}{2}$

There are two points on the unit circle with cosine $-\frac{1}{2}$.

They correspond to angles $\frac{2\pi}{3}$ and $\frac{4\pi}{3}$.

$-\pi \leqslant x \leqslant \pi$

$\therefore \quad -3\pi \leqslant 3x \leqslant 3\pi$

So, $3x = -\frac{8\pi}{3}, -\frac{4\pi}{3}, -\frac{2\pi}{3}, \frac{2\pi}{3}, \frac{4\pi}{3}, \frac{8\pi}{3}$

$\therefore \quad x = -\frac{8\pi}{9}, -\frac{4\pi}{9}, -\frac{2\pi}{9}, \frac{2\pi}{9}, \frac{4\pi}{9}, \frac{8\pi}{9}$

$x = -\frac{1}{2}$

**i** $\sin\left(4(x - \frac{\pi}{4})\right) = 0, \quad 0 \leqslant x \leqslant \pi$

There are two points on the unit circle with sine 0.

They correspond to angles 0 and $\pi$.

$0 \leqslant x \leqslant \pi$

$\therefore \quad -\frac{\pi}{4} \leqslant x - \frac{\pi}{4} \leqslant \frac{3\pi}{4}$

$\therefore \quad -\pi \leqslant 4(x - \frac{\pi}{4}) \leqslant 3\pi$

So, $4(x - \frac{\pi}{4}) = -\pi, 0, \pi, 2\pi, 3\pi$

$\therefore \quad x - \frac{\pi}{4} = -\frac{\pi}{4}, 0, \frac{\pi}{4}, \frac{\pi}{2}, \frac{3\pi}{4}$

$\therefore \quad x = 0, \frac{\pi}{4}, \frac{\pi}{2}, \frac{3\pi}{4}, \pi$

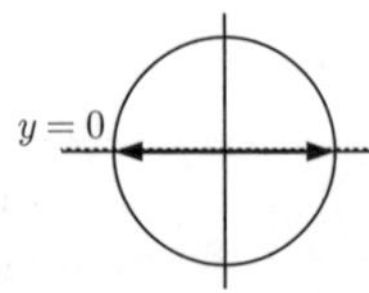


**j** $2\sin\left(2(x - \frac{\pi}{3})\right) = -\sqrt{3}, \quad 0 \leqslant x \leqslant 2\pi$

$\therefore \quad \sin\left(2(x - \frac{\pi}{3})\right) = -\frac{\sqrt{3}}{2}$

There are two points on the unit circle with sine $-\frac{\sqrt{3}}{2}$.

They correspond to angles $\frac{4\pi}{3}$ and $\frac{5\pi}{3}$.

$0 \leqslant x \leqslant 2\pi$

$\therefore \quad -\frac{\pi}{3} \leqslant x - \frac{\pi}{3} \leqslant \frac{5\pi}{3}$

$\therefore \quad -\frac{2\pi}{3} \leqslant 2(x - \frac{\pi}{3}) \leqslant \frac{10\pi}{3}$

So, $2(x - \frac{\pi}{3}) = -\frac{2\pi}{3}, -\frac{\pi}{3}, \frac{4\pi}{3}, \frac{5\pi}{3}, \frac{10\pi}{3}$

$\therefore \quad x - \frac{\pi}{3} = -\frac{\pi}{3}, -\frac{\pi}{6}, \frac{2\pi}{3}, \frac{5\pi}{6}, \frac{5\pi}{3}$

$\therefore \quad x = 0, \frac{\pi}{6}, \pi, \frac{7\pi}{6}, 2\pi$

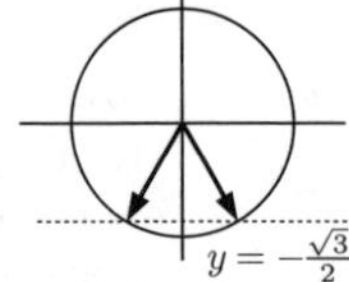

**7** $\tan x = \sqrt{3}, \quad 0 \leqslant x \leqslant 2\pi$

There are two points on the unit circle with tangent $\sqrt{3}$. They correspond to angles $\frac{\pi}{3}$ and $\frac{4\pi}{3}$.

**a** $0 \leqslant x \leqslant 2\pi$

$\therefore\ -\frac{\pi}{6} \leqslant x - \frac{\pi}{6} \leqslant \frac{11\pi}{6}$

$\therefore\ x - \frac{\pi}{6} = \frac{\pi}{3}, \frac{4\pi}{3}$

$\therefore\ x = \frac{\pi}{2}, \frac{3\pi}{2}$

**b** $0 \leqslant x \leqslant 2\pi$

$\therefore\ 0 \leqslant 4x \leqslant 8\pi$

$\therefore\ 4x = \frac{\pi}{3}, \frac{4\pi}{3}, \frac{7\pi}{3}, \frac{10\pi}{3}, \frac{13\pi}{3}, \frac{16\pi}{3}, \frac{19\pi}{3}, \frac{22\pi}{3}$

$\therefore\ x = \frac{\pi}{12}, \frac{\pi}{3}, \frac{7\pi}{12}, \frac{5\pi}{6}, \frac{13\pi}{12}, \frac{4\pi}{3}, \frac{19\pi}{12}, \frac{11\pi}{6}$

**c** $\tan^2 x = 3, \quad 0 \leqslant x \leqslant 2\pi$

$\therefore\ \tan x = \pm\sqrt{3}$

$\therefore\ x = \frac{\pi}{3}, \frac{2\pi}{3}, \frac{4\pi}{3}, \frac{5\pi}{3}$

**8** **a** The zeros of $y = \sin 2x$ are the solutions of $\sin 2x = 0$ $\{0° \leqslant x \leqslant 180°\}$

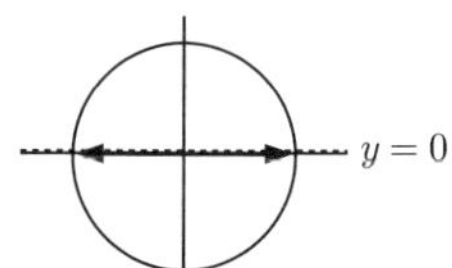


$0° \leqslant x \leqslant 180° \quad \therefore\ 0 \leqslant 2x \leqslant 360°$

So, $2x = 0°, 180°, 360°$

$\therefore\ x = 0°, 90°, 180°$

**b** The zeros of $y = \sin(x - \frac{\pi}{4})$ are the solutions of $\sin(x - \frac{\pi}{4}) = 0$ $\{0 \leqslant x \leqslant 3\pi\}$

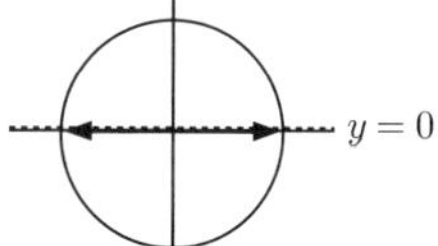


$0 \leqslant x \leqslant 3\pi \quad \therefore\ -\frac{\pi}{4} \leqslant x - \frac{\pi}{4} \leqslant \frac{11\pi}{4}$

So, $x - \frac{\pi}{4} = 0, \pi, 2\pi$

$\therefore\ x = \frac{\pi}{4}, \frac{5\pi}{4}, \frac{9\pi}{4}$

**9** **a**

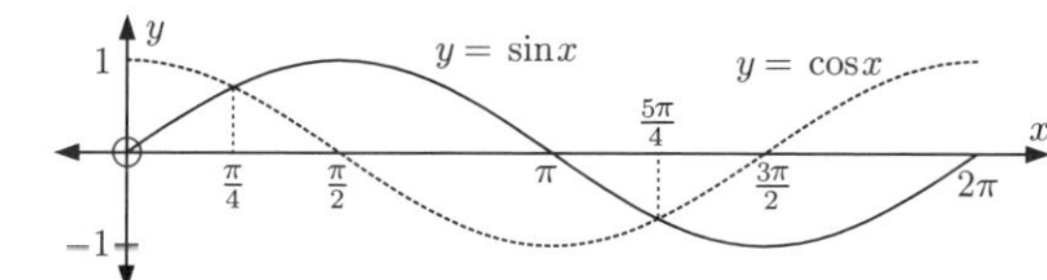


**b** $x = \frac{\pi}{4}$ or $\frac{5\pi}{4}$

**10** **a** $\sin x = -\cos x, \quad 0 \leqslant x \leqslant 2\pi$

$\therefore\ \frac{\sin x}{\cos x} = -1$

$\therefore\ \tan x = -1$

On the domain $0 \leqslant x \leqslant 2\pi$,

$\therefore\ x = \frac{3\pi}{4}, \frac{7\pi}{4}$

**b** $\sin(3x) = \cos(3x), \quad 0 \leqslant x \leqslant 2\pi$

$\therefore\ \frac{\sin(3x)}{\cos(3x)} = 1$

$\therefore\ \tan(3x) = 1$

The two points on the unit circle with tangent 1 correspond to angles $\frac{\pi}{4}$ and $\frac{5\pi}{4}$.

$0 \leqslant x \leqslant 2\pi \quad \therefore\ 0 \leqslant 3x \leqslant 6\pi$

So, $3x = \frac{\pi}{4}, \frac{5\pi}{4}, \frac{9\pi}{4}, \frac{13\pi}{4}, \frac{17\pi}{4}, \frac{21\pi}{4}$

$\therefore\ x = \frac{\pi}{12}, \frac{5\pi}{12}, \frac{3\pi}{4}, \frac{13\pi}{12}, \frac{17\pi}{12}, \frac{7\pi}{4}$

**c** $\sin(2x) = \sqrt{3}\cos(2x), \quad 0 \leqslant x \leqslant 2\pi$

$\therefore\ \frac{\sin(2x)}{\cos(2x)} = \sqrt{3}$

$\therefore\ \tan(2x) = \sqrt{3}$

The two points on the unit circle with tangent $\sqrt{3}$ correspond to angles $\frac{\pi}{3}$ and $\frac{4\pi}{3}$.

$0 \leqslant x \leqslant 2\pi \quad \therefore\ 0 \leqslant 2x \leqslant 4\pi$

So, $2x = \frac{\pi}{3}, \frac{4\pi}{3}, \frac{7\pi}{3}, \frac{10\pi}{3}$

$\therefore\ x = \frac{\pi}{6}, \frac{2\pi}{3}, \frac{7\pi}{6}, \frac{5\pi}{3}$

## EXERCISE 11B

**1** **a** $H(t) = 10\sin\left(\frac{\pi}{50}(t-25)\right) + 12$ metres

After 50 seconds, $t = 50$ and $H(50) = 10\sin\left(\frac{\pi}{50}(50-25)\right) + 12 = 22$ metres.

So, the green light will be 22 metres high.

**b** A full circle is completed in one period.

$$\text{period} = \frac{2\pi}{\frac{\pi}{50}} = 2\pi \times \frac{50}{\pi} = 100 \text{ seconds}$$

So, it takes 100 seconds for the wheel to complete a full circle.

**c** We need to solve $H(t) = 16$ so $10\sin\left(\frac{\pi}{50}(t-25)\right) + 12 = 16$.

3 minutes $= 3 \times 60 = 180$ seconds

Using technology to find the intercepts of $y_1 = 10\sin\left(\frac{\pi}{50}(t-25)\right) + 12$ and $y_2 = 16$ for $0 \leqslant t \leqslant 180$, $t \approx 31.5, 68.5, 132, 168$

So, in the first three minutes, the green light is 16 metres above the ground at 31.5, 68.5, 132, and 168 seconds.

**2** **a** $P(t) = 7500 + 3000\sin(\frac{\pi t}{8})$, $0 \leqslant t \leqslant 12$

**i** $P(0) = 7500 + 3000\sin 0$
$= 7500 + 0$
$= 7500$ grasshoppers

**ii** $P(5) = 7500 + 3000\sin(\frac{5\pi}{8})$
$\approx 10\,271.63$
$\approx 10\,300$ grasshoppers

**b** The greatest value of $P(t)$ occurs when $\sin\left(\frac{\pi t}{8}\right) = 1$, so the greatest population is $7500 + 3000 = 10\,500$ grasshoppers.

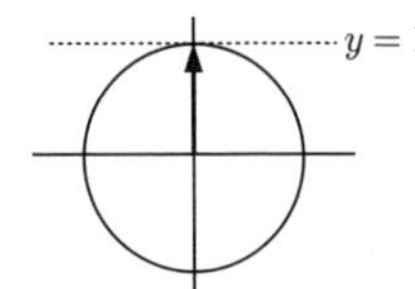


The point on the unit circle with sine 1 corresponds to angle $\frac{\pi}{2}$.

$0 \leqslant t \leqslant 12 \quad \therefore \quad 0 \leqslant \frac{\pi t}{8} \leqslant \frac{3\pi}{2}$

So $\frac{\pi t}{8} = \frac{\pi}{2}$

$\therefore \quad t = 4$

So the greatest population occurs after 4 weeks.

**c** **i** When $P(t) = 9000$,

$7500 + 3000\sin\left(\frac{\pi t}{8}\right) = 9000$

$\therefore \quad 3000\sin\left(\frac{\pi t}{8}\right) = 1500$

$\therefore \quad \sin\left(\frac{\pi t}{8}\right) = \frac{1}{2}$

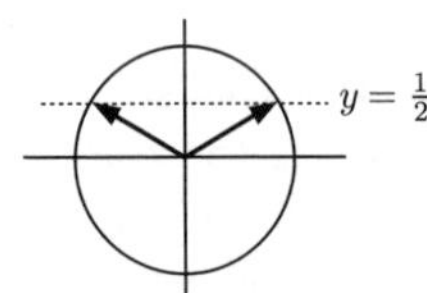


The points on the unit circle with sine $\frac{1}{2}$ correspond to angles $\frac{\pi}{6}$ and $\frac{5\pi}{6}$.

$0 \leqslant t \leqslant 12 \quad \therefore \quad 0 \leqslant \frac{\pi t}{8} \leqslant \frac{3\pi}{2}$

So $\frac{\pi t}{8} = \frac{\pi}{6}, \frac{5\pi}{6}$

$\therefore \quad t = \frac{8}{6}, \frac{40}{6}$

$\therefore \quad t = 1\frac{1}{3}$ or $6\frac{2}{3}$

So, the population is 9000 at $1\frac{1}{3}$ weeks and $6\frac{2}{3}$ weeks.

**ii** When $P(t) = 6000$,

$7500 + 3000\sin\left(\frac{\pi t}{8}\right) = 6000$

$\therefore \quad 3000\sin\left(\frac{\pi t}{8}\right) = -1500$

$\therefore \quad \sin\left(\frac{\pi t}{8}\right) = -\frac{1}{2}$

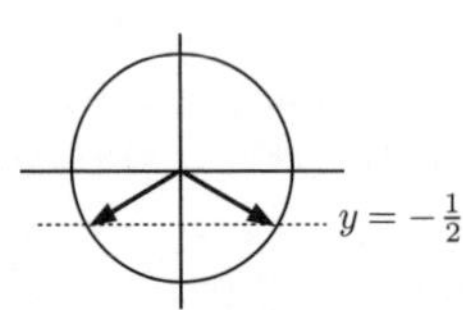


The points on the unit circle with sine $-\frac{1}{2}$ correspond to angles $\frac{7\pi}{6}$ and $\frac{11\pi}{6}$.

$0 \leqslant t \leqslant 12 \quad \therefore \quad 0 \leqslant \frac{\pi t}{8} \leqslant \frac{3\pi}{2}$

So $\frac{\pi t}{8} = \frac{7\pi}{6}$

$\therefore \quad t = \frac{56}{6}$

$\therefore \quad t = 9\frac{1}{3}$

So, the population is 6000 at $9\frac{1}{3}$ weeks.

**d** If $P(t) > 10\,000$, then

$7500 + 3000\sin\left(\frac{\pi t}{8}\right) > 10\,000$

$\therefore\ 3000\sin\left(\frac{\pi t}{8}\right) > 2500$

$\therefore\ \sin\left(\frac{\pi t}{8}\right) > \frac{5}{6}$

Solving $\sin\left(\frac{\pi t}{8}\right) = \frac{5}{6}$ using technology

$t \approx 2.51$ or $5.49$ So, $2.51 \leqslant t \leqslant 5.49$ weeks.

$y = \frac{5}{6}$

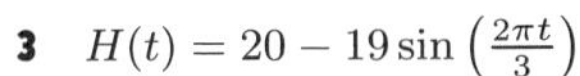

**3** $H(t) = 20 - 19\sin\left(\frac{2\pi t}{3}\right)$

**a** $H(0) = 20 - 19(0)$
$= 20$ m

So, at time $t = 0$, the light is 20 m above the ground.

**b** $H$ is lowest when $\sin\left(\frac{2\pi t}{3}\right) = 1$

$\therefore\ \dfrac{2\pi t}{3} = \frac{\pi}{2} + k2\pi$

$\therefore\ \dfrac{2t}{3} = \frac{1}{2} + k2$

$\therefore\ t = \frac{3}{4} + k3$

$\therefore\ t = \frac{3}{4}$ min $\{$as $k = 0\}$

$y = 1$

**c** period $= \dfrac{2\pi}{\frac{2\pi}{3}} = 3$ min

$\therefore$ one revolution takes 3 minutes

**d**

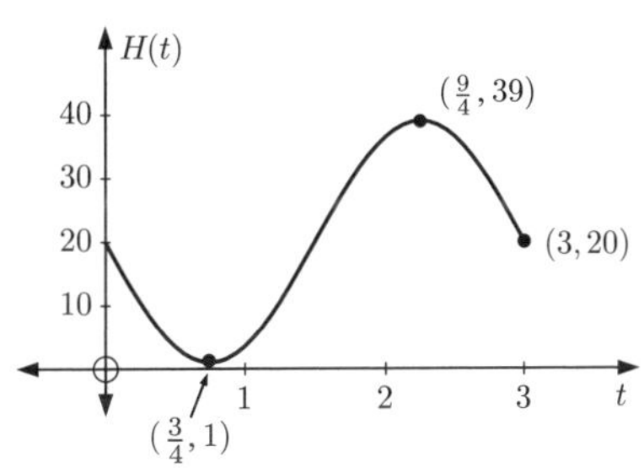


**4** $P(t) = 400 + 250\sin\left(\frac{\pi t}{2}\right)$ years

**a** $P(0) = 400 + 250(0)$
$= 400$ water buffalo

**b** **i** $P(\frac{1}{2}) = 400 + 250\sin\left(\dfrac{\pi(\frac{1}{2})}{2}\right)$
$= 400 + 250\sin(\frac{\pi}{4})$
$= 400 + 250 \times \frac{1}{\sqrt{2}}$
$\approx 577$ water buffalo

**ii** $P(2) = 400 + 250\sin\pi$
$= 400 + 250(0)$
$= 400$ water buffalo

**c** $P(1) = 400 + 250\sin(\frac{\pi}{2})$
$= 400 + 250 \times 1$
$= 650$ water buffalo

This is the maximum herd size.

**d** $P(t)$ is smallest when $\sin\left(\frac{\pi t}{2}\right) = -1$ and is $400 - 250 = 150$ water buffalo.

It occurs when $\dfrac{\pi t}{2} = \dfrac{3\pi}{2} + k2\pi$

$\therefore\ \dfrac{t}{2} = \frac{3}{2} + k2$

$\therefore\ t = 3 + 4k$

So, the first time is after 3 years.

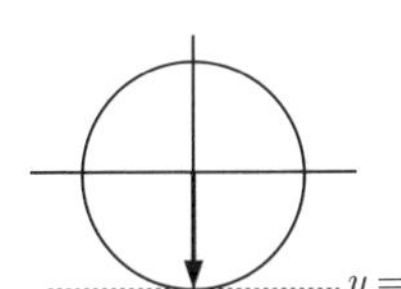

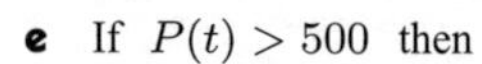

**e** If $P(t) > 500$ then

$400 + 250\sin\left(\frac{\pi t}{2}\right) > 500$

$\therefore \quad 250\sin\left(\frac{\pi t}{2}\right) > 100$

$\therefore \quad \sin\left(\frac{\pi t}{2}\right) > \frac{2}{5}$

$\sin\left(\dfrac{\pi t}{2}\right) = \frac{2}{5}$ when

$\dfrac{\pi t}{2} \approx 0.4115$ or $\pi - 0.4115$

$\therefore \quad t \approx 0.262$ or $1.74$

So, for $\sin\left(\frac{\pi t}{2}\right) > \frac{2}{5}$, $0.262 < t < 1.74$

$\therefore$ the herd first exceeded 500 when $t \approx 0.262$ years.

**5** **a** The period is 4 seconds.

$\therefore \quad \dfrac{2\pi}{b} = 4$

$\therefore \quad b = \frac{\pi}{2}$

Amplitude is 3

$\therefore \quad a = 3$

$d = 1 + 3 = 4$

$c = 0$

$\therefore \quad H(t) = 3\cos(\frac{\pi}{2}(t - 0)) + 4$ metres

$\therefore \quad H(t) = 3\cos(\frac{\pi}{2}t) + 4$ metres

*Check:* When $t = 0$, $H(0) = 3\cos 0 + 4 = 7$ ✓

**b** X enters the water when $H(t) = 2$

$\therefore \quad 3\cos\left(\frac{\pi t}{2}\right) + 4 = 2$

$\therefore \quad \cos\left(\frac{\pi t}{2}\right) = -\frac{2}{3}$

Using technology, $t \approx 1.46$ seconds

**6** $C(t) = 9.2\sin(\frac{\pi}{7}(t - 4)) + 107.8$ cents $\text{L}^{-1}$

**a** **i** 107.8 is the median value. Values are between $107.8 - 9.2$ and $107.8 + 9.2$

$= 98.6$ cents $\text{L}^{-1}$ (min.) and $117.0$ cents $\text{L}^{-1}$ (max.)

$\therefore$ the statement is true.

**ii** period $= \dfrac{2\pi}{\frac{\pi}{7}} = 14$ days $\therefore$ true

**b** $C(7) = 9.2\sin(\frac{\pi}{7}(3)) + 107.8 \approx 116.8$ cents $\text{L}^{-1}$

**c** When $C(t) = \$1.10\ \text{L}^{-1}$ then $9.2\sin(\frac{\pi}{7}(t - 4)) + 107.8 = 110$

$\therefore \quad \sin(\frac{\pi}{7}(t - 4)) = \dfrac{2.2}{9.2} \approx 0.239\,13$

$\therefore \quad \frac{\pi}{7}(t - 4) \approx 0.2415,\ \pi - 0.2415,\ 2\pi + 0.2415,\ 3\pi - 0.2415$

$\approx 0.2415,\ 2.9001,\ 6.5247,\ 9.1833$

$\therefore \quad t - 4 \approx 0.538,\ 6.462,\ 14.538,\ 20.462$

$\therefore \quad t \approx 4.538,\ 10.462,\ 18.538,\ 24.462$

So, the price is \$1.10 per litre on the 5th, 11th, 19th, and 25th days.

**d** The minimum cost per litre is $-9.2 + 107.8 = 98.6$ cents $\text{L}^{-1}$

when $\sin\frac{\pi}{7}(t - 4) = -1$

$\therefore \quad \frac{\pi}{7}(t - 4) = \frac{3\pi}{2}$

$\therefore \quad \dfrac{t - 4}{7} = \frac{3}{2}$

$\therefore \quad 2t - 8 = 21$

$\therefore \quad 2t = 29$

$\therefore \quad t = 14.5 \pm 14k$

{period is 14 days}

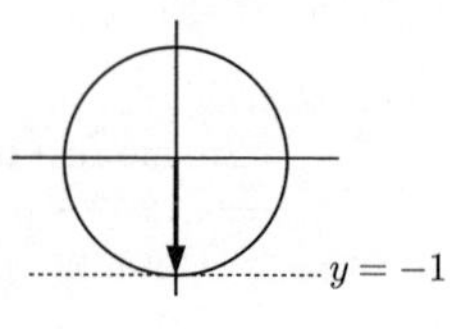


So, the minimum occurred on the 1st day and the 15th day.

## EXERCISE 11C.1

**1** **a** $\sin\theta + \sin\theta$
$= 2\sin\theta$

**b** $2\cos\theta + \cos\theta$
$= 3\cos\theta$

**c** $3\sin\theta - \sin\theta$
$= 2\sin\theta$

**d** $3\sin\theta - 2\sin\theta$
$= \sin\theta$

**e** $\tan\theta - 3\tan\theta$
$= -2\tan\theta$

**f** $2\cos^2\theta - 5\cos^2\theta$
$= -3\cos^2\theta$

**2** **a** $3\sin^2\theta + 3\cos^2\theta$
$= 3(\sin^2\theta + \cos^2\theta)$
$= 3(1)$
$= 3$

**b** $-2\sin^2\theta - 2\cos^2\theta$
$= -2(\sin^2\theta + \cos^2\theta)$
$= -2(1)$
$= -2$

**c** $-\cos^2\theta - \sin^2\theta$
$= -(\cos^2\theta + \sin^2\theta)$
$= -(1)$
$= -1$

**d** $3 - 3\sin^2\theta$
$= 3(1 - \sin^2\theta)$
$= 3\cos^2\theta$

**e** $4 - 4\cos^2\theta$
$= 4(1 - \cos^2\theta)$
$= 4\sin^2\theta$

**f** $\cos^3\theta + \cos\theta\sin^2\theta$
$= \cos\theta(\cos^2\theta + \sin^2\theta)$
$= \cos\theta(1)$
$= \cos\theta$

**g** $\cos^2\theta - 1$
$= 1 - \sin^2\theta - 1$
$= -\sin^2\theta$

**h** $\sin^2\theta - 1$
$= 1 - \cos^2\theta - 1$
$= -\cos^2\theta$

**i** $2\cos^2\theta - 2$
$= -2(1 - \cos^2\theta)$
$= -2\sin^2\theta$

**j** $\dfrac{1 - \sin^2\theta}{\cos^2\theta}$
$= \dfrac{\cos^2\theta}{\cos^2\theta}$
$= 1$

**k** $\dfrac{1 - \cos^2\theta}{\sin\theta}$
$= \dfrac{\sin^2\theta}{\sin\theta}$
$= \sin\theta$

**l** $\dfrac{\cos^2\theta - 1}{-\sin\theta}$
$= \dfrac{1 - \sin^2\theta - 1}{-\sin\theta}$
$= \dfrac{-\sin^2\theta}{-\sin\theta}$
$= \sin\theta$

**3** **a** $3\tan x - \dfrac{\sin x}{\cos x}$
$= 3\tan x - \tan x$
$= 2\tan x$

**b** $\dfrac{\sin^2 x}{\cos^2 x}$
$= \left(\dfrac{\sin x}{\cos x}\right)^2$
$= \tan^2 x$

**c** $\tan x\cos x$
$= \dfrac{\sin x}{\cos x} \times \cos x$
$= \sin x$

**d** $\dfrac{\sin x}{\tan x}$
$= \sin x \div \dfrac{\sin x}{\cos x}$
$= \sin x \times \dfrac{\cos x}{\sin x}$
$= \cos x$

**e** $3\sin x + 2\cos x\tan x$
$= 3\sin x + 2\cos x\dfrac{\sin x}{\cos x}$
$= 3\sin x + 2\sin x$
$= 5\sin x$

**f** $\dfrac{2\tan x}{\sin x}$
$= 2\left(\dfrac{\sin x}{\cos x}\right) \div \dfrac{\sin x}{1}$
$= \dfrac{2\cancel{\sin x}}{\cos x} \times \dfrac{1}{\cancel{\sin x}_1}$
$= \dfrac{2}{\cos x}$

**4** **a** $(1 + \sin\theta)^2$
$= 1 + 2\sin\theta + \sin^2\theta$

**b** $(\sin\alpha - 2)^2$
$= \sin^2\alpha - 4\sin\alpha + 4$

**c** $(\tan\alpha - 1)^2$
$= \tan^2\alpha - 2\tan\alpha + 1$

**d** $(\sin\alpha + \cos\alpha)^2$
$= \sin^2\alpha + 2\sin\alpha\cos\alpha + \cos^2\alpha$
$= 1 + 2\sin\alpha\cos\alpha$

**e** $(\sin\beta - \cos\beta)^2$
$= \sin^2\beta - 2\sin\beta\cos\beta + \cos^2\beta$
$= 1 - 2\sin\beta\cos\beta$

**f** $-(2 - \cos\alpha)^2$
$= -[4 - 4\cos\alpha + \cos^2\alpha]$
$= -4 + 4\cos\alpha - \cos^2\alpha$

**5** $(\sin x + \tan x)(\sin x - \tan x) = \sin^2 x - \tan^2 x$

## EXERCISE 11C.2

**1** **a** $1-\sin^2\theta$
$=(1+\sin\theta)(1-\sin\theta)$

**b** $\sin^2\alpha-\cos^2\alpha$
$=(\sin\alpha+\cos\alpha)(\sin\alpha-\cos\alpha)$

**c** $\tan^2\alpha-1$
$=(\tan\alpha+1)(\tan\alpha-1)$

**d** $2\sin^2\beta-\sin\beta$
$=\sin\beta(2\sin\beta-1)$

**e** $2\cos\phi+3\cos^2\phi$
$=\cos\phi(2+3\cos\phi)$

**f** $3\sin^2\theta-6\sin\theta$
$=3\sin\theta(\sin\theta-2)$

**g** $\tan^2\theta+5\tan\theta+6$
$=(\tan\theta+2)(\tan\theta+3)$

**h** $2\cos^2\theta+7\cos\theta+3$
$=(2\cos\theta+1)(\cos\theta+3)$

**i** $6\cos^2\alpha-\cos\alpha-1$
$=(3\cos\alpha+1)(2\cos\alpha-1)$

**2** **a** $\dfrac{1-\sin^2\alpha}{1-\sin\alpha}$
$=\dfrac{(1+\sin\alpha)\cancel{(1-\sin\alpha)}}{\cancel{1-\sin\alpha}\ _1}$
$=1+\sin\alpha$

**b** $\dfrac{\tan^2\beta-1}{\tan\beta+1}$
$=\dfrac{\cancel{(\tan\beta+1)}(\tan\beta-1)}{\cancel{\tan\beta+1}\ _1}$
$=\tan\beta-1$

**c** $\dfrac{\cos^2\phi-\sin^2\phi}{\cos\phi+\sin\phi}$
$=\dfrac{\cancel{(\cos\phi+\sin\phi)}(\cos\phi-\sin\phi)}{\cancel{\cos\phi+\sin\phi}\ _1}$
$=\cos\phi-\sin\phi$

**d** $\dfrac{\cos^2\phi-\sin^2\phi}{\cos\phi-\sin\phi}$
$=\dfrac{(\cos\phi+\sin\phi)\cancel{(\cos\phi-\sin\phi)}}{\cancel{\cos\phi-\sin\phi}\ _1}$
$=\cos\phi+\sin\phi$

**e** $\dfrac{\sin\alpha+\cos\alpha}{\sin^2\alpha-\cos^2\alpha}$
$=\dfrac{\cancel{\sin\alpha+\cos\alpha}}{\cancel{(\sin\alpha+\cos\alpha)}(\sin\alpha-\cos\alpha)}$
$=\dfrac{1}{\sin\alpha-\cos\alpha}$

**f** $\dfrac{3-3\sin^2\theta}{6\cos\theta}=\dfrac{3(1-\sin^2\theta)}{6\cos\theta}$
$=\dfrac{^1\cancel{3\cos^2\theta}\ \cos\theta}{_2\cancel{6\cos\theta}}$
$=\dfrac{\cos\theta}{2}$

**3** **a** $(\cos\theta+\sin\theta)^2+(\cos\theta-\sin\theta)^2$
$=\cos^2\theta+\cancel{2\cos\theta\sin\theta}+\sin^2\theta$
$\quad+\cos^2\theta-\cancel{2\cos\theta\sin\theta}+\sin^2\theta$
$=2\cos^2\theta+2\sin^2\theta$
$=2(\cos^2\theta+\sin^2\theta)$
$=2(1)=2$

**b** $(2\sin\theta+3\cos\theta)^2+(3\sin\theta-2\cos\theta)^2$
$=4\sin^2\theta+\cancel{12\sin\theta\cos\theta}+9\cos^2\theta$
$\quad+9\sin^2\theta-\cancel{12\sin\theta\cos\theta}+4\cos^2\theta$
$=13\sin^2\theta+13\cos^2\theta$
$=13(\sin^2\theta+\cos^2\theta)$
$=13(1)=13$

**c** $(1-\cos\theta)\left(1+\dfrac{1}{\cos\theta}\right)$
$=1+\dfrac{1}{\cos\theta}-\cos\theta-1$
$=\dfrac{1}{\cos\theta}-\cos\theta$
$=\dfrac{1}{\cos\theta}-\cos\theta\left(\dfrac{\cos\theta}{\cos\theta}\right)$
$=\dfrac{1-\cos^2\theta}{\cos\theta}$
$=\dfrac{\sin^2\theta}{\cos\theta}=\tan\theta\sin\theta$

**d** $\left(1+\dfrac{1}{\sin\theta}\right)(\sin\theta-\sin^2\theta)$
$=\cancel{\sin\theta}-\sin^2\theta+1-\cancel{\sin\theta}$
$=1-\sin^2\theta$
$=\cos^2\theta$

**e**
$$\begin{aligned}
&\frac{\sin\theta}{1+\cos\theta}+\frac{1+\cos\theta}{\sin\theta}\\
&=\frac{\sin^2\theta+(1+\cos\theta)(1+\cos\theta)}{\sin\theta(1+\cos\theta)}\\
&=\frac{\sin^2\theta+1+2\cos\theta+\cos^2\theta}{\sin\theta(1+\cos\theta)}\\
&=\frac{1+1+2\cos\theta}{\sin\theta(1+\cos\theta)}\\
&=\frac{2\cancel{(1+\cos\theta)}}{\sin\theta\cancel{(1+\cos\theta)}}\\
&=\frac{2}{\sin\theta}
\end{aligned}$$

**f**
$$\begin{aligned}
&\frac{\sin\theta}{1-\cos\theta}-\frac{\sin\theta}{1+\cos\theta}\\
&=\frac{\sin\theta(1+\cos\theta)-\sin\theta(1-\cos\theta)}{(1-\cos\theta)(1+\cos\theta)}\\
&=\frac{\cancel{\sin\theta}+\sin\theta\cos\theta-\cancel{\sin\theta}+\sin\theta\cos\theta}{1-\cos^2\theta}\\
&=\frac{2\sin\theta\cos\theta}{\sin^2\theta}\\
&=\frac{2\cancel{\sin\theta}^{\,1}\cos\theta}{\cancel{\sin\theta}\sin\theta}\\
&=\frac{2\cos\theta}{\sin\theta}\\
&=\frac{2}{\tan\theta}
\end{aligned}$$

**g**
$$\begin{aligned}
&\frac{1}{1-\sin\theta}+\frac{1}{1+\sin\theta}\\
&=\frac{1+\sin\theta+1-\sin\theta}{(1-\sin\theta)(1+\sin\theta)}\\
&=\frac{2}{1-\sin^2\theta}\\
&=\frac{2}{\cos^2\theta}
\end{aligned}$$

## EXERCISE 11D

**1** **a** $\sin 2\theta = 2\sin\theta\cos\theta$
$= 2(\frac{4}{5})(\frac{3}{5})$
$= \frac{24}{25}$

**b** $\cos 2\theta = \cos^2\theta - \sin^2\theta$
$= \frac{9}{25} - \frac{16}{25}$
$= -\frac{7}{25}$

**c** $\tan 2\theta = \dfrac{\sin 2\theta}{\cos 2\theta}$
$= \dfrac{\frac{24}{25}}{-\frac{7}{25}}$
$= -\frac{24}{7}$

**2** **a** $\cos 2A = 2\cos^2 A - 1$
$= 2(\frac{1}{3})^2 - 1$
$= 2\times\frac{1}{9} - 1$
$= \frac{2}{9} - 1$
$= -\frac{7}{9}$

**b** $\cos 2\phi = 1 - 2\sin^2\phi$
$= 1 - 2(-\frac{2}{3})^2$
$= 1 - 2(\frac{4}{9})$
$= 1 - \frac{8}{9}$
$= \frac{1}{9}$

**3** **a** $\sin\alpha = -\frac{2}{3}$
$\alpha$ is in Q3
$\therefore\ \cos\alpha < 0$

$\cos^2\alpha + \sin^2\alpha = 1$
$\therefore\ \cos^2\alpha + \frac{4}{9} = 1$
$\therefore\ \cos^2\alpha = \frac{5}{9}$
$\therefore\ \cos\alpha = -\frac{\sqrt{5}}{3}$

**b** $\sin 2\alpha = 2\sin\alpha\cos\alpha$
$= 2\left(-\frac{2}{3}\right)\left(-\frac{\sqrt{5}}{3}\right)$
$= \frac{4\sqrt{5}}{9}$

**4** **a** $\cos\beta = \frac{2}{5}$
$\beta$ is in Q4
$\therefore\ \sin\beta < 0$

$\cos^2\beta + \sin^2\beta = 1$
$\therefore\ \frac{4}{25} + \sin^2\beta = 1$
$\therefore\ \sin^2\beta = \frac{21}{25}$
$\therefore\ \sin\beta = -\frac{\sqrt{21}}{5}$

**b** $\sin 2\beta = 2\sin\beta\cos\beta$
$= 2\left(-\frac{\sqrt{21}}{5}\right)\left(\frac{2}{5}\right)$
$= -\frac{4\sqrt{21}}{25}$

**5** $\alpha$ is acute $\therefore$ $\cos\alpha$ and $\sin\alpha$ are positive

**a**
$$\cos 2\alpha = 2\cos^2\alpha - 1$$
$$\therefore\ -\tfrac{7}{9} = 2\cos^2\alpha - 1$$
$$\therefore\ 2\cos^2\alpha = \tfrac{2}{9}$$
$$\therefore\ \cos^2\alpha = \tfrac{1}{9}$$
$$\therefore\ \cos\alpha = \tfrac{1}{3}$$

**b**
$$\sin\alpha = \sqrt{1-\cos^2\alpha}$$
$$= \sqrt{1-\tfrac{1}{9}}$$
$$= \sqrt{\tfrac{8}{9}}$$
$$= \tfrac{2\sqrt{2}}{3}$$

**6**
$$\left[\cos(\tfrac{\pi}{12}) + \sin(\tfrac{\pi}{12})\right]^2$$
$$= \cos^2(\tfrac{\pi}{12}) + 2\cos(\tfrac{\pi}{12})\sin(\tfrac{\pi}{12}) + \sin^2(\tfrac{\pi}{12})$$
$$= 1 + 2\cos(\tfrac{\pi}{12})\sin(\tfrac{\pi}{12})$$
$$= 1 + \sin(\tfrac{\pi}{6}) \qquad \{\sin 2A = 2\cos A\sin A\}$$
$$= 1 + \tfrac{1}{2}$$
$$= \tfrac{3}{2}$$

**7** **a**
$$2\sin\alpha\cos\alpha$$
$$= \sin 2\alpha$$

**b**
$$4\cos\alpha\sin\alpha$$
$$= 2(2\sin\alpha\cos\alpha)$$
$$= 2\sin 2\alpha$$

**c**
$$\sin\alpha\cos\alpha$$
$$= \tfrac{1}{2}(2\sin\alpha\cos\alpha)$$
$$= \tfrac{1}{2}\sin 2\alpha$$

**d**
$$2\cos^2\beta - 1$$
$$= \cos 2\beta$$

**e**
$$1 - 2\cos^2\phi$$
$$= -(2\cos^2\phi - 1)$$
$$= -\cos 2\phi$$

**f**
$$1 - 2\sin^2 N$$
$$= \cos 2N$$

**g**
$$2\sin^2 M - 1$$
$$= -(1 - 2\sin^2 M)$$
$$= -\cos 2M$$

**h**
$$\cos^2\alpha - \sin^2\alpha$$
$$= \cos 2\alpha$$

**i**
$$\sin^2\alpha - \cos^2\alpha$$
$$= -(\cos^2\alpha - \sin^2\alpha)$$
$$= -\cos 2\alpha$$

**j**
$$2\sin 2A\cos 2A$$
$$= \sin 2(2A)$$
$$= \sin 4A$$

**k**
$$2\cos 3\alpha\sin 3\alpha$$
$$= \sin 2(3\alpha)$$
$$= \sin 6\alpha$$

**l**
$$2\cos^2 4\theta - 1$$
$$= \cos 2(4\theta)$$
$$= \cos 8\theta$$

**m**
$$1 - 2\cos^2 3\beta$$
$$= -(2\cos^2 3\beta - 1)$$
$$= -\cos 2(3\beta)$$
$$= -\cos 6\beta$$

**n**
$$1 - 2\sin^2 5\alpha$$
$$= \cos 2(5\alpha)$$
$$= \cos 10\alpha$$

**o**
$$2\sin^2 3D - 1$$
$$= -(1 - 2\sin^2 3D)$$
$$= -\cos 2(3D)$$
$$= -\cos 6D$$

**p**
$$\cos^2 2A - \sin^2 2A$$
$$= \cos 2(2A)$$
$$= \cos 4A$$

**q**
$$\cos^2(\tfrac{\alpha}{2}) - \sin^2(\tfrac{\alpha}{2})$$
$$= \cos 2(\tfrac{\alpha}{2})$$
$$= \cos\alpha$$

**r**
$$2\sin^2 3P - 2\cos^2 3P$$
$$= -2[\cos^2 3P - \sin^2 3P]$$
$$= -2\cos 2(3P)$$
$$= -2\cos 6P$$

**8** **a**
$$(\sin\theta + \cos\theta)^2$$
$$= \sin^2\theta + 2\sin\theta\cos\theta + \cos^2\theta$$
$$= \sin^2\theta + \cos^2\theta + 2\sin\theta\cos\theta$$
$$= 1 + \sin 2\theta$$

**b**
$$\cos^4\theta - \sin^4\theta$$
$$= (\cos^2\theta + \sin^2\theta)(\cos^2\theta - \sin^2\theta)$$
$$= 1 \times \cos 2\theta$$
$$= \cos 2\theta$$

**9** **a** $\sin 2x + \sin x = 0$

$\therefore\ 2\sin x\cos x + \sin x = 0$

$\therefore\ \sin x(2\cos x + 1) = 0$

$\therefore\ \sin x = 0$ or $\cos x = -\frac{1}{2}$

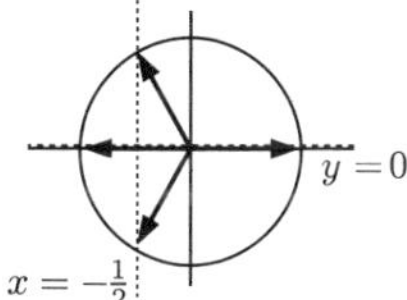


$\therefore\ x = 0, \frac{2\pi}{3}, \pi, \frac{4\pi}{3}, 2\pi$

**b** $\sin 2x - 2\cos x = 0$

$\therefore\ 2\sin x\cos x - 2\cos x = 0$

$\therefore\ 2\cos x(\sin x - 1) = 0$

$\therefore\ \cos x = 0$ or $\sin x = 1$

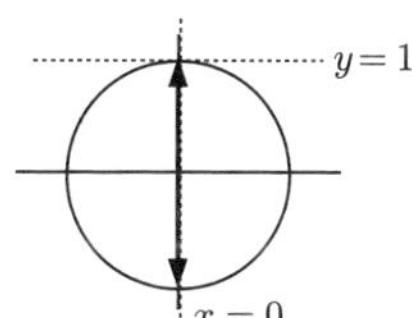


$\therefore\ x = \frac{\pi}{2}, \frac{3\pi}{2}$

**c** $\sin 2x + 3\sin x = 0$

$\therefore\ 2\sin x\cos x + 3\sin x = 0$

$\therefore\ \sin x(2\cos x + 3) = 0$

$\therefore\ \sin x = 0$ or $\cos x = -\frac{3}{2}$

impossible

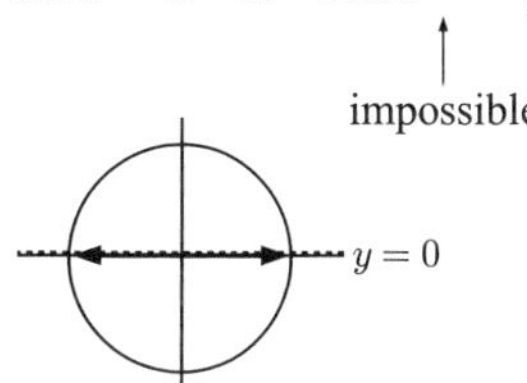


$\therefore\ x = 0, \pi, 2\pi$

**10** **a** $\frac{1}{2} - \frac{1}{2}\cos 2\theta$

$= \frac{1}{2} - \frac{1}{2}(1 - 2\sin^2\theta)$

$= \frac{1}{2} - \frac{1}{2} + \sin^2\theta$

$= \sin^2\theta$

**b** $\frac{1}{2} + \frac{1}{2}\cos 2\theta$

$= \frac{1}{2} + \frac{1}{2}(2\cos^2\theta - 1)$

$= \frac{1}{2} + \cos^2\theta - \frac{1}{2}$

$= \cos^2\theta$

## EXERCISE 11E

**1** **a** $2\sin^2 x + \sin x = 0$

$\therefore\ \sin x(2\sin x + 1) = 0$

$\therefore\ \sin x = 0$ or $-\frac{1}{2}$

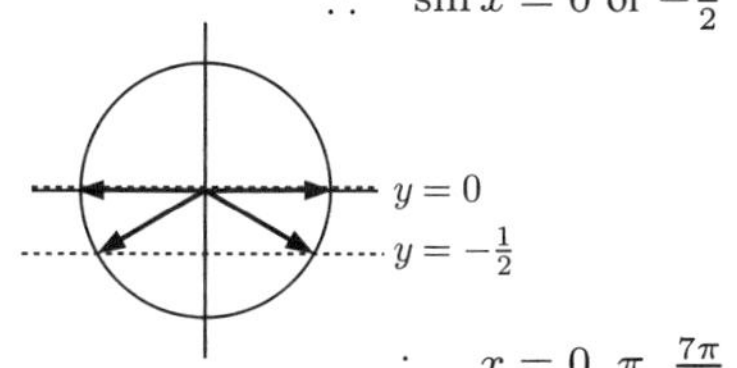


$\therefore\ x = 0, \pi, \frac{7\pi}{6}, \frac{11\pi}{6}, 2\pi$

**b** $2\cos^2 x = \cos x$

$\therefore\ 2\cos^2 x - \cos x = 0$

$\therefore\ \cos x(2\cos x - 1) = 0$

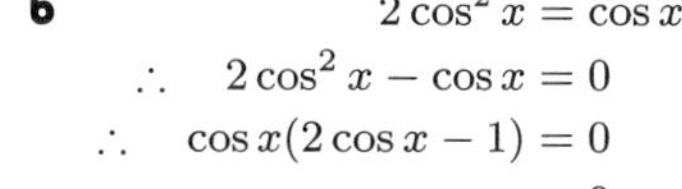

$\therefore\ \cos x = 0$ or $\frac{1}{2}$

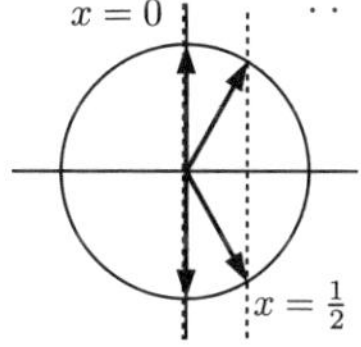


$\therefore\ x = \frac{\pi}{3}, \frac{\pi}{2}, \frac{3\pi}{2}, \frac{5\pi}{3}$

**c** $2\cos^2 x + \cos x - 1 = 0$

$\therefore\ (2\cos x - 1)(\cos x + 1) = 0$

$\therefore\ \cos x = \frac{1}{2}$ or $-1$

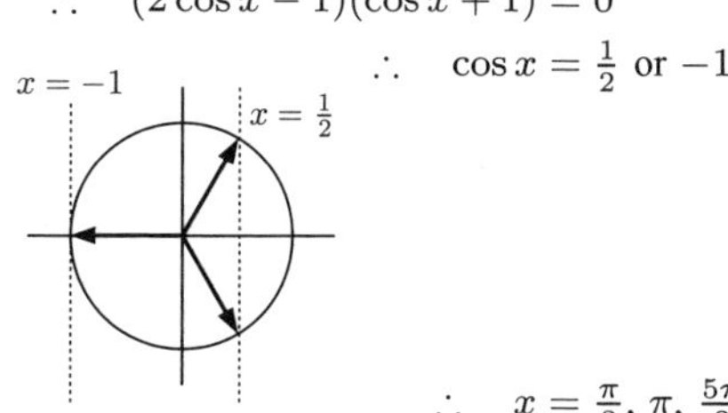


$\therefore\ x = \frac{\pi}{3}, \pi, \frac{5\pi}{3}$

**d** $2\sin^2 x + 3\sin x + 1 = 0$

$\therefore\ (2\sin x + 1)(\sin x + 1) = 0$

$\therefore\ \sin x = -\frac{1}{2}$ or $-1$

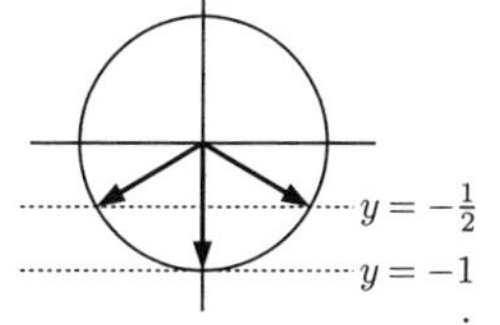


$\therefore\ x = \frac{7\pi}{6}, \frac{3\pi}{2}, \frac{11\pi}{6}$

**e**
$$\sin^2 x = 2 - \cos x$$
$$\therefore\ 1 - \cos^2 x = 2 - \cos x$$
$$\therefore\ \cos^2 x - \cos x + 1 = 0$$
$$\text{where } \Delta = (-1)^2 - 4(1)(1) = 1 - 4 = -3$$
$\therefore$ no real solutions exist

**2** **a**
$$\cos 2x - \cos x = 0$$
$$\therefore\ (2\cos^2 x - 1) - \cos x = 0$$
$$\therefore\ 2\cos^2 x - \cos x - 1 = 0$$
$$\therefore\ (2\cos x + 1)(\cos x - 1) = 0$$
$$\therefore\ \cos x = -\tfrac{1}{2} \text{ or } 1$$

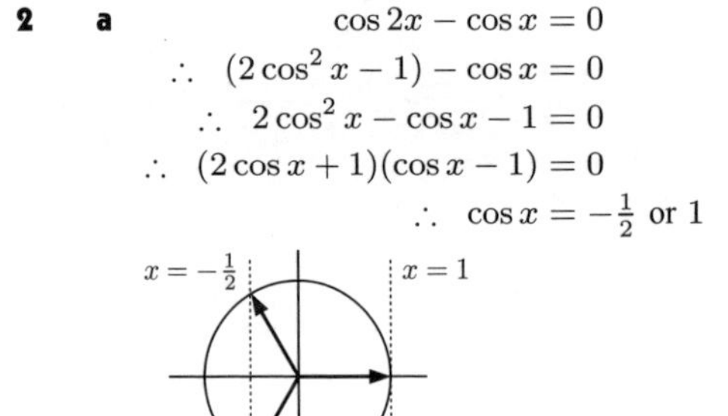


$$\therefore\ x = 0,\ \tfrac{2\pi}{3},\ \tfrac{4\pi}{3},\ 2\pi$$

**b**
$$\cos 2x + 3\cos x = 1$$
$$\therefore\ (2\cos^2 x - 1) + 3\cos x = 1$$
$$\therefore\ 2\cos^2 x + 3\cos x - 2 = 0$$
$$\therefore\ (2\cos x - 1)(\cos x + 2) = 0$$
$$\therefore\ \cos x = \tfrac{1}{2} \quad \{-1 \leqslant \cos x \leqslant 1\}$$

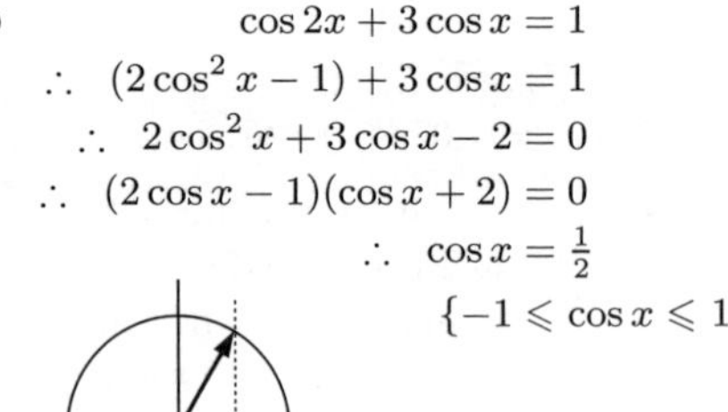


$$\therefore\ x = \tfrac{\pi}{3},\ \tfrac{5\pi}{3}$$

**c**
$$\cos 2x + \sin x = 0$$
$$\therefore\ (1 - 2\sin^2 x) + \sin x = 0$$
$$\therefore\ -2\sin^2 x + \sin x + 1 = 0$$
$$\therefore\ 2\sin^2 x - \sin x - 1 = 0$$
$$\therefore\ (2\sin x + 1)(\sin x - 1) = 0$$
$$\therefore\ \sin x = -\tfrac{1}{2} \text{ or } 1$$
$$\therefore\ x = \tfrac{\pi}{2},\ \tfrac{7\pi}{6},\ \tfrac{11\pi}{6}$$

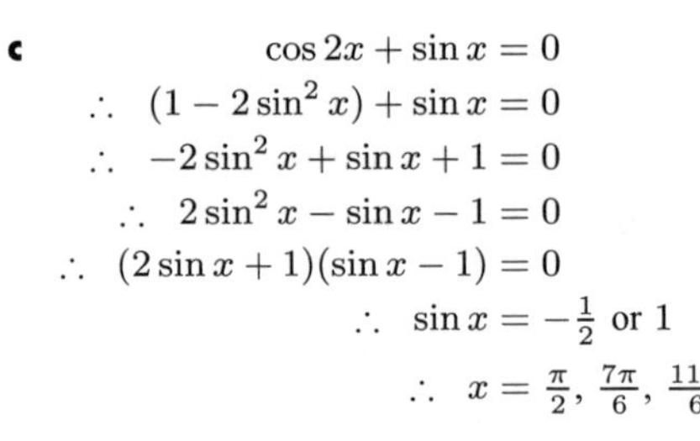

$y = 1$

$y = -\frac{1}{2}$

**d**
$$\sin 4x = \sin 2x$$
$$\therefore\ 2\sin 2x \cos 2x = \sin 2x$$
$$\therefore\ 2\sin 2x \cos 2x - \sin 2x = 0$$
$$\therefore\ \sin 2x(2\cos 2x - 1) = 0$$
$$\therefore\ \sin 2x = 0 \quad \text{or} \quad \cos 2x = \tfrac{1}{2}$$
$$\therefore\ 2x = 0,\ \tfrac{\pi}{3},\ \pi,\ \tfrac{5\pi}{3},\ 2\pi,\ \tfrac{7\pi}{3},\ 3\pi,\ \tfrac{11\pi}{3},\ 4\pi \quad \{0 \leqslant 2x \leqslant 4\pi\}$$
$$\therefore\ x = 0,\ \tfrac{\pi}{6},\ \tfrac{\pi}{2},\ \tfrac{5\pi}{6},\ \pi,\ \tfrac{7\pi}{6},\ \tfrac{3\pi}{2},\ \tfrac{11\pi}{6},\ 2\pi$$

$y = 0$

$x = \frac{1}{2}$

**e** $\sin x + \cos x = \sqrt{2}$

Squaring both sides we get:
$$\sin^2 x + 2\sin x\cos x + \cos^2 x = 2$$
$$\therefore\ \sin 2x + 1 = 2$$
$$\therefore\ \sin 2x = 1$$
$$\therefore\ 2x = \tfrac{\pi}{2},\ \tfrac{5\pi}{2} \quad \{0 \leqslant 2x \leqslant 4\pi\}$$
$$\therefore\ x = \tfrac{\pi}{4},\ \tfrac{5\pi}{4}$$

$y = 1$

Since we squared the original equation, we must check our answers.

$\sin\frac{\pi}{4} + \cos\frac{\pi}{4} = \frac{1}{\sqrt{2}} + \frac{1}{\sqrt{2}} = \frac{2}{\sqrt{2}} = \sqrt{2}$ ✓

$\sin\frac{5\pi}{4} + \cos\frac{5\pi}{4} = -\frac{1}{\sqrt{2}} + (-\frac{1}{\sqrt{2}}) = -\frac{2}{\sqrt{2}} = -\sqrt{2}$ ✗

$\therefore\ x = \frac{\pi}{4}$ is the only solution

**f**
$$2\cos^2 x = 3\sin x$$
$$\therefore\ 2(1-\sin^2 x) = 3\sin x$$
$$\therefore\ 2\sin^2 x + 3\sin x - 2 = 0$$
$$\therefore\ (2\sin x - 1)(\sin x + 2) = 0$$
$$\therefore\ \sin x = \tfrac{1}{2} \quad \{-1 \leqslant \sin x \leqslant 1\}$$
$$\therefore\ x = \tfrac{\pi}{6}, \tfrac{5\pi}{6}$$

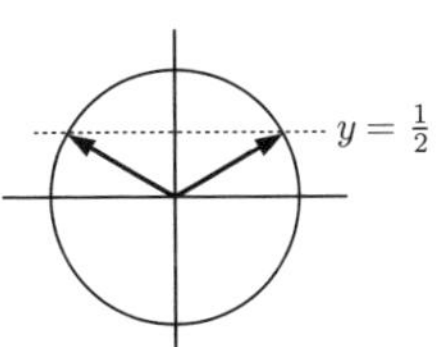


## REVIEW SET 11A

**1**

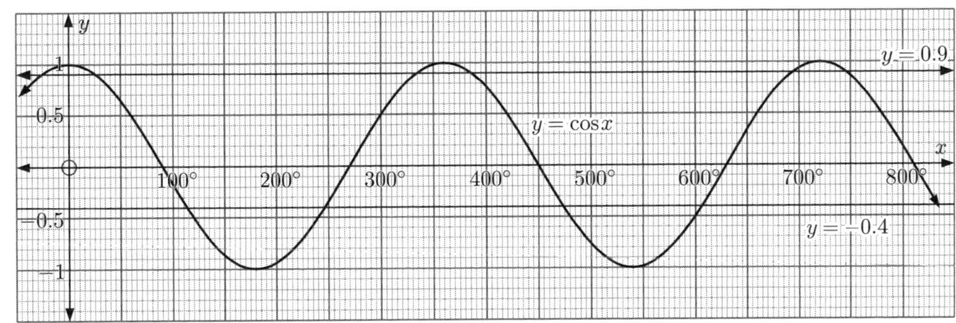


**a** When $\cos x = -0.4$, $0° \leqslant x \leqslant 800°$, $x \approx 115°, 245°, 475°, 605°$

**b** When $\cos x = 0.9$, $0° \leqslant x \leqslant 600°$, $x \approx 25°, 335°, 385°$

**2** **a** $2\sin x = -1$, $0 \leqslant x \leqslant 4\pi$

$\therefore\ \sin x = -\frac{1}{2}$

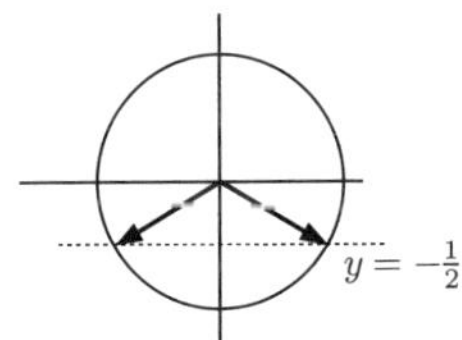


The points on the unit circle with sine $-\frac{1}{2}$ correspond to angles $\frac{7\pi}{6}$ and $\frac{11\pi}{6}$.

$\therefore\ x = \frac{7\pi}{6}, \frac{11\pi}{6}, \frac{19\pi}{6}, \frac{23\pi}{6}$

**b** $\sqrt{2}\sin x - 1 = 0$, $-2\pi \leqslant x \leqslant 2\pi$

$\therefore\ \sin x = \frac{1}{\sqrt{2}}$

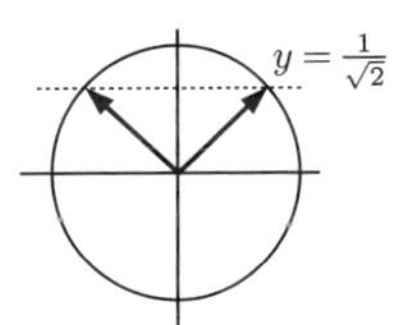


The points on the unit circle with sine $\frac{1}{\sqrt{2}}$ correspond to angles $\frac{\pi}{4}$ and $\frac{3\pi}{4}$.

$\therefore\ x = -\frac{7\pi}{4}, -\frac{5\pi}{4}, \frac{\pi}{4}, \frac{3\pi}{4}$

**3** **a** $2\sin 3x + \sqrt{3} = 0$, $0 \leqslant x \leqslant 2\pi$

$\therefore\ \sin 3x = -\frac{\sqrt{3}}{2}$



The points on the unit circle with sine $-\frac{\sqrt{3}}{2}$ correspond to angles $\frac{4\pi}{3}$ and $\frac{5\pi}{3}$.

$0 \leqslant x \leqslant 2\pi \quad \therefore\ 0 \leqslant 3x \leqslant 6\pi$

So $3x = \frac{4\pi}{3}, \frac{5\pi}{3}, \frac{10\pi}{3}, \frac{11\pi}{3}, \frac{16\pi}{3}, \frac{17\pi}{3}$

So, the $x$-intercepts are $\frac{4\pi}{9}, \frac{5\pi}{9}, \frac{10\pi}{9}, \frac{11\pi}{9}, \frac{16\pi}{9}, \frac{17\pi}{9}$.

**b** $\sqrt{2}\sin(x + \frac{\pi}{4}) = 0$, $0 \leqslant x \leqslant 3\pi$

$\therefore\ \sin(x + \frac{\pi}{4}) = 0$

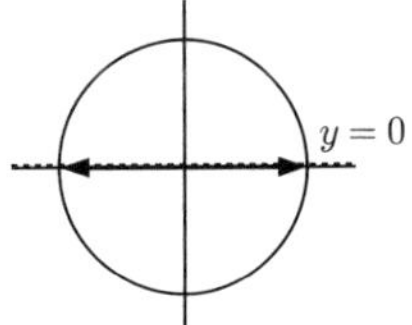


The points on the unit circle with sine 0 correspond to angles 0 and $\pi$.

$0 \leqslant x \leqslant 3\pi \quad \therefore\ \frac{\pi}{4} \leqslant x + \frac{\pi}{4} \leqslant \frac{13\pi}{4}$

So $x + \frac{\pi}{4} = \pi, 2\pi, 3\pi$

So, the $x$-intercepts are $\frac{3\pi}{4}, \frac{7\pi}{4}, \frac{11\pi}{4}$.

**4** $\sqrt{2}\cos(x+\frac{\pi}{4})-1=0, \quad 0 \leqslant x \leqslant 4\pi$

$\therefore \quad \cos(x+\frac{\pi}{4}) = \frac{1}{\sqrt{2}}$

The points on the unit circle with cosine $\frac{1}{\sqrt{2}}$ correspond to angles $\frac{\pi}{4}$ and $\frac{7\pi}{4}$.

$x = \frac{1}{\sqrt{2}}$

$0 \leqslant x \leqslant 4\pi \quad \therefore \quad \frac{\pi}{4} \leqslant x+\frac{\pi}{4} \leqslant \frac{17\pi}{4}$

So $x+\frac{\pi}{4} = \frac{\pi}{4}, \frac{7\pi}{4}, \frac{9\pi}{4}, \frac{15\pi}{4}, \frac{17\pi}{4}$

$\therefore \quad x = 0, \frac{3\pi}{2}, 2\pi, \frac{7\pi}{2}, 4\pi$

**5** **a** $\dfrac{1-\cos^2\theta}{1+\cos\theta}$

$= \dfrac{(1+\cos\theta)(1-\cos\theta)}{1+\cos\theta}$

$= 1-\cos\theta$

**b** $\dfrac{\sin\alpha-\cos\alpha}{\sin^2\alpha-\cos^2\alpha}$

$= \dfrac{\sin\alpha-\cos\alpha}{(\sin\alpha+\cos\alpha)(\sin\alpha-\cos\alpha)}$

$= \dfrac{1}{\sin\alpha+\cos\alpha}$

**c** $\dfrac{4\sin^2\alpha-4}{8\cos\alpha}$

$= \dfrac{-4(1-\sin^2\alpha)}{8\cos\alpha}$

$= \dfrac{-4\cos^2\alpha}{8\cos\alpha}$

$= \dfrac{-\cos\alpha}{2}$

**6**

S A T C $\alpha$

$\cos^2\alpha+\sin^2\alpha = 1$

$\therefore \quad \cos^2\alpha + \frac{9}{16} = 1$

$\therefore \quad \cos^2\alpha = \frac{7}{16}$

$\therefore \quad \cos\alpha = \pm\frac{\sqrt{7}}{4}$

But in Q3, $\cos\alpha < 0$

$\therefore \quad \cos\alpha = -\frac{\sqrt{7}}{4}$

$\sin 2\alpha = 2\sin\alpha\cos\alpha$

$= 2(-\frac{3}{4})(-\frac{\sqrt{7}}{4})$

$= \frac{3\sqrt{7}}{8}$

**7** $\dfrac{\sin 2\alpha-\sin\alpha}{\cos 2\alpha-\cos\alpha+1} = \dfrac{2\sin\alpha\cos\alpha-\sin\alpha}{2\cos^2\alpha-1-\cos\alpha+1}$

$= \dfrac{\sin\alpha(2\cos\alpha-1)}{\cos\alpha(2\cos\alpha-1)} = \dfrac{\sin\alpha}{\cos\alpha} = \tan\alpha$

## REVIEW SET 11B

**1** **a** $\sin x = 0.382, \quad 0 \leqslant x \leqslant 8$

$\therefore \quad x \approx 0.392, 2.75, 6.68$ {using technology}

**b** $\tan\left(\frac{x}{2}\right) = -0.458, \quad 0 \leqslant x \leqslant 8$

$\therefore \quad x \approx 5.42$ {using technology}

**2** **a** $\cos x = 0.4379, \quad 0 \leqslant x \leqslant 10$

$\therefore \quad x \approx 1.12, 5.17, 7.40$ {using technology}

**b** $\cos(x-2.4) = -0.6014, \quad 0 \leqslant x \leqslant 6$

$\therefore \quad x \approx 0.184, 4.62$ {using technology}

**3** **a** $\sin 2A = 2\sin A\cos A$

$= 2(\frac{5}{13})(\frac{12}{13})$

$= \frac{120}{169}$

**b** $\cos 2A = \cos^2 A - \sin^2 A$

$= (\frac{12}{13})^2 - (\frac{5}{13})^2$

$= \frac{144-25}{169}$

$= \frac{119}{169}$

**c** $\tan 2A = \dfrac{\sin 2A}{\cos 2A}$

$= \dfrac{\frac{120}{169}}{\frac{119}{169}}$

$= \frac{120}{119}$

**4** **a** **i** $\tan x = 4$

$\therefore \quad x \approx 1.33, 4.47, 7.61$

**ii** $\tan\left(\frac{x}{4}\right) = 4$

$\therefore \quad x \approx 5.30$

**iii** $\tan(x-1.5) = 4$

$\therefore \quad x \approx 2.83, 5.97, 9.11$

**b** **i** $\tan(x + \frac{\pi}{6}) = -\sqrt{3}, \quad -\pi \leqslant x \leqslant \pi$

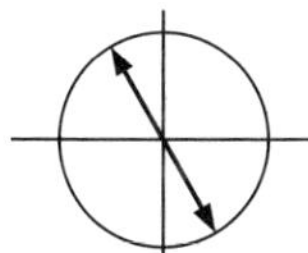

The points on the unit circle with tangent $-\sqrt{3}$ correspond to angles $\frac{2\pi}{3}$ and $\frac{5\pi}{3}$.

$-\pi \leqslant x \leqslant \pi \quad \therefore \quad -\frac{5\pi}{6} \leqslant x + \frac{\pi}{6} \leqslant \frac{7\pi}{6}$

So $\quad x + \frac{\pi}{6} = -\frac{\pi}{3}, \frac{2\pi}{3}$

$\therefore \quad x = -\frac{\pi}{2}, \frac{\pi}{2}$

**ii** $\tan 2x = -\sqrt{3}, \quad -\pi \leqslant x \leqslant \pi$

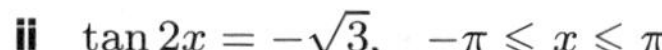

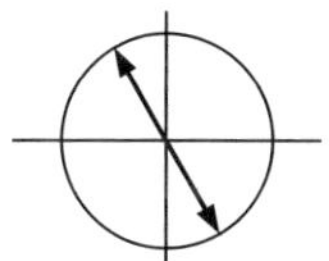

The points on the unit circle with tangent $-\sqrt{3}$ correspond to angles $\frac{2\pi}{3}$ and $\frac{5\pi}{3}$.

$-\pi \leqslant x \leqslant \pi \quad \therefore \quad -2\pi \leqslant x \leqslant 2\pi$

So $\quad 2x = -\frac{4\pi}{3}, -\frac{\pi}{3}, \frac{2\pi}{3}, \frac{5\pi}{3}$

$\therefore \quad x = -\frac{2\pi}{3}, -\frac{\pi}{6}, \frac{\pi}{3}, \frac{5\pi}{6}$

**iii** $\tan^2 x - 3 = 0, \quad -\pi \leqslant x \leqslant \pi$

$\therefore \quad \tan x = \pm\sqrt{3}$

The points on the unit circle with tangent $\pm\sqrt{3}$ correspond to angles $\frac{\pi}{3}, \frac{2\pi}{3}, \frac{4\pi}{3}, \frac{5\pi}{3}$.

$\therefore \quad x = -\frac{2\pi}{3}, -\frac{\pi}{3}, \frac{\pi}{3}, \frac{2\pi}{3}$

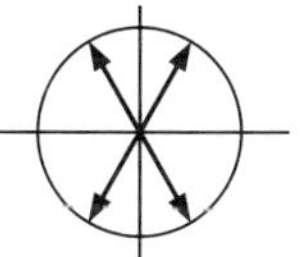

**c** $3\tan(x - 1.2) = -2$

$\therefore \quad x \approx 0.612, 3.75, 6.90$

**5** Using technology:

**a** $x \approx 1.27, 5.02$ **b** $x = \frac{\pi}{12}, \frac{\pi}{4}, \frac{3\pi}{4}, \frac{11\pi}{12}, \frac{17\pi}{12}, \frac{19\pi}{12}$

$\approx 0.262, 0.785, 2.36, 2.88, 4.45, 4.97$

**c** $x \approx 1.09, 2.05$

**6** $P(t) = 5 + 2\sin\left(\frac{\pi t}{3}\right), \quad 0 \leqslant t \leqslant 8,$ where $P(t)$ is in thousands of water beetles.

**a** $P(0) = 5 + 2\sin 0$

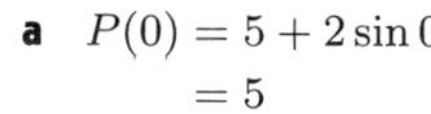

$= 5$

So, 5000 water beetles.

**b** Smallest $P = 5 + 2(-1) = 3$

Largest $P = 5 + 2(1) = 7$

$\therefore$ smallest is 3000 water beetles
largest is 7000 water beetles

**c** If population is $> 6000$,

then $\quad P(t) > 6$

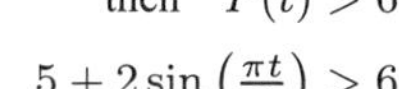

$\therefore \quad 5 + 2\sin\left(\frac{\pi t}{3}\right) > 6$

$\therefore \quad 2\sin\left(\frac{\pi t}{3}\right) > 1$

$\therefore \quad \sin\left(\frac{\pi t}{3}\right) > \frac{1}{2}$

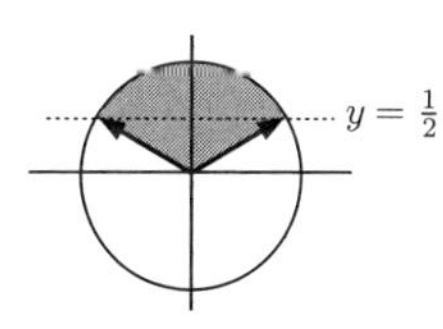


The points on the unit circle with sine $\frac{1}{2}$ correspond to angles $\frac{\pi}{6}$ and $\frac{5\pi}{6}$.

$0 \leqslant t \leqslant 8 \quad \therefore \quad 0 \leqslant \frac{\pi t}{3} \leqslant \frac{8\pi}{3}$

So $\quad \frac{\pi}{6} < \frac{\pi t}{3} < \frac{5\pi}{6}, \quad \frac{13\pi}{6} < \frac{\pi t}{3} \leqslant \frac{8\pi}{3}$

$\therefore \quad \frac{1}{2} < t < \frac{5}{2}, \quad \frac{13}{2} < t \leqslant 8$

$\therefore \quad 0.5 < t < 2.5, \quad 6.5 < t \leqslant 8$

**7** $3\cos x + \sin 2x = 1, \quad 0 \leqslant x \leqslant 10$

$\therefore \quad x \approx 1.37, 5.44, 7.65 \quad$ {using technology}

## REVIEW SET 11C

**1**

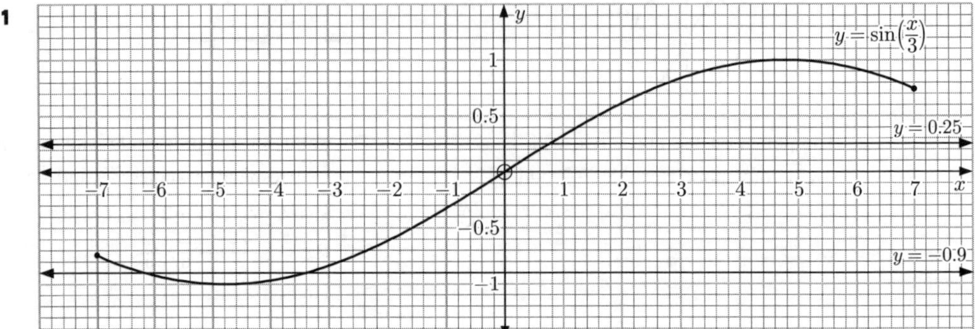


**a** $\sin\left(\frac{x}{3}\right) = -0.9, \quad -7 \leqslant x \leqslant 7$

$\therefore \; x \approx -6.1, \, -3.4$

**b** $\sin\left(\frac{x}{3}\right) = \frac{1}{4}, \quad -7 \leqslant x \leqslant 7$

$\therefore \; x \approx 0.8$

**2** **a** $\sin^2 x - \sin x - 2 = 0, \;\; 0 \leqslant x \leqslant 2\pi$

$\therefore \; (\sin x - 2)(\sin x + 1) = 0$

$\therefore \; \sin x = 2$ or $-1$

But $\sin x$ values lie between $-1$ and 1 inclusive.

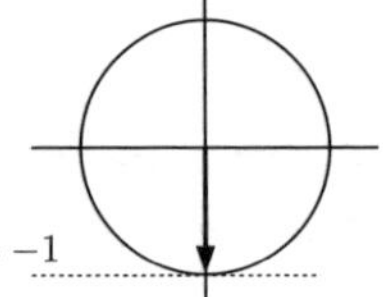


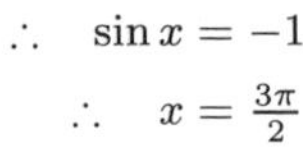

$\therefore \; \sin x = -1$

$\therefore \; x = \frac{3\pi}{2}$

**b** $4\sin^2 x = 1, \;\; 0 \leqslant x \leqslant 2\pi$

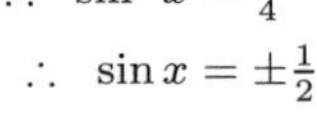

$\therefore \; \sin^2 x = \frac{1}{4}$

$\therefore \; \sin x = \pm\frac{1}{2}$

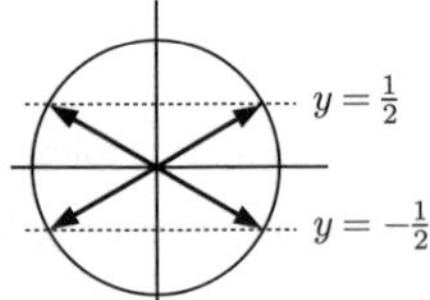


$\therefore \; x = \frac{\pi}{6}, \frac{5\pi}{6}, \frac{7\pi}{6}, \frac{11\pi}{6}$

**3** **a** $\tan(x - \frac{\pi}{3}) = \frac{1}{\sqrt{3}}, \quad 0 \leqslant x \leqslant 4\pi$

The points on the unit circle with tangent $\frac{1}{\sqrt{3}}$ correspond to angles $\frac{\pi}{6}$ and $\frac{7\pi}{6}$.

$0 \leqslant x \leqslant 4\pi$

$\therefore \; -\frac{\pi}{3} \leqslant x - \frac{\pi}{3} \leqslant \frac{11\pi}{3}$

So $x - \frac{\pi}{3} = \frac{\pi}{6}, \frac{7\pi}{6}, \frac{13\pi}{6}, \frac{19\pi}{6}$

$\therefore \; x = \frac{\pi}{2}, \frac{3\pi}{2}, \frac{5\pi}{2}, \frac{7\pi}{2}$

**b** $\cos(x + \frac{2\pi}{3}) = \frac{1}{2}, \quad -2\pi \leqslant x \leqslant 2\pi$

The points on the unit circle with cosine $\frac{1}{2}$ correspond to angles $\frac{\pi}{3}$ and $\frac{5\pi}{3}$.

$x = \frac{1}{2}$

$-2\pi \leqslant x \leqslant 2\pi$

$\therefore \; -\frac{4\pi}{3} \leqslant x + \frac{2\pi}{3} \leqslant \frac{8\pi}{3}$

So $x + \frac{2\pi}{3} = -\frac{\pi}{3}, \frac{\pi}{3}, \frac{5\pi}{3}, \frac{7\pi}{3}$

$\therefore \; x = -\pi, -\frac{\pi}{3}, \pi, \frac{5\pi}{3}$

**4** **a**

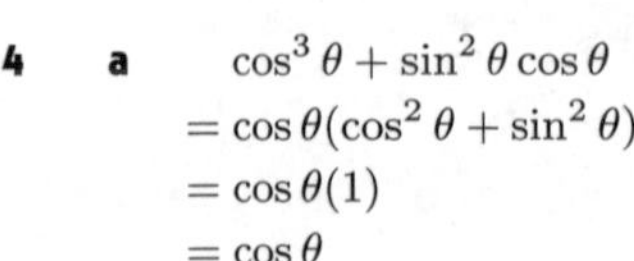

$\cos^3\theta + \sin^2\theta\cos\theta$

$= \cos\theta(\cos^2\theta + \sin^2\theta)$

$= \cos\theta(1)$

$= \cos\theta$

**b** $\frac{\cos^2\theta - 1}{\sin\theta} = \frac{-(1-\cos^2\theta)}{\sin\theta}$

$= -\frac{\sin^2\theta}{\sin\theta}$

$= -\sin\theta$

**c** $5 - 5\sin^2\theta$

$= 5(1 - \sin^2\theta)$

$= 5\cos^2\theta$

**d** $\frac{\sin^2\theta - 1}{\cos\theta} = -\frac{(1-\sin^2\theta)}{\cos\theta}$

$= -\frac{\cos^2\theta}{\cos\theta}$

$= -\cos\theta$

**5** **a** $(2\sin\alpha - 1)^2$
$= 4\sin^2\alpha - 4\sin\alpha + 1$

**b** $(\cos\alpha - \sin\alpha)^2$
$= \cos^2\alpha - 2\sin\alpha\cos\alpha + \sin^2\alpha$
$= \cos^2\alpha + \sin^2\alpha - 2\sin\alpha\cos\alpha$
$= 1 - \sin 2\alpha$

**6** **a** $\dfrac{\cos\theta}{1+\sin\theta} + \dfrac{1+\sin\theta}{\cos\theta}$

$= \dfrac{\cos^2\theta + (1+\sin\theta)^2}{(1+\sin\theta)\cos\theta}$

$= \dfrac{\cos^2\theta + 1 + 2\sin\theta + \sin^2\theta}{(1+\sin\theta)\cos\theta}$

$= \dfrac{2 + 2\sin\theta}{(1+\sin\theta)\cos\theta}$ $\{\cos^2\theta + \sin^2\theta = 1\}$

$= \dfrac{2\cancel{(1+\sin\theta)}}{\cancel{(1+\sin\theta)}\cos\theta}$

$= \dfrac{2}{\cos\theta}$

**b** $\left(1 + \dfrac{1}{\cos\theta}\right)\left(\cos\theta - \cos^2\theta\right)$
$= \cancel{\cos\theta} - \cos^2\theta + 1 - \cancel{\cos\theta}$
$= 1 - \cos^2\theta$
$= \sin^2\theta$

**7** $\tan\theta = -\frac{2}{3}$, $\frac{\pi}{2} < \theta < \pi$

S A T C θ

$\therefore \dfrac{\sin\theta}{\cos\theta} = -\frac{2}{3}$

$\therefore \sin\theta = -2k$, $\cos\theta = 3k$
but $\cos^2\theta + \sin^2\theta = 1$
$\therefore 9k^2 + 4k^2 = 1$
$\therefore 13k^2 = 1$
$\therefore k = \pm\frac{1}{\sqrt{13}}$

But in Q2,
$\sin\theta > 0$, $\cos\theta < 0$
$\therefore k = -\frac{1}{\sqrt{13}}$
$\therefore \sin\theta = \frac{2}{\sqrt{13}}$, $\cos\theta = -\frac{3}{\sqrt{13}}$

# Chapter 12
## VECTORS

### EXERCISE 12A.1

**1** **a** 135° 30 N

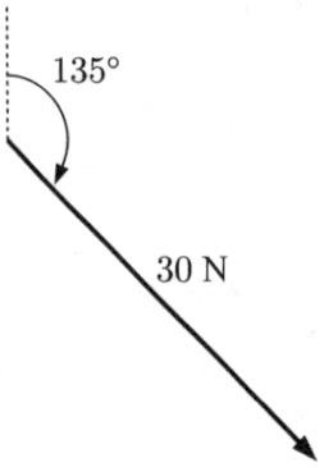

**b** 25 m s$^{-1}$

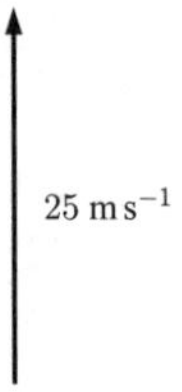

**c** ground 30° 30 cm min$^{-1}$

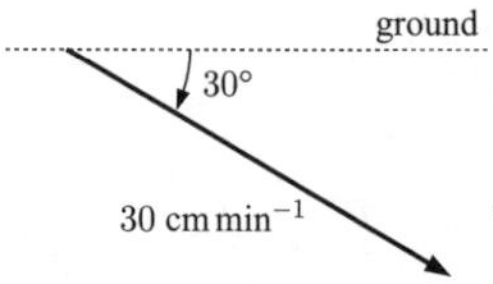

**d** 10° 50 m s$^{-1}$

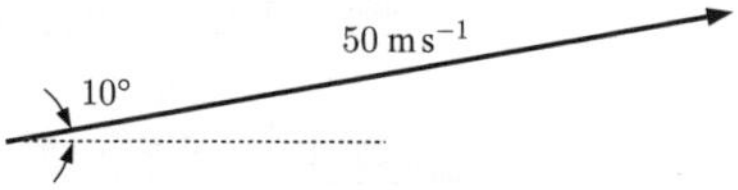

**2** **a** 100 m s$^{-1}$

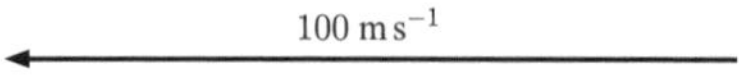

**b** 45° 75 m s$^{-1}$

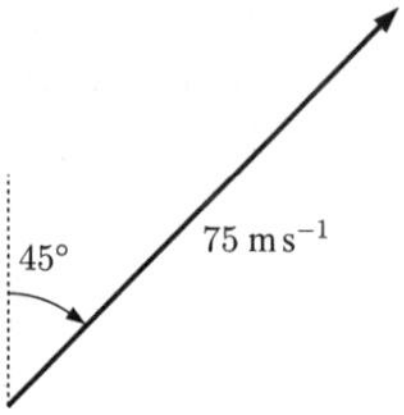

**3** **a** 30 N N 45°

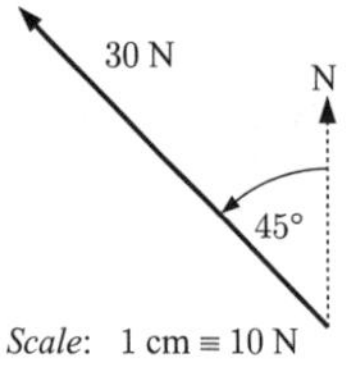

*Scale*: 1 cm ≡ 10 N

**b** 36 m s$^{-1}$

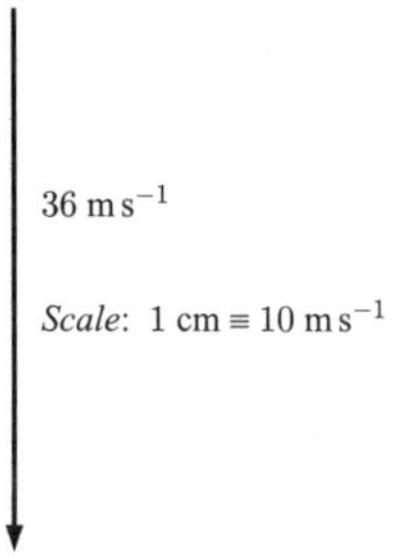

*Scale*: 1 cm ≡ 10 m s$^{-1}$

**c** $y$ 4 units 15° $x$

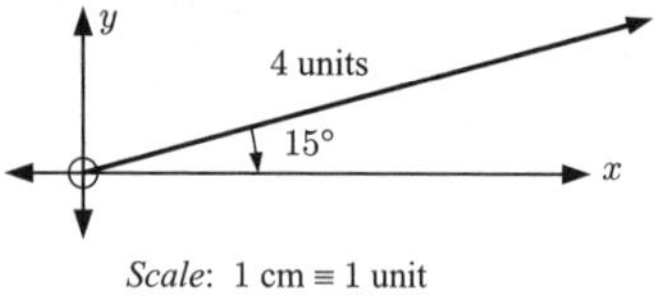

*Scale*: 1 cm ≡ 1 unit

**d** *Scale*: 1 cm ≡ 30 km h$^{-1}$

150 km h$^{-1}$ 8°

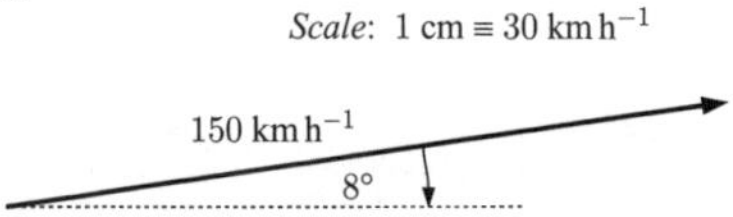

### EXERCISE 12A.2

**1** **a** If they are equal in magnitude, they have the same length. These are **p**, **q**, **s**, and **t**.

**b** Those parallel are **p**, **q**, **r**, and **t**.

**c** Those in the same direction are: **p** and **r**, **q** and **t**.

**d** To be equal they must have the same direction and be equal in length $\therefore$ $\mathbf{q} = \mathbf{t}$.

**e** **p** and **q** are negatives (equal length, but opposite direction). Likewise, **p** and **t** are negatives. We write $\mathbf{p} = -\mathbf{q}$ and $\mathbf{p} = -\mathbf{t}$.

**2** **a** True, as they have the same length and direction.

**b** True, as they are sides of an equilateral triangle.

**c** False, as they do not have the same direction.

**d** False, as they have opposite directions.

**e** True, as they have the same length and direction.

**f** False, as they do not have the same direction.

**3** **a** **i** $\overrightarrow{BC}$ is the vector which originates at B and terminates at C.

**ii** $\overrightarrow{ED} = \overrightarrow{AB}$, as they have the same length and direction.

**b** **i** $\overrightarrow{FE}$ and $\overrightarrow{BC}$ are negatives of $\overrightarrow{EF}$, as they both have the same length but opposite direction.

**ii** All sides of the hexagon are equal in length

$\therefore$ the vectors with the same length as $\overrightarrow{ED}$ are
$\overrightarrow{DE}$, $\overrightarrow{EF}$, $\overrightarrow{FE}$, $\overrightarrow{FA}$, $\overrightarrow{AF}$, $\overrightarrow{AB}$, $\overrightarrow{BA}$, $\overrightarrow{BC}$, $\overrightarrow{CB}$, $\overrightarrow{CD}$, and $\overrightarrow{DC}$.

**c** The vector $\overrightarrow{FC}$ is parallel to $\overrightarrow{AB}$ and twice its length.
$\overrightarrow{CF}$ is also parallel to $\overrightarrow{AB}$ and twice its length (but in the opposite direction).

## EXERCISE 12B.1

**1** **a**

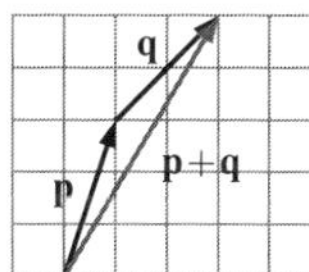


**b**

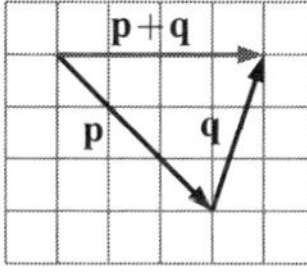


**c**

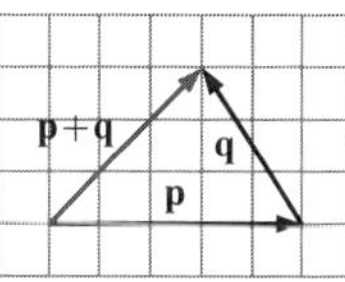


**d**

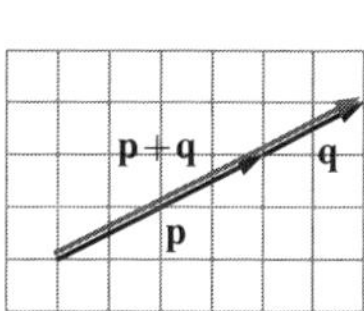


**e**

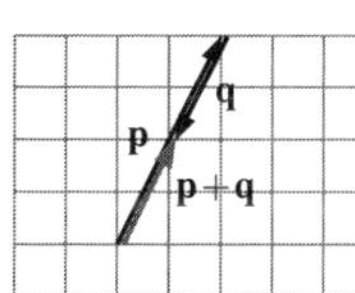


**f**

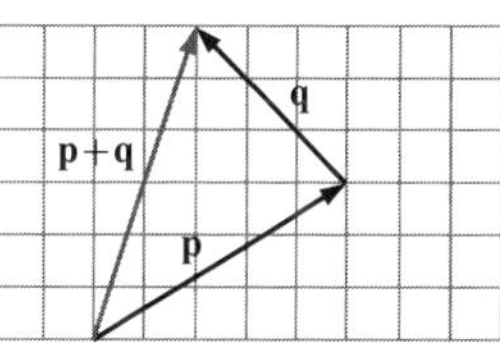


**2** **a** $\overrightarrow{AB} + \overrightarrow{BC}$
$= \overrightarrow{AC}$

**b** $\overrightarrow{BC} + \overrightarrow{CD}$
$= \overrightarrow{BD}$

**c** $\overrightarrow{AB} + \overrightarrow{BA}$
$= \overrightarrow{AA}$
$= \mathbf{0}$

**d** $\overrightarrow{AB} + \overrightarrow{BC} + \overrightarrow{CD}$
$= \overrightarrow{AC} + \overrightarrow{CD}$
$= \overrightarrow{AD}$

**e** $\overrightarrow{AC} + \overrightarrow{CB} + \overrightarrow{BD}$
$= \overrightarrow{AB} + \overrightarrow{BD}$
$= \overrightarrow{AD}$

**f** $\overrightarrow{BC} + \overrightarrow{CA} + \overrightarrow{AB}$
$= \overrightarrow{BA} + \overrightarrow{AB}$
$= \overrightarrow{BB}$
$= \mathbf{0}$

**3** **a** **i**

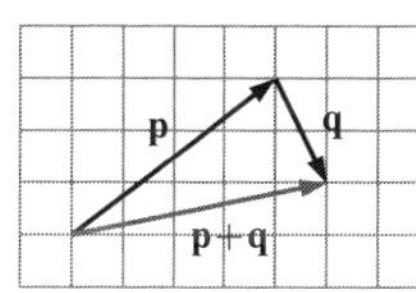


**ii**

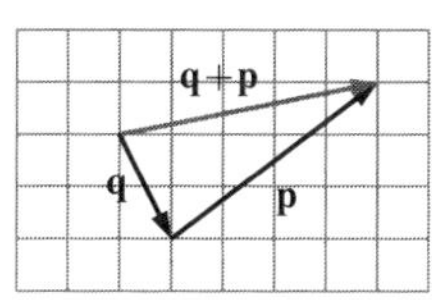


**b** yes

**4** $\overrightarrow{PS} = \overrightarrow{PR} + \overrightarrow{RS}$
$= (\mathbf{a} + \mathbf{b}) + \mathbf{c}$

But $\overrightarrow{PS} = \overrightarrow{PQ} + \overrightarrow{QS}$
$= \mathbf{a} + (\mathbf{b} + \mathbf{c})$

$\therefore$ $(\mathbf{a} + \mathbf{b}) + \mathbf{c} = \mathbf{a} + (\mathbf{b} + \mathbf{c})$ {as both are equal to $\overrightarrow{PS}$}

**5** **a** *Scale*: 1 cm = 100 km h$^{-1}$

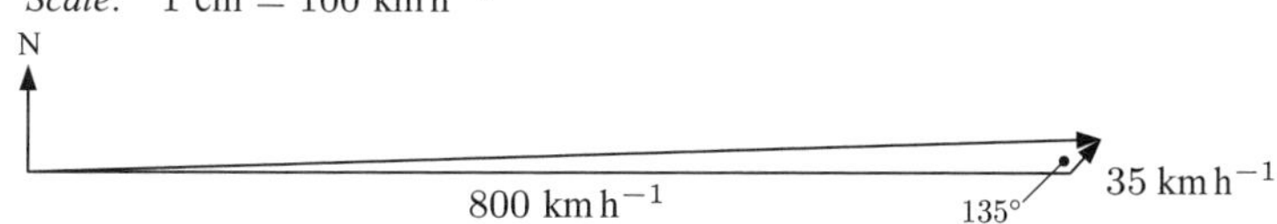


**b** We need to perform vector addition to find the effect of the wind on the aeroplane.

**c** Measuring the length of the resulting vector, we get 82.5 mm, or 8.25 cm.

$\therefore$ the resulting speed of the plane is $8.25 \times 100 = 825 \text{ km h}^{-1}$.

Using a protractor to measure the angle between 'true north' and the resulting vector, we get 88°.

$\therefore$ the direction of the aeroplane is 88° east of north.

## EXERCISE 12B.2

**1** **a**

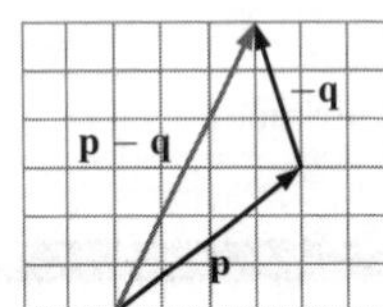


**b**

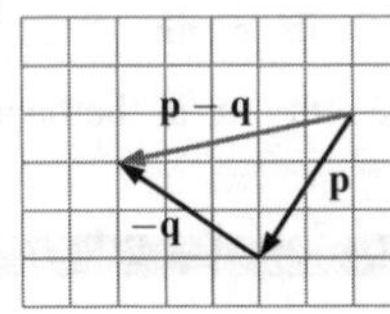


**c**

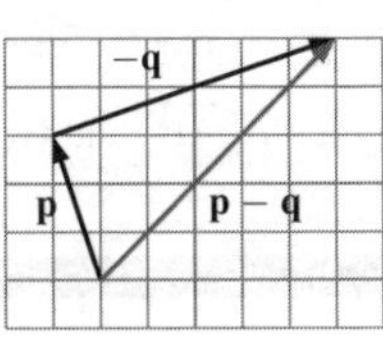


**d**

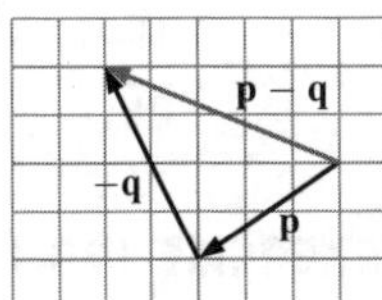


**2** **a**

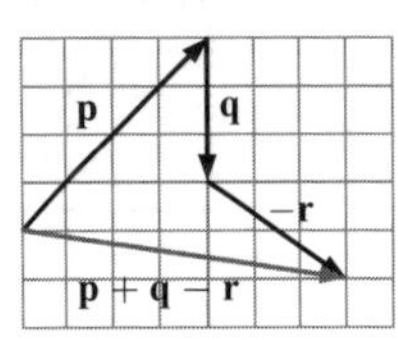


**b**

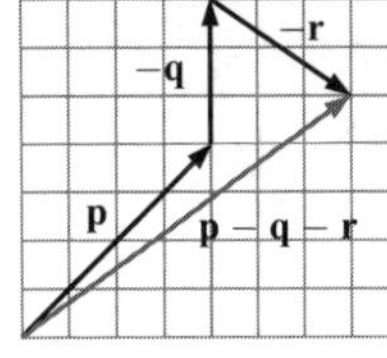


**c**

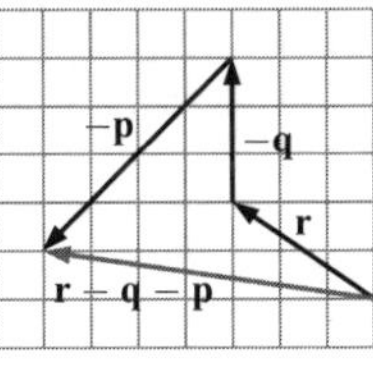


**3** **a** $\overrightarrow{AC} + \overrightarrow{CB}$
$= \overrightarrow{AB}$

**b** $\overrightarrow{AD} - \overrightarrow{BD}$
$= \overrightarrow{AD} + \overrightarrow{DB}$
$= \overrightarrow{AB}$

**c** $\overrightarrow{AC} + \overrightarrow{CA}$
$= \overrightarrow{AA}$
$= \mathbf{0}$

**d** $\overrightarrow{AB} + \overrightarrow{BC} + \overrightarrow{CD}$
$= \overrightarrow{AC} + \overrightarrow{CD}$
$= \overrightarrow{AD}$

**e** $\overrightarrow{BA} - \overrightarrow{CA} + \overrightarrow{CB}$
$= \overrightarrow{BA} + \overrightarrow{AC} + \overrightarrow{CB}$
$= \overrightarrow{BC} + \overrightarrow{CB}$
$= \overrightarrow{BB}$
$= \mathbf{0}$

**f** $\overrightarrow{AB} - \overrightarrow{CB} - \overrightarrow{DC}$
$= \overrightarrow{AB} + \overrightarrow{BC} + \overrightarrow{CD}$
$= \overrightarrow{AC} + \overrightarrow{CD}$
$= \overrightarrow{AD}$

## EXERCISE 12B.3

**1** **a** $\mathbf{t} = \mathbf{r} + \mathbf{s}$ **b** $\mathbf{r} = -\mathbf{s} - \mathbf{t}$ **c** $\mathbf{r} = -\mathbf{p} - \mathbf{q} - \mathbf{s}$

**d** $\mathbf{r} = \mathbf{q} - \mathbf{p} + \mathbf{s}$ **e** $\mathbf{p} = \mathbf{t} + \mathbf{s} + \mathbf{r} - \mathbf{q}$ **f** $\mathbf{p} = -\mathbf{u} + \mathbf{t} + \mathbf{s} - \mathbf{r} - \mathbf{q}$

**2** **a** **i** $\overrightarrow{OB} = \overrightarrow{OA} + \overrightarrow{AB}$
$= \mathbf{r} + \mathbf{s}$

**ii** $\overrightarrow{CA} = \overrightarrow{CB} + \overrightarrow{BA}$
$= -\overrightarrow{BC} - \overrightarrow{AB}$
$= -\mathbf{t} - \mathbf{s}$

**iii** $\overrightarrow{OC} = \overrightarrow{OA} + \overrightarrow{AB} + \overrightarrow{BC}$
$= \mathbf{r} + \mathbf{s} + \mathbf{t}$

**b** **i** $\overrightarrow{AD} = \overrightarrow{AB} + \overrightarrow{BD}$
$= \mathbf{p} + \mathbf{q}$

**ii** $\overrightarrow{BC} = \overrightarrow{BD} + \overrightarrow{DC}$
$= \mathbf{q} + \mathbf{r}$

**iii** $\overrightarrow{AC} = \overrightarrow{AB} + \overrightarrow{BD} + \overrightarrow{DC}$
$= \mathbf{p} + \mathbf{q} + \mathbf{r}$

## EXERCISE 12B.4

**1** **a**

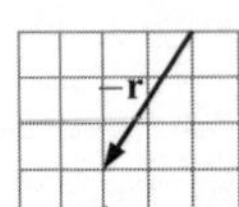


**b**

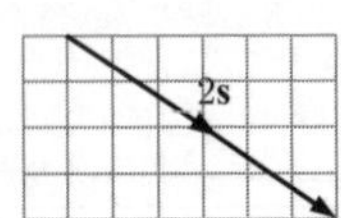


**c**

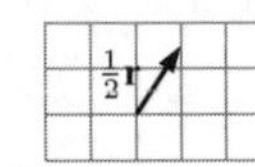


**d**

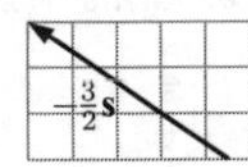

**e**

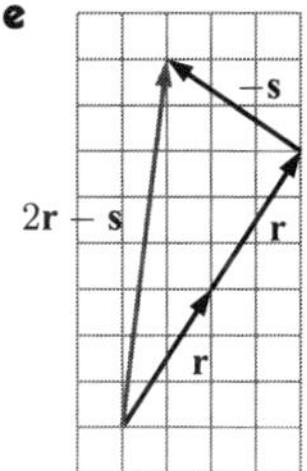


**f**

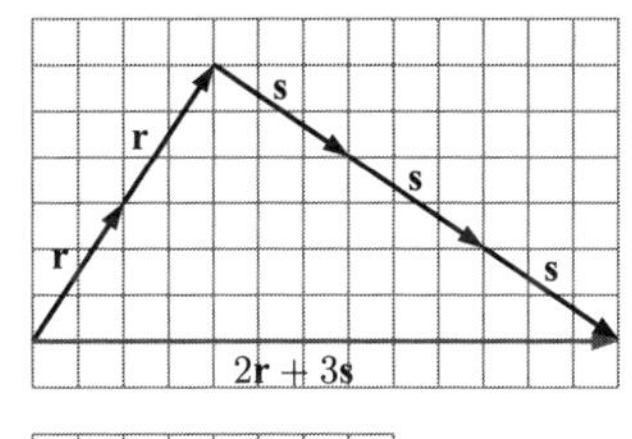


**g**

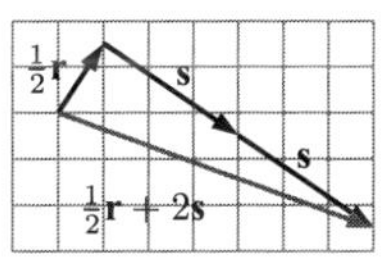


**h**

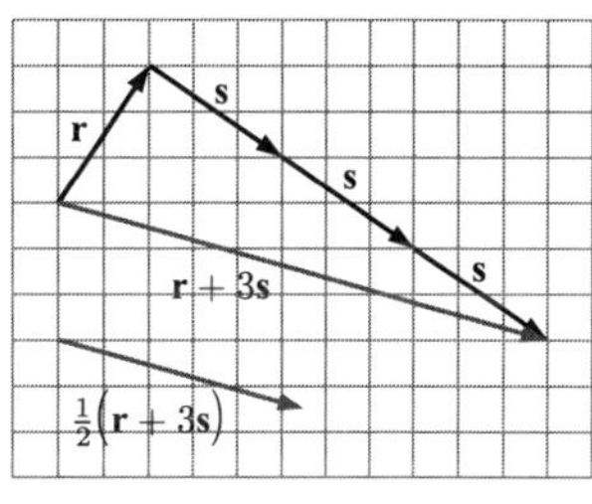


**2** **a**

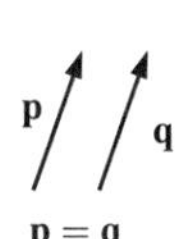


$\mathbf{p} = \mathbf{q}$

**b** p q $\mathbf{p} = -\mathbf{q}$

**c** p q $\mathbf{p} = 2\mathbf{q}$

**d** $\mathbf{p} = \frac{1}{3}\mathbf{q}$ q p

**e** p $\mathbf{p} = -3\mathbf{q}$ q

**3** **a**

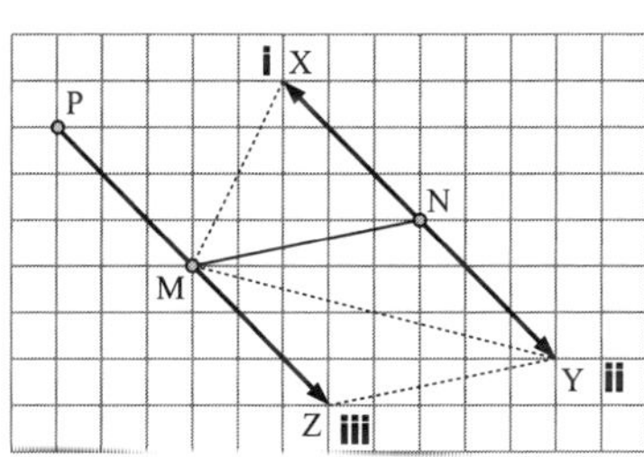


**b** a parallelogram

**4**

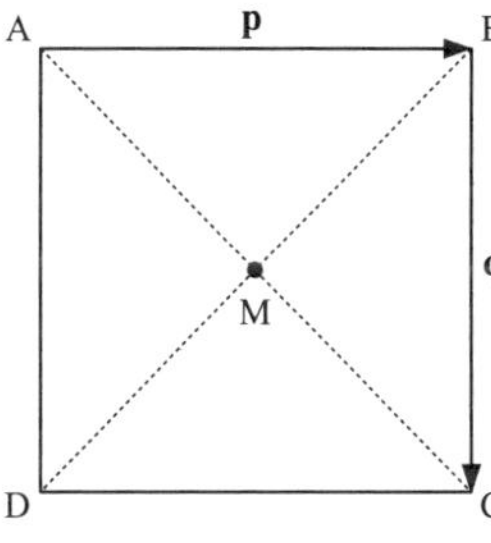


**a** $\overrightarrow{CD} = -\overrightarrow{AB}$
$= -\mathbf{p}$

**b** $\overrightarrow{AC} = \overrightarrow{AB} + \overrightarrow{BC}$
$= \mathbf{p} + \mathbf{q}$

**c** $\overrightarrow{AM} = \frac{1}{2}\overrightarrow{AC}$
$= \frac{1}{2}(\mathbf{p} + \mathbf{q})$
{using part **b**}

**d** $\overrightarrow{BD} = \overrightarrow{BC} + \overrightarrow{CD}$
$= \mathbf{q} + (-\mathbf{p})$ {using part **a**}
$= \mathbf{q} - \mathbf{p}$
and $\overrightarrow{BM} = \frac{1}{2}\overrightarrow{BD}$
$= \frac{1}{2}(\mathbf{q} - \mathbf{p})$

**5** **a** $\overrightarrow{PX} = \overrightarrow{QR}$
$= \mathbf{b}$

**b** $\overrightarrow{PS} = 2\overrightarrow{PX}$
$= 2\mathbf{b}$
{using part **a**}

**c** $\overrightarrow{QX} = \overrightarrow{QR} + \overrightarrow{RX}$
$= \mathbf{b} + (-\mathbf{a})$
$= \mathbf{b} - \mathbf{a}$

**d** $\overrightarrow{RS} = \overrightarrow{QX}$
$= \mathbf{b} - \mathbf{a}$
{using part **c**}

## EXERCISE 12C

**1** **a** $\begin{pmatrix} 7 \\ 3 \end{pmatrix}$, $7\mathbf{i} + 3\mathbf{j}$ **b** $\begin{pmatrix} -6 \\ 0 \end{pmatrix}$, $-6\mathbf{i}$ **c** $\begin{pmatrix} 2 \\ -5 \end{pmatrix}$, $2\mathbf{i} - 5\mathbf{j}$

**d** $\begin{pmatrix} 0 \\ 6 \end{pmatrix}$, $6\mathbf{j}$ **e** $\begin{pmatrix} -6 \\ 3 \end{pmatrix}$, $-6\mathbf{i} + 3\mathbf{j}$ **f** $\begin{pmatrix} -5 \\ -5 \end{pmatrix}$, $-5\mathbf{i} - 5\mathbf{j}$

**2** **a** $\begin{pmatrix} 3 \\ 4 \end{pmatrix} = 3\mathbf{i} + 4\mathbf{j}$ **b** $\begin{pmatrix} 2 \\ 0 \end{pmatrix} = 2\mathbf{i}$ **c** $\begin{pmatrix} 2 \\ -5 \end{pmatrix} = 2\mathbf{i} - 5\mathbf{j}$ **d** $\begin{pmatrix} -1 \\ -3 \end{pmatrix} = -\mathbf{i} - 3\mathbf{j}$

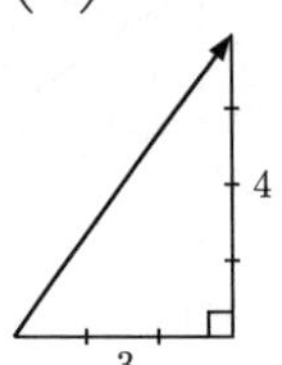


2

2
−5

−1
−3

**3** **a** $\overrightarrow{BA} = \begin{pmatrix} -4 \\ -1 \end{pmatrix} = -4\mathbf{i} - \mathbf{j}$ **b** $\overrightarrow{BC} = \begin{pmatrix} -1 \\ -5 \end{pmatrix} = -\mathbf{i} - 5\mathbf{j}$ **c** $\overrightarrow{DC} = \begin{pmatrix} 2 \\ 0 \end{pmatrix} = 2\mathbf{i}$

**d** $\overrightarrow{AC} = \begin{pmatrix} 3 \\ -4 \end{pmatrix} = 3\mathbf{i} - 4\mathbf{j}$ **e** $\overrightarrow{CA} = \begin{pmatrix} -3 \\ 4 \end{pmatrix} = -3\mathbf{i} + 4\mathbf{j}$ **f** $\overrightarrow{DB} = \begin{pmatrix} 3 \\ 5 \end{pmatrix} = 3\mathbf{i} + 5\mathbf{j}$

**4** **a** $\mathbf{i} + 2\mathbf{j} = \begin{pmatrix} 1 \\ 2 \end{pmatrix}$ **b** $-\mathbf{i} + 3\mathbf{j} = \begin{pmatrix} -1 \\ 3 \end{pmatrix}$ **c** $-5\mathbf{j} = \begin{pmatrix} 0 \\ -5 \end{pmatrix}$ **d** $4\mathbf{i} - 2\mathbf{j} = \begin{pmatrix} 4 \\ -2 \end{pmatrix}$

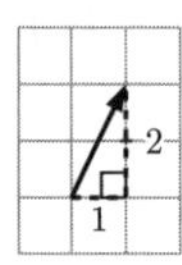


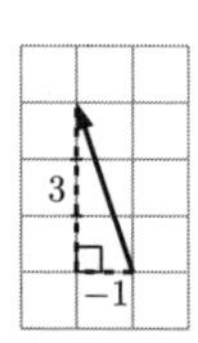


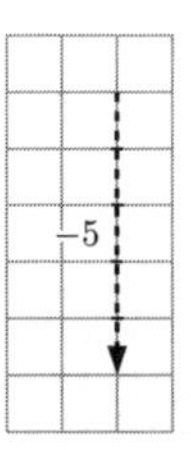


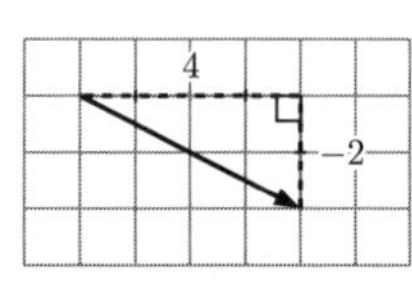


**5** The zero vector in component form is $\begin{pmatrix} 0 \\ 0 \end{pmatrix}$.

## EXERCISE 12D

**1** **a** $\left|\begin{pmatrix} 3 \\ 4 \end{pmatrix}\right| = \sqrt{3^2 + 4^2}$
$= \sqrt{9 + 16}$
$= \sqrt{25} = 5$ units

**b** $\left|\begin{pmatrix} -4 \\ 3 \end{pmatrix}\right| = \sqrt{(-4)^2 + 3^2}$
$= \sqrt{16 + 9}$
$= \sqrt{25} = 5$ units

**c** $\left|\begin{pmatrix} 2 \\ 0 \end{pmatrix}\right| = \sqrt{2^2 + 0^2}$
$= \sqrt{4}$
$= 2$ units

**d** $\left|\begin{pmatrix} -2 \\ 2 \end{pmatrix}\right| = \sqrt{(-2)^2 + 2^2}$
$= \sqrt{4 + 4}$
$= \sqrt{8}$ units

**e** $\left|\begin{pmatrix} 0 \\ -3 \end{pmatrix}\right| = \sqrt{0^2 + (-3)^2}$
$= \sqrt{9}$
$= 3$ units

**2** **a** As $\mathbf{i} + \mathbf{j} = \begin{pmatrix} 1 \\ 1 \end{pmatrix}$,
$|\mathbf{i} + \mathbf{j}| = \sqrt{1^2 + 1^2}$
$= \sqrt{2}$ units

**b** As $5\mathbf{i} - 12\mathbf{j} = \begin{pmatrix} 5 \\ -12 \end{pmatrix}$,
$|5\mathbf{i} - 12\mathbf{j}| = \sqrt{5^2 + (-12)^2}$
$= \sqrt{25 + 144}$
$= \sqrt{169} = 13$ units

**c** As $-\mathbf{i} + 4\mathbf{j} = \begin{pmatrix} -1 \\ 4 \end{pmatrix}$,
$|-\mathbf{i} + 4\mathbf{j}| = \sqrt{(-1)^2 + 4^2}$
$= \sqrt{1 + 16}$
$= \sqrt{17}$ units

**d** As $3\mathbf{i} = \begin{pmatrix} 3 \\ 0 \end{pmatrix}$,
$|3\mathbf{i}| = \sqrt{3^2 + 0^2}$
$= \sqrt{9}$
$= 3$ units

**e** As $k\mathbf{j} = \begin{pmatrix} 0 \\ k \end{pmatrix}$,
$|k\mathbf{j}| = \sqrt{0^2 + k^2}$
$= \sqrt{k^2}$
$= |k|$ units

**3** **a** $\sqrt{0^2+(-1)^2}=1$

$\therefore \begin{pmatrix}0\\-1\end{pmatrix}$ is a unit vector.

**b** $\sqrt{(-\frac{1}{\sqrt{2}})^2+(\frac{1}{\sqrt{2}})^2}=\sqrt{\frac{1}{2}+\frac{1}{2}}=1$

$\therefore \begin{pmatrix}-\frac{1}{\sqrt{2}}\\\frac{1}{\sqrt{2}}\end{pmatrix}$ is a unit vector.

**c** $\sqrt{(\frac{2}{3})^2+(\frac{1}{3})^2}=\sqrt{\frac{4}{9}+\frac{1}{9}}=\frac{\sqrt{5}}{3}$

$\therefore \begin{pmatrix}\frac{2}{3}\\\frac{1}{3}\end{pmatrix}$ is not a unit vector.

**d** $\sqrt{(-\frac{3}{5})^2+(-\frac{4}{5})^2}=\sqrt{\frac{9}{25}+\frac{16}{25}}=1$

$\therefore \begin{pmatrix}-\frac{3}{5}\\-\frac{4}{5}\end{pmatrix}$ is a unit vector.

**e** $\sqrt{(\frac{2}{7})^2+(-\frac{5}{7})^2}=\sqrt{\frac{4}{49}+\frac{25}{49}}=\frac{\sqrt{29}}{7}$

$\therefore \begin{pmatrix}\frac{2}{7}\\-\frac{5}{7}\end{pmatrix}$ is not a unit vector.

**4** **a** length $=1$

$\therefore \sqrt{0^2+k^2}=1$

$\therefore k^2=1$

$\therefore k=\pm1$

**b** length $=1$

$\therefore \sqrt{k^2+0}=1$

$\therefore k^2=1$

$\therefore k=\pm1$

**c** length $=1$

$\therefore \sqrt{k^2+1}=1$

$\therefore k^2+1=1$

$\therefore k^2=0$

$\therefore k=0$

**d** length $=1$

$\therefore \sqrt{k^2+k^2}=1$

$\therefore 2k^2=1$

$\therefore k^2=\frac{1}{2}$

$\therefore k=\pm\sqrt{\frac{1}{2}}=\pm\frac{1}{\sqrt{2}}$

**e** length $=1$

$\therefore \sqrt{\left(\frac{1}{2}\right)^2+k^2}=1$

$\therefore \frac{1}{4}+k^2=1$

$\therefore k^2=\frac{3}{4}$

$\therefore k=\pm\sqrt{\frac{3}{4}}=\pm\frac{\sqrt{3}}{2}$

**5** If $|\mathbf{v}|=\sqrt{73}$ units then $\sqrt{8^2+p^2}=\sqrt{73}$

$\therefore 64+p^2=73$

$\therefore p^2=9 \quad \therefore p=\pm3$

## EXERCISE 12E

**1** **a** $\mathbf{a}+\mathbf{b}=\begin{pmatrix}-3\\2\end{pmatrix}+\begin{pmatrix}1\\4\end{pmatrix}=\begin{pmatrix}-2\\6\end{pmatrix}$

**b** $\mathbf{b}+\mathbf{a}=\begin{pmatrix}1\\4\end{pmatrix}+\begin{pmatrix}-3\\2\end{pmatrix}=\begin{pmatrix}-2\\6\end{pmatrix}$

**c** $\mathbf{b}+\mathbf{c}=\begin{pmatrix}1\\4\end{pmatrix}+\begin{pmatrix}-2\\-5\end{pmatrix}=\begin{pmatrix}-1\\-1\end{pmatrix}$

**d** $\mathbf{c}+\mathbf{b}=\begin{pmatrix}-2\\-5\end{pmatrix}+\begin{pmatrix}1\\4\end{pmatrix}=\begin{pmatrix}-1\\-1\end{pmatrix}$

**e** $\mathbf{a}+\mathbf{c}=\begin{pmatrix}-3\\2\end{pmatrix}+\begin{pmatrix}-2\\-5\end{pmatrix}=\begin{pmatrix}-5\\-3\end{pmatrix}$

**f** $\mathbf{c}+\mathbf{a}=\begin{pmatrix}-2\\-5\end{pmatrix}+\begin{pmatrix}-3\\2\end{pmatrix}=\begin{pmatrix}-5\\-3\end{pmatrix}$

**g** $\mathbf{a}+\mathbf{a}=\begin{pmatrix}-3\\2\end{pmatrix}+\begin{pmatrix}-3\\2\end{pmatrix}=\begin{pmatrix}-6\\4\end{pmatrix}$

**h** $\mathbf{b}+\mathbf{a}+\mathbf{c}=\begin{pmatrix}1\\4\end{pmatrix}+\begin{pmatrix}-3\\2\end{pmatrix}+\begin{pmatrix}-2\\-5\end{pmatrix}$

$=\begin{pmatrix}-2\\6\end{pmatrix}+\begin{pmatrix}-2\\-5\end{pmatrix}=\begin{pmatrix}-4\\1\end{pmatrix}$

**2** **a** $\mathbf{p}-\mathbf{q}=\begin{pmatrix}-4\\2\end{pmatrix}-\begin{pmatrix}-1\\-5\end{pmatrix}=\begin{pmatrix}-3\\7\end{pmatrix}$

**b** $\mathbf{q}-\mathbf{r}=\begin{pmatrix}-1\\-5\end{pmatrix}-\begin{pmatrix}3\\-2\end{pmatrix}=\begin{pmatrix}-4\\-3\end{pmatrix}$

**c** $\mathbf{p}+\mathbf{q}-\mathbf{r}$

$=\begin{pmatrix}-4\\2\end{pmatrix}+\begin{pmatrix}-1\\-5\end{pmatrix}-\begin{pmatrix}3\\-2\end{pmatrix}=\begin{pmatrix}-8\\-1\end{pmatrix}$

**d** $\mathbf{p}-\mathbf{q}-\mathbf{r}$

$=\begin{pmatrix}-4\\2\end{pmatrix}-\begin{pmatrix}-1\\-5\end{pmatrix}-\begin{pmatrix}3\\-2\end{pmatrix}=\begin{pmatrix}-6\\9\end{pmatrix}$

**e** $\mathbf{q}-\mathbf{r}-\mathbf{p}=\begin{pmatrix}-1\\-5\end{pmatrix}-\begin{pmatrix}3\\-2\end{pmatrix}-\begin{pmatrix}-4\\2\end{pmatrix}$
$=\begin{pmatrix}0\\-5\end{pmatrix}$

**f** $\mathbf{r}+\mathbf{q}-\mathbf{p}=\begin{pmatrix}3\\-2\end{pmatrix}+\begin{pmatrix}-1\\-5\end{pmatrix}-\begin{pmatrix}-4\\2\end{pmatrix}$
$=\begin{pmatrix}6\\-9\end{pmatrix}$

**3** **a** $\mathbf{a}+\mathbf{0}=\begin{pmatrix}a_1\\a_2\end{pmatrix}+\begin{pmatrix}0\\0\end{pmatrix}$
$=\begin{pmatrix}a_1+0\\a_2+0\end{pmatrix}=\begin{pmatrix}a_1\\a_2\end{pmatrix}=\mathbf{a}$

**b** $\mathbf{a}-\mathbf{a}=\begin{pmatrix}a_1\\a_2\end{pmatrix}-\begin{pmatrix}a_1\\a_2\end{pmatrix}$
$=\begin{pmatrix}a_1-a_1\\a_2-a_2\end{pmatrix}=\begin{pmatrix}0\\0\end{pmatrix}=\mathbf{0}$

**4** **a** $-3\mathbf{p}$
$=-3\begin{pmatrix}1\\5\end{pmatrix}$
$=\begin{pmatrix}-3\\-15\end{pmatrix}$

**b** $\frac{1}{2}\mathbf{q}$
$=\frac{1}{2}\begin{pmatrix}-2\\4\end{pmatrix}$
$=\begin{pmatrix}-1\\2\end{pmatrix}$

**c** $2\mathbf{p}+\mathbf{q}$
$=2\begin{pmatrix}1\\5\end{pmatrix}+\begin{pmatrix}-2\\4\end{pmatrix}$
$=\begin{pmatrix}2\\10\end{pmatrix}+\begin{pmatrix}-2\\4\end{pmatrix}$
$=\begin{pmatrix}0\\14\end{pmatrix}$

**d** $\mathbf{p}-2\mathbf{q}$
$=\begin{pmatrix}1\\5\end{pmatrix}-2\begin{pmatrix}-2\\4\end{pmatrix}$
$=\begin{pmatrix}1\\5\end{pmatrix}-\begin{pmatrix}-4\\8\end{pmatrix}$
$=\begin{pmatrix}5\\-3\end{pmatrix}$

**e** $\mathbf{p}-\frac{1}{2}\mathbf{r}=\begin{pmatrix}1\\5\end{pmatrix}-\begin{pmatrix}-\frac{3}{2}\\-\frac{1}{2}\end{pmatrix}$
$=\begin{pmatrix}\frac{5}{2}\\\frac{11}{2}\end{pmatrix}$

**f** $2\mathbf{p}+3\mathbf{r}=2\begin{pmatrix}1\\5\end{pmatrix}+3\begin{pmatrix}-3\\-1\end{pmatrix}$
$=\begin{pmatrix}2\\10\end{pmatrix}+\begin{pmatrix}-9\\-3\end{pmatrix}$
$=\begin{pmatrix}-7\\7\end{pmatrix}$

**g** $2\mathbf{q}-3\mathbf{r}=2\begin{pmatrix}-2\\4\end{pmatrix}-3\begin{pmatrix}-3\\-1\end{pmatrix}$
$=\begin{pmatrix}-4\\8\end{pmatrix}-\begin{pmatrix}-9\\-3\end{pmatrix}$
$=\begin{pmatrix}5\\11\end{pmatrix}$

**h** $2\mathbf{p}-\mathbf{q}+\frac{1}{3}\mathbf{r}=\begin{pmatrix}2\\10\end{pmatrix}-\begin{pmatrix}-2\\-4\end{pmatrix}+\begin{pmatrix}-1\\-\frac{1}{3}\end{pmatrix}$
$=\begin{pmatrix}3\\\frac{17}{3}\end{pmatrix}$

**5** **a**

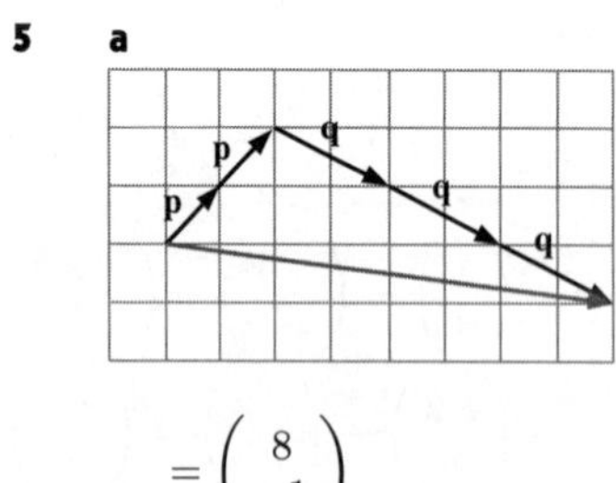


$=\begin{pmatrix}8\\-1\end{pmatrix}$

**b**

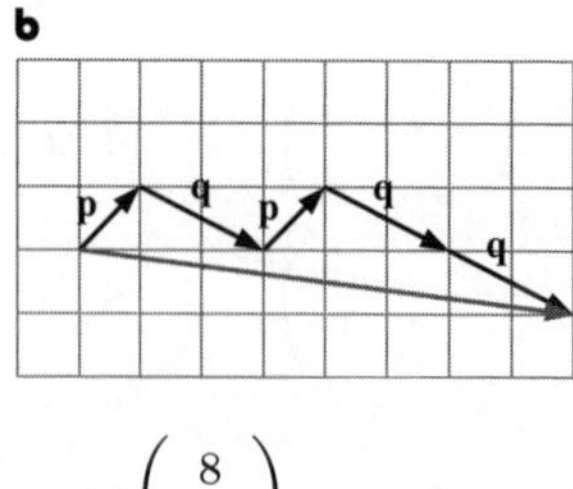


$=\begin{pmatrix}8\\-1\end{pmatrix}$

**c**

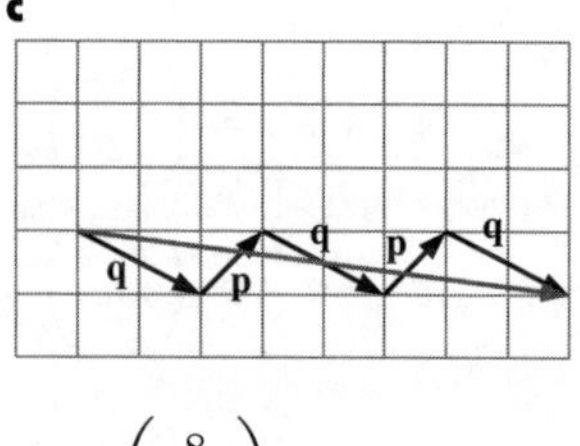


$=\begin{pmatrix}8\\-1\end{pmatrix}$

The vector expressions are equal, as each consists of 2 **p**s and 3 **q**s. Each expression is equal to $2\mathbf{p}+3\mathbf{q}$.

**6** **a** $|\mathbf{r}|=\sqrt{2^2+3^2}$
$=\sqrt{13}$ units

**b** $|\mathbf{s}|=\sqrt{(-1)^2+4^2}$
$=\sqrt{17}$ units

**c**
$$\begin{aligned} &\mathbf{r}+\mathbf{s} \\ &= \begin{pmatrix} 2 \\ 3 \end{pmatrix} + \begin{pmatrix} -1 \\ 4 \end{pmatrix} \\ &= \begin{pmatrix} 1 \\ 7 \end{pmatrix} \\ \therefore\ &|\mathbf{r}+\mathbf{s}| \\ &= \sqrt{1^2+7^2} \\ &= \sqrt{50} \text{ units} \\ &= 5\sqrt{2} \text{ units} \end{aligned}$$

**d**
$$\begin{aligned} &\mathbf{r}-\mathbf{s} \\ &= \begin{pmatrix} 2 \\ 3 \end{pmatrix} - \begin{pmatrix} -1 \\ 4 \end{pmatrix} \\ &= \begin{pmatrix} 3 \\ -1 \end{pmatrix} \\ \therefore\ &|\mathbf{r}-\mathbf{s}| \\ &= \sqrt{3^2+(-1)^2} \\ &= \sqrt{10} \text{ units} \end{aligned}$$

**e**
$$\begin{aligned} &\mathbf{s}-2\mathbf{r} \\ &= \begin{pmatrix} -1 \\ 4 \end{pmatrix} - \begin{pmatrix} 4 \\ 6 \end{pmatrix} \\ &= \begin{pmatrix} -5 \\ -2 \end{pmatrix} \\ \therefore\ &|\mathbf{s}-2\mathbf{r}| \\ &= \sqrt{(-5)^2+(-2)^2} \\ &= \sqrt{29} \text{ units} \end{aligned}$$

**7** **a**
$$\begin{aligned} |\mathbf{p}| &= \sqrt{1^2+3^2} \\ &= \sqrt{10} \text{ units} \end{aligned}$$

**b**
$$\begin{aligned} 2\mathbf{p} &= \begin{pmatrix} 2 \\ 6 \end{pmatrix} \\ \therefore\ |2\mathbf{p}| &= \sqrt{2^2+6^2} \\ &= \sqrt{4+36} \\ &= \sqrt{40} \\ &= 2\sqrt{10} \text{ units} \end{aligned}$$

**c**
$$\begin{aligned} -2\mathbf{p} &= \begin{pmatrix} -2 \\ -6 \end{pmatrix} \\ \therefore\ |-2\mathbf{p}| &= \sqrt{(-2)^2+(-6)^2} \\ &= \sqrt{4+36} \\ &= \sqrt{40} \\ &= 2\sqrt{10} \text{ units} \end{aligned}$$

**d**
$$\begin{aligned} 3\mathbf{p} &= \begin{pmatrix} 3 \\ 9 \end{pmatrix} \\ \therefore\ |3\mathbf{p}| &= \sqrt{3^2+9^2} \\ &= \sqrt{9+81} \\ &= \sqrt{90} \\ &= 3\sqrt{10} \text{ units} \end{aligned}$$

**e**
$$\begin{aligned} -3\mathbf{p} &= \begin{pmatrix} -3 \\ -9 \end{pmatrix} \\ \therefore\ |-3\mathbf{p}| &= \sqrt{(-3)^2+(-9)^2} \\ &= \sqrt{9+81} \\ &= \sqrt{90} \\ &= 3\sqrt{10} \text{ units} \end{aligned}$$

**f**
$$\begin{aligned} |\mathbf{q}| &= \sqrt{(-2)^2+4^2} \\ &= \sqrt{4+16} \\ &= \sqrt{20} \\ &= 2\sqrt{5} \text{ units} \end{aligned}$$

**g**
$$\begin{aligned} 4\mathbf{q} &= \begin{pmatrix} -8 \\ 16 \end{pmatrix} \\ \therefore\ |4\mathbf{q}| &= \sqrt{(-8)^2+16^2} \\ &= \sqrt{64+256} \\ &= \sqrt{320} \\ &= 8\sqrt{5} \text{ units} \end{aligned}$$

**h**
$$\begin{aligned} 4\mathbf{q} &= \begin{pmatrix} 8 \\ -16 \end{pmatrix} \\ \therefore\ |-4\mathbf{q}| &= \sqrt{8^2+(-16)^2} \\ &= \sqrt{64+256} \\ &= \sqrt{320} \\ &= 8\sqrt{5} \text{ units} \end{aligned}$$

**i**
$$\begin{aligned} \tfrac{1}{2}\mathbf{q} &= \begin{pmatrix} -1 \\ 2 \end{pmatrix} \\ \therefore\ \left|\tfrac{1}{2}\mathbf{q}\right| &= \sqrt{(-1)^2+2^2} \\ &= \sqrt{5} \text{ units} \end{aligned}$$

**j**
$$\begin{aligned} -\tfrac{1}{2}\mathbf{q} &= \begin{pmatrix} 1 \\ -2 \end{pmatrix} \\ \therefore\ \left|-\tfrac{1}{2}\mathbf{q}\right| &= \sqrt{1^2+(-2)^2} \\ &= \sqrt{5} \text{ units} \end{aligned}$$

**8**
$$\begin{aligned} k\mathbf{x} &= \mathbf{a} \\ \therefore\ k\begin{pmatrix} x_1 \\ x_2 \end{pmatrix} &= \begin{pmatrix} a_1 \\ a_2 \end{pmatrix} \\ \therefore\ kx_1 = a_1 &\quad \text{and} \quad kx_2 = a_2 \\ \therefore\ x_1 = \tfrac{1}{k}a_1 &\quad \text{and} \quad x_2 = \tfrac{1}{k}a_2 \\ \therefore\ \begin{pmatrix} x_1 \\ x_2 \end{pmatrix} &= \begin{pmatrix} \frac{1}{k}a_1 \\ \frac{1}{k}a_2 \end{pmatrix} = \tfrac{1}{k}\begin{pmatrix} a_1 \\ a_2 \end{pmatrix} \end{aligned}$$

and so $\mathbf{x} = \frac{1}{k}\mathbf{a}$

**9**
$$\begin{aligned} k\mathbf{v} &= \begin{pmatrix} kv_1 \\ kv_2 \end{pmatrix} \\ \therefore\ |k\mathbf{v}| &= \sqrt{(kv_1)^2+(kv_2)^2} \\ &= \sqrt{k^2{v_1}^2+k^2{v_2}^2} \\ &= \sqrt{k^2({v_1}^2+{v_2}^2)} \\ &= \sqrt{k^2}\sqrt{{v_1}^2+{v_2}^2} \\ &= |k|\sqrt{{v_1}^2+{v_2}^2} \\ &= |k|\,|\mathbf{v}| \end{aligned}$$

## EXERCISE 12F

**1** **a** $\overrightarrow{AB} = \begin{pmatrix} b_1 - a_1 \\ b_2 - a_2 \end{pmatrix} = \begin{pmatrix} 4-2 \\ 7-3 \end{pmatrix} = \begin{pmatrix} 2 \\ 4 \end{pmatrix}$

**b** $\overrightarrow{AB} = \begin{pmatrix} b_1 - a_1 \\ b_2 - a_2 \end{pmatrix} = \begin{pmatrix} 1-3 \\ 4--1 \end{pmatrix} = \begin{pmatrix} -2 \\ 5 \end{pmatrix}$

**c** $\overrightarrow{AB} = \begin{pmatrix} b_1 - a_1 \\ b_2 - a_2 \end{pmatrix} = \begin{pmatrix} 1--2 \\ 4-7 \end{pmatrix} = \begin{pmatrix} 3 \\ -3 \end{pmatrix}$

**d** $\overrightarrow{AB} = \begin{pmatrix} b_1 - a_1 \\ b_2 - a_2 \end{pmatrix} = \begin{pmatrix} 3-2 \\ 0-5 \end{pmatrix} = \begin{pmatrix} 1 \\ -5 \end{pmatrix}$

**e** $\overrightarrow{AB} = \begin{pmatrix} b_1 - a_1 \\ b_2 - a_2 \end{pmatrix} = \begin{pmatrix} 6-0 \\ -1-4 \end{pmatrix} = \begin{pmatrix} 6 \\ -5 \end{pmatrix}$

**f** $\overrightarrow{AB} = \begin{pmatrix} b_1 - a_1 \\ b_2 - a_2 \end{pmatrix} = \begin{pmatrix} 0--1 \\ 0--3 \end{pmatrix} = \begin{pmatrix} 1 \\ 3 \end{pmatrix}$

**2** **a** Let B have coordinates $(b_1, b_2)$.

$\therefore \overrightarrow{AB} = \begin{pmatrix} b_1 - 1 \\ b_2 - 4 \end{pmatrix}$

$\therefore \begin{pmatrix} b_1 - 1 \\ b_2 - 4 \end{pmatrix} = \begin{pmatrix} 3 \\ -2 \end{pmatrix}$

$\therefore b_1 - 1 = 3$ and $b_2 - 4 = -2$

$\therefore b_1 = 4$ and $b_2 = 2$

$\therefore$ B has coordinates (4, 2).

**b** Let C have coordinates $(c_1, c_2)$.

$\therefore \overrightarrow{CA} = \begin{pmatrix} 1 - c_1 \\ 4 - c_2 \end{pmatrix}$

$\therefore \begin{pmatrix} 1 - c_1 \\ 4 - c_2 \end{pmatrix} = \begin{pmatrix} -1 \\ 2 \end{pmatrix}$

$\therefore 1 - c_1 = -1$ and $4 - c_2 = 2$

$\therefore c_1 = 2$ and $c_2 = 2$

$\therefore$ C has coordinates (2, 2).

**3** **a** $\overrightarrow{PC} = \begin{pmatrix} 1-(-1) \\ 2-1 \end{pmatrix} = \begin{pmatrix} 2 \\ 1 \end{pmatrix}$

**b** Let Q have coordinates $(q_1, q_2)$.

$\therefore \overrightarrow{CQ} = \begin{pmatrix} q_1 - 1 \\ q_2 - 2 \end{pmatrix}$

But $\overrightarrow{CQ} = \overrightarrow{PC}$

$\therefore \begin{pmatrix} q_1 - 1 \\ q_2 - 2 \end{pmatrix} = \begin{pmatrix} 2 \\ 1 \end{pmatrix}$

$\therefore q_1 - 1 = 2$ and $q_2 - 2 = 1$

$\therefore q_1 = 3$ and $q_2 = 3$

$\therefore$ Q has coordinates (3, 3).

**4** **a** $\overrightarrow{AB} = \begin{pmatrix} 6-1 \\ 5-4 \end{pmatrix} = \begin{pmatrix} 5 \\ 1 \end{pmatrix}$

**b** $\overrightarrow{CD} = -\overrightarrow{AB} = -\begin{pmatrix} 5 \\ 1 \end{pmatrix} = \begin{pmatrix} -5 \\ -1 \end{pmatrix}$

**c** Let D have coordinates $(d_1, d_2)$.

$\therefore \overrightarrow{CD} = \begin{pmatrix} d_1 - 4 \\ d_2 - (-1) \end{pmatrix} = \begin{pmatrix} d_1 - 4 \\ d_2 + 1 \end{pmatrix}$

But $\overrightarrow{CD} = \begin{pmatrix} -5 \\ -1 \end{pmatrix}$

$\therefore \begin{pmatrix} d_1 - 4 \\ d_2 + 1 \end{pmatrix} = \begin{pmatrix} -5 \\ -1 \end{pmatrix}$

$\therefore d_1 - 4 = -5$ and $d_2 + 1 = -1$

$\therefore d_1 = -1$ and $d_2 = -2$

$\therefore$ D has coordinates (−1, −2).

**5** **a** $\overrightarrow{AB} = \begin{pmatrix} 3-(-1) \\ k-3 \end{pmatrix} = \begin{pmatrix} 4 \\ k-3 \end{pmatrix}$

Since A and B are 5 units apart,

$|\overrightarrow{AB}| = 5$ units and $|\overrightarrow{AB}| = \sqrt{4^2 + (k-3)^2} = \sqrt{16 + (k-3)^2}$

**b** $|\overrightarrow{AB}| = 5$

$\therefore \sqrt{16+(k-3)^2} = 5$

$\therefore 16 + k^2 - 6k + 9 = 25$

$\therefore k^2 - 6k = 0$

$\therefore k(k-6) = 0$

$\therefore k = 0$ or $k - 6 = 0$

$\therefore k = 0$ or $6$

**c**

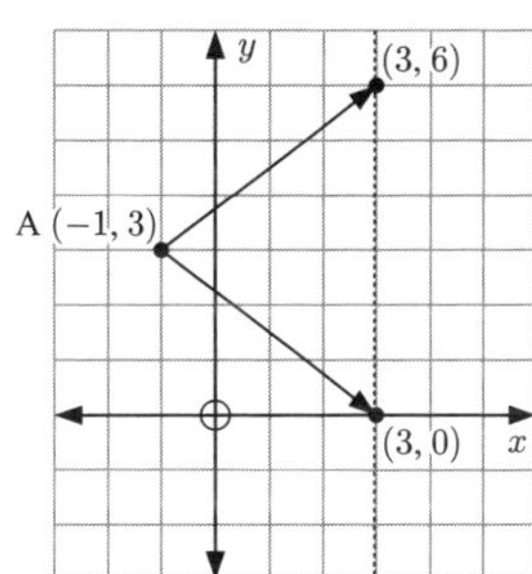


**6** **a** $\overrightarrow{AB} = \begin{pmatrix} 3-1 \\ 5-2 \end{pmatrix} = \begin{pmatrix} 2 \\ 3 \end{pmatrix}$

$\overrightarrow{AC} = \begin{pmatrix} 4-1 \\ -1-2 \end{pmatrix} = \begin{pmatrix} 3 \\ -3 \end{pmatrix}$

**b** $\overrightarrow{BC} = \overrightarrow{BA} + \overrightarrow{AC} = -\overrightarrow{AB} + \overrightarrow{AC}$

**c** $\overrightarrow{BC} = -\overrightarrow{AB} + \overrightarrow{AC} = -\begin{pmatrix} 2 \\ 3 \end{pmatrix} + \begin{pmatrix} 3 \\ -3 \end{pmatrix} = \begin{pmatrix} -2 \\ -3 \end{pmatrix} + \begin{pmatrix} 3 \\ -3 \end{pmatrix} = \begin{pmatrix} -2+3 \\ -3-3 \end{pmatrix} = \begin{pmatrix} 1 \\ -6 \end{pmatrix}$

**d** $\overrightarrow{BC} = \begin{pmatrix} 4-3 \\ -1-5 \end{pmatrix} = \begin{pmatrix} 1 \\ -6 \end{pmatrix}$ ✓

**7** **a** $\overrightarrow{AC} = \overrightarrow{AB} + \overrightarrow{BC} = -\overrightarrow{BA} + \overrightarrow{BC} = -\begin{pmatrix} 2 \\ -3 \end{pmatrix} + \begin{pmatrix} -3 \\ 1 \end{pmatrix} = \begin{pmatrix} -5 \\ 4 \end{pmatrix}$

**b** $\overrightarrow{CB} = \overrightarrow{CA} + \overrightarrow{AB} = \begin{pmatrix} 2 \\ -1 \end{pmatrix} + \begin{pmatrix} -1 \\ 3 \end{pmatrix} = \begin{pmatrix} 1 \\ 2 \end{pmatrix}$

**c** $\overrightarrow{SP} = \overrightarrow{SR} + \overrightarrow{RQ} + \overrightarrow{QP} = -\overrightarrow{RS} + \overrightarrow{RQ} - \overrightarrow{PQ} = -\begin{pmatrix} -3 \\ 2 \end{pmatrix} + \begin{pmatrix} 2 \\ 1 \end{pmatrix} - \begin{pmatrix} -1 \\ 4 \end{pmatrix} = \begin{pmatrix} 6 \\ -5 \end{pmatrix}$

**8** **a** M is $\left(\dfrac{3+-1}{2}, \dfrac{6+2}{2}\right)$

$\therefore$ M is $(1, 4)$

**b** $\overrightarrow{CA} = \begin{pmatrix} 3--4 \\ 6-1 \end{pmatrix} = \begin{pmatrix} 7 \\ 5 \end{pmatrix}$

$\overrightarrow{CM} = \begin{pmatrix} 1--4 \\ 4-1 \end{pmatrix} = \begin{pmatrix} 5 \\ 3 \end{pmatrix}$

$\overrightarrow{CB} = \begin{pmatrix} -1--4 \\ 2-1 \end{pmatrix} = \begin{pmatrix} 3 \\ 1 \end{pmatrix}$

**c** $\frac{1}{2}\overrightarrow{CA} + \frac{1}{2}\overrightarrow{CB} = \frac{1}{2}\begin{pmatrix} 7 \\ 5 \end{pmatrix} + \frac{1}{2}\begin{pmatrix} 3 \\ 1 \end{pmatrix} = \begin{pmatrix} 5 \\ 3 \end{pmatrix}$ which is $\overrightarrow{CM}$

## EXERCISE 12G

**1** **a**

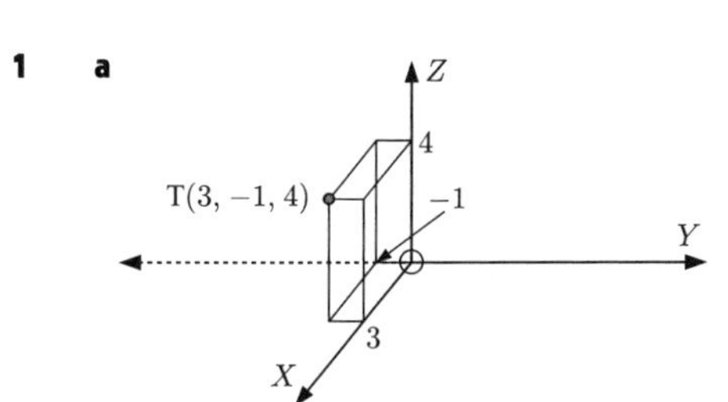


**b** $\overrightarrow{OT} = \begin{pmatrix} 3 \\ -1 \\ 4 \end{pmatrix}$

**c** $OT = \sqrt{(3-0)^2 + (-1-0)^2 + (4-0)^2} = \sqrt{9+1+16} = \sqrt{26}$ units

**2** **a**

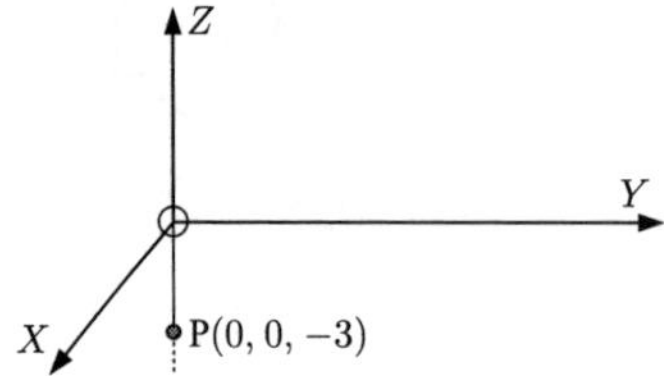


$\text{OP} = \sqrt{0^2 + 0^2 + (-3)^2} = 3$ units

**b**

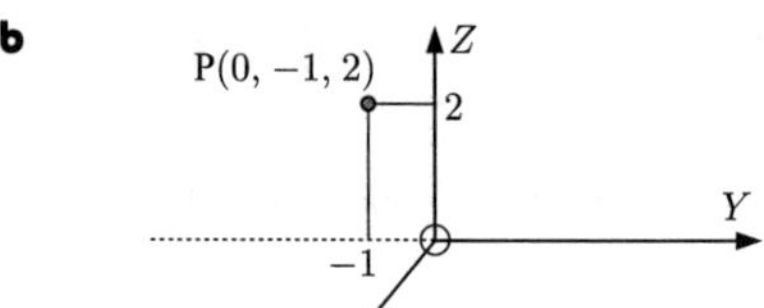


$\text{OP} = \sqrt{0^2 + (-1)^2 + 2^2} = \sqrt{5}$ units

**c**

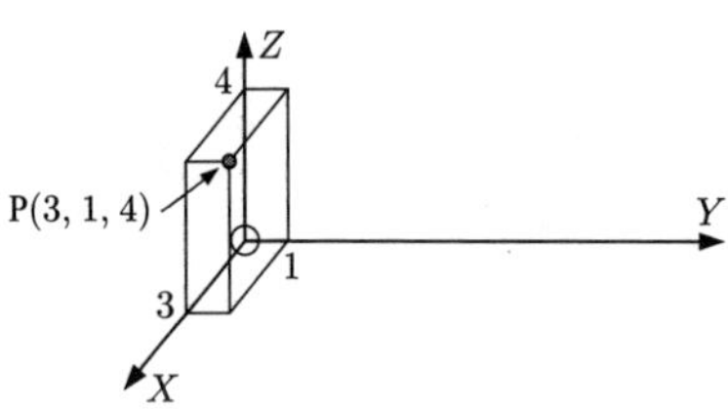


$\text{OP} = \sqrt{3^2 + 1^2 + 4^2} = \sqrt{26}$ units

**d**

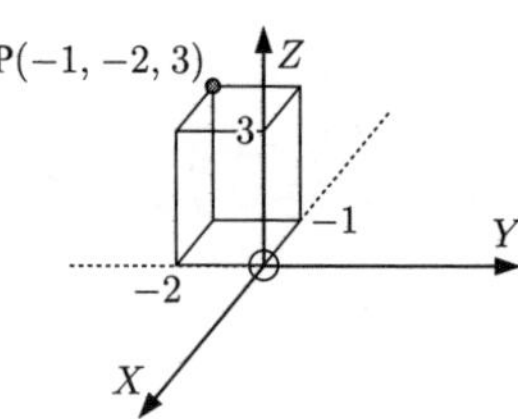


$\text{OP} = \sqrt{(-1)^2 + (-2)^2 + 3^2} = \sqrt{14}$ units

**3** **a** $\overrightarrow{\text{AB}} = \begin{pmatrix} 1-(-3) \\ 0-1 \\ -1-2 \end{pmatrix} = \begin{pmatrix} 4 \\ -1 \\ -3 \end{pmatrix}$, $\quad \overrightarrow{\text{BA}} = \begin{pmatrix} -3-1 \\ 1-0 \\ 2-(-1) \end{pmatrix} = \begin{pmatrix} -4 \\ 1 \\ 3 \end{pmatrix}$

**b** $|\overrightarrow{\text{AB}}| = \sqrt{4^2 + (-1)^2 + (-3)^2} = \sqrt{26}$ units, $\quad |\overrightarrow{\text{BA}}| = \sqrt{(-4)^2 + 1^2 + 3^2} = \sqrt{26}$ units

**4** $\overrightarrow{\text{OA}} = \begin{pmatrix} 3 \\ 1 \\ 0 \end{pmatrix}$, $\quad \overrightarrow{\text{OB}} = \begin{pmatrix} -1 \\ 1 \\ 2 \end{pmatrix}$, $\quad \overrightarrow{\text{AB}} = \begin{pmatrix} -1-3 \\ 1-1 \\ 2-0 \end{pmatrix} = \begin{pmatrix} -4 \\ 0 \\ 2 \end{pmatrix}$

**5** **a** The position vector of M relative to N

$$= \overrightarrow{\text{NM}} = \begin{pmatrix} 4-(-1) \\ -2-2 \\ -1-0 \end{pmatrix} = \begin{pmatrix} 5 \\ -4 \\ -1 \end{pmatrix}$$

**b** The position vector of N relative to M

$$= \overrightarrow{\text{MN}} = \begin{pmatrix} -1-4 \\ 2-(-2) \\ 0-(-1) \end{pmatrix} = \begin{pmatrix} -5 \\ 4 \\ 1 \end{pmatrix}$$

**c** $\text{MN} = \sqrt{(-5)^2 + 4^2 + 1^2} = \sqrt{25+16+1} = \sqrt{42}$ units

**6** **a** The position vector of A relative to O

$$= \overrightarrow{\text{OA}} = \begin{pmatrix} -1 \\ 2 \\ 5 \end{pmatrix}$$

$$\begin{aligned} \therefore \quad \text{OA} &= \sqrt{(-1)^2 + 2^2 + 5^2} \\ &= \sqrt{1+4+25} \\ &= \sqrt{30} \text{ units} \end{aligned}$$

**b** The position vector of B relative to A

$$= \overrightarrow{\text{AB}} = \begin{pmatrix} 2-(-1) \\ 0-2 \\ 3-5 \end{pmatrix} = \begin{pmatrix} 3 \\ -2 \\ -2 \end{pmatrix}$$

$$\begin{aligned} \therefore \quad \text{AB} &= \sqrt{3^2 + (-2)^2 + (-2)^2} \\ &= \sqrt{9+4+4} \\ &= \sqrt{17} \text{ units} \end{aligned}$$

**c** The position vector of C relative to A

$$= \overrightarrow{\text{AC}} = \begin{pmatrix} -3-(-1) \\ 1-2 \\ 0-5 \end{pmatrix} = \begin{pmatrix} -2 \\ -1 \\ -5 \end{pmatrix}$$

$$\begin{aligned} \therefore \quad \text{AC} &= \sqrt{(-2)^2 + (-1)^2 + (-5)^2} \\ &= \sqrt{4+1+25} \\ &= \sqrt{30} \text{ units} \end{aligned}$$

**d** The position vector of B relative to C

$$= \overrightarrow{\text{CB}} = \begin{pmatrix} 2-(-3) \\ 0-1 \\ 3-0 \end{pmatrix} = \begin{pmatrix} 5 \\ -1 \\ 3 \end{pmatrix}$$

$$\begin{aligned} \therefore \quad \text{CB} &= \sqrt{5^2 + (-1)^2 + 3^2} \\ &= \sqrt{25+1+9} \\ &= \sqrt{35} \text{ units} \end{aligned}$$

**e** Triangle ABC has $\text{AC} = \sqrt{30}$ units, $\text{BC} = \sqrt{35}$ units, and $\text{AB} = \sqrt{17}$ units.
All the side lengths are different, and $\left(\sqrt{17}\right)^2 + \left(\sqrt{30}\right)^2 \neq \left(\sqrt{35}\right)^2$.
$\therefore$ triangle ABC is scalene, and not right angled.

**7**

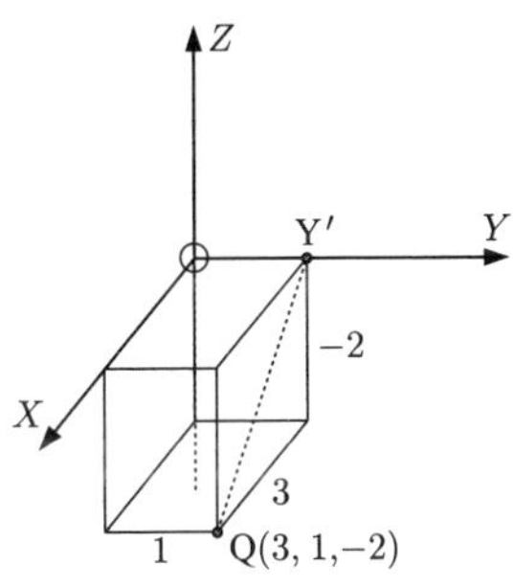


**a** The distance from Q to the $Y$-axis is the distance from Q to Y′(0, 1, 0).

$$\therefore \quad QY' = \sqrt{(3-0)^2+(1-1)^2+(-2-0)^2} = \sqrt{9+4} = \sqrt{13} \text{ units}$$

**b** The distance from Q to the origin is

$$QO = \sqrt{(3-0)^2+(1-0)^2+(-2-0)^2} = \sqrt{9+1+4} = \sqrt{14} \text{ units}$$

**c** The distance from Q to the $ZOY$ plane is the distance from Q to $(0, 1, -2)$, which is 3 units.

**8** P(0, 4, 4), Q(2, 6, 5), R(1, 4, 3)

$$PQ = \sqrt{(2-0)^2+(6-4)^2+(5-4)^2} = \sqrt{4+4+1} = 3$$

$$PR = \sqrt{(1-0)^2+(4-4)^2+(3-4)^2} = \sqrt{1+0+1} = \sqrt{2}$$

$$QR = \sqrt{(1-2)^2+(4-6)^2+(3-5)^2} = \sqrt{1+4+4} = 3$$

$\therefore$ PQ = QR and so △PQR is isosceles.

**9** **a** A(0, 0, 3), B(2, 8, 1), C(−9, 6, 18)

$$AB = \sqrt{(2-0)^2+(8-0)^2+(1-3)^2} = \sqrt{4+64+4} = \sqrt{72}$$

$$AC = \sqrt{(-9-0)^2+(6-0)^2+(18-3)^2} = \sqrt{81+36+225} = \sqrt{342}$$

$$BC = \sqrt{(-9-2)^2+(6-8)^2+(18-1)^2} = \sqrt{121+4+289} = \sqrt{414}$$

Since $BC^2 = AB^2 + AC^2$, △ABC is right angled.

**b** A(1, 0, −3), B(2, 2, 0), C(4, 6, 6)

$$AB = \sqrt{(2-1)^2+(2-0)^2+(0--3)^2} = \sqrt{1+4+9} = \sqrt{14}$$

$$AC = \sqrt{(4-1)^2+(6-0)^2+(6--3)^2} = \sqrt{9+36+81} = \sqrt{126} = 3\sqrt{14}$$

$$BC = \sqrt{(4-2)^2+(6-2)^2+(6-0)^2} = \sqrt{4+16+36} = \sqrt{56} = 2\sqrt{14}$$

Since AB + BC = AC, the points A, B, and C lie on a straight line, so they do not form a triangle.

**10** **a**
$$\overrightarrow{AB} = \begin{pmatrix} 6-5 \\ 12-6 \\ 9-(-2) \end{pmatrix} = \begin{pmatrix} 1 \\ 6 \\ 11 \end{pmatrix}$$

and so $|\overrightarrow{AB}| = \sqrt{1^2+6^2+11^2} = \sqrt{1+36+121} = \sqrt{158}$ units

$$\overrightarrow{BC} = \begin{pmatrix} 2-6 \\ 4-12 \\ 2-9 \end{pmatrix} = \begin{pmatrix} -4 \\ -8 \\ -7 \end{pmatrix}$$

and so $|\overrightarrow{BC}| = \sqrt{(-4)^2+(-8)^2+(-7)^2} = \sqrt{16+64+49} = \sqrt{129}$ units

$$\overrightarrow{AC} = \begin{pmatrix} 2-5 \\ 4-6 \\ 2-(-2) \end{pmatrix} = \begin{pmatrix} -3 \\ -2 \\ 4 \end{pmatrix}$$

and so $|\overrightarrow{AC}| = \sqrt{(-3)^2+(-2)^2+4^2} = \sqrt{9+4+16} = \sqrt{29}$ units

Now, $\left(\sqrt{29}\right)^2 + \left(\sqrt{129}\right)^2 = 29+129 = 158 = \left(\sqrt{158}\right)^2$

So, $AC^2 + BC^2 = AB^2$

$\therefore$ triangle ABC is right angled with the right angle at C.

**b** 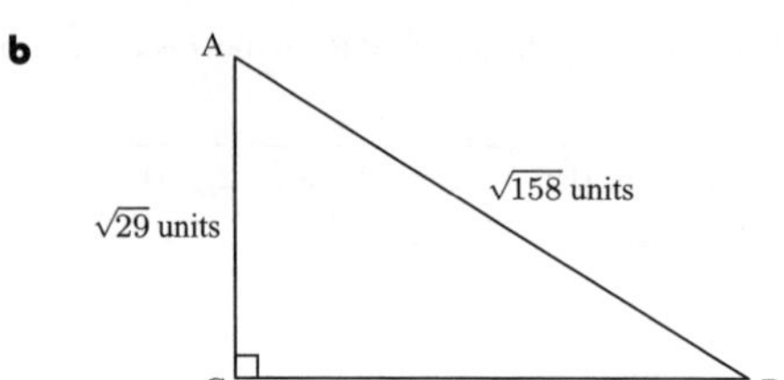


$$\begin{aligned}\text{Area} &= \tfrac{1}{2} \times \text{base} \times \text{height} \\ &= \tfrac{1}{2} \times \sqrt{29} \times \sqrt{129} \\ &\approx 30.6 \text{ units}^2\end{aligned}$$

**11** B; A $(-2, 1, 3)$; C$(-1, 2, 4)$

If B is $(a, b, c)$ then $\dfrac{a-2}{2} = -1, \quad \dfrac{b+1}{2} = 2, \quad \dfrac{c+3}{2} = 4$

$\therefore \quad a = 0, \quad b = 3, \quad c = 5$

$\therefore \quad$ B is $(0, 3, 5)$

$$\begin{aligned} r = \text{AC} &= \sqrt{(-1 - -2)^2 + (2-1)^2 + (4-3)^2} \\ &= \sqrt{1+1+1} \\ &= \sqrt{3} \text{ units}\end{aligned}$$

**12** **a** $(0, y, 0)$ for any $y$

**b** The distance between $(0, y, 0)$ and B$(-1, -1, 2)$ is $\sqrt{(-1)^2 + (-1-y)^2 + 2^2}$.

$\therefore \quad \sqrt{1 + (y+1)^2 + 4} = \sqrt{14}$

$\therefore \quad (y+1)^2 = 9$

$\therefore \quad y + 1 = \pm 3$

$\therefore \quad y = -1 \pm 3$

$\therefore \quad y = -4$ or $2 \qquad \therefore \quad$ the two points are $(0, -4, 0)$ and $(0, 2, 0)$.

**13** **a** $\begin{pmatrix} a-4 \\ b-3 \\ c+2 \end{pmatrix} = \begin{pmatrix} 1 \\ 3 \\ -4 \end{pmatrix}$

$\therefore \quad \begin{cases} a - 4 = 1 \\ b - 3 = 3 \\ c + 2 = -4 \end{cases}$

$\therefore \quad a = 5, \quad b = 6, \quad c = -6$

**b** $\begin{pmatrix} a-5 \\ b-2 \\ c+3 \end{pmatrix} = \begin{pmatrix} 3-a \\ 2-b \\ 5-c \end{pmatrix}$

$\therefore \quad \begin{cases} a - 5 = 3 - a \\ b - 2 = 2 - b \\ c + 3 = 5 - c \end{cases}$

$\therefore \quad 2a = 8, \quad 2b = 4, \quad 2c = 2$

$\therefore \quad a = 4, \quad b = 2, \quad c = 1$

**14** **a** length $= 1$

$\therefore \quad \sqrt{\left(\tfrac{1}{4}\right) + k^2 + \tfrac{1}{16}} = 1$

$\therefore \quad \sqrt{k^2 + \tfrac{5}{16}} = 1$

$\therefore \quad k^2 = \tfrac{11}{16}$

$\therefore \quad k = \pm\tfrac{\sqrt{11}}{4}$

**b** length $= 1$

$\therefore \quad \sqrt{k^2 + \tfrac{4}{9} + \tfrac{1}{9}} = 1$

$\therefore \quad \sqrt{k^2 + \tfrac{5}{9}} = 1$

$\therefore \quad k^2 = \tfrac{4}{9}$

$\therefore \quad k = \pm\tfrac{2}{3}$

**15** A$(-1, 3, 4)$, B$(2, 5, -1)$, C$(-1, 2, -2)$, D$(r, s, t)$

**a** If $\overrightarrow{\text{AC}} = \overrightarrow{\text{BD}}$ then $\begin{pmatrix} -1-(-1) \\ 2-3 \\ -2-4 \end{pmatrix} = \begin{pmatrix} r-2 \\ s-5 \\ t+1 \end{pmatrix}$

$\therefore \quad r - 2 = 0, \quad s - 5 = -1,$ and $t + 1 = -6 \qquad \therefore \quad r = 2, \quad s = 4,$ and $t = -7$

**b** If $\overrightarrow{\text{AB}} = \overrightarrow{\text{DC}}$ then $\begin{pmatrix} 2-(-1) \\ 5-3 \\ -1-4 \end{pmatrix} = \begin{pmatrix} -1-r \\ 2-s \\ -2-t \end{pmatrix}$

$\therefore \quad -1 - r = 3, \quad 2 - s = 2,$ and $-2 - t = -5 \qquad \therefore \quad r = -4, \quad s = 0,$ and $t = 3$

**16** **a** $\overrightarrow{AB} = \begin{pmatrix} 3-1 \\ -3-2 \\ 2-3 \end{pmatrix} = \begin{pmatrix} 2 \\ -5 \\ -1 \end{pmatrix}$ and $\overrightarrow{DC} = \begin{pmatrix} 7-5 \\ -4-1 \\ 5-6 \end{pmatrix} = \begin{pmatrix} 2 \\ -5 \\ -1 \end{pmatrix}$.

**b** ABCD is a parallelogram since its opposite sides are parallel and equal in length.

**17** **a** Suppose S is at $(x, y, z)$. $\overrightarrow{PQ} = \overrightarrow{SR}$ {opposite sides are parallel and equal in length}

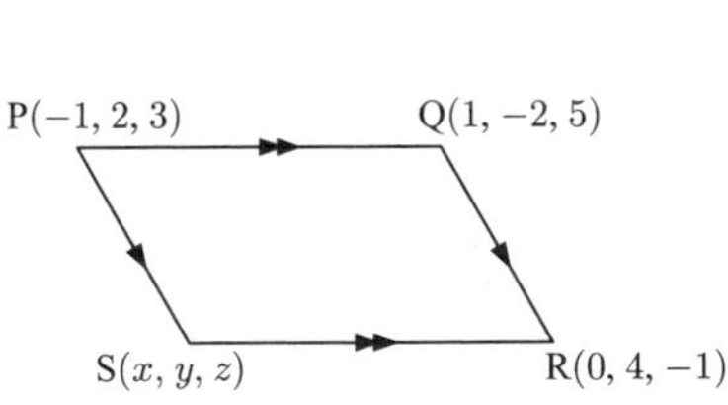


$$\therefore \begin{pmatrix} 1-(-1) \\ -2-2 \\ 5-3 \end{pmatrix} = \begin{pmatrix} 0-x \\ 4-y \\ -1-z \end{pmatrix}$$

$$\therefore \begin{pmatrix} 2 \\ -4 \\ 2 \end{pmatrix} = \begin{pmatrix} -x \\ 4-y \\ -1-z \end{pmatrix}$$

$\therefore\ -x = 2 \qquad 4-y=-4 \qquad -1-z=2$

$\therefore\ x=-2 \qquad y=8 \qquad z=-3$

$\therefore$ S is at $(-2, 8, -3)$.

**b** The midpoint of [PR] is $\left(\frac{-1+0}{2}, \frac{2+4}{2}, \frac{3+(-1)}{2}\right)$ which is $(-\frac{1}{2}, 3, 1)$.

The midpoint of [QS] is $\left(\frac{1+(-2)}{2}, \frac{-2+8}{2}, \frac{5+(-3)}{2}\right)$ which is $(-\frac{1}{2}, 3, 1)$.

So, [PR] and [QS] have the same midpoint. ✓

## EXERCISE 12H

**1** **a** $2\mathbf{x} = \mathbf{q}$

$\therefore\ \frac{1}{2}(2\mathbf{x}) = \frac{1}{2}\mathbf{q}$

$\therefore\ \mathbf{x} = \frac{1}{2}\mathbf{q}$

**b** $\frac{1}{2}\mathbf{x} = \mathbf{n}$

$\therefore\ 2(\frac{1}{2}\mathbf{x}) = 2\mathbf{n}$

$\therefore\ \mathbf{x} = 2\mathbf{n}$

**c** $-3\mathbf{x} = \mathbf{p}$

$\therefore\ 3\mathbf{x} = -\mathbf{p}$

$\therefore\ \frac{1}{3}(3\mathbf{x}) = -\frac{1}{3}\mathbf{p}$

$\therefore\ \mathbf{x} = -\frac{1}{3}\mathbf{p}$

**d** $\mathbf{q} + 2\mathbf{x} = \mathbf{r}$

$\therefore\ 2\mathbf{x} = \mathbf{r} - \mathbf{q}$

$\therefore\ \mathbf{x} = \frac{1}{2}(\mathbf{r} - \mathbf{q})$

**e** $4\mathbf{s} - 5\mathbf{x} = \mathbf{t}$

$\therefore\ -5\mathbf{x} = \mathbf{t} - 4\mathbf{s}$

$\therefore\ 5\mathbf{x} = 4\mathbf{s} - \mathbf{t}$

$\therefore\ \mathbf{x} = \frac{1}{5}(4\mathbf{s} - \mathbf{t})$

**f** $4\mathbf{m} - \frac{1}{3}\mathbf{x} = \mathbf{n}$

$\therefore\ 4\mathbf{m} - \mathbf{n} = \frac{1}{3}\mathbf{x}$

$\therefore\ \mathbf{x} = 3(4\mathbf{m} - \mathbf{n})$

**2** **a** $2\mathbf{a} + \mathbf{x} = \mathbf{b}$

$\therefore\ \mathbf{x} = \mathbf{b} - 2\mathbf{a}$

$$= \begin{pmatrix} 2 \\ -2 \\ 1 \end{pmatrix} - 2\begin{pmatrix} -1 \\ 2 \\ 3 \end{pmatrix}$$

$$= \begin{pmatrix} 2 \\ -2 \\ 1 \end{pmatrix} - \begin{pmatrix} -2 \\ 4 \\ 6 \end{pmatrix}$$

$$\therefore\ \mathbf{x} = \begin{pmatrix} 4 \\ -6 \\ -5 \end{pmatrix}$$

**b** $3\mathbf{x} - \mathbf{a} = 2\mathbf{b}$

$\therefore\ 3\mathbf{x} = \mathbf{a} + 2\mathbf{b}$

$\therefore\ \mathbf{x} = \frac{1}{3}(\mathbf{a} + 2\mathbf{b})$

$$= \frac{1}{3}\left[\begin{pmatrix} -1 \\ 2 \\ 3 \end{pmatrix} + 2\begin{pmatrix} 2 \\ -2 \\ 1 \end{pmatrix}\right]$$

$$= \frac{1}{3}\left[\begin{pmatrix} -1 \\ 2 \\ 3 \end{pmatrix} + \begin{pmatrix} 4 \\ -4 \\ 2 \end{pmatrix}\right]$$

$$\therefore\ \mathbf{x} = \frac{1}{3}\begin{pmatrix} 3 \\ -2 \\ 5 \end{pmatrix} = \begin{pmatrix} 1 \\ -\frac{2}{3} \\ \frac{5}{3} \end{pmatrix}$$

**c** $2\mathbf{b} - 2\mathbf{x} = -\mathbf{a}$

$\therefore\ \mathbf{a} + 2\mathbf{b} = 2\mathbf{x}$

$\therefore\ \mathbf{x} = \frac{1}{2}(\mathbf{a} + 2\mathbf{b}) = \frac{1}{2}\begin{pmatrix} 3 \\ -2 \\ 5 \end{pmatrix}$ {using **b**}

$$= \begin{pmatrix} \frac{3}{2} \\ -1 \\ \frac{5}{2} \end{pmatrix}$$

**3** $\overrightarrow{AB} = \overrightarrow{AO} + \overrightarrow{OB} = -\overrightarrow{OA} + \overrightarrow{OB} = -\begin{pmatrix} -2 \\ -1 \\ 1 \end{pmatrix} + \begin{pmatrix} 1 \\ 3 \\ -1 \end{pmatrix} = \begin{pmatrix} 2 \\ 1 \\ -1 \end{pmatrix} + \begin{pmatrix} 1 \\ 3 \\ -1 \end{pmatrix} = \begin{pmatrix} 3 \\ 4 \\ -2 \end{pmatrix}$

$\therefore \quad |\overrightarrow{AB}| = \sqrt{3^2 + 4^2 + (-2)^2} = \sqrt{9 + 16 + 4} = \sqrt{29}$ units

**4** **a** $\overrightarrow{AB} = \begin{pmatrix} 3-(-1) \\ -2-3 \\ 1-2 \end{pmatrix} = \begin{pmatrix} 4 \\ -5 \\ -1 \end{pmatrix}$

$= 4\mathbf{i} - 5\mathbf{j} - \mathbf{k}$

**b** $|\overrightarrow{AB}| = \sqrt{4^2 + (-5)^2 + (-1)^2}$

$= \sqrt{16 + 25 + 1}$

$= \sqrt{42}$ units

**5** **a** $|\mathbf{a}| = \sqrt{1^2 + 0^2 + 3^2}$

$= \sqrt{1 + 9}$

$= \sqrt{10}$ units

**b** $|\mathbf{b}| = \sqrt{(-2)^2 + 1^2 + 1^2}$

$= \sqrt{4 + 1 + 1}$

$= \sqrt{6}$ units

**c** $2|\mathbf{a}| = 2\sqrt{10}$ units {using part **a**}

**d** $2\mathbf{a} = 2\begin{pmatrix} 1 \\ 0 \\ 3 \end{pmatrix} = \begin{pmatrix} 2 \\ 0 \\ 6 \end{pmatrix}$

$\therefore \quad |2\mathbf{a}| = \sqrt{2^2 + 0^2 + 6^2}$

$= \sqrt{4 + 36}$

$= \sqrt{40}$

$= \sqrt{4}\sqrt{10}$

$= 2\sqrt{10}$ units

**e** $-3|\mathbf{b}| = -3\sqrt{6}$ units {using part **b**}

**f** $-3\mathbf{b} = -3\begin{pmatrix} -2 \\ 1 \\ 1 \end{pmatrix} = \begin{pmatrix} 6 \\ -3 \\ -3 \end{pmatrix}$

$\therefore \quad |-3\mathbf{b}| = \sqrt{6^2 + (-3)^2 + (-3)^2}$

$= \sqrt{36 + 9 + 9}$

$= \sqrt{54}$

$= \sqrt{9}\sqrt{6}$

$= 3\sqrt{6}$ units

**g** $\mathbf{a} + \mathbf{b} = \begin{pmatrix} 1 \\ 0 \\ 3 \end{pmatrix} + \begin{pmatrix} -2 \\ 1 \\ 1 \end{pmatrix}$

$= \begin{pmatrix} 1-2 \\ 0+1 \\ 3+1 \end{pmatrix}$

$= \begin{pmatrix} -1 \\ 1 \\ 4 \end{pmatrix}$

$\therefore \quad |\mathbf{a} + \mathbf{b}| = \sqrt{(-1)^2 + 1^2 + 4^2}$

$= \sqrt{1 + 1 + 16}$

$= \sqrt{18}$

$= \sqrt{9}\sqrt{2}$ units $= 3\sqrt{2}$ units

**h** $\mathbf{a} - \mathbf{b} = \begin{pmatrix} 1 \\ 0 \\ 3 \end{pmatrix} - \begin{pmatrix} -2 \\ 1 \\ 1 \end{pmatrix}$

$= \begin{pmatrix} 1-(-2) \\ 0-1 \\ 3-1 \end{pmatrix}$

$= \begin{pmatrix} 3 \\ -1 \\ 2 \end{pmatrix}$

$\therefore \quad |\mathbf{a} - \mathbf{b}| = \sqrt{3^2 + (-1)^2 + 2^2}$

$= \sqrt{9 + 1 + 4}$

$= \sqrt{14}$ units

**6** $\overrightarrow{AC} = \overrightarrow{AB} + \overrightarrow{BC}$

$= (\mathbf{i} - \mathbf{j} + \mathbf{k}) + (-2\mathbf{i} + \mathbf{j} - 3\mathbf{k})$

$= -\mathbf{i} - 2\mathbf{k}$

**7** A(2, 1, −2), B(0, 3, −4), C(1, −2, 1), D(−2, −3, 2)

$\overrightarrow{AC} = \begin{pmatrix} 1-2 \\ -2-1 \\ 1-(-2) \end{pmatrix} = \begin{pmatrix} -1 \\ -3 \\ 3 \end{pmatrix}$

$\overrightarrow{BD} = \begin{pmatrix} -2-0 \\ -3-3 \\ 2-(-4) \end{pmatrix} = \begin{pmatrix} -2 \\ -6 \\ 6 \end{pmatrix} = 2\begin{pmatrix} -1 \\ -3 \\ 3 \end{pmatrix} = 2\overrightarrow{AC}$

**8** $\overrightarrow{AB} = \begin{pmatrix} 2 - -1 \\ 3 - 5 \\ -3 - 2 \end{pmatrix} = \begin{pmatrix} 3 \\ -2 \\ -5 \end{pmatrix}$ $\therefore$ C is $(2+3,\ 3-2,\ -3-5)$, or $(5,\ 1,\ -8)$,
D is $(5+3,\ 1-2,\ -8-5)$, or $(8,\ -1,\ -13)$,
E is $(8+3,\ -1-2,\ -13-5)$, or $(11,\ -3,\ -18)$.

**9**

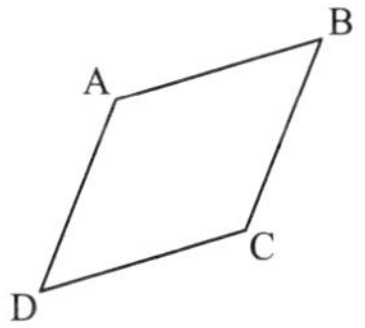


**a** $\overrightarrow{AB} = \begin{pmatrix} 4-3 \\ 2 - -1 \end{pmatrix} = \begin{pmatrix} 1 \\ 3 \end{pmatrix}$

$\overrightarrow{DC} = \begin{pmatrix} -1 - -2 \\ 4-1 \end{pmatrix} = \begin{pmatrix} 1 \\ 3 \end{pmatrix}$

Now $\overrightarrow{AB} = \overrightarrow{DC}$
$\therefore$ sides [AB] and [DC] are equal in length and parallel.
This is sufficient to deduce that ABCD is a parallelogram.

**b** $\overrightarrow{AB} = \begin{pmatrix} -1-5 \\ 2-0 \\ 4-3 \end{pmatrix} = \begin{pmatrix} -6 \\ 2 \\ 1 \end{pmatrix}$

$\overrightarrow{DC} = \begin{pmatrix} 4-10 \\ -3 - -5 \\ 6-5 \end{pmatrix} = \begin{pmatrix} -6 \\ 2 \\ 1 \end{pmatrix}$

So $\overrightarrow{AB} = \overrightarrow{DC}$
$\therefore$ sides [AB] and [DC] are equal in length and parallel.
This is sufficient to deduce that ABCD is a parallelogram.

**c** $\overrightarrow{AB} = \begin{pmatrix} 1-2 \\ 4 - -3 \\ -1-2 \end{pmatrix} = \begin{pmatrix} -1 \\ 7 \\ -3 \end{pmatrix}$

$\overrightarrow{DC} = \begin{pmatrix} -2 - -1 \\ 6 - -1 \\ -2-2 \end{pmatrix} = \begin{pmatrix} -1 \\ 7 \\ -4 \end{pmatrix}$

So, $\overrightarrow{AB} \neq \overrightarrow{DC}$
$\therefore$ ABCD cannot be a parallelogram.

**10** **a** Let D be $(a,\ b)$.

Now $\overrightarrow{CD} = \overrightarrow{BA}$

$\therefore \begin{pmatrix} a-8 \\ b - -2 \end{pmatrix} = \begin{pmatrix} 3-2 \\ 0 - -1 \end{pmatrix}$

$\therefore \begin{pmatrix} a-8 \\ b+2 \end{pmatrix} = \begin{pmatrix} 1 \\ 1 \end{pmatrix}$

$\therefore$ $a = 9,\ b = -1$

So, D is $(9,\ -1)$.

**b** Let R be $(a,\ b,\ c)$.

Now $\overrightarrow{SR} = \overrightarrow{PQ}$

$\therefore \begin{pmatrix} a-4 \\ b-0 \\ c-7 \end{pmatrix} = \begin{pmatrix} -2 - -1 \\ 5-4 \\ 2-3 \end{pmatrix}$

$\therefore \begin{pmatrix} a-4 \\ b \\ c-7 \end{pmatrix} = \begin{pmatrix} -1 \\ 1 \\ -1 \end{pmatrix}$

$\therefore$ $a = 3,\ b = 1,\ c = 6$

So, R is $(3,\ 1,\ 6)$.

**c** Let X be $(a,\ b,\ c)$.

Now $\overrightarrow{WX} = \overrightarrow{ZY}$

$\therefore \begin{pmatrix} a - -1 \\ b-5 \\ c-8 \end{pmatrix} = \begin{pmatrix} 3-0 \\ -2-4 \\ -2-6 \end{pmatrix}$

$\therefore \begin{pmatrix} a+1 \\ b-5 \\ c-8 \end{pmatrix} = \begin{pmatrix} 3 \\ -6 \\ -8 \end{pmatrix}$

$\therefore$ $a = 2,\ b = -1,\ c = 0$

So, X is $(2,\ -1,\ 0)$.

**11** **a** $\overrightarrow{BD} = \frac{1}{2}\overrightarrow{OA}$
$= \frac{1}{2}\mathbf{a}$

**b** $\overrightarrow{AB} = \overrightarrow{AO} + \overrightarrow{OB}$
$= -\mathbf{a} + \mathbf{b}$
$= \mathbf{b} - \mathbf{a}$

**c** $\overrightarrow{BA} = -\overrightarrow{AB}$
$= -(\mathbf{b} - \mathbf{a})$
$= -\mathbf{b} + \mathbf{a}$ or $\mathbf{a} - \mathbf{b}$

**d** $\overrightarrow{OD} = \overrightarrow{OB} + \overrightarrow{BD}$
$= \mathbf{b} + \frac{1}{2}\mathbf{a}$

**e** $\overrightarrow{AD} = \overrightarrow{AO} + \overrightarrow{OD}$
$= -\mathbf{a} + \mathbf{b} + \frac{1}{2}\mathbf{a}$
$= -\frac{1}{2}\mathbf{a} + \mathbf{b}$ or $\mathbf{b} - \frac{1}{2}\mathbf{a}$

**f** $\overrightarrow{DA} = -\overrightarrow{AD}$
$= \frac{1}{2}\mathbf{a} - \mathbf{b}$

**12** **a** $\overrightarrow{AD} = \overrightarrow{AB} + \overrightarrow{BD} = \begin{pmatrix} -1 \\ 3 \\ 2 \end{pmatrix} + \begin{pmatrix} 0 \\ 2 \\ -3 \end{pmatrix} = \begin{pmatrix} -1 \\ 5 \\ -1 \end{pmatrix}$

**b** $\overrightarrow{CB} = \overrightarrow{CA} + \overrightarrow{AB} = -\overrightarrow{AC} + \overrightarrow{AB} = -\begin{pmatrix} 2 \\ -1 \\ 4 \end{pmatrix} + \begin{pmatrix} -1 \\ 3 \\ 2 \end{pmatrix} = \begin{pmatrix} -2 \\ 1 \\ -4 \end{pmatrix} + \begin{pmatrix} -1 \\ 3 \\ 2 \end{pmatrix} = \begin{pmatrix} -3 \\ 4 \\ -2 \end{pmatrix}$

**c** $\overrightarrow{CD} = \overrightarrow{CB} + \overrightarrow{BD} = \begin{pmatrix} -3 \\ 4 \\ -2 \end{pmatrix} + \begin{pmatrix} 0 \\ 2 \\ -3 \end{pmatrix}$ {using **b**} $= \begin{pmatrix} -3 \\ 6 \\ -5 \end{pmatrix}$

**13** **a** $\mathbf{a}+\mathbf{b}=\begin{pmatrix}2\\-1\\1\end{pmatrix}+\begin{pmatrix}1\\2\\-3\end{pmatrix}=\begin{pmatrix}3\\1\\-2\end{pmatrix}$

**b** $\mathbf{a}-\mathbf{b}=\begin{pmatrix}2\\-1\\1\end{pmatrix}-\begin{pmatrix}1\\2\\-3\end{pmatrix}=\begin{pmatrix}1\\-3\\4\end{pmatrix}$

**c** $\mathbf{b}+2\mathbf{c}=\begin{pmatrix}1\\2\\-3\end{pmatrix}+2\begin{pmatrix}0\\1\\-3\end{pmatrix}=\begin{pmatrix}1\\2\\-3\end{pmatrix}+\begin{pmatrix}0\\2\\-6\end{pmatrix}=\begin{pmatrix}1\\4\\-9\end{pmatrix}$

**d** $\mathbf{c}-\frac{1}{2}\mathbf{a}=\begin{pmatrix}0\\1\\-3\end{pmatrix}-\frac{1}{2}\begin{pmatrix}2\\-1\\1\end{pmatrix}=\begin{pmatrix}0\\1\\-3\end{pmatrix}-\begin{pmatrix}1\\-\frac{1}{2}\\\frac{1}{2}\end{pmatrix}=\begin{pmatrix}-1\\\frac{3}{2}\\-\frac{7}{2}\end{pmatrix}$

**e** $\mathbf{a}-\mathbf{b}-\mathbf{c}=\begin{pmatrix}2\\-1\\1\end{pmatrix}-\begin{pmatrix}1\\2\\-3\end{pmatrix}-\begin{pmatrix}0\\1\\-3\end{pmatrix}=\begin{pmatrix}1\\-4\\7\end{pmatrix}$

**f** $2\mathbf{b}-\mathbf{c}+\mathbf{a}=2\begin{pmatrix}1\\2\\-3\end{pmatrix}-\begin{pmatrix}0\\1\\-3\end{pmatrix}+\begin{pmatrix}2\\-1\\1\end{pmatrix}=\begin{pmatrix}2\\4\\-6\end{pmatrix}-\begin{pmatrix}0\\1\\-3\end{pmatrix}+\begin{pmatrix}2\\-1\\1\end{pmatrix}=\begin{pmatrix}4\\2\\-2\end{pmatrix}$

**14** **a** $|\mathbf{a}|=\sqrt{(-1)^2+1^2+3^2}$
$=\sqrt{11}$ units

**b** $|\mathbf{b}|=\sqrt{1^2+(-3)^2+2^2}$
$=\sqrt{14}$ units

**c** $|\mathbf{b}+\mathbf{c}|=\left|\begin{pmatrix}1\\-3\\2\end{pmatrix}+\begin{pmatrix}-2\\2\\4\end{pmatrix}\right|=\left|\begin{pmatrix}-1\\-1\\6\end{pmatrix}\right|$
$=\sqrt{(-1)^2+(-1)^2+6^2}$
$=\sqrt{1+1+36}$
$=\sqrt{38}$ units

**d** $|\mathbf{a}-\mathbf{c}|=\left|\begin{pmatrix}-1\\1\\3\end{pmatrix}-\begin{pmatrix}-2\\2\\4\end{pmatrix}\right|=\left|\begin{pmatrix}1\\-1\\-1\end{pmatrix}\right|$
$=\sqrt{1^2+(-1)^2+(-1)^2}$
$=\sqrt{3}$ units

**e** $|\mathbf{a}|\mathbf{b}=\sqrt{11}\begin{pmatrix}1\\-3\\2\end{pmatrix}=\begin{pmatrix}\sqrt{11}\\-3\sqrt{11}\\2\sqrt{11}\end{pmatrix}$

**f** $\frac{1}{|\mathbf{a}|}\mathbf{a}=\frac{1}{\sqrt{11}}\begin{pmatrix}-1\\1\\3\end{pmatrix}=\begin{pmatrix}-\frac{1}{\sqrt{11}}\\\frac{1}{\sqrt{11}}\\\frac{3}{\sqrt{11}}\end{pmatrix}$

**15** **a** $2\begin{pmatrix}1\\0\\3a\end{pmatrix}=\begin{pmatrix}b\\c-1\\2\end{pmatrix}$

$\therefore \begin{pmatrix}2\\0\\6a\end{pmatrix}=\begin{pmatrix}b\\c-1\\2\end{pmatrix}$

$\therefore\ 2=b,\quad 0=c-1,\quad \text{and}\quad 6a=2$
$\therefore\ a=\frac{1}{3},\quad b=2,\quad \text{and}\quad c=1$

**b** $a\begin{pmatrix}1\\1\\0\end{pmatrix}+b\begin{pmatrix}2\\0\\-1\end{pmatrix}+c\begin{pmatrix}0\\1\\1\end{pmatrix}=\begin{pmatrix}-1\\3\\3\end{pmatrix}$

$\therefore \begin{pmatrix}a\\a\\0\end{pmatrix}+\begin{pmatrix}2b\\0\\-b\end{pmatrix}+\begin{pmatrix}0\\c\\c\end{pmatrix}=\begin{pmatrix}-1\\3\\3\end{pmatrix}$

So, $a+2b=-1$ .... (1)
$a+c=3$
$\therefore\ c=3-a$ .... (2)
$-b+c=3$
$\therefore\ c=b+3$ .... (3)

Substituting (2) into (3), we get $3-a=b+3$
$\therefore\ -a=b$

Substituting into (1), we get
$a+2(-a)=-1$
$\therefore\ -a=-1$
$\therefore\ a=1$
$\therefore\ b=-1$
and $c=-1+3=2$ {using (3)}
$\therefore\ a=1,\quad b=-1,\quad \text{and}\quad c=2$

**c** $a\begin{pmatrix}2\\-3\\1\end{pmatrix}+b\begin{pmatrix}1\\7\\2\end{pmatrix}=\begin{pmatrix}7\\-19\\2\end{pmatrix}$

$\therefore \begin{pmatrix}2a\\-3a\\a\end{pmatrix}+\begin{pmatrix}b\\7b\\2b\end{pmatrix}=\begin{pmatrix}7\\-19\\2\end{pmatrix}$

$\therefore \begin{pmatrix}2a+b\\-3a+7b\\a+2b\end{pmatrix}=\begin{pmatrix}7\\-19\\2\end{pmatrix}$

So, $2a+b=7$

$\therefore \quad b=7-2a \quad .... (1)$

$-3a+7b=-19 \quad .... (2)$

$a+2b=2 \quad .... (3)$

Substituting (1) into (3), we get

$a+2(7-2a)=2$

$\therefore \quad a+14-4a=2$

$\therefore \quad -3a=-12$

$\therefore \quad a=4$

and so $b=7-2(4)=-1$

$\therefore \quad a=4, \quad b=-1$

*Check*: $-3(4)+7(-1)=-12-7=-19$ ✓

## EXERCISE 12I

**1** Since **a** and **b** are parallel, then $\mathbf{b}=k\mathbf{a}$. $\therefore \begin{pmatrix}-6\\r\\s\end{pmatrix}=k\begin{pmatrix}2\\-1\\3\end{pmatrix}=\begin{pmatrix}2k\\-k\\3k\end{pmatrix}$

$\therefore \quad 2k=-6, \ r=-k, \ s=3k \qquad \therefore \quad k=-3, \ r=3, \ s=-9$

**2** If $\begin{pmatrix}3\\-1\\2\end{pmatrix}$ and $\begin{pmatrix}a\\2\\b\end{pmatrix}$ are parallel, then $\begin{pmatrix}a\\2\\b\end{pmatrix}=k\begin{pmatrix}3\\-1\\2\end{pmatrix}$.

$\therefore \quad a=3k, \ 2=-k, \ b=2k \qquad \therefore \quad k=-2, \ a=-6, \text{ and } b=-4$

**3** **a** $\overrightarrow{AB}=3\overrightarrow{CD}$ means that $\overrightarrow{AB}$ is parallel to $\overrightarrow{CD}$ and 3 times its length.

**b** $\overrightarrow{RS}=-\frac{1}{2}\overrightarrow{KL}$ means that $\overrightarrow{RS}$ is parallel to $\overrightarrow{KL}$, half its length, and in the opposite direction.

**c** 

$\overrightarrow{AB}=2\overrightarrow{BC}$ means that A, B, and C are collinear and the length of $\overrightarrow{AB}$ is twice the length of $\overrightarrow{BC}$.

**4** $\overrightarrow{OP}=\begin{pmatrix}3\\2\\-1\end{pmatrix}, \ \overrightarrow{OQ}=\begin{pmatrix}1\\4\\-3\end{pmatrix}, \ \overrightarrow{OR}=\begin{pmatrix}2\\-1\\2\end{pmatrix}, \ \overrightarrow{OS}=\begin{pmatrix}-1\\-2\\3\end{pmatrix}$

**a** $\overrightarrow{PR}=\overrightarrow{PO}+\overrightarrow{OR}=-\begin{pmatrix}3\\2\\-1\end{pmatrix}+\begin{pmatrix}2\\-1\\2\end{pmatrix}=\begin{pmatrix}-1\\-3\\3\end{pmatrix}$

$\overrightarrow{QS}=\overrightarrow{QO}+\overrightarrow{OS}=-\begin{pmatrix}1\\4\\-3\end{pmatrix}+\begin{pmatrix}-1\\-2\\3\end{pmatrix}=\begin{pmatrix}-2\\-6\\6\end{pmatrix}=2\begin{pmatrix}-1\\-3\\3\end{pmatrix}=2\overrightarrow{PR}$ and so [QS] || [PR].

**b** Since $\overrightarrow{QS}=2\overrightarrow{PR}$, $|\overrightarrow{QS}|=2|\overrightarrow{PR}|$, and so [QS] is twice as long as [PR].

**5** **a** The vector in the same direction as **a** and twice its length is $2\mathbf{a}$.

$$2\mathbf{a} = 2\begin{pmatrix} 2 \\ 4 \end{pmatrix} = \begin{pmatrix} 4 \\ 8 \end{pmatrix}$$

**b** The vector in the opposite direction to **a** and half its length is $-\frac{1}{2}\mathbf{a}$.

$$-\tfrac{1}{2}\mathbf{a} = -\tfrac{1}{2}\begin{pmatrix} 2 \\ 4 \end{pmatrix} = \begin{pmatrix} -\frac{2}{2} \\ -\frac{4}{2} \end{pmatrix} = \begin{pmatrix} -1 \\ -2 \end{pmatrix}$$

**6** **a** $\mathbf{i} + 2\mathbf{j}$ has length $\sqrt{1^2 + 2^2} = \sqrt{5}$ units $\therefore$ unit vector $= \frac{1}{\sqrt{5}}(\mathbf{i} + 2\mathbf{j})$

**b** $2\mathbf{i} - 3\mathbf{k}$ has length $\sqrt{2^2 + 0^2 + (-3)^2} = \sqrt{4+9} = \sqrt{13}$ units

$\therefore$ unit vector is $\frac{1}{\sqrt{13}}(2\mathbf{i} - 3\mathbf{k})$

**c** $2\mathbf{i} - 2\mathbf{j} + \mathbf{k}$ has length $\sqrt{2^2 + (-2)^2 + 1^2} = \sqrt{9} = 3$ units

$\therefore$ unit vector is $\frac{1}{3}(2\mathbf{i} - 2\mathbf{j} + \mathbf{k})$

**7** **a** $\begin{pmatrix} 2 \\ -1 \end{pmatrix}$ has length $\sqrt{2^2 + (-1)^2} = \sqrt{5}$ units

$\therefore$ the unit vector in the same direction is $\frac{1}{\sqrt{5}}\begin{pmatrix} 2 \\ -1 \end{pmatrix}$

$\therefore$ the vector of length 3 units in the same direction is $\frac{3}{\sqrt{5}}\begin{pmatrix} 2 \\ -1 \end{pmatrix} = \begin{pmatrix} \frac{6}{\sqrt{5}} \\ -\frac{3}{\sqrt{5}} \end{pmatrix}$

**b** $\begin{pmatrix} -1 \\ -4 \end{pmatrix}$ has length $\sqrt{(-1)^2 + (-4)^2} = \sqrt{17}$ units

$\therefore$ the unit vector in the opposite direction is $-\frac{1}{\sqrt{17}}\begin{pmatrix} -1 \\ -4 \end{pmatrix} = \frac{1}{\sqrt{17}}\begin{pmatrix} 1 \\ 4 \end{pmatrix}$

$\therefore$ the vector of length 2 units in the opposite direction is $\frac{2}{\sqrt{17}}\begin{pmatrix} 1 \\ 4 \end{pmatrix} = \begin{pmatrix} \frac{2}{\sqrt{17}} \\ \frac{8}{\sqrt{17}} \end{pmatrix}$

**8** **a** $\overrightarrow{AB}$ is a vector in the same direction as $\begin{pmatrix} 1 \\ -1 \end{pmatrix}$ with length 4 units.

Now, $\begin{pmatrix} 1 \\ -1 \end{pmatrix}$ has length $\sqrt{1^2 + (-1)^2} = \sqrt{2}$ units

$\therefore$ the unit vector in the same direction is $\frac{1}{\sqrt{2}}\begin{pmatrix} 1 \\ -1 \end{pmatrix}$

$\therefore$ the vector of length 4 units in the same direction is $\frac{4}{\sqrt{2}}\begin{pmatrix} 1 \\ -1 \end{pmatrix} = 2\sqrt{2}\begin{pmatrix} 1 \\ -1 \end{pmatrix}$

$$= \begin{pmatrix} 2\sqrt{2} \\ -2\sqrt{2} \end{pmatrix}$$

$\therefore$ $\overrightarrow{AB} = \begin{pmatrix} 2\sqrt{2} \\ -2\sqrt{2} \end{pmatrix}$.

**b** $\overrightarrow{OA} = \begin{pmatrix} 3 \\ 2 \end{pmatrix}$, $\overrightarrow{AB} = \begin{pmatrix} 2\sqrt{2} \\ -2\sqrt{2} \end{pmatrix}$

Now $\overrightarrow{OB} = \overrightarrow{OA} + \overrightarrow{AB}$

$$\therefore \overrightarrow{OB} = \begin{pmatrix} 3 \\ 2 \end{pmatrix} + \begin{pmatrix} 2\sqrt{2} \\ -2\sqrt{2} \end{pmatrix}$$

$$= \begin{pmatrix} 3 + 2\sqrt{2} \\ 2 - 2\sqrt{2} \end{pmatrix}$$

**c** If $\overrightarrow{OB} = \begin{pmatrix} 3 + 2\sqrt{2} \\ 2 - 2\sqrt{2} \end{pmatrix}$, then the coordinates of B are $(3 + 2\sqrt{2}, 2 - 2\sqrt{2})$.

**9** **a** $|\mathbf{a}| = \sqrt{2^2 + (-1)^2 + (-2)^2}$
$= \sqrt{4+1+4}$
$= 3$ units

$\therefore$ the vectors of length 1 unit parallel to $\mathbf{a}$ are $\pm\frac{1}{3}\mathbf{a}$.

$\therefore$ the vectors are $\begin{pmatrix} \frac{2}{3} \\ -\frac{1}{3} \\ -\frac{2}{3} \end{pmatrix}$ and $\begin{pmatrix} -\frac{2}{3} \\ \frac{1}{3} \\ \frac{2}{3} \end{pmatrix}$.

**b** $|\mathbf{b}| = \sqrt{(-2)^2 + (-1)^2 + 2^2}$
$= \sqrt{4+1+4}$
$= 3$ units

$\therefore$ the vectors of length 2 units parallel to $\mathbf{b}$ are $\pm\frac{2}{3}\mathbf{b}$.

$\therefore$ the vectors are $\begin{pmatrix} -\frac{4}{3} \\ -\frac{2}{3} \\ \frac{4}{3} \end{pmatrix}$ and $\begin{pmatrix} \frac{4}{3} \\ \frac{2}{3} \\ -\frac{4}{3} \end{pmatrix}$.

**10** **a** $\begin{pmatrix} -1 \\ 4 \\ 1 \end{pmatrix}$ has length $\sqrt{(-1)^2 + 4^2 + 1^2} = \sqrt{18} = 3\sqrt{2}$ units

$\therefore$ the unit vector in the same direction is $\frac{1}{3\sqrt{2}}\begin{pmatrix} -1 \\ 4 \\ 1 \end{pmatrix}$

$\therefore$ the vector of length 6 units in the same direction is $\frac{6}{3\sqrt{2}}\begin{pmatrix} -1 \\ 4 \\ 1 \end{pmatrix} = \sqrt{2}\begin{pmatrix} -1 \\ 4 \\ 1 \end{pmatrix} = \begin{pmatrix} -\sqrt{2} \\ 4\sqrt{2} \\ \sqrt{2} \end{pmatrix}$

**b** $\begin{pmatrix} -1 \\ -2 \\ -2 \end{pmatrix}$ has length $\sqrt{(-1)^2 + (-2)^2 + (-2)^2} = \sqrt{9} = 3$ units

$\therefore$ the unit vector in the opposite direction is $-\frac{1}{3}\begin{pmatrix} -1 \\ -2 \\ -2 \end{pmatrix} = \frac{1}{3}\begin{pmatrix} 1 \\ 2 \\ 2 \end{pmatrix}$

$\therefore$ the vector of length 5 units in the opposite direction is $\frac{5}{3}\begin{pmatrix} 1 \\ 2 \\ 2 \end{pmatrix} = \begin{pmatrix} \frac{5}{3} \\ \frac{10}{3} \\ \frac{10}{3} \end{pmatrix}$

## EXERCISE 12J

**1** **a** $\mathbf{q} \bullet \mathbf{p}$
$= \begin{pmatrix} -1 \\ 5 \end{pmatrix} \bullet \begin{pmatrix} 3 \\ 2 \end{pmatrix}$
$= -1(3) + 5(2)$
$= -3 + 10$
$= 7$

**b** $\mathbf{q} \bullet \mathbf{r}$
$= \begin{pmatrix} -1 \\ 5 \end{pmatrix} \bullet \begin{pmatrix} -2 \\ 4 \end{pmatrix}$
$= -1(-2) + 5(4)$
$= 2 + 20$
$= 22$

**c** $\mathbf{q} \bullet (\mathbf{p} + \mathbf{r})$
$= \begin{pmatrix} -1 \\ 5 \end{pmatrix} \bullet \left[\begin{pmatrix} 3 \\ 2 \end{pmatrix} + \begin{pmatrix} -2 \\ 4 \end{pmatrix}\right]$
$= \begin{pmatrix} -1 \\ 5 \end{pmatrix} \bullet \begin{pmatrix} 1 \\ 6 \end{pmatrix}$
$= -1(1) + 5(6)$
$= -1 + 30 = 29$

**d** $3\mathbf{r} \bullet \mathbf{q}$
$= 3\begin{pmatrix} -2 \\ 4 \end{pmatrix} \bullet \begin{pmatrix} -1 \\ 5 \end{pmatrix}$
$= \begin{pmatrix} -6 \\ 12 \end{pmatrix} \bullet \begin{pmatrix} -1 \\ 5 \end{pmatrix}$
$= -6(-1) + 12(5)$
$= 6 + 60 = 66$

**e** $2\mathbf{p} \bullet 2\mathbf{p}$
$= 2\begin{pmatrix} 3 \\ 2 \end{pmatrix} \bullet 2\begin{pmatrix} 3 \\ 2 \end{pmatrix}$
$= \begin{pmatrix} 6 \\ 4 \end{pmatrix} \bullet \begin{pmatrix} 6 \\ 4 \end{pmatrix}$
$= 6(6) + 4(4)$
$= 36 + 16 = 52$

**f** $\mathbf{i} \bullet \mathbf{p}$
$= \begin{pmatrix} 1 \\ 0 \end{pmatrix} \bullet \begin{pmatrix} 3 \\ 2 \end{pmatrix}$
$= 1(3) + 0(2)$
$= 3 + 0$
$= 3$

**g** $\mathbf{q} \bullet \mathbf{j} = \begin{pmatrix} -1 \\ 5 \end{pmatrix} \bullet \begin{pmatrix} 0 \\ 1 \end{pmatrix}$
$= -1(0) + 5(1)$
$= 0 + 5$
$= 5$

**h** $\mathbf{i} \bullet \mathbf{i} = \begin{pmatrix} 1 \\ 0 \end{pmatrix} \bullet \begin{pmatrix} 1 \\ 0 \end{pmatrix}$
$= 1(1) + 0(0)$
$= 1 + 0$
$= 1$

**2** **a** $\mathbf{a} \bullet \mathbf{b} = \begin{pmatrix} 2 \\ 1 \\ 3 \end{pmatrix} \bullet \begin{pmatrix} -1 \\ 1 \\ 1 \end{pmatrix}$
$= 2(-1) + 1(1) + 3(1)$
$= -2 + 1 + 3$
$= 2$

**b** $\mathbf{b} \bullet \mathbf{a} = \begin{pmatrix} -1 \\ 1 \\ 1 \end{pmatrix} \bullet \begin{pmatrix} 2 \\ 1 \\ 3 \end{pmatrix}$
$= (-1)(2) + 1(1) + 1(3)$
$= -2 + 1 + 3$
$= 2$

**c** $|\mathbf{a}|^2 = \left(\sqrt{2^2 + 1^2 + 3^2}\right)^2$
$= 14$

**d** $\mathbf{a} \bullet \mathbf{a} = \begin{pmatrix} 2 \\ 1 \\ 3 \end{pmatrix} \bullet \begin{pmatrix} 2 \\ 1 \\ 3 \end{pmatrix}$
$= 2(2) + 1(1) + 3(3)$
$= 14$

**e** $\mathbf{a} \bullet (\mathbf{b} + \mathbf{c})$
$= \begin{pmatrix} 2 \\ 1 \\ 3 \end{pmatrix} \bullet \left[\begin{pmatrix} -1 \\ 1 \\ 1 \end{pmatrix} + \begin{pmatrix} 0 \\ -1 \\ 1 \end{pmatrix}\right]$
$= \begin{pmatrix} 2 \\ 1 \\ 3 \end{pmatrix} \bullet \begin{pmatrix} -1 \\ 0 \\ 2 \end{pmatrix}$
$= 2(-1) + 1(0) + 3(2) = 4$

**f** $\mathbf{a} \bullet \mathbf{b} + \mathbf{a} \bullet \mathbf{c}$
$= 2 + \begin{pmatrix} 2 \\ 1 \\ 3 \end{pmatrix} \bullet \begin{pmatrix} 0 \\ -1 \\ 1 \end{pmatrix}$ {using **a**}
$= 2 + 2(0) + 1(-1) + 3(1)$
$= 4$

**3** **a** $\mathbf{p} \bullet \mathbf{q}$
$= \begin{pmatrix} 3 \\ -1 \\ 2 \end{pmatrix} \bullet \begin{pmatrix} -2 \\ 1 \\ 3 \end{pmatrix}$
$= 3(-2) + (-1)(1) + 2(3)$
$= -6 - 1 + 6$
$= -1$

**b** If the angle between $\mathbf{p}$ and $\mathbf{q}$ is $\theta$, then

$$\cos\theta = \frac{\mathbf{p} \bullet \mathbf{q}}{|\mathbf{p}||\mathbf{q}|} = \frac{-1}{\sqrt{3^2 + (-1)^2 + 2^2}\sqrt{(-2)^2 + 1^2 + 3^2}}$$
$$= \frac{-1}{\sqrt{14}\sqrt{14}}$$
$$\therefore \quad \theta = \cos^{-1}(-\tfrac{1}{14}) \approx 94.1^\circ$$

**4** **a** If the angle between $\mathbf{m}$ and $\mathbf{n}$ is $\theta$, then

$$\cos\theta = \frac{\mathbf{m} \bullet \mathbf{n}}{|\mathbf{m}||\mathbf{n}|} = \frac{2(-1) + (-1)(3) + (-1)(2)}{\sqrt{2^2 + (-1)^2 + (-1)^2}\sqrt{(-1)^2 + 3^2 + 2^2}}$$
$$= \frac{-7}{\sqrt{6}\sqrt{14}} = -\frac{7}{\sqrt{84}}$$
$$\therefore \quad \theta = \cos^{-1}\left(-\tfrac{7}{\sqrt{84}}\right) \approx 140^\circ$$

**b** $\mathbf{m} = 2\mathbf{j} - \mathbf{k} = \begin{pmatrix} 0 \\ 2 \\ -1 \end{pmatrix}$ and $\mathbf{n} = \mathbf{i} + 2\mathbf{k} = \begin{pmatrix} 1 \\ 0 \\ 2 \end{pmatrix}$

If the angle between $\mathbf{m}$ and $\mathbf{n}$ is $\theta$, then

$$\cos\theta = \frac{\mathbf{m} \bullet \mathbf{n}}{|\mathbf{m}||\mathbf{n}|} = \frac{(0)(1) + (2)(0) + (-1)(2)}{\sqrt{0^2 + 2^2 + (-1)^2}\sqrt{1^2 + 0^2 + 2^2}}$$
$$= \frac{-2}{\sqrt{5}\sqrt{5}} = -\frac{2}{5}$$
$$\therefore \quad \theta = \cos^{-1}\left(-\tfrac{2}{5}\right) \approx 114^\circ$$

**5** **a** $(\mathbf{i} + \mathbf{j} - \mathbf{k}) \bullet (2\mathbf{j} + \mathbf{k})$
$= \begin{pmatrix} 1 \\ 1 \\ -1 \end{pmatrix} \bullet \begin{pmatrix} 0 \\ 2 \\ 1 \end{pmatrix}$
$= 1(0) + 1(2) - 1(1) = 1$

**b** $\mathbf{i} \bullet \mathbf{i}$
$= \begin{pmatrix} 1 \\ 0 \\ 0 \end{pmatrix} \bullet \begin{pmatrix} 1 \\ 0 \\ 0 \end{pmatrix}$
$= 1$

**c** $\mathbf{i} \bullet \mathbf{j}$
$= \begin{pmatrix} 1 \\ 0 \\ 0 \end{pmatrix} \bullet \begin{pmatrix} 0 \\ 1 \\ 0 \end{pmatrix}$
$= 0$

**6** **a** $\mathbf{p} \bullet \mathbf{q} = |\mathbf{p}||\mathbf{q}|\cos\theta$
$= 2 \times 5 \times \cos 60°$
$= 5$

**b** $\mathbf{p} \bullet \mathbf{q} = |\mathbf{p}||\mathbf{q}|\cos\theta$
$= 6 \times 3 \times \cos 120°$
$= -9$

**7** **a** **i** If $\mathbf{v}$ and $\mathbf{w}$ are parallel, then they are either in the same direction (so the angle between them is $0°$) or in opposite directions (so the angle between them is $180°$).

If the angle between them is $0°$, then $\mathbf{v} \bullet \mathbf{w} = |\mathbf{v}||\mathbf{w}|\cos 0°$
$= 3 \times 4 \times 1$
$= 12$

If the angle between them is $180°$, then $\mathbf{v} \bullet \mathbf{w} = |\mathbf{v}||\mathbf{w}|\cos 180°$
$= 3 \times 4 \times -1$
$= -12$

**ii** $\mathbf{v} \bullet \mathbf{w} = |\mathbf{v}||\mathbf{w}|\cos 60°$
$= 3 \times 4 \times \frac{1}{2}$
$= 6$

**b** **i** $\mathbf{a}$ and $\mathbf{b}$ are not perpendicular as their dot product is not equal to 0.

**ii** $\mathbf{a} \bullet \mathbf{b} = |\mathbf{a}||\mathbf{b}|\cos\theta$

If $\mathbf{a}$ and $\mathbf{b}$ are parallel, then $\theta = 0$ or $180$
$\therefore \cos\theta = \pm 1$

$\therefore -12 = |\mathbf{a}| \times 1 \times \pm 1$
$\therefore |\mathbf{a}| = \pm 12$ but $|\mathbf{a}| > 0$
$\therefore |\mathbf{a}| = 12$ units

{**Note:** This means that $\cos\theta$ must be $-1$, so the angle between $\mathbf{a}$ and $\mathbf{b}$ is $180°$
$\therefore$ $\mathbf{a}$ and $\mathbf{b}$ are in opposite directions.}

**c** **i** $\mathbf{c} \bullet \mathbf{d} = |\mathbf{c}||\mathbf{d}|\cos\theta$
$\therefore 5 = |\sqrt{5}||\sqrt{5}|\cos\theta$
$\therefore 5 = 5\cos\theta$
$\therefore \cos\theta = 1$
$\therefore \theta = 0°$
So, $\mathbf{c} = \mathbf{d}$

**ii** $\mathbf{c} \bullet \mathbf{d} = |\mathbf{c}||\mathbf{d}|\cos\theta$
$\therefore -5 = |\sqrt{5}||\sqrt{5}|\cos\theta$
$\therefore -5 = 5\cos\theta$
$\therefore \cos\theta = -1$
$\therefore \theta = 180°$
So, $\mathbf{c} = -\mathbf{d}$

**8** **a** P has coordinates $(\cos\theta, \sin\theta)$.

**b** $\overrightarrow{BP} = \overrightarrow{BO} + \overrightarrow{OP}$

$= \begin{pmatrix} 1 \\ 0 \end{pmatrix} + \begin{pmatrix} \cos\theta \\ \sin\theta \end{pmatrix}$

$= \begin{pmatrix} 1 + \cos\theta \\ \sin\theta \end{pmatrix} = \begin{pmatrix} \cos\theta + 1 \\ \sin\theta \end{pmatrix}$

$\overrightarrow{AP} = \overrightarrow{AO} + \overrightarrow{OP}$

$= \begin{pmatrix} -1 \\ 0 \end{pmatrix} + \begin{pmatrix} \cos\theta \\ \sin\theta \end{pmatrix}$

$= \begin{pmatrix} -1 + \cos\theta \\ \sin\theta \end{pmatrix} = \begin{pmatrix} \cos\theta - 1 \\ \sin\theta \end{pmatrix}$

**c** $\overrightarrow{AP} \bullet \overrightarrow{BP} = (\cos\theta + 1)(\cos\theta - 1) + \sin^2\theta$
$= \cos^2\theta - 1 + \sin^2\theta$
$= 1 - 1 \quad \{\cos^2\theta + \sin^2\theta = 1\}$
$= 0$

**d** $\overrightarrow{AP} \bullet \overrightarrow{BP} = 0$, which mean $\overrightarrow{AP}$ and $\overrightarrow{BP}$ are perpendicular.

Now, triangle APB is in a semi-circle, and the angle at P is $90°$.

So, we have deduced that the angle in a semi-circle is a right angle.

**9** $\mathbf{a} \bullet (\mathbf{b} + \mathbf{c})$

$= \begin{pmatrix} a_1 \\ a_2 \\ a_3 \end{pmatrix} \bullet \left[ \begin{pmatrix} b_1 \\ b_2 \\ b_3 \end{pmatrix} + \begin{pmatrix} c_1 \\ c_2 \\ c_3 \end{pmatrix} \right]$

$= \begin{pmatrix} a_1 \\ a_2 \\ a_3 \end{pmatrix} \bullet \begin{pmatrix} b_1 + c_1 \\ b_2 + c_2 \\ b_3 + c_3 \end{pmatrix}$

$= a_1(b_1 + c_1) + a_2(b_2 + c_2) + a_3(b_3 + c_3)$

$= a_1b_1 + a_1c_1 + a_2b_2 + a_2c_2 + a_3b_3 + a_3c_3$

$= (a_1b_1 + a_2b_2 + a_3b_3) + (a_1c_1 + a_2c_2 + a_3c_3)$

$= \mathbf{a} \bullet \mathbf{b} + \mathbf{a} \bullet \mathbf{c}$

$\therefore \quad \mathbf{p} \bullet (\mathbf{c} + \mathbf{d}) = \mathbf{p} \bullet \mathbf{c} + \mathbf{p} \bullet \mathbf{d}$

If we let $\mathbf{p} = \mathbf{a} + \mathbf{b}$,

then $(\mathbf{a} + \mathbf{b}) \bullet (\mathbf{c} + \mathbf{d})$

$= \mathbf{p} \bullet (\mathbf{c} + \mathbf{d})$

$= \mathbf{p} \bullet \mathbf{c} + \mathbf{p} \bullet \mathbf{d}$

$= (\mathbf{a} + \mathbf{b}) \bullet \mathbf{c} + (\mathbf{a} + \mathbf{b}) \bullet \mathbf{d}$

$= \mathbf{c} \bullet (\mathbf{a} + \mathbf{b}) + \mathbf{d} \bullet (\mathbf{a} + \mathbf{b})$

$= \mathbf{c} \bullet \mathbf{a} + \mathbf{c} \bullet \mathbf{b} + \mathbf{d} \bullet \mathbf{a} + \mathbf{d} \bullet \mathbf{b}$

$= \mathbf{a} \bullet \mathbf{c} + \mathbf{a} \bullet \mathbf{d} + \mathbf{b} \bullet \mathbf{c} + \mathbf{b} \bullet \mathbf{d}$

**10** **a** $\begin{pmatrix} 3 \\ t \end{pmatrix} \bullet \begin{pmatrix} -2 \\ 1 \end{pmatrix} = 0$

$\therefore \quad -6 + t = 0$

$\therefore \quad t = 6$

**b** $\begin{pmatrix} t \\ t+2 \end{pmatrix} \bullet \begin{pmatrix} 3 \\ -4 \end{pmatrix} = 0$

$\therefore \quad 3t - 4(t+2) = 0$

$\therefore \quad 3t - 4t - 8 = 0$

$\therefore \quad -t = 8$

$\therefore \quad t = -8$

**c** $\begin{pmatrix} t \\ t+2 \end{pmatrix} \bullet \begin{pmatrix} 2-3t \\ t \end{pmatrix} = 0$

$\therefore \quad 2t - 3t^2 + t^2 + 2t = 0$

$\therefore \quad -2t^2 + 4t = 0$

$\therefore \quad t^2 - 2t = 0$

$\therefore \quad t(t-2) = 0$

$\therefore \quad t = 0$ or $2$

**11** **a** If $\mathbf{p} \parallel \mathbf{q}$ then $\begin{pmatrix} 3 \\ t \end{pmatrix} = k \begin{pmatrix} -2 \\ 1 \end{pmatrix}$ where $k \neq 0$

$\therefore \quad 3 = -2k$ and $t = k$

$\therefore \quad k = -\frac{3}{2}$ and $t = -\frac{3}{2}$

**b** If $\mathbf{r} \parallel \mathbf{s}$ then $\begin{pmatrix} t \\ t+2 \end{pmatrix} = k \begin{pmatrix} 3 \\ -4 \end{pmatrix}$

where $k \neq 0$

$\therefore \quad t = 3k$ and $t + 2 = -4k$

$\therefore \quad t + 2 = -4\left(\frac{t}{3}\right)$

$\therefore \quad 3t + 6 = -4t$

$\therefore \quad 7t = -6$

$\therefore \quad t = -\frac{6}{7}$

**c** If $\mathbf{a} \parallel \mathbf{b}$ then $\begin{pmatrix} t \\ t+2 \end{pmatrix} = k \begin{pmatrix} 2-3t \\ t \end{pmatrix}$

$\therefore \quad t = k(2-3t)$ and $t + 2 = kt$

$\therefore \quad \dfrac{t}{2-3t} = \dfrac{t+2}{t}$ {equating $k$s}

$\therefore \quad t^2 = (t+2)(2-3t)$

$\therefore \quad t^2 = 2t - 3t^2 + 4 - 6t$

$\therefore \quad 4t^2 + 4t - 4 = 0$

$\therefore \quad t^2 + t - 1 = 0$

which has $\Delta = 1^2 - 4(1)(-1) = 5$

$\therefore \quad t = \dfrac{-1 \pm \sqrt{5}}{2}$

**12** **a** $\begin{pmatrix} 1 \\ 1 \\ 5 \end{pmatrix} \bullet \begin{pmatrix} 2 \\ 3 \\ -1 \end{pmatrix} = 1(2) + 1(3) + 5(-1) = 0$

$\therefore \quad \begin{pmatrix} 1 \\ 1 \\ 5 \end{pmatrix}$ and $\begin{pmatrix} 2 \\ 3 \\ -1 \end{pmatrix}$ are perpendicular.

**b** $\mathbf{a} \bullet \mathbf{b}$

$= \begin{pmatrix} 3 \\ 1 \\ 2 \end{pmatrix} \bullet \begin{pmatrix} -1 \\ 1 \\ 1 \end{pmatrix}$

$= 3(-1) + 1(1) + 2(1)$

$= 0$

$\mathbf{b} \bullet \mathbf{c}$

$= \begin{pmatrix} -1 \\ 1 \\ 1 \end{pmatrix} \bullet \begin{pmatrix} 1 \\ 5 \\ -4 \end{pmatrix}$

$= (-1)(1) + 1(5) + 1(-4)$

$= 0$

$\mathbf{a} \bullet \mathbf{c}$

$= \begin{pmatrix} 3 \\ 1 \\ 2 \end{pmatrix} \bullet \begin{pmatrix} 1 \\ 5 \\ -4 \end{pmatrix}$

$= (3)(1) + 1(5) + 2(-4)$

$= 0$

$\therefore \quad \mathbf{a}$, $\mathbf{b}$, and $\mathbf{c}$ are mutually perpendicular.

**c** **i** $\begin{pmatrix} 3 \\ -1 \\ t \end{pmatrix} \bullet \begin{pmatrix} 2t \\ -3 \\ -4 \end{pmatrix} = 0$

$\therefore\ 3(2t) + (-1)(-3) + t(-4) = 0$

$\therefore\ 6t + 3 - 4t = 0$

$\therefore\ 2t + 3 = 0$

$\therefore\ t = -\frac{3}{2}$

**ii** $\begin{pmatrix} 3 \\ t \\ -2 \end{pmatrix} \bullet \begin{pmatrix} 1-t \\ -3 \\ 4 \end{pmatrix} = 0$

$\therefore\ 3(1-t) + t(-3) + (-2)4 = 0$

$\therefore\ 3 - 3t - 3t - 8 = 0$

$\therefore\ -6t = 5$

$\therefore\ t = -\frac{5}{6}$

**13** **a** We have three points: A(−2, 1), B(−2, 5), C(3, 1).

Then $\overrightarrow{AB} = \begin{pmatrix} 0 \\ 4 \end{pmatrix}$, $\overrightarrow{AC} = \begin{pmatrix} 5 \\ 0 \end{pmatrix}$, and $\overrightarrow{BC} = \begin{pmatrix} 5 \\ -4 \end{pmatrix}$

Now $\overrightarrow{AB} \bullet \overrightarrow{AC} = \begin{pmatrix} 0 \\ 4 \end{pmatrix} \bullet \begin{pmatrix} 5 \\ 0 \end{pmatrix} = 0 + 0 = 0$

$\therefore$ $\overrightarrow{AB}$ is perpendicular to $\overrightarrow{AC}$ and so $\triangle$ABC is right angled at A.

**b** We have three points: A(4, 7), B(1, 2), C(−1, 6)

Then $\overrightarrow{AB} = \begin{pmatrix} -3 \\ -5 \end{pmatrix}$, $\overrightarrow{AC} = \begin{pmatrix} -5 \\ -1 \end{pmatrix}$, and $\overrightarrow{BC} = \begin{pmatrix} -2 \\ 4 \end{pmatrix}$

Now $\overrightarrow{AB} \bullet \overrightarrow{AC} = \begin{pmatrix} -3 \\ -5 \end{pmatrix} \bullet \begin{pmatrix} -5 \\ -1 \end{pmatrix} = 15 + 5 = 20$

$\overrightarrow{AB} \bullet \overrightarrow{BC} = \begin{pmatrix} -3 \\ -5 \end{pmatrix} \bullet \begin{pmatrix} -2 \\ 4 \end{pmatrix} = 6 + (-20) = -14$

$\overrightarrow{AC} \bullet \overrightarrow{BC} = \begin{pmatrix} -5 \\ -1 \end{pmatrix} \bullet \begin{pmatrix} -2 \\ 4 \end{pmatrix} = 10 + (-4) = 6$

$\therefore$ none of the sides are perpendicular to each other and so $\triangle$ABC is not right angled.

**c** We have three points: A(2, −2), B(5, 7), C(−1, −1)

Then $\overrightarrow{AB} = \begin{pmatrix} 3 \\ 9 \end{pmatrix}$, $\overrightarrow{AC} = \begin{pmatrix} -3 \\ 1 \end{pmatrix}$, and $\overrightarrow{BC} = \begin{pmatrix} -6 \\ -8 \end{pmatrix}$

Now $\overrightarrow{AB} \bullet \overrightarrow{AC} = \begin{pmatrix} 3 \\ 9 \end{pmatrix} \bullet \begin{pmatrix} -3 \\ 1 \end{pmatrix} = -9 + 9 = 0$

$\therefore$ $\overrightarrow{AB}$ is perpendicular to $\overrightarrow{AC}$ and so $\triangle$ABC is right angled at A.

**d** We have three points: A(10, 1), B(5, 2), C(7, 4)

Then $\overrightarrow{AB} = \begin{pmatrix} -5 \\ 1 \end{pmatrix}$, $\overrightarrow{AC} = \begin{pmatrix} -3 \\ 3 \end{pmatrix}$, and $\overrightarrow{BC} = \begin{pmatrix} 2 \\ 2 \end{pmatrix}$

Now $\overrightarrow{AC} \bullet \overrightarrow{BC} = \begin{pmatrix} -3 \\ 3 \end{pmatrix} \bullet \begin{pmatrix} 2 \\ 2 \end{pmatrix} = -6 + 6 = 0$

$\therefore$ $\overrightarrow{AC}$ is perpendicular to $\overrightarrow{BC}$ and so $\triangle$ABC is right angled at C.

**14** We have three points: A(5, 1, 2), B(6, −1, 0), C(3, 2, 0)

Then $\overrightarrow{AB} = \begin{pmatrix} 1 \\ -2 \\ -2 \end{pmatrix}$, $\overrightarrow{AC} = \begin{pmatrix} -2 \\ 1 \\ -2 \end{pmatrix}$, and $\overrightarrow{BC} = \begin{pmatrix} -3 \\ 3 \\ 0 \end{pmatrix}$

Now $\overrightarrow{AB} \bullet \overrightarrow{AC} = \begin{pmatrix} 1 \\ -2 \\ -2 \end{pmatrix} \bullet \begin{pmatrix} -2 \\ 1 \\ -2 \end{pmatrix} = (-2) + (-2) + 4 = 0$

$\therefore$ $\overrightarrow{AB}$ is perpendicular to $\overrightarrow{AC}$ and so $\triangle$ABC is right angled at A.

**15** **a** A(2, 4, 2) B(−1, 2, 3) D(0, 5, 5) C(−3, 3, 6)

$\overrightarrow{AB} = \begin{pmatrix} -3 \\ -2 \\ 1 \end{pmatrix}$, $\overrightarrow{BC} = \begin{pmatrix} -2 \\ 1 \\ 3 \end{pmatrix}$

$\overrightarrow{DC} = \begin{pmatrix} -3 \\ -2 \\ 1 \end{pmatrix}$, $\overrightarrow{AD} = \begin{pmatrix} -2 \\ 1 \\ 3 \end{pmatrix}$

$\therefore$ $\overrightarrow{AB}$ is parallel to $\overrightarrow{DC}$ and $\overrightarrow{BC}$ is parallel to $\overrightarrow{AD}$.

$\therefore$ ABCD is a parallelogram.

**b** $|\overrightarrow{AB}| = \sqrt{14}$ units and $|\overrightarrow{BC}| = \sqrt{14}$ units $\therefore$ ABCD is a rhombus.

**c** $\overrightarrow{AC} \bullet \overrightarrow{BD} = \begin{pmatrix} -5 \\ -1 \\ 4 \end{pmatrix} \bullet \begin{pmatrix} 1 \\ 3 \\ 2 \end{pmatrix} = (-5)(1) + (-1)(3) + 4(2) = 0$

$\therefore$ $\overrightarrow{AC}$ is perpendicular to $\overrightarrow{BD}$ which illustrates that the diagonals of a rhombus are perpendicular.

**16** **a** $\begin{pmatrix} 5 \\ 2 \end{pmatrix} \bullet \begin{pmatrix} -2 \\ 5 \end{pmatrix} = -10 + 10 = 0$, so $\begin{pmatrix} -2 \\ 5 \end{pmatrix}$ is one such vector.

$\therefore$ required vectors have form $k\begin{pmatrix} -2 \\ 5 \end{pmatrix}$, $k \neq 0$. **Note:** $k\begin{pmatrix} 2 \\ -5 \end{pmatrix}$, $k \neq 0$ is also acceptable.

**b** $\begin{pmatrix} -1 \\ -2 \end{pmatrix} \bullet \begin{pmatrix} 2 \\ -1 \end{pmatrix} = -2 + 2 = 0$, so $\begin{pmatrix} 2 \\ -1 \end{pmatrix}$ is one such vector.

$\therefore$ required vectors have form $k\begin{pmatrix} 2 \\ -1 \end{pmatrix}$, $k \neq 0$.

**c** $\begin{pmatrix} 3 \\ -1 \end{pmatrix} \bullet \begin{pmatrix} 1 \\ 3 \end{pmatrix} = 3 - 3 = 0$, so $\begin{pmatrix} 1 \\ 3 \end{pmatrix}$ is one such vector.

$\therefore$ required vectors have form $k\begin{pmatrix} 1 \\ 3 \end{pmatrix}$, $k \neq 0$.

**d** $\begin{pmatrix} -4 \\ 3 \end{pmatrix} \bullet \begin{pmatrix} 3 \\ 4 \end{pmatrix} = -12 + 12 = 0$, so $\begin{pmatrix} 3 \\ 4 \end{pmatrix}$ is one such vector.

$\therefore$ required vectors have form $k\begin{pmatrix} 3 \\ 4 \end{pmatrix}$, $k \neq 0$.

**e** $\begin{pmatrix} 2 \\ 0 \end{pmatrix} \bullet \begin{pmatrix} 0 \\ 1 \end{pmatrix} = 0 + 0 = 0$, so $\begin{pmatrix} 0 \\ 1 \end{pmatrix}$ is one such vector.

$\therefore$ required vectors have form $k\begin{pmatrix} 0 \\ 1 \end{pmatrix}$, $k \neq 0$.

**17** Suppose $\begin{pmatrix} a \\ b \\ c \end{pmatrix}$ is perpendicular to $\begin{pmatrix} 1 \\ 2 \\ -1 \end{pmatrix}$. $\therefore$ $\begin{pmatrix} a \\ b \\ c \end{pmatrix} \bullet \begin{pmatrix} 1 \\ 2 \\ -1 \end{pmatrix} = 0$

So, to find a vector perpendicular to $\begin{pmatrix} 1 \\ 2 \\ -1 \end{pmatrix}$, we pick two non-zero integer values for $a$ and $b$, then solve for $c$.

For example, if $a = 1$, $b = 2$

then $\begin{pmatrix} 1 \\ 2 \\ c \end{pmatrix} \bullet \begin{pmatrix} 1 \\ 2 \\ -1 \end{pmatrix} = 0$

$\therefore\ 1 + 4 - c = 0$

$\therefore\ 5 - c = 0$

$\therefore\ c = 5$

So, the vector $\begin{pmatrix} 1 \\ 2 \\ 5 \end{pmatrix}$ is perpendicular to $\begin{pmatrix} 1 \\ 2 \\ -1 \end{pmatrix}$.

Repeating this process with a different value of $a$ (or $b$) will give another vector which is perpendicular to $\begin{pmatrix} 1 \\ 2 \\ -1 \end{pmatrix}$.

**18** Given A(3, 0, 1), B(−3, 1, 2), and C(−2, 1, −1),

$\overrightarrow{BC} = \begin{pmatrix} 1 \\ 0 \\ -3 \end{pmatrix}$ and $\overrightarrow{BA} = \begin{pmatrix} 6 \\ -1 \\ -1 \end{pmatrix}$

$$\therefore \quad \cos\theta = \frac{\overrightarrow{BC} \bullet \overrightarrow{BA}}{|\overrightarrow{BC}||\overrightarrow{BA}|}$$

$$= \frac{\begin{pmatrix} 1 \\ 0 \\ -3 \end{pmatrix} \bullet \begin{pmatrix} 6 \\ -1 \\ -1 \end{pmatrix}}{\sqrt{1+9}\sqrt{36+1+1}}$$

$$= \frac{6+0+3}{\sqrt{10}\sqrt{38}} = \frac{9}{\sqrt{380}}$$

$$\therefore \quad \theta \approx 62.5°$$

If $\overrightarrow{BA}$ and $\overrightarrow{CB}$ are used we would find the exterior angle of the triangle at B, which is 117.5°.

**19**

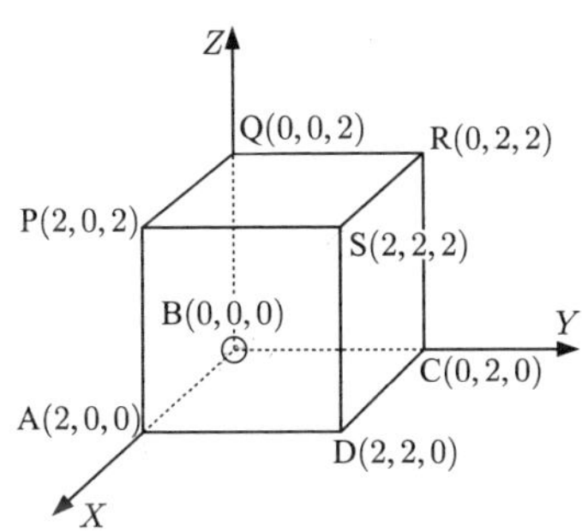



**a** Suppose the origin is at B.

Now $\overrightarrow{BA} = \begin{pmatrix} 2 \\ 0 \\ 0 \end{pmatrix}$ and $\overrightarrow{BS} = \begin{pmatrix} 2 \\ 2 \\ 2 \end{pmatrix}$

$$\therefore \quad \overrightarrow{BA} \bullet \overrightarrow{BS} = \begin{pmatrix} 2 \\ 0 \\ 0 \end{pmatrix} \bullet \begin{pmatrix} 2 \\ 2 \\ 2 \end{pmatrix} = 4+0+0 = 4$$

$$\therefore \quad \cos A\widehat{B}S = \frac{4}{\sqrt{4+0+0}\sqrt{4+4+4}}$$

$$= \frac{4}{2 \times 2\sqrt{3}} = \frac{1}{\sqrt{3}}$$

$$\therefore \quad A\widehat{B}S \approx 54.7°$$

**b** Consider vectors away from B.

$\overrightarrow{BR} = \begin{pmatrix} 0 \\ 2 \\ 2 \end{pmatrix}$ and $\overrightarrow{BP} = \begin{pmatrix} 2 \\ 0 \\ 2 \end{pmatrix}$

$$\therefore \quad \overrightarrow{BR} \bullet \overrightarrow{BP} = \begin{pmatrix} 0 \\ 2 \\ 2 \end{pmatrix} \bullet \begin{pmatrix} 2 \\ 0 \\ 2 \end{pmatrix} = 0+0+4 = 4$$

$$\therefore \quad \cos R\widehat{B}P = \frac{4}{\sqrt{0+4+4}\sqrt{4+0+4}}$$

$$= \frac{4}{\sqrt{8} \times \sqrt{8}}$$

$$= \tfrac{1}{2} \text{ and so } R\widehat{B}P = 60°$$

**c** $\overrightarrow{BP} = \begin{pmatrix} 2 \\ 0 \\ 2 \end{pmatrix}$ and $\overrightarrow{BS} = \begin{pmatrix} 2 \\ 2 \\ 2 \end{pmatrix}$

$$\therefore \quad \overrightarrow{BP} \bullet \overrightarrow{BS} = \begin{pmatrix} 2 \\ 0 \\ 2 \end{pmatrix} \bullet \begin{pmatrix} 2 \\ 2 \\ 2 \end{pmatrix}$$

$$= 4+0+4 = 8$$

$$\therefore \quad \cos P\widehat{B}S = \frac{8}{\sqrt{4+4}\sqrt{4+4+4}}$$

$$= \frac{8}{\sqrt{96}}$$

$$\therefore \quad P\widehat{B}S \approx 35.3°$$

**20** Suppose the origin is at N.

**a**

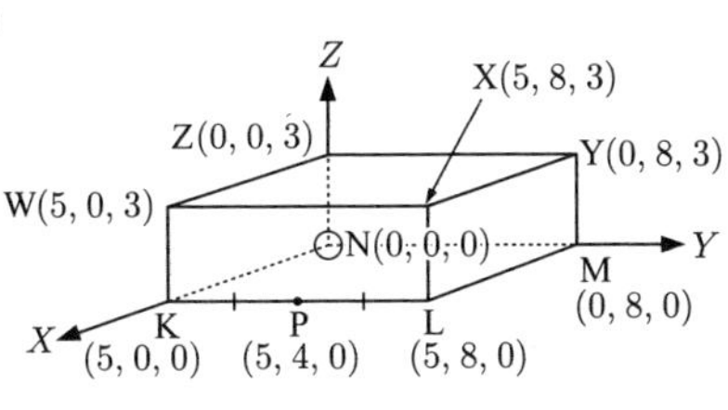


$\overrightarrow{NY} = \begin{pmatrix} 0 \\ 8 \\ 3 \end{pmatrix}$ and $\overrightarrow{NX} = \begin{pmatrix} 5 \\ 8 \\ 3 \end{pmatrix}$

$$\overrightarrow{NY} \bullet \overrightarrow{NX} = \begin{pmatrix} 0 \\ 8 \\ 3 \end{pmatrix} \bullet \begin{pmatrix} 5 \\ 8 \\ 3 \end{pmatrix} = 0+64+9 = 73$$

$$\therefore \quad \cos Y\widehat{N}X = \frac{73}{\sqrt{64+9}\sqrt{25+64+9}}$$

$$= \frac{73}{\sqrt{73}\sqrt{98}} = \sqrt{\frac{73}{98}}$$

$$\therefore \quad Y\widehat{N}X \approx 30.3°$$

**b** $\overrightarrow{NY} = \begin{pmatrix} 0 \\ 8 \\ 3 \end{pmatrix}$ and $\overrightarrow{NP} = \begin{pmatrix} 5 \\ 4 \\ 0 \end{pmatrix}$

$$\overrightarrow{NY} \bullet \overrightarrow{NP} = \begin{pmatrix} 0 \\ 8 \\ 3 \end{pmatrix} \bullet \begin{pmatrix} 5 \\ 4 \\ 0 \end{pmatrix} = 0 + 32 + 0 = 32$$

$$\therefore \cos \widehat{YNP} = \frac{32}{\sqrt{64+9}\sqrt{25+16}} = \frac{32}{\sqrt{73}\sqrt{41}}$$

$$\therefore \widehat{YNP} \approx 54.2^\circ$$

**21** **a** M is the midpoint of [BC]. $\therefore$ M is at $\left(\frac{2+1}{2}, \frac{2+3}{2}, \frac{2+1}{2}\right)$, which is $\left(\frac{3}{2}, \frac{5}{2}, \frac{3}{2}\right)$.

**b** Now $\overrightarrow{MD} = \begin{pmatrix} \frac{3}{2} \\ -\frac{1}{2} \\ -\frac{3}{2} \end{pmatrix}$ and $\overrightarrow{MA} = \begin{pmatrix} \frac{1}{2} \\ -\frac{3}{2} \\ -\frac{1}{2} \end{pmatrix}$

M, D, A, $\theta$

$$\therefore \cos\theta = \frac{\overrightarrow{MD} \bullet \overrightarrow{MA}}{|\overrightarrow{MD}||\overrightarrow{MA}|} = \frac{\begin{pmatrix} \frac{3}{2} \\ -\frac{1}{2} \\ -\frac{3}{2} \end{pmatrix} \bullet \begin{pmatrix} \frac{1}{2} \\ -\frac{3}{2} \\ -\frac{1}{2} \end{pmatrix}}{\sqrt{\frac{9}{4}+\frac{1}{4}+\frac{9}{4}}\sqrt{\frac{1}{4}+\frac{9}{4}+\frac{1}{4}}}$$

$$\therefore \cos\theta = \frac{\frac{3}{4}+\frac{3}{4}+\frac{3}{4}}{\sqrt{\frac{19}{4}}\sqrt{\frac{11}{4}}} = \frac{\frac{9}{4}}{\frac{\sqrt{209}}{4}} = \frac{9}{\sqrt{209}} \quad \text{and so} \quad \theta \approx 51.5^\circ$$

**22** **a** $\begin{pmatrix} 2 \\ t \\ t-2 \end{pmatrix} \bullet \begin{pmatrix} t \\ 3 \\ t \end{pmatrix} = 0$

$\therefore\ 2t + 3t + t(t-2) = 0$
$\therefore\ 5t + t^2 - 2t = 0$
$\therefore\ t^2 + 3t = 0$
$\therefore\ t(t+3) = 0$ and so $t = 0$ or $t = -3$

**b** Given that $\mathbf{a} = \begin{pmatrix} 1 \\ 2 \\ 3 \end{pmatrix}$, $\mathbf{b} = \begin{pmatrix} 2 \\ 2 \\ r \end{pmatrix}$, and $\mathbf{c} = \begin{pmatrix} s \\ t \\ 1 \end{pmatrix}$ are mutually perpendicular,

$\mathbf{a} \bullet \mathbf{b} = 0$, $\mathbf{b} \bullet \mathbf{c} = 0$, and $\mathbf{a} \bullet \mathbf{c} = 0$

$\therefore \begin{pmatrix} 1 \\ 2 \\ 3 \end{pmatrix} \bullet \begin{pmatrix} 2 \\ 2 \\ r \end{pmatrix} = 0$ $\quad\therefore\ 2 + 4 + 3r = 0$
$\therefore\ 3r = -6$
$\therefore\ r = -2$

and $\begin{pmatrix} 2 \\ 2 \\ -2 \end{pmatrix} \bullet \begin{pmatrix} s \\ t \\ 1 \end{pmatrix} = 0$ $\quad\therefore\ 2s + 2t - 2 = 0$
$\therefore\ s + t = 1$ .... (1)

and $\begin{pmatrix} 1 \\ 2 \\ 3 \end{pmatrix} \bullet \begin{pmatrix} s \\ t \\ 1 \end{pmatrix} = 0$ $\quad\therefore\ s + 2t + 3 = 0$
$\therefore\ s + 2t = -3$ .... (2)

(2) – (1) gives $t = -4$ and so $s = 5$

$\therefore\ r = -2$, $s = 5$, and $t = -4$

**23** **a** Let $\theta$ be the angle between the vectors $\mathbf{i} = \begin{pmatrix} 1 \\ 0 \\ 0 \end{pmatrix}$ and $\begin{pmatrix} 1 \\ 2 \\ 3 \end{pmatrix}$.

Then $\cos\theta = \dfrac{\begin{pmatrix} 1 \\ 0 \\ 0 \end{pmatrix} \bullet \begin{pmatrix} 1 \\ 2 \\ 3 \end{pmatrix}}{\left|\begin{pmatrix} 1 \\ 0 \\ 0 \end{pmatrix}\right|\left|\begin{pmatrix} 1 \\ 2 \\ 3 \end{pmatrix}\right|} = \dfrac{1}{\sqrt{1}\sqrt{1+4+9}} = \dfrac{1}{\sqrt{14}}$ and so $\theta \approx 74.5^\circ$.

**b** Let $\theta$ be the angle between the vectors $\mathbf{j} = \begin{pmatrix} 0 \\ 1 \\ 0 \end{pmatrix}$ and $\begin{pmatrix} -1 \\ 1 \\ 3 \end{pmatrix}$.

Then $\cos\theta = \dfrac{\begin{pmatrix} 0 \\ 1 \\ 0 \end{pmatrix} \bullet \begin{pmatrix} -1 \\ 1 \\ 3 \end{pmatrix}}{\sqrt{1}\sqrt{1+1+9}} = \dfrac{1}{\sqrt{11}}$ and so $\theta \approx 72.5^\circ$.

## REVIEW SET 12A

**1** **a**

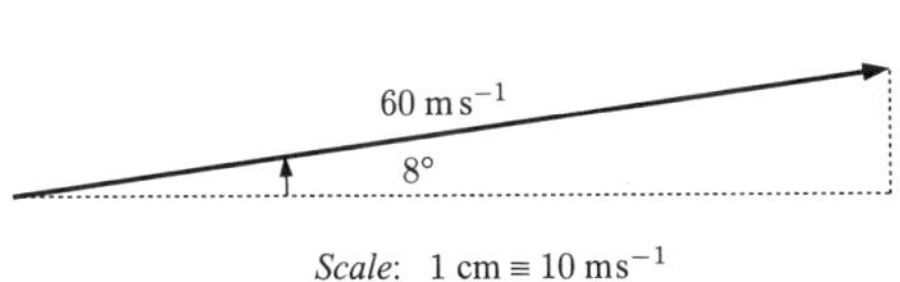


**b**

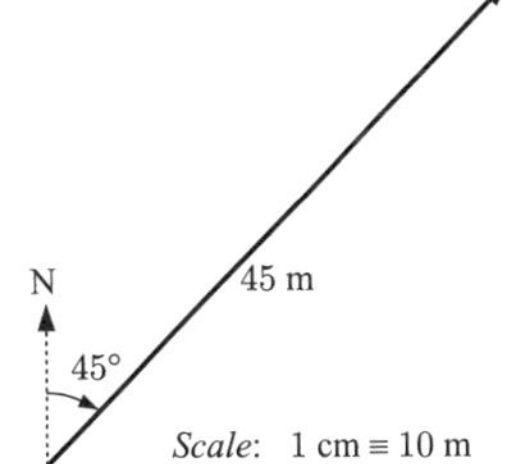


**2** **a** $\overrightarrow{AB} - \overrightarrow{CB} = \overrightarrow{AB} + \overrightarrow{BC} = \overrightarrow{AC}$

**b** $\overrightarrow{AB} + \overrightarrow{BC} - \overrightarrow{DC} = \overrightarrow{AC} + \overrightarrow{CD} = \overrightarrow{AD}$

**3** **a** $\mathbf{q} = \mathbf{p} + \mathbf{r}$

**b** $\mathbf{l} = \mathbf{k} - \mathbf{j} + \mathbf{n} - \mathbf{m}$

**4** $\overrightarrow{SP} = \overrightarrow{SR} + \overrightarrow{RQ} + \overrightarrow{QP}$
$= -\overrightarrow{RS} + \overrightarrow{RQ} - \overrightarrow{PQ}$
$= -\begin{pmatrix} 2 \\ -3 \end{pmatrix} + \begin{pmatrix} -1 \\ 2 \end{pmatrix} - \begin{pmatrix} -4 \\ 1 \end{pmatrix} = \begin{pmatrix} 1 \\ 4 \end{pmatrix}$

**5** **a** $\overrightarrow{BC} = 2\overrightarrow{OA} = 2\mathbf{p}$

Now $\overrightarrow{AC} = \overrightarrow{AO} + \overrightarrow{OB} + \overrightarrow{BC}$
$= -\mathbf{p} + \mathbf{q} + 2\mathbf{p}$
$= \mathbf{p} + \mathbf{q}$

**b** $\overrightarrow{OM} = \overrightarrow{OA} + \overrightarrow{AM}$
$= \mathbf{p} + \frac{1}{2}\overrightarrow{AC}$
$= \mathbf{p} + \frac{1}{2}(\mathbf{p} + \mathbf{q})$
$= \frac{3}{2}\mathbf{p} + \frac{1}{2}\mathbf{q}$

**6** The vectors are parallel, so $\begin{pmatrix} -12 \\ -20 \\ 2 \end{pmatrix} = k\begin{pmatrix} 3 \\ m \\ n \end{pmatrix}$ $\therefore$ $3k = -12$, $km = -20$, $kn = 2$
$\therefore$ $k = -4$, $m = 5$, $n = -\frac{1}{2}$

**7** $\overrightarrow{CB} = \overrightarrow{CA} + \overrightarrow{AB} = -\overrightarrow{AC} + \overrightarrow{AB} = \begin{pmatrix} 6 \\ -1 \\ 3 \end{pmatrix} + \begin{pmatrix} 2 \\ -7 \\ 4 \end{pmatrix} = \begin{pmatrix} 8 \\ -8 \\ 7 \end{pmatrix}$

**8** **a** $\mathbf{p} \bullet \mathbf{q} = \begin{pmatrix} 3 \\ -2 \end{pmatrix} \bullet \begin{pmatrix} -1 \\ 5 \end{pmatrix}$
$= -3 + (-10)$
$= -13$

**b** $\mathbf{p} - \mathbf{r} = \begin{pmatrix} 3 \\ -2 \end{pmatrix} - \begin{pmatrix} -3 \\ 4 \end{pmatrix} = \begin{pmatrix} 6 \\ -6 \end{pmatrix}$

$\therefore$ $\mathbf{q} \bullet (\mathbf{p} - \mathbf{r}) = \begin{pmatrix} -1 \\ 5 \end{pmatrix} \bullet \begin{pmatrix} 6 \\ -6 \end{pmatrix} = -6 - 30$
$= -36$

**9** [Figure: parallelogram with vertices X, Z, Y, W]

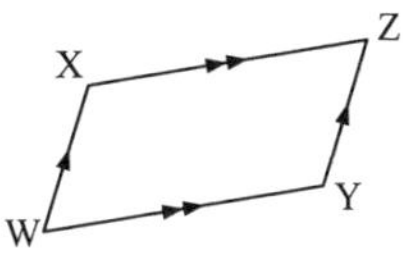

$\overrightarrow{WY} = \begin{pmatrix} 3 - -3 \\ 4 - -1 \end{pmatrix} = \begin{pmatrix} 6 \\ 5 \end{pmatrix}$

$\overrightarrow{XZ} = \begin{pmatrix} 4 - -2 \\ 10 - 5 \end{pmatrix} = \begin{pmatrix} 6 \\ 5 \end{pmatrix}$

So, $\overrightarrow{WY} = \overrightarrow{XZ}$

$\therefore$ [WY] is parallel to [XZ] and they are equal in length. This is sufficient to deduce that WYZX is a parallelogram.

**10** $\overrightarrow{AB} = \begin{pmatrix} -1 - 2 \\ 4 - 3 \end{pmatrix} = \begin{pmatrix} -3 \\ 1 \end{pmatrix}$

$\overrightarrow{AC} = \begin{pmatrix} 3 - 2 \\ k - 3 \end{pmatrix} = \begin{pmatrix} 1 \\ k - 3 \end{pmatrix}$

Now $\overrightarrow{AB} \bullet \overrightarrow{AC} = 0$ {as $B\widehat{A}C = 90^\circ$}

$\therefore$ $\begin{pmatrix} -3 \\ 1 \end{pmatrix} \bullet \begin{pmatrix} 1 \\ k - 3 \end{pmatrix} = 0$

$\therefore$ $-3 + k - 3 = 0$

$\therefore$ $k = 6$

**11** One vector perpendicular to $\begin{pmatrix} -4 \\ 5 \end{pmatrix}$ is $\begin{pmatrix} 5 \\ 4 \end{pmatrix}$ as the dot product $= -20 + 20 = 0$

$\therefore$ all vectors have the form $t\begin{pmatrix} 5 \\ 4 \end{pmatrix}$, $t \neq 0$.

**12**

**a** **i**
$$\begin{aligned} &\overrightarrow{OB} \\ &= \overrightarrow{OA} + \overrightarrow{AB} \\ &= \overrightarrow{OA} + \overrightarrow{OC} \\ &= \mathbf{p} + \mathbf{q} \end{aligned}$$

**ii**
$$\begin{aligned} &\overrightarrow{OM} \\ &= \overrightarrow{OA} + \overrightarrow{AM} \\ &= \overrightarrow{OA} + \tfrac{1}{2}\overrightarrow{AC} \\ &= \mathbf{p} + \tfrac{1}{2}(\overrightarrow{AO} + \overrightarrow{OC}) \\ &= \mathbf{p} + \tfrac{1}{2}(-\mathbf{p} + \mathbf{q}) \\ &= \mathbf{p} - \tfrac{1}{2}\mathbf{p} + \tfrac{1}{2}\mathbf{q} \\ &= \tfrac{1}{2}\mathbf{p} + \tfrac{1}{2}\mathbf{q} \end{aligned}$$

**b** We notice that $\overrightarrow{OM} = \frac{1}{2}\overrightarrow{OB}$

$\therefore$ [OM] ∥ [OB] and $\text{OM} = \frac{1}{2}\text{OB}$

So, O, M, and B are collinear (as O is common) and hence M is the midpoint of [OB].

**13** **a**
$$\begin{aligned} \mathbf{a} \bullet \mathbf{b} &= |\mathbf{a}||\mathbf{b}|\cos\theta \\ &= 2 \times 4 \times \cos 120° \\ &= 8(-\tfrac{1}{2}) \\ &= -4 \end{aligned}$$

**b**
$$\begin{aligned} \mathbf{b} \bullet \mathbf{c} &= |\mathbf{b}||\mathbf{c}|\cos\theta \\ &= 4 \times 5 \times \cos 60° \\ &= 20(\tfrac{1}{2}) \\ &= 10 \end{aligned}$$

**c**
$$\begin{aligned} \mathbf{a} \bullet \mathbf{c} &= |\mathbf{a}||\mathbf{c}|\cos\theta \\ &= 2 \times 5 \times \cos 180° \\ &= 10(-1) \\ &= -10 \end{aligned}$$

**14** $\overrightarrow{JK} = \begin{pmatrix} 6 \\ -3 \\ -3 \end{pmatrix}$, $\overrightarrow{JL} = \begin{pmatrix} a+4 \\ b-1 \\ -1 \end{pmatrix}$

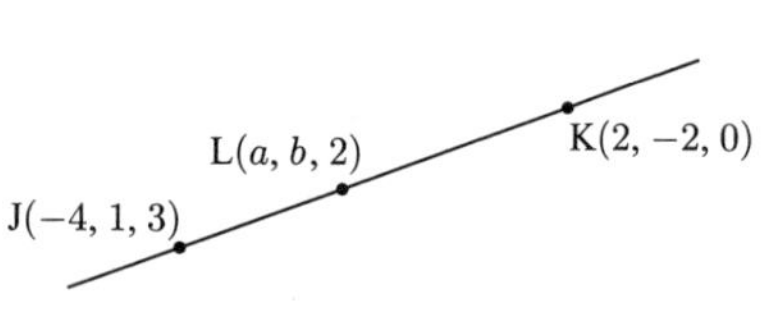


If J, K, and L are collinear then $\overrightarrow{JK} \parallel \overrightarrow{JL}$

$\therefore \begin{pmatrix} 6 \\ -3 \\ -3 \end{pmatrix} = k\begin{pmatrix} a+4 \\ b-1 \\ -1 \end{pmatrix}$ for some $k \neq 0$

$\therefore k = 3$

$\therefore a + 4 = 2$ and $b - 1 = -1$

$\therefore a = -2$ and $b = 0$

**15**

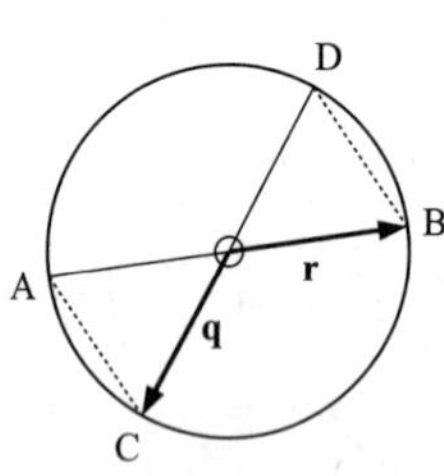


**a**
$$\begin{aligned} &\overrightarrow{DB} \\ &= \overrightarrow{DO} + \overrightarrow{OB} \\ &= \overrightarrow{OC} + \overrightarrow{OB} \\ &= \mathbf{q} + \mathbf{r} \end{aligned}$$

**b**
$$\begin{aligned} &\overrightarrow{AC} \\ &= \overrightarrow{AO} + \overrightarrow{OC} \\ &= \overrightarrow{OB} + \overrightarrow{OC} \\ &= \mathbf{r} + \mathbf{q} \end{aligned}$$

We see that $\overrightarrow{DB} = \overrightarrow{AC}$

$\therefore$ [DB] is parallel to [AC] and equal in length.

**16** **a** $\begin{pmatrix} 2-t \\ 3 \\ t \end{pmatrix} \bullet \begin{pmatrix} t \\ 4 \\ t+1 \end{pmatrix} = 0$

$\therefore (2-t)t + 12 + t(t+1) = 0$

$\therefore 2t - t^2 + 12 + t^2 + t = 0$

$\therefore 3t + 12 = 0$

$\therefore t = -4$

**b** $\overrightarrow{KL} = \begin{pmatrix} -7 \\ 1 \\ 3 \end{pmatrix}$, $\overrightarrow{KM} = \begin{pmatrix} -2 \\ -2 \\ -1 \end{pmatrix}$, $\overrightarrow{LM} = \begin{pmatrix} 5 \\ -3 \\ -4 \end{pmatrix}$

Now $\overrightarrow{KM} \bullet \overrightarrow{LM} = \begin{pmatrix} -2 \\ -2 \\ -1 \end{pmatrix} \bullet \begin{pmatrix} 5 \\ -3 \\ -4 \end{pmatrix}$

$= -2(5) - 2(-3) - 1(-4) = 0$

$\therefore$ [KM] and [LM] are perpendicular.

$\therefore$ triangle KLM is right angled at M.

## REVIEW SET 12B

**1** **a**

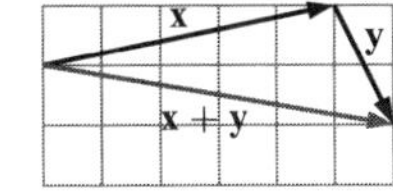


**b**

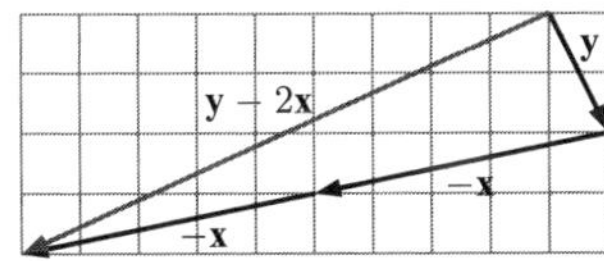


**2** $\overrightarrow{AB} = \begin{pmatrix} 4-(-2) \\ 0-(-1) \\ -1-3 \end{pmatrix} = \begin{pmatrix} 6 \\ 1 \\ -4 \end{pmatrix}$, $\overrightarrow{AC} = \begin{pmatrix} -2-(-2) \\ 1-(-1) \\ -4-3 \end{pmatrix} = \begin{pmatrix} 0 \\ 2 \\ -7 \end{pmatrix}$,

$\overrightarrow{BC} = \begin{pmatrix} -2-4 \\ 1-0 \\ -4-(-1) \end{pmatrix} = \begin{pmatrix} -6 \\ 1 \\ -3 \end{pmatrix}$

$\therefore$ $AB = \sqrt{6^2+1^2+(-4)^2} = \sqrt{53}$ units  $AC = \sqrt{0^2+2^2+(-7)^2} = \sqrt{53}$ units  $BC = \sqrt{(-6)^2+1^2+(-3)^2} = \sqrt{46}$ units

$\therefore$ AB = AC, so ABC is an isosceles triangle.

**3** **a** $|\mathbf{s}| = \sqrt{(-3)^2+2^2} = \sqrt{13}$ units

**b** $\mathbf{r}+\mathbf{s} = \begin{pmatrix} 4 \\ 1 \end{pmatrix} + \begin{pmatrix} -3 \\ 2 \end{pmatrix} = \begin{pmatrix} 1 \\ 3 \end{pmatrix}$

$\therefore$ $|\mathbf{r}+\mathbf{s}| = \sqrt{1^2+3^2} = \sqrt{10}$ units

**c** $2\mathbf{s}-\mathbf{r} = 2\begin{pmatrix} -3 \\ 2 \end{pmatrix} - \begin{pmatrix} 4 \\ 1 \end{pmatrix} = \begin{pmatrix} -10 \\ 3 \end{pmatrix}$

$\therefore$ $|2\mathbf{s}-\mathbf{r}| = \sqrt{(-10)^2+3^2} = \sqrt{109}$ units

**4** $r\begin{pmatrix} -2 \\ 1 \end{pmatrix} + s\begin{pmatrix} 3 \\ -4 \end{pmatrix} = \begin{pmatrix} 13 \\ -24 \end{pmatrix}$

$\therefore$ $\begin{pmatrix} -2r+3s \\ r-4s \end{pmatrix} = \begin{pmatrix} 13 \\ -24 \end{pmatrix}$

$\therefore$ $-2r+3s = 13$
$r-4s = -24$ .... (1)

$\therefore$ $-2r+3s = 13$
$2r-8s = -48$ $\{2\times(1)\}$
adding $-5s = -35$
$\therefore$ $s = 7$

Now using (1), $r-4(7) = -24$
$\therefore$ $r = -24+28$
$\therefore$ $r = 4$ and $s = 7$

**5** **a** $\overrightarrow{PQ} = \begin{pmatrix} -4-2 \\ 4-3 \\ 2--1 \end{pmatrix} = \begin{pmatrix} -6 \\ 1 \\ 3 \end{pmatrix}$

**b** $PQ = |\overrightarrow{PQ}| = \sqrt{36+1+9} = \sqrt{46}$ units

**c** The midpoint is at $\left(\frac{2+-4}{2}, \frac{3+4}{2}, \frac{-1+2}{2}\right)$ which is $(-1, \frac{7}{2}, \frac{1}{2})$ or $(-1, 3\frac{1}{2}, \frac{1}{2})$.

**6** A (4, 2, −1), B (−1, 5, 2), C (3, −3, c)

$\overrightarrow{BA} = \begin{pmatrix} 4--1 \\ 2-5 \\ -1-2 \end{pmatrix} = \begin{pmatrix} 5 \\ -3 \\ -3 \end{pmatrix}$

$\overrightarrow{BC} = \begin{pmatrix} 3--1 \\ -3-5 \\ c-2 \end{pmatrix} = \begin{pmatrix} 4 \\ -8 \\ c-2 \end{pmatrix}$

But $\overrightarrow{BA} \bullet \overrightarrow{BC} = 0$
$\therefore$ $20+24-3(c-2) = 0$
$\therefore$ $44 = 3(c-2)$
$\therefore$ $3c-6 = 44$
$\therefore$ $3c = 50$
$\therefore$ $c = \frac{50}{3}$

**7** $\mathbf{a}-3\mathbf{x} = \mathbf{b}$ $\therefore$ $\mathbf{a}-\mathbf{b} = 3\mathbf{x}$ $\therefore$ $\mathbf{x} = \frac{1}{3}(\mathbf{a}-\mathbf{b}) = \frac{1}{3}\begin{pmatrix} 3 \\ -5 \\ -2 \end{pmatrix} = \begin{pmatrix} 1 \\ -\frac{5}{3} \\ -\frac{2}{3} \end{pmatrix}$

**8** If the angle is $\theta$ then $\begin{pmatrix} 3 \\ 1 \\ -2 \end{pmatrix} \bullet \begin{pmatrix} 2 \\ 5 \\ 1 \end{pmatrix} = \sqrt{9+1+4}\sqrt{4+25+1}\cos\theta$

$$\therefore \quad 6+5-2 = \sqrt{14}\sqrt{30}\cos\theta$$

$$\therefore \quad \frac{9}{\sqrt{14\times 30}} = \cos\theta$$

$$\therefore \quad \theta \approx 64.0°$$

**9** Let Q$(0, 0, z)$ be a point on the $Z$-axis.

$$\text{PQ} = \sqrt{4^2+(-2)^2+(z-5)^2} = 6$$

$$\therefore \quad 16+4+(z-5)^2 = 36$$

$$\therefore \quad (z-5)^2 = 16$$

$$\therefore \quad z-5 = \pm 4$$

$\therefore \quad z = 1$ or $9$ $\quad \therefore$ Q is $(0, 0, 1)$ or $(0, 0, 9)$.

**10** Since they are perpendicular, $\begin{pmatrix} 3 \\ 3-2t \end{pmatrix} \bullet \begin{pmatrix} t^2+t \\ -2 \end{pmatrix} = 0$

$$\therefore \quad 3(t^2+t) - 2(3-2t) = 0$$

$$\therefore \quad 3t^2+3t-6+4t = 0$$

$$\therefore \quad 3t^2+7t-6 = 0$$

$$\therefore \quad (3t-2)(t+3) = 0$$

$$\therefore \quad t = \tfrac{2}{3} \text{ or } -3$$

**11** **a** $\mathbf{u} \bullet \mathbf{v}$

$$= \begin{pmatrix} -4 \\ 2 \\ 1 \end{pmatrix} \bullet \begin{pmatrix} -1 \\ 3 \\ -2 \end{pmatrix}$$

$$= -4(-1) + 2(3) + 1(-2)$$

$$= 8$$

**b** If $\theta$ is the angle between $\mathbf{u}$ and $\mathbf{v}$ then

$$\cos\theta = \frac{\mathbf{u}\bullet\mathbf{v}}{|\mathbf{u}||\mathbf{v}|}$$

$$= \frac{8}{\sqrt{(-4)^2+2^2+1^2}\sqrt{(-1)^2+3^2+(-2)^2}}$$

$$= \frac{8}{\sqrt{21}\sqrt{14}}$$

$$\therefore \quad \theta \approx 62.2°$$

**12**

**a** $\overrightarrow{AC} = \overrightarrow{AO} + \overrightarrow{OC} = -\mathbf{p} + \mathbf{r} = \mathbf{r} - \mathbf{p}$

$\overrightarrow{BC} = \overrightarrow{BO} + \overrightarrow{OC} = -\mathbf{q} + \mathbf{r} = \mathbf{r} - \mathbf{q}$

**b** [AP] $\perp$ [BC] and [BQ] $\perp$ [AC]

$\therefore \ \mathbf{p} \perp \mathbf{r} - \mathbf{q}$ $\qquad \therefore \ \mathbf{q} \perp (\mathbf{r} - \mathbf{p})$

$\therefore \ \mathbf{p} \bullet (\mathbf{r} - \mathbf{q}) = 0$ $\qquad \therefore \ \mathbf{q} \bullet (\mathbf{r} - \mathbf{p}) = 0$

$\therefore \ \mathbf{p}\bullet\mathbf{r} - \mathbf{p}\bullet\mathbf{q} = 0$ $\qquad \therefore \ \mathbf{q}\bullet\mathbf{r} - \mathbf{q}\bullet\mathbf{p} = 0$

$\therefore \ \mathbf{p}\bullet\mathbf{r} = \mathbf{p}\bullet\mathbf{q}$ $\qquad \therefore \ \mathbf{q}\bullet\mathbf{r} = \mathbf{p}\bullet\mathbf{q}$

$$\therefore \quad \mathbf{q}\bullet\mathbf{r} = \mathbf{p}\bullet\mathbf{q} = \mathbf{p}\bullet\mathbf{r}$$

**c** $\mathbf{r} \bullet \overrightarrow{AB} = \mathbf{r} \bullet (-\mathbf{p} + \mathbf{q})$

$= -\mathbf{r}\bullet\mathbf{p} + \mathbf{r}\bullet\mathbf{q}$

$= -\mathbf{p}\bullet\mathbf{q} + \mathbf{p}\bullet\mathbf{q}$ {from **b**}

$= 0$ and so $\mathbf{r} \perp \overrightarrow{AB}$ $\quad \therefore$ [OC] $\perp$ [AB]

**13** $|3\mathbf{i} - 2\mathbf{j} + \mathbf{k}| = \sqrt{3^2+(-2)^2+1^2} = \sqrt{14}$

$\therefore$ a unit vector in the direction $3\mathbf{i} - 2\mathbf{j} + \mathbf{k}$ is $\frac{1}{\sqrt{14}}(3\mathbf{i} - 2\mathbf{j} + \mathbf{k})$

$\therefore$ a vector 4 units long and parallel to $3\mathbf{i} - 2\mathbf{j} + \mathbf{k}$ is $\pm\frac{4}{\sqrt{14}}(3\mathbf{i} - 2\mathbf{j} + \mathbf{k})$.

**14** M is $\left(\frac{-2+2}{2}, \frac{1+5}{2}, \frac{-3-1}{2}\right)$ or $(0, 3, -2)$.

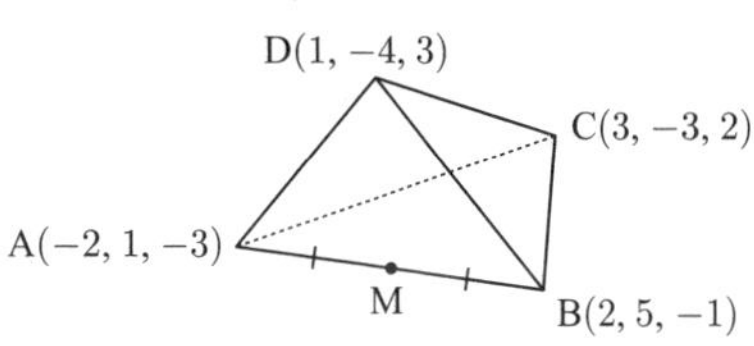


$\therefore \overrightarrow{MD} = \begin{pmatrix} 1 \\ -7 \\ 5 \end{pmatrix}, \quad \overrightarrow{MC} = \begin{pmatrix} 3 \\ -6 \\ 4 \end{pmatrix}$

$\therefore \quad \overrightarrow{MD} \bullet \overrightarrow{MC} = |\overrightarrow{MD}|\,|\overrightarrow{MC}| \cos\theta$

$\therefore \quad 3 + 42 + 20 = \sqrt{1+49+25}\sqrt{9+36+16}\cos\theta$

$\therefore \quad 65 = \sqrt{75}\sqrt{61}\cos\theta$

$\therefore \quad \theta \approx 16.1°$

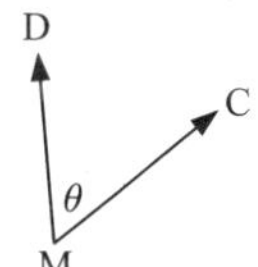


**15** **a** $\sqrt{k^2 + (\frac{1}{\sqrt{2}})^2 + (-k)^2} = 1$

$\therefore \quad k^2 + \frac{1}{2} + k^2 = 1$

$\therefore \quad 2k^2 = \frac{1}{2}$

$\therefore \quad k^2 = \frac{1}{4}$

$\therefore \quad k = \pm\frac{1}{2}$

**b** $\begin{pmatrix} 3 \\ 2 \\ -1 \end{pmatrix}$ has length $\sqrt{3^2 + 2^2 + (-1)^2} = \sqrt{14}$ units

$\therefore$ a unit vector in the opposite direction is $-\frac{1}{\sqrt{14}}\begin{pmatrix} 3 \\ 2 \\ -1 \end{pmatrix}$

$\therefore$ a vector of length 5 units in the opposite direction

is $-\frac{5}{\sqrt{14}}\begin{pmatrix} 3 \\ 2 \\ -1 \end{pmatrix}$.

## REVIEW SET 12C

**1** **a** $\overrightarrow{PR} + \overrightarrow{RQ}$
$= \overrightarrow{PQ}$

**b** $\overrightarrow{PS} + \overrightarrow{SQ} + \overrightarrow{QR}$
$= \overrightarrow{PQ} + \overrightarrow{QR}$
$= \overrightarrow{PR}$

**2** **a** $\mathbf{m} - \mathbf{n} + \mathbf{p} = \begin{pmatrix} 6 \\ -3 \\ 1 \end{pmatrix} - \begin{pmatrix} 2 \\ 3 \\ -4 \end{pmatrix} + \begin{pmatrix} -1 \\ 3 \\ 6 \end{pmatrix} = \begin{pmatrix} 3 \\ -3 \\ 11 \end{pmatrix}$

**b** $2\mathbf{n} - 3\mathbf{p} = 2\begin{pmatrix} 2 \\ 3 \\ -4 \end{pmatrix} - 3\begin{pmatrix} -1 \\ 3 \\ 6 \end{pmatrix} = \begin{pmatrix} 4 \\ 6 \\ -8 \end{pmatrix} - \begin{matrix} -3 \\ 9 \\ 18 \end{matrix} = \begin{pmatrix} 7 \\ -3 \\ -26 \end{pmatrix}$

**c** $\mathbf{m} + \mathbf{p} = \begin{pmatrix} 6 \\ -3 \\ 1 \end{pmatrix} + \begin{pmatrix} -1 \\ 3 \\ 6 \end{pmatrix} = \begin{pmatrix} 5 \\ 0 \\ 7 \end{pmatrix}$ $\quad \therefore \quad |\mathbf{m} + \mathbf{p}| = \sqrt{25 + 0 + 49}$
$= \sqrt{74}$ units

**3** **a** If $\overrightarrow{AB} = \frac{1}{2}\overrightarrow{CD}$ then [AB] ∥ [CD] and [AB] is half the length of [CD].

**b** If $\overrightarrow{AB} = 2\overrightarrow{AC}$ then [AB] ∥ [AC] and AB = 2AC

$\therefore$ A, B and C are collinear and AB = 2AC.

So, C is the midpoint of [AB].

**4** **a** $\overrightarrow{PQ} = \begin{pmatrix} -1-2 \\ 7 - -5 \\ 9-6 \end{pmatrix} = \begin{pmatrix} -3 \\ 12 \\ 3 \end{pmatrix}$

**b** $PQ = \sqrt{(-3)^2 + 12^2 + 3^2}$
$= \sqrt{162}$ units

**c**

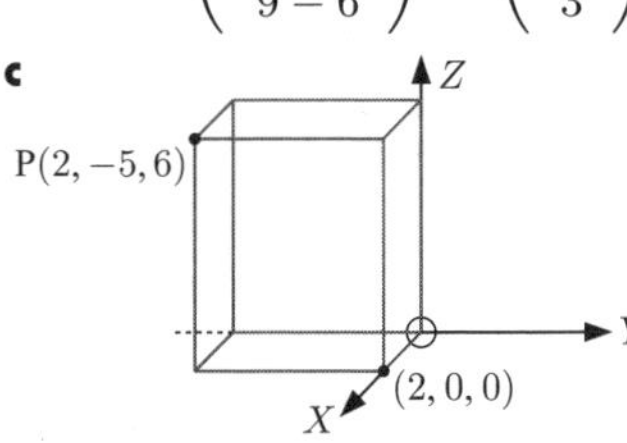


$\therefore$ distance

$= \sqrt{(2-2)^2 + (0 - -5)^2 + (0-6)^2}$

$= \sqrt{0 + 25 + 36}$

$= \sqrt{61}$ units

**5** **a** $\overrightarrow{OQ} = \overrightarrow{OR} + \overrightarrow{RQ} = \mathbf{r} + \mathbf{q}$

**b** $\overrightarrow{PQ} = \overrightarrow{PO} + \overrightarrow{OR} + \overrightarrow{RQ} = -\mathbf{p} + \mathbf{r} + \mathbf{q}$

**c** $\overrightarrow{ON} = \overrightarrow{OR} + \overrightarrow{RN} = \mathbf{r} + \frac{1}{2}\mathbf{q}$

**d**
$$\begin{aligned}\overrightarrow{MN} &= \overrightarrow{MQ} + \overrightarrow{QN}\\ &= \tfrac{1}{2}\overrightarrow{PQ} + \tfrac{1}{2}\overrightarrow{QR}\\ &= \tfrac{1}{2}(-\mathbf{p}+\mathbf{r}+\mathbf{q}) + \tfrac{1}{2}(-\mathbf{q})\\ &= -\tfrac{1}{2}\mathbf{p} + \tfrac{1}{2}\mathbf{r} + \tfrac{1}{2}\mathbf{q} - \tfrac{1}{2}\mathbf{q}\\ &= -\tfrac{1}{2}\mathbf{p} + \tfrac{1}{2}\mathbf{r}\end{aligned}$$

**6** **a**
$$\begin{aligned}\mathbf{p} - 3\mathbf{x} &= \mathbf{0}\\ \therefore\ 3\mathbf{x} &= \mathbf{p}\\ \therefore\ \mathbf{x} &= \tfrac{1}{3}\mathbf{p}\\ &= \tfrac{1}{3}\begin{pmatrix}-3\\1\\2\end{pmatrix}\\ \therefore\ \mathbf{x} &= \begin{pmatrix}-1\\ \frac{1}{3}\\ \frac{2}{3}\end{pmatrix}\end{aligned}$$

**b**
$$\begin{aligned}2\mathbf{q} - \mathbf{x} &= \mathbf{r}\\ \therefore\ \mathbf{r} + \mathbf{x} &= 2\mathbf{q}\\ \therefore\ \mathbf{x} &= 2\mathbf{q} - \mathbf{r}\\ &= 2\begin{pmatrix}2\\-4\\1\end{pmatrix} - \begin{pmatrix}3\\2\\0\end{pmatrix}\\ \therefore\ \mathbf{x} &= \begin{pmatrix}4\\-8\\2\end{pmatrix} - \begin{pmatrix}3\\2\\0\end{pmatrix} = \begin{pmatrix}4-3\\-8-2\\2-0\end{pmatrix} = \begin{pmatrix}1\\-10\\2\end{pmatrix}\end{aligned}$$

**7** Since $\mathbf{v}$ is parallel to $\mathbf{w}$, the angle $\theta$ between $\mathbf{v}$ and $\mathbf{w}$ is either $0°$ or $180°$.

Now,
$$\begin{aligned}\mathbf{v} \bullet \mathbf{w} &= |\mathbf{v}|\,|\mathbf{w}|\cos\theta\\ &= 3\times 2\times\cos 0° \quad\text{or}\quad 3\times 2\times\cos 180°\\ &= 6(1) \quad\text{or}\quad 6(-1)\\ &= \pm 6\end{aligned}$$

**8** As the vectors are perpendicular,
$$\begin{aligned}\begin{pmatrix}-4\\t+2\\t\end{pmatrix} \bullet \begin{pmatrix}t\\1+t\\-3\end{pmatrix} &= 0\\ \therefore\ -4t + (t+2)(1+t) - 3t &= 0\\ \therefore\ -4t + t + t^2 + 2 + 2t - 3t &= 0\\ \therefore\ t^2 - 4t + 2 &= 0\\ \therefore\ t &= \frac{4 \pm \sqrt{16 - 4(1)(2)}}{2}\\ \therefore\ t &= \frac{4\pm\sqrt{8}}{2} = 2 \pm \sqrt{2}\end{aligned}$$

**9**
$$\overrightarrow{MK} = \begin{pmatrix}3-4\\1-1\\4-3\end{pmatrix} = \begin{pmatrix}-1\\0\\1\end{pmatrix} \qquad \overrightarrow{LK} = \begin{pmatrix}3--2\\1-1\\4-3\end{pmatrix} = \begin{pmatrix}5\\0\\1\end{pmatrix}$$

$$\overrightarrow{ML} = \begin{pmatrix}-2-4\\1-1\\3-3\end{pmatrix} = \begin{pmatrix}-6\\0\\0\end{pmatrix} \qquad \overrightarrow{LM} = \begin{pmatrix}4--2\\1-1\\3-3\end{pmatrix} = \begin{pmatrix}6\\0\\0\end{pmatrix}$$

$$\begin{aligned}\overrightarrow{MK} \bullet \overrightarrow{ML} &= |\overrightarrow{MK}|\,|\overrightarrow{ML}|\cos\widehat{M}\\ \therefore\ 6+0+0 &= \sqrt{1+0+1}\sqrt{36+0+0}\cos\widehat{M}\\ \therefore\ 6 &= \sqrt{2}\times 6\cos\widehat{M}\\ \therefore\ \cos\widehat{M} &= \tfrac{1}{\sqrt{2}}\\ \therefore\ \widehat{M} &= 45°\end{aligned}$$

$$\begin{aligned}\therefore\ \overrightarrow{LK} \bullet \overrightarrow{LM} &= |\overrightarrow{LK}|\,|\overrightarrow{LM}|\cos\widehat{L}\\ \therefore\ 30+0+0 &= \sqrt{25+0+1}\sqrt{36+0+0}\cos\widehat{L}\\ \therefore\ 30 &= \sqrt{26}\times 6\cos\widehat{L}\\ \therefore\ \tfrac{5}{\sqrt{26}} &= \cos\widehat{L}\\ \therefore\ \widehat{L} &\approx 11.3°\end{aligned}$$

and $\widehat{K} \approx 180° - 45° - 11.3° \approx 123.7°$

**10** **a** $\begin{pmatrix}\frac{5}{13}\\k\end{pmatrix}$ is a unit vector if
$$\begin{aligned}\sqrt{\left(\tfrac{5}{13}\right)^2 + k^2} &= 1\\ \therefore\ \tfrac{25}{169} + k^2 &= 1\\ \therefore\ k^2 &= \tfrac{144}{169}\\ \therefore\ k &= \pm\tfrac{12}{13}\end{aligned}$$

**b** $\begin{pmatrix}k\\k\\k\end{pmatrix}$ is a unit vector if
$$\begin{aligned}\sqrt{k^2+k^2+k^2} &= 1\\ \therefore\ 3k^2 &= 1\\ \therefore\ k^2 &= \tfrac{1}{3}\\ \therefore\ k &= \pm\tfrac{1}{\sqrt{3}}\end{aligned}$$

**11** If D is the origin, (DA) the $X$-axis, (DC) the $Y$-axis, and (DE) the $Z$-axis, then A is (4, 0, 0), C is (0, 8, 0), and G is (4, 8, 5).

$$\overrightarrow{AG} = \begin{pmatrix} 4-4 \\ 8-0 \\ 5-0 \end{pmatrix} = \begin{pmatrix} 0 \\ 8 \\ 5 \end{pmatrix}, \quad \overrightarrow{AC} = \begin{pmatrix} 0-4 \\ 8-0 \\ 0-0 \end{pmatrix} = \begin{pmatrix} -4 \\ 8 \\ 0 \end{pmatrix}$$

If we let $\widehat{GAC}$ be $\theta$ then $\begin{pmatrix} 0 \\ 8 \\ 5 \end{pmatrix} \bullet \begin{pmatrix} -4 \\ 8 \\ 0 \end{pmatrix} = \sqrt{0+64+25}\sqrt{16+64+0}\cos\theta$

$$\therefore \quad 0+64+0 = \sqrt{89}\sqrt{80}\cos\theta$$

$$\therefore \quad \cos\theta = \frac{64}{\sqrt{89 \times 80}} \quad \text{and so} \quad \theta \approx 40.7°$$

$$\therefore \quad \widehat{GAC} \approx 40.7°$$

**12** LHS $= \mathbf{p} \bullet (\mathbf{q} - \mathbf{r})$

$$= \begin{pmatrix} 3 \\ -2 \end{pmatrix} \bullet \left[\begin{pmatrix} -2 \\ 5 \end{pmatrix} - \begin{pmatrix} 1 \\ -3 \end{pmatrix}\right]$$

$$= \begin{pmatrix} 3 \\ -2 \end{pmatrix} \bullet \begin{pmatrix} -3 \\ -8 \end{pmatrix}$$

$$= -9 - 16 = -25$$

RHS $= \mathbf{p} \bullet \mathbf{q} - \mathbf{p} \bullet \mathbf{r}$

$$= \begin{pmatrix} 3 \\ -2 \end{pmatrix} \bullet \begin{pmatrix} -2 \\ 5 \end{pmatrix} - \begin{pmatrix} 3 \\ -2 \end{pmatrix} \bullet \begin{pmatrix} 1 \\ -3 \end{pmatrix}$$

$$= (-6-10) - (3+6)$$

$$= -16 - 9$$

$$= -25 \qquad \therefore \text{ LHS} = \text{RHS} \ \checkmark$$

**13** $\overrightarrow{MP} \bullet \overrightarrow{PT} = 0$

Now, $\begin{pmatrix} 5 \\ -1 \end{pmatrix} \bullet \begin{pmatrix} 1 \\ 5 \end{pmatrix} = 5 + (-5) = 0$

$\therefore$ $\overrightarrow{PT}$ has the form $k\begin{pmatrix} 1 \\ 5 \end{pmatrix}$, where $k$ is a scalar.

Also, $|\overrightarrow{MP}| = |\overrightarrow{PT}|$

$$\therefore \quad \sqrt{5^2 + (-1)^2} = \sqrt{k^2 + (5k)^2}$$

$$\therefore \quad \sqrt{26} = \sqrt{26k^2}$$

$$= |k|\sqrt{26}$$

$$\therefore \quad k = \pm 1$$

So $\overrightarrow{PT} = \begin{pmatrix} 1 \\ 5 \end{pmatrix}$ or $\begin{pmatrix} -1 \\ -5 \end{pmatrix}$

Finally, $\overrightarrow{OT} = \overrightarrow{OM} + \overrightarrow{MP} + \overrightarrow{PT}$

$$= \begin{pmatrix} -2 \\ 4 \end{pmatrix} + \begin{pmatrix} 5 \\ -1 \end{pmatrix} + \begin{pmatrix} 1 \\ 5 \end{pmatrix}$$

or $\begin{pmatrix} -2 \\ 4 \end{pmatrix} + \begin{pmatrix} 5 \\ -1 \end{pmatrix} + \begin{pmatrix} -1 \\ -5 \end{pmatrix}$

$$= \begin{pmatrix} 4 \\ 8 \end{pmatrix} \text{ or } \begin{pmatrix} 2 \\ -2 \end{pmatrix}$$

**14** **a** $\mathbf{p} \bullet \mathbf{q}$

$$= \begin{pmatrix} 2 \\ -1 \\ 4 \end{pmatrix} \bullet \begin{pmatrix} -1 \\ -4 \\ 2 \end{pmatrix}$$

$$= 2(-1) - 1(-4) + 4(2)$$

$$= 10$$

**b** If $\theta$ is the angle between $\mathbf{p}$ and $\mathbf{q}$ then

$$\cos\theta = \frac{\mathbf{p} \bullet \mathbf{q}}{|\mathbf{p}||\mathbf{q}|} = \frac{10}{\sqrt{2^2 + (-1)^2 + 4^2}\sqrt{(-1)^2 + (-4)^2 + 2^2}}$$

$$= \frac{10}{\sqrt{21}\sqrt{21}}$$

$$\therefore \quad \theta \approx 61.6°$$

**15** $\mathbf{u} = \begin{pmatrix} 2 \\ 1 \end{pmatrix}, \quad \mathbf{v} = \begin{pmatrix} 0 \\ 3 \end{pmatrix}$

$$\cos\theta = \frac{\mathbf{u} \bullet \mathbf{v}}{|\mathbf{u}||\mathbf{v}|} = \frac{\begin{pmatrix} 2 \\ 1 \end{pmatrix} \bullet \begin{pmatrix} 0 \\ 3 \end{pmatrix}}{\sqrt{2^2 + 1^2}\sqrt{3^2}}$$

$$= \frac{3}{\sqrt{5}\sqrt{9}} = \frac{1}{\sqrt{5}}$$

Now $\sin^2\theta + \cos^2\theta = 1$

$$\therefore \quad \sin^2\theta + \tfrac{1}{5} = 1$$

$$\therefore \quad \sin^2\theta = \tfrac{4}{5}$$

$$\therefore \quad \sin\theta = \pm\tfrac{2}{\sqrt{5}}$$

But $\theta$ is acute, so $\sin\theta = \frac{2}{\sqrt{5}}$.

# Chapter 13

## VECTOR APPLICATIONS

### EXERCISE 13A

**1** **a**

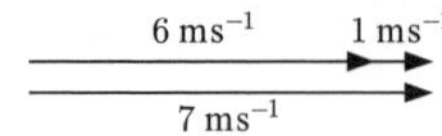


If the athlete is assisted by a wind of 1 m s$^{-1}$ his speed will be 7 m s$^{-1}$.

**b**

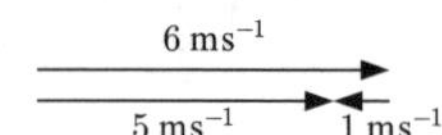


If the athlete runs into a head wind of 1 m s$^{-1}$ his speed will be 5 m s$^{-1}$.

**2** **a**

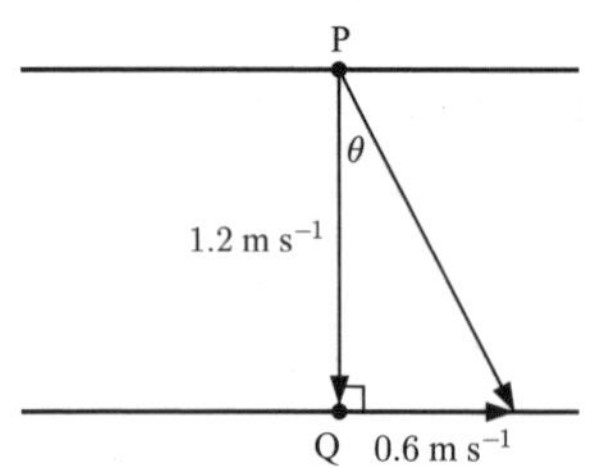


$$(\text{actual speed})^2 = (\text{swimming speed})^2 + (\text{current})^2 = 1.2^2 + 0.6^2 = 1.8$$

$\therefore$ actual speed $= \sqrt{1.8} \approx 1.34$ m s$^{-1}$

$\tan\theta = \dfrac{0.6}{1.2}$

$\therefore \theta \approx 26.6°$

$\therefore$ Mary's actual velocity is approximately 1.34 m s$^{-1}$ in the direction 26.6° to the left of her intended line.

**b** **i**

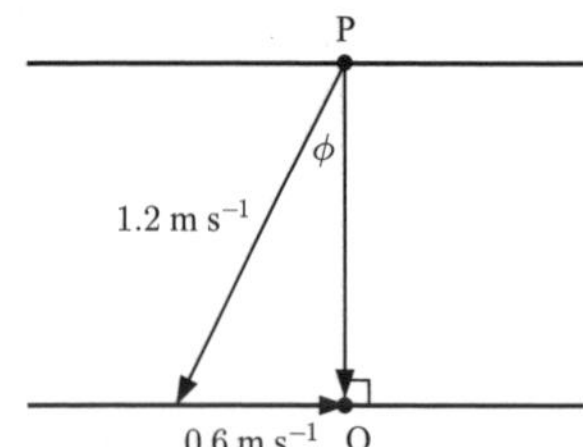


Mary needs to aim to the right of Q so the current will correct her direction.

$\sin\phi = \dfrac{0.6}{1.2}$

$\therefore \phi = 30°$

$\therefore$ Mary should aim to swim 30° to the right of Q.

**ii**

$$(\text{swimming speed})^2 = (\text{actual speed})^2 + (\text{current})^2$$

$$\therefore (\text{actual speed})^2 = 1.2^2 - 0.6^2 = 1.08$$

$\therefore$ Mary's actual speed $= \sqrt{1.08} \approx 1.04$ m s$^{-1}$

**3**

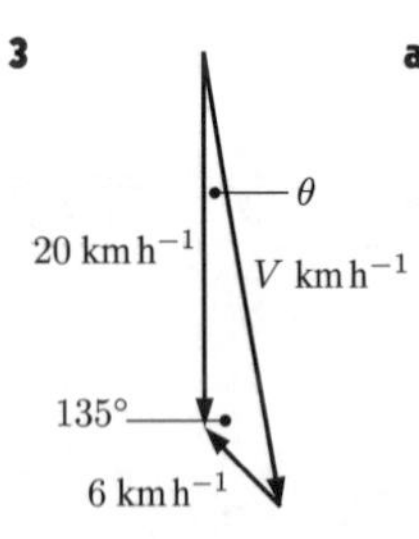


**a** Using the cosine rule,

$V^2 = 20^2 + 6^2 - 2 \times 20 \times 6 \times \cos 135°$

$\therefore V \approx 24.6$

$\therefore$ the equivalent speed in still water is 24.6 km h$^{-1}$.

**b** Using the sine rule,

$\dfrac{\sin\theta}{6} \approx \dfrac{\sin 135°}{24.61}$

$\therefore \theta \approx \sin^{-1}\left(\dfrac{6 \times \sin 135°}{24.61}\right)$

$\therefore \theta \approx 9.93°$

$\therefore$ the boat should head 9.93° east of south.

**4**

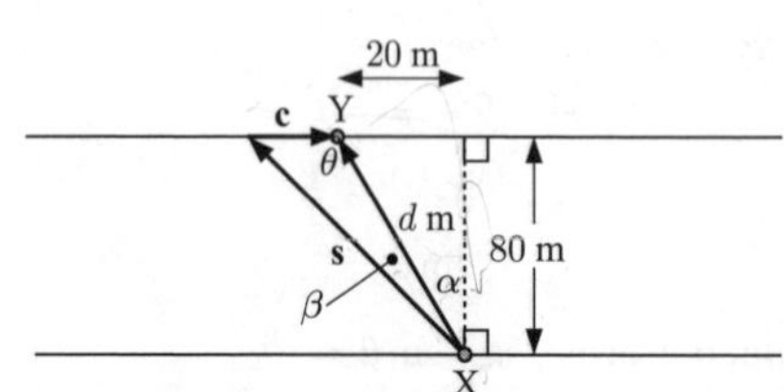


**a** $d^2 = 80^2 + 20^2$ {Pythagoras}

$\therefore d = \sqrt{80^2 + 20^2}$ $\{d > 0\}$

$\therefore d \approx 82.5$

$\therefore$ the distance from X to Y is about 82.5 m.

**b** $\alpha = \tan^{-1}(\frac{20}{80}) \approx 14.04°$

$\therefore\ \theta \approx 90° + 14.04°$ {exterior angle of △}

$\therefore\ \theta \approx 104.04°$

In $t$ seconds, Stephanie can swim $1.8t$ metres, and the current will move $0.3t$ metres.

$\therefore\ |\mathbf{s}| = 1.8t$ and $|\mathbf{c}| = 0.3t$

Using the sine rule,

$$\frac{\sin\beta}{0.3t} = \frac{\sin\theta}{1.8t}$$

$$\beta \approx \sin^{-1}\left(\frac{0.3 \times \sin 104.04°}{1.8}\right)$$

$\therefore\ \beta \approx 9.31°$

$\therefore\ \alpha + \beta \approx 23.3°$

$\therefore$ Stephanie should head 23.3° to the left of the perpendicular across the river.

**c** $\tan(\alpha+\beta) = \dfrac{20 + 0.3t}{80}$

$\therefore\ 20 + 0.3t \approx 80\tan(23.34°)$

$\therefore\ t \approx \dfrac{80\tan(23.34°) - 20}{0.3}$

$\therefore\ t \approx 48.4$

$\therefore$ Stephanie will take 48.4 seconds to cross the river.

**5**

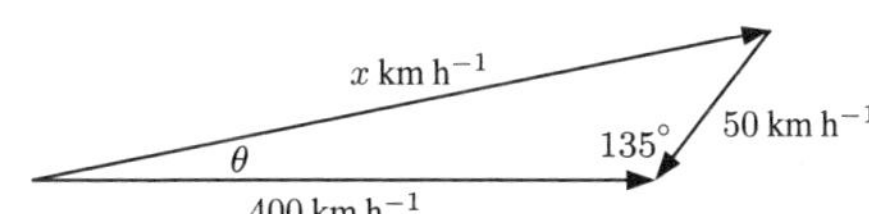


**a** Using the cosine rule,

$x^2 = 50^2 + 400^2 - 2 \times 50 \times 400 \cos 135°$

$\therefore\ x \approx 436.79$

The aeroplane should fly so that its speed in still air would be 437 km h$^{-1}$.
The wind slows the aeroplane down to 400 km h$^{-1}$.

**b** Using the sine rule,

$$\frac{\sin\theta}{50} \approx \frac{\sin 135°}{436.79}$$

$\therefore\ \theta \approx 4.64°$

The aeroplane should head 4.64° north of east.

## EXERCISE 13B

**1** **a** **i** $\begin{pmatrix} x \\ y \end{pmatrix} = \begin{pmatrix} 3 \\ -4 \end{pmatrix} + t\begin{pmatrix} 1 \\ 4 \end{pmatrix},\quad t \in \mathbb{R}$

**ii** $x = 3 + t,\quad y = -4 + 4t,\quad t \in \mathbb{R}$

**iii** $t = x - 3 = \dfrac{y+4}{4}$

$\therefore\ 4x - 12 = y + 4$

$\therefore\ 4x - y = 16$

**b** **i** If the line has direction vector $\mathbf{b}$ perpendicular to $\begin{pmatrix} 5 \\ 2 \end{pmatrix}$, then

$\mathbf{b} \bullet \begin{pmatrix} 5 \\ 2 \end{pmatrix} = 0$

$\therefore\ \mathbf{b} = \begin{pmatrix} -2 \\ 5 \end{pmatrix}$ is a reasonable choice

$\therefore\ \begin{pmatrix} x \\ y \end{pmatrix} = \begin{pmatrix} 5 \\ 2 \end{pmatrix} + t\begin{pmatrix} -2 \\ 5 \end{pmatrix},\quad t \in \mathbb{R}$

**ii** $x = 5 - 2t,\quad y = 2 + 5t,\quad t \in \mathbb{R}$

**iii** $t = \dfrac{x-5}{-2} = \dfrac{y-2}{5}$

$\therefore\ 5x - 25 = -2y + 4$

$\therefore\ 5x + 2y = 29$

**c** **i** $\begin{pmatrix} x \\ y \end{pmatrix} = \begin{pmatrix} -6 \\ 0 \end{pmatrix} + t\begin{pmatrix} 3 \\ 7 \end{pmatrix},\quad t \in \mathbb{R}$

**ii** $x = -6 + 3t,\quad y = 7t,\quad t \in \mathbb{R}$

**iii** $t = \dfrac{x+6}{3} = \dfrac{y}{7}$

$\therefore\ 7x + 42 = 3y$

$\therefore\ 7x - 3y = -42$

**d** **i** Take $(-1, 11)$ as our fixed point,

so $\mathbf{a} = \begin{pmatrix} -1 \\ 11 \end{pmatrix}$.

The direction vector $\mathbf{b} = \begin{pmatrix} -3-(-1) \\ 12-11 \end{pmatrix}$

$= \begin{pmatrix} -2 \\ 1 \end{pmatrix}$

$\therefore \begin{pmatrix} x \\ y \end{pmatrix} = \begin{pmatrix} -1 \\ 11 \end{pmatrix} + t\begin{pmatrix} -2 \\ 1 \end{pmatrix}, \quad t \in \mathbb{R}$

**ii** $x = -1 - 2t, \quad y = 11 + t, \quad t \in \mathbb{R}$

**iii** $t = \dfrac{x+1}{-2} = y - 11$

$\therefore \ x + 1 = -2y + 22$

$\therefore \ x + 2y = 21$

**2** **a** $x = -1 + 2t, \quad y = 4 - t, \quad t \in \mathbb{R}$

**b** When $t = 0$, $x = -1 + 2(0) = -1$ and $y = 4 - 0 = 4$ $\quad\therefore$ the point is $(-1, 4)$.

When $t = 1$, $x = -1 + 2(1) = 1$ and $y = 4 - 1 = 3$ $\quad\therefore$ the point is $(1, 3)$.

When $t = 3$, $x = -1 + 2(3) = 5$ and $y = 4 - 3 = 1$ $\quad\therefore$ the point is $(5, 1)$.

When $t = -1$, $x = -1 + 2(-1) = -3$ and $y = 4 - (-1) = 5$ $\quad\therefore$ the point is $(-3, 5)$.

When $t = -4$, $x = -1 + 2(-4) = -9$ and $y = 4 - (-4) = 8$ $\quad\therefore$ the point is $(-9, 8)$.

**3** **a** If $t + 2 = 3$ and $1 - 3t = -2$,

then $t = 1$ and $-3t = -3$

$\therefore \ t = 1$

Since $t = 1$ in each case,

$(3, -2)$ lies on the line.

**b** If $(k, 4)$ lies on $x = 1 - 2t, \quad y = 1 + t$, then

$k = 1 - 2t$ and $4 = 1 + t$

$\therefore \ t = 3$

and $k = 1 - 6$

$\therefore \ k = -5$

**4** **a** When $t = 1$, $\mathbf{r} = \begin{pmatrix} 1 \\ 5 \end{pmatrix} + 1 \times \begin{pmatrix} -1 \\ 3 \end{pmatrix} = \begin{pmatrix} 1-1 \\ 5+3 \end{pmatrix} = \begin{pmatrix} 0 \\ 8 \end{pmatrix}$

$\therefore$ the point is $(0, 8)$.

**b** $\begin{pmatrix} 1 \\ -3 \end{pmatrix}$ is a non-zero scalar multiple of $\begin{pmatrix} -1 \\ 3 \end{pmatrix}$, so it could also be used to describe the direction of the line.

**c** The line passes through point $(0, 8)$ and has direction vector $\begin{pmatrix} 1 \\ -3 \end{pmatrix}$.

$\therefore \ \mathbf{r} = \begin{pmatrix} 0 \\ 8 \end{pmatrix} + s\begin{pmatrix} 1 \\ -3 \end{pmatrix}$ is an alternative vector equation for the line.

**5** **a** **i** $\begin{pmatrix} x \\ y \\ z \end{pmatrix} = \begin{pmatrix} 1 \\ 3 \\ -7 \end{pmatrix} + t\begin{pmatrix} 2 \\ 1 \\ 3 \end{pmatrix}, \quad t \in \mathbb{R}$

**ii** $x = 1 + 2t, \quad y = 3 + t, \quad z = -7 + 3t, \quad t \in \mathbb{R}$

**b** **i** $\begin{pmatrix} x \\ y \\ z \end{pmatrix} = \begin{pmatrix} 0 \\ 1 \\ 2 \end{pmatrix} + t\begin{pmatrix} 1 \\ 1 \\ -2 \end{pmatrix}, \quad t \in \mathbb{R}$

**ii** $x = t, \quad y = 1 + t, \quad z = 2 - 2t, \quad t \in \mathbb{R}$

**c** **i** Since the line is parallel to the $X$-axis, it has direction vector $\begin{pmatrix} 1 \\ 0 \\ 0 \end{pmatrix}$

$\therefore \begin{pmatrix} x \\ y \\ z \end{pmatrix} = \begin{pmatrix} -2 \\ 2 \\ 1 \end{pmatrix} + t\begin{pmatrix} 1 \\ 0 \\ 0 \end{pmatrix}, \quad t \in \mathbb{R}$

**ii** $x = -2 + t, \quad y = 2, \quad z = 1, \quad t \in \mathbb{R}$

**d** **i** $\begin{pmatrix} x \\ y \\ z \end{pmatrix} = \begin{pmatrix} 0 \\ 2 \\ -1 \end{pmatrix} + t\begin{pmatrix} 2 \\ -1 \\ 3 \end{pmatrix}, \quad t \in \mathbb{R}$

**ii** $x = 2t, \quad y = 2 - t, \quad z = -1 + 3t, \quad t \in \mathbb{R}$

**6** **a** $\overrightarrow{AB} = \begin{pmatrix} -1-1 \\ 3-2 \\ 2-1 \end{pmatrix} = \begin{pmatrix} -2 \\ 1 \\ 1 \end{pmatrix} \quad \therefore \begin{pmatrix} x \\ y \\ z \end{pmatrix} = \begin{pmatrix} 1 \\ 2 \\ 1 \end{pmatrix} + t\begin{pmatrix} -2 \\ 1 \\ 1 \end{pmatrix}, \quad t \in \mathbb{R}$

**b** $\overrightarrow{CD} = \begin{pmatrix} 3-0 \\ 1-1 \\ -1-3 \end{pmatrix} = \begin{pmatrix} 3 \\ 0 \\ -4 \end{pmatrix} \quad \therefore \begin{pmatrix} x \\ y \\ z \end{pmatrix} = \begin{pmatrix} 0 \\ 1 \\ 3 \end{pmatrix} + t\begin{pmatrix} 3 \\ 0 \\ -4 \end{pmatrix}, \quad t \in \mathbb{R}$

**c** $\overrightarrow{EF} = \begin{pmatrix} 1-1 \\ -1-2 \\ 5-5 \end{pmatrix} = \begin{pmatrix} 0 \\ -3 \\ 0 \end{pmatrix} \quad \therefore \begin{pmatrix} x \\ y \\ z \end{pmatrix} = \begin{pmatrix} 1 \\ 2 \\ 5 \end{pmatrix} + t\begin{pmatrix} 0 \\ -3 \\ 0 \end{pmatrix}, \quad t \in \mathbb{R}$

**d** $\overrightarrow{GH} = \begin{pmatrix} 5-0 \\ -1-1 \\ 3--1 \end{pmatrix} = \begin{pmatrix} 5 \\ -2 \\ 4 \end{pmatrix} \quad \therefore \begin{pmatrix} x \\ y \\ z \end{pmatrix} = \begin{pmatrix} 0 \\ 1 \\ -1 \end{pmatrix} + t\begin{pmatrix} 5 \\ -2 \\ 4 \end{pmatrix}, \quad t \in \mathbb{R}$

**7** Given $x = 1 - t, \quad y = 3 + t, \quad z = 3 - 2t$:

**a** The line meets the $XOY$ plane when $z = 0 \quad \therefore\ 3 - 2t = 0$

$\therefore\ t = \frac{3}{2}$

Then $x = 1 - \frac{3}{2} = -\frac{1}{2}$ and $y = 3 + \frac{3}{2} = \frac{9}{2}$, so the point is $\left(-\frac{1}{2}, \frac{9}{2}, 0\right)$.

**b** The line meets the $YOZ$ plane when $x = 0 \quad \therefore\ 1 - t = 0$

$\therefore\ t = 1$

Then $y = 3 + 1 = 4$ and $z = 3 - 2 = 1$, so the point is $(0, 4, 1)$.

**c** The line meets the $XOZ$ plane when $y = 0 \quad \therefore\ 3 + t = 0$

$\therefore\ t = -3$

Then $x = 1 - (-3) = 4$ and $z = 3 - 2(-3) = 9$, so the point is $(4, 0, 9)$.

**8** Given a line with equations $x = 2 - t, \quad y = 3 + 2t$ and $z = 1 + t$,

the distance to the point $(1, 0, -2)$ is $\sqrt{(2-t-1)^2 + (3+2t-0)^2 + (1+t+2)^2}$.

But this distance $= 5\sqrt{3}$ units

$$\therefore\ \sqrt{(1-t)^2 + (3+2t)^2 + (t+3)^2} = 5\sqrt{3}$$
$$\therefore\ (1-t)^2 + (3+2t)^2 + (t+3)^2 = 75$$
$$\therefore\ 1 - 2t + t^2 + 9 + 12t + 4t^2 + t^2 + 6t + 9 = 75$$
$$\therefore\ 6t^2 + 16t - 56 = 0$$
$$\therefore\ 3t^2 + 8t - 28 = 0$$
$$\therefore\ (3t + 14)(t - 2) = 0$$
$$\therefore\ t = -\tfrac{14}{3} \text{ or } t = 2$$

When $t = 2$, the point is $(0, 7, 3)$, and when $t = -\frac{14}{3}$, the point is $\left(\frac{20}{3}, -\frac{19}{3}, -\frac{11}{3}\right)$.

## EXERCISE 13C

**1** $L_1$ has direction vector $\begin{pmatrix} 12 \\ 5 \end{pmatrix}$ and $L_2$ has direction vector $\begin{pmatrix} 3 \\ -4 \end{pmatrix}$.

If $\theta$ is the angle between them, $\cos\theta = \dfrac{\left|\begin{pmatrix} 12 \\ 5 \end{pmatrix} \bullet \begin{pmatrix} 3 \\ -4 \end{pmatrix}\right|}{\sqrt{144+25}\sqrt{9+16}} = \dfrac{|36 + (-20)|}{13 \times 5} = \frac{16}{65}$

$\therefore\ \theta = \cos^{-1}\left(\frac{16}{65}\right) \approx 75.7°$

**2** Line 1 has direction vector $\begin{pmatrix} 5 \\ -2 \end{pmatrix}$ and line 2 has direction vector $\begin{pmatrix} 4 \\ 10 \end{pmatrix}$

and $\begin{pmatrix} 5 \\ -2 \end{pmatrix} \bullet \begin{pmatrix} 4 \\ 10 \end{pmatrix} = 20 + (-20) = 0$

$\therefore$ the lines are perpendicular.

**3** $L_1$ has direction vector $\begin{pmatrix} 4 \\ -3 \end{pmatrix}$ and $L_2$ has direction vector $\begin{pmatrix} 5 \\ 4 \end{pmatrix}$.

If $\theta$ is the angle between them, $\cos\theta = \dfrac{\left|\begin{pmatrix} 4 \\ -3 \end{pmatrix} \bullet \begin{pmatrix} 5 \\ 4 \end{pmatrix}\right|}{\sqrt{16+9}\sqrt{25+16}} = \dfrac{|20+(-12)|}{\sqrt{25\times 41}} = \dfrac{8}{\sqrt{25\times 41}}$

$$\therefore \quad \theta = \cos^{-1}\left(\frac{8}{\sqrt{25\times 41}}\right) \approx 75.5^\circ$$

$\therefore$ the required angle measures $75.5^\circ$.

**4** **a** Line 1 has direction vector $\begin{pmatrix} 3 \\ -16 \\ 7 \end{pmatrix}$ and line 2 has direction vector $\begin{pmatrix} 3 \\ 8 \\ -5 \end{pmatrix}$.

If $\theta$ is the angle between them,

$$\cos\theta = \frac{\left|\begin{pmatrix} 3 \\ -16 \\ 7 \end{pmatrix} \bullet \begin{pmatrix} 3 \\ 8 \\ -5 \end{pmatrix}\right|}{\sqrt{9+256+49}\sqrt{9+64+25}} = \frac{|9-128-35|}{\sqrt{314}\sqrt{98}} = \frac{154}{\sqrt{314\times 98}}$$

$\therefore \quad \theta \approx 28.6^\circ$

**b** Since $L_1 \perp L_3$, $\begin{pmatrix} 3 \\ -16 \\ 7 \end{pmatrix} \bullet \begin{pmatrix} 0 \\ -3 \\ x \end{pmatrix} = 0$

$$\therefore \quad 48 + 7x = 0$$

$$\therefore \quad x = -\tfrac{48}{7}$$

**5** **a** $x - y = 3$ has gradient $+\frac{1}{1}$ and so has direction vector $\begin{pmatrix} 1 \\ 1 \end{pmatrix}$.

$3x + 2y = 11$ has gradient $-\frac{3}{2}$ and so has direction vector $\begin{pmatrix} 2 \\ -3 \end{pmatrix}$.

$\therefore \begin{pmatrix} 1 \\ 1 \end{pmatrix} \bullet \begin{pmatrix} 2 \\ -3 \end{pmatrix} = \sqrt{1+1}\sqrt{4+9}\cos\theta$

$\therefore \quad 2 - 3 = \sqrt{2}\sqrt{13}\cos\theta$

$\therefore \quad \frac{-1}{\sqrt{26}} = \cos\theta$

$\therefore \quad \theta \approx 101.3^\circ$

$\therefore$ the angle is $180^\circ - 101.3^\circ \approx 78.7^\circ$

**b** $y = x + 2$ has gradient $1 = \frac{1}{1}$ and so has direction vector $\begin{pmatrix} 1 \\ 1 \end{pmatrix}$.

$y = 1 - 3x$ has gradient $-3 = \frac{-3}{1}$ and so has direction vector $\begin{pmatrix} 1 \\ -3 \end{pmatrix}$.

$\therefore \begin{pmatrix} 1 \\ 1 \end{pmatrix} \bullet \begin{pmatrix} 1 \\ -3 \end{pmatrix} = \sqrt{1+1}\sqrt{1+9}\cos\theta$

$\therefore \quad 1 - 3 = \sqrt{2}\sqrt{10}\cos\theta$

$\therefore \quad \frac{-2}{\sqrt{20}} = \cos\theta$

$\therefore \quad \theta \approx 116.6^\circ$

$\therefore$ the angle is $180^\circ - 116.6^\circ \approx 63.4^\circ$

**c** $y+x=7$ has gradient $-1=\frac{-1}{1}$ and so has direction vector $\begin{pmatrix}1\\-1\end{pmatrix}$.

$x-3y+2=0$ has gradient $\frac{1}{3}$ and so has direction vector $\begin{pmatrix}3\\1\end{pmatrix}$.

$$\therefore \begin{pmatrix}1\\-1\end{pmatrix}\bullet\begin{pmatrix}3\\1\end{pmatrix}=\sqrt{1+1}\sqrt{9+1}\cos\theta$$
$$\therefore 3-1=\sqrt{2}\sqrt{10}\cos\theta$$
$$\therefore \frac{2}{\sqrt{20}}=\cos\theta$$
$$\therefore \theta\approx 63.4°$$

$\therefore$ the angle is $63.4°$.

**d** $y=2-x$ has gradient $-1=\frac{-1}{1}$ and so has direction vector $\begin{pmatrix}1\\-1\end{pmatrix}$.

$x-2y=7$ has gradient $\frac{1}{2}$ and so has direction vector $\begin{pmatrix}2\\1\end{pmatrix}$.

$$\therefore \begin{pmatrix}1\\-1\end{pmatrix}\bullet\begin{pmatrix}2\\1\end{pmatrix}=\sqrt{1+1}\sqrt{4+1}\cos\theta$$
$$\therefore 2-1=\sqrt{2}\sqrt{5}\cos\theta$$
$$\therefore \frac{1}{\sqrt{10}}=\cos\theta$$
$$\therefore \theta\approx 71.6°$$

$\therefore$ the angle is $71.6°$.

## EXERCISE 13D

**1** **a** $x(0)=1$ and $y(0)=2$,

$\therefore$ the initial position is $(1, 2)$

**b**

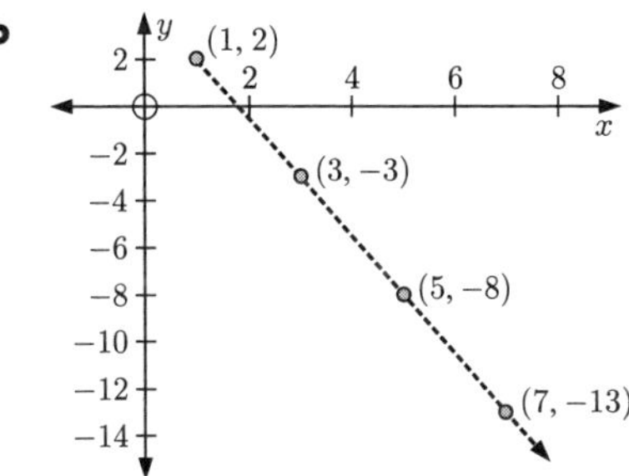


**c** The velocity vector is $\begin{pmatrix}2\\-5\end{pmatrix}$.

**d** In 1 second, the $x$-step is 2 and $y$-step is $-5$, which is a distance of $\sqrt{2^2+(-5)^2}=\sqrt{29}$ cm

$\therefore$ the speed is $\sqrt{29}\ \text{cm s}^{-1}$.

**2** **a** $\begin{pmatrix}x\\y\end{pmatrix}=\begin{pmatrix}2\\3\end{pmatrix}+t\begin{pmatrix}4\\-5\end{pmatrix},\quad t\geqslant 0$

**b** 90 minutes = 1.5 hours

When $t=1.5$, $x=2+4(1.5)=8$

and $y=3-5(1.5)=-4.5$

$\therefore$ the boat is at $(8, -4.5)$ after 90 minutes.

**c** When the boat reaches the point $(5, -0.75)$,

$2+4t=5$ and $3-5t=-0.75$

$\therefore 4t=3$ $\qquad -5t=-3.75$

$\therefore t=0.75$ $\qquad t=0.75$

It will take 0.75 hours = 45 minutes for the boat to reach point $(5, -0.75)$.

**3** **a**

$$\mathbf{r}=\mathbf{a}+t\mathbf{b}$$

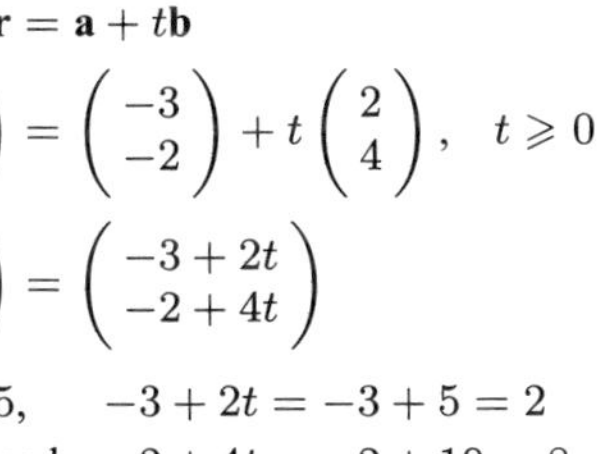

$$\therefore \begin{pmatrix}x\\y\end{pmatrix}=\begin{pmatrix}-3\\-2\end{pmatrix}+t\begin{pmatrix}2\\4\end{pmatrix},\quad t\geqslant 0$$
$$\therefore \begin{pmatrix}x\\y\end{pmatrix}=\begin{pmatrix}-3+2t\\-2+4t\end{pmatrix}$$

**b** At $t=2.5$, $-3+2t=-3+5=2$

and $-2+4t=-2+10=8$

So, the position vector is $\begin{pmatrix}2\\8\end{pmatrix}$.

**c** **i** When the car is due north, $x=0$.

$\therefore -3+2t=0$

$\therefore t=1.5$ seconds

**ii** When the car is due west, $y=0$.

$\therefore -2+4t=0$

$\therefore t=0.5$ seconds

**d**

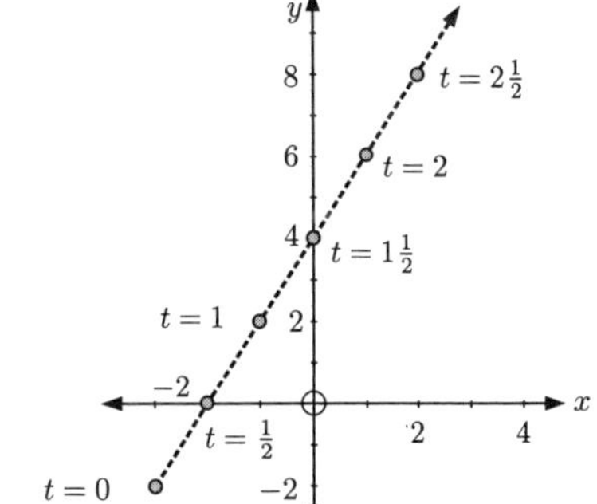

**4** **a** **i** When $t = 0$, $\begin{pmatrix} x \\ y \end{pmatrix} = \begin{pmatrix} -4 \\ 3 \end{pmatrix}$ **ii** The velocity vector is $\begin{pmatrix} 12 \\ 5 \end{pmatrix}$.

$\therefore$ the object is at $(-4, 3)$.

**iii** The speed is $\sqrt{12^2 + 5^2} = 13 \text{ m s}^{-1}$

**b** **i** When $t = 0$, $\begin{pmatrix} x \\ y \\ z \end{pmatrix} = \begin{pmatrix} 3 \\ 0 \\ 4 \end{pmatrix}$ **ii** The velocity vector is $\begin{pmatrix} 2 \\ -1 \\ -2 \end{pmatrix}$.

$\therefore$ the object is at $(3, 0, 4)$.

**iii** The speed is $\sqrt{2^2 + (-1)^2 + (-2)^2} = 3 \text{ m s}^{-1}$

**5** **a** $\begin{pmatrix} 4 \\ -3 \end{pmatrix}$ has length $\sqrt{4^2 + (-3)^2} = 5$

$\therefore 30\begin{pmatrix} 4 \\ -3 \end{pmatrix}$ has length 150

$\therefore$ the velocity vector is $\begin{pmatrix} 120 \\ -90 \end{pmatrix}$.

**b** $2\mathbf{i} + \mathbf{j} = \begin{pmatrix} 2 \\ 1 \end{pmatrix}$ has length $\sqrt{2^2 + 1^2} = \sqrt{5}$

$\therefore 10\sqrt{5}\begin{pmatrix} 2 \\ 1 \end{pmatrix}$ has length 50

$\therefore$ the velocity vector is $\begin{pmatrix} 20\sqrt{5} \\ 10\sqrt{5} \end{pmatrix}$.

**6** $-2\mathbf{i} + 5\mathbf{j} - 14\mathbf{k}$ has length $\sqrt{(-2)^2 + 5^2 + (-14)^2} = \sqrt{4 + 25 + 196} = \sqrt{225} = 15$

$\therefore 6\begin{pmatrix} -2 \\ 5 \\ -14 \end{pmatrix}$ has length 90, so the velocity vector is $\begin{pmatrix} -12 \\ 30 \\ -84 \end{pmatrix}$.

**7** Yacht A: $\begin{pmatrix} x_A \\ y_A \end{pmatrix} = \begin{pmatrix} 4 \\ 5 \end{pmatrix} + t\begin{pmatrix} 1 \\ -2 \end{pmatrix}$ Yacht B: $\begin{pmatrix} x_B \\ y_B \end{pmatrix} = \begin{pmatrix} 1 \\ -8 \end{pmatrix} + t\begin{pmatrix} 2 \\ 3 \end{pmatrix}$, $t \geqslant 0$

**a** When $t = 0$, $\begin{pmatrix} x_A \\ y_A \end{pmatrix} = \begin{pmatrix} 4 \\ 5 \end{pmatrix}$ $\therefore$ A is at $(4, 5)$

and $\begin{pmatrix} x_B \\ y_B \end{pmatrix} = \begin{pmatrix} 1 \\ -8 \end{pmatrix}$ $\therefore$ B is at $(1, -8)$.

**b** For A, the velocity vector is $\begin{pmatrix} 1 \\ -2 \end{pmatrix}$, and for B it is $\begin{pmatrix} 2 \\ 1 \end{pmatrix}$.

**c** Speed of A $= \sqrt{1^2 + (-2)^2} = \sqrt{5} \text{ km h}^{-1}$. Speed of B $= \sqrt{2^2 + 1^2} = \sqrt{5} \text{ km h}^{-1}$.

**d** A has direction vector $\begin{pmatrix} 1 \\ -2 \end{pmatrix}$ and B has direction vector $\begin{pmatrix} 2 \\ 1 \end{pmatrix}$.

Since $\begin{pmatrix} 1 \\ -2 \end{pmatrix} \bullet \begin{pmatrix} 2 \\ 1 \end{pmatrix} = 2 - 2 = 0$, the paths of the yachts are at right angles to each other.

**8** **a** P's torpedo has position $\begin{pmatrix} x_1 \\ y_1 \end{pmatrix} = \begin{pmatrix} -5 \\ 4 \end{pmatrix} + t\begin{pmatrix} 3 \\ -1 \end{pmatrix}$ and at $t = 0$, the time is 1:34 pm

$\therefore x_1(t) = -5 + 3t, \quad y_1(t) = 4 - t.$

**b** Speed $= \sqrt{3^2 + (-1)^2} = \sqrt{10} \text{ km min}^{-1}$

**c** Q fires its torpedo after $a$ minutes.

$\therefore$ at time $t$, its torpedo has travelled for $(t - a)$ minutes.

$\therefore \begin{pmatrix} x_2 \\ y_2 \end{pmatrix} = \begin{pmatrix} 15 \\ 7 \end{pmatrix} + (t - a)\begin{pmatrix} -4 \\ -3 \end{pmatrix}$, $t \geqslant a$

$\therefore x_2(t) = 15 - 4(t - a)$ and $y_2(t) = 7 - 3(t - a)$

$\therefore$ position is $Q(15 - 4(t - a), 7 - 3(t - a))$.

**d** They meet when $x_1(t) = x_2(t)$ and $y_1(t) = y_2(t)$

$\therefore \quad -5+3t = 15-4(t-a)$ and $4-t = 7-3(t-a)$

$\therefore \quad 7t-4a = 20$ .... (1) and $2t-3a = 3$ .... (2)

Solving simultaneously,

$$\begin{array}{rll} & 21t - 12a = 60 & \{3\times(1)\} \\ & -8t + 12a = -12 & \{-4\times(2)\} \\ \hline \text{adding} & 13t \qquad = 48 & \end{array}$$

$\therefore \quad t = \frac{48}{13}$ and $7\left(\frac{48}{13}\right) - 4a = 20$

$\therefore \quad t \approx 3.6923$ $\qquad \therefore \quad 5.8462 = 4a$

$\therefore \quad t \approx 3$ min $41.54$ sec $\qquad \therefore \quad a \approx 1.4615 \approx 1$ min $27.7$ sec

So, as $a \approx 1.4615$, Q fired at 1:35:28 pm, and the explosion occurred at 1:37:42 pm.

**9** **a** $\overrightarrow{AB} = \begin{pmatrix} 3-6 \\ 10-9 \\ 2.5-3 \end{pmatrix} = \begin{pmatrix} -3 \\ 1 \\ -0.5 \end{pmatrix}$

**b** $|\overrightarrow{AB}| = \sqrt{(-3)^2+1^2+(-0.5)^2}$

$= \sqrt{10.25}$ km

The helicopter travels $\sqrt{10.25}$ km in 10 minutes.

$\therefore$ the helicopter's speed is

$6 \times \sqrt{10.25} \approx 19.2 \text{ km h}^{-1}$.

**c** $\begin{pmatrix} x \\ y \\ z \end{pmatrix} = \begin{pmatrix} 6 \\ 9 \\ 3 \end{pmatrix} + t\begin{pmatrix} -3 \\ 1 \\ -0.5 \end{pmatrix}, \quad t \in \mathbb{R}$

**d** If $z = 0$, $3 + (-0.5)t = 0$

$\therefore \quad t = 6$

$t = 1$ represents 10 minutes, so $t = 6$ represents 60 minutes.

$\therefore$ the helicopter lands on the helipad after 1 hour.

## EXERCISE 13E

**1** **a** Let N be the point on the line closest to P.

N has coordinates $(2+t, 3+2t)$ for some $t \in \mathbb{R}$.

$\overrightarrow{PN}$ is $\begin{pmatrix} 2+t-3 \\ 3+2t-2 \end{pmatrix} = \begin{pmatrix} t-1 \\ 2t+1 \end{pmatrix}$.

Now $\overrightarrow{PN} \bullet \begin{pmatrix} 1 \\ 2 \end{pmatrix} = 0$, as $\begin{pmatrix} 1 \\ 2 \end{pmatrix}$ is the direction vector of the line.

$\therefore \quad \begin{pmatrix} t-1 \\ 2t+1 \end{pmatrix} \bullet \begin{pmatrix} 1 \\ 2 \end{pmatrix} = 0$

$\therefore \quad (t-1) + 2(2t+1) = 0$

$\therefore \quad t-1+4t+2 = 0$

$\therefore \quad 5t = -1$

$\therefore \quad t = -\frac{1}{5}$

Thus $\overrightarrow{PN} = \begin{pmatrix} -\frac{1}{5}-1 \\ -\frac{2}{5}+1 \end{pmatrix} = \begin{pmatrix} -\frac{6}{5} \\ \frac{3}{5} \end{pmatrix}$

$= \frac{3}{5}\begin{pmatrix} -2 \\ 1 \end{pmatrix}$

and $|\overrightarrow{PN}| = \frac{3}{5}\left|\begin{pmatrix} -2 \\ 1 \end{pmatrix}\right| = \frac{3}{5}\sqrt{(-2)^2+1^2}$

$= \frac{3}{5}\sqrt{5}$

So the shortest distance from P to the line is $\frac{3}{5}\sqrt{5}$ units.

**b** Let N be the point on the line closest to Q.

N has coordinates $(t, 1-t)$ for some $t \in \mathbb{R}$.

$\overrightarrow{QN}$ is $\begin{pmatrix} t-(-1) \\ 1-t-1 \end{pmatrix} = \begin{pmatrix} t+1 \\ -t \end{pmatrix}$.

Now $\overrightarrow{QN} \bullet \begin{pmatrix} 1 \\ -1 \end{pmatrix} = 0$, as $\begin{pmatrix} 1 \\ -1 \end{pmatrix}$ is the direction vector of the line.

$\therefore \quad \begin{pmatrix} t+1 \\ -t \end{pmatrix} \bullet \begin{pmatrix} 1 \\ -1 \end{pmatrix} = 0$

$\therefore \quad (t+1) + (-1)(-t) = 0$

$\therefore \quad t+1+t = 0$

$\therefore \quad 2t = -1$

$\therefore \quad t = -\frac{1}{2}$

Thus $\overrightarrow{QN} = \begin{pmatrix} -\frac{1}{2}+1 \\ -\left(-\frac{1}{2}\right) \end{pmatrix} = \begin{pmatrix} \frac{1}{2} \\ \frac{1}{2} \end{pmatrix}$

$= \frac{1}{2}\begin{pmatrix} 1 \\ 1 \end{pmatrix}$

and $|\overrightarrow{QN}| = \frac{1}{2}\left|\begin{pmatrix} 1 \\ 1 \end{pmatrix}\right| = \frac{1}{2}\sqrt{1^2+1^2}$

$= \frac{1}{2}\sqrt{2}$

So the shortest distance from Q to the line is $\frac{1}{2}\sqrt{2}$ units.

**c** Let N be the point on the line closest to R.
N has coordinates $(2+s,\ 3-s)$ for some $s \in \mathbb{R}$.

$\overrightarrow{RN}$ is $\begin{pmatrix} 2+s-(-3) \\ 3-s-(-1) \end{pmatrix} = \begin{pmatrix} s+5 \\ 4-s \end{pmatrix}$.

Now $\overrightarrow{RN} \bullet \begin{pmatrix} 1 \\ -1 \end{pmatrix} = 0$, as $\begin{pmatrix} 1 \\ -1 \end{pmatrix}$ is the direction vector of the line.

$$\therefore \begin{pmatrix} s+5 \\ 4-s \end{pmatrix} \bullet \begin{pmatrix} 1 \\ -1 \end{pmatrix} = 0$$

$$\therefore (s+5)+(-1)(4-s)=0$$
$$\therefore s+5-4+s=0$$
$$\therefore 2s=-1$$
$$\therefore s=-\tfrac{1}{2}$$

Thus $\overrightarrow{RN} = \begin{pmatrix} -\frac{1}{2}+5 \\ 4-(-\frac{1}{2}) \end{pmatrix} = \begin{pmatrix} \frac{9}{2} \\ \frac{9}{2} \end{pmatrix}$

$$= \tfrac{9}{2}\begin{pmatrix} 1 \\ 1 \end{pmatrix}$$

and $|\overrightarrow{RN}| = \frac{9}{2}\left|\begin{pmatrix} 1 \\ 1 \end{pmatrix}\right| = \frac{9}{2}\sqrt{1^2+1^2}$

$$= \tfrac{9}{2}\sqrt{2}$$

So the shortest distance from R to the line is $\frac{9}{2}\sqrt{2}$ units.

**d** Let N be the point on the line closest to S.
N has coordinates $(2+3t,\ 5-7t)$ for some $t \in \mathbb{R}$.

$\overrightarrow{SN}$ is $\begin{pmatrix} 2+3t-5 \\ 5-7t-(-2) \end{pmatrix} = \begin{pmatrix} 3t-3 \\ 7-7t \end{pmatrix}$.

Now $\overrightarrow{SN} \bullet \begin{pmatrix} 3 \\ -7 \end{pmatrix} = 0$, as $\begin{pmatrix} 3 \\ -7 \end{pmatrix}$ is the direction vector of the line.

$$\therefore \begin{pmatrix} 3t-3 \\ 7-7t \end{pmatrix} \bullet \begin{pmatrix} 3 \\ -7 \end{pmatrix} = 0$$

$$\therefore 3(3t-3)-7(7-7t)=0$$
$$\therefore 9t-9-49+49t=0$$
$$\therefore 58t=-58$$
$$\therefore t=1$$

Thus $\overrightarrow{SN} = \begin{pmatrix} 3(1)-3 \\ 7-7(1) \end{pmatrix} = \begin{pmatrix} 0 \\ 0 \end{pmatrix}$

$\therefore\ |\overrightarrow{SN}| = 0$

So S actually lies on the line, and the shortest distance is 0 units.

**2** **a** $6\mathbf{i} - 6\mathbf{j}$

**b** The length of $\begin{pmatrix} -3 \\ 4 \end{pmatrix} = \sqrt{(-3)^2+4^2} = \sqrt{25} = 5$

As the speed is 10 km h$^{-1}$, the liner has velocity vector $2\begin{pmatrix} -3 \\ 4 \end{pmatrix} = \begin{pmatrix} -6 \\ 8 \end{pmatrix}$.

$\therefore$ the liner has position $\begin{pmatrix} x \\ y \end{pmatrix} = \begin{pmatrix} 6 \\ -6 \end{pmatrix} + t\begin{pmatrix} -6 \\ 8 \end{pmatrix} = \begin{pmatrix} 6-6t \\ -6+8t \end{pmatrix}$, $t \geqslant 0$, $t$ in hours.

**c** The liner is due east when $y=0$

$$\therefore -6+8t=0$$
$$\therefore \text{at } t=\tfrac{3}{4} \text{ hours}$$

**d** The liner L is nearest the fishing boat O when $\overrightarrow{OL} \perp \binom{-3}{4}$

(0, 0)
L(6 – 6t, –6 + 8t)
$\binom{-3}{4}$

$$\therefore \overrightarrow{OL} \bullet \begin{pmatrix} -3 \\ 4 \end{pmatrix} = 0$$

$$\therefore \begin{pmatrix} 6-6t \\ -6+8t \end{pmatrix} \bullet \begin{pmatrix} -3 \\ 4 \end{pmatrix} = 0$$

$$\therefore (-18+18t)+(-24+32t)=0$$
$$\therefore 50t=42$$
$$\therefore t=0.84 \text{ hours} = 50.4 \text{ minutes}$$

and when $t=0.84$, $\overrightarrow{OL} = \begin{pmatrix} 6-6(0.84) \\ -6+8(0.84) \end{pmatrix} = \begin{pmatrix} 0.96 \\ 0.72 \end{pmatrix}$

$\therefore$ the liner is closest to the fishing boat after 0.84 hours or 50.4 minutes, when it is at (0.96, 0.72).

**3** **a** $|\mathbf{b}| = \sqrt{(-3)^2 + (-1)^2} = \sqrt{10}$

As the speed is $40\sqrt{10}$ km h$^{-1}$, the velocity vector is $40\begin{pmatrix}-3\\-1\end{pmatrix} = \begin{pmatrix}-120\\-40\end{pmatrix}$.

**b** $\begin{pmatrix}x\\y\end{pmatrix} = \begin{pmatrix}200\\100\end{pmatrix} + t\begin{pmatrix}-120\\-40\end{pmatrix}$, $t \geqslant 0$ $\{t = 0$ at 12:00 noon$\}$

**c** At 1:00 pm, $t = 1$ and $\begin{pmatrix}x\\y\end{pmatrix} = \begin{pmatrix}200-120\\100-40\end{pmatrix} = \begin{pmatrix}80\\60\end{pmatrix}$

$\therefore$ the aircraft is at (80, 60).

**d** The distance from O(0, 0) to $P_1(80, 60)$ is $\left|\begin{pmatrix}80\\60\end{pmatrix}\right| = \sqrt{80^2 + 60^2} = 100$ km,

which is when it becomes visible to radar. {within 100 km of O(0, 0)}

**e**

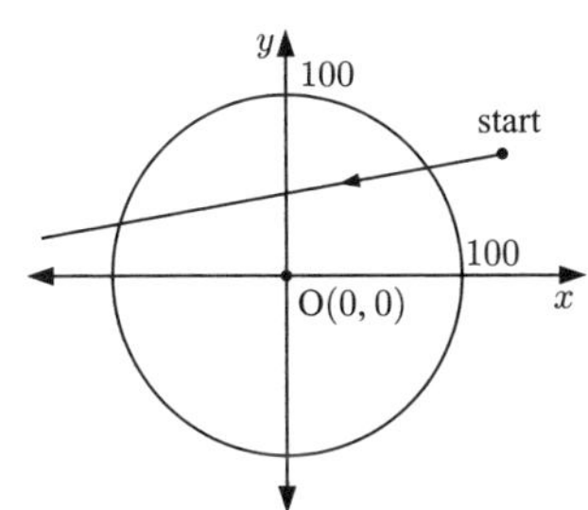


A general point on the path is $P(200 - 120t, 100 - 40t)$.

Now $\overrightarrow{OP} = \begin{pmatrix}200-120t\\100-40t\end{pmatrix}$,

and for the closest point $\overrightarrow{OP} \bullet \begin{pmatrix}-3\\-1\end{pmatrix} = 0$

$\therefore \quad -3(200 - 120t) - 1(100 - 40t) = 0$

$\therefore \quad -700 + 400t = 0$

$\therefore \quad t = \frac{7}{4} = 1\frac{3}{4}$ hours

$\therefore$ the time when the aircraft is closest is 1:45 pm, and

at this time $\overrightarrow{OP} = \begin{pmatrix}200 - 120(\frac{7}{4})\\100 - 40(\frac{7}{4})\end{pmatrix} = \begin{pmatrix}-10\\30\end{pmatrix}$

$\therefore \quad d_{\min} = \sqrt{(-10)^2 + 30^2} \approx 31.6$ km

**f** It disappears from radar when $|\overrightarrow{OP}| = 100$ and $t > 1\frac{3}{4}$

$\therefore \quad \sqrt{(200 - 120t)^2 + (100 - 40t)^2} = 100$

$\therefore \quad 40\,000 - 48\,000t + 14\,400t^2 + 10\,000 - 8000t + 1600t^2 = 10\,000$

$\therefore \quad 16\,000t^2 - 56\,000t + 40\,000 = 0$

$\therefore \quad 16t^2 - 56t + 40 = 0 \qquad \{\div 1000\}$

$\therefore \quad 2t^2 - 7t + 5 = 0 \qquad \{\div 8\}$

$\therefore \quad (2t - 5)(t - 1) = 0$

$\therefore \quad t = \frac{5}{2} \qquad \{\text{as } t > 1\frac{3}{4}\}$

So, the aircraft disappears from the radar screen $2\frac{1}{2}$ hours after noon, or at 2:30 pm.

**4** **a** At A, $y = 0$

$\therefore \quad 2x = 36$

$\therefore \quad x = 18$

At B, $x = 0$

$\therefore \quad 3y = 36$

$\therefore \quad y = 12$

So A is (18, 0) and B is (0, 12).

**b** $2x + 3y = 36$

$\therefore \quad 3y = 36 - 2x$

$\therefore \quad y = \dfrac{36 - 2x}{3}$

$\therefore$ any point R on the railway track can be written $R\left(x, \dfrac{36 - 2x}{3}\right)$.

**c** $\overrightarrow{PR} = \begin{pmatrix}x - 4\\\frac{36-2x}{3} - 0\end{pmatrix} = \begin{pmatrix}x - 4\\\frac{36-2x}{3}\end{pmatrix}$,

$\overrightarrow{AB} = \begin{pmatrix}0 - 18\\12 - 0\end{pmatrix} = \begin{pmatrix}-18\\12\end{pmatrix}$

**d** The point closest to the railway track is R such that $\overrightarrow{PR} \perp \overrightarrow{AB}$.

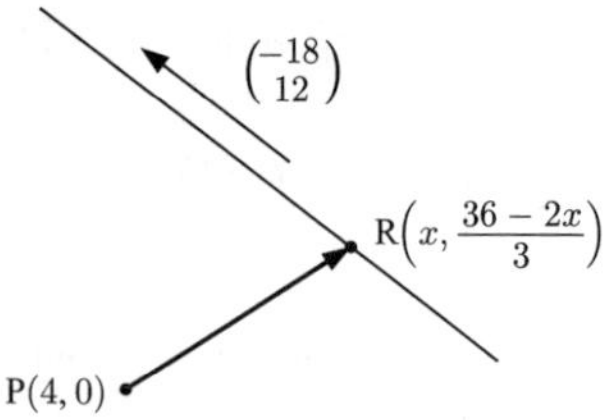

$$\therefore \quad \overrightarrow{PR} \bullet \overrightarrow{AB} = 0$$

$$\therefore \quad \begin{pmatrix} x-4 \\ \frac{36-2x}{3} \end{pmatrix} \bullet \begin{pmatrix} -18 \\ 12 \end{pmatrix} = 0$$

$$\therefore \quad -18(x-4) + 4(36-2x) = 0$$

$$\therefore \quad -18x + 72 + 144 - 8x = 0$$

$$\therefore \quad 26x = 216$$

$$\therefore \quad x = \tfrac{108}{13}$$

$\begin{pmatrix} -18 \\ 12 \end{pmatrix}$ R$\left(x, \frac{36-2x}{3}\right)$ P(4, 0)

Now when $x = \frac{108}{13}$, $\dfrac{36-2x}{3} = 12 - \frac{2}{3}x = 12 - \frac{2}{3}\left(\frac{108}{13}\right) = \frac{84}{13}$. So R is $\left(\frac{108}{13}, \frac{84}{13}\right)$.

$$|\overrightarrow{PR}| = \sqrt{\left(\tfrac{108}{13} - 4\right)^2 + \left(\tfrac{84}{13} - 0\right)^2} = \sqrt{\tfrac{784}{13}} \approx 7.77 \text{ km}$$

The closest point on the track to the camp is $\left(\frac{108}{13}, \frac{84}{13}\right)$, a distance of 7.77 km.

**5** For A, $x_A(t) = 3 - t$, $y_A(t) = 2t - 4$ For B, $x_B(t) = 4 - 3t$, $y_B(t) = 3 - 2t$

**a** When $t = 0$, $x_A(0) = 3$, $y_A(0) = -4$ and $x_B(0) = 4$, $y_B(0) = 3$

$\therefore$ A is at (3, −4). $\therefore$ B is at (4, 3).

**b** The velocity vector of A is $\begin{pmatrix} -1 \\ 2 \end{pmatrix}$ and the velocity vector of B is $\begin{pmatrix} -3 \\ -2 \end{pmatrix}$.

**c** If the angle is $\theta$, $\begin{pmatrix} -1 \\ 2 \end{pmatrix} \bullet \begin{pmatrix} -3 \\ -2 \end{pmatrix} = \sqrt{1+4}\sqrt{9+4}\cos\theta$

$$\therefore \quad 3 - 4 = \sqrt{5}\sqrt{13}\cos\theta$$

$$\therefore \quad \frac{-1}{\sqrt{65}} = \cos\theta \quad \text{and so} \quad \theta \approx 97.1^\circ$$

$\therefore$ the acute angle between the paths is $\approx 82.9^\circ$.

**d** If $D$ is the distance between them, then

$$D = \sqrt{[(4-3t)-(3-t)]^2 + [(3-2t)-(2t-4)]^2}$$
$$= \sqrt{[1-2t]^2 + [7-4t]^2}$$
$$= \sqrt{1 - 4t + 4t^2 + 49 - 56t + 16t^2}$$
$$= \sqrt{20t^2 - 60t + 50}$$

Under the square root we have a quadratic in $t$, so $D$ is a minimum when

$$t = -\frac{b}{2a} = \tfrac{60}{40} = 1\tfrac{1}{2}$$

$\therefore$ $t = 1.5$ hours

$\therefore$ the boats are closest after $1\frac{1}{2}$ hours.

**6** **a** The direction vector of the line is $\mathbf{b} = \begin{pmatrix} 3 \\ 2 \\ 1 \end{pmatrix}$.

Let the point $(3, 0, -1)$ be P, and $A(2+3t, -1+2t, 4+t)$ be any point on the line.

$$\therefore \quad \overrightarrow{PA} = \begin{pmatrix} 2+3t-3 \\ -1+2t-0 \\ 4+t-(-1) \end{pmatrix} = \begin{pmatrix} -1+3t \\ -1+2t \\ 5+t \end{pmatrix}$$

Now $\overrightarrow{PA}$ and $\mathbf{b}$ are perpendicular, so $\overrightarrow{PA} \bullet \mathbf{b} = 0$.

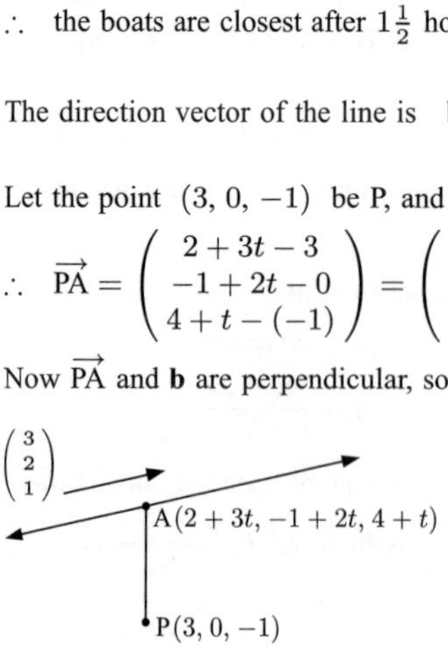

$\begin{pmatrix} 3 \\ 2 \\ 1 \end{pmatrix}$ A$(2+3t, -1+2t, 4+t)$ P$(3, 0, -1)$

$$\therefore \quad \begin{pmatrix} -1+3t \\ -1+2t \\ 5+t \end{pmatrix} \bullet \begin{pmatrix} 3 \\ 2 \\ 1 \end{pmatrix} = 0$$

$$\therefore \quad -3 + 9t - 2 + 4t + 5 + t = 0$$

$$\therefore \quad 14t = 0$$

$$\therefore \quad t = 0$$

Substituting $t = 0$ into the parametric equations, we obtain the foot of the perpendicular (2, −1, 4).

**b** When $t = 0$, $\overrightarrow{PA} = \begin{pmatrix} -1 \\ -1 \\ 5 \end{pmatrix}$, so $PA = \sqrt{1+1+25}$
$= \sqrt{27}$ units

$\therefore$ the shortest distance from the point to the line is $\sqrt{27}$ units.

**7** **a** The line has direction vector $\mathbf{b} = \begin{pmatrix} 2 \\ 3 \\ 1 \end{pmatrix}$.

Let the point $(1, 1, 3)$ be P and $A(1 + 2t, -1 + 3t, 2 + t)$ be any point on the line.

$\therefore \overrightarrow{PA} = \begin{pmatrix} 1+2t-1 \\ -1+3t-1 \\ 2+t-3 \end{pmatrix} = \begin{pmatrix} 2t \\ -2+3t \\ -1+t \end{pmatrix}$.

Now $\overrightarrow{PA}$ and $\mathbf{b}$ are perpendicular, so $\overrightarrow{PA} \bullet \mathbf{b} = 0$

$\begin{pmatrix} 2 \\ 3 \\ 1 \end{pmatrix}$ P(1, 1, 3) A$(1 + 2t, -1 + 3t, 2 + t)$

$\therefore \begin{pmatrix} 2t \\ -2+3t \\ -1+t \end{pmatrix} \bullet \begin{pmatrix} 2 \\ 3 \\ 1 \end{pmatrix} = 0$

$\therefore 4t - 6 + 9t - 1 + t = 0$

$\therefore 14t = 7$

$\therefore t = \frac{1}{2}$

Substituting $t = \frac{1}{2}$ into the parametric equations, we obtain the foot of the perpendicular $(2, \frac{1}{2}, \frac{5}{2})$.

**b** When $t = \frac{1}{2}$, $\overrightarrow{PA} = \begin{pmatrix} 1 \\ -2+\frac{3}{2} \\ -1+\frac{1}{2} \end{pmatrix} = \begin{pmatrix} 1 \\ -\frac{1}{2} \\ -\frac{1}{2} \end{pmatrix}$

$\therefore PA = \sqrt{1 + \frac{1}{4} + \frac{1}{4}} = \sqrt{\frac{3}{2}}$ units

$\therefore$ the shortest distance from the point to the line is $\sqrt{\frac{3}{2}}$ units.

## EXERCISE 13F

**1** **a**

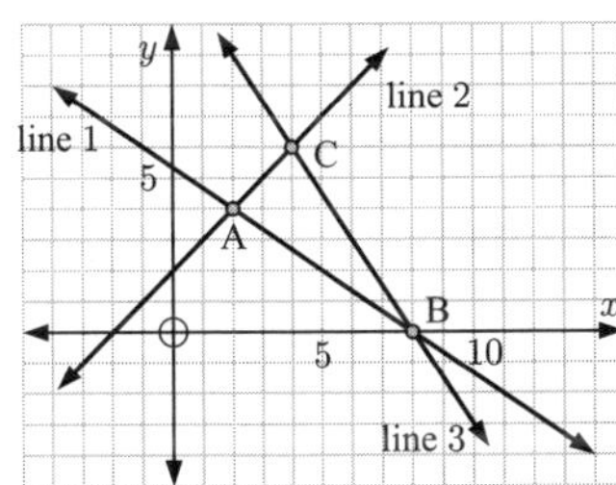


**b** A is (2, 4), B is (8, 0), C is (4, 6)

**c** $BC = \sqrt{(4-8)^2 + (6-0)^2} = \sqrt{16+36}$
$= \sqrt{52}$ units

$AB = \sqrt{(8-2)^2 + (0-4)^2} = \sqrt{36+16}$
$= \sqrt{52}$ units

$\therefore BC = AB$ and so $\triangle ABC$ is isosceles.

**d** Line 1 and Line 2 meet at A.

$\therefore \begin{pmatrix} -1 \\ 6 \end{pmatrix} + r\begin{pmatrix} 3 \\ -2 \end{pmatrix} = \begin{pmatrix} 0 \\ 2 \end{pmatrix} + s\begin{pmatrix} 1 \\ 1 \end{pmatrix}$

$\therefore \begin{pmatrix} 3r - s \\ -2r - s \end{pmatrix} = \begin{pmatrix} 1 \\ -4 \end{pmatrix}$

$\therefore 3r - s = 1$
and $2r + s = 4$
Adding, $5r = 5$ $\therefore r = 1$

$\therefore \begin{pmatrix} x \\ y \end{pmatrix} = \begin{pmatrix} -1 \\ 6 \end{pmatrix} + \begin{pmatrix} 3 \\ -2 \end{pmatrix} = \begin{pmatrix} 2 \\ 4 \end{pmatrix}$ ✓

Line 2 and Line 3 meet at C.

$\therefore \begin{pmatrix} 0 \\ 2 \end{pmatrix} + s\begin{pmatrix} 1 \\ 1 \end{pmatrix} = \begin{pmatrix} 10 \\ -3 \end{pmatrix} + t\begin{pmatrix} -2 \\ 3 \end{pmatrix}$

$\therefore \begin{pmatrix} s + 2t \\ s - 3t \end{pmatrix} = \begin{pmatrix} 10 \\ -5 \end{pmatrix}$

$\therefore s + 2t = 10$
$-s + 3t = 5$
Adding, $5t = 15$ $\therefore t = 3$

$\therefore \begin{pmatrix} x \\ y \end{pmatrix} = \begin{pmatrix} 10 \\ -3 \end{pmatrix} + 3\begin{pmatrix} -2 \\ 3 \end{pmatrix} = \begin{pmatrix} 4 \\ 6 \end{pmatrix}$ ✓

Line 1 and Line 3 meet at B.

$$\therefore \begin{pmatrix} -1 \\ 6 \end{pmatrix} + r\begin{pmatrix} 3 \\ -2 \end{pmatrix} = \begin{pmatrix} 10 \\ -3 \end{pmatrix} + t\begin{pmatrix} -2 \\ 3 \end{pmatrix}$$

$$\therefore \begin{pmatrix} 3r + 2t \\ -2r - 3t \end{pmatrix} = \begin{pmatrix} 11 \\ -9 \end{pmatrix}$$

$$\therefore \quad 3r + 2t = 11 \quad .... (1)$$
$$-2r - 3t = -9 \quad .... (2)$$

$$\therefore \quad 9r + 6t = 33 \quad \{3 \times (1)\}$$
$$-4r - 6t = -18 \quad \{2 \times (2)\}$$

Adding, $5r = 15$

$$\therefore \quad r = 3$$

So, $\begin{pmatrix} x \\ y \end{pmatrix} = \begin{pmatrix} -1 \\ 6 \end{pmatrix} + 3\begin{pmatrix} 3 \\ -2 \end{pmatrix} = \begin{pmatrix} 8 \\ 0 \end{pmatrix}$ ✓

**2** **a**

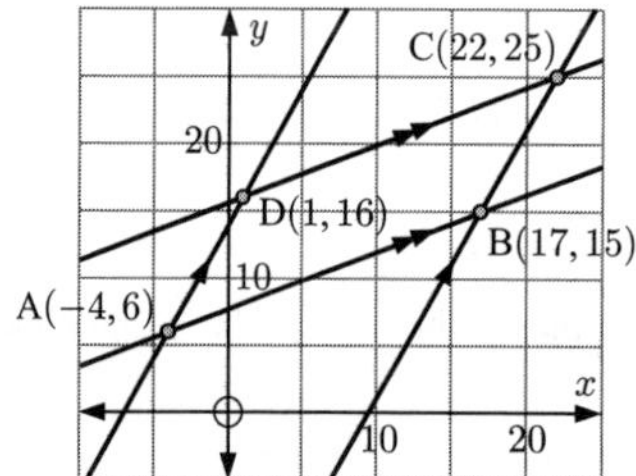


**b** A(−4, 6), B(17, 15), C(22, 25), D(1, 16)

**c** Lines (AB) and (AD) meet at A.

$$\therefore \begin{pmatrix} -4 \\ 6 \end{pmatrix} + r\begin{pmatrix} 7 \\ 3 \end{pmatrix} = \begin{pmatrix} -4 \\ 6 \end{pmatrix} + s\begin{pmatrix} 1 \\ 2 \end{pmatrix}$$

$$\therefore \begin{pmatrix} 7r - s \\ 3r - 2s \end{pmatrix} = \begin{pmatrix} 0 \\ 0 \end{pmatrix}$$

$$\therefore \quad 7r - s = 0 \quad .... (1)$$
and $3r - 2s = 0$
$$-14r + 2s = 0 \quad \{-2 \times (1)\}$$

Adding, $-11r = 0$

$$\therefore \quad r = 0$$

$$\therefore \begin{pmatrix} x \\ y \end{pmatrix} = \begin{pmatrix} -4 \\ 6 \end{pmatrix} \quad ✓$$

Lines (AB) and (CB) meet at B.

$$\therefore \begin{pmatrix} -4 \\ 6 \end{pmatrix} + r\begin{pmatrix} 7 \\ 3 \end{pmatrix} = \begin{pmatrix} 22 \\ 25 \end{pmatrix} + u\begin{pmatrix} -1 \\ -2 \end{pmatrix}$$

$$\therefore \begin{pmatrix} 7r + u \\ 3r + 2u \end{pmatrix} = \begin{pmatrix} 26 \\ 19 \end{pmatrix}$$

$$\therefore \quad 7r + u = 26 \quad .... (1)$$
and $3r + 2u = 19$
$$-14r - 2u = -52 \quad \{-2 \times (1)\}$$

Adding, $-11r = -33$

$$\therefore \quad r = 3$$

$$\therefore \begin{pmatrix} x \\ y \end{pmatrix} = \begin{pmatrix} -4 \\ 6 \end{pmatrix} + 3\begin{pmatrix} 7 \\ 3 \end{pmatrix} = \begin{pmatrix} 17 \\ 15 \end{pmatrix} \quad ✓$$

Lines (CD) and (CB) meet at C.

$$\therefore \begin{pmatrix} 22 \\ 25 \end{pmatrix} + t\begin{pmatrix} -7 \\ -3 \end{pmatrix} = \begin{pmatrix} 22 \\ 25 \end{pmatrix} + u\begin{pmatrix} -1 \\ -2 \end{pmatrix}$$

$$\therefore \begin{pmatrix} -7t + u \\ -3t + 2u \end{pmatrix} = \begin{pmatrix} 0 \\ 0 \end{pmatrix}$$

$$\therefore \quad -7t + u = 0 \quad .... (1)$$
and $-3t + 2u = 0$
$$14t - 2u = 0 \quad \{-2 \times (1)\}$$

Adding, $11t = 0$

$$\therefore \quad t = 0$$

$$\therefore \begin{pmatrix} x \\ y \end{pmatrix} = \begin{pmatrix} 22 \\ 25 \end{pmatrix} \quad ✓$$

Lines (AD) and (CD) meet at D.

$$\therefore \begin{pmatrix} -4 \\ 6 \end{pmatrix} + s\begin{pmatrix} 1 \\ 2 \end{pmatrix} = \begin{pmatrix} 22 \\ 25 \end{pmatrix} + t\begin{pmatrix} -7 \\ -3 \end{pmatrix}$$

$$\therefore \begin{pmatrix} s + 7t \\ 2s + 3t \end{pmatrix} = \begin{pmatrix} 26 \\ 19 \end{pmatrix}$$

$$\therefore \quad s + 7t = 26 \quad .... (1)$$
and $2s + 3t = 19$
$$-2s - 14t = -52 \quad \{-2 \times (1)\}$$

Adding, $-11t = -33$

$$\therefore \quad t = 3$$

$$\therefore \begin{pmatrix} x \\ y \end{pmatrix} = \begin{pmatrix} 22 \\ 25 \end{pmatrix} + 3\begin{pmatrix} -7 \\ -3 \end{pmatrix} = \begin{pmatrix} 1 \\ 16 \end{pmatrix} \quad ✓$$

**3** **a** Lines (AB) and (AC) meet at A.

$$\therefore \begin{pmatrix} 0 \\ 2 \end{pmatrix} + r\begin{pmatrix} 2 \\ 1 \end{pmatrix} = \begin{pmatrix} 0 \\ 5 \end{pmatrix} + t\begin{pmatrix} 1 \\ -1 \end{pmatrix}$$

$$\therefore \begin{pmatrix} 2r - t \\ r + t \end{pmatrix} = \begin{pmatrix} 0 \\ 3 \end{pmatrix}$$

$$\therefore \quad 2r - t = 0$$
$$r + t = 3$$
$$\text{Adding,} \quad 3r = 3$$
$$\therefore \quad r = 1$$

$$\therefore \begin{pmatrix} x \\ y \end{pmatrix} = \begin{pmatrix} 0 \\ 2 \end{pmatrix} + \begin{pmatrix} 2 \\ 1 \end{pmatrix} = \begin{pmatrix} 2 \\ 3 \end{pmatrix}$$

$\therefore$ A is (2, 3)

Lines (AB) and (BC) meet at B.

$$\therefore \begin{pmatrix} 0 \\ 2 \end{pmatrix} + r\begin{pmatrix} 2 \\ 1 \end{pmatrix} = \begin{pmatrix} 8 \\ 6 \end{pmatrix} + s\begin{pmatrix} -1 \\ -2 \end{pmatrix}$$

$$\therefore \begin{pmatrix} 2r + s \\ r + 2s \end{pmatrix} = \begin{pmatrix} 8 \\ 4 \end{pmatrix}$$

$$\therefore \quad -4r - 2s = -16$$
$$r + 2s = 4$$
$$\text{Adding,} \quad -3r = -12$$
$$\therefore \quad r = 4$$

$$\therefore \begin{pmatrix} x \\ y \end{pmatrix} = \begin{pmatrix} 0 \\ 2 \end{pmatrix} + 4\begin{pmatrix} 2 \\ 1 \end{pmatrix} = \begin{pmatrix} 8 \\ 6 \end{pmatrix}$$

$\therefore$ B is (8, 6)

Lines (BC) and (AC) meet at C.

$$\therefore \begin{pmatrix} 8 \\ 6 \end{pmatrix} + s\begin{pmatrix} -1 \\ -2 \end{pmatrix} = \begin{pmatrix} 0 \\ 5 \end{pmatrix} + t\begin{pmatrix} 1 \\ -1 \end{pmatrix}$$

$$\therefore \begin{pmatrix} -s - t \\ -2s + t \end{pmatrix} = \begin{pmatrix} -8 \\ -1 \end{pmatrix}$$

$$\therefore \quad -s - t = -8$$
$$-2s + t = -1$$
$$\text{Adding,} \quad -3s = -9$$
$$\therefore \quad s = 3$$

$$\therefore \begin{pmatrix} x \\ y \end{pmatrix} = \begin{pmatrix} 8 \\ 6 \end{pmatrix} + 3\begin{pmatrix} -1 \\ -2 \end{pmatrix} = \begin{pmatrix} 5 \\ 0 \end{pmatrix}$$

$\therefore$ C is (5, 0)

**b** A(2, 3), B(8, 6), C(5, 0)

$$\text{AB} = \sqrt{(8-2)^2 + (6-3)^2} = \sqrt{36+9} = \sqrt{45}$$

$$\text{BC} = \sqrt{(5-8)^2 + (0-6)^2} = \sqrt{9+36} = \sqrt{45}$$

$$\text{AC} = \sqrt{(5-2)^2 + (0-3)^2} = \sqrt{9+9} = \sqrt{18}$$

The two equal sides are [AB] and [BC] and they have length $\sqrt{45}$ units. [AC] has length $\sqrt{18}$ units.

**4** **a** Lines (QP) and (PR) meet at P.

$$\therefore \begin{pmatrix} 3 \\ -1 \end{pmatrix} + r\begin{pmatrix} 14 \\ 10 \end{pmatrix} = \begin{pmatrix} 0 \\ 18 \end{pmatrix} + t\begin{pmatrix} 5 \\ -7 \end{pmatrix}$$

$$\therefore \begin{pmatrix} 14r - 5t \\ 10r + 7t \end{pmatrix} = \begin{pmatrix} -3 \\ 19 \end{pmatrix}$$

$$\therefore \quad 14r - 5t = -3 \quad .... (1)$$
$$10r + 7t = 19 \quad .... (2)$$

$$\therefore \quad 98r - 35t = -21 \quad \{7 \times (1)\}$$
$$50r + 35t = 95 \quad \{5 \times (2)\}$$
$$\text{Adding,} \quad 148r = 74$$
$$\therefore \quad r = \tfrac{1}{2}$$

$$\therefore \begin{pmatrix} x \\ y \end{pmatrix} = \begin{pmatrix} 3 \\ -1 \end{pmatrix} + \tfrac{1}{2}\begin{pmatrix} 14 \\ 10 \end{pmatrix} = \begin{pmatrix} 10 \\ 4 \end{pmatrix}$$

$\therefore$ P is (10, 4)

Lines (QR) and (PR) meet at R.

$$\therefore \begin{pmatrix} 3 \\ -1 \end{pmatrix} + s\begin{pmatrix} 17 \\ -9 \end{pmatrix} = \begin{pmatrix} 0 \\ 18 \end{pmatrix} + t\begin{pmatrix} 5 \\ -7 \end{pmatrix}$$

$$\therefore \begin{pmatrix} 17s - 5t \\ -9s + 7t \end{pmatrix} = \begin{pmatrix} -3 \\ 19 \end{pmatrix}$$

$$\therefore \quad 17s - 5t = -3 \quad .... (1)$$
$$-9s + 7t = 19 \quad .... (2)$$

$$\therefore \quad 119s - 35t = -21 \quad \{7 \times (1)\}$$
$$-45s + 35t = 95 \quad \{5 \times (2)\}$$
$$\text{Adding,} \quad 74s = 74$$
$$\therefore \quad s = 1$$

$$\therefore \begin{pmatrix} x \\ y \end{pmatrix} = \begin{pmatrix} 3 \\ -1 \end{pmatrix} + \begin{pmatrix} 17 \\ -9 \end{pmatrix} = \begin{pmatrix} 20 \\ -10 \end{pmatrix}$$

$\therefore$ R is (20, −10)

Lines (QP) and (PR) meet at Q.

$$\begin{pmatrix} 3 \\ -1 \end{pmatrix} + r\begin{pmatrix} 14 \\ 10 \end{pmatrix} = \begin{pmatrix} 3 \\ -1 \end{pmatrix} + s\begin{pmatrix} 17 \\ -9 \end{pmatrix}$$

$$\therefore \quad r\begin{pmatrix} 14 \\ 10 \end{pmatrix} = s\begin{pmatrix} 17 \\ -9 \end{pmatrix}$$

$$\therefore \quad r = s = 0$$

$$\text{So,} \quad \begin{pmatrix} x \\ y \end{pmatrix} = \begin{pmatrix} 3 \\ -1 \end{pmatrix}$$

$\therefore$ Q is (3, −1)

**b** $\overrightarrow{PQ} = \begin{pmatrix} 3-10 \\ -1-4 \end{pmatrix} = \begin{pmatrix} -7 \\ -5 \end{pmatrix}$

$\overrightarrow{PR} = \begin{pmatrix} 20-10 \\ -10-4 \end{pmatrix} = \begin{pmatrix} 10 \\ -14 \end{pmatrix}$

and $\overrightarrow{PQ} \bullet \overrightarrow{PR} = -70 + 70 = 0$

**c** [PQ] $\perp$ [PR] $\therefore$ $\widehat{QPR} = 90°$

**d** Area $= \frac{1}{2} \mid \overrightarrow{PQ} \mid \mid \overrightarrow{PR} \mid$

$= \frac{1}{2}\sqrt{49+25}\sqrt{100+196}$

$= 74 \text{ units}^2$

**5** **a** Lines (AB) and (AD) meet at A.

$$\therefore \quad \begin{pmatrix} 2 \\ 5 \end{pmatrix} + r\begin{pmatrix} 4 \\ 1 \end{pmatrix} = \begin{pmatrix} 3 \\ 1 \end{pmatrix} + u\begin{pmatrix} -3 \\ 12 \end{pmatrix}$$

$$\therefore \quad \begin{pmatrix} 4r+3u \\ r-12u \end{pmatrix} = \begin{pmatrix} 1 \\ -4 \end{pmatrix}$$

$\therefore$ $4r + 3u = 1$
$r - 12u = -4$ .... (1)

$\therefore$ $4r + 3u = 1$
$-4r + 48u = 16$ $\{-4 \times (1)\}$

Adding, $51u = 17$

$\therefore$ $u = \frac{1}{3}$

$$\therefore \quad \begin{pmatrix} x \\ y \end{pmatrix} = \begin{pmatrix} 3 \\ 1 \end{pmatrix} + \frac{1}{3}\begin{pmatrix} -3 \\ 12 \end{pmatrix} = \begin{pmatrix} 2 \\ 5 \end{pmatrix}$$

$\therefore$ A is (2, 5)

Lines (AB) and (BC) meet at B.

$$\therefore \quad \begin{pmatrix} 2 \\ 5 \end{pmatrix} + r\begin{pmatrix} 4 \\ 1 \end{pmatrix} = \begin{pmatrix} 18 \\ 9 \end{pmatrix} + s\begin{pmatrix} -8 \\ 32 \end{pmatrix}$$

$$\therefore \quad \begin{pmatrix} 4r+8s \\ r-32s \end{pmatrix} = \begin{pmatrix} 16 \\ 4 \end{pmatrix}$$

$\therefore$ $4r + 8s = 16$ .... (1)
$r - 32s = 4$ .... (2)

$\therefore$ $r + 2s = 4$ $\{(1) \div 4\}$
$-r + 32s = -4$ $\{-1 \times (2)\}$

Adding, $34s = 0$

$\therefore$ $s = 0$

$$\therefore \quad \begin{pmatrix} x \\ y \end{pmatrix} = \begin{pmatrix} 18 \\ 9 \end{pmatrix}$$

$\therefore$ B is (18, 9)

Lines (BC) and (CD) meet at C.

$$\therefore \quad \begin{pmatrix} 18 \\ 9 \end{pmatrix} + s\begin{pmatrix} -8 \\ 32 \end{pmatrix} = \begin{pmatrix} 14 \\ 25 \end{pmatrix} + t\begin{pmatrix} -8 \\ -2 \end{pmatrix}$$

$$\therefore \quad \begin{pmatrix} -8s+8t \\ 32s+2t \end{pmatrix} = \begin{pmatrix} -4 \\ 16 \end{pmatrix}$$

$\therefore$ $-8s + 8t = -4$ .... (1)
$32s + 2t = 16$

$\therefore$ $2s - 2t = 1$ $\{(1) \div -4\}$
$32s + 2t = 16$

Adding, $34s = 17$

$\therefore$ $s = \frac{1}{2}$

$$\therefore \quad \begin{pmatrix} x \\ y \end{pmatrix} = \begin{pmatrix} 18 \\ 9 \end{pmatrix} + \frac{1}{2}\begin{pmatrix} -8 \\ 32 \end{pmatrix} = \begin{pmatrix} 14 \\ 25 \end{pmatrix}$$

$\therefore$ C is (14, 25)

Lines (CD) and (AD) meet at D.

$$\therefore \quad \begin{pmatrix} 14 \\ 25 \end{pmatrix} + t\begin{pmatrix} -8 \\ -2 \end{pmatrix} = \begin{pmatrix} 3 \\ 1 \end{pmatrix} + u\begin{pmatrix} -3 \\ 12 \end{pmatrix}$$

$$\therefore \quad \begin{pmatrix} -8t+3u \\ -2t-12u \end{pmatrix} = \begin{pmatrix} -11 \\ -24 \end{pmatrix}$$

$\therefore$ $-8t + 3u = -11$ .... (1)
$-2t - 12u = -24$ .... (2)

$\therefore$ $16t - 6u = 22$ $\{(-2) \times (1)\}$
$t + 6u = 12$ $\{(2) \div -2\}$

Adding, $17t = 34$

$\therefore$ $t = 2$

$$\therefore \quad \begin{pmatrix} x \\ y \end{pmatrix} = \begin{pmatrix} 14 \\ 25 \end{pmatrix} + 2\begin{pmatrix} -8 \\ -2 \end{pmatrix} = \begin{pmatrix} -2 \\ 21 \end{pmatrix}$$

$\therefore$ D is (−2, 21)

**b** $\overrightarrow{AC} = \begin{pmatrix} 14-2 \\ 25-5 \end{pmatrix} = \begin{pmatrix} 12 \\ 20 \end{pmatrix}$

$\overrightarrow{DB} = \begin{pmatrix} 18--2 \\ 9-21 \end{pmatrix} = \begin{pmatrix} 20 \\ -12 \end{pmatrix}$

**i** $|\overrightarrow{AC}| = \sqrt{12^2+20^2} = \sqrt{544}$ units

**ii** $|\overrightarrow{DB}| = \sqrt{20^2+(-12)^2} = \sqrt{544}$ units

**iii** $\overrightarrow{AC} \bullet \overrightarrow{DB} = 240 - 240 = 0$

**c** The diagonals are perpendicular and equal in length, and as their midpoints are the same (at (8, 15)), ABCD is a square.

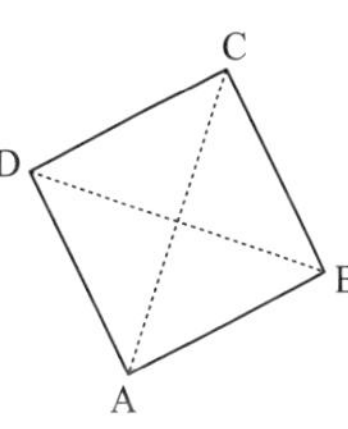


## EXERCISE 13G

**1** **a** Line 1 has direction vector $\begin{pmatrix} 2 \\ -1 \\ 1 \end{pmatrix}$ and line 2 has direction vector $\begin{pmatrix} 3 \\ -1 \\ 2 \end{pmatrix}$.

As one vector is not a scalar multiple of the other, the lines are not parallel.

Now $1+2t = -2+3s$ $\quad 2-t = 3-s \quad 3+t = 1+2s$

$\therefore\ 2t-3s = -3$ .... (1) $\quad \therefore\ -t+s = 1$ .... (2) $\quad \therefore\ t-2s=-2$ .... (3)

Solving (2) and (3) simultaneously:

$$\begin{aligned} -t+s &= 1 \\ t-2s &= -2 \\ \hline -s &= -1 \quad \therefore\ s = 1 \text{ and } t = 0 \end{aligned}$$

and in (1), LHS $= 2t-3s = 2(0)-3(1) = -3$ ✓

$\therefore\ s=1,\ t=0$ satisfies all three equations

$\therefore$ the two lines meet at (1, 2, 3) {using $t=0$ or $s=1$}

The acute angle between the lines has $\cos\theta = \dfrac{|6+1+2|}{\sqrt{4+1+1}\,\sqrt{9+1+4}} = \dfrac{9}{\sqrt{84}}$

and so $\theta \approx 10.9°$

**b** Line 1 has direction vector $\begin{pmatrix} 2 \\ -12 \\ 12 \end{pmatrix}$ and line 2 has direction vector $\begin{pmatrix} 4 \\ 3 \\ -1 \end{pmatrix}$.

As one vector is not a scalar multiple of the other, the lines are not parallel.

Now $-1+2t = 4s-3 \quad 2-12t = 3s+2 \quad 4+12t = -s-1$

$\therefore\ 2t-4s=-2 \quad -12t-3s = 0 \quad 12t+s=-5$ .... (3)

$\therefore\ t-2s = -1$ .... (1) $\quad s = -4t$ .... (2)

Solving (1) and (2) simultaneously: $t-2(-4t) = -1$

$\therefore\ 9t = -1$

$\therefore\ t = -\frac{1}{9}$ and so $s = \frac{4}{9}$

In (3), $12t+s = 12\left(-\frac{1}{9}\right)+\frac{4}{9} = -\frac{12}{9}+\frac{4}{9} = -\frac{8}{9}$, which is not $-5$.

Since the system is inconsistent, the lines do not intersect, so the lines are skew.

The acute angle between the lines has $\cos\theta = \dfrac{|8-36-12|}{\sqrt{292}\,\sqrt{26}} = \dfrac{40}{\sqrt{7592}}$ and so $\theta \approx 62.7°$.

**c** Line 1 has direction vector $\begin{pmatrix} 6 \\ 8 \\ 2 \end{pmatrix}$ and line 2 has direction vector $\begin{pmatrix} 3 \\ 4 \\ 1 \end{pmatrix}$.

As $\begin{pmatrix} 6 \\ 8 \\ 2 \end{pmatrix} = 2\begin{pmatrix} 3 \\ 4 \\ 1 \end{pmatrix}$ the two lines are parallel. Hence, $\theta = 0°$.

To see if the lines are coincident, try to find a shared point.

The point on line 1 where $t=1$ is (6, 11, 1).

The unique point on line 2 with $z$-coordinate 1 is the point where $1+s=1 \ \therefore\ s=0$.

This point is (2, 0, 1). Since $(6, 11, 1) \neq (2, 0, 1)$ the lines are not coincident.

**d** In line 1 let $x = 2 - y = z + 2 = t$, so $x = t$, $y = 2 - t$, and $z = t - 2$, $t \in \mathbb{R}$.

Line 1 has direction vector $\begin{pmatrix} 1 \\ -1 \\ 1 \end{pmatrix}$ and line 2 has direction vector $\begin{pmatrix} 3 \\ -2 \\ 2 \end{pmatrix}$.

As one vector is not a scalar multiple of the other, the lines are not parallel.

Now $t = 1 + 3s$ .... (1) $\quad 2 - t = -2 - 2s$ $\quad -2 + t = 2s + \frac{1}{2}$

$-t + 2s = -4$ .... (2) $\quad t - 2s = 2\frac{1}{2}$ .... (3)

Solving (1) and (2) simultaneously, $-(1 + 3s) + 2s = -4$

$\therefore\ -1 - 3s + 2s = -4$

$\therefore\ -s = -3$

$\therefore\ s = 3$ and so $t = 1 + 3(3) = 10$

Substituting in (3), $t - 2s = 10 - 2(3) = 4 \neq 2\frac{1}{2}$

Since the system is inconsistent, the lines do not meet. $\therefore$ they are skew.

The acute angle between the lines has $\cos\theta = \dfrac{|3 + 2 + 2|}{\sqrt{1 + 1 + 1}\sqrt{9 + 4 + 4}} = \dfrac{7}{\sqrt{3}\sqrt{17}}$

$\therefore\ \theta \approx 11.4°$

**e** Line 1 has direction vector $\begin{pmatrix} 1 \\ -1 \\ 2 \end{pmatrix}$ and line 2 has direction vector $\begin{pmatrix} 3 \\ -2 \\ 1 \end{pmatrix}$.

As one vector is not a scalar multiple of the other, the lines are not parallel.

$1 + t = 2 + 3s$ $\quad 2 - t = 3 - 2s$ $\quad 3 + 2t = s - 5$

$t - 3s = 1$ .... (1) $\quad -t + 2s = 1$ .... (2) $\quad 2t - s = -8$ .... (3)

Solving (1) and (2) simultaneously, $t - 3s = 1$

$-t + 2s = 1$

Adding, $-s = 2$

$\therefore\ s = -2$ and $t - 3(-2) = 1$ $\therefore\ t = -5$

Checking in (3), $2t - s = 2(-5) - (-2) = -10 + 2 = -8$ ✓

Since $s = -2$, $t = -5$ satisfies all three equations, the lines meet.

They meet at $x = 1 + (-5)$, $y = 2 - (-5)$, $z = 3 + 2(-5)$, or at $(-4, 7, -7)$.

The acute angle between the lines has $\cos\theta = \dfrac{|3 + 2 + 2|}{\sqrt{1 + 1 + 4}\sqrt{9 + 4 + 1}} = \dfrac{7}{\sqrt{84}}$

and so $\theta \approx 40.2°$

**f** Line 1 has direction vector $\begin{pmatrix} -2 \\ 1 \\ 0 \end{pmatrix}$ and line 2 has direction vector $\begin{pmatrix} 4 \\ -2 \\ 0 \end{pmatrix}$.

Now $\begin{pmatrix} 4 \\ -2 \\ 0 \end{pmatrix} = -2\begin{pmatrix} -2 \\ 1 \\ 0 \end{pmatrix}$, so the lines are parallel and hence $\theta = 0°$.

All points on line 1 have $z$-coordinate 5 and all points on line 2 have $z$-coordinate 3.

$\therefore$ the lines are not coincident.

**g** Line 1 has direction vector $\begin{pmatrix} 2 \\ -1 \\ 3 \end{pmatrix}$ and line 2 has direction vector $\begin{pmatrix} -4 \\ 2 \\ -6 \end{pmatrix}$.

As $\begin{pmatrix} -4 \\ 2 \\ -6 \end{pmatrix} = -2\begin{pmatrix} 2 \\ -1 \\ 3 \end{pmatrix}$, the two lines are parallel. Hence $\theta = 0°$.

The point on line 1 where $t = 1$ is $(3, -1, 4)$.

The unique point on line 2 with $x$-coordinate 3 is the point where $3 - 4s = 3$ $\therefore\ s = 0$.

This point is $(3, -1, 4)$.

Lines 1 and 2 are parallel and share the point $(3, -1, 4)$. $\therefore$ they are coincident.

## REVIEW SET 13A

**1** **a** The vector equation is

$$\begin{pmatrix} x \\ y \end{pmatrix} = \begin{pmatrix} -6 \\ 3 \end{pmatrix} + t\begin{pmatrix} 4 \\ -3 \end{pmatrix}$$

**b** The parametric equations are
$x = -6 + 4t, \quad y = 3 - 3t, \quad t \in \mathbb{R}$

**c** $$\frac{x+6}{4} = \frac{y-3}{-3} = t$$

$\therefore \quad -3x - 18 = 4y - 12$

So, the Cartesian equation is $3x + 4y = -6$.

**2** $(-3, m)$ lies on the line, so $\begin{pmatrix} -3 \\ m \end{pmatrix} = \begin{pmatrix} 18 \\ -2 \end{pmatrix} + \begin{pmatrix} -7t \\ 4t \end{pmatrix}$

$\therefore \quad -3 = 18 - 7t \quad$ and $\quad m = -2 + 4t$

$\therefore \quad 7t = 21$

$\therefore \quad t = 3 \quad$ and so $\quad m = -2 + 4(3) = 10$

**3** **a** When $t = 1$, $\mathbf{r} = \begin{pmatrix} 3 \\ -3 \end{pmatrix} + 1\begin{pmatrix} 2 \\ 5 \end{pmatrix} = \begin{pmatrix} 3+2 \\ -3+5 \end{pmatrix} = \begin{pmatrix} 5 \\ 2 \end{pmatrix}$

$\therefore$ the point is (5, 2).

**b** $\begin{pmatrix} 4 \\ 10 \end{pmatrix}$ is a non-zero scalar multiple of $\begin{pmatrix} 2 \\ 5 \end{pmatrix}$, so it could also be used to describe the direction of the line.

**c** The line passes through point (5, 2) and has direction vector $\begin{pmatrix} 4 \\ 10 \end{pmatrix}$.

$\therefore \quad \mathbf{r} = \begin{pmatrix} 5 \\ 2 \end{pmatrix} + s\begin{pmatrix} 4 \\ 10 \end{pmatrix}$, $s \in \mathbb{R}$ is an alternative vector equation for the line.

**4** P(2, 0, 1), Q(3, 4, −2), R(−1, 3, 2)

**a** $\overrightarrow{PQ} = \begin{pmatrix} 1 \\ 4 \\ -3 \end{pmatrix}$

Since $\overrightarrow{PQ} = \begin{pmatrix} 1 \\ 4 \\ -3 \end{pmatrix}$ and P is at (2, 0, 1), the line has parametric equations

$x = 2 + t, \quad y = 0 + 4t, \quad z = 1 - 3t$

$\therefore \quad x = 2 + t, \quad y = 4t, \quad z = 1 - 3t, \quad t$ in $\mathbb{R}$

**b** $\overrightarrow{QR} = \begin{pmatrix} -4 \\ -1 \\ 4 \end{pmatrix}$

$|\overrightarrow{PQ}| = \sqrt{1^2 + 4^2 + (-3)^2} = \sqrt{26}$

$|\overrightarrow{QR}| = \sqrt{(-4)^2 + (-1)^2 + 4^2} = \sqrt{33}$

$$\overrightarrow{PQ} \bullet \overrightarrow{QR} = \begin{pmatrix} 1 \\ 4 \\ -3 \end{pmatrix} \bullet \begin{pmatrix} -4 \\ -1 \\ 4 \end{pmatrix} = 1 \times (-4) + 4 \times (-1) + (-3) \times 4$$
$$= -20$$

If $\theta = P\widehat{Q}R$, then $\cos\theta = \dfrac{|\overrightarrow{PQ} \bullet \overrightarrow{QR}|}{|\overrightarrow{PQ}||\overrightarrow{QR}|} = \dfrac{|-20|}{\sqrt{26}\sqrt{33}} = \dfrac{20}{\sqrt{26}\sqrt{33}}$

**5** **a** Lines (AB) and (AC) meet at A.

$$\therefore \begin{pmatrix} 4 \\ -1 \end{pmatrix} + t\begin{pmatrix} 1 \\ 3 \end{pmatrix} = \begin{pmatrix} -1 \\ 0 \end{pmatrix} + u\begin{pmatrix} 3 \\ 1 \end{pmatrix}$$

$$\therefore \begin{pmatrix} 4+t \\ -1+3t \end{pmatrix} = \begin{pmatrix} -1+3u \\ u \end{pmatrix}$$

$$\therefore\ t - 3u = -5 \quad .... (1)$$
$$3t - u = 1$$

$$\therefore\ -3t + 9u = 15 \quad \{-3 \times (1)\}$$
$$3t - u = 1$$

Adding, $8u = 16$

$$\therefore\ u = 2$$

So, $\begin{pmatrix} x \\ y \end{pmatrix} = \begin{pmatrix} -1 \\ 0 \end{pmatrix} + 2\begin{pmatrix} 3 \\ 1 \end{pmatrix} = \begin{pmatrix} 5 \\ 2 \end{pmatrix}$

$\therefore$ A is (5, 2).

Lines (AB) and (BC) meet at B.

$$\therefore \begin{pmatrix} 4 \\ -1 \end{pmatrix} + t\begin{pmatrix} 1 \\ 3 \end{pmatrix} = \begin{pmatrix} 7 \\ 4 \end{pmatrix} + s\begin{pmatrix} 1 \\ -1 \end{pmatrix}$$

$$\therefore \begin{pmatrix} 4+t \\ -1+3t \end{pmatrix} = \begin{pmatrix} 7+s \\ 4-s \end{pmatrix}$$

$$\therefore\ t - s = 3$$
$$3t + s = 5$$

Adding, $4t = 8$

$$\therefore\ t = 2$$

So, $\begin{pmatrix} x \\ y \end{pmatrix} = \begin{pmatrix} 4 \\ -1 \end{pmatrix} + 2\begin{pmatrix} 1 \\ 3 \end{pmatrix} = \begin{pmatrix} 6 \\ 5 \end{pmatrix}$

$\therefore$ B is (6, 5).

Lines (BC) and (AC) meet at C.

$$\therefore \begin{pmatrix} 7 \\ 4 \end{pmatrix} + s\begin{pmatrix} 1 \\ -1 \end{pmatrix} = \begin{pmatrix} -1 \\ 0 \end{pmatrix} + u\begin{pmatrix} 3 \\ 1 \end{pmatrix}$$

$$\therefore \begin{pmatrix} 7+s \\ 4-s \end{pmatrix} = \begin{pmatrix} -1+3u \\ u \end{pmatrix}$$

$$\therefore\ s - 3u = -8$$
$$-s - u = -4$$

Adding, $-4u = -12$

$$\therefore\ u = 3$$

So, $\begin{pmatrix} x \\ y \end{pmatrix} = \begin{pmatrix} -1 \\ 0 \end{pmatrix} + 3\begin{pmatrix} 3 \\ 1 \end{pmatrix} = \begin{pmatrix} 8 \\ 3 \end{pmatrix}$ $\quad\therefore$ C is (8, 3).

**b** $\overrightarrow{AB} = \begin{pmatrix} 1 \\ 3 \end{pmatrix}$, so $|\overrightarrow{AB}| = \sqrt{1+9} = \sqrt{10}$ units

$\overrightarrow{BC} = \begin{pmatrix} 2 \\ -2 \end{pmatrix}$, so $|\overrightarrow{BC}| = \sqrt{4+4} = \sqrt{8}$ units

$\overrightarrow{AC} = \begin{pmatrix} 3 \\ 1 \end{pmatrix}$, so $|\overrightarrow{AC}| = \sqrt{9+1} = \sqrt{10}$ units

**c** Triangle ABC is isosceles.

**6** **a**

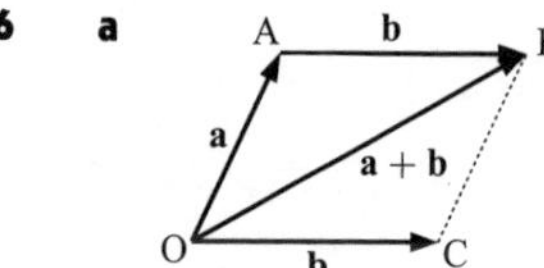


As **a** and **b** are unit vectors, OABC is a rhombus.
But the angles of a rhombus are bisected by its diagonals,
so **a** + **b** bisects the angle between vector **a** and vector **b**.

**b**

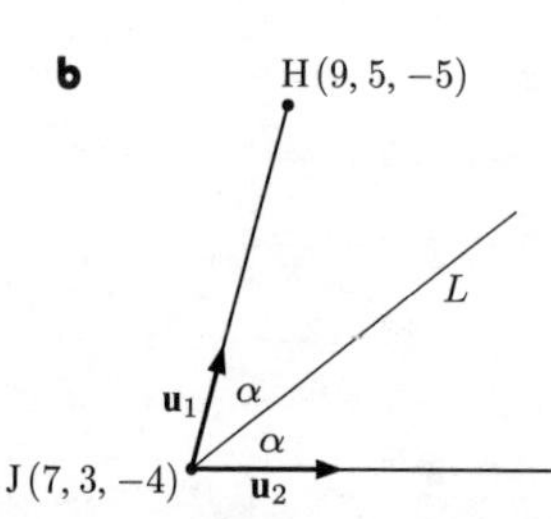


$\overrightarrow{JH} = \begin{pmatrix} 9-7 \\ 5-3 \\ -5--4 \end{pmatrix} = \begin{pmatrix} 2 \\ 2 \\ -1 \end{pmatrix}$, $\overrightarrow{JK} = \begin{pmatrix} 1-7 \\ 0-3 \\ 2--4 \end{pmatrix} = \begin{pmatrix} -6 \\ -3 \\ 6 \end{pmatrix}$

Using **a**, we write unit vectors $\mathbf{u}_1$ and $\mathbf{u}_2$ in the direction of $\overrightarrow{JH}$ and $\overrightarrow{JK}$ respectively.

$\therefore \quad \mathbf{u}_1 = \dfrac{1}{\sqrt{4+4+1}}\begin{pmatrix}2\\2\\-1\end{pmatrix} = \begin{pmatrix}\frac{2}{3}\\\frac{2}{3}\\-\frac{1}{3}\end{pmatrix}$ and $\mathbf{u}_2 = \dfrac{1}{\sqrt{36+9+36}}\begin{pmatrix}-6\\-3\\6\end{pmatrix} = \begin{pmatrix}-\frac{2}{3}\\-\frac{1}{3}\\\frac{2}{3}\end{pmatrix}$

and $\mathbf{u}_1 + \mathbf{u}_2 = \begin{pmatrix}0\\\frac{1}{3}\\\frac{1}{3}\end{pmatrix}$, which bisects $\widehat{\text{HJK}}$, by **a**.

$\therefore$ the equation of the line $L$ is $\begin{pmatrix}x\\y\\z\end{pmatrix} = \begin{pmatrix}7\\3\\-4\end{pmatrix} + t\begin{pmatrix}0\\\frac{1}{3}\\\frac{1}{3}\end{pmatrix}$, $t \in \mathbb{R}$

**c** $\overrightarrow{\text{HK}} = \begin{pmatrix}1-9\\0-5\\2--5\end{pmatrix} = \begin{pmatrix}-8\\-5\\7\end{pmatrix}$ so (HK) has equation $\begin{pmatrix}x\\y\\z\end{pmatrix} = \begin{pmatrix}1\\0\\2\end{pmatrix} + s\begin{pmatrix}-8\\-5\\7\end{pmatrix}$, $s \in \mathbb{R}$.

This line meets $L$ where

$7 = 1 - 8s, \qquad 3 + \dfrac{t}{3} = -5s, \qquad \text{and} \qquad -4 + \dfrac{t}{3} = 2 + 7s \quad .... (*)$

$\therefore \; 8s = -6$

$\therefore \; s = -\frac{3}{4}$ and so $3 + \dfrac{t}{3} = \frac{15}{4}$

$\therefore \; \dfrac{t}{3} = \frac{3}{4}$

$\therefore \; t = \frac{9}{4}$

In $(*)$, $\text{LHS} = -4 + \dfrac{t}{3} = -4 + \frac{3}{4} = -\frac{13}{4}$ $\qquad \text{RHS} = 2 + 7s = 2 + 7\left(-\frac{3}{4}\right) = \frac{8}{4} - \frac{21}{4} = -\frac{13}{4}$ ✓

$\therefore \; s = -\frac{3}{4}, \; t = \frac{9}{4}$ satisfy all 3 equations.

So, $\begin{pmatrix}x\\y\\z\end{pmatrix} = \begin{pmatrix}1\\0\\2\end{pmatrix} - \frac{3}{4}\begin{pmatrix}-8\\-5\\7\end{pmatrix} = \begin{pmatrix}1+6\\0+\frac{15}{4}\\2-\frac{21}{4}\end{pmatrix} = \begin{pmatrix}7\\3\frac{3}{4}\\-3\frac{1}{4}\end{pmatrix}$

$\therefore \; L$ meets (HK) at $(7, 3\frac{3}{4}, -3\frac{1}{4})$.

**7** If A is $(3, -1, -2)$ and $B(5, 3, -4)$ then $\overrightarrow{\text{AB}} = \begin{pmatrix}5-3\\3--1\\-4--2\end{pmatrix} = \begin{pmatrix}2\\4\\-2\end{pmatrix} = 2\begin{pmatrix}1\\2\\-1\end{pmatrix}$

$\therefore$ the line has equation $\begin{pmatrix}x\\y\\z\end{pmatrix} = \begin{pmatrix}3\\-1\\-2\end{pmatrix} + t\begin{pmatrix}1\\2\\-1\end{pmatrix}$, $t \in \mathbb{R}$

and it meets $x^2 + y^2 + z^2 = 26$ where

$$(3+t)^2 + (-1+2t)^2 + (-2-t)^2 = 26$$

$$\therefore \; 9 + 6t + t^2 + 1 - 4t + 4t^2 + 4 + 4t + t^2 - 26 = 0$$

$$\therefore \; 6t^2 + 6t - 12 = 0$$

$$\therefore \; t^2 + t - 2 = 0$$

$$\therefore \; (t+2)(t-1) = 0$$

$$\therefore \; t = -2 \text{ or } 1$$

$\therefore \; \begin{pmatrix}x\\y\\z\end{pmatrix} = \begin{pmatrix}3\\-1\\-2\end{pmatrix} - 2\begin{pmatrix}1\\2\\-1\end{pmatrix}$ or $\begin{pmatrix}3\\-1\\-2\end{pmatrix} + \begin{pmatrix}1\\2\\-1\end{pmatrix}$

$\therefore$ the line meets the sphere at $(1, -5, 0)$ and $(4, 1, -3)$.

**8** **a**

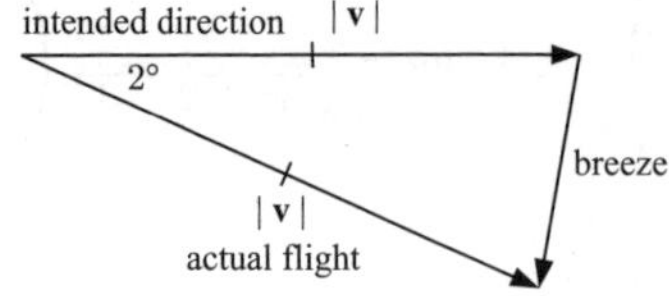


**b** **i**

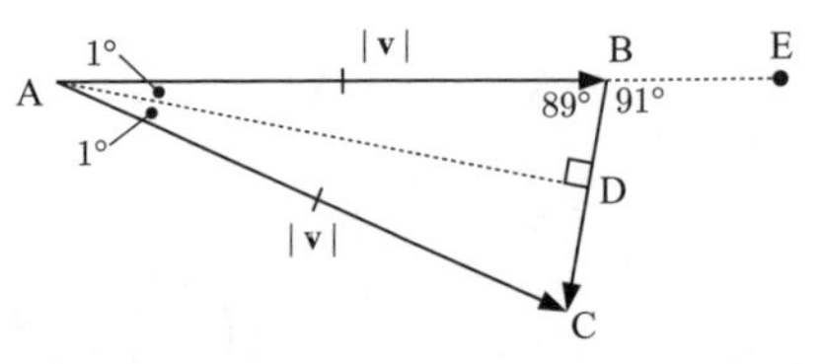


Triangle ABC is isosceles.

$\therefore$ line (AD) that meets the base at the midpoint, D, bisects the angle at A and is perpendicular to the base.

The triangle ABD is right angled at D and angle DAB $= 1°$.

$\therefore$ angle ABD $= 180 - 90 - 1 = 89°$.

If line (AB) is extended to E, then angle DBE $= 180 - 89 = 91°$.

Line (AB) is the arrow's intended path and line (BC) is the breeze, so the breeze is 91° to the intended direction of the arrow.

**ii** The speed of the breeze is the length of (BC) $= 2\times \mid \overrightarrow{BD} \mid$.

Using the sine rule, $\dfrac{\mid \overrightarrow{BD} \mid}{\sin 1°} = \dfrac{\mid \mathbf{v} \mid}{\sin 90°}$

$$\therefore \mid \overrightarrow{BD} \mid = \frac{\mid \mathbf{v} \mid \sin 1°}{1} = \mid \mathbf{v} \mid \sin 1°$$

$$\therefore \mid \overrightarrow{BC} \mid = 2 \mid \mathbf{v} \mid \sin 1°$$

So the speed of the breeze is $2 \mid \mathbf{v} \mid \sin 1°$.

## REVIEW SET 13B

**1** The vector equation is $\begin{pmatrix} x \\ y \end{pmatrix} = \begin{pmatrix} 0 \\ 8 \end{pmatrix} + t \begin{pmatrix} 5 \\ 4 \end{pmatrix}, \quad t \in \mathbb{R}$

**2** **a** **i** The yacht is initially at $(-6, 10)$, so its initial position vector is $\begin{pmatrix} -6 \\ 10 \end{pmatrix}$ or $-6\mathbf{i} + 10\mathbf{j}$.

**ii** $-\mathbf{i} - 3\mathbf{j}$ has length $\sqrt{(-1)^2 + (-3)^2} = \sqrt{10}$

$\therefore$ $5(-\mathbf{i} - 3\mathbf{j})$ has length $5\sqrt{10}$

$\therefore$ the velocity vector is $-5\mathbf{i} - 15\mathbf{j}$

**iii** $\begin{pmatrix} x \\ y \end{pmatrix} = \begin{pmatrix} -6 \\ 10 \end{pmatrix} + t \begin{pmatrix} -5 \\ -15 \end{pmatrix}$ $\therefore$ the position vector is $-6\mathbf{i} + 10\mathbf{j} + t(-5\mathbf{i} - 15\mathbf{j})$

$= (-6 - 5t)\mathbf{i} + (10 - 15t)\mathbf{j}, \quad t \geqslant 0$

**b** Let P be the point on the yacht's path when it is closest to the beacon.

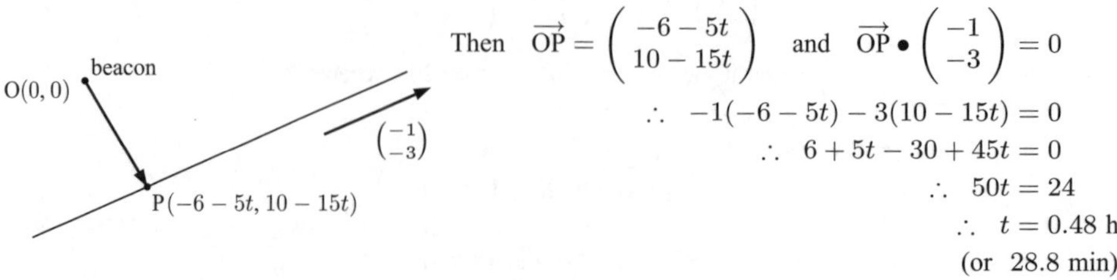


Then $\overrightarrow{OP} = \begin{pmatrix} -6 - 5t \\ 10 - 15t \end{pmatrix}$ and $\overrightarrow{OP} \bullet \begin{pmatrix} -1 \\ -3 \end{pmatrix} = 0$

$\therefore \ -1(-6 - 5t) - 3(10 - 15t) = 0$

$\therefore \ 6 + 5t - 30 + 45t = 0$

$\therefore \ 50t = 24$

$\therefore \ t = 0.48$ h

(or 28.8 min)

**c** When $t = 0.48$, $\overrightarrow{OP} = \begin{pmatrix} -6 - 5(0.48) \\ 10 - 15(0.48) \end{pmatrix} = \begin{pmatrix} -8.4 \\ 2.8 \end{pmatrix}$

and $\mid \overrightarrow{OP} \mid = \sqrt{(-8.4)^2 + (2.8)^2} \approx 8.85$ km

As the closest distance is 8.85 km and the radius is 8 km, the yacht will miss the reef.

**3** **a** **i** $\begin{pmatrix} x \\ y \end{pmatrix} = \begin{pmatrix} 2 \\ -3 \end{pmatrix} + t\begin{pmatrix} 4 \\ -1 \end{pmatrix}, \quad t \in \mathbb{R}$ **ii** $x = 2 + 4t, \quad y = -3 - t, \quad t \in \mathbb{R}$

**b** **i** The line has direction vector $\begin{pmatrix} 5-(-1) \\ -2-6 \\ 0-3 \end{pmatrix} = \begin{pmatrix} 6 \\ -8 \\ -3 \end{pmatrix}$

$$\therefore \begin{pmatrix} x \\ y \\ z \end{pmatrix} = \begin{pmatrix} -1 \\ 6 \\ 3 \end{pmatrix} + t\begin{pmatrix} 6 \\ -8 \\ -3 \end{pmatrix}, \quad t \in \mathbb{R}$$

**ii** $x = -1 + 6t, \quad y = 6 - 8t, \quad z = 3 - 3t, \quad t \in \mathbb{R}$

**4**

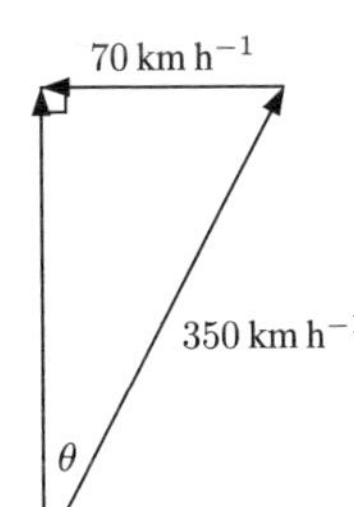


**a** $\sin\theta = \dfrac{70}{350}$

$\therefore \theta \approx 11.5°$

So the pilot should face the plane $11.5°$ east of north.

**b** $x^2 + 70^2 = 350^2$

$\therefore x^2 = 350^2 - 70^2$

$\therefore x \approx 343 \text{ km h}^{-1}$

So the speed of the plane will be $343 \text{ km h}^{-1}$.

**5** $L_1$ has direction vector $\mathbf{b}_1 = \begin{pmatrix} 5-0 \\ -2-3 \end{pmatrix} = \begin{pmatrix} 5 \\ -5 \end{pmatrix}$

$L_2$ has direction vector $\mathbf{b}_2 = \begin{pmatrix} -6-(-2) \\ 7-4 \end{pmatrix} = \begin{pmatrix} -4 \\ 3 \end{pmatrix}$

If the angle between the lines is $\theta$,

$$\cos\theta = \frac{|\,\mathbf{b}_1 \bullet \mathbf{b}_2\,|}{|\,\mathbf{b}_1\,|\,|\,\mathbf{b}_2\,|} = \frac{|-20-15|}{\sqrt{25+25}\sqrt{16+9}} = \frac{35}{\sqrt{50}\times 5}$$

$$\therefore \theta \approx 8.13°$$

$\therefore$ the angle between $L_1$ and $L_2$ is about $8.13°$.

**6** **a** $\begin{pmatrix} x_1(t) \\ y_1(t) \end{pmatrix} = \begin{pmatrix} 2 \\ 4 \end{pmatrix} + t\begin{pmatrix} 1 \\ -3 \end{pmatrix}$ where $t \geqslant 0$. When $t = 0$, the time is 2:17 pm.

$\therefore x_1(t) = 2 + t, \quad y_1(t) = 4 - 3t, \quad t \geqslant 0$

**b** After time $t$ has passed, submarine Y18's torpedo has been moving for time $(t-2)$.

$\therefore x_2(t) = 11 - (t-2), \qquad y_2(t) = 3 + a(t-2)$

$\therefore x_2(t) = 13 - t \qquad y_2(t) = [3-2a] + at, \quad t \geqslant 2$

**c** They meet where $2 + t = 13 - t$ and $4 - 3t = [3-2a] + at$

$\therefore 2t = 11$

$\therefore t = \frac{11}{2}$ $\therefore$ the time would be 2:17 pm plus $5\frac{1}{2}$ min, or 2:22:30 pm

**d** When $t = \frac{11}{2}$,

$4 - 3\left(\frac{11}{2}\right) = [3-2a] + a\left(\frac{11}{2}\right)$

$\therefore -\frac{25}{2} = 3 + \dfrac{7a}{2}$

$\therefore -25 = 6 + 7a$

$\therefore 7a = -31$

$\therefore a = -\frac{31}{7}$

Y18's torpedo has velocity vector $\begin{pmatrix} -1 \\ -\frac{31}{7} \end{pmatrix}$

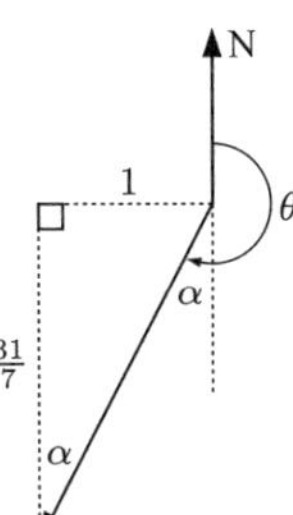


with speed $= \sqrt{(-1)^2 + \left(-\frac{31}{7}\right)^2}$

$\approx 4.54$ units per minute

$\tan\alpha = \dfrac{1}{\frac{31}{7}} = \frac{7}{31}$

$\therefore \alpha = \tan^{-1}\left(\frac{7}{31}\right) \approx 12.7°$

$\therefore$ the direction is $180° + \alpha° \approx 192.7°$

So, the torpedo has speed 4.54 units per minute and direction $7°$ west of south.

**7** **a** Line 1 has direction vector $\begin{pmatrix} 1 \\ 2 \\ -1 \end{pmatrix}$ and line 2 has direction vector $\begin{pmatrix} 4 \\ 1 \\ -2 \end{pmatrix}$.

As one vector is not a scalar multiple of the other, the lines are not parallel.

Now, $2+t=-8+4s$ .... (1) $\quad -1+2t=s$ .... (2) $\quad 3-t=7-2s$ .... (3)

Substituting (2) into (1), $2+t=-8+4(-1+2t)$

$\therefore\ 2+t=-8-4+8t$

$\therefore\ 7t=14$

$\therefore\ t=2$

$\therefore\ s=-1+2(2)=3$

In (3), LHS $=3-2=1$ $\qquad$ RHS $=7-2(3)=1$ ✓

$\therefore$ $s=3$, $t=2$ satisfies all three equations.

$\therefore$ the lines meet at $(4, 3, 1)$ {substituting $t=2$ into line 1}

The angle $\theta$ between the lines has

$$\cos\theta = \frac{\left|\begin{pmatrix} 1 \\ 2 \\ -1 \end{pmatrix} \bullet \begin{pmatrix} 4 \\ 1 \\ -2 \end{pmatrix}\right|}{\sqrt{1+4+1}\sqrt{16+1+4}} = \frac{|4+2+2|}{\sqrt{6}\sqrt{21}} = \frac{8}{3\sqrt{14}}$$

$\therefore\ \theta \approx 44.5°$

**b** Line 1 has direction vector $\begin{pmatrix} 1 \\ -2 \\ 3 \end{pmatrix}$ and line 2 has direction vector $\begin{pmatrix} -1 \\ 3 \\ 1 \end{pmatrix}$.

As one vector is not a scalar multiple of the other, the lines are not parallel.

Now, $3+t=2-s$ $\qquad 5-2t=1+3s$ $\qquad -1+3t=4+s$

$\therefore\ t+s=-1$ .... (1) $\qquad 2t+3s=4$ .... (2) $\qquad 3t-s=5$ .... (3)

Solving (1) and (3) simultaneously:

$$\begin{aligned} t+s&=-1 \\ 3t-s&=5 \\ \text{Adding,}\quad 4t\quad &=4 \end{aligned}$$

$\therefore\ t=1 \qquad \therefore\ s=-2$

In (2), LHS $=2(1)+3(-2)=-4$ ✗

$\therefore$ the system of equations is inconsistent and so the lines are skew.

The angle $\theta$ between them has

$$\cos\theta = \frac{\left|\begin{pmatrix} 1 \\ -2 \\ 3 \end{pmatrix} \bullet \begin{pmatrix} -1 \\ 3 \\ 1 \end{pmatrix}\right|}{\sqrt{1+4+9}\sqrt{1+9+1}} = \frac{|-1-6+3|}{\sqrt{14}\sqrt{11}} = \frac{4}{\sqrt{154}}$$

$\therefore\ \theta \approx 71.2°$

## REVIEW SET 13C

**1** The direction vector is $\begin{pmatrix} 3 \\ -1 \end{pmatrix}$ which has length $\sqrt{3^2+(-1)^2}=\sqrt{10}$ units

$\therefore\ 2\sqrt{10}\begin{pmatrix} 3 \\ -1 \end{pmatrix}$ has length 20. So, the velocity vector is $\begin{pmatrix} 6\sqrt{10} \\ -2\sqrt{10} \end{pmatrix}$ or $2\sqrt{10}(3\mathbf{i}-\mathbf{j})$.

**2** **a** $x(0)=-4$ and $y(0)=3$, so the initial position is $(-4, 3)$.

**b** $x(4)=-4+8(4)=28$ and $y(4)=3+6(4)=27$, so at $t=4$ the position is $(28, 27)$.

**c** The velocity vector is $\begin{pmatrix} 8 \\ 6 \end{pmatrix}$. $\qquad$ **d** The speed is $\sqrt{8^2+6^2}=10\ \text{m s}^{-1}$.

**3** **a** (KL) has direction vector $\begin{pmatrix} 5 \\ -2 \end{pmatrix}$ and (MN) has direction vector $\begin{pmatrix} -5 \\ 2 \end{pmatrix}$.

Now $\begin{pmatrix} 5 \\ -2 \end{pmatrix} = -\begin{pmatrix} -5 \\ 2 \end{pmatrix}$, so (KL) $\parallel$ (MN).

**b** $\overrightarrow{KL} = a\begin{pmatrix} 5 \\ -2 \end{pmatrix}$, $\overrightarrow{NK} = b\begin{pmatrix} 4 \\ 10 \end{pmatrix}$, $\overrightarrow{MN} = c\begin{pmatrix} -5 \\ 2 \end{pmatrix}$ {for some constants $a$, $b$, $c$}

$\therefore$ $\overrightarrow{KL} \bullet \overrightarrow{NK} = ab(20 - 20) = 0$ and $\overrightarrow{NK} \bullet \overrightarrow{MN} = bc(-20 + 20) = 0$

$\therefore$ (NK) is perpendicular to both (KL) and (MN).

**c** (KL) and (NK) meet at K.

$$\therefore \begin{pmatrix} 2 \\ 19 \end{pmatrix} + p\begin{pmatrix} 5 \\ -2 \end{pmatrix} = \begin{pmatrix} 3 \\ 7 \end{pmatrix} + r\begin{pmatrix} 4 \\ 10 \end{pmatrix}$$

$$\therefore \begin{pmatrix} 5p - 4r \\ -2p - 10r \end{pmatrix} = \begin{pmatrix} 1 \\ -12 \end{pmatrix}$$

$$\begin{aligned} \therefore\ 5p - 4r &= 1 \quad .... (1) \\ 2p + 10r &= 12 \quad .... (2) \\ \therefore\ 25p - 20r &= 5 \quad \{5 \times (1)\} \\ 4p + 20r &= 24 \quad \{2 \times (2)\} \\ \text{Adding,}\quad 29p &= 29 \\ \therefore\ p &= 1 \end{aligned}$$

and $\begin{pmatrix} x \\ y \end{pmatrix} = \begin{pmatrix} 2 \\ 19 \end{pmatrix} + \begin{pmatrix} 5 \\ -2 \end{pmatrix} = \begin{pmatrix} 7 \\ 17 \end{pmatrix}$

$\therefore$ K is (7, 17).

(KL) and (ML) meet at L.

$$\therefore \begin{pmatrix} 2 \\ 19 \end{pmatrix} + p\begin{pmatrix} 5 \\ -2 \end{pmatrix} = \begin{pmatrix} 33 \\ -5 \end{pmatrix} + q\begin{pmatrix} -11 \\ 16 \end{pmatrix}$$

$$\therefore \begin{pmatrix} 5p + 11q \\ -2p - 16q \end{pmatrix} = \begin{pmatrix} 31 \\ -24 \end{pmatrix}$$

$$\begin{aligned} \therefore\ 5p + 11q &= 31 \quad .... (1) \\ -2p - 16q &= -24 \quad .... (2) \\ \therefore\ 10p + 22q &= 62 \quad \{2 \times (1)\} \\ -10p - 80q &= -120 \quad \{5 \times (2)\} \\ \text{Adding,}\quad -58q &= -58 \\ \therefore\ q &= 1 \end{aligned}$$

and $\begin{pmatrix} x \\ y \end{pmatrix} = \begin{pmatrix} 33 \\ -5 \end{pmatrix} + \begin{pmatrix} -11 \\ 16 \end{pmatrix} = \begin{pmatrix} 22 \\ 11 \end{pmatrix}$

$\therefore$ L is (22, 11).

(ML) and (MN) meet at M.

$$\therefore \begin{pmatrix} 33 \\ -5 \end{pmatrix} + q\begin{pmatrix} -11 \\ 16 \end{pmatrix} = \begin{pmatrix} 43 \\ -9 \end{pmatrix} + s\begin{pmatrix} -5 \\ 2 \end{pmatrix}$$

$$\therefore \begin{pmatrix} -11q + 5s \\ 16q - 2s \end{pmatrix} = \begin{pmatrix} 10 \\ -4 \end{pmatrix}$$

$$\begin{aligned} \therefore\ -11q + 5s &= 10 \quad .... (1) \\ 16q - 2s &= -4 \quad .... (2) \\ \therefore\ -22q + 10s &= 20 \quad \{2 \times (1)\} \\ 80q - 10s &= -20 \quad \{5 \times (2)\} \\ \text{Adding,}\quad 58q &= 0 \end{aligned}$$

$\therefore$ $q = 0$ and so $\begin{pmatrix} x \\ y \end{pmatrix} = \begin{pmatrix} 33 \\ -5 \end{pmatrix}$

$\therefore$ M is (33, –5).

(NK) and (MN) meet at N.

$$\therefore \begin{pmatrix} 3 \\ 7 \end{pmatrix} + r\begin{pmatrix} 4 \\ 10 \end{pmatrix} = \begin{pmatrix} 43 \\ -9 \end{pmatrix} + s\begin{pmatrix} -5 \\ 2 \end{pmatrix}$$

$$\therefore \begin{pmatrix} 4r + 5s \\ 10r - 2s \end{pmatrix} = \begin{pmatrix} 40 \\ -16 \end{pmatrix}$$

$$\begin{aligned} \therefore\ 4r + 5s &= 40 \quad .... (1) \\ 10r - 2s &= -16 \quad .... (2) \\ \therefore\ 8r + 10s &= 80 \quad \{2 \times (1)\} \\ 50r - 10s &= -80 \quad \{5 \times (2)\} \\ \text{Adding,}\quad 58r &= 0 \end{aligned}$$

$\therefore$ $r = 0$ and so $\begin{pmatrix} x \\ y \end{pmatrix} = \begin{pmatrix} 3 \\ 7 \end{pmatrix}$

$\therefore$ N is (3, 7).

**d**

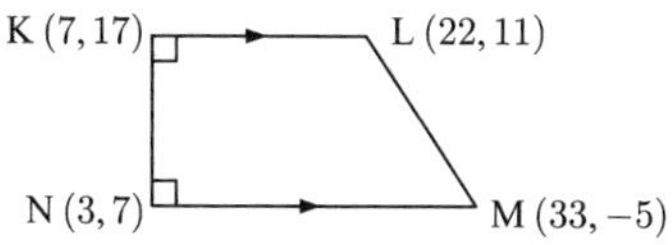

K (7, 17)
L (22, 11)
N (3, 7)
M (33, –5)

$$\begin{aligned} \text{KL} &= \sqrt{(22-7)^2 + (11-17)^2} \\ &= \sqrt{225 + 36} \\ &= \sqrt{261} \text{ units} \end{aligned}$$

$$\begin{aligned} \text{NM} &= \sqrt{(33-3)^2 + (-5-7)^2} \\ &= \sqrt{900 + 144} \\ &= \sqrt{1044} \text{ units} \end{aligned}$$

$$\begin{aligned} \text{KN} &= \sqrt{(7-3)^2 + (17-7)^2} \\ &= \sqrt{16 + 100} \\ &= \sqrt{116} \text{ units} \end{aligned}$$

$$\therefore \text{ area} = \left(\frac{\sqrt{261} + \sqrt{1044}}{2}\right) \times \sqrt{116} = 261 \text{ units}^2$$

**4** $L_1$ has direction vector $\mathbf{b}_1 = \begin{pmatrix} -4 \\ 3 \end{pmatrix}$, $L_2$ has direction vector $\mathbf{b}_2 = \begin{pmatrix} 5 \\ -12 \end{pmatrix}$

If $\theta$ is the angle between them,

$$\cos\theta = \frac{|\mathbf{b}_1 \bullet \mathbf{b}_2|}{|\mathbf{b}_1||\mathbf{b}_2|} = \frac{|-20-36|}{\sqrt{16+9}\sqrt{25+144}} = \frac{56}{5\times 13}$$

$$\therefore \ \theta \approx 30.5°$$

$\therefore$ the angle between $L_1$ and $L_2$ is about $30.5°$.

**5** **a** $$\overrightarrow{AB} = \begin{pmatrix} 0-3 \\ 2--1 \\ -2-1 \end{pmatrix} = \begin{pmatrix} -3 \\ 3 \\ -3 \end{pmatrix}$$

$\therefore \ |\overrightarrow{AB}| = \sqrt{9+9+9} = \sqrt{27}$ units

**b** $$\begin{pmatrix} x \\ y \\ z \end{pmatrix} = \begin{pmatrix} 0 \\ 2 \\ -2 \end{pmatrix} + t\begin{pmatrix} -3 \\ 3 \\ -3 \end{pmatrix}, \quad t \in \mathbb{R}$$

$$\therefore \ \begin{pmatrix} x \\ y \\ z \end{pmatrix} = \begin{pmatrix} 0 \\ 2 \\ -2 \end{pmatrix} + \lambda\begin{pmatrix} -1 \\ 1 \\ -1 \end{pmatrix}, \quad \text{where } \lambda = 3t$$

$\therefore \ \mathbf{r} = 2\mathbf{j} - 2\mathbf{k} + \lambda(-\mathbf{i}+\mathbf{j}-\mathbf{k}), \quad$ where $\lambda \in \mathbb{R}$

A lies on the line $\mathbf{r}$ when $\lambda = -3$ and B lies on $\mathbf{r}$ when $\lambda = 0$.

$\therefore$ the line between A and B is the same line as $\mathbf{r}$, so it can be described by $\mathbf{r}$.

**c** The line with equation $t(\mathbf{i}+\mathbf{j}+\mathbf{k})$ has direction vector $\begin{pmatrix} 1 \\ 1 \\ 1 \end{pmatrix}$.

$$\therefore \ \overrightarrow{AB} \bullet \begin{pmatrix} 1 \\ 1 \\ 1 \end{pmatrix} = \begin{pmatrix} -3 \\ 3 \\ -3 \end{pmatrix} \bullet \begin{pmatrix} 1 \\ 1 \\ 1 \end{pmatrix}$$
$$= -3+3-3$$
$$= -3$$

$$\left|\begin{pmatrix} 1 \\ 1 \\ 1 \end{pmatrix}\right| = \sqrt{1^2+1^2+1^2} = \sqrt{3}$$

$$\therefore \ \cos\theta = \frac{|-3|}{3\sqrt{3}\times\sqrt{3}} = \frac{3}{9} = \frac{1}{3}$$

$$\therefore \ \theta \approx 70.5°$$

$\therefore$ the angle between the two lines is about $70.5°$.

**6** **a** Road A has direction vector $\begin{pmatrix} 15--9 \\ -16-2 \end{pmatrix} = \begin{pmatrix} 24 \\ -18 \end{pmatrix} = 6\begin{pmatrix} 4 \\ -3 \end{pmatrix}$.

So, Road A has equation $\begin{pmatrix} x \\ y \end{pmatrix} = \begin{pmatrix} -9 \\ 2 \end{pmatrix} + t\begin{pmatrix} 4 \\ -3 \end{pmatrix}$, $t \in \mathbb{R}$.

Road B has direction vector $\begin{pmatrix} 21-6 \\ 18--18 \end{pmatrix} = \begin{pmatrix} 15 \\ 36 \end{pmatrix} = 3\begin{pmatrix} 5 \\ 12 \end{pmatrix}$

So, Road B has equation $\begin{pmatrix} x \\ y \end{pmatrix} = \begin{pmatrix} 6 \\ -18 \end{pmatrix} + s\begin{pmatrix} 5 \\ 12 \end{pmatrix}$, $s \in \mathbb{R}$.

**b** Let $A(-9+4t, 2-3t)$ be any point on Road A.

$\therefore \overrightarrow{HA} = \begin{pmatrix} -9+4t-4 \\ 2-3t-11 \end{pmatrix} = \begin{pmatrix} -13+4t \\ -9-3t \end{pmatrix}$

The closest point to H(4, 11) on Road A is such that $\overrightarrow{HA} \perp \begin{pmatrix} 4 \\ -3 \end{pmatrix}$.

$$\therefore \begin{pmatrix} -13+4t \\ -9-3t \end{pmatrix} \bullet \begin{pmatrix} 4 \\ -3 \end{pmatrix} = 0$$

$$\therefore -52+16t+27+9t = 0$$

$$\therefore 25t = 25$$

$$\therefore t = 1$$

So A is $(-9+4, 2-3)$ or $(-5, -1)$.

$$\therefore \overrightarrow{HA} = \begin{pmatrix} -13+4 \\ -9-3 \end{pmatrix} = \begin{pmatrix} -9 \\ -12 \end{pmatrix}$$

$$\therefore |\overrightarrow{HA}| = \sqrt{81+144} = \sqrt{225} = 15 \text{ km}$$

Now, let $B(6+5s, -18+12s)$ be any point on Road B.

$$\therefore \overrightarrow{HB} = \begin{pmatrix} 6+5s-4 \\ -18+12s-11 \end{pmatrix} = \begin{pmatrix} 2+5s \\ -29+12s \end{pmatrix}$$

The closest point to H(4, 11) on Road B is such that $\overrightarrow{HB} \perp \begin{pmatrix} 5 \\ 12 \end{pmatrix}$.

$$\therefore \begin{pmatrix} 2+5s \\ -29+12s \end{pmatrix} \bullet \begin{pmatrix} 5 \\ 12 \end{pmatrix} = 0$$

$$\therefore 10+25s-348+144s = 0$$

$$\therefore 169s = 338$$

$$\therefore s = 2$$

So B is $(6+5(2), -18+12(2))$ or $(16, 6)$.

$$\therefore \overrightarrow{HB} = \begin{pmatrix} 2+5(2) \\ -29+12(2) \end{pmatrix} = \begin{pmatrix} 12 \\ -5 \end{pmatrix}$$

$$\therefore |\overrightarrow{HB}| = \sqrt{144+25} = \sqrt{169} = 13 \text{ km}$$

The hiker should head toward Road B, a distance of 13 km.

**7** **a** $\overrightarrow{AB} = \begin{pmatrix} 2-4 \\ 1-2 \\ 5--1 \end{pmatrix} = \begin{pmatrix} -2 \\ -1 \\ 6 \end{pmatrix}$ and $\overrightarrow{AC} = \begin{pmatrix} 9-4 \\ 4-2 \\ 1--1 \end{pmatrix} = \begin{pmatrix} 5 \\ 2 \\ 2 \end{pmatrix}$

$$\therefore \overrightarrow{AB} \bullet \overrightarrow{AC} = (-2)(5)+(-1)(2)+(6)(2) = -10-2+12 = 0$$

$$\therefore \overrightarrow{AB} \perp \overrightarrow{AC}$$

**b** **i** The equation is $\begin{pmatrix} x \\ y \\ z \end{pmatrix} = \begin{pmatrix} 4 \\ 2 \\ -1 \end{pmatrix} + t\begin{pmatrix} -2 \\ -1 \\ 6 \end{pmatrix}, \quad t \in \mathbb{R}$

or $x = 4-2t, \quad y = 2-t, \quad z = -1+6t, \quad t \in \mathbb{R}$

**ii** The equation is $\begin{pmatrix} x \\ y \\ z \end{pmatrix} = \begin{pmatrix} 4 \\ 2 \\ -1 \end{pmatrix} + s\begin{pmatrix} 5 \\ 2 \\ 2 \end{pmatrix}, \quad s \in \mathbb{R}$

or $x = 4+5s, \quad y = 2+2s, \quad z = -1+2s, \quad s \in \mathbb{R}$

# Chapter 14
## INTRODUCTION TO DIFFERENTIAL CALCULUS

### EXERCISE 14A

**1** **a** As $x \to 3$, $x+4 \to 7$
$\therefore \lim_{x\to 3} (x+4) = 7$

**b** As $x \to -1$, $5-2x \to 7$
$\therefore \lim_{x\to -1} (5-2x) = 7$

**c** As $x \to 4$, $3x-1 \to 11$
$\therefore \lim_{x\to 4} (3x-1) = 11$

**d** As $x \to 2$, $5x^2-3x+2 \to 5(4)-3(2)+2 = 16$
$\therefore \lim_{x\to 2} (5x^2-3x+2) = 16$

**e** As $h \to 0$, $h^2 \to 0$ and $1-h \to 1$
$\therefore \lim_{h\to 0} h^2(1-h) = 0 \times 1 = 0$

**f** As $x \to 0$, $x^2+5 \to 5$
$\therefore \lim_{x\to 0} (x^2+5) = 5$

**2** **a** $\lim_{x\to 0} 5 = 5$ **b** $\lim_{h\to 2} 7 = 7$ **c** $\lim_{x\to 0} c = c$ (when $c$ is a constant)

**3** **a** $\lim_{x\to 1} \dfrac{x^2-3x}{x}$
$= \lim_{x\to 1} \dfrac{x(x-3)}{x}$
$= \lim_{x\to 1} (x-3)$ since $x \neq 0$
$= -2$

**b** $\lim_{h\to 2} \dfrac{h^2+5h}{h}$
$= \lim_{h\to 2} \dfrac{h(h+5)}{h}$
$= \lim_{h\to 2} (h+5)$ since $h \neq 0$
$= 7$

**c** $\dfrac{x-1}{x+1}$ can be made as close as we like to $-1$ by making $x$ sufficiently close to 0.
$\therefore \lim_{x\to 0} \dfrac{x-1}{x+1} = -1$

**d** $\lim_{x\to 0} \dfrac{x}{x}$
$= \lim_{x\to 0} 1$ since $x \neq 0$
$= 1$

**4** **a** $\lim_{x\to 0} \dfrac{x^2-3x}{x}$
$= \lim_{x\to 0} \dfrac{x(x-3)}{x}$
$= \lim_{x\to 0} (x-3)$ since $x \neq 0$
$= -3$

**b** $\lim_{x\to 0} \dfrac{x^2+5x}{x}$
$= \lim_{x\to 0} \dfrac{x(x+5)}{x}$
$= \lim_{x\to 0} (x+5)$ since $x \neq 0$
$= 5$

**c** $\lim_{x\to 0} \dfrac{2x^2-x}{x}$
$= \lim_{x\to 0} \dfrac{x(2x-1)}{x}$
$= \lim_{x\to 0} (2x-1)$ since $x \neq 0$
$= -1$

**d** $\lim_{h\to 0} \dfrac{2h^2+6h}{h}$
$= \lim_{h\to 0} \dfrac{2h(h+3)}{h}$
$= \lim_{h\to 0} 2(h+3)$ since $h \neq 0$
$= 6$

**e** $\lim_{h\to 0} \dfrac{3h^2-4h}{h}$
$= \lim_{h\to 0} \dfrac{h(3h-4)}{h}$
$= \lim_{h\to 0} (3h-4)$ since $h \neq 0$
$= -4$

**f** $\lim_{h\to 0} \dfrac{h^3-8h}{h}$
$= \lim_{h\to 0} \dfrac{h(h^2-8)}{h}$
$= \lim_{h\to 0} (h^2-8)$ since $h \neq 0$
$= -8$

**g** $\lim\limits_{x\to 1} \dfrac{x^2-x}{x-1}$

$= \lim\limits_{x\to 1} \dfrac{x(x-1)}{x-1}$

$= \lim\limits_{x\to 1} x$ since $x \neq 1$

$= 1$

**h** $\lim\limits_{x\to 2} \dfrac{x^2-2x}{x-2}$

$= \lim\limits_{x\to 2} \dfrac{x(x-2)}{x-2}$

$= \lim\limits_{x\to 2} x$ since $x \neq 2$

$= 2$

**i** $\lim\limits_{x\to 3} \dfrac{x^2-x-6}{x-3}$

$= \lim\limits_{x\to 3} \dfrac{(x+2)(x-3)}{x-3}$

$= \lim\limits_{x\to 3} (x+2)$ since $x \neq 3$

$= 5$

## EXERCISE 14B

**1** **a** **i** As $x\to 0^-$, $f(x)\to -\infty$
As $x\to 0^+$, $f(x)\to \infty$
As $x\to \infty$, $f(x)\to 0^+$
As $x\to -\infty$, $f(x)\to 0^-$
The vertical asymptote is $x=0$.
The horizontal asymptote is $y=0$.

**ii** $\lim\limits_{x\to -\infty} f(x) = 0$,

$\lim\limits_{x\to \infty} f(x) = 0$

**b** **i** As $x\to -3^-$, $f(x)\to \infty$
As $x\to -3^+$, $f(x)\to -\infty$
As $x\to \infty$, $f(x)\to 3^-$
As $x\to -\infty$, $f(x)\to 3^+$
The vertical asymptote is $x=-3$.
The horizontal asymptote is $y=3$.

**ii** $\lim\limits_{x\to -\infty} f(x) = 3$,

$\lim\limits_{x\to \infty} f(x) = 3$

**c** **i** As $x\to -\frac{2}{3}^-$, $f(x)\to -\infty$
As $x\to -\frac{2}{3}^+$, $f(x)\to \infty$
As $x\to \infty$, $f(x)\to -\frac{2}{3}^+$
As $x\to -\infty$, $f(x)\to -\frac{2}{3}^-$
The vertical asymptote is $x=-\frac{2}{3}$.
The horizontal asymptote is $y=-\frac{2}{3}$.

**ii** $\lim\limits_{x\to -\infty} f(x) = -\frac{2}{3}$,

$\lim\limits_{x\to \infty} f(x) = -\frac{2}{3}$

**d** **i** As $x\to 1^-$, $f(x)\to \infty$
As $x\to 1^+$, $f(x)\to -\infty$
As $x\to \infty$, $f(x)\to -1^-$
As $x\to -\infty$, $f(x)\to -1^+$
The vertical asymptote is $x=1$.
The horizontal asymptote is $y=-1$.

**ii** $\lim\limits_{x\to -\infty} f(x) = -1$,

$\lim\limits_{x\to \infty} f(x) = -1$

**e** **i** Since there are no real values of $x$ that make $x^2+1=0$, $f(x)$ is defined for all $x\in\mathbb{R}$.
$\therefore$ there are no vertical asymptotes.
As $x\to \infty$, $f(x)\to 1^-$
As $x\to -\infty$, $f(x)\to 1^-$
The horizontal asymptote is $y=1$.

**ii** $\lim\limits_{x\to -\infty} f(x) = 1$,

$\lim\limits_{x\to \infty} f(x) = 1$

**f** **i** Since there are no real values of $x$ that make $x^2+1=0$, $f(x)$ is defined for all $x\in\mathbb{R}$.
$\therefore$ there are no vertical asymptotes.
As $x\to \infty$, $f(x)\to 0^+$
As $x\to -\infty$, $f(x)\to 0^-$
The horizontal asymptote is $y=0$.

**ii** $\lim\limits_{x\to -\infty} f(x) = 0$,

$\lim\limits_{x\to \infty} f(x) = 0$

**2** **a**

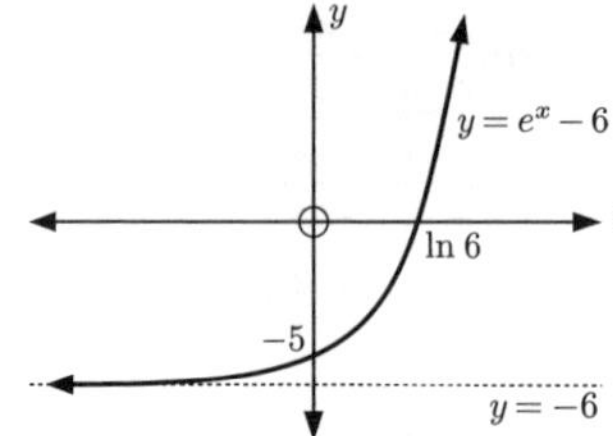


**b** **i** As $x \to -\infty$, $e^x - 6 \to -6^+$

$\therefore \lim_{x \to -\infty} (e^x - 6) = -6$

$\therefore$ the function has horizontal asymptote $y = -6$.

**ii** As $x \to \infty$, $e^x - 6 \to \infty$

$\therefore \lim_{x \to \infty} (e^x - 6)$ does not exist.

**3** We sketch the graph of $y = 2e^{-x} - 3$:

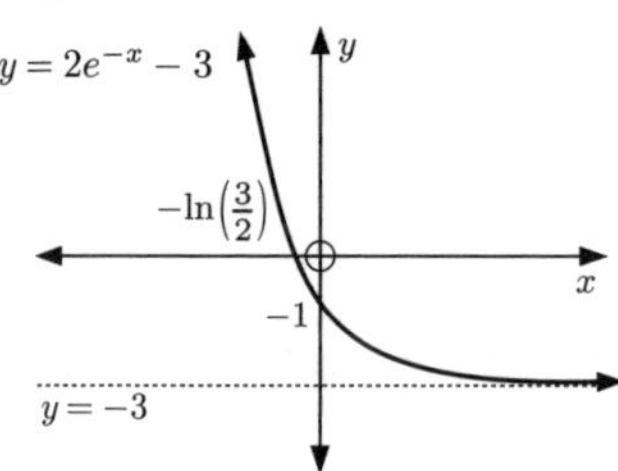


As $x \to -\infty$, $2e^{-x} - 3 \to \infty$

$\therefore \lim_{x \to -\infty} (2e^{-x} - 3)$ does not exist.

As $x \to \infty$, $2e^{-x} - 3 \to -3^+$

$\therefore \lim_{x \to \infty} (2e^{-x} - 3) = -3$.

## EXERCISE 14C

**1** **a**

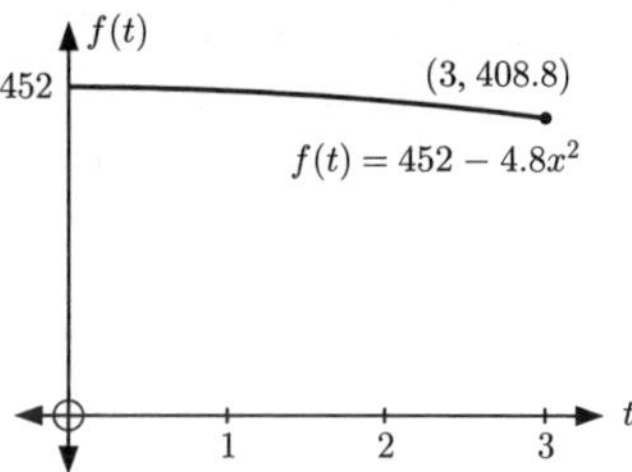


**b** The graph of $f(t)$ is not a straight line, so the jumper is not travelling with constant speed.

**c** If the *altitude* of the jumper is given by $f(t) = 452 - 4.8t^2$,
then the *distance* covered by the jumper $d(t) = 452 - f(t)$

$\therefore d(t) = 452 - (452 - 4.8t^2)$

$\therefore d(t) = 452 - 452 + 4.8t^2$

$\therefore d(t) = 4.8t^2$

**i** We choose a fixed point F on $d(t)$ when $t = 0$ seconds. This is the point $(0, 0)$.
We then choose another point M on the curve, for example the point $(1, 4.8)$.
The average speed in the interval $0 \leqslant t \leqslant 1$ is $\frac{4.8 - 0}{1 - 0} = 4.8 \text{ m s}^{-1}$.

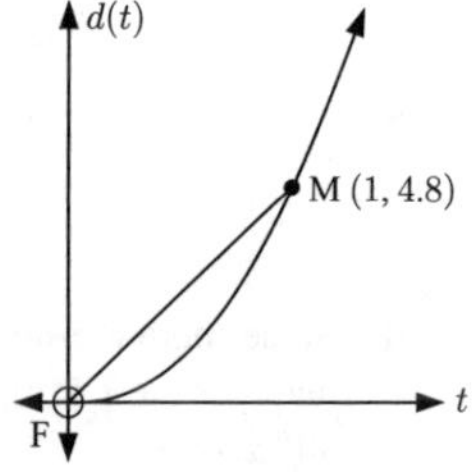


We repeat this process, moving M closer to F each time, and get the following results:

| $t$ | gradient of [FM] |
|---|---|
| 1 | 4.8 |
| 0.5 | 2.4 |
| 0.1 | 0.48 |
| 0.01 | 0.048 |
| 0.001 | 0.0048 |

So, as M approaches F, the gradient of [FM] approaches 0.
$\therefore$ the speed of the jumper at $t = 0$ seconds is $0 \text{ m s}^{-1}$.

**ii** We now choose point F on $d(t)$ when $t = 1$ second. This is the point $(1,\ 4.8)$.
We then choose another point M on the curve, for example the point $(2,\ 19.2)$.
The average speed in the interval $1 \leqslant t \leqslant 2$ is $\dfrac{19.2 - 4.8}{2 - 1} = 14.4\ \text{m s}^{-1}$.

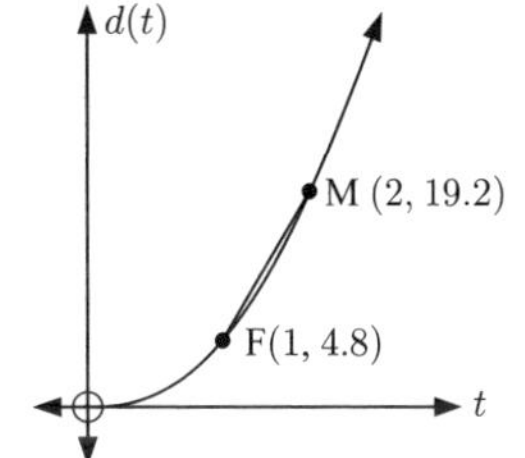


We repeat this process, moving M closer to F each time, and get the following results:

| $t$ | gradient of [FM] |
|---|---|
| 2 | 14.4 |
| 1.5 | 12 |
| 1.1 | 10.08 |
| 1.01 | 9.648 |
| 1.001 | 9.6048 |
| 1.0001 | 9.600 48 |

| $t$ | gradient of [FM] |
|---|---|
| 0 | 4.8 |
| 0.5 | 7.2 |
| 0.9 | 9.12 |
| 0.99 | 9.552 |
| 0.999 | 9.5952 |
| 0.9999 | 9.599 52 |

So, as M approaches F (from either direction), the gradient of [FM] approaches 9.6 .
$\therefore$ the speed of the jumper at $t = 1$ second is $9.6\ \text{m s}^{-1}$.

**iii** We now choose point F on $d(t)$ when $t = 2$ seconds. This is the point $(2,\ 19.2)$.
We then choose another point M on the curve, for example the point $(3,\ 43.2)$.
The average speed in the interval $2 \leqslant t \leqslant 3$ is $\dfrac{43.2 - 19.2}{3 - 2} = 24\ \text{m s}^{-1}$.

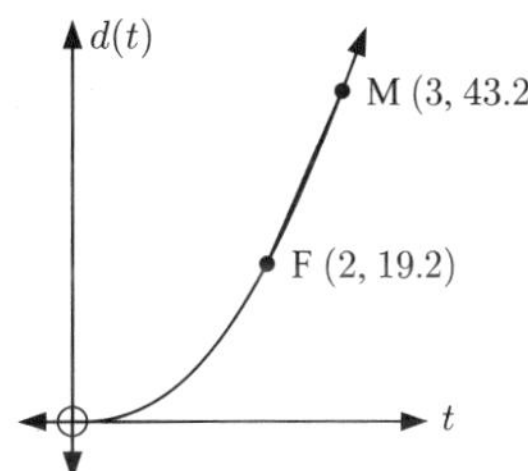


We repeat this process, moving M closer to F each time, and get the following results:

| $t$ | gradient of [FM] |
|---|---|
| 3 | 24 |
| 2.5 | 21.6 |
| 2.1 | 19.68 |
| 2.01 | 19.248 |
| 2.001 | 19.2048 |
| 2.0001 | 19.200 48 |

| $t$ | gradient of [FM] |
|---|---|
| 1 | 14.4 |
| 1.5 | 16.8 |
| 1.9 | 18.72 |
| 1.99 | 19.152 |
| 1.999 | 19.1952 |
| 1.9999 | 19.199 52 |

So, as M approaches F (from either direction), the gradient of [FM] approaches 19.2 .
$\therefore$ the speed of the jumper at $t = 2$ seconds is $19.2\ \text{m s}^{-1}$.

**iv** We choose a fixed point F on $d(t)$ when $t = 3$ seconds. This is the point $(3,\ 43.2)$.
We then choose another point M on the curve, for example the point $(2,\ 19.2)$.
The gradient of [MF] is

$$\frac{43.2 - 19.2}{3 - 2} = 24\ \text{m s}^{-1}.$$

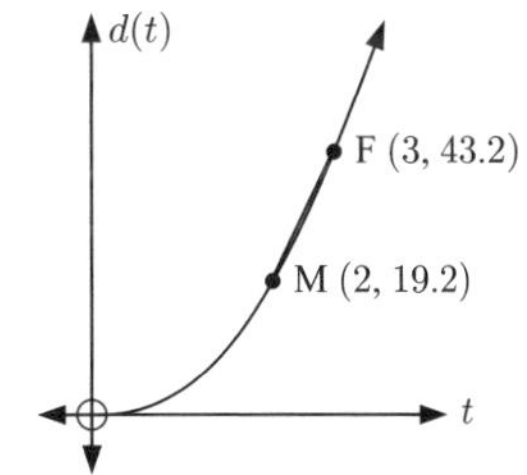


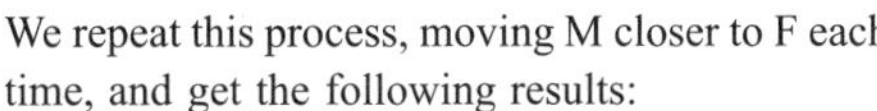
We repeat this process, moving M closer to F each time, and get the following results:

| $t$ | gradient of [MF] |
|---|---|
| 2 | 24 |
| 2.5 | 26.4 |
| 2.9 | 28.32 |
| 2.99 | 28.752 |
| 2.999 | 28.7952 |
| 2.9999 | 28.799 52 |

So, as M approaches F, the gradient of [MF] approaches 28.8 .

$\therefore$ the speed of the jumper at $t = 3$ seconds is $28.8\ \text{m s}^{-1}$.

**2** **a** Suppose A is the point (2, 4) and B is a point on $y = x^2$ with coordinates $(x, x^2)$.
The chord [AB] has gradient

$$\frac{x^2 - 4}{x - 2} \quad \left(\text{or } \frac{4 - x^2}{2 - x}\right).$$

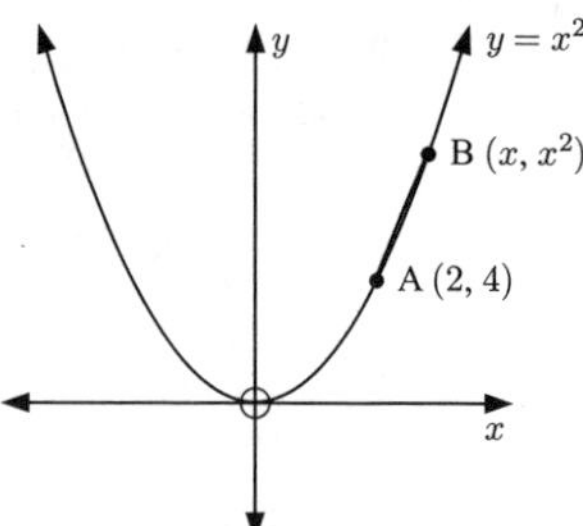


As B moves closer to A (from either side), we get the following results:

| $x$ | Point B | Gradient of [AB] |
|---|---|---|
| 0 | (0, 0) | 2 |
| 1 | (1, 1) | 3 |
| 1.5 | (1.5, 2.25) | 3.5 |
| 1.9 | (1.9, 3.61) | 3.9 |
| 1.99 | (1.99, 3.9601) | 3.99 |
| 1.999 | (1.999, 3.996 001) | 3.999 |

| $x$ | Point B | Gradient of [AB] |
|---|---|---|
| 5 | (5, 25) | 7 |
| 3 | (3, 9) | 5 |
| 2.5 | (2.5, 6.25) | 4.5 |
| 2.1 | (2.1, 4.41) | 4.1 |
| 2.01 | (2.01, 4.0401) | 4.01 |
| 2.001 | (2.001, 4.004 001) | 4.001 |

So, as B approaches A, the gradient of [AB] approaches 4.
$\therefore$ the gradient of the tangent to $y = x^2$ at the point (2, 4) is 4.

**b** $\displaystyle\lim_{x \to 2} \frac{x^2 - 4}{x - 2}$

$= \displaystyle\lim_{x \to 2} \frac{(x+2)(x-2)}{x - 2}$

$= \displaystyle\lim_{x \to 2} (x + 2)$ since $x \neq 2$

$= 4$

This is the gradient of the tangent to $y = x^2$ at the point where $x = 2$.

## EXERCISE 14D

**1** **a** $f(2) = 3$

**b** $f'(2)$ is the gradient of the tangent to $f(x)$ at the point where $x = 2$.
Since $f(x)$ is a straight line, this is the same as the gradient of $f(x)$ itself.
$f(x)$ is a horizontal line, and hence has gradient 0.
$\therefore$ $f'(2) = 0$

**2** **a** $f(0) = 4$

**b** $f'(0)$ is the gradient of the tangent to $f(x)$ at the point where $x = 0$.
Since $f(x)$ is a straight line, this is the same as the gradient of $f(x)$ itself.
$f(x)$ goes through (0, 4) and (4, 0), so it has gradient $= \dfrac{0 - 4}{4 - 0} = -1$
$\therefore$ $f'(0) = -1$

**3** The graph shows the tangent to the curve $y = f(x)$ at the point where $x = 2$.

The tangent passes through (0, 1) and (4, 5), so its gradient is $f'(2) = \dfrac{5 - 1}{4 - 0} = 1$.

The equation of the tangent is $\dfrac{y - 1}{x - 0} = 1$

$\therefore$ $y = x + 1$

When $x = 2$, $y = 3$, so the point of contact is (2, 3).
$\therefore$ $f(2) = 3$ and $f'(2) = 1$.

## EXERCISE 14E

**1** **a** **i**
$$\begin{aligned} f(x) &= x \\ \therefore\ f'(x) &= \lim_{h\to 0} \frac{f(x+h)-f(x)}{h} \\ &= \lim_{h\to 0} \frac{(x+h)-x}{h} \\ &= \lim_{h\to 0} \frac{h}{h} \\ &= \lim_{h\to 0} 1 \qquad \{\text{as } h \neq 0\} \\ &= 1 \end{aligned}$$

**ii**
$$\begin{aligned} f(x) &= 5 \\ \therefore\ f'(x) &= \lim_{h\to 0} \frac{f(x+h)-f(x)}{h} \\ &= \lim_{h\to 0} \frac{5-5}{h} \\ &= \lim_{h\to 0} \frac{0}{h} \\ &= \lim_{h\to 0} 0 \qquad \{\text{as } h \neq 0\} \\ &= 0 \end{aligned}$$

**iii**
$$\begin{aligned} f(x) &= x^3 \\ \therefore\ f'(x) &= \lim_{h\to 0} \frac{f(x+h)-f(x)}{h} \\ &= \lim_{h\to 0} \frac{(x+h)^3-x^3}{h} \\ &= \lim_{h\to 0} \frac{\cancel{x^3}+3x^2h+3xh^2+h^3-\cancel{x^3}}{h} \\ &= \lim_{h\to 0} \frac{3x^2h+3xh^2+h^3}{h} \\ &= \lim_{h\to 0} 3x^2+3xh+h^2 \qquad \{\text{as } h \neq 0\} \\ &= 3x^2 \end{aligned}$$

**iv**
$$\begin{aligned} f(x) &= x^4 \\ \therefore\ f'(x) &= \lim_{h\to 0} \frac{f(x+h)-f(x)}{h} \\ &= \lim_{h\to 0} \frac{(x+h)^4-x^4}{h} \\ &= \lim_{h\to 0} \frac{\cancel{x^4}+4x^3h+6x^2h^2+4xh^3+h^4-\cancel{x^4}}{h} \\ &= \lim_{h\to 0} \frac{4x^3h+6x^2h^2+4xh^3+h^4}{h} \\ &= \lim_{h\to 0} 4x^3+6x^2h+4xh^2+h^3 \qquad \{\text{as } h \neq 0\} \\ &= 4x^3 \end{aligned}$$

**b** From **a**, we predict that if $f(x) = x^n$, $f'(x) = nx^{n-1}$, $n \in \mathbb{N}$.

**2** **a**
$$\begin{aligned} & f(x) = 2x+5 \\ & \therefore\ f'(x) \\ &= \lim_{h\to 0} \frac{f(x+h)-f(x)}{h} \\ &= \lim_{h\to 0} \frac{(2(x+h)+5)-(2x+5)}{h} \\ &= \lim_{h\to 0} \frac{\cancel{2x}+2h+\cancel{5}-\cancel{2x}-\cancel{5}}{h} \\ &= \lim_{h\to 0} \frac{2h}{h} \\ &= \lim_{h\to 0} 2 \qquad \{\text{as } h \neq 0\} \\ &= 2 \end{aligned}$$

**b**
$$\begin{aligned} & f(x) = x^2-3x \\ & \therefore\ f'(x) \\ &= \lim_{h\to 0} \frac{f(x+h)-f(x)}{h} \\ &= \lim_{h\to 0} \frac{[(x+h)^2-3(x+h)]-[x^2-3x]}{h} \\ &= \lim_{h\to 0} \frac{\cancel{x^2}+2xh+h^2-\cancel{3x}-3h-\cancel{x^2}+\cancel{3x}}{h} \\ &= \lim_{h\to 0} \frac{2xh+h^2-3h}{h} \\ &= \lim_{h\to 0} 2x+h-3 \qquad \{\text{as } h \neq 0\} \\ &= 2x-3 \end{aligned}$$

**c** $f(x) = -x^2 + 5x - 3$

$$\begin{aligned} \therefore\ f'(x) &= \lim_{h\to 0} \frac{f(x+h)-f(x)}{h} \\ &= \lim_{h\to 0} \frac{[-(x+h)^2 + 5(x+h) - 3] - [-x^2 + 5x - 3]}{h} \\ &= \lim_{h\to 0} \frac{\cancel{-x^2} - 2xh - h^2 + \cancel{5x} + 5h \cancel{-3} + \cancel{x^2} - \cancel{5x} \cancel{+3}}{h} \\ &= \lim_{h\to 0} \frac{-2xh - h^2 + 5h}{h} \\ &= \lim_{h\to 0} -2x + 5 - h \qquad \{\text{as } h \neq 0\} \\ &= -2x + 5 \end{aligned}$$

**3** **a** $y = f(x) = 4 - x$

$$\begin{aligned} \therefore\ \frac{dy}{dx} & \\ &= \lim_{h\to 0} \frac{f(x+h)-f(x)}{h} \\ &= \lim_{h\to 0} \frac{[4-(x+h)]-[4-x]}{h} \\ &= \lim_{h\to 0} \frac{\cancel{4} - \cancel{x} - h - \cancel{4} + \cancel{x}}{h} \\ &= \lim_{h\to 0} \frac{-h}{h} \\ &= \lim_{h\to 0} -1 \qquad \{\text{as } h \neq 0\} \\ &= -1 \end{aligned}$$

**b** $y = f(x) = 2x^2 + x - 1$

$$\begin{aligned} \therefore\ \frac{dy}{dx} & \\ &= \lim_{h\to 0} \frac{f(x+h)-f(x)}{h} \\ &= \lim_{h\to 0} \frac{[2(x+h)^2 + (x+h) - 1] - [2x^2 + x - 1]}{h} \\ &= \lim_{h\to 0} \frac{\cancel{2x^2} + 4xh + 2h^2 + \cancel{x} + h \cancel{-1} - \cancel{2x^2} - \cancel{x} + \cancel{1}}{h} \\ &= \lim_{h\to 0} \frac{4xh + 2h^2 + h}{h} \\ &= \lim_{h\to 0} 4x + 1 + 2h \qquad \{\text{as } h \neq 0\} \\ &= 4x + 1 \end{aligned}$$

**c** $y = f(x) = x^3 - 2x^2 + 3$

$$\begin{aligned} \therefore\ \frac{dy}{dx} &= \lim_{h\to 0} \frac{f(x+h)-f(x)}{h} \\ &= \lim_{h\to 0} \frac{[(x+h)^3 - 2(x+h)^2 + 3] - [x^3 - 2x^2 + 3]}{h} \\ &= \lim_{h\to 0} \frac{\cancel{x^3} + 3x^2h + 3xh^2 + h^3 - \cancel{2x^2} - 4xh - 2h^2 \cancel{+3} - \cancel{x^3} + \cancel{2x^2} \cancel{-3}}{h} \\ &= \lim_{h\to 0} \frac{3x^2h + 3xh^2 + h^3 - 4xh - 2h^2}{h} \\ &= \lim_{h\to 0} 3x^2 + 3xh + h^2 - 4x - 2h \qquad \{\text{as } h \neq 0\} \\ &= 3x^2 - 4x \end{aligned}$$

**4** **a** $f(x) = x^3$ $\quad \therefore\ f'(2) = \lim_{h\to 0} \frac{f(2+h)-f(2)}{h}$ where $f(2) = 2^3 = 8$

$$\begin{aligned} &= \lim_{h\to 0} \frac{(2+h)^3 - 8}{h} \\ &= \lim_{h\to 0} \frac{\cancel{8} + 12h + 6h^2 + h^3 - \cancel{8}}{h} \\ &= \lim_{h\to 0} \frac{12h + 6h^2 + h^3}{h} \\ &= \lim_{h\to 0} 12 + 6h + h^2 \qquad \{\text{as } h \neq 0\} \\ &= 12 \end{aligned}$$

**b** $f(x) = x^4$ $\quad \therefore \quad f'(3) = \lim_{h \to 0} \frac{f(3+h) - f(3)}{h}$ where $f(3) = 3^4 = 81$

$$= \lim_{h \to 0} \frac{(3+h)^4 - 81}{h}$$

$$= \lim_{h \to 0} \frac{\cancel{81} + 108h + 54h^2 + 12h^3 + h^4 - \cancel{81}}{h}$$

$$= \lim_{h \to 0} \frac{108h + 54h^2 + 12h^3 + h^4}{h}$$

$$= \lim_{h \to 0} 108 + 54h + 12h^2 + h^3 \qquad \{\text{as } h \neq 0\}$$

$$= 108$$

**5** **a** $f(x) = 3x + 5$

We need to find $f'(-2)$.

$$f'(-2) = \lim_{h \to 0} \frac{f(-2+h) - f(-2)}{h}$$

where $f(-2) = 3(-2) + 5 = -1$

$$= \lim_{h \to 0} \frac{[3(-2+h) + 5] - [-1]}{h}$$

$$= \lim_{h \to 0} \frac{\cancel{-6} + 3h + \cancel{5} + \cancel{1}}{h}$$

$$= \lim_{h \to 0} \frac{3h}{h}$$

$$= \lim_{h \to 0} 3 \qquad \{\text{as } h \neq 0\}$$

$$= 3$$

$\therefore$ the gradient of the tangent to $f(x) = 3x + 5$ at $x = -2$ is 3.

**b** $f(x) = 5 - 2x^2$

We need to find $f'(3)$.

$$f'(3) = \lim_{h \to 0} \frac{f(3+h) - f(3)}{h}$$

where $f(3) = 5 - 2(3)^2 = -13$

$$= \lim_{h \to 0} \frac{[5 - 2(3+h)^2] - [-13]}{h}$$

$$= \lim_{h \to 0} \frac{5 - 2(9 + 6h + h^2) + 13}{h}$$

$$= \lim_{h \to 0} \frac{\cancel{5} - \cancel{18} - 12h - 2h^2 + \cancel{13}}{h}$$

$$= \lim_{h \to 0} \frac{-12h - 2h^2}{h}$$

$$= \lim_{h \to 0} \frac{2h(6 + h)}{h}$$

$$= \lim_{h \to 0} -2(6+h) \qquad \{\text{as } h \neq 0\}$$

$$= -12$$

**c** $f(x) = x^2 + 3x - 4$

We need to find $f'(3)$.

$f'(3) = \lim_{h \to 0} \frac{f(3+h) - f(3)}{h}$ where $f(3) = 3^2 + 3(3) - 4 = 14$

$$= \lim_{h \to 0} \frac{[(3+h)^2 + 3(3+h) - 4] - 14}{h}$$

$$= \lim_{h \to 0} \frac{\cancel{9} + 6h + h^2 + \cancel{9} + 3h - \cancel{4} - \cancel{14}}{h}$$

$$= \lim_{h \to 0} \frac{9h + h^2}{h}$$

$$= \lim_{h \to 0} \frac{h(9+h)}{h}$$

$$= \lim_{h \to 0} (9+h) \qquad \{\text{as } h \neq 0\}$$

$$= 9$$

**d** $f(x) = 5 - 2x - 3x^2$

We need to find $f'(-2)$.

$$\begin{aligned} f'(-2) &= \lim_{h \to 0} \frac{f(-2+h) - f(-2)}{h} \quad \text{where} \quad f(-2) = 5 - 2(-2) - 3(-2)^2 = -3 \\ &= \lim_{h \to 0} \frac{[5 - 2(-2+h) - 3(-2+h)^2] - [-3]}{h} \\ &= \lim_{h \to 0} \frac{5 + 4 - 2h - 3(4 - 4h + h^2) + 3}{h} \\ &= \lim_{h \to 0} \frac{\cancel{5} + \cancel{4} - 2h - \cancel{12} + 12h - 3h^2 + \cancel{3}}{h} \\ &= \lim_{h \to 0} \frac{10h - 3h^2}{h} \\ &= \lim_{h \to 0} \frac{h(10 - 3h)}{h} \\ &= \lim_{h \to 0} (10 - 3h) \qquad \{\text{as } h \neq 0\} \\ &= 10 \end{aligned}$$

**6** **a**

$$\begin{aligned} y &= x^3 - 3x \\ \therefore \frac{dy}{dx} &= \lim_{h \to 0} \frac{f(x+h) - f(x)}{h} \\ &= \lim_{h \to 0} \frac{[(x+h)^3 - 3(x+h)] - [x^3 - 3x]}{h} \\ &= \lim_{h \to 0} \frac{\cancel{x^3} + 3x^2h + 3xh^2 + h^3 - \cancel{3x} - 3h - \cancel{x^3} + \cancel{3x}}{h} \\ &= \lim_{h \to 0} \frac{3x^2h + 3xh^2 + h^3 - 3h}{h} \\ &= \lim_{h \to 0} \frac{h(3x^2 + 3xh + h^2 - 3)}{h} \\ &= \lim_{h \to 0} (3x^2 + 3xh + h^2 - 3) \qquad \{\text{as } h \neq 0\} \\ &= 3x^2 - 3 \end{aligned}$$

**b** The tangent has zero gradient when $f'(x) = 0$

$\therefore\ 3x^2 - 3 = 0$

$\therefore\ 3x^2 = 3$

$\therefore\ x^2 = 1$

$\therefore\ x = \pm 1$

When $x = -1$, $y = (-1)^3 - 3(-1) = 2$ When $x = 1$, $y = (1)^3 - 3(1) = -2$

So, the points on the graph at which the tangent has zero gradient are $(-1, 2)$ and $(1, -2)$.

## REVIEW SET 14A

**1** **a** We can make $6x - 7$ as close as we like to $-1$ by making $x$ sufficiently close to 1.

$\therefore\ \lim_{x \to 1} (6x - 7) = -1$

**b**
$$\begin{aligned} \lim_{h \to 0} \frac{2h^2 - h}{h} &= \lim_{h \to 0} \frac{h(2h - 1)}{h} \\ &= \lim_{h \to 0} (2h - 1) \quad \{\text{as } h \neq 0\} \\ &= -1 \end{aligned}$$

**c**
$$\begin{aligned} \lim_{x \to 4} \frac{x^2 - 16}{x - 4} &= \lim_{x \to 4} \frac{(x+4)(x-4)}{x - 4} \\ &= \lim_{x \to 4} (x + 4) \quad \{\text{as } x \neq 0\} \\ &= 8 \end{aligned}$$

**2** **a** $f(x) = x^2 + 2x$

$$\begin{aligned}
\therefore\ & f'(x) \\
&= \lim_{h\to 0} \frac{f(x+h) - f(x)}{h} \\
&= \lim_{h\to 0} \frac{[(x+h)^2 + 2(x+h)] - [x^2 + 2x]}{h} \\
&= \lim_{h\to 0} \frac{\cancel{x^2} + 2xh + h^2 + \cancel{2x} + 2h - \cancel{x^2} - \cancel{2x}}{h} \\
&= \lim_{h\to 0} \frac{2xh + h^2 + 2h}{h} \\
&= \lim_{h\to 0} \frac{h(2x + h + 2)}{h} \\
&= \lim_{h\to 0} (2x + h + 2) \qquad \{\text{as } h \neq 0\} \\
&= 2x + 2
\end{aligned}$$

**b** $y = f(x) = 4 - 3x^2$

$$\begin{aligned}
\therefore\ & \frac{dy}{dx} \\
&= \lim_{h\to 0} \frac{f(x+h) - f(x)}{h} \\
&= \lim_{h\to 0} \frac{[4 - 3(x+h)^2] - [4 - 3x^2]}{h} \\
&= \lim_{h\to 0} \frac{4 - 3(x^2 + 2xh + h^2) - 4 + 3x^2}{h} \\
&= \lim_{h\to 0} \frac{\cancel{4} - \cancel{3x^2} - 6xh - 3h^2 - \cancel{4} + \cancel{3x^2}}{h} \\
&= \lim_{h\to 0} \frac{-3h(2x + h)}{h} \\
&= \lim_{h\to 0} -3(2x + h) \qquad \{\text{as } h \neq 0\} \\
&= -6x
\end{aligned}$$

**3** $f(x) = 5x - x^2$

$$\begin{aligned}
f'(1) &= \lim_{h\to 0} \frac{f(1+h) - f(1)}{h} \quad \text{where} \quad f(1) = 5(1) - (1)^2 = 4 \\
&= \lim_{h\to 0} \frac{[5(1+h) - (1+h)^2] - 4}{h} \\
&= \lim_{h\to 0} \frac{5 + 5h - (1 + 2h + h^2) - 4}{h} \\
&= \lim_{h\to 0} \frac{\cancel{5} + 5h - \cancel{1} - 2h - h^2 - \cancel{4}}{h} \\
&= \lim_{h\to 0} \frac{3h - h^2}{h} \\
&= \lim_{h\to 0} \frac{h(3 - h)}{h} \\
&= \lim_{h\to 0} (3 - h) \qquad \{\text{as } h \neq 0\} \\
&= 3
\end{aligned}$$

**4** **a** $f(t) = 452 - 4.8t^2$

$$\begin{aligned}
\therefore\ & f'(t) \\
&= \lim_{h\to 0} \frac{f(t+h) - f(t)}{h} \\
&= \lim_{h\to 0} \frac{[452 - 4.8(t+h)^2] - [452 - 4.8t^2]}{h} \\
&= \lim_{h\to 0} \frac{\cancel{452} - 4.8(t^2 + 2th + h^2) - \cancel{452} + 4.8t^2}{h} \\
&= \lim_{h\to 0} \frac{-\cancel{4.8t^2} - 9.6th - 4.8h^2 + \cancel{4.8t^2}}{h} \\
&= \lim_{h\to 0} \frac{h(-9.6t - 4.8h)}{h} \\
&= \lim_{h\to 0} (-9.6t - 4.8h) \qquad \{\text{as } h \neq 0\} \\
&= -9.6t \text{ m s}^{-1}
\end{aligned}$$

**b** To find the speed of the jumper at $t = 2$ seconds, we need to find $f'(2)$.

Now $f'(t) = -9.6t$ {from **a**}

$$\begin{aligned}
\therefore\ f'(2) &= -9.6 \times 2 \\
&= -19.2
\end{aligned}$$

$\therefore$ the speed of the jumper at $t = 2$ seconds is $19.2 \text{ m s}^{-1}$.

(The $-$ sign indicates the jumper is moving downwards.)

## REVIEW SET 14B

**1** **a** We sketch the graph of $y = \dfrac{x-7}{3x+2}$:

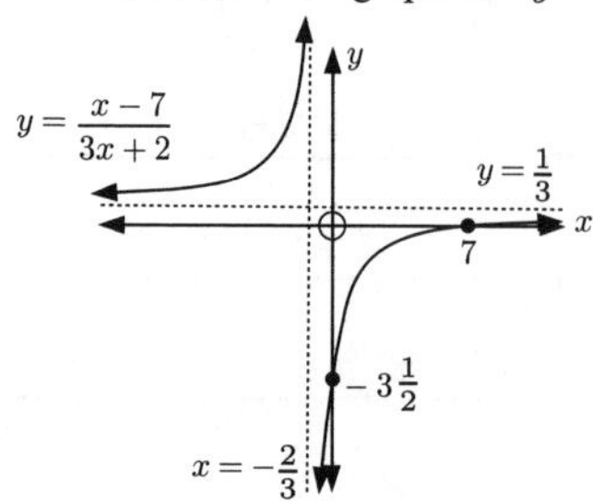


As $x \to -\frac{2}{3}^-$, $y \to \infty$

As $x \to -\frac{2}{3}^+$, $y \to -\infty$

As $x \to \infty$, $y \to \frac{1}{3}^-$

As $x \to -\infty$, $y \to \frac{1}{3}^+$

The vertical asymptote is $x = -\frac{2}{3}$.

The horizontal asymptote is $y = \frac{1}{3}$.

**b** $\lim\limits_{x \to -\infty} \left(\dfrac{x-7}{3x+2}\right) = \frac{1}{3}$, $\lim\limits_{x \to \infty} \left(\dfrac{x-7}{3x+2}\right) = \frac{1}{3}$

**2** **a**

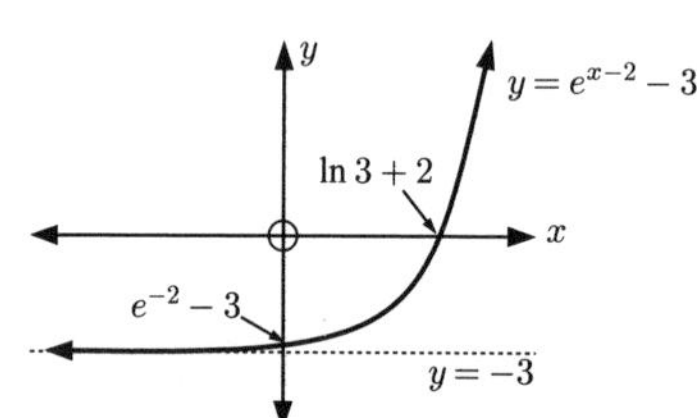


**b** $\lim\limits_{x \to -\infty} (e^{x-2} - 3) = -3$,

$\lim\limits_{x \to \infty} (e^{x-2} - 3)$ does not exist.

**c** The horizontal asymptote is $y = -3$.

**3** **a**
$$\begin{aligned}\frac{f(x+h)-f(x)}{h} &= \frac{2(x+h)^2 - 2x^2}{h} \\ &= \frac{2(x^2+2xh+h^2) - 2x^2}{h} \\ &= \frac{\cancel{2x^2} + 4xh + 2h^2 - \cancel{2x^2}}{h} \\ &= \frac{h(4x+2h)}{h} \\ &= 4x + 2h \quad \text{provided } h \neq 0\end{aligned}$$

**b** If $x = 3$ then
$$\begin{aligned}\frac{f(3+h)-f(3)}{h} &= 4(3) + 2h \quad \{\text{using } \mathbf{a}\} \\ &= 12 + 2h\end{aligned}$$

**i** When $h = 0.1$,
$$\begin{aligned}\frac{f(3+h)-f(3)}{h} &= 12 + 2(0.1) \\ &= 12 + 0.2 \\ &= 12.2\end{aligned}$$

**ii** When $h = 0.01$,
$$\begin{aligned}\frac{f(3+h)-f(3)}{h} &= 12 + 2(0.01) \\ &= 12 + 0.02 \\ &= 12.02\end{aligned}$$

**iii**
$$\begin{aligned}\lim_{h \to 0} \frac{f(3+h)-f(3)}{h} &= \lim_{h \to 0} 12 + 2h \\ &= 12\end{aligned}$$

**c** The gradient of the tangent to $y = 2x^2$ at the point (3, 18) is 12.

**4** $\lim\limits_{x \to -\infty} \left(\dfrac{2x+3}{4-x}\right) = -2$, $\lim\limits_{x \to \infty} \left(\dfrac{2x+3}{4-x}\right) = -2$

## REVIEW SET 14C

**1** **a** $\lim\limits_{h \to 0} \dfrac{h^3 - 3h}{h}$
$= \lim\limits_{h \to 0} \dfrac{h(h^2 - 3)}{h}$
$= \lim\limits_{h \to 0} h^2 - 3 \quad \{\text{as } h \neq 0\}$
$= -3$

**b** $\lim\limits_{x \to 1} \dfrac{3x^2 - 3x}{x - 1}$
$= \lim\limits_{x \to 1} \dfrac{3x(x - 1)}{x - 1}$
$= \lim\limits_{x \to 1} 3x \quad \{\text{as } x \neq 1\}$
$= 3$

**c** $\lim\limits_{x \to 2} \dfrac{x^2 - 3x + 2}{2 - x}$
$= \lim\limits_{x \to 2} \dfrac{(x - 1)(x - 2)}{-(x - 2)}$
$= \lim\limits_{x \to 2} -(x - 1) \quad \{\text{as } x \neq 2\}$
$= -1$

**2** **a**

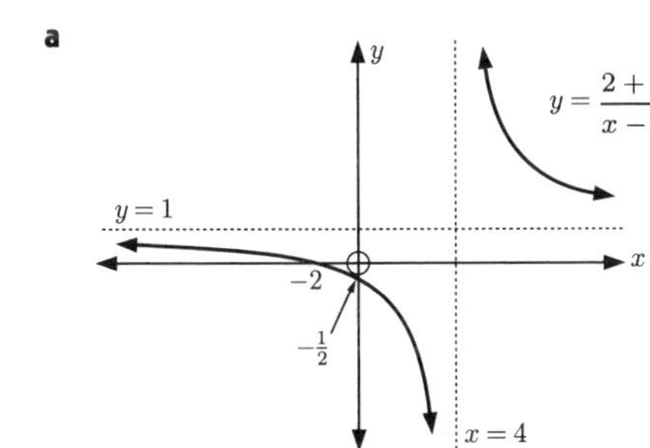


**b** As $x \to 4^-$, $y \to -\infty$
As $x \to 4^+$, $y \to \infty$
As $x \to \infty$, $y \to 1^+$
As $x \to -\infty$, $y \to 1^-$
The vertical asymptote is $x = 4$.
The horizontal asymptote is $y = 1$.

**c** $\lim\limits_{x \to -\infty} \dfrac{2 + x}{x - 4} = 1$, $\lim\limits_{x \to \infty} \dfrac{2 + x}{x - 4} = 1$

**3** $f(x) = x^4 - 2x$

$f'(1) = \lim\limits_{h \to 0} \dfrac{f(1 + h) - f(1)}{h}$ where $f(1) = 1^4 - 2(1) = -1$
$= \lim\limits_{h \to 0} \dfrac{[(1 + h)^4 - 2(1 + h)] - [-1]}{h}$
$= \lim\limits_{h \to 0} \dfrac{\not{1} + 4h + 6h^2 + 4h^3 + h^4 - \not{2} - 2h + \not{1}}{h}$
$= \lim\limits_{h \to 0} \dfrac{h^4 + 4h^3 + 6h^2 + 2h}{h}$
$= \lim\limits_{h \to 0} \dfrac{h(h^3 + 4h^2 + 6h + 2)}{h}$
$= \lim\limits_{h \to 0} (h^3 + 4h^2 + 6h + 2) \quad \{\text{as } h \neq 0\}$
$= 2$

**4** **a** $y = 2x^2 - 1$
$\therefore \dfrac{dy}{dx} = \lim\limits_{h \to 0} \dfrac{[2(x + h)^2 - 1] - [2x^2 - 1]}{h}$
$= \lim\limits_{h \to 0} \dfrac{2(x^2 + 2hx + h^2) - 1 - 2x^2 + 1}{h}$
$= \lim\limits_{h \to 0} \dfrac{\not{2x^2} + 4hx + 2h^2 - \not{1} - \not{2x^2} + \not{1}}{h}$
$= \lim\limits_{h \to 0} \dfrac{4hx + 2h^2}{h}$
$= \lim\limits_{h \to 0} \dfrac{h(4x + 2h)}{h}$
$= \lim\limits_{h \to 0} 4x + 2h \quad \{\text{as } h \neq 0\}$
$= 4x$

**b** The gradient of the tangent to $y = 2x^2 - 1$ at the point where $x = 4$ is $4 \times 4 = 16$.

**c** If the gradient of the tangent is equal to $-12$, then $4x = -12$
$\therefore x = -3$

# Chapter 15
## RULES OF DIFFERENTIATION

### EXERCISE 15A

**1** **a** $f(x) = x^3$
$\therefore\ f'(x) = 3x^2$

**b** $f(x) = 2x^3$
$\therefore\ f'(x) = 2(3x^2) = 6x^2$

**c** $f(x) = 7x^2$
$\therefore\ f'(x) = 7(2x) = 14x$

**d** $f(x) = 6\sqrt{x} = 6x^{\frac{1}{2}}$
$\therefore\ f'(x) = 6\left(\frac{1}{2}x^{-\frac{1}{2}}\right) = \dfrac{3}{\sqrt{x}}$

**e** $f(x) = 3\sqrt[3]{x} = 3x^{\frac{1}{3}}$
$\therefore\ f'(x) = 3\left(\frac{1}{3}x^{-\frac{2}{3}}\right) = \dfrac{1}{\sqrt[3]{x^2}}$

**f** $f(x) = x^2 + x$
$\therefore\ f'(x) = 2x + 1$

**g** $f(x) = 4 - 2x^2$
$\therefore\ f'(x) = 0 - 2(2x) = -4x$

**h** $f(x) = x^2 + 3x - 5$
$\therefore\ f'(x) = 2x + 3 - 0 = 2x + 3$

**i** $f(x) = \frac{1}{2}x^4 - 6x^2$
$\therefore\ f'(x) = \frac{1}{2}(4x^3) - 6(2x) = 2x^3 - 12x$

**j** $f(x) = \dfrac{3x-6}{x} = 3 - 6x^{-1}$
$\therefore\ f'(x) = 0 - 6(-1x^{-2}) = \dfrac{6}{x^2}$

**k** $f(x) = \dfrac{2x-3}{x^2} = \dfrac{2x}{x^2} - \dfrac{3}{x^2} = 2x^{-1} - 3x^{-2}$
$\therefore\ f'(x) = -2x^{-2} + 6x^{-3} = -\dfrac{2}{x^2} + \dfrac{6}{x^3}$

**l** $f(x) = \dfrac{x^3+5}{x} = x^2 + 5x^{-1}$
$\therefore\ f'(x) = 2x - 5x^{-2} = 2x - \dfrac{5}{x^2}$

**m** $f(x) = \dfrac{x^3+x-3}{x} = x^2 + 1 - 3x^{-1}$
$\therefore\ f'(x) = 2x + 0 + 3x^{-2} = 2x + \dfrac{3}{x^2}$

**n** $f(x) = \dfrac{1}{\sqrt{x}} = x^{-\frac{1}{2}}$
$\therefore\ f'(x) = -\frac{1}{2}x^{-\frac{3}{2}} = -\dfrac{1}{2x\sqrt{x}}$

**o** $f(x) = (2x-1)^2 = 4x^2 - 4x + 1$
$\therefore\ f'(x) = 8x - 4$

**p** $f(x) = (x+2)^3 = x^3 + 3x^2(2) + 3x(2^2) + 2^3 = x^3 + 6x^2 + 12x + 8$
$\therefore\ f'(x) = 3x^2 + 12x + 12$

**2** **a** $y = 2.5x^3 - 1.4x^2 - 1.3$
$\therefore\ \dfrac{dy}{dx} = 7.5x^2 - 2.8x$

**b** $y = \pi x^2$
$\therefore\ \dfrac{dy}{dx} = 2\pi x$

**c** $y = \dfrac{1}{5x^2} = \frac{1}{5}x^{-2}$
$\therefore\ \dfrac{dy}{dx} = -\frac{2}{5}x^{-3} = -\dfrac{2}{5x^3}$

**d** $y = 100x$
$\therefore\ \dfrac{dy}{dx} = 100$

**e** $y = 10(x+1) = 10x + 10$
$\therefore\ \dfrac{dy}{dx} = 10$

**f** $y = 4\pi x^3$
$\therefore\ \dfrac{dy}{dx} = 12\pi x^2$

**3** **a** $\frac{d}{dx}(6x+2)$
$= 6$

**b** $\frac{d}{dx}(x\sqrt{x})$
$= \frac{d}{dx}(x^{\frac{3}{2}})$
$= \frac{3}{2}x^{\frac{1}{2}}$
$= \frac{3\sqrt{x}}{2}$

**c** $\frac{d}{dx}(5-x)^2$
$= \frac{d}{dx}(25-10x+x^2)$
$= -10+2x$
$= 2x-10$

**d** $\frac{d}{dx}\left(\frac{6x^2-9x^4}{3x}\right)$
$= \frac{d}{dx}(2x-3x^3)$
$= 2-9x^2$

**e** $\frac{d}{dx}((x+1)(x-2))$
$= \frac{d}{dx}(x^2-x-2)$
$= 2x-1$

**f** $\frac{d}{dx}\left(\frac{1}{x^2}+6\sqrt{x}\right)$
$= \frac{d}{dx}\left(x^{-2}+6x^{\frac{1}{2}}\right)$
$= -2x^{-3}+3x^{-\frac{1}{2}}$
$= -\frac{2}{x^3}+\frac{3}{\sqrt{x}}$

**g** $\frac{d}{dx}\left(4x-\frac{1}{4x}\right)$
$= \frac{d}{dx}\left(4x-\frac{1}{4}x^{-1}\right)$
$= 4+\frac{1}{4}x^{-2}$
$= 4+\frac{1}{4x^2}$

**h** $\frac{d}{dx}(x(x+1)(2x-5))$
$= \frac{d}{dx}\left(x(2x^2-3x-5)\right)$
$= \frac{d}{dx}\left(2x^3-3x^2-5x\right)$
$= 6x^2-6x-5$

**4** **a** Consider $y = x^2$ when $x = 2$
Now $\frac{dy}{dx} = 2x$
$\therefore$ when $x = 2$,
$\frac{dy}{dx} = 2(2) = 4$
$\therefore$ the tangent has gradient 4.

**b** Consider $y = \frac{8}{x^2}$ at the point $(9, \frac{8}{81})$
Now $y = 8x^{-2}$
$\therefore \frac{dy}{dx} = -16x^{-3} = -\frac{16}{x^3}$
$\therefore$ at $(9, \frac{8}{81})$, $x = 9$ and so $\frac{dy}{dx} = -\frac{16}{729}$
$\therefore$ the tangent has gradient $-\frac{16}{729}$.

**c** Consider $y = 2x^2-3x+7$ when $x = -1$
Now $\frac{dy}{dx} = 4x-3$
$\therefore$ when $x = -1$,
$\frac{dy}{dx} = 4(-1)-3 = -7$
$\therefore$ the tangent has gradient $-7$.

**d** Consider $y = \frac{2x^2-5}{x}$ at the point $(2, \frac{3}{2})$
Now $y = 2x-5x^{-1}$
$\therefore \frac{dy}{dx} = 2+5x^{-2} = 2+\frac{5}{x^2}$
$\therefore$ at $(2, \frac{3}{2})$, $x = 2$ and so $\frac{dy}{dx} = 2+\frac{5}{4}$
$= \frac{13}{4}$
$\therefore$ the tangent has gradient $\frac{13}{4}$.

**e** Consider $y = \frac{x^2-4}{x^2}$ at the point $(4, \frac{3}{4})$
Now $y = 1-4x^{-2}$
$\therefore \frac{dy}{dx} = 0+8x^{-3} = \frac{8}{x^3}$
$\therefore$ at $(4, \frac{3}{4})$, $x = 4$ and so
$\frac{dy}{dx} = \frac{8}{4^3} = \frac{1}{8}$
$\therefore$ the tangent has gradient $\frac{1}{8}$.

**f** Consider $y = \frac{x^3-4x-8}{x^2}$ when $x = -1$
Now $y = x-4x^{-1}-8x^{-2}$
$\therefore \frac{dy}{dx} = 1+4x^{-2}+16x^{-3}$
$= 1+\frac{4}{x^2}+\frac{16}{x^3}$
$\therefore$ when $x = -1$,
$\frac{dy}{dx} = 1+4-16 = -11$
$\therefore$ the tangent has gradient $-11$.

**5** $f(x) = x^2 + (b+1)x + 2c, \quad f(2) = 4, \quad \text{and} \quad f'(-1) = 2$

$\therefore \ f'(x) = 2x + (b+1)$

But $f'(-1) = 2$, so $2(-1) + b + 1 = 2$

$\therefore \ -1 + b = 2$

$\therefore \ b = 3$

So, $f(x) = x^2 + (3+1)x + 2c$

$= x^2 + 4x + 2c$

But $f(2) = 4$, so $2^2 + 4(2) + 2c = 4$

$\therefore \ 2c = -8$

$\therefore \ c = -4$

**6** **a** $f(x) = 4\sqrt{x} + x = 4x^{\frac{1}{2}} + x$

$\therefore \ f'(x) = 4\left(\frac{1}{2}x^{-\frac{1}{2}}\right) + 1$

$= \dfrac{2}{\sqrt{x}} + 1$

**b** $f(x) = \sqrt[3]{x} = x^{\frac{1}{3}}$

$\therefore \ f'(x) = \frac{1}{3}x^{-\frac{2}{3}}$

$= \dfrac{1}{3\sqrt[3]{x^2}}$

**c** $f(x) = -\dfrac{2}{\sqrt{x}} = -2x^{-\frac{1}{2}}$

$\therefore \ f'(x) = -2\left(-\frac{1}{2}x^{-\frac{3}{2}}\right)$

$= x^{-\frac{3}{2}}$

$= \dfrac{1}{x\sqrt{x}}$

**d** $f(x) = 2x - \sqrt{x} = 2x - x^{\frac{1}{2}}$

$\therefore \ f'(x) = 2 - \frac{1}{2}x^{-\frac{1}{2}}$

$= 2 - \dfrac{1}{2\sqrt{x}}$

**e** $f(x) = \dfrac{4}{\sqrt{x}} - 5 = 4x^{-\frac{1}{2}} - 5$

$\therefore \ f'(x) = 4\left(-\frac{1}{2}x^{-\frac{3}{2}}\right)$

$= -2x^{-\frac{3}{2}} = -\dfrac{2}{x\sqrt{x}}$

**f** $f(x) = 3x^2 - x\sqrt{x} = 3x^2 - x^{\frac{3}{2}}$

$\therefore \ f'(x) = 6x - \frac{3}{2}x^{\frac{1}{2}}$

$= 6x - \frac{3}{2}\sqrt{x}$

**g** $f(x) = \dfrac{5}{x^2\sqrt{x}} = 5x^{-\frac{5}{2}}$

$\therefore \ f'(x) = 5\left(-\frac{5}{2}x^{-\frac{7}{2}}\right)$

$= -\frac{25}{2}x^{-\frac{7}{2}}$

$= \dfrac{-25}{2x^3\sqrt{x}}$

**h** $f(x) = 2x - \dfrac{3}{x\sqrt{x}} = 2x - 3x^{-\frac{3}{2}}$

$\therefore \ f'(x) = 2 - 3\left(-\frac{3}{2}x^{-\frac{5}{2}}\right)$

$= 2 + \frac{9}{2}x^{-\frac{5}{2}}$

$= 2 + \dfrac{9}{2x^2\sqrt{x}}$

**7** **a** $y = 4x - \dfrac{3}{x} = 4x - 3x^{-1} \quad \therefore \quad \dfrac{dy}{dx} = 4 + 3x^{-2} = 4 + \dfrac{3}{x^2}$

$\dfrac{dy}{dx}$ is the gradient function of $y = 4x - \dfrac{3}{x}$ from which the gradient at any point can be found.

**b** $S = 2t^2 + 4t$ m $\quad \therefore \quad \dfrac{dS}{dt} = 4t + 4 \ \text{m s}^{-1}$

$\dfrac{dS}{dt}$ is the instantaneous rate of change in position at time $t$. It is the velocity function.

**c** $C = 1785 + 3x + 0.002x^2$ dollars.

$\dfrac{dC}{dx} = 3 + 0.002(2x) = 3 + 0.004x$ dollars per toaster

$\dfrac{dC}{dx}$ is the instantaneous rate of change in cost as the number of toasters changes.

## EXERCISE 15B.1

**1** **a** $g(x) = x^2, \quad f(x) = 2x + 7$
$\therefore \quad g(f(x)) = g(2x + 7) = (2x + 7)^2$

**b** $g(x) = 2x + 7, \quad f(x) = x^2$
$g(f(x)) = g(x^2) = 2x^2 + 7$

**c** $g(x) = \sqrt{x}, \quad f(x) = 3 - 4x$
$g(f(x)) = g(3 - 4x) = \sqrt{3 - 4x}$

**d** $g(x) = 3 - 4x, \quad f(x) = \sqrt{x}$
$g(f(x)) = g(\sqrt{x}) = 3 - 4\sqrt{x}$

**e** $g(x) = \dfrac{2}{x}, \quad f(x) = x^2 + 3$
$g(f(x)) = g(x^2 + 3) = \dfrac{2}{x^2 + 3}$

**f** $g(x) = x^2 + 3, \quad f(x) = \dfrac{2}{x}$
$g(f(x)) = g\left(\dfrac{2}{x}\right) = \left(\dfrac{2}{x}\right)^2 + 3 = \dfrac{4}{x^2} + 3$

**2** **a** $g(f(x)) = (3x + 10)^3 \quad \therefore \quad g(x) = x^3, \quad f(x) = 3x + 10$

**b** $g(f(x)) = \dfrac{1}{2x + 4} \quad \therefore \quad g(x) = \dfrac{1}{x}, \quad f(x) = 2x + 4$

**c** $g(f(x)) = \sqrt{x^2 - 3x} \quad \therefore \quad g(x) = \sqrt{x}, \quad f(x) = x^2 - 3x$

**d** $g(f(x)) = \dfrac{10}{(3x - x^2)^3} \quad \therefore \quad g(x) = \dfrac{10}{x^3}, \quad f(x) = 3x - x^2$

{other answers are possible for **2**}

## EXERCISE 15B.2

**1** **a** $\dfrac{1}{(2x - 1)^2}$
$= (2x - 1)^{-2}$
$= u^{-2}$
where $u = 2x - 1$

**b** $\sqrt{x^2 - 3x}$
$= (x^2 - 3x)^{\frac{1}{2}}$
$= u^{\frac{1}{2}}$
where $u = x^2 - 3x$

**c** $\dfrac{2}{\sqrt{2 - x^2}}$
$= 2(2 - x^2)^{-\frac{1}{2}}$
$= 2u^{-\frac{1}{2}}$
where $u = 2 - x^2$

**d** $\sqrt[3]{x^3 - x^2}$
$= (x^3 - x^2)^{\frac{1}{3}}$
$= u^{\frac{1}{3}}$
where $u = x^3 - x^2$

**e** $\dfrac{4}{(3 - x)^3}$
$= 4(3 - x)^{-3}$
$= 4u^{-3}$
where $u = 3 - x$

**f** $\dfrac{10}{x^2 - 3}$
$= 10(x^2 - 3)^{-1}$
$= 10u^{-1}$
where $u = x^2 - 3$

**2** **a** $y = (4x - 5)^2$
$\therefore \quad y = u^2$ where $u = 4x - 5$

Now $\dfrac{dy}{dx} = \dfrac{dy}{du}\dfrac{du}{dx}$
$= 2u(4)$
$= 8u$
$= 8(4x - 5)$

**b** $y = \dfrac{1}{5 - 2x}$
$\therefore \quad y = u^{-1}$ where $u = 5 - 2x$

Now $\dfrac{dy}{dx} = \dfrac{dy}{du}\dfrac{du}{dx}$
$= -u^{-2}(-2)$
$= 2u^{-2}$
$= 2\,(5 - 2x)^{-2}$

**c** $y = \sqrt{3x - x^2}$
$\therefore \quad y = u^{\frac{1}{2}}$ where $u = 3x - x^2$

Now $\dfrac{dy}{dx} = \dfrac{dy}{du}\dfrac{du}{dx}$
$= \frac{1}{2}u^{-\frac{1}{2}}(3 - 2x)$
$= \frac{1}{2}\left(3x - x^2\right)^{-\frac{1}{2}}(3 - 2x)$

**d** $y = (1 - 3x)^4$
$\therefore \quad y = u^4$ where $u = 1 - 3x$

Now $\dfrac{dy}{dx} = \dfrac{dy}{du}\dfrac{du}{dx}$
$= 4u^3(-3)$
$= -12u^3$
$= -12(1 - 3x)^3$

**e** $y = 6(5-x)^3$

$\therefore\ y = 6u^3$ where $u = 5-x$

Now $\dfrac{dy}{dx} = \dfrac{dy}{du}\dfrac{du}{dx}$

$= 18u^2(-1)$

$= -18u^2$

$= -18(5-x)^2$

**f** $y = \sqrt[3]{2x^3 - x^2}$

$\therefore\ y = u^{\frac{1}{3}}$ where $u = 2x^3 - x^2$

Now $\dfrac{dy}{dx} = \dfrac{dy}{du}\dfrac{du}{dx}$

$= \frac{1}{3}u^{-\frac{2}{3}}(6x^2 - 2x)$

$= \frac{1}{3}\left(2x^3 - x^2\right)^{-\frac{2}{3}}(6x^2 - 2x)$

**g** $y = \dfrac{6}{(5x-4)^2}$

$\therefore\ y = 6u^{-2}$ where $u = 5x - 4$

Now $\dfrac{dy}{dx} = \dfrac{dy}{du}\dfrac{du}{dx}$

$= -12u^{-3}(5)$

$= -60\,(5x-4)^{-3}$

**h** $y = \dfrac{4}{3x - x^2}$

$\therefore\ y = 4u^{-1}$ where $u = 3x - x^2$

Now $\dfrac{dy}{dx} = \dfrac{dy}{du}\dfrac{du}{dx}$

$= -4u^{-2}(3-2x)$

$= -4\left(3x - x^2\right)^{-2}(3-2x)$

**i** $y = 2\left(x^2 - \dfrac{2}{x}\right)^3$

$\therefore\ y = 2u^3$ where $u = x^2 - 2x^{-1}$

Now $\dfrac{dy}{dx} = \dfrac{dy}{du}\dfrac{du}{dx}$

$= 6u^2(2x + 2x^{-2})$

$= 6\left(x^2 - \dfrac{2}{x}\right)^2\left(2x + \dfrac{2}{x^2}\right)$

**3** **a** $y = \sqrt{1-x^2}$ at $x = \frac{1}{2}$

$\therefore\ y = \sqrt{u}$ where $u = 1 - x^2$

Now $\dfrac{dy}{dx} = \dfrac{dy}{du}\dfrac{du}{dx} = \frac{1}{2}u^{-\frac{1}{2}}(-2x)$

$= \dfrac{-x}{\sqrt{u}}$

$= \dfrac{-x}{\sqrt{1-x^2}}$

At $x = \frac{1}{2}$, $\dfrac{dy}{dx} = \dfrac{-\frac{1}{2}}{\sqrt{1-\frac{1}{4}}} = -\frac{1}{2}\left(\frac{2}{\sqrt{3}}\right)$

$\therefore$ gradient of tangent $= -\frac{1}{\sqrt{3}}$

**b** $y = (3x+2)^6$ at $x = -1$

$\therefore\ y = u^6$ where $u = 3x + 2$

Now $\dfrac{dy}{dx} = \dfrac{dy}{du}\dfrac{du}{dx}$

$= 6u^5(3)$

$= 18u^5$

$= 18(3x+2)^5$

At $x = -1$, $\dfrac{dy}{dx} = 18(-1)^5$

$\therefore$ gradient of tangent $= -18$

**c** $y = \dfrac{1}{(2x-1)^4}$ at $x = 1$

$\therefore\ y = u^{-4}$ where $u = 2x - 1$

Now $\dfrac{dy}{dx} = \dfrac{dy}{du}\dfrac{du}{dx} = -4u^{-5}(2)$

$= \dfrac{-8}{u^5}$

$= \dfrac{-8}{(2x-1)^5}$

At $x = 1$, $\dfrac{dy}{dx} = \dfrac{-8}{1^5}$

$\therefore$ gradient of tangent $= -8$

**d** $y = 6 \times \sqrt[3]{1-2x}$ at $x = 0$

$\therefore\ y = 6u^{\frac{1}{3}}$ where $u = 1 - 2x$

Now $\dfrac{dy}{dx} = \dfrac{dy}{du}\dfrac{du}{dx} = 6(\frac{1}{3})u^{-\frac{2}{3}}(-2)$

$= 2u^{-\frac{2}{3}}(-2)$

$= \dfrac{-4}{\sqrt[3]{u^2}}$

$= \dfrac{-4}{\sqrt[3]{(1-2x)^2}}$

At $x = 0$, $\dfrac{dy}{dx} = \dfrac{-4}{\sqrt[3]{1^2}}$

$\therefore$ gradient of tangent $= -4$

**e** $y = \dfrac{4}{x+2\sqrt{x}}$ at $x = 4$

$\therefore\ y = 4u^{-1}$ where $u = x + 2x^{\frac{1}{2}}$

$$\text{Now}\quad \frac{dy}{dx} = \frac{dy}{du}\frac{du}{dx}$$
$$= -4u^{-2}(1 + x^{-\frac{1}{2}})$$
$$= -\frac{4}{u^2}\left(1+\frac{1}{\sqrt{x}}\right)$$
$$= \frac{-4}{(x+2\sqrt{x})^2}\left(1+\frac{1}{\sqrt{x}}\right)$$

At $x = 4$, $\dfrac{dy}{dx} = \dfrac{-4}{(4+4)^2}(1+\frac{1}{2}) = -\frac{6}{64}$

$\therefore$ gradient of tangent $= -\frac{3}{32}$

**f** $y = \left(x+\dfrac{1}{x}\right)^3$ at $x = 1$

$\therefore\ y = u^3$ where $u = x + x^{-1}$

$$\text{Now}\quad \frac{dy}{dx} = \frac{dy}{du}\frac{du}{dx}$$
$$= 3u^2(1 - x^{-2})$$
$$= 3\left(x+\frac{1}{x}\right)^2\left(1-\frac{1}{x^2}\right)$$

At $x = 1$, $\dfrac{dy}{dx} = 3(1+1)^2(1-1)$

$\therefore$ gradient of tangent $= 0$

**4** If $f(x) = (2x-b)^a$

then $f'(x) = a(2x-b)^{a-1} \times 2$
$= 2a(2x-b)^{a-1}$

but $f'(x) = 24x^2 - 24x + 6$
$= 6(4x^2 - 4x + 1)$
$= 6(2x-1)^2$

Solving by inspection, we get $a - 1 = 2$ and $b = 1$

$\therefore\ a = 3$ and $b = 1$

(**Note:** We can also use $2a = 6$ when solving for $a$.)

**5** $y = \dfrac{a}{\sqrt{1+bx}} = a\,(1+bx)^{-\frac{1}{2}}$

When $x = 3$, $y = 1$

$\therefore\ 1 = \dfrac{a}{\sqrt{1+3b}}$

$\therefore\ a = \sqrt{1+3b}$ .... (1)

$\dfrac{dy}{dx} = -\frac{1}{2}a\,(1+bx)^{-\frac{3}{2}} \times b$

When $x = 3$, $\dfrac{dy}{dx} = -\frac{1}{8}$

$\therefore\ -\frac{1}{8} = -\frac{1}{2}ab\,(1+3b)^{-\frac{3}{2}}$

$\therefore\ a = \dfrac{1}{4b}\,(1+3b)^{\frac{3}{2}}$ .... (2)

$= \dfrac{1}{4b}\,(1+3b)\,\sqrt{1+3b}$

Equating the RHS of (1) and (2) we get:

$$\sqrt{1+3b} = \frac{1}{4b}\,(1+3b)\,\sqrt{1+3b}$$

$$\therefore\ 1 = \frac{1}{4b}\,(1+3b) \qquad \{\text{since } b \neq 0,\ \sqrt{1+3b} \neq 0\}$$

$\therefore\ 4b = 1 + 3b$

$\therefore\ b = 1$

$\therefore\ a = \sqrt{1+3(1)}$
$= \sqrt{4}$
$= 2$

So, $a = 2$ and $b = 1$.

**6** **a** $y = x^3 \quad \therefore \quad \dfrac{dy}{dx} = 3x^2$

$x = y^{\frac{1}{3}} \quad \therefore \quad \dfrac{dx}{dy} = \frac{1}{3}y^{-\frac{2}{3}}$

$$\begin{aligned}\frac{dy}{dx}\frac{dx}{dy} &= 3x^2\left(\tfrac{1}{3}\right)y^{-\frac{2}{3}}\\ &= x^2(y)^{-\frac{2}{3}}\\ &= x^2(x^3)^{-\frac{2}{3}} \quad \{\text{substituting } y = x^3\}\\ &= x^2(x^{-2})\\ &= x^0\\ &= 1 \quad \text{as required}\end{aligned}$$

**b** We know that $\dfrac{dy}{du}\dfrac{du}{dx} = \dfrac{dy}{dx}$ {chain rule}

Letting $x = y$, $\dfrac{dy}{du}\dfrac{du}{dy} = \dfrac{dy}{dy}$

$\therefore \quad \dfrac{dy}{du}\dfrac{du}{dy} = 1$

Letting $u = x$, $\dfrac{dy}{dx}\dfrac{dx}{dy} = 1$

## EXERCISE 15C

**1** **a** $f(x) = x(x-1)$ is the product of

$u = x$ and $v = x - 1$

$\therefore \quad u' = 1$ and $v' = 1$

Now $f'(x) = u'v + uv'$ {product rule}

$$\begin{aligned}&= 1 \times (x-1) + x \times 1\\ &= x - 1 + x\\ &= 2x - 1\end{aligned}$$

**b** $f(x) = 2x(x+1)$ is the product of

$u = 2x$ and $v = x + 1$

$\therefore \quad u' = 2$ and $v' = 1$

Now $f'(x) = u'v + uv'$ {product rule}

$$\begin{aligned}&= 2(x+1) + 2x \times 1\\ &= 2x + 2 + 2x\\ &= 4x + 2\end{aligned}$$

**c** $f(x) = x^2\sqrt{x+1}$ is the product of $u = x^2$ and $v = (x+1)^{\frac{1}{2}}$

$\therefore \quad u' = 2x$ and $v' = \frac{1}{2}(x+1)^{-\frac{1}{2}}$

Now $f'(x) = u'v + uv'$ {product rule}

$$\begin{aligned}&= 2x(x+1)^{\frac{1}{2}} + x^2 \times \tfrac{1}{2}(x+1)^{-\frac{1}{2}}\\ &= 2x(x+1)^{\frac{1}{2}} + \tfrac{1}{2}x^2(x+1)^{-\frac{1}{2}}\end{aligned}$$

**2** **a** $y = x^2(2x-1)$ is the product of

$u = x^2$ and $v = 2x - 1$

$\therefore \quad u' = 2x$ and $v' = 2$

Now $\dfrac{dy}{dx} = u'v + uv'$ {product rule}

$$\begin{aligned}\therefore \quad \frac{dy}{dx} &= 2x(2x-1) + x^2(2)\\ &= 2x(2x-1) + 2x^2\end{aligned}$$

**b** $y = 4x(2x+1)^3$ is the product of

$u = 4x$ and $v = (2x+1)^3$

$\therefore \quad u' = 4$ and $v' = 3(2x+1)^2 \times 2$

$= 6(2x+1)^2$

Now $\dfrac{dy}{dx} = u'v + uv'$ {product rule}

$\therefore \quad \dfrac{dy}{dx} = 4(2x+1)^3 + 24x(2x+1)^2$

**c** $y = x^2\sqrt{3-x}$ is the product of

$u = x^2$ and $v = (3-x)^{\frac{1}{2}}$

$\therefore \quad u' = 2x$ and $v' = \frac{1}{2}(3-x)^{-\frac{1}{2}}(-1)$

$= -\frac{1}{2}(3-x)^{-\frac{1}{2}}$

Now $\dfrac{dy}{dx} = u'v + uv'$ {product rule}

$$\begin{aligned}\therefore \quad \frac{dy}{dx} &= 2x(3-x)^{\frac{1}{2}} + x^2\left[-\tfrac{1}{2}(3-x)^{-\frac{1}{2}}\right]\\ &= 2x(3-x)^{\frac{1}{2}} - \tfrac{1}{2}x^2(3-x)^{-\frac{1}{2}}\end{aligned}$$

**d** $y = \sqrt{x}(x-3)^2$ is the product of

$u = x^{\frac{1}{2}}$ and $v = (x-3)^2$

$\therefore \quad u' = \frac{1}{2}x^{-\frac{1}{2}}$ and $v' = 2(x-3)^1$

Now $\dfrac{dy}{dx} = u'v + uv'$ {product rule}

$\therefore \quad \dfrac{dy}{dx} = \frac{1}{2}x^{-\frac{1}{2}}(x-3)^2 + 2\sqrt{x}(x-3)$

**e** $y = 5x^2(3x^2-1)^2$ is the product of $u = 5x^2$ and $v = (3x^2-1)^2$

$\therefore\ u' = 10x$ and $v' = 2(3x^2-1)^1(6x) = 12x(3x^2-1)$

Now $\dfrac{dy}{dx} = u'v + uv'$ {product rule}

$$\therefore\ \frac{dy}{dx} = 10x(3x^2-1)^2 + 5x^2(12x)(3x^2-1) = 10x(3x^2-1)^2 + 60x^3(3x^2-1)$$

**f** $y = \sqrt{x}(x-x^2)^3$ is the product of $u = x^{\frac{1}{2}}$ and $v = (x-x^2)^3$

$\therefore\ u' = \frac{1}{2}x^{-\frac{1}{2}}$ and $v' = 3(x-x^2)^2(1-2x)$

Now $\dfrac{dy}{dx} = u'v + uv'$ {product rule}

$$\therefore\ \frac{dy}{dx} = \tfrac{1}{2}x^{-\frac{1}{2}}(x-x^2)^3 + 3\sqrt{x}(x-x^2)^2(1-2x)$$

**3** **a** $y = x^4(1-2x)^2$ is the product of $u = x^4$ and $v = (1-2x)^2$

$\therefore\ u' = 4x^3$ and $v' = 2(1-2x)^1(-2) = -4(1-2x)$

Now $\dfrac{dy}{dx} = u'v + uv'$ {product rule}

$$\therefore\ \frac{dy}{dx} = 4x^3(1-2x)^2 - 4x^4(1-2x)$$

At $x=-1$, $\dfrac{dy}{dx} = 4(-1)^3(3)^2 - 4(-1)^4(3) = -48$ $\quad\therefore$ gradient of tangent $= -48$

**b** $y = \sqrt{x}(x^2-x+1)^2$ is the product of $u = x^{\frac{1}{2}}$ and $v = (x^2-x+1)^2$

$\therefore\ u' = \frac{1}{2}x^{-\frac{1}{2}}$ and $v' = 2(x^2-x+1)(2x-1)$

Now $\dfrac{dy}{dx} = u'v + uv'$ {product rule}

$$\therefore\ \frac{dy}{dx} = \tfrac{1}{2}x^{-\frac{1}{2}}(x^2-x+1)^2 + 2\sqrt{x}(x^2-x+1)(2x-1)$$

At $x=4$, $\dfrac{dy}{dx} = \frac{1}{2}(4)^{-\frac{1}{2}}(13)^2 + 2\sqrt{4}(13)(7) = 406\frac{1}{4}$ $\quad\therefore$ gradient of tangent $= 406\frac{1}{4}$

**c** $y = x\sqrt{1-2x}$ is the product of $u = x$ and $v = (1-2x)^{\frac{1}{2}}$

$\therefore\ u' = 1$ and $v' = \frac{1}{2}(1-2x)^{-\frac{1}{2}}(-2) = -(1-2x)^{-\frac{1}{2}}$

Now $\dfrac{dy}{dx} = u'v + uv'$ {product rule}

$$\therefore\ \frac{dy}{dx} = \sqrt{1-2x} - \frac{x}{\sqrt{1-2x}}$$

At $x=-4$, $\dfrac{dy}{dx} = \sqrt{9} - \dfrac{(-4)}{\sqrt{9}} = 3 + \frac{4}{3} = \frac{13}{3}$ $\quad\therefore$ gradient of tangent $= \frac{13}{3}$

**d** $y = x^3\sqrt{5-x^2}$ is the product of $u = x^3$ and $v = (5-x^2)^{\frac{1}{2}}$

$\therefore\ u' = 3x^2$ and $v' = \frac{1}{2}(5-x^2)^{-\frac{1}{2}}(-2x) = -x(5-x^2)^{-\frac{1}{2}}$

Now $\dfrac{dy}{dx} = u'v + uv'$ {product rule}

$$\therefore\ \frac{dy}{dx} = 3x^2\sqrt{5-x^2} - \frac{x^4}{\sqrt{5-x^2}}$$

At $x=1$, $\dfrac{dy}{dx} = 3(1)^2\sqrt{4} - \dfrac{1}{\sqrt{4}} = 6 - \frac{1}{2} = \frac{11}{2}$ $\quad\therefore$ gradient of tangent $= \frac{11}{2}$

**4** **a** $y=\sqrt{x}(3-x)^2$ is the product of

$u=x^{\frac{1}{2}}$ and $v=(3-x)^2$

$\therefore\ u'=\frac{1}{2}x^{-\frac{1}{2}}$ and $v'=2(3-x)^1(-1)$
$=-2(3-x)$

Now $\frac{dy}{dx}=u'v+uv'$ {product rule}

$$\begin{aligned}\therefore\ \frac{dy}{dx}&=\frac{1}{2\sqrt{x}}(3-x)^2-2\sqrt{x}(3-x)\\&=\frac{(3-x)^2-(2\sqrt{x})(2\sqrt{x})(3-x)}{2\sqrt{x}}\\&=\frac{(3-x)\left[(3-x)-4x\right]}{2\sqrt{x}}\\&=\frac{(3-x)(3-5x)}{2\sqrt{x}}\quad\text{as required}\end{aligned}$$

**b** Tangents are horizontal when their gradients are 0.

$\frac{dy}{dx}=0$ when $(3-x)(3-5x)=0$

$\therefore\ 3-x=0$ or $3-5x=0$

$\therefore\ x=3$ or $x=\frac{3}{5}$

**c** $\frac{dy}{dx}$ is undefined when $2\sqrt{x}\leqslant 0$

$\therefore\ \sqrt{x}\leqslant 0$

$\therefore\ x\leqslant 0$

**d** $\frac{dy}{dx}$ is undefined when $x\leqslant 0$ {from **c**}

But when $x=0$, $y=\sqrt{0}(3-0)^2$
$=0$

So, when $x=0$, $y$ is defined but $\frac{dy}{dx}$ is not.

**e** As we approach the point $x=0$ from the right, the curve has steeper and steeper gradient and approaches vertical.

**5** $y=-2x^2(x+4)$ is the product of $u=-2x^2$ and $v=x+4$

$\therefore\ u'=-4x$ and $v'=1$

Now $\frac{dy}{dx}=u'v+uv'$ {product rule}

$$\begin{aligned}&=-4x(x+4)-2x^2\times 1\\&=-4x^2-16x-2x^2\\&=-6x^2-16x\end{aligned}$$

If $\frac{dy}{dx}=10$, $-6x^2-16x=10$

$\therefore\ 6x^2+16x+10=0$

$\therefore\ 3x^2+8x+5=0$

$\therefore\ (3x+5)(x+1)=0$

$\therefore\ x=-\frac{5}{3}$ and $x=-1$

## EXERCISE 15D

**1** **a** $y=\frac{1+3x}{2-x}$ is a quotient where

$u=1+3x$ and $v=2-x$

$\therefore\ u'=3$ and $v'=-1$

Now $\frac{dy}{dx}=\frac{u'v-uv'}{v^2}$ {quotient rule}

$$\begin{aligned}\therefore\ \frac{dy}{dx}&=\frac{3(2-x)-(1+3x)(-1)}{(2-x)^2}\\&=\frac{7}{(2-x)^2}\end{aligned}$$

**b** $y=\frac{x^2}{2x+1}$ is a quotient where

$u=x^2$ and $v=2x+1$

$\therefore\ u'=2x$ and $v'=2$

Now $\frac{dy}{dx}=\frac{u'v-uv'}{v^2}$ {quotient rule}

$$\begin{aligned}\therefore\ \frac{dy}{dx}&=\frac{2x(2x+1)-x^2(2)}{(2x+1)^2}\\&=\frac{2x(2x+1)-2x^2}{(2x+1)^2}\end{aligned}$$

**c** $y=\frac{x}{x^2-3}$ is a quotient where

$u=x$ and $v=x^2-3$

$\therefore\ u'=1$ and $v'=2x$

Now $\frac{dy}{dx}=\frac{u'v-uv'}{v^2}$ {quotient rule}

$$\begin{aligned}\therefore\ \frac{dy}{dx}&=\frac{1(x^2-3)-x(2x)}{(x^2-3)^2}\\&=\frac{(x^2-3)-2x^2}{(x^2-3)^2}\end{aligned}$$

**d** $y=\frac{\sqrt{x}}{1-2x}$ is a quotient where

$u=x^{\frac{1}{2}}$ and $v=1-2x$

$\therefore\ u'=\frac{1}{2}x^{-\frac{1}{2}}$ and $v'=-2$

Now $\frac{dy}{dx}=\frac{u'v-uv'}{v^2}$ {quotient rule}

$$\begin{aligned}\therefore\ \frac{dy}{dx}&=\frac{\frac{1}{2}x^{-\frac{1}{2}}(1-2x)-\sqrt{x}(-2)}{(1-2x)^2}\\&=\frac{\frac{1}{2}x^{-\frac{1}{2}}(1-2x)+2\sqrt{x}}{(1-2x)^2}\end{aligned}$$

**e** $y = \dfrac{x^2-3}{3x-x^2}$ is a quotient where $u = x^2-3$ and $v = 3x - x^2$
$\therefore\ u' = 2x$ and $v' = 3-2x$

Now $\dfrac{dy}{dx} = \dfrac{u'v-uv'}{v^2}$ {quotient rule}

$\therefore\ \dfrac{dy}{dx} = \dfrac{2x(3x-x^2)-(x^2-3)(3-2x)}{(3x-x^2)^2}$

**f** $y = \dfrac{x}{\sqrt{1-3x}}$ is a quotient where $u = x$ and $v = (1-3x)^{\frac{1}{2}}$
$\therefore\ u' = 1$ and $v' = -\frac{3}{2}(1-3x)^{-\frac{1}{2}}$

Now $\dfrac{dy}{dx} = \dfrac{u'v-uv'}{v^2}$ {quotient rule}

$\therefore\ \dfrac{dy}{dx} = \dfrac{(1-3x)^{\frac{1}{2}} - x\left(-\frac{3}{2}(1-3x)^{-\frac{1}{2}}\right)}{1-3x}$

$= \dfrac{(1-3x)^{\frac{1}{2}} + \frac{3}{2}x(1-3x)^{-\frac{1}{2}}}{1-3x}$

**2** **a** $y = \dfrac{x}{1-2x}$ is a quotient where
$u = x$ and $v = 1-2x$
$\therefore\ u' = 1$ and $v' = -2$

Now $\dfrac{dy}{dx} = \dfrac{u'v-uv'}{v^2}$ {quotient rule}

$\therefore\ \dfrac{dy}{dx} = \dfrac{1(1-2x)-x(-2)}{(1-2x)^2}$

$= \dfrac{1}{(1-2x)^2}$

At $x=1$, $\dfrac{dy}{dx} = \dfrac{1}{(1-2)^2} = \dfrac{1}{(-1)^2} = 1$

$\therefore$ the gradient of the tangent $= 1$

**b** $y = \dfrac{x^3}{x^2+1}$ is a quotient where
$u = x^3$ and $v = x^2+1$
$\therefore\ u' = 3x^2$ and $v' = 2x$

Now $\dfrac{dy}{dx} = \dfrac{u'v-uv'}{v^2}$ {quotient rule}

$\therefore\ \dfrac{dy}{dx} = \dfrac{3x^2(x^2+1)-x^3(2x)}{(x^2+1)^2}$

$= \dfrac{x^4+3x^2}{(x^2+1)^2}$

At $x=-1$, $\dfrac{dy}{dx} = \dfrac{1+3}{(1+1)^2} = \dfrac{4}{4} = 1$

$\therefore$ the gradient of the tangent $= 1$

**c** $y = \dfrac{\sqrt{x}}{2x+1}$ is a quotient where
$u = x^{\frac{1}{2}}$ and $v = 2x+1$
$\therefore\ u' = \frac{1}{2}x^{-\frac{1}{2}}$ and $v' = 2$

Now $\dfrac{dy}{dx} = \dfrac{u'v-uv'}{v^2}$ {quotient rule}

$= \dfrac{\frac{1}{2\sqrt{x}}(2x+1)-\sqrt{x}(2)}{(2x+1)^2}$

At $x=4$, $\dfrac{dy}{dx} = \dfrac{\frac{9}{4}-4}{81} = \dfrac{\left(\frac{9}{4}-4\right)}{81} \times \dfrac{4}{4}$

$= \dfrac{9-16}{324}$

$\therefore$ the gradient of the tangent $= -\frac{7}{324}$

**d** $y = \dfrac{x^2}{\sqrt{x^2+5}}$ is a quotient where
$u = x^2$ and $v = (x^2+5)^{\frac{1}{2}}$
$\therefore\ u' = 2x$ and $v' = \frac{1}{2}(x^2+5)^{-\frac{1}{2}}(2x)$
$= x(x^2+5)^{-\frac{1}{2}}$

Now $\dfrac{dy}{dx} = \dfrac{u'v-uv'}{v^2}$ {quotient rule}

$= \dfrac{2x\sqrt{x^2+5} - x^2\left(\dfrac{x}{\sqrt{x^2+5}}\right)}{(x^2+5)}$

At $x=-2$, $\dfrac{dy}{dx} = \dfrac{-4(3)-4\left(\frac{-2}{3}\right)}{9}$

$= \dfrac{\left(-12+\frac{8}{3}\right)}{9} \times \dfrac{3}{3}$

$= \dfrac{-36+8}{27}$

$\therefore$ the gradient of the tangent $= -\frac{28}{27}$

**3** **a** $y = \dfrac{2\sqrt{x}}{1-x}$ is a quotient where $u = 2x^{\frac{1}{2}}$ and $v = 1 - x$

$\therefore\ u' = x^{-\frac{1}{2}}$ and $v' = -1$

Now $\dfrac{dy}{dx} = \dfrac{u'v - uv'}{v^2}$ {quotient rule}

$$\therefore\ \frac{dy}{dx} = \frac{\frac{1}{\sqrt{x}}(1-x) - 2\sqrt{x}(-1)}{(1-x)^2} \times \left(\frac{\sqrt{x}}{\sqrt{x}}\right)$$

$$= \frac{(1-x) + 2x}{\sqrt{x}(1-x)^2}$$

$$= \frac{x+1}{\sqrt{x}(1-x)^2} \quad \text{as required}$$

**b** **i** $\dfrac{dy}{dx} = 0$ when $x + 1 = 0$ $\therefore\ x = -1.$

However $\dfrac{dy}{dx}$ is not defined for $x \leqslant 0$ because of the $\sqrt{x}$ term. Hence $\dfrac{dy}{dx}$ never equals 0.

**ii** $\dfrac{dy}{dx}$ is undefined when $x \leqslant 0$ and when $x = 1$.

**4** **a** $y = \dfrac{x^2 - 3x + 1}{x+2}$ is a quotient where $u = x^2 - 3x + 1$ and $v = x + 2$

$\therefore\ u' = 2x - 3$ and $v' = 1$

Now $\dfrac{dy}{dx} = \dfrac{u'v - uv'}{v^2}$ {quotient rule}

$$\therefore\ \frac{dy}{dx} = \frac{(2x-3)(x+2) - (x^2 - 3x + 1)(1)}{(x+2)^2}$$

$$= \frac{2x^2 + 4x - 3x - 6 - x^2 + 3x - 1}{(x+2)^2}$$

$$= \frac{x^2 + 4x - 7}{(x+2)^2} \quad \text{as required}$$

**b** **i** $\dfrac{dy}{dx} = 0$ when $x^2 + 4x - 7 = 0$ $\therefore\ x = \dfrac{-4 \pm \sqrt{44}}{2} = -2 \pm \sqrt{11}$

**ii** $\dfrac{dy}{dx}$ is undefined when $(x+2)^2 = 0$

$\therefore\ x = -2$

**c** $\dfrac{dy}{dx}$ is zero when the tangent to the function is horizontal. This occurs at the function's turning points or points of horizontal inflection.

$\dfrac{dy}{dx}$ is undefined at vertical asymptotes of the function.

## EXERCISE 15E

**1** **a** $f(x) = e^{4x}$

$\therefore\ f'(x) = 4e^{4x}$

**b** $f(x) = e^x + 3$

$\therefore\ f'(x) = e^x + 0$

$= e^x$

**c** $f(x) = \exp(-2x)$

$= e^{-2x}$

$\therefore\ f'(x) = -2e^{-2x}$

**d** $f(x) = e^{\frac{x}{2}}$

$\therefore\ f'(x) = \frac{1}{2}e^{\frac{x}{2}}$

**e** $f(x) = 2e^{-\frac{x}{2}}$

$\therefore\ f'(x) = 2e^{-\frac{x}{2}}\left(-\frac{1}{2}\right)$

$= -e^{-\frac{x}{2}}$

**f** $f(x) = 1 - 2e^{-x}$

$\therefore\ f'(x) = 0 - 2e^{-x}(-1)$

$= 2e^{-x}$

**g** $f(x) = 4e^{\frac{x}{2}} - 3e^{-x}$

$\therefore\ f'(x) = 4e^{\frac{x}{2}}\left(\frac{1}{2}\right) - 3e^{-x}(-1)$

$= 2e^{\frac{x}{2}} + 3e^{-x}$

**h** $f(x) = \dfrac{e^x + e^{-x}}{2} = \frac{1}{2}(e^x + e^{-x})$

$\therefore\ f'(x) = \frac{1}{2}(e^x + e^{-x}(-1))$

$= \dfrac{e^x - e^{-x}}{2}$

**i** $f(x) = e^{-x^2}$

$\therefore\ f'(x) = e^{-x^2}(-2x)$

$= -2xe^{-x^2}$

**j** $f(x) = e^{\frac{1}{x}}$

$\therefore\ f'(x) = e^{\frac{1}{x}}\left(-\dfrac{1}{x^2}\right)$

**k** $f(x) = 10\left(1 + e^{2x}\right)$

$= 10 + 10e^{2x}$

$\therefore\ f'(x) = 0 + 10e^{2x}\,(2)$

$= 20e^{2x}$

**l** $f(x) = 20\left(1 - e^{-2x}\right)$

$= 20 - 20e^{-2x}$

$\therefore\ f'(x) = 0 - 20e^{-2x}\,(-2)$

$= 40e^{-2x}$

**m** $f(x) = e^{2x+1}$

$\therefore\ f'(x) = e^{2x+1}(2)$

$= 2e^{2x+1}$

**n** $f(x) = e^{\frac{x}{4}}$

$\therefore\ f'(x) = e^{\frac{x}{4}}\left(\frac{1}{4}\right)$

$= \frac{1}{4}e^{\frac{x}{4}}$

**o** $f(x) = e^{1-2x^2}$

$\therefore\ f'(x) = e^{1-2x^2}(-4x)$

$= -4xe^{1-2x^2}$

**p** $f(x) = e^{-0.02x}$

$\therefore\ f'(x) = e^{-0.02x} \times (-0.02)$

$= -0.02e^{-0.02x}$

**2** **a** $f(x) = xe^x$

$\therefore\ f'(x) = 1e^x + xe^x$ {product rule}

$= e^x + xe^x$

**b** $f(x) = x^3e^{-x}$

$\therefore\ f'(x) = 3x^2e^{-x} + x^3(-e^{-x})$

{product rule}

$= 3x^2e^{-x} - x^3e^{-x}$

**c** $f(x) = \dfrac{e^x}{x}$

$\therefore\ f'(x) = \dfrac{e^x x - e^x\,(1)}{x^2}$ {quotient rule}

$= \dfrac{xe^x - e^x}{x^2}$

**d** $f(x) = \dfrac{x}{e^x}$

$\therefore\ f'(x) = \dfrac{1e^x - xe^x}{(e^x)^2}$ {quotient rule}

$= \dfrac{e^x(1-x)}{(e^x)^2} = \dfrac{1-x}{e^x}$

**e** $f(x) = x^2e^{3x}$

$\therefore\ f'(x) = 2xe^{3x} + 3x^2e^{3x}$ {product rule}

**f** $f(x) = \dfrac{e^x}{\sqrt{x}}$

$\therefore\ f'(x) = \dfrac{e^x\sqrt{x} - \dfrac{e^x}{2\sqrt{x}}}{(\sqrt{x})^2}$ {quotient rule}

$= \dfrac{xe^x - \frac{1}{2}e^x}{x\sqrt{x}}$

**g** $f(x) = \sqrt{x}e^{-x}$

$\therefore\ f'(x) = \frac{1}{2}x^{-\frac{1}{2}}e^{-x} - x^{\frac{1}{2}}e^{-x}$

{product rule}

**h** $f(x) = \dfrac{e^x + 2}{e^{-x} + 1}$

$\therefore\ f'(x) = \dfrac{e^x(e^{-x} + 1) - (e^x + 2)\left(-e^{-x}\right)}{(e^{-x} + 1)^2}$

{quotient rule}

$= \dfrac{1 + e^x + 1 + 2e^{-x}}{(e^{-x} + 1)^2}$

$= \dfrac{e^x + 2 + 2e^{-x}}{(e^{-x} + 1)^2}$

**3** **a**
$$\begin{aligned} f(x) &= (e^x+2)^4 \\ &= u^4 \text{ where } u = e^x+2 \\ \frac{dy}{dx} &= \frac{dy}{du}\frac{du}{dx} \quad \{\text{chain rule}\} \\ &= 4u^3(e^x) \\ \therefore\ f'(x) &= 4(e^x+2)^3(e^x) \\ &= 4e^x(e^x+2)^3 \\ \therefore\ f'(0) &= 4\left(e^0+2\right)^3\left(e^0\right) \\ &= 108 \end{aligned}$$
$\therefore$ gradient of tangent $= 108$

**b**
$$\begin{aligned} f(x) &= \frac{1}{2-e^{-x}} \\ &= u^{-1} \text{ where } u = 2-e^{-x} \\ \frac{dy}{dx} &= \frac{dy}{du}\frac{du}{dx} \quad \{\text{chain rule}\} \\ &= -u^{-2}(e^{-x}) \\ \therefore\ f'(x) &= -\frac{e^{-x}}{(2-e^{-x})^2} \\ \therefore\ f'(0) &= -\frac{e^0}{(2-e^0)^2} \\ &= -1 \end{aligned}$$
$\therefore$ gradient of tangent $= -1$

**c**
$$\begin{aligned} f(x) &= \sqrt{e^{2x}+10} \\ &= u^{\frac{1}{2}} \text{ where } u = e^{2x}+10 \\ \frac{dy}{dx} &= \frac{dy}{du}\frac{du}{dx} \quad \{\text{chain rule}\} \\ &= \tfrac{1}{2}u^{-\frac{1}{2}}(2e^{2x}) \\ \therefore\ f'(x) &= \frac{e^{2x}}{\sqrt{e^{2x}+10}} \\ \therefore\ f'(\ln 3) &= \frac{e^{2\ln 3}}{\sqrt{e^{2\ln 3}+10}} = \frac{9}{\sqrt{19}} \end{aligned}$$
$\therefore$ gradient of tangent $= \frac{9}{\sqrt{19}}$

**4** $f(x) = e^{kx}+x \quad \therefore\ f'(x) = ke^{kx}+1$

Now $f'(0) = -8$, so $ke^0+1 = -8$
$$\therefore\ k \times 1 = -9$$
$$\therefore\ k = -9$$

**5** **a**
$$\begin{aligned} y &= 2^x \\ &= (e^{\ln 2})^x \\ &= e^{x\ln 2} \\ \therefore\ \frac{dy}{dx} &= e^{x\ln 2} \times \ln 2 \\ &= 2^x \ln 2 \end{aligned}$$

**b**
$$\begin{aligned} y &= b^x \\ &= (e^{\ln b})^x \\ &= e^{x\ln b} \\ \therefore\ \frac{dy}{dx} &= e^{x\ln b} \times \ln b \\ &= b^x \times \ln b \end{aligned}$$

**6** $f(x) = x^2e^{-x}$ is the product of $u = x^2$ and $v = e^{-x}$
$\therefore\ u' = 2x$ and $v' = -e^{-x}$

$$\begin{aligned} \therefore\ f'(x) &= u'v+uv' \\ &= 2x\left(e^{-x}\right)+x^2\left(-e^{-x}\right) \\ &= 2xe^{-x}-x^2e^{-x} \end{aligned}$$

Now $f'(x) = 0$ when $2xe^{-x}-x^2e^{-x} = 0$
$$\therefore\ xe^{-x}(2-x) = 0$$
$$\therefore\ x = 0 \text{ or } 2-x = 0 \quad \{e^{-x} > 0 \text{ for all } x\}$$
$$\therefore\ x = 0 \text{ or } x = 2$$

$f(0) = 0^2e^0$ and $f(2) = 2^2e^{-2}$
$= 0 \qquad\qquad = 4e^{-2} \left(\text{or } \frac{4}{e^2}\right)$

So, the possible coordinates of P are $(0, 0)$ and $\left(2, \frac{4}{e^2}\right)$.

## EXERCISE 15F

**1** **a** $y = \ln(7x)$
$\therefore\ y = \ln 7 + \ln x$
$\therefore\ \dfrac{dy}{dx} = 0 + \dfrac{1}{x} = \dfrac{1}{x}$

*or*

$y = \ln(7x)$
$\therefore\ \dfrac{dy}{dx} = \dfrac{7}{7x} \begin{matrix} \leftarrow f'(x) \\ \leftarrow f(x) \end{matrix}$
$= \dfrac{1}{x}$

**b** $y = \ln(2x+1)$
$\therefore\ \dfrac{dy}{dx} = \dfrac{2}{2x+1} \begin{matrix} \leftarrow f'(x) \\ \leftarrow f(x) \end{matrix}$

**c** $y = \ln(x - x^2)$
$\therefore\ \dfrac{dy}{dx} = \dfrac{1-2x}{x-x^2} \begin{matrix} \leftarrow f'(x) \\ \leftarrow f(x) \end{matrix}$

**d** $y = 3 - 2\ln x$
$\therefore\ \dfrac{dy}{dx} = 0 - 2\left(\dfrac{1}{x}\right)$
$= -\dfrac{2}{x}$

**e** $y = x^2 \ln x$
$\therefore\ \dfrac{dy}{dx} = 2x\ln x + x^2\left(\dfrac{1}{x}\right)$
$= 2x\ln x + x$

**f** $y = \dfrac{\ln x}{2x}$
$\therefore\ \dfrac{dy}{dx} = \dfrac{\left(\dfrac{1}{x}\right)2x - \ln x \times 2}{(2x)^2}$
$= \dfrac{2 - 2\ln x}{4x^2}$
$= \dfrac{1 - \ln x}{2x^2}$

**g** $y = e^x \ln x$
$\therefore\ \dfrac{dy}{dx} = e^x \ln x + \dfrac{e^x}{x}$

**h** $y = (\ln x)^2$
$\therefore\ \dfrac{dy}{dx} = 2(\ln x)^1\left(\dfrac{1}{x}\right)$
$= \dfrac{2\ln x}{x}$

**i** $y = \sqrt{\ln x} = (\ln x)^{\frac{1}{2}}$
$\therefore\ \dfrac{dy}{dx} = \frac{1}{2}(\ln x)^{-\frac{1}{2}}\left(\dfrac{1}{x}\right)$
$= \dfrac{1}{2x\sqrt{\ln x}}$

**j** $y = e^{-x}\ln x$
$\therefore\ \dfrac{dy}{dx} = -e^{-x}\ln x + e^{-x}\left(\dfrac{1}{x}\right)$
$= \dfrac{e^{-x}}{x} - e^{-x}\ln x$

**k** $y = \sqrt{x}\ln(2x)$
$\therefore\ \dfrac{dy}{dx} = \dfrac{1}{2\sqrt{x}}\ln(2x) + \sqrt{x}\left(\dfrac{1}{x}\right)$
$= \dfrac{\ln(2x)}{2\sqrt{x}} + \dfrac{1}{\sqrt{x}}$

**l** $y = \dfrac{2\sqrt{x}}{\ln x}$
$\therefore\ \dfrac{dy}{dx} = \dfrac{\dfrac{1}{\sqrt{x}}\ln x - 2\sqrt{x}\left(\dfrac{1}{x}\right)}{(\ln x)^2}$
$= \dfrac{\dfrac{1}{\sqrt{x}}\ln x - \dfrac{2}{\sqrt{x}}}{(\ln x)^2}$
$= \dfrac{\ln x - 2}{\sqrt{x}(\ln x)^2}$

**m** $y = 3 - 4\ln(1-x)$
$\therefore\ \dfrac{dy}{dx} = -\dfrac{4}{1-x} \times -1$
$= \dfrac{4}{1-x}$

**n** $y = x\ln(x^2+1)$
$\therefore\ \dfrac{dy}{dx} = \ln(x^2+1) + x\,\dfrac{2x}{x^2+1}$
$= \ln(x^2+1) + \dfrac{2x^2}{x^2+1}$

**2** **a** $y = x\ln 5$

$\therefore \dfrac{dy}{dx} = \ln 5$

**b** $y = \ln(x^3) = 3\ln x$

$\therefore \dfrac{dy}{dx} = 3\left(\dfrac{1}{x}\right) = \dfrac{3}{x}$

**c** $y = \ln(x^4 + x)$

$\therefore \dfrac{dy}{dx} = \dfrac{4x^3+1}{x^4+x}$

**d** $y = \ln(10 - 5x)$

$\therefore \dfrac{dy}{dx} = \dfrac{-5}{10-5x} = \dfrac{1}{x-2}$

**e** $y = [\ln(2x+1)]^3$

$\therefore \dfrac{dy}{dx} = 3[\ln(2x+1)]^2 \times \dfrac{2}{2x+1}$

$= \dfrac{6}{2x+1}[\ln(2x+1)]^2$

**f** $y = \dfrac{\ln(4x)}{x}$

$\therefore \dfrac{dy}{dx} = \dfrac{\left(\frac{4}{4x}\right)x - \ln(4x) \times 1}{x^2}$

$= \dfrac{1-\ln(4x)}{x^2}$

**g** $y = \ln\left(\dfrac{1}{x}\right)$

$= -\ln x$

$\therefore \dfrac{dy}{dx} = -\dfrac{1}{x}$

**h** $y = \ln(\ln x)$

$\therefore \dfrac{dy}{dx} = \dfrac{\frac{1}{x}}{\ln x}$

$= \dfrac{1}{x\ln x}$

**i** $y = \dfrac{1}{\ln x} = (\ln x)^{-1}$

$\therefore \dfrac{dy}{dx} = -1(\ln x)^{-2} \times \dfrac{1}{x}$

$= \dfrac{-1}{x(\ln x)^2}$

**3** **a** $y = \ln\sqrt{1-2x}$

$= \ln(1-2x)^{\frac{1}{2}}$

$= \frac{1}{2}\ln(1-2x)$

$\therefore \dfrac{dy}{dx} = \frac{1}{2} \times \dfrac{-2}{1-2x}$

$= \dfrac{-1}{1-2x}$

**b** $y = \ln\left(\dfrac{1}{2x+3}\right)$

$= -\ln(2x+3)$

$\therefore \dfrac{dy}{dx} = -\dfrac{2}{2x+3}$

**c** $y = \ln(e^x\sqrt{x})$

$= \ln e^x + \ln x^{\frac{1}{2}}$

$= \ln e^x + \frac{1}{2}\ln x$

$= x + \frac{1}{2}\ln x$

$\therefore \dfrac{dy}{dx} = 1 + \frac{1}{2}\left(\dfrac{1}{x}\right)$

$= 1 + \dfrac{1}{2x}$

**d** $y = \ln(x\sqrt{2-x})$

$= \ln x + \ln(2-x)^{\frac{1}{2}}$

$= \ln x + \frac{1}{2}\ln(2-x)$

$\therefore \dfrac{dy}{dx} = \dfrac{1}{x} + \frac{1}{2}\left(\dfrac{-1}{2-x}\right)$

$= \dfrac{1}{x} - \dfrac{1}{2(2-x)}$

**e** $y = \ln\left(\dfrac{x+3}{x-1}\right)$

$= \ln(x+3) - \ln(x-1)$

$\therefore \dfrac{dy}{dx} = \dfrac{1}{x+3} - \dfrac{1}{x-1}$

**f** $y = \ln\left(\dfrac{x^2}{3-x}\right)$

$= \ln x^2 - \ln(3-x)$

$= 2\ln x - \ln(3-x)$

$\therefore \dfrac{dy}{dx} = \dfrac{2}{x} - \dfrac{-1}{3-x} = \dfrac{2}{x} + \dfrac{1}{3-x}$

**g** $f(x) = \ln\left((3x-4)^3\right)$

$= 3\ln(3x-4)$

$\therefore f'(x) = 3 \times \dfrac{3}{3x-4} = \dfrac{9}{3x-4}$

**h** $f(x) = \ln\left(x(x^2+1)\right)$

$= \ln x + \ln(x^2+1)$

$\therefore f'(x) = \dfrac{1}{x} + \dfrac{2x}{x^2+1}$

**i** $f(x) = \ln\left(\dfrac{x^2+2x}{x-5}\right)$

$= \ln(x^2+2x) - \ln(x-5)$

$\therefore f'(x) = \dfrac{2x+2}{x^2+2x} - \dfrac{1}{x-5}$

**4** $y = x\ln x$ is the product of $u = x$ and $v = \ln x$

$\therefore\ u' = 1$ and $v' = \dfrac{1}{x}$

$$\text{Now}\quad \frac{dy}{dx} = u'v + uv' \quad \{\text{product rule}\}$$
$$= 1\ln x + x \times \frac{1}{x}$$
$$= \ln x + 1$$

At $x = e$, $\dfrac{dy}{dx} = \ln e + 1 = 1 + 1 = 2$

$\therefore$ gradient of tangent $= 2$

**5** $f(x) = a\ln(2x+b)$

Now $f(e) = 3$, $\therefore\ 3 = a\ln(2e+b)$

$$\therefore\ a = \frac{3}{\ln(2e+b)}$$

$$f'(x) = a \times \frac{2}{2x+b}$$

Now $f'(e) = \dfrac{6}{e}$ $\therefore\ \dfrac{6}{e} = \dfrac{2a}{2e+b}$

$$\therefore\ 6(2e+b) = 2ae$$
$$\therefore\ 3(2e+b) = ae$$
$$\therefore\ a = \frac{3(2e+b)}{e}$$
$$= \frac{6e+3b}{e}$$
$$= 6 + \frac{3b}{e}$$

Using technology we get $a = 3$, $b = -e$.

## EXERCISE 15G

**1** **a** $y = \sin(2x)$

$$\therefore\ \frac{dy}{dx} = \cos(2x)\,\frac{d}{dx}(2x)$$
$$= 2\cos(2x)$$

**b** $y = \sin x + \cos x$

$$\therefore\ \frac{dy}{dx} = \cos x - \sin x$$

**c** $y = \cos(3x) - \sin x$

$$\therefore\ \frac{dy}{dx} = -\sin(3x) \times 3 - \cos x$$
$$= -3\sin(3x) - \cos x$$

**d** $y = \sin(x+1)$

$$\therefore\ \frac{dy}{dx} = \cos(x+1)\,\frac{d}{dx}(x+1)$$
$$= 1\cos(x+1)$$
$$= \cos(x+1)$$

**e** $y = \cos(3-2x)$

$$\therefore\ \frac{dy}{dx} = -\sin(3-2x) \times -2$$
$$= 2\sin(3-2x)$$

**f** $y = \tan(5x)$

$$\therefore\ \frac{dy}{dx} = \frac{1}{\cos^2(5x)} \times 5$$
$$= \frac{5}{\cos^2(5x)}$$

**g** $y = \sin\left(\frac{x}{2}\right) - 3\cos x$

$$\therefore\ \frac{dy}{dx} = \tfrac{1}{2}\cos\left(\tfrac{x}{2}\right) + 3\sin x$$

**h** $y = 3\tan(\pi x)$

$$\therefore\ \frac{dy}{dx} = 3 \times \frac{1}{\cos^2(\pi x)} \times \pi$$
$$= \frac{3\pi}{\cos^2(\pi x)}$$

**i** $y = 4\sin x - \cos(2x)$

$$\therefore\ \frac{dy}{dx} = 4\cos x + \sin(2x) \times 2$$
$$= 4\cos x + 2\sin(2x)$$

**2** **a** $y = x^2 + \cos x$

$\therefore \dfrac{dy}{dx} = 2x - \sin x$

**b** $y = \tan x - 3\sin x$

$\therefore \dfrac{dy}{dx} = \dfrac{1}{\cos^2 x} - 3\cos x$

**c** $y = e^x \cos x$

$\therefore \dfrac{dy}{dx} = e^x \cos x + e^x(-\sin x)$

$= e^x \cos x - e^x \sin x$

**d** $y = e^{-x} \sin x$

$\therefore \dfrac{dy}{dx} = -e^{-x} \sin x + e^{-x} \cos x$

**e** $y = \ln(\sin x)$

$\therefore \dfrac{dy}{dx} = \dfrac{\cos x}{\sin x}$

**f** $y = e^{2x} \tan x$

$\therefore \dfrac{dy}{dx} = 2e^{2x} \tan x + \dfrac{e^{2x}}{\cos^2 x}$

**g** $y = \sin(3x)$

$\therefore \dfrac{dy}{dx} = 3\cos(3x)$

**h** $y = \cos(\frac{x}{2})$

$\therefore \dfrac{dy}{dx} = -\frac{1}{2}\sin(\frac{x}{2})$

**i** $y = 3\tan(2x)$

$\therefore \dfrac{dy}{dx} = \dfrac{3}{\cos^2(2x)} \times 2$

$= \dfrac{6}{\cos^2(2x)}$

**j** $y = x\cos x$

$\therefore \dfrac{dy}{dx} = 1 \times \cos x + x(-\sin x)$

$= \cos x - x\sin x$

**k** $y = \dfrac{\sin x}{x}$

$\therefore \dfrac{dy}{dx} = \dfrac{(\cos x)(x) - \sin x \times 1}{x^2}$

$= \dfrac{x\cos x - \sin x}{x^2}$

**l** $y = x\tan x$

$\therefore \dfrac{dy}{dx} = 1 \times \tan x + x \times \dfrac{1}{\cos^2 x}$

$= \tan x + \dfrac{x}{\cos^2 x}$

**3** **a** $y = \sin(x^2)$

$\therefore \dfrac{dy}{dx} = 2x\cos(x^2)$

**b** $y = \cos\left(\sqrt{x}\right) = \cos(x^{\frac{1}{2}})$

$\therefore \dfrac{dy}{dx} = -\sin(x^{\frac{1}{2}}) \times \frac{1}{2}x^{-\frac{1}{2}}$

$= -\dfrac{1}{2\sqrt{x}}\sin(\sqrt{x})$

**c** $y = \sqrt{\cos x} = (\cos x)^{\frac{1}{2}}$

$\therefore \dfrac{dy}{dx} = \frac{1}{2}(\cos x)^{-\frac{1}{2}} \times (-\sin x)$

$= -\dfrac{\sin x}{2\sqrt{\cos x}}$

**d** $y = \sin^2 x = (\sin x)^2$

$\therefore \dfrac{dy}{dx} = 2\sin x\cos x$

**e** $y = \cos^3 x = (\cos x)^3$

$\therefore \dfrac{dy}{dx} = 3\cos^2 x \times (-\sin x)$

$= -3\sin x\cos^2 x$

**f** $y = \cos x\sin(2x)$

$\therefore \dfrac{dy}{dx} = (-\sin x)\sin(2x) + \cos x(2\cos(2x))$

$= -\sin x\sin(2x) + 2\cos x\cos(2x)$

**g** $y = \cos(\cos x)$

$\therefore \dfrac{dy}{dx} = -\sin(\cos x) \times (-\sin x)$

$= \sin x\sin(\cos x)$

**h** $y = \cos^3(4x) = (\cos(4x))^3$

$\therefore \dfrac{dy}{dx} = 3(\cos(4x))^2 \times (-4\sin(4x))$

$= -12\sin(4x)\cos^2(4x)$

**i** $y = \dfrac{1}{\sin x} = (\sin x)^{-1}$

$\therefore \dfrac{dy}{dx} = -1(\sin x)^{-2} \times \cos x$

$= -\dfrac{\cos x}{\sin^2 x}$

**j** $y = \dfrac{1}{\cos(2x)} = (\cos(2x))^{-1}$

$\therefore \dfrac{dy}{dx} = -1(\cos(2x))^{-2} \times (-2\sin(2x))$

$= \dfrac{2\sin(2x)}{\cos^2(2x)}$

**k** $y = \dfrac{2}{\sin^2(2x)} = 2(\sin(2x))^{-2}$

$\therefore \dfrac{dy}{dx} = -4(\sin(2x))^{-3} \times 2\cos(2x)$

$= -\dfrac{8\cos(2x)}{\sin^3(2x)}$

**l** $y = \dfrac{8}{\tan^3\left(\frac{x}{2}\right)} = 8\left[\tan\left(\frac{x}{2}\right)\right]^{-3}$

$\therefore \dfrac{dy}{dx} = -24\left[\tan\left(\frac{x}{2}\right)\right]^{-4} \times \frac{1}{2} \times \dfrac{1}{\cos^2\left(\frac{x}{2}\right)}$

$= \dfrac{-12}{\cos^2\left(\frac{x}{2}\right)\tan^4\left(\frac{x}{2}\right)}$

**4** **a** $f(x) = \sin^3 x$

$= [\sin x]^3$

$= u^3$ where $u = \sin x$

$f'(x) = \dfrac{dy}{dx} = \dfrac{dy}{du}\dfrac{du}{dx}$ {chain rule}

$\therefore f'(x) = 3u^2 \times \dfrac{du}{dx}$

$= 3\sin^2 x \cos x$

$\therefore f'\left(\frac{2\pi}{3}\right) = 3\sin^2\left(\frac{2\pi}{3}\right)\cos\left(\frac{2\pi}{3}\right)$

$= -\frac{9}{8}$

**b** $f(x) = \cos x \sin x$ is the product of

$u = \cos x$ and $v = \sin x$

$\therefore u' = -\sin x$ and $v' = \cos x$

Now $f'(x) = u'v + uv'$

$= -\sin x \sin x + \cos x \cos x$

$= \cos^2 x - \sin^2 x$

$\therefore f'\left(\frac{\pi}{4}\right) = \cos^2\left(\frac{\pi}{4}\right) - \sin^2\left(\frac{\pi}{4}\right)$

$= 0$

## EXERCISE 15H

**1** **a** $f(x) = 3x^2 - 6x + 2$

$\therefore f'(x) = 6x - 6$

$\therefore f''(x) = 6$

**b** $f(x) = \dfrac{2}{\sqrt{x}} - 1 = 2x^{-\frac{1}{2}} - 1$

$\therefore f'(x) = -x^{-\frac{3}{2}}$

$f''(x) = \frac{3}{2}x^{-\frac{5}{2}}$

$= \dfrac{3}{2x^{\frac{5}{2}}}$

**c** $f(x) = 2x^3 - 3x^2 - x + 5$

$\therefore f'(x) = 6x^2 - 6x - 1$

$\therefore f''(x) = 12x - 6$

**d** $f(x) = \dfrac{2-3x}{x^2} = 2x^{-2} - 3x^{-1}$

$\therefore f'(x) = -4x^{-3} + 3x^{-2}$

$\therefore f''(x) = 12x^{-4} - 6x^{-3}$

$= \dfrac{12-6x}{x^4}$

**e** $f(x) = (1-2x)^3$

$\therefore f'(x) = 3(1-2x)^2(-2)$

$= -6(1-2x)^2$

$\therefore f''(x) = -12(1-2x)^1(-2)$

$= 24(1-2x) = 24 - 48x$

**f** $f(x) = \dfrac{x+2}{2x-1}$ is a quotient with $u = x + 2$ and $v = 2x - 1$

$\therefore u' = 1$ and $v' = 2$

$\therefore f'(x) = \dfrac{1(2x-1) - 2(x+2)}{(2x-1)^2}$ {quotient rule}

$= \dfrac{-5}{(2x-1)^2}$

$= -5(2x-1)^{-2}$

$\therefore f''(x) = 10(2x-1)^{-3}(2) = \dfrac{20}{(2x-1)^3}$

**2** **a** $y = x - x^3$

$\therefore \dfrac{dy}{dx} = 1 - 3x^2$

$\therefore \dfrac{d^2y}{dx^2} = -6x$

**b** $y = x^2 - \dfrac{5}{x^2}$

$= x^2 - 5x^{-2}$

$\therefore \dfrac{dy}{dx} = 2x + 10x^{-3}$

$\therefore \dfrac{d^2y}{dx^2} = 2 - 30x^{-4}$

$= 2 - \dfrac{30}{x^4}$

**c** $y = 2 - \dfrac{3}{\sqrt{x}}$

$= 2 - 3x^{-\frac{1}{2}}$

$\therefore \dfrac{dy}{dx} = \frac{3}{2}x^{-\frac{3}{2}}$

$\therefore \dfrac{d^2y}{dx^2} = -\frac{9}{4}x^{-\frac{5}{2}}$

**d** $y = \dfrac{4-x}{x}$

$= 4x^{-1} - 1$

$\therefore \dfrac{dy}{dx} = -4x^{-2}$

$\therefore \dfrac{d^2y}{dx^2} = 8x^{-3}$

$= \dfrac{8}{x^3}$

**e** $y = (x^2 - 3x)^3$

$\therefore \dfrac{dy}{dx} = 3(x^2 - 3x)^2(2x - 3)$

$= (6x - 9)(x^2 - 3x)^2$ which is a product where

$u = 6x - 9$ and $v = (x^2 - 3x)^2$

$\therefore u' = 6$ and $v' = 2(x^2 - 3x)^1(2x - 3)$

$\therefore \dfrac{d^2y}{dx^2} = 6(x^2 - 3x)^2 + (6x - 9)(2)(x^2 - 3x)(2x - 3)$

$= 6(x^2 - 3x)\left[(x^2 - 3x) + (2x - 3)^2\right]$

$= 6(x^2 - 3x)(x^2 - 3x + 4x^2 - 12x + 9)$

$= 6(x^2 - 3x)(5x^2 - 15x + 9)$

**f** $y = x^2 - x + \dfrac{1}{1-x}$

$= x^2 - x + (1-x)^{-1}$

$\therefore \dfrac{dy}{dx} = 2x - 1 + (-1)(1-x)^{-2}(-1)$

$= 2x - 1 + (1-x)^{-2}$

$\therefore \dfrac{d^2y}{dx^2} = 2 - 2(1-x)^{-3}(-1)$

$= 2 + \dfrac{2}{(1-x)^3}$

**3** **a** $f(x) = x^3 - 2x + 5$

$\therefore f(2) = (2)^3 - 2(2) + 5$

$= 9$

**b** $f(x) = x^3 - 2x + 5$

$\therefore f'(x) = 3x^2 - 2$

$\therefore f'(2) = 3(2)^2 - 2$

$= 10$

**c** $f'(x) = 3x^2 - 2$ {from **b**}

$\therefore f''(x) = 6x$

$\therefore f''(2) = 6(2)$

$= 12$

**d** $f''(x) = 6x$ {from **c**}

$\therefore f^{(3)}(x) = 6$

$\therefore f^{(3)}(2) = 6$

**4** $y = Ae^{kx}$

**a** $\dfrac{dy}{dx} = Ae^{kx}(k)$

$= k(Ae^{kx})$

$= ky$

**b** $\dfrac{d^2y}{dx^2} = \dfrac{d}{dx}kAe^{kx}$ {from **a**}

$= k^2Ae^{kx}$

$= k^2y$

**c** $\dfrac{d^3y}{dx^3} = \dfrac{d}{dx}k^2Ae^{kx}$ {from **b**}

$= k^3Ae^{kx}$

$= k^3y$

**5** **a** $f(x) = 2x^3 - 6x^2 + 5x + 1$

$\therefore f'(x) = 6x^2 - 12x + 5$

$\therefore f''(x) = 12x - 12$

So, $f''(x) = 0$ when $12x - 12 = 0$

$\therefore 12x = 12$

$\therefore x = 1$

**b** $f(x) = \dfrac{x}{x^2+2}$ is a quotient where $u = x$ and $v = x^2 + 2$

$\therefore\ u' = 1$ and $v' = 2x$

$$\therefore\ f'(x) = \frac{1(x^2+2) - 2x^2}{(x^2+2)^2} \quad \{\text{quotient rule}\}$$
$$= \frac{2 - x^2}{(x^2+2)^2}$$

This is another quotient, this time with $u = 2 - x^2$ and $v = (x^2+2)^2$

$\therefore\ u' = -2x$ and $v' = 2(x^2+2)(2x)$

$$\therefore\ f''(x) = \frac{-2x(x^2+2)^2 - 4x(x^2+2)(2-x^2)}{(x^2+2)^4}$$
$$= \frac{-2x(x^2+2)[x^2+2+2(2-x^2)]}{(x^2+2)^4}$$
$$= \frac{-2x[-x^2+6]}{(x^2+2)^3}$$
$$= \frac{2x[x^2-6]}{(x^2+2)^3}$$

So, $f''(x) = 0$ when $2x[x^2-6] = 0$

$\therefore\ x = 0$ or $x^2 - 6 = 0$

$\therefore\ x = 0$ or $x = \pm\sqrt{6}$

**6** $f(x) = 2x^3 - x$

$\therefore\ f'(x) = 6x^2 - 1$

$\therefore\ f''(x) = 12x$

By substituting the various values of $x$ into these three functions, we can fill in the table as follows:

| $x$ | $-1$ | $0$ | $1$ |
|---|---|---|---|
| $f(x)$ | $-$ | $0$ | $+$ |
| $f'(x)$ | $+$ | $-$ | $+$ |
| $f''(x)$ | $-$ | $0$ | $+$ |

**7** $f(x) = \frac{2}{3}\sin 3x$

$$\therefore\ f'(x) = \tfrac{2}{3} \times (\cos 3x) \times 3$$
$$= 2\cos 3x$$
$$\therefore\ f''(x) = 2 \times (-\sin 3x) \times 3$$
$$= -6\sin 3x$$
$$\therefore\ f^{(3)}(x) = -6 \times (\cos 3x) \times 3$$
$$= -18\cos 3x$$
$$\therefore\ f^{(3)}\left(\tfrac{2\pi}{9}\right) = -18\cos 3\left(\tfrac{2\pi}{9}\right)$$
$$= -18\cos\left(\tfrac{2\pi}{3}\right)$$
$$= 9$$

**8** **a** $f(x) = 2\sin^3 x - 3\sin x$

$$= 2(\sin x)^3 - 3\sin x$$
$$\therefore\ f'(x) = 2 \times 3(\sin x)^2 \times (\cos x) - 3\cos x$$
$$= -3\cos x(1 - 2\sin^2 x)$$
$$= -3\cos x\cos 2x$$

**b** $f''(x) = -3(-\sin x \times \cos 2x + \cos x \times (-2)\sin 2x)$

$$= 3\sin x\cos 2x + 6\cos x\sin 2x$$

**9** **a** $y = -\ln x$

$$\therefore\ \frac{dy}{dx} = -1 \times \frac{1}{x}$$
$$= -x^{-1}$$
$$\therefore\ \frac{d^2y}{dx^2} = -\left(-x^{-2}\right)$$
$$= x^{-2} = \frac{1}{x^2}$$

**b** $y = x\ln x$ is the product of $u = x$ and $v = \ln x$

$\therefore\ u' = 1$ and $v' = \dfrac{1}{x}$

Now $\dfrac{dy}{dx} = u'v + uv'$ {product rule}

$$= 1 \times \ln x + x \times \frac{1}{x}$$
$$= \ln x + 1$$
$$\therefore\ \frac{d^2y}{dx^2} = \frac{1}{x}$$

**c** $y = (\ln x)^2$

$\therefore \dfrac{dy}{dx} = 2(\ln x)\left(\dfrac{1}{x}\right) = \dfrac{2\ln x}{x}$ which is a quotient with $u = 2\ln x$ and $v = x$

$\therefore u' = \dfrac{2}{x}$ and $v' = 1$

$$\therefore \frac{d^2y}{dx^2} = \frac{u'v - uv'}{v^2} \quad \{\text{quotient rule}\}$$
$$= \frac{\frac{2}{x} \times x - 2\ln x \times 1}{x^2}$$
$$= \frac{2 - 2\ln x}{x^2} = \frac{2}{x^2}(1 - \ln x)$$

**10** **a** $f(x) = x^2 - \dfrac{1}{x}$

$$\therefore f(1) = (1)^2 - \frac{1}{1} = 1 - 1 = 0$$

**b** $f(x) = x^2 - \dfrac{1}{x}$

$$= x^2 - x^{-1}$$
$$\therefore f'(x) = 2x - \left(-x^{-2}\right) = 2x + x^{-2} = 2x + \frac{1}{x^2}$$
$$\therefore f'(1) = 2(1) + \frac{1}{1^2} = 2 + 1 = 3$$

**c** $f'(x) = 2x + x^{-2}$ {from **b**}

$$\therefore f''(x) = 2 - 2x^{-3} = 2 - \frac{2}{x^3}$$
$$\therefore f''(1) = 2 - \frac{2}{1^3} = 2 - 2 = 0$$

**d** $f''(x) = 2 - 2x^{-3}$ {from **c**}

$$\therefore f^{(3)}(x) = 6x^{-4} = \frac{6}{x^4}$$
$$\therefore f^{(3)}(1) = \frac{6}{1^4} = 6$$

**11** $y = 2e^{3x} + 5e^{4x}$ $\quad \therefore \dfrac{dy}{dx} = 6e^{3x} + 20e^{4x}$ and $\dfrac{d^2y}{dx^2} = 18e^{3x} + 80e^{4x}$

Now
$$\frac{d^2y}{dx^2} - 7\frac{dy}{dx} + 12y = \left(18e^{3x} + 80e^{4x}\right) - 7\left(6e^{3x} + 20e^{4x}\right) + 12\left(2e^{3x} + 5e^{4x}\right)$$
$$= 18e^{3x} + 80e^{4x} - 42e^{3x} - 140e^{4x} + 24e^{3x} + 60e^{4x}$$
$$= e^{3x}[18 - 42 + 24] + e^{4x}[80 - 140 + 60]$$
$$= e^{3x}(0) + e^{4x}(0)$$
$$= 0$$
$$\therefore \frac{d^2y}{dx^2} - 7\frac{dy}{dx} + 12y = 0$$

**12** If $y = \sin(2x + 3)$, then $\dfrac{dy}{dx} = 2\cos(2x+3)$ and $\dfrac{d^2y}{dx^2} = -4\sin(2x+3)$

$$\therefore \frac{d^2y}{dx^2} + 4y = -4\sin(2x+3) + 4\sin(2x+3) = 0$$

**13**
$$y = \sin x$$
$$\therefore \frac{dy}{dx} = \cos x$$
$$\therefore \frac{d^2y}{dx^2} = -\sin x$$
$$\therefore \frac{d^3y}{dx^3} = -\cos x$$
$$\therefore \frac{d^4y}{dx^4} = -(-\sin x) = \sin x = y$$

**14** If $y = 2\sin x + 3\cos x$, then $y' = 2\cos x - 3\sin x$ and $y'' = -2\sin x - 3\cos x$

$\therefore\ y'' + y = -2\sin x - 3\cos x + 2\sin x + 3\cos x = 0$

## REVIEW SET 15A

**1** **a** $f(x) = 7 + x - 3x^2$

$\therefore\ f(3) = 7 + 3 - 3(3)^2 = -17$

**b** $f'(x) = 1 - 6x$

$\therefore\ f'(3) = 1 - 6(3) = -17$

**c** $f''(x) = -6$

$\therefore\ f''(3) = -6$

**2** **a** $y = 3x^2 - x^4$

$\therefore\ \dfrac{dy}{dx} = 6x - 4x^3$

**b** $y = \dfrac{x^3 - x}{x^2} = x - x^{-1}$

$\therefore\ \dfrac{dy}{dx} = 1 + x^{-2} = 1 + \dfrac{1}{x^2}$

**3** $f(x) = \dfrac{x}{\sqrt{x^2+1}}$ is a quotient with $u = x$ and $v = \left(x^2+1\right)^{\frac{1}{2}}$

$\therefore\ u' = 1$ and $v' = \frac{1}{2}\left(x^2+1\right)^{-\frac{1}{2}} \times 2x$

$= x\left(x^2+1\right)^{-\frac{1}{2}}$

Now $f'(x) = \dfrac{u'v - uv'}{v^2}$ {quotient rule}

$$= \frac{1 \times \left(x^2+1\right)^{\frac{1}{2}} - x \times x\left(x^2+1\right)^{-\frac{1}{2}}}{x^2+1}$$

$$= \frac{\sqrt{x^2+1} - \frac{x^2}{\sqrt{x^2+1}}}{x^2+1}$$

$$= \frac{(x^2+1) - x^2}{(x^2+1)\sqrt{x^2+1}}$$

$$= \frac{1}{(x^2+1)\sqrt{x^2+1}}$$

The tangent to $f(x)$ has gradient 1 when $f'(x) = 1$

$$\therefore\ \frac{1}{(x^2+1)\sqrt{x^2+1}} = 1$$

$\therefore\ \left(x^2+1\right)^{\frac{3}{2}} = 1$

$\therefore\ x^2 + 1 = 1$

$\therefore\ x^2 = 0$

$\therefore\ x = 0$

and $f(0) = \dfrac{0}{\sqrt{0^2+1}} = 0$

$\therefore$ the tangent to $f(x) = \dfrac{x}{\sqrt{x^2+1}}$ has gradient 1 at the point (0, 0).

**4** **a** $y = e^{x^3+2}$

$= e^u$ where $u = x^3 + 2$

$\therefore\ \dfrac{dy}{dx} = \dfrac{dy}{du}\dfrac{du}{dx}$ {chain rule}

$= e^u \times 3x^2$

$= 3x^2 e^u$

$= 3x^2 e^{x^3+2}$

**b** $y = \ln\left(\dfrac{x+3}{x^2}\right)$

$= \ln(x+3) - \ln(x^2)$

$\therefore\ \dfrac{dy}{dx} = \dfrac{1}{x+3} - \dfrac{2x}{x^2}$

$= \dfrac{1}{x+3} - \dfrac{2}{x}$

**5** $y = 3e^x - e^{-x}$

$\therefore \dfrac{dy}{dx} = 3e^x - \left(-e^{-x}\right)$
$= 3e^x + e^{-x}$

$\therefore \dfrac{d^2y}{dx^2} = 3e^x + \left(-e^{-x}\right)$
$= 3e^x - e^{-x}$
$= y$

**6** **a** $\dfrac{d}{dx}(\sin(5x)\ln x) = \dfrac{d}{dx}(\sin(5x))\ln x + \sin(5x)\dfrac{d}{dx}(\ln x)$ {product rule}
$= 5\cos(5x)\ln x + \dfrac{\sin(5x)}{x}$

**b** $\dfrac{d}{dx}(\sin x\cos(2x)) = \dfrac{d}{dx}(\sin x)\cos(2x) + \sin x\dfrac{d}{dx}(\cos(2x))$ {product rule}
$= \cos x\cos(2x) + \sin x(-2\sin(2x))$
$= \cos x\cos(2x) - 2\sin x\sin(2x)$

**c** $\dfrac{d}{dx}\left(e^{-2x}\tan x\right) = \dfrac{d}{dx}\left(e^{-2x}\right)\tan x + e^{-2x}\dfrac{d}{dx}(\tan x)$ {product rule}
$= -2e^{-2x}\tan x + \dfrac{e^{-2x}}{\cos^2 x}$

**7** $y = \sin^2 x$
$= (\sin x)^2 = u^2$ where $u = \sin x$
$\therefore \dfrac{du}{dx} = \cos x$

Now $\dfrac{dy}{dx} = \dfrac{dy}{du}\dfrac{du}{dx}$ {chain rule}
$= 2u\dfrac{du}{dx}$
$= 2\sin x\cos x$

When $x = \frac{\pi}{3}$, $\dfrac{dy}{dx} = 2\sin\left(\frac{\pi}{3}\right)\cos\left(\frac{\pi}{3}\right)$
$= \frac{\sqrt{3}}{2}$

$\therefore$ gradient of tangent $= \frac{\sqrt{3}}{2}$

**8** **a** $f(x) = (x^2+3)^4$
$\therefore f'(x) = 4(x^2+3)^3(2x)$
$= 8x(x^2+3)^3$

**b** $g(x) = \dfrac{\sqrt{x+5}}{x^2}$ is a quotient with
$u = (x+5)^{\frac{1}{2}}$ and $v = x^2$
$\therefore u' = \frac{1}{2}(x+5)^{-\frac{1}{2}}$ and $v' = 2x$
$\therefore g'(x) = \dfrac{\frac{1}{2}(x+5)^{-\frac{1}{2}}(x^2) - (x+5)^{\frac{1}{2}}(2x)}{x^4}$
$= \dfrac{\frac{1}{2}x(x+5)^{-\frac{1}{2}} - 2(x+5)^{\frac{1}{2}}}{x^3}$

**9** **a** $f(x) = 3x^2 - \dfrac{1}{x} = 3x^2 - x^{-1}$
$\therefore f'(x) = 6x + x^{-2}$
$\therefore f''(x) = 6 - 2x^{-3} = 6 - \dfrac{2}{x^3}$
$\therefore f''(2) = 6 - \dfrac{2}{2^3} = \dfrac{23}{4}$

**b** $f(x) = \sqrt{x} = x^{\frac{1}{2}}$
$\therefore f'(x) = \frac{1}{2}x^{-\frac{1}{2}}$
$\therefore f''(x) = -\frac{1}{4}x^{-\frac{3}{2}}$
$\therefore f''(2) = -\frac{1}{4}(2^{-\frac{3}{2}})$
$= -\dfrac{1}{4\sqrt{2^3}} = -\dfrac{1}{8\sqrt{2}}$

**10** $y = \left(1 - \frac{1}{3}x\right)^3$

$\therefore \frac{dy}{dx} = 3\left(1 - \frac{1}{3}x\right)^2\left(-\frac{1}{3}\right)$

$= -\left(1 - \frac{1}{3}x\right)^2$

$\therefore \frac{d^2y}{dx^2} = -2\left(1 - \frac{1}{3}x\right)\left(-\frac{1}{3}\right)$

$= \frac{2}{3}\left(1 - \frac{1}{3}x\right)$

$= \frac{2}{3} - \frac{2}{9}x$

$\therefore \frac{d^3y}{dx^3} = -\frac{2}{9}$

## REVIEW SET 15B

**1** **a** $y = 5x - 3x^{-1}$

$\therefore \frac{dy}{dx} = 5 + 3x^{-2}$

**b** $y = (3x^2 + x)^4$

$\therefore \frac{dy}{dx} = 4(3x^2 + x)^3(6x + 1)$

**c** $y = (x^2 + 1)(1 - x^2)^3$ is a product with $u = x^2 + 1$ and $v = (1 - x^2)^3$

$\therefore u' = 2x$ and $v' = 3(1 - x^2)^2(-2x)$

$= -6x(1 - x^2)^2$

$\therefore \frac{dy}{dx} = 2x(1 - x^2)^3 + (x^2 + 1) \times -6x(1 - x^2)^2$ {product rule}

$= 2x(1 - x^2)^3 - 6x(x^2 + 1)(1 - x^2)^2$

**2** $y = 2x^3 + 3x^2 - 10x + 3$

$\therefore \frac{dy}{dx} = 6x^2 + 6x - 10$

The gradient of the tangent is 2 where $6x^2 + 6x - 10 = 2$

$\therefore 6x^2 + 6x - 12 = 0$

$\therefore 6(x^2 + x - 2) = 0$

$\therefore 6(x + 2)(x - 1) = 0$

$\therefore x = -2$ or $1$

$\therefore y = 2(-2)^3 + 3(-2)^2 - 10(-2) + 3$
$= -16 + 12 + 20 + 3$
$= 19$

or $y = 2(1)^3 + 3(1)^2 - 10(1) + 3$
$= 2 + 3 - 10 + 3$
$= -2$

So, the gradient of the tangent is 2 at $(-2, 19)$ and $(1, -2)$.

**3** $y = \sqrt{5 - 4x} = (5 - 4x)^{\frac{1}{2}}$

**a** $\frac{dy}{dx} = \frac{1}{2}(5 - 4x)^{-\frac{1}{2}}(-4)$

$= -2(5 - 4x)^{-\frac{1}{2}}$

**b** $\frac{d^2y}{dx^2} = -2(-\frac{1}{2})(5 - 4x)^{-\frac{3}{2}}(-4)$

$= -4(5 - 4x)^{-\frac{3}{2}}$

**4** **a**

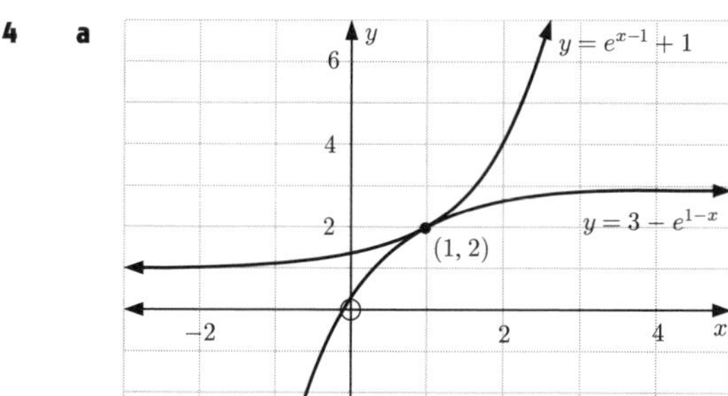


**b** Using technology, the point of intersection is $(1, 2)$.

**c** If $y = e^{x-1} + 1$

then $\frac{dy}{dx} = e^{x-1}$

When $x = 1$, $\frac{dy}{dx} = e^{1-1}$

$= e^0$

$= 1$

$\therefore$ gradient of tangent $= 1$

If $y = 3 - e^{1-x}$

then $\frac{dy}{dx} = -e^{1-x} \times -1$

$= e^{1-x}$

When $x = 1$, $\frac{dy}{dx} = e^{1-1}$

$= e^0$

$= 1$

$\therefore$ gradient of tangent $= 1$

So, the tangents to each curve at $(1, 2)$ both have a gradient of 1.

**d** If the tangents of each curve at $(1, 2)$ have the same gradient, then they are in fact the same line. That is, the two curves have a *common tangent* at $(1, 2)$.

**5** **a** $y = \ln(x^3 - 3x)$

$\therefore \frac{dy}{dx} = \frac{3x^2 - 3}{x^3 - 3x}$

**b** $y = \frac{e^x}{x^2}$

$\therefore \frac{dy}{dx} = \frac{e^x(x^2) - e^x(2x)}{x^4}$ {quotient rule}

$= \frac{e^x(x-2)}{x^3}$

**6** $f(x) = 2x^4 - 4x^3 - 9x^2 + 4x + 7$

$\therefore f'(x) = 8x^3 - 12x^2 - 18x + 4$

$\therefore f''(x) = 24x^2 - 24x - 18$

So, $f''(x) = 0$ where $24x^2 - 24x - 18 = 0$

$\therefore 4x^2 - 4x - 3 = 0$

$\therefore (2x+1)(2x-3) = 0$

$\therefore x = -\frac{1}{2}$ or $x = \frac{3}{2}$

**7** **a** $f(x) = x - \cos x$

$\therefore f(\pi) = \pi - \cos\pi$

$= \pi - (-1)$

$= \pi + 1$

**b** $f(x) = x - \cos x$

$\therefore f'(x) = 1 - (-\sin x)$

$= 1 + \sin x$

$\therefore f'\left(\frac{\pi}{2}\right) = 1 + \sin\left(\frac{\pi}{2}\right)$

$= 1 + 1 = 2$

**c** $f'(x) = 1 + \sin x$ {from **b**}

$\therefore f''(x) = \cos x$

$\therefore f''\left(\frac{3\pi}{4}\right) = \cos\left(\frac{3\pi}{4}\right)$

$= -\frac{1}{\sqrt{2}}$ $\left(\text{or } -\frac{\sqrt{2}}{2}\right)$

**8** $y = 3\sin bx - a\cos 2x$

$\therefore \frac{dy}{dx} = 3 \times (\cos bx) \times b - a \times (-\sin 2x) \times 2$

$= 3b\cos bx + 2a\sin 2x$

$\therefore \frac{d^2y}{dx^2} = 3b \times (-\sin bx) \times b + 2a \times (\cos 2x) \times 2$

$= -3b^2\sin bx + 4a\cos 2x$

Now $y + \frac{d^2y}{dx^2} = 6\cos 2x$

$\therefore 3\sin bx - a\cos 2x + 4a\cos 2x - 3b^2\sin bx = 6\cos 2x$

$\therefore 3\sin bx - 3b^2\sin bx + 3a\cos 2x = 6\cos 2x$

$\therefore (3 - 3b^2)\sin bx + 3a\cos 2x = 6\cos 2x$

$\therefore 3 - 3b^2 = 0$ and $3a = 6$

$\therefore 3(1 - b^2) = 0$ and $a = 2$

$\therefore 3(1+b)(1-b) = 0$

$\therefore 1 + b = 0$ or $1 - b = 0$

$\therefore b = -1$ or $1$

So, $a = 2$ and $b = -1$ or $1$.

**9** **a** $\frac{d}{dx}(10x - \sin(10x)) = 10 - 10\cos(10x)$

**b** $\frac{d}{dx}\left(\ln\left(\frac{1}{\cos x}\right)\right) = \frac{1}{\left(\frac{1}{\cos x}\right)} \times \frac{d}{dx}\left(\frac{1}{\cos x}\right)$ {chain rule}

$= \cos x \times \frac{d}{dx}\left((\cos x)^{-1}\right)$

$= \cos x \times \left(-(\cos x)^{-2} \times (-\sin x)\right)$

$= \frac{\cos x \sin x}{\cos^2 x} = \tan x$

**c** $\frac{d}{dx}(\sin(5x)\ln(2x)) = \frac{d}{dx}(\sin(5x))\ln(2x) + \sin(5x)\frac{d}{dx}(\ln(2x))$ {product rule}

$= 5\cos(5x)\ln(2x) + \sin(5x) \times \frac{2}{2x}$

$= 5\cos(5x)\ln(2x) + \frac{\sin(5x)}{x}$

## REVIEW SET 15C

**1** **a** $y = x^3\sqrt{1-x^2}$ is a product where $u = x^3$ and $v = \left(1-x^2\right)^{\frac{1}{2}}$

$\therefore\ u' = 3x^2$ and $v' = \frac{1}{2}\left(1-x^2\right)^{-\frac{1}{2}}(-2x)$

$= -x\left(1-x^2\right)^{-\frac{1}{2}}$

$\therefore\ \frac{dy}{dx} = 3x^2\left(1-x^2\right)^{\frac{1}{2}} + x^3 \times -x\left(1-x^2\right)^{-\frac{1}{2}}$ {product rule}

$= 3x^2\left(1-x^2\right)^{\frac{1}{2}} - x^4\left(1-x^2\right)^{-\frac{1}{2}}$

**b** $y = \frac{x^2-3x}{\sqrt{x+1}}$ is a quotient where $u = x^2 - 3x$ and $v = (x+1)^{\frac{1}{2}}$

$\therefore\ u' = 2x - 3$ and $v' = \frac{1}{2}(x+1)^{-\frac{1}{2}}$

$\therefore\ \frac{dy}{dx} = \frac{(2x-3)(x+1)^{\frac{1}{2}} - \frac{1}{2}(x^2-3x)(x+1)^{-\frac{1}{2}}}{x+1}$ {quotient rule}

**2** **a** $y = 3x^4 - \frac{2}{x} = 3x^4 - 2x^{-1}$

$\therefore\ \frac{dy}{dx} = 12x^3 + 2x^{-2}$

$\therefore\ \frac{d^2y}{dx^2} = 36x^2 - 4x^{-3}$

$= 36x^2 - \frac{4}{x^3}$

**b** $y = x^3 - x + \frac{1}{\sqrt{x}} = x^3 - x + x^{-\frac{1}{2}}$

$\therefore\ \frac{dy}{dx} = 3x^2 - 1 - \frac{1}{2}x^{-\frac{3}{2}}$

$\therefore\ \frac{d^2y}{dx^2} = 6x + \frac{3}{4}x^{-\frac{5}{2}}$

**3** $y = xe^x$ is the product of $u = x$ and $v = e^x$

$\therefore\ u' = 1$ and $v' = e^x$

Now $\frac{dy}{dx} = u'v + uv'$ {product rule}

$= 1 \times e^x + x \times e^x$

$= e^x + xe^x$

$= (1+x)e^x$

If $\frac{dy}{dx} = 2e$, then $(1+x)e^x = 2e$

Solving by inspection, we get $x = 1$. When $x = 1$, $y = 1 \times e^1 = e$.

$\therefore$ the gradient of $y = xe^x$ is $2e$ at the point $(1, e)$.

**4** **a** $$f(x) = \ln(e^x + 3)$$
$$\therefore\ f'(x) = \frac{e^x}{e^x + 3}$$

**b** $$f(x) = \ln\left[\frac{(x+2)^3}{x}\right]$$
$$= \ln(x+2)^3 - \ln x$$
$$= 3\ln(x+2) - \ln x$$
$$\therefore\ f'(x) = \frac{3}{x+2} - \frac{1}{x}$$

**5** $$y = \left(x - \frac{1}{x}\right)^4$$
$$= \left(x - x^{-1}\right)^4$$
$$\therefore\ \frac{dy}{dx} = 4\left(x - x^{-1}\right)^3(1 + x^{-2})$$
$$= 4\left(x - \frac{1}{x}\right)^3\left(1 + \frac{1}{x^2}\right)$$
When $x = 1$, $$\frac{dy}{dx} = 4\left(1 - \frac{1}{1}\right)^3\left(1 + \frac{1}{1^2}\right)$$
$$= 4 \times 0 \times 2$$
$$= 0$$

**6** **a** $$f(x) = x^{\frac{1}{2}}\cos(4x)$$
$$\therefore\ f'(x) = \tfrac{1}{2}x^{-\frac{1}{2}}\cos(4x) + x^{\frac{1}{2}}(-4\sin(4x)) \qquad \{\text{product rule}\}$$
$$= \tfrac{1}{2}x^{-\frac{1}{2}}\cos(4x) - 4x^{\frac{1}{2}}\sin(4x)$$
and $$f''(x) = -\tfrac{1}{4}x^{-\frac{3}{2}}\cos(4x) + \tfrac{1}{2}x^{-\frac{1}{2}}(-4\sin(4x)) - \left[2x^{-\frac{1}{2}}\sin(4x) + 4x^{\frac{1}{2}} \times 4\cos(4x)\right]$$
$$= -\tfrac{1}{4}x^{-\frac{3}{2}}\cos(4x) - 4x^{-\frac{1}{2}}\sin(4x) - 16x^{\frac{1}{2}}\cos(4x)$$

**b** $$f'(x) = \tfrac{1}{2}x^{-\frac{1}{2}}\cos(4x) - 4x^{\frac{1}{2}}\sin(4x) \qquad \{\text{from } \mathbf{a}\}$$
$$\therefore\ f'\left(\tfrac{\pi}{16}\right) = \tfrac{1}{2}\left(\tfrac{\pi}{16}\right)^{-\frac{1}{2}}\cos\left(\tfrac{\pi}{4}\right) - 4\left(\tfrac{\pi}{16}\right)^{\frac{1}{2}}\sin\left(\tfrac{\pi}{4}\right) \approx -0.455$$
$$f''(x) = -\tfrac{1}{4}x^{-\frac{3}{2}}\cos(4x) - 4x^{-\frac{1}{2}}\sin(4x) - 16x^{\frac{1}{2}}\cos(4x)$$
$$\therefore\ f''\left(\tfrac{\pi}{8}\right) = -\tfrac{1}{4}\left(\tfrac{\pi}{8}\right)^{-\frac{3}{2}}\cos\left(\tfrac{\pi}{2}\right) - 4\left(\tfrac{\pi}{8}\right)^{-\frac{1}{2}}\sin\left(\tfrac{\pi}{2}\right) - 16\left(\tfrac{\pi}{8}\right)^{\frac{1}{2}}\cos\left(\tfrac{\pi}{2}\right)$$
$$= 0 - 4\left(\tfrac{\pi}{8}\right)^{-\frac{1}{2}}(1) - 0 \qquad \{\text{since } \cos\left(\tfrac{\pi}{2}\right) = 0\}$$
$$\approx -6.38$$

**7** $$y = 3\sin 2x + 2\cos 2x$$
$$\therefore\ \frac{dy}{dx} = 3 \times (\cos 2x) \times 2 + 2 \times (-\sin 2x) \times 2$$
$$= 6\cos 2x - 4\sin 2x$$
$$\therefore\ \frac{d^2y}{dx^2} = 6 \times (-\sin 2x) \times 2 - 4 \times (\cos 2x) \times 2$$
$$= -12\sin 2x - 8\cos 2x$$
$$\therefore\ 4y + \frac{d^2y}{dx^2} = 4(3\sin 2x + 2\cos 2x) + (-12\sin 2x - 8\cos 2x)$$
$$= 12\sin 2x + 8\cos 2x - 12\sin 2x - 8\cos 2x$$
$$= 0$$

**8** **a** $$f(x) = \frac{6x}{3 + x^2}$$
Now, $f(x) = -\frac{1}{2}$ when $$\frac{6x}{3 + x^2} = -\frac{1}{2}$$
$$\therefore\ 12x = -(3 + x^2)$$
$$\therefore\ x^2 + 12x + 3 = 0$$
$$\therefore\ x = \frac{-12 \pm \sqrt{12^2 - 4 \times 1 \times 3}}{2}$$
$$= \frac{-12 \pm \sqrt{132}}{2} \approx -11.7 \text{ or } -0.255$$

**b** $f(x) = \dfrac{6x}{3+x^2}$ is a quotient where $u = 6x$ and $v = 3 + x^2$

$\therefore\ u' = 6$ and $v' = 2x$

$$\begin{aligned}\therefore\ f'(x) &= \frac{u'v - uv'}{v^2} \quad \{\text{quotient rule}\}\\ &= \frac{6(3+x^2) - 6x(2x)}{(3+x^2)^2}\\ &= \frac{18 + 6x^2 - 12x^2}{(3+x)^2}\\ &= \frac{18 - 6x^2}{(3+x^2)^2}\end{aligned}$$

Now, $f'(x) = 0$ when $\dfrac{18-6x^2}{(3+x^2)^2} = 0$

$$\begin{aligned}\therefore\ 18 - 6x^2 &= 0 \quad \{\text{since } (3+x^2)^2 > 0 \text{ for all } x \in \mathbb{R}\}\\ \therefore\ 6(3 - x^2) &= 0\\ \therefore\ x^2 &= 3\\ \therefore\ x &= \pm\sqrt{3}\\ &\approx -1.73 \text{ or } 1.73\end{aligned}$$

**c** $f'(x) = \dfrac{18-6x^2}{(3+x^2)^2}$ is a quotient where $u = 18 - 6x^2$ and $v = (3+x^2)^2$

$\therefore\ u' = -12x$ and $v' = 2(3+x^2) \times 2x$

$= 4x(3+x^2)$

$$\begin{aligned}f''(x) &= \frac{u'v - uv'}{v^2} \quad \{\text{quotient rule}\}\\ &= \frac{-12x(3+x^2)^2 - (18-6x^2) \times 4x(3+x^2)}{(3+x^2)^4}\\ &= \frac{(3+x^2)[-12x(3+x^2) - 4x(18-6x^2)]}{(3+x^2)^4}\\ &= \frac{-36x - 12x^3 - 72x + 24x^3}{(3+x^2)^3}\\ &= \frac{12x^3 - 108x}{(3+x^2)^3}\end{aligned}$$

Now, $f''(x) = 0$ when $\dfrac{12x^3 - 108x}{(3+x^2)^3} = 0$

$$\begin{aligned}\therefore\ 12x^3 - 108x &= 0 \quad \{\text{since } 3 + x^2 > 0 \text{ for all } x \in \mathbb{R}\}\\ \therefore\ 12x(x^2 - 9) &= 0\\ \therefore\ 12x(x+3)(x-3) &= 0\\ \therefore\ x &= 0, -3, \text{ or } 3\end{aligned}$$

**9** **a** $f(x) = -10\sin 2x \cos 2x,\quad 0 \leqslant x \leqslant \pi$

$\therefore\ f(x) = -5\sin 4x \quad \{2\sin A\cos A = \sin 2A\}$

**b** $f'(x) = -20\cos 4x$

If $f'(x) = 0,\quad -20\cos 4x = 0$

$$\begin{aligned}\therefore\ \cos 4x &= 0\\ \therefore\ 4x &= \tfrac{\pi}{2} + n\pi, \quad n \text{ any integer}\\ \therefore\ x &= \tfrac{\pi}{8} + \tfrac{n\pi}{4}\end{aligned}$$

So, for the domain $0 \leqslant x \leqslant \pi$, $x = \frac{\pi}{8}, \frac{3\pi}{8}, \frac{5\pi}{8}, \frac{7\pi}{8}$

# Chapter 16

## PROPERTIES OF CURVES

### EXERCISE 16A

**1** **a** We seek the tangent to $y = x - 2x^2 + 3$ at $x = 2$.

When $x = 2$, $y = 2 - 2(2)^2 + 3 = -3$

$\therefore$ the point of contact is $(2, -3)$.

Now $\dfrac{dy}{dx} = 1 - 4x$, so at $x = 2$,

$\dfrac{dy}{dx} = 1 - 8 = -7$

$\therefore$ the tangent has equation

$$\frac{y-(-3)}{x-2} = -7$$

$\therefore$ $y + 3 = -7(x - 2)$

$\therefore$ $y = -7x + 14 - 3$

$\therefore$ $y = -7x + 11$

**b** We seek the tangent to

$y = \sqrt{x} + 1 = x^{\frac{1}{2}} + 1$ at $x = 4$.

When $x = 4$, $y = \sqrt{4} + 1 = 3$

$\therefore$ the point of contact is $(4, 3)$.

Now $\dfrac{dy}{dx} = \dfrac{1}{2\sqrt{x}}$, so at $x = 4$,

$\dfrac{dy}{dx} = \dfrac{1}{2\sqrt{4}} = \frac{1}{4}$

$\therefore$ the tangent has equation

$$\frac{y-3}{x-4} = \frac{1}{4}$$

$\therefore$ $4y - 12 = x - 4$

$\therefore$ $4y = x + 8$

**c** We seek the tangent to $y = x^3 - 5x$ at $x = 1$.

When $x = 1$, $y = 1^3 - 5(1) = -4$

$\therefore$ the point of contact is $(1, -4)$.

Now $\dfrac{dy}{dx} = 3x^2 - 5$, so at $x = 1$,

$\dfrac{dy}{dx} = 3 - 5 = -2$

$\therefore$ the tangent has equation

$$\frac{y-(-4)}{x-1} = -2$$

$\therefore$ $y + 4 = -2x + 2$

$\therefore$ $y = -2x - 2$

**d** We seek the tangent to $y = \dfrac{4}{\sqrt{x}}$ at $(1, 4)$.

Now $y = \dfrac{4}{\sqrt{x}} = 4x^{-\frac{1}{2}}$

$\therefore$ $\dfrac{dy}{dx} = -2x^{-\frac{3}{2}}$ so at $x = 1$,

$\dfrac{dy}{dx} = -2\left(1^{-\frac{3}{2}}\right) = -2$

$\therefore$ the tangent has equation

$$\frac{y-4}{x-1} = -2$$

$\therefore$ $y - 4 = -2x + 2$

$\therefore$ $y = -2x + 6$

**e** We seek the tangent to

$y = \dfrac{3}{x} - \dfrac{1}{x^2} = 3x^{-1} - x^{-2}$ at $(-1, -4)$.

Now $\dfrac{dy}{dx} = -3x^{-2} + 2x^{-3}$

$= -\dfrac{3}{x^2} + \dfrac{2}{x^3}$ so at $(1, -4)$,

$\dfrac{dy}{dx} = -\dfrac{3}{(-1)^2} + \dfrac{2}{(-1)^3}$

$= -3 - 2$

$= -5$

$\therefore$ the tangent has equation

$$\frac{y-(-4)}{x-(-1)} = -5$$

$\therefore$ $y + 4 = -5x - 5$

$\therefore$ $y = -5x - 9$

**f** We seek the tangent to

$y = 3x^2 - \dfrac{1}{x} = 3x^2 - x^{-1}$ at $x = -1$.

When $x = -1$, $y = 3(-1)^2 - \frac{1}{(-1)} = 4$

$\therefore$ the point of contact is $(-1, 4)$.

Now $\dfrac{dy}{dx} = 6x + x^{-2}$

$= 6x + \dfrac{1}{x^2}$ so at $x = -1$,

$\dfrac{dy}{dx} = 6(-1) + \dfrac{1}{(-1)^2} = -5$

$\therefore$ the tangent has equation

$$\frac{y-4}{x-(-1)} = -5$$

$\therefore$ $y - 4 = -5x - 5$

$\therefore$ $y = -5x - 1$

**2** **a** We seek the normal to $y = x^2$ at $(3, 9)$.

Now $\dfrac{dy}{dx} = 2x$ so at $x = 3$,

$\dfrac{dy}{dx} = 2(3) = 6 = \frac{6}{1}$

$\therefore$ the normal at (3, 9) has gradient $-\frac{1}{6}$, so the equation of the normal is

$$\frac{y-9}{x-3} = -\tfrac{1}{6}$$

$$\therefore \quad 6y - 54 = -x + 3$$

$$\therefore \quad 6y = -x + 57$$

**b** We seek the normal to $y = x^3 - 5x + 2$ at $x = -2$.

When $x = -2$, $y = (-2)^3 - 5(-2) + 2$
$= 4$

and so the point of contact is $(-2, 4)$.

Now $\dfrac{dy}{dx} = 3x^2 - 5$ so at $x = -2$,

$\dfrac{dy}{dx} = 3(-2)^2 - 5 = 7$

$\therefore$ the normal at $(-2, 4)$ has gradient $-\frac{1}{7}$, so the equation of the normal is

$$\frac{y-4}{x-(-2)} = -\tfrac{1}{7}$$

$$\therefore \quad 7y - 28 = -(x+2)$$

$$\therefore \quad 7y = -x + 26$$

**c** We seek the normal to $y = \dfrac{5}{\sqrt{x}} - \sqrt{x}$ at (1, 4).

Now $y = 5x^{-\frac{1}{2}} - x^{\frac{1}{2}}$

$\therefore \dfrac{dy}{dx} = -\frac{5}{2}x^{-\frac{3}{2}} - \frac{1}{2}x^{-\frac{1}{2}}$ so at $x = 1$,

$\dfrac{dy}{dx} = -\frac{5}{2}\left(1^{-\frac{3}{2}}\right) - \frac{1}{2}\left(1^{-\frac{1}{2}}\right)$

$= -\frac{5}{2} - \frac{1}{2} = -3$

$\therefore$ the normal at (1, 4) has gradient $\frac{1}{3}$, so the equation of the normal is

$$\frac{y-4}{x-1} = \tfrac{1}{3}$$

$$\therefore \quad 3y - 12 = x - 1$$

$$\therefore \quad 3y = x + 11$$

**d** We seek the normal to $y = 8\sqrt{x} - \dfrac{1}{x^2}$ at $x = 1$.

When $x = 1$, $y = 8\sqrt{1} - \dfrac{1}{1^2} = 7$

$\therefore$ the point of contact is (1, 7).

Now $y = 8\sqrt{x} - \dfrac{1}{x^2} = 8x^{\frac{1}{2}} - x^{-2}$

$\therefore \dfrac{dy}{dx} = 4x^{-\frac{1}{2}} + 2x^{-3}$ so at $x = 1$,

$\dfrac{dy}{dx} = 4 + 2 = 6$

$\therefore$ the normal at (1, 7) has gradient $-\frac{1}{6}$, so the equation of the normal is

$$\frac{y-7}{x-1} = -\tfrac{1}{6}$$

$$\therefore \quad 6y - 42 = -x + 1$$

$$\therefore \quad 6y = -x + 43$$

**3** **a** $y = 2x^3 + 3x^2 - 12x + 1$

$\therefore \dfrac{dy}{dx} = 6x^2 + 6x - 12$

Horizontal tangents have gradient $= 0$

so $6x^2 + 6x - 12 = 0$

$\therefore x^2 + x - 2 = 0$

$\therefore (x+2)(x-1) = 0$

$\therefore x = -2$ or $x = 1$

Now at $x = -2$,

$y = 2(-2)^3 + 3(-2)^2 - 12(-2) + 1$
$= 21$

and at $x = 1$,

$y = 2(1)^3 + 3(1)^2 - 12(1) + 1$
$= -6$

$\therefore$ the points of contact are $(-2, 21)$ and $(1, -6)$

$\therefore$ the tangents are $y = 21$ and $y = -6$.

**b** Now $y = 2\sqrt{x} + \dfrac{1}{\sqrt{x}} = 2x^{\frac{1}{2}} + x^{-\frac{1}{2}}$

$\therefore \dfrac{dy}{dx} = x^{-\frac{1}{2}} - \frac{1}{2}x^{-\frac{3}{2}} = \dfrac{1}{\sqrt{x}} - \dfrac{1}{2x\sqrt{x}}$

Horizontal tangents have gradient $= 0$

$\therefore \dfrac{1}{\sqrt{x}} - \dfrac{1}{2x\sqrt{x}} = 0$

$\therefore \dfrac{2x-1}{2x\sqrt{x}} = 0$

$\therefore 2x - 1 = 0$

$\therefore x = \frac{1}{2}$

Now at $x = \frac{1}{2}$,

$y = 2\sqrt{\frac{1}{2}} + \dfrac{1}{\sqrt{\frac{1}{2}}} = \dfrac{2\left(\frac{1}{2}\right) + 1}{\sqrt{\frac{1}{2}}} = \dfrac{2}{\sqrt{\frac{1}{2}}}$
$= 2\sqrt{2}$

$\therefore$ the only horizontal tangent touches at the curve at $(\frac{1}{2}, 2\sqrt{2})$.

**c** Now $y = 2x^3 + kx^2 - 3$

$\therefore \dfrac{dy}{dx} = 6x^2 + 2kx$

When $x = 2$, $\dfrac{dy}{dx} = 4$

$\therefore 6(2)^2 + 2k(2) = 4$

$\therefore 24 + 4k = 4$

$\therefore 4k = -20$

$\therefore k = -5$

**d** Now $y = 1 - 3x + 12x^2 - 8x^3$

$\therefore \dfrac{dy}{dx} = -3 + 24x - 24x^2$

When $x = 1$, $\dfrac{dy}{dx} = -3 + 24 - 24 = -3$

$\therefore$ the tangent at (1, 2) has gradient $-3$

The tangents to the curve have gradient $-3$ when $-3 + 24x - 24x^2 = -3$

$\therefore 24x^2 - 24x = 0$

$\therefore 24x(x-1) = 0$

$\therefore$ when $x = 0$ or $x = 1$

So the other $x$-value for which the tangent to the curve has gradient $-3$ is $x = 0$, and when $x = 0$, $y = 1 - 0 + 0 - 0 = 1$

$\therefore$ the tangent to the curve at (0, 1) is parallel to the tangent at (1, 2).

This tangent has equation $\dfrac{y-1}{x-0} = -3$ or $y = -3x + 1$.

**4** **a** Now $y = x^2 + ax + b$

$\therefore \dfrac{dy}{dx} = 2x + a$

At $x = 1$, $\dfrac{dy}{dx} = 2 + a$

$\therefore$ the gradient of the tangent to the curve at $x = 1$ will be $2 + a$.

However the equation of the tangent is $2x + y = 6$ or $y = -2x + 6$ and so the gradient of the tangent is $-2$.

$\therefore 2 + a = -2$

$\therefore a = -4$

So, the curve is $y = x^2 - 4x + b$.

We also know that the tangent contacts the curve when $x = 1$.

$\therefore 1^2 - 4(1) + b = -2(1) + 6$

$\therefore 1 - 4 + b = 4$

$\therefore b = 7$

$\therefore a = -4, \quad b = 7$

**b** Now $y = a\sqrt{x} + \dfrac{b}{\sqrt{x}} = ax^{\frac{1}{2}} + bx^{-\frac{1}{2}}$

$\therefore \dfrac{dy}{dx} = \dfrac{a}{2}x^{-\frac{1}{2}} - \dfrac{b}{2}x^{-\frac{3}{2}}$

$\therefore$ at $x = 4$, $\dfrac{dy}{dx} = \dfrac{a}{2}\left(4^{-\frac{1}{2}}\right) - \dfrac{b}{2}\left(4^{-\frac{3}{2}}\right)$

$= \dfrac{a}{2}\left(\dfrac{1}{2}\right) - \dfrac{b}{2}\left(\dfrac{1}{8}\right)$

$= \dfrac{a}{4} - \dfrac{b}{16}$

$\therefore$ the gradient of the tangent to the curve at $x = 4$ will be $\dfrac{a}{4} - \dfrac{b}{16} = \dfrac{4a - b}{16}$

However the equation of the *normal* is $4x + y = 22$ or $y = -4x + 22$.

$\therefore$ the normal has gradient $-4$.

$\therefore$ the tangent has gradient $\frac{1}{4}$, and so

$\dfrac{4a - b}{16} = \frac{1}{4}$

$\therefore 4a - b = 4$

$\therefore b = 4a - 4$ .... (1)

Also, at $x = 4$ the normal line intersects the curve.

$\therefore a\sqrt{4} + \dfrac{b}{\sqrt{4}} = -4(4) + 22$

$\therefore 2a + \dfrac{b}{2} = 6$

Consequently, $2a + \dfrac{4a - 4}{2} = 6$ {using (1)}

$\therefore 2a + 2a - 2 = 6$

$\therefore 4a = 8$

$\therefore a = 2$

and so $b = 4(2) - 4 = 4$ {from (1)}

**c** $y = 2x^2 - 1$

$\therefore \dfrac{dy}{dx} = 4x$

$\therefore$ at the point where $x = a$, $\dfrac{dy}{dx} = 4a$

$\therefore$ the gradient of the tangent at the point where $x = a$ is $4a$.

Also, at $x = a$, $y = 2a^2 - 1$.

$\therefore$ the tangent has equation

$\dfrac{y - (2a^2 - 1)}{x - a} = 4a$

$\therefore y - 2a^2 + 1 = 4a(x - a)$

$\therefore y - 2a^2 + 1 = 4ax - 4a^2$

$\therefore 4ax - y = 2a^2 + 1$

**5** **a** $y=\sqrt{2x+1}$

When $x=4$, $y=\sqrt{2(4)+1}=3$,

so the point of contact is $(4, 3)$.

Now $\dfrac{dy}{dx}=\frac{1}{2}(2x+1)^{-\frac{1}{2}}(2)=\dfrac{1}{\sqrt{2x+1}}$

$\therefore$ at $x=4$, $\dfrac{dy}{dx}=\dfrac{1}{\sqrt{2(4)+1}}=\frac{1}{3}$

$\therefore$ the tangent has equation $\dfrac{y-3}{x-4}=\frac{1}{3}$

or $3y=x+5$.

**b** $y=\dfrac{1}{2-x}=(2-x)^{-1}$

$\therefore$ at $x=-1$, $y=\dfrac{1}{2-(-1)}=\frac{1}{3}$

So the point of contact is $(-1, \frac{1}{3})$.

Now $\dfrac{dy}{dx}=-1(2-x)^{-2}(-1)=\dfrac{1}{(2-x)^2}$

$\therefore$ at $x=-1$, $\dfrac{dy}{dx}=\dfrac{1}{(2-(-1))^2}=\frac{1}{9}$

$\therefore$ the tangent has equation

$$\frac{y-\frac{1}{3}}{x-(-1)}=\frac{1}{9}$$

$\therefore$ $9y-3=x+1$

$\therefore$ $9y=x+4$

**c** We seek the tangent to $f(x)=\dfrac{x}{1-3x}$ at $(-1, -\frac{1}{4})$.

$f(x)$ is a quotient where

$u=x$ and $v=1-3x$

$\therefore$ $u'=1$ and $v'=-3$

Now $f'(x)=\dfrac{u'v-uv'}{v^2}$ {quotient rule}

$\therefore$ $f'(x)=\dfrac{1(1-3x)-x(-3)}{(1-3x)^2}$

$=\dfrac{1}{(1-3x)^2}$

$\therefore$ $f'(-1)=\dfrac{1}{(1-3(-1))^2}=\frac{1}{16}$

$\therefore$ the tangent has equation

$$\frac{y-\left(-\frac{1}{4}\right)}{x-(-1)}=\frac{1}{16}$$

$\therefore$ $16y+4=x+1$

$\therefore$ $16y=x-3$

**d** We seek the tangent to $f(x)=\dfrac{x^2}{1-x}$ at $(2, -4)$.

$f(x)$ is a quotient where

$u=x^2$ and $v=1-x$

$\therefore$ $u'=2x$ and $v'=-1$

Now $f'(x)=\dfrac{u'v-uv'}{v^2}$ {quotient rule}

$\therefore$ $f'(x)=\dfrac{2x(1-x)-x^2(-1)}{(1-x)^2}$

$=\dfrac{2x-2x^2+x^2}{(1-x)^2}=\dfrac{2x-x^2}{(1-x)^2}$

$\therefore$ $f'(2)=\dfrac{2(2)-2^2}{(1-2)^2}=\dfrac{4-4}{1}=0$

As the tangent has gradient 0, it is horizontal.

$\therefore$ its equation is $y=c$

Since the contact point is $(2, -4)$, the tangent has equation $y=-4$.

**6** **a** We seek the normal to $y=\dfrac{1}{(x^2+1)^2}$ at $(1, \frac{1}{4})$.

As $y=(x^2+1)^{-2}$,

$\dfrac{dy}{dx}=-2(x^2+1)^{-3}(2x)=\dfrac{-4x}{(x^2+1)^3}$

$\therefore$ at $x=1$, $\dfrac{dy}{dx}=\dfrac{-4}{(1+1)^3}=\frac{-4}{8}=-\frac{1}{2}$

$\therefore$ the normal at $(1, \frac{1}{4})$ has gradient 2.

So the equation of the normal is

$$\frac{y-\frac{1}{4}}{x-1}=2$$

$\therefore$ $y-\frac{1}{4}=2x-2$

$\therefore$ $y=2x-\frac{7}{4}$

**b** $y=\dfrac{1}{\sqrt{3-2x}}$

$\therefore$ at $x=-3$, $y=\dfrac{1}{\sqrt{3-2(-3)}}=\frac{1}{3}$

$\therefore$ the point of contact is $(-3, \frac{1}{3})$

Now $y=(3-2x)^{-\frac{1}{2}}$

$\therefore$ $\dfrac{dy}{dx}=-\frac{1}{2}(3-2x)^{-\frac{3}{2}}(-2)=(3-2x)^{-\frac{3}{2}}$

$\therefore$ at $x=-3$, $\dfrac{dy}{dx}=(3-2(-3))^{-\frac{3}{2}}$

$=9^{-\frac{3}{2}}=3^{-3}=\frac{1}{27}$

$\therefore$ the normal at $(-3, \frac{1}{3})$ has gradient $-27$.

So the equation of the normal is

$$\frac{y-\frac{1}{3}}{x-(-3)}=-27$$

$\therefore$ $y-\frac{1}{3}=-27(x+3)$

$\therefore$ $y=-27x-\frac{242}{3}$

**c** $f(x) = \sqrt{x}(1-x)^2$

Since $f(4) = \sqrt{4}(1-4)^2 = 18$,

the point of contact is $(4, 18)$.

Now $f(x)$ is a product where

$u = x^{\frac{1}{2}}$ and $v = (1-x)^2$

$\therefore\ u' = \frac{1}{2}x^{-\frac{1}{2}}$ and $v' = 2(1-x)(-1)$
$= -2(1-x)$

Now $f'(x) = u'v + uv'$ {product rule}

$\therefore\ f'(x) = \frac{1}{2}x^{-\frac{1}{2}}(1-x)^2 - x^{\frac{1}{2}}2(1-x)$

$\therefore\ f'(4) = \frac{1}{2\sqrt{4}}(1-4)^2 - \sqrt{4}(2)(1-4)$
$= \frac{1}{4}(9) - 2(2)(-3) = \frac{57}{4}$

$\therefore$ the normal at (4, 18) has gradient $-\frac{4}{57}$.

So, the equation of the normal is

$$\frac{y-18}{x-4} = -\frac{4}{57}$$

$\therefore\ 57(y-18) = -4(x-4)$

$\therefore\ 57y = -4x + 1042$

**d** $f(x) = \dfrac{x^2-1}{2x+3}$

Since $f(-1) = \dfrac{(-1)^2-1}{2(-1)+3} = \dfrac{0}{1} = 0$,

the point of contact is $(-1, 0)$.

Now $f(x)$ is a quotient where

$u = x^2 - 1$ and $v = 2x + 3$

$\therefore\ u' = 2x$ and $v' = 2$

Now $f'(x) = \dfrac{u'v - uv'}{v^2}$
$= \dfrac{2x(2x+3) - (x^2-1)(2)}{(2x+3)^2}$

$\therefore\ f'(-1)$
$= \dfrac{2(-1)(-2+3) - ((-1)^2-1)(2)}{(2(-1)+3)^2}$
$= \dfrac{-2(1) - (0)(2)}{(1)^2} = -2$

$\therefore$ the normal at $(-1, 0)$ has gradient $\frac{1}{2}$.

So, the equation of the normal is

$$\frac{y-0}{x-(-1)} = \frac{1}{2}$$

or $2y = x + 1$

**7** The tangent has equation $3x + y = 5$ or $y = -3x + 5$

$\therefore$ the tangent has gradient $-3$ .... (1)

Also, at $x = -1$, $y = -3(-1) + 5 = 8$

$\therefore$ the tangent contacts the curve at $(-1, 8)$ .... (2)

Now $y = a(1-bx)^{\frac{1}{2}}$, so $\dfrac{dy}{dx} = \frac{1}{2}a(1-bx)^{-\frac{1}{2}}(-b)$

$\therefore\ -3 = \frac{1}{2}a(1+b)^{-\frac{1}{2}}(-b)$ {using (1)}

$\therefore\ 6 = \dfrac{ab}{\sqrt{1+b}}$ .... (3)

Using (2), $(-1, 8)$ must lie on the curve $y = a\sqrt{1-bx}$

$\therefore\ 8 = a\sqrt{1+b}$ .... (4)

$\therefore\ \dfrac{6\sqrt{1+b}}{b} = \dfrac{8}{\sqrt{1+b}}$ {equating $a$ s in (3) and (4)}

$\therefore\ 6(1+b) = 8b$

$\therefore\ 6 + 6b = 8b$

$\therefore\ 6 = 2b$

$\therefore\ b = 3$ and $a = \frac{8}{\sqrt{4}} = 4$

**8** **a** $f(x) = e^{-x}$

$\therefore\ f(1) = e^{-1}$

$\therefore$ the point of contact is $\left(1, \dfrac{1}{e}\right)$.

Now $f'(x) = -e^{-x}$

$\therefore\ f'(1) = -e^{-1} = -\dfrac{1}{e}$

So, the gradient of the tangent is $-\dfrac{1}{e}$

$\therefore$ the tangent has equation $\dfrac{y - \frac{1}{e}}{x-1} = -\dfrac{1}{e}$

$\therefore\ e\left(y - \dfrac{1}{e}\right) = -(x-1)$

$\therefore\ ey - 1 = -x + 1$

$\therefore\ x + ey = 2$

or $y = -\dfrac{1}{e}x + \dfrac{2}{e}$

**b** $y = \ln(2 - x)$

so when $x = -1$, $y = \ln 3$

$\therefore$ the point of contact is $(-1, \ln 3)$.

Now $\dfrac{dy}{dx} = \dfrac{-1}{2-x}$

$\therefore$ when $x = -1$, $\dfrac{dy}{dx} = -\dfrac{1}{2+1} = -\frac{1}{3}$

So, the gradient of the tangent is $-\frac{1}{3}$.

$\therefore$ the tangent has equation $\dfrac{y - \ln 3}{x+1} = -\frac{1}{3}$

$\therefore$ $3(y - \ln 3) = -(x+1)$

$\therefore$ $3y - 3\ln 3 = -x - 1$

$\therefore$ $x + 3y = 3\ln 3 - 1$

**c** $y = \ln\sqrt{x}$
$= \ln x^{\frac{1}{2}}$
$= \frac{1}{2}\ln x$

$\therefore$ when $y = -1$, $-1 = \frac{1}{2}\ln x$

$\therefore$ $\ln x = -2$

$\therefore$ $x = e^{-2}$

$\therefore$ $x = \dfrac{1}{e^2}$ $\qquad\therefore$ the point of contact is $\left(\dfrac{1}{e^2}, -1\right)$

Now $\dfrac{dy}{dx} = \frac{1}{2}\,\dfrac{1}{x} = \dfrac{1}{2x}$, so at the point of contact, $\dfrac{dy}{dx} = \dfrac{1}{2e^{-2}} = \dfrac{e^2}{2}$

$\therefore$ the tangent has gradient $\dfrac{e^2}{2}$ and the normal has gradient $-\dfrac{2}{e^2}$

$\therefore$ the normal has equation $\dfrac{y+1}{x - \dfrac{1}{e^2}} = -\dfrac{2}{e^2}$

$\therefore$ $e^2(y+1) = -2\left(x - \dfrac{1}{e^2}\right)$

$\therefore$ $e^2y + e^2 = -2x + \dfrac{2}{e^2}$

$\therefore$ $2x + e^2y = \dfrac{2}{e^2} - e^2$ or $y = -\dfrac{2}{e^2}x + \dfrac{2}{e^4} - 1$

**9** $y = \dfrac{\cos x}{1 + \sin x}$ $\qquad\therefore$

$$\frac{dy}{dx} = \frac{(-\sin x)(1+\sin x) - \cos x(\cos x)}{(1+\sin x)^2}$$
$$= \frac{-\sin x - \sin^2 x - \cos^2 x}{(1+\sin x)^2}$$
$$= \frac{-1 - \sin x}{(1+\sin x)^2} \qquad \{\sin^2 x + \cos^2 x = 1\}$$
$$= -\frac{(1+\sin x)}{(1+\sin x)^2}$$
$$= \frac{-1}{1+\sin x}$$

Since $\dfrac{-1}{1+\sin x}$ never equals 0, there are no horizontal tangents.

**10** **a** $y = \sin x$ $\quad\therefore$ $\dfrac{dy}{dx} = \cos x$

When $x = 0$, $\dfrac{dy}{dx} = \cos 0 = 1$

$\therefore$ the tangent has equation $\dfrac{y-0}{x-0} = 1$

or $y = x$

**b** $y = \tan x$ $\quad\therefore$ $\dfrac{dy}{dx} = \dfrac{1}{\cos^2 x}$

When $x = 0$, $\dfrac{dy}{dx} = \dfrac{1}{\cos^2 0} = 1$

$\therefore$ the tangent has equation $\dfrac{y-0}{x-0} = 1$

or $y = x$

**c** $y = \cos x \quad \therefore \quad \dfrac{dy}{dx} = -\sin x$

When $x = \frac{\pi}{6}$, $y = \frac{\sqrt{3}}{2}$

and $\dfrac{dy}{dx} = -\sin\left(\frac{\pi}{6}\right) = -\frac{1}{2}$

So, the normal has gradient 2,

and its equation is $\dfrac{y - \frac{\sqrt{3}}{2}}{x - \frac{\pi}{6}} = 2$

$\therefore \quad y - \frac{\sqrt{3}}{2} = 2x - \frac{\pi}{3}$

$\therefore \quad 2x - y = \frac{\pi}{3} - \frac{\sqrt{3}}{2}$

**d** $y = \dfrac{1}{\sin(2x)} = (\sin(2x))^{-1}$

$\therefore \quad \dfrac{dy}{dx} = -1(\sin(2x))^{-2} \times 2\cos(2x)$

$= -\dfrac{2\cos(2x)}{(\sin(2x))^2}$

When $x = \frac{\pi}{4}$, $y = 1$

and $\dfrac{dy}{dx} = -\dfrac{2\cos(\frac{\pi}{2})}{\left(\sin(\frac{\pi}{2})\right)^2} = 0$

$\therefore$ the gradient of the normal is undefined, so the normal is $x = \frac{\pi}{4}$.

**11** **a** Consider the tangent to $y = x^3$ at $x = 2$.
When $x = 2$, $y = 2^3 = 8$ so the point of contact is (2, 8)

Now $\dfrac{dy}{dx} = 3x^2$ and so at $x = 2$,

$\dfrac{dy}{dx} = 3(2)^2 = 12$

$\therefore$ the tangent at (2, 8) has gradient 12 and

its equation is $\dfrac{y-8}{x-2} = 12$

$\therefore \quad y - 8 = 12x - 24$

$\therefore \quad y = 12x - 16$

$\therefore$ the tangent meets the curve where

$12x - 16 = x^3$

$\therefore \quad x^3 - 12x + 16 = 0$

Because the tangent touches the curve at $x = 2$, there must be a repeated solution at this point.

$\therefore$ $(x-2)^2$ must be a factor of this cubic

$\therefore$ $(x-2)^2(x+4) = 0$

$\therefore$ the tangent meets the curve again when $x = -4$.

When $x = -4$, $y = (-4)^3 = -64$

$\therefore$ the tangent meets the curve again at $(-4, -64)$.

**b** Consider the tangent to
$y = -x^3 + 2x^2 + 1$ at $x = -1$.

When $x = -1$, $y = -(-1)^3 + 2(-1)^2 + 1$
$= 4$

and so the point of contact is $(-1, 4)$.

Now $\dfrac{dy}{dx} = -3x^2 + 4x$ and so at $x = -1$,

$\dfrac{dy}{dx} = -3(-1)^2 + 4(-1) = -7$

$\therefore$ the tangent at $(-1, 4)$ has gradient $-7$

and its equation is $\dfrac{y-4}{x-(-1)} = -7$

$\therefore \quad y - 4 = -7(x+1)$

$\therefore \quad y = -7x - 3$

$\therefore$ the tangent meets the curve where

$-7x - 3 = -x^3 + 2x^2 + 1$

$\therefore \quad x^3 - 2x^2 - 7x - 4 = 0$

Because the tangent touches the curve at $x = -1$, there must be a repeated solution at this point.

$\therefore$ $(x+1)^2$ must be a factor of this cubic

$\therefore$ $(x+1)^2(x-4) = 0$

$\therefore$ the tangent meets the curve again when $x = 4$.

When $x = 4$, $y = -(4)^3 + 2(4)^2 + 1$
$= -64 + 32 + 1 = -31$

$\therefore$ the tangent meets the curve again at $(4, -31)$.

**12** **a** $f(x) = x^2 + \dfrac{4}{x^2}$

$\therefore \quad f'(x) = 2x - 2 \times \dfrac{4}{x^3}$

$\therefore \quad f'(x) = 2x - \dfrac{8}{x^3}$

**b** Horizontal tangents have gradient 0, so

$2x - \dfrac{8}{x^3} = 0$

$\therefore \quad 2x^4 = 8$

$\therefore \quad x^4 = 4$

$\therefore \quad x = \pm\sqrt{2}$

**c** When $x = -\sqrt{2}$,

$f(-\sqrt{2}) = \left(-\sqrt{2}\right)^2 + \dfrac{4}{\left(-\sqrt{2}\right)^2} = 2 + \dfrac{4}{2} = 4$

$\therefore$ the horizontal tangent at $(-\sqrt{2}, 4)$ is $y = 4$.

When $x = \sqrt{2}$,

$f(\sqrt{2}) = \left(\sqrt{2}\right)^2 + \dfrac{4}{\left(\sqrt{2}\right)^2} = 2 + \dfrac{4}{2} = 4$

$\therefore$ the horizontal tangent at $(\sqrt{2}, 4)$ is $y = 4$.

$\therefore$ the tangents are the same line because they have the same equation.

**13** $y = x^2e^x$ so when $x = 1,\ y = e$

$\therefore$ the point of contact is $(1, e)$.

Now $\dfrac{dy}{dx} = 2xe^x + x^2e^x$

$\therefore$ when $x = 1,\ \dfrac{dy}{dx} = 2e + e = 3e$

$\therefore$ the tangent has equation $\dfrac{y-e}{x-1} = 3e$

$\therefore\ y - e = 3ex - 3e$

$\therefore\ y - 3ex = -2e$

$\therefore\ 3ex - y = 2e$

The tangent cuts the $x$-axis when

$y = 0$

$\therefore\ 3ex = 2e$

$\therefore\ x = \frac{2}{3}$

and the $y$-axis when

$x = 0$

$\therefore\ -y = 2e$

$\therefore\ y = -2e$

So, A is $(\frac{2}{3}, 0)$ and B is $(0, -2e)$.

**14** **a** Consider the tangent to $y = x^2 - x + 9$ at $x = a$.

When $x = a,\ y = a^2 - a + 9$, so the point of contact is $(a, a^2 - a + 9)$.

Now $\dfrac{dy}{dx} = 2x - 1$ and so at $x = a,\ \dfrac{dy}{dx} = 2a - 1$

$\therefore$ the gradient of the tangent at $(a, a^2 - a + 9)$ is $2a - 1$

$\therefore$ the equation of the tangent is $\dfrac{y - (a^2 - a + 9)}{x - a} = 2a - 1$

$\therefore\ y - (a^2 - a + 9) = (2a - 1)(x - a)$

$\therefore\ y = (2a - 1)x - 2a^2 + a + a^2 - a + 9$

$\therefore\ y = (2a - 1)x - a^2 + 9$ .... (1)

But this tangent passes through $(0, 0)$, so $0 = a^2 - 9$

$\therefore\ (a + 3)(a - 3) = 0$

$\therefore\ a = \pm 3$

$\therefore$ the tangents are: At $a = 3,\ y = (2(3) - 1)x - 3^2 + 9$ {from (1)}

$\therefore\ y = 5x$, with contact at $(3, 15)$.

At $a = -3,\ y = (2(-3) - 1)x - (-3)^2 + 9$ {from (1)}

$\therefore\ y = -7x$, with contact at $(-3, 21)$.

**b** Let $(a, a^3)$ lie on $y = x^3$.

Now $\dfrac{dy}{dx} = 3x^2$, so at $x = a,\ \dfrac{dy}{dx} = 3a^2$

$\therefore$ the gradient of the tangent at $(a, a^3)$ is $3a^2$

$\therefore$ the equation of the tangent is $\dfrac{y - a^3}{x - a} = 3a^2$ or $y - a^3 = (3a^2)(x - a)$

But this tangent passes through $(-2, 0)$, so $0 - a^3 = 3a^2(-2 - a)$

$\therefore\ -a^3 = -6a^2 - 3a^3$

$\therefore\ 2a^3 + 6a^2 = 0$

$\therefore\ 2a^2(a + 3) = 0$

$\therefore\ a = 0$ or $-3$

If $a = 0$, the tangent equation is $y = 0$, with contact point $(0, 0)$.

If $a = -3$, the tangent equation is $y - (-27) = 27(x + 3)$

$\therefore\ y = 27x + 54$, with contact point $(-3, -27)$.

**c** Let $(a, \sqrt{a})$ lie on $y = \sqrt{x}$.

Now $\dfrac{dy}{dx} = \frac{1}{2}x^{-\frac{1}{2}} = \dfrac{1}{2\sqrt{x}}$, so at $x = a,\ \dfrac{dy}{dx} = \dfrac{1}{2\sqrt{a}}$

$\therefore$ the gradient of the tangent at $(a, \sqrt{a})$ is $\dfrac{1}{2\sqrt{a}}$

and the gradient of the normal at this point is $-2\sqrt{a}$.

$\therefore$ the normal has equation $\dfrac{y-\sqrt{a}}{x-a}=-2\sqrt{a}$

or $y-\sqrt{a}=-2\sqrt{a}(x-a)$.

But this normal passes through $(4, 0)$, so $0-\sqrt{a}=-2\sqrt{a}(4-a)$

$\therefore\ 2\sqrt{a}(4-a)-\sqrt{a}=0$

$\therefore\ \sqrt{a}(8-2a-1)=0$

$\therefore\ \sqrt{a}(7-2a)=0$

$\therefore\ a=0$ or $\frac{7}{2}$

But $a=0$ is the endpoint of the function, so there is no normal here.

When $a=\frac{7}{2}$, $\quad y-\sqrt{\frac{7}{2}}=-2\sqrt{\frac{7}{2}}\left(x-\frac{7}{2}\right)$

$\therefore\ \sqrt{2}y-\sqrt{7}=-2\sqrt{7}\left(x-\frac{7}{2}\right)$

$\therefore\ \sqrt{2}y+2\sqrt{7}x=7\sqrt{7}+\sqrt{7}$

$\therefore\ \sqrt{2}y+2\sqrt{7}x=8\sqrt{7}$

$\therefore\ y=-\sqrt{14}x+4\sqrt{14}$ with contact point $\left(\frac{7}{2}, \sqrt{\frac{7}{2}}\right)$.

**15** $y=e^x$ so when $x=a$, $y=e^a$

$\therefore$ the point of contact is $(a, e^a)$.

Now $\dfrac{dy}{dx}=e^x$

$\therefore$ at the point $(a, e^a)$, $\dfrac{dy}{dx}=e^a$

$\therefore$ the tangent has equation $\dfrac{y-e^a}{x-a}=e^a$

or $y-e^a=e^a(x-a)$ .... $(*)$

$\therefore\ y=e^ax+e^a-ae^a$

Since the tangent passes through the origin, $(0, 0)$ must satisfy $(*)$

$\therefore\ 0-e^a=e^a(0-a)$

$\therefore\ -e^a=-ae^a$

$\therefore\ e^a(a-1)=0$

$\therefore\ a=1 \quad$ {as $e^a>0$}

So the equation of the tangent is

$y-e=ex-e$ or $y=ex$.

**16** **a**

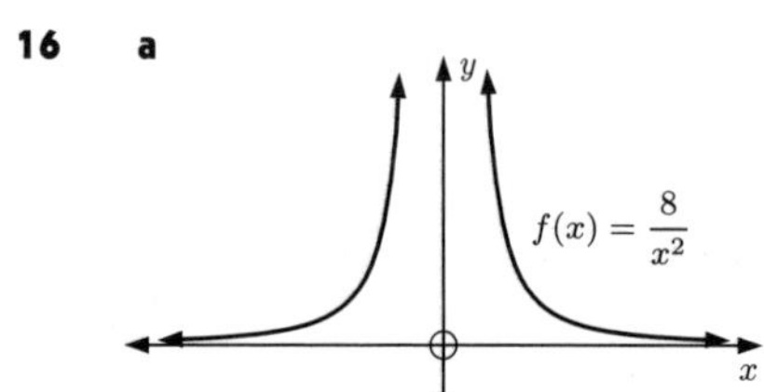


**b** Let $\left(a, \dfrac{8}{a^2}\right)$ lie on $f(x)=\dfrac{8}{x^2}=8x^{-2}$

Now $f'(x)=-16x^{-3}=-\dfrac{16}{x^3}$

$\therefore\ f'(a)=-\dfrac{16}{a^3}$

$\therefore$ the gradient of the tangent at $\left(a, \dfrac{8}{a^2}\right)$ is $-\dfrac{16}{a^3}$

$\therefore$ the equation of the tangent is $\dfrac{y-\frac{8}{a^2}}{x-a}=-\dfrac{16}{a^3}$

$\therefore\ a^3y-8a=-16x+16a$

$\therefore\ 16x+a^3y=24a$

**c** The tangent cuts the $x$-axis when $y=0$

$\therefore\ 16x=24a$

$\therefore\ x=\frac{3}{2}a$

$\therefore$ A is $(\frac{3}{2}a, 0)$.

The tangent cuts the $y$-axis when $x=0$

$\therefore\ a^3y=24a$

$\therefore\ y=\dfrac{24}{a^2}$

$\therefore$ B is $\left(0, \dfrac{24}{a^2}\right)$.

**d** Area of triangle OAB

$=\left|\frac{1}{2}\times\left(\frac{3}{2}a\right)\times\left(\dfrac{24}{a^2}\right)\right|$

$=\dfrac{18}{|a|}$ units$^2$

As $a\to\infty$, $\dfrac{18}{a}\to 0$

$\therefore$ area $\to 0$

**17** $y = 3e^{-x}$ and $y = 2 + e^x$ meet when $3e^{-x} = 2 + e^x$

$$\therefore \quad 3 = 2e^x + e^{2x} \quad \{\times e^x\}$$
$$\therefore \quad e^{2x} + 2e^x - 3 = 0$$
$$\therefore \quad (e^x + 3)(e^x - 1) = 0$$
$$\therefore \quad e^x = -3 \text{ or } 1$$
$$\therefore \quad e^x = 1 \text{ and so } x = 0 \quad \{\text{as } e^x > 0\}$$

Now when $x = 0$, $y = 3e^0 = 3$, so the graphs meet at $(0, 3)$.

For $y = 2 + e^x$, $\dfrac{dy}{dx} = e^x$,

so at the point $(0, 3)$, $\dfrac{dy}{dx} = e^0 = 1$

$\therefore$ the gradient of the tangent at this point is 1

$\therefore$ the tangent has direction vector $\begin{pmatrix} 1 \\ 1 \end{pmatrix}$

For $y = 3e^{-x}$, $\dfrac{dy}{dx} = -3e^{-x}$,

so at the point $(0, 3)$, $\dfrac{dy}{dx} = -3$

$\therefore$ the gradient of the tangent at this point is $-3$

$\therefore$ the tangent has direction vector $\begin{pmatrix} 1 \\ -3 \end{pmatrix}$

If $\theta$ is the acute angle between the tangents, then $\cos\theta = \dfrac{|1(1) + 1(-3)|}{\sqrt{1^2+1^2}\sqrt{1^2+(-3)^2}} = \dfrac{|-2|}{\sqrt{2}\sqrt{10}} = \dfrac{2}{\sqrt{20}}$

$$\therefore \quad \theta \approx 63.43°$$

**18** **a** $y = ax^2, a > 0$ touches $y = \ln x$ when $ax^2 = \ln x$

If the curves touch when $x = b$ then $ab^2 = \ln b$ .... (1)

Now for $y = ax^2$, $\dfrac{dy}{dx} = 2ax$ and for $y = \ln x$, $\dfrac{dy}{dx} = \dfrac{1}{x}$

$\therefore$ when $x = b$, $\dfrac{dy}{dx} = 2ab$ $\therefore$ when $x = b$, $\dfrac{dy}{dx} = \dfrac{1}{b}$

Since the curves touch each other, they share a common tangent. $\therefore \dfrac{1}{b} = 2ab$ .... (2)

**b** Now $ab^2 = \frac{1}{2}$ {from (2)}

and $ab^2 = \ln b$ {from (1)}

$\therefore \ln b = \frac{1}{2}$

$\therefore b = e^{\frac{1}{2}} = \sqrt{e}$

When $x = b = \sqrt{e}$, $y = \ln x = \ln e^{\frac{1}{2}} = \frac{1}{2}$

$\therefore$ the point of contact is $(\sqrt{e}, \frac{1}{2})$.

**c** $a = \dfrac{1}{2b^2}$ {from (2)}

$\therefore a = \dfrac{1}{2(\sqrt{e})^2} = \dfrac{1}{2e}$

**d** The tangent has gradient $\dfrac{1}{b} = \dfrac{1}{\sqrt{e}}$ and passes through $(\sqrt{e}, \frac{1}{2})$

$\therefore$ the tangent is $\dfrac{y - \frac{1}{2}}{x - \sqrt{e}} = \dfrac{1}{\sqrt{e}}$ $\therefore y - \frac{1}{2} = \dfrac{1}{\sqrt{e}}\left(x - \sqrt{e}\right)$

$$\therefore \quad y - \tfrac{1}{2} = \frac{1}{\sqrt{e}}x - 1$$
$$\therefore \quad y = e^{-\frac{1}{2}}x - \tfrac{1}{2}$$

## EXERCISE 16B

**1** **a**

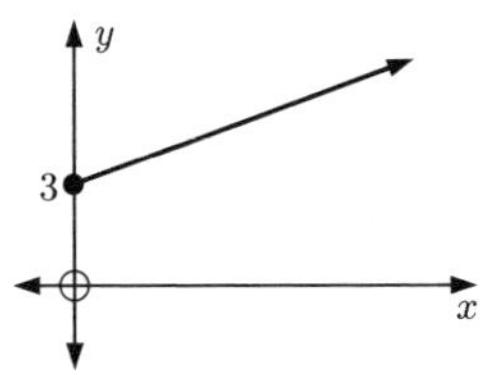


**i** $x \geqslant 0$ **ii** never

**b**

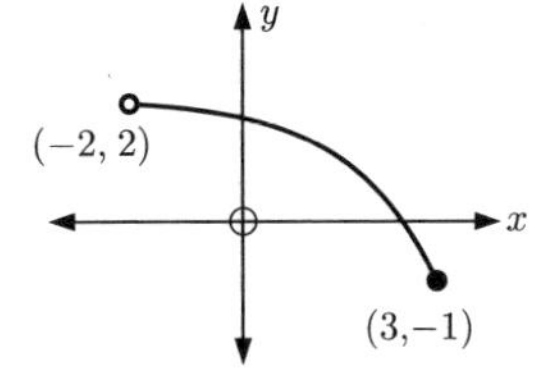


**i** never **ii** $-2 < x \leqslant 3$

**c**

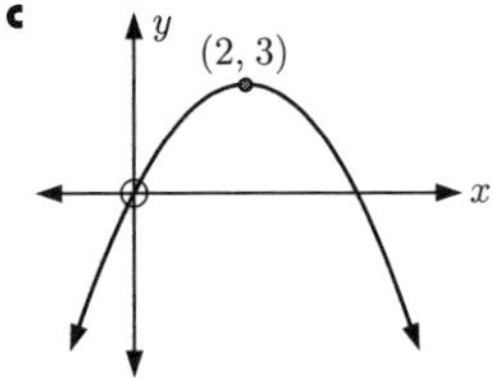


**i** $x \leqslant 2$ **ii** $x \geqslant 2$

**d**

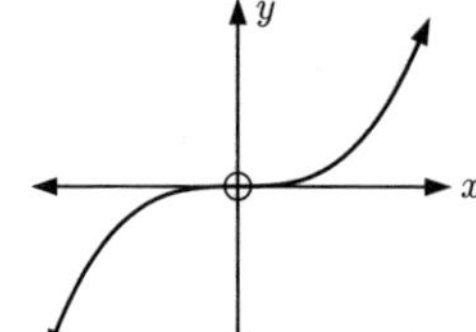


**i** all real $x$ **ii** never

**e**

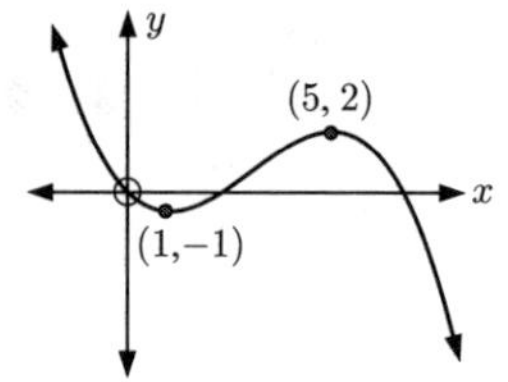


**i** $1 \leqslant x \leqslant 5$
**ii** $x \leqslant 1,\ x \geqslant 5$

**f**

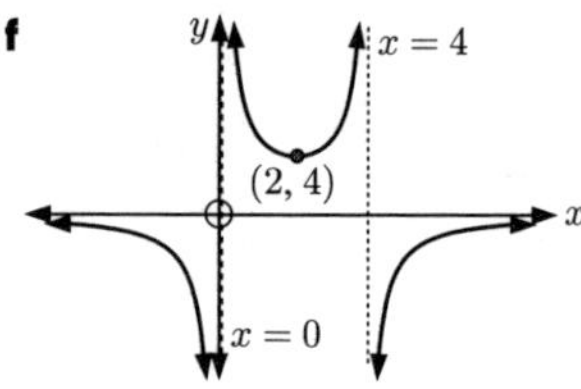


**i** $2 \leqslant x < 4,\quad x > 4$
**ii** $x < 0,\quad 0 < x \leqslant 2$

**2** **a** $f(x) = x^2,\quad f'(x) = 2x$
Sign diagram of $f'(x)$: $-$ | 0 | $+$
increasing when $x \geqslant 0$,
decreasing when $x \leqslant 0$

**b** $f(x) = -x^3,\quad f'(x) = -3x^2$
Sign diagram of $f'(x)$: $-$ | 0 | $-$
decreasing for all $x$

**c** $f(x) = 2x^2 + 3x - 4,\quad f'(x) = 4x + 3$
Sign diagram of $f'(x)$: $-$ | $-\frac{3}{4}$ | $+$
increasing when $x \geqslant -\frac{3}{4}$,
decreasing when $x \leqslant -\frac{3}{4}$

**d** $f(x) = \sqrt{x} = x^{\frac{1}{2}}$,
$f'(x) = \frac{1}{2}x^{-\frac{1}{2}} = \dfrac{1}{2\sqrt{x}}$
Sign diagram of $f'(x)$: (undefined) | 0 | $+$
$f(x)$ is only defined when $x \geqslant 0$
increasing when $x \geqslant 0$, never decreasing

**e** $f(x) = \dfrac{2}{\sqrt{x}} = 2x^{-\frac{1}{2}}$
$f'(x) = -x^{-\frac{3}{2}} = \dfrac{-1}{x\sqrt{x}}$
Sign diagram of $f'(x)$: (undefined) | 0 | $-$
$f(x)$ is only defined for $x > 0$
never increasing, decreasing when $x > 0$

**f** $f(x) = x^3 - 6x^2,\quad f'(x) = 3x^2 - 12x$
$= 3x(x-4)$
Sign diagram of $f'(x)$: $+$ | 0 | $-$ | 4 | $+$
increasing when $x \leqslant 0$ or $x \geqslant 4$,
decreasing when $0 \leqslant x \leqslant 4$

**g** $f(x) = e^x,\quad f'(x) = e^x$

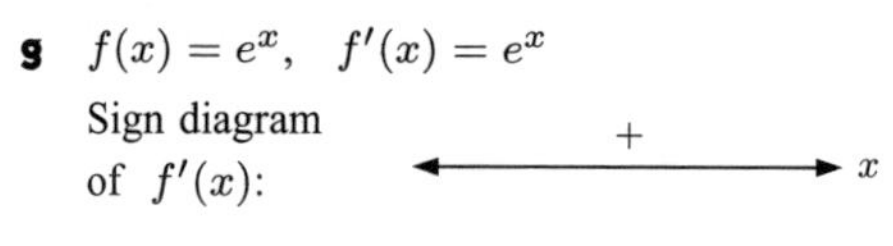


$f(x)$ is increasing for all $x$

**h** $f(x) = \ln x,\quad f'(x) = \dfrac{1}{x}$
Sign diagram of $f'(x)$: (undefined) | 0 | $+$
$f(x)$ is only defined when $x > 0$
increasing when $x > 0$, never decreasing

**i** $f(x) = -2x^3 + 4x$
$f'(x) = -6x^2 + 4$
$= -2(3x^2 - 2)$

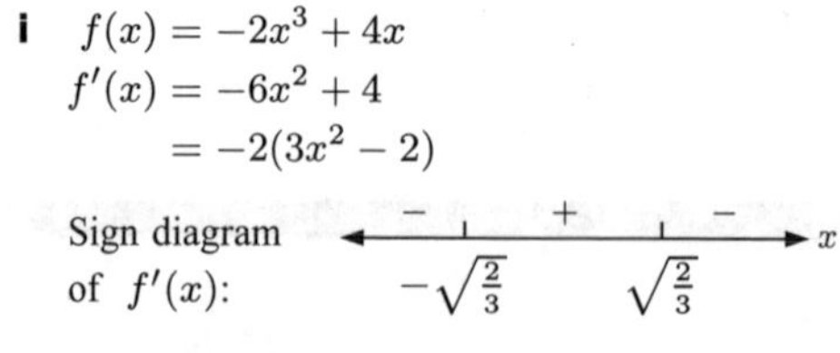


Sign diagram of $f'(x)$: $-$ | $-\sqrt{\frac{2}{3}}$ | $+$ | $\sqrt{\frac{2}{3}}$ | $-$

increasing for $-\sqrt{\frac{2}{3}} \leqslant x \leqslant \sqrt{\frac{2}{3}}$,
decreasing for $x \leqslant -\sqrt{\frac{2}{3}}$ or $x \geqslant \sqrt{\frac{2}{3}}$

**j** $f(x) = -4x^3 + 15x^2 + 18x + 3$
$f'(x) = -12x^2 + 30x + 18$
$= -6(2x^2 - 5x - 3)$
$= -6(2x+1)(x-3)$
Sign diagram of $f'(x)$: $-$ | $-\frac{1}{2}$ | $+$ | 3 | $-$
increasing when $-\frac{1}{2} \leqslant x \leqslant 3$,
decreasing when $x \leqslant -\frac{1}{2}$ or $x \geqslant 3$

**k** $f(x) = 3 + e^{-x}$, $f'(x) = -e^{-x}$

Sign diagram of $f'(x)$: $-$ for all $x$

$f(x)$ is decreasing for all $x$.

**l** $f(x) = xe^x$, $f'(x) = e^x + xe^x = e^x(1+x)$

Sign diagram of $f'(x)$: $-$ | $-1$ | $+$

increasing when $x \geqslant -1$
decreasing when $x \leqslant -1$.

**m** $f(x) = 3x^4 - 16x^3 + 24x^2 - 2$,

$f'(x) = 12x^3 - 48x^2 + 48x$
$= 12x(x^2 - 4x + 4)$
$= 12x(x-2)^2$

Sign diagram of $f'(x)$: $-$ | $0$ | $+$ | $2$ | $+$

increasing when $x \geqslant 0$,
decreasing when $x \leqslant 0$

**n** $f(x) = 2x^3 + 9x^2 + 6x - 7$,

$f'(x) = 6x^2 + 18x + 6$
$= 6(x^2 + 3x + 1)$

$f'(x) = 0$ when $x = \frac{-3 \pm \sqrt{9-4}}{2} = \frac{-3 \pm \sqrt{5}}{2}$

Sign diagram of $f'(x)$: $+$ | $\frac{-3-\sqrt{5}}{2}$ | $-$ | $\frac{-3+\sqrt{5}}{2}$ | $+$

increasing for $x \leqslant \frac{-3-\sqrt{5}}{2}$ or $x \geqslant \frac{-3+\sqrt{5}}{2}$,
decreasing for $\frac{-3-\sqrt{5}}{2} \leqslant x \leqslant \frac{-3+\sqrt{5}}{2}$

**o** $f(x) = x^3 - 6x^2 + 3x - 1$,

$f'(x) = 3x^2 - 12x + 3$
$= 3(x^2 - 4x + 1)$

$f'(x) = 0$ when $x = \frac{4 \pm \sqrt{16-4}}{2} = 2 \pm \sqrt{3}$

Sign diagram of $f'(x)$: $+$ | $2-\sqrt{3}$ | $-$ | $2+\sqrt{3}$ | $+$

increasing when $x \leqslant 2 - \sqrt{3}$
or $x \geqslant 2 + \sqrt{3}$,
decreasing when $2 - \sqrt{3} \leqslant x \leqslant 2 + \sqrt{3}$

**p** $f(x) = x - 2\sqrt{x} = x - 2x^{\frac{1}{2}}$

$f'(x) = 1 - x^{-\frac{1}{2}} = 1 - \dfrac{1}{\sqrt{x}} = \dfrac{\sqrt{x} - 1}{\sqrt{x}}$

Sign diagram of $f'(x)$: (undefined for $x < 0$) $0$ | $-$ | $1$ | $+$

increasing when $x \geqslant 1$,
decreasing when $0 \leqslant x \leqslant 1$

**3** **a** $f(x) = \dfrac{4x}{x^2+1}$ is a quotient with

$u = 4x$ and $v = x^2 + 1$
$\therefore$ $u' = 4$ and $v' = 2x$

$$\therefore \quad f'(x) = \frac{4(x^2+1) - 4x \times 2x}{(x^2+1)^2} = \frac{4x^2 + 4 - 8x^2}{(x^2+1)^2} = \frac{4 - 4x^2}{(x^2+1)^2} = \frac{-4(x^2-1)}{(x^2+1)^2} = \frac{-4(x+1)(x-1)}{(x^2+1)^2}$$

Sign diagram of $f'(x)$:

$-$ | $-1$ | $+$ | $1$ | $-$

**b** $f(x)$ is increasing for $-1 \leqslant x \leqslant 1$,
decreasing for $x \leqslant -1$ and $x \geqslant 1$

**4** **a** $f(x) = \dfrac{4x}{(x-1)^2}$ is a quotient with

$u = 4x$ and $v = (x-1)^2$
$\therefore$ $u' = 4$ and $v' = 2(x-1)$

$$\therefore \quad f'(x) = \frac{4(x-1)^2 - 8x(x-1)}{(x-1)^4} = \frac{4(x-1)[(x-1) - 2x]}{(x-1)^4} = \frac{4(-1-x)}{(x-1)^3} = \frac{-4(x+1)}{(x-1)^3}$$

Sign diagram of $f'(x)$:

**b** $f(x)$ is increasing for $-1 \leqslant x < 1$,
decreasing for $x \leqslant -1$ and $x > 1$

**5** **a** $f(x) = \dfrac{-x^2+4x-7}{x-1}$ is a quotient with $u = -x^2+4x-7$ and $v = x-1$

$\therefore$ $u' = -2x+4$ and $v' = 1$

$$\therefore \quad f'(x) = \frac{(-2x+4)(x-1)-(-x^2+4x-7)(1)}{(x-1)^2}$$
$$= \frac{-2x^2+6x-4+x^2-4x+7}{(x-1)^2}$$
$$= \frac{-x^2+2x+3}{(x-1)^2}$$
$$= \frac{-(x^2-2x-3)}{(x-1)^2}$$
$$= \frac{-(x+1)(x-3)}{(x-1)^2}$$

Sign diagram of $f'(x)$:

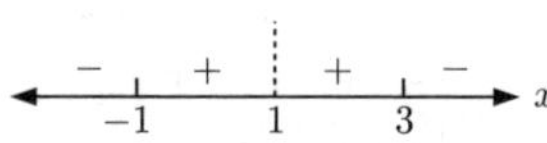


**b** $f(x)$ is increasing for $-1 \leqslant x < 1$ and $1 < x \leqslant 3$, and decreasing for $x \leqslant -1$ and $x \geqslant 3$.

**6** **a** $f(x) = \dfrac{x^3}{x^2-1}$ is a quotient with $u = x^3$ and $v = x^2-1$

$\therefore$ $u' = 3x^2$ and $v' = 2x$

$$\therefore \quad f'(x) = \frac{3x^2(x^2-1)-x^3 \times 2x}{(x^2-1)^2}$$
$$= \frac{3x^4-3x^2-2x^4}{(x^2-1)^2}$$
$$= \frac{x^2(x^2-3)}{(x^2-1)^2}$$
$$= \frac{x^2(x+\sqrt{3})(x-\sqrt{3})}{(x^2-1)^2}$$

Sign diagram of $f'(x)$:

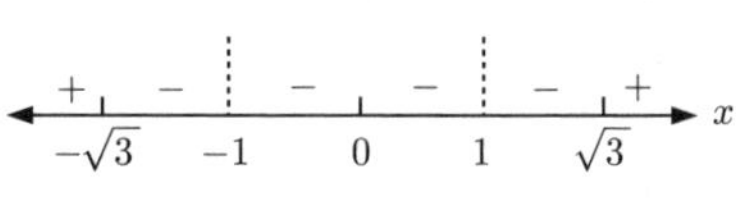


$\therefore$ $f(x)$ is increasing for $x \leqslant -\sqrt{3}$ and $x \geqslant \sqrt{3}$, and decreasing for $-\sqrt{3} \leqslant x < -1$, $-1 < x < 1$, and $1 < x \leqslant \sqrt{3}$.

**b** $f(x) = e^{-x^2}$

$\therefore$ $f'(x) = -2xe^{-x^2}$

Sign diagram of $f'(x)$:



$\therefore$ $f(x)$ is increasing for $x \leqslant 0$ and decreasing for $x \geqslant 0$.

**c** $f(x) = x^2 + \dfrac{4}{x-1} = x^2 + 4(x-1)^{-1}$

$$\therefore \quad f'(x) = 2x - 4(x-1)^{-2} \times 1$$
$$= 2x - \frac{4}{(x-1)^2}$$
$$= \frac{2x(x-1)^2-4}{(x-1)^2}$$
$$= \frac{2x(x^2-2x+1)-4}{(x-1)^2}$$
$$= \frac{2x^3-4x^2+2x-4}{(x-1)^2}$$
$$= \frac{(x-2)(2x^2+2)}{(x-1)^2}$$

Sign diagram of $f'(x)$:

$\therefore$ $f(x)$ is increasing for $x \geqslant 2$, and decreasing for $x < 1$ and $1 < x \leqslant 2$.

**d** $f(x) = \dfrac{e^{-x}}{x}$ is a quotient with $u = e^{-x}$ and $v = x$

$\therefore\ u' = -e^{-x}$ and $v' = 1$

$$\therefore\ f'(x) = \frac{-e^{-x}x - e^{-x} \times 1}{x^2} = \frac{-e^{-x}(x+1)}{x^2}$$

Sign diagram of $f'(x)$:

$\therefore$ $f(x)$ is increasing for $x \leqslant -1$, and decreasing for $-1 \leqslant x < 0$ and $x > 0$.

## EXERCISE 16C

**1** **a** A is a *local maximum*, B is a *stationary inflection*, C is a *local minimum*.

**b** $f'(x)$ has sign diagram:

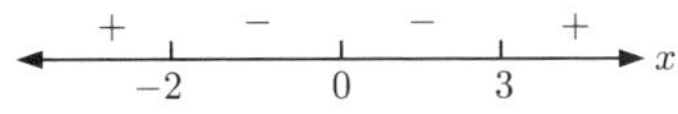


**c** **i** $f(x)$ is increasing for $x \leqslant -2$ and $x \geqslant 3$ **ii** $f(x)$ is decreasing for $-2 \leqslant x \leqslant 3$

**d** $f(x)$ has sign diagram:

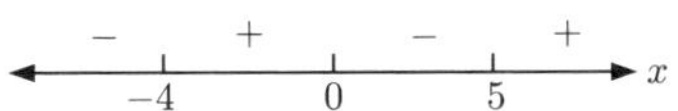


**2** **a** $f(x) = x^2 - 2$ $\therefore\ f'(x) = 2x$

with sign diagram: – + 0 $x$

Now $f(0) = -2$,
so there is a local minimum at $(0, -2)$.

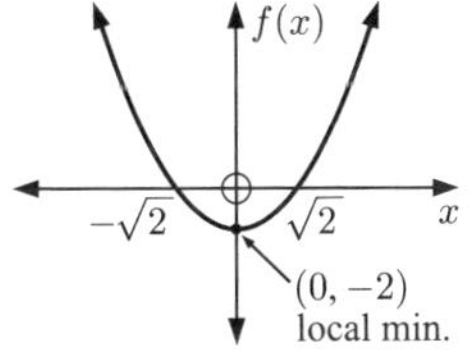


**b** $f(x) = x^3 + 1$ $\therefore\ f'(x) = 3x^2$

with sign diagram:

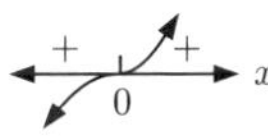


Now $f(0) = 1$,
so there is a stationary inflection at $(0, 1)$.

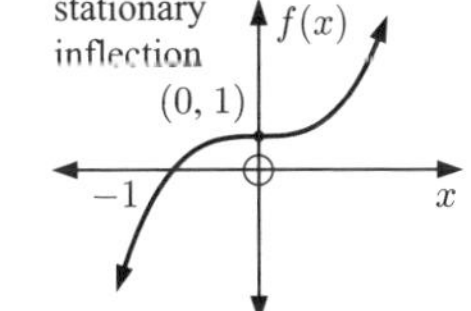


**c** $f(x) = x^3 - 3x + 2$

$\therefore\ f'(x) = 3x^2 - 3$
$= 3(x^2 - 1)$
$= 3(x+1)(x-1)$

with sign diagram: + – + –1 1 $x$

Now $f(-1) = 4$, $f(1) = 0$,
so there is a local maximum at $(-1, 4)$,
and a local minimum at $(1, 0)$.

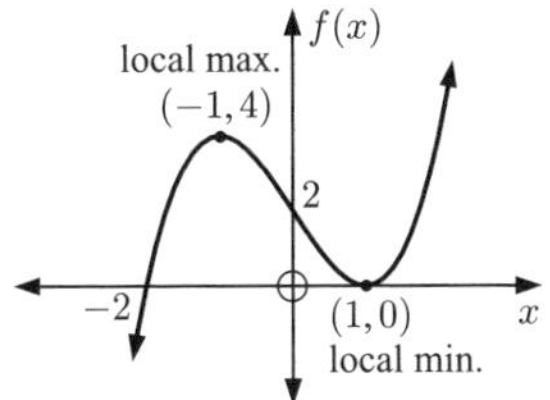


**d** $f(x) = x^4 - 2x^2$

$\therefore\ f'(x) = 4x^3 - 4x$
$= 4x(x^2 - 1)$
$= 4x(x+1)(x-1)$

with sign diagram: – + – + –1 0 1 $x$

Now $f(-1) = -1$, $f(1) = -1$, $f(0) = 0$,
so there are local minima at $(-1, -1)$ and
$(1, -1)$, and a local maximum at $(0, 0)$.

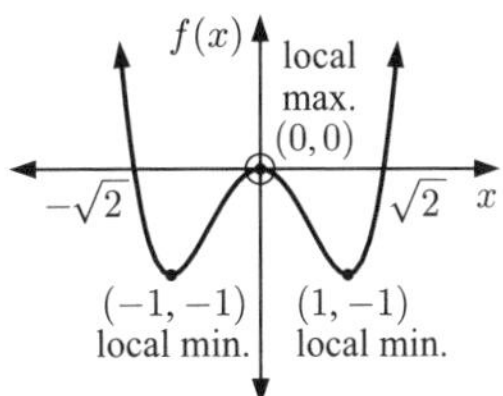

**e** $f(x) = x^3 - 6x^2 + 12x + 1$

$\therefore \quad f'(x) = 3x^2 - 12x + 12$
$= 3(x^2 - 4x + 4)$
$= 3(x-2)^2$

with sign diagram:

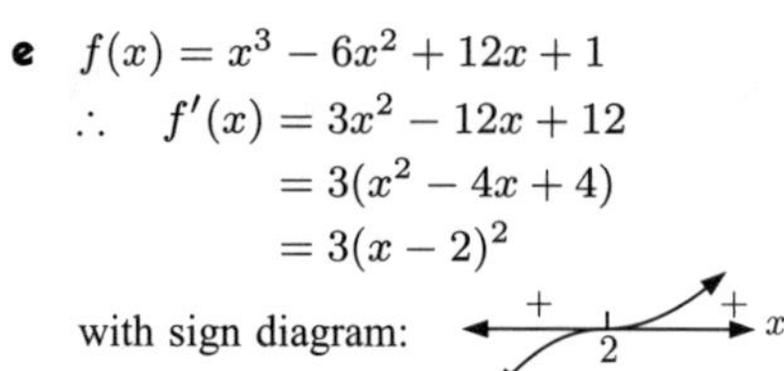

Now $f(2) = 9$, so there is a stationary inflection at (2, 9).

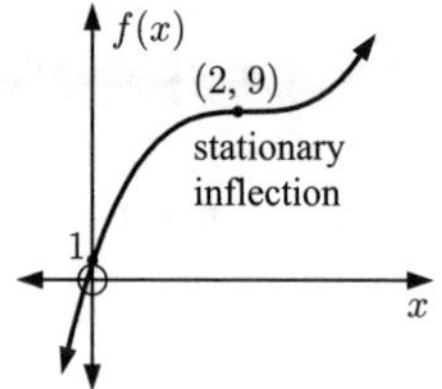


**f** $f(x) = \sqrt{x} + 2$

$\therefore \quad f'(x) = \frac{1}{2}x^{-\frac{1}{2}}$
$= \dfrac{1}{2\sqrt{x}} \neq 0$

with sign diagram: (shaded for $x < 0$; + for $x > 0$)

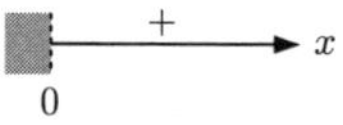

$\therefore$ no stationary points.

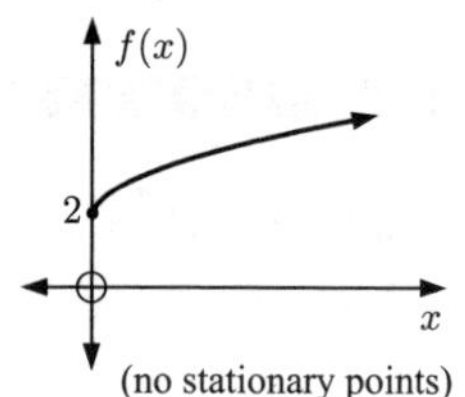


**g** $f(x) = x - \sqrt{x}$

$\therefore \quad f'(x) = 1 - \frac{1}{2}x^{-\frac{1}{2}}$
$= 1 - \dfrac{1}{2\sqrt{x}} = \dfrac{2\sqrt{x} - 1}{2\sqrt{x}}$

with sign diagram:

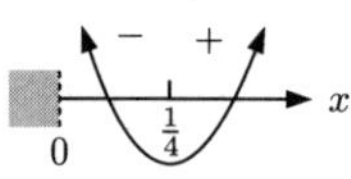


$f(x)$ is defined for all $x \geqslant 0$

Now $f(\frac{1}{4}) = -\frac{1}{4}$, so there is a local minimum at $(\frac{1}{4}, -\frac{1}{4})$.

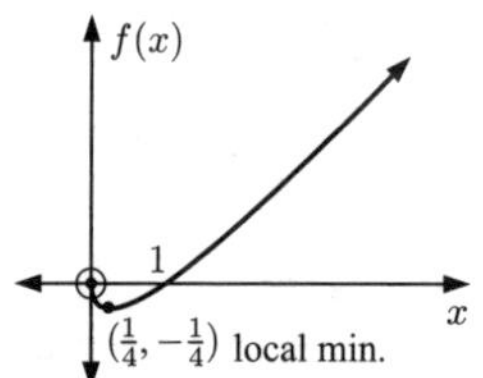


**h** $f(x) = x^4 - 6x^2 + 8x - 3$

$\therefore \quad f'(x) = 4x^3 - 12x + 8$
$= 4(x^3 - 3x + 2)$
$= 4(x-1)(x^2 + x - 2)$
$= 4(x-1)(x+2)(x-1)$

with sign diagram: (− for $x < -2$, + for $-2 < x < 1$, + for $x > 1$)

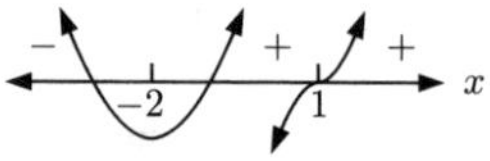

Now $f(-2) = -27$, $f(1) = 0$, so there is a local minimum at $(-2, -27)$, and a stationary inflection at $(1, 0)$.

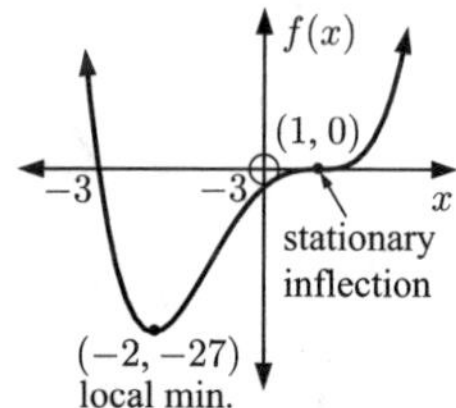


**i** $f(x) = 1 - x\sqrt{x} = 1 - x^{\frac{3}{2}}$

$\therefore \quad f'(x) = -\frac{3}{2}x^{\frac{1}{2}} = \dfrac{-3\sqrt{x}}{2}$

with sign diagram: (shaded for $x < 0$; − for $x > 0$)

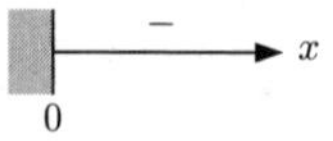

$f(x)$ is only defined when $x \geqslant 0$

Now $f(0) = 1$, so there is a local maximum at (0, 1).

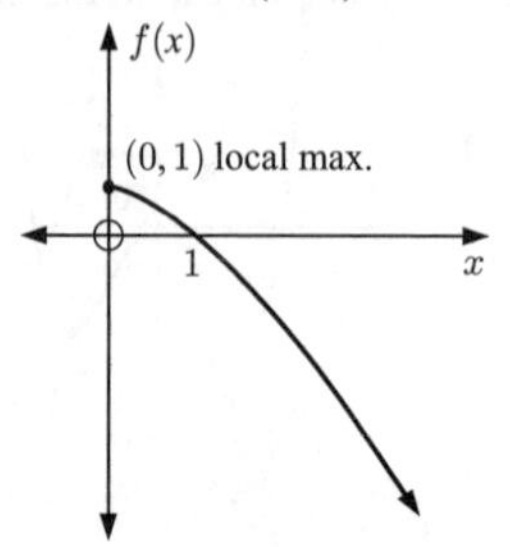


**j** $f(x) = x^4 - 2x^2 - 8$

$\therefore \quad f'(x) = 4x^3 - 4x$
$= 4x(x^2 - 1) = 4x(x+1)(x-1)$

with sign diagram: (− for $x < -1$, + for $-1 < x < 0$, − for $0 < x < 1$, + for $x > 1$)

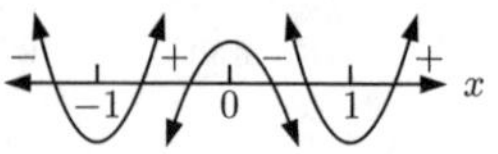

Now $f(-1) = -9$, $f(1) = -9$, $f(0) = -8$, so there are local minima at $(-1, -9)$ and $(1, -9)$, and a local maximum at $(0, -8)$.

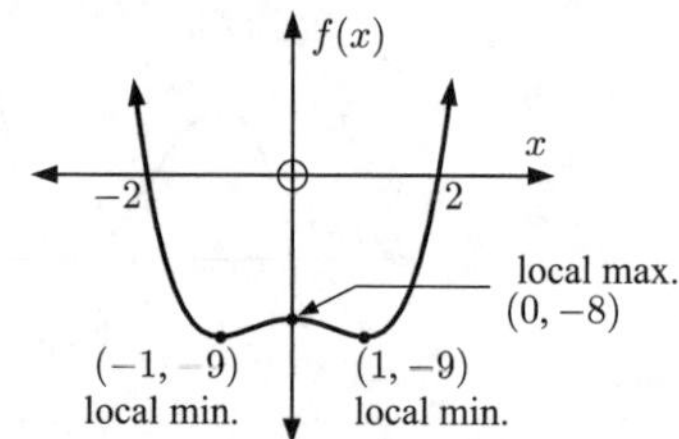

**3** $f(x) = ax^2 + bx + c, \quad a \neq 0$

$\therefore \quad f'(x) = 2ax + b$

$f(x)$ has a stationary point when $f'(x) = 0$

$\therefore \quad x = -\dfrac{b}{2a}$

There is a local maximum when $a < 0$  and there is a local minimum when $a > 0$ 

**4** **a** $y = xe^{-x}$

$\therefore \quad \dfrac{dy}{dx} = 1e^{-x} - xe^{-x}$ {product rule}

$= e^{-x}(1 - x)$

$= \dfrac{1 - x}{e^x}$ which has sign diagram:

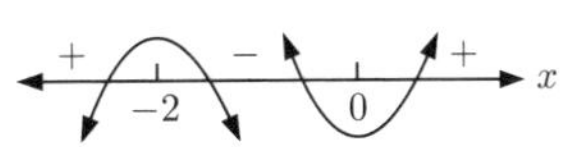


When $x = 1$, $y = 1e^{-1} = \dfrac{1}{e}$, so we have a local maximum at $\left(1, \dfrac{1}{e}\right)$.

**b** $y = x^2e^x$

$\therefore \quad \dfrac{dy}{dx} = 2xe^x + x^2e^x$ {product rule}

$= xe^x(2 + x)$ which has sign diagram:

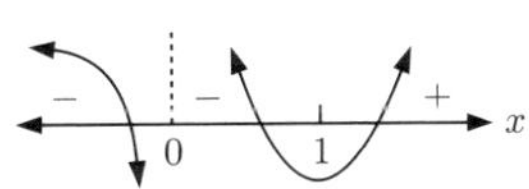


When $x = -2$, $y = 4e^{-2}$, and when $x = 0$, $y = 0$.

So, we have a local maximum at $\left(-2, \dfrac{4}{e^2}\right)$, and a local minimum at $(0, 0)$.

**c** $y = \dfrac{e^x}{x}$

$\therefore \quad \dfrac{dy}{dx} = \dfrac{e^x x - e^x\,(1)}{x^2}$ {quotient rule}

$= \dfrac{e^x(x - 1)}{x^2}$ which has sign diagram:

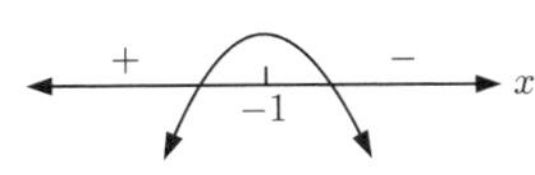


When $x = 1$, $y = \dfrac{e^1}{1} = e$, so we have a local minimum at $(1, e)$.

**d** $y = e^{-x}(x + 2)$

$\therefore \quad \dfrac{dy}{dx} = -e^{-x}(x + 2) + e^{-x}$ {product rule}

$= e^{-x}(-x - 2 + 1)$

$= e^{-x}(-x - 1)$ which has sign diagram:

+ −
−1 $x$

When $x = -1$, $y = e(-1 + 2) = e$, so we have a local maximum at $(-1, e)$.

**5** $f(x) = 2x^3 + ax^2 - 24x + 1$

$\therefore \quad f'(x) = 6x^2 + 2ax - 24$

But $f'(-4) = 0$, so $96 - 8a - 24 = 0$

$\therefore \quad 72 = 8a$

$\therefore \quad a = 9$

**6** **a** $f(x) = x^3 + ax + b$

$\therefore \quad f'(x) = 3x^2 + a$

But $f'(-2) = 0$

$\therefore \quad 3(-2)^2 + a = 0$

$\therefore \quad 12 + a = 0$

$\therefore \quad a = -12$

Also, $f(-2) = 3$

$\therefore \quad (-2)^3 - 12(-2) + b = 3$

$\therefore \quad -8 + 24 + b = 3$

$\therefore \quad b = -13$

**b** Now $f(x) = x^3 - 12x - 13$

$\therefore \ f'(x) = 3x^2 - 12$

$= 3(x^2 - 4)$

$= 3(x + 2)(x - 2)$ with sign diagram:

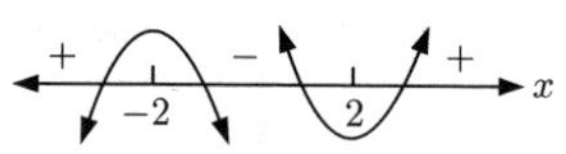


Now $f(2) = -29$, so there is a local minimum at $(2, -29)$ and a local maximum at $(-2, 3)$.

**7** **a** $f(x)$ is defined when $\ln x$ is defined $\therefore$ $f(x)$ is defined for $x > 0$

**b** $f'(x) = \ln x + \dfrac{x}{x}$ {product rule}

$= \ln x + 1$

which is 0 when $\ln x = -1$

$\therefore \ x = e^{-1}$

Sign diagram of $f'(x)$ is:

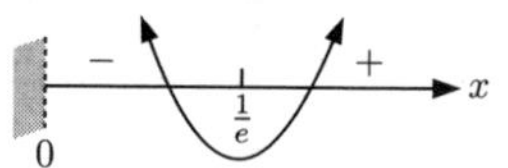


So, there is a local minimum at $\left(\frac{1}{e}, \frac{1}{e}\ln\frac{1}{e}\right)$

$\therefore$ the global minimum value of $f(x)$ is $\frac{1}{e}\ln e^{-1} = -\frac{1}{e}$

**8** **a** If $f(x) = \sin x$ then $f'(x) = \cos x$

Stationary points occur when $f'(x) = 0$,

which is when $x = \frac{\pi}{2}, \frac{3\pi}{2}$

Sign diagram for $f'(x)$ is:

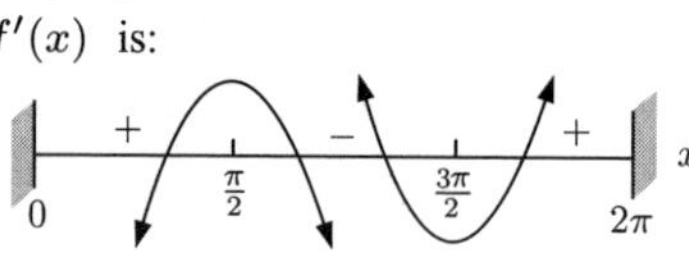


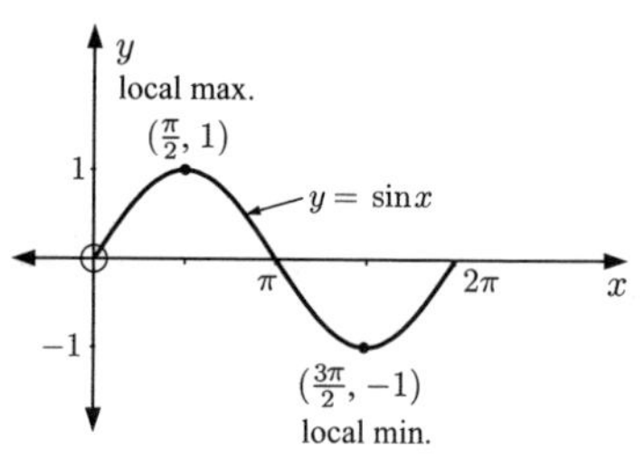


There is a local maximum at $(\frac{\pi}{2}, 1)$

and a local minimum at $(\frac{3\pi}{2}, -1)$.

**b** If $f(x) = \cos(2x)$ then $f'(x) = -2\sin(2x)$

$\therefore \ f'(x) = 0$ when $-2\sin(2x) = 0$

$\therefore \ \sin(2x) = 0$

$\therefore \ 2x = k\pi$ for any integer $k$

$\therefore \ x = \dfrac{k\pi}{2}$

On the domain $0 \leqslant x \leqslant 2\pi$, $f'(x) = 0$

when $x = 0, \frac{\pi}{2}, \pi, \frac{3\pi}{2}$, and $2\pi$.

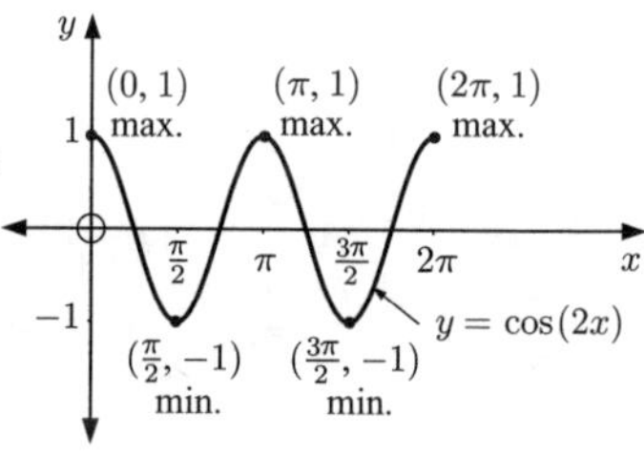


Sign diagram for $f'(x)$ is:

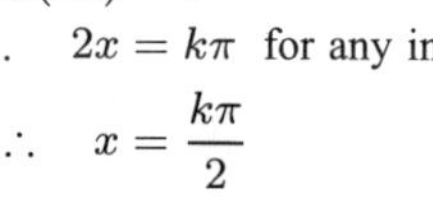


There are local maxima at $(0, 1)$, $(\pi, 1)$, $(2\pi, 1)$ and local minima at $(\frac{\pi}{2}, -1)$, $(\frac{3\pi}{2}, -1)$.

**c** If $f(x) = \sin^2 x$ then $f'(x) = 2\sin x\cos x = \sin(2x)$

$\therefore \ f'(x) = 0$ when $\sin(2x) = 0$

Using **b**, we know that on the domain $0 \leqslant x \leqslant 2\pi$

$f'(x) = 0$ when $x = 0, \frac{\pi}{2}, \pi, \frac{3\pi}{2}$, and $2\pi$.

Sign diagram for $f'(x)$ is:

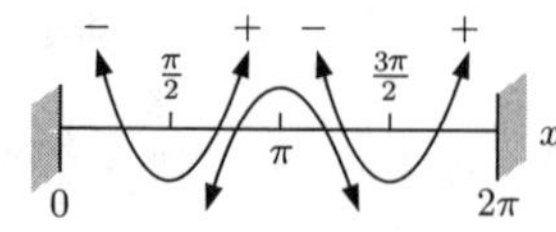


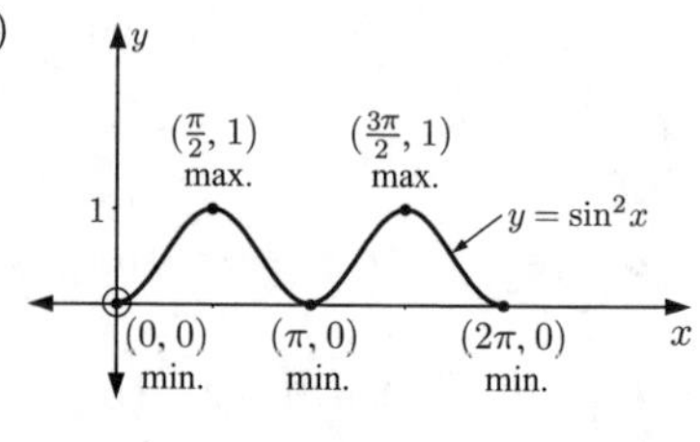


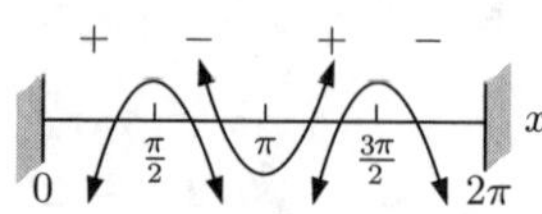

There are local mimima at $(0, 0)$, $(\pi, 0)$, $(2\pi, 0)$ and local maxima at $(\frac{\pi}{2}, 1)$, $(\frac{3\pi}{2}, 1)$.

**d** If $f(x) = e^{\sin x}$ then $f'(x) = e^{\sin x} \times \cos x$

$\therefore$ $f'(x) = 0$ when $\cos x\, e^{\sin x} = 0$

$\therefore$ $\cos x = 0$ $\{e^{\sin x} > 0$ for all $x\}$

$\therefore$ $x = \frac{\pi}{2} + k\pi$, $k$ an integer

On the domain $0 \leqslant x \leqslant 2\pi$, $f'(x) = 0$ when $x = \frac{\pi}{2}, \frac{3\pi}{2}$.

Sign diagram for $f'(x)$ is:

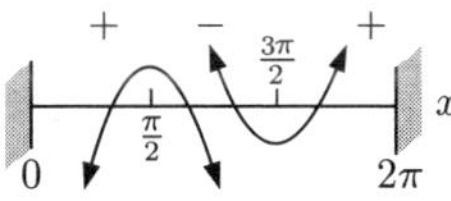

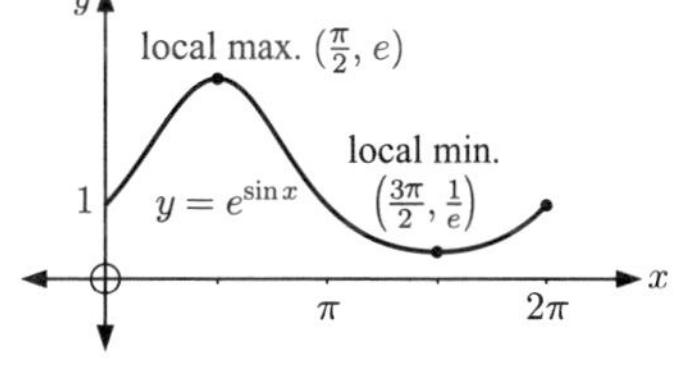


There is a local maximum at $(\frac{\pi}{2}, e)$ and a local minimum at $(\frac{3\pi}{2}, \frac{1}{e})$.

**e** If $f(x) = \sin(2x) + 2\cos x$ then $f'(x) = 2\cos(2x) - 2\sin x$

$\therefore$ $f'(x) = 0$ when $2\cos(2x) - 2\sin x = 0$

$\therefore$ $2(1 - 2\sin^2 x) - 2\sin x = 0$

$\therefore$ $-2(2\sin^2 x + \sin x - 1) = 0$

$\therefore$ $-2(2\sin x - 1)(\sin x + 1) = 0$

$\therefore$ when $\sin x = \frac{1}{2}$ or $\sin x = -1$

On the domain $0 \leqslant x \leqslant 2\pi$, when $x = \frac{\pi}{6}, \frac{5\pi}{6}, \frac{3\pi}{2}$.

Sign diagram of $f'(x)$:

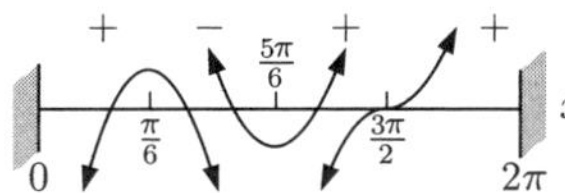

$f(\frac{\pi}{6}) = \sin(\frac{2\pi}{6}) + 2\cos(\frac{\pi}{6})$

$= \frac{\sqrt{3}}{2} + 2 \times \frac{\sqrt{3}}{2} = \frac{3\sqrt{3}}{2}$

$f(\frac{5\pi}{6}) = \sin(\frac{10\pi}{6}) + 2\cos(\frac{5\pi}{6})$

$= -\frac{\sqrt{3}}{2} + 2(-\frac{\sqrt{3}}{2}) = -\frac{3\sqrt{3}}{2}$

$f(\frac{3\pi}{2}) = \sin(3\pi) + 2\cos(\frac{3\pi}{2})$

$= 0 + 2 \times 0 = 0$

$\therefore$ there is a local maximum at $(\frac{\pi}{6}, \frac{3\sqrt{3}}{2})$, a local minimum at $(\frac{5\pi}{6}, -\frac{3\sqrt{3}}{2})$ and a stationary point of inflection at $(\frac{3\pi}{2}, 0)$.

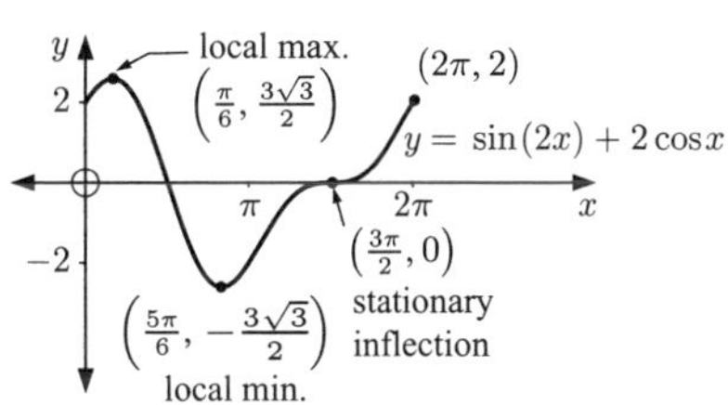


**9** Let the cubic polynomial be

$P(x) = ax^3 + bx^2 + cx + d$

$\therefore$ $P'(x) = 3ax^2 + 2bx + c$ .... (1)

Now (0, 2) lies on $P(x)$, so $P(0) = 2$

$\therefore$ $a(0) + b(0) + c(0) + d = 2$

$\therefore$ $d = 2$

The tangent at (0, 2) is $y = 9x + 2$, so

$P'(0) = 9$

$\therefore$ $3a(0) + 2b(0) + c = 9$

$\therefore$ $c = 9$ .... (2)

There is a stationary point at $(-1, -7)$, so $P'(-1) = 0$

$\therefore$ $3a(-1)^2 + 2b(-1) + c = 0$ {using (1)}

$\therefore$ $3a - 2b + c = 0$

So, using (2), $3a - 2b = -9$ .... (3)

Finally, $(-1, -7)$ lies on $P(x)$

$\therefore$ $a(-1)^3 + b(-1)^2 + c(-1) + d = -7$

$\therefore$ $-a + b - 9 + 2 = -7$

$\therefore$ $b - a = 0$

$\therefore$ $a = b$

So, using (3), $3a - 2a = -9$

$\therefore$ $a = -9$

$\therefore$ $a = b = -9$

$\therefore$ $P(x) = -9x^3 - 9x^2 + 9x + 2$

**10** **a** $f(x) = x^3 - 12x - 2$, for $-3 \leqslant x \leqslant 5$

$\therefore \quad f'(x) = 3x^2 - 12$

$= 3(x+2)(x-2)$

which is 0 when $x = -2$ or 2

| $x$ | $-3$ | $-2$ | $2$ | $5$ |
|---|---|---|---|---|
| $f(x)$ | 7 | 14 | $-18$ | 63 |

$\therefore$ the greatest value is 63 when $x = 5$, and the least value is $-18$ when $x = 2$.

**b** $f(x) = 4 - 3x^2 + x^3$, for $-2 \leqslant x \leqslant 3$

$\therefore \quad f'(x) = -6x + 3x^2$

$= 3x(x-2)$

which is 0 when $x = 0$ or 2

| $x$ | $-2$ | $0$ | $2$ | $3$ |
|---|---|---|---|---|
| $f(x)$ | $-16$ | 4 | 0 | 4 |

$\therefore$ greatest value is 4 when $x = 0$ or $x = 3$, least value is $-16$ when $x = -2$.

**11** $y = 4e^{-x} \sin x$

$\therefore \quad \frac{dy}{dx} = -4e^{-x} \sin x + 4e^{-x} \cos x$

$\therefore$ stationary points occur when $-4e^{-x} \sin x + 4e^{-x} \cos x = 0$

$\therefore \quad 4e^{-x}(\cos x - \sin x) = 0$

$\therefore \quad \cos x - \sin x = 0 \quad \{e^{-x} > 0 \text{ for all } x\}$

$\therefore \quad \sin x = \cos x$

$\therefore \quad \tan x = 1$

$\therefore \quad x = \frac{\pi}{4} + k\pi, \quad k$ an integer

Sign diagram of $\frac{dy}{dx}$ is:

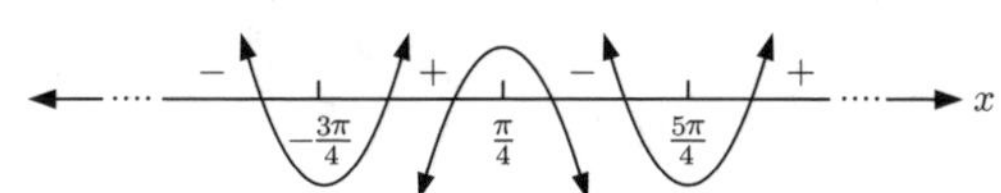


$\therefore \quad y = 4e^{-x} \sin x$ has a local maximum when $x = \frac{\pi}{4}$.

**12** Consider $f(x) = \frac{\ln x}{x}$

$\therefore \quad f'(x) = \frac{\left(\frac{1}{x}\right) x - \ln x(1)}{x^2} = \frac{1 - \ln x}{x^2}$

$\therefore \quad f'(x) = 0$ when $1 - \ln x = 0$

$\therefore \quad \ln x = 1$

$\therefore \quad x = e$

Sign diagram of $f'(x)$ is:

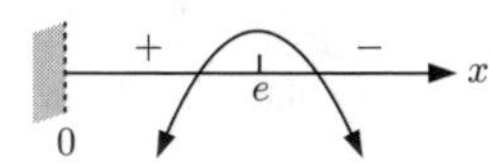


Now $f(e) = \frac{\ln e}{e} = \frac{1}{e}$

$\therefore$ there is a local maximum at $\left(e, \frac{1}{e}\right)$

$\therefore \quad f(x) \leqslant \frac{1}{e}$ for all $x$, and so $\frac{\ln x}{x} \leqslant \frac{1}{e}$ for all $x > 0$

**13** **a** $f(x) = x - \ln x$

$\therefore \quad f'(x) = 1 - \frac{1}{x} = \frac{x-1}{x}$ and the sign diagram of $f'(x)$ is:

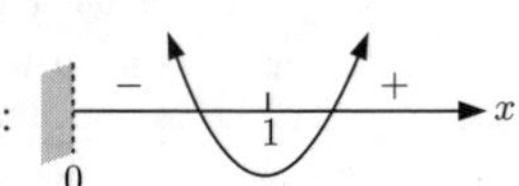


$\therefore \quad f(x)$ has a local minimum at $(1, 1 - \ln 1)$ or $(1, 1)$. This is the only turning point.

**b** $f(x) \geqslant 1$ for all $x > 0$

$\therefore \quad x - \ln x \geqslant 1$

$\therefore \quad \ln x \leqslant x - 1$ for all $x > 0$

## EXERCISE 16D.1

**1** **a**

| Point | $f(x)$ | $f'(x)$ | $f''(x)$ |
|---|---|---|---|
| A | + | − | + |
| B | − | 0 | + |
| C | + | + | 0 |
| D | + | 0 | − |
| E | 0 | − | − |

**b** The turning points of $y = f(x)$ are point B, a local minimum, and point D, a local maximum.

**c** The inflection point of $y = f(x)$ is point C, a non-stationary point of inflection.

**2** **a**

$f(x) = x^2 + 3$

$\therefore\ f'(x) = 2x$

$\therefore\ f''(x) = 2$

Since $f''(x) \neq 0$,
no points of inflection exist.

**b**

$f(x) = 2 - x^3$

$\therefore\ f'(x) = -3x^2$

$\therefore\ f''(x) = -6x$

Now $f''(x) = 0$ when $x = 0$,
and $f'(0) = 0$

$\therefore$ there is a stationary inflection at (0, 2).

**c**

$f(x) = x^3 - 6x^2 + 9x + 1$

$\therefore\ f'(x) = 3x^2 - 12x + 9$
$= 3(x^2 - 4x + 3)$
$= 3(x-3)(x-1)$

and $f''(x) = 6x - 12 = 6(x-2)$

Now $f''(x) = 0$ when $x = 2$
and $f'(2) \neq 0$

$\therefore$ there is a non-stationary inflection at $(2, f(2))$ which is (2, 3).

**d**

$f(x) = x^3 + 6x^2 + 12x + 5$

$\therefore\ f'(x) = 3x^2 + 12x + 12$
$= 3(x^2 + 4x + 4)$
$= 3(x+2)^2$

and $f''(x) = 6x + 12 = 6(x+2)$

Now $f''(x) = 0$ when $x = -2$
and $f'(-2) = 0$

$\therefore$ there is a stationary inflection at $(-2, f(-2))$ which is $(-2, -3)$.

**e**

$f(x) = -3x^4 - 8x^3 + 2$

$\therefore\ f'(x) = -12x^3 - 24x^2$
$= -12x^2(x+2)$

and $f''(x) = -36x^2 - 48x$
$= -12x(3x+4)$

$\therefore$ there is a stationary inflection at (0, 2), and a non-stationary inflection at $(-\frac{4}{3}, f(-\frac{4}{3}))$, which is $\left(-\frac{4}{3}, \frac{310}{27}\right)$.

**f**

$f(x) = 3 - \dfrac{1}{\sqrt{x}} = 3 - x^{-\frac{1}{2}}$

$\therefore\ f'(x) = \frac{1}{2}x^{-\frac{3}{2}}$

and $f''(x) = -\frac{3}{4}x^{-\frac{5}{2}} = \dfrac{-3}{4x^2\sqrt{x}}$

Now $f''(x) \neq 0$ for all $x$

$\therefore$ there are no points of inflection.

**3** **a** $f(x) = x^2$

$\therefore \quad f'(x) = 2x$ which has sign diagram:

and $f''(x) = 2$

**i** There is a local minimum at (0, 0).

**ii** There are no points of inflection as $f''(x) \neq 0$.

**iii** $f(x)$ is increasing when $x \geqslant 0$, and decreasing when $x \leqslant 0$.

**iv** $f(x)$ is concave up for all $x$ as $f''(x) > 0$ for all $x$.

**v**

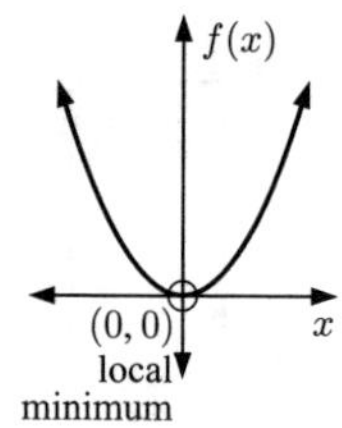


**b** $f(x) = x^3$

$\therefore \quad f'(x) = 3x^2$ which has sign diagram:

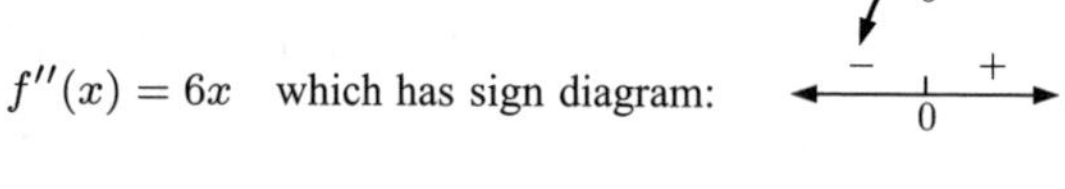

and $f''(x) = 6x$ which has sign diagram:

**v**

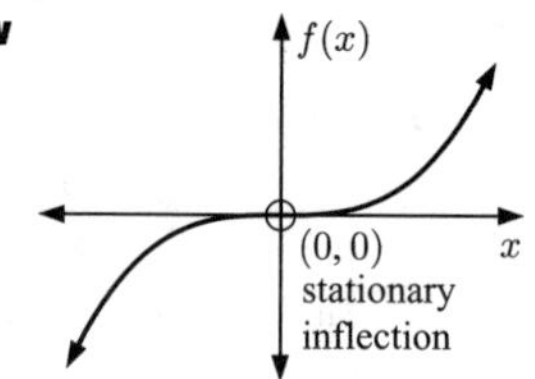


**i** A stationary inflection at (0, 0).

**ii** A stationary inflection at (0, 0).

**iii** $f(x)$ is increasing for all real $x$.

**iv** $f(x)$ is concave up when $x \geqslant 0$, and concave down when $x \leqslant 0$.

**c** $f(x) = \sqrt{x}$

$\therefore \quad f'(x) = \frac{1}{2}x^{-\frac{1}{2}} = \dfrac{1}{2\sqrt{x}}$ which has sign diagram:

and $f''(x) = -\frac{1}{4}x^{-\frac{3}{2}} = \dfrac{-1}{4x\sqrt{x}}$ which has sign diagram:

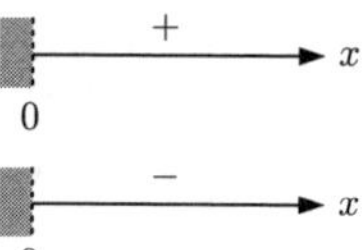

**i** There are no stationary points as $f'(x) \neq 0$.

**ii** There are no points of inflection as $f''(x) \neq 0$.

**iii** $f(x)$ is increasing for all $x \geqslant 0$, never decreasing.

**iv** $f(x)$ is concave down for all $x \geqslant 0$ as $f''(x) < 0$ for all $x > 0$, never concave up.

**v**

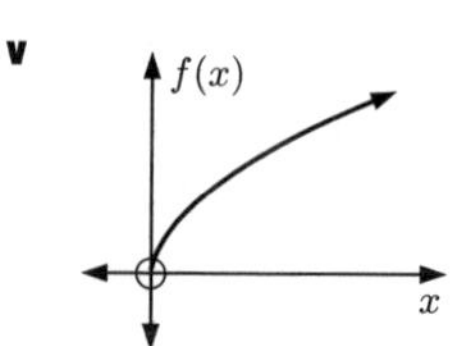


**d** $f(x) = x^3 - 3x^2 - 24x + 1$

$\therefore \quad f'(x) = 3x^2 - 6x - 24$

$= 3(x^2 - 2x - 8)$

$= 3(x-4)(x+2)$ which has sign diagram:

and $f''(x) = 6x - 6$

$= 6(x-1)$ which has sign diagram:

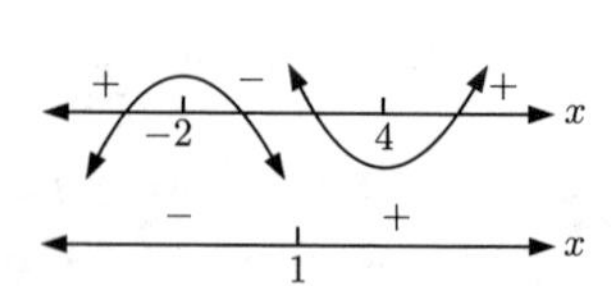

**i** $f(-2) = 29$, $f(4) = -79$, so there is a local maximum at (−2, 29), and a local minimum at (4, −79).

**ii** $f(1) = -25$, so there is a non-stationary inflection at (1, −25).

**iii** $f(x)$ is increasing for $x \leqslant -2$ and $x \geqslant 4$, and decreasing for $-2 \leqslant x \leqslant 4$.

**iv** $f(x)$ is concave down for $x \leqslant 1$, and concave up for $x \geqslant 1$.

**v**

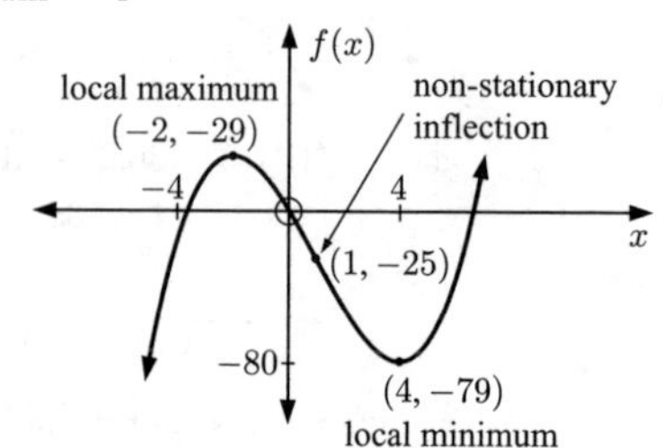

**e** $f(x) = 3x^4 + 4x^3 - 2$

$\therefore$ $f'(x) = 12x^3 + 12x^2$

$= 12x^2(x+1)$ which has sign diagram:

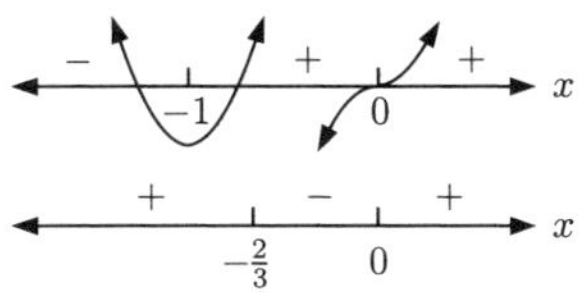


and $f''(x) = 36x^2 + 24x$

$= 12x(3x+2)$ which has sign diagram:

**i** There is a local minimum at $(-1, f(-1))$ which is $(-1, -3)$, and a stationary inflection at $(0, -2)$.

**ii** There is a non-stationary inflection at $(-\frac{2}{3}, f(-\frac{2}{3}))$ which is $(-\frac{2}{3}, -\frac{70}{27})$, and a stationary inflection at $(0, -2)$.

**iii** $f(x)$ is increasing for $x \geqslant -1$, and decreasing for $x \leqslant -1$.

**iv** $f(x)$ is concave down for $-\frac{2}{3} \leqslant x \leqslant 0$, and concave up for $x \leqslant -\frac{2}{3}$ and $x \geqslant 0$.

**v**

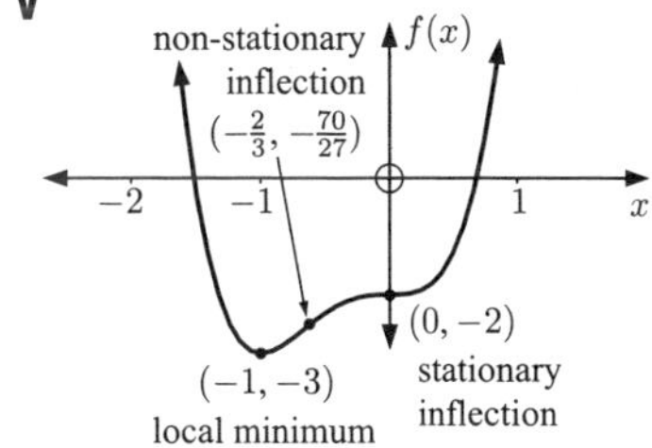


**f** $f(x) = (x-1)^4$

$\therefore$ $f'(x) = 4(x-1)^3$ which has sign diagram:

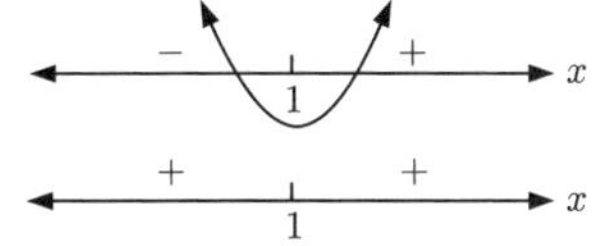


and $f''(x) = 12(x-1)^2$ which has sign diagram:

**i** There is a local minimum at $(1, 0)$.

**ii** Since there is no sign change in $f''(x)$ at $x = 1$, there are no points of inflection.

**iii** $f(x)$ is increasing for $x \geqslant 1$, and decreasing for $x \leqslant 1$.

**iv** $f(x)$ is concave up for all $x$.

**v**

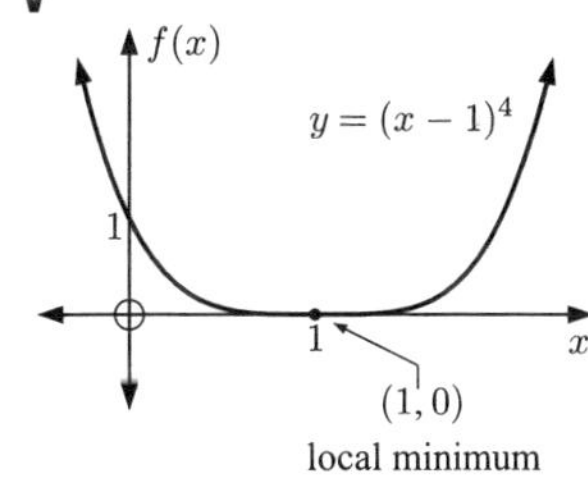


**g** $f(x) = x^4 - 4x^2 + 3$

$f'(x) = 4x^3 - 8x = 4x(x^2-2)$

$= 4x(x+\sqrt{2})(x-\sqrt{2})$ which has sign diagram:

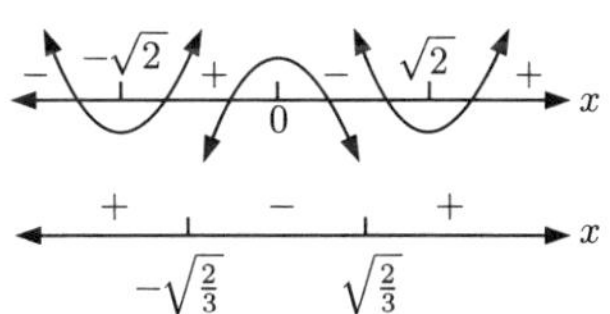


$f''(x) = 12x^2 - 8 = 4(3x^2-2)$

$= 4(\sqrt{3}x+\sqrt{2})(\sqrt{3}x-\sqrt{2})$ which has sign diagram:

**i** There is a local maximum at $(0, 3)$, and $f(-\sqrt{2}) = f(\sqrt{2}) = -1$, so there are local minima at $(\sqrt{2}, -1)$ and $(-\sqrt{2}, -1)$.

**ii** $f\left(\sqrt{\frac{2}{3}}\right) = f\left(-\sqrt{\frac{2}{3}}\right) = \frac{7}{9}$, so there are non-stationary inflections at $(\sqrt{\frac{2}{3}}, \frac{7}{9})$ and $(-\sqrt{\frac{2}{3}}, \frac{7}{9})$.

**iii** $f(x)$ is increasing for $-\sqrt{2} \leqslant x \leqslant 0$ and $x \geqslant \sqrt{2}$, and decreasing for $x \leqslant -\sqrt{2}$ and $0 \leqslant x \leqslant \sqrt{2}$.

**iv** $f(x)$ is concave down for $-\sqrt{\frac{2}{3}} \leqslant x \leqslant \sqrt{\frac{2}{3}}$, and concave up for $x \leqslant -\sqrt{\frac{2}{3}}$ and $x \geqslant \sqrt{\frac{2}{3}}$.

**v**

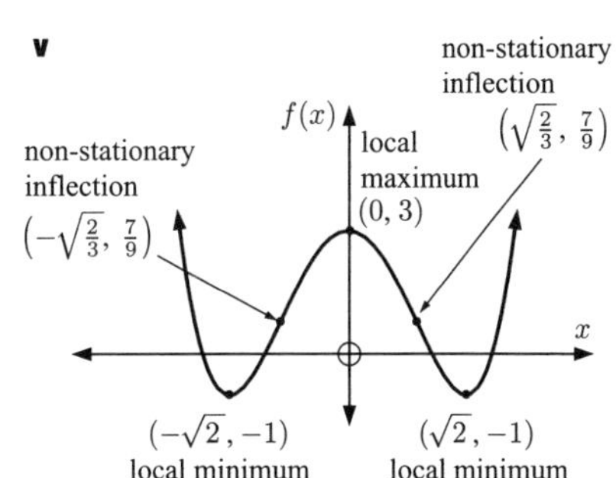

**h** $f(x) = 3 - \dfrac{4}{\sqrt{x}} = 3 - 4x^{-\frac{1}{2}}, \quad x > 0$

$\therefore \quad f'(x) = 2x^{-\frac{3}{2}} = \dfrac{2}{x\sqrt{x}}$ with sign diagram:

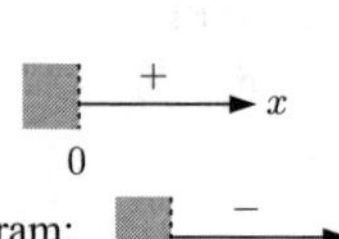


and $f''(x) = -3x^{-\frac{5}{2}} = -\dfrac{3}{x^2\sqrt{x}}$ with sign diagram:

−

$x$

0

**v**

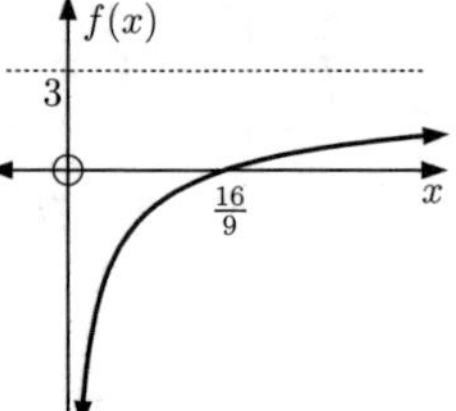


**i** There are no stationary points as $f'(x) \neq 0$.

**ii** There are no points of inflection as $f''(x) \neq 0$.

**iii** $f(x)$ is increasing for all $x > 0$ and never decreasing.

**iv** $f(x)$ is concave down for all $x > 0$ and never concave up.

**4** **a** Consider $f(x) = e^{2x} - 3$

$f(x)$ cuts the $x$-axis at A when $f(x) = 0$

$\therefore \quad e^{2x} - 3 = 0$

$\therefore \quad e^{2x} = 3$

$\therefore \quad 2x = \ln 3$

$\therefore \quad x = \dfrac{\ln 3}{2}$

$\therefore$ A is $\left(\frac{\ln 3}{2}, 0\right)$

$f(x)$ cuts the $y$-axis at B when $x = 0$

$\therefore \quad f(0) = e^{2\times 0} - 3$

$= e^0 - 3$

$= -2$

$\therefore$ B is $(0, -2)$

**b** $f(x) = e^{2x} - 3$

$\therefore \quad f'(x) = 2e^{2x}$

Since $e^{2x} > 0$ for all $x$, $f'(x) > 0$ for all $x$, and hence $f(x)$ is increasing for all $x$.

**c** $f''(x) = 4e^{2x}$, which is always $> 0$.

$\therefore \quad f(x)$ is concave up for all $x$.

**d**

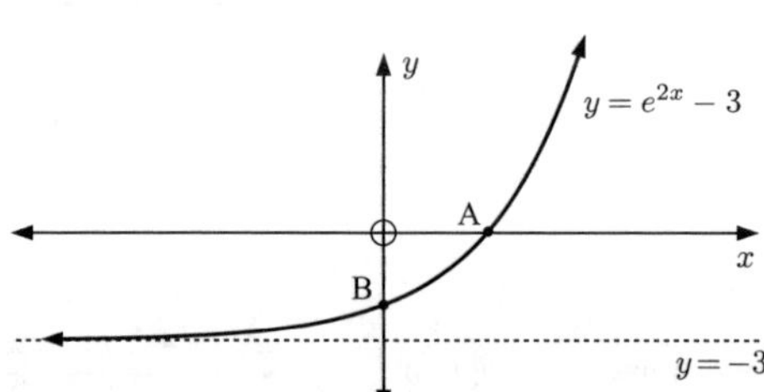


**e** As $x \to -\infty$, $e^{2x} \to 0$, so $e^{2x} - 3 \to -3^+$

$\therefore \quad y = -3$ is a horizontal asymptote.

**5** **a** The $x$-intercepts occur when $y = 0$

For $f(x) = e^x - 3$, $\quad e^x - 3 = 0$

$\therefore \quad e^x = 3$

$\therefore \quad x = \ln 3$

and for $g(x) = 3 - \dfrac{5}{e^x}$, $\quad 3 - \dfrac{5}{e^x} = 0$

$\therefore \quad \dfrac{3e^x - 5}{e^x} = 0$

$\therefore \quad 3e^x - 5 = 0$

$\therefore \quad e^x = \frac{5}{3}$

$\therefore \quad x = \ln(\frac{5}{3})$

$\therefore \quad f(x)$ has $x$-intercept $\ln 3$

and $g(x)$ has $x$-intercept $\ln(\frac{5}{3})$.

The $y$-intercepts occur when $x = 0$

Now $f(0) = e^0 - 3 = -2$ and $g(0) = 3 - \dfrac{5}{e^0} = 3 - 5 = -2$

$\therefore$ both $f(x)$ and $g(x)$ have $y$-intercept $-2$.

**b** As $x \to \infty$, $f(x) \to \infty$ As $x \to \infty$, $g(x) \to 3^-$

$x \to -\infty$, $f(x) \to -3^+$ $x \to -\infty$, $g(x) \to -\infty$

**c** $f(x)$ and $g(x)$ meet when

$$e^x - 3 = 3 - 5e^{-x}$$
$$\therefore \quad e^{2x} - 3e^x = 3e^x - 5 \quad \{\times e^x\}$$
$$\therefore \quad e^{2x} - 6e^x + 5 = 0$$
$$\therefore \quad (e^x - 5)(e^x - 1) = 0$$
$$\therefore \quad e^x = 5 \text{ or } 1$$
$$\therefore \quad x = \ln 5 \text{ or } 0$$

Now $f(\ln 5) = e^{\ln 5} - 3 = 5 - 3 = 2$

and $f(0) = -2$

$\therefore$ $f(x)$ and $g(x)$ meet at $(\ln 5,\ 2)$ and $(0,\ -2)$.

**d**

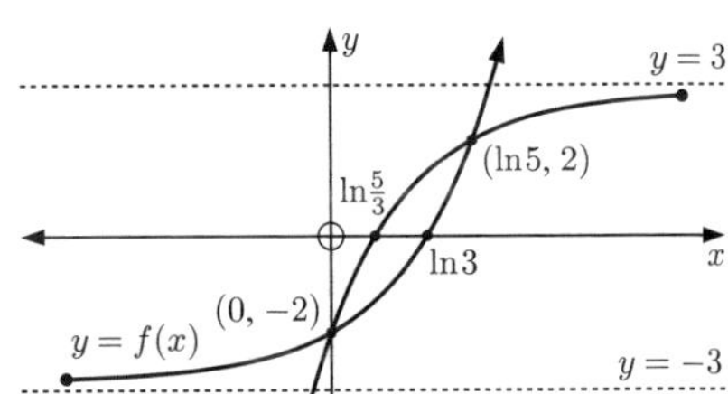


**6** **a** Consider $y = e^x - 3e^{-x}$

It cuts the $x$-axis at P when $y = 0$

$$\therefore \quad e^x - 3e^{-x} = 0$$
$$\therefore \quad e^{2x} - 3 = 0 \quad \{\times e^x\}$$
$$\therefore \quad e^{2x} = 3$$
$$\therefore \quad 2x = \ln 3$$
$$\therefore \quad x = \tfrac{1}{2}\ln 3$$

It cuts the $y$-axis at Q when $x = 0$

$$\therefore \quad y = e^0 - 3e^0 = 1 - 3 = -2$$

$\therefore$ P is $(\frac{1}{2}\ln 3,\ 0)$ and Q is $(0,\ -2)$.

**b**
$$\frac{dy}{dx} = e^x + 3e^{-x} = e^x + \frac{3}{e^x}$$

Since $e^x > 0$ for all $x$,

$\dfrac{dy}{dx} > 0$ for all $x$

$\therefore$ the function is increasing for all $x$

**c**
$$\frac{dy}{dx} = e^x + 3e^{-x}$$
$$\therefore \quad \frac{d^2y}{dx^2} = e^x - 3e^{-x} = y$$

Above the $x$-axis $y > 0$ $\therefore$ $\dfrac{d^2y}{dx^2} > 0$

$\therefore$ the function is concave up

Below the $x$-axis $y < 0$ $\therefore$ $\dfrac{d^2y}{dx^2} < 0$

$\therefore$ the function is concave down

$\therefore$ a non-stationary inflection occurs when $y = 0$

**d**

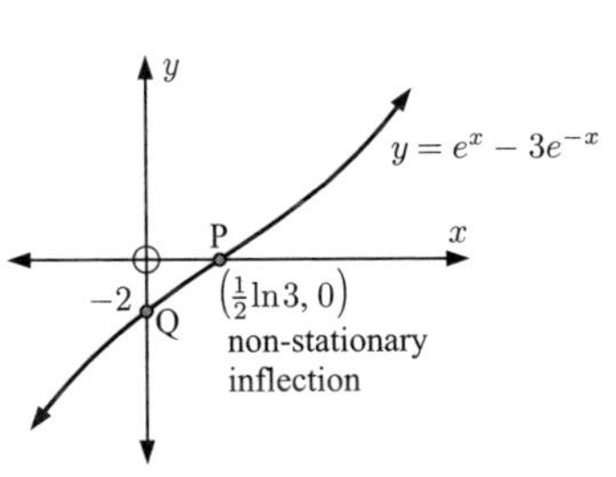


**7** $f(x) = \ln(2x - 1) - 3$

**a** $f(x) = 0$ when $\ln(2x - 1) = 3$

$$\therefore \quad 2x - 1 = e^3$$
$$\therefore \quad 2x = e^3 + 1$$
$$\therefore \quad x = \frac{e^3 + 1}{2} \approx 10.5$$

$\therefore$ the $x$-intercept is $\dfrac{e^3+1}{2}$

**b** $f(0)$ cannot be found as $\ln(-1)$ is not defined. $\therefore$ there is no $y$-intercept.

**c** $f'(x) = \dfrac{2}{2x-1}$ $\therefore$ $f'(1) = \dfrac{2}{2-1} = 2$ $\therefore$ gradient of tangent $= 2$

**d** $\ln(2x - 1)$ has meaning provided $2x - 1 > 0$ $\therefore$ $2x > 1$ and so $x > \frac{1}{2}$

$\therefore$ the domain of $f$ is $\{x \mid x > \frac{1}{2}\}$

**e** $f'(x) = 2(2x-1)^{-1}$

$\therefore\ f''(x) = -2(2x-1)^{-2}(2)$

$= \dfrac{-4}{(2x-1)^2},\quad x > \frac{1}{2}$

$\therefore$ provided $x > \frac{1}{2}$, $f''(x) < 0$

$\therefore$ $f(x)$ is concave down for all $x$ in the domain of $f$.

**f**

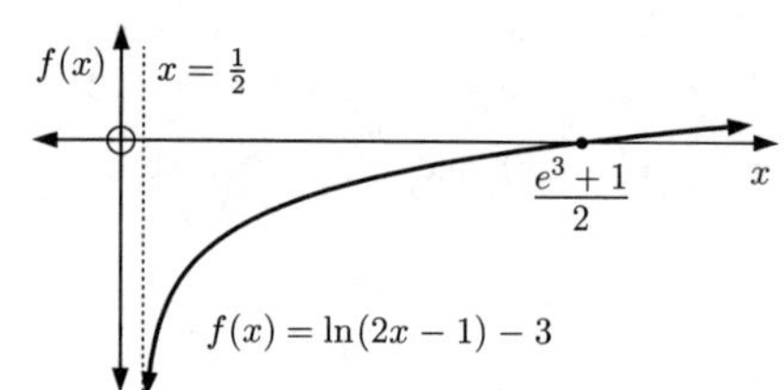


**8** **a** $f(x) = \ln x$ is defined for all $x > 0$.

**b** $f'(x) = \dfrac{1}{x}$ which is $> 0$ for all $x > 0$

$\therefore$ $f(x)$ is increasing on $x > 0$; its gradient is always positive.

$f''(x) = -x^{-2} = \dfrac{-1}{x^2}$ which is $< 0$ for all $x > 0$ $\quad\therefore$ $f(x)$ is concave down on $x > 0$.

**c**

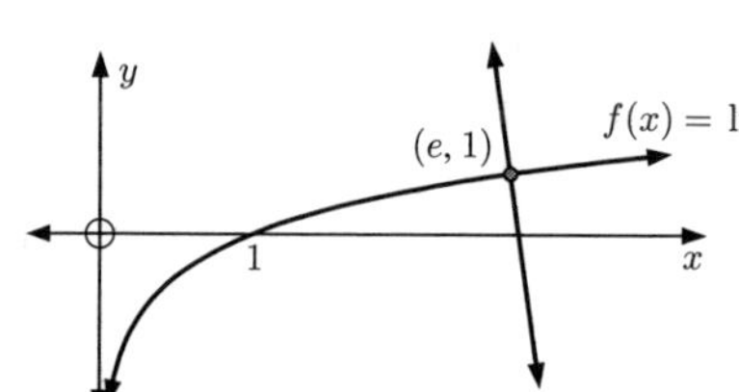


At $y = 1$, $1 = \ln x$

$\therefore\ x = e^1 = e$

$\therefore$ the point of contact is $(e, 1)$

Now $\dfrac{dy}{dx} = \dfrac{1}{x}$

$\therefore$ at $(e, 1)$, $\dfrac{dy}{dx} = \dfrac{1}{e}$

$\therefore$ the gradient of the tangent is $\dfrac{1}{e}$, and the gradient of the normal is $-e$

$\therefore$ the equation of the normal is $\dfrac{y-1}{x-e} = -e$ $\quad\therefore\ y - 1 = -e(x-e)$

$\therefore\ y - 1 = -ex + e^2$

$\therefore\ y = -ex + 1 + e^2$

**9** Consider $f(x) = \dfrac{e^x}{x}$.

**a** $e^x \neq 0$ for all $x$, so $f(x) \neq 0$ and there is no $x$-intercept.

$f(0) = \dfrac{e^0}{0}$ is undefined, so there is also no $y$-intercept.

**b** As $x \to +\infty$ $f(x) \to \infty$, and as $x \to -\infty$, $f(x) \to 0^-$

(As $x \to 0^+$, $y \to +\infty$, and as $x \to 0^-$, $y \to -\infty$
$\therefore$ $x = 0$ is a vertical asymptote.)

**c** Using the quotient rule, $f'(x) = \dfrac{e^x x - e^x(1)}{x^2} = \dfrac{e^x(x-1)}{x^2}$

with sign diagram:

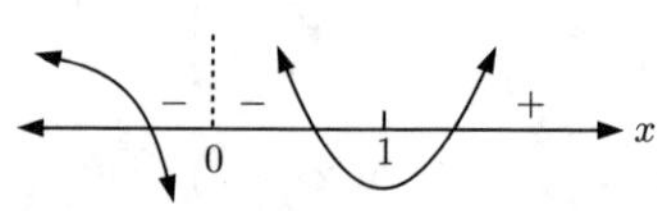


$f(1) = \dfrac{e^1}{1} = e$, so there is a local minimum at $(1, e)$.

**d** Using the product and quotient rules,

$f''(x) = \dfrac{[e^x(x-1) + e^x]x^2 - e^x(x-1)2x}{x^4}$

$= \dfrac{x^3e^x - \cancel{x^2e^x} + \cancel{x^2e^x} - 2x^2e^x + 2xe^x}{x^4}$

$= \dfrac{e^x(x^2 - 2x + 2)}{x^3}$ with sign diagram:

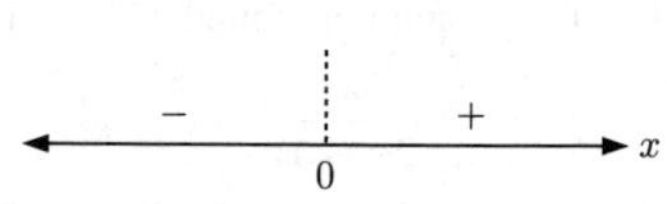


**i** $f(x)$ is concave up for $x > 0$. **ii** $f(x)$ is concave down for $x < 0$.

**e**

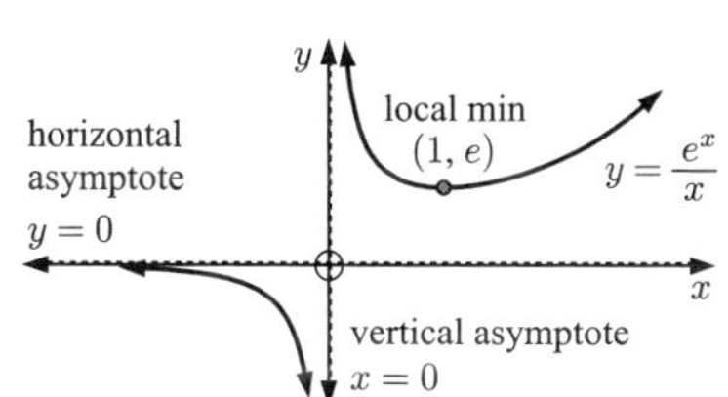


**f** Now $f'(x) = \dfrac{e^x(x-1)}{x^2}$

$\therefore \quad f'(-1) = \dfrac{e^{-1}(-1-1)}{(-1)^2} = -\dfrac{2}{e}$

$\therefore$ the gradient of the tangent is $= -\dfrac{2}{e}$

When $x = -1$, $y = \dfrac{e^{-1}}{-1} = -\dfrac{1}{e}$

$\therefore$ the equation of tangent is $\dfrac{y - \left(-\frac{1}{e}\right)}{x - (-1)} = -\dfrac{2}{e}$ $\quad \therefore \quad \dfrac{y + \frac{1}{e}}{x+1} = -\dfrac{2}{e}$

$\therefore \quad e\left(y + \dfrac{1}{e}\right) = -2(x+1)$

$ey + 1 = -2x - 2$

$\therefore \quad ey = -2x - 3$

**10** **a** $f(x) = \frac{1}{\sqrt{2\pi}} e^{-\frac{1}{2}x^2}$

$\therefore \quad f'(x) = \frac{1}{\sqrt{2\pi}} e^{-\frac{1}{2}x^2}(-x)$

$= \dfrac{-x}{\sqrt{2\pi}} e^{-\frac{1}{2}x^2}$

$\therefore \quad f'(x) = 0$ when $x = 0$

$f'(x)$ has sign diagram:

Now $f(0) = \frac{1}{\sqrt{2\pi}}$

so there is a local maximum at $\left(0, \frac{1}{\sqrt{2\pi}}\right)$.

The function is increasing for $x \leqslant 0$ and decreasing for $x \geqslant 0$

**b** $f'(x) = \dfrac{-x}{\sqrt{2\pi}} e^{-\frac{1}{2}x^2} = \frac{1}{\sqrt{2\pi}}\left(-xe^{-\frac{1}{2}x^2}\right)$

$\therefore \quad f''(x) = \frac{1}{\sqrt{2\pi}}\left((-1)\, e^{-\frac{1}{2}x^2} + (-x)\, e^{-\frac{1}{2}x^2}\,(-x)\right)$ {product rule}

$= \frac{1}{\sqrt{2\pi}} e^{-\frac{1}{2}x^2}\left(x^2 - 1\right)$

$= \frac{1}{\sqrt{2\pi}} e^{-\frac{1}{2}x^2}\,(x+1)\,(x-1)$ which has sign diagram: + −1 − 1 + $x$

Now $f(1) = \frac{1}{\sqrt{2\pi}} e^{-\frac{1}{2}} = \frac{1}{\sqrt{2e\pi}}$ and $f(-1) = \frac{1}{\sqrt{2e\pi}}$

$\therefore$ there are non-stationary points of inflection at $\left(1, \frac{1}{\sqrt{2e\pi}}\right)$ and $\left(-1, \frac{1}{\sqrt{2e\pi}}\right)$.

**c** As $x \to \infty$, $\quad e^{-\frac{1}{2}x^2} \to 0^+$

$\therefore \quad f(x) \to 0^+$

As $x \to -\infty$, $\quad e^{-\frac{1}{2}x^2} \to 0^+$

$\therefore \quad f(x) \to 0^+$

**d**

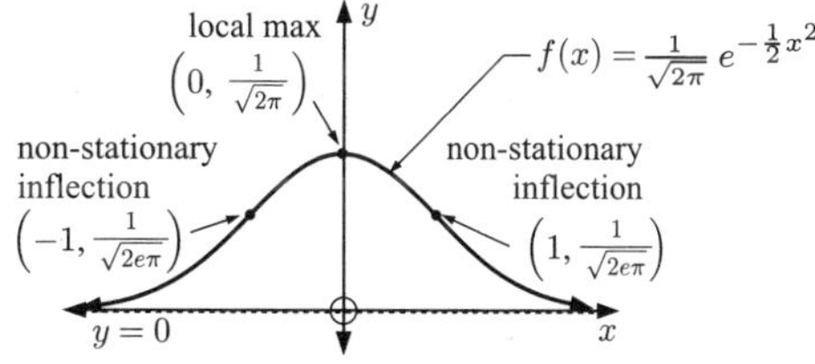


## EXERCISE 16D.2

**1** **a** $f(x)$ is quadratic, so $f'(x)$ will be linear and $f''(x)$ will be constant.

$f(x)$ is decreasing for $x \leqslant 1$ and increasing for $x \geqslant 1$

$\therefore \quad f'(x) \leqslant 0$ for $x \leqslant 1$ and $f'(x) \geqslant 0$ for $x \geqslant 1$

$\therefore \quad f'(x)$ is an increasing linear function which cuts the $x$-axis at 1.

As $f'(x)$ is increasing, $f''(x) > 0$.

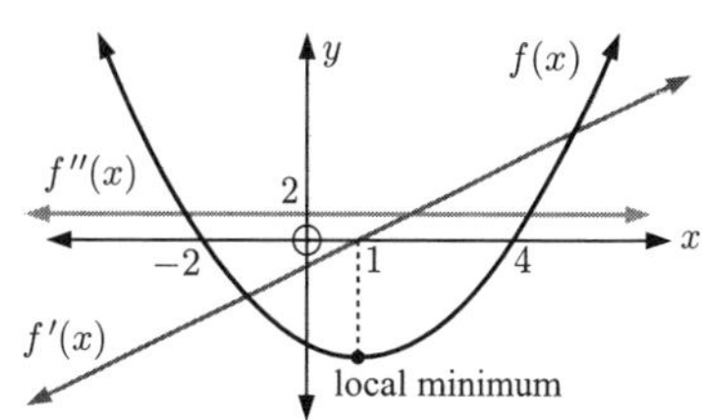

**b** $f(x)$ is cubic, so $f'(x)$ will be quadratic and $f''(x)$ will be linear.
$f(x)$ has turning points at $x \approx \pm 1$
$\therefore$ $f'(x)$ cuts the $x$-axis at these points.
$f(x)$ has a non-stationary inflection point at $x = 0$
$\therefore$ $f'(x)$ has a turning point at $x = 0$, and $f''(0) = 0$.
$f(x)$ is concave down for $x \leqslant 0$ and concave up for $x \geqslant 0$
$\therefore$ $f'(x)$ is decreasing for $x \leqslant 0$ and increasing for $x \geqslant 0$
and $f''(x) \leqslant 0$ for $x \leqslant 0$ and $\geqslant 0$ for $x \geqslant 0$.

**c** $f(x)$ is cubic, so $f'(x)$ will be quadratic and $f''(x)$ will be linear.
$f(x)$ has turning points at $x \approx 1$ and $x = 3$
$\therefore$ $f'(x)$ cuts the $x$-axis at these points.
$f(x)$ has a non-stationary inflection point at $x \approx 2$
$\therefore$ $f'(x)$ has a turning point at $x \approx 2$, and $f''(2) = 0$
$f(x)$ is concave down for $x \leqslant 2$ and concave up for $x \geqslant 2$
$\therefore$ $f'(x)$ is decreasing for $x \leqslant 2$ and increasing for $x \geqslant 2$
and $f''(x) \leqslant 0$ for $x \leqslant 2$ and $\geqslant 0$ for $x \geqslant 2$.

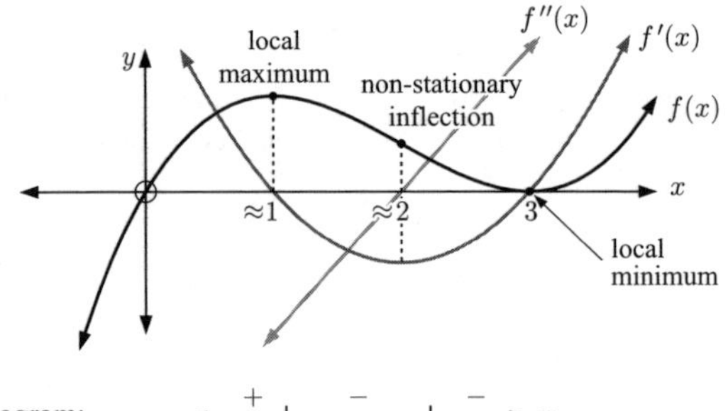


**2** **a** $f'(x)$ has sign diagram: $+$ | $-3$ | $-$ | $1$ | $-$ $x$

$\therefore$ $f(x)$ is increasing for $x \leqslant -3$, and decreasing for $x \geqslant -3$
$\therefore$ $f(x)$ has a local maximum at $x = -3$
$f'(x)$ has a turning point at $x \approx -1.7$.
At this point, $f''(x) = 0$, but $f'(x) \neq 0$
$\therefore$ $f(x)$ has a non-stationary inflection point here.
$f'(x)$ has another turning point at $x = 1$.
At this point, $f''(x) = 0$ and $f'(x) = 0$
$\therefore$ $f(x)$ has a stationary inflection point at $x = 1$.
A possible graph of $f(x)$ is shown alongside:

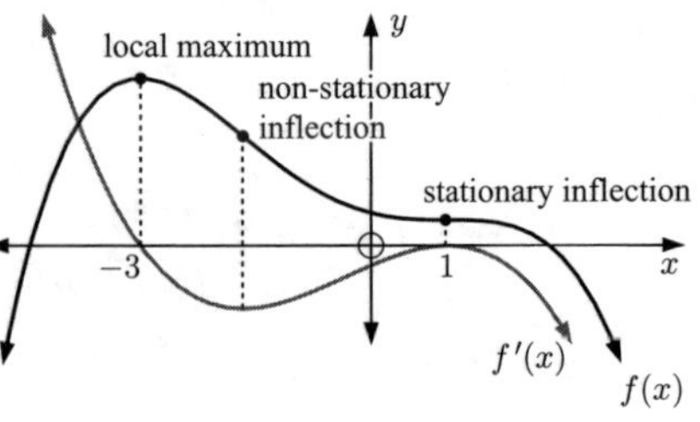


**b** $f'(x)$ has sign diagram:

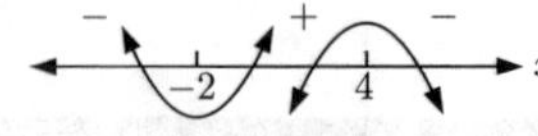


$\therefore$ $f(x)$ has a local minimum at $x = -2$ and a local maximum at $x = 4$
$f'(x)$ has a turning point at $x \approx 1$.
At this point, $f''(x) = 0$, but $f'(x) \neq 0$
$\therefore$ $f(x)$ has a non-stationary inflection point at $x \approx 1$.
A possible graph of $f(x)$ is shown alongside:

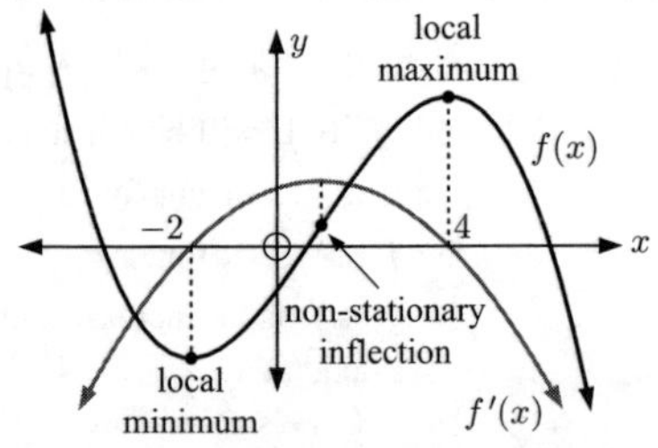

## REVIEW SET 16A

**1** Consider $y = -2x^2$. When $x = -1$, $y = -2(-1)^2 = -2$, so the point of contact is $(-1, -2)$.

Now $\dfrac{dy}{dx} = -4x$

$\therefore$ at $x = -1$, $\dfrac{dy}{dx} = -4(-1) = 4$

$\therefore$ the tangent has equation $\dfrac{y - (-2)}{x - (-1)} = 4$ or $y = 4x + 2$

**2** Consider $y = \dfrac{1 - 2x}{x^2}$. When $x = 1$, $y = \dfrac{1 - 2(1)}{1^2} = -1$, so the point of contact is $(1, -1)$.

Since $y = \dfrac{1}{x^2} - \dfrac{2}{x}$, $\dfrac{dy}{dx} = -2x^{-3} + 2x^{-2} = -\dfrac{2}{x^3} + \dfrac{2}{x^2}$

$\therefore$ at $x = 1$, $\dfrac{dy}{dx} = -2 + 2 = 0$

So, the tangent is a horizontal line, and the normal must be a vertical line of the form $x = k$.
As the normal passes through $(1, -1)$, its equation must be $x = 1$.

**3** **a** The vertical asymptote is $x + 3 = 0$ or $x = -3$.

**b** When $y = 0$, $\dfrac{3x - 2}{x + 3} = 0$

$\therefore$ $3x - 2 = 0$

$\therefore$ $x = \frac{2}{3}$

$\therefore$ the $x$-intercept is $\frac{2}{3}$.

When $x = 0$, $f(0) = \frac{-2}{3}$

$\therefore$ the $y$-intercept is $-\frac{2}{3}$.

**c** $f'(x) = \dfrac{3(x + 3) - (3x - 2)(1)}{(x + 3)^2}$

$= \dfrac{3x + 9 - 3x + 2}{(x + 3)^2}$

$= \dfrac{11}{(x + 3)^2}$

$f'(x)$ has sign diagram: $+$ | $+$ with critical value $-3$ on the $x$ axis.

**d** $f'(x) \neq 0$ for any $x$, so $f(x)$ has no stationary points.

**4** $y = e^{-x^2}$ so when $x = 1$, $y = e^{-1} = \dfrac{1}{e}$

$\therefore$ the point of contact is $\left(1, \dfrac{1}{e}\right)$

Now $\dfrac{dy}{dx} = -2xe^{-x^2}$

$\therefore$ when $x = 1$, $\dfrac{dy}{dx} = -2e^{-1}$

$\therefore$ the gradient of the tangent is $-\dfrac{2}{e}$

and the gradient of the normal is $\dfrac{e}{2}$

$\therefore$ the equation of the normal is $\dfrac{y - \frac{1}{e}}{x - 1} = \dfrac{e}{2}$

$\therefore$ $2\left(y - \dfrac{1}{e}\right) = e(x - 1)$

$\therefore$ $2y - \dfrac{2}{e} = ex - e$

$\therefore$ $2y = ex + \dfrac{2}{e} - e$

$\therefore$ $y = \dfrac{e}{2}x + \dfrac{1}{e} - \dfrac{e}{2}$

**5** $y = x \tan x$

$\therefore$ $\dfrac{dy}{dx} = 1 \times \tan x + x \times \left(\dfrac{1}{\cos^2 x}\right)$

$= \tan x + \dfrac{x}{\cos^2 x}$

Now $\cos \frac{\pi}{4} = \frac{1}{\sqrt{2}}$ and $\tan \frac{\pi}{4} = 1$

$\therefore$ at $x = \frac{\pi}{4}$, $y = \frac{\pi}{4}$,

and $\dfrac{dy}{dx} = 1 + \dfrac{\frac{\pi}{4}}{\left(\frac{1}{\sqrt{2}}\right)^2} = 1 + \frac{\pi}{2}$

$\therefore$ the equation of the tangent is

$\dfrac{y - \frac{\pi}{4}}{x - \frac{\pi}{4}} = 1 + \frac{\pi}{2}$

$\therefore$ $y - \frac{\pi}{4} = (1 + \frac{\pi}{2})(x - \frac{\pi}{4})$

$= x - \frac{\pi}{4} + \frac{\pi}{2}x - \frac{\pi^2}{8}$

$\therefore$ $y = (1 + \frac{\pi}{2})x - \frac{\pi^2}{8}$

$\therefore$ $2y = (2 + \pi)x - \frac{\pi^2}{4}$

$\therefore$ $(2 + \pi)x - 2y = \frac{\pi^2}{4}$ as required

**6** $y = \dfrac{ax+b}{\sqrt{x}} = a\sqrt{x} + \dfrac{b}{\sqrt{x}} = ax^{\frac{1}{2}} + bx^{-\frac{1}{2}}$

$\therefore \quad \dfrac{dy}{dx} = \dfrac{a}{2}x^{-\frac{1}{2}} - \dfrac{b}{2}x^{-\frac{3}{2}} = \dfrac{a}{2\sqrt{x}} - \dfrac{b}{2x\sqrt{x}}$

The equation of the tangent at $x = 1$ is $2x - y = 1$ or $y = 2x - 1$

so the gradient of the tangent is 2

$\therefore$ at $x = 1$, $\dfrac{dy}{dx} = \dfrac{a}{2} - \dfrac{b}{2} = 2$

$\therefore \quad a - b = 4$

$\therefore \quad a = b + 4$ .... (1)

Also at $x = 1$, the tangent touches the curve

$\therefore \quad \dfrac{a(1)+b}{\sqrt{1}} = 2(1) - 1$

$\therefore \quad a + b = 1$

$\therefore \quad b + 4 + b = 1$ {using (1)}

$\therefore \quad 2b = -3$

$\therefore \quad b = -\frac{3}{2}$ and $a = 4 - \frac{3}{2} = \frac{5}{2}$

**7** $f(x) = 4\ln(2x)$, P$(1, 4\ln 2)$

$\therefore \quad f'(x) = 4 \times \dfrac{2}{2x}$

$= \dfrac{4}{x}$

$\therefore$ at $x = 1$, $f'(1) = \dfrac{4}{1} = 4$

$\therefore$ the tangent has equation

$\dfrac{y - 4\ln 2}{x - 1} = 4$

$\therefore \quad y - 4\ln 2 = 4x - 4$

$\therefore \quad y = 4x + 4\ln 2 - 4$

**8** **a** $f(x) = \dfrac{e^x}{x-1}$

Now $f(0) = \dfrac{e^0}{-1} = -1$ so the $y$-intercept is $-1$.

**b** $f(x)$ is defined for all $x \neq 1$.

**c** $f'(x) = \dfrac{e^x(x-1) - e^x(1)}{(x-1)^2}$ {quotient rule}

$= \dfrac{e^x(x-2)}{(x-1)^2}$ and has sign diagram:

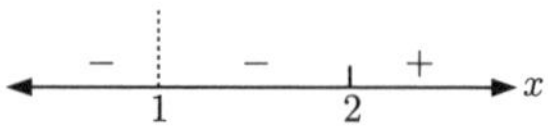


$\therefore \quad f'(x) \leqslant 0$ for $x < 1$ and $1 < x \leqslant 2$ and $f'(x) \geqslant 0$ for $x \geqslant 2$

$\therefore \quad f(x)$ is decreasing for $x < 1$ and $1 < x \leqslant 2$, and increasing for $x \geqslant 2$.

$$f''(x) = \frac{[e^x(x-2) + e^x(1)](x-1)^2 - e^x(x-2)[2(x-1)^1(1)]}{(x-1)^4} \quad \text{\{product and quotient rules\}}$$

$$= \frac{[e^x(x-2+1)(x-1)^2] - 2e^x(x-2)(x-1)}{(x-1)^4}$$

$$= \frac{e^x(x-1)(x-1)^2 - 2e^x(x-2)(x-1)}{(x-1)^4}$$

$$= \frac{e^x(x-1)\left[(x-1)^2 - 2(x-2)\right]}{(x-1)^4}$$

$$= \frac{e^x(x-1)\left[x^2 - 2x + 1 - 2x + 4\right]}{(x-1)^4}$$

$$= \frac{e^x(x^2 - 4x + 5)}{(x-1)^3} \quad \text{where the quadratic term has } \Delta < 0$$

The sign diagram of $f''(x)$ is:

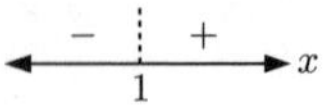


$\therefore \quad f''(x) > 0$ for $x > 1$

and $f''(x) < 0$ for $x < 1$.

$\therefore \quad f(x)$ is concave down for all $x < 1$

and concave up for all $x > 1$.

**d**

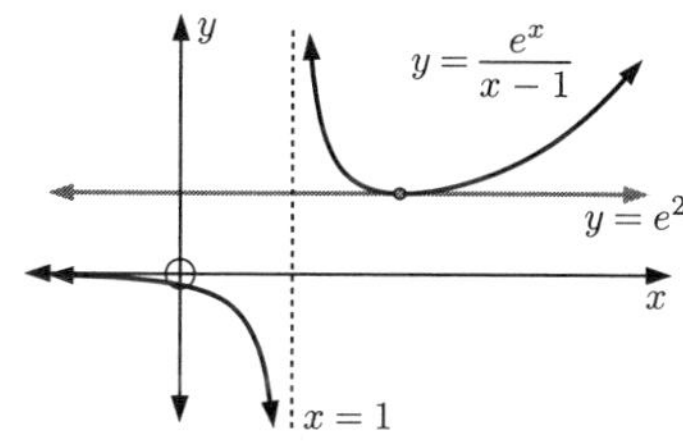


**e** Now $f(2) = \dfrac{e^2}{2-1} = e^2$

Using **c**, we have a local minimum at $(2,\ e^2)$

$\therefore$ the tangent at $x = 2$ is horizontal and is $y = e^2$.

**9** $y = 2x^3 + ax + b \qquad \therefore\ \dfrac{dy}{dx} = 6x^2 + a$

Now as the gradient at $(-2,\ 33)$ is 10,

$$\text{at } x = -2,\quad \frac{dy}{dx} = 10$$
$$\therefore\ 10 = 6(-2)^2 + a$$
$$\therefore\ a = -14$$
$$\therefore\ y = 2x^3 - 14x + b$$

Then, since $(-2,\ 33)$ lies on the curve,

$$\text{at } x = -2,\quad y = 33$$
$$\therefore\ 2(-2)^3 - 14(-2) + b = 33$$
$$\therefore\ -16 + 28 + b = 33$$
$$\therefore\ b = 21$$

**10** $y = \dfrac{a}{(x+2)^2} = a(x+2)^{-2}$

The gradient of the line (AB) is

$$\frac{y_2 - y_1}{x_2 - x_1} = \frac{8-4}{0-2} = \frac{4}{-2} = -2$$

$\therefore$ the equation of the tangent is

$$\frac{y-8}{x-0} = -2 \quad \text{or} \quad y = -2x + 8$$

Now $\dfrac{dy}{dx} = -2a(x+2)^{-3}$, so for the given tangent, $-2a(x+2)^{-3} = -2$

$$\therefore\ \frac{a}{(x+2)^3} = 1$$
$$\therefore\ a = (x+2)^3 \quad .... (1)$$

The line (AB) meets the curve where

$$-2x + 8 = \frac{a}{(x+2)^2}$$
$$\therefore\ -2x + 8 = \frac{(x+2)^3}{(x+2)^2} \quad \{\text{using (1)}\}$$
$$\therefore\ -2x + 8 = x + 2$$
$$\therefore\ -3x = -6$$
$$\therefore\ x = 2$$

and so $a = (2+2)^3 = 64$

**11** $y = \dfrac{5}{\sqrt{x}} = 5x^{-\frac{1}{2}}$

$$\therefore\ \frac{dy}{dx} = -\tfrac{5}{2}x^{-\frac{3}{2}}$$

$\therefore$ the gradient of the tangent at the point $(1,\ 5)$ is $-\frac{5}{2}(1)^{-\frac{3}{2}} = -\frac{5}{2}$

$\therefore$ the equation of the tangent is

$$\frac{y-5}{x-1} = -\frac{5}{2}$$
$$\therefore\ y - 5 = -\tfrac{5}{2}x + \tfrac{5}{2}$$
$$\therefore\ y = -\tfrac{5}{2}x + \tfrac{15}{2}$$

Now, P and Q are the $y$- and $x$-intercepts, so:

P: $y = -\frac{5}{2}(0) + \frac{15}{2} = \frac{15}{2}$

Q: $0 = -\frac{5}{2}x + \frac{15}{2}$

$$\therefore\ \tfrac{5}{2}x = \tfrac{15}{2}$$
$$\therefore\ x = 3$$

So P is $(0,\ 7.5)$ and Q is $(3,\ 0)$.

**12** At $x = A$, $f'(x) = 0$ and $f''(x) = 0$

$\therefore$ $f(x)$ has a stationary inflection point at $x = A$.

At $x = B$, $f''(x) = 0$ but $f'(x) \neq 0$

$\therefore$ $f(x)$ has a non-stationary inflection point at $x = B$.

$f'(x)$ is above the $x$-axis for $x \leqslant C$, and below the $x$-axis for $x \geqslant C$

$\therefore$ $f(x)$ is increasing for $x \leqslant C$ and decreasing for $x \geqslant C$, so $f(x)$ has a local maximum at $x = C$.

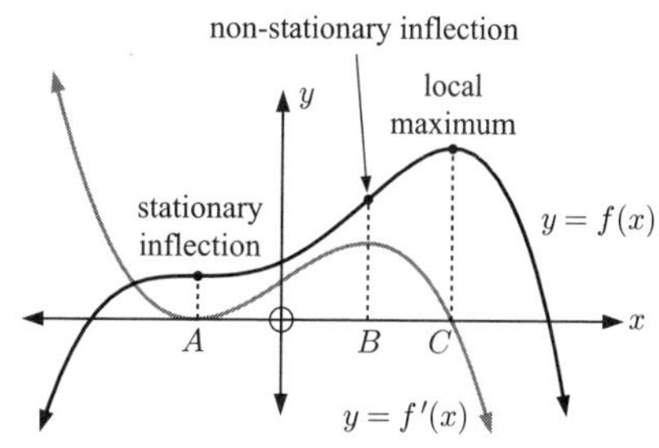

**13** $y = \ln(x^2+3)$ $\quad \therefore \quad \dfrac{dy}{dx} = \dfrac{2x}{x^2+3}$

When $x = 0$, $\dfrac{dy}{dx} = 0$ so the gradient of the tangent at this point is 0.

But when $x = 0$, $y = \ln(0+3) = \ln 3$

$\therefore$ the tangent is $y = \ln 3$ which does not cut the $x$-axis.

## REVIEW SET 16B

**1** $y = x^3 - 3x^2 - 9x + 2 \quad \therefore \quad \dfrac{dy}{dx} = 3x^2 - 6x - 9$

Horizontal tangents occur when $\dfrac{dy}{dx} = 0$

$$\therefore \quad 3x^2 - 6x - 9 = 0$$
$$\therefore \quad x^2 - 2x - 3 = 0$$
$$\therefore \quad (x-3)(x+1) = 0$$
$$\therefore \quad x = 3 \text{ or } x = -1$$

When $x = 3$, the horizontal tangent has equation $y = -25$.

When $x = -1$, the horizontal tangent has equation $y = 7$.

**2** Consider the tangent to $y = x^2\sqrt{1-x}$ at $x = -3$.

When $x = -3$, $y = (-3)^2\sqrt{1-(-3)} = 9\sqrt{4} = 18 \quad \{y \geqslant 0\}$

$\therefore$ the point of contact is $(-3, 18)$.

Also, $y = x^2\sqrt{1-x}$ is a product with $u = x^2$ and $v = (1-x)^{\frac{1}{2}}$

$\therefore \quad u' = 2x$ and $v' = \frac{1}{2}(1-x)^{-\frac{1}{2}}(-1)$

$$\therefore \quad \frac{dy}{dx} = 2x(1-x)^{\frac{1}{2}} - x^2\left(\tfrac{1}{2}\right)(1-x)^{-\frac{1}{2}}$$

$$\therefore \quad \text{at } x=-3, \quad \frac{dy}{dx} = 2(-3)(1-(-3))^{\frac{1}{2}} - (-3)^2\left(\tfrac{1}{2}\right)(1-(-3))^{-\frac{1}{2}}$$
$$= -6(2) - 9\left(\tfrac{1}{2}\right)\left(\tfrac{1}{2}\right)$$
$$= -\tfrac{57}{4}$$

$\therefore$ the tangent at $(-3, 18)$ has equation $\dfrac{y-18}{x-(-3)} = -\frac{57}{4}$

$$\therefore \quad 4y - 72 = -57x - 171$$
$$\therefore \quad 4y = -57x - 99$$

Now when $x = 0$, $y = -\frac{99}{4}$ and when $y = 0$, $x = -\frac{99}{57}$

$\therefore$ the area of $\triangle OAB = \frac{1}{2}\left(\frac{99}{4}\right)\left(\frac{99}{57}\right) = \frac{3267}{152} \approx 21.5$ units$^2$

**3** **a** $f(x) = x^3 + ax, \quad a < 0$

$\therefore \quad f'(x) = 3x^2 + a$

$f(x)$ has a turning point at $x = \sqrt{2}$, so $f'(\sqrt{2}) = 0$

$$\therefore \quad 3\left(\sqrt{2}\right)^2 + a = 0$$
$$\therefore \quad a = -6$$

**b** $f'(x) = 3x^2 - 6 = 3(x^2-2) = 3(x+\sqrt{2})(x-\sqrt{2})$

$\therefore \quad f'(x)$ has sign diagram:

+ | $-\sqrt{2}$ | − | $\sqrt{2}$ | + → $x$

$\therefore \quad f(x)$ has a local maximum at $\left(-\sqrt{2}, (-\sqrt{2})^3 - 6(-\sqrt{2})\right)$ or $\left(-\sqrt{2}, 4\sqrt{2}\right)$,

and a local minimum at $\left(\sqrt{2}, \left(\sqrt{2}\right)^3 - 6\sqrt{2}\right)$ or $\left(\sqrt{2}, -4\sqrt{2}\right)$.

**c**

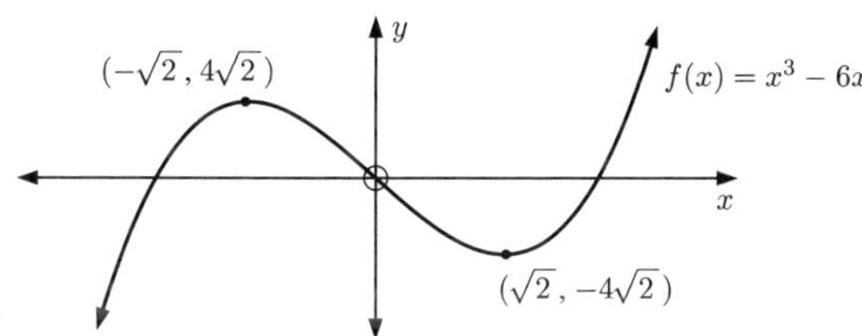


**4** $\quad f(x) = e^{4x} + px + q$

$\therefore \quad f'(x) = 4e^{4x} + p$

At the point where $x = 0$, the tangent to $f(x)$ has equation $y = 5x - 7$, so $f'(0) = 5$

$\therefore \quad 4e^0 + p = 5$

$\therefore \quad p = 1$

The tangent meets $f(x)$ when $x = 0$ and $y = 5(0) - 7 = -7$, so $(0, -7)$ must lie on $f(x)$ too.

$\therefore \quad e^{4(0)} + p(0) + q = -7$

$\therefore \quad 1 + q = -7$

$\therefore \quad q = -8$

**5** Consider the tangent to $y = 2x^3 + 4x - 1$ at $(1, 5)$.

$\dfrac{dy}{dx} = 6x^2 + 4 \qquad \therefore$ at $x = 1$, $\dfrac{dy}{dx} = 6(1)^2 + 4 = 10$

$\therefore$ the tangent has equation $\dfrac{y-5}{x-1} = 10$ or $y = 10x - 5$

Now the tangent meets the curve again where $10x - 5 = 2x^3 + 4x - 1$

$\therefore \quad 2x^3 - 6x + 4 = 0$

$\therefore \quad x^3 - 3x + 2 = 0$

We know that $(x-1)^2$ is a factor since the line is tangent to the curve at $x = 1$.

Consequently, $x^3 - 3x + 2 = (x-1)^2(x+2) = 0$ {since the constant term is 2}

Thus $x = -2$ is the other solution and when $x = -2$, $y = 2(-2)^3 + 4(-2) - 1 = -25$

$\therefore$ the tangent meets the curve again at $(-2, -25)$.

**6** Consider $y = 4(ax+1)^{-2}$.

When $x = 0$, $y = 4(0+1)^{-2} = 4$, so the point of contact is $(0, 4)$.

Now $\dfrac{dy}{dx} = -8(ax+1)^{-3}(a) = \dfrac{-8a}{(ax+1)^3} \qquad \therefore$ at $x = 0$, $\dfrac{dy}{dx} = -8a$

$\therefore$ the tangent has equation $\dfrac{y-4}{x-0} = -8a$ or $y - 4 = -8ax$

This tangent passes through $(1, 0)$, so $0 - 4 = -8a(1) \qquad \therefore \quad a = \frac{1}{2}$

**7** $f(x) = e^x - x$

**a** $f'(x) = e^x - 1$

so $f'(x) = 0$ when $e^x = 1$

$\therefore \quad x = 0$

Sign diagram of $f'(x)$ is: $-$ | $0$ | $+$ $\;x$

Now $f(0) = e^0 - 0 = 1$

$\therefore$ there is a local minimum at $(0, 1)$.

**b** As $x \to \infty$, $e^x \to \infty$ faster than $x$

$\therefore \quad f(x) \to \infty$

**c** 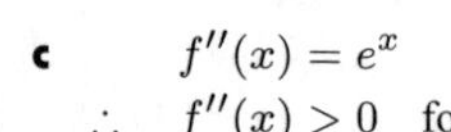

$f''(x) = e^x$

$\therefore \quad f''(x) > 0 \quad$ for all $x$

(sign diagram: $+$ for all $x$)

$\therefore \quad f(x)$ is concave up for all $x$

**d** (graph of $f(x) = e^x - x$, with local minimum at $(0, 1)$)

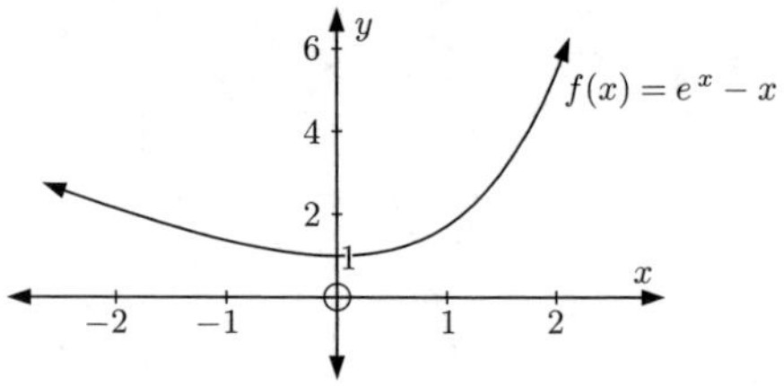

**e** Since a local minimum exists at $(0, 1)$,

$f(x) \geqslant 1$ for all $x$

$\therefore \quad e^x - x \geqslant 1$

$\therefore \quad e^x \geqslant x + 1$ for all $x$

**8** 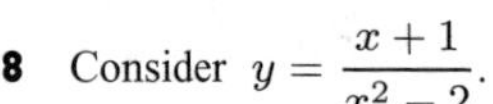

Consider $y = \dfrac{x+1}{x^2-2}$.

When $x = 1$, $\quad y = \dfrac{1+1}{1^2-2} = -2$

$\therefore$ the point of contact is $(1, -2)$.

$y = \dfrac{x+1}{x^2-2}$ is a quotient with

$u = x + 1 \quad$ and $\quad v = x^2 - 2$

$\therefore \quad u' = 1 \quad$ and $\quad v' = 2x$

$$\frac{dy}{dx} = \frac{1(x^2-2) - (x+1)(2x)}{(x^2-2)^2} \quad \{\text{quotient rule}\}$$

$$\therefore \text{ at } x = 1, \quad \frac{dy}{dx} = \frac{(1-2) - 2(1+1)}{(1-2)^2}$$

$$= \frac{-1-4}{1} = -5$$

$\therefore$ the normal at $(1, -2)$ has gradient $\frac{1}{5}$.

So the normal has equation $\quad \dfrac{y-(-2)}{x-1} = \frac{1}{5}$

$\therefore \quad 5y + 10 = x - 1$

$\therefore \quad y = \frac{1}{5}x - \frac{11}{5} \quad$ (or $\quad x - 5y = 11$)

**9**

$$y = 2 - \frac{7}{1+2x} = 2 - 7(1+2x)^{-1}$$

$$\therefore \quad \frac{dy}{dx} = 7(1+2x)^{-2} \times 2 \quad \{\text{chain rule}\}$$

$$= \frac{14}{(1+2x)^2}$$

The tangent is horizontal when the gradient $\dfrac{dy}{dx} = 0$.

But $\dfrac{14}{(1+2x)^2}$ is never 0, so

$y = 2 - \dfrac{7}{1+2x}$ has no horizontal tangents.

**10** Let $\quad g(x) = ax^2 + bx + c$.

$g(x)$ has $y$-intercept $(0, 3)$, so $\quad g(0) = a(0)^2 + b(0) + c = 3$

$\therefore \quad c = 3$

$\therefore \quad g(x) = ax^2 + bx + 3$

The point $(2, 7)$ lies on $g(x)$, so $\quad g(2) = a(2)^2 + b(2) + 3 = 7$

$\therefore \quad 4a + 2b = 4 \quad$ .... (1)

Also, $\quad g'(x) = 2ax + b$

$\therefore \quad g'(2) = 2a(2) + b = 4a + b$

$\therefore$ the gradient of the tangent to $g(x)$ at $(2, 7)$ is $4a + b$.

But, the tangent at $(2, 7)$ passes through $(0, 11)$, so the gradient $= \dfrac{7-11}{2-0} = -2$

$\therefore \quad 4a + b = -2 \quad$ .... (2)

Solving (1) and (2) simultaneously,

$$\begin{aligned} 4a + 2b &= 4 \\ 4a + b &= -2 \\ \hline \text{subtracting:} \quad b &= 6 \end{aligned}$$

Using (2), $\quad 4a + 6 = -2, \quad$ so $\quad a = -2$

So, $\quad g(x) = -2x^2 + 6x + 3$

**11** **a** $f(x) = \sqrt{\cos x}, \quad 0 \leqslant x \leqslant 2\pi$
$f(x)$ is defined when $\cos x \geqslant 0$,
which is when $0 \leqslant x \leqslant \frac{\pi}{2}$
and $\frac{3\pi}{2} \leqslant x \leqslant 2\pi$.

**b** $f(x) = (\cos x)^{\frac{1}{2}}$

$\therefore \quad f'(x) = \frac{1}{2}(\cos x)^{-\frac{1}{2}}(-\sin x)$

$= \dfrac{-\sin x}{2\sqrt{\cos x}}$

$\therefore \quad f'(x) = 0$ when $-\sin x = 0$

For $0 \leqslant x \leqslant 2\pi$, this is when $x = 0, \pi, 2\pi$.

Sign diagram for $f'(x)$ is:

$f(x)$ is increasing for $\frac{3\pi}{2} \leqslant x \leqslant 2\pi$
and decreasing for $0 \leqslant x \leqslant \frac{\pi}{2}$.

**c**

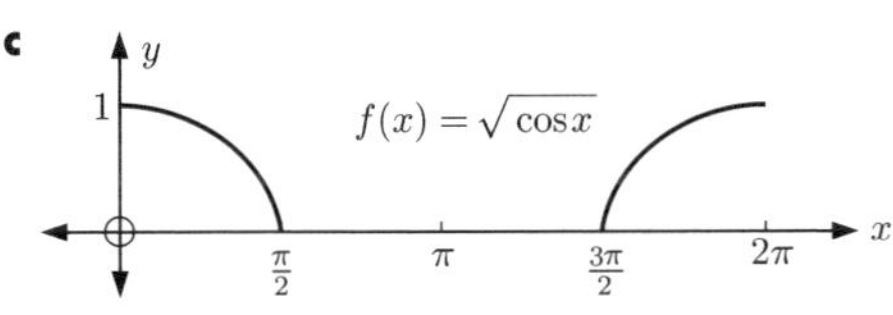


**12** $f(x)$ has a turning point at $x = 0$
$\therefore \quad f'(0) = 0$
$f(x)$ is increasing for $x \geqslant 0$,
except at the asymptote,
so $f'(x)$ is positive for $x \geqslant 0$.
$f(x)$ is decreasing for $x \leqslant 0$,
except at the asymptote,
so $f'(x)$ is negative for $x \leqslant 0$.
As $x \to \pm\infty$, $f(x)$ becomes
closer to horizontal so $f'(x) \to 0$.

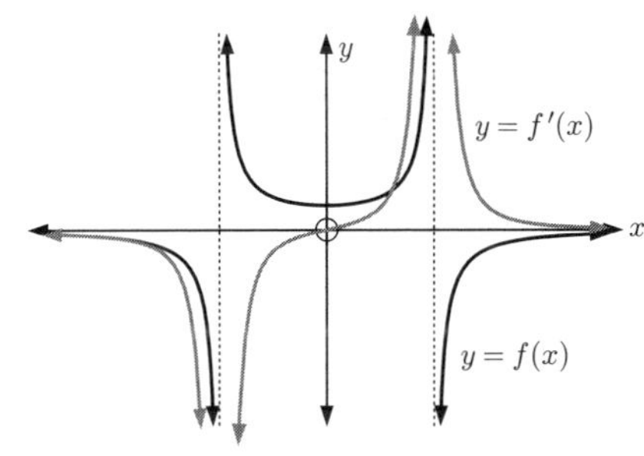


**13** **a**

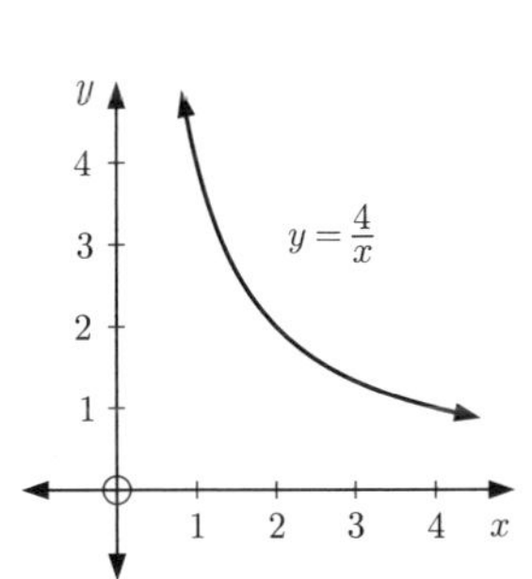


**b** For $f(x) = \dfrac{4}{x} = 4x^{-1}$,

$f'(x) = -4x^{-2} = -\dfrac{4}{x^2}$ and $f'(k) = -\dfrac{4}{k^2}, \quad k > 0$

$\therefore$ the gradient of the tangent to $f(x)$ at $\left(k, \dfrac{4}{k}\right)$ is $-\dfrac{4}{k^2}$

$\therefore$ the equation of the tangent is $\dfrac{y - \frac{4}{k}}{x - k} = -\dfrac{4}{k^2}$

$\therefore \quad yk^2 - 4k = -4x + 4k$

$\therefore \quad k^2 y = -4x + 8k$

$\therefore \quad y = -\dfrac{4}{k^2}x + \dfrac{8}{k}$

**c** $y = -\dfrac{4}{k^2}x + \dfrac{8}{k}$ cuts the $x$-axis when $y = 0$

$\therefore \quad -\dfrac{4}{k^2}x + \dfrac{8}{k} = 0$

$\therefore \quad \dfrac{4}{k^2}x = \dfrac{8}{k}$

$\therefore \quad x = 2k \qquad \therefore$ A is at $(2k, 0)$

$y = -\dfrac{4}{k^2}x + \dfrac{8}{k}$ cuts the $y$-axis when $x = 0$

$\therefore \quad y = \dfrac{8}{k} \qquad \therefore$ B is at $\left(0, \dfrac{8}{k}\right)$

**d** Area of triangle OAB $= \frac{1}{2}(2k)\left(\dfrac{8}{k}\right) = 8$ units$^2$

**e** The gradient of the tangent to $f(x)$ at $\left(k, \frac{4}{k}\right)$ is $-\frac{4}{k^2}$

$\therefore$ the gradient of the normal to $f(x)$ at $\left(k, \frac{4}{k}\right)$ is $\frac{k^2}{4}$

$\therefore$ the equation of the normal is $\frac{y - \frac{4}{k}}{x - k} = \frac{k^2}{4}$

$$\therefore \quad 4y - \frac{16}{k} = k^2x - k^3$$

$$\therefore \quad 4ky - k^3x = 16 - k^4$$

This normal passes through $(1, 1)$, so $4k - k^3 = 16 - k^4$

$$\therefore \quad k^4 - k^3 + 4k - 16 = 0$$

$$\therefore \quad (k-2)(k+2)(k^2 - k + 4) = 0 \quad \{\text{using technology}\}$$

$$\therefore \quad k = \pm 2$$

But $k > 0$, so $k = 2$

## REVIEW SET 16C

**1** Consider the normal to the curve $y = \frac{1}{\sqrt{x}}$ at $x = 4$.

When $x = 4$, $y = \frac{1}{\sqrt{4}} = \frac{1}{2}$, so the point of contact is $(4, \frac{1}{2})$.

Now $\frac{dy}{dx} = -\frac{1}{2}x^{-\frac{3}{2}}$ $\quad\therefore$ at $x = 4$, $\frac{dy}{dx} = -\frac{1}{2}\left(4^{-\frac{3}{2}}\right) = -\frac{1}{2}\left(\frac{1}{8}\right) = -\frac{1}{16}$

$\therefore$ the normal at $(4, \frac{1}{2})$ has gradient 16.

So the equation is $\frac{y - \frac{1}{2}}{x - 4} = 16$

$$\therefore \quad y - \tfrac{1}{2} = 16x - 64$$

$$\therefore \quad y = 16x - \tfrac{127}{2}$$

**2** **a** The tangent shown on the graph passes through $(0, 5)$ and $(5, 0)$.

$\therefore$ the gradient of the tangent is

$\frac{0-5}{5-0} = -1$, so $f'(3) = -1$.

Also, since the tangent passes through $(0, 5)$,

it has equation $\frac{y-5}{x-0} = -1$

$$\therefore \quad y - 5 = -x$$

$$\therefore \quad y = -x + 5$$

So when $x = 3$, $y = -3 + 5 = 2$

$\therefore$ the point of contact is $(3, 2)$, and hence $f(3) = 2$.

**b** $f(x)$ has the form $f(x) = ax^2 + bx + c$

The $y$-intercept is 14 $\therefore$ $f(0) = 14$.

$$\therefore \quad a(0)^2 + b(0) + c = 14$$

$$\therefore \quad c = 14$$

$f(3) = 2$

$$\therefore \quad a(3)^2 + b(3) + 14 = 2$$

$$\therefore \quad 9a + 3b = -12 \quad .... (1)$$

$f'(3) = -1$

and $f'(x) = 2ax + b$

$$\therefore \quad 2a(3) + b = -1$$

$$\therefore \quad 6a + b = -1$$

$$\therefore \quad b = -6a - 1 \quad .... (2)$$

Substituting (2) into (1),

$$9a + 3(-6a - 1) = -12$$

$$\therefore \quad 9a - 18a - 3 = -12$$

$$\therefore \quad -9a = -9$$

$$\therefore \quad a = 1$$

Using (2), $b = -6(1) - 1$

$$\therefore \quad b = -7$$

So, $f(x) = x^2 - 7x + 14$

**3** $y = x^3 + ax + b \qquad \therefore \quad \dfrac{dy}{dx} = 3x^2 + a$

$\therefore \quad \text{at} \quad x = 1, \quad \dfrac{dy}{dx} = 3 + a$

The equation of the tangent at $x = 1$ is $y = 2x$, so the gradient is 2.

$\therefore \quad 3 + a = 2$ and so $a = -1$

Also at $x = 1$, the tangent touches the curve.

$$\therefore \quad x^3 + ax + b = 2x \text{ when } x = 1$$
$$\therefore \quad (1)^3 + (-1)(1) + b = 2(1)$$
$$\therefore \quad 1 - 1 + b = 2$$
$$\therefore \quad b = 2$$

**4** **a** $y = x^3 + ax^2 - 4x + 3 \qquad \therefore \quad \dfrac{dy}{dx} = 3x^2 + 2ax - 4$

The tangent at $x = 1$ is parallel to $y = 3x$, so when $x = 1$, $\dfrac{dy}{dx} = 3$

$$\therefore \quad 3 = 3(1)^2 + 2a(1) - 4$$
$$\therefore \quad 2a = 4$$
$$\therefore \quad a = 2$$

When $x = 1$, $y = 1^3 + 2(1)^2 - 4(1) + 3 = 2$

The contact point is $(1, 2)$ and since the gradient is 3, the tangent at $(1, 2)$ has equation

$$\frac{y-2}{x-1} = 3 \qquad \therefore \quad y - 2 = 3x - 3$$
$$\therefore \quad y = 3x - 1$$

**b** The tangent meets the curve where $x^3 + 2x^2 - 4x + 3 = 3x - 1$

$$\therefore \quad x^3 + 2x^2 - 7x + 4 = 0$$

Since the line touches the curve at $x = 1$, $(x-1)^2$ must be a factor.

Consequently, $x^3 + 2x^2 - 7x + 4 = (x-1)^2(x+4) = 0$ {since the constant term is 4}

$\therefore$ the curve cuts the tangent when $x = 4$.

When $x = -4$, $y = (-4)^3 + 2(-4)^2 - 4(-4) + 3 = -13$

$\therefore$ the curve cuts the tangent at $(-4, -13)$.

**5**

$$y = \ln(x^4 + 3)$$
$$\therefore \quad \frac{dy}{dx} = \frac{4x^3}{x^4 + 3}$$

$\therefore$ when $x = 1$, $\dfrac{dy}{dx} = \dfrac{4(1)^3}{1^4 + 3} = 1$ and $y = \ln(1^4 + 3) = \ln 4$

$\therefore$ the tangent has equation $\dfrac{y - \ln 4}{x - 1} = 1$ or $y = x - 1 + \ln 4$

Now when $x = 0$, $y = \ln 4 - 1$, so the tangent cuts the $y$-axis at $(0, \ln 4 - 1)$.

**6** **a**

$$f(x) = 2x^3 - 3x^2 - 36x + 7$$
$$\therefore \quad f'(x) = 6x^2 - 6x - 36$$
$$= 6(x^2 - x - 6)$$
$$= 6(x-3)(x+2)$$

with sign diagram:

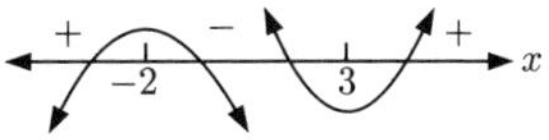

Now $f(-2) = 51$, $f(3) = -74$, so there is a local maximum at $(-2, 51)$, and a local minimum at $(3, -74)$.

$$f''(x) = 12x - 6$$
$$= 6(2x - 1)$$

with sign diagram:

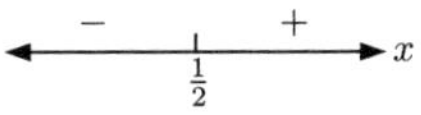

Now $f(\frac{1}{2}) = -\frac{23}{2}$, so there is a non-stationary inflection at $(\frac{1}{2}, -\frac{23}{2})$.

**b** $f(x)$ is increasing when $x \leqslant -2$ or $x \geqslant 3$, and decreasing when $-2 \leqslant x \leqslant 3$.

**c** $f(x)$ is concave up when $x \geqslant \frac{1}{2}$, and concave down when $x \leqslant \frac{1}{2}$.

**d**

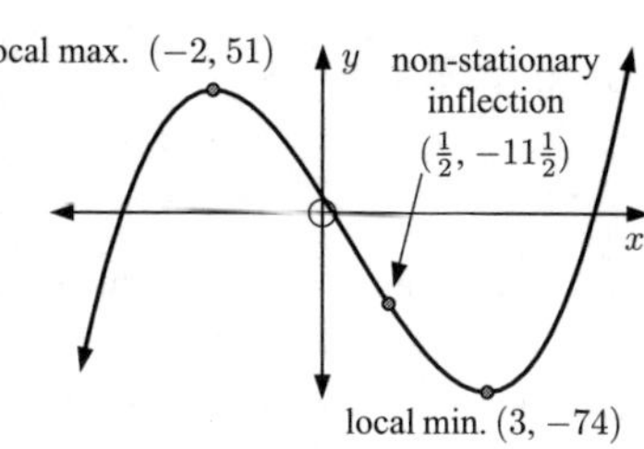


**7** Consider the normal to $f(x) = \dfrac{3x}{1+x}$ at $(2, 2)$.

$f(x)$ is a quotient with $u = 3x$ and $v = 1 + x$

$\therefore\ u' = 3$ and $v' = 1$

$$\therefore\ f'(x) = \frac{3(1+x) - 1(3x)}{(1+x)^2} = \frac{3}{(1+x)^2} \quad \{\text{quotient rule}\}$$

$\therefore\ f'(2) = \frac{3}{9} = \frac{1}{3}$

$\therefore$ the normal at $(2, 2)$ has gradient $-3$

So, the equation of the normal is $\dfrac{y-2}{x-2} = -3$

$\therefore\ y - 2 = -3(x - 2)$

$\therefore\ y = -3x + 8$

When $x = 0$, $y = 8$ and when $y = 0$, $x = \frac{8}{3}$

$\therefore$ B and C are at $(0, 8)$ and $(\frac{8}{3}, 0)$,

and the distance BC $= \sqrt{\left(0 - \frac{8}{3}\right)^2 + (8-0)^2} = \sqrt{\frac{64}{9} + 64} = \sqrt{\frac{640}{9}} = \frac{8\sqrt{10}}{3}$ units

**8** $f(x) = x^3 - 4x^2 + 4x$
$= x(x^2 - 4x + 4)$
$= x(x-2)^2$

**a** $f(0) = 0$, so the $y$-intercept is 0.
$f(x)$ cuts the $x$-axis when $y = 0$ $\quad \therefore\ x(x-2)^2 = 0$
$\therefore\ x = 0$ or $2$

$\therefore$ the $x$-intercepts are 0 and 2.

**b** $f'(x) = 3x^2 - 8x + 4$
$= (3x - 2)(x - 2)$

which is 0 when $x = \frac{2}{3}$ or 2

Sign diagram of $f'(x)$ is:

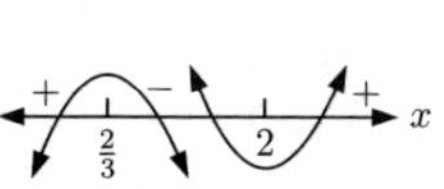


Now $f(\frac{2}{3}) = \frac{32}{27}$, so there is a local maximum at $(\frac{2}{3}, \frac{32}{27})$, and a local minimum at $(2, 0)$.

$f''(x) = 6x - 8 = 2(3x - 4)$

Sign diagram of $f''(x)$ is:

Now $f(\frac{4}{3}) = \frac{16}{27}$, so there is a non-stationary inflection at $(\frac{4}{3}, \frac{16}{27})$.

**c**

$y$ local maximum $(\frac{2}{3}, \frac{32}{27})$ non-stationary inflection $(\frac{4}{3}, \frac{16}{27})$ $y = f(x)$ axis intercept $(0, 0)$ local minimum $(2, 0)$ $x$

**9** **a** $y = \dfrac{1}{\sin x} = (\sin x)^{-1}$

$\therefore\ \dfrac{dy}{dx} = -(\sin x)^{-2}(\cos x)$
$= -\dfrac{\cos x}{\sin^2 x}$

When $x = \frac{\pi}{3}$, $y = \dfrac{1}{\sin(\frac{\pi}{3})} = \frac{2}{\sqrt{3}}$

and $\dfrac{dy}{dx} = -\dfrac{\cos(\frac{\pi}{3})}{\sin^2(\frac{\pi}{3})} = -\dfrac{\frac{1}{2}}{(\frac{\sqrt{3}}{2})^2} = -\frac{2}{3}$

$\therefore$ the tangent has equation $\dfrac{y - \frac{2}{\sqrt{3}}}{x - \frac{\pi}{3}} = -\frac{2}{3}$ which is $3y - 2\sqrt{3} = -2x + \frac{2\pi}{3}$

or $2x + 3y = 2\sqrt{3} + \frac{2\pi}{3}$

**b** $y = \cos(\frac{x}{2})$

$\therefore \dfrac{dy}{dx} = -\frac{1}{2}\sin(\frac{x}{2})$

When $x = \frac{\pi}{2}$, $y = \cos(\frac{\pi}{4}) = \frac{1}{\sqrt{2}}$

and $\dfrac{dy}{dx} = -\frac{1}{2}\sin(\frac{\pi}{4}) = -\frac{1}{2\sqrt{2}}$

$\therefore$ the normal has gradient $2\sqrt{2}$, and its equation is $\dfrac{y - \frac{1}{\sqrt{2}}}{x - \frac{\pi}{2}} = 2\sqrt{2}$

$$\therefore\ y - \tfrac{1}{\sqrt{2}} = 2\sqrt{2}x - \pi\sqrt{2}$$

$$\therefore\ y - 2\sqrt{2}x = \tfrac{1}{\sqrt{2}} - \pi\sqrt{2}$$

or $\sqrt{2}y - 4x = 1 - 2\pi$

**10** $f(x) = 3x^3 + ax^2 + b$

$\therefore\ f'(x) = 9x^2 + 2ax$

Since the tangent at $(-2, 14)$ has gradient 0,

$$f'(-2) = 0$$
$$\therefore\ 36 - 4a = 0$$
$$\therefore\ a = 9$$

As the point $(-2, 14)$ lies on the curve,

$$14 = 3(-2)^3 + 9(-2)^2 + b$$
$$\therefore\ b = 14 + 24 - 36$$
$$\therefore\ b = 2$$

$\therefore\ f'(x) = 9x^2 + 18x$

$\therefore\ f''(x) = 18x + 18$ and so $f''(-2) = -36 + 18 = -18$

**11** The curves $y = \sqrt{3x+1}$ and $y = \sqrt{5x - x^2}$ meet when $\sqrt{3x+1} = \sqrt{5x - x^2}$

Squaring both sides, $3x + 1 = 5x - x^2$

$$\therefore\ x^2 - 2x + 1 = 0$$
$$\therefore\ (x-1)^2 = 0$$
$$\therefore\ x = 1$$

When $x = 1$, $y = \sqrt{3+1} = 2$, so the curves meet at $(1, 2)$.

Now for $y = \sqrt{3x+1} = (3x+1)^{\frac{1}{2}}$

$$\frac{dy}{dx} = \tfrac{1}{2}(3x+1)^{-\frac{1}{2}}(3)$$

$\therefore$ at $(1, 2)$, $\dfrac{dy}{dx} = \dfrac{3}{2(3+1)^{\frac{1}{2}}} = \frac{3}{4}$

*Check*: $y = \sqrt{5x - x^2} = (5x - x^2)^{\frac{1}{2}}$

$$\frac{dy}{dx} = \tfrac{1}{2}(5x - x^2)^{-\frac{1}{2}}(5 - 2x) = \frac{5 - 2x}{2\sqrt{5x - x^2}}$$

$\therefore$ at $(1, 2)$, $\dfrac{dy}{dx} = \dfrac{5-2}{2\sqrt{5-1}} = \frac{3}{4}$ ✓

$\therefore$ the curves have a common tangent at their point of intersection.

The equation of the common tangent at $(1, 2)$ is $\dfrac{y-2}{x-1} = \frac{3}{4}$

$$\therefore\ 4(y-2) = 3(x-1)$$
$$\therefore\ 4y = 3x + 5$$

**12** **a** $f(x) = x + \ln x$ is defined when $x > 0$

**b** $f'(x) = 1 + \dfrac{1}{x} = \dfrac{x+1}{x}$ which has sign diagram:

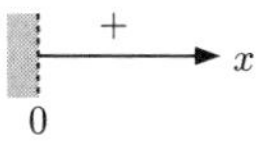


$\therefore\ f(x)$ is increasing for all $x > 0$.

$f''(x) = -\dfrac{1}{x^2}$ which has sign diagram:

$\therefore\ f(x)$ is concave down for all $x > 0$.

**c**

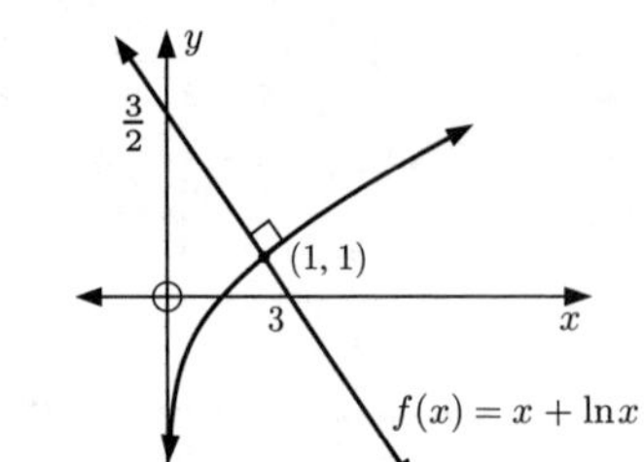


**d** $f(1) = 1 + \ln(1) = 1$

$\therefore$ $(1, 1)$ is the point of contact.

$f'(1) = \dfrac{1+1}{1} = 2$

$\therefore$ the tangent at $x = 1$ has gradient 2, so the normal has gradient $-\frac{1}{2}$

$\therefore$ the normal has equation $\dfrac{y-1}{x-1} = -\frac{1}{2}$

$\therefore$ $2y - 2 = -x + 1$

$\therefore$ $x + 2y = 3$

**13** At $x = B$, $f''(x) = 0$ but $f'(x) \neq 0$

$\therefore$ $f(x)$ has a non-stationary inflection point at $x = B$.

$f'(x)$ is above the $x$-axis for $x \leqslant A$ and $x \geqslant C$, and below the $x$-axis for $A \leqslant x \leqslant C$

$\therefore$ $f(x)$ is increasing for $x \leqslant A$, decreasing for $A \leqslant x \leqslant C$, then increasing for $x \geqslant C$

$\therefore$ $f(x)$ has a local maximum at $x = A$ and a local minimum at $x = C$.

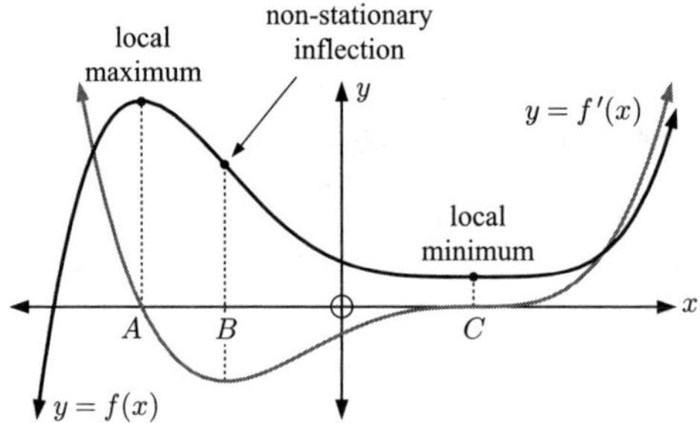

# Chapter 17
## APPLICATIONS OF DIFFERENTIAL CALCULUS

### EXERCISE 17A.1

**1** **a** $s(t) = t^2 + 3t - 2, \quad t \geqslant 0$

Average velocity $= \dfrac{s(t_2) - s(t_1)}{t_2 - t_1}$

$= \dfrac{s(3) - s(1)}{3 - 1}$

$= \dfrac{16 - 2}{2}$

$= 7 \text{ m s}^{-1}$

**b** Average velocity $= \dfrac{s(t_2) - s(t_1)}{t_2 - t_1}$

$= \dfrac{s(1+h) - s(1)}{(1+h) - 1}$

$= \dfrac{(1+h)^2 + 3(1+h) - 2 - 2}{h}$

$= \dfrac{2h + h^2 + 3h}{h}$

$= (h + 5) \text{ m s}^{-1}, \quad h \neq 0$

**c** $\lim\limits_{h \to 0} \dfrac{s(1+h) - s(1)}{h}$

$= \lim\limits_{h \to 0} 5 + h$

$= 5 \text{ m s}^{-1}$

This is the instantaneous velocity at $t = 1$ second, or $s'(1)$.

**d** Average velocity

$= \dfrac{s(t_2) - s(t_1)}{t_2 - t_1}$

$= \dfrac{s(t+h) - s(t)}{(t+h) - t}$

$= \dfrac{\left[(t+h)^2 + 3(t+h) - 2\right] - \left[t^2 + 3t - 2\right]}{h}$

$= \dfrac{2ht + h^2 + 3h}{h}$

$= (2t + h + 3) \text{ m s}^{-1}, \quad h \neq 0$

Now $\lim\limits_{h \to 0} \dfrac{s(t+h) - s(t)}{h} = \lim\limits_{h \to 0} (2t + h + 3)$

$= (2t + 3) \text{ m s}^{-1}$

This is the instantaneous velocity at $t$ seconds.

**2** **a** $s(t) = 5 - 2t^2$ cm

Average velocity $= \dfrac{s(t_2) - s(t_1)}{t_2 - t_1}$

$= \dfrac{s(5) - s(2)}{5 - 2}$

$= \dfrac{(-45) - (-3)}{3}$

$= -14 \text{ cm s}^{-1}$

**b** Average velocity $= \dfrac{s(t_2) - s(t_1)}{t_2 - t_1}$

$= \dfrac{s(2+h) - s(2)}{(2+h) - 2}$

$= \dfrac{5 - 2(2+h)^2 + 3}{h}$

$= \dfrac{-8h - 2h^2}{h}$

$= (-8 - 2h) \text{ cm s}^{-1}, h \neq 0$

**c** $\lim\limits_{h \to 0} \dfrac{s(2+h) - s(2)}{h} = \lim\limits_{h \to 0} (-8 - 2h)$

$= -8 \text{ cm s}^{-1}$

This is the instantaneous velocity when $t = 2$ seconds, or $s'(2)$.

**d** $\lim\limits_{h \to 0} \dfrac{s(t+h) - s(t)}{h}$

$= \lim\limits_{h \to 0} \dfrac{\left[5 - 2(t+h)^2\right] - \left[5 - 2t^2\right]}{h}$

$= \lim\limits_{h \to 0} \dfrac{-4th - 2h^2}{h}$

$= \lim\limits_{h \to 0} (-4t - 2h)$

$= -4t \text{ cm s}^{-1}$

This is the instantaneous velocity at $t$ seconds.

**3** $v(t) = 2\sqrt{t} + 3 \text{ cm s}^{-1}, \; t \geqslant 0$

**a** Average acceleration

$$= \frac{v(t_2) - v(t_1)}{t_2 - t_1} = \frac{v(4) - v(1)}{4 - 1} = \frac{7 - 5}{3} = \tfrac{2}{3} \text{ cm s}^{-2}$$

**b** Average acceleration

$$= \frac{v(t_2) - v(t_1)}{t_2 - t_1} = \frac{v(1+h) - v(1)}{(1+h) - 1} = \frac{[2\sqrt{1+h} + 3] - [2\sqrt{1} + 3]}{h} = \frac{2\sqrt{1+h} - 2}{h} \text{ cm s}^{-2}$$

**c**

$$\lim_{h \to 0} \frac{v(1+h) - v(1)}{(1+h) - 1}$$
$$= \lim_{h \to 0} \frac{2\sqrt{1+h} - 2}{h}$$
$$= \lim_{h \to 0} \frac{2\left(\sqrt{1+h} - 1\right)}{h} \times \frac{\sqrt{1+h} + 1}{\sqrt{1+h} + 1}$$
$$= \lim_{h \to 0} \frac{2(1 + h - 1)}{h\left(\sqrt{1+h} + 1\right)}$$
$$= \lim_{h \to 0} \frac{2h}{h\left(\sqrt{1+h} + 1\right)}$$
$$= \lim_{h \to 0} \frac{2}{\sqrt{1+h} + 1}$$
$$= \tfrac{2}{2}$$
$$= 1 \text{ cm s}^{-2}$$

This is the instantaneous acceleration when $t = 1$ second.

**d**

$$\lim_{h \to 0} \frac{v(t+h) - v(t)}{h}$$
$$= \lim_{h \to 0} \frac{2\sqrt{t+h} - 2\sqrt{t}}{h}$$
$$= \lim_{h \to 0} \frac{2(\sqrt{t+h} - \sqrt{t})}{h} \times \frac{\sqrt{t+h} + \sqrt{t}}{\sqrt{t+h} + t}$$
$$= \lim_{h \to 0} \frac{2h}{h(\sqrt{t+h} + \sqrt{t})}$$
$$= \frac{2}{2\sqrt{t}}$$
$$= \frac{1}{\sqrt{t}} \text{ cm s}^{-2}$$

This is the instantaneous acceleration at $t$ seconds.

**4** **a** This is the instantaneous velocity at $t = 4$ seconds.

**b** This is the instantaneous acceleration at $t = 4$ seconds.

## EXERCISE 17A.2

**1** **a** $s(t) = t^2 - 4t + 3$ cm, $t \geqslant 0$ $\quad \therefore \; v(t) = 2t - 4 \text{ cm s}^{-1}$ and $a(t) = 2 \text{ cm s}^{-2}$.

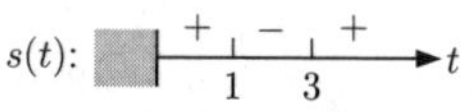

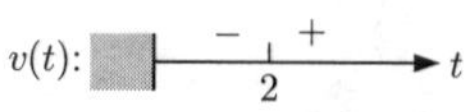

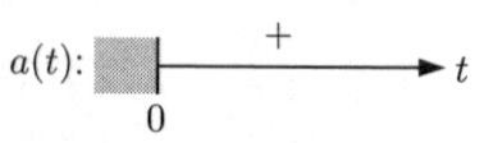

$s(t)$: + | − | + at $t$ = 1, 3

$v(t)$: − | + at $t$ = 2

$a(t)$: + from $t$ = 0

**b** When $t = 0$, $s(0) = 3$ cm, $v(0) = -4 \text{ cm s}^{-1}$, $a(0) = 2 \text{ cm s}^{-2}$

$\therefore$ the object is 3 cm right of O and is moving to the left with a velocity of 4 $\text{cm s}^{-1}$ and slowing down, its acceleration being 2 $\text{cm s}^{-2}$ to the right.

**c** When $t = 2$, $s(2) = -1$ cm, $v(2) = 0 \text{ cm s}^{-1}$, $a(2) = 2 \text{ cm s}^{-2}$

$\therefore$ the object is 1 cm left of O, instantaneously stationary and accelerating to the right at 2 $\text{cm s}^{-2}$.

**d** The object reverses direction when $v(t) = 0$, which occurs at $t = 2$ seconds.
At $t = 2$, the particle is 1 cm left of O.

**e**

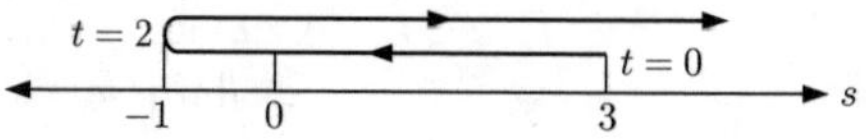

$t = 2$
$t = 0$
$-1$ $0$ $3$ $s$

**f** Speed decreases when $v(t)$ and $a(t)$ have opposite signs, which is when $0 \leqslant t \leqslant 2$.

**2** $s(t) = 98t - 4.9t^2$ m, $t \geqslant 0$

**a** $v(t) = 98 - 9.8t \text{ m s}^{-1}$
$a(t) = -9.8 \text{ m s}^{-2}$

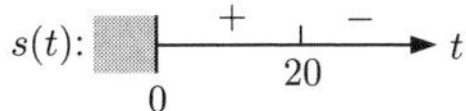


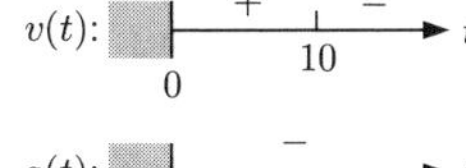


**b** When $t = 0$, $s(0) = 0$ m, $v(0) = 98 \text{ m s}^{-1}$ skyward

**c** When $t = 5$, $s(5) = 367.5$ m, $v(5) = 49 \text{ m s}^{-1}$, $a(5) = -9.8 \text{ m s}^{-2}$

The stone is 367.5 m above the ground and moving skyward at $49 \text{ m s}^{-1}$. Its speed is decreasing.

When $t = 12$, $s(12) = 470.4$ m, $v(12) = -19.6 \text{ m s}^{-1}$, $a(12) = -9.8 \text{ m s}^{-2}$

The stone is 470.4 m above the ground and moving groundward at $19.6 \text{ m s}^{-1}$. Its speed is increasing.

**d** The maximum height is reached when $v(t) = 0 \text{ m s}^{-1}$

$\therefore \ 98 - 9.8t = 0$
$\therefore \ 9.8t = 98$
$\therefore \ t = 10$ seconds

$\therefore$ the maximum height is
$s(10) = 98(10) - 4.9\,(100)$
$= 980 - 490$
$= 490$ m

**e** The stone is at ground level when $s(t) = 0$ which is when $98t - 4.9t^2 = 0$
$\therefore \ 4.9t(20 - t) = 0$
$\therefore \ t = 0$ or 20 seconds
$\therefore$ it hits the ground after 20 seconds.

**3** $s(t) = 1.2 + 28.1t - 4.9t^2$ metres

**a** When released, $t = 0$ and $s(0) = 1.2$ m $\therefore$ it is released 1.2 m above the ground.

**b** $s'(t) = 28.1 - 9.8t \text{ m s}^{-1}$ is the instantaneous velocity of the ball at the time $t$ seconds after release.

**c** When $s'(t) = 0$, $28.1 - 9.8t = 0$ $\therefore \ t = \dfrac{28.1}{9.8} \approx 2.87$ seconds

So, after 2.87 seconds the ball has reached its maximum height, and is instantaneously at rest.

**d** $s(2.867) = 1.2 + 28.1 \times 2.867 - 4.9 \times 2.867^2 \approx 41.5$ m

So, the maximum height reached is about 41.5 m.

**e** **i** $s'(0) = 28.1 \text{ m s}^{-1}$ **ii** $s'(2) = 28.1 - 19.6 = 8.5 \text{ m s}^{-1}$ **iii** $s'(5) = 28.1 - 49 = -20.9 \text{ m s}^{-1}$
$\therefore$ speed $= 20.9 \text{ m s}^{-1}$

If $s'(t) \geqslant 0$, the ball is travelling upwards. If $s'(t) \leqslant 0$, the ball is travelling downwards.

**f** $s(t) = 0$ when $1.2 + 28.1t - 4.9t^2 = 0$
$\therefore \ 4.9t^2 - 28.1t - 1.2 = 0$

$$\therefore \ t = \frac{28.1 \pm \sqrt{28.1^2 - 4(4.9)(-1.2)}}{9.8} \approx -0.0424 \text{ or } 5.777$$

But $t > 0$, so the ball hits the ground after 5.78 seconds.

**g** $s''(t) = -9.8 \text{ m s}^{-2}$ and is the rate of change in $s'(t)$

$\therefore$ the instantaneous acceleration is constant at $-9.8 \text{ m s}^{-2}$ for the entire motion.

**4** **a** $s(t) = bt - 4.9t^2$
$s'(t) = b - 9.8t$
$\therefore \ s'(0) = b$
$\therefore$ the initial velocity is $b \text{ m s}^{-1}$ upwards.

**b** Since $s(14.2) = 0$,
$b(14.2) - 4.9(14.2)^2 = 0$
$\therefore \ 14.2\,[b - 4.9 \times 14.2] = 0$
$\therefore \ b = 4.9 \times 14.2$
$\therefore \ b = 69.58$
$\therefore$ the initial velocity is $69.6 \text{ m s}^{-1}$.

**5** **a** $s(t) = 12t - 2t^3 - 1$ cm, $t \geqslant 0$

$\therefore$ $v(t) = 12 - 6t^2$ cm s$^{-1}$

and $a(t) = -12t$ cm s$^{-2}$

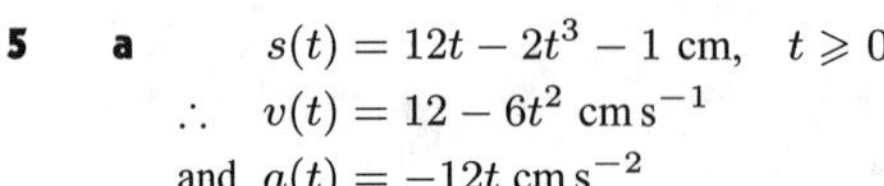

$v(t)$: $+$ $\sqrt{2}$ $-$ $t$

$a(t)$: $-$ $t$, $0$

**b** When $t = 0$, $s(0) = -1$ cm
$v(0) = 12$ cm s$^{-1}$
$a(0) = 0$ cm s$^{-2}$

The particle was 1 cm left of O and was moving right at a constant speed of 12 cm s$^{-1}$.

**c** The particle reverses direction when $v(t) = 0$ which is when $12 - 6t^2 = 0$

$\therefore$ $t^2 = 2$

$\therefore$ $t = \sqrt{2}$ $\{t > 0\}$

When $t = \sqrt{2}$, $s(\sqrt{2}) = 12\sqrt{2} - 2(2\sqrt{2}) - 1$
$= 8\sqrt{2} - 1$

$\therefore$ the particle is $(8\sqrt{2} - 1)$ cm to the right of O.

**d** **i** From the sign diagrams in **a**, the speed increases for $t \geqslant \sqrt{2}$ seconds.

**ii** The velocity of the particle never increases $\{a(t) \leqslant 0\}$.

**6** **a** $x(t) = t^3 - 9t^2 + 24t$ m, $t \geqslant 0$

$v(t) = 3t^2 - 18t + 24$
$= 3(t^2 - 6t + 8)$
$= 3(t - 4)(t - 2)$ m s$^{-1}$

and $a(t) = 6t - 18$
$= 6(t - 3)$ m s$^{-2}$

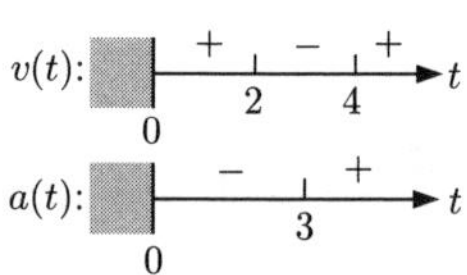


**b** The particle reverses direction when $v(t) = 0$, which occurs at $t = 2$ and $t = 4$ seconds.

$x(2) = 8 - 36 + 48$ m
$= 20$ m

and $x(4) = 64 - 144 + 96$
$= 16$ m

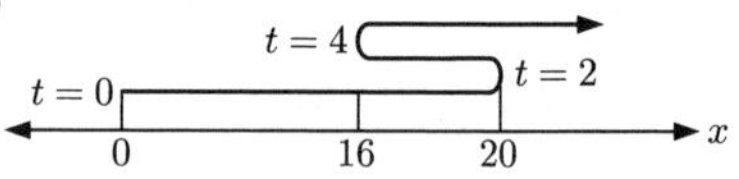


**c** **i** The speed decreases when $v(t)$ and $a(t)$ have the opposite sign, which is when $0 \leqslant t \leqslant 2$ and $3 \leqslant t \leqslant 4$.

**ii** The velocity decreases when $a(t) \leqslant 0$, which is when $0 \leqslant t \leqslant 3$.

**d** When $t = 5$, $x(5) = 5^3 - 9(5)^2 + 24(5)$
$= 125 - 225 + 120$
$= 20$ m

$\therefore$ distance travelled $= 20 + 4 + 4$ m
$= 28$ m

**7** **a** $s(t) = 100t + 200e^{-\frac{t}{5}}$ cm, $t \geqslant 0$

$v(t) = 100 - 40e^{-\frac{t}{5}}$ cm s$^{-1}$

$a(t) = 8e^{-\frac{t}{5}}$ cm s$^{-2}$

**b** When $t = 0$, $s(0) = 200$ cm (right of the origin)
$v(0) = 60$ cm s$^{-1}$
$a(0) = 8$ cm s$^{-2}$

**c** As $t \to \infty$, $e^{-\frac{t}{5}} \to 0$,

$\therefore$ $v(t) \to 100$ cm s$^{-1}$ (below)

**d**

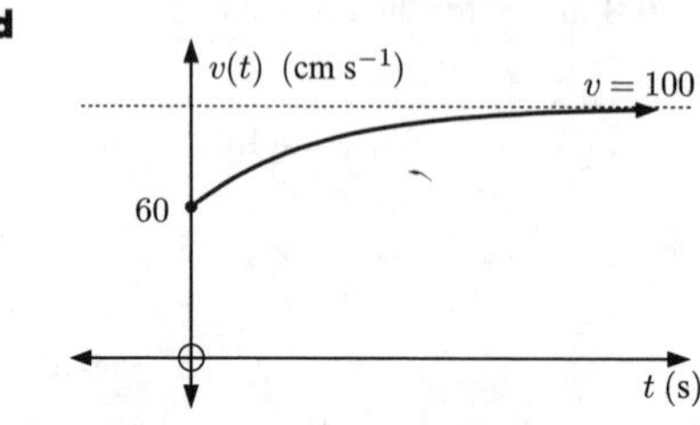


**e** When $v(t) = 80$ cm s$^{-1}$,

$100 - 40e^{-\frac{t}{5}} = 80$

$\therefore$ $-40e^{-\frac{t}{5}} = -20$

$\therefore$ $e^{-\frac{t}{5}} = 0.5$

$\therefore$ $-\frac{t}{5} = \ln 0.5$

$\therefore$ $t = -5\ln 0.5 \approx 3.47$ s

**8** $x(t) = 1 - 2\cos t$ cm

$\therefore\ v(t) = x'(t) = 2\sin t$

$\therefore\ a(t) = v'(t) = 2\cos t$

**a** When $t = 0$,

$x(0) = 1 - 2\cos 0$
$= -1$ cm

$v(0) = 2\sin 0$
$= 0$ cm s$^{-1}$

$a(0) = 2\cos 0$
$= 2$ cm s$^{-2}$

**b** When $t = \frac{\pi}{4}$,

$x(\frac{\pi}{4}) = 1 - \frac{2}{\sqrt{2}}$
$= 1 - \sqrt{2}$ cm

$v(\frac{\pi}{4}) = \frac{2}{\sqrt{2}} = \sqrt{2}$ cm s$^{-1}$

$a(\frac{\pi}{4}) = \frac{2}{\sqrt{2}} = \sqrt{2}$ cm s$^{-2}$

The particle is $(\sqrt{2} - 1)$ cm left of the origin, moving right at $\sqrt{2}$ cm s$^{-1}$ with increasing speed.

**c** We need to look for the points where the velocity equals zero.

If $v(t) = 2\sin t = 0$

then $\sin t = 0$

$\therefore\ t = \pi \quad (0 < t < 2\pi)$

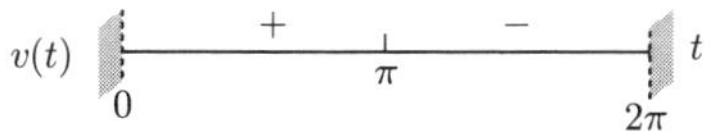


The particle reverses direction when $t = \pi$.

At $t = \pi$, $x(\pi) = 3$ cm.

**d** The particle's speed is increasing when $v(t) = 2\sin t$ and $a(t) = 2\cos t$ have the same sign.

If $a(t) = 2\cos t = 0$

then $\cos t = 0$

$t = \frac{\pi}{2}, \frac{3\pi}{2} \quad (0 \leqslant t \leqslant 2\pi)$

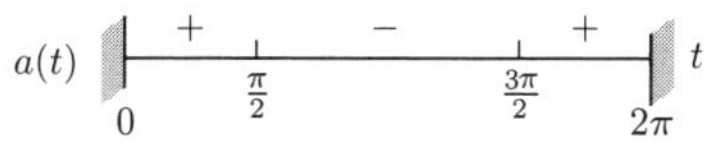


$\therefore$ the particle's speed is increasing when $0 \leqslant t \leqslant \frac{\pi}{2}$ and $\pi \leqslant t \leqslant \frac{3\pi}{2}$.

**9** **a** Let the equation be $s(t) = at^2 + bt + c$

$\therefore\ v(t) = 2at + b$

and $a(t) = 2a = g$ {gravitational acceleration}

$\therefore\ a = \frac{1}{2}g$ and $v(t) = gt + b$

But when $t = 0$, $v(0) = g \times 0 + b = b$

$\therefore$ the initial velocity is $b$

$\therefore\ v(t) = v(0) + gt$ as required

**b** Now when $t = 0$, $s(0) = 0$

$\therefore\ a \times 0^2 + b \times 0 + c = 0$

$\therefore\ c = 0$

and so $s(t) = \left(\frac{1}{2}g\right)t^2 + v(0)t$

$\therefore\ s(t) = v(0) \times t + \frac{1}{2}gt^2$ as required

## EXERCISE 17B

**1** $P(t) = 2t^2 - 12t + 118$ thousand dollars, $t \geqslant 0$

**a** $P(0) = \$118\,000$ is the current annual profit

**b** $\dfrac{dP}{dt} = 4t - 12$ thousand dollars per year

**c** $\dfrac{dP}{dt}$ is the rate of change in profit with time.

**d** **i** The profit decreases when $\dfrac{dP}{dt} \leqslant 0$, which occurs when $4t - 12 \leqslant 0$

$\therefore\ 4t \leqslant 12$

$\therefore\ t \leqslant 3$

But $t \geqslant 0$, so $0 \leqslant t \leqslant 3$ years.

**ii** The profit increases on the previous year when $\dfrac{dP}{dt} \geqslant 0$, which is for $t > 3$ years.

**e** The profit function is a quadratic with $a > 0$ $\therefore$ the shape is $\cup$

So, a minimum profit occurs when $\dfrac{dP}{dt} = 0$, which is when $t = 3$ years

and $P(3) = 18 - 36 + 118 = 100$ thousand dollars or \$100 000.

**f** When $t = 4$, $\dfrac{dP}{dt} = 4$ thousand dollars per year.

So, the profit is increasing at \$4000 per year after 4 years.

When $t = 10$, $\dfrac{dP}{dt} = 28$ thousand dollars per year.

So, the profit is increasing at \$28 000 per year after 10 years.

When $t = 25$, $\dfrac{dP}{dt} = 88$ thousand dollars per year.

So, the profit is increasing at \$88 000 per year after 25 years.

**2** $V = 200(50 - t)^2$ m$^3$

**a** average rate on $0 \leqslant t \leqslant 5$

$$= \frac{V(5) - V(0)}{5 - 0}$$

$$= \frac{200(45)^2 - 200(50)^2}{5}$$

$= -19\,000$ m$^3$ per minute

$\therefore$ leaving at 19 000 m$^3$ per minute

**b** $V'(t) = 400(50 - t)^1 \times (-1)$

$\therefore$ $V'(5) = 400 \times 45 \times -1$

$= -18\,000$ m$^3$ per minute

$\therefore$ leaving at 18 000 m$^3$ per minute

**3** $Q = 100 - 10\sqrt{t}$, $t \geqslant 0$

**a** **i** At $t = 0$, $Q = 100$ units

**ii** At $t = 25$, $Q = 50$ units

**iii** At $t = 100$, $Q = 0$ units

**b** $\dfrac{dQ}{dt} = -5t^{-\frac{1}{2}} = -\dfrac{5}{\sqrt{t}}$

**i** At $t = 25$, $\dfrac{dQ}{dt} = -1$ unit per year

$\therefore$ decreasing at 1 unit per year

**ii** At $t = 50$, $\dfrac{dQ}{dt} = -\frac{5}{\sqrt{50}}$

$= -\frac{1}{\sqrt{2}}$ units per year

$\therefore$ decreasing at $\frac{1}{\sqrt{2}}$ units per year

**c** $\dfrac{dQ}{dt} = -\dfrac{5}{\sqrt{t}}$ $\therefore$ the skin *loses* the chemical at the rate $R = \dfrac{5}{\sqrt{t}} = 5t^{-\frac{1}{2}}$ units per year.

Now $\dfrac{dR}{dt} = -\frac{5}{2}t^{-\frac{3}{2}} = -\dfrac{5}{2t\sqrt{t}}$

Since $2t\sqrt{t} > 0$ for all $t > 0$, $\dfrac{dR}{dt} < 0$ for all $t > 0$.

$\therefore$ the rate at which the skin loses the chemical is decreasing for all $t > 0$.

**4** $H = 20 - \dfrac{97.5}{t+5}$ m, $t \geqslant 0$

**a** At planting, $t = 0$ $\therefore$ $H(0) = 20 - \dfrac{97.5}{0+5} = 0.5$ m

**b** $H(4) = 20 - \dfrac{97.5}{4+5} \approx 9.17$ m

$H(8) = 20 - \dfrac{97.5}{8+5} = 12.5$ m

$H(12) = 20 - \dfrac{97.5}{12+5} \approx 14.3$ m

**c** Now $\dfrac{dH}{dt} = 97.5(t+5)^{-2} = \dfrac{97.5}{(t+5)^2}$

When $t = 0$, $\dfrac{dH}{dt} = \frac{97.5}{25} = 3.9$ m year$^{-1}$

When $t = 5$, $\dfrac{dH}{dt} = \frac{97.5}{100} = 0.975$ m year$^{-1}$

When $t = 10$, $\dfrac{dH}{dt} = \frac{97.5}{225} \approx 0.433$ m year$^{-1}$

**d** Now $\dfrac{dH}{dt} = \dfrac{97.5}{(t+5)^2}$

Since $(t+5)^2 > 0$ for all $t \geqslant 0$, $\dfrac{dH}{dt} > 0$ for all $t \geqslant 0$

$\therefore$ the height of the tree is always increasing, which means that the tree is always growing.

**5** **a** $C(x) = 0.0003x^3 + 0.02x^2 + 4x + 2250$

$\therefore$ $C'(x) = 0.0009x^2 + 0.04x + 4$ dollars per pair

**b** $C'(220) = 0.0009(220)^2 + 0.04(220) + 4 = \$56.36$ per pair

This estimates the cost of making the 221st pair of jeans if 220 pairs are currently being made.

**c** $C(221) - C(220) \approx \$7348.98 - \$7292.40 \approx \$56.58$

This is the actual cost to make the extra pair of jeans (221 instead of 220).

**d** $C''(x) = 0.0018x + 0.04$

$C''(x) = 0$ when $0.0018x + 0.04 = 0$

$$\therefore \quad x = -\frac{0.04}{0.0018} \approx -22.2$$

This is the point when the rate of change is a minimum. However, it is out of the bounds of our model, as we cannot make a negative quantity of jeans.

**6** **a** $C(v) = \frac{1}{5}v^2 + 200\,000v^{-1}$ euros

**i** At $v = 50$ km h$^{-1}$, $C = \frac{1}{5}(50)^2 + \dfrac{200\,000}{50} = €4500$

**ii** At $v = 100$ km h$^{-1}$, $C = \frac{1}{5}(100)^2 + \dfrac{200\,000}{100} = €4000$

**b** $\dfrac{dC}{dv} = \frac{2}{5}v - 200\,000v^{-2} = \frac{2}{5}v - \dfrac{200\,000}{v^2}$

**i** At $v = 30$ km h$^{-1}$,

$$\frac{dC}{dv} = \tfrac{2}{5}(30) - \frac{200\,000}{30^2}$$
$$\approx -€210.22 \text{ per km h}^{-1}$$

So, a decrease of €210.22 per km h$^{-1}$.

**ii** At $v = 90$ km h$^{-1}$,

$$\frac{dC}{dv} = \tfrac{2}{5}(90) - \frac{200\,000}{90^2}$$
$$\approx €11.31 \text{ per km h}^{-1}$$

So, an increase of €11.31 per km h$^{-1}$.

**c** The cost is a minimum when $\dfrac{dC}{dv} = 0$, which occurs when $\frac{2}{5}v - \dfrac{200\,000}{v^2} = 0$

$$\therefore \quad \tfrac{2}{5}v = \frac{200\,000}{v^2}$$
$$\therefore \quad v^3 = 500\,000$$
$$\therefore \quad v \approx 79.4 \text{ km h}^{-1}$$

**7** **a** $V = 50\,000\left(1 - \dfrac{t}{80}\right)^2$, $0 \leqslant t \leqslant 80$

$\therefore$ $\dfrac{dV}{dt} = 2 \times 50\,000\left(1 - \dfrac{t}{80}\right)^1 \times \left(-\frac{1}{80}\right) = -1250\left(1 - \dfrac{t}{80}\right)$ L min$^{-1}$



**b** The outflow was fastest when $t = 0$, when the tap was first opened.

**c** $\dfrac{dV}{dt} = -1250 + \frac{1250}{80}t$ $\quad\therefore\quad$ $\dfrac{d^2V}{dt^2} = \frac{1250}{80} = \frac{125}{8}$ L min$^{-2}$

Since $\dfrac{d^2V}{dt^2}$ is constant and positive, $\dfrac{dV}{dt}$ is constantly increasing

$\therefore$ the outflow is decreasing at a constant rate.

**8** $y = \frac{1}{10}x(x-2)(x-3) = \frac{1}{10}(x^3 - 5x^2 + 6x)$

**a** When $y = 0$, $x = 0$, 2 or 3

$\therefore$ the lake is between 2 and 3 km from the shoreline.

**b** $\dfrac{dy}{dx} = \frac{1}{10}(3x^2 - 10x + 6)$
$= \frac{3}{10}x^2 - x + \frac{3}{5}$

When $x = \frac{1}{2}$, $\dfrac{dy}{dx} = \frac{7}{40}$ $\therefore$ land is sloping upwards.

When $x = 1\frac{1}{2}$, $\dfrac{dy}{dx} = -\frac{9}{40}$ $\therefore$ land is sloping downwards.

This means the top of the hill is between $x = \frac{1}{2}$ and $x = 1\frac{1}{2}$.

**c** The deepest point of the lake occurs when the slope of the land is 0, which is when $\dfrac{dy}{dx} = 0$

$\therefore$ $\frac{1}{10}(3x^2 - 10x + 6) = 0$

$\therefore$ $3x^2 - 10x + 6 = 0$

$\therefore$ $x = \dfrac{10 \pm \sqrt{100-72}}{6} = \dfrac{5 \pm \sqrt{7}}{3}$

but it must be the value between 2 and 3 km, so $x = \dfrac{5+\sqrt{7}}{3} \approx 2.55$ km from the sea

The depth at this point is $y(2.549) \approx \frac{1}{10}(2.549)(0.549)(-0.451)$
$\approx -0.06311$ km
$\approx 63.1$ m

**9** **a** $W = 20e^{-kt}$ so when $t = 50$ hours, $W = 10$ g

$\therefore$ $20e^{-50k} = 10$

$\therefore$ $e^{-50k} = \frac{1}{2}$

$\therefore$ $-50k = \ln\frac{1}{2} = -\ln 2$ $\therefore$ $k = \frac{1}{50}\ln 2 \approx 0.0139$

**b** **i** When $t = 0$,
$W = 20e^0$
$= 20$ g

**ii** When $t = 24$,
$W = 20e^{-24k}$
$= 20e^{-24\frac{\ln 2}{50}}$
$\approx 14.3$ g

**iii** When $t = 1$ week
$= 7 \times 24$ hours
$= 168$ hours
$W = 20e^{-168\frac{\ln 2}{50}}$
$\approx 1.95$ g

**c** When $W = 1$ g, $20e^{-\frac{\ln 2}{50}\times t} = 1$

$\therefore$ $e^{-\frac{\ln 2}{50}\times t} = 0.05$

$\therefore$ $-\frac{\ln 2}{50} \times t = \ln 0.05$

$\therefore$ $t = \dfrac{-50\ln 0.05}{\ln 2} = 216.096\,404\,7$
$\approx 216$ hours or 9 days and 6 minutes

**d** $\dfrac{dW}{dt}$
$= 20e^{-kt}(-k)$
$= \left(-20\dfrac{\ln 2}{50}\right) \times e^{-\frac{\ln 2}{50}t}$

**i** When $t = 100$ hours,
$\dfrac{dW}{dt} = \left(\dfrac{-20\ln 2}{50}\right)e^{-2\ln 2}$
$\approx -0.0693$ g h$^{-1}$

**ii** When $t = 1000$ hours,
$\dfrac{dW}{dt} = \left(\dfrac{-20\ln 2}{50}\right)e^{-20\ln 2}$
$\approx -2.64 \times 10^{-7}$ g h$^{-1}$

**e** $\dfrac{dW}{dt} = -k(20e^{-kt}) = -kW$ $\therefore$ $\dfrac{dW}{dt} \propto W$

**10** $T = 5 + 95e^{-kt}$ °C

**a** $T = 20$°C when $t = 15$

$\therefore\ 20 = 5 + 95e^{-15k}$

$\therefore\ 15 = 95e^{-15k}$

$\therefore\ e^{15k} = \frac{95}{15}$

$\therefore\ 15k = \ln\left(\frac{19}{3}\right)$

$\therefore\ k = \frac{1}{15}\ln\left(\frac{19}{3}\right) \approx 0.123$

**b** When $t = 0$,

$T = 5 + 95e^0$

$= 5 + 95$

$= 100$°C

**c** $\dfrac{dT}{dt} = 0 + 95e^{-kt}(-k)$

$= -(95e^{-kt})k$

$= c(T-5)$ where $c = -k$

$\therefore\ c \approx -0.123$

**d** $\dfrac{dT}{dt} = -95e^{-kt} \times k \approx -11.6902e^{-0.1231t}$

**i** When $t = 0$, $\dfrac{dT}{dt} \approx -11.69$, so the temperature is decreasing at 11.7°C min$^{-1}$.

**ii** When $t = 10$, $\dfrac{dT}{dt} \approx -11.6902e^{-1.231} \approx -3.415$,

so the temperature is decreasing at 3.42°C min$^{-1}$.

**iii** When $t = 20$, $\dfrac{dT}{dt} \approx -11.6902e^{-2.461} \approx -0.998$,

so the temperature is decreasing at 0.998°C min$^{-1}$.

**11** $H(t) = 20\ln(3t+2) + 30$ cm, $t \geqslant 0$

**a** The shrubs were planted when $t = 0$. $H(0) = 20\ln(2) + 30 \approx 43.9$ cm

**b** When $H = 1$ m $= 100$ cm,

$20\ln(3t+2) + 30 = 100$

$\therefore\ 20\ln(3t+2) = 70$

$\therefore\ \ln(3t+2) = 3.5$

$\therefore\ 3t + 2 = e^{3.5}$

$\therefore\ 3t = e^{3.5} - 2$

$\therefore\ t = \dfrac{e^{3.5}-2}{3}$ years

$\therefore\ t \approx 10.4$ years

**c** $\dfrac{dH}{dt} = 20 \times \dfrac{3}{(3t+2)} = \dfrac{60}{3t+2}$ cm year$^{-1}$

**i** When $t = 3$, $\dfrac{dH}{dt} = \frac{60}{11} \approx 5.4545$

$\therefore$ it is growing at 5.45 cm year$^{-1}$

**ii** When $t = 10$, $\dfrac{dH}{dt} = \frac{60}{32} = 1.875$

$\therefore$ it is growing at 1.88 cm year$^{-1}$

**12** **a** $A = s(1 - e^{-kt})$, $t \geqslant 0$

When $t = 0$, $A = s(1 - e^0)$

$= s(1-1) = 0$

**b** **i** When $t = 3$, $A = 5$, and $s = 10$,

$5 = 10(1 - e^{-3k})$

$\therefore\ 0.5 = 1 - e^{-3k}$

$\therefore\ e^{-3k} = 0.5$

$\therefore\ e^{3k} = 2$

$\therefore\ 3k = \ln 2$

$\therefore\ k = \dfrac{\ln 2}{3} \approx 0.231$

**ii** $\dfrac{dA}{dt} = ske^{-kt}$

$\therefore$ when $t = 5$ and $s = 10$,

$\dfrac{dA}{dt} = 10\left(\frac{1}{3}\ln 2\right)\left(e^{-\frac{5}{3}\ln 2}\right)$

$\approx 0.728$ litres per hour

**c** $A = s\left(1 - e^{-kt}\right)$

$\therefore\ A = s - se^{-kt}$

$\therefore\ A - s = -se^{-kt}$

Now, $\dfrac{dA}{dt} = ske^{-kt}$

$= k\left(se^{-kt}\right)$

$= -k\left(-se^{-kt}\right)$

$= -k(A-s)$

$\therefore\ \dfrac{dA}{dt} \propto (A-s)$

**13** Triangle PQR has area $A = \frac{1}{2} \times 6 \times 7 \times \sin\theta$

$\therefore\ A = 21\sin\theta$ cm$^2$

$\therefore\ \dfrac{dA}{d\theta} = 21\cos\theta$ cm$^2$ per radian

When $\theta = 45° = \frac{\pi}{4}$, $\dfrac{dA}{d\theta} = 21\cos(\frac{\pi}{4})$ cm$^2$ per radian

$= 21 \times \frac{1}{\sqrt{2}}$ cm$^2$ per radian

$\approx \frac{21}{\sqrt{2}}$ cm$^2$ per radian

**14** $d = 9.3 + 6.8\cos(0.507t)$ m

$\therefore\ \dfrac{dd}{dt} = -6.8\sin(0.507t) \times 0.507$

$= -3.4476\sin(0.507t)$

**a** When $t = 8$, $\dfrac{dd}{dt} = -3.4476\sin(0.507 \times 8)$

$\approx 2.73$

$\therefore$ the rate of change in the depth of water is 2.73 m per hour.

**b** When $t = 8$, $\dfrac{dd}{dt} \approx 2.73 > 0$

$\therefore$ the tide is rising.

**15** **a** $V(t) = 340\sin(100\pi t)$

$\therefore\ V'(t) = 340\cos(100\pi t) \times 100\pi$

$= 34\,000\pi\cos(100\pi t)$

When $t = 0.01$,

$V'(0.01) = 34\,000\pi \times \cos\pi$

$= -34\,000\pi$ units per second

**b** When $V(t)$ is a maximum,
$V'(t)$ must be 0 units per second.

**16** **a** The distance from $A(-x, 0)$ to $P(\cos t, \sin t)$ is fixed at 2 m.

$\cos t = \dfrac{OQ}{1} = OQ$

$\therefore\ (\cos t + x)^2 + \sin^2 t = 2^2$ {Pythagoras in triangle APQ}

$\therefore\ (\cos t + x)^2 = 4 - \sin^2 t$

$\therefore\ x + \cos t = \pm\sqrt{4 - \sin^2 t}$

$\therefore$ since $x > 0$, $x = \sqrt{4 - \sin^2 t} - \cos t$



**b** Now $\dfrac{dx}{dt} = \frac{1}{2}(4 - \sin^2 t)^{-\frac{1}{2}}(-2\sin t\cos t) + \sin t$

$= \dfrac{-\sin t\cos t}{\sqrt{4 - \sin^2 t}} + \sin t$

**i** When $t = 0$,
$\sin t = 0$ and $\cos t = 1$

$\therefore\ \dfrac{dx}{dt} = 0 + 0$

$= 0$ m s$^{-1}$

**ii** When $t = \frac{\pi}{2}$,
$\sin t = 1$ and $\cos t = 0$

$\therefore\ \dfrac{dx}{dt} = 0 + \sin(\frac{\pi}{2})$

$= 1$ m s$^{-1}$

**iii** When $t = \frac{2\pi}{3}$,
$\sin t = \frac{\sqrt{3}}{2}$ and $\cos t = -\frac{1}{2}$

$\therefore\ \dfrac{dx}{dt} = \dfrac{-\frac{\sqrt{3}}{2}(-\frac{1}{2})}{\sqrt{4 - \frac{3}{4}}} + \dfrac{\sqrt{3}}{2}$

$\approx 1.11$ m s$^{-1}$

## EXERCISE 17C

**1** $C(x) = 720 + 4x + 0.02x^2$

$p(x) = 15 - 0.002x$

Revenue $R(x) = xp(x) = 15x - 0.002x^2$

Profit $P(x) = \text{revenue} - \text{cost}$

$= (15x - 0.002x^2) - (720 + 4x + 0.02x^2)$

$= -0.022x^2 + 11x - 720$

$\therefore \quad P'(x) = -0.044x + 11$

Now $P'(x) = 0$ when $x = \dfrac{11}{0.044} = 250$

$\therefore$ as $P''(x) = -0.044 < 0$, $P$ is maximised when 250 items are produced.

**2** **a**

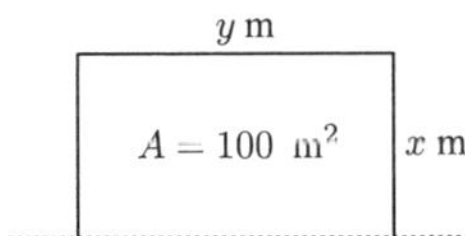


Now $xy = 100$

$\therefore \quad y = \dfrac{100}{x}$

$\therefore \quad L = 2x + y$

$\therefore \quad L = 2x + \dfrac{100}{x}$

**b**

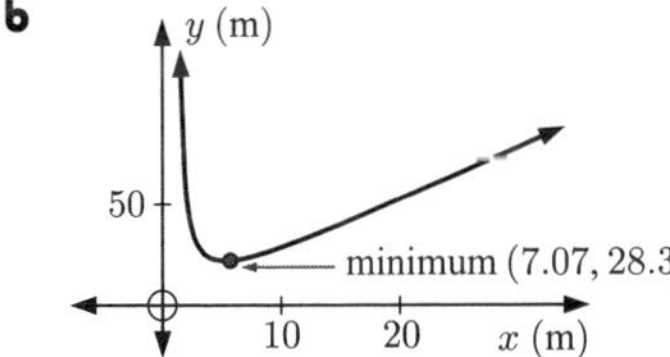


**c** $\dfrac{dL}{dx} = 2 - 100x^{-2} = 2 - \dfrac{100}{x^2}$

which is 0 when $\dfrac{100}{x^2} - 2$

$\therefore \quad x^2 = 50$

$\therefore \quad x = \sqrt{50} \quad \{x > 0\}$

$\dfrac{d^2L}{dt^2} = 200x^{-3} = \dfrac{200}{x^3} > 0$ for $x > 0$

$\therefore \quad L_{\text{min}} = 2\sqrt{50} + \dfrac{100}{\sqrt{50}}$

$= 2\sqrt{50} + 2\sqrt{50}$

$= 4\sqrt{50}$

$= 20\sqrt{2}$ m when $x = 5\sqrt{2}$ m

$\therefore$ min $L \approx 28.3$ m when $x \approx 7.07$ m

**d**

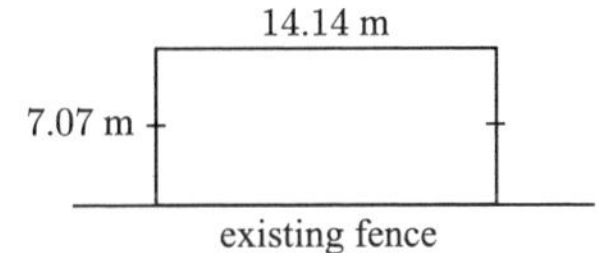


**3** Suppose $x$ fittings are produced daily.

$\therefore \quad C(x) = 1000 + 2x + \dfrac{5000}{x}$

$= 1000 + 2x + 5000x^{-1}$ euros

$\therefore \quad C'(x) = 2 - \dfrac{5000}{x^2}$

Now $C'(x) = 0$ when $x^2 = 2500$

$\therefore \quad x = 50 \quad \{\text{as } x > 0\}$

Also, $C''(x) = 10\,000x^{-3} = \dfrac{10\,000}{x^3}$

which $> 0$ when $x > 0$.

$\therefore$ the cost is minimised when 50 fittings are produced.

**4** $C(x) = \frac{1}{4}x^2 + 8x + 20$

$p(x) = 23 - \frac{1}{2}x$

Revenue $R(x) = xp(x) = 23x - \frac{1}{2}x^2$

Profit $P(x) = \text{revenue} - \text{cost}$

$= (23x - \frac{1}{2}x^2) - (\frac{1}{4}x^2 + 8x + 20)$

$= -\frac{3}{4}x^2 + 15x - 20$

$\therefore \quad P'(x) = -\frac{3}{2}x + 15$

Now $P'(x) = 0$ when $x = \dfrac{15}{\frac{3}{2}} = 10$

$\therefore$ as $P''(x) = -\frac{3}{2} < 0$, $P$ is maximised when 10 blankets per day are produced.

**5** Cost per hour $= \dfrac{v^2}{10} + 22$

Now, cost per km $= \dfrac{\text{cost per hour}}{\text{km per hour}}$

$\therefore \quad C(v) = \dfrac{\frac{v^2}{10} + 22}{v} = 0.1v + 22v^{-1}$

$\therefore \quad C'(v) = 0.1 - 22v^{-2}$

Now $C'(v) = 0$ when $0.1 = \dfrac{22}{v^2}$

$\therefore \quad v^2 = 220$

$\therefore \quad v \approx 14.8 \text{ km h}^{-1}$

**6** **a** $A(t) = t\ln t + 1, \quad 0 < t \leqslant 5$

$\therefore \quad A'(t) = \ln t + t \times \dfrac{1}{t} + 0$ {product rule}

$= \ln t + 1$

$\therefore \quad A'(t) = 0$ when $\ln t = -1$

$\therefore \quad t = e^{-1}$

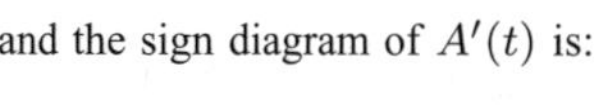
and the sign diagram of $A'(t)$ is:

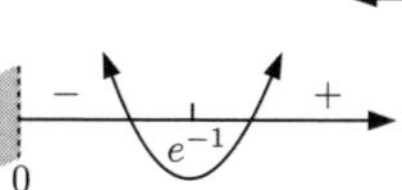


$\therefore \quad A(t)$ is a minimum when $t = \dfrac{1}{e}$

$\approx 0.3679$ years

$\approx 4.41$ months

$\therefore$ the child's memorising ability is a minimum at 4.41 months old.

**b**

$A(t)$

$(5, 5\ln 5 + 1)$

minimum

$(e^{-1}, 0.632)$

$t$ (years)

**7** $C(x) = 0.0007x^3 - 0.1796x^2 + 14.663x + 160$ for $50 \leqslant x \leqslant 150$

$C'(x) = 0.0021x^2 - 0.3592x + 14.663$

$C'(x) = 0$ when $0.0021x^2 - 0.3592x + 14.663 = 0$

Using technology, $x \approx 103.74$ or $x \approx 67.30$

| $x$ | 50 | 67.30 | 103.74 | 150 |
|---|---|---|---|---|
| $C(x)$ | 531.65 | 546.73 | 529.80 | 680.95 |

$\therefore$ the maximum hourly cost is \$680.95 when 150 hinges are made per hour. The minimum hourly cost is \$529.80 when 104 hinges are made per hour.

**8** **a** Inner length of box $= 2x$ cm

**b** Volume $= 200 \text{ cm}^3$

$\therefore \quad x \times 2x \times h = 200$

$2x^2h = 200$

$\therefore \quad x^2h = 100$

**c** From **b**, $h = \dfrac{100}{x^2}$

Area of inner surface is

$A(x) = 2(2x \times x) + 2(2x \times h) + 2(x \times h)$

$= 4x^2 + 4xh + 2xh$

$= 4x^2 + 6xh$

$= 4x^2 + \dfrac{600}{x} \text{ cm}^2$

**d**

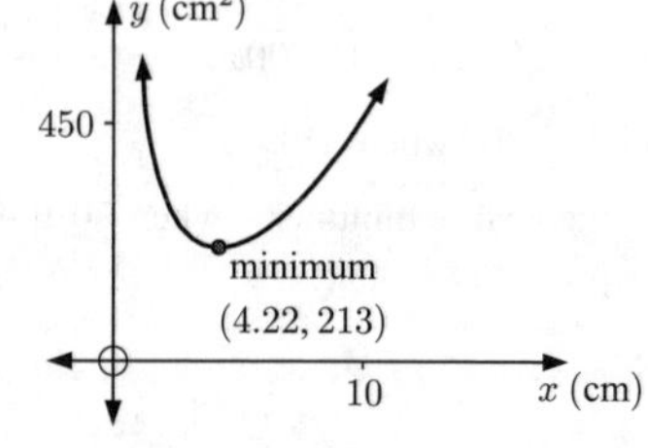

**e** $A(x) = 4x^2 + 600x^{-1}$

$\therefore \quad A'(x) = 8x - 600x^{-2}$

$= 8x - \dfrac{600}{x^2}$

$\therefore \quad A'(x) = 0 \quad \text{when} \quad 8x = \dfrac{600}{x^2}$

$\therefore \quad 8x^3 = 600$

$\therefore \quad x^3 = 75$

$\therefore \quad x \approx 4.217 \text{ cm}$

$A''(x) = 8 + 1200x^{-3}$

$= 8 + \dfrac{1200}{x^3}$

$\therefore \quad A''(x) > 0 \qquad \{\text{as } x > 0\}$

$\therefore$ area is minimised when $x \approx 4.22$ cm

$A_{\min} \approx 4(4.217)^2 + \dfrac{600}{(4.217)}$

$\approx 213 \text{ cm}^2$

**f** Height $h \approx \dfrac{100}{(4.217)^2}$

$\approx 5.62$ cm

5.62 cm
8.43 cm
4.22 cm

**9** $C(x) = 4\ln x + \left(\dfrac{30-x}{10}\right)^2, \quad x \geqslant 10$

$\therefore \quad C'(x) = \dfrac{4}{x} + 2\left(\dfrac{30-x}{10}\right)\left(-\frac{1}{10}\right)$

$= \dfrac{4}{x} - \dfrac{30-x}{50}$

$= \dfrac{200 - x(30-x)}{50x}$

$= \dfrac{200 - 30x + x^2}{50x}$

$= \dfrac{(x-10)(x-20)}{50x}$

$C'(x)$ has sign diagram:

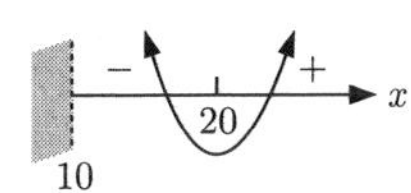


$\therefore$ the minimum cost occurs when $x = 20$ or when 20 kettles per day are produced.

**10** Let coordinates of D be $(x, 0)$ where $x > 0$.

$\therefore$ the coordinates of C are $(x, e^{-x^2})$.

$\therefore$ area ABCD $= 2xe^{-x^2}$

$\therefore \quad \dfrac{dA}{dx} = 2e^{-x^2} + 2xe^{-x^2}(-2x) \quad \{\text{product rule}\}$

$= 2e^{-x^2}\left(1 - 2x^2\right)$

$= 2e^{-x^2}(1 + \sqrt{2}x)(1 - \sqrt{2}x)$

$\dfrac{dA}{dx}$ has sign diagram:

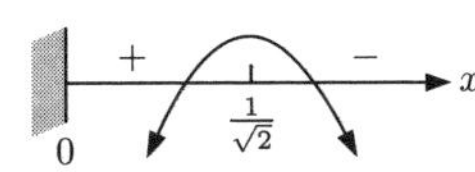


$\therefore$ the area is a maximum when $x = \frac{1}{\sqrt{2}}$

and so C is $\left(\frac{1}{\sqrt{2}}, e^{-\frac{1}{2}}\right)$.

**11** **a** Volume of can $= \pi r^2 h$

$\therefore \quad 1000 = \pi r^2 h \quad$ (in cm)

$\therefore \quad h = \dfrac{1000}{\pi r^2}$ cm

**b** Opening the can up we get:

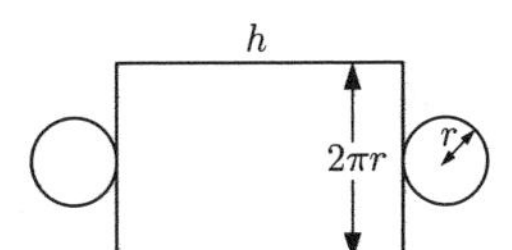


$\therefore \quad A(r) = \pi r^2 + \pi r^2 + 2\pi r h$

$= 2\pi r^2 + 2\pi r h$

$= 2\pi r^2 + \dfrac{2000}{r}$ cm$^2$

**c**

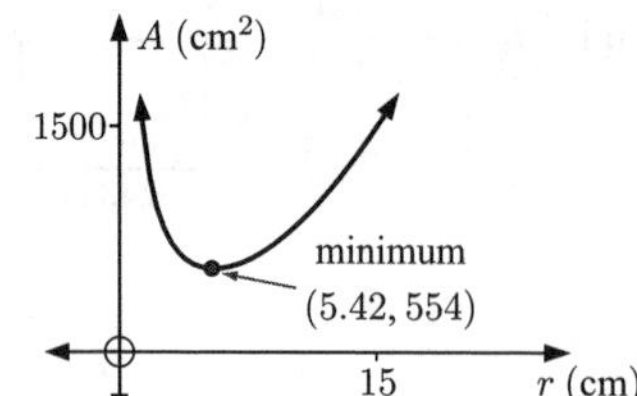


**d** $A(r) = 2\pi r^2 + 2000r^{-1}$

$A'(r) = 4\pi r - 2000r^{-2} = 4\pi r - \dfrac{2000}{r^2}$

So, $A'(r) = 0$ when $4\pi r = \dfrac{2000}{r^2}$

$r^3 = \dfrac{2000}{4\pi}$

$r = \sqrt[3]{\dfrac{500}{\pi}} \approx 5.42$ cm

**e**

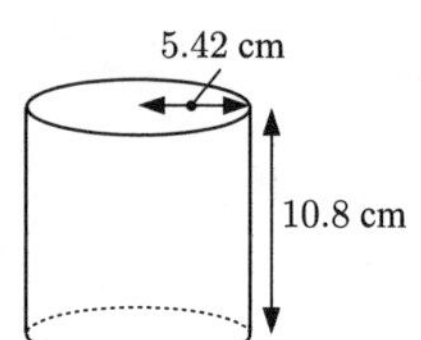


$A''(r) = 4\pi + 4000r^{-3} = 4\pi + \dfrac{4000}{r^3}$

and as $r > 0$, $A''(r) > 0$

$\therefore$ area is a minimum when $r \approx 5.42$ cm

and $h = \dfrac{1000}{\pi r^2} \approx 10.8$ cm

$A_{\min} = 2\pi r^2 + 2\pi rh \approx 554$ cm$^2$

**12**

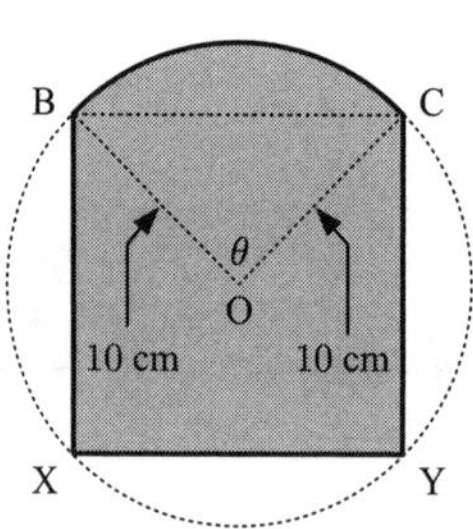


**a** Using the cosine rule in $\triangle$BCO,

$\text{BC}^2 = 10^2 + 10^2 - 2 \times 10 \times 10\cos\theta$

$\therefore$ $\text{BC} = \sqrt{200 - 200\cos\theta}$

$\therefore$ $\text{XY} = \sqrt{200 - 200\cos\theta}$ also

Now $\text{BY}^2 = \text{BX}^2 + \text{XY}^2$ {Pythagoras}

$\therefore$ $400 = \text{BX}^2 + (200 - 200\cos\theta)$

$\therefore$ $\text{BX}^2 = 200 + 200\cos\theta$

$\therefore$ $\text{BX} = \sqrt{200 + 200\cos\theta}$

The shaded area is equal to the area of the sector plus $\frac{3}{4}$ of the area of BCYX.

$\therefore$ $A = \frac{1}{2}(10)^2\theta + \frac{3}{4}[\text{BX} \times \text{BC}]$

$= 50\theta + \frac{3}{4}\sqrt{200 + 200\cos\theta}\sqrt{200 - 200\cos\theta}$

$= 50\theta + \frac{3}{4} \times 200\sqrt{1 + \cos\theta}\sqrt{1 - \cos\theta}$

$= 50\theta + 150\sqrt{1 - \cos^2\theta}$

$= 50\theta + 150\sin\theta$

$= 50(\theta + 3\sin\theta)$ as required

**b** $\dfrac{dA}{d\theta} = 50 + 150\cos\theta = 50(1 + 3\cos\theta)$

which is zero when $\cos\theta = -\frac{1}{3}$

$\therefore$ $\theta \approx 1.91$

The sign diagram of $\dfrac{dA}{d\theta}$ is:

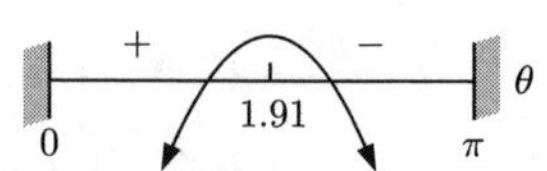


Since $0 < \theta < \pi$, $A$ is maximised when $\theta \approx 1.91$.

When $\theta \approx 1.91$, $A \approx 237$ $\therefore$ the maximum area is 237 cm$^2$.

**13** **a**

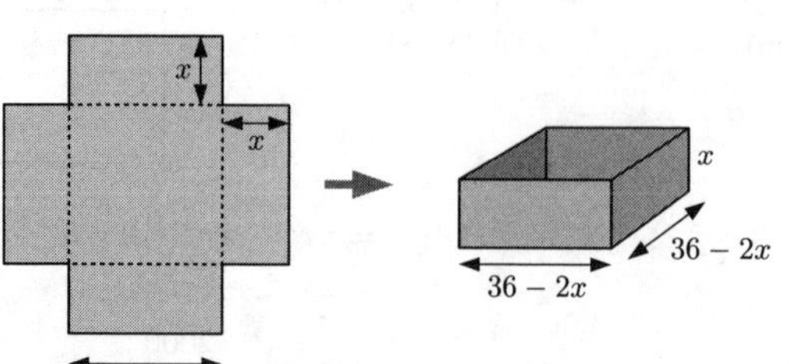


The volume of the container is

$V = lbd$

$= x(36 - 2x)(36 - 2x)$

$\therefore$ $V = x(36 - 2x)^2$ cm$^3$

**b** The container will have the greatest capacity when $V(x)$ is maximised.

Using the product rule, $V'(x) = (36 - 2x)^2 - 4x(36 - 2x)$
$= (36 - 2x)[(36 - 2x) - 4x]$
$= (36 - 2x)(36 - 6x)$

$\therefore \quad V'(x) = 0$ when $x = 6$ or $x = 18$

Sign diagram of $V'(x)$ is:

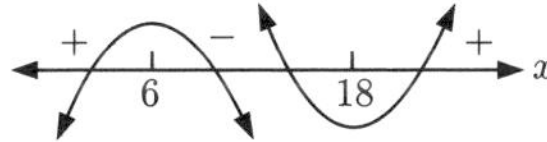


$\therefore$ the volume is maximised when $x = 6$ cm $\{0 \leqslant x < 18\}$

So, 6 cm × 6 cm squares should be cut out to maximise the capacity.

**14** **a** $P = 2\pi r + 2l$

$\therefore \quad 400 = 2\pi(x) + 2l$

$\therefore \quad 200 = \pi x + l$

$\therefore \quad l = 200 - \pi x$

Now clearly $x \geqslant 0$ and $l \geqslant 0$

$\therefore \quad \pi x \leqslant 200$

$\therefore \quad x \leqslant \dfrac{200}{\pi}$

So, $0 \leqslant x \leqslant \dfrac{200}{\pi}$

or $0 \leqslant x \leqslant 63.7$

**b** Area of shaded rectangle $A = 2xl$ m$^2$

$\therefore \quad A = 2x(200 - \pi x)$ m$^2$
$= 400x - 2\pi x^2$ m$^2$

Now $\dfrac{dA}{dx} = 400 - 4\pi x$

which is 0 when $4\pi x = 400$

$\therefore \quad x = \dfrac{100}{\pi} \approx 31.83$ m

and so $l = 200 - \pi\left(\dfrac{100}{\pi}\right)$
$= 100$ m

So, the maximum area of the rectangle is

$2 \times \dfrac{100}{\pi} \times 100 \approx 6366$ m$^2$.

**15**
$$P(t) = \frac{50\,000}{1 + 1000e^{-0.5t}}, \quad 0 \leqslant t \leqslant 25$$
$$= 50\,000(1 + 1000e^{-0.5t})^{-1}$$
$$\therefore \quad P'(t) = -50\,000(1 + 1000e^{-0.5t})^{-2}\left(-500e^{-0.5t}\right)$$
$$= 2.5 \times 10^7 e^{-0.5t}(1 + 1000e^{-0.5t})^{-2}$$

The wasp population is growing the fastest when $\dfrac{dP}{dt}$ is a maximum.

Using technology, the graph of $P'(t)$ can be drawn and the maximum obtained.

The maximum occurs when $t \approx 13.8$ weeks.

**16**
$$E(t) = 750te^{-1.5t}$$
$$\therefore \quad E'(t) = 750e^{-1.5t} + 750t \times -1.5e^{-1.5t}$$
$$= 750e^{-1.5t} - 1125te^{-1.5t}$$
$$= e^{-1.5t}(750 - 1125t)$$

Now $E'(t) = 0$ when $750 - 1125t = 0$ $\{$since $e^{-1.5t} > 0$ for all $t \in \mathbb{R}\}$

$\therefore \quad 1125t = 750$

$\therefore \quad t = \frac{2}{3}$ hours (or 40 minutes)

Sign diagram for $E'(t)$: + on $(0, \frac{2}{3})$, − for $t > \frac{2}{3}$

So, the drug is most effective 40 minutes after the injection.

**17** **a** $\text{AB} = x$ m

$\therefore$ $\text{BC} = (24 - x)$ m $\qquad \therefore$ $D(x) = \sqrt{x^2 + (24 - x)^2}$ {Pythagoras}

**b**
$$\begin{aligned}[D(x)]^2 &= x^2 + (24 - x)^2 \\ &= x^2 + 576 - 48x + x^2 \\ &= 2x^2 - 48x + 576\end{aligned}$$

$\therefore$ $\dfrac{d[D(x)]^2}{dx} = 4x - 48$

$\therefore$ $\dfrac{d[D(x)]^2}{dx} = 0$ when $x = 12$

Sign diagram for $\dfrac{d[D(x)]^2}{dx}$:

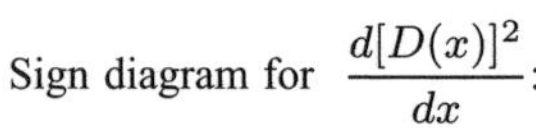

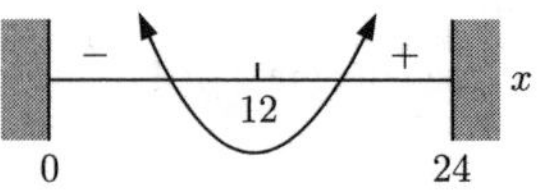


**c** When $\text{AB} = \text{BC} = 12$ m, $D(x)$ is a minimum,
and the minimum $D(x) = 12\sqrt{2}$ m $\approx 17.0$ m.

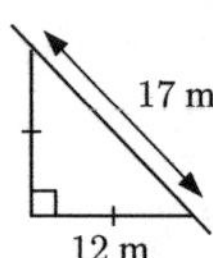


**18** **a**

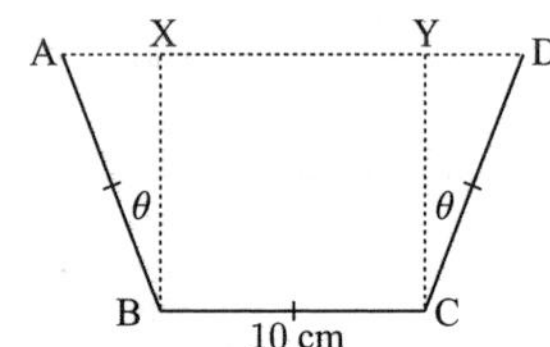


The triangles have height $10\cos\theta$ and width $10\sin\theta$.

$\therefore$ area $A$
$$\begin{aligned}&= \text{area of } \triangle\text{s} + \text{area of rectangle} \\ &= 2 \times \tfrac{1}{2} \times 10\cos\theta \times 10\sin\theta + 10 \times 10\cos\theta \\ &= 100\sin\theta\cos\theta + 100\cos\theta \\ &= 100\cos\theta(1 + \sin\theta)\end{aligned}$$

**b**
$$\begin{aligned}\frac{dA}{d\theta} &= 100(-\sin\theta(1 + \sin\theta) + \cos\theta \times \cos\theta) \\ &= 100(-\sin\theta - \sin^2\theta + \cos^2\theta) \\ &= 100(-\sin\theta - \sin^2\theta + 1 - \sin^2\theta) \\ &= -100(2\sin^2\theta + \sin\theta - 1) \\ &= -100(2\sin\theta - 1)(\sin\theta + 1)\end{aligned}$$

$\therefore$ $\dfrac{dA}{d\theta} = 0$ when $2\sin\theta - 1 = 0$ or $\sin\theta + 1 = 0$

$\therefore$ $\sin\theta = \frac{1}{2}$ or $\sin\theta = -1$

**c** The maximum carrying capacity occurs when $A$ is maximised.

Using **b**, $\dfrac{dA}{d\theta} = 0$ when $\theta = \frac{\pi}{6}, \frac{5\pi}{6}$, or $\frac{3\pi}{2}$.

But $0 \leqslant \theta \leqslant \frac{\pi}{2}$, so the sign diagram for $\dfrac{dA}{d\theta}$ is:

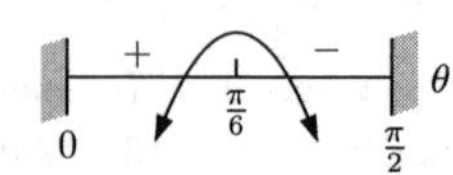


So, the maximum area occurs when $\theta = \frac{\pi}{6} = 30°$

When $\theta = 30°$, 
$$\begin{aligned}A &= 100\cos 30°(1 + \sin 30°) \\ &= 100 \times \tfrac{\sqrt{3}}{2} \times \tfrac{3}{2} \\ &= 75\sqrt{3} \\ &\approx 130 \text{ cm}^2\end{aligned}$$

$\therefore$ the maximum cross-sectional area is 130 cm$^2$.

**19** **a**
$$\begin{aligned}\text{Arc AC} &= \frac{\theta}{360} \times (2\pi r_{\text{sector}}) \\ &= \frac{\theta}{360}(2 \times \pi \times 10) \\ &= \frac{\theta\pi}{18}\end{aligned}$$

**b** Now arc AC forms the base of the cone.

$\therefore$ $2\pi r = \dfrac{\theta\pi}{18}$ {from **a**}

$\therefore$ $r = \dfrac{\theta}{36}$

**c** Height of cone $= \sqrt{10^2 - r^2}$ {Pythagoras}

$\therefore \quad h = \sqrt{100 - \left(\dfrac{\theta}{36}\right)^2}$

**d** $V = \frac{1}{3}\pi r^2 h$

$= \frac{1}{3}\pi \left(\dfrac{\theta}{36}\right)^2 \sqrt{100 - \left(\dfrac{\theta}{36}\right)^2}$

**e**

$V$ (cm$^3$)
500
(294, 403)
360
$\theta$ (°)

**f** $V(\theta) = \frac{1}{3}\pi \left(\dfrac{\theta}{36}\right)^2 \sqrt{100 - \left(\dfrac{\theta}{36}\right)^2}$

$$= \frac{\pi\theta^2}{3 \times 36^2}\sqrt{\frac{100 \times 36^2 - \theta^2}{36^2}}$$

$$= \frac{\pi\theta^2}{3888} \times \frac{1}{36}\sqrt{129\,600 - \theta^2}$$

$$= \frac{\pi\theta^2}{139\,968}\sqrt{129\,600 - \theta^2}$$

Now $V'(\theta) = \dfrac{2\pi\theta}{139\,968}\left(129\,600 - \theta^2\right)^{\frac{1}{2}} + \dfrac{\pi\theta^2}{139\,968}\left(\tfrac{1}{2}\right)\left(129\,600 - \theta^2\right)^{-\frac{1}{2}}(-2\theta)$ {product rule}

$$= \frac{\pi\theta}{139\,968}\left(\frac{2\sqrt{129\,600 - \theta^2}}{1} - \frac{\theta^2}{\sqrt{129\,600 - \theta^2}}\right)$$

$$= \frac{\pi\theta}{139\,968}\left(\frac{2(129\,600 - \theta^2) - \theta^2}{\sqrt{129\,600 - \theta^2}}\right)$$

and $V'(\theta) = 0$ when $\theta = 0$ or $2(129\,600 - \theta^2) = \theta^2$

$259\,200 - 2\theta^2 = \theta^2$

$\therefore \quad 3\theta^2 = 259\,200$

$\therefore \quad \theta = \sqrt{86\,400}$ {as $\theta > 0$}

$\therefore \quad \theta \approx 293.9°$

Sign diagram of $V'(\theta)$ is:

+ −
0° 293.9° 360° $\theta$

$\therefore$ maximum $V$ occurs when $\theta \approx 294°$

**20** **a** Consider each boat's position $t$ hours after 1:00 pm.

PA $= 12t$ and QB $= 8t$

$\therefore$ PB $= 100 - 8t$

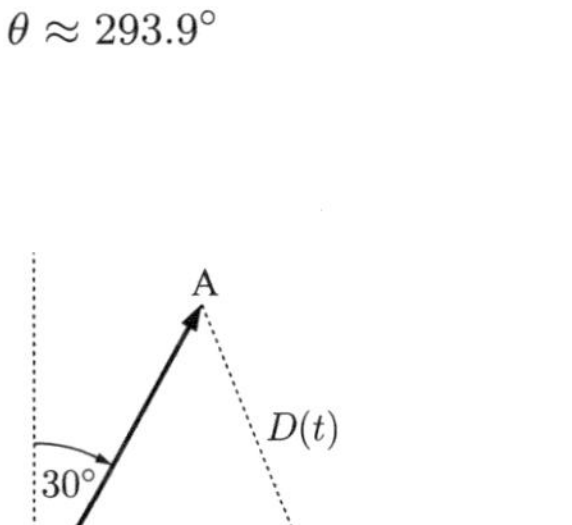


Using the cosine rule in $\triangle$PAB,

$$D(t)^2 = \text{AP}^2 + \text{BP}^2 - 2\text{AP} \times \text{BP}\cos 60°$$
$$= (12t)^2 + (100 - 8t)^2 - 2(12t)(100 - 8t)\tfrac{1}{2}$$
$$= 144t^2 + (100 - 8t)^2 - 12t(100 - 8t)$$
$$= 144t^2 + 10\,000 - 1600t + 64t^2 - 1200t + 96t^2$$
$$= 304t^2 - 2800t + 10\,000$$

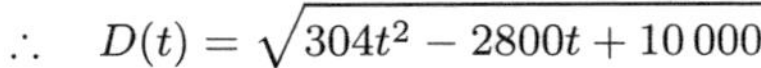

$\therefore \quad D(t) = \sqrt{304t^2 - 2800t + 10\,000}$

**b** Now $\dfrac{d[D(t)]^2}{dt} = 608t - 2800$

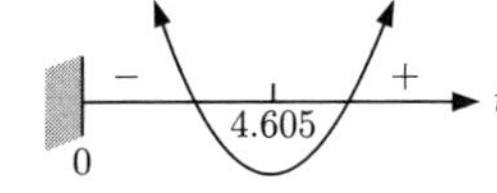


$\therefore \quad \dfrac{d[D(t)]^2}{dt} = 0$ when $t = \dfrac{2800}{608} \approx 4.605\,26$

$\therefore \quad D(t)$ is a minimum when $t \approx 4.605\,26$ hours after 1:00 pm

and $[D(t)]^2_{\text{min}} \approx 304\,(4.6053)^2 - 2800\,(4.6053) + 10\,000$

$\therefore \quad [D(t)]^2_{\text{min}} \approx 3550$ km$^2$

**c** The ships are closest when $t = 4.605\,26$ hours which occurs when the time is 4 hours 36 minutes after 1:00 pm.
So, the ships are closest at approximately 5:36 pm.

**21**

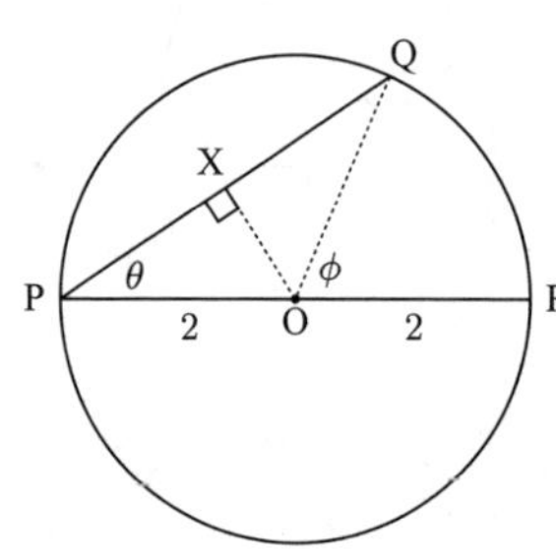


$\dfrac{\text{PX}}{2} = \cos\theta \quad \therefore \quad \text{PQ} = 2\text{PX} = 4\cos\theta$

$\therefore$ the time taken to row from P to Q is $\dfrac{4\cos\theta}{3}$ hours

Now $\phi = 2\theta$ {angle at the centre}

But, arc length $\text{QR}_{\text{arc}} = 2\phi$

$\therefore \quad \text{QR}_{\text{arc}} = 4\theta$

and thc time taken to walk from Q to R is $\dfrac{4\theta}{5}$

$\therefore$ the total time from P to R, $T = \frac{4}{3}\cos\theta + \dfrac{4\theta}{5}$

$\therefore \quad \dfrac{dT}{d\theta} = -\frac{4}{3}\sin\theta + \frac{4}{5}$

$\therefore \quad \dfrac{dT}{d\theta} = 0$ when $-\frac{4}{3}\sin\theta = -\frac{4}{5}$

$\therefore \quad \sin\theta = \frac{3}{5}$

$\therefore \quad \theta \approx 0.6435$ radians

$\therefore \quad \theta \approx 36.87°$

and the sign diagram of $\dfrac{dT}{d\theta}$ is:

+ − $\theta$ 0° 36.87° 90°

So the maximum time occurs when $\theta \approx 36.9°$

and the maximum time is $\frac{4}{3}\cos 0.6435 + \frac{4}{5} \times 0.6435$

$\approx 1.581$ hours

$\approx 1$ hour 34 min 53 s

**22**

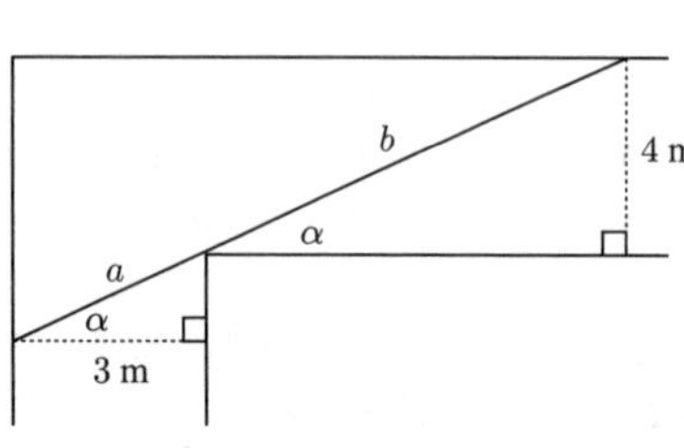


$\cos\alpha = \dfrac{3}{a}$ and $\sin\alpha = \dfrac{4}{b}$

$\therefore \quad a = \dfrac{3}{\cos\alpha}$ and $b = \dfrac{4}{\sin\alpha}$

Now $L = a + b$

$\therefore \quad L = \dfrac{3}{\cos\alpha} + \dfrac{4}{\sin\alpha}$

$= 3(\cos\alpha)^{-1} + 4(\sin\alpha)^{-1}$

$$\therefore \quad \frac{dL}{d\alpha} = -3(\cos\alpha)^{-2} \times (-\sin\alpha) - 4(\sin\alpha)^{-2} \times \cos\alpha$$

$$= \frac{3\sin\alpha}{\cos^2\alpha} - \frac{4\cos\alpha}{\sin^2\alpha}$$

$$= \frac{3\sin^3\alpha - 4\cos^3\alpha}{\cos^2\alpha\sin^2\alpha}$$

$\therefore \quad \dfrac{dL}{d\alpha} = 0$

when $3\sin^3\alpha - 4\cos^3\alpha = 0$

$\therefore \quad 3\sin^3\alpha = 4\cos^3\alpha$

$\therefore \quad \tan^3\alpha = \frac{4}{3}$

$\therefore \quad \tan\alpha = \sqrt[3]{\frac{4}{3}}$

$\therefore \quad \alpha \approx 47.74°$

Sign diagram of $\dfrac{dL}{d\alpha}$ is:

− + $\theta$ 0° 47.74° 90°

$\therefore$ $L$ is minimised when $\alpha \approx 47.74°$ and $L = \dfrac{3}{\cos\alpha} + \dfrac{4}{\sin\alpha} \approx 9.87$ m

So, the maximum possible length of the X-ray screen is 9.87 m.

## REVIEW SET 17A

**1** **a** $s(t) = 2t^3 - 9t^2 + 12t - 5$ cm, $t \geqslant 0$

$v(t) = 6t^2 - 18t + 12$

$= 6(t^2 - 3t + 2)$

$= 6(t-2)(t-1)$ cm s$^{-1}$

and $a(t) = 12t - 18$

$= 6(2t-3)$ cm s$^{-2}$

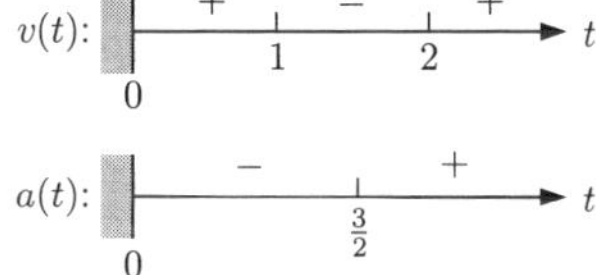


**b** When $t = 0$, $s(0) = -5$ cm

$v(0) = 12$ cm s$^{-1}$

$a(0) = -18$ cm s$^{-2}$

Initially, the particle is 5 cm to the left of O, moving at 12 cm s$^{-1}$ towards the origin and decreasing in speed.

**c** When $t = 2$, $s(2) = -1$ cm

$v(2) = 0$ cm s$^{-1}$

$a(2) = 6$ cm s$^{-2}$

When $t = 2$, the particle is 1 cm to the left of O, instantaneously at rest and increasing in speed towards O.

**d** The particle changes direction when $t = 1$ and $t = 2$, at $s(1) = 0$ cm, $s(2) = -1$ cm.

**e** $t = 0$, $t = 2$, $t = 1$; $-5$, $-1$, $0$, $s$

**f** The speed is increasing when $1 \leqslant t \leqslant \frac{3}{2}$ and $t \geqslant 2$

$\{v(t)$ and $a(t)$ have the same sign$\}$

**2** **a** Now if OD $= x$, the coordinates of C are $(x, k - x^2)$.

$\therefore$ the area of ABCD $= 2x \times (k - x^2)$

$\therefore$ $A = 2kx - 2x^3$, $x > 0$

**b** Now $\dfrac{dA}{dx} = 2k - 6x^2$

But $\dfrac{dA}{dx} = 0$ when AD $= 2\sqrt{3}$, and this occurs when $x = \sqrt{3}$

$\therefore$ $2k - 6(\sqrt{3})^2 = 0$

$\therefore$ $2k - 18 = 0$

$\therefore$ $2k = 18$

$\therefore$ $k = 9$

*Check:* $\dfrac{dA}{dx} = 18 - 6x^2$

$= 6(3 - x^2)$

$= 6(\sqrt{3} + x)(\sqrt{3} - x)$

**3** **a** $x(t) = 3 + \sin(2t)$ cm, $t \geqslant 0$ s

$v(t) = x'(t) = 0 + 2\cos(2t)$ cm s$^{-1}$

$a(t) = v'(t) = -4\sin(2t)$ cm s$^{-2}$

$\therefore$ $x(0) = 3$ cm

$v(0) = 2$ cm s$^{-1}$

$a(0) = 0$ cm s$^{-2}$

$\therefore$ initially the particle is 3 cm right of O, moving right at a speed of 2 cm s$^{-1}$.

**b** $x'(t) = 0$ when $2\cos(2t) = 0$

$\therefore$ $\cos(2t) = 0$

$\therefore$ $2t = \frac{\pi}{2} + k\pi$

For the interval $0 \leqslant t \leqslant \pi$, $t = \frac{\pi}{4}$ or $\frac{3\pi}{4}$

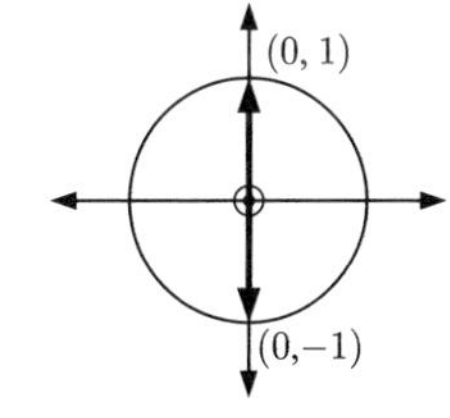


$+$, $-$, $+$; $0$, $\frac{\pi}{4}$, $\frac{3\pi}{4}$, $\pi$, $t$

$\therefore$ the particle reverses direction at $t = \frac{\pi}{4}$ s, $\frac{3\pi}{4}$ s.

**c** $x(0) = 3$, $x\left(\frac{\pi}{4}\right) = 3 + \sin\left(\frac{\pi}{2}\right) = 4$,

$x\left(\frac{3\pi}{4}\right) = 3 + \sin\left(\frac{3\pi}{2}\right) = 3 - 1 = 2$,

$x(\pi) = 3 + \sin(2\pi) = 3$

$\therefore$ the total distance travelled $= 1 + 2 + 1 = 4$ cm.

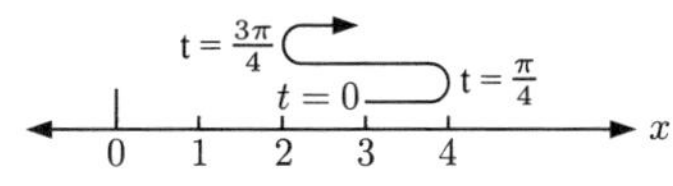

**4** Suppose the sheet is bent $x$ cm from each end. To maximise the water carried we need to maximise the area of cross-section.

$$A = x(24-2x), \quad 0 \leqslant x \leqslant 12$$
$$= 24x - 2x^2$$
$$\therefore \quad \frac{dA}{dx} = 24 - 4x$$

So, $\frac{dA}{dx} = 0$ when $x = 6$, and $\frac{dA}{dx}$ has sign diagram:

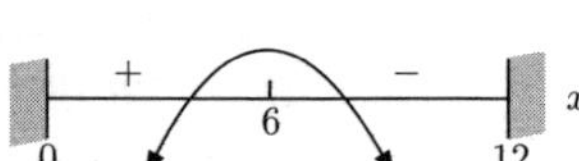


The maximum water is held when $x = 6$ cm

$\therefore$ the bends must be made 6 cm from each end.

**5** **a** $s(t) = 2t - \frac{4}{t+1} = 2t - 4(t+1)^{-1}$

$$\therefore \quad v(t) = 2 + 4(t+1)^{-2}$$
$$= 2 + \frac{4}{(t+1)^2} \text{ cm s}^{-1}$$

$v(t)$: + (for $t > 0$)

$$\therefore \quad a(t) = -8(t+1)^{-3}$$
$$= -\frac{8}{(t+1)^3} \text{ cm s}^{-2}$$

**b** $s(1) = 2(1) - \frac{4}{(1+1)} = 0$ cm

$v(1) = 2 + \frac{4}{(1+1)^2} = 3 \text{ cm s}^{-1}$

$a(1) = -\frac{8}{(1+1)^3} = -1 \text{ cm s}^{-2}$

$\therefore$ the particle is at the origin and is moving to the right with velocity 3 cm s$^{-1}$ and slowing down, its acceleration being 1 cm s$^{-2}$ to the left.

**c** $v(t) = 2 + \frac{4}{(t+1)^2} = \frac{2(t+1)^2 + 4}{(t+1)^2}$

$\therefore$ $v(t) \neq 0$ for any real $t$, so the particle never changes direction.

**d**

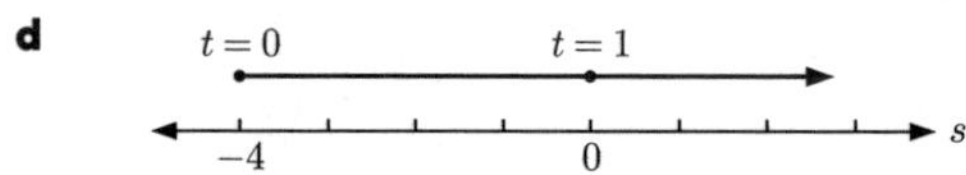


**e** **i** The velocity is never increasing {acceleration is negative for all $t > 0$}.

**ii** The speed is never increasing, as $v(t)$ and $a(t)$ have different signs for all $t > 0$.

**6** When the box is manufactured its base is $(2k - 2x)$ by $(k - 2x)$ and its height is $x$ cm.

$$\therefore \quad V = x(2k-2x)(k-2x)$$
$$\therefore \quad V = x(2k^2 - 4kx - 2xk + 4x^2)$$
$$= 2k^2x - 6kx^2 + 4x^3$$
$$\therefore \quad \frac{dV}{dx} = 2k^2 - 12kx + 12x^2$$
$$= 2(6x^2 - 6kx + k^2)$$

So, $\frac{dV}{dx} = 0$ when
$$x = \frac{6k \pm \sqrt{36k^2 - 4(6)k^2}}{12}$$
$$= \frac{6k \pm k\sqrt{12}}{12}$$
$$= \frac{k}{2} \pm \frac{k}{\sqrt{12}}$$
$$= \frac{k}{2} - \frac{k}{2\sqrt{3}} \quad \{\text{as } x \leqslant \frac{k}{2}\}$$
$$= \frac{k}{2}\left(1 - \tfrac{1}{\sqrt{3}}\right)$$

The sign diagram of $\frac{dV}{dx}$ is:

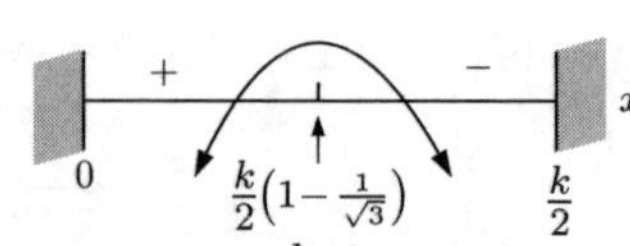


$\therefore$ the maximum capacity occurs when $x = \frac{k}{2}\left(1 - \frac{1}{\sqrt{3}}\right)$.

**7** **a** $s(t) = 30 + \cos(\pi t)$ cm, $t \geqslant 0$

$\therefore\ v(t) = s'(t) = -\pi \sin(\pi t)$

So, $v(0) = 0 \text{ cm s}^{-1}$, $v(\frac{1}{2}) = -\pi \text{ cm s}^{-1}$,

$v(1) = 0 \text{ cm s}^{-1}$, $v(\frac{3}{2}) = \pi \text{ cm s}^{-1}$,

$v(2) = 0 \text{ cm s}^{-1}$

Sign diagram of $v(t)$ is:

− + − + .... $t$ (0, 1, 2, 3)

**b** The cork is falling when $v(t) \leqslant 0$, which is for $0 \leqslant t \leqslant 1$, $2 \leqslant t \leqslant 3$, $4 \leqslant t \leqslant 5$, ....

$\therefore$ the cork is falling for $2n \leqslant t \leqslant 2n+1$, $n \in \{0, 1, 2, 3, ....\}$

## REVIEW SET 17B

**1** $H(t) = 60 + 40\ln(2t+1)$ cm, $t \geqslant 0$

**a** When first planted, $t = 0$ $\therefore\ H(0) = 60 + 40\ln(1) = 60 + 40(0) = 60$ cm.

**b** **i** When $H(t) = 150$ cm,

$\therefore\ 60 + 40\ln(2t+1) = 150$

$\therefore\ 40\ln(2t+1) = 90$

$\therefore\ \ln(2t+1) = \frac{90}{40} = 2.25$

$\therefore\ 2t + 1 = e^{2.25}$

$\therefore\ 2t = e^{2.25} - 1$

$\therefore\ t = \frac{1}{2}(e^{2.25} - 1)$

$\therefore\ t \approx 4.24$ years

**ii** When $H(t) = 300$ cm,

$\therefore\ 60 + 40\ln(2t+1) = 300$

$\therefore\ 40\ln(2t+1) = 240$

$\therefore\ \ln(2t+1) = 6$

$\therefore\ 2t + 1 = e^6$

$\therefore\ 2t = e^6 - 1$

$\therefore\ t = \frac{1}{2}(e^6 - 1)$

$\therefore\ t \approx 201$ years

**c** $H'(t) = 40\left(\dfrac{2}{2t+1}\right) = \dfrac{80}{2t+1}$ cm per year

**i** When $t = 2$, $H'(2) = \frac{80}{5} = 16$ cm per year

**ii** When $t = 20$, $H'(20) = \frac{80}{41} \approx 1.95$ cm per year

**2** $s(t) = 80e^{-\frac{t}{10}} - 40t$ metres, $t \geqslant 0$

**a** $v(t) = s'(t) = -8e^{-\frac{t}{10}} - 40 \text{ m s}^{-1}$

$a(t) = v'(t) = \frac{4}{5}e^{-\frac{t}{10}} \text{ m s}^{-2}$

**b** When $t = 0$, $s(0) = 80$ m

$v(0) = -48 \text{ m s}^{-1}$

$a(0) = 0.8 \text{ m s}^{-2}$

**c** As $t \to \infty$, $e^{-\frac{t}{10}} \to 0$ $\therefore\ v(t) \to -40 \text{ m s}^{-1}$ (below)

**d**

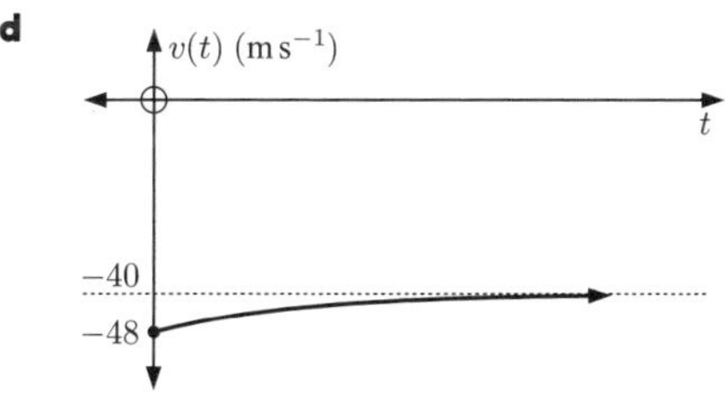


**e** When $v(t) = -44 \text{ m s}^{-1}$

$\therefore\ -8e^{-\frac{t}{10}} - 40 = -44$

$\therefore\ -8e^{-\frac{t}{10}} = -4$

$\therefore\ e^{-\frac{t}{10}} = 0.5$

$\therefore\ -\frac{t}{10} = \ln 0.5$

$\therefore\ t = -10\ln(2^{-1})$

$\therefore\ t = 10\ln 2$ seconds

**3** $C(v) = \dfrac{v^2}{30} + \dfrac{9000}{v}$ dollars per hour

**a** **i** For $t = 2$ hours at $v = 45 \text{ km h}^{-1}$,

$\text{cost} = \left(\dfrac{45^2}{30} + \dfrac{9000}{45}\right) \times 2$ dollars

$= \$535.00$

**ii** For $t = 5$ hours at $v = 64 \text{ km h}^{-1}$,

$\text{cost} = \left(\dfrac{64^2}{30} + \dfrac{9000}{64}\right) \times 5$ dollars

$\approx \$1385.79$

**b** $C'(v) = \dfrac{2v}{30} - 9000v^{-2} = \dfrac{v}{15} - \dfrac{9000}{v^2}$

**i** For $v = 50\text{ km h}^{-1}$

$\therefore \quad C'(50) = \dfrac{50}{15} - \dfrac{9000}{50^2}$

$\approx -\$0.267 \text{ per km h}^{-1}$

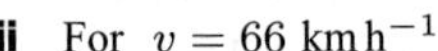

**ii** For $v = 66\text{ km h}^{-1}$

$\therefore \quad C'(66) = \dfrac{66}{15} - \dfrac{9000}{66^2}$

$= \$2.33 \text{ per km h}^{-1}$

**c** Now $C'(v) = \dfrac{v}{15} - \dfrac{9000}{v^2} = \dfrac{v^3 - 135\,000}{15v^2}$

$\therefore \quad C'(v) = 0$ when $v^3 = 135\,000$

$\therefore \quad v \approx 51.3$

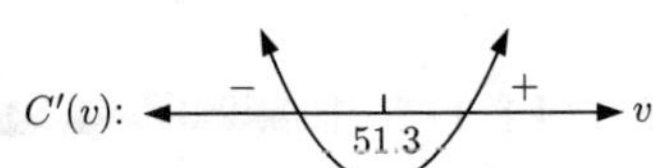


$\therefore$ the minimum cost occurs when $v \approx 51.3\text{ km h}^{-1}$

**4** **a** $x(t) = 3t - \sqrt{t+1} = 3t - (t+1)^{\frac{1}{2}}$ cm, $t \geqslant 0$

$\therefore \quad v(t) = 3 - \frac{1}{2}(t+1)^{-\frac{1}{2}} = 3 - \dfrac{1}{2\sqrt{t+1}}$

since $t \geqslant 0$, $\sqrt{t+1}$ exists and is $> 0$

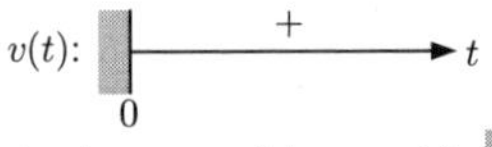

v(t): + (for $t > 0$)

and $a(t) = \frac{1}{4}(t+1)^{-\frac{3}{2}} = \dfrac{1}{4(t+1)^{\frac{3}{2}}}$ which is always positive

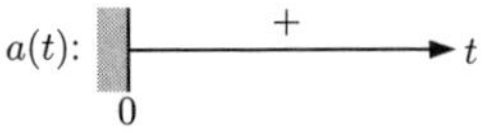

a(t): + (for $t > 0$)

**b** $x(0) = 3(0) - \sqrt{0+1} = -1$ cm

$v(0) = 3 - \dfrac{1}{2\sqrt{0+1}} = 2.5\text{ cm s}^{-1}$

$a(0) = \dfrac{1}{4(0+1)^{\frac{3}{2}}} = 0.25\text{ cm s}^{-2}$

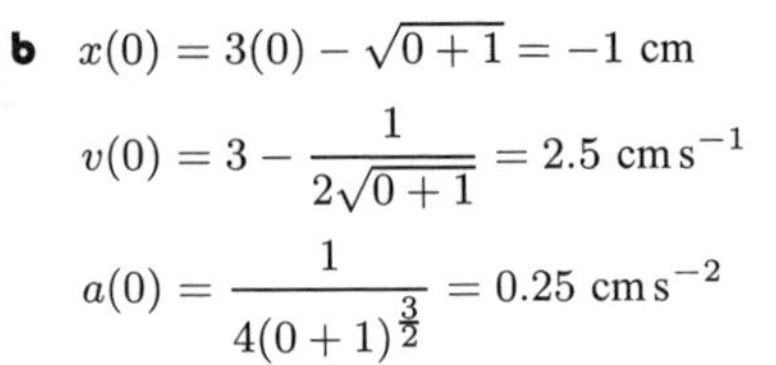

The particle is 1 cm to the left of the origin, is travelling to the right at $2.5\text{ cm s}^{-1}$, and accelerating at $0.25\text{ cm s}^{-2}$.

**c** $x(8) = 3(8) - \sqrt{8+1} = 21$ cm

$v(8) = 3 - \dfrac{1}{2\sqrt{8+1}} \approx 2.83\text{ cm s}^{-1}$

$a(8) = \dfrac{1}{4(8+1)^{\frac{3}{2}}} \approx 0.009\,26\text{ cm s}^{-2}$

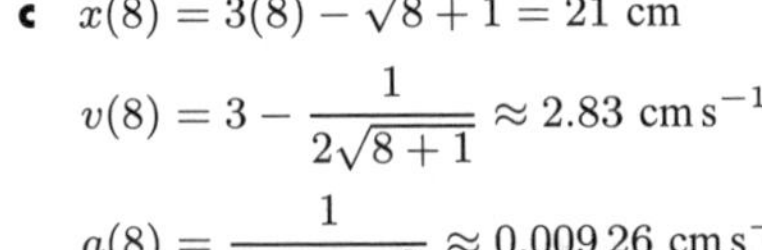

The particle is 21 cm to the right of the origin, is travelling to the right at $2.83\text{ cm s}^{-1}$, and accelerating at $0.009\,26\text{ cm s}^{-2}$.

**d** Since $v(t)$ is $> 0$ for all $t \geqslant 0$, the particle never changes direction.

**e** $v(t)$ and $a(t)$ have the same sign for all $t \geqslant 0$, so the speed of the particle is always increasing.

$\therefore$ the speed of the particle is never decreasing.

**5** **a** At time $t = 0$, $V = 20\,000e^{-0.4\times 0}$

$= 20\,000$ dollars

$\therefore$ the purchase price of the car was \$20 000.

**b** $V' = -0.4(20\,000)e^{-0.4t}$

$= -8000e^{-0.4t}$

At time $t = 10$, $V' = -8000e^{-0.4\times 10}$

$\approx -146.53 \text{ dollars year}^{-1}$

$\therefore$ after 10 years, the car is decreasing in value at \$146.53 per year.

**6** $P(x) = I(x) - C(x)$

$= \left[200\ln\left(1 + \dfrac{x}{100}\right) + 1000\right] - \left[(x-100)^2 + 200\right]$

$= 200\ln(1 + 0.01x) - (x-100)^2 + 800$

$$\therefore \quad \frac{dP}{dx} = 200\left(\frac{0.01}{1+0.01x}\right) - 2(x-100)^1$$
$$= \frac{2}{1+0.01x} - \frac{2(x-100)}{1}$$
$$= \frac{2-2(x-100)(1+0.01x)}{1+0.01x}$$
$$= \frac{2-2(x+0.01x^2-100-x)}{1+0.01x}$$
$$= \frac{2-0.02x^2+200}{1+0.01x}$$
$$= \frac{202-0.02x^2}{1+0.01x}$$

$\therefore \quad \frac{dP}{dx} = 0$ when $0.02x^2 = 202$

$\therefore \quad x^2 = 10\,100$

$\therefore \quad x = \sqrt{10\,100} \quad \{x > 0\}$

$\therefore \quad x \approx 100.49$

and the sign diagram of $\frac{dP}{dx}$ is: $+$ | 100.49 | $-$ ($x$)

$\therefore$ the maximum profit occurs when $x \approx 100.49$

Now $P(100) \approx \$938.63$ and $P(101) \approx \$938.63$

$\therefore$ the maximum daily profit is $938.63 when 100 or 101 shirts are made.

**7** **a** $P = 200$ m

But $P = 2x + 2y + \pi x$

$\therefore \quad 200 = 2x + 2y + \pi x$

$\therefore \quad 2y = 200 - 2x - \pi x$

$\therefore \quad y = 100 - x - \frac{\pi}{2}x$

**b** Area of lawn $= 2x \times y + \frac{1}{2}\pi x^2$

$= 2x\left[100 - x - \frac{\pi}{2}x\right] + \frac{1}{2}\pi x^2$

$= 200x - 2x^2 - \pi x^2 + \frac{1}{2}\pi x^2$

$\therefore \quad A = 200x - 2x^2 - \frac{1}{2}\pi x^2$

**c** $A = 200x - 2x^2 - \frac{1}{2}\pi x^2 = 200x - (2 + \frac{\pi}{2})x^2$ m$^2$

$\therefore \quad \frac{dA}{dx} = 200 - 2\left(2 + \frac{\pi}{2}\right)x = 200 - (4+\pi)x$

$\therefore \quad \frac{dA}{dx} = 0$ when $(4+\pi)x = 200 \quad \therefore \quad x = \frac{200}{4+\pi}$

and the sign diagram for $\frac{dA}{dx}$ is: 0 | $+$ | $\frac{200}{4+\pi}$ | $-$ ($x$)

28.0 m

56.0 m

$\therefore$ the maximum area occurs when $x = \frac{200}{4+\pi} \approx 28.0$ m

and $y = 100 - x - \frac{\pi}{2}x \approx 28.0$ m

## REVIEW SET 17C

**1** **a** Volume $= lbd$

$\therefore \quad x^2 y = 1$

$\therefore \quad y = \frac{1}{x^2}, \quad x > 0$

**b** area $= x^2 + 4xy$

$\therefore \quad$ cost $= (x^2 + 4xy) \times 2$

$\therefore \quad C = 2x^2 + 8xy$

$= 2x^2 + \frac{8}{x}$ dollars $\quad$ {using **a**}

**c** $\frac{dC}{dx} = 4x - 8x^{-2}$

$= 4x - \frac{8}{x^2}$

$= \frac{4(x^3-2)}{x^2}$

So, $\frac{dC}{dx} = 0$ when $x = \sqrt[3]{2}$ m

$\frac{dC}{dx}$ has sign diagram: $-$ | $\sqrt[3]{2}$ | $+$ ($x$)

The minimum cost is when $x = \sqrt[3]{2} \approx 1.26$ m

$\therefore \quad y = \frac{1}{x^2} \approx 0.630$

and the box is 1.26 m by 1.26 m by 0.630 m.

**2** **a** $s(t) = 15t - \dfrac{60}{(t+1)^2}$ cm, $t \geqslant 0$

$= 15t - 60(t+1)^{-2}$ cm

$\therefore\ v(t) = 15 + 120(t+1)^{-3}$ cm s$^{-1}$

$\therefore\ a(t) = -360(t+1)^{-4}$ cm s$^{-2}$

**b** When $t = 3$, $s(t) = 41.25$ cm

$v(t) \approx 16.88$ cm s$^{-1}$

$a(t) \approx -1.41$ cm s$^{-2}$

The particle is 41.25 cm right of O, travelling right at 16.88 cm s$^{-1}$, and is slowing down (decelerating) at 1.41 cm s$^{-2}$.

**c** $v(t) = 15 + \dfrac{120}{(t+1)^3}$ cm s$^{-1}$

$v(t) = 0$ when $15 + \dfrac{120}{(t+1)^3} = 0$

$\therefore\ 15(t+1)^3 + 120 = 0$

$\therefore\ (t+1)^3 = -8$

$\therefore\ t = -3$

$a(t) = -360(t+1)^{-4} = \dfrac{-360}{(t+1)^4}$ cm s$^{-2}$

where $(t+1)^4$ is always positive. $\therefore\ a(t) < 0$ for all $t > 0$

Since $v(t) > 0$ and $a(t) < 0$ for all $t > 0$, $v(t)$ is always decreasing.

$\therefore$ the particle's speed is never increasing.

**3** Let the coordinates of B be $(x, 0)$, so the coordinates of A are $(x, e^{-2x})$.

$\therefore$ the area OBAC is $A = xe^{-2x}$

$$\therefore\ \frac{dA}{dx} = (1)e^{-2x} + x(-2e^{-2x}) \quad \{\text{product rule}\}$$

$$= e^{-2x}(1-2x)$$

$$= \frac{1-2x}{e^{2x}}$$

and has sign diagram:

So, the maximum area occurs when $x = \frac{1}{2}$ and $y = e^{-2(\frac{1}{2})} = e^{-1} = \frac{1}{e}$

$\therefore$ the coordinates of A are $\left(\frac{1}{2}, \frac{1}{e}\right)$.

**4** **a** The tree was $H(0) = 6\left(1 - \frac{2}{3}\right) = 2$ metres tall when first planted.

**b** $t = 3$: $H(3) = 6\left(1 - \frac{2}{3+3}\right) = 4$ metres

$t = 6$: $H(6) = 6\left(1 - \frac{2}{6+3}\right) = 4\frac{2}{3}$ metres

$t = 9$: $H(9) = 6\left(1 - \frac{2}{9+3}\right) = 5$ metres

**c** $H(t) = 6\left(1 - \dfrac{2}{t+3}\right)$

$= 6 - 12(t+3)^{-1}$

$\therefore\ H'(t) = 12(t+3)^{-2}$

$= \dfrac{12}{(t+3)^2}$

So, $t = 0$: $H'(0) = \dfrac{12}{3^2} = \frac{4}{3}$ m year$^{-1}$

$t = 3$: $H'(3) = \dfrac{12}{6^2} = \frac{1}{3}$ m year$^{-1}$

$t = 6$: $H'(6) = \dfrac{12}{9^2} = \frac{4}{27}$ m year$^{-1}$

$t = 9$: $H'(9) = \dfrac{12}{12^2} = \frac{1}{12}$ m year$^{-1}$

**d** $H'(t) = \dfrac{12}{(t+3)^2}$,

and $(t+3)^2 > 0$ for all $t \geqslant 0$

$\therefore\ \dfrac{12}{(t+3)^2} > 0$

$\therefore\ H'(t) > 0$ for all $t \geqslant 0$

This means that the height of the tree is always increasing.

**e**

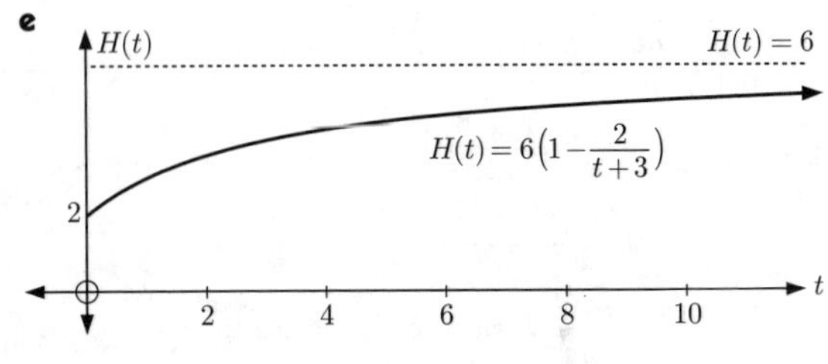

**5** **a** $s(t) = 25t - 10\ln t$ cm, $t \geqslant 1$

$\therefore\ v(t) = 25 - \dfrac{10}{t}$ cm min$^{-1}$

$\therefore\ a(t) = 10t^{-2} = \dfrac{10}{t^2}$ cm min$^{-2}$

**b** When $t = e$,

$s(e) = 25e - 10\ln e = (25e - 10)$ cm
$\approx 58.0$ cm

$v(e) = \left(25 - \dfrac{10}{e}\right)$ cm min$^{-1}$
$\approx 21.3$ cm min$^{-1}$

$a(e) = \dfrac{10}{e^2}$ cm min$^{-2}$ $\approx 1.35$ cm min$^{-2}$

**c** As $t \to \infty$, $\dfrac{10}{t} \to 0$

$\therefore\ v(t) \to 25$ cm min$^{-1}$ (below)

**d** $v(t)$ (cm min$^{-1}$) against $t$ (min): curve starting at $(1, 15)$, increasing towards the asymptote $v = 25$. Axis marks: 5, 15, 25; 4, 8, 12.

**e** When $v(t) = 20$ cm min$^{-1}$,

$25 - \dfrac{10}{t} = 20$

$\therefore\ \dfrac{10}{t} = 5$

$\therefore\ t = 2$ minutes

**6** **a** AC $= 2x$ m

Now ABC is an isosceles triangle.

$\therefore$ XC $= x$

But BC$^2$ = BX$^2$ + XC$^2$ {Pythagoras}

$\therefore\ 2500 = \text{BX}^2 + x^2$

$\therefore\ \text{BX} = \sqrt{2500 - x^2}$

$\therefore\ A(x) = \frac{1}{2}(2x)\sqrt{2500 - x^2} = x\sqrt{2500 - x^2}$

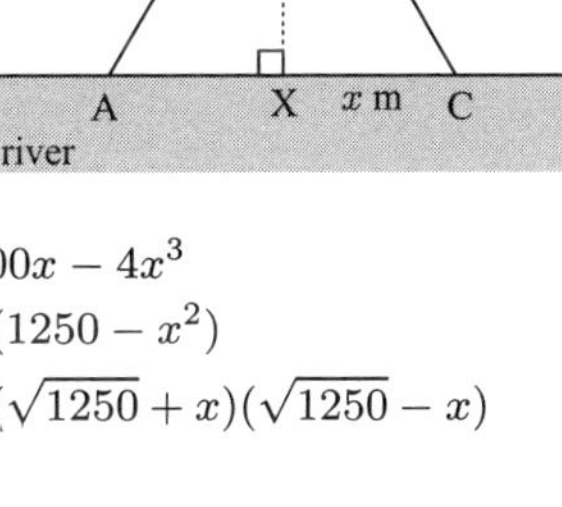


**b** Now $[A(x)]^2 = x^2(2500 - x^2)$

$\therefore\ A^2 = 2500x^2 - x^4$

$\therefore\ \dfrac{d[(A(x)]^2}{dx} = 5000x - 4x^3$
$= 4x(1250 - x^2)$
$= 4x(\sqrt{1250} + x)(\sqrt{1250} - x)$

Sign diagram for $\dfrac{d\,[A(x)]^2}{dx}$ is: $+$ on $(0, 25\sqrt{2})$, $-$ for $x > 25\sqrt{2}$

$\therefore$ the maximum area occurs when $x = 25\sqrt{2}$ m $\approx 35.4$ m

The corresponding maximum area $= \sqrt{1250} \times \sqrt{1250} = 1250$ m$^2$.

**7** **a** $\sin\theta = \dfrac{\text{NA}}{x} = \dfrac{1}{x}$

$\therefore\ \dfrac{1}{x^2} = \sin^2\theta$

$\therefore$ at A, $I = \dfrac{\sqrt{8}\cos\theta}{x^2} = \sqrt{8}\cos\theta\sin^2\theta$

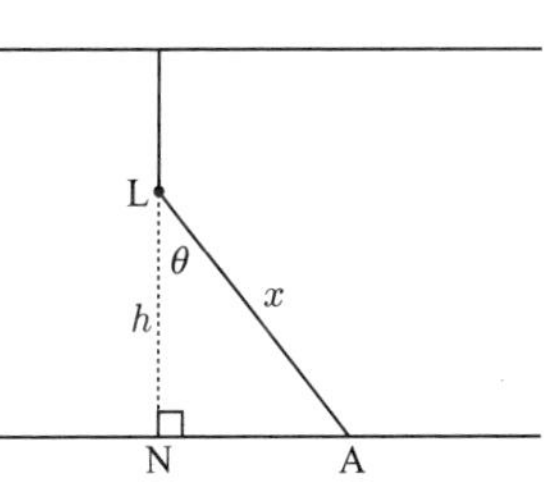


**b** $\dfrac{dI}{d\theta} = \sqrt{8}(-\sin\theta)\sin^2\theta + \sqrt{8}\cos\theta(2\sin\theta\cos\theta)$
$= \sqrt{8}\sin\theta[2\cos^2\theta - \sin^2\theta]$
$= \sqrt{8}\sin\theta[2(1 - \sin^2\theta) - \sin^2\theta]$
$= \sqrt{8}\sin\theta[2 - 3\sin^2\theta]$

$\dfrac{dI}{d\theta} = 0$ when $\sin\theta = \sqrt{\frac{2}{3}}$, $0 < \theta < \frac{\pi}{2}$

and the sign diagram of $\dfrac{dI}{d\theta}$ is:

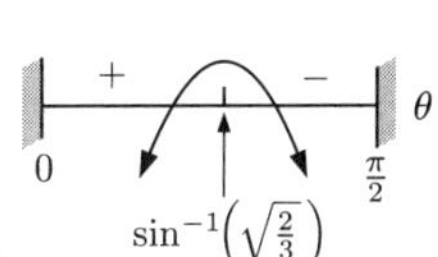


$\therefore$ the maximum illumination at A is obtained when $\sin\theta = \sqrt{\frac{2}{3}}$

$\therefore\ x = \dfrac{1}{\sin\theta} = \sqrt{\frac{3}{2}}$ and $h = \sqrt{x^2 - \text{NA}^2} = \sqrt{\frac{3}{2} - 1} = \frac{1}{\sqrt{2}}$

$\therefore$ the bulb is $\frac{1}{\sqrt{2}}$ m above the floor.

# Chapter 18
## INTEGRATION

### EXERCISE 18A.1

**1** **a** **i**

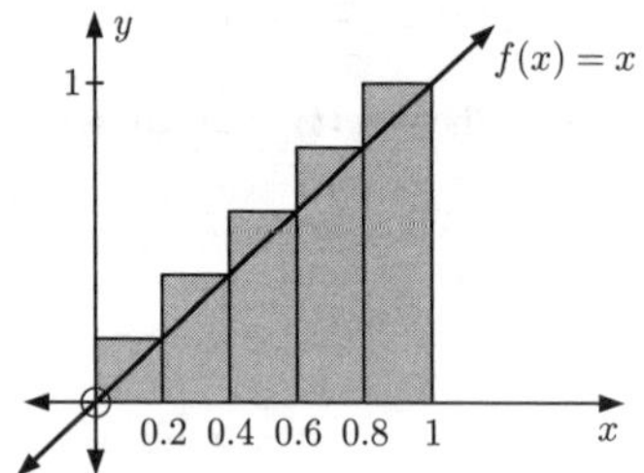


The rectangles are $\frac{1}{5} = 0.2$ units wide.

$A_U = 0.2 \times f(0.2) + 0.2 \times f(0.4) + 0.2 \times f(0.6) + 0.2 \times f(0.8) + 0.2 \times f(1)$
$= 0.2 \times 0.2 + 0.2 \times 0.4 + 0.2 \times 0.6 + 0.2 \times 0.8 + 0.2 \times 1$
$= 0.6$ units$^2$

**ii**

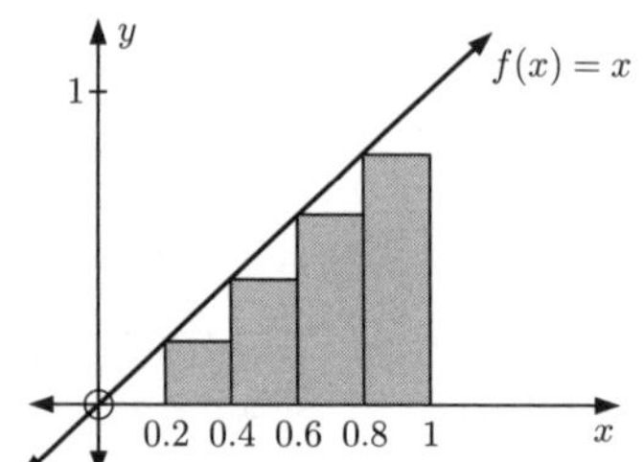


$A_L = 0.2 \times f(0) + 0.2 \times f(0.2) + 0.2 \times f(0.4) + 0.2 \times f(0.6) + 0.2 \times f(0.8)$
$= 0.2 \times 0 + 0.2 \times 0.2 + 0.2 \times 0.4 + 0.2 \times 0.6 + 0.2 \times 0.8$
$= 0.4$ units$^2$

**b** The area between $y = x$ and the $x$-axis from $x = 0$ to $x = 1$ is a triangle.

$\therefore$ area $= \frac{1}{2} \times$ base $\times$ height
$= \frac{1}{2} \times 1 \times 1$
$= \frac{1}{2}$ unit$^2$

$\therefore$ $A_L <$ area $< A_U$, and both $A_L$ and $A_U$ are within 0.1 unit$^2$, or 20%, of the actual area.

**2** The rectangles are $\frac{2}{6} = \frac{1}{3}$ units wide.

**a**

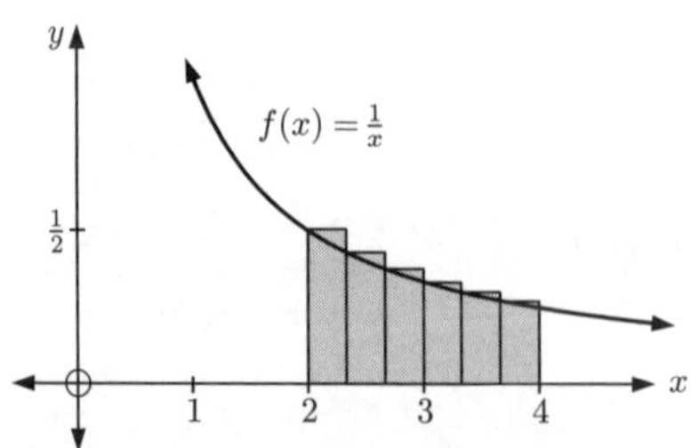


$A_U = \frac{1}{3}f(2) + \frac{1}{3}f(\frac{7}{3}) + \frac{1}{3}f(\frac{8}{3}) + \frac{1}{3}f(3) + \frac{1}{3}f(\frac{10}{3}) + \frac{1}{3}f(\frac{11}{3})$
$= \frac{1}{3} \times \frac{1}{2} + \frac{1}{3} \times \frac{3}{7} + \frac{1}{3} \times \frac{3}{8} + \frac{1}{3} \times \frac{1}{3} + \frac{1}{3} \times \frac{3}{10} + \frac{1}{3} \times \frac{3}{11}$
$\approx 0.737$ units$^2$

**b**

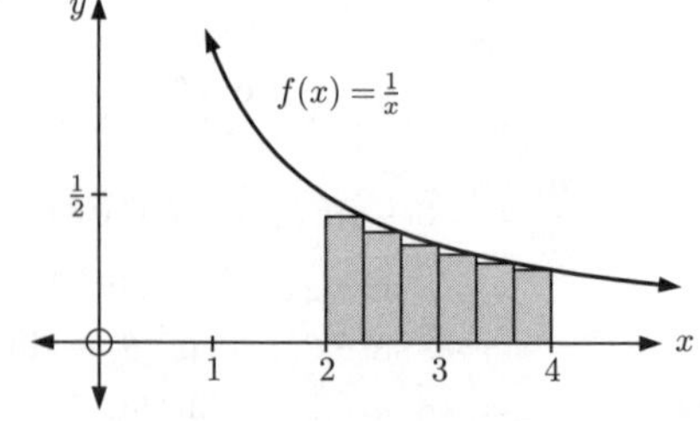


$A_L = \frac{1}{3}f(\frac{7}{3}) + \frac{1}{3}f(\frac{8}{3}) + \frac{1}{3}f(3) + \frac{1}{3}f(\frac{10}{3}) + \frac{1}{3}f(\frac{11}{3}) + \frac{1}{3}f(4)$
$= \frac{1}{3} \times \frac{3}{7} + \frac{1}{3} \times \frac{3}{8} + \frac{1}{3} \times \frac{1}{3} + \frac{1}{3} \times \frac{3}{10} + \frac{1}{3} \times \frac{3}{11} + \frac{1}{3} \times \frac{1}{4}$
$\approx 0.653$ units$^2$

**3** Using provided software,

| $n$ | $A_L$ | $A_U$ |
|---|---|---|
| 10 | 2.1850 | 2.4850 |
| 25 | 2.2736 | 2.3936 |
| 50 | 2.3034 | 2.3634 |
| 100 | 2.3184 | 2.3484 |
| 500 | 2.3303 | 2.3363 |

$A_L$ and $A_U$ converge to $\frac{7}{3}$

**4** **a** **i**

| $n$ | $A_L$ | $A_U$ |
|---|---|---|
| 5 | 0.160 00 | 0.360 00 |
| 10 | 0.202 50 | 0.302 50 |
| 50 | 0.240 10 | 0.260 10 |
| 100 | 0.245 03 | 0.255 03 |
| 500 | 0.249 00 | 0.251 00 |
| 1000 | 0.249 50 | 0.250 50 |
| 10 000 | 0.249 95 | 0.250 05 |

**ii**

| $n$ | $A_L$ | $A_U$ |
|---|---|---|
| 5 | 0.400 00 | 0.600 00 |
| 10 | 0.450 00 | 0.550 00 |
| 50 | 0.490 00 | 0.510 00 |
| 100 | 0.495 00 | 0.505 00 |
| 500 | 0.499 00 | 0.501 00 |
| 1000 | 0.499 50 | 0.500 50 |
| 10 000 | 0.499 95 | 0.500 05 |

**iii**

| $n$ | $A_L$ | $A_U$ |
|---|---|---|
| 5 | 0.549 74 | 0.749 74 |
| 10 | 0.610 51 | 0.710 51 |
| 50 | 0.656 10 | 0.676 10 |
| 100 | 0.661 46 | 0.671 46 |
| 500 | 0.665 65 | 0.667 65 |
| 1000 | 0.666 16 | 0.667 16 |
| 10 000 | 0.666 62 | 0.666 72 |

**iv**

| $n$ | $A_L$ | $A_U$ |
|---|---|---|
| 5 | 0.618 67 | 0.818 67 |
| 10 | 0.687 40 | 0.787 40 |
| 50 | 0.738 51 | 0.758 51 |
| 100 | 0.744 41 | 0.754 41 |
| 500 | 0.748 93 | 0.750 93 |
| 1000 | 0.749 47 | 0.750 47 |
| 10 000 | 0.749 95 | 0.750 05 |

**b** **i** $A_L$ and $A_U$ converge to $0.25 = \frac{1}{4} = \frac{1}{3+1}$

**ii** $A_L$ and $A_U$ converge to $0.5 = \frac{1}{2} = \frac{1}{1+1}$

**iii** $A_L$ and $A_U$ converge to $0.6\overline{6} = \frac{2}{3} = \frac{1}{\frac{1}{2}+1}$

**iv** $A_L$ and $A_U$ converge to $0.75 = \frac{3}{4} = \frac{1}{\frac{1}{3}+1}$

**c** From **b**, it appears that the area between the graph of $y = x^a$ and the $x$-axis for $0 \leqslant x \leqslant 1$ is $\dfrac{1}{a+1}$.

**5** **a**

| $n$ | *Rational bounds for* $\pi$ |
|---|---|
| 10 | $2.9045 < \pi < 3.3045$ |
| 50 | $3.0983 < \pi < 3.1783$ |
| 100 | $3.1204 < \pi < 3.1604$ |
| 200 | $3.1312 < \pi < 3.1512$ |
| 1000 | $3.1396 < \pi < 3.1436$ |
| 10 000 | $3.1414 < \pi < 3.1418$ |

**b** $3\frac{10}{71} < \pi < 3\frac{1}{7}$ is approximately
$3.1408 < \pi < 3.1429$
From **a**, this is a better approximation than our estimate using $n = 10, 50, 100, 200, 1000$.
Only $n = 10\,000$ gives us a better estimate.

## EXERCISE 18A.2

**1** **a**

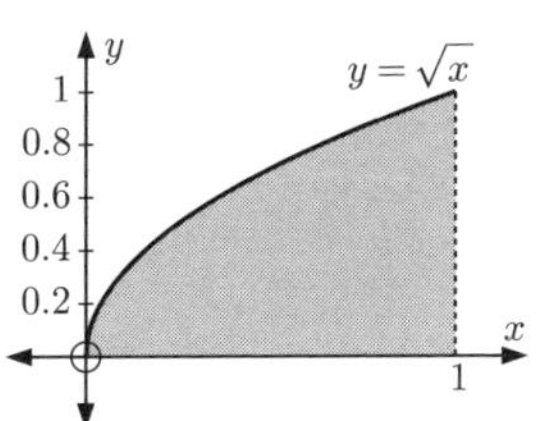


**b**

| $n$ | $A_L$ | $A_U$ |
|---|---|---|
| 5 | 0.5497 | 0.7497 |
| 10 | 0.6105 | 0.7105 |
| 50 | 0.6561 | 0.6761 |
| 100 | 0.6615 | 0.6715 |
| 500 | 0.6656 | 0.6676 |

**c** $\int_0^1 \sqrt{x}\, dx \approx 0.67$

**2** **a**

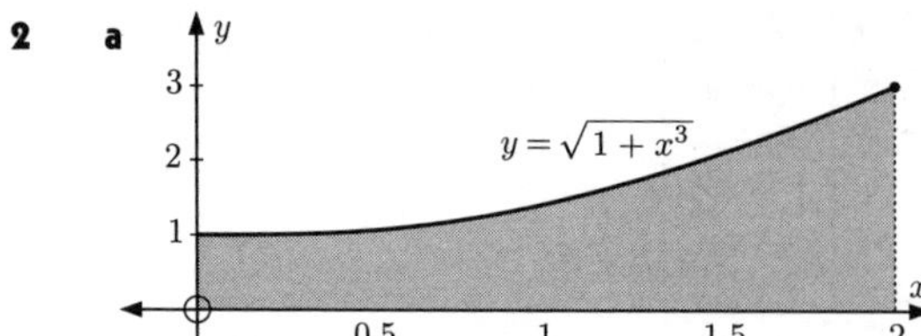


**c**

| $n$ | $A_L$ | $A_U$ |
|---|---|---|
| 50 | 3.2016 | 3.2816 |
| 100 | 3.2214 | 3.2614 |
| 500 | 3.2373 | 3.2453 |

**d** $\int_0^2 \sqrt{1+x^3}\,dx \approx 3.24$

**b** The rectangles will have width $\dfrac{2-0}{n} = \dfrac{2}{n}$.

The lower rectangle sum will be

$$A_L = \frac{2}{n} \times \sqrt{1+{x_0}^3} + \frac{2}{n} \times \sqrt{1+{x_1}^3} + .... + \frac{2}{n} \times \sqrt{1+x_{n-1}^3}$$

$$= \frac{2}{n} \sum_{i=0}^{n-1} \sqrt{1+{x_i}^3}$$

The upper rectangle sum will be

$$A_U = \frac{2}{n}\sqrt{1+{x_1}^3} + \frac{2}{n}\sqrt{1+{x_2}^3} + .... + \frac{2}{n}\sqrt{1+x_n^3}$$

$$= \frac{2}{n} \sum_{i=1}^{n} \sqrt{1+{x_i}^3}$$

**3** **a**

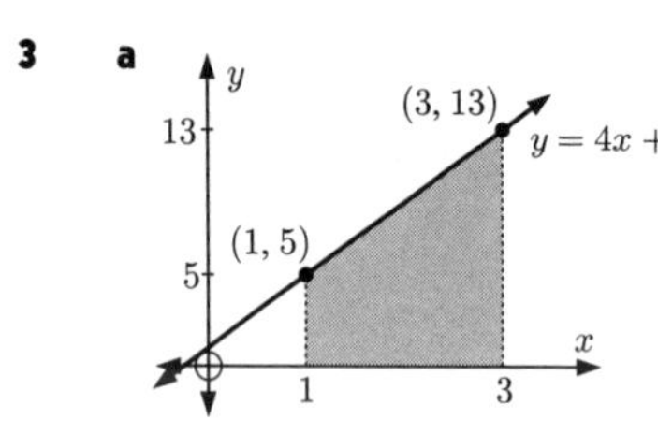


$\int_1^3 (1+4x)\ dx$

= area of the shaded trapezium

$= \left(\dfrac{5+13}{2}\right) \times 2$

$= 18$

**b**

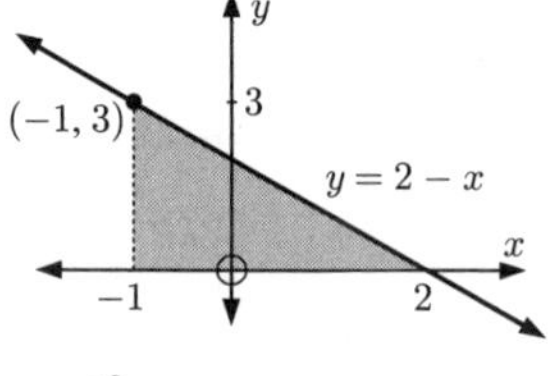


$\int_{-1}^2 (2-x)\ dx$

= area of shaded triangle

$= \frac{1}{2}(3 \times 3)$

$= 4.5$

**c**

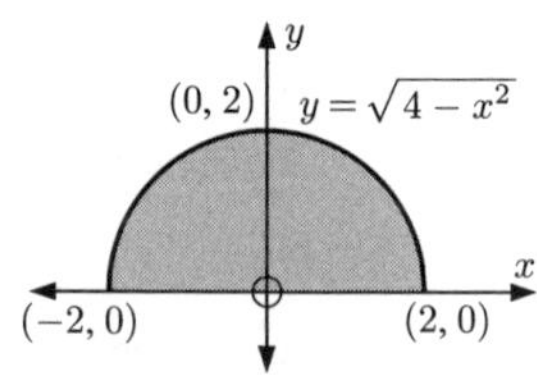


$\int_{-2}^2 \sqrt{4-x^2}\ dx$

= area of semi-circle, radius 2

$= \frac{1}{2}\left(\pi \times 2^2\right)$

$= 2\pi$

## EXERCISE 18B

**1** **a** **i**

$\dfrac{d}{dx}(x^2) = 2x$

$\therefore\ \dfrac{d}{dx}\left(\frac{1}{2}x^2\right) = x$

$\therefore$ the antiderivative of $x$ is $\frac{1}{2}x^2$ or $\dfrac{x^2}{2}$

**ii**

$\dfrac{d}{dx}(x^3) = 3x^2$

$\therefore\ \dfrac{d}{dx}\left(\frac{1}{3}x^3\right) = x^2$

$\therefore$ the antiderivative of $x^2$ is $\frac{1}{3}x^3$ or $\dfrac{x^3}{3}$

**iii**

$\dfrac{d}{dx}(x^6) = 6x^5$

$\therefore\ \dfrac{d}{dx}\left(\frac{1}{6}x^6\right) = x^5$

$\therefore$ the antiderivative of $x^5$ is $\frac{1}{6}x^6$ or $\dfrac{x^6}{6}$

**iv**

$\dfrac{d}{dx}(x^{-1}) = -x^{-2}$

$\therefore\ \dfrac{d}{dx}(-x^{-1}) = x^{-2}$

$\therefore$ the antiderivative of $x^{-2}$ is $-x^{-1}$ or $-\dfrac{1}{x}$

**v** $\frac{d}{dx}(x^{-3}) = -3x^{-4}$

$\therefore \frac{d}{dx}\left(-\frac{1}{3}x^{-3}\right) = x^{-4}$

$\therefore$ the antiderivative of $x^{-4}$ is $-\frac{1}{3}x^{-3} = -\frac{1}{3x^3}$

**vi** $\frac{d}{dx}\left(x^{\frac{4}{3}}\right) = \frac{4}{3}x^{\frac{1}{3}}$

$\therefore \frac{d}{dx}\left(\frac{3}{4}x^{\frac{4}{3}}\right) = x^{\frac{1}{3}}$

$\therefore$ the antiderivative of $x^{\frac{1}{3}}$ is $\frac{3}{4}x^{\frac{4}{3}}$

**vii** $\frac{d}{dx}(x^{\frac{1}{2}}) = \frac{1}{2}x^{-\frac{1}{2}}$

$\therefore \frac{d}{dx}(2x^{\frac{1}{2}}) = x^{-\frac{1}{2}}$

$\therefore$ the antiderivative of $x^{-\frac{1}{2}}$ is $2x^{\frac{1}{2}} = 2\sqrt{x}$

**b** The antiderivative of $x^n$ is $\frac{x^{n+1}}{n+1}$ $(n \neq -1)$.

**2** **a** **i** $\frac{d}{dx}(e^{2x}) = 2e^{2x}$

$\therefore \frac{d}{dx}\left(\frac{1}{2}e^{2x}\right) = e^{2x}$

$\therefore$ the antiderivative of $e^{2x}$ is $\frac{1}{2}e^{2x}$

**ii** $\frac{d}{dx}(e^{5x}) = 5e^{5x}$

$\therefore \frac{d}{dx}\left(\frac{1}{5}e^{5x}\right) = e^{5x}$

$\therefore$ the antiderivative of $e^{5x}$ is $\frac{1}{5}e^{5x}$

**iii** $\frac{d}{dx}(e^{\frac{1}{2}x}) = \frac{1}{2}e^{\frac{1}{2}x}$

$\therefore \frac{d}{dx}\left(2e^{\frac{1}{2}x}\right) = e^{\frac{1}{2}x}$

$\therefore$ the antiderivative of $e^{\frac{1}{2}x}$ is $2e^{\frac{1}{2}x}$

**iv** $\frac{d}{dx}(e^{0.01x}) = 0.01e^{0.01x}$

$\therefore \frac{d}{dx}\left(100e^{0.01x}\right) = e^{0.01x}$

$\therefore$ the antiderivative of $e^{0.01x}$ is $100e^{0.01x}$

**v** $\frac{d}{dx}(e^{\pi x}) = \pi e^{\pi x}$

$\therefore \frac{d}{dx}\left(\frac{1}{\pi}e^{\pi x}\right) = e^{\pi x}$

$\therefore$ the antiderivative of $e^{\pi x}$ is $\frac{1}{\pi}e^{\pi x}$

**vi** $\frac{d}{dx}\left(e^{\frac{x}{3}}\right) = \frac{1}{3}e^{\frac{x}{3}}$

$\therefore \frac{d}{dx}\left(3e^{\frac{x}{3}}\right) = e^{\frac{x}{3}}$

$\therefore$ the antiderivative of $e^{\frac{x}{3}}$ is $3e^{\frac{x}{3}}$

**b** The antiderivative of $e^{kx}$ is $\frac{1}{k}e^{kx}$.

**3** **a** $\frac{d}{dx}(x^3 + x^2) = 3x^2 + 2x$

$\therefore \frac{d}{dx}\left(2x^3 + 2x^2\right) = 6x^2 + 4x$

$\therefore$ the antiderivative of $6x^2 + 4x$ is $2x^3 + 2x^2$

**b** $\frac{d}{dx}(e^{3x+1}) = 3e^{3x+1}$

$\therefore \frac{d}{dx}\left(\frac{1}{3}e^{3x+1}\right) = e^{3x+1}$

$\therefore$ the antiderivative of $e^{3x+1}$ is $\frac{1}{3}e^{3x+1}$

**c** $\frac{d}{dx}(x\sqrt{x}) = \frac{d}{dx}(x^{\frac{3}{2}}) = \frac{3}{2}x^{\frac{1}{2}}$
$= \frac{3}{2}\sqrt{x}$

$\therefore \frac{d}{dx}\left(\frac{2}{3}x\sqrt{x}\right) = \sqrt{x}$

$\therefore$ the antiderivative of $\sqrt{x}$ is $\frac{2}{3}x\sqrt{x}$

**d** $\frac{d}{dx}((2x+1)^4) = 4(2x+1)^3 \times 2$
$= 8(2x+1)^3$

$\therefore \frac{d}{dx}\left(\frac{1}{8}(2x+1)^4\right) = (2x+1)^3$

$\therefore$ the antiderivative of $(2x+1)^3$ is $\frac{1}{8}(2x+1)^4$

## EXERCISE 18C

**1** **a** $f(x) = x^3$ has antiderivative $F(x) = \frac{x^4}{4}$

$\therefore$ area $= \int_0^1 x^3\,dx$
$= F(1) - F(0)$
$= \frac{1}{4} - 0 = \frac{1}{4}$ units$^2$

**b** $f(x) = x^2$ has antiderivative $F(x) = \frac{x^3}{3}$

$\therefore$ area $= \int_1^2 x^2\,dx$
$= F(2) - F(1)$
$= \frac{8}{3} - \frac{1}{3} = 2\frac{1}{3}$ units$^2$

**c** $f(x) = \sqrt{x} = x^{\frac{1}{2}}$ has antiderivative $F(x) = \dfrac{x^{\frac{3}{2}}}{\frac{3}{2}} = \frac{2}{3}x\sqrt{x}$

$\therefore$ area $= \int_0^1 \sqrt{x}\,dx$
$= F(1) - F(0)$
$= \frac{2}{3} \times 1\sqrt{1} - 0 = \frac{2}{3}$ units$^2$

**2** **a** $\int_a^a f(x)\,dx = F(a) - F(a) = 0$
$\int_a^a f(x)\,dx =$ area of the strip between $x = a$ and $x = a$.
This strip has 0 width, so its area $= 0$.

**b** The antiderivative of $c$ is $cx$.
$\therefore$ $\int_a^b c\,dx = F(b) - F(a)$
$= cb - ca$
$= c(b - a)$

**c** $\int_b^a f(x)\,dx = F(a) - F(b)$
$= -[F(b) - F(a)]$
$= -\int_a^b f(x)\,dx$

**d** If $\dfrac{d}{dx}F(x) = f(x)$ then
$\dfrac{d}{dx}\,c\,F(x) = c\,f(x)$
$\therefore$ $\int_a^b c\,f(x)\,dx = c\,F(b) - c\,F(a)$
$= c[F(b) - F(a)]$
$= c\int_a^b f(x)\,dx$

**e** $\int_a^b (f(x) + g(x))\,dx$
$= [F(b) + G(b)] - [F(a) + G(a)]$
$= [F(b) - F(a)] + [G(b) - G(a)]$
$= \int_a^b f(x)\,dx + \int_a^b g(x)\,dx$

**3** **a** $f(x) = x^3$ has antiderivative $F(x) = \dfrac{x^4}{4}$
$\therefore$ area $= \int_1^2 x^3\,dx$
$= F(2) - F(1)$
$= \frac{16}{4} - \frac{1}{4}$
$= 3\frac{3}{4}$ units$^2$

**b** $f(x) = x^2 + 3x + 2$ has antiderivative
$F(x) = \dfrac{x^3}{3} + \dfrac{3x^2}{2} + 2x$
$\therefore$ area $= \int_1^3 (x^2 + 3x + 2)\,dx$
$= F(3) - F(1)$
$= (\frac{27}{3} + \frac{27}{2} + 6) - (\frac{1}{3} + \frac{3}{2} + 2)$
$= 24\frac{2}{3}$ units$^2$

**c** $f(x) = \sqrt{x} = x^{\frac{1}{2}}$ has antiderivative
$F(x) = \dfrac{x^{\frac{3}{2}}}{\frac{3}{2}} = \frac{2}{3}x\sqrt{x}$
$\therefore$ area $= \int_1^2 \sqrt{x}\,dx$
$= F(2) - F(1)$
$= \frac{2}{3}\,2\sqrt{2} - \frac{2}{3}\,1\sqrt{1}$
$= \frac{4\sqrt{2}}{3} - \frac{2}{3}$
$= \frac{-2+4\sqrt{2}}{3}$ units$^2$

**d** $f(x) = e^x$ has antiderivative $F(x) = e^x$
$\therefore$ area $= \int_0^{1.5} e^x\,dx$
$= F(1.5) - F(0)$
$= e^{1.5} - e^0$
$= e^{1.5} - 1$
$\approx 3.48$ units$^2$

**e** $f(x) = \dfrac{1}{\sqrt{x}} = x^{-\frac{1}{2}}$ has antiderivative
$F(x) = \dfrac{x^{\frac{1}{2}}}{\frac{1}{2}} = 2\sqrt{x}$
$\therefore$ area $= \int_1^4 \dfrac{1}{\sqrt{x}}\,dx$
$= F(4) - F(1)$
$= 2\sqrt{4} - 2\sqrt{1} = 2$ units$^2$

**f** $f(x) = x^3 + 2x^2 + 7x + 4$ has antiderivative
$F(x) = \dfrac{x^4}{4} + \dfrac{2x^3}{3} + \dfrac{7x^2}{2} + 4x$
$\therefore$ area $= \int_1^{1.25} \left(x^3 + 2x^2 + 7x + 4\right)\,dx$
$= F(1.25) - F(1)$
$= [12.381\,18 - 8.416\,67]$
$\approx 3.96$ units$^2$

**4** Using technology, area $= \int_0^3 \sqrt{9-x^2}\, dx \approx 7.07$ units$^2$

*Check*: The area is a quarter circle with radius 3 units.

$\therefore$ area $= \frac{1}{4}\pi r^2$

$= \frac{1}{4} \times \pi \times 3^2 = \frac{9}{4}\pi \approx 7.07$ units$^2$ ✓

**5** **a** If $\dfrac{d}{dx}F(x) = f(x)$ then $\dfrac{d}{dx}(-F(x)) = -f(x)$

$$\therefore \int_a^b (-f(x))\, dx = -F(b) - (-F(a))$$
$$= -(F(b) - F(a))$$
$$= -\int_a^b f(x)\, dx$$

**b** Since $y = -f(x)$ is a reflection of $y = f(x)$ in the $x$-axis, then

shaded area = area between the $x$-axis and $y = -f(x)$ from $x = a$ to $x = b$

$= \int_a^b (-f(x))\, dx$

$= -\int_a^b f(x)\, dx$ {using **a**}

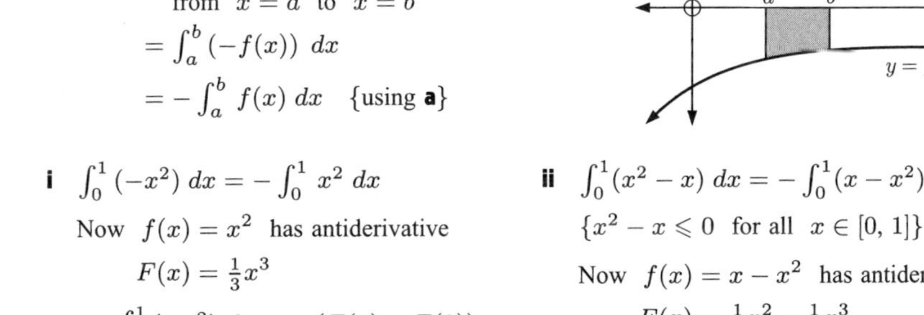


**c** **i** $\int_0^1 (-x^2)\, dx = -\int_0^1 x^2\, dx$

Now $f(x) = x^2$ has antiderivative

$F(x) = \frac{1}{3}x^3$

$\therefore \int_0^1 (-x^2)\, dx = -(F(1) - F(0))$

$= -\left(\frac{1}{3} - 0\right)$

$= -\frac{1}{3}$

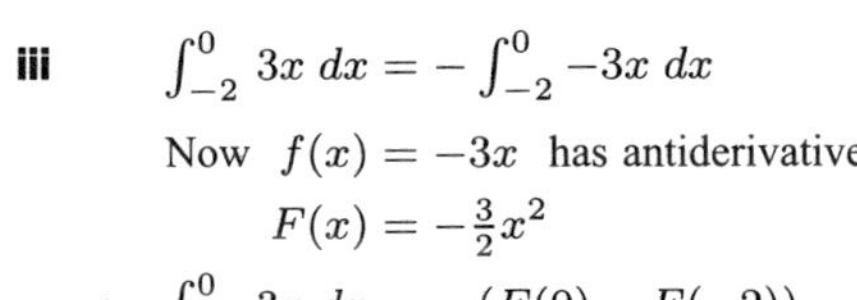


The shaded region has area $\frac{1}{3}$ units$^2$.

**ii** $\int_0^1 (x^2 - x)\, dx = -\int_0^1 (x - x^2)\, dx$

$\{x^2 - x \leqslant 0$ for all $x \in [0, 1]\}$

Now $f(x) = x - x^2$ has antiderivative

$F(x) = \frac{1}{2}x^2 - \frac{1}{3}x^3$

$\therefore \int_0^1 (x^2 - x)\, dx = -(F(1) - F(0))$

$= -\left(\frac{1}{2} - \frac{1}{3} - (0 - 0)\right)$

$= -\frac{1}{6}$

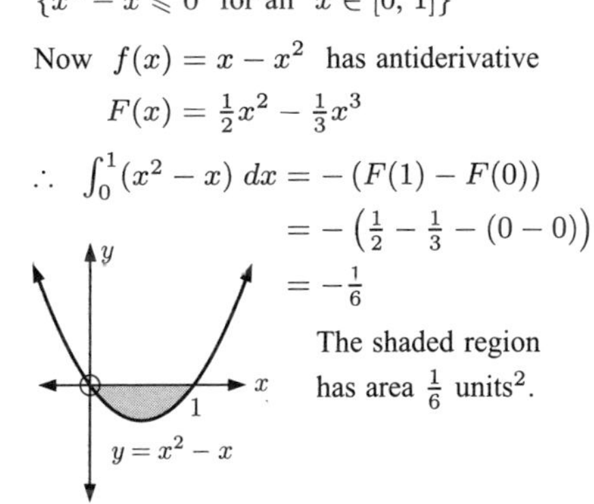


The shaded region has area $\frac{1}{6}$ units$^2$.

**iii** $\int_{-2}^0 3x\, dx = -\int_{-2}^0 -3x\, dx$

Now $f(x) = -3x$ has antiderivative

$F(x) = -\frac{3}{2}x^2$

$\therefore \int_{-2}^0 3x\, dx = -(F(0) - F(-2))$

$= -(0 - (-6))$

$= -6$

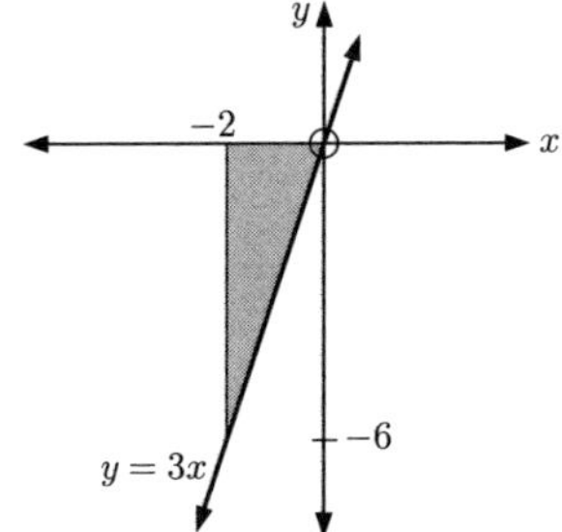


The shaded region has area 6 units$^2$.

**d** $\int_0^2 \left(-\sqrt{4-x^2}\right) dx = -\int_0^2 \sqrt{4-x^2}\, dx$

Now $f(x) = \sqrt{4-x^2}$ is the top half of a circle with radius 2 units and centre (0, 0).

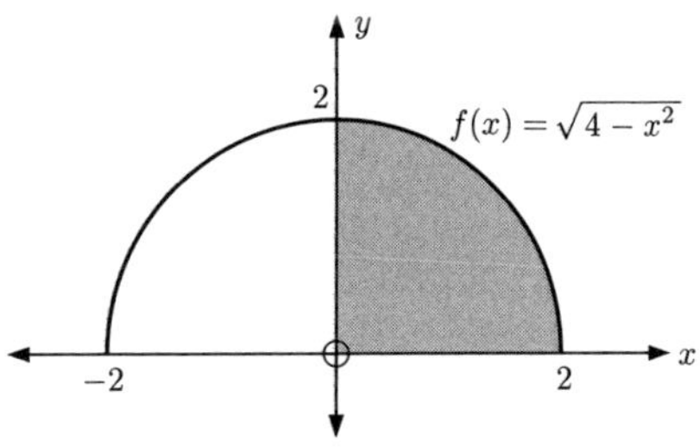


$\therefore \int_0^2 \left(-\sqrt{4-x^2}\right) dx = -\int_0^2 \sqrt{4-x^2}\, dx$

$= -$(shaded area)

$= -\frac{1}{4} \times \pi \times 2^2$

$= -\pi$

## EXERCISE 18D

**1** If $y = x^7$ then $\frac{dy}{dx} = 7x^6$

$\therefore \int 7x^6\,dx = x^7 + c$

$\therefore 7\int x^6\,dx = x^7 + c$

$\therefore \int x^6\,dx = \frac{1}{7}x^7 + c$

**2** If $y = x^3 + x^2$ then $\frac{dy}{dx} = 3x^2 + 2x$

$\therefore \int (3x^2 + 2x)\,dx = x^3 + x^2 + c$

**3** If $y = e^{2x+1}$ then $\frac{dy}{dx} = 2e^{2x+1}$

$\therefore \int 2e^{2x+1}\,dx = e^{2x+1} + c$

$\therefore 2\int e^{2x+1}\,dx = e^{2x+1} + c$

$\therefore \int e^{2x+1}\,dx = \frac{1}{2}e^{2x+1} + c$

**4** If $y = (2x+1)^4$

then $\frac{dy}{dx} = 4(2x+1)^3 \times 2 = 8(2x+1)^3$

$\therefore \int 8(2x+1)^3\,dx = (2x+1)^4 + c$

$\therefore 8\int (2x+1)^3\,dx = (2x+1)^4 + c$

$\therefore \int (2x+1)^3\,dx = \frac{1}{8}(2x+1)^4 + c$

**5** If $y = x\sqrt{x} = x^{\frac{3}{2}}$

then $\frac{dy}{dx} = \frac{3}{2}x^{\frac{1}{2}} = \frac{3}{2}\sqrt{x}$

$\therefore \int \frac{3}{2}\sqrt{x}\,dx = x\sqrt{x} + c$

$\therefore \frac{3}{2}\int \sqrt{x}\,dx = x\sqrt{x} + c$

$\therefore \int \sqrt{x}\,dx = \frac{2}{3}x\sqrt{x} + c$

**6** If $y = \frac{1}{\sqrt{x}} = x^{-\frac{1}{2}}$

then $\frac{dy}{dx} = -\frac{1}{2}x^{-\frac{3}{2}} = -\frac{1}{2x\sqrt{x}}$

$\therefore \int -\frac{1}{2}\left(\frac{1}{x\sqrt{x}}\right)dx = \frac{1}{\sqrt{x}} + c$

$\therefore -\frac{1}{2}\int \frac{1}{x\sqrt{x}}\,dx = \frac{1}{\sqrt{x}} + c$

$\therefore \int \frac{1}{x\sqrt{x}}\,dx = -\frac{2}{\sqrt{x}} + c$

**7** If $y = \cos 2x$

then $\frac{dy}{dx} = -2\sin 2x$

$\therefore \int -2\sin 2x\,dx = \cos 2x + c$

$\therefore -2\int \sin 2x\,dx = \cos 2x + c$

$\therefore \int \sin 2x\,dx = -\frac{1}{2}\cos 2x + c$

**8** If $y = \sin(1-5x)$

then $\frac{dy}{dx} = -5\cos(1-5x)$

$\therefore \int -5\cos(1-5x)\,dx = \sin(1-5x) + c$

$\therefore -5\int \cos(1-5x)\,dx = \sin(1-5x) + c$

$\therefore \int \cos(1-5x)\,dx = -\frac{1}{5}\sin(1-5x) + c$

**9** $\frac{d}{dx}\left[(x^2-x)^3\right] = 3(x^2-x)^2(2x-1)$

$\therefore \int 3(2x-1)(x^2-x)^2\,dx = (x^2-x)^3 + c$

$\therefore 3\int (2x-1)(x^2-x)^2\,dx = (x^2-x)^3 + c$

$\therefore \int (2x-1)(x^2-x)^2\,dx = \frac{1}{3}(x^2-x)^3 + c$

**10** Suppose $F(x)$ is the antiderivative of $f(x)$ and $G(x)$ is the antiderivative of $g(x)$.

$\therefore \frac{d}{dx}(F(x) + G(x)) = f(x) + g(x)$

$\therefore \int [f(x) + g(x)]\,dx$

$= F(x) + G(x) + c$

$= (F(x) + c_1) + (G(x) + c_2)$

$= \int f(x)\,dx + \int g(x)\,dx$

**11** $y = \sqrt{1-4x} = (1-4x)^{\frac{1}{2}}$

$\therefore \frac{dy}{dx} = \frac{1}{2}(1-4x)^{-\frac{1}{2}}(-4)$

$= \frac{-2}{\sqrt{1-4x}}$

$\therefore \int \frac{-2}{\sqrt{1-4x}}\,dx = \sqrt{1-4x} + c$

$\therefore -2\int \frac{1}{\sqrt{1-4x}}\,dx = \sqrt{1-4x} + c$

$\therefore \int \frac{1}{\sqrt{1-4x}}\,dx = -\frac{1}{2}\sqrt{1-4x} + c$

**12** $\dfrac{d}{dx}(\ln(5-3x+x^2)) = \dfrac{2x-3}{5-3x+x^2}$

Now $5-3x+x^2>0$ for all $x$,
as $a>0$ and $\Delta=-11<0$

$\therefore \displaystyle\int \frac{2x-3}{5-3x+x^2}\,dx = \ln(5-3x+x^2)+c$

$\therefore \displaystyle\int \frac{4x-6}{5-3x+x^2}\,dx = 2\ln(5-3x+x^2)+c$

## EXERCISE 18E.1

**1** **a** $\int(x^4-x^2-x+2)\,dx$
$=\dfrac{x^5}{5}-\dfrac{x^3}{3}-\dfrac{x^2}{2}+2x+c$

**b** $\int(\sqrt{x}+e^x)\,dx$
$=\int(x^{\frac{1}{2}}+e^x)\,dx$
$=\dfrac{x^{\frac{3}{2}}}{\frac{3}{2}}+e^x+c$
$=\frac{2}{3}x^{\frac{3}{2}}+e^x+c$

**c** $\displaystyle\int\left(3e^x-\frac{1}{x}\right)dx$
$=3e^x-\ln|x|+c$

**d** $\displaystyle\int\left(x\sqrt{x}-\frac{2}{x}\right)dx$
$=\displaystyle\int\left(x^{\frac{3}{2}}-\frac{2}{x}\right)dx$
$=\dfrac{x^{\frac{5}{2}}}{\frac{5}{2}}-2\ln|x|+c$
$=\frac{2}{5}x^{\frac{5}{2}}-2\ln|x|+c$

**e** $\displaystyle\int\left(\frac{1}{x\sqrt{x}}+\frac{4}{x}\right)dx$
$=\displaystyle\int\left(x^{-\frac{3}{2}}+\frac{4}{x}\right)dx$
$=\dfrac{x^{-\frac{1}{2}}}{-\frac{1}{2}}+4\ln|x|+c$
$=-2x^{-\frac{1}{2}}+4\ln|x|+c$

**f** $\int(\frac{1}{2}x^3-x^4+x^{\frac{1}{3}})\,dx$
$=\frac{1}{2}\dfrac{x^4}{4}-\dfrac{x^5}{5}+\dfrac{x^{\frac{4}{3}}}{\frac{4}{3}}+c$
$=\frac{1}{8}x^4-\frac{1}{5}x^5+\frac{3}{4}x^{\frac{4}{3}}+c$

**g** $\displaystyle\int\left(x^2+\frac{3}{x}\right)dx$
$=\frac{1}{3}x^3+3\ln|x|+c$

**h** $\displaystyle\int\left(\frac{1}{2x}+x^2-e^x\right)dx$
$=\frac{1}{2}\ln|x|+\frac{1}{3}x^3-e^x+c$

**i** $\displaystyle\int\left(5e^x+\frac{1}{3}x^3-\frac{4}{x}\right)dx$
$=5e^x+\frac{1}{3}\dfrac{x^4}{4}-4\ln|x|+c$
$=5e^x+\frac{1}{12}x^4-4\ln|x|+c$

**2** **a** $\int(3\sin x-2)\,dx$
$=-3\cos x-2x+c$

**b** $\int(4x-2\cos x)\,dx$
$=2x^2-2\sin x+c$

**c** $\int(\sin x-2\cos x+e^x)\,dx$
$=-\cos x-2\sin x+e^x+c$

**d** $\int(x^2\sqrt{x}-10\sin x)\,dx$
$=\int(x^{\frac{5}{2}}-10\sin x)\,dx$
$=\frac{2}{7}x^{\frac{7}{2}}+10\cos x+c$
$=\frac{2}{7}x^3\sqrt{x}+10\cos x+c$

**e** $\displaystyle\int\left(\frac{x(x-1)}{3}+\cos x\right)dx$
$=\displaystyle\int\left(\frac{1}{3}x^2-\frac{1}{3}x+\cos x\right)dx$
$=\dfrac{x^3}{9}-\dfrac{x^2}{6}+\sin x+c$

**f** $\int(-\sin x+2\sqrt{x})\,dx$
$=\int(-\sin x+2x^{\frac{1}{2}})\,dx$
$=\cos x+\frac{4}{3}x^{\frac{3}{2}}+c$
$=\cos x+\frac{4}{3}x\sqrt{x}+c$

**3** **a** $\int(x^2+3x-2)\,dx$
$=\frac{1}{3}x^3+\frac{3}{2}x^2-2x+c$

**b** $\displaystyle\int\left(\sqrt{x}-\frac{1}{\sqrt{x}}\right)dx$
$=\displaystyle\int\left(x^{\frac{1}{2}}-x^{-\frac{1}{2}}\right)dx$
$=\dfrac{x^{\frac{3}{2}}}{\frac{3}{2}}-\dfrac{x^{\frac{1}{2}}}{\frac{1}{2}}+c$
$=\frac{2}{3}x^{\frac{3}{2}}-2x^{\frac{1}{2}}+c$

**c** $\displaystyle\int\left(2e^x-\frac{1}{x^2}\right)dx$
$=\int\left(2e^x-x^{-2}\right)dx$
$=2e^x-\dfrac{x^{-1}}{-1}+c$
$=2e^x+\dfrac{1}{x}+c$

**d** $\int \left(\frac{1-4x}{x\sqrt{x}}\right) dx$

$= \int \left(\frac{1}{x\sqrt{x}} - \frac{4}{\sqrt{x}}\right) dx$

$= \int (x^{-\frac{3}{2}} - 4x^{-\frac{1}{2}})\, dx$

$= \frac{x^{-\frac{1}{2}}}{-\frac{1}{2}} - \frac{4x^{\frac{1}{2}}}{\frac{1}{2}} + c$

$= -2x^{-\frac{1}{2}} - 8x^{\frac{1}{2}} + c$

**e** $\int (2x+1)^2\, dx$

$= \int (4x^2 + 4x + 1)\, dx$

$= \frac{4x^3}{3} + \frac{4x^2}{2} + x + c$

$= \frac{4}{3}x^3 + 2x^2 + x + c$

**f** $\int \frac{x^2 + x - 3}{x}\, dx$

$= \int \left(x + 1 - \frac{3}{x}\right) dx$

$= \frac{1}{2}x^2 + x - 3\ln|x| + c$

**g** $\int \frac{2x-1}{\sqrt{x}}\, dx$

$= \int \left(2x^{\frac{1}{2}} - x^{-\frac{1}{2}}\right) dx$

$= \frac{2x^{\frac{3}{2}}}{\frac{3}{2}} - \frac{x^{\frac{1}{2}}}{\frac{1}{2}} + c$

$= \frac{4}{3}x^{\frac{3}{2}} - 2x^{\frac{1}{2}} + c$

**h** $\int \frac{x^2 - 4x + 10}{x^2\sqrt{x}}\, dx$

$= \int \left(\frac{x^2}{x^2\sqrt{x}} - \frac{4x}{x^2\sqrt{x}} + \frac{10}{x^2\sqrt{x}}\right) dx$

$= \int \left(x^{-\frac{1}{2}} - 4x^{-\frac{3}{2}} + 10x^{-\frac{5}{2}}\right) dx$

$= \frac{x^{\frac{1}{2}}}{\frac{1}{2}} - \frac{4x^{-\frac{1}{2}}}{-\frac{1}{2}} + \frac{10x^{-\frac{3}{2}}}{-\frac{3}{2}} + c$

$= 2x^{\frac{1}{2}} + 8x^{-\frac{1}{2}} - \frac{20}{3}x^{-\frac{3}{2}} + c$

**i** $\int (x+1)^3\, dx$

$= \int (x^3 + 3x^2 + 3x + 1)\, dx$

$= \frac{1}{4}x^4 + x^3 + \frac{3}{2}x^2 + x + c$

**4** **a** $\int (\sqrt{x} + \frac{1}{2}\cos x)\, dx$

$= \int (x^{\frac{1}{2}} + \frac{1}{2}\cos x)\, dx$

$= \frac{2}{3}x^{\frac{3}{2}} + \frac{1}{2}\sin x + c$

**b** $\int (2e^t - 4\sin t)\, dt$

$= 2e^t + 4\cos t + c$

**c** $\int \left(3\cos t - \frac{1}{t}\right) dt$

$= 3\sin t - \ln|t| + c$

**5** **a** $\frac{dy}{dx} = 6$

$\therefore\ y = \int 6\, dx$

$\therefore\ y = 6x + c$

**b** $\frac{dy}{dx} = 4x^2$

$\therefore\ y = \int 4x^2\, dx$

$\therefore\ y = \frac{4}{3}x^3 + c$

**c** $\frac{dy}{dx} = 5\sqrt{x} - x^2 = 5x^{\frac{1}{2}} - x^2$

$\therefore\ y = \int (5x^{\frac{1}{2}} - x^2)\, dx$

$\therefore\ y = \frac{10}{3}x^{\frac{3}{2}} - \frac{1}{3}x^3 + c$

$\therefore\ y = \frac{10}{3}x\sqrt{x} - \frac{1}{3}x^3 + c$

**d** $\frac{dy}{dx} = \frac{1}{x^2} = x^{-2}$

$\therefore\ y = \int x^{-2}\, dx$

$\therefore\ y = \frac{x^{-1}}{-1} + c$

$\therefore\ y = -\frac{1}{x} + c$

**e** $\frac{dy}{dx} = 2e^x - 5$

$\therefore\ y = \int (2e^x - 5)\, dx$

$\therefore\ y = 2e^x - 5x + c$

**f** $\frac{dy}{dx} = 4x^3 + 3x^2$

$\therefore\ y = \int (4x^3 + 3x^2)\, dx$

$= \frac{4x^4}{4} + \frac{3x^3}{3} + c$

$\therefore\ y = x^4 + x^3 + c$

**6** **a** $\frac{dy}{dx} = (1-2x)^2$

$\therefore\ y = \int (1-2x)^2\, dx$

$= \int (1 - 4x + 4x^2)\, dx$

$= x - \frac{4x^2}{2} + \frac{4x^3}{3} + c$

$= x - 2x^2 + \frac{4}{3}x^3 + c$

**b** $\frac{dy}{dx} = \sqrt{x} - \frac{2}{\sqrt{x}}$

$= x^{\frac{1}{2}} - 2x^{-\frac{1}{2}}$

$\therefore\ y = \int (x^{\frac{1}{2}} - 2x^{-\frac{1}{2}})\, dx$

$= \frac{x^{\frac{3}{2}}}{\frac{3}{2}} - \frac{2x^{\frac{1}{2}}}{\frac{1}{2}} + c$

$= \frac{2}{3}x^{\frac{3}{2}} - 4\sqrt{x} + c$

**c** $\frac{dy}{dx} = \frac{x^2 + 2x - 5}{x^2}$

$= 1 + 2x^{-1} - 5x^{-2}$

$\therefore\ y = \int (1 + 2x^{-1} - 5x^{-2})\, dx$

$= x + 2\ln|x| - \frac{5x^{-1}}{-1} + c$

$= x + 2\ln|x| + \frac{5}{x} + c$

**7** **a** $f'(x) = x^3 - 5\sqrt{x} + 3$
$= x^3 - 5x^{\frac{1}{2}} + 3$
$\therefore\ f(x) = \int (x^3 - 5x^{\frac{1}{2}} + 3)\ dx$
$= \frac{1}{4}x^4 - \frac{10}{3}x^{\frac{3}{2}} + 3x + c$
$= \frac{1}{4}x^4 - \frac{10}{3}x\sqrt{x} + 3x + c$

**b** $f'(x) = 2\sqrt{x}(1 - 3x)$
$= 2x^{\frac{1}{2}} - 6x^{\frac{3}{2}}$
$\therefore\ f(x) = \int (2x^{\frac{1}{2}} - 6x^{\frac{3}{2}})\ dx$
$= \dfrac{2x^{\frac{3}{2}}}{\frac{3}{2}} - \dfrac{6x^{\frac{5}{2}}}{\frac{5}{2}} + c$
$= \frac{4}{3}x^{\frac{3}{2}} - \frac{12}{5}x^{\frac{5}{2}} + c$

**c** $f'(x) = 3e^x - \dfrac{4}{x}$
$\therefore\ f(x) = \displaystyle\int \left(3e^x - \frac{4}{x}\right) dx$
$= 3e^x - 4\ln|x| + c$

## EXERCISE 18E.2

**1** **a** $f'(x) = 2x - 1$
$\therefore\ f(x) = \int (2x - 1)\ dx$
$= \dfrac{2x^2}{2} - x + c$
$= x^2 - x + c$
But $f(0) = 3$, so $0 - 0 + c = 3$
$\therefore\ c = 3$
$\therefore\ f(x) = x^2 - x + 3$

**b** $f'(x) = 3x^2 + 2x$
$\therefore\ f(x) = \int (3x^2 + 2x)\ dx$
$= \dfrac{3x^3}{3} + \dfrac{2x^2}{2} + c$
$= x^3 + x^2 + c$
But $f(2) = 5$, so $8 + 4 + c = 5$
$\therefore\ c = -7$
$\therefore\ f(x) = x^3 + x^2 - 7$

**c** $f'(x) = e^x + \dfrac{1}{\sqrt{x}} = e^x + x^{-\frac{1}{2}}$
$\therefore\ f(x) = \int (e^x + x^{-\frac{1}{2}})\ dx$
$= e^x + 2x^{\frac{1}{2}} + c$
But $f(1) = 1$, so $e^1 + 2 + c = 1$
$\therefore\ c = -1 - e$
$\therefore\ f(x) = e^x + 2\sqrt{x} - 1 - e$

**d** $f'(x) = x - \dfrac{2}{\sqrt{x}} = x - 2x^{-\frac{1}{2}}$
$\therefore\ f(x) = \int (x - 2x^{-\frac{1}{2}})\ dx$
$= \dfrac{x^2}{2} - \dfrac{2x^{\frac{1}{2}}}{\frac{1}{2}} + c$
$= \frac{1}{2}x^2 - 4\sqrt{x} + c$
But $f(1) = 2$, so $\frac{1}{2} - 4 + c = 2$
$\therefore\ c = \frac{11}{2}$
$\therefore\ f(x) = \frac{1}{2}x^2 - 4\sqrt{x} + \frac{11}{2}$

**2** **a** $f'(x) = x^2 - 4\cos x$
$\therefore\ f(x) = \int (x^2 - 4\cos x)\ dx$
$= \dfrac{x^3}{3} - 4\sin x + c$
But $f(0) = 3$
$\therefore\ 0 - 4\sin(0) + c = 3$
$\therefore\ c = 3$
$\therefore\ f(x) = \dfrac{x^3}{3} - 4\sin x + 3$

**b** $f'(x) = 2\cos x - 3\sin x$
$\therefore\ f(x) = \int (2\cos x - 3\sin x)\ dx$
$= 2\sin x + 3\cos x + c$
But $f(\frac{\pi}{4}) = \frac{1}{\sqrt{2}}$
$\therefore\ 2\sin\frac{\pi}{4} + 3\cos\frac{\pi}{4} + c = \frac{1}{\sqrt{2}}$
$\therefore\ 2(\frac{1}{\sqrt{2}}) + 3(\frac{1}{\sqrt{2}}) + c = \frac{1}{\sqrt{2}}$
$\therefore\ c = -\frac{4}{\sqrt{2}}$
$\therefore\ c = -2\sqrt{2}$
$\therefore\ f(x) = 2\sin x + 3\cos x - 2\sqrt{2}$

**3** **a** Given:

$f''(x) = 2x + 1, \quad f'(1) = 3, \quad f(2) = 7$

$\therefore \; f'(x) = \int (2x+1)\,dx$

$= \dfrac{2x^2}{2} + x + c$

$= x^2 + x + c$

But $f'(1) = 3$ so $1 + 1 + c = 3$

$\therefore \; c = 1$

$\therefore \; f'(x) = x^2 + x + 1$

Then $f(x) = \int (x^2 + x + 1)\,dx$

$= \dfrac{x^3}{3} + \dfrac{x^2}{2} + x + k$

But $f(2) = 7$, so $\frac{8}{3} + 2 + 2 + k = 7$

$\therefore \; k = 7 - 4 - \frac{8}{3}$

$\therefore \; k = \frac{1}{3}$

$\therefore \; f(x) = \frac{1}{3}x^3 + \frac{1}{2}x^2 + x + \frac{1}{3}$

**b** Given: $f''(x) = 15\sqrt{x} + \dfrac{3}{\sqrt{x}}$,

$f'(1) = 12, \quad f(0) = 5$

Now $f''(x) = 15x^{\frac{1}{2}} + 3x^{-\frac{1}{2}}$

$\therefore \; f'(x) = \dfrac{15x^{\frac{3}{2}}}{\frac{3}{2}} + \dfrac{3x^{\frac{1}{2}}}{\frac{1}{2}} + c$

$= 10x^{\frac{3}{2}} + 6x^{\frac{1}{2}} + c$

But $f'(1) = 12$ so $10 + 6 + c = 12$

$\therefore \; c = -4$

$\therefore \; f'(x) = 10x^{\frac{3}{2}} + 6x^{\frac{1}{2}} - 4$

$\therefore \; f(x) = \int (10x^{\frac{3}{2}} + 6x^{\frac{1}{2}} - 4)\,dx$

$= \dfrac{10x^{\frac{5}{2}}}{\frac{5}{2}} + \dfrac{6x^{\frac{3}{2}}}{\frac{3}{2}} - 4x + k$

$= 4x^{\frac{5}{2}} + 4x^{\frac{3}{2}} - 4x + k$

But $f(0) = 5$, so $k = 5$

$\therefore \; f(x) = 4x^{\frac{5}{2}} + 4x^{\frac{3}{2}} - 4x + 5$

**c** Given: $f''(x) = \cos x$, $f'(\frac{\pi}{2}) = 0$ and $f(0) = 3$

Now $f'(x) = \int \cos x\,dx = \sin x + c$

But $f'(\frac{\pi}{2}) = 0$ so $\sin(\frac{\pi}{2}) + c = 0$

$\therefore \; c = -1$

$\therefore \; f'(x) = \sin x - 1$

So, $f(x) = \int (\sin x - 1)\,dx$

$= -\cos x - x + k$

But $f(0) = 3$ so $-\cos 0 - 0 + k = 3$

$\therefore \; -1 + k = 3$

$\therefore \; k = 4$

So, $f(x) = -\cos x - x + 4$

**d** Given: $f''(x) = 2x$ and that $(1, 0)$ and $(0, 5)$ lie on the curve

Now $f'(x) = \int 2x\,dx = \dfrac{2x^2}{2} + c = x^2 + c$

$\therefore \; f(x) = \int (x^2 + c)\,dx = \dfrac{x^3}{3} + cx + k$

But $f(0) = 5$ so $0 + 0 + k = 5$ and so $k = 5$

and $f(1) = 0$ so $\frac{1}{3} + c + 5 = 0$ and so $c = -5\frac{1}{3}$

$\therefore \; f(x) = \frac{1}{3}x^3 - \frac{16}{3}x + 5$

## EXERCISE 18F

**1** **a** $\int (2x+5)^3\,dx$

$= \frac{1}{2} \times \dfrac{(2x+5)^4}{4} + c$

$= \frac{1}{8}(2x+5)^4 + c$

**b** $\displaystyle\int \frac{1}{(3-2x)^2}\,dx$

$= \int (3-2x)^{-2}\,dx$

$= \frac{1}{-2} \times \dfrac{(3-2x)^{-1}}{-1} + c$

$= \dfrac{1}{2(3-2x)} + c$

**c** $\displaystyle\int \frac{4}{(2x-1)^4}\,dx$

$= \int 4(2x-1)^{-4}\,dx$

$= 4(\frac{1}{2}) \times \dfrac{(2x-1)^{-3}}{-3} + c$

$= \dfrac{-2}{3(2x-1)^3} + c$

**d** $\int (4x-3)^7\,dx$
$= \frac{1}{4} \times \frac{(4x-3)^8}{8} + c$
$= \frac{1}{32}(4x-3)^8 + c$

**e** $\int \sqrt{3x-4}\,dx$
$= \int (3x-4)^{\frac{1}{2}}\,dx$
$= \frac{1}{3} \times \frac{(3x-4)^{\frac{3}{2}}}{\frac{3}{2}} + c$
$= \frac{2}{9}(3x-4)^{\frac{3}{2}} + c$

**f** $\int \frac{10}{\sqrt{1-5x}}\,dx$
$= \int 10(1-5x)^{-\frac{1}{2}}\,dx$
$= 10(\frac{1}{-5}) \times \frac{(1-5x)^{\frac{1}{2}}}{\frac{1}{2}} + c$
$= -4\sqrt{1-5x} + c$

**g** $\int 3(1-x)^4\,dx$
$= 3\int (1-x)^4\,dx$
$= 3(\frac{1}{-1}) \times \frac{(1-x)^5}{5} + c$
$= -\frac{3}{5}(1-x)^5 + c$

**h** $\int \frac{4}{\sqrt{3-4x}}\,dx$
$= \int 4(3-4x)^{-\frac{1}{2}}\,dx$
$= 4(\frac{1}{-4}) \times \frac{(3-4x)^{\frac{1}{2}}}{\frac{1}{2}} + c$
$= -2\sqrt{3-4x} + c$

**2** **a** $\int \sin(3x)\,dx$
$= -\frac{1}{3}\cos(3x) + c$

**b** $\int (2\cos(-4x) + 1)\,dx$
$= 2 \times (\frac{1}{-4})\sin(-4x) + x + c$
$= -\frac{1}{2}\sin(-4x) + x + c$

**c** $\int 3\cos\left(\frac{x}{2}\right)dx$
$= 6\sin\left(\frac{x}{2}\right) + c$

**d** $\int (3\sin(2x) - e^{-x})\,dx$
$= -\frac{3}{2}\cos(2x) + e^{-x} + c$

**e** $\int 2\sin(2x + \frac{\pi}{6})\,dx$
$= -\frac{2}{2}\cos(2x + \frac{\pi}{6}) + c$
$= -\cos(2x + \frac{\pi}{6}) + c$

**f** $\int -3\cos(\frac{\pi}{4} - x)\,dx$
$= -3 \times (-1)\sin(\frac{\pi}{4} - x) + c$
$= 3\sin(\frac{\pi}{4} - x) + c$

**g** $\int (\cos(2x) + \sin(2x))\,dx$
$= \frac{1}{2}\sin(2x) - \frac{1}{2}\cos(2x) + c$

**h** $\int (2\sin(3x) + 5\cos(4x))\,dx$
$= -\frac{2}{3}\cos(3x) + \frac{5}{4}\sin(4x) + c$

**i** $\int \left(\frac{1}{2}\cos(8x) - 3\sin x\right)dx$
$= \frac{1}{2}(\frac{1}{8})\sin(8x) + 3\cos x + c$
$= \frac{1}{16}\sin(8x) + 3\cos x + c$

**3** **a** $\frac{dy}{dx} = \sqrt{2x-7} = (2x-7)^{\frac{1}{2}}$
$\therefore\ y = \frac{1}{2} \times \frac{(2x-7)^{\frac{3}{2}}}{\frac{3}{2}} + c$
$= \frac{1}{3}(2x-7)^{\frac{3}{2}} + c$

But $y = 11$ when $x = 8$
$\therefore\ \frac{1}{3}(16-7)^{\frac{3}{2}} + c = 11$
$\therefore\ \frac{1}{3}(27) + c = 11$
$\therefore\ 9 + c = 11$ and so $c = 2$
$\therefore\ y = \frac{1}{3}(2x-7)^{\frac{3}{2}} + 2$

**b** $f(x)$ has gradient function $f'(x) = \frac{4}{\sqrt{1-x}} = 4(1-x)^{-\frac{1}{2}}$
$\therefore\ f(x) = 4(\frac{1}{-1}) \times \frac{(1-x)^{\frac{1}{2}}}{\frac{1}{2}} + c$
$= -8\sqrt{1-x} + c$

But $y = -11$ when $x = -3$
$\therefore\ -8\sqrt{1-(-3)} + c = -11$
$\therefore\ -8\sqrt{4} + c = -11$
$\therefore\ -16 + c = -11$ and so $c = 5$
$\therefore\ f(x) = 5 - 8\sqrt{1-x}$

Now $f(-8) = 5 - 8\sqrt{1-(-8)} = 5 - 8(3) = -19$, so the point is $(-8, -19)$.

**4** **a** $\int \cos^2 x \, dx$
$= \int (\frac{1}{2} + \frac{1}{2}\cos(2x)) \, dx$
$= \frac{1}{2}x + \frac{1}{4}\sin(2x) + c$

**b** $\int \sin^2 x \, dx$
$= \int (\frac{1}{2} - \frac{1}{2}\cos(2x)) \, dx$
$= \frac{1}{2}x - \frac{1}{4}\sin(2x) + c$

**c** $\int (1 + \cos^2(2x)) \, dx$
$= \int (1 + \frac{1}{2} + \frac{1}{2}\cos(4x)) \, dx$
$= \int (\frac{3}{2} + \frac{1}{2}\cos(4x)) \, dx$
$= \frac{3}{2}x + \frac{1}{8}\sin(4x) + c$

**d** $\int (3 - \sin^2(3x)) \, dx$
$= \int (3 - (\frac{1}{2} - \frac{1}{2}\cos(6x))) \, dx$
$= \int (\frac{5}{2} + \frac{1}{2}\cos(6x)) \, dx$
$= \frac{5}{2}x + \frac{1}{12}\sin(6x) + c$

**e** $\int \frac{1}{2}\cos^2(4x) \, dx$
$= \int \frac{1}{2}(\frac{1}{2} + \frac{1}{2}\cos(8x)) \, dx$
$= \int (\frac{1}{4} + \frac{1}{4}\cos(8x)) \, dx$
$= \frac{1}{4}x + \frac{1}{32}\sin(8x) + c$

**f** $\int (1 + \cos x)^2 \, dx$
$= \int (1 + 2\cos x + \cos^2 x) \, dx$
$= \int (1 + 2\cos x + \frac{1}{2} + \frac{1}{2}\cos(2x)) \, dx$
$= \int \left(\frac{3}{2} + 2\cos x + \frac{1}{2}\cos(2x)\right) dx$
$= \frac{3}{2}x + 2\sin x + \frac{1}{4}\sin(2x) + c$

**5** **a** $\int 3(2x-1)^2 \, dx$
$= 3\int (2x-1)^2 \, dx$
$= 3(\frac{1}{2})\,\dfrac{(2x-1)^3}{3} + c$
$= \frac{1}{2}(2x-1)^3 + c$

**b** $\int (x^2 - x)^2 \, dx$
$= \int \left(x^4 - 2x^3 + x^2\right) dx$
$= \dfrac{x^5}{5} - \dfrac{2x^4}{4} + \dfrac{x^3}{3} + c$
$= \frac{1}{5}x^5 - \frac{1}{2}x^4 + \frac{1}{3}x^3 + c$

**c** $\int (1 - 3x)^3 \, dx$
$= (\frac{1}{-3})\,\dfrac{(1-3x)^4}{4} + c$
$= -\frac{1}{12}(1-3x)^4 + c$

**d** $\int (1 - x^2)^2 \, dx$
$= \int (1 - 2x^2 + x^4) \, dx$
$= x - \frac{2}{3}x^3 + \frac{1}{5}x^5 + c$

**e** $\int 4\sqrt{5-x} \, dx$
$= 4\int (5-x)^{\frac{1}{2}} \, dx$
$= 4(\frac{1}{-1})\,\dfrac{(5-x)^{\frac{3}{2}}}{\frac{3}{2}} + c$
$= -\frac{8}{3}(5-x)^{\frac{3}{2}} + c$

**f** $\int (x^2 + 1)^3 \, dx$
$= \int (x^6 + 3x^4 + 3x^2 + 1) \, dx$
$= \dfrac{x^7}{7} + \dfrac{3x^5}{5} + \dfrac{3x^3}{3} + x + c$
$= \frac{1}{7}x^7 + \frac{3}{5}x^5 + x^3 + x + c$

**6** **a** $\int \left(2e^x + 5e^{2x}\right) dx$
$= 2e^x + 5(\frac{1}{2})e^{2x} + c$
$= 2e^x + \frac{5}{2}e^{2x} + c$

**b** $\int (3e^{5x-2}) \, dx$
$= 3(\frac{1}{5})e^{5x-2} + c$
$= \frac{3}{5}e^{5x-2} + c$

**c** $\int (e^{7-3x}) \, dx$
$= \frac{1}{-3}e^{7-3x} + c$
$= -\frac{1}{3}e^{7-3x} + c$

**d** $\displaystyle\int \frac{1}{2x-1} \, dx$
$= \frac{1}{2}\ln|2x-1| + c$

**e** $\displaystyle\int \frac{5}{1-3x} \, dx$
$= 5\displaystyle\int \frac{1}{1-3x} \, dx$
$= 5(\frac{1}{-3})\ln|1-3x| + c$
$= -\frac{5}{3}\ln|1-3x| + c$

**f** $\displaystyle\int \left(e^{-x} - \frac{4}{2x+1}\right) dx$
$= \frac{1}{-1}e^{-x} - 4(\frac{1}{2})\ln|2x+1| + c$
$= -e^{-x} - 2\ln|2x+1| + c$

**g** $\int (e^x + e^{-x})^2 \, dx$
$= \int \left(e^{2x} + 2 + e^{-2x}\right) dx$
$= \frac{1}{2}e^{2x} + 2x + (\frac{1}{-2})e^{-2x} + c$
$= \frac{1}{2}e^{2x} + 2x - \frac{1}{2}e^{-2x} + c$

**h** $\int (e^{-x}+2)^2\,dx$
$= \int \left(e^{-2x}+4e^{-x}+4\right) dx$
$= \frac{1}{-2}e^{-2x}+4(\frac{1}{-1})e^{-x}+4x+c$
$= -\frac{1}{2}e^{-2x}-4e^{-x}+4x+c$

**i** $\int \left(x-\frac{5}{1-x}\right) dx$
$= \frac{x^2}{2}-5(\frac{1}{-1})\ln|1-x|+c$
$= \frac{1}{2}x^2+5\ln|1-x|+c$

**7** **a** $\frac{dy}{dx}=(1-e^x)^2$
$= 1-2e^x+e^{2x}$
$\therefore\ y=x-2e^x+\frac{1}{2}e^{2x}+c$

**b** $\frac{dy}{dx}=1-2x+\frac{3}{x+2}$
$\therefore\ y=x-\frac{2x^2}{2}+3\ln|x+2|+c$
$= x-x^2+3\ln|x+2|+c$

**c** $\frac{dy}{dx}=e^{-2x}+\frac{4}{2x-1}$
$\therefore\ y=\frac{1}{-2}e^{-2x}+4(\frac{1}{2})\ln|2x-1|+c$
$= -\frac{1}{2}e^{-2x}+2\ln|2x-1|+c$

**8** Differentiating Tracy's answer gives
$\frac{d}{dx}\left(\frac{1}{4}\ln(4x)+c\right)=\frac{1}{4}\left(\frac{1}{4x}\right)\times 4+0,\ \ x>0$
$= \frac{1}{4x},\ \ x>0$

Differentiating Nadine's answer gives
$\frac{d}{dx}\left(\frac{1}{4}\ln(x)+c\right)=\frac{1}{4}\left(\frac{1}{x}\right)+0,\ \ x>0$
$= \frac{1}{4x},\ \ x>0$

Both answers give the correct derivative and both are correct. This result occurs because $\ln(4x)=\ln 4+\ln x$. Their answers differ by a constant which is accounted for by $c$.

**9** Given: $f'(x)=p\sin(\frac{1}{2}x)$, $f(0)=1$ and $f(2\pi)=0$
$\therefore\ f(x)=-2p\cos(\frac{1}{2}x)+c$
But $f(0)=1$, so $-2p\cos(0)+c=1$
$\therefore\ -2p+c=1$
$\therefore\ c=1+2p$ .... (1)
Also, $f(2\pi)=0$, so $-2p\cos(\pi)+c=0$
$\therefore\ 2p+c=0$
$\therefore\ 2p+1+2p=0$ {using (1)}
$\therefore\ p=-\frac{1}{4}$
$\therefore\ c=\frac{1}{2}$ {from (1)}
$\therefore\ f(x)=\frac{1}{2}\cos(\frac{1}{2}x)+\frac{1}{2}$

**10** $g''(x)=-\sin 2x$
Integrating both sides with respect to $x$, we get $g'(x)=\frac{1}{2}\cos 2x+c$, $c$ some constant.
So, $g'(\pi)=\frac{1}{2}\cos(2\pi)+c$
$= \frac{1}{2}+c$
and $g'(-\pi)=\frac{1}{2}\cos(-2\pi)+c$
$= \frac{1}{2}+c$
$= g'(\pi)$
$\therefore$ the gradients of the tangents to $y=g(x)$ at $x=\pi$ and $x=-\pi$ are equal.

**11** **a** $f'(x)=2e^{-2x}$
$\therefore\ f(x)=2(\frac{1}{-2})e^{-2x}+c$
$= -e^{-2x}+c$
But $f(0)=3$ so $-e^0+c=3$
$\therefore\ c=4$
$\therefore\ f(x)=-e^{-2x}+4$

**b** $f'(x)=2x-\frac{2}{1-x}$
$\therefore\ f(x)=\frac{2x^2}{2}-\frac{2}{-1}\ln|1-x|+c$
$= x^2+2\ln|1-x|+c$
But $f(-1)=3$ so $1+2\ln|2|+c=3$
$\therefore\ c=2-2\ln 2$
$\therefore\ f(x)=x^2+2\ln|1-x|+2-2\ln 2$

**c**
$$f'(x) = \sqrt{x} + \tfrac{1}{2}e^{-4x}$$
$$= x^{\frac{1}{2}} + \tfrac{1}{2}e^{-4x}$$
$$\therefore\ f(x) = \frac{x^{\frac{3}{2}}}{\frac{3}{2}} + \tfrac{1}{2}(\tfrac{1}{-4})e^{-4x} + c$$
$$= \tfrac{2}{3}x^{\frac{3}{2}} - \tfrac{1}{8}e^{-4x} + c$$

But $f(1) = 0$
$$\therefore\ \tfrac{2}{3} - \tfrac{1}{8}e^{-4} + c = 0$$
$$\therefore\ c = \tfrac{1}{8}e^{-4} - \tfrac{2}{3}$$
$$\therefore\ f(x) = \tfrac{2}{3}x^{\frac{3}{2}} - \tfrac{1}{8}e^{-4x} + \tfrac{1}{8}e^{-4} - \tfrac{2}{3}$$

**12**
$$(\sin x + \cos x)^2 = \sin^2 x + 2\sin x \cos x + \cos^2 x$$
$$= 1 + \sin 2x$$
$$\therefore\ \int (\sin x + \cos x)^2\,dx = \int (1 + \sin 2x)\,dx$$
$$= x - \tfrac{1}{2}\cos(2x) + c$$

**13**
$$(\cos x + 1)^2 = \cos^2 x + 2\cos x + 1$$
$$= (\tfrac{1}{2} + \tfrac{1}{2}\cos 2x) + 2\cos x + 1$$
$$= \tfrac{1}{2}\cos 2x + 2\cos x + \tfrac{3}{2}$$
$$\therefore\ \int (\cos x + 1)^2\,dx = \int (\tfrac{1}{2}\cos 2x + 2\cos x + \tfrac{3}{2})\,dx$$
$$= \tfrac{1}{4}\sin 2x + 2\sin x + \tfrac{3}{2}x + c$$

## EXERCISE 18G

**1** **a** $u = x^3 + 1,\quad \dfrac{du}{dx} = 3x^2$
$$\therefore\ \int 3x^2(x^3+1)^4\,dx$$
$$= \int u^4 \frac{du}{dx}\,dx$$
$$= \int u^4\,du$$
$$= \tfrac{1}{5}u^5 + c$$
$$= \tfrac{1}{5}(x^3+1)^5 + c$$

**b** $u = x^3 + 1,\quad \dfrac{du}{dx} = 3x^2$
$$\therefore\ \int x^2 e^{x^3+1}\,dx$$
$$= \tfrac{1}{3}\int (3x^2)e^{x^3+1}\,dx$$
$$= \tfrac{1}{3}\int e^u \frac{du}{dx}\,dx$$
$$= \tfrac{1}{3}\int e^u\,du$$
$$= \tfrac{1}{3}e^u + c$$
$$= \tfrac{1}{3}e^{x^3+1} + c$$

**c** $u = \sin x,\quad \dfrac{du}{dx} = \cos x$
$$\therefore\ \int \sin^4 x \cos x\,dx$$
$$= \int u^4 \frac{du}{dx}\,dx$$
$$= \int u^4\,du$$
$$= \frac{u^5}{5} + c$$
$$= \tfrac{1}{5}\sin^5 x + c$$

**2** **a**
$$\int 4x^3(2+x^4)^3\,dx$$
$$= \int u^3 \frac{du}{dx}\,dx \quad \{u = 2 + x^4,\ \frac{du}{dx} = 4x^3\}$$
$$= \int u^3\,du$$
$$= \frac{u^4}{4} + c$$
$$= \tfrac{1}{4}(2+x^4)^4 + c$$

**b**
$$\int \frac{2x}{\sqrt{x^2+3}}\,dx$$
$$= \int \left(\left(x^2+3\right)^{-\frac{1}{2}} \times 2x\right)dx$$
$$= \int u^{-\frac{1}{2}} \frac{du}{dx}\,dx \quad \{u = x^2 + 3,\ \frac{du}{dx} = 2x\}$$
$$= \int u^{-\frac{1}{2}}\,du$$
$$= \frac{u^{\frac{1}{2}}}{\frac{1}{2}} + c$$
$$= 2\sqrt{u} + c$$
$$= 2\sqrt{x^2+3} + c$$

**c**

$$\begin{aligned}
&\int \frac{x}{(1-x^2)^5}\,dx\\
&= -\tfrac{1}{2}\int (1-x^2)^{-5} \times (-2x)\,dx\\
&= -\tfrac{1}{2}\int u^{-5}\,\frac{du}{dx}\,dx \quad \{u = 1-x^2,\\
&= -\tfrac{1}{2}\int u^{-5}\,du \qquad \frac{du}{dx} = -2x\}\\
&= -\tfrac{1}{2}\,\frac{u^{-4}}{-4} + c\\
&= \frac{1}{8(1-x^2)^4} + c
\end{aligned}$$

**d**

$$\begin{aligned}
&\int \sqrt{x^3+x}\,(3x^2+1)\,dx\\
&= \int \sqrt{u}\,\frac{du}{dx}\,dx \quad \{u = x^3+x,\\
&= \int u^{\frac{1}{2}}\,du \qquad \frac{du}{dx} = 3x^2+1\}\\
&= \frac{u^{\frac{3}{2}}}{\frac{3}{2}} + c\\
&= \tfrac{2}{3}u^{\frac{3}{2}} + c\\
&= \tfrac{2}{3}(x^3+x)^{\frac{3}{2}} + c
\end{aligned}$$

**3** **a**

$$\begin{aligned}
&\int -2e^{1-2x}\,dx\\
&= \int e^u\,\frac{du}{dx}\,dx \quad \{u = 1-2x, \quad \frac{du}{dx} = -2\}\\
&= \int e^u\,du\\
&= e^u + c\\
&= e^{1-2x} + c
\end{aligned}$$

**b**

$$\begin{aligned}
&\int 2xe^{x^2}\,dx\\
&= \int e^u\,\frac{du}{dx}\,dx \quad \{u = x^2, \quad \frac{du}{dx} = 2x\}\\
&= \int e^u\,du\\
&= e^u + c\\
&= e^{x^2} + c
\end{aligned}$$

**c**

$$\begin{aligned}
\int \frac{e^{\sqrt{x}}}{\sqrt{x}}\,dx &= 2\int \frac{e^{\sqrt{x}}}{2\sqrt{x}}\,dx\\
&= 2\int e^u\,\frac{du}{dx}\,dx \quad \{u = \sqrt{x},\\
&= 2\int e^u\,du \qquad \frac{du}{dx} = \frac{1}{2\sqrt{x}}\}\\
&= 2e^u + c\\
&= 2e^{\sqrt{x}} + c
\end{aligned}$$

**d**

$$\begin{aligned}
&\int (2x-1)e^{x-x^2}\,dx\\
&= -\int (1-2x)e^{x-x^2}\,dx\\
&= -\int e^u\,\frac{du}{dx}\,dx \quad \{u = x-x^2,\\
&= -\int e^u\,du \qquad \frac{du}{dx} = 1-2x\}\\
&= -e^u + c\\
&= -e^{x-x^2} + c
\end{aligned}$$

**4** **a** Let $u = x^2+1, \quad \frac{du}{dx} = 2x$

$$\begin{aligned}
\therefore \int \frac{2x}{x^2+1}\,dx &= \int \frac{1}{x^2+1}\,(2x)\,dx\\
&= \int \frac{1}{u}\,\frac{du}{dx}\,dx\\
&= \int \frac{1}{u}\,du\\
&= \ln|u| + c\\
&= \ln\left|x^2+1\right| + c
\end{aligned}$$

**b** Let $u = 2-x^2, \quad \frac{du}{dx} = -2x$

$$\begin{aligned}
\therefore \int \frac{x}{2-x^2}\,dx &= -\tfrac{1}{2}\int \frac{1}{2-x^2}\,(-2x)\,dx\\
&= -\tfrac{1}{2}\int \frac{1}{u}\,\frac{du}{dx}\,dx\\
&= -\tfrac{1}{2}\int \frac{1}{u}\,du\\
&= -\tfrac{1}{2}\ln|u| + c\\
&= -\tfrac{1}{2}\ln\left|2-x^2\right| + c
\end{aligned}$$

**c** Let $u = x^2-3x, \quad \frac{du}{dx} = 2x-3$

$$\begin{aligned}
\therefore \int \frac{2x-3}{x^2-3x}\,dx &= \int \frac{1}{x^2-3x}\,(2x-3)\,dx\\
&= \int \frac{1}{u}\,\frac{du}{dx}\,dx\\
&= \int \frac{1}{u}\,du\\
&= \ln|u| + c\\
&= \ln\left|x^2-3x\right| + c
\end{aligned}$$

**d** Let $u = x^3-x, \quad \frac{du}{dx} = 3x^2-1$

$$\begin{aligned}
\therefore \int \frac{6x^2-2}{x^3-x}\,dx &= 2\int \frac{1}{x^3-3x}\,(3x^2-1)\,dx\\
&= 2\int \frac{1}{u}\,\frac{du}{dx}\,dx\\
&= 2\int \frac{1}{u}\,du\\
&= 2\ln|u| + c\\
&= 2\ln\left|x^3-x\right| + c
\end{aligned}$$

**5** **a** Let $u = \cos x$, $\dfrac{du}{dx} = -\sin x$

$$\therefore \int \frac{\sin x}{\sqrt{\cos x}}\,dx = -\int \frac{-\sin x}{\sqrt{\cos x}}\,dx$$
$$= -\int u^{-\frac{1}{2}} \frac{du}{dx}\,dx$$
$$= -\int u^{-\frac{1}{2}}\,du$$
$$= -\frac{u^{\frac{1}{2}}}{\frac{1}{2}} + c$$
$$= -2(\cos x)^{\frac{1}{2}} + c$$

**b** Let $u = \cos x$, $\dfrac{du}{dx} = -\sin x$

$$\therefore \int \frac{\sin x}{\cos x}\,dx = -\int \frac{-\sin x}{\cos x}\,dx$$
$$= -\int \frac{1}{u}\frac{du}{dx}\,dx$$
$$= -\int \frac{1}{u}\,du = -\ln|u| + c$$
$$= -\ln|\cos x| + c$$

**c** Let $u = \sin x$, $\dfrac{du}{dx} = \cos x$

$$\therefore \int \sqrt{\sin x}\cos x\,dx = \int u^{\frac{1}{2}}\frac{du}{dx}\,dx$$
$$= \int u^{\frac{1}{2}}\,du$$
$$= \tfrac{2}{3}u^{\frac{3}{2}} + c$$
$$= \tfrac{2}{3}(\sin x)^{\frac{3}{2}} + c$$

**d** Let $u = x^2$, $\dfrac{du}{dx} = 2x$

$$\therefore \int x\sin(x^2)\,dx = \tfrac{1}{2}\int (2x)\sin(x^2)\,dx$$
$$= \tfrac{1}{2}\int \sin u \frac{du}{dx}\,dx$$
$$= \tfrac{1}{2}\int \sin u\,du$$
$$= \tfrac{1}{2}(-\cos u) + c$$
$$= -\tfrac{1}{2}\cos(x^2) + c$$

## EXERCISE 18H

**1** **a** $\int_1^4 \sqrt{x}\,dx = \int_1^4 x^{\frac{1}{2}}\,dx$
$= \left[\frac{2}{3}x^{\frac{3}{2}}\right]_1^4$
$= \frac{2}{3}(8) - \frac{2}{3}(1)$
$= \frac{14}{3} \approx 4.67$

$\int_1^4 (-\sqrt{x})\,dx = \int_1^4 \left(-x^{\frac{1}{2}}\right)\,dx$
$= \left[-\frac{2}{3}x^{\frac{3}{2}}\right]_1^4$
$= -\frac{2}{3}(8) - (-\frac{2}{3}(1))$
$= -\frac{14}{3} \approx -4.67$

**b** $\int_0^1 x^7\,dx = \left[\frac{1}{8}x^8\right]_0^1$
$= \frac{1}{8} - 0 = \frac{1}{8}$

$\int_0^1 (-x^7)\,dx = \left[-\frac{1}{8}x^8\right]_0^1$
$= -\frac{1}{8} - 0 = -\frac{1}{8}$

Property: $\int_a^b [-f(x)]\,dx = -\int_a^b f(x)\,dx$

**2** **a** $\int_0^1 x^2\,dx$
$= \left[\frac{1}{3}x^3\right]_0^1$
$= \frac{1}{3} - 0$
$= \frac{1}{3}$

**b** $\int_1^2 x^2\,dx$
$= \left[\frac{1}{3}x^3\right]_1^2$
$= \frac{1}{3}(8) - \frac{1}{3}(1)$
$= \frac{7}{3}$

**c** $\int_0^2 x^2\,dx$
$= \left[\frac{1}{3}x^3\right]_0^2$
$= \frac{1}{3}(8) - 0$
$= \frac{8}{3}$

**d** $\int_0^1 3x^2\,dx$
$= \left[x^3\right]_0^1$
$= 1 - 0$
$= 1$

Properties: $\int_a^b f(x)\,dx + \int_b^c f(x)\,dx = \int_a^c f(x)\,dx$
$\int_a^b c\,f(x)\,dx = c\int_a^b f(x)\,dx$, $c$ a constant

**3** **a** $\int_0^2 (x^3 - 4x)\,dx$
$= \left[\frac{1}{4}x^4 - 2x^2\right]_0^2$
$= \left[\frac{1}{4}(16) - 2(4)\right] - [0 - 0]$
$= -4$

**b** $\int_2^3 (x^3 - 4x)\,dx$
$= \left[\frac{1}{4}x^4 - 2x^2\right]_2^3$
$= \left[\frac{1}{4}(81) - 2(9)\right] - \left[\frac{1}{4}(16) - 2(4)\right]$
$= \frac{25}{4} = 6.25$

**c** $\int_0^3 (x^3 - 4x)\, dx = \left[\frac{1}{4}x^4 - 2x^2\right]_0^3$
$= \left[\frac{1}{4}(81) - 2(9)\right] - [0 - 0]$
$= \frac{9}{4} = 2.25$

**4** **a** $\int_0^1 x^2\, dx = \left[\frac{1}{3}x^3\right]_0^1$
$= \frac{1}{3}(1) - 0$
$= \frac{1}{3}$

**b** $\int_0^1 \sqrt{x}\, dx = \int_0^1 x^{\frac{1}{2}}\, dx$
$= \left[\frac{2}{3}x^{\frac{3}{2}}\right]_0^1$
$= \frac{2}{3}(1) - 0$
$= \frac{2}{3}$

**c** $\int_0^1 (x^2 + \sqrt{x})\, dx$
$= \int_0^1 (x^2 + x^{\frac{1}{2}})\, dx$
$= \left[\frac{1}{3}x^3 + \frac{2}{3}x^{\frac{3}{2}}\right]_0^1$
$= \left[\frac{1}{3}(1) + \frac{2}{3}(1)\right] - [0 + 0]$
$= 1$

Property: $\int_a^b f(x)\, dx + \int_a^b g(x)\, dx = \int_a^b [f(x) + g(x)]\, dx$

**5** **a** $\int_0^1 x^3\, dx = \left[\frac{x^4}{4}\right]_0^1$
$= \frac{1}{4} - 0$
$= \frac{1}{4}$

**b** $\int_0^2 (x^2 - x)\, dx = \left[\frac{x^3}{3} - \frac{x^2}{2}\right]_0^2$
$= (\frac{8}{3} - 2) - (0 - 0)$
$= \frac{2}{3}$

**c** $\int_0^1 e^x\, dx = [e^x]_0^1$
$= e^1 - e^0$
$= e - 1$
$\approx 1.72$

**d** $\int_0^{\frac{\pi}{6}} \cos x\, dx$
$= [\sin x]_0^{\frac{\pi}{6}}$
$= \sin\frac{\pi}{6} - \sin 0$
$= \frac{1}{2}$

**e** $\int_1^4 \left(x - \frac{3}{\sqrt{x}}\right) dx$
$= \int_1^4 (x - 3x^{-\frac{1}{2}})\, dx$
$= \left[\frac{x^2}{2} - \frac{3x^{\frac{1}{2}}}{\frac{1}{2}}\right]_1^4$
$= \left[\frac{x^2}{2} - 6\sqrt{x}\right]_1^4$
$= \left[\frac{16}{2} - 12\right] - (\frac{1}{2} - 6)$
$= 1\frac{1}{2}$

**f** $\int_4^9 \frac{x - 3}{\sqrt{x}}\, dx$
$= \int_4^9 (x^{\frac{1}{2}} - 3x^{-\frac{1}{2}})\, dx$
$= \left[\frac{x^{\frac{3}{2}}}{\frac{3}{2}} - \frac{3x^{\frac{1}{2}}}{\frac{1}{2}}\right]_4^9$
$= \left[\frac{2}{3}x^{\frac{3}{2}} - 6x^{\frac{1}{2}}\right]_4^9$
$= \left[\frac{2}{3}(27) - 6(3)\right] - \left[\frac{2}{3}(8) - 6(2)\right]$
$= (18 - 18) - (\frac{16}{3} - 12)$
$= 6\frac{2}{3}$

**g** $\int_1^3 \frac{1}{x}\, dx = [\ln|x|]_1^3$
$= \ln 3 - \ln 1$
$= \ln 3 - 0$
$= \ln 3$
$\approx 1.10$

**h** $\int_{\frac{\pi}{3}}^{\frac{\pi}{2}} \sin x\, dx = \left[-\cos x\right]_{\frac{\pi}{3}}^{\frac{\pi}{2}}$
$= -\cos\frac{\pi}{2} + \cos\frac{\pi}{3}$
$= \frac{1}{2}$

**i** $\int_1^2 (e^{-x} + 1)^2\, dx$
$= \int_1^2 (e^{-2x} + 2e^{-x} + 1)\, dx$
$= \left[(\frac{1}{-2})e^{-2x} + 2(\frac{1}{-1})e^{-x} + x\right]_1^2$
$= \left[-\frac{e^{-2x}}{2} - 2e^{-x} + x\right]_1^2$
$= \left(-\frac{e^{-4}}{2} - 2e^{-2} + 2\right) - \left(-\frac{e^{-2}}{2} - 2e^{-1} + 1\right)$
$\approx 1.52$

**j** $\int_2^6 \frac{1}{\sqrt{2x - 3}}\, dx$
$= \int_2^6 (2x - 3)^{-\frac{1}{2}}\, dx$
$= \left[\frac{1}{2}\,\frac{(2x - 3)^{\frac{1}{2}}}{\frac{1}{2}}\right]_2^6$
$= \left[\sqrt{2x - 3}\right]_2^6$
$= \sqrt{9} - \sqrt{1}$
$= 2$

**k** $\int_0^1 e^{1-x}\,dx = \left[(\frac{1}{-1})e^{1-x}\right]_0^1$
$= \left(\frac{e^0}{-1}\right) - \left(\frac{e^1}{-1}\right)$
$= e - 1$
$\approx 1.72$

**l** $\int_0^{\frac{\pi}{6}} \sin(3x)\,dx = \left[-\frac{1}{3}\cos(3x)\right]_0^{\frac{\pi}{6}}$
$= -\frac{1}{3}[\cos\frac{\pi}{2} - \cos 0]$
$= -\frac{1}{3}[0-1]$
$= \frac{1}{3}$

**6** $\int_0^{\frac{\pi}{4}} \cos^2 x\,dx = \int_0^{\frac{\pi}{4}} (\frac{1}{2} + \frac{1}{2}\cos(2x))\,dx$
$= \left[\frac{x}{2} + \frac{1}{4}\sin(2x)\right]_0^{\frac{\pi}{4}}$
$= [\frac{\pi}{8} + \frac{1}{4}\sin\frac{\pi}{2}] - 0$
$= \frac{\pi}{8} + \frac{1}{4}$

**7** $\int_0^{\frac{\pi}{2}} \sin^2 x\,dx = \int_0^{\frac{\pi}{2}} (\frac{1}{2} - \frac{1}{2}\cos(2x))\,dx$
$= \left[\frac{x}{2} - \frac{1}{4}\sin(2x)\right]_0^{\frac{\pi}{2}}$
$= [\frac{\pi}{4} - \frac{1}{4}\sin\pi] - 0$
$= \frac{\pi}{4}$

**8** Using technology:

**a** $\int_1^3 \ln x\,dx \approx 1.30$ **b** $\int_{-1}^1 e^{-x^2}\,dx \approx 1.49$ **c** $\displaystyle\int_{\frac{\pi}{4}}^{\frac{\pi}{6}} \sin(\sqrt{x})\,dx \approx -0.189$

**9** $\dfrac{4x+1}{x-1} = \dfrac{4x-4+1+4}{x-1}$
$= \dfrac{4(x-1)+5}{x-1}$
$= \dfrac{4(x-1)}{x-1} + \dfrac{5}{x-1}$
$= 4 + \dfrac{5}{x-1}$ as required

$\therefore \displaystyle\int_3^5 \frac{4x+1}{x-1}\,dx = \int_3^5 4 + \frac{5}{x-1}\,dx$
$= [4x + 5\ln|x-1|]_3^5$
$= 4(5) + 5\ln|5-1| - (4(3) + 5\ln|3-1|)$
$= 20 + 5\ln 4 - 12 - 5\ln 2$
$= 20 + 5\ln 2^2 - 12 - 5\ln 2$
$= 8 + 10\ln 2 - 5\ln 2$
$= 8 + 5\ln 2$

**10** **a** $\displaystyle\int_m^{-2} \frac{1}{4-x}\,dx = \ln\frac{3}{2}$
$\therefore [-\ln|4-x|]_m^{-2} = \ln\frac{3}{2}$
$\therefore -\ln|4--2| + \ln|4-m| = \ln\frac{3}{2}$
$\therefore \ln|4-m| - \ln 6 = \ln\frac{3}{2}$
$\therefore \ln\left|\frac{4-m}{6}\right| = \ln\frac{3}{2}$
$\therefore \left|\frac{4-m}{6}\right| = \frac{3}{2}$
$\therefore \frac{4-m}{6} = \pm\frac{3}{2}$
$\therefore 4-m = \pm 9$
$\therefore m = 4 \pm 9$
$\therefore m = -5$ or $13$

However, the solution $m = 13$ is invalid, since the vertical asymptote $x = 4$ lies between $-2$ and $13$.

$\therefore m = -5$ is the only valid answer.

**b** $\displaystyle\int_m^{2m} (2x-1)\,dx = 4$
$\therefore \left[x^2 - x\right]_m^{2m} = 4$
$\therefore (2m)^2 - 2m - (m^2 - m) = 4$
$\therefore 4m^2 - 2m - m^2 + m = 4$
$\therefore 3m^2 - m - 4 = 0$
$\therefore (3m-4)(m+1) = 0$
$\therefore m = \frac{4}{3}$ or $-1$

**11** **a** $\int_0^3 f(x)\,dx =$ area between $f(x)$ and the $x$-axis from $x = 0$ to $x = 3$
$= 2 + 3 + 1.5 = 6.5$

**b** $\int_3^7 f(x)\,dx = -(\text{area between } f(x) \text{ and the } x\text{-axis from } x=3 \text{ to } x=7)$
$= -\left(\frac{3}{2}+3+\frac{5}{2}+2\right) = -9$

**c** $\int_2^4 f(x)\,dx = (\text{area between } f(x) \text{ and the } x\text{-axis from } x=2 \text{ to } x=3)$
$-(\text{area between } f(x) \text{ and the } x\text{-axis from } x=3 \text{ to } x=4)$
$= 1.5 - 1.5 = 0$

**d** $\int_0^7 f(x)\,dx = (\text{area between } f(x) \text{ and the } x\text{-axis from } x=0 \text{ to } x=3)$
$-(\text{area between } f(x) \text{ and the } x\text{-axis from } x=3 \text{ to } x=7)$
$= 6.5 - 9 = -2.5$

**12** **a** $\int_0^4 f(x)\,dx$
= area of semi-circle with radius 2
$= \frac{1}{2}\pi(2)^2 = 2\pi$

**b** $\int_4^6 f(x)\,dx$
$= -(\text{area of 2 by 2 rectangle})$
$= -(2\times 2) = -4$

**c** $\int_6^8 f(x)\,dx$
= area of semi-circle with radius 1
$= \frac{1}{2}\pi(1)^2 = \frac{\pi}{2}$

**d** $\int_0^8 f(x)\,dx$
$= \int_0^4 f(x)\,dx + \int_4^6 f(x)\,dx + \int_6^8 f(x)\,dx$
$= 2\pi + (-4) + \frac{\pi}{2} = \frac{5\pi}{2} - 4$

**13** **a** $\int_2^4 f(x)\,dx + \int_4^7 f(x)\,dx$
$= \int_2^7 f(x)\,dx$

**b** $\int_1^3 g(x)\,dx + \int_3^8 g(x)\,dx + \int_8^9 g(x)\,dx$
$= \int_1^9 g(x)\,dx$

**14** **a** $\int_1^3 f(x)\,dx + \int_3^6 f(x)\,dx = \int_1^6 f(x)\,dx$
$\therefore\ \int_3^6 f(x)\,dx = \int_1^6 f(x)\,dx - \int_1^3 f(x)\,dx$
$= (-3) - 2$
$= -5$

**b** $\int_0^2 f(x)\,dx + \int_2^4 f(x)\,dx + \int_4^6 f(x)\,dx = \int_0^6 f(x)\,dx$
$\therefore\ \int_2^4 f(x)\,dx = \int_0^6 f(x)\,dx - \int_4^6 f(x)\,dx - \int_0^2 f(x)\,dx$
$= (7) - (-2) - (5)$
$= 4$

**15** **a** $\int_1^{-1} f(x)\,dx = -\int_{-1}^1 f(x)\,dx$
$= -(-4)$
$= 4$

**b** $\int_{-1}^1 (2+f(x))\,dx = \int_{-1}^1 2\,dx + \int_{-1}^1 f(x)\,dx$
$= [2x]_{-1}^1 + (-4)$
$= (2-(-2)) - 4$
$= 0$

**c** $\int_{-1}^1 2\,f(x)\,dx = 2\int_{-1}^1 f(x)\,dx$
$= 2(-4)$
$= -8$

**d** $\int_{-1}^1 k\,f(x)\,dx = 7$
$\therefore\ k\int_{-1}^1 f(x)\,dx = 7$
$\therefore\ k(-4) = 7$
$\therefore\ k = -\frac{7}{4}$

**16** $\int_2^3 (g'(x)-1)\,dx = \int_2^3 g'(x)\,dx + \int_2^3 -1\,dx$
$= [g(x)]_2^3 + [-x]_2^3$
$= (g(3)-g(2)) + (-3-(-2))$
$= 5 - 4 - 1$
$= 0$

## REVIEW SET 18A

**1** **a** $\int_0^4 f(x)\,dx = \text{area of semi-circle with radius 2}$
$= \frac{1}{2} \times \pi \times 2^2$
$= 2\pi \text{ units}^2$

**b** $\int_4^6 f(x)\,dx = \text{area of square}$
$= 2 \times 2$
$= 4 \text{ units}^2$

**2** **a** $\displaystyle\int \frac{4}{\sqrt{x}}\,dx = 4\int x^{-\frac{1}{2}}\,dx$
$\displaystyle= 4\,\frac{x^{\frac{1}{2}}}{\frac{1}{2}} + c = 8\sqrt{x} + c$

**b** $\displaystyle\int \frac{3}{1-2x}\,dx = 3\int \frac{1}{1-2x}\,dx$
$= 3(\frac{1}{-2})\ln|1-2x| + c$
$= -\frac{3}{2}\ln|1-2x| + c$

**c** $\int \sin(4x-5)\,dx = -\frac{1}{4}\cos(4x-5) + c$

**d** $\int e^{4-3x}\,dx = \frac{1}{-3}e^{4-3x} + c$
$= -\frac{1}{3}e^{4-3x} + c$

**3** **a** $\int_{-5}^{-1} \sqrt{1-3x}\,dx = \int_{-5}^{-1}(1-3x)^{\frac{1}{2}}\,dx$
$\displaystyle= \left[\frac{1}{-3} \times \frac{(1-3x)^{\frac{3}{2}}}{\frac{3}{2}}\right]_{-5}^{-1}$
$= -\frac{2}{9}\left[(1-3x)^{\frac{3}{2}}\right]_{-5}^{-1}$
$= -\frac{2}{9}\left(4^{\frac{3}{2}} - 16^{\frac{3}{2}}\right)$
$= -\frac{2}{9}(8-64) = 12\frac{4}{9}$

**b** $\int_0^{\frac{\pi}{2}} \cos\left(\frac{x}{2}\right)\,dx = \left[2\sin\left(\frac{x}{2}\right)\right]_0^{\frac{\pi}{2}}$
$= 2\sin(\frac{\pi}{4}) - 2\sin(0)$
$= 2\left(\frac{1}{\sqrt{2}}\right) - 2(0)$
$= \sqrt{2}$

**4** $y = \sqrt{x^2-4} = (x^2-4)^{\frac{1}{2}}$
$\therefore \displaystyle\frac{dy}{dx} = \frac{1}{2}\left(x^2-4\right)^{-\frac{1}{2}} \times 2x$
$\displaystyle= \frac{x}{(x^2-4)^{\frac{1}{2}}}$
$\displaystyle= \frac{x}{\sqrt{x^2-4}}$
$\therefore \displaystyle\int \frac{x}{\sqrt{x^2-4}}\,dx = \sqrt{x^2-4} + c$

**5** $\int_0^b \cos x\,dx = \frac{1}{\sqrt{2}}, \quad 0 < b < \pi$
$\therefore \quad [\sin x]_0^b = \frac{1}{\sqrt{2}}$
$\therefore \quad \sin b - \sin 0 = \frac{1}{\sqrt{2}}$
$\therefore \quad \sin b = \frac{1}{\sqrt{2}}$
$\therefore \quad b = \frac{\pi}{4}, \frac{3\pi}{4} \quad \{0 < b < \pi\}$

**6** $\int (2-\cos x)^2\,dx$
$= \int (4 - 4\cos x + \cos^2 x)\,dx$
$= \int (4 - 4\cos x + \frac{1}{2} + \frac{1}{2}\cos(2x))\,dx$
$= \frac{9}{2}x - 4\sin x + \frac{1}{4}\sin(2x) + c$

**7** $\displaystyle\frac{d}{dx}(3x^2+x)^3 = 3(3x^2+x)^2(6x+1)$
$\therefore \int 3(3x^2+x)^2(6x+1)\,dx = (3x^2+x)^3 + c$
$\therefore 3\int (3x^2+x)^2(6x+1)\,dx = (3x^2+x)^3 + c$
$\therefore \int (3x^2+x)^2(6x+1)\,dx = \frac{1}{3}(3x^2+x)^3 + c$

**8** **a** $\int_1^4 (f(x)+1)\,dx$
$= \int_1^4 f(x)\,dx + \int_1^4 1\,dx$
$= 3 + [x]_1^4$
$= 3 + (4-1) = 6$

**b** $\int_1^2 f(x)\,dx - \int_4^2 f(x)\,dx$
$= \int_1^2 f(x)\,dx + \int_2^4 f(x)\,dx$
$= \int_1^4 f(x)\,dx$
$= 3$

**9** If $u(x) = \sin x$, $\dfrac{du}{dx} = \cos x$

$$\therefore \int e^{\sin x} \cos x \, dx = \int e^u \frac{du}{dx} \, dx$$
$$= \int e^u \, du$$
$$= e^u + c$$
$$= e^{\sin x} + c$$

**10** Given: $f''(x) = 2\sin(2x)$, $f'(\frac{\pi}{2}) = 0$, and $f(0) = 3$

Now $f'(x) = \int 2\sin(2x) \, dx$
$= -\cos(2x) + c$

But $f'(\frac{\pi}{2}) = 0$, so $-\cos(\pi) + c = 0$
$\therefore \; -(-1) + c = 0$
$\therefore \; c = -1$

$\therefore \; f'(x) = -\cos(2x) - 1$

$\therefore \; f(x) = \int (-\cos(2x) - 1) \, dx$
$= -\frac{1}{2}\sin(2x) - x + k$

But $f(0) = 3$, so $-\frac{1}{2}\sin(0) - 0 + k = 3$
$\therefore \; k = 3$

so $f(x) = -\frac{1}{2}\sin(2x) - x + 3$

$\therefore \; f(\frac{\pi}{2}) = -\frac{1}{2}\sin(\pi) - \frac{\pi}{2} + 3$
$= 3 - \frac{\pi}{2}$

**11** $\int_0^{\frac{\pi}{6}} \sin^2\left(\frac{x}{2}\right) dx$
$= \int_0^{\frac{\pi}{6}} \left(\frac{1}{2} - \frac{1}{2}\cos x\right) dx$
$= \left[\frac{1}{2}x - \frac{1}{2}\sin x\right]_0^{\frac{\pi}{6}}$
$= \frac{\pi}{12} - \frac{1}{2}\left(\frac{1}{2}\right) - 0 + 0$
$= \frac{\pi}{12} - \frac{1}{4}$

**12** **a** If $u(x) = x^2 + 1$, then $\dfrac{du}{dx} = 2x$

$$\therefore \int 2x(x^2+1)^3 \, dx = \int \frac{du}{dx} u^3 \, dx$$
$$= \int u^3 \, du$$
$$= \frac{u^4}{4} + c$$
$$= \tfrac{1}{4}(x^2+1)^4 + c$$

**b** **i** $\int_0^1 2x(x^2+1)^3 \, dx$
$= \left[\dfrac{(x^2+1)^4}{4}\right]_0^1$ {using **a**}
$= \dfrac{(1^2+1)^4}{4} - \dfrac{(0+1)^4}{4}$
$= \frac{16}{4} - \frac{1}{4}$
$= \frac{15}{4}$

**ii** $\int_{-1}^2 -x(1+x^2)^3 \, dx$
$= \int_{-1}^2 -\frac{1}{2} \times 2x(1+x^2)^3 \, dx$
$= -\frac{1}{2}\int_{-1}^2 2x(x^2+1)^3 \, dx$
$= -\frac{1}{2}\left[\dfrac{(x^2+1)^4}{4}\right]_{-1}^2$
$= -\frac{1}{2}\left[\dfrac{(4+1)^4}{4} - \dfrac{(1+1)^4}{4}\right]$
$= -\frac{1}{2}\left(\frac{625}{4} - \frac{16}{4}\right)$
$= -\frac{609}{8}$

## REVIEW SET 18B

**1** **a**

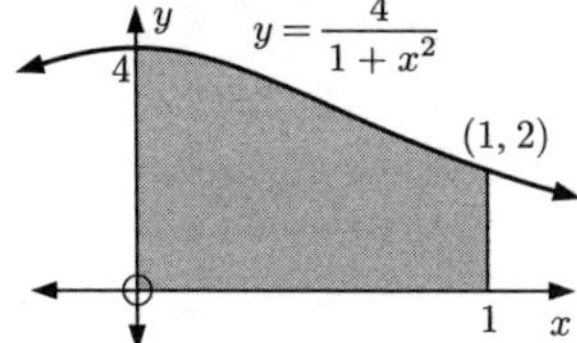


lower rectangles      upper rectangles

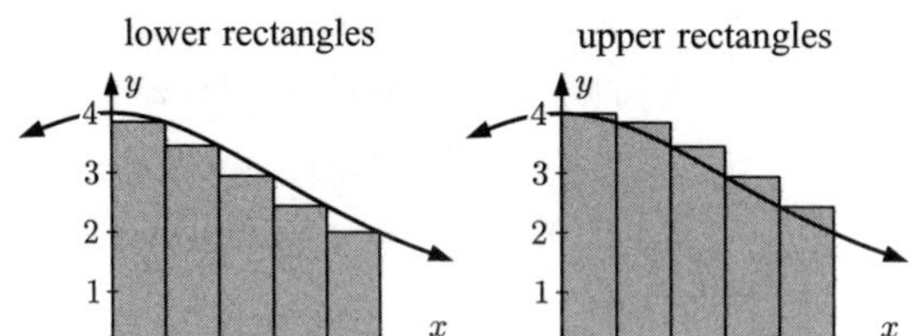


**b**

| $n$ | $A_L$ | $A_U$ |
|---|---|---|
| 5 | 2.9349 | 3.3349 |
| 50 | 3.1215 | 3.1615 |
| 100 | 3.1316 | 3.1516 |
| 500 | 3.1396 | 3.1436 |

**c** $\displaystyle\int_0^1 \frac{4}{1+x^2}\,dx \approx 3.1416$

(the average of $A_L$ and $A_U$ for $n = 500$). This value agrees with $\pi$ to 4 decimal places.

**2** **a**
$$\frac{dy}{dx} = (x^2-1)^2$$
$$\therefore\quad y = \int (x^2-1)^2\,dx$$
$$= \int (x^4 - 2x^2 + 1)\,dx$$
$$= \tfrac{1}{5}x^5 - \tfrac{2}{3}x^3 + x + c$$

**b**
$$\frac{dy}{dx} = 400 - 20e^{-\frac{x}{2}}$$
$$\therefore\quad y = \int (400 - 20e^{-\frac{x}{2}})\,dx$$
$$= 400x - \frac{20e^{-\frac{x}{2}}}{-\frac{1}{2}} + c$$
$$= 400x + 40e^{-\frac{x}{2}} + c$$

**3** Using technology: **a** $\int_{-2}^{0} 4e^{-x^2}\,dx \approx 3.528$ **b** $\displaystyle\int_0^1 \frac{10x}{\sqrt{3x+1}}\,dx \approx 2.963$

**4**
$$\frac{d}{dx}(\ln x)^2 = 2(\ln x)^1\left(\frac{1}{x}\right)$$
$$= \frac{2\ln x}{x}$$
$$\therefore\quad \int \frac{2\ln x}{x}\,dx = (\ln x)^2 + c$$
$$\therefore\quad \int \frac{\ln x}{x}\,dx = \tfrac{1}{2}(\ln x)^2 + c$$

**5** Given: $f''(x) = 18x + 10,\quad f(0) = -1,\quad f(1) = 13$
$$f'(x) = \int (18x+10)\,dx$$
$$= 9x^2 + 10x + c$$
$$\therefore\quad f(x) = 3x^3 + 5x^2 + cx + d$$
But $f(0) = -1$ so $d = -1$
$$\therefore\quad f(x) = 3x^3 + 5x^2 + cx - 1$$
And $f(1) = 13$ so $3 + 5 + c - 1 = 13$
$$\therefore\quad c + 7 = 13$$
$$\therefore\quad c = 6$$
$$\therefore\quad f(x) = 3x^3 + 5x^2 + 6x - 1$$

**6**
$$\int_0^a e^{1-2x}\,dx = \frac{e}{4}$$
$$\therefore\quad \left[\tfrac{1}{-2}e^{1-2x}\right]_0^a = \frac{e}{4}$$
$$\therefore\quad (-\tfrac{1}{2}e^{1-2a}) - (-\tfrac{1}{2}e^1) = \frac{e}{4}$$
$$\therefore\quad -\tfrac{1}{2}e^{1-2a} + \frac{e}{2} = \frac{e}{4}$$
$$\therefore\quad \tfrac{1}{2}e^{1-2a} = \frac{e}{4}$$
$$\therefore\quad e^{1-2a} = \frac{e}{2}$$
$$\therefore\quad 1 - 2a = \ln\left(\frac{e}{2}\right) = \ln e - \ln 2$$
$$\therefore\quad 1 - 2a = 1 - \ln 2$$
$$\therefore\quad 2a = \ln 2$$
$$\therefore\quad a = \tfrac{1}{2}\ln 2$$
$$\therefore\quad a = \ln 2^{\frac{1}{2}}$$
$$\therefore\quad a = \ln\sqrt{2}$$

**7** Using technology: **a** $\displaystyle\int_3^4 \frac{x}{\sqrt{2x+1}}\,dx \approx 1.236\,17$ **b** $\int_0^1 x^2 e^{x+1}\,dx \approx 1.952\,49$

**8** **a** $f''(x) = 3x^2 + 2x$

$\therefore\ f'(x) = \dfrac{3x^3}{3} + \dfrac{2x^2}{2} + c$

$= x^3 + x^2 + c$

$\therefore\ f(x) = \dfrac{x^4}{4} + \dfrac{x^3}{3} + cx + d$

But $f(0) = 3$ so $d = 3$

$\therefore\ f(x) = \dfrac{x^4}{4} + \dfrac{x^3}{3} + cx + 3$

Also, $f(2) = 3$ so $4 + \frac{8}{3} + 2c + 3 = 3$

$\therefore\ \frac{20}{3} = -2c$

$\therefore\ c = -\frac{10}{3}$

$\therefore\ f(x) = \frac{1}{4}x^4 + \frac{1}{3}x^3 - \frac{10}{3}x + 3$

**b** Now $f'(2) = 2^3 + 2^2 - \frac{10}{3}$

$= 12 - \frac{10}{3}$

$= \frac{26}{3}$

$\therefore$ the normal has gradient $-\frac{3}{26}$

$\therefore$ equation is $\dfrac{y-3}{x-2} = -\frac{3}{26}$

$\therefore\ y - 3 = -\frac{3}{26}(x-2)$

$\therefore\ y = -\frac{3}{26}x + \frac{6}{26} + 3$

or $3x + 26y = 84$

**9** **a** $(e^x + 2)^3$

$= (e^x)^3 + 3(e^x)^2(2) + 3(e^x)(2)^2 + (2)^3$

$= e^{3x} + 6e^{2x} + 12e^x + 8$

**b** $\int_0^1 (e^x+2)^3\,dx$

$= \left[\frac{1}{3}e^{3x} + 3e^{2x} + 12e^x + 8x\right]_0^1$

$= \left(\frac{1}{3}e^3 + 3e^2 + 12e + 8\right) - \left(\frac{1}{3} + 3 + 12\right)$

$= \frac{1}{3}e^3 + 3e^2 + 12e - 7\frac{1}{3}$ $(\approx 54.1)$

**c** Using technology, $\int_0^1 (e^x+2)^3\,dx \approx 54.1$

**10** $\int \sin^5 x \cos x\,dx$

$= \int (\sin x)^5 \cos x\,dx$

$= \int u^5 \frac{du}{dx}\,dx$ $\{u = \sin x,\ \frac{du}{dx} = \cos x\}$

$= \int u^5\,du$

$= \dfrac{u^6}{6} + c$

$= \dfrac{\sin^6 x}{6} + c$

$\therefore\ \displaystyle\int_{\frac{\pi}{4}}^{\frac{\pi}{3}} \sin^5 x \cos x\,dx$

$= \left[\dfrac{\sin^6 x}{6}\right]_{\frac{\pi}{4}}^{\frac{\pi}{3}}$

$= \frac{1}{6}\left(\left(\frac{\sqrt{3}}{2}\right)^6 - \left(\frac{1}{\sqrt{2}}\right)^6\right)$

$= \frac{19}{384}$

## REVIEW SET 18C

**1** **a** $\displaystyle\int \left(2e^{-x} - \frac{1}{x} + 3\right)dx$

$= -2e^{-x} - \ln|x| + 3x + c$

**b** $\displaystyle\int \left(\sqrt{x} - \frac{1}{\sqrt{x}}\right)^2 dx$

$= \displaystyle\int \left(x - 2 + \frac{1}{x}\right)dx$

$= \frac{1}{2}x^2 - 2x + \ln|x| + c$

**c** $\int (3 + e^{2x-1})^2\,dx$

$= \int (9 + 6e^{2x-1} + e^{4x-2})\,dx$

$= 9x + 3e^{2x-1} + \frac{1}{4}e^{4x-2} + c$

**2** $f'(x) = x^2 - 3x + 2$

$\therefore \quad f(x) = \dfrac{x^3}{3} - \dfrac{3x^2}{2} + 2x + c$

But $f(1) = 3$

so $\frac{1}{3} - \frac{3}{2} + 2 + c = 3$

$\therefore \quad c = 1 - \frac{1}{3} + 1\frac{1}{2}$

$\therefore \quad c = 2\frac{1}{6}$

$\therefore \quad f(x) = \frac{1}{3}x^3 - \frac{3}{2}x^2 + 2x + 2\frac{1}{6}$

**3** $\displaystyle\int_2^3 \frac{1}{\sqrt{3x-4}}\,dx$

$= \int_2^3 (3x-4)^{-\frac{1}{2}}\,dx$

$= \left[\frac{1}{3}\dfrac{(3x-4)^{\frac{1}{2}}}{\frac{1}{2}}\right]_2^3$

$= \left[\frac{2}{3}\sqrt{3x-4}\right]_2^3$

$= \frac{2}{3}\sqrt{5} - \frac{2}{3}\sqrt{2}$

$= \frac{2}{3}(\sqrt{5} - \sqrt{2})$

**4** $\int_0^{\frac{\pi}{3}} \cos^2\left(\frac{x}{2}\right)\,dx$

$= \int_0^{\frac{\pi}{3}} \left(\frac{1}{2} + \frac{1}{2}\cos x\right)\,dx$

$= \left[\frac{1}{2}x + \frac{1}{2}\sin x\right]_0^{\frac{\pi}{3}}$

$= \frac{\pi}{6} + \frac{1}{2}\left(\frac{\sqrt{3}}{2}\right) - 0 - 0$

$= \frac{\pi}{6} + \frac{\sqrt{3}}{4}$

**5** $\dfrac{d}{dx}(e^{-2x}\sin x) = -2e^{-2x}\sin x + e^{-2x}\cos x$ {product rule}

$= e^{-2x}(\cos x - 2\sin x)$

$\therefore \quad \int_0^{\frac{\pi}{2}} e^{-2x}(\cos x - 2\sin x)\,dx = \left[e^{-2x}\sin x\right]_0^{\frac{\pi}{2}}$

$= e^{-\pi}(1) - e^0(0) = e^{-\pi}$

**6** If $n \neq -1$, $\int (2x+3)^n\,dx = \frac{1}{2}\dfrac{(2x+3)^{n+1}}{n+1} + c = \dfrac{1}{2(n+1)}(2x+3)^{n+1} + c$

If $n = -1$, $\int (2x+3)^{-1}\,dx = \displaystyle\int \frac{1}{2x+3}\,dx = \frac{1}{2}\ln|2x+3| + c$

So, $\int (2x+3)^n\,dx = \begin{cases} \dfrac{1}{2(n+1)}(2x+3)^{n+1} + c & \text{if } n \neq -1 \\ \frac{1}{2}\ln|2x+3| + c & \text{if } n = -1 \end{cases}$

**7** $f'(x) = 2\sqrt{x} + \dfrac{a}{\sqrt{x}}$

$= 2x^{\frac{1}{2}} + ax^{-\frac{1}{2}}$

$\therefore \quad f(x) = \frac{4}{3}x^{\frac{3}{2}} + 2ax^{\frac{1}{2}} + c$

$= \dfrac{4x\sqrt{x}}{3} + 2a\sqrt{x} + c$

Now $f(0) = 2$ so $c = 2$

$\therefore \quad f(x) = \dfrac{4x\sqrt{x}}{3} + 2a\sqrt{x} + 2$

Also, $f(1) = 4$ so $\frac{4}{3} + 2a + 2 = 4$

$\therefore \quad 2a = \frac{2}{3}$

$\therefore \quad a = \frac{1}{3}$

$\therefore \quad f'(x) = 2\sqrt{x} + \dfrac{1}{3\sqrt{x}} = \dfrac{6x+1}{3\sqrt{x}}$

Now $f(x)$ is only defined for $x > 0$,
so $f'(x) > 0$ for all $x$ in the domain.
$\therefore$ the function has no stationary points.

**8** $\displaystyle\int_a^{2a} (x^2 + ax + 2)\,dx = \frac{73a}{2}$

$\therefore \quad \left[\dfrac{x^3}{3} + \dfrac{ax^2}{2} + 2x\right]_a^{2a} = \dfrac{73a}{2}$

$\therefore \quad \left(\dfrac{8a^3}{3} + \dfrac{a}{2}(4a^2) + 4a\right) - \left(\dfrac{a^3}{3} + \dfrac{a^3}{2} + 2a\right) = \dfrac{73a}{2}$

$\dfrac{8a^3}{3} + 2a^3 + 4a - \dfrac{a^3}{3} - \dfrac{a^3}{2} - 2a = \dfrac{73a}{2}$

$\therefore \quad 16a^3 + 12a^3 + 24a - 2a^3 - 3a^3 - 12a = 219a$

$\therefore \quad 23a^3 - 207a = 0$

$\therefore \quad 23a(a^2 - 9) = 0$

$\therefore \quad 23a(a+3)(a-3) = 0$

$\therefore \quad a = 0$ or $a = \pm 3$

**9** **a**

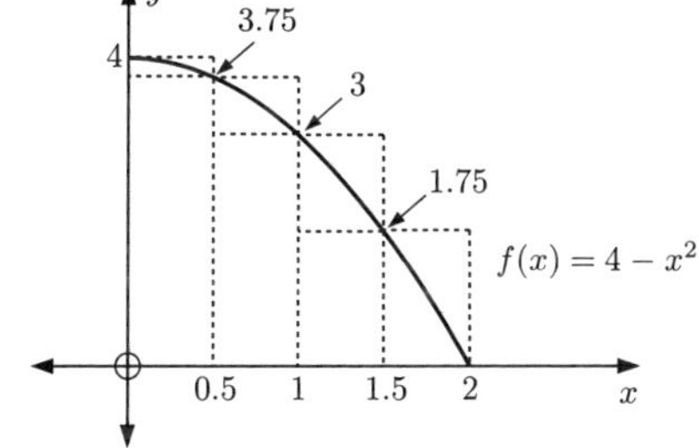


$$\begin{aligned} A_U &= 0.5\,[f(0)+f(0.5)+f(1)+f(1.5)] \\ &= 0.5(4+3.75+3+1.75) \\ &= 6.25 \end{aligned}$$

$$\begin{aligned} A_L &= 0.5\,[f(0.5)+f(1)+f(1.5)+f(2)] \\ &= 0.5(3.75+3+1.75+0) \\ &= 4.25 \end{aligned}$$

$\therefore\quad 4.25 < \int_0^2 (4-x^2)\,dx < 6.25$

$\therefore\quad A = 4.25 = \frac{17}{4},\quad B = 6.25 = \frac{25}{4}$

**b** an estimate of $\int_0^2 (4-x^2)\,dx \approx \dfrac{A+B}{2} \approx \frac{42}{8} \approx \frac{21}{4}$

**10** **a**

$$\begin{aligned} &\int \frac{2x}{\sqrt{x^2-5}}\,dx \\ &= \int 2x(x^2-5)^{-\frac{1}{2}}\,dx \\ &= \int u^{-\frac{1}{2}}\frac{du}{dx}\,dx \quad \left\{u = x^2-5,\quad \frac{du}{dx} = 2x\right\} \\ &= \int u^{-\frac{1}{2}}\,du \\ &= \frac{u^{\frac{1}{2}}}{\frac{1}{2}} + c \\ &= 2\sqrt{u} + c \\ &= 2\sqrt{x^2-5} + c \end{aligned}$$

**b**

$$\begin{aligned} &\int 3x^2\sqrt{x^3-1}\,dx \\ &= \int 3x^2(x^3-1)^{\frac{1}{2}}\,dx \\ &= \int u^{\frac{1}{2}}\frac{du}{dx}\,dx \quad \left\{u = x^3-1,\quad \frac{du}{dx} = 3x^2\right\} \\ &= \int u^{\frac{1}{2}}\,du \\ &= \frac{u^{\frac{3}{2}}}{\frac{3}{2}} + c \\ &= \tfrac{2}{3}u\sqrt{u} + c \\ &= \tfrac{2}{3}(x^3-1)\sqrt{x^3-1} + c \end{aligned}$$

$$\begin{aligned} \therefore\quad &\int_1^2 3x^2\sqrt{x^3-1}\,dx \\ &= \tfrac{2}{3}\left[(x^3-1)\sqrt{x^3-1}\right]_1^2 \\ &= \tfrac{2}{3}\left((8-1)\sqrt{8-1} - 0\right) \\ &= \tfrac{2}{3} \times 7\sqrt{7} \\ &= \tfrac{14\sqrt{7}}{3} \end{aligned}$$

# Chapter 19

## APPLICATIONS OF INTEGRATION

### EXERCISE 19A

**1** **a**

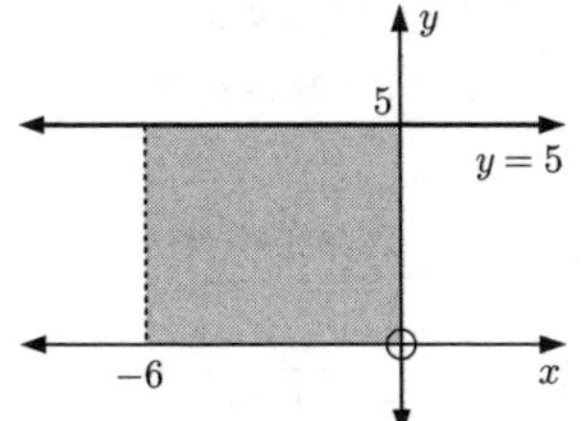


**i** $\text{Area} = 5 \times 6$
$= 30 \text{ units}^2$

**ii** $\text{Area} = \int_{-6}^{0} 5\, dx$
$= [5x]_{-6}^{0}$
$= 5(0) - 5(-6)$
$= 30 \text{ units}^2$

**b**

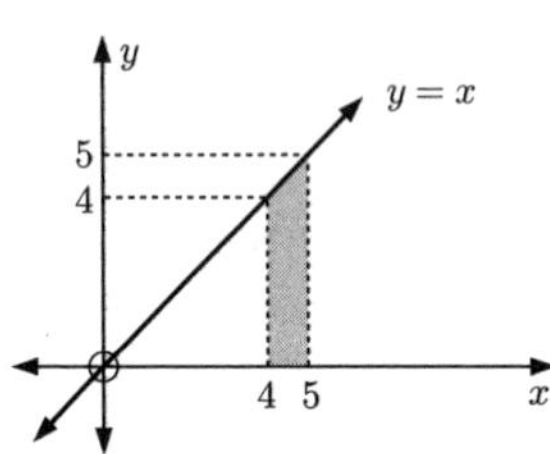


**i** $\text{Area} = \left(\dfrac{4+5}{2}\right) \times 1$
$= \frac{9}{2} \text{ units}^2$

**ii** $\text{Area} = \int_{4}^{5} x\, dx$
$= \left[\frac{1}{2}x^2\right]_4^5$
$= \frac{1}{2}(25) - \frac{1}{2}(16)$
$= \frac{9}{2} \text{ units}^2$

**c**

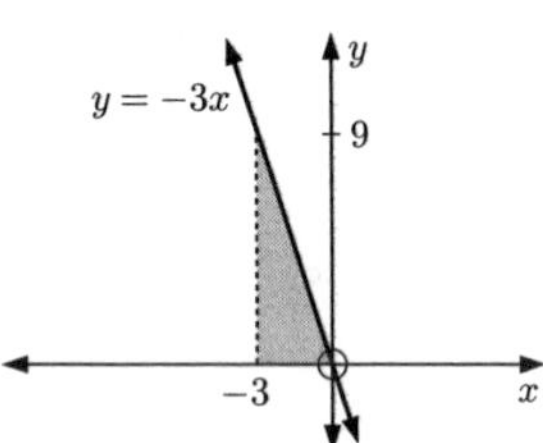


**i** $\text{Area} = \frac{1}{2} \times 3 \times 9$
$= \frac{27}{2} \text{ units}^2$

**ii** $\text{Area} = \int_{-3}^{0} (-3x)\, dx$
$= \left[-\frac{3}{2}x^2\right]_{-3}^{0}$
$= -\frac{3}{2}(0) - (-\frac{3}{2})(9)$
$= \frac{27}{2} \text{ units}^2$

**d**

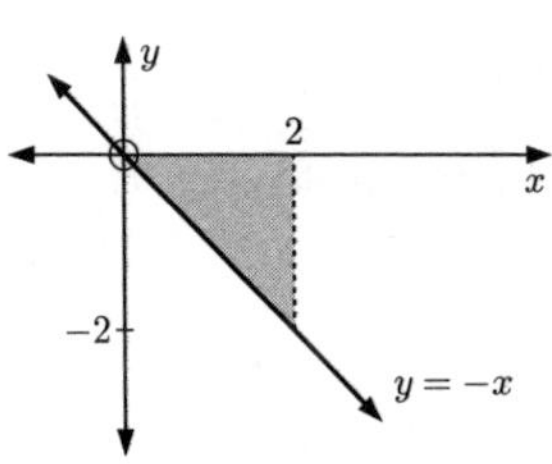


**i** $\text{Area} = \frac{1}{2} \times 2 \times 2$
$= 2 \text{ units}^2$

**ii** $\text{Area} = -\int_0^2 -x\, dx$
$= -\left[-\frac{1}{2}x^2\right]_0^2$
$= -\left(-\frac{1}{2}(4) - (-\frac{1}{2})(0)\right)$
$= 2 \text{ units}^2$

**2** **a**

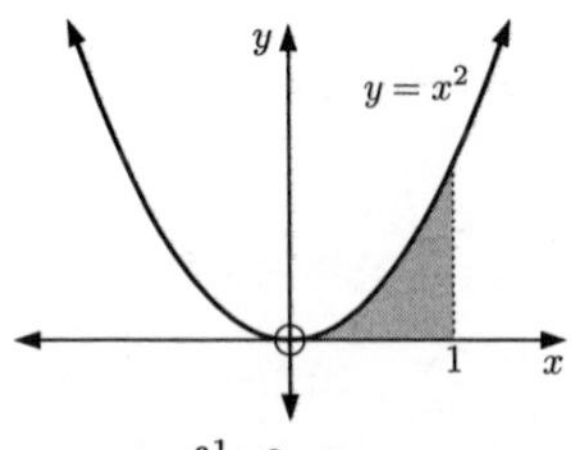


$\text{Area} = \int_0^1 x^2\, dx$
$= \left[\dfrac{x^3}{3}\right]_0^1$
$= \frac{1}{3} - 0$
$= \frac{1}{3} \text{ units}^2$

**b**

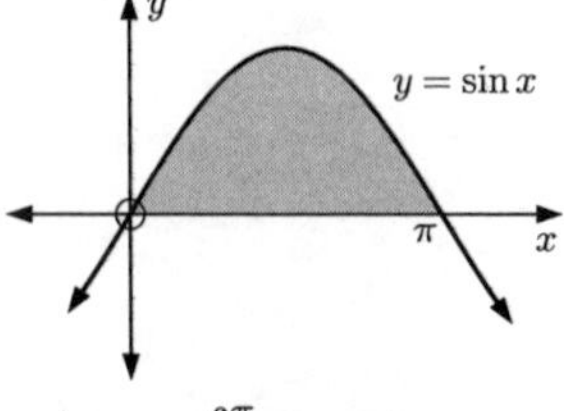


$\text{Area} = \int_0^{\pi} \sin x\, dx$
$= [-\cos x]_0^{\pi}$
$= -\cos \pi - (-\cos 0)$
$= 2 \text{ units}^2$

**c**

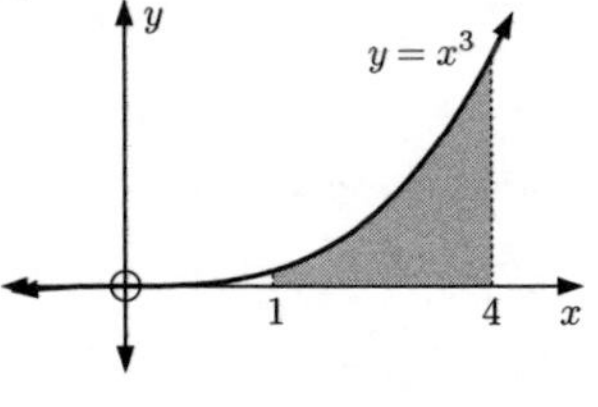


$\text{Area} = \int_1^4 x^3\, dx$
$= \left[\dfrac{x^4}{4}\right]_1^4$
$= \frac{256}{4} - \frac{1}{4}$
$= 63\frac{3}{4} \text{ units}^2$

**d**

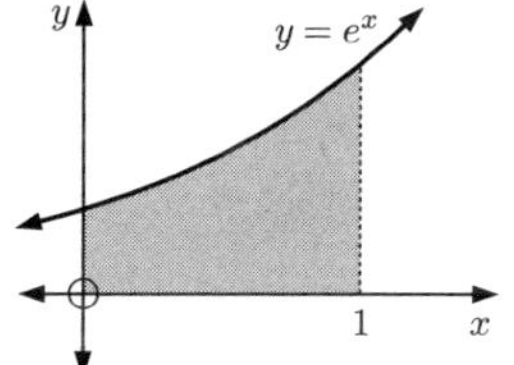


$\text{Area} = \int_0^1 e^x \, dx$

$= [e^x]_0^1$

$= (e - 1) \text{ units}^2$

**e** The graph cuts the $x$-axis at $y = 0$.

$\therefore \quad 6 + x - x^2 = 0$

$\therefore \quad (3 - x)(2 + x) = 0$

$\therefore \quad x = 3 \text{ or } -2$

The $x$-intercepts are 3 and $-2$.

$\text{Area} = \int_{-2}^{3} (6 + x - x^2) \, dx$

$= \left[6x + \dfrac{x^2}{2} - \dfrac{x^3}{3}\right]_{-2}^{3}$

$= (18 + \frac{9}{2} - 9) - (-12 + 2 + \frac{8}{3})$

$= 20\frac{5}{6} \text{ units}^2$

**f**

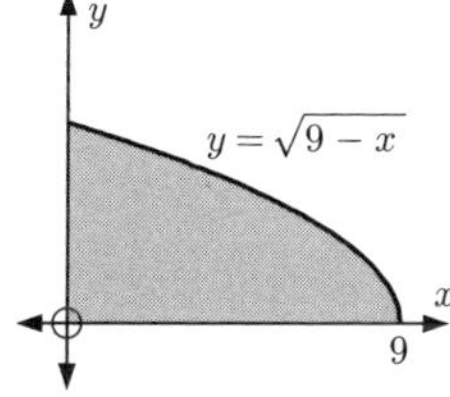


$\text{Area} = \int_0^9 (9 - x)^{\frac{1}{2}} \, dx$

$= \left[\left(\frac{1}{-1}\right) \dfrac{(9 - x)^{\frac{3}{2}}}{\frac{3}{2}}\right]_0^9$

$= -\frac{2}{3}\left[(9 - x)^{\frac{3}{2}}\right]_0^9$

$= -\frac{2}{3}(0 - 27)$

$= 18 \text{ units}^2$

**g**

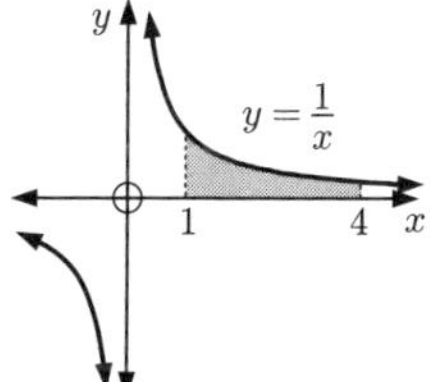


$\text{Area} = \displaystyle\int_1^4 \frac{1}{x} \, dx$

$= [\ln x]_1^4 \quad \{x > 0\}$

$= \ln 4 - \ln 1$

$= \ln 4 - 0$

$= \ln 4 \text{ units}^2$

**h**

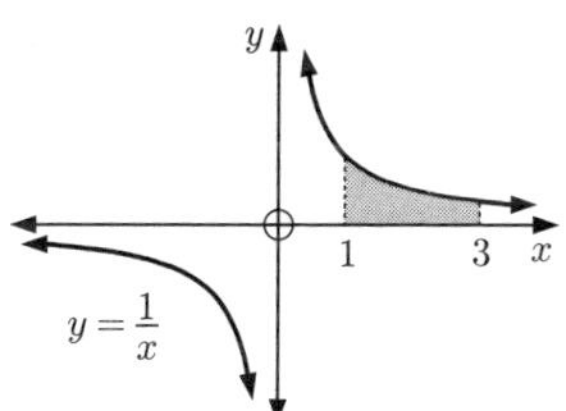


$\text{Area} = \displaystyle\int_1^3 \frac{1}{x} \, dx$

$= [\ln x]_1^3 \quad \{x > 0\}$

$= \ln 3 - \ln 1$

$= \ln 3 - 0$

$= \ln 3 \text{ units}^2$

**i**

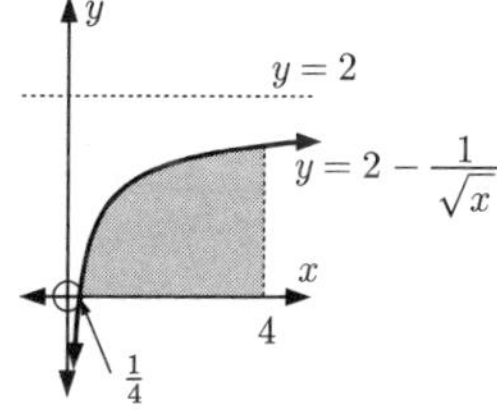


$\text{Area} = \displaystyle\int_{\frac{1}{4}}^{4} \left(2 - \frac{1}{\sqrt{x}}\right) dx$

$= \displaystyle\int_{\frac{1}{4}}^{4} (2 - x^{-\frac{1}{2}}) \, dx$

$= \left[2x - \dfrac{x^{\frac{1}{2}}}{\frac{1}{2}}\right]_{\frac{1}{4}}^{4}$

$= \left[2x - 2\sqrt{x}\right]_{\frac{1}{4}}^{4}$

$= (8 - 4) - (\frac{1}{2} - 1)$

$= 4\frac{1}{2} \text{ units}^2$

**j**

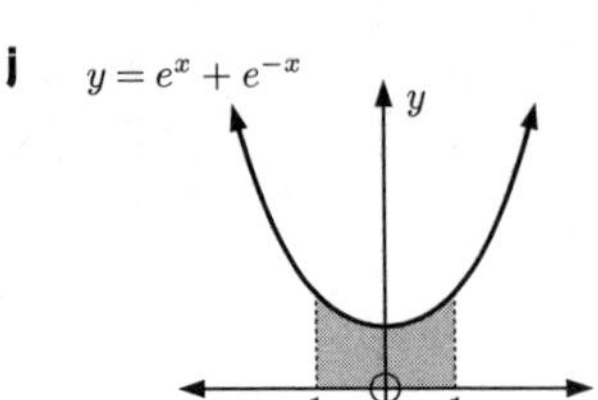


$$\begin{aligned}\text{Area} &= \int_{-1}^{1} (e^x + e^{-x})\,dx \\ &= \left[e^x - e^{-x}\right]_{-1}^{1} \\ &= (e - e^{-1}) - (e^{-1} - e) \\ &= \left(2e - \frac{2}{e}\right) \text{ units}^2\end{aligned}$$

**3**

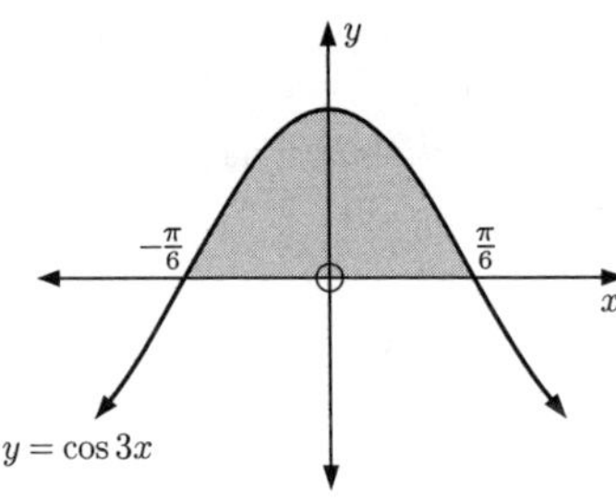


$y = \cos 3x$ has zeros at $\begin{cases} \frac{\pi}{6} \\ -\frac{\pi}{6} \end{cases} + \frac{2k\pi}{3}$, $k$ an integer

$$\begin{aligned}\therefore\ \text{area} &= \int_{-\frac{\pi}{6}}^{\frac{\pi}{6}} \cos 3x\,dx \\ &= \left[\tfrac{1}{3}\sin 3x\right]_{-\frac{\pi}{6}}^{\frac{\pi}{6}} \\ &= \tfrac{1}{3}\left(\sin(\tfrac{\pi}{2}) - \sin(-\tfrac{\pi}{2})\right) \\ &= \tfrac{1}{3}\left(1 - (-1)\right) \\ &= \tfrac{2}{3} \text{ units}^2\end{aligned}$$

**4** **a**

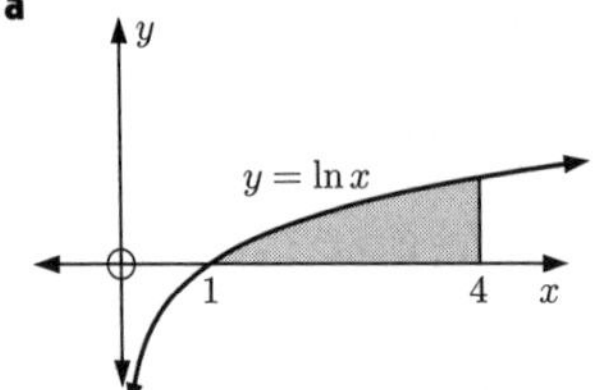


Area $= \int_1^4 \ln x\,dx$
$\approx 2.55$ units$^2$

**b**

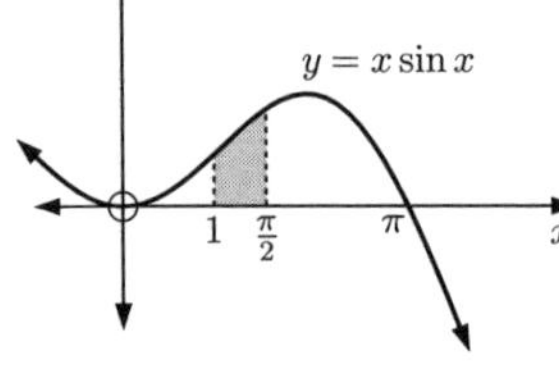


Area $= \int_1^{\frac{\pi}{2}} x\sin x\,dx$
$\approx 0.699$ units$^2$

**c**

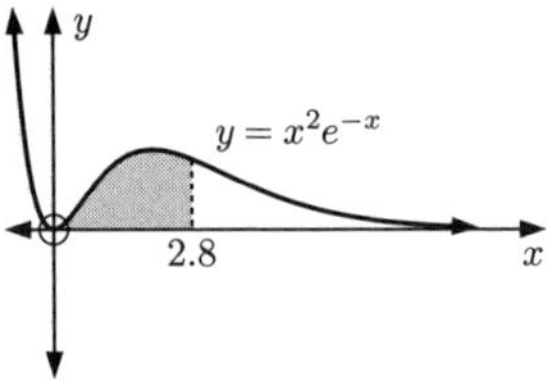

Area $= \int_0^{2.8} x^2 e^{-x}\,dx$
$\approx 1.06$ units$^2$

## EXERCISE 19B

**1** **a** The curve cuts the $x$-axis when $y = 0$.

$\therefore\ x^2 + x - 2 = 0$
$\therefore\ (x+2)(x-1) = 0$
$\therefore\ x = -2$ or $1$
$\therefore$ the $x$-intercepts are $-2$ and $1$

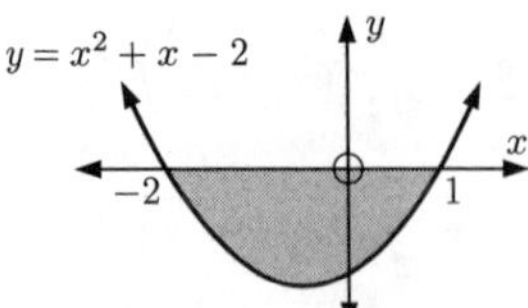


$$\begin{aligned}\text{Area} &= \int_{-2}^{1} [0 - (x^2 + x - 2)]\,dx \\ &= \int_{-2}^{1} (-x^2 - x + 2)\,dx \\ &= \left[-\frac{x^3}{3} - \frac{x^2}{2} + 2x\right]_{-2}^{1} \\ &= (-\tfrac{1}{3} - \tfrac{1}{2} + 2) - (\tfrac{8}{3} - 2 - 4) \\ &= 4\tfrac{1}{2} \text{ units}^2\end{aligned}$$

**b** The curve cuts the $x$-axis at $(0, 0)$.

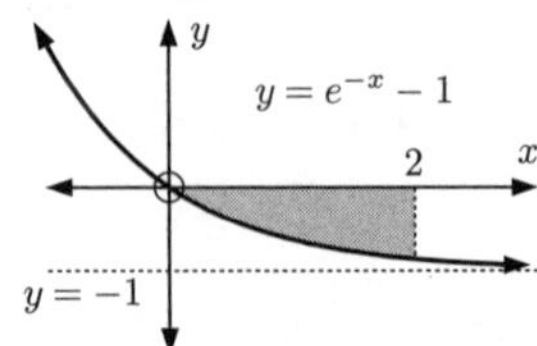


$$\begin{aligned}\text{Area} &= \int_0^2 [0 - (e^{-x} - 1)]\,dx \\ &= \int_0^2 (1 - e^{-x})\,dx \\ &= \left[x + e^{-x}\right]_0^2 \\ &= \left(2 + e^{-2}\right) - \left(0 + e^0\right) \\ &= (1 + e^{-2}) \text{ units}^2\end{aligned}$$

**c** The curve cuts the $x$-axis when $y = 0$.

$\therefore \quad 3x^2 - 8x + 4 = 0$

$\therefore \quad (3x-2)(x-2) = 0$

$\therefore \quad x = 2$ or $\frac{2}{3}$

$\therefore$ the $x$-intercepts are 2 and $\frac{2}{3}$.

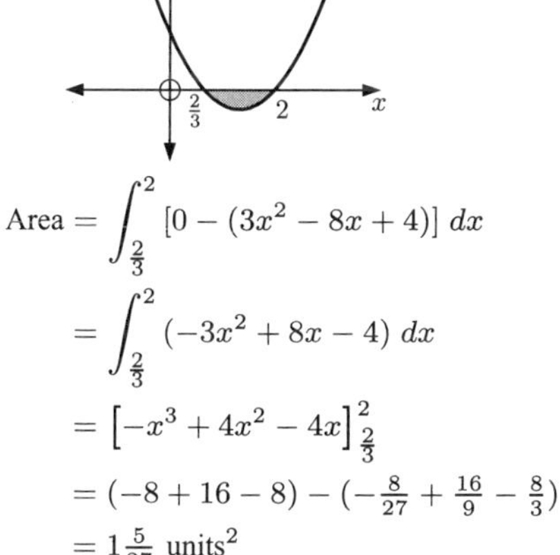


$$\text{Area} = \int_{\frac{2}{3}}^{2} [0 - (3x^2 - 8x + 4)]\, dx$$

$$= \int_{\frac{2}{3}}^{2} (-3x^2 + 8x - 4)\, dx$$

$$= \left[-x^3 + 4x^2 - 4x\right]_{\frac{2}{3}}^{2}$$

$$= (-8 + 16 - 8) - (-\tfrac{8}{27} + \tfrac{16}{9} - \tfrac{8}{3})$$

$$= 1\tfrac{5}{27} \text{ units}^2$$

**d**

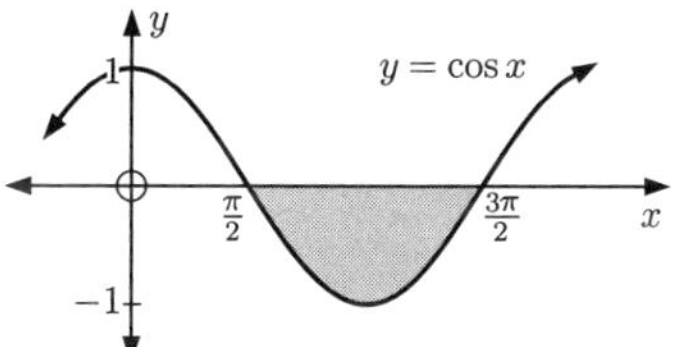


$$\text{Area} = \int_{\frac{\pi}{2}}^{\frac{3\pi}{2}} [0 - \cos x]\, dx$$

$$= \int_{\frac{\pi}{2}}^{\frac{3\pi}{2}} -\cos x \, dx$$

$$= [-\sin x]_{\frac{\pi}{2}}^{\frac{3\pi}{2}}$$

$$= -\sin(\tfrac{3\pi}{2}) - \left(-\sin(\tfrac{\pi}{2})\right)$$

$$= -(-1) - (-1)$$

$$= 2 \text{ units}^2$$

**e** The curve cuts the $x$-axis when $y = 0$.

$\therefore \quad x^3 - 4x = 0$

$\therefore \quad x(x^2 - 4) = 0$

$\therefore \quad x(x+2)(x-2) = 0$

$\therefore$ the $x$-intercepts are 0 and $\pm 2$

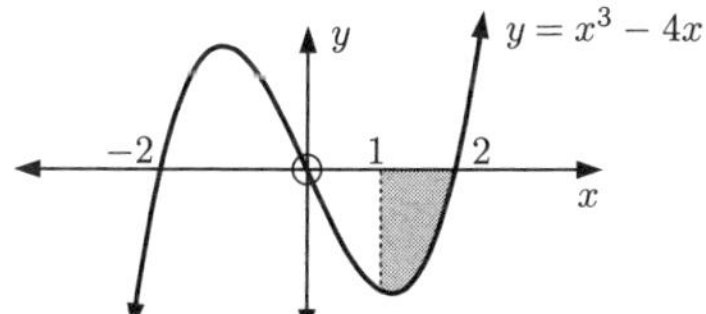


$$\text{Area} = \int_1^2 [0 - (x^3 - 4x)]\, dx$$

$$= \int_1^2 (-x^3 + 4x)\, dx$$

$$= \left[-\frac{x^4}{4} + 2x^2\right]_1^2$$

$$= (-4 + 8) - (-\tfrac{1}{4} + 2)$$

$$= 2\tfrac{1}{4} \text{ units}^2$$

**f** $y = \sin x - 1$ is the graph of $\sin x$ translated vertically $-1$ unit downwards.

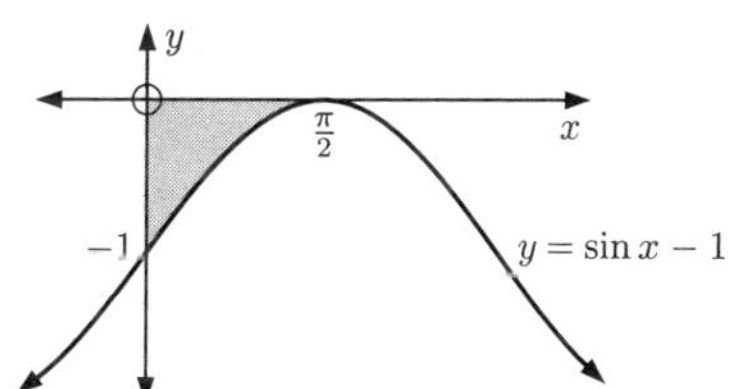


$$\text{Area} = \int_0^{\frac{\pi}{2}} [0 - (\sin x - 1)]\, dx$$

$$= \int_0^{\frac{\pi}{2}} (1 - \sin x)\, dx$$

$$= [x + \cos x]_0^{\frac{\pi}{2}}$$

$$= (\tfrac{\pi}{2} + \cos\tfrac{\pi}{2}) - (0 + \cos 0)$$

$$= (\tfrac{\pi}{2} - 1) \text{ units}^2$$

**g** The curve cuts the $x$-axis when $y = 0$.

$\therefore \quad \sin^2 x = 0$

$\therefore \quad \sin x = 0$

$\therefore \quad x = 0 + k\pi, \quad k$ an integer

So, the first two non-negative $x$-intercepts are 0, $\pi$.

$$\text{Area} = \int_0^{\pi} \left[\sin^2 x - 0\right] dx$$

$$= \int_0^{\pi} \left(\tfrac{1}{2} - \tfrac{1}{2}\cos 2x\right) dx$$

$$= \left[\tfrac{1}{2}x - \tfrac{1}{4}\sin 2x\right]_0^{\pi}$$

$$= (\tfrac{1}{2}(\pi) - \tfrac{1}{4}\sin(2\pi)) - (\tfrac{1}{2}(0) - \tfrac{1}{4}\sin(0))$$

$$= \tfrac{\pi}{2} \text{ units}^2$$

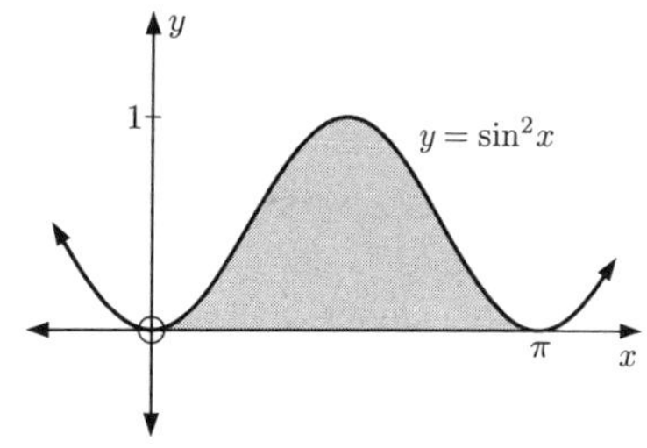

**2** $y = x^2 - 2x$ meets $y = 3$

when $x^2 - 2x = 3$

$\therefore \quad x^2 - 2x - 3 = 0$

$\therefore \quad (x-3)(x+1) = 0$

$\therefore \quad x = 3$ or $-1$

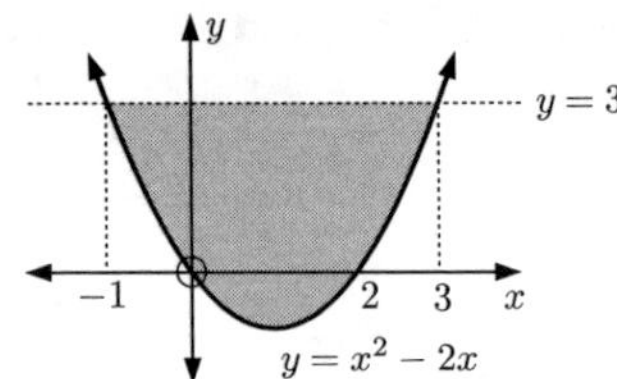


$$\begin{aligned} A &= \int_{-1}^{3} [3 - (x^2 - 2x)]\, dx \\ &= \int_{-1}^{3} (3 + 2x - x^2)\, dx \\ &= \left[3x + x^2 - \frac{x^3}{3}\right]_{-1}^{3} \\ &= (9 + 9 - 9) - (-3 + 1 + \tfrac{1}{3}) \\ &= 10\tfrac{2}{3} \text{ units}^2 \end{aligned}$$

**3** **a**

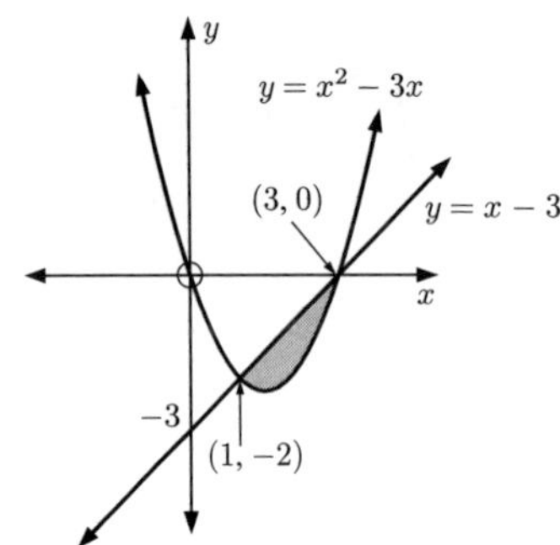


**b** The graphs meet where $x - 3 = x^2 - 3x$

$\therefore \quad x^2 - 3x - x + 3 = 0$

$\therefore \quad x^2 - 4x + 3 = 0$

$\therefore \quad (x-1)(x-3) = 0$

$\therefore \quad x = 1$ or $3$

$\therefore$ the graphs meet at $(1, -2)$ and $(3, 0)$.

**c**
$$\begin{aligned} \text{Area} &= \int_{1}^{3} [(x-3) - (x^2 - 3x)]\, dx \\ &= \int_{1}^{3} (-3 + 4x - x^2)\, dx \\ &= \left[-3x + 2x^2 - \frac{x^3}{3}\right]_{1}^{3} \\ &= (-9 + 18 - 9) - (-3 + 2 - \tfrac{1}{3}) \\ &= 1\tfrac{1}{3} \text{ units}^2 \end{aligned}$$

**4** $y = \sqrt{x}$ meets $y = x^2$ where $\sqrt{x} = x^2$

$\therefore \quad x = x^4$

$\therefore \quad x^4 - x = 0$

$\therefore \quad x(x^3 - 1) = 0$

$\therefore \quad x(x-1)(x^2 + x + 1) = 0$

$\therefore \quad x = 0$ or $1$

The factor $(x^2 + x + 1)$ has no real root since $\Delta = -3$ which is $< 0$.

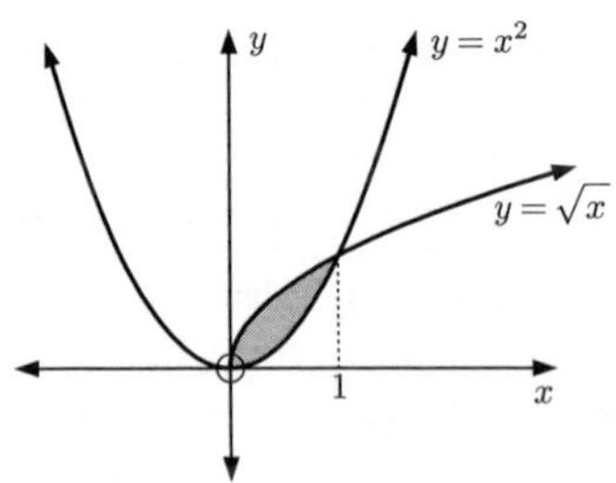


$$\begin{aligned} \text{Area} &= \int_{0}^{1} \left(\sqrt{x} - x^2\right) dx \\ &= \int_{0}^{1} (x^{\frac{1}{2}} - x^2)\, dx \\ &= \left[\tfrac{2}{3}x^{\frac{3}{2}} - \frac{x^3}{3}\right]_{0}^{1} \\ &= \tfrac{2}{3} - \tfrac{1}{3} \\ &= \tfrac{1}{3} \text{ unit}^2 \end{aligned}$$

**5** **a** $y = e^x - 1$ has no vertical asymptotes.

As $x \to \infty$, $e^x - 1 \to \infty$

As $x \to -\infty$, $e^x \to 0$

so $e^x - 1 \to -1^+$

$\therefore \quad y = -1$ is a horizontal asymptote.

$y = 0$ when $e^x - 1 = 0$

$\therefore \quad e^x = 1$

$\therefore \quad x = 0$

$\therefore \quad x$-intercept is $(0, 0)$.

This is also the $y$-intercept.

$y = 2 - 2e^{-x}$ has no vertical asymptotes.

As $x \to \infty$, $e^{-x} \to 0$

so $2 - 2e^{-x} \to 2^-$

$\therefore \quad y = 2$ is a horizontal asymptote.

$y = 0$ when $2 - 2c^{-x} - 0$

$\therefore \quad e^{-x} = 1$

$\therefore \quad x = 0$

$\therefore \quad x$-intercept is $(0, 0)$.

This is also the $y$-intercept.

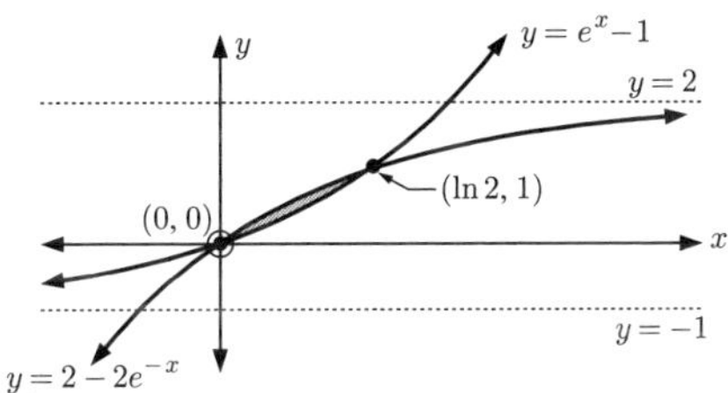


**b** $y = e^x - 1$ meets $y = 2 - 2e^{-x}$

where $e^x - 1 = 2 - 2e^{-x}$

$\therefore \; e^{2x} - e^x = 2e^x - 2 \quad \{\times\, e^x\}$

$\therefore \; e^{2x} - 3e^x + 2 = 0$

$\therefore \; (e^x - 1)(e^x - 2) = 0$

$\therefore \; e^x = 1$ or $2$

$\therefore \; x = 0$ or $\ln 2$

$\therefore$ the graphs meet at $(0, 0)$ and $(\ln 2, 1)$.

**c** $A = \int_0^{\ln 2} [(2 - 2e^{-x}) - (e^x - 1)]\, dx$

$= \int_0^{\ln 2} (3 - e^x - 2e^{-x})\, dx$

$= \left[3x - e^x + 2e^{-x}\right]_0^{\ln 2}$

$= (3\ln 2 - 2 + 1) - (0 - 1 + 2)$

$= 3\ln 2 - 2$

$\approx 0.0794$ units$^2$

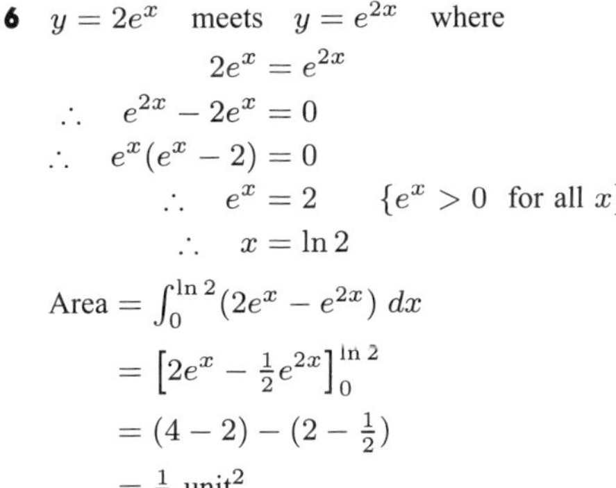

**6** $y = 2e^x$ meets $y = e^{2x}$ where

$2e^x = e^{2x}$

$\therefore \; e^{2x} - 2e^x = 0$

$\therefore \; e^x(e^x - 2) = 0$

$\therefore \; e^x = 2 \quad \{e^x > 0$ for all $x\}$

$\therefore \; x = \ln 2$

Area $= \int_0^{\ln 2} (2e^x - e^{2x})\, dx$

$= \left[2e^x - \frac{1}{2}e^{2x}\right]_0^{\ln 2}$

$= (4 - 2) - (2 - \frac{1}{2})$

$= \frac{1}{2}$ unit$^2$

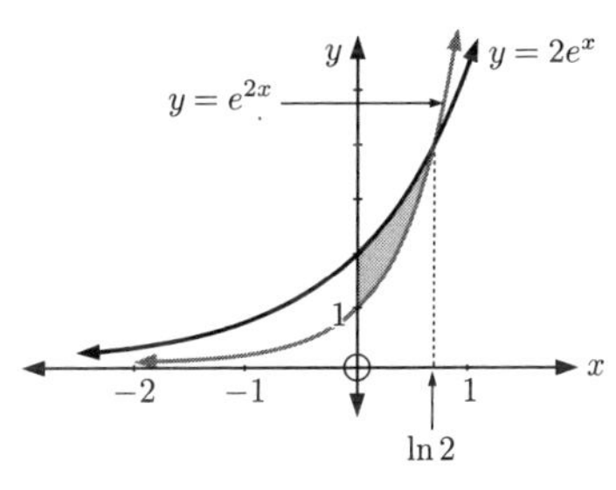


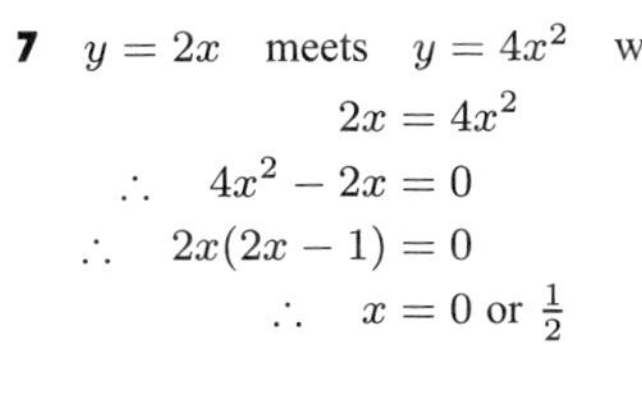

**7** $y = 2x$ meets $y = 4x^2$ where

$2x = 4x^2$

$\therefore \; 4x^2 - 2x = 0$

$\therefore \; 2x(2x - 1) = 0$

$\therefore \; x = 0$ or $\frac{1}{2}$

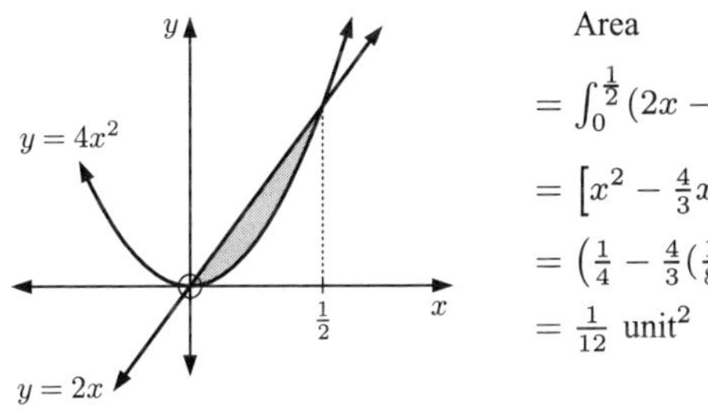


Area

$= \int_0^{\frac{1}{2}} (2x - 4x^2)\, dx$

$= \left[x^2 - \frac{4}{3}x^3\right]_0^{\frac{1}{2}}$

$= \left(\frac{1}{4} - \frac{4}{3}\left(\frac{1}{8}\right)\right) - (0 - 0)$

$= \frac{1}{12}$ unit$^2$

**8** **a** Now $x^2 + y^2 = 9 \quad \therefore \; y^2 = 9 - x^2$

$\therefore \; y = \pm\sqrt{9 - x^2}$

In the upper half of the circle all $y$-values are $\geqslant 0$

$\therefore \; y = +\sqrt{9 - x^2}$ is the required equation.

**b** The shaded area is $A$ where $A = \int_0^3 \sqrt{9 - x^2}\, dx$

This is a quarter of the area of a circle with radius 3 units.

$\therefore \; A = \frac{1}{4}(\pi \times 3^2) = \frac{9\pi}{4} \approx 7.07$ units$^2$

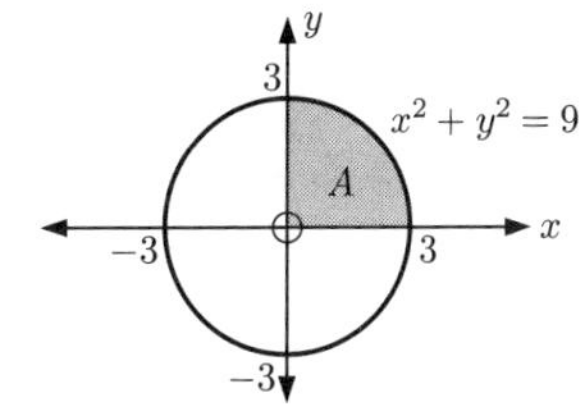

**9** **a** $f(x) = x^3 - 9x$
$= x(x^2 - 9)$
$= x(x+3)(x-3)$

$\therefore$ $y = f(x)$ cuts the $x$-axis at $0, \pm 3$

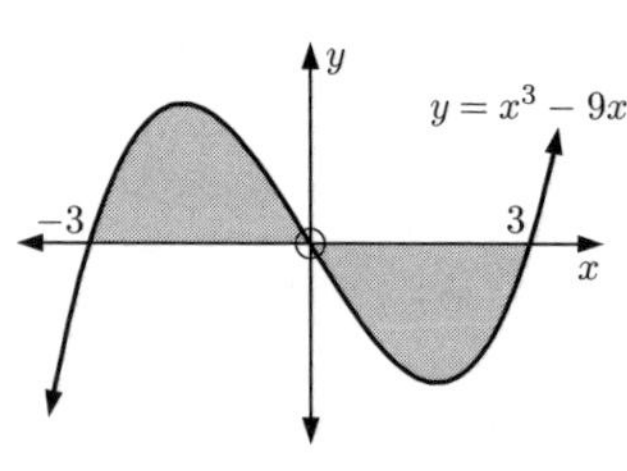


Area $= \int_{-3}^{0} (x^3 - 9x)\, dx + \int_0^3 [0 - (x^3 - 9x)]\, dx$

$= \left[\frac{x^4}{4} - \frac{9x^2}{2}\right]_{-3}^{0} + \left[-\frac{x^4}{4} + \frac{9x^2}{2}\right]_0^3$

$= 0 - (\frac{81}{4} - \frac{81}{2}) + (-\frac{81}{4} + \frac{81}{2}) - 0$

$= 40\frac{1}{2}$ units$^2$

**b** $f(x) = -x(x-2)(x-4)$
$= -x^3 + 6x^2 - 8x$

$\therefore$ $y = f(x)$ cuts the $x$-axis at 0, 2, and 4

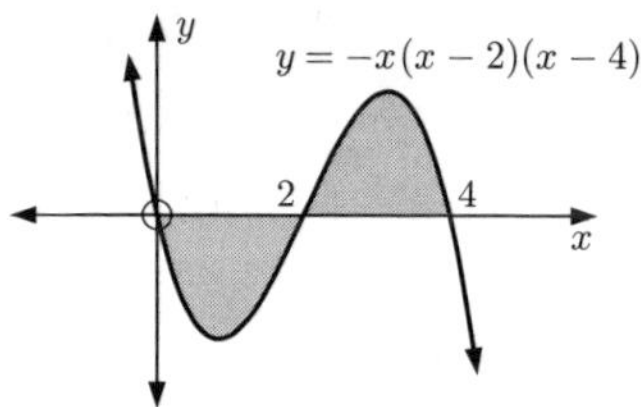


Area $= \int_0^2 [0 - (-x^3 + 6x^2 - 8x)]\, dx$
$+ \int_2^4 (-x^3 + 6x^2 - 8x)\, dx$

$= \int_0^2 (x^3 - 6x^2 + 8x)\, dx + \int_2^4 (-x^3 + 6x^2 - 8x)\, dx$

$= \left[\frac{x^4}{4} - 2x^3 + 4x^2\right]_0^2 + \left[-\frac{x^4}{4} + 2x^3 - 4x^2\right]_2^4$

$= ([4 - 16 + 16] - 0) + ([-64 + 128 - 64] - [-4 + 16 - 16])$

$= 8$ units$^2$

**c** $f(x) = x^4 - 5x^2 + 4$
$= (x^2 - 1)(x^2 - 4)$
$= (x+1)(x-1)(x+2)(x-2)$

$\therefore$ $y = f(x)$ cuts the $x$-axis at $\pm 1, \pm 2$

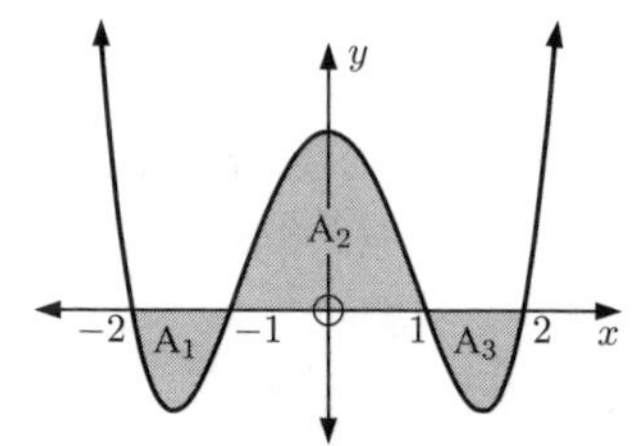


$A_1 = \int_{-2}^{-1} [0 - (x^4 - 5x^2 + 4)]\, dx$

$= \int_{-2}^{-1} (-x^4 + 5x^2 - 4)\, dx$

$= \left[-\frac{x^5}{5} + \frac{5x^3}{3} - 4x\right]_{-2}^{-1}$

$= (\frac{1}{5} - \frac{5}{3} + 4) - (\frac{32}{5} - \frac{40}{3} + 8)$

$= \frac{22}{15}$ units$^2$

$A_2 = \int_{-1}^{1} (x^4 - 5x^2 + 4)\, dx$

$= \left[\frac{x^5}{5} - \frac{5x^3}{3} + 4x\right]_{-1}^{1}$

$= (\frac{1}{5} - \frac{5}{3} + 4) - (-\frac{1}{5} + \frac{5}{3} - 4)$

$= \frac{76}{15}$ units$^2$

By symmetry, $A_3 = A_1$ $\therefore$ area $= \frac{22}{15} + \frac{76}{15} + \frac{22}{15} = \frac{120}{15} = 8$ units$^2$

**10** **a** $y = \sin(2x)$ is the curve $C_1$ and $y = \sin x$ is the curve $C_2$.

**b** The curves meet when $\sin(2x) = \sin x$

$\therefore$ $2\sin x \cos x - \sin x = 0$

$\therefore$ $\sin x(2\cos x - 1) = 0$

$\therefore$ $\sin x = 0$ or $\cos x = \frac{1}{2}$

$\therefore$ $x = 0 + k\pi$ or $x = \begin{cases} \frac{\pi}{3} \\ \frac{5\pi}{3} \end{cases} + 2k\pi$, $k$ an integer

$\therefore$ the $x$-coordinate of A $= \frac{\pi}{3}$

{smallest positive solution}

$\therefore$ A is at $(\frac{\pi}{3}, \frac{\sqrt{3}}{2})$

**c** $\text{Area} = \int_0^{\frac{\pi}{3}} (\sin(2x) - \sin x)\, dx + \int_{\frac{\pi}{3}}^{\pi} (\sin x - \sin(2x))\, dx$

$= \left[-\frac{1}{2}\cos(2x) + \cos x\right]_0^{\frac{\pi}{3}} + \left[-\cos x + \frac{1}{2}\cos(2x)\right]_{\frac{\pi}{3}}^{\pi}$

$= \left(-\frac{1}{2}\cos\frac{2\pi}{3} + \cos\frac{\pi}{3}\right) - \left(-\frac{1}{2}\cos 0 + \cos 0\right) + \left(-\cos\pi + \frac{1}{2}\cos 2\pi\right)$
$\quad - \left(-\cos\frac{\pi}{3} + \frac{1}{2}\cos\frac{2\pi}{3}\right)$

$= (\frac{1}{4} + \frac{1}{2}) - (-\frac{1}{2} + 1) + (1 + \frac{1}{2}) - (-\frac{1}{2} - \frac{1}{4})$

$= 2\frac{1}{2}$ units$^2$

**11**

**a** $\int_1^7 f(x)\, dx$ only gives us the correct area provided that $f(x)$ is positive on the interval $1 \leqslant x \leqslant 7$. But $f(x)$ is not positive for $3 \leqslant x \leqslant 5$,

so $\int_1^7 f(x)\, dx = A_1 - A_2 + A_3$

which is *not* the shaded area.

**b** shaded area

$= \int_1^3 f(x)\, dx + \int_3^5 [0 - f(x)]\, dx + \int_5^7 f(x)\, dx$

$= \int_1^3 f(x)\, dx - \int_3^5 f(x)\, dx + \int_5^7 f(x)\, dx$

**12** **a** $y = \cos(2x)$ is the curve $C_2$ and $y = \cos^2 x$ is the curve $C_1$.

**b** Point A lies on $y = \cos(2x)$. When $x = 0$, $y = \cos 0 = 1$. $\therefore$ A is at $(0, 1)$.

Point B lies on $y = \cos(2x)$. When $x = \frac{\pi}{4}$, $y = \cos\frac{\pi}{2} = 0$. $\therefore$ B is at $(\frac{\pi}{4}, 0)$.

Point C lies on $y = \cos^2 x$. When $x = \frac{\pi}{2}$, $y = \cos^2\frac{\pi}{2} = 0$. $\therefore$ C is at $(\frac{\pi}{2}, 0)$.

Point D lies on $y = \cos(2x)$. When $x = \frac{3\pi}{4}$, $y = \cos\frac{3\pi}{2} = 0$. $\therefore$ D is at $(\frac{3\pi}{4}, 0)$.

Point E lies where the curves meet. Now $\cos(2\pi) = \cos^2\pi = 1$. $\therefore$ E is at $(\pi, 1)$.

**c** $A = \int_0^{\pi} (\cos^2 x - \cos(2x))\, dx$

$= \int_0^{\pi} \left(\frac{1}{2} + \frac{1}{2}\cos(2x) - \cos(2x)\right) dx$

$= \int_0^{\pi} \left(\frac{1}{2} - \frac{1}{2}\cos(2x)\right) dx$

$= \left[\frac{x}{2} - \frac{1}{4}\sin(2x)\right]_0^{\pi} = (\frac{\pi}{2} - 0) - (0 - 0) = \frac{\pi}{2}$ units$^2$

**13** **a** The graphs meet when $e^{-x^2} = x^2 - 1$

$\therefore$ $x = \pm 1.1307$ {technology}

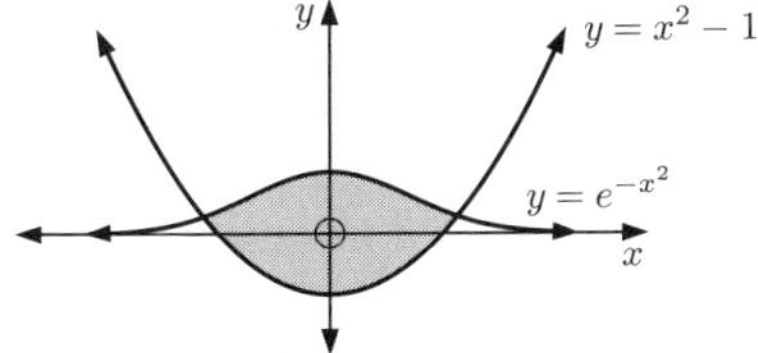


$\therefore$ area $= \int_{-1.1307}^{1.1307} [e^{-x^2} - (x^2 - 1)]\, dx$

$\approx 2.88$ units$^2$ {technology}

**b** The graphs meet when $x^x = 4x - \frac{1}{10}x^4$

$\therefore$ $x \approx 0.1832$ or $2.2696$ {technology}

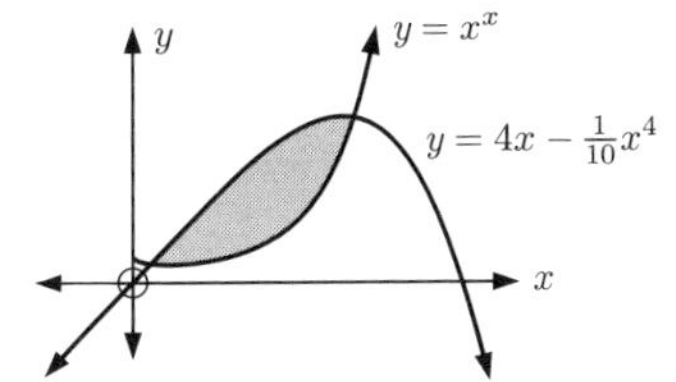


$\therefore$ area $= \int_{0.1832}^{2.2696} (4x - \frac{1}{10}x^4 - x^x)\, dx$

$\approx 4.97$ units$^2$ {technology}

**14** Area $= \displaystyle\int_1^k \frac{1}{1+2x}\,dx = 0.2$ units$^2$

$\therefore \left[\frac{1}{2}\ln(1+2x)\right]_1^k = 0.2, \quad 1+2x > 0$

$\therefore [\ln(1+2x)]_1^k = 0.4$

$\therefore \ln(1+2k) - \ln 3 = 0.4$

{since $k \geqslant 1$, $1+2x>0$ for all $x$ in the shaded region}

$\therefore \ln\left(\dfrac{1+2k}{3}\right) = 0.4$

$\therefore \dfrac{1+2k}{3} = e^{0.4}$

$\therefore 1+2k = 3e^{0.4}$

$\therefore k = \dfrac{3e^{0.4}-1}{2} \approx 1.7377$

**15** Area $= \int_0^b \sqrt{x}\,dx$

$\therefore \int_0^b x^{\frac{1}{2}}\,dx = 1$

$\therefore \left[\frac{2}{3}x^{\frac{3}{2}}\right]_0^b = 1$

$\therefore \frac{2}{3}b\sqrt{b} - 0 = 1$

$\therefore b\sqrt{b} = \frac{3}{2}$

$\therefore b^{\frac{3}{2}} = 1.5$

$\therefore b = (1.5)^{\frac{2}{3}} \approx 1.3104$

**16** By symmetry, the area bounded by $x=0$ and $x=a$ is $\frac{1}{2}(6a)$ units$^2$.

$\therefore \int_0^a (x^2+2)\,dx = 3a$

$\therefore \left[\dfrac{x^3}{3} + 2x\right]_0^a = 3a$

$\therefore \dfrac{a^3}{3} + 2a - 0 = 3a$

$\therefore a^3 + 6a = 9a$

$\therefore a^3 - 3a = 0$

$\therefore a(a^2-3) = 0$

$\therefore a = 0$ or $\pm\sqrt{3}$ $\quad \therefore a = \sqrt{3}$ {as $a > 0$}

## EXERCISE 19C.1

**1**

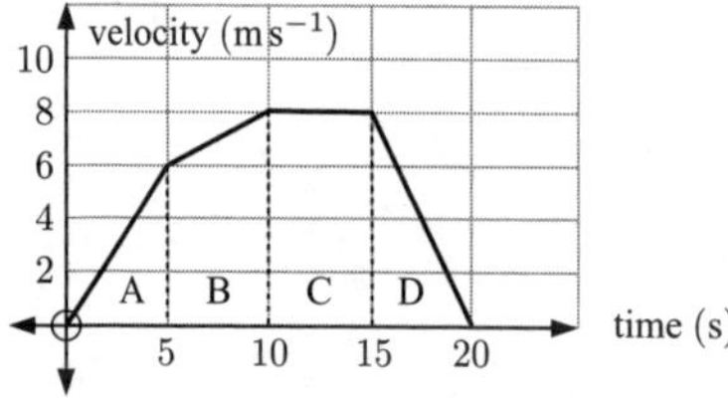


Total distance travelled

$= $ area A + area B + area C + area D

$= \frac{1}{2}(5 \times 6) + \left(\dfrac{6+8}{2}\right)5 + 5 \times 8 + \frac{1}{2}(5 \times 8)$

$= 15 + 35 + 40 + 20$

$= 110$ m

**2**

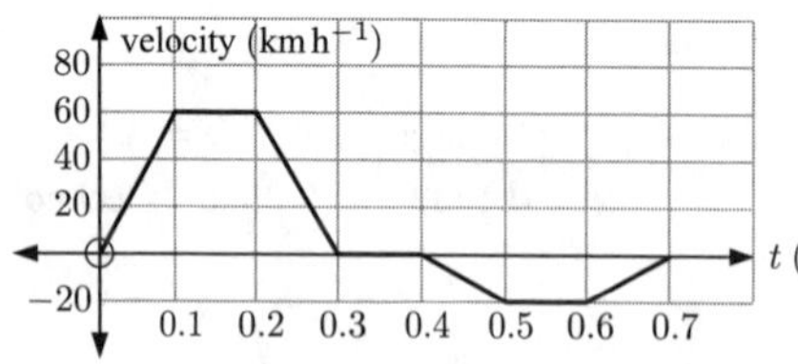


**a** **i** The graph above the $t$-axis indicates that the velocity is positive and the car is travelling forwards.

**ii** The graph below the $t$-axis indicates that the velocity is negative and the car is travelling backwards (opposite direction).

**b** Total distance travelled = area above the $t$-axis + area below the $t$-axis

$= \left(\frac{0.1}{2} + 0.1 + \frac{0.1}{2}\right)60 + \left(\frac{0.1}{2} + 0.1 + \frac{0.1}{2}\right)20$

$= 12 + 4$

$= 16$ km

**c** Final displacement = area above the $t$-axis − area below the $t$-axis

$= 12 - 4$

$= 8$ km from the starting point (on the positive side)

**3** **a**

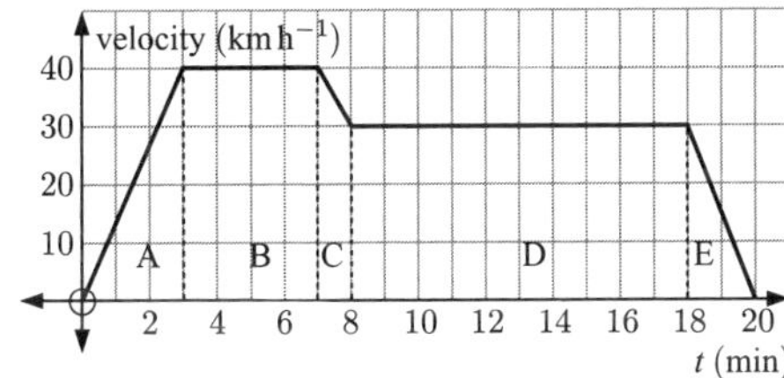


**b** Total distance travelled

$= \text{area A} + \text{area B} + \text{area C} + \text{area D} + \text{area E}$

$= \frac{1}{60}\left[\frac{1}{2}(3 \times 40) + (40 \times 4) + \left(\frac{40+30}{2}\right)1 + (10 \times 30) + \frac{1}{2}(2 \times 30)\right]$

{the factor $\frac{1}{60}$ accounts for the fact that the times are in minutes while the speeds are in $\text{km h}^{-1}$}

$= \frac{1}{60}[60 + 160 + 35 + 300 + 30]$

$= 9.75$ km

## EXERCISE 19C.2

**1** **a**

$$\begin{aligned} s(t) &= \int (1-2t)\,dt \\ &= t - 2\left(\frac{t^2}{2}\right) + c \\ &= t - t^2 + c \end{aligned}$$

But $s(0) = 2$

$\therefore\ 0 - 0^2 + c = 2$

$\therefore\ c = 2$

$\therefore\ s(t) = t - t^2 + 2$ cm

**b** The particle changes direction when

$v(t) = 0$

$\therefore\ 1 - 2t = 0$

$\therefore\ t = \frac{1}{2}$ s

Now $s(\frac{1}{2}) = \frac{1}{2} - (\frac{1}{2})^2 + 2 = 2\frac{1}{4}$ cm

and $s(1) = 1 - 1 + 2 = 2$ cm

$\therefore$ motion diagram is:

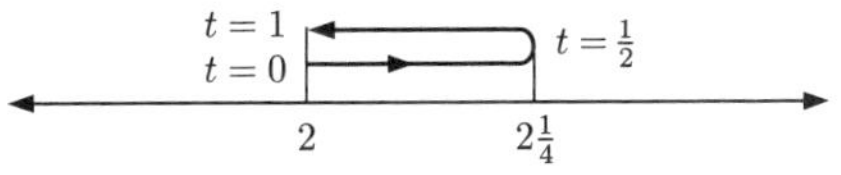

$t = 1$, $t = 0$ at $2$; $t = \frac{1}{2}$ at $2\frac{1}{4}$

$\therefore$ total distance travelled $= \frac{1}{4} + \frac{1}{4}$

$= \frac{1}{2}$ cm

**c** Displacement $= s(1) - s(0)$

$= 2 - 2$

$= 0$ cm

**2** **a**

$$\begin{aligned} s(t) &= \int (t^2 - t - 2)\,dt \\ &= \tfrac{1}{3}t^3 - \tfrac{1}{2}t^2 - 2t + c \end{aligned}$$

But $s(0) = 0$, $\therefore\ c = 0$

$\therefore\ s(t) = \frac{1}{3}t^3 - \frac{1}{2}t^2 - 2t$ cm

**b** P changes direction when $v(t) = 0$

$\therefore\ t^2 - t - 2 = 0$

$\therefore\ (t-2)(t+1) = 0$

$\therefore$ P changes direction when $t = 2$ (since $t \geqslant 0$)

Now $s(2) = \frac{2^3}{3} - \frac{2^2}{2} - 2(2) = -\frac{10}{3}$

and $s(3) = \frac{3^3}{3} - \frac{3^2}{2} - 2(3) = -\frac{3}{2}$

$\therefore$ motion diagram is:

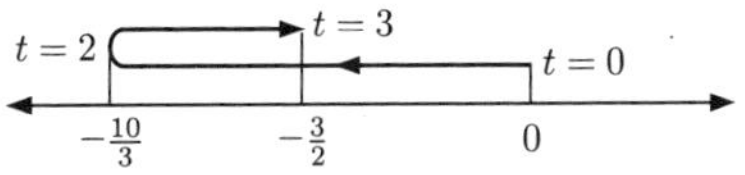

$t = 2$ at $-\frac{10}{3}$; $t = 3$ at $-\frac{3}{2}$; $t = 0$ at $0$

$\therefore$ total distance travelled $= \frac{10}{3} + (\frac{10}{3} - \frac{3}{2})$

$= 5\frac{1}{6}$ cm

**c** Displacement

$= s(3) - s(0)$

$= -\frac{3}{2} - 0$

$= -\frac{3}{2}$ cm ($1\frac{1}{2}$ cm left of its starting point)

**3** **a**

$$s(t) = \int (32 + 4t)\, dt$$
$$= 32t + 4\left(\frac{t^2}{2}\right) + c$$
$$\therefore\ s(t) = 32t + 2t^2 + c$$

But $s(0) = 16$

$$\therefore\ 0 + 0 + c = 16$$
$$\therefore\ c = 16$$
$$\therefore\ s(t) = 32t + 2t^2 + 16 \text{ m}$$

**b** We check for any changes in direction.

These occur when $v(t) = 0$

$$\therefore\ 32 + 4t = 0$$
$$\therefore\ 4t = -32$$
$$\therefore\ t = -8$$

But $0 \leqslant t \leqslant t_1$, so the object does not change direction on the interval.

$$\therefore\ \text{displacement} = s(t_1) - s(0)$$
$$= \int_0^{t_1} (32 + 4t)\, dt$$

**c** $a(t) = v'(t)$

$= 4$

$\therefore$ the object is travelling with constant acceleration of 4 $\text{m}\,\text{s}^{-2}$.

**4**

$$s(t) = \int (\cos(2t))\, dt$$
$$= \tfrac{1}{2}\sin(2t) + c$$

But $s(\frac{\pi}{4}) = 1$

$$\therefore\ \tfrac{1}{2}\sin(\tfrac{\pi}{2}) + c = 1$$
$$\therefore\ c + \tfrac{1}{2} = 1$$
$$\therefore\ c = \tfrac{1}{2}$$
$$\therefore\ s(t) = \tfrac{1}{2}\sin(2t) + \tfrac{1}{2}$$

$$\therefore\ s(\tfrac{\pi}{3}) = \tfrac{1}{2}\sin(\tfrac{2\pi}{3}) + \tfrac{1}{2}$$
$$= \tfrac{1}{2} \times \tfrac{\sqrt{3}}{2} + \tfrac{1}{2}$$
$$= \tfrac{\sqrt{3}}{4} + \tfrac{1}{2}$$
$$= \tfrac{\sqrt{3}+2}{4} \text{ m}$$

**5** $x'(t) = 16t - 4t^3$ units $\text{s}^{-1}$, $t \geqslant 0$

$= 4t(4 - t^2)$

$= 4t(2 + t)(2 - t)$ which has sign diagram:

$\therefore$ a direction reversal occurs at $t = 2$.

Now $x(t) = \int (16t - 4t^3)\, dt = 8t^2 - t^4 + c$

**a** $x(0) = c$

$x(2) = 32 - 16 + c = c + 16$

$x(3) = 72 - 81 + c = c - 9$

$\therefore$ motion diagram for $0 \leqslant t \leqslant 3$ is:

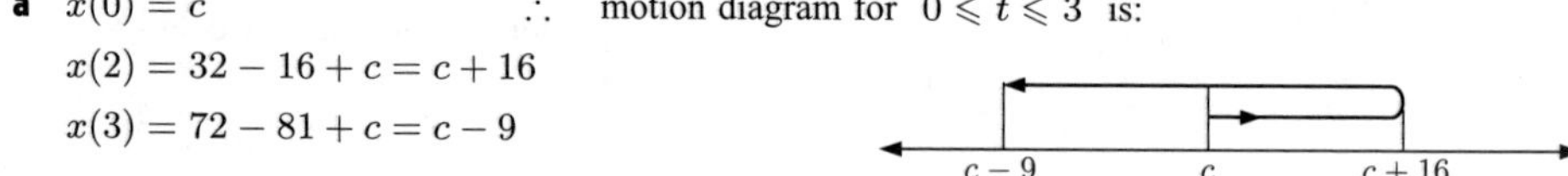


$\therefore$ total distance travelled $= (c + 16 - c) + (c + 16 - [c - 9])$

$= 41$ units

**b** $x(1) = 7 + c = c + 7$ $\therefore$ motion diagram for $1 \leqslant t \leqslant 3$ is:

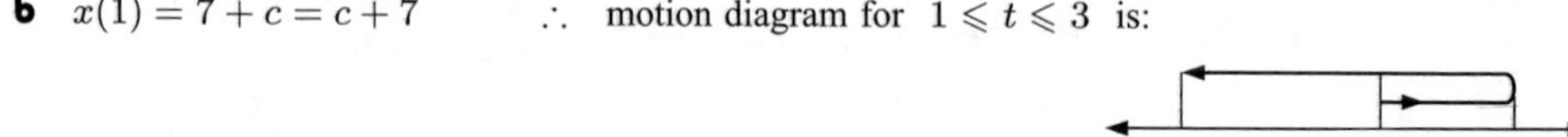

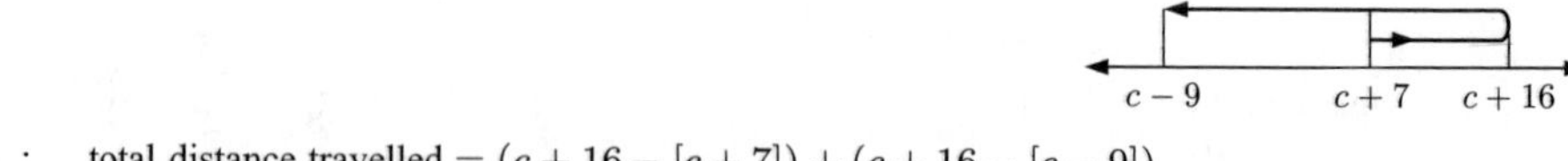


$\therefore$ total distance travelled $= (c + 16 - [c + 7]) + (c + 16 - [c - 9])$

$= 34$ units

**6** **a** $v(t) = \cos t$ $\text{m}\,\text{s}^{-1}$, $t \geqslant 0$

$\therefore$ $v(t)$ has sign diagram:

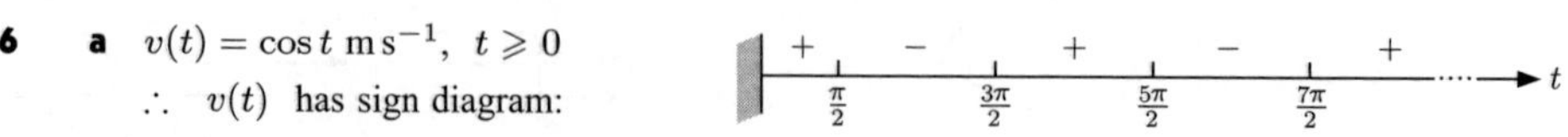


$\therefore$ a direction reversal occurs at $t = \frac{\pi}{2}, \frac{3\pi}{2}, \frac{5\pi}{2}, \frac{7\pi}{2}, \ldots$

$s(t) = \int \cos t\, dt = \sin t + c$

$\therefore \quad s(0) = c$

$s\left(\frac{\pi}{2}\right) = c + 1$

$s\left(\frac{3\pi}{2}\right) = c - 1$

$s\left(\frac{5\pi}{2}\right) = c + 1$

$s\left(\frac{7\pi}{2}\right) = c - 1$

The motion diagram is:

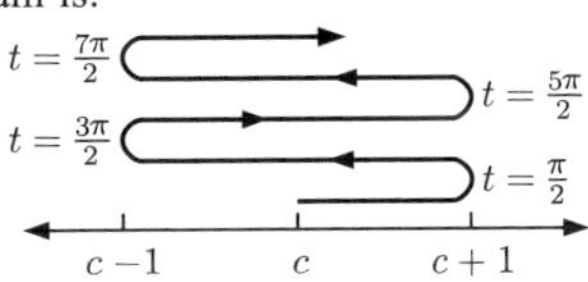


$\therefore$ the particle oscillates between the points $(c - 1)$ and $(c + 1)$.

**b** distance $= (c + 1) - (c - 1)$

$= 2$ m

**7** $v(t) = 50 - 10e^{-0.5t}$ m s$^{-1}$, $t \geqslant 0$

**a** $v(0) = 50 - \dfrac{10}{e^0} = 50 - 10 = 40$ m s$^{-1}$

**b** $v(3) = 50 - \dfrac{10}{e^{1.5}} \approx 47.8$ m s$^{-1}$

**c** The velocity reaches 45 m s$^{-1}$

when $45 = 50 - 10e^{-0.5t}$

$\therefore \quad 10e^{-\frac{t}{2}} = 5$

$\therefore \quad e^{\frac{t}{2}} = 2$

$\therefore \quad \dfrac{t}{2} = \ln 2$

$\therefore \quad t = 2\ln 2 \approx 1.39$ seconds

**d** $v(t) = 50 - \dfrac{10}{e^{\frac{t}{2}}}$

As $t \to \infty$, $\dfrac{10}{e^{\frac{t}{2}}} \to 0^+$

$\therefore \quad v(t) \to 50^-$

**e** $a(t) = v'(t)$

$= 10e^{-0.5t}(-0.5)$

$= 5e^{-0.5t}$ m s$^{-2}$

$\therefore$ $a(t) > 0$ for all $t$ $\{e^x > 0$ for all $x\}$

$\therefore$ the acceleration is always positive

**f**

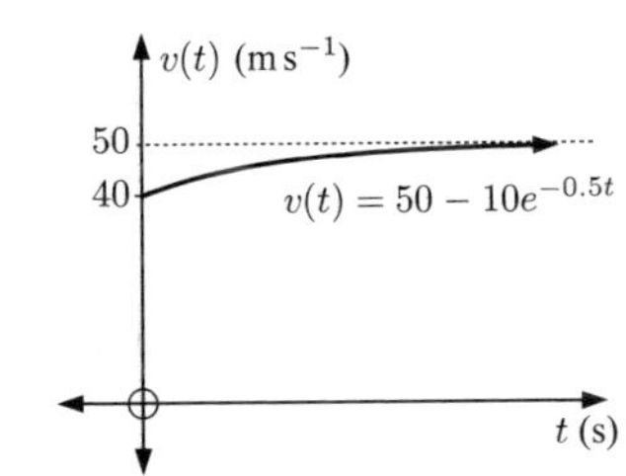


**g** total distance travelled

$= \int_0^3 (50 - 10e^{-0.5t})\, dt$

$= \left[50t + 20e^{-0.5t}\right]_0^3$

$= 150 + 20e^{-1.5} - 20$

$\approx 134.5$ m

**8** **a** $v(t) = \displaystyle\int \frac{-1}{(t+1)^2}\, dt$

$= \int -(t+1)^{-2}\, dt$

$= (t+1)^{-1} + c$

But $v(0) = 0$

$\therefore \quad \dfrac{1}{0+1} + c = 0$

$\therefore \quad c + 1 = 0$

$\therefore \quad c = -1$

$\therefore \quad v(t) = \dfrac{1}{t+1} - 1$ m s$^{-1}$

**b** $s(t) = \displaystyle\int \left(\frac{1}{t+1} - 1\right) dt$

$= \ln|t+1| - t + c$

But $s(0) = 0$

$\therefore \quad \ln 1 - 0 + c = 0$

$\therefore \quad c = 0$

$\therefore \quad s(t) = \ln|t+1| - t$ m

**c** $s(2) = \ln 3 - 2$ m $\approx -0.901$ m

$v(2) = \dfrac{1}{2+1} - 1 = -\frac{2}{3}\ \text{m s}^{-1}$

$a(2) = \dfrac{-1}{(2+1)^2} = -\frac{1}{9}\ \text{m s}^{-2}$

The object is approximately 0.901 m to the left of the origin, travelling left at $\frac{2}{3}$ m s$^{-1}$, with acceleration $-\frac{1}{9}$ m s$^{-2}$.

**9** **a**
$$v(t) = \int \left(\frac{t}{10} - 3\right) dt$$
$$= \tfrac{1}{10}\left(\frac{t^2}{2}\right) - 3t + c$$
$$= \tfrac{1}{20}t^2 - 3t + c \ \text{m s}^{-1}$$

But $v(0) = 45$

$\therefore\ \frac{1}{20}(0)^2 - 3(0) + c = 45$

$\therefore\ c = 45$

$\therefore\ v(t) = \frac{1}{20}t^2 - 3t + 45\ \text{m s}^{-1}$

**b**
$$\int_0^{60} (\tfrac{1}{20}t^2 - 3t + 45)\, dt$$
$$= \left[\tfrac{1}{60}t^3 - \tfrac{3}{2}t^2 + 45t\right]_0^{60}$$
$$= \tfrac{1}{60}(60)^3 - \tfrac{3}{2}(60)^2 + 45(60)$$
$$= 900$$

The train travels a total of 900 m in the first 60 seconds.

**10** $a(t) = 4e^{-\frac{t}{20}}\ \text{m s}^{-2}$

$\therefore\ v(t) = \int 4e^{-\frac{t}{20}}\, dt$

$= 4\dfrac{1}{-\frac{1}{20}}e^{-\frac{t}{20}} + c$

$= -80e^{-\frac{t}{20}} + c$

Now $v(0) = 20\ \text{m s}^{-1}$

$\therefore\ c = 100$

$\therefore\ v(t) = 100 - 80e^{-\frac{t}{20}}\ \text{m s}^{-1}$

**a** As $t \to \infty$, $e^{-\frac{t}{20}} \to 0^+$ $\quad\therefore\ v(t) \to 100^-\ \text{m s}^{-1}$

$\therefore$ the object approaches a limiting velocity of $100\ \text{m s}^{-1}$.

**b** The total distance travelled $= \int_0^{10} (100 - 80e^{-\frac{t}{20}})\, dt \qquad \{v(t) > 0 \text{ for } 0 \leqslant t \leqslant 10\}$

$$= \left[100t + 1600e^{-\frac{t}{20}}\right]_0^{10}$$
$$= \left(1000 + 1600e^{-\frac{1}{2}}\right) - (0 + 1600)$$
$$\approx 370.4 \text{ m}$$

## EXERCISE 19D.1

**1** **a**

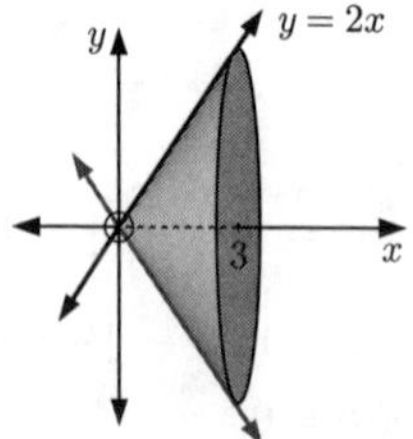


Volume $= \pi \int_0^3 (2x)^2\, dx$

$= 4\pi \int_0^3 x^2\, dx$

$= 4\pi \left[\frac{1}{3}x^3\right]_0^3$

$= 4\pi(9 - 0)$

$= 36\pi$ units$^3$

**b**

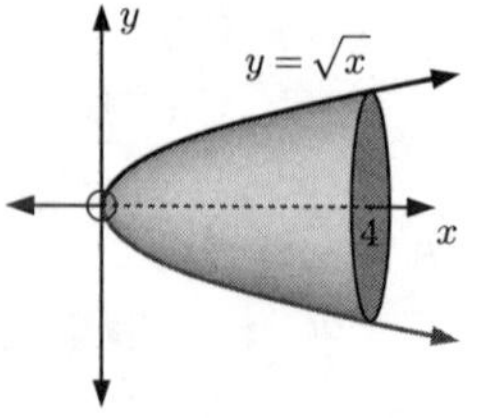


Volume $= \pi \int_0^4 (\sqrt{x})^2\, dx$

$= \pi \int_0^4 x\, dx$

$= \pi \left[\frac{1}{2}x^2\right]_0^4$

$= \pi(8 - 0)$

$= 8\pi$ units$^3$

**c**

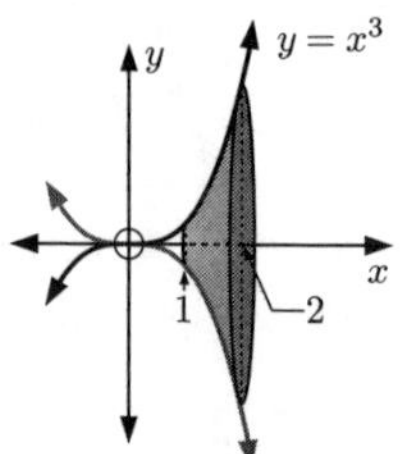


Volume $= \pi \int_1^2 (x^3)^2\, dx$

$= \pi \int_1^2 x^6\, dx$

$= \pi \left[\frac{1}{7}x^7\right]_1^2$

$= \pi \left(\frac{128}{7} - \frac{1}{7}\right)$

$= \frac{127\pi}{7}$ units$^3$

**d**

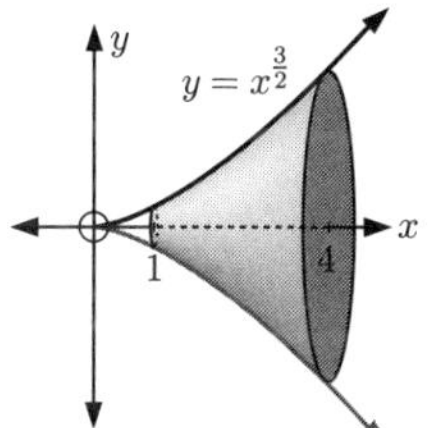


$$\begin{aligned}\text{Volume} &= \pi \int_1^4 (x^{\frac{3}{2}})^2\, dx \\ &= \pi \int_1^4 x^3\, dx \\ &= \pi \left[\tfrac{1}{4}x^4\right]_1^4 \\ &= \pi\left(\tfrac{256}{4} - \tfrac{1}{4}\right) \\ &= \tfrac{255\pi}{4} \text{ units}^3\end{aligned}$$

**e**

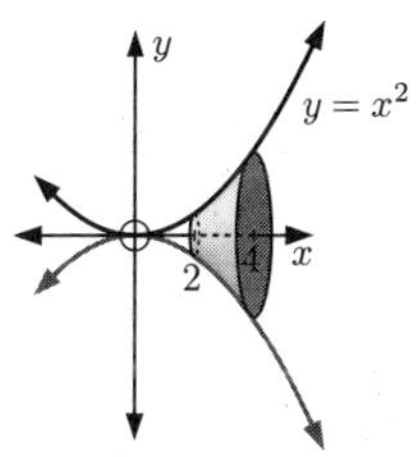


$$\begin{aligned}\text{Volume} &= \pi \int_2^4 (x^2)^2\, dx \\ &= \pi \int_2^4 x^4\, dx \\ &= \pi \left[\tfrac{1}{5}x^5\right]_2^4 \\ &= \pi\left(\tfrac{1024}{5} - \tfrac{32}{5}\right) \\ &= \tfrac{992\pi}{5} \text{ units}^3\end{aligned}$$

**f**

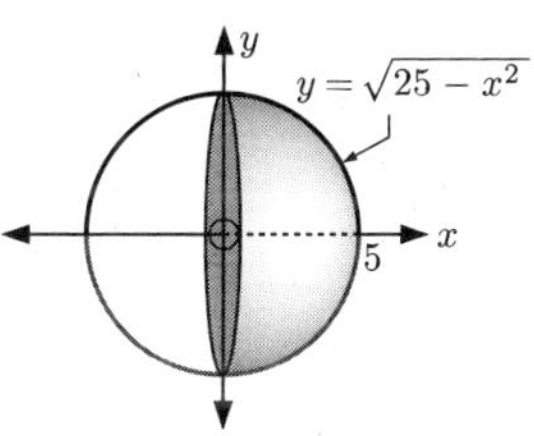


$$\begin{aligned}\text{Volume} &= \pi \int_0^5 (25 - x^2)\, dx \\ &= \pi \left[25x - \frac{x^3}{3}\right]_0^5 \\ &= \pi\left(125 - \tfrac{125}{3}\right) \\ &= \pi\left(\tfrac{2}{3}\right)125 \\ &= \tfrac{250\pi}{3} \text{ units}^3\end{aligned}$$

**g**

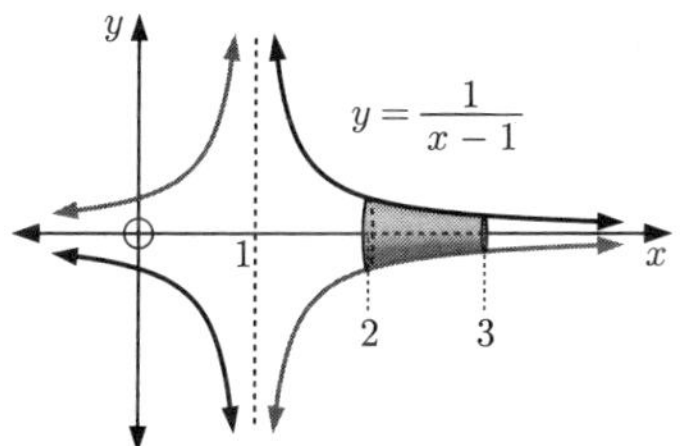


$$\begin{aligned}\text{Volume} &= \pi \int_2^3 \left(\frac{1}{x-1}\right)^2 dx \\ &= \pi \int_2^3 (x-1)^{-2}\, dx \\ &= \pi \left[-\frac{1}{x-1}\right]_2^3 \\ &= \pi\left(-\tfrac{1}{2} + 1\right) \\ &= \tfrac{\pi}{2} \text{ units}^3\end{aligned}$$

**h**

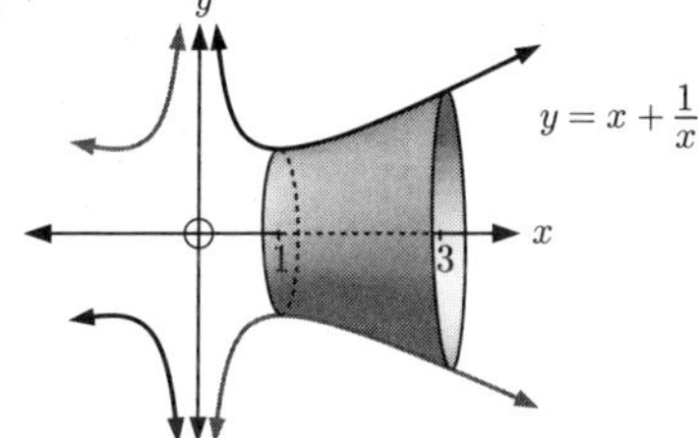


$$\begin{aligned}\text{Volume} &= \pi \int_1^3 \left(x + \frac{1}{x}\right)^2 dx \\ &= \pi \int_1^3 \left(x^2 + 2 + x^{-2}\right) dx \\ &= \pi \left[\frac{x^3}{3} + 2x - \frac{1}{x}\right]_1^3 \\ &= \pi\left[9 + 6 - \tfrac{1}{3} - \left(\tfrac{1}{3} + 2 - 1\right)\right] \\ &= \tfrac{40\pi}{3} \text{ units}^3\end{aligned}$$

**2** **a** $\text{Volume} = \pi \displaystyle\int_1^3 \left(\frac{x^3}{x^2+1}\right)^2 dx$

$\approx 5.926\pi$ {using technology}

$\approx 18.6$ units$^3$

**b** $\text{Volume} = \pi \int_0^2 \left(e^{\sin x}\right)^2 dx$

$\approx 9.613\pi$ {using technology}

$\approx 30.2$ units$^3$

**3** **a**

$$\begin{aligned}V &= \pi \int_0^6 \left(\frac{x}{2} + 4\right)^2 dx \\ &= \pi \int_0^6 \left(\tfrac{1}{4}x^2 + 4x + 16\right) dx \\ &= \pi \left[\frac{x^3}{12} + \frac{4x^2}{2} + 16x\right]_0^6 \\ &= \pi(18 + 72 + 96) - 0 \\ &= 186\pi \text{ units}^3\end{aligned}$$

**b**

$$\begin{aligned}V &= \pi \int_1^2 (x^2+3)^2\, dx \\ &= \pi \int_1^2 \left(x^4 + 6x^2 + 9\right) dx \\ &= \pi \left[\frac{x^5}{5} + \frac{6x^3}{3} + 9x\right]_1^2 \\ &= \pi\left[\left(\tfrac{32}{5} + 16 + 18\right) - \left(\tfrac{1}{5} + 2 + 9\right)\right] \\ &= \pi\left(\tfrac{146}{5}\right) \\ &= \tfrac{146\pi}{5} \text{ units}^3\end{aligned}$$

**c** $V = \pi \int_0^4 (e^x)^2 \, dx$
$= \pi \int_0^4 e^{2x} \, dx$
$= \pi \left[\frac{1}{2}e^{2x}\right]_0^4$
$= \pi \left(\frac{1}{2}e^8 - \frac{1}{2}\right)$
$= \frac{\pi}{2}(e^8 - 1)$ units$^3$

**4** **a** If we take a vertical slice of the bowl, we get a circle.

**b** Volume of revolution
$= \pi \int_a^b y^2 \, dx$
$= \pi \int_0^4 \left(4\sqrt{x}\right)^2 \, dx$
$= \int_0^4 \pi \left(4\sqrt{x}\right)^2 \, dx$

**c** Capacity $= \int_0^4 \pi \times 16x \, dx$
$= \int_0^4 16\pi x \, dx$
$= \left[8\pi x^2\right]_0^4$
$= 8\pi \times 16$
$= 128\pi$ units$^3$
$\approx 402$ units$^3$

**5** **a** Volume $= \pi \int_5^8 y^2 \, dx$
$= \pi \int_5^8 (64 - x^2) \, dx$
$= \pi \left[64x - \dfrac{x^3}{3}\right]_5^8$
$= \pi \left[\left(512 - \frac{512}{3}\right) - \left(320 - \frac{125}{3}\right)\right]$
$= 63\pi$ units$^3$

**b** $63\pi$ cm$^3$ $\approx 198$ cm$^3$

**6** **a** a cone of base radius $r$ and height $h$

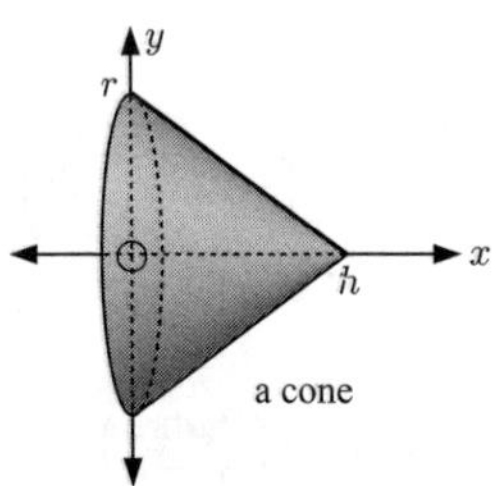


**b** [AB] has gradient $= \dfrac{r-0}{0-h} = -\dfrac{r}{h}$

$\therefore$ its equation is $y = -\left(\dfrac{r}{h}\right)x + r$

**c** $V = \pi \displaystyle\int_0^h \left(\frac{-r}{h}x + r\right)^2 dx$
$= \pi r^2 \displaystyle\int_0^h \left(-\frac{x}{h} + 1\right)^2 dx$
$= \pi r^2 \displaystyle\int_0^h \left(\frac{x^2}{h^2} - \frac{2x}{h} + 1\right) dx$
$= \pi r^2 \left[\dfrac{x^3}{3h^2} - \dfrac{2x^2}{2h} + x\right]_0^h$
$= \pi r^2 \left[\left(\dfrac{h}{3} - h + h\right) - 0\right]$
$= \frac{1}{3}\pi r^2 h$ units$^3$

**7** **a** a sphere of radius $r$

**b** $V = \pi \int_{-r}^r y^2 \, dx = 2\pi \int_0^r (r^2 - x^2) \, dx$
$= 2\pi \left[r^2 x - \dfrac{x^3}{3}\right]_0^r$
$= 2\pi(r^3 - \dfrac{r^3}{3} - 0)$
$= 2\pi \times \frac{2}{3}r^3$
$= \frac{4}{3}\pi r^3$ units$^3$

**8** **a**

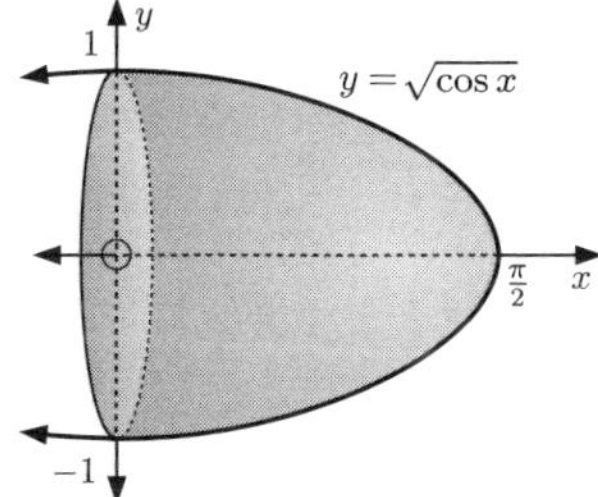


$$\text{Volume} = \pi\int_0^{\frac{\pi}{2}} \left(\sqrt{\cos x}\right)^2\ dx$$
$$= \pi\int_0^{\frac{\pi}{2}} \cos x\ dx$$
$$= \pi\Big[\sin x\Big]_0^{\frac{\pi}{2}}$$
$$= \pi\left(\sin\left(\tfrac{\pi}{2}\right) - \sin 0\right)$$
$$= \pi \text{ units}^3$$

**b**

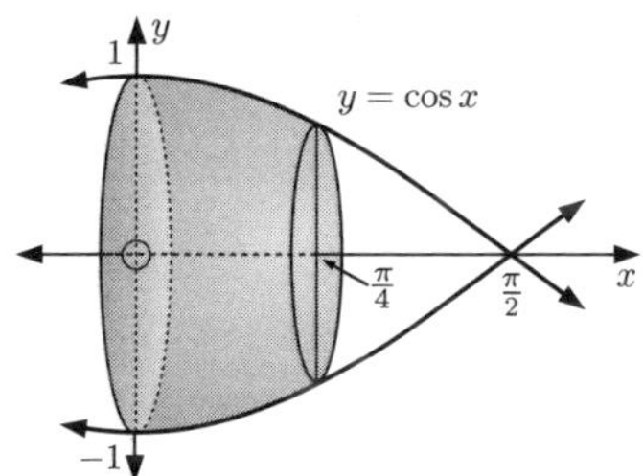


$$\text{Volume} = \pi\int_0^{\frac{\pi}{4}} (\cos x)^2\ dx$$
$$= \pi\int_0^{\frac{\pi}{4}} \cos^2 x\ dx$$
$$= \pi\int_0^{\frac{\pi}{4}} \left(\tfrac{1}{2} + \tfrac{1}{2}\cos(2x)\right)\ dx$$
$$= \pi\left[\tfrac{1}{2}x + \tfrac{1}{4}\sin(2x)\right]_0^{\frac{\pi}{4}}$$
$$= \pi\left(\tfrac{\pi}{8} + \tfrac{1}{4}\sin\left(\tfrac{\pi}{2}\right) - 0\right)$$
$$= \tfrac{\pi^2}{8} + \tfrac{\pi}{4} \text{ units}^3$$

**9** **a**

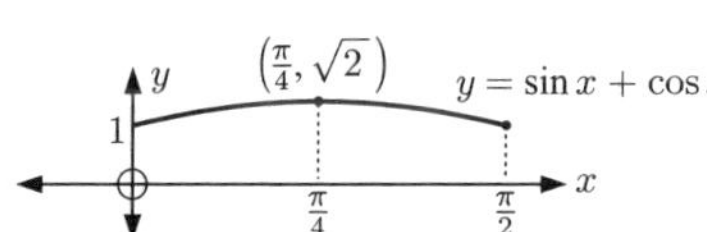


**b** Volume

$$= \pi\int_0^{\frac{\pi}{4}} (\sin x + \cos x)^2\ dx$$
$$= \pi\int_0^{\frac{\pi}{4}} \left(\sin^2 x + 2\sin x\cos x + \cos^2 x\right)\ dx$$
$$= \pi\int_0^{\frac{\pi}{4}} (1 + \sin(2x))\ dx$$
$$= \pi\left[x - \tfrac{1}{2}\cos(2x)\right]_0^{\frac{\pi}{4}}$$
$$= \pi\left[\left(\tfrac{\pi}{4} - \tfrac{1}{2}\cos\left(\tfrac{\pi}{2}\right)\right) - \left(0 - \tfrac{1}{2}\cos 0\right)\right]$$
$$= \pi\left(\tfrac{\pi}{4} + \tfrac{1}{2}\right) \text{ units}^3$$

## EXERCISE 19D.2

**1** **a** The graphs meet where $4 - x^2 = 3$

$\therefore\ x^2 = 1$

$\therefore\ x = \pm 1$

$\therefore$ A is at $(-1, 3)$ and B is at $(1, 3)$.

**b**
$$V = \pi\int_{-1}^{1} \left((4-x^2)^2 - 3^2\right)\ dx$$
$$= \pi\int_{-1}^{1} \left(16 - 8x^2 + x^4 - 9\right)\ dx$$
$$= \pi\int_{-1}^{1} \left(x^4 - 8x^2 + 7\right)\ dx$$
$$= \pi\left[\frac{x^5}{5} - \frac{8x^3}{3} + 7x\right]_{-1}^{1}$$
$$= \pi\left(\tfrac{1}{5} - \tfrac{8}{3} + 7 - \left(\tfrac{-1}{5} - \tfrac{-8}{3} - 7\right)\right)$$
$$= \tfrac{136\pi}{15} \text{ units}^3$$

**2** **a** The graphs meet where $e^{\frac{x}{2}} = e$

$\therefore\ e^{\frac{x}{2}} = e^1$

$\therefore\ \dfrac{x}{2} = 1$

$\therefore\ x = 2$

$\therefore$ A is at $(2, e)$.

**b**
$$V = \pi\int_0^2 \left(e^2 - \left(e^{\frac{x}{2}}\right)^2\right)\ dx$$
$$= \pi\int_0^2 \left(e^2 - e^x\right)\ dx$$
$$= \pi\left[e^2x - e^x\right]_0^2$$
$$= \pi\left[(2e^2 - e^2) - (0 - 1)\right]$$
$$= \pi(e^2 + 1) \text{ units}^3$$

**3** **a** The graphs meet where $x = \frac{1}{x}$

$\therefore \; x^2 = 1$

$\therefore \; x = \pm 1$

$\therefore \; x = 1 \quad \{\text{as } x > 0\}$

$\therefore$ A is at (1, 1).

**b** $V = \pi \int_1^2 \left(x^2 - \left(\frac{1}{x}\right)^2\right) dx$

$= \pi \int_1^2 \left(x^2 - x^{-2}\right) dx$

$= \pi \left[\frac{x^3}{3} - \frac{x^{-1}}{-1}\right]_1^2$

$= \pi \left[\left(\frac{8}{3} + \frac{1}{2}\right) - \left(\frac{1}{3} + 1\right)\right]$

$= \frac{11\pi}{6}$ units$^3$

**4** **a** The curves meet where $\sqrt{x-4} = 1$

$\therefore \; x - 4 = 1$

$\therefore \; x = 5$

$\therefore$ A is at (5, 1).

**b** $V = \pi \int_5^8 \left(\left(\sqrt{x-4}\right)^2 - 1^2\right) dx$

$= \pi \int_5^8 (x - 4 - 1)\, dx$

$= \pi \int_5^8 (x - 5)\, dx$

$= \pi \left[\frac{x^2}{2} - 5x\right]_5^8$

$= \pi \left[(32 - 40) - \left(\frac{25}{2} - 25\right)\right]$

$= \frac{9\pi}{2}$ units$^3$

**5** The shaded area $= \int_1^\infty \frac{1}{x}\, dx$

$= \lim_{t\to\infty} \int_1^t \frac{1}{x}\, dx$

$= \lim_{t\to\infty} [\ln(x)]_1^t, \; x > 0$

$= \lim_{t\to\infty} \ln t$, which is infinite

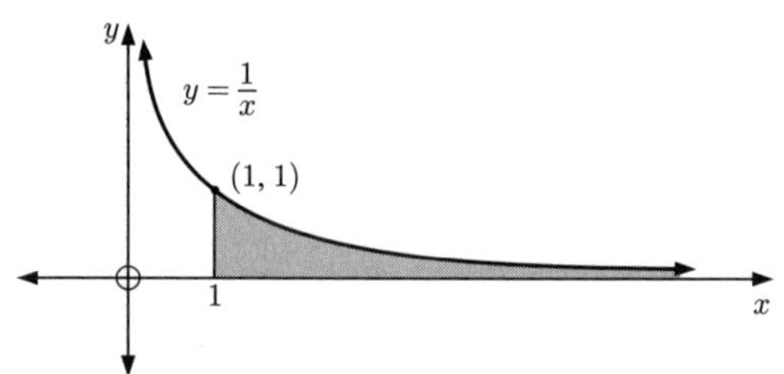


The volume of revolution $= \pi \int_1^\infty \left(\frac{1}{x}\right)^2 dx$

$= \pi \lim_{t\to\infty} \int_1^t x^{-2}\, dx$

$= \pi \lim_{t\to\infty} \left[-\frac{1}{x}\right]_1^t$

$= \pi \lim_{t\to\infty} \left(-\frac{1}{t} + 1\right)$

$= \pi$, which is finite

## REVIEW SET 19A

**1** shaded area $= \int_a^b [f(x) - g(x)]\, dx + \int_b^c [g(x) - f(x)]\, dx + \int_c^d [f(x) - g(x)]\, dx$

**2**

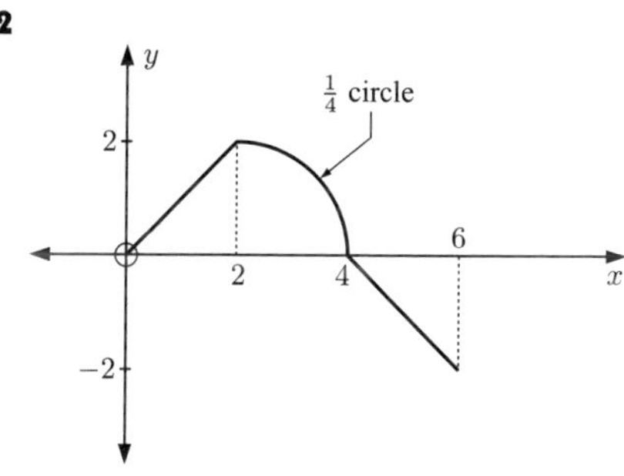


**a** $\int_0^4 f(x)\,dx = \text{area of triangle} + \text{area of } \frac{1}{4} \text{ circle}$
$= \frac{1}{2}(2 \times 2) + \frac{1}{4}\pi(2)^2$
$= (2 + \pi) \text{ units}^2$

**b** $\int_4^6 f(x)\,dx = -\text{area of triangle below } x\text{-axis}$
$= -\frac{1}{2}(2 \times 2)$
$= -2 \text{ units}^2$

**c** $\int_0^6 f(x)\,dx = \int_0^4 f(x)\,dx + \int_4^6 f(x)\,dx$
$= (2 + \pi) + (-2)$
$= \pi \text{ units}^2$

**3** $\int_{-1}^3 f(x)\,dx$ gives us the correct area only if $f(x)$ is non-negative on the interval $-1 \leqslant x \leqslant 3$.

In this case $f(x)$ is negative for $1 < x < 3$, so $\int_{-1}^3 f(x)\,dx$ does not provide the correct answer.

(The shaded area which is below the $x$-axis is given by $\int_1^3 [0 - f(x)]\,dx = -\int_1^3 f(x)\,dx$.)

**4** $y = k$ meets $y = x^2$ where $x^2 = k$ $\quad \therefore \ x = \pm\sqrt{k}$

By symmetry, $\int_0^{\sqrt{k}} (k - x^2)\,dx = \frac{1}{2} \times 5\frac{1}{3} = \frac{1}{2} \times \frac{16}{3}$

$$\therefore \ \left[kx - \frac{x^3}{3}\right]_0^{\sqrt{k}} = \frac{8}{3}$$

$$\therefore \ k\sqrt{k} - \frac{k\sqrt{k}}{3} = \frac{8}{3}$$

$$\therefore \ \tfrac{2}{3}k\sqrt{k} = \tfrac{8}{3}$$

$$\therefore \ k\sqrt{k} = 4$$

$$\therefore \ k^{\frac{3}{2}} = 4$$

$$\therefore \ k = 4^{\frac{2}{3}} = \sqrt[3]{16}$$

**5**

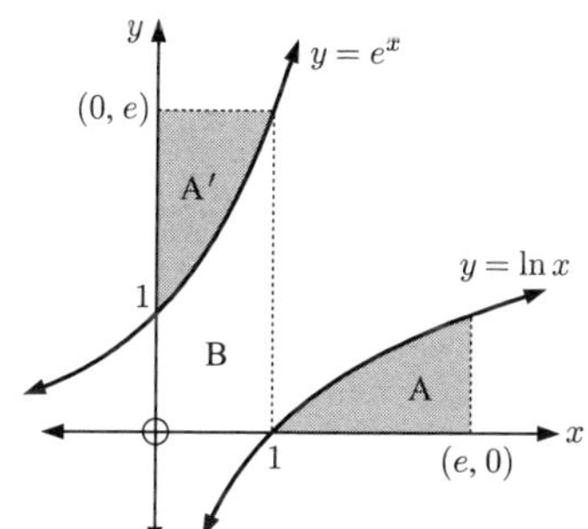


$y = e^x$ and $y = \ln x$ are inverse functions, so they are symmetrical about $y = x$

$\therefore$ area A = area A′

But area A′ + area B = area of rectangle

$\therefore$ area A + area B $= e \times 1 = e$

Since area A $= \int_1^e \ln x\,dx$

and area B $= \int_0^1 e^x\,dx$,

$\int_0^1 e^x\,dx + \int_1^e \ln x\,dx = e$

**6** $y = x^2 + 4x + 1$ meets $y = 3x + 3$ where

$$x^2 + 4x + 1 = 3x + 3$$
$$\therefore \ x^2 + x - 2 = 0$$
$$\therefore \ (x + 2)(x - 1) = 0$$
$$\therefore \ x = -2 \text{ or } 1$$

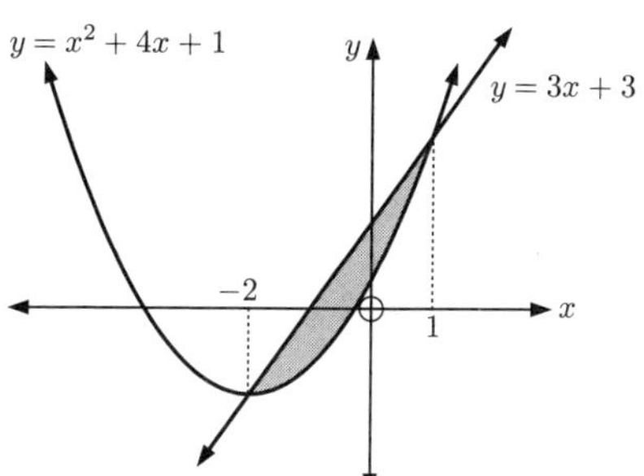


$$\therefore \ \text{area} = \int_{-2}^1 [(3x + 3) - (x^2 + 4x + 1)]\,dx$$
$$= \int_{-2}^1 (-x^2 - x + 2)\,dx$$
$$= \left[-\frac{x^3}{3} - \frac{x^2}{2} + 2x\right]_{-2}^1$$
$$= \left(-\tfrac{1}{3} - \tfrac{1}{2} + 2\right) - \left(\tfrac{8}{3} - 2 - 4\right)$$
$$= -\tfrac{1}{3} - \tfrac{1}{2} + 2 - \tfrac{8}{3} + 2 + 4$$
$$= 4\tfrac{1}{2} \text{ units}^2$$

**7** **a** $v(t) = t^2 - 6t + 8 \text{ m s}^{-1}, \quad t \geqslant 0$

$= (t-4)(t-2)$ which has sign diagram:

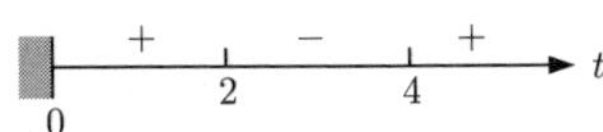


**b** Now $s(t) = \int (t^2 - 6t + 8)\, dt$

$= \dfrac{t^3}{3} - 3t^2 + 8t + c$

$\therefore \quad s(0) = c$

$s(2) = c + 6\frac{2}{3}$

$s(4) = c + 5\frac{1}{3}$

$s(5) = c + 6\frac{2}{3}$

the motion diagram is:

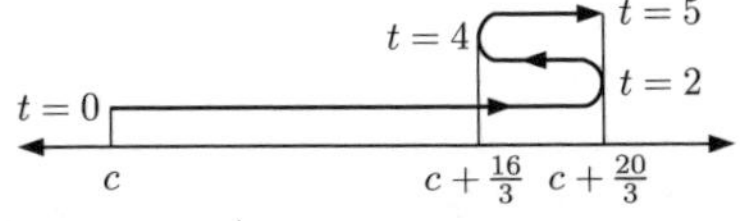


The particle moves in the positive direction initially, then at $t = 2$, $6\frac{2}{3}$ m from its starting point, it changes direction. It changes direction again at $t = 4$, $5\frac{1}{3}$ m from its starting point. When $t = 5$ it is $6\frac{2}{3}$ m from its starting point.

**c** After 5 seconds, the particle is $6\frac{2}{3}$ m to the right of its starting point.

**d** The total distance travelled $= (c + \frac{20}{3} - c) + [(c + \frac{20}{3}) - (c + \frac{16}{3})] + [(c + \frac{20}{3}) - (c + \frac{16}{3})]$

$= 9\frac{1}{3}$ m

**8** Consider $y = 4e^x - 1$.

The $x$-intercept occurs when $y = 0$

$\therefore \quad 4e^x - 1 = 0$

$\therefore \quad e^x = \frac{1}{4}$

$\therefore \quad x = \ln \frac{1}{4} < 0$

$y = 4e^x - 1$ is the graph of $y = e^x$ with a vertical stretch of factor 4 and a vertical translation of $-1$.

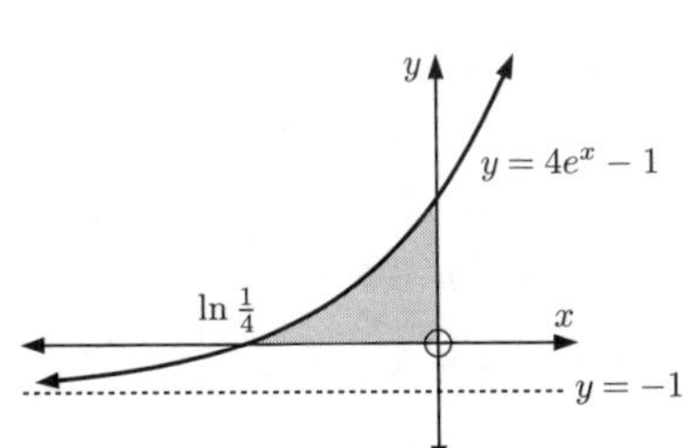


$$\text{Area} = \int_{\ln \frac{1}{4}}^{0} (4e^x - 1)\, dx$$

$$= [4e^x - x]_{\ln \frac{1}{4}}^{0}$$

$$= \left(4e^0 - 0\right) - \left(4e^{\ln \frac{1}{4}} - \ln \tfrac{1}{4}\right)$$

$$= 4 - 0 - 4(\tfrac{1}{4}) + \ln \tfrac{1}{4}$$

$$= 3 + \ln \tfrac{1}{4} \text{ units}^2$$

$$= (3 - \ln 4) \text{ units}^2$$

**9** The slope of the straight line is $\dfrac{0-8}{4-0} = \frac{-8}{4} = -2$

$\therefore$ the straight line has equation $y = -2x + 8$

$\therefore$ the volume of revolution

$= \pi \int_0^2 \left(x^2\right)^2\, dx + \pi \int_2^4 (-2x + 8)^2\, dx$

$= \pi \int_0^2 x^4\, dx + \pi \int_2^4 (4x^2 - 32x + 64)\, dx$

$= \pi \left[\frac{1}{5}x^5\right]_0^2 + \pi \left[\frac{4}{3}x^3 - 16x^2 + 64x\right]_2^4$

$= \pi \left(\frac{1}{5}(2)^5 - 0\right) + \pi \left[\frac{4}{3}(4)^3 - 16(4)^2 + 64(4) - \left(\frac{4}{3}(2)^3 - 16(2)^2 + 64(2)\right)\right]$

$= \pi \times \frac{32}{5} + \pi \left(\frac{256}{3} - \frac{224}{3}\right)$

$= \frac{32\pi}{5} + \frac{32\pi}{3}$

$= \frac{256\pi}{15}$ as required

## REVIEW SET 19B

**1** **a** $a(t) = v'(t)$

$\therefore\ a(t) = 2 - 6t\ \ \text{m s}^{-2}$

**b** $s(t) = \int (2t - 3t^2)\, dt$

$\therefore\ s(t) = t^2 - t^3 + c\ \ \text{m}$

**c** Change in displacement after two seconds

$= s(2) - s(0)$

$= 2^2 - 2^3 + c - (0^2 - 0^3 + c)$

$= 4 - 8 + c - c$

$= -4$ m (4 m to the left)

**2** **a** $f(x) = \dfrac{x}{1+x^2}$ $\therefore\ f'(x) = \dfrac{1(1+x^2) - x(2x)}{(1+x^2)^2}$ {quotient rule}

$= \dfrac{1 + x^2 - 2x^2}{(1+x^2)^2}$

$= \dfrac{1 - x^2}{(1+x^2)^2}$

$= \dfrac{(1+x)(1-x)}{(1+x^2)^2}$

which has sign diagram:

$-$ $+$ $-$ ; $-1$, $1$, $x$

$\therefore$ there is a local minimum at $(-1, -\frac{1}{2})$ and a local maximum at $(1, \frac{1}{2})$.

**b** As $x \to \infty$, $f(x) \to 0^+$.

As $x \to -\infty$, $f(x) \to 0^-$.

**c**

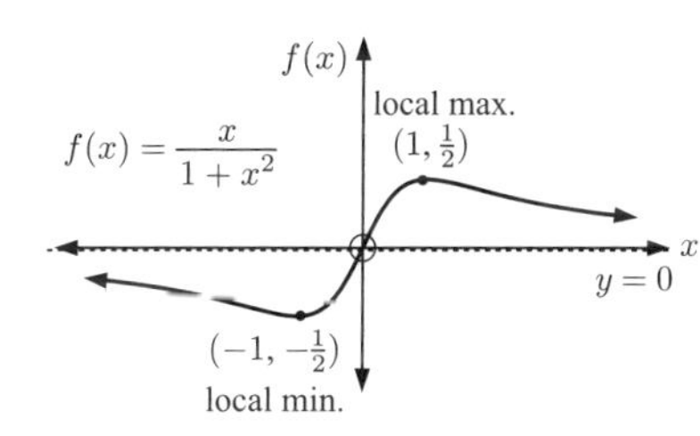


**d** Area $= \displaystyle\int_{-2}^{0} \left[0 - \frac{x}{1+x^2}\right] dx$

$= \displaystyle\int_{-2}^{0} \frac{-x}{1+x^2}\, dx$

$\approx 0.805$ units$^2$

**3** $v(t) = \sin t$ which has sign diagram:

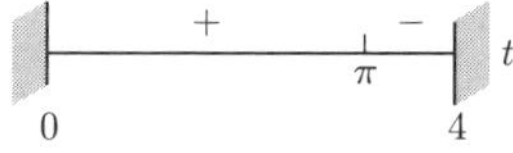


Now $s(t) = \int \sin t\, dt$

$= -\cos t + c$ metres

$\therefore\ s(0) = -1 + c$

$s(\pi) = 1 + c$

$s(4) = -\cos 4 + c \approx c + 0.654$

motion diagram:

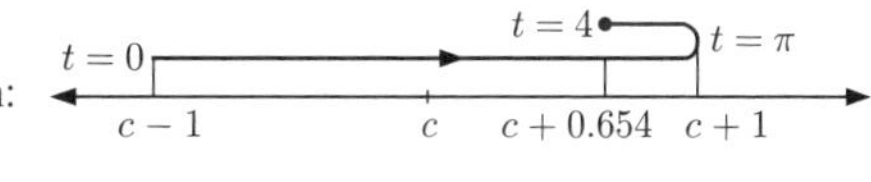


$\therefore$ total distance travelled $= [(c+1) - (c-1)] + [(c+1) - (c+0.654)]$

$\approx 2.35$ m

**4** $v(t) = \dfrac{100}{(t+2)^2} = 100(t+2)^{-2}\ \text{m s}^{-1}$

**a** At $t = 0$, $v(0) = \dfrac{100}{2^2} = 25\ \text{m s}^{-1}$

At $t = 3$, $v(3) = \dfrac{100}{5^2} = 4\ \text{m s}^{-1}$

**b** As $t \to \infty$, $v(t) \to 0^+$

**c**

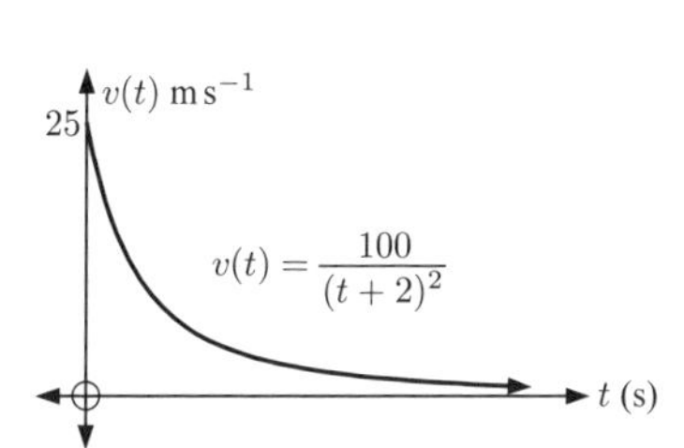

**d** As $v(t)$ is always positive, the boat is always travelling forwards.

$$s(t) = \int v(t)\,dt$$
$$= \int 100(t+2)^{-2}\,dt$$
$$= -100(t+2)^{-1} + c$$
$$= \frac{-100}{t+2} + c$$

$\therefore\ s(0) = c - 50$ m

$\therefore$ when the boat has travelled 30 m,

$s(t) = c - 20$ m

$$\therefore\ c - 20 = \frac{-100}{t+2} + c$$
$$\therefore\ \frac{-100}{t+2} = -20$$
$$\therefore\ t + 2 = 5$$
$$\therefore\ t = 3 \text{ seconds}$$

**e** $$a(t) = v'(t)$$
$$= -200(t+2)^{-3}$$
$$= \frac{-200}{(t+2)^3}\ \text{m s}^{-2}, \quad t \geqslant 0$$

**f** $$\frac{dv}{dt} = \frac{-200}{(t+2)^3} = -\tfrac{1}{5}\frac{1000}{(t+2)^3}$$
$$= -\tfrac{1}{5}\left(\frac{100}{(t+2)^2}\right)^{\frac{3}{2}}$$
$$= -\tfrac{1}{5}\,v^{\frac{3}{2}}$$
$$\therefore\ \frac{dv}{dt} = -kv^{\frac{3}{2}} \quad \text{where} \quad k = \tfrac{1}{5}$$

**5** **a** The graphs meet when $\cos 2x = e^{3x}$

Using technology, $x = 0$ and $x \approx -0.7292$

**b** Shaded area $\approx \int_{-0.7292}^{0} (\cos 2x - e^{3x})\,dx$

$\approx 0.2009$ units$^2$ {using technology}

**6**
$$\int_0^m \sin x\,dx = \tfrac{1}{2}$$
$$\therefore\ [-\cos x]_0^m = \tfrac{1}{2}$$
$$\therefore\ -\cos m + \cos 0 = \tfrac{1}{2}$$
$$\therefore\ \cos m = \tfrac{1}{2}$$
$$\therefore\ m = \tfrac{\pi}{3} \quad \{0 < m < \tfrac{\pi}{2}\}$$

**7** **a** The graphs meet where

$x^2 = \sin x$

$\therefore\ x = 0$ or $\approx 0.8767$ {using technology}

$\therefore\ a \approx 0.8767$

**b** area $\approx \int_0^{0.8767} (\sin x - x^2)\,dx$

$\approx 0.1357$ units$^2$ {using technology}

**8** **a** $y = \cos(2x)$ meets the $x$-axis where $2x = \frac{\pi}{2}$, or $x = \frac{\pi}{4}$.

$$\therefore\ V = \pi \int_{\frac{\pi}{16}}^{\frac{\pi}{4}} \cos^2(2x)\,dx = \pi \int_{\frac{\pi}{16}}^{\frac{\pi}{4}} \left(\tfrac{1}{2} + \tfrac{1}{2}\cos(4x)\right) dx$$
$$= \pi\left[\tfrac{1}{2}x + \tfrac{1}{8}\sin(4x)\right]_{\frac{\pi}{16}}^{\frac{\pi}{4}}$$
$$= \pi\left[\left(\tfrac{\pi}{8} + \tfrac{1}{8}\sin\pi\right) - \left(\tfrac{\pi}{32} + \tfrac{1}{8}\sin\left(\tfrac{\pi}{4}\right)\right)\right]$$
$$= \pi\left(\tfrac{\pi}{8} - \tfrac{\pi}{32} - \tfrac{1}{8}\left(\tfrac{1}{\sqrt{2}}\right)\right)$$
$$= \pi\left(\tfrac{3\pi}{32} - \tfrac{1}{8\sqrt{2}}\right) \text{ units}^3$$

**b** $V = \pi \int_0^2 (e^{-x} + 4)^2 \, dx$

$= \pi \int_0^2 \left(e^{-2x} + 8e^{-x} + 16\right) \, dx$

$= \pi \left[\frac{1}{-2}e^{-2x} + \frac{8}{-1}e^{-x} + 16x\right]_0^2$

$= \pi \left[\left(-\frac{1}{2}e^{-4} - 8e^{-2} + 32\right) - \left(-\frac{1}{2} - 8\right)\right]$

$= \pi \left(\frac{81}{2} - \frac{1}{2e^4} - \frac{8}{e^2}\right)$ units$^3$

$\approx 124$ units$^3$

## REVIEW SET 19C

**1** **a** $a(t) = 6t - 30 \text{ cm s}^{-2}$

$v(t) = \int (6t - 30) \, dt$

$= 3t^2 - 30t + c$

But $v(0) = 27$

$\therefore \ 0 - 0 + c = 27$

$\therefore \ c = 27$

$\therefore \ v(t) = 3t^2 - 30t + 27 \text{ cm s}^{-1}$

**b** Displacement after 6 seconds

$= \int_0^6 (3t^2 - 30t + 27) \, dt$

$= \left[t^3 - 15t^2 + 27t\right]_0^6$

$= 6^3 - 15(6)^2 + 27(6) - 0$

$= -162$ cm

(162 cm to the left of the origin)

**2** **a**

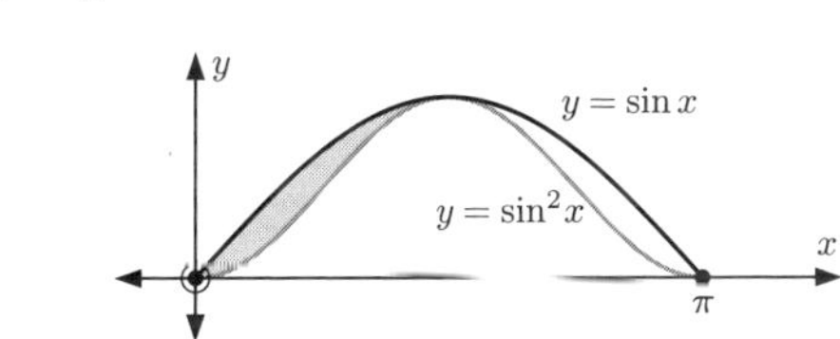


**b** Area $= \int_0^{\frac{\pi}{2}} (\sin x - \sin^2 x) \, dx$

$= \int_0^{\frac{\pi}{2}} (\sin x - (\frac{1}{2} - \frac{1}{2}\cos 2x)) \, dx$

$= \int_0^{\frac{\pi}{2}} (\sin x + \frac{1}{2}\cos 2x - \frac{1}{2}) \, dx$

$= \left[-\cos x + \frac{1}{4}\sin 2x - \frac{1}{2}x\right]_0^{\frac{\pi}{2}}$

$= \left(0 + \frac{1}{4}(0) - \frac{\pi}{4}\right) - (-1 + 0 - 0)$

$= \left(1 - \frac{\pi}{4}\right)$ units$^2$

**3** The area between $x = 0$ and $x = a$ is 2 units$^2$.

$\therefore \ \int_0^a e^x \, dx = 2$

$\therefore \ [e^x]_0^a = 2$

$\therefore \ e^a - e^0 = 2$

$\therefore \ e^a = 3$

$\therefore \ a = \ln 3$

The area between $x = a = \ln 3$ and $x = b$ is 2 units$^2$.

$\therefore \ \int_{\ln 3}^b e^x \, dx = 2$

$\therefore \ [e^x]_{\ln 3}^b = 2$

$\therefore \ e^b - e^{\ln 3} = 2$

$\therefore \ e^b - 3 = 2$

$\therefore \ e^b = 5$

$\therefore \ b = \ln 5$

**4**

Required area = area of $\triangle$ – area under sine curve

$= \frac{1}{2}\pi \times \pi - \int_0^\pi \sin x \, dx$

$= \frac{\pi^2}{2} - [-\cos x]_0^\pi$

$= \frac{\pi^2}{2} - [-\cos \pi + \cos 0]$

$= \left(\frac{\pi^2}{2} - 2\right)$ units$^2$

**5** The graphs meet when $\frac{2}{\pi}x = \sin x$

$\therefore\ x = -\frac{\pi}{2},\ 0,\ \frac{\pi}{2}$ {using technology}

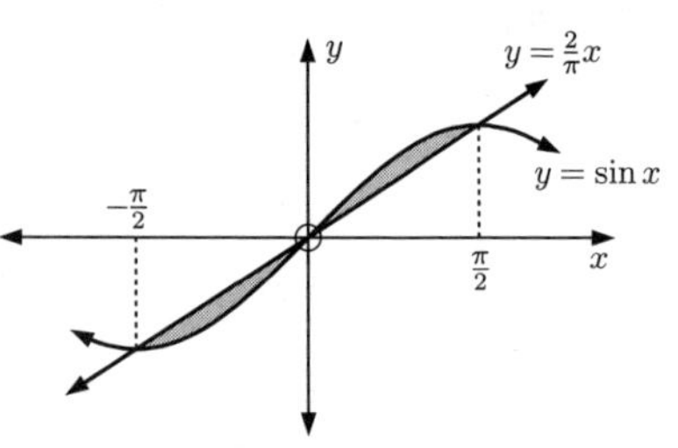


$$\therefore\ \text{area} = \int_{-\frac{\pi}{2}}^{0} \left(\tfrac{2}{\pi}x - \sin x\right) dx + \int_{0}^{\frac{\pi}{2}} \left(\sin x - \tfrac{2}{\pi}x\right) dx$$

$$= \left[\frac{x^2}{\pi} + \cos x\right]_{-\frac{\pi}{2}}^{0} + \left[-\cos x - \frac{x^2}{\pi}\right]_{0}^{\frac{\pi}{2}}$$

$$= (0+1) - \left(\tfrac{\pi}{4} + 0\right) + \left(0 - \tfrac{\pi}{4}\right) - (-1 - 0)$$

$$= \left(2 - \tfrac{\pi}{2}\right) \text{ units}^2$$

**6** The coordinates of B are $(2,\ 4+k)$

$\therefore$ area rectangle OABC $= 2 \times (4+k)$

$= 8 + 2k$

$\therefore$ since the two shaded regions are equal in area, each area is $4+k$ units$^2$.

$$\therefore\ \int_0^2 (x^2 + k)\, dx = 4 + k$$

$$\therefore\ \left[\frac{x^3}{3} + kx\right]_0^2 = 4 + k$$

$$\therefore\ \tfrac{8}{3} + 2k = 4 + k$$

$$\therefore\ k = 4 - \tfrac{8}{3}$$

$$\therefore\ k = \tfrac{4}{3}$$

**7** **a**

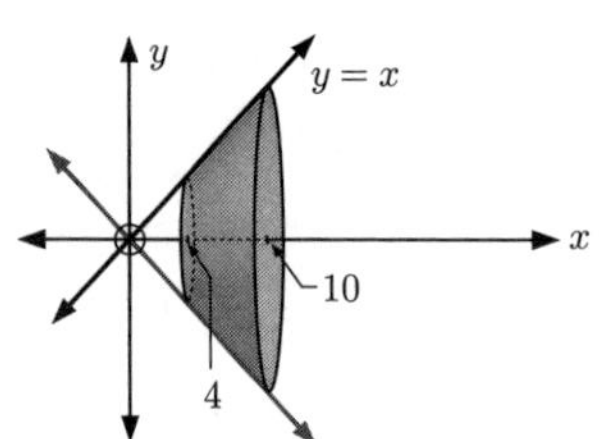


$$V = \pi \int_4^{10} x^2\, dx$$

$$= \pi \left[\frac{x^3}{3}\right]_4^{10}$$

$$= \pi \left(\tfrac{1000}{3} - \tfrac{64}{3}\right)$$

$$= \tfrac{936\pi}{3}$$

$$= 312\pi \text{ units}^3$$

**b**

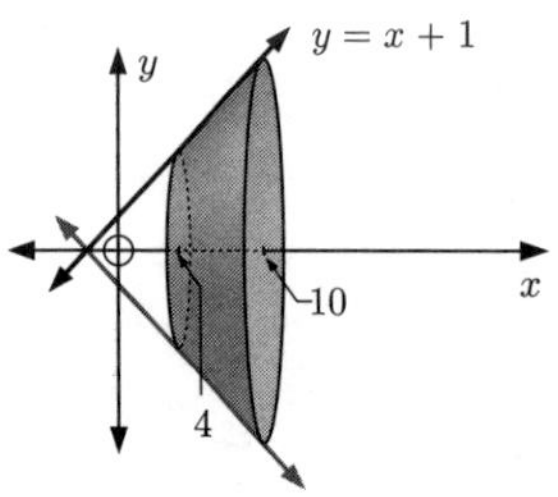


$$V = \pi \int_4^{10} (x+1)^2\, dx$$

$$= \pi \left[\frac{(x+1)^3}{3}\right]_4^{10}$$

$$= \pi \left(\tfrac{11^3}{3} - \tfrac{5^3}{3}\right)$$

$$= \tfrac{1206\pi}{3} = 402\pi \text{ units}^3$$

**c**

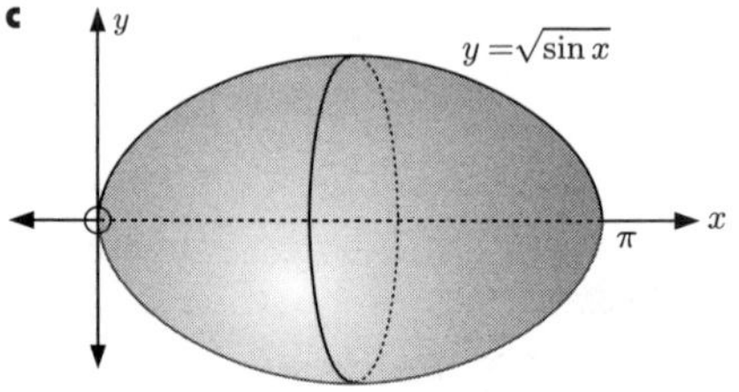

$$V = \pi \int_0^{\pi} \left(\sqrt{\sin x}\right)^2 dx$$

$$= \pi \int_0^{\pi} \sin x\, dx$$

$$= \pi \left[-\cos x\right]_0^{\pi}$$

$$= \pi\,(1 - (-1))$$

$$= 2\pi \text{ units}^3$$

**d**

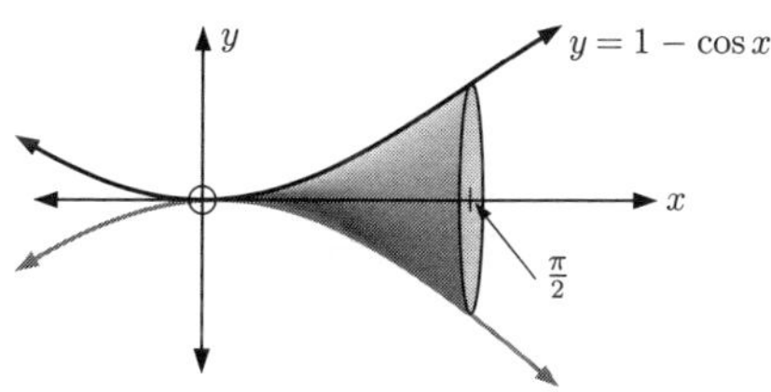


$V = \pi \int_0^{\frac{\pi}{2}} (1 - \cos x)^2 \, dx$

$= \pi \int_0^{\frac{\pi}{2}} (1 - 2\cos x + \cos^2 x) \, dx$

$= \pi \int_0^{\frac{\pi}{2}} (1 - 2\cos x + \frac{1}{2} + \frac{1}{2}\cos 2x) \, dx$

$= \pi \left[\frac{3}{2}x - 2\sin x + \frac{1}{4}\sin 2x\right]_0^{\frac{\pi}{2}}$

$= \pi \left[\left(\frac{3}{2}(\frac{\pi}{2}) - 2\sin(\frac{\pi}{2}) + \frac{1}{4}\sin\pi\right) - \left(\frac{3}{2}(0) - 2\sin 0 + \frac{1}{4}\sin 0\right)\right]$

$= \pi \left(\frac{3\pi}{4} - 2\right)$ units$^3$

$= \pi \left(\frac{3\pi - 8}{4}\right)$ units$^3$

**8** **a** $V = \frac{1}{3}\pi r^2 h$

$= \frac{1}{3}\pi \times 4^2 \times 8$

$= \frac{1}{3}\pi \times 128$

$= \frac{128\pi}{3}$ units$^3$

**b**

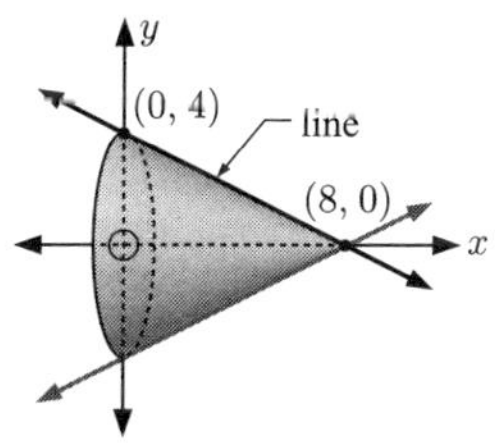


gradient $= \frac{0-4}{8-0} = -\frac{1}{2}$

$\therefore$ the line has equation $y = -\frac{1}{2}x + 4$

$\therefore \quad V = \pi \int_0^8 \left(-\frac{1}{2}x + 4\right)^2 dx$

$= \pi \int_0^8 \left(\frac{x^2}{4} - 4x + 16\right) dx$

$= \pi \left[\frac{x^3}{12} - \frac{4x^2}{2} + 16x\right]_0^8$

$= \pi \left(\frac{128}{3} - 128 + 128 - 0\right)$

$= \frac{128\pi}{3}$ units$^3$ ✓

**9** $y = \sin x$ and $y = \cos x$ meet where $\sin x = \cos x$

$\therefore \quad \frac{\sin x}{\cos x} = 1$

$\therefore \quad \tan x = 1$

$\therefore \quad x = \frac{\pi}{4}$

Hence $V = \pi \int_0^{\frac{\pi}{4}} (\cos^2 x - \sin^2 x) \, dx$

$= \pi \int_0^{\frac{\pi}{4}} \cos(2x) \, dx$

$= \pi \left[\frac{1}{2}\sin(2x)\right]_0^{\frac{\pi}{4}}$

$= \pi \left(\frac{1}{2}\sin\left(\frac{\pi}{2}\right) - \frac{1}{2}\sin 0\right)$

$= \pi \left(\frac{1}{2}(1) - 0\right)$

$= \frac{\pi}{2}$ units$^3$

# Chapter 20
## DESCRIPTIVE STATISTICS

### EXERCISE 20A

**1** **a** Heights can take any value from 170 cm to 205 cm, including decimal values such as 181.37 cm. The 'height' variable can take any real number between 170 and 205.

**b** 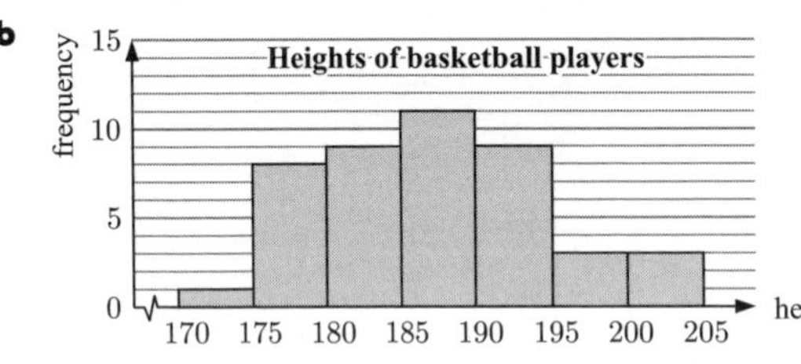


**c** The modal class is the class occurring most often. This is $185 \leqslant H < 190$ cm.

**d** The distribution is slightly positively skewed, as there is more of a 'tail' to the right.

**2** **a** The data is continuous numerical. Actual time is continuous and could be measured to the nearest second, millisecond, and so on. After it has been rounded to the nearest minute, it becomes discrete numerical data.

**b**

| *Time* (min) | *Tally* | *Frequency* |
|---|---|---|
| 0 - 9 | ~~\|\|\|\|~~ \| | 6 |
| 10 - 19 | ~~\|\|\|\|~~ ~~\|\|\|\|~~ ~~\|\|\|\|~~ ~~\|\|\|\|~~ ~~\|\|\|\|~~ \| | 26 |
| 20 - 29 | ~~\|\|\|\|~~ ~~\|\|\|\|~~ \|\|\| | 13 |
| 30 - 39 | ~~\|\|\|\|~~ \|\|\|\| | 9 |
| 40 - 49 | ~~\|\|\|\|~~ \| | 6 |

**c** The distribution is positively skewed, or skewed to the high end.

**d** The travelling time modal class was between 10 and 19 min, if considering classes. The mode is actually 10, as 10 occurs most frequently.

**3** **a** The data is discrete numerical, so a column graph should be used.

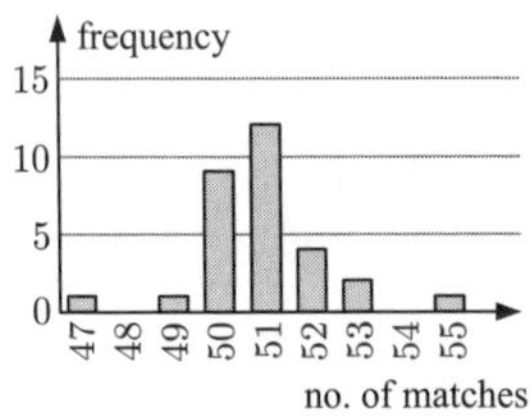


**b** The data is continuous, so a frequency histogram should be used.

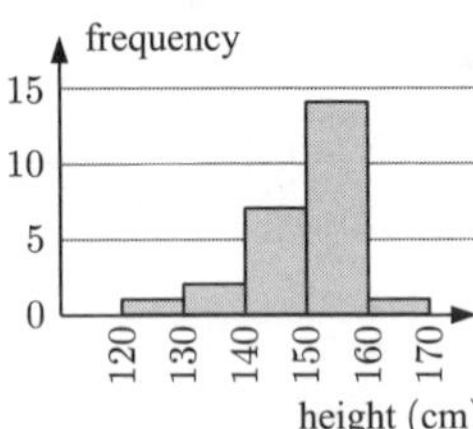


**4** **a** Number which are $\geqslant 400$ mm is $14 + 6 = 20$ seedlings.

**b** $12 + 18 + 42 + 28 + 14 + 6 = 120$ seedlings have been sampled.

$$\therefore \ \% \text{ between 349 and 400} = \frac{42+28}{120} \times 100\% = \frac{70}{120} \times 100\% \approx 58.3\%$$

**c** **i** Number less than 400 mm

$$= \frac{12+18+42+28}{120} \times 1462 = \frac{100}{120} \times 1462 \approx 1218 \text{ seedlings}$$

**ii** Number between 374 mm and 425 mm

$$= \frac{28+14}{120} \times 1462 = \frac{42}{120} \times 1462 \approx 512 \text{ seedlings}$$

## EXERCISE 20B.1

**1** **a** **i** mean $= \dfrac{2+3+3+3+4+\ldots.+9+9}{23}$
$= \frac{129}{23}$
$\approx 5.61$

**ii** median = 12th score (when in order)
$= 6$

**iii** mode $= 6$ (6 occurs most often)

**b** **i** mean $= \dfrac{10+12+12+15+\ldots.+20+21}{15}$
$= \frac{245}{15}$
$\approx 16.3$

**ii** median = 8th score (when in order)
$= 17$

**iii** mode $= 18$

**c** **i** mean $= \dfrac{22.4+24.6+21.8+\ldots.+23.5}{11}$
$= \frac{273}{11}$
$\approx 24.8$

**ii** median = 6th score (when in order)
$= 24.9$

**iii** mode $= 23.5$

**2** **a** mean of set A $= \dfrac{3+4+4+5+\ldots.+10}{13}$
$\approx 6.46$

mean of set B $= \dfrac{3+4+4+5+\ldots.+15}{13}$
$\approx 6.85$

**b** median of set A = 7th score = 7 median of set B = 7th score = 7

**c** The data sets are the same except for the last value, and the last value of set A is less than that of set B. So, the mean of set A is less than that of set B.

**d** The middle value of both data sets is the same, so the median is the same.

**3** **a** mean $= \dfrac{23\,000+46\,000+23\,000+\ldots.+32\,000}{10} = \$29\,300$

median = middle score when in order of size $= \dfrac{\$23\,000+\$24\,000}{2} = \$23\,500$

mode $= \$23\,000$

**b** The mode is unsatisfactory because it is the lowest salary. It does not take the higher values into account.

**c** The median is too close to the lower end of the distribution since the data is positively skewed. So the median is not a satisfactory measure of the middle.

**4** **a** mean $= \dfrac{3+1+0+0+\ldots.+1+0+0}{31} = \frac{99}{31} \approx 3.19$

median = 16th score (when in order) $= 0$

mode $= 0$ (most frequently occurring score)

**b** The median is not in the centre, as the data is very positively skewed.

**c** The mode is the lowest value. It does not take the higher values into account.

**d** Yes, 21 and 42. **e** No, as this would ignore actual valid data.

**5** **a** mean $= \dfrac{43+55+41+37}{4} = \frac{176}{4} = 44$ points **b** another 44 points

**c** **i** new mean $= \dfrac{176+25}{5} = 40.2$ points

**ii** It will increase the new mean to 40.3 points as 41 points is greater than the old mean of 40.2 points.

$\left\{\dfrac{5\times 40.2+41}{6} \approx 40.3\right\}$

**6** mean $= \dfrac{\text{total}}{12}$ $\therefore$ $\$15\,467 = \dfrac{\text{total}}{12}$ $\therefore$ total $= \$15\,467 \times 12 = \$185\,604$

**7** mean $= \dfrac{\text{total}}{12}$ $\quad \therefore \quad 262 = \dfrac{\text{total}}{12}$ $\quad \therefore \quad$ total $= 262 \times 12 = 3144$ km

**8** $\overline{x} = \dfrac{\sum\limits_{i=1}^{n} x_i}{n} \quad \therefore \quad 11.6 = \dfrac{\sum\limits_{i=1}^{10} x_i}{10} \quad \therefore \quad \sum\limits_{i=1}^{10} x_i = 11.6 \times 10 = 116$

**9** Total for first 14 matches $= 14 \times 16.5$ goals $= 231$ goals

$$\therefore \quad \text{new average} = \frac{231 + 21 + 24}{16} = \frac{276}{16} = 17.25 \text{ goals per game}$$

**10** **a** mean selling price $= \dfrac{146\,400 + 127\,600 + 211\,000 + .... + 162\,500}{10} = \$163\,770$

median selling price $= \dfrac{\text{5th} + \text{6th}}{2} = \dfrac{146\,400 + 148\,000}{2} = \$147\,200$

These figures differ by \$16 570. There are more selling prices at the lower end of the market.

**b** **i** Use the mean as it tends to inflate the average house value of that district.

**ii** Use the median as you want to buy at the lowest price possible.

**11** $\dfrac{5 + 9 + 11 + 12 + 13 + 14 + 17 + x}{8} = 12$

$$\therefore \quad \frac{81 + x}{8} = 12$$
$$\therefore \quad 81 + x = 96$$
$$\therefore \quad x = 15$$

**12** $\dfrac{3 + 0 + a + a + 4 + a + 6 + a + 3}{9} = 4$

$$\therefore \quad \frac{4a + 16}{9} = 4$$
$$\therefore \quad 4a + 16 = 36$$
$$\therefore \quad 4a = 20$$
$$\therefore \quad a = 5$$

**13** $\dfrac{29 + 36 + 32 + 38 + 35 + 34 + 39 + x}{8} = 35$

$$\therefore \quad \frac{243 + x}{8} = 35$$
$$\therefore \quad 243 + x = 280$$
$$\therefore \quad x = 37$$

So, her 8th result was 37.

**14** Total for first 10 measurements $= 10 \times 15.7$
$= 157$

Total for next 20 measurements $= 20 \times 14.3$
$= 286$

$$\therefore \quad \text{mean} = \frac{157 + 286}{30} \approx 14.8$$

**15** If there are 9 measurements with a median of 12, then 12 must be one of the unknown measurements.
So, the measurements are 7, 9, 11, 12, 13, 14, 17, 19, and $a$.

$$\text{mean} = \frac{7 + 9 + 11 + 12 + 13 + 14 + 17 + 19 + a}{9} = \frac{102 + a}{9}$$
$$\therefore \quad \frac{102 + a}{9} = 12$$
$$\therefore \quad 102 + a = 108$$
$$\therefore \quad a = 6$$

So, the other measurements are 6 and 12.

**16** Scores were 5 7 9 9 10 $a$ $b$ where $a \leqslant b$ say.

$$\text{mean} = \frac{5 + 7 + 9 + 9 + 10 + a + b}{7} = 8$$
$$\therefore \quad \frac{40 + a + b}{7} = 8$$
$$\therefore \quad 40 + a + b = 56$$
$$\therefore \quad a + b = 16 \qquad \{a \leqslant 12,\ b \leqslant 12\}$$

Possibilities are:

| $a$ | 5 | 6 | 7 | 8 |
|---|---|---|---|---|
| $b$ | 11 | 10 | 9 | 8 |
| | ✗ | ✗ | ✓ | ✗ |

$a=8$: reject as modes are 8 and 9
$a=5$: reject as modes are 9 and 10
reject as modes are 5 and 9

So, the missing results are 7 and 9.

## EXERCISE 20B.2

**1** **a** The mode is 1 head, as this is the result which occurs most often.

**b** The median is the average of the 15th and 16th scores

$= \dfrac{1+1}{2} = 1$ head

**c**

| $x$ | $f$ | $fx$ |
|---|---|---|
| 0 | 4 | 0 |
| 1 | 12 | 12 |
| 2 | 11 | 22 |
| 3 | 3 | 9 |
| $\sum$ | 30 | 43 |

mean $= \dfrac{\sum fx}{\sum f}$

$= \dfrac{43}{30}$

$\approx 1.43$ heads

**2** **a** **i**

| $x$ | $f$ | $fx$ |
|---|---|---|
| 0 | 5 | 0 |
| 1 | 8 | 8 |
| 2 | 13 | 26 |
| 3 | 8 | 24 |
| 4 | 6 | 24 |
| 5 | 3 | 15 |
| 6 | 3 | 18 |
| 7 | 2 | 14 |
| 8 | 1 | 8 |
| 9 | 0 | 0 |
| 10 | 0 | 0 |
| 11 | 1 | 11 |
| $\sum$ | 50 | 148 |

mean $= \dfrac{\sum fx}{\sum f}$

$= \dfrac{148}{50}$

$= 2.96$ phone calls

**ii** median

$=$ average of 25th and 26th scores (when in order)

$= \dfrac{2+2}{2}$ $\left\{\begin{array}{l}13 \text{ scores are 1 or 0} \\ 26 \text{ scores are 2, 1, or 0}\end{array}\right\}$

$= 2$ phone calls

**iii** mode $= 2$ phone calls {occurs most often}

**b**

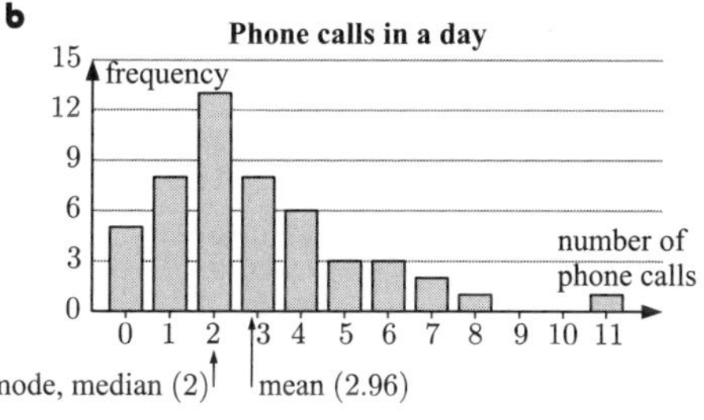


**c** The distribution is positively skewed. 11 is an outlier.

**d** The mean takes into account the larger numbers of phone calls.

**e** The mean, as it best represents all the data.

**3** **a** **i** mode $= 49$ matches {occurs most often}

**ii** median $=$ average of 15th and 16th values (when in order)

$= \dfrac{49+49}{2} = 49$ matches {9 are 47 or 48 and the next 11 are 49}

**iii**

| $x$ | $f$ | $fx$ |
|---|---|---|
| 47 | 5 | 235 |
| 48 | 4 | 192 |
| 49 | 11 | 539 |
| 50 | 6 | 300 |
| 51 | 3 | 153 |
| 52 | 1 | 52 |
| $\sum$ | 30 | 1471 |

mean $= \dfrac{\sum fx}{\sum f}$

$= \dfrac{1471}{30}$

$\approx 49.0$ matches

**b** No, as they claim the average is 50 matches per box.

**c** The sample of only 30 is not large enough. The company could have won its case by arguing that a larger sample would have found an average of 50 matches per box.

**4** **a** **i**

| $x$ | $f$ | $fx$ |
|---|---|---|
| 1 | 5 | 5 |
| 2 | 28 | 56 |
| 3 | 15 | 45 |
| 4 | 8 | 32 |
| 5 | 2 | 10 |
| 6 | 1 | 6 |
| $\sum$ | 59 | 154 |

$$\text{mean} = \frac{\sum fx}{\sum f} = \frac{154}{59} \approx 2.61 \text{ children}$$

**ii** mode = 2 children {occurs most often}

**iii** median = 30th score = 2 children

**b** This school has more children per family (2.61) than the average Australian family (2.2).

**c** Positive as the higher values are more spread out.

**d** The mean is higher than the mode and median.

**5** **a**

| *Donation* ($x$) | *Frequency* ($f$) | $fx$ |
|---|---|---|
| 1 | 7 | 7 |
| 2 | 9 | 18 |
| 3 | 2 | 6 |
| 4 | 4 | 16 |
| 5 | 8 | 40 |
| *Total* | 30 | 87 |

**b** Total number of donations $= 7 + 9 + 2 + 4 + 8$
$= 30$

**c** **i** $\text{mean} = \dfrac{\sum fx}{\sum f} = \dfrac{87}{30} = 2.9$

$\therefore$ the mean donation is \$2.90.

**ii** median = average of the 15th and 16th values (when in order)

$= \dfrac{2+2}{2} = 2$

$\therefore$ the median donation is \$2.

**iii** mode = \$2 {occurs most often}

**d** The mode can be found easily using the graph only, as it is the value with the tallest column.

**6** **a**

$$\text{mean} = \frac{\sum fx}{\sum f}$$

$$\therefore\ 4.45 = \frac{1\times 0 + 2\times 2 + 3\times 3 + 4\times 5 + 5\times x + 6\times 4 + 7\times 1}{0+2+3+5+x+4+1}$$

$$\therefore\ 4.45 = \frac{64+5x}{15+x}$$

$$\therefore\ 4.45(15+x) = 64+5x$$

$$\therefore\ 2.75 = 0.55x$$

$$\therefore\ x = 5$$

**b** From a total of $2+3+5+5+4+1 = 20$ students, $5+5+4+1 = 15$ students scored 4 or more.

$\therefore$ $\frac{15}{20} = 75\%$ of the students passed.

**7** **a** **Without fertiliser**

| | |
|---|---|
| 2 | \|\| |
| 3 | 卌 卌 \| |
| 4 | 卌 卌 卌 \|\|\|\| |
| 5 | 卌 卌 卌 卌 卌 \|\|\|\| |
| 6 | 卌 卌 卌 卌 卌 卌 卌 卌 卌 卌 \| |
| 7 | 卌 卌 卌 卌 卌 |
| 8 | 卌 卌 \|\| |
| 9 | \| |

| $x$ | $f$ | $fx$ | $cf$ |
|---|---|---|---|
| 2 | 2 | 4 | 2 |
| 3 | 11 | 33 | 13 |
| 4 | 19 | 76 | 32 |
| 5 | 29 | 145 | 61 |
| 6 | 51 | 306 | 112 |
| 7 | 25 | 175 | 137 |
| 8 | 12 | 96 | 149 |
| 9 | 1 | 9 | 150 |

**i** mean $= \dfrac{\sum fx}{\sum f} = \dfrac{844}{150} \approx 5.63$ peas/pod  **ii** mode = 6 peas/pod {occurs most often}

**iii** median = average of 75th and 76th scores $= \dfrac{6+6}{2} = 6$ peas/pod

**b** **With fertiliser**

| | |
|---|---|
| 3 | \|\|\|\| |
| 4 | 卌 卌 \|\|\| |
| 5 | 卌 卌 \| |
| 6 | 卌 卌 卌 卌 卌 \|\|\| |
| 7 | 卌 卌 卌 卌 卌 卌 卌 卌 卌 \|\| |
| 8 | 卌 卌 卌 卌 卌 \|\| |
| 9 | 卌 卌 \|\|\|\| |
| 10 | \|\|\|\| |
| 11 | \| |
| 13 | \| |

| $x$ | $f$ | $fx$ | $cf$ |
|---|---|---|---|
| 3 | 4 | 12 | 4 |
| 4 | 13 | 52 | 17 |
| 5 | 11 | 55 | 28 |
| 6 | 28 | 168 | 56 |
| 7 | 47 | 329 | 103 |
| 8 | 27 | 216 | 130 |
| 9 | 14 | 126 | 144 |
| 10 | 4 | 40 | 148 |
| 11 | 1 | 11 | 149 |
| 13 | 1 | 13 | 150 |

**i** mean $= \dfrac{\sum fx}{\sum f} = \dfrac{1022}{150} \approx 6.81$ peas/pod  **ii** mode = 7 peas/pod {occurs most often}

**iii** median = average of 75th and 76th scores $= \dfrac{7+7}{2} = 7$ peas/pod

**c** The mean best represents the centre for this data.

**d** Yes, as a mean of 6.81 peas per pod is significantly greater than a mean of 5.63 peas per pod.

**Note:** The total yield of the crop may not have improved as, for example, the number of pods per plant may have decreased when using the fertiliser.

**8** **a** mean birth mass $= \dfrac{75+70+80+ .... +83}{8} = \dfrac{567}{8} \approx 70.9$ grams

**b** mean after 2 weeks $= \dfrac{210+200+200+ .... +230}{8} = \dfrac{1681}{8} \approx 210$ grams

**c** mean increase $\approx (210.13 - 70.88)$ grams $\approx 139$ grams

**9** The 31 scores in order are: {15 scores below 10}, 10.1, 10.4, 10.7, 10.9, {12 scores above 11}

Median = 16th score (when in order) = 10.1 cm

**10** **a** **Brand A**

| $x$ | $f$ | $fx$ |
|---|---|---|
| 46 | 1 | 46 |
| 47 | 2 | 94 |
| 48 | 3 | 144 |
| 49 | 7 | 343 |
| 50 | 10 | 500 |
| 51 | 20 | 1020 |
| 52 | 15 | 780 |
| 53 | 3 | 159 |
| 55 | 1 | 55 |
| $\sum$ | 62 | 3141 |

mean $= \dfrac{\sum fx}{\sum f}$

$= \dfrac{3141}{62}$

$\approx 50.7$

**Brand B**

| $x$ | $f$ | $fx$ |
|---|---|---|
| 48 | 3 | 144 |
| 49 | 17 | 833 |
| 50 | 30 | 1500 |
| 51 | 7 | 357 |
| 52 | 2 | 104 |
| 53 | 1 | 53 |
| 54 | 1 | 54 |
| $\sum$ | 61 | 3045 |

mean $= \dfrac{\sum fx}{\sum f}$

$= \dfrac{3045}{61}$

$\approx 49.9$

**b** Based on average contents, the C.P.S. should not prosecute either manufacturer. To the nearest toothpick, the average contents for A is 51 and for B is 50.

**11** **a** **i** median salary $= \dfrac{10\text{th} + 11\text{th}}{2}$ (when in order) $= \dfrac{35\,000 + 28\,000}{2}$ $= €31\,500$

**ii** modal salary $= €28\,000$ {occurs most often}

**iii**

| $x$ | $f$ | $fx$ |
|---|---|---|
| 50 000 | 1 | 50 000 |
| 42 000 | 3 | 126 000 |
| 35 000 | 6 | 210 000 |
| 28 000 | 10 | 280 000 |
| $\sum$ | 20 | 666 000 |

mean $= \dfrac{\sum fx}{\sum f}$ $= \dfrac{666\,000}{20}$ $= €33\,300$

**b** The mode, as it is the most commonly occurring value.

## EXERCISE 20B.3

**1**

| *midpoint* ($x$) | $f$ | $fx$ |
|---|---|---|
| 4.5 | 2 | 9 |
| 14.5 | 5 | 72.5 |
| 24.5 | 7 | 171.5 |
| 34.5 | 27 | 931.5 |
| 44.5 | 9 | 400.5 |
| $\sum$ | 50 | 1585 |

$\therefore$ mean result $\approx \dfrac{1585}{50}$ $\approx 31.7$

**2**

| *midpoint* ($x$) | $f$ | $fx$ |
|---|---|---|
| 2500 | 4 | 10 000 |
| 3500 | 4 | 14 000 |
| 4500 | 9 | 40 500 |
| 5500 | 14 | 77 000 |
| 6500 | 23 | 149 500 |
| 7500 | 16 | 120 000 |
| $\sum$ | 70 | 411 000 |

**a** 70

**b** $\approx 411\,000$ litres $\approx 411$ kL

**c** mean $\approx \dfrac{\sum fx}{\sum f}$ $\approx \dfrac{411\,000}{70}$ $\approx 5870$ litres

**3** **a** $5 + 10 + 25 + 40 + 10 + 15 + 10 + 10 = 125$ people

**b**

| *midpoint* ($x$) | *frequency* ($f$) | $fx$ |
|---|---|---|
| 85 | 5 | 425 |
| 95 | 10 | 950 |
| 105 | 25 | 2625 |
| 115 | 40 | 4600 |
| 125 | 10 | 1250 |
| 135 | 15 | 2025 |
| 145 | 10 | 1450 |
| 155 | 10 | 1550 |
| $\sum$ | 125 | 14 875 |

mean $\approx \dfrac{\sum fx}{\sum f}$ $\approx \dfrac{14\,875}{125}$ $\approx 119$ marks

**c** $\frac{15}{125} = \frac{3}{25}$ scored $< 100$

**d** There are $15 + 10 + 10 = 35$ people who scored more than 130 for the test.

$\therefore$ % who scored more than 130 $= \dfrac{35}{125} \times 100\%$ $= 28\%$

## EXERCISE 20C

**1** **a** 2 3 3 3 4 4 4 5 5 5 5 6 6 6 6 6 7 7 8 8 8 9 9 ($n = 23$)

min = 2, $Q_1$ = 4, median = 6, $Q_3$ = 7, max = 9

**i** median = 6

**ii** $Q_1 = 4$, $Q_3 = 7$

**iii** range $= 9 - 2$ $= 7$

**iv** IQR $= Q_3 - Q_1$ $= 3$

**b** 10 12 12 14 15 15 16 16 17 18 18 18 18 19 20 21 22 24 $(n = 18)$

min: 10, $Q_1$: 15, median: 17 18, $Q_3$: 19, max: 24

**i** median $= 17.5$

**ii** $Q_1 = 15$
$Q_3 = 19$

**iii** range
$= 24 - 10$
$= 14$

**iv** IQR
$= Q_3 - Q_1$
$= 4$

**c** 21.8 22.4 23.5 23.5 24.6 24.9 25.0 25.3 26.1 26.4 29.5 $(n = 11)$

min: 21.8, $Q_1$: 23.5, median: 24.9, $Q_3$: 26.1, max: 29.5

**i** median $= 24.9$

**ii** $Q_1 = 23.5$
$Q_3 = 26.1$

**iii** range
$= 29.5 - 21.8$
$= 7.7$

**iv** IQR
$= Q_3 - Q_1$
$= 2.6$

**2** 0 0 0 0.8 1.4 1.5 1.6 1.9 2.1 2.2 2.7 3.0 3.4 3.6 3.8 3.8 4.5 4.8 5.2 5.2

min: 0, $Q_1$: 1.4 1.5, median: 2.2 2.7, $Q_3$: 3.8 3.8, max: 5.2

**a** median $= \dfrac{2.2 + 2.7}{2}$
$= 2.45$

$Q_1 = 1.45$
$Q_3 = 3.8$

**b** range $= 5.2 - 0$
$= 5.2$

IQR $= Q_3 - Q_1$
$= 3.8 - 1.45$
$= 2.35$

**c** **i** The median $= 2.45$ min, so "50% of the waiting times were greater than 2.45 min."

**ii** $Q_3 = 3.8$ min, so "75% of the waiting times were less than or equal to 3.8 min."

**iii** The minimum waiting time was 0 minutes and the maximum waiting time was 5.2 minutes. The waiting times were spread over 5.2 minutes.

**3** 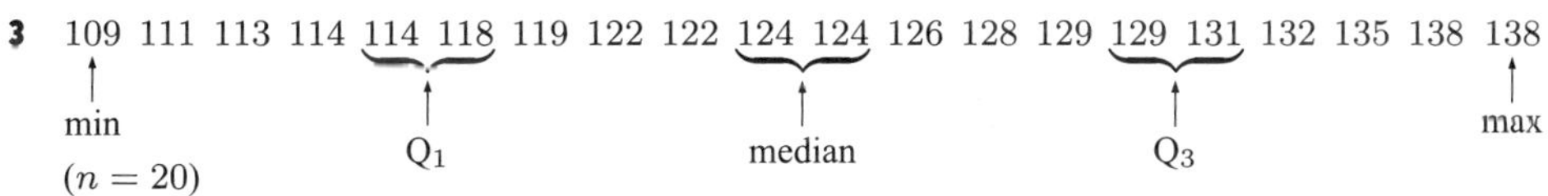


**a** **i** median $= 124$ cm

**ii** $Q_1 = 116$ cm
$Q_3 = 130$ cm

**b** **i** .... 124 cm tall

**ii** .... 130 cm tall

**c** **i** range $= 138 - 109$
$= 29$ cm

**ii** IQR $= Q_3 - Q_1$
$= 14$ cm

**d** the IQR $= 14$ cm, so .... over 14 cm

**4**

| | *Measure* | Median | Mode | Range | Interquartile range |
|---|---|---|---|---|---|
| **a** | *Value* | $9 + 2 = 11$ | $7 + 2 = 9$ | 13 | 6 |
| **b** | *Value* | $9 \times 2 = 18$ | $7 \times 2 = 14$ | $13 \times 2 = 26$ | $6 \times 2 = 12$ |

**5** See **Exercise 20B.2**, solution to question **7**.

**a** **Without fertiliser**

**i** range $= 9 - 2 = 7$ **ii** median $= 6$ **iii** lower quartile $=$ 38th score $= 5$

**iv** upper quartile $=$ 113th score $= 7$ **v** interquartile range $= 7 - 5 = 2$

**b** **With fertiliser**

**i** range $= 13 - 3 = 10$ **ii** median $= 7$ **iii** lower quartile $=$ 38th score $= 6$

**iv** upper quartile $=$ 113th score $= 8$ **v** interquartile range $= 8 - 6 = 2$

## EXERCISE 20D

**1** **a** **i** median $= 35$ **ii** max. value $= 78$ **iii** min. value $= 13$

**iv** $Q_3 = 53$ **v** $Q_1 = 26$

**b** **i** range $= 78 - 13 = 65$ **ii** $\text{IQR} = Q_3 - Q_1 = 53 - 26 = 27$

**2** **a** **i** highest mark was 98, lowest mark was 25 **ii** .... the median which is 70
**iii** .... $Q_3$ which is 85 **iv** .... $Q_1 = 55$ and $Q_3 = 85$
**b** range $= 98 - 25 = 73$ **c** $\text{IQR} = Q_3 - Q_1 = 85 - 55 = 30$

**3** **a** **i** 3 4 5 5 5 6 6 6 7 7 8 8 9 10

min (3), $Q_1$ (5), median (between 6 and 6), $Q_3$ (8), max (10)

So, min $= 3$, $Q_1 = 5$, median $= 6$, $Q_3 = 8$, max $= 10$

**ii**

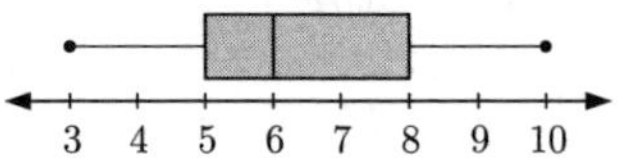


**iii** range $= 10 - 3$
$= 7$

**iv** $\text{IQR} = Q_3 - Q_1$
$= 8 - 5$
$= 3$

**b** **i** 0 1 2 3 4 5 6 6 7 7 7 8 8 8 8 8 8 9 9

min (0), $Q_1$ (4), median (7), $Q_3$ (8), max (9)

So, min $= 0$, $Q_1 = 4$, median $= 7$, $Q_3 = 8$, max $= 9$

**ii**

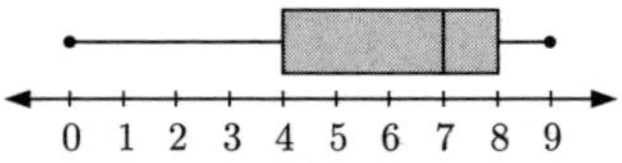


**iii** range $= 9 - 0$
$= 9$

**iv** $\text{IQR} = Q_3 - Q_1$
$= 8 - 4$
$= 4$

**4** **a**

| *Statistic* | *Year 9* | *Year 12* |
|---|---|---|
| minimum | 1 | 6 |
| $Q_1$ | 5 | 10 |
| median | 7.5 | 14 |
| $Q_3$ | 10 | 16 |
| maximum | 12 | 17.5 |

**b** For the Year 9 group
**i** range $= 12 - 1$
$= 11$
**ii** $\text{IQR} = 10 - 5$
$= 5$

For the Year 12 group
**i** range $= 17.5 - 6$
$= 11.5$
**ii** $\text{IQR} = 16 - 10$
$= 6$

**c** **i** True, as indicated by the median.
**ii** There is not enough information to tell if this is true. We do not know if there are any data values between 5 and 6 in the Year 9 boxplot.

**5** 2 3 3 4 4 4 4 5 5 5 5 5 5 5 6 6 6 6 6 6 7 7 7 7 8 8 8 9 9 9 10 12 13

min (2), $Q_1$ (between 5 and 5), median (6), $Q_3$ (between 8 and 8), max (13)

**a** median $= 6$, $Q_1 = 5$, $Q_3 = 8$ **b** $\text{IQR} = 8 - 5 = 3$

**c** Lower boundary $= Q_1 - 1.5 \times \text{IQR}$
$= 5 - 1.5 \times 3$
$= 0.5$

Upper boundary $= Q_3 + 1.5 \times \text{IQR}$
$= 8 + 1.5 \times 3$
$= 12.5$

So, 13 is an outlier.

$\therefore$ the boxplot is:

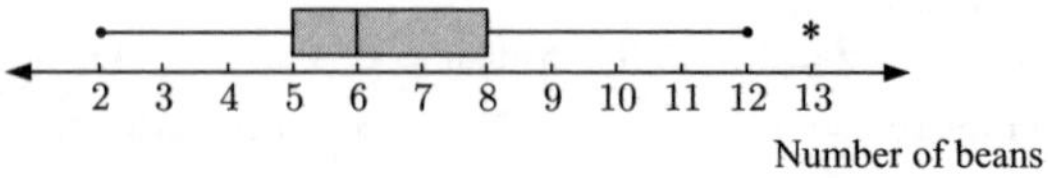


**6** **a** $\text{IQR} =$ upper quartile $-$ lower quartile
$= 43.5 - 31.5$
$= 12$

**b** Lower boundary $= Q_1 - 1.5 \times \text{IQR}$
$= 31.5 - 1.5 \times 12$
$= 13.5$

Upper boundary $= Q_3 + 1.5 \times \text{IQR}$
$= 43.5 + 1.5 \times 12$
$= 61.5$

**c** 13.2 and 65 would be outliers.

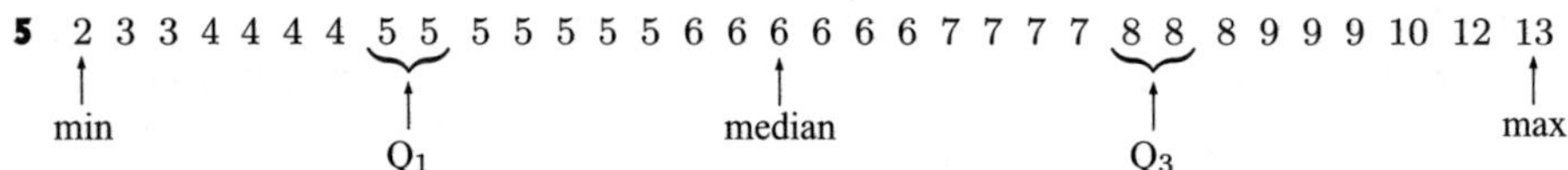

**7** **a** 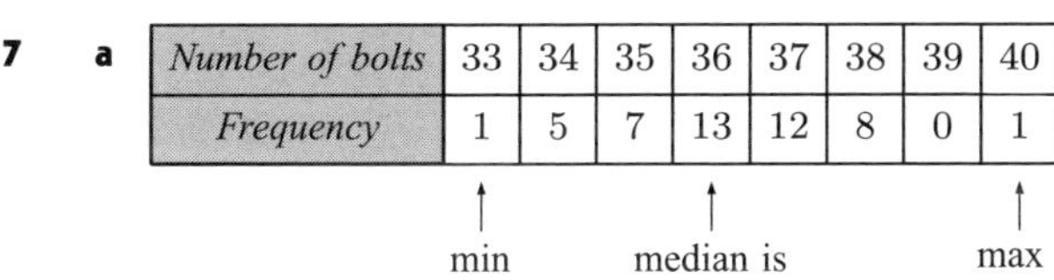

| *Number of bolts* | 33 | 34 | 35 | 36 | 37 | 38 | 39 | 40 |
|---|---|---|---|---|---|---|---|---|
| *Frequency* | 1 | 5 | 7 | 13 | 12 | 8 | 0 | 1 |



There are 47 scores

$\therefore$ median = 24th

$\left(\frac{47+1}{2} = 24\right)$

13 scores are 35 or less

26 scores are 36 or less

$\therefore$ median is 36

$Q_1 = 12\text{th} \quad \left(\frac{23+1}{2} = 12\right)$

$= 35$

$Q_3 = 36\text{th}$

$= 37$

So, min = 33, $Q_1 = 35$, median = 36, $Q_3 = 37$, max = 40

**b** **i** range = 40 − 33 = 7

**ii** IQR = 37 − 35 = 2

**c** 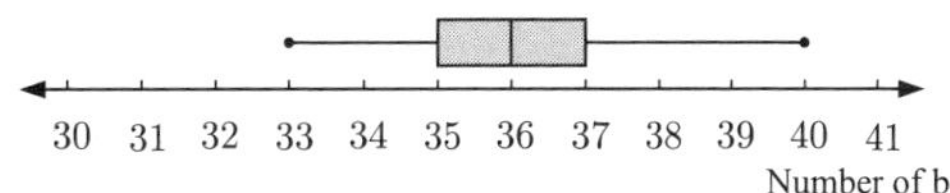



**d** Lower boundary $= Q_1 - 1.5 \times \text{IQR}$
$= 35 - 1.5 \times 2$
$= 32$

Upper boundary $= Q_3 + 1.5 \times \text{IQR}$
$= 37 + 1.5 \times 2$
$= 40$

Since all data values are on or within these boundaries, there are no outliers.

**8** 3 5 6 7 7 8 8 9 9 9 10 10 10 11 11 12 12 13 13 13 14 14 16 18 22

(min = 3, $Q_1$ at 8 8, median = 10, $Q_3$ at 13 13, max = 22)

**a** median = 10

$Q_1 = \frac{8+8}{2}$
$= 8$

$Q_3 = \frac{13+13}{2}$
$= 13$

**b** $\text{IQR} = Q_3 - Q_1$
$= 13 - 8$
$= 5$

**c** Lower boundary $= Q_1 - 1.5 \times \text{IQR}$
$= 8 - 1.5 \times 5$
$= 0.5$

Upper boundary $= Q_3 + 1.5 \times \text{IQR}$
$= 13 + 1.5 \times 5$
$= 20.5$

**d** 22 is an outlier, as it is greater than 20.5.

**e** 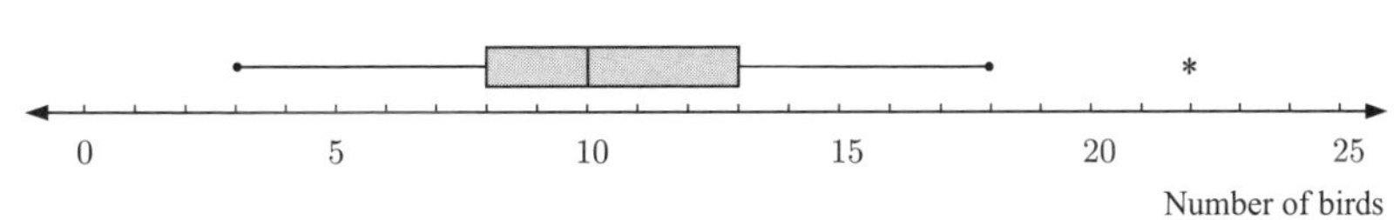



**9** The data displayed in graph **A** has a minimum of 2 and a maximum of 8. There are no outliers.

$\therefore$ graph **A** matches boxplot **I**.

The data displayed in graph **B** has a minimum of 1 and a maximum of 13. 13 is clearly an outlier, and there are no outliers at the lower end of the data set.

$\therefore$ graph **B** matches boxplot **IV**.

The data displayed in graph **C** has a minimum of 1 and a maximum of 13. 1 and 13 are clear outliers.

$\therefore$ graph **C** matches boxplot **III**.

The data displayed in graph **D** has a minimum of 1 and a maximum of 8. There are no outliers.

$\therefore$ graph **D** matches boxplot **II**.

**10** **a** discrete

**c**

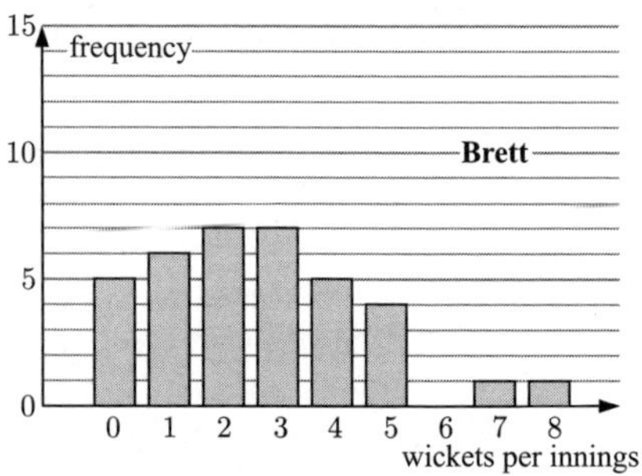


**d** There are no outliers for Shane. Brett has outliers of 7 and 8 which must not be removed.

**e** Shane's distribution is reasonably symmetrical. Brett's distribution is positively skewed.

**f** Shane has a higher mean ($\approx$ 2.89 wickets) compared with Brett ($\approx$ 2.67 wickets). Shane has a higher median (3 wickets) compared with Brett (2.5 wickets). Shane's modal number of wickets is 3 (14 times) compared with Brett, who has two modal values of 2 and 3 (7 times each).

**g** Shane's range is 6 wickets, compared with Brett's range of 8 wickets. Shane's IQR is 2 wickets, compared with Brett's IQR of 3 wickets. Brett's wicket taking shows greater spread or variability.

**h**

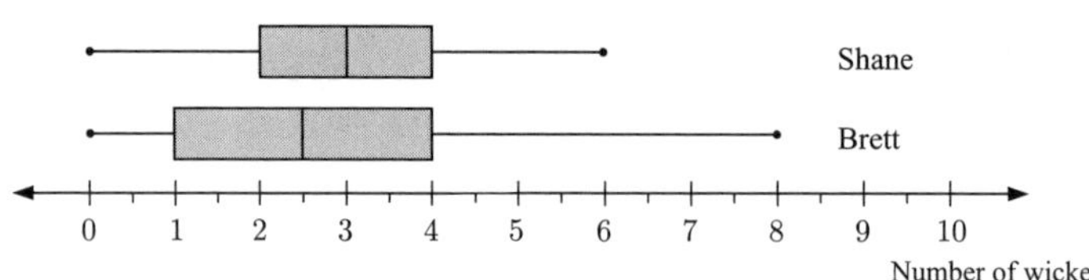


**i** Generally, Shane takes more wickets than Brett and is a more consistent bowler.

**11** **a** continuous

**c** For the 'new type' globes, 191 hours could be considered an outlier. However, it could be a genuine piece of data, so we will include it in the analysis.

**d** The mean and median are $\approx 25\%$ and $\approx 19\%$ higher for the 'new type' of globe compared with the 'old type'.

The range is higher for the 'new type' of globe (but has been affected by the 191 hours).

The IQR for each type of globe is almost the same.

| | *Old type* | *New type* |
|---|---|---|
| Mean | 107 | 134 |
| Median | 110.5 | 132 |
| Mode | 87, 111, 113 | 131, 133 |
| Range | 56 | 84 |
| IQR | 19 | 18.5 |

**e**

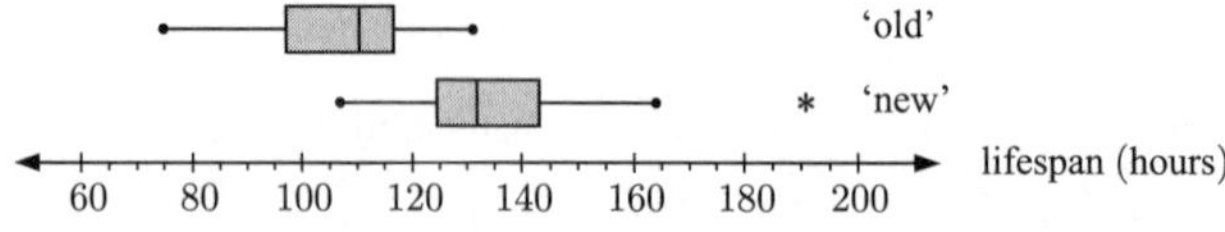


**f** For the 'old type' of globe, the data is bunched to the right of the median, hence the distribution is negatively skewed. For the 'new type' of globe, the data is bunched to the left of the median, hence the distribution is slightly positively skewed.

**g** The manufacturer's claim, that the 'new type' of globe has a 20% longer life than the 'old type' seems to be backed up by the 25% higher mean life and 19.5% higher median life.

## EXERCISE 20E

**1**

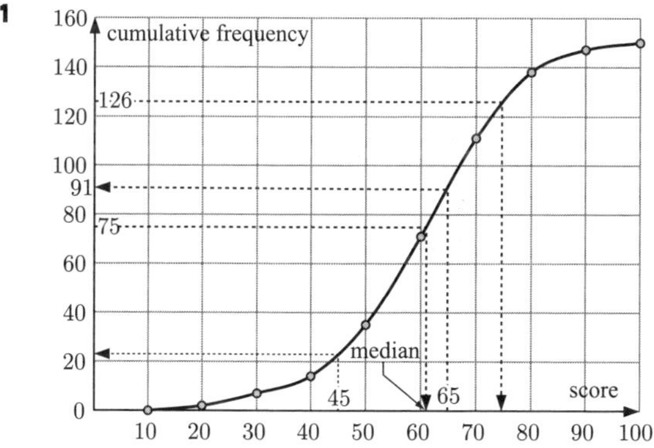


**a** Total frequency $= 150$
$\therefore$ median $\approx 61$ {from the graph}

**b** When the score $= 65$,
CF $\approx 91$ {from the graph}
$\therefore$ about 91 students scored less than 65.

**c** $\approx 36 + 40 \approx 76$ students scored between 50 and 70.

**d** For a pass mark of 45, CF $\approx 23.5$
$\therefore$ 23 or 24 students failed the exam.

**e** 84% of 150 $= 126$
So, for a CF of 126, the score value is 75.
So, the minimum credit mark would be 75.

**2** **a**

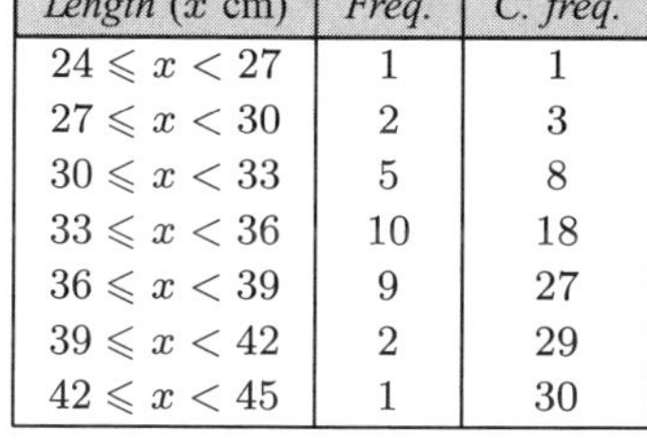

| *Length* ($x$ cm) | *Freq.* | *C. freq.* |
|---|---|---|
| $24 \leqslant x < 27$ | 1 | 1 |
| $27 \leqslant x < 30$ | 2 | 3 |
| $30 \leqslant x < 33$ | 5 | 8 |
| $33 \leqslant x < 36$ | 10 | 18 |
| $36 \leqslant x < 39$ | 9 | 27 |
| $39 \leqslant x < 42$ | 2 | 29 |
| $42 \leqslant x < 45$ | 1 | 30 |

**b**

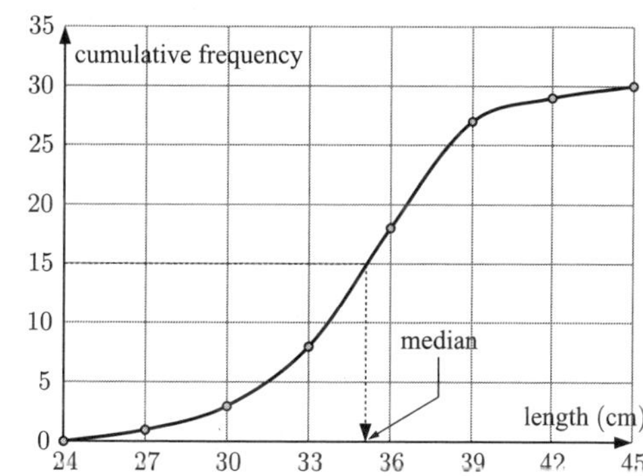


**c** median $\approx 35$ cm

**d** There are 30 data values.
So, the median is the average of the 15th and 16th scores (when in order).
In order they are:

24 27 28 30 31 31 32 32 33 33 33 33 34 34 34 35 35 35 36 .... and so on.

median $= \dfrac{34+35}{2} = 34.5$

So, the median from the graph is a good approximation.

**3** **a** For $h = 5$, CF $\approx 9$ $\therefore$ 9 seedlings have height 5 cm or less

**b** For $h = 8$, CF $\approx 43$ $\therefore$ % taller than 8 cm $= \dfrac{60-43}{60} \times 100\% \approx 28.3\%$

**c** The approximate median occurs at CF $= 30$, $\therefore$ median $\approx 7$ cm.

**d** $\text{IQR} = Q_3 - Q_1$
$= (h \text{ when CF} = 45) - (h \text{ when CF} = 15)$
$\approx 8.3 - 5.9$
$\approx 2.4$ cm

**e** 90th percentile occurs when CF $= 90\%$ of $60 = 54$
$\therefore$ 90th percentile $= 10$
This means that 90% of the seedlings have a height of 10 cm or less.

**4**

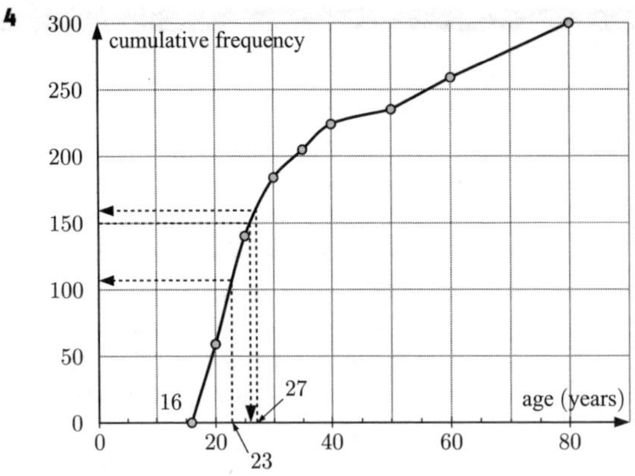


Total frequency = 300

**a** median ≈ 26 years {from the graph}

**b** when the age = 23, CF ≈ 108

and $\frac{108}{300} \times 100\% \approx 36\%$

**c** **i** when age is 27, CF ≈ 158

$\therefore$ $\text{P(age} \leqslant 27) = \frac{158}{300} = 0.527$

**ii** when age is 26 or less, CF ≈ 150 {**a**}

when age is 27 or less, CF ≈ 158 {**c i**}

$\therefore$ 8 were 27 years old

$\therefore$ $\text{P(aged 27)} \approx \frac{8}{300} \approx 0.0267$

**5** **a** The lower quartile occurs when CF = 25% of 80 = 20 $\therefore$ $Q_1 = 27$ min

**b** The median occurs when CF = 50% of 80 = 40 $\therefore$ median = 29 min

**c** The upper quartile occurs when CF = 75% of 80 = 60 $\therefore$ $Q_3 \approx 31.3$ min

**d** $\text{IQR} = Q_3 - Q_1 \approx 31.3 - 27 \approx 4.3$ min

**e** For the 40th percentile, CF = 40% of 80 = 32

When CF = 32, $x \approx 28.2$ So, the 40th percentile is about 28 min 10 sec.

**f** From the cumulative frequency graph we can obtain the cumulative frequency table:

| *Time t* (mins) | $21 \leqslant t < 24$ | $24 \leqslant t < 27$ | $27 \leqslant t < 30$ | $30 \leqslant t < 33$ | $33 \leqslant t < 36$ |
|---|---|---|---|---|---|
| *Cumulative frequency* | 5 | 20 | 50 | 70 | 80 |

We can then complete the table as follows:

| *Time t* (mins) | $21 \leqslant t < 24$ | $24 \leqslant t < 27$ | $27 \leqslant t < 30$ | $30 \leqslant t < 33$ | $33 \leqslant t < 36$ |
|---|---|---|---|---|---|
| *Number of competitors* | $5 - 0 = 5$ | $20 - 5 = 15$ | $50 - 20 = 30$ | $70 - 50 = 20$ | $80 - 70 = 10$ |

**6**

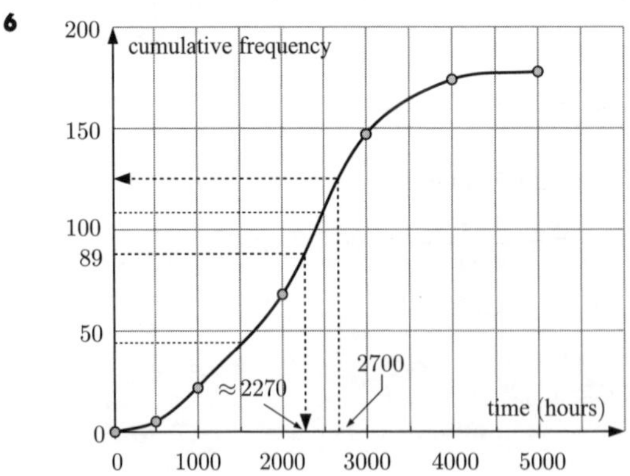


Total frequency = 178

**a** median ≈ 2270 hours

**b** For a life of 2700 hours CF ≈ 126

and $\frac{126}{178} \times 100\% \approx 71\%$

So, about 71% have a life $\leqslant$ 2700 h

**c** For a life of 1500 h, CF ≈ 42

For a life of 2500 h, CF ≈ 107.5

and $107.5 - 42 = 65.5$

$\therefore$ ≈ 65 or 66 had a life between 1500 and 2500 hours.

## EXERCISE 20F.1

**1** **a** Looking at the graphs, Sample A appears to have the wider spread.

**b** *Sample A*:

| $x$ | $f$ | $fx$ |
|---|---|---|
| 4 | 1 | 4 |
| 5 | 2 | 10 |
| 6 | 3 | 18 |
| 7 | 4 | 28 |
| 8 | 5 | 40 |
| 9 | 4 | 36 |
| 10 | 3 | 30 |
| 11 | 2 | 22 |
| 12 | 1 | 12 |
| $\sum$ | 25 | 200 |

$\therefore$ mean $= \frac{200}{25} = 8$

*Sample B*:

| $x$ | $f$ | $fx$ |
|---|---|---|
| 6 | 2 | 12 |
| 7 | 6 | 42 |
| 8 | 9 | 72 |
| 9 | 6 | 54 |
| 10 | 2 | 20 |
| $\sum$ | 25 | 200 |

$\therefore$ mean $= \frac{200}{25} = 8$

**c** *Sample A*:

| $x$ | $x-\overline{x}$ | $(x-\overline{x})^2$ | $f$ | $f(x-\overline{x})^2$ |
|---|---|---|---|---|
| 4 | $-4$ | 16 | 1 | 16 |
| 5 | $-3$ | 9 | 2 | 18 |
| 6 | $-2$ | 4 | 3 | 12 |
| 7 | $-1$ | 1 | 4 | 4 |
| 8 | 0 | 0 | 5 | 0 |
| 9 | 1 | 1 | 4 | 4 |
| 10 | 2 | 4 | 3 | 12 |
| 11 | 3 | 9 | 2 | 18 |
| 12 | 4 | 16 | 1 | 16 |
| | | | $\sum$ | 100 |

$\therefore \quad s = \sqrt{\frac{\sum(x-\overline{x})^2}{n}} = \sqrt{\frac{100}{25}} = 2$

*Sample B*:

| $x$ | $x-\overline{x}$ | $(x-\overline{x})^2$ | $f$ | $f(x-\overline{x})^2$ |
|---|---|---|---|---|
| 6 | $-2$ | 4 | 2 | 8 |
| 7 | $-1$ | 1 | 6 | 6 |
| 8 | 0 | 0 | 9 | 0 |
| 9 | 1 | 1 | 6 | 6 |
| 10 | 2 | 4 | 2 | 8 |
| | | | $\sum$ | 28 |

$\therefore \quad s = \sqrt{\frac{\sum(x-\overline{x})^2}{n}} = \sqrt{\frac{28}{25}} \approx 1.06$

The standard deviation is higher for Sample A.
$\therefore$ Sample A has a greater spread.

**d** The standard deviation is calculated using all data values, not just two. (Range only uses maximum and minimum; IQR only uses the upper and lower quartiles.)

**2** **a** *Andrew*: $\overline{x} = \dfrac{23 + 17 + .... + 28 + 32}{8} = 25$

| $x$ | $x-\overline{x}$ | $(x-\overline{x})^2$ |
|---|---|---|
| 23 | $-2$ | 4 |
| 17 | $-8$ | 64 |
| 31 | 6 | 36 |
| 25 | 0 | 0 |
| 25 | 0 | 0 |
| 19 | $-6$ | 36 |
| 28 | 3 | 9 |
| 32 | 7 | 49 |
| | $\sum$ | 198 |

$\therefore \quad s = \sqrt{\frac{\sum(x-\overline{x})^2}{n}} = \sqrt{\frac{198}{8}} \approx 4.97$

*Brad*: $\overline{x} = \dfrac{9 + 29 + .... + 38 + 43}{8} = 30.5$

| $x$ | $x-\overline{x}$ | $(x-\overline{x})^2$ |
|---|---|---|
| 9 | $-21.5$ | 462.25 |
| 29 | $-1.5$ | 2.25 |
| 41 | 10.5 | 110.25 |
| 26 | $-4.5$ | 20.25 |
| 14 | $-16.5$ | 272.25 |
| 44 | 13.5 | 182.25 |
| 38 | 7.5 | 56.25 |
| 43 | 12.5 | 156.25 |
| | $\sum$ | 1262 |

$\therefore \quad s = \sqrt{\frac{\sum(x-\overline{x})^2}{n}} = \sqrt{\frac{1262}{8}} \approx 12.6$

**b** Andrew, as he has the smaller standard deviation.

**3** **a** Rockets have mean $= \dfrac{0 + 10 + 1 + 9 + 11 + 0 + 8 + 5 + 6 + 7}{10} = \dfrac{57}{10} = 5.7$ runs

Bullets have mean $= \dfrac{4 + 3 + 4 + 1 + 4 + 11 + 7 + 6 + 12 + 5}{10} = \dfrac{57}{10} = 5.7$ runs

Rockets' range $= 11 - 0 = 11$ runs  Bullets' range $= 12 - 1 = 11$ runs

**b** We suspect the Rockets, as they have two zeros.

**c** *Rockets*

| $x$ | $(x-\overline{x})^2$ |
|---|---|
| 0 | $(-5.7)^2$ |
| 10 | $(4.3)^2$ |
| 1 | $(-4.7)^2$ |
| 9 | $(3.3)^2$ |
| 11 | $(5.3)^2$ |
| 0 | $(-5.7)^2$ |
| 8 | $(2.3)^2$ |
| 5 | $(-0.7)^2$ |
| 6 | $(0.3)^2$ |
| 7 | $(1.3)^2$ |
| | 152.1 |

$s = \sqrt{\dfrac{\sum(x-\overline{x})^2}{n}}$

$= \sqrt{\dfrac{152.1}{10}}$

$= 3.9$ runs ↑ greater variability

*Bullets*

| $x$ | $(x-\overline{x})^2$ |
|---|---|
| 4 | $(-1.7)^2$ |
| 3 | $(-2.7)^2$ |
| 4 | $(-1.7)^2$ |
| 1 | $(-4.7)^2$ |
| 4 | $(-1.7)^2$ |
| 11 | $(5.3)^2$ |
| 7 | $(1.3)^2$ |
| 6 | $(0.3)^2$ |
| 12 | $(6.3)^2$ |
| 5 | $(-0.7)^2$ |
| | 108.1 |

$s = \sqrt{\dfrac{\sum(x-\overline{x})^2}{n}}$

$= \sqrt{\dfrac{108.1}{10}}$

$\approx 3.29$ runs

**d** The standard deviation, as it takes all values into account, not just the lowest and highest.

**4** **a** We suspect variability in standard deviation since the factors may change every day.

**b** **i** sample mean **ii** sample standard deviation

**c** A low standard deviation would mean less variability in the volume of soft drink per can.

**5** **a**

| $x$ | $(x-\overline{x})^2$ |
|---|---|
| 79 | $10^2$ |
| 64 | $(-5)^2$ |
| 59 | $(-10)^2$ |
| 71 | $2^2$ |
| 68 | $(-1)^2$ |
| 68 | $(-1)^2$ |
| 74 | $5^2$ |
| 483 | 256 |

$\overline{x} = \dfrac{\sum x}{n}$

$= \dfrac{483}{7} = 69$ kg

$s = \sqrt{\dfrac{\sum(x-\overline{x})^2}{n}}$

$= \sqrt{\dfrac{256}{7}}$

$\approx 6.05$ kg

**b**

| $x$ | $(x-\overline{x})^2$ |
|---|---|
| 89 | $10^2$ |
| 74 | $(-5)^2$ |
| 69 | $(-10)^2$ |
| 81 | $2^2$ |
| 78 | $(-1)^2$ |
| 78 | $(-1)^2$ |
| 84 | $5^2$ |
| 553 | 256 |

$\overline{x} = \dfrac{\sum x}{n}$

$= \dfrac{553}{7}$

$= 79$ kg

$s \approx 6.05$ kg

**c** The distribution has simply shifted by 10 kg. The mean increases by 10 kg and the standard deviation remains the same. In general, changing the values by a constant does not affect standard deviation.

**6** **a**

| $x$ | $(x-\overline{x})^2$ |
|---|---|
| 0.8 | $(-0.21)^2$ |
| 1.1 | $(0.09)^2$ |
| 1.2 | $(0.19)^2$ |
| 0.9 | $(-0.11)^2$ |
| 1.2 | $(0.19)^2$ |
| 1.2 | $(0.19)^2$ |
| 0.9 | $(-0.11)^2$ |
| 0.7 | $(-0.31)^2$ |
| 1.0 | $(-0.01)^2$ |
| 1.1 | $(0.09)^2$ |
| 10.1 | 0.289 |

$\overline{x} = \dfrac{\sum x}{n}$

$= \dfrac{10.1}{10}$

$= 1.01$ kg

$s = \sqrt{\dfrac{\sum(x-\overline{x})^2}{n}}$

$= \sqrt{\dfrac{0.289}{10}}$

$= 0.17$ kg

**b**

| $x$ | $(x-\overline{x})^2$ |
|---|---|
| 1.6 | $(-0.42)^2$ |
| 2.2 | $(0.18)^2$ |
| 2.4 | $(0.38)^2$ |
| 1.8 | $(-0.22)^2$ |
| 2.4 | $(0.38)^2$ |
| 2.4 | $(0.38)^2$ |
| 1.8 | $(-0.22)^2$ |
| 1.4 | $(-0.62)^2$ |
| 2.0 | $(-0.02)^2$ |
| 2.2 | $(0.18)^2$ |
| 20.2 | 1.156 |

$\overline{x} = \dfrac{\sum x}{n}$

$= \dfrac{20.2}{10}$

$= 2.02$ kg

$s = \sqrt{\dfrac{\sum(x-\overline{x})^2}{n}}$

$= \sqrt{\dfrac{1.156}{10}}$

$= 0.34$ kg

**c** Doubling the values doubles the mean and the standard deviation.

**7** **a** $\overline{x} = \dfrac{0.8+0.6+0.7+0.8+0.4+2.8}{6}$

$\approx 1.017$

| $x$ | $(x-\overline{x})^2$ |
|---|---|
| 0.8 | $(-0.217)^2$ |
| 0.6 | $(-0.417)^2$ |
| 0.7 | $(-0.317)^2$ |
| 0.8 | $(-0.217)^2$ |
| 0.4 | $(-0.617)^2$ |
| 2.8 | $(1.783)^2$ |
| $\sum$ | 3.928 |

$\therefore \quad s = \sqrt{\dfrac{\sum(x-\overline{x})^2}{n}} \approx \sqrt{\dfrac{3.928}{6}}$

$\approx 0.809$

**b** $\overline{x} = \dfrac{0.8+0.6+0.7+0.8+0.4}{5}$

$= 0.66$

| $x$ | $(x-\overline{x})^2$ |
|---|---|
| 0.8 | $(0.14)^2$ |
| 0.6 | $(-0.06)^2$ |
| 0.7 | $(0.04)^2$ |
| 0.8 | $(0.14)^2$ |
| 0.4 | $(-0.26)^2$ |
| $\sum$ | 0.112 |

$\therefore \quad s = \sqrt{\dfrac{\sum(x-\overline{x})^2}{n}} = \sqrt{\dfrac{0.112}{5}}$

$\approx 0.150$

**c** The extreme value greatly increases the standard deviation.

## EXERCISE 20F.2

**1** **a** $s = \sqrt{\text{variance}}$

$= \sqrt{45.9}$ kg

$\approx 6.77$ kg

We estimate the standard deviation $\sigma$ by $s \approx 6.77$ kg.

**b** Unbiased estimate of $\mu$ is $\overline{x} = 93.8$ kg

**2** **a** Using technology, $\overline{x} \approx 77.5$ g and $s_n \approx 7.44$ g

**b** Unbiased estimate of $\mu$ is $\overline{x} \approx 77.5$ g

Estimate of $\sigma$ is $s_n \approx 7.44$ g

## EXERCISE 20F.3

**1** **a**

| $x$ | $f$ | $fx$ | $x-\overline{x}$ | $f(x-\overline{x})^2$ |
|---|---|---|---|---|
| 0 | 14 | 0 | $-1.7241$ | 41.62 |
| 1 | 18 | 18 | $-0.7241$ | 9.44 |
| 2 | 13 | 26 | 0.2759 | 0.99 |
| 3 | 5 | 15 | 1.2759 | 8.14 |
| 4 | 3 | 12 | 2.2759 | 15.54 |
| 5 | 2 | 10 | 3.2759 | 21.46 |
| 6 | 2 | 12 | 4.2759 | 36.57 |
| 7 | 1 | 7 | 5.2759 | 27.83 |
| $\sum$ | 58 | 100 | | 161.59 |

$\overline{x} = \dfrac{\sum fx}{\sum f}$

$= \dfrac{100}{58}$

$\approx 1.72$ children

$s = \sqrt{\dfrac{\sum f(x-\overline{x})^2}{\sum f}}$

$\approx \sqrt{\dfrac{161.59}{58}}$

$\approx 1.67$ children

**b** Estimate of $\mu$ is $\overline{x} \approx 1.72$ children. Estimate of $\sigma$ is $s \approx 1.67$ children.

**2** **a**

| $x$ | $f$ | $fx$ | $x-\overline{x}$ | $f(x-\overline{x})^2$ |
|---|---|---|---|---|
| 11 | 2 | 22 | $-3.48$ | 24.22 |
| 12 | 1 | 12 | $-2.48$ | 6.150 |
| 13 | 4 | 52 | $-1.48$ | 8.762 |
| 14 | 5 | 70 | $-0.48$ | 1.152 |
| 15 | 6 | 90 | 0.52 | 1.622 |
| 16 | 4 | 64 | 1.52 | 9.242 |
| 17 | 2 | 34 | 2.52 | 12.70 |
| 18 | 1 | 18 | 3.52 | 12.39 |
| $\sum$ | 25 | 362 | | 76.24 |

$\overline{x} = \dfrac{\sum fx}{\sum f}$

$= \dfrac{362}{25}$

$\approx 14.5$ years

$s = \sqrt{\dfrac{\sum f(x-\overline{x})^2}{\sum f}}$

$= \sqrt{\dfrac{76.24}{25}}$

$\approx 1.75$ years

**b** Estimate of $\mu$ is $\overline{x} \approx 14.5$ years. Estimate of $\sigma$ is $s \approx 1.75$ years

**3** **a**

| $x$ | $f$ | $fx$ | $x-\overline{x}$ | $f(x-\overline{x})^2$ |
|---|---|---|---|---|
| 33 | 1 | 33 | −4.2708 | 18.24 |
| 35 | 5 | 175 | −2.2708 | 25.78 |
| 36 | 7 | 252 | −1.2708 | 11.30 |
| 37 | 13 | 481 | −0.2708 | 0.95 |
| 38 | 12 | 456 | 0.7292 | 6.38 |
| 39 | 8 | 312 | 1.7292 | 23.92 |
| 40 | 2 | 80 | 2.7292 | 14.90 |
| $\sum$ | 48 | 1789 | | 101.47 |

$$\overline{x} = \frac{\sum fx}{\sum f} = \frac{1789}{48} \approx 37.3 \text{ toothpicks}$$

$$s = \sqrt{\frac{\sum f(x-\overline{x})^2}{\sum f}} = \sqrt{\frac{101.47}{48}} \approx 1.45 \text{ toothpicks}$$

**b** Estimate of $\mu$ is $\overline{x} \approx 37.3$ toothpicks
Estimate of $\sigma$ is $s \approx 1.45$ toothpicks

**4** **a**

| *Midpoint* ($x$) | $f$ | $fx$ | $f(x-\overline{x})^2$ |
|---|---|---|---|
| 41 | 1 | 41 | 52.80 |
| 43 | 1 | 43 | 27.74 |
| 45 | 3 | 135 | 32.01 |
| 47 | 7 | 329 | 11.23 |
| 49 | 11 | 539 | 5.91 |
| 51 | 5 | 255 | 37.35 |
| 53 | 2 | 106 | 44.80 |
| $\sum$ | 30 | 1448 | 211.87 |

$$\overline{x} = \frac{\sum fx}{\sum f} = \frac{1448}{30} \approx 48.3 \text{ cm}$$

$$s = \sqrt{\frac{\sum f(x-\overline{x})^2}{\sum f}} \approx \sqrt{\frac{211.87}{30}} \approx 2.66 \text{ cm}$$

**b** Estimate of $\mu$ is $\overline{x} \approx 48.3$ cm
Estimate of $\sigma$ is $s \approx 2.66$ cm

**5** **a**

| *Midpoint* ($x$) | $f$ | $fx$ | $f(x-\overline{x})^2$ |
|---|---|---|---|
| 364.995 | 17 | 6204.9 | 10 881.53 |
| 374.995 | 38 | 14 249.8 | 8895.42 |
| 384.995 | 47 | 18 094.8 | 1320.23 |
| 394.995 | 57 | 22 514.7 | 1259.13 |
| 404.995 | 18 | 7289.9 | 3889.62 |
| 414.995 | 10 | 4150.0 | 6100.9 |
| 424.995 | 10 | 4250.0 | 12 040.9 |
| 434.995 | 3 | 1305.0 | 5994.27 |
| $\sum$ | 200 | 78 059 | 50 382 |

$$\overline{x} = \frac{\sum fx}{\sum f} = \frac{78\,059}{200} \approx \$390.30$$

$$s = \sqrt{\frac{\sum f(x-\overline{x})^2}{\sum f}} = \sqrt{\frac{50\,382}{200}} \approx \$15.87$$

**b** Estimate of $\mu$ is $\overline{x} \approx \$390.30$
Estimate of $\sigma$ is $s \approx \$15.87$

## REVIEW SET 20A

**1** **a** **i** There are 30 colonies.
$\therefore$ median = average of 15th and 16th colonies
$= \frac{3.1+3.2}{2} = 3.15$ cm

**ii** range $= 4.9 - 0.4 = 4.5$ cm

**b**

| *Diameter* | *Tally* | *Frequency* |
|---|---|---|
| 0.0 - 0.9 | \|\|\| | 3 |
| 1.0 - 1.9 | \|\|\|\| | 4 |
| 2.0 - 2.9 | ~~\|\|\|\|~~ \| | 6 |
| 3.0 - 3.9 | ~~\|\|\|\|~~ ~~\|\|\|\|~~ \|\| | 12 |
| 4.0 - 4.9 | ~~\|\|\|\|~~ | 5 |

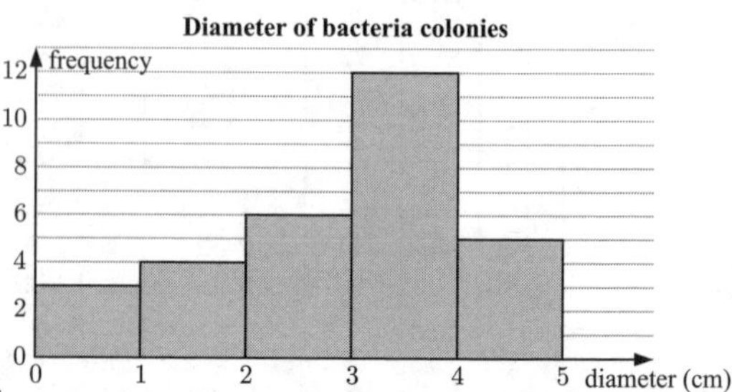


**c** The distribution is slightly negatively skewed.

**2** If the mode is 6 then one of the unknown numbers must be 6.
Suppose the other unknown number is $x$.

$$\therefore \quad \frac{4+6+9+6+3+x}{6} = 6$$

$$\therefore \quad 28 + x = 36$$

$$\therefore \quad x = 8$$

Since $a > b$, $a = 8$ and $b = 6$.

**3** **a** We first organise the data into tables:

*Girls*:

| *Time $x$ (s)* | $f$ | $fx$ |
|---|---|---|
| 33 | 1 | 33 |
| 34 | 3 | 102 |
| 35 | 5 | 175 |
| 36 | 4 | 144 |
| 37 | 4 | 148 |
| 38 | 1 | 38 |
| 39 | 1 | 39 |
| 40 | 0 | 0 |
| 41 | 1 | 41 |
| *Total* | 20 | 720 |

*Boys*:

| *Time $x$ (s)* | $f$ | $fx$ |
|---|---|---|
| 32 | 1 | 32 |
| 33 | 4 | 132 |
| 34 | 5 | 170 |
| 35 | 6 | 210 |
| 36 | 3 | 108 |
| 37 | 1 | 37 |
| *Total* | 20 | 689 |

Both boys and girls have 20 member squads, so the median is the average of the 10th and 11th swimmer.

*Girls*: median $= \dfrac{36+36}{2}$ $= 36$ s

mean $= \dfrac{\sum fx}{\sum f}$ $= \dfrac{720}{20}$ $= 36$ s

The tallest column on the *Girls* histogram is the '35' column.

$\therefore$ the modal class is 34.5 - 35.5 s.

*Boys*: median $= \dfrac{34+35}{2}$ $= 34.5$ s

mean $= \dfrac{\sum fx}{\sum f}$ $= \dfrac{689}{20}$ $= 34.45$ s

The tallest column on the *Boys* histogram is the '35' column.

$\therefore$ the modal class is 34.5 - 35.5 s.

So,

| *Distribution* | *Girls* | *Boys* |
|---|---|---|
| shape | positively skewed | approximately symmetrical |
| median | 36 s | 34.5 s |
| mean | 36 s | 34.45 s |
| modal class | 34.5 - 35.5 s | 34.5 - 35.5 s |

**b** The girls' distribution is positively skewed and the boys' distribution is approximately symmetrical. The median and mean swim times for boys are both about 1.5 seconds lower than for girls. Despite this, the distributions have the same modal class because of the skewness in the girls' distribution. The analysis supports the conjecture that boys generally swim faster than girls with less spread of times.

**4** 11 12 12 13 14 14 15 15 15 16 17 17 18

min: 11, $Q_1 = 12.5$, median: 15, $Q_3 = 16.5$, max: 18

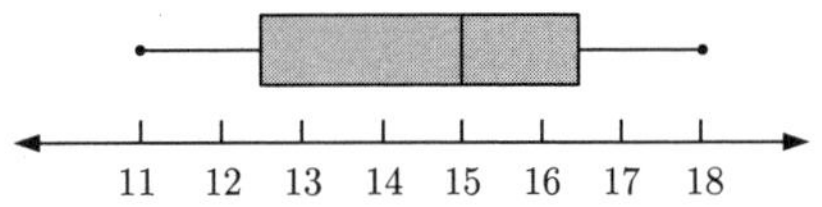

**5** **a** median = score for CF of 40
$\approx 58.5$ seconds

**b** IQR = (score for CF of 60) – (score for CF of 20)
$\approx 61.5 - 55.5$
$\approx 6$ seconds

**c** Time corresponding to the top 10% of runners = score for a CF of 8
$\approx 53$ seconds

**6** **a**

| *Score* | *Frequency* | *Cumulative frequency* |
|---|---|---|
| 6 | 2 | 2 |
| 7 | 4 | $m = 2 + 4 = 6$ |
| 8 | 7 | 13 |
| 9 | $p = 25 - 13 = 12$ | 25 |
| 10 | 5 | 30 |

$\therefore \quad m = 6, \quad p = 12$

**b** The mode is the most frequently occurring score.
$\therefore$ mode = 9
There are 30 values, so the median is the average of the 15th and 16th values.

$\therefore$ median $= \dfrac{9+9}{2}$
$= 9$

Range = maximum – minimum
$= 10 - 6$
$= 4$

So,

| *Measure* | mode | median | range |
|---|---|---|---|
| *Value* | 9 | 9 | 4 |

**c** mean $= \dfrac{\sum fx}{\sum f}$
$= \dfrac{254}{30}$
$= \dfrac{127}{15}$

**7** **a** When $t = 20$, CF $\approx 108$ and when $t = 10$, CF $\approx 20$.
So, approximately $108 - 20 \approx 88$ students spent between 10 and 20 minutes travelling to school.

**b** If 30% of students spent more than $m$ minutes, 70% of students spent less than $m$ minutes.
70% of 200 students = 140 students.
When CF = 140, $t \approx 24$ minutes $\quad \therefore \quad m \approx 24$.

**c**

| *Time t* (min) | $5 \leqslant t < 10$ | $10 \leqslant t < 15$ | $15 \leqslant t < 20$ | $20 \leqslant t < 25$ |
|---|---|---|---|---|
| *Cumulative frequency* | 20 | 60 | 108 | 150 |
| *Frequency* | 20 | $60 - 20 = 40$ | $108 - 60 = 48$ | $150 - 108 = 42$ |

| *Time t* (min) | $25 \leqslant t < 30$ | $30 \leqslant t < 35$ | $35 \leqslant t < 40$ |
|---|---|---|---|
| *Cumulative frequency* | 178 | 195 | 200 |
| *Frequency* | $178 - 150 = 28$ | $195 - 178 = 17$ | $200 - 195 = 5$ |

So,

| *Time t* (min) | $5 \leqslant t < 10$ | $10 \leqslant t < 15$ | $15 \leqslant t < 20$ | $20 \leqslant t < 25$ |
|---|---|---|---|---|
| *Frequency* | 20 | 40 | 48 | 42 |

| *Time t* (min) | $25 \leqslant t < 30$ | $30 \leqslant t < 35$ | $35 \leqslant t < 40$ |
|---|---|---|---|
| *Frequency* | 28 | 17 | 5 |

## REVIEW SET 20B

**1** **a** highest = 97.5 m, lowest = 64.6 m

**b** The range = 97.5 − 64.6 = 32.9
So, if intervals of length 5 are used we need about 7 of them.
We choose $60 \leqslant d < 65$, $65 \leqslant d < 70$, $70 \leqslant d < 75$, and so on.

**c** **Distances thrown by Thabiso**

| *Distance* (m) | *Tally* | *Frequency* ($f$) |
|---|---|---|
| $60 \leqslant d < 65$ | \| | 1 |
| $65 \leqslant d < 70$ | \|\|\| | 3 |
| $70 \leqslant d < 75$ | ~~\|\|\|\|~~ | 5 |
| $75 \leqslant d < 80$ | \|\| | 2 |
| $80 \leqslant d < 85$ | ~~\|\|\|\|~~ \|\|\| | 8 |
| $85 \leqslant d < 90$ | ~~\|\|\|\|~~ \| | 6 |
| $90 \leqslant d < 95$ | \|\|\| | 3 |
| $95 \leqslant d < 100$ | \|\| | 2 |
| | Total | 30 |

**d** **Frequency histogram displaying the distance Thabiso throws a baseball**

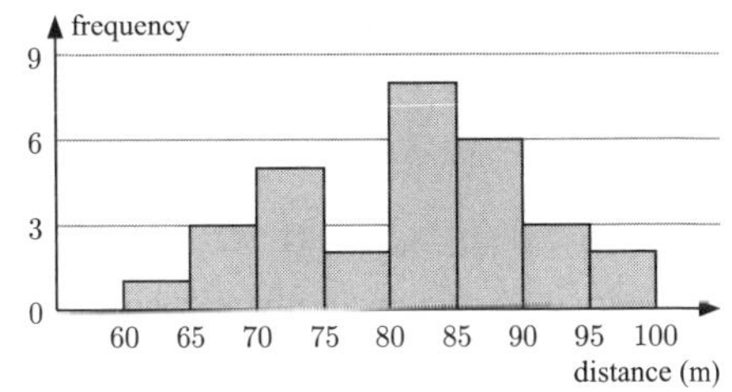


**e** Using technology:
**i** $\overline{x} \approx 81.1$ m **ii** median $\approx 83.1$ m

**2** **a** mean $= \dfrac{(k-2)+k+(k+3)+(k+3)}{4}$
$= \dfrac{4k+4}{4}$
$= \dfrac{\not{4}(k+1)}{\not{4}}$
$= k+1$ as required

**b** The members are now
$k$, $k+2$, $k+5$, and $k+5$.
$\therefore$ new mean
$= \dfrac{k+(k+2)+(k+5)+(k+5)}{4}$
$= \dfrac{4k+12}{4}$
$= \dfrac{\not{4}(k+3)}{\not{4}}$
$= k+3$

**3** **a**

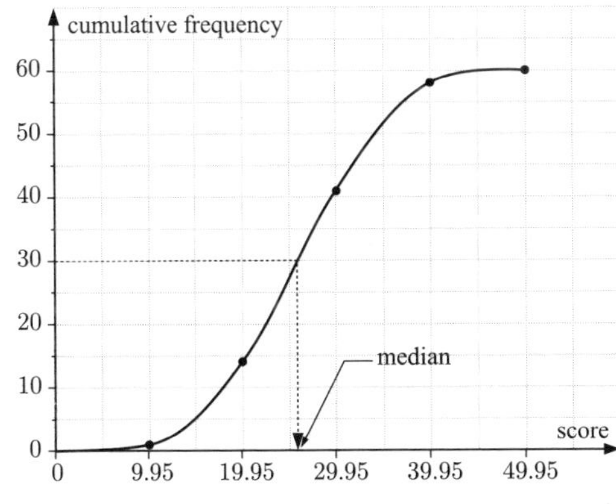


**b** median $\approx 25.9$ (see graph)

**c** IQR $= Q_3 - Q_1$
= (score for CF of 45) − (score for CF of 15)
$\approx 32.0 - 20.0$
$\approx 12.0$

**d**

| *Scores* | $f$ | midpt $x$ | $fx$ | $x-\overline{x}$ | $f(x-\overline{x})^2$ |
|---|---|---|---|---|---|
| $0 \leqslant x < 10$ | 1 | 5 | 5 | −21 | 441 |
| $10 \leqslant x < 20$ | 13 | 15 | 195 | −11 | 1573 |
| $20 \leqslant x < 30$ | 27 | 25 | 675 | −1 | 27 |
| $30 \leqslant x < 40$ | 17 | 35 | 595 | 9 | 1377 |
| $40 \leqslant x < 50$ | 2 | 45 | 90 | 19 | 722 |
| $\sum$ | 60 | | 1560 | | |

$\overline{x} = \dfrac{\sum fx}{\sum f}$
$= \dfrac{1560}{60}$
$= 26$

$s = \sqrt{\dfrac{\sum f(x-\overline{x})^2}{\sum f}}$
$= \sqrt{\dfrac{4140}{60}}$
$\approx 8.31$

**4** **a**

| *Litres* ($x$) | $f$ | $fx$ | $f(x-\overline{x})^2$ |
|---|---|---|---|
| 17.5 | 5 | 87.5 | 1299.38 |
| 22.5 | 13 | 292.5 | 1607.71 |
| 27.5 | 17 | 467.5 | 636.87 |
| 32.5 | 29 | 942.5 | 36.42 |
| 37.5 | 27 | 1012.5 | 406.32 |
| 42.5 | 18 | 765 | 1419.16 |
| 47.5 | 7 | 332.5 | 1348.45 |
| $\sum$ | 116 | 3900 | 6754.31 |

$$\overline{x} = \frac{\sum fx}{\sum f} = \frac{3900}{116} \approx 33.6 \text{ litres}$$

$$s = \sqrt{\frac{\sum f(x-\overline{x})^2}{\sum f}} \approx \sqrt{\frac{6754.31}{116}} \approx 7.63 \text{ litres}$$

**b** Estimate of $\mu$ is $\overline{x} \approx 33.6$ litres

Estimate of $\sigma$ is $s \approx 7.63$ litres

**5** Using technology: **a** **i** 101.5 **ii** 98 **iii** 105.5 **b** 7.5 **c** $\overline{x} = 100.2, \ s \approx 7.59$

**6** **a** Using technology, $\overline{x} \approx 49.6, \ s \approx 1.60$.

**b** This does not justify the claim. A larger sample is needed.

**7** **a** Using technology, $s \approx 11.65$

**b** Using technology, $Q_1 = 117$ and $Q_3 = 130$.

**c** IQR $= 130 - 117$ $= 13$

Lower boundary $= Q_1 - 1.5 \times \text{IQR}$ $= 117 - 1.5 \times 13$ $= 97.5$

Upper boundary $= Q_3 + 1.5 \times \text{IQR}$ $= 130 + 1.5 \times 13$ $= 149.5$

So, 93 is an outlier.

**d**

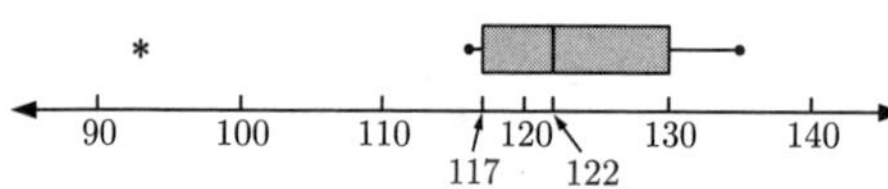

 **8** **a**

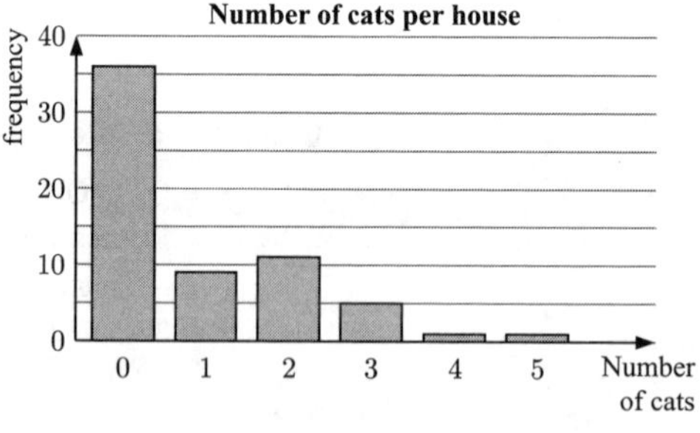


**b** The distribution is positively skewed.

**c** **i** The mode is 0 cats.

**ii**

| *Number of cats* | *Frequency* | $fx$ |
|---|---|---|
| 0 | 36 | 0 |
| 1 | 9 | 9 |
| 2 | 11 | 22 |
| 3 | 5 | 15 |
| 4 | 1 | 4 |
| 5 | 1 | 5 |
| *Total* | 63 | 55 |

$$\overline{x} = \frac{\sum fx}{\sum f} = \frac{55}{63} \approx 0.873$$

**iii** There are 63 values, so the median is the $\frac{63+1}{2} = 32$nd value.

$\therefore$ median $= 0$ cats.

**d** The mean, as it suggests that some people have cats. (The mode and median are both 0.)

## REVIEW SET 20C

**1** **a** The data is discrete.

**c** No, as we do not know each individual data value, only the intervals they fall in.

**b**

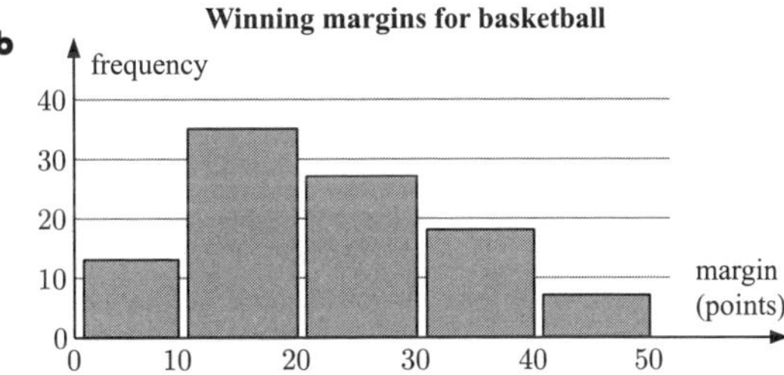


**2** **a**

| *Score* | *f* | *Product* |
|---|---|---|
| 2 | 3 | 6 |
| 5 | 2 | 10 |
| $x$ | 4 | $4x$ |
| $x+6$ | 1 | $x+6$ |
| *Total* | 10 | $5x+22$ |

mean $= 5.7$

$\therefore \dfrac{5x+22}{10} = 5.7$

$\therefore 5x+22 = 57$

$\therefore 5x = 35$

$\therefore x = 7$

**b** The data set is:

| *Score* $x$ | $f$ | $x-\overline{x}$ | $(x-\overline{x})^2$ | $f(x-\overline{x})^2$ |
|---|---|---|---|---|
| 2 | 3 | $-3.7$ | 13.69 | 41.07 |
| 5 | 2 | $-0.7$ | 0.49 | 0.98 |
| 7 | 4 | 1.3 | 1.69 | 6.76 |
| 13 | 1 | 7.3 | 53.29 | 53.29 |
| $\sum f$ | 10 | | $\sum f(x-\overline{x})^2$ | 102.1 |

Variance $= \dfrac{\sum f(x-\overline{x})^2}{\sum f}$

$= \dfrac{102.1}{10}$

$= 10.21$

$\therefore s^2 \approx 10.2$

**3** Use technology or

| *Midpoint* $(x)$ | $f$ | $fx$ |
|---|---|---|
| 274.5 | 14 | 3843 |
| 324.5 | 34 | 11 033 |
| 374.5 | 68 | 25 466 |
| 424.5 | 72 | 30 564 |
| 474.5 | 54 | 25 623 |
| 524.5 | 23 | 12 063.5 |
| 574.5 | 7 | 4021.5 |
| $\sum$ | 272 | 112 614 |

$\overline{x} = \dfrac{\sum fx}{\sum f}$

$= \dfrac{112\,614}{272}$

$\approx 414$ customers

**4** **a** Reading from the boxplots:

| | $A$ | $B$ |
|---|---|---|
| Min | 11 | 11.2 |
| $Q_1$ | 11.6 | 12 |
| Median | 12 | 12.6 |
| $Q_3$ | 12.6 | 13.2 |
| Max | 13 | 13.8 |

**b** **i** range of $A$
$= 13 - 11$
$= 2$

range of $B$
$= 13.8 - 11.2$
$= 2.6$

**ii** IQR of $A$
$= 12.6 - 11.6$
$= 1$

IQR of $B$
$= 13.2 - 12$
$= 1.2$

**c** **i** The members of squad $A$ generally ran faster because their median time is lower.

**ii** The times in squad $B$ are more varied because their range and IQR are higher.

**5** **a** Using technology with $x$ values 74.995, 84.995, 94.995, and so on,
$\overline{x} \approx$ €103.51 and $s \approx$ €19.40

**b** Estimate of $\mu$ is $\overline{x} \approx$ €103.51
Estimate of $\sigma$ is $s \approx$ €19.40

**6** **a** No, it will not be the same. Extreme values have less effect on the standard deviation of a larger population.

**b** **i** The mean would be used. **ii** The standard deviation would be used.

**c** A low standard deviation means that the weight of biscuits in each packet is, on average, close to 250 g.

**7** Let the number of marks be $x$.

**a** When $x = 45$, $\text{CF} \approx 120$

$\therefore$ about 120 students scored 45 marks or less.

**b** When $\text{CF} = 400$, $x \approx 65$

$\therefore$ the median mark was about 65 marks.

**c** When $\text{CF} = 200$, $x \approx 54$ and when $\text{CF} = 600$, $x \approx 75$

$\therefore$ the middle 50% of results lie between 54 and 75 marks.

**d** $\text{IQR} \approx 75 - 54 \approx 21$ marks

**e** When $x = 55$, $\text{CF} \approx 215$

$\therefore$ about $\frac{215}{800} \approx 27\%$ of students scored less than 55

$\therefore$ about 73% of students scored 55 or more

**f** 10% of 800 students = 80 students

When $\text{CF} = 800 - 80 = 720$, $x \approx 82$

$\therefore$ a score of 82 marks is required for a 'distinction'.

# Chapter 21

## LINEAR MODELLING

### EXERCISE 21A

**1** **a** A scatter diagram consists of points plotted on a set of axes where the independent variable is placed on the horizontal axis and the dependent variable on the vertical axis.

**b** Correlation refers to the relationship or association between two variables.

**c** Positive correlation describes the relationship when increasing the independent variable generally results in the dependent variable increasing.

**d** Negative correlation describes the relationship when increasing the independent variable generally results in the dependent variable decreasing.

**e** An outlier is a data point that does not fit the general trend of the data and is isolated from the main body of data.

**2** **a** **i** no correlation **ii** zero **iii** non-linear **iv** no outliers

**b** **i** positive correlation **ii** weak **iii** roughly linear **iv** no outliers

**c** **i** negative correlation **ii** moderate **iii** non-linear **iv** one outlier

**d** **i** positive correlation **ii** moderate **iii** linear **iv** no outliers

**e** **i** negative correlation **ii** strong **iii** linear **iv** one outlier

**f** **i** positive correlation **ii** moderate **iii** non-linear **iv** no outliers

**3** **a**

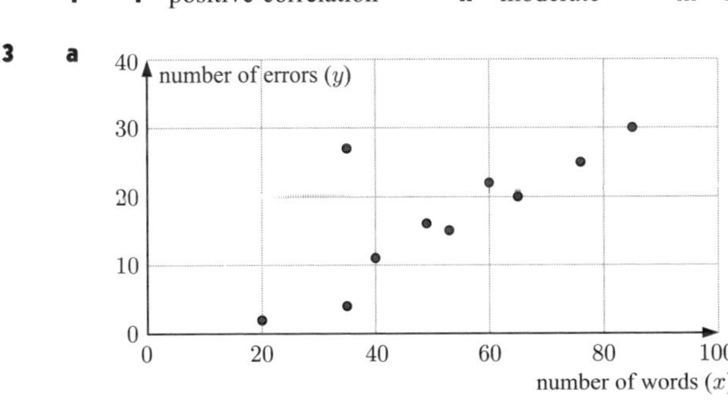


**b** **i** Student C is best described as slow but accurate.

**ii** Student G is best described as fast but inaccurate.

**iii** Student I is best described as an outlier.

**c** There is a strong, positive correlation between the two variables.

**d** The data is approximately linear.

**4** **a**

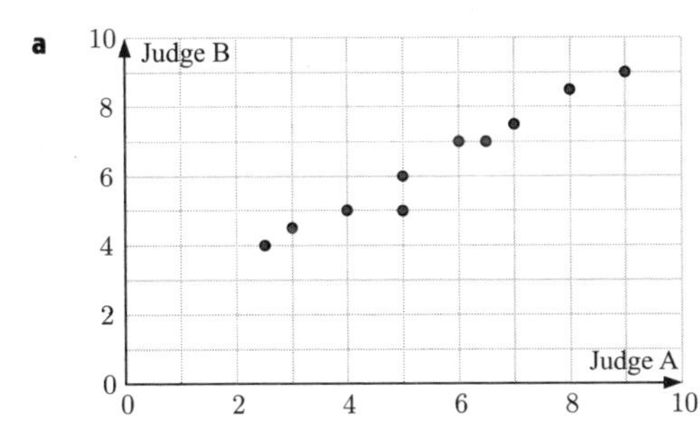


**b** There appears to be a **strong, positive** correlation between Judge A's scores and Judge B's scores. This means that as Judge A's scores increase, Judge B's scores **increase**.

**c** No, it is not reasonable. The judges are able to make decisions independently of one another.

### EXERCISE 21B

**1** **a** **B** **b** **A** **c** **D** **d** **C** **e** **E**

**2** **a**

| | $x$ | $y$ | $x-\overline{x}$ | $y-\overline{y}$ | $(x-\overline{x})(y-\overline{y})$ | $(x-\overline{x})^2$ | $(y-\overline{y})^2$ |
|---|---|---|---|---|---|---|---|
| | 0 | 1 | $-3.5$ | $-1.75$ | 6.125 | 12.25 | 3.0625 |
| | 2 | 2 | $-1.5$ | $-0.75$ | 1.125 | 2.25 | 0.5625 |
| | 4 | 3 | 0.5 | 0.25 | 0.125 | 0.25 | 0.0625 |
| | 8 | 5 | 4.5 | 2.25 | 10.125 | 20.25 | 5.0625 |
| Totals: | 14 | 11 | | | 17.5 | 35 | 8.75 |

There are four pairs of data values, so $n = 4$.

$$\therefore\ \overline{x} = \frac{\sum x}{n} = \frac{14}{4} = 3.5$$

$$\therefore\ \overline{y} = \frac{\sum y}{n} = \frac{11}{4} = 2.75$$

$$r = \frac{\sum(x-\overline{x})(y-\overline{y})}{\sqrt{\sum(x-\overline{x})^2\sum(y-\overline{y})^2}} = \frac{17.5}{\sqrt{35\times 8.75}} = 1$$

**b**

| | $x$ | $y$ | $x-\overline{x}$ | $y-\overline{y}$ | $(x-\overline{x})(y-\overline{y})$ | $(x-\overline{x})^2$ | $(y-\overline{y})^2$ |
|---|---|---|---|---|---|---|---|
| | 0 | 5 | $-2.5$ | 2.5 | $-6.25$ | 6.25 | 6.25 |
| | 1 | 4 | $-1.5$ | 1.5 | $-2.25$ | 2.25 | 2.25 |
| | 4 | 1 | 1.5 | $-1.5$ | $-2.25$ | 2.25 | 2.25 |
| | 5 | 0 | 2.5 | $-2.5$ | $-6.25$ | 6.25 | 6.25 |
| Totals: | 10 | 10 | | | $-17$ | 17 | 17 |

There are four pairs of data values, so $n = 4$.

$$\therefore\ \overline{x} = \frac{\sum x}{n} = \frac{10}{4} = 2.5$$

$$\therefore\ \overline{y} = \frac{\sum y}{n} = \frac{10}{4} = 2.5$$

$$r = \frac{\sum(x-\overline{x})(y-\overline{y})}{\sqrt{\sum(x-\overline{x})^2\sum(y-\overline{y})^2}} = \frac{-17}{\sqrt{17\times 17}} = -1$$

**c**

| | $x$ | $y$ | $x-\overline{x}$ | $y-\overline{y}$ | $(x-\overline{x})(y-\overline{y})$ | $(x-\overline{x})^2$ | $(y-\overline{y})^2$ |
|---|---|---|---|---|---|---|---|
| | 1 | 1 | $-1$ | $-1$ | 1 | 1 | 1 |
| | 1 | 3 | $-1$ | 1 | $-1$ | 1 | 1 |
| | 3 | 1 | 1 | $-1$ | $-1$ | 1 | 1 |
| | 3 | 3 | 1 | 1 | 1 | 1 | 1 |
| Totals: | 8 | 8 | | | 0 | 4 | 4 |

There are four pairs of data values, so $n = 4$.

$$\therefore\ \overline{x} = \frac{\sum x}{n} = \frac{8}{4} = 2$$

$$\therefore\ \overline{y} = \frac{\sum y}{n} = \frac{8}{4} = 2$$

$$r = \frac{\sum(x-\overline{x})(y-\overline{y})}{\sqrt{\sum(x-\overline{x})^2\sum(y-\overline{y})^2}} = \frac{0}{\sqrt{4\times 4}} = 0$$

**3** **a**

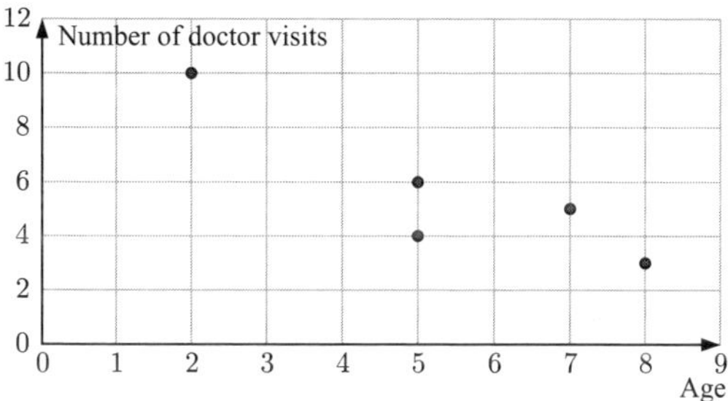


**b**

| | $x$ | $y$ | $x-\overline{x}$ | $y-\overline{y}$ | $(x-\overline{x})(y-\overline{y})$ | $(x-\overline{x})^2$ | $(y-\overline{y})^2$ |
|---|---|---|---|---|---|---|---|
| | 2 | 10 | $-3.4$ | 4.4 | $-14.96$ | 11.56 | 19.36 |
| | 5 | 6 | $-0.4$ | 0.4 | $-0.16$ | 0.16 | 0.16 |
| | 7 | 5 | 1.6 | $-0.6$ | 0.96 | 2.56 | 0.36 |
| | 5 | 4 | $-0.4$ | $-1.6$ | 0.64 | 0.16 | 2.56 |
| | 8 | 3 | 2.6 | $-2.6$ | $-6.76$ | 6.76 | 6.76 |
| Totals: | 27 | 28 | | | $-22.2$ | 21.2 | 29.2 |

There are five pairs of data values, so $n = 5$.

$$\therefore \quad \overline{x} = \frac{\sum x}{n} = \frac{27}{5} = 5.4$$

$$\therefore \quad \overline{y} = \frac{\sum y}{n} = \frac{28}{5} = 5.6$$

$$r = \frac{\sum(x-\overline{x})(y-\overline{y})}{\sqrt{\sum(x-\overline{x})^2 \sum(y-\overline{y})^2}} = \frac{-22.2}{\sqrt{21.2 \times 29.2}} \approx -0.892$$

**c** There is a strong, negative correlation between *age* and *number of doctor visits*.

**4** **a**

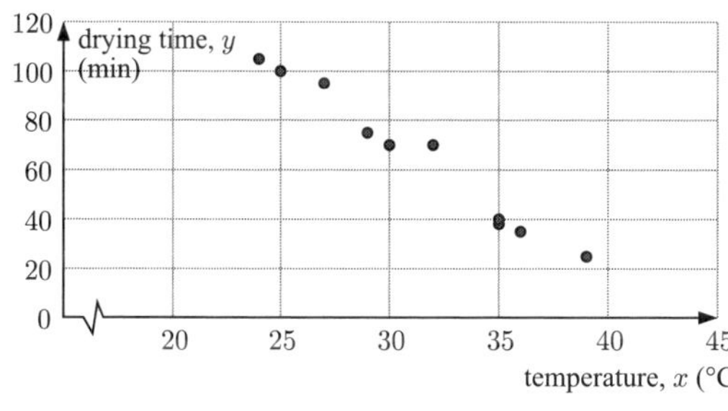


**b** Using technology, $r \approx -0.987$

**c** There is a very strong, negative correlation between *temperature* and *drying time*.

**5** **a**

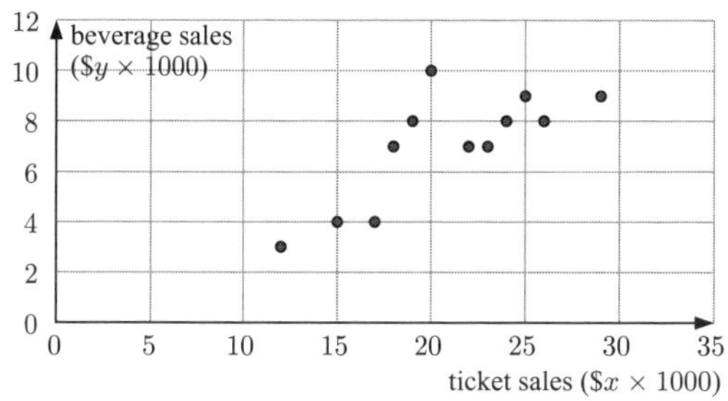


**b** Using technology, $r \approx 0.776$

**c** There is a moderate, positive correlation between *ticket sales* and *beverage sales*.

**6** **a**

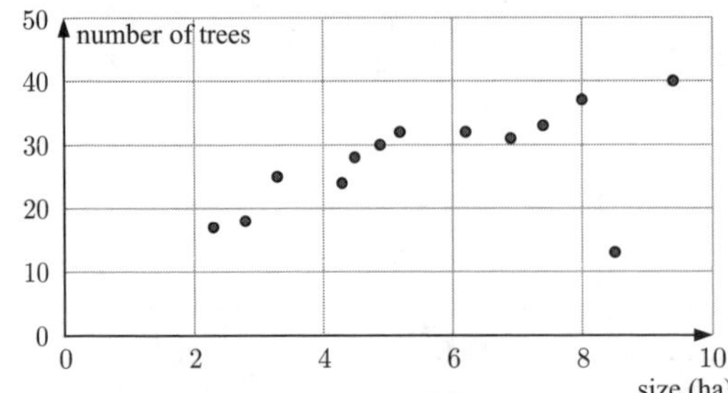


**b** The value of $r$ is expected to be positive, since more trees can be planted in a larger space.

**c** Using technology, $r \approx 0.520$

**d** Yes, the point $(8.5,\ 13)$ is an outlier.

**e** Using technology, $r \approx 0.943$

## EXERCISE 21C

**1** **a** **i** The mean of the $x$-values is

$$\overline{x} = \frac{1+3+4+5+6+8}{6} = \frac{27}{6} = 4.5$$

The mean of the $y$-values is

$$\overline{y} = \frac{2+3+3+4+5+7}{6} = \frac{24}{6} = 4$$

So, the mean point is $(4.5,\ 4)$.

$\therefore$ the scatter diagram and line of best fit are:

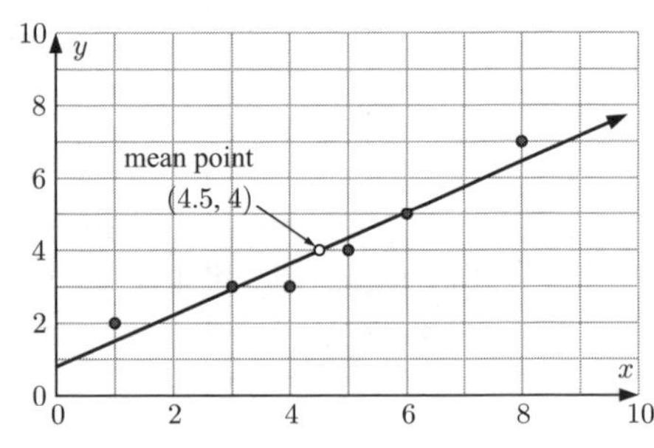


**ii** The $y$-intercept of the line of best fit is $\approx 0.8$.

The gradient of the line of best fit $\approx \dfrac{4-0.8}{4.5-0}$

$\approx 0.7$

$\therefore$ the line of best fit is $y \approx 0.7x + 0.8$

**b** **i** The mean of the $x$-values is

$$\overline{x} = \frac{13+18+7+1+12+6+15+4+17+3}{10} = \frac{96}{10} = 9.6$$

The mean of the $y$-values is

$$\overline{y} = \frac{10+6+17+18+13+14+6+15+5+14}{10} = \frac{118}{10} = 11.8$$

So, the mean point is $(9.6,\ 11.8)$.

$\therefore$ the scatter diagram and line of best fit are:

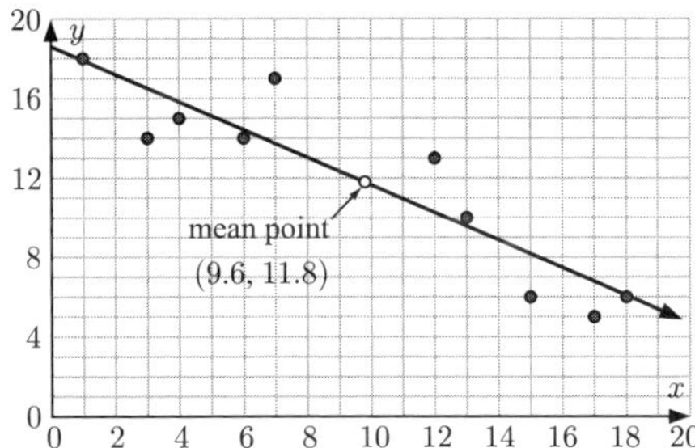


**ii** The $y$-intercept of the line of best fit is $\approx 18.6$.

The gradient of the line of best fit $\approx \dfrac{11.8 - 18.6}{9.6 - 0}$

$\approx -0.7$

$\therefore$ the line of best fit is $y \approx -0.7x + 18.6$

**c** **i** The mean of the $x$-values is

$$\overline{x} = \frac{11 + 7 + 16 + 4 + .... + 9 + 18 + 5 + 12}{16}$$

$$= \frac{157}{16}$$

$$\approx 9.8$$

The mean of the $y$-values is

$$\overline{y} = \frac{16 + 12 + 32 + 5 + .... + 15 + 34 + 6 + 26}{16}$$

$$= \frac{282}{16}$$

$$\approx 17.6$$

So, the mean point is (9.8, 17.6).

$\therefore$ the scatter diagram and line of best fit are:

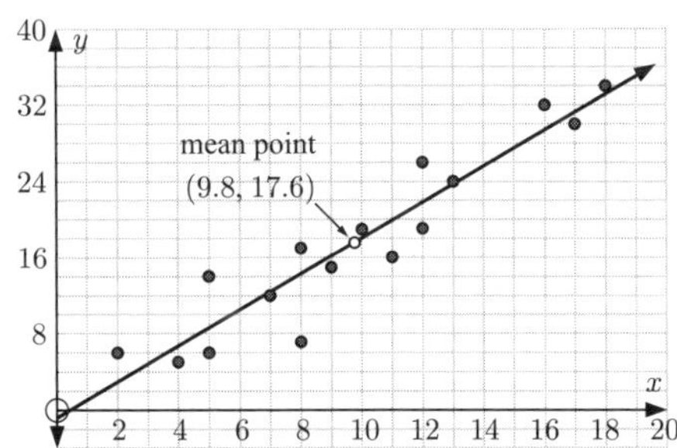


**ii** The points (9.8, 17.6) and (2, 3) lie on the line of best fit.

The gradient is $\dfrac{17.6 - 3}{9.8 - 2} \approx 1.9$, so the line has form $y \approx 1.9x + c$.

If (2, 3) lies on the line then $3 \approx 1.9 \times 2 + c$

$\therefore c + 3.8 \approx 3$

$\therefore c \approx -0.8$

So, the line of best fit is $y \approx 1.9x - 0.8$.

**2** **a, c**

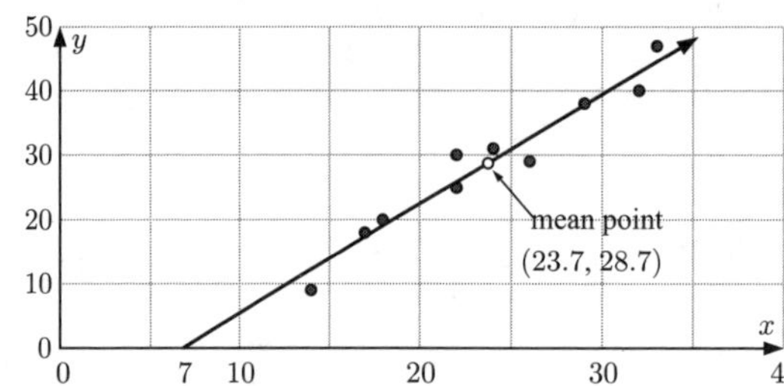


**b** There is a very strong, positive correlation between the *maximum temperature* and *number of car break-ins*.

**d** The points $(7, 0)$ and $(23.7, 28.7)$ lie on the line of best fit.

The gradient is $\dfrac{28.7 - 0}{23.7 - 7} \approx 1.7$, so the line has form $y \approx 1.7x + c$.

If $(7, 0)$ lies on the line then $0 \approx 1.7 \times 7 + c$

$\therefore \; c \approx -11.9$

$\therefore \; c \approx -12$

So, the line of best fit is $y \approx 1.7x - 12$.

**e** If $x = 25$, then $y \approx 1.7 \times 25 - 12$

$\approx 30.5$

$\approx 31$ {round to nearest whole number}

So, we would expect about 31 break-ins to happen on a 25°C day.

**3** **a, d**

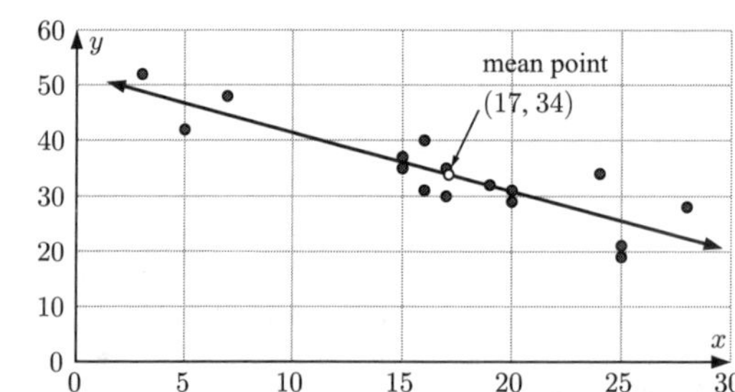


**b** Using technology, $r \approx -0.878$

**c** There is a strong, negative correlation between the *number of speed cameras* and the *number of car accidents*.

**d** The mean of the $x$-values is:

$$\overline{x} = \frac{7 + 15 + 20 + .... + 25 + 15 + 19}{16} = \frac{272}{16} = 17$$

The mean of the $y$-values is:

$$\overline{y} = \frac{48 + 35 + 31 + .... + 21 + 37 + 32}{16} = \frac{544}{16} = 34$$

$\therefore$ the mean point is (17, 34).

**e** If we extend the line of best fit so that it cuts the $y$-axis, we see that the $y$-intercept $\approx 52$.
This means that we would expect a city with no speed cameras to have approximately 52 car accidents.

## EXERCISE 21D

**1** **a** Using technology, the equation of the least squares regression line is $y \approx 0.712x + 0.797$.

**b** Using technology, the equation of the least squares regression line is $y \approx -0.707x + 18.6$.

**c** Using technology, the equation of the least squares regression line is $y \approx 1.88x - 0.797$.

**2** Using technology, the equation of the least squares regression line is $y \approx 1.72x - 12.2$.
When $x = 25$, $y \approx 1.72 \times 25 - 12.2 \approx 30.8$
$\therefore$ we would expect about 31 break-ins. This matches our previous answer.

**3** **a**

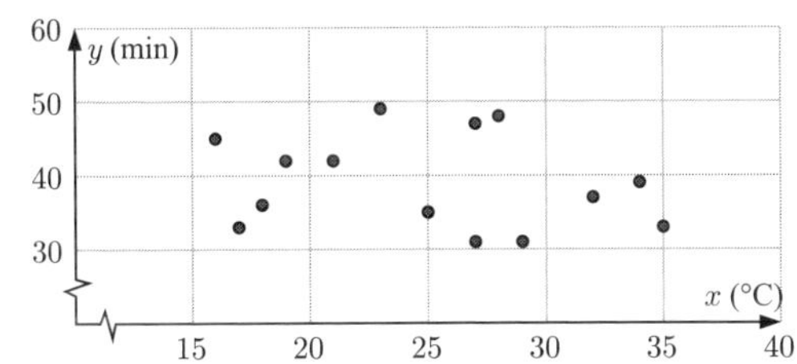


**b** Using technology, $r \approx -0.219$

**c** There is a weak, negative correlation between *temperature* and *time*.

**d** No, as there is only a weak, negative correlation between the variables.

**4** **a** Using technology, $r \approx -0.924$

**b** There is a strong, negative correlation between *petrol price* and the *number of customers*.

**c** Using technology, the equation of the least squares regression line is $y \approx -4.27x + 489$.

## EXERCISE 21E

**1** **a** Using technology, the equation of the least squares regression line is $y \approx -5.75x + 245$.

**b** When $x = 28$, $y \approx -5.75 \times 28 + 245$

$\therefore \; y \approx 84$

So, it will take about 84 minutes for Jill's clothes to dry on a 28°C day.

**c** Since we are interpolating, and the variables are strongly correlated, the estimate is quite reliable.

**2** **a** Using technology, the equation of the least squares regression line is $y \approx 0.350x - 0.293$.

**b** **i** When $x = 35$, $y \approx 0.350 \times 35 - 0.293$

$\therefore \; y \approx 11.957$ thousand dollars

So, the predicted beverage sales are $\approx$ \$12 000.

**ii** The variables are only moderately correlated, and we are extrapolating, so the prediction is not very reliable.

**3** **a**

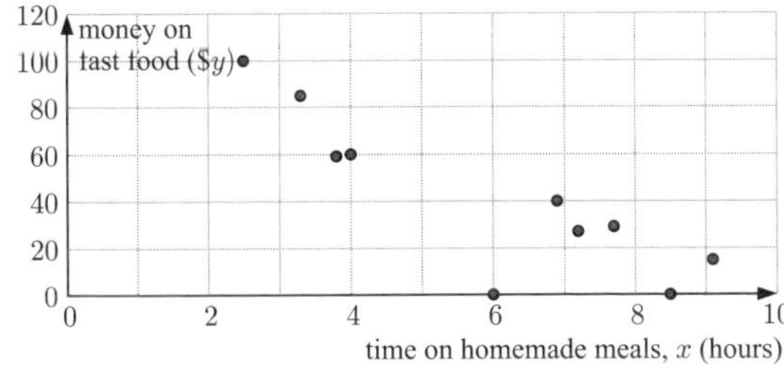


**b** Using technology, $r \approx -0.868$

**c** Using technology, the equation of the least squares regression line is $y \approx -12.654\,22x + 116.159\,902$, which we round to $y \approx -12.7x + 116$.

**d** gradient $\approx -12.7$. For each extra hour spent on homemade meals, a family spends about \$12.70 less each week on fast food.

$y$-intercept $\approx 116$. If no time is spent on homemade meals, a family will spend \$116 each week on fast food.

**e** When $x = 5$, $y \approx -12.654\,22 \times 5 + 116.159\,902$

$\therefore \; y \approx 52.89$

So, the family spends about \$52.89 on fast food each week. It is reasonably reliable as it is an interpolation and the variables are strongly correlated.

**4** **a**

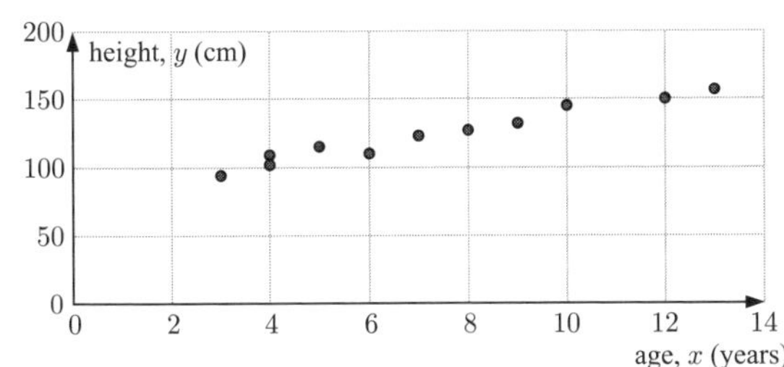


**b** Using technology, the least squares regression line is $y \approx 5.979\,806\,13x + 79.966\,882$, which we round to $y \approx 5.98x + 80.0$

**c** When $y = 140$, $5.98x + 80.0 = 140$
$\therefore\ 5.98x = 60$
$\therefore\ x \approx 10.0$

We would expect children to reach a height of 140 cm when they were 10 years old.

**d** gradient $\approx 5.98$. Every year, a child grows about 5.98 cm taller.

**e** When $x = 20$, $y \approx 5.98 \times 20 + 80.0$
$\approx 199.6$
$\therefore\ y \approx 200$

We would expect a 20 year old to be 200 cm tall. This prediction is not very reliable as it is an extrapolation well beyond the upper pole. Most children finish growing taller before 20 years.

**5** **a**

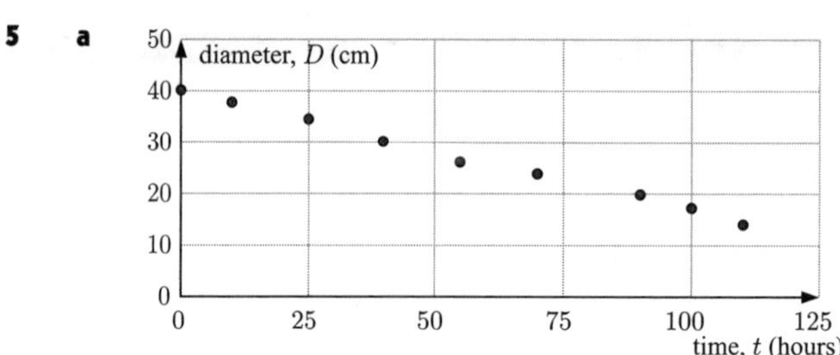


**b** There is a very strong, negative correlation between *time* and *diameter*.

**c** Using technology, the equation of the least squares regression line is
$D \approx -0.231\,994\,7t + 39.966\,374\,3$, which we round to $D \approx -0.232t + 40.0$.

**d** **i** When $t = 80$,
$D \approx 21.4$
$\therefore$ the diameter of the balloon is 21.4 cm after 80 hours.

**ii** When $D = 0$,
$-0.232t + 40.0 = 0$
$\therefore\ -0.232t = -40.0$
$\therefore\ t \approx 172$
It will take about 172 hours for the balloon to completely deflate.

**e** The prediction in part **d i** is more likely to be reliable, as it is an interpolation and the prediction in part **d ii** was extrapolated.

**6** **a**

| $t$ | 1 | 2 | 3 | 4 | 5 |
|---|---|---|---|---|---|
| $M$ | 3.6 | 5.7 | 9.1 | 14.6 | 23.3 |
| $\ln M$ | 1.28 | 1.74 | 2.21 | 2.68 | 3.15 |

**b**

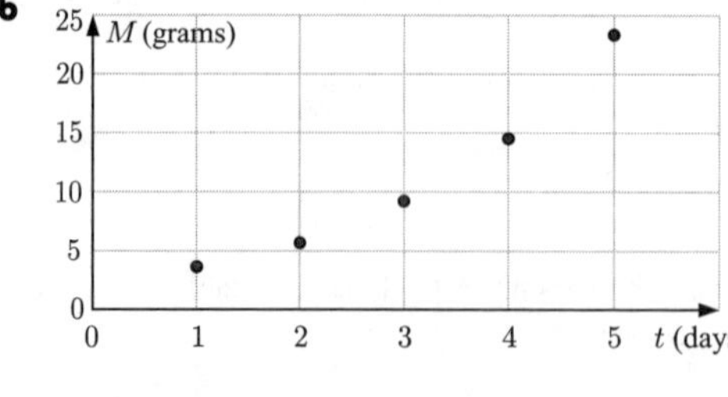


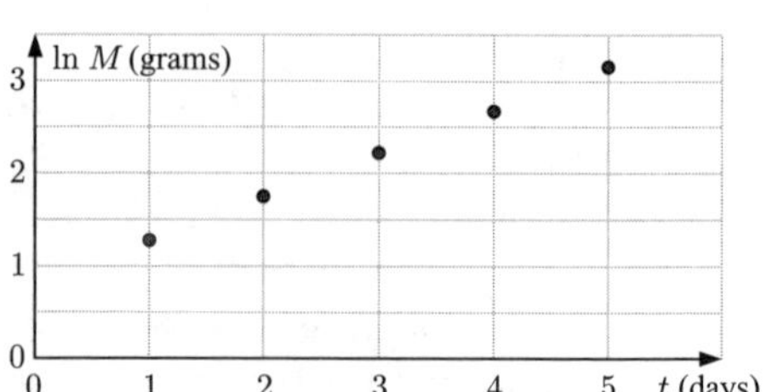


The scatter diagram of $\ln M$ against $t$ is linear.

**c** Using technology, $\ln M \approx 0.467\,559\,43t + 0.809\,151\,54$
which we round to $\ln M \approx 0.468t + 0.809$

**d** $\ln M \approx 0.468t + 0.809$
$\therefore\ e^{\ln M} \approx e^{0.468t + 0.809}$
$\therefore\ M \approx e^{0.809} \times e^{0.468t}$
$\therefore\ M \approx 2.25 \times 1.60^t$ as required

**e** When $t = 0$, $M \approx 2.25 \times 1.60^0$
$\approx 2.25$

There was originally about 2.25 grams of bacteria in the culture.

**7** **a** As $t$ increases, $Q$ decreases.

**b** The graph is not linear.

**c**

| $\sqrt{t}$ | 1.41 | 3.16 | 4.47 | 5.48 | 7.07 | 8.06 | 8.94 |
|---|---|---|---|---|---|---|---|
| $Q$ | 119 | 98 | 82 | 70 | 50 | 40 | 30 |

**d** $Q = m\sqrt{t} + c$ is a likely model for the data, as the graph of $Q$ against $\sqrt{t}$ is approximately linear.

**e** Using technology, $m \approx -11.880\,144$ and $c \approx 135.373\,507$

$\therefore\ m \approx -11.9$ and $c \approx 135$

**f** **i** When $t = 0$, $Q \approx -11.880\,144\sqrt{0} + 135.373\,507$

$\therefore\ Q \approx 135$

There is approximately 135 mg of the chemical in a newly born baby.

**ii** When $t = 25$, $Q \approx -11.880\,144\sqrt{25} + 135.373\,507$

$\therefore\ Q \approx 76.0$

There is approximately 76.0 mg of the chemical in a 25 year old.

**g** The answer in part **f ii** is more likely to be reliable as it is an interpolation, and the answer in part **f i** was an extrapolation.

**8** **a**

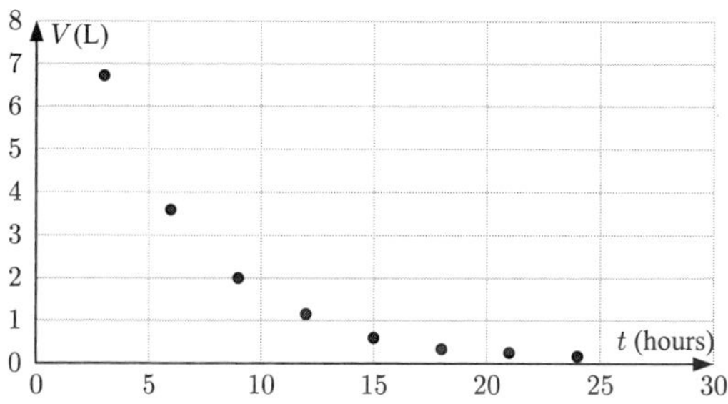


**b** **i**

| $t$ | 3 | 6 | 9 | 12 | 15 | 18 | 21 | 24 |
|---|---|---|---|---|---|---|---|---|
| $\ln V$ | 1.90 | 1.28 | 0.693 | 0.0953 | −0.511 | −1.14 | −1.71 | −2.30 |

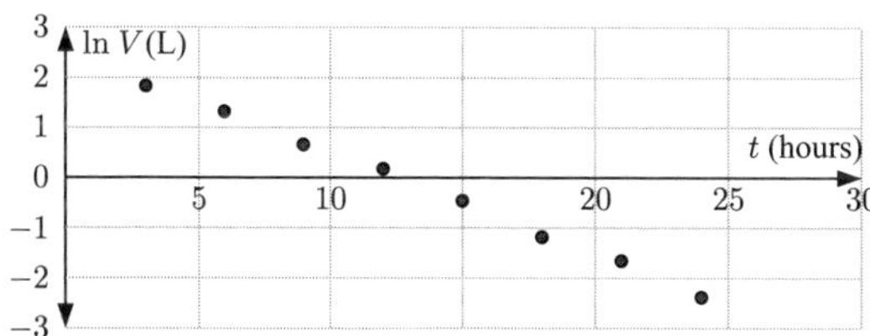


**ii**

| $\ln t$ | 1.10 | 1.79 | 2.20 | 2.48 | 2.71 | 2.89 | 3.04 | 3.18 |
|---|---|---|---|---|---|---|---|---|
| $\ln V$ | 1.90 | 1.28 | 0.693 | 0.0953 | −0.511 | −1.14 | −1.71 | −2.30 |

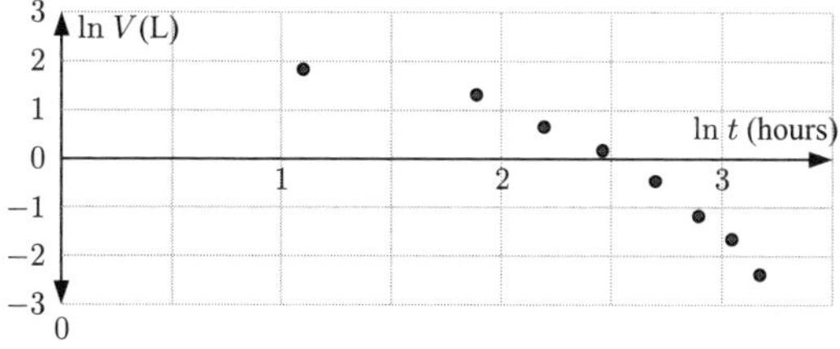

**c** The graph of $\ln V$ against $t$ is approximately linear.

$\therefore$ the model has the form $\ln V = mt + c$

Using technology, $\ln V \approx -0.200\,457\,8t + 2.494\,163\,51$

$\therefore \quad e^{\ln V} \approx e^{-0.200\,457\,8t + 2.494\,163\,51}$

$\therefore \quad V \approx e^{2.494\,163\,51} \times e^{-0.200\,457\,8t}$

$\therefore \quad V \approx 12.1 \times (0.818)^t$

**d** When $t = 5$, $V \approx 12.1 \times (0.818)^5$

$\therefore \quad V \approx 4.43$

There is approximately 4.43 L of water remaining in the bird bath after 5 hours.

**e** When $t = 0$, $V \approx 12.1 \times (0.818)^0$

$\therefore \quad V \approx 12.1$

When $t = 10$, $V \approx 12.1 \times (0.818)^{10}$

$\therefore \quad V \approx 1.622\,999\,578$

and $12.1 \text{ L} - 1.622\,999\,578 \text{ L}$

$\approx 10.5 \text{ L}$

So, approximately 10.5 L of water has evaporated after 10 hours.

## REVIEW SET 21A

**1** **a** **i** There is moderate, negative correlation between the variables.

**ii** The relationship is not linear.

**b** **i** There is strong, positive correlation between the variables.

**ii** The relationship appears to be linear.

**2** **a, c**

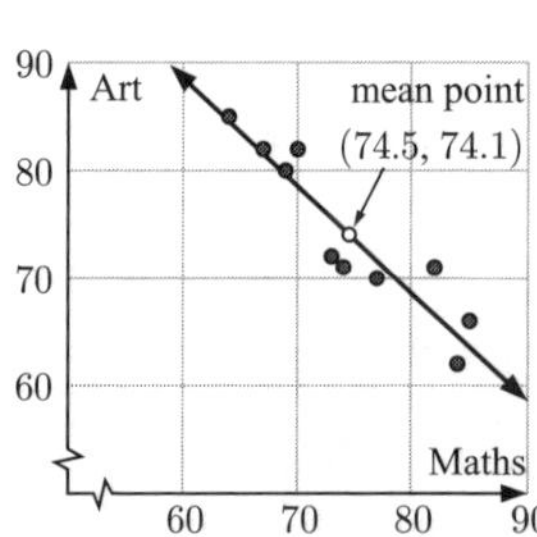


**b** There is strong, negative, linear correlation between Mathematics and Art marks.

**3** **a**

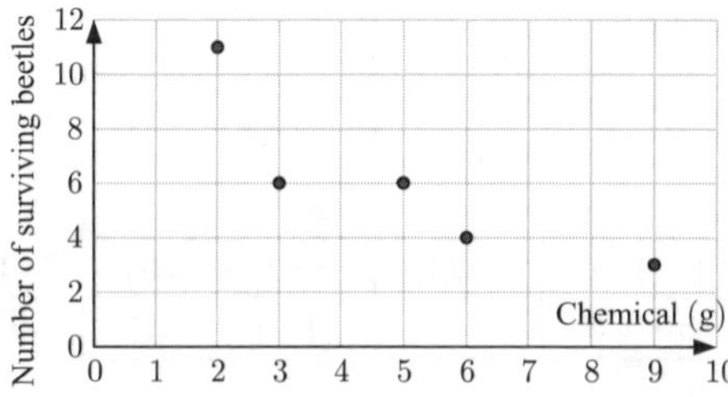


**b** There is moderate, negative correlation between the *quantity of chemical* and the *number of surviving beetles.*

**4** **a, b**

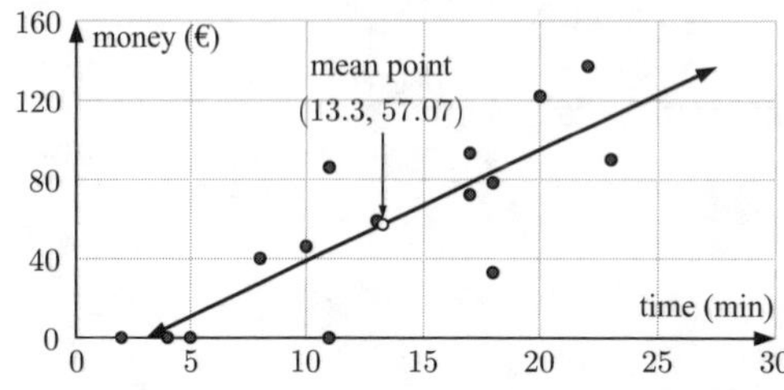


**c** There is a moderate, positive, linear correlation between *time in the store* and *money spent.*

**5** **a, b**

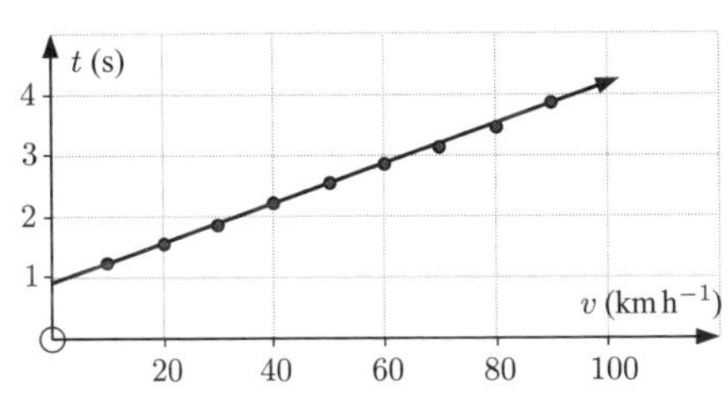


**c** **i** When $v = 55$,
$t = 0.03 \times 55 + 0.9$
$\therefore \ t = 2.55$
The stopping time for a speed of 55 km h$^{-1}$ is about 2.55 seconds.

**ii** When $v = 110$,
$t = 0.03 \times 110 + 0.9$
$\therefore \ t = 4.2$
The stopping time for a speed of 110 km h$^{-1}$ is about 4.2 seconds.

**d** The estimate in part **c i** is more likely to be reliable, as it is an interpolation.

## REVIEW SET 21B

**1** **a**

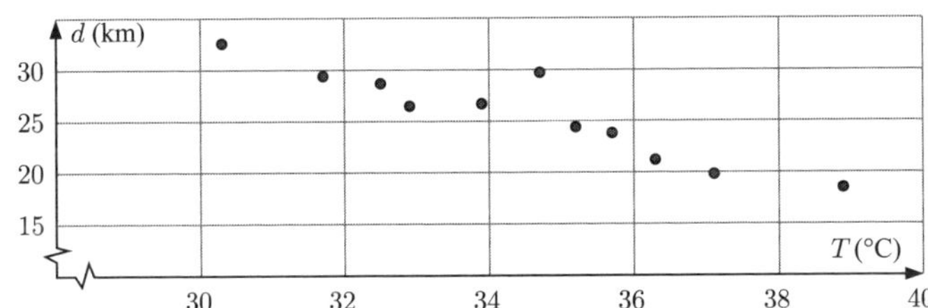


**b** $r \approx -0.928$ {using technology}
There is a strong, negative, linear correlation between the *temperature* and the *number of kilometres ridden*.

**c** $d \approx -1.64T + 82.3$ {using technology}

**d** If Thomas does not ride at all, $d = 0$

$\therefore \ 0 \approx -1.644T + 82.25$

$\therefore \ 1.644T \approx 82.25$

$\therefore \ T \approx \dfrac{82.25}{1.644} \approx 50.0°\text{C}$

Note that this is not a reliable extrapolation. Fifty degrees lies well outside the poles of the data set (the closest data value we have is for 38.9 degrees) and it may be that Thomas stops riding at some other temperature.

**2** **a**

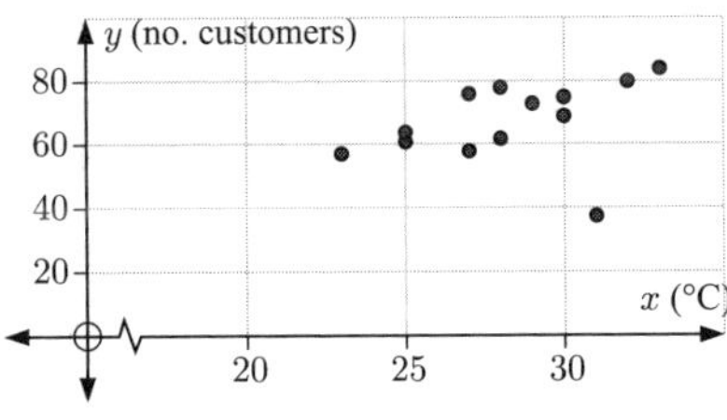


**b** Using technology, $r \approx 0.314$

**c** Yes, the point (31, 35) is an outlier.

**d** Using technology, $r \approx 0.801$

**e** There is a moderate, positive correlation between the *number of customers* and the *noon temperature* at the garden centre.

**3** **a, d**

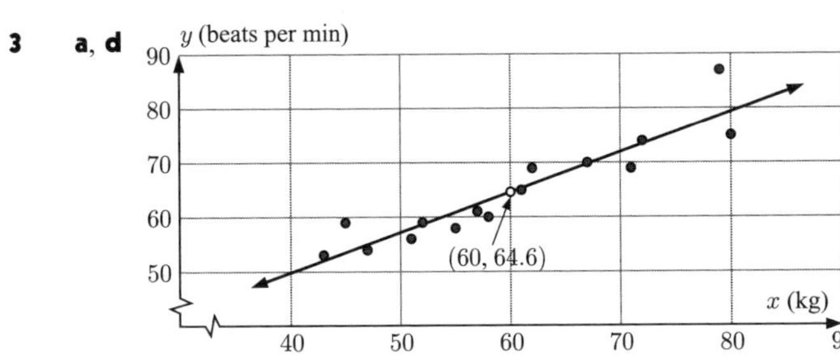


**b** $r \approx 0.929$
$\therefore$ there is a strong, positive correlation between *weight* and *pulse rate*.

**c** $\overline{x} = \dfrac{61+52+47+....+55}{15} = 60, \quad \overline{y} = \dfrac{65+59+54+....+58}{15} = 64.6,$

so $(\overline{x}, \overline{y}) = (60, 64.6)$.

**e** Using technology, the equation of the line of best fit is $y \approx 0.738x + 20.3$.

**f** When $x = 65$, $\quad y \approx 0.738 \times 65 + 20.3$

$\therefore \quad y \approx 68.27$

So, the pulse rate for a 65 kg student is about 68 beats per minute. This is a reliable estimate, as it is an interpolation on data with a strong correlation.

**4** **a** The independent variable is the number of waterings, $n$.

**b** $f \approx 34.0n + 19.3$ {using technology}

**c**

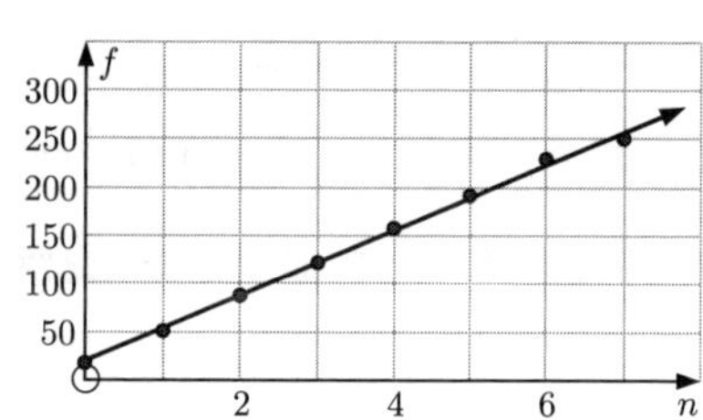


**d** **i** If $n = 2.5$, $f \approx 34.0 \times 2.5 + 19.3 \approx 104$
If $n = 10$, $f \approx 34.0 \times 10 + 19.3 \approx 359$

**ii** The case $n = 10$ is unreliable as it lies outside the poles and over-watering could be a problem. The case $n = 2.5$ is a much more reliable estimate.

**5** **a**

$(50, 4.4)$ is an outlier.

**b** Using technology:

**i** With the outlier: $r \approx 0.798$

**ii** Without the outlier: $r \approx 0.993$

**c** Using technology:

**i** With the outlier, the least squares regression line is $y \approx 0.0672x + 2.22$.

**ii** Without the outlier, the least squares regression line is $y \approx 0.119x + 1.32$.

**d** We should use the line in **ii**, which does not include the outlier.

**e** The outlier may have been caused by over-fertilisation, which can make plants sick or even kill them.

## REVIEW SET 21C

**1** **a** There is a moderate, positive correlation between *hours of study* and *marks obtained.*

**b** The number of marks is greater than 50, so the outlier appears to be an error. It should be discarded.

**2** **a**

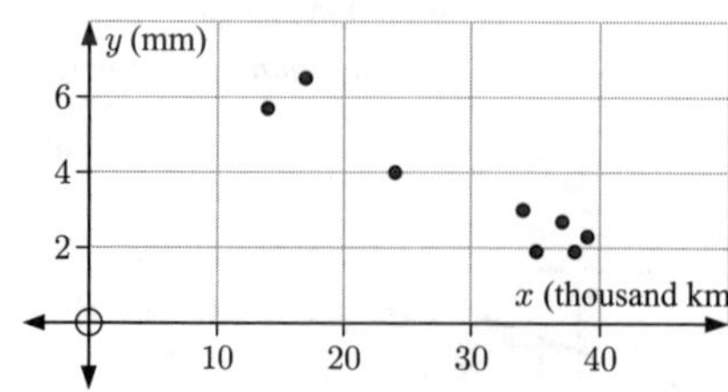


**b** $r \approx -0.951$

**c** There is a very strong, negative correlation between the *tread depth* and the *number of kilometres travelled.*

**3** **a, e**

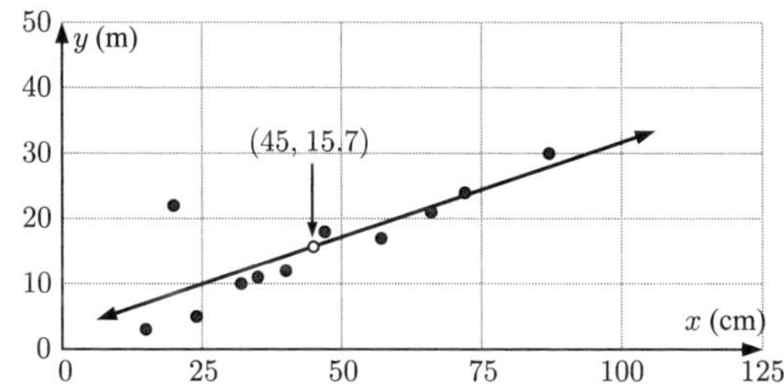


**b** The point (20, 22) is an outlier.

**c** The outlier tree is unusually tall and thin.

**d** $\overline{x} = \dfrac{35 + 47 + 72 + .... + 32}{11} = 45, \qquad \overline{y} = \dfrac{11 + 18 + 24 + .... + 10}{11} = 15.7,$

so $(\overline{x}, \overline{y}) = (45, 15.7)$.

**f** Using technology, the equation of the line of best fit is $y \approx 0.289\,030\,32x + 2.720\,908\,06$, which we round to $y \approx 0.289x + 2.72$.

**g** When $x = 120$, $y \approx 0.289 \times 120 + 2.72$

$\therefore \quad y \approx 37.4$

So, the height of a tree with width 120 cm will be about 37 m. This is an extrapolation, and so may not be reliable.

**4** **a**

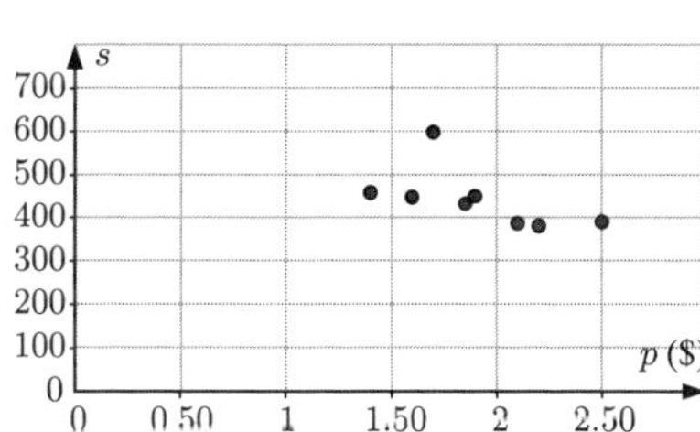


**b** Yes, the point (1.7, 597) is an outlier. It should not be deleted as there is no evidence that it is a mistake.

**c** $s \approx -116p + 665$ {using technology}

**d** 50 cents is a long way outside the poles, so the prediction would therefore be unreliable.

**5** **a**

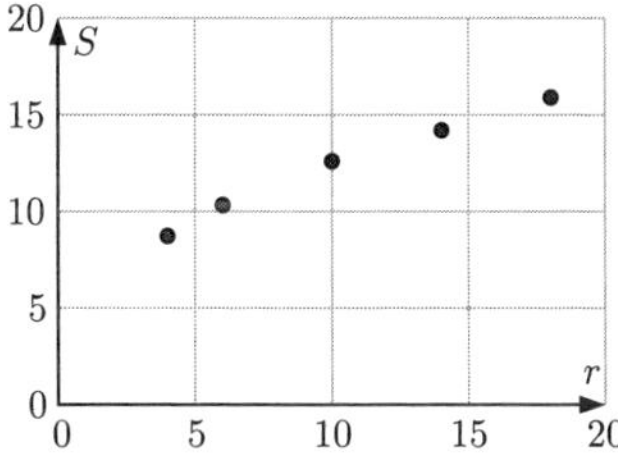


**b**

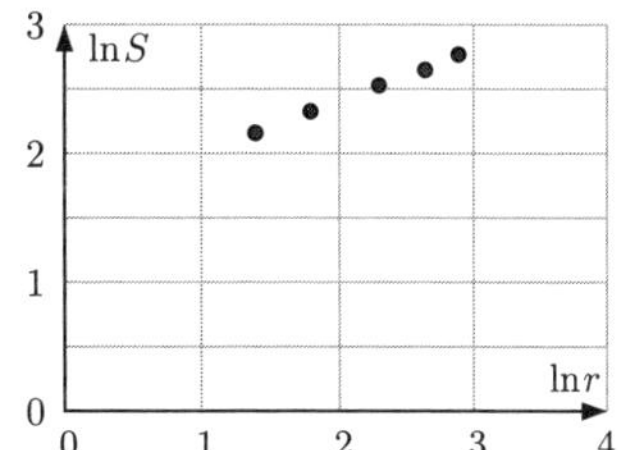


**c** Using technology, $\ln S \approx 0.395\,835\,26 \times \ln r + 1.618\,110\,33$

$\therefore \quad e^{\ln S} \approx e^{0.395\,835\,26 \ln r + 1.618\,110\,33}$

$\therefore \quad S \approx (e^{\ln r})^{0.395\,835\,26} \times e^{1.618\,110\,33}$

$\therefore \quad S \approx 5.043\,550\,66 \times r^{0.395\,835\,26}$

$\therefore \quad S \approx 5.04 \times r^{0.396}$

**d** When $r = 8$, $S \approx 5.04 \times 8^{0.396}$

$\therefore \quad S \approx 11.5$

The maximum speed of the canoe if there are 8 rowers is about $11.5$ km h$^{-1}$.

# Chapter 22
## PROBABILITY

### EXERCISE 22A

**1** **a** P(inside a square) $= \frac{113}{145} \approx 0.779$ **b** P(on a line) $= \frac{32}{145} \approx 0.221$

**2** Total frequency $= 17 + 38 + 19 + 4 = 78$

**a** P(20 to 39 seconds) $= \frac{38}{78} \approx 0.487$ **b** P(> 60 seconds) $= \frac{4}{78} \approx 0.051$

**c** P(between 20 and 59 seconds inclusive) $= \frac{38+19}{78} \approx 0.731$

**3**

| *Calls/day* | *No. of days* |
|---|---|
| 0 | 2 |
| 1 | 7 |
| 2 | 11 |
| 3 | 8 |
| 4 | 7 |
| 5 | 4 |
| 6 | 3 |
| 7 | 0 |
| 8 | 1 |

**a** Survey lasted $2+7+11+8+7+4+3+0+1$
$= 43$ days

**b** **i** P(0 calls) $\approx \frac{2}{43} \approx 0.0465$ **ii** P($\geqslant$ 5 calls) $\approx \frac{4+3+0+1}{43} \approx 0.186$ **iii** P(< 3 calls) $\approx \frac{2+7+11}{43} \approx 0.465$

**4** Total frequency
$= 37 + 81 + 48 + 17 + 6 + 1$
$= 190$

**a** P(4 days gap) $\approx \frac{17}{190} \approx 0.0895$ **b** P(at least 4 days gap) $\approx \frac{17+6+1}{190} \approx 0.126$

### EXERCISE 22B

**1** **a** {A, B, C, D}

**b** {BB, BG, GB, GG}

**c** {ABCD, ABDC, ACBD, ACDB, ADBC, ADCB, BACD, BADC, BCAD, BCDA, BDAC, BDCA, CABD, CADB, CBAD, CBDA, CDAB, CDBA, DABC, DACB, DBAC, DBCA, DCAB, DCBA}

**d** {GGG, GGB, GBG, BGG, GBB, BGB, BBG, BBB}

**2** **a** **b** 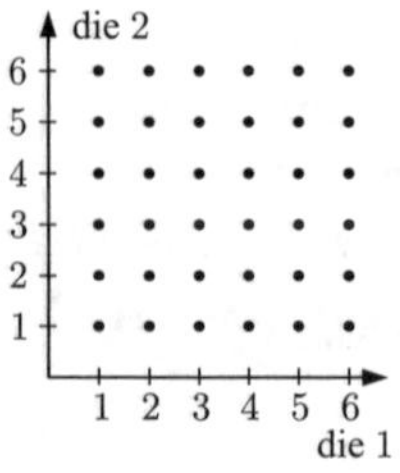 **c** 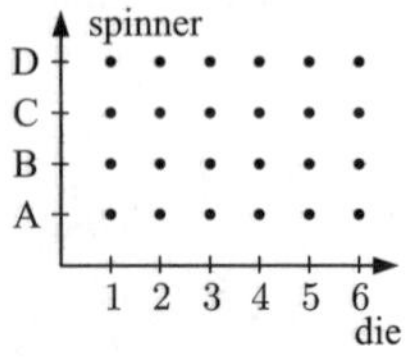 **d**

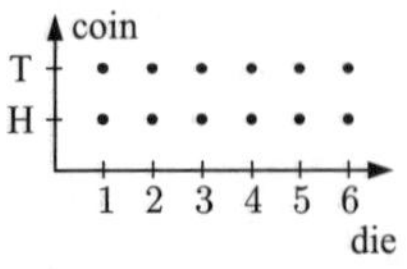

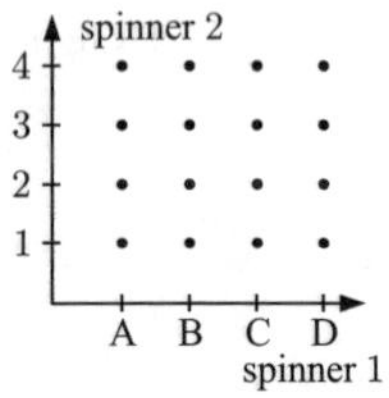

coin
T
H
1 2 3 4 5 6
die

die 2
6
5
4
3
2
1
1 2 3 4 5 6
die 1

spinner
D
C
B
A
1 2 3 4 5 6
die

spinner 2
4
3
2
1
A B C D
spinner 1

**3** **a** 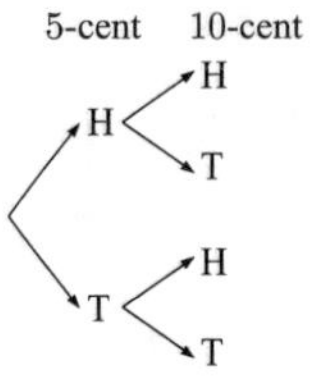 **b** 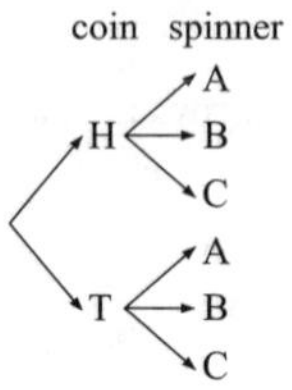 **c** 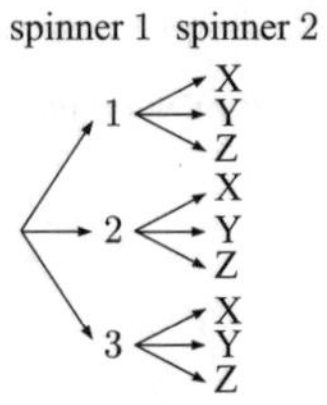 **d**

5-cent 10-cent
H H T
T H T

coin spinner
H A B C
T A B C

spinner 1 spinner 2
1 X Y Z
2 X Y Z
3 X Y Z

draw 1 draw 2
P P B W
B P B W
W P B W

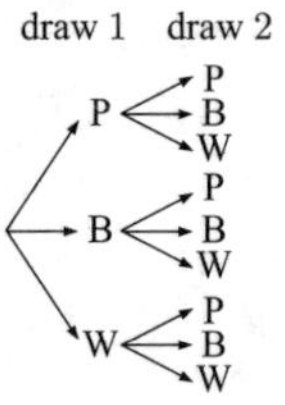

## EXERCISE 22C.1

**1** Total number of marbles $= 5 + 3 + 7 = 15$

**a** $\text{P(red)} = \frac{3}{15} = \frac{1}{5}$

**b** $\text{P(green)} = \frac{5}{15} = \frac{1}{3}$

**c** $\text{P(blue)} = \frac{7}{15}$

**d** $\text{P(not red)} = \dfrac{5+7}{15} = \frac{12}{15}$ or $\frac{4}{5}$

**e** $\text{P(neither green nor blue)} = \text{P(red)} = \frac{1}{5}$

**f** $\text{P(green or red)} = \dfrac{5+3}{15} = \frac{8}{15}$

**2** **a** 8 are brown and so 4 are white.

**b** **i** $\text{P(brown)} = \frac{8}{12} = \frac{2}{3}$

**ii** $\text{P(white)} = \frac{4}{12} = \frac{1}{3}$

**3** **a** P(multiple of 4)

$= \text{P}(4, 8, 12, 16, 20, 24, 28, 32, 36)$

$= \frac{9}{36}$

$= \frac{1}{4}$

**b** P(between 6 and 9 inclusive)

$= \text{P}(6, 7, 8 \text{ or } 9)$

$= \frac{4}{36}$

$= \frac{1}{9}$

**c** $\text{P}(> 20)$

$= \text{P}(21, 22, 23, 24, \ldots., 35, 36)$

$= \dfrac{36-20}{36}$

$= \frac{16}{36}$

$= \frac{4}{9}$

**d** P(9)

$= \frac{1}{36}$

**e** P(multiple of 13)

$= \text{P}(13 \text{ or } 26)$

$= \frac{2}{36}$

$= \frac{1}{18}$

**f** P(odd multiple of 3)

$= \text{P}(3, 9, 15, 21, 27, \text{ or } 33)$

$= \frac{6}{36}$

$= \frac{1}{6}$

**g** P(multiple of 4 and 6)

$= \text{P(multiple of 12)}$

$= \text{P}(12, 24, 36)$

$= \frac{3}{36}$

$= \frac{1}{12}$

**h** P(multiple of 4 or 6)

$= \text{P}(4, 6, 8, 12, 16, 18, 20, 24, 28, 30, 32, 36)$

$= \frac{12}{36}$

$= \frac{1}{3}$

**4** **a** P(on Tuesday)

$= \frac{1}{7}$

**b** P(on a weekend)

$= \frac{2}{7}$

**c** P(in July)

$= \dfrac{4 \times 31}{365 \times 3 + 366}$ {over a 4 year period}

$= \frac{124}{1461}$

**d** P(in January or February)

$= \dfrac{4 \times 31 + 3 \times 28 + 1 \times 29}{3 \times 365 + 1 \times 366}$ {over a 4 year period}

$= \frac{237}{1461} \quad (= \frac{79}{487})$

**5** **a** Let A denote Antti, K denote Kai, and N denote Neda.
Possible orders are: {AKN, ANK, KAN, KNA, NAK, NKA}

**b** **i** P(A in middle)
$= \frac{2}{6}$
$= \frac{1}{3}$

**ii** P(A at left end)
$= \frac{2}{6}$
$= \frac{1}{3}$

**iii** P(A does not sit at right end)
$= 1 -$ P(A at right end)
$= 1 - \frac{2}{6}$
$= \frac{4}{6} \quad (= \frac{2}{3})$

**iv** P(K and N are together) $= \frac{4}{6} = \frac{2}{3}$

**6** Let G denote 'a girl' and B denote 'a boy'.

**a** Possible orders are: {GGG, GGB, GBG, BGG, GBB, BGB, BBG, BBB}

**b** **i** P(all boys) = P(BBB) $= \frac{1}{8}$

**ii** P(all girls) = P(GGG) $= \frac{1}{8}$

**iii** P(boy, then girl, then girl)
= P(BGG)
$= \frac{1}{8}$

**iv** P(2 girls and a boy)
= P(GGB or GBG or BGG)
$= \frac{3}{8}$

**v** P(girl is eldest)
= P(GGG or GBG or GBB or GGB)
$= \frac{4}{8} = \frac{1}{2}$

**vi** P(at least one boy)
$= \frac{7}{8}$ {all except GGG}

**7** **a** {ABCD, ABDC, ACBD, ACDB, ADBC, ADCB, BACD, BADC, BCAD, BCDA, BDAC, BDCA, CABD, CADB, CBAD, CBDA, CDAB, CDBA, DABC, DACB, DBAC, DBCA, DCAB, DCBA}

**b** **i** P(A sits on one end) $= \frac{12}{24} = \frac{1}{2}$

**ii** P(B sits on one of the two middle seats) $= \frac{12}{24} = \frac{1}{2}$

**iii** P(A and B are together) $= \frac{12}{24} = \frac{1}{2}$

**iv** P(A, B, and C are together) $= \frac{12}{24} = \frac{1}{2}$

## EXERCISE 22C.2

**1**

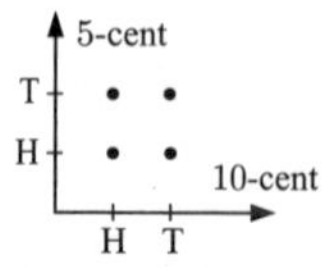

**a** P(2 heads) $= \frac{1}{4}$

**b** P(2 tails) $= \frac{1}{4}$

**c** P(exactly 1 head)
= P(HT or TH)
$= \frac{2}{4}$ or $\frac{1}{2}$

**d** P(at least one H)
= P(HT or TH or HH)
$= \frac{3}{4}$

**2** **a**

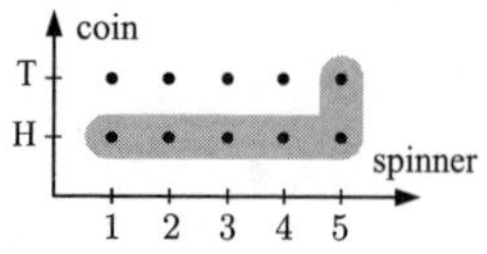

**b** There are $2 \times 5 = 10$ possible outcomes.

**c** **i** P(T and 3)
$= \frac{1}{10}$

**ii** P(H and even)
= P(H2 or H4)
$= \frac{2}{10}$ or $\frac{1}{5}$

**iii** P(an odd)
= P(H1, T1, H3, T3, H5, T5)
$= \frac{6}{10} = \frac{3}{5}$

**iv** P(H or 5)
$= \frac{6}{10}$
$= \frac{3}{5}$ {shaded}

**3** **a** P(two 3s)
$= \text{P}((3, 3))$
$= \frac{1}{36}$

**b** P(5 and a 6)
$= \text{P}((5, 6), (6, 5))$
$= \frac{2}{36}$
$= \frac{1}{18}$

**c** P(5 or a 6)
$= \frac{20}{36}$
$= \frac{5}{9}$

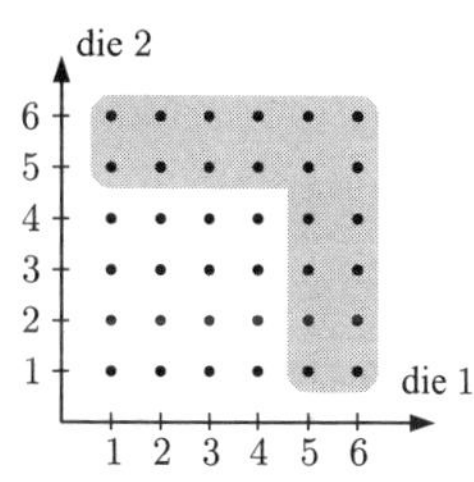

**d** P(at least one 6)
$= \frac{11}{36}$

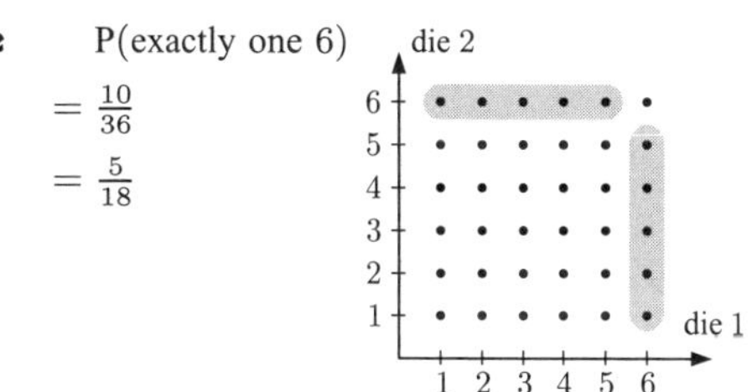

**e** P(exactly one 6)
$= \frac{10}{36}$
$= \frac{5}{18}$

**f** P(no sixes)
$= \frac{25}{36}$

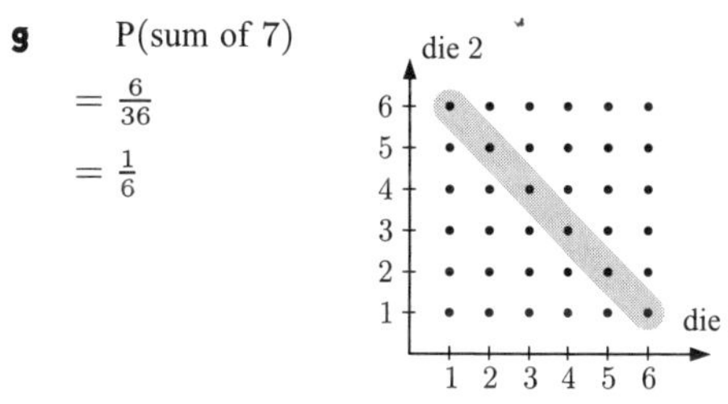

**g** P(sum of 7)
$= \frac{6}{36}$
$= \frac{1}{6}$

**h** P(sum > 8)
$= \frac{10}{36}$
$= \frac{5}{18}$

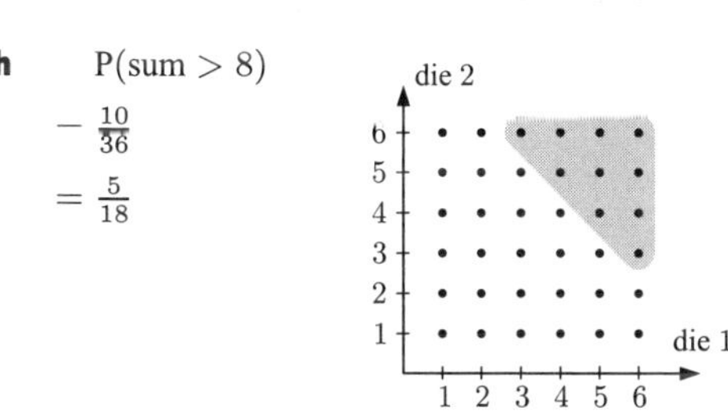

**i** P(sum of 7 or 11)
$= \dfrac{6+2}{36}$
$= \frac{2}{9}$

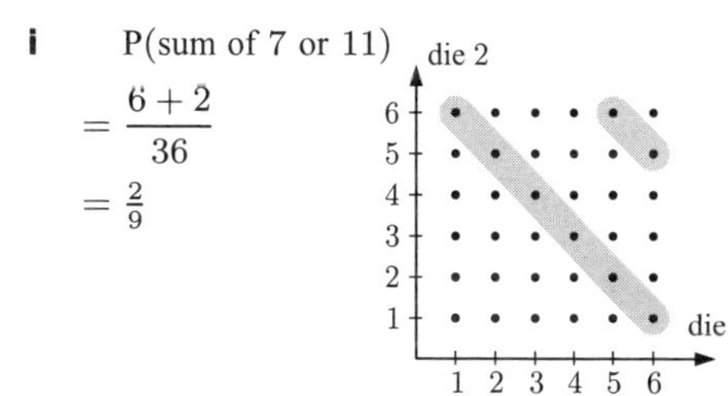

**j** P(sum no more than 8)
$= \text{P}(\text{sum} \leqslant 8)$
$= \frac{26}{36}$
$= \frac{13}{18}$

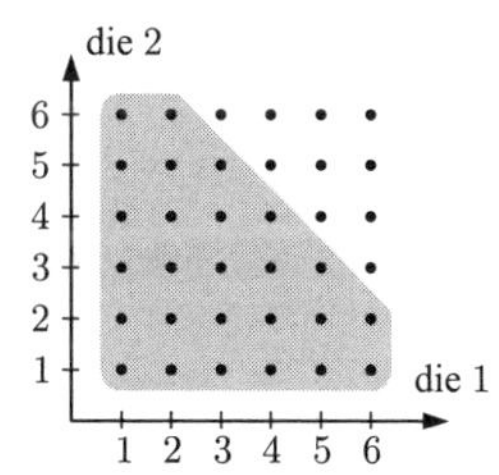

## EXERCISE 22D

**1** We extend the table to include the totals:

| | Employed | Unemployed | Total |
|---|---|---|---|
| Attended university | 225 | 34 | 259 |
| Did not attend university | 197 | 81 | 278 |
| Total | 422 | 115 | 537 |

**a** 259 out of the 537 adults surveyed attended university.

$\therefore$ P(attended university) $\approx \frac{259}{537} \approx 0.482$

**b** 197 out of the 537 adults surveyed did not attend university and are currently employed.

$\therefore$ P(did not attend university and is currently employed) $\approx \frac{197}{537} \approx 0.367$

**c** 115 out of the 537 adults surveyed were unemployed.

$\therefore$ P(unemployed) $\approx \frac{115}{537} \approx 0.214$

**d** Of the 259 adults who attended university, 225 are currently employed.

$\therefore$ P(employed given that they attended university) $\approx \frac{225}{259} \approx 0.869$

**e** Of the 115 unemployed adults, 34 attended university.

$\therefore$ P(attended university given that they are currently unemployed) $\approx \frac{34}{115} \approx 0.296$

**2** We extend the table to include the totals:

| | Adult | Child | Total |
|---|---|---|---|
| Season ticket holder | 1824 | 779 | 2603 |
| Not a season ticket holder | 3247 | 1660 | 4907 |
| Total | 5071 | 2439 | 7510 |

**a** Total match attendance was 7510.

**b** **i** P(child) $= \frac{2439}{7510} \approx 0.325$ **ii** P(not a season ticket holder) $= \frac{4907}{7510} \approx 0.653$

**iii** P(adult season ticket holder) $= \frac{1824}{7510} \approx 0.243$

**3** We extend the table to include the totals:

| | Single | Double | Family | Total |
|---|---|---|---|---|
| Peak season | 125 | 220 | 98 | 443 |
| Off-peak season | 248 | 192 | 152 | 592 |
| Total | 373 | 412 | 250 | 1035 |

**a** P(peak season) $= \frac{443}{1035} \approx 0.428$

**b** P(single room in the off-peak season) $= \frac{248}{1035} \approx 0.240$

**c** P(single room or double room) $= \frac{373+412}{1035} = \frac{785}{1035} \approx 0.758$

**d** Of the 592 off-peak season bookings, 152 were for family rooms.

$\therefore$ P(family room given it was in the off-peak season) $= \frac{152}{592} \approx 0.257$

**e** $412 + 250 = 662$ bookings were not for a single room. Of these, $220 + 98 = 318$ were in the peak season.

$\therefore$ P(peak season given it was not a single room) $= \frac{318}{662} \approx 0.480$

## EXERCISE 22E.1

**1** **a** P(rains on any one day)

$= \frac{6}{7}$

**b** P(rains on 2 successive days)

$=$ P(R and R)

$= \frac{6}{7} \times \frac{6}{7}$

$= \frac{36}{49}$

**c** P(rains on 3 successive days)

$=$ P(R and R and R)

$= \frac{6}{7} \times \frac{6}{7} \times \frac{6}{7}$ or $\frac{216}{343}$

**2** **a** P(H, then H, then H)

$=$ P(H and H and H)

$= \frac{1}{2} \times \frac{1}{2} \times \frac{1}{2}$

$= \frac{1}{8}$

**b** P(T, then H, then T)

$=$ P(T and H and T)

$= \frac{1}{2} \times \frac{1}{2} \times \frac{1}{2}$

$= \frac{1}{8}$

**3** Let $A$ be the event of photocopier A malfunctioning and $B$ be the event of photocopier B malfunctioning.

**a** P(both malfunction)
$= \text{P}(A \text{ and } B)$
$= 0.08 \times 0.12$
$= 0.0096$

**b** P(both work)
$= \text{P}(A' \text{ and } B')$
$= 0.92 \times 0.88$
$= 0.8096$

**4** **a** P(they will be happy)
= P(B, then G, then B, then G)
= P(B and G and B and G)
$= \frac{1}{2} \times \frac{1}{2} \times \frac{1}{2} \times \frac{1}{2}$
$= \frac{1}{16}$

**b** P(they will be unhappy)
= 1 − P(they will be happy)
$= 1 - \frac{1}{16}$
$= \frac{15}{16}$

**5** Let $J$ be the event of Jiri hitting the target and $B$ be the event of Benita hitting the target.

**a** P(both hit)
$= \text{P}(JB)$
$= 0.7 \times 0.8$
$= 0.56$

**b** P(both miss)
$= \text{P}(J'B')$
$= 0.3 \times 0.2$
$= 0.06$

**c** P(J hits and B misses)
$= \text{P}(JB')$
$= 0.7 \times 0.2$
$= 0.14$

**d** P(B hits and J misses)
$= \text{P}(BJ')$
$= 0.8 \times 0.3$
$= 0.24$

**6** Let $H$ be the event the archer hits the bullseye. $\therefore$ $\text{P}(H) = \frac{2}{5}$, $\text{P}(H') = \frac{3}{5}$

**a** P(3 hits)
$= \text{P}(HHH)$
$= \frac{2}{5} \times \frac{2}{5} \times \frac{2}{5}$
$= \frac{8}{125}$

**b** P(2 hits then a miss)
$= \text{P}(HHH')$
$= \frac{2}{5} \times \frac{2}{5} \times \frac{3}{5}$
$= \frac{12}{125}$

**c** P(all misses)
$= \text{P}(H'H'H')$
$= \frac{3}{5} \times \frac{3}{5} \times \frac{3}{5}$
$= \frac{27}{125}$

## EXERCISE 22E.2

**1** **a** P(all strawberry creams)
= P(SSS)
$= \frac{8}{12} \times \frac{7}{11} \times \frac{6}{10}$
$= \frac{14}{55}$

**b** P(none is a strawberry cream)
= P(S′S′S′)
$= \frac{4}{12} \times \frac{3}{11} \times \frac{2}{10}$
$= \frac{1}{55}$

**2** **a** P(both red)
= P(RR)
$= \frac{7}{10} \times \frac{6}{9}$
$= \frac{7}{15}$

**b** P(GR)
$= \frac{3}{10} \times \frac{7}{9}$
$= \frac{7}{30}$

**c** P(a green and a red)
= P(GR or RG)
$= \frac{3}{10} \times \frac{7}{9} + \frac{7}{10} \times \frac{3}{9}$
$= \frac{7}{15}$

**3** **a** P(wins first prize) $= \frac{3}{100}$

**b** P(wins 1st and 2nd)
= P(WW)
$= \frac{3}{100} \times \frac{2}{99}$
$\approx 0.000\,606$

**c** P(wins all 3)
= P(WWW)
$= \frac{3}{100} \times \frac{2}{99} \times \frac{1}{98}$
$\approx 0.000\,006\,18$

**d** P(wins none of them)
= P(W′W′W′)
$= \frac{97}{100} \times \frac{96}{99} \times \frac{95}{98}$
$\approx 0.912$

**4** **a** P(does not contain captain)
$= \text{P}(\text{C}'\text{C}'\text{C}')$
$= \frac{6}{7} \times \frac{5}{6} \times \frac{4}{5}$
$= \frac{4}{7}$

**b** P(does not contain captain or vice captain)
$= \text{P}(\text{OOO})$ {O ≡ other}
$= \frac{5}{7} \times \frac{4}{6} \times \frac{3}{5}$
$= \frac{2}{7}$

**5** **a** P(two boys) = P(first selected is a boy *and* second selected is a boy)
= P(first selected is a boy) × P(second selected is a boy)
$= \frac{5}{7} \times \frac{4}{6}$
$= \frac{20}{42} = \frac{10}{21}$

**b** P(eldest two students) = P(either of the two eldest students *and* the remaining student)
= P(either of the two eldest students) × P(the remaining student)
$= \frac{2}{7} \times \frac{1}{6}$
$= \frac{2}{42} = \frac{1}{21}$

## EXERCISE 22F

**1** **a**

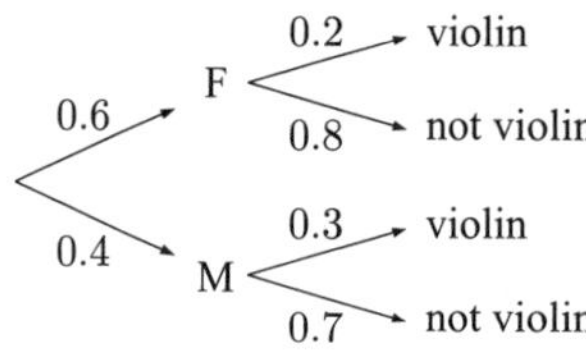


**b** **i** P(male and not violin) = 0.4 × 0.7
= 0.28

**ii** P(plays the violin)
= P(F and V) + P(M and V)
= 0.6 × 0.2 + 0.4 × 0.3
= 0.24

**2** **a**

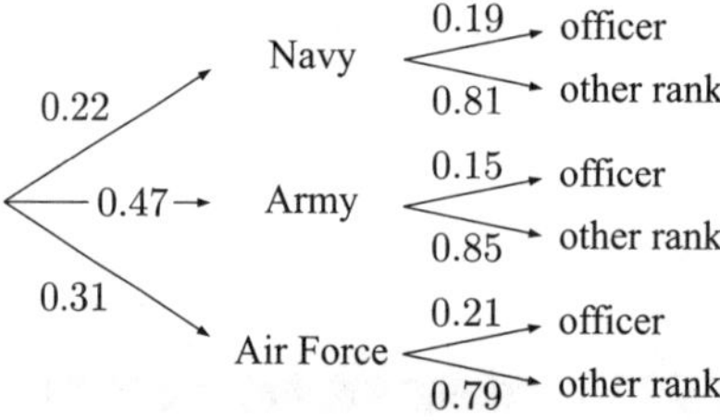


**b** **i** P(officer) = P(N and O) + P(A and O) + P(AF and O) {where O represents officer}
= 0.22 × 0.19 + 0.47 × 0.15 + 0.31 × 0.21
= 0.1774 ≈ 0.177

**ii** P(not an officer in the navy)
= P((N and O)′)
= 1 − P(N and O)
= 1 − 0.22 × 0.19
= 0.9582 ≈ 0.958

**iii** P(not an army or air force officer)
= 1 − (P(army or air force officer))
= 1 − (P(A and O) + P(AF and O))
= 1 − (0.47 × 0.15 + 0.31 × 0.21)
= 0.8644 ≈ 0.864

**3** **a**

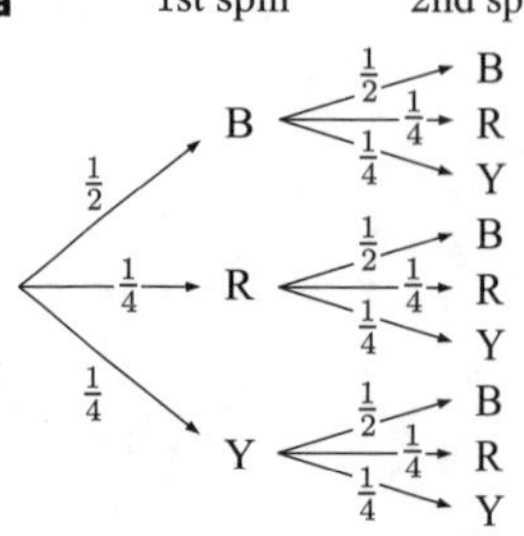


**b** P(both black)
= P(BB)
$= \frac{1}{2} \times \frac{1}{2}$
$= \frac{1}{4}$

**c** P(both yellow)
= P(YY)
$= \frac{1}{4} \times \frac{1}{4}$
$= \frac{1}{16}$

**d** P(both different)

$= \text{P(BR or BY or RB or RY or YB or YR)}$

$= \frac{1}{2} \times \frac{1}{4} + \frac{1}{2} \times \frac{1}{4} + \frac{1}{4} \times \frac{1}{2} + \frac{1}{4} \times \frac{1}{4} + \frac{1}{4} \times \frac{1}{2} + \frac{1}{4} \times \frac{1}{4}$

$= \frac{4}{8} + \frac{2}{16}$

$= \frac{5}{8}$

**e** P(B appears on either spin)

$= \text{P(BB or BR or BY or RB or YB)}$

$= \frac{1}{2} \times \frac{1}{2} + \frac{1}{2} \times \frac{1}{4} + \frac{1}{2} \times \frac{1}{4} + \frac{1}{4} \times \frac{1}{2} + \frac{1}{4} \times \frac{1}{2}$

$= 4(\frac{1}{8}) + \frac{1}{4}$

$= \frac{3}{4}$

**4**

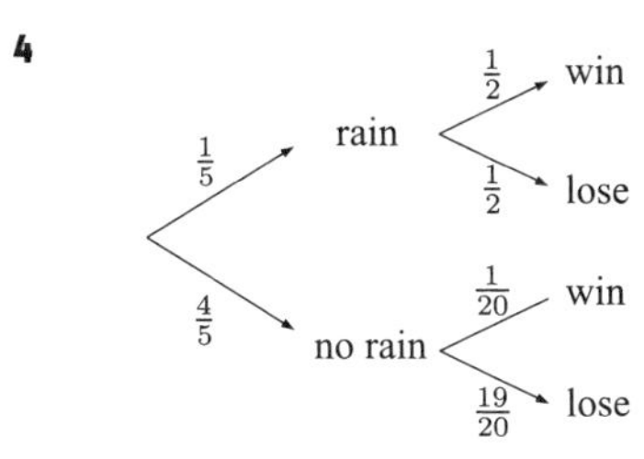


P(Mudlark wins)

$= \text{P(rain and win or no rain and win)}$

$= \frac{1}{5} \times \frac{1}{2} + \frac{4}{5} \times \frac{1}{20}$

$= \frac{1}{10} + \frac{4}{100}$

$= \frac{14}{100}$

$= \frac{7}{50}$

**5**

0.4 → Machine A: 0.05 spoiled, 0.95 not spoiled
0.6 → Machine B: 0.02 spoiled, 0.98 not spoiled

P(next is spoiled)

$= \text{P(from A and spoiled or from B and spoiled)}$

$= 0.4 \times 0.05 + 0.6 \times 0.02$

$= 0.020 + 0.012$

$= 0.032 \quad (3.2\%)$

**6**

A: 2W 3R  B: 3W 1R

$\frac{1}{2}$ → A: $\frac{2}{5}$ W, $\frac{3}{5}$ R
$\frac{1}{2}$ → B: $\frac{3}{4}$ W, $\frac{1}{4}$ R

P(red)

$= \text{P(A and red or B and red)}$

$= \frac{1}{2} \times \frac{3}{5} + \frac{1}{2} \times \frac{1}{4}$

$= \frac{3}{10} + \frac{1}{8}$

$= \frac{17}{40}$

**7** There are 7 teams above Tottenham and 12 teams below Tottenham.

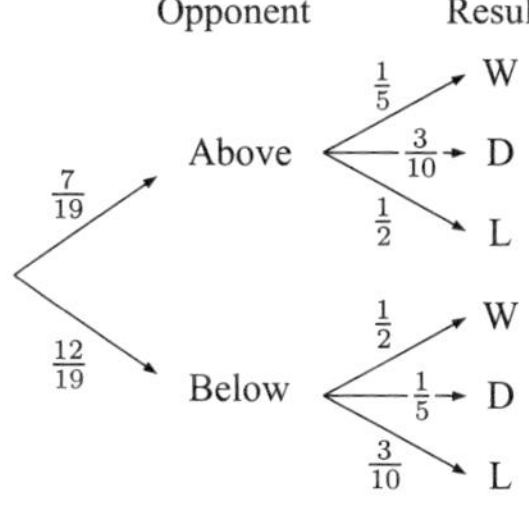


$\therefore$ P(Draw)

$= \frac{7}{19} \times \frac{3}{10} + \frac{12}{19} \times \frac{1}{5}$

$= \frac{21}{190} + \frac{24}{190}$

$= \frac{9}{38}$

**8**

$\frac{3}{6}$ → A: $\frac{3}{5}$ R, $\frac{2}{5}$ Bl
$\frac{2}{6}$ → B: $\frac{4}{5}$ R, $\frac{1}{5}$ Bl
$\frac{1}{6}$ → C: $\frac{2}{5}$ R, $\frac{3}{5}$ Bl

**a** P(blue) $= \text{P(A and Bl or B and Bl or C and Bl)}$

$= \frac{3}{6} \times \frac{2}{5} + \frac{2}{6} \times \frac{1}{5} + \frac{1}{6} \times \frac{3}{5}$

$= \frac{11}{30}$

**b** P(red) $= 1 - \text{P(blue)}$

$= 1 - \frac{11}{30}$

$= \frac{19}{30}$

## EXERCISE 22G

**1**

2P
5G

**a** P(different colours)
$= \text{P(PG or GP)}$
$= \frac{2}{7} \times \frac{5}{7} + \frac{5}{7} \times \frac{2}{7}$
$= \frac{20}{49}$

**b** P(different colours)
$= \text{P(PG or GP)}$
$= \frac{2}{7} \times \frac{5}{6} + \frac{5}{7} \times \frac{2}{6}$
$= \frac{20}{42}$ or $\frac{10}{21}$

**2** **a**

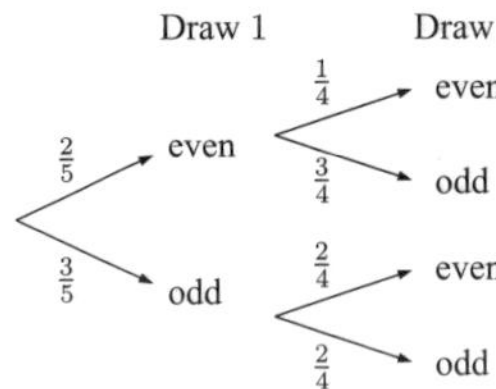


**b** **i** P(both odd)
$= \text{P(odd and odd)}$
$= \frac{3}{5} \times \frac{2}{4}$
$= \frac{3}{10}$

**ii** P(both even)
$= \text{P(even and even)}$
$= \frac{2}{5} \times \frac{1}{4}$
$= \frac{1}{10}$

**iii** P(one odd and other even)
$= 1 - \text{P(both odd)} - \text{P(both even)}$
$= 1 - \frac{3}{10} - \frac{1}{10}$
$= \frac{6}{10}$
$= \frac{3}{5}$

**3**

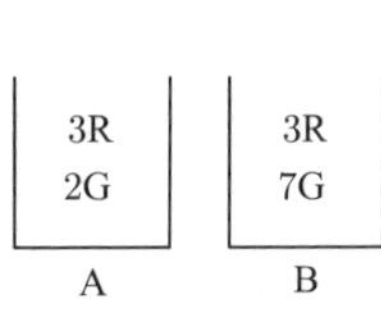


A: $\frac{4}{6}$; R $\frac{3}{5}$ (R $\frac{2}{4}$, G $\frac{2}{4}$); G $\frac{2}{5}$ (R $\frac{3}{4}$, G $\frac{1}{4}$)
B: $\frac{2}{6}$; R $\frac{3}{10}$ (R $\frac{2}{9}$, G $\frac{7}{9}$); G $\frac{7}{10}$ (R $\frac{3}{9}$, G $\frac{6}{9}$)

**a** P(both green)
$= \text{P(AGG or BGG)}$
$= \frac{4}{6} \times \frac{2}{5} \times \frac{1}{4} + \frac{2}{6} \times \frac{7}{10} \times \frac{6}{9}$
$= \frac{1}{15} + \frac{7}{45}$
$= \frac{10}{45}$
$= \frac{2}{9}$

**b** P(different in colour)
$= 1 - \text{P(both green)} - \text{P(both red)}$
$= 1 - \frac{2}{9} - \text{P(ARR or BRR)}$
$= \frac{7}{9} - (\frac{4}{6} \times \frac{3}{5} \times \frac{2}{4} + \frac{2}{6} \times \frac{3}{10} \times \frac{2}{9})$
$= \frac{7}{9} - (\frac{1}{5} + \frac{1}{45})$
$= \frac{5}{9}$

**4**

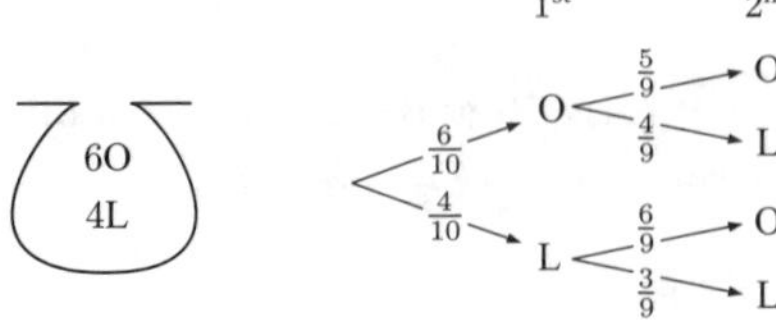


**a** **i** P(both O)
$= \frac{6}{10} \times \frac{5}{9}$
$= \frac{1}{3}$

**ii** P(both L)
$= \frac{4}{10} \times \frac{3}{9}$
$= \frac{2}{15}$

**iii** P(OL)
$= \frac{6}{10} \times \frac{4}{9}$
$= \frac{4}{15}$

**iv** P(LO)
$= \frac{4}{10} \times \frac{6}{9}$
$= \frac{4}{15}$

**b** $\frac{1}{3}+\frac{2}{15}+\frac{4}{15}+\frac{4}{15}$
$=\frac{5}{15}+\frac{2}{15}+\frac{4}{15}+\frac{4}{15}$
$=\frac{15}{15}$ which is 1

The answer must be 1 as the four categories **i**, **ii**, **iii**, **iv** are all the possibilities that could occur.

**5** **a**

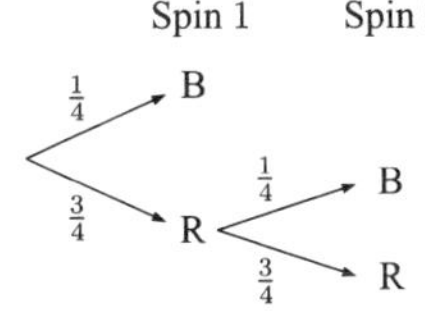


**b** P(blue)
$=$ P(B) + P(RB)
$=\frac{1}{4}+\frac{3}{4}\times\frac{1}{4}$
$=\frac{7}{16}$

**6**

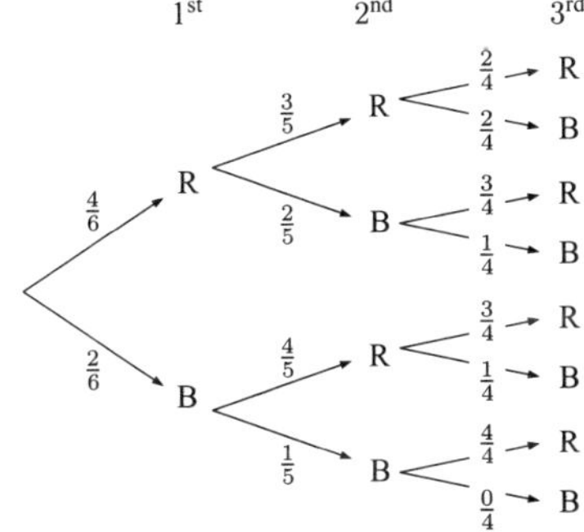


**a** P(all red)
$=$ P(RRR)
$=\frac{4}{6}\times\frac{3}{5}\times\frac{2}{4}$
$=\frac{1}{5}$

**b** P(only two are red)
$=$ P(RRB or RBR or BRR)
$=\frac{4}{6}\times\frac{3}{5}\times\frac{2}{4}+\frac{4}{6}\times\frac{2}{5}\times\frac{3}{4}+\frac{2}{6}\times\frac{4}{5}\times\frac{3}{4}$
$=3\times\left(\frac{24}{6\times5\times4}\right)$
$=\frac{3}{5}$

**c** P(at least two are red)
$=$ P(all red or only two are red)
$=\frac{1}{5}+\frac{3}{5}$ {from **a** and **b**}
$=\frac{4}{5}$

**7**

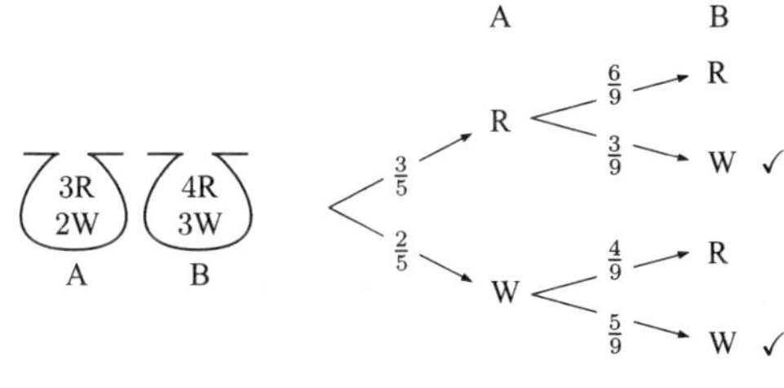


P(marble from B is W)
$=$ P(RW or WW) {paths ticked}
$=\frac{3}{5}\times\frac{3}{9}+\frac{2}{5}\times\frac{5}{9}$
$=\frac{19}{45}$

**8**

1st draw 2nd draw

$\frac{2}{100}$ W: $\frac{1}{99}$ W, $\frac{98}{99}$ L

$\frac{98}{100}$ L: $\frac{2}{99}$ W, $\frac{97}{99}$ L

**a** P(wins both)
$=$ P(WW)
$=\frac{2}{100}\times\frac{1}{99}$
$\approx 0.000\,202$

**b** P(wins neither)
$=$ P(LL)
$=\frac{98}{100}\times\frac{97}{99}$
$\approx 0.960$

**c** P(wins at least one prize) $=1-$ P(wins neither)
$=1-\frac{98}{100}\times\frac{97}{99}$
$\approx 0.0398$

**9**

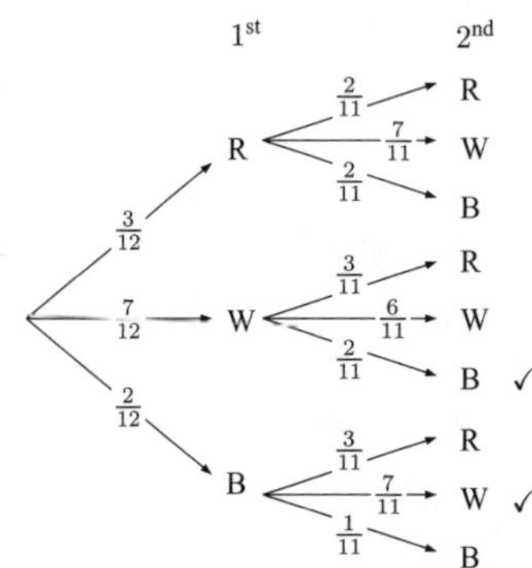


P(one white and one black)

$= \text{P(WB or BW)}$ {paths ticked}

$= \frac{7}{12} \times \frac{2}{11} + \frac{2}{12} \times \frac{7}{11}$

$= \frac{7}{33}$

**10** There are $(n+7)$ markers in total.

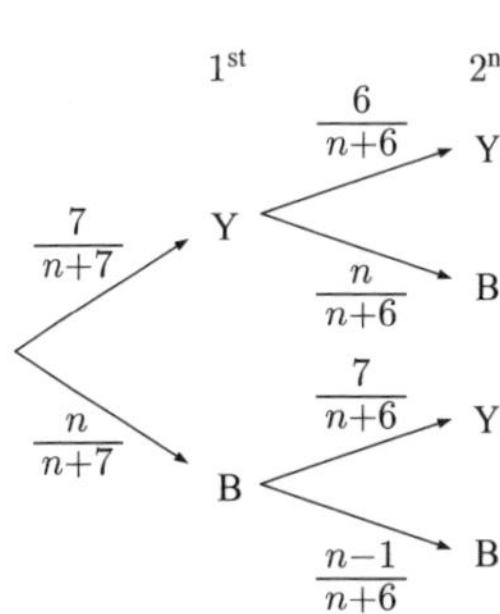


$\text{P(YY)} = \frac{3}{13}$

$\therefore \quad \frac{7}{n+7} \times \frac{6}{n+6} = \frac{3}{13}$

$\therefore \quad \frac{42}{n^2+13n+42} = \frac{3}{13}$

$\therefore \quad 546 = 3n^2 + 39n + 126$

$\therefore \quad 3(n^2 + 13n - 140) = 0$

$\therefore \quad 3(n-7)(n+20) = 0$

$\therefore \quad n = 7 \qquad \{n \geqslant 0\}$

$\therefore$ there are 7 blue markers in the bag to start with.

## EXERCISE 22H.1

**1** **a** $A = \{1, 2, 3, 6\}, \quad B = \{2, 4, 6, 8, 10\}$

**b**

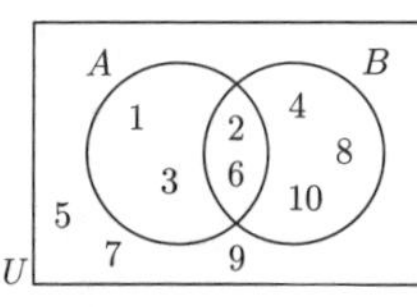


**c** **i** $n(A) = 4$

**ii** $A \cup B = \{1, 2, 3, 4, 6, 8, 10\}$

**iii** $A \cap B = \{2, 6\}$

**2** **a**

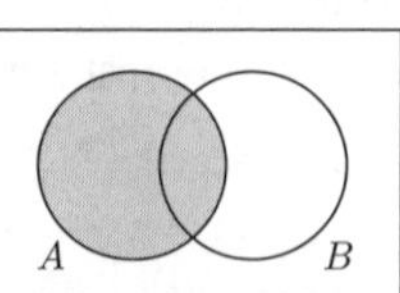


**b**

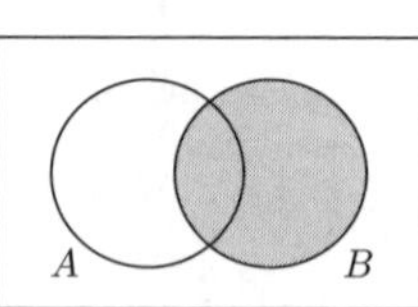


**c**

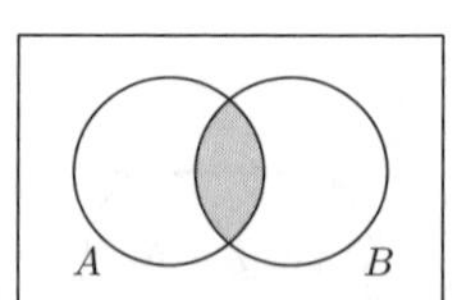


**d**

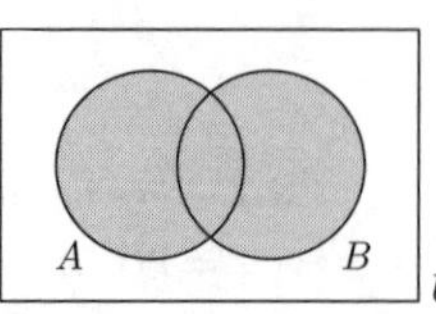


**e**

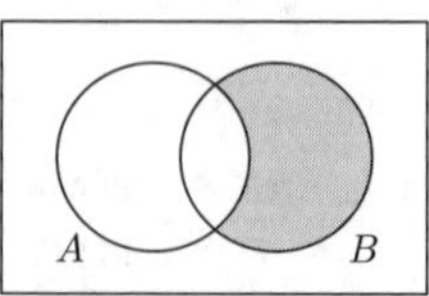


**f**

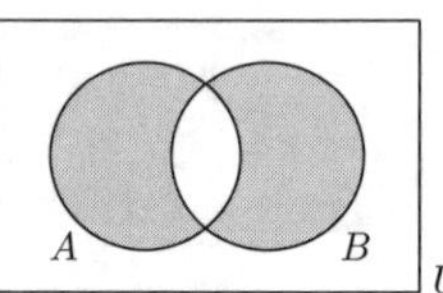


**3** **a**

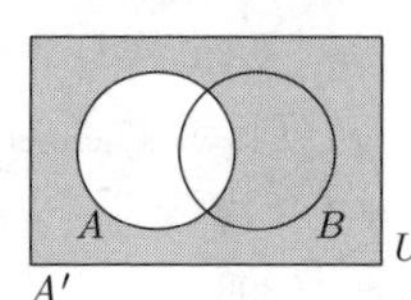


$A'$

**b**

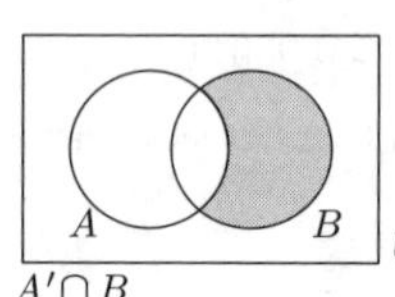


$A' \cap B$

**c**

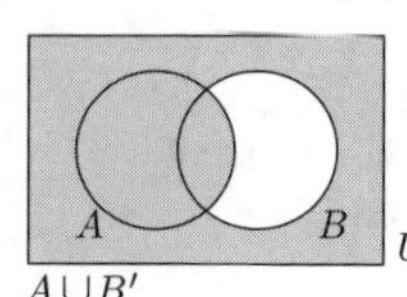


$A \cup B'$

**d**

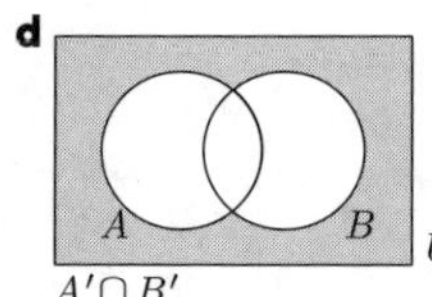


$A' \cap B'$

**4** **a**

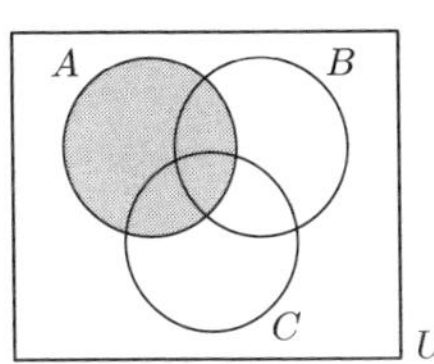


**b**

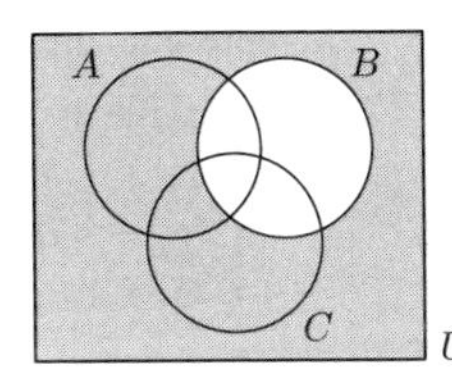


**c**

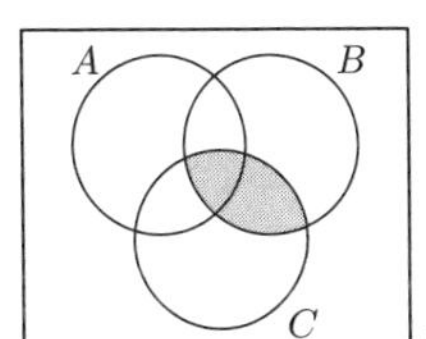


**d**

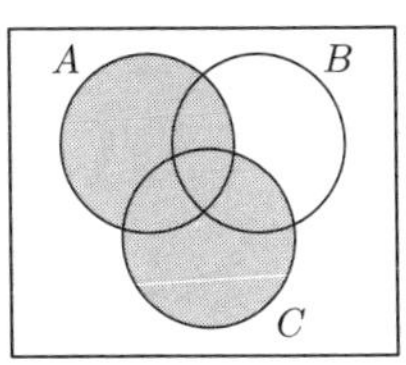


**e**

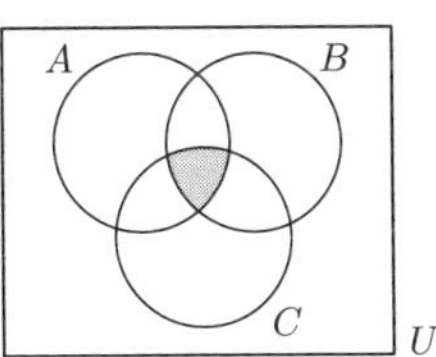


**f**

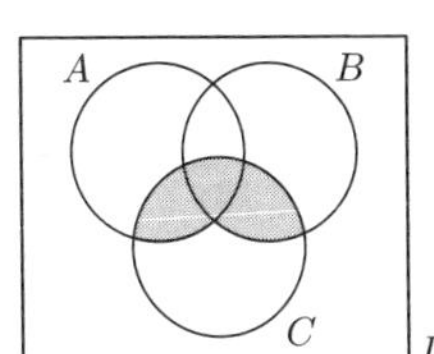


**5** **a** Total number in the class $= 3 + 5 + 17 + 4 = 29$

**b** Number who study both $= 17$ {the intersection}

**c** Number who study at least one $= 5 + 17 + 4 = 26$ {the union}

**d** Number who study only Chemistry $= 5$

**6** **a** Total number in the survey $= 37 + 9 + 15 + 4 = 65$

**b** Number who liked both $= 9$ {the intersection}

**c** Number who liked neither $= 4$

**d** Number who liked exactly one $= 37 + 15 = 52$

**7** **a**

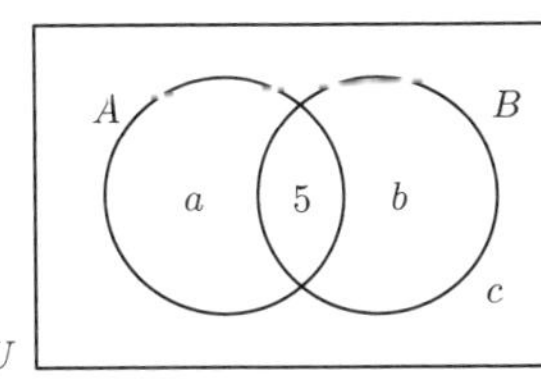


$a + 5 = 11$ {since $n(A) = 11$}

$\therefore\ a = 6$

$6 + 5 + b = 12$ {since $n(A \cup B) = 12$}

$\therefore\ b = 1$

$6 + c = 8$ {since $n(B') = 8$}

$\therefore\ c = 2$

$\therefore$ the completed Venn diagram is:

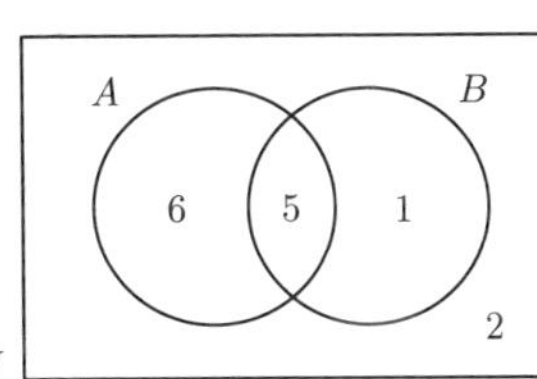


**b** **i** $n(U) = 6 + 5 + 1 + 2 = 14$

$$\therefore\ P(A \cup B) = \frac{n(A \cup B)}{n(U)} = \frac{6 + 5 + 1}{14} = \frac{12}{14} = \frac{6}{7}$$

**ii** $$P(A') = \frac{n(A')}{n(U)} = \frac{1 + 2}{14} = \frac{3}{14}$$

**8**

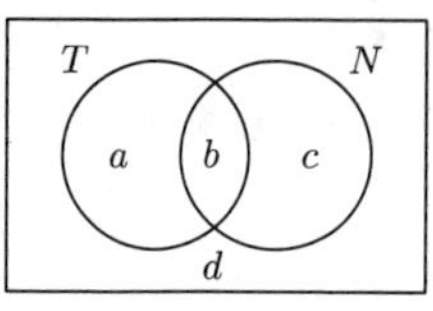


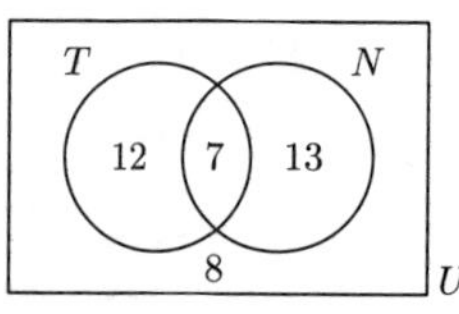


$T$ represents those playing tennis
$N$ represents those playing netball

$$\therefore \begin{cases} a+b+c+d=40 \\ a+b=19 \\ b+c=20 \\ d=8 \end{cases}$$

So, $a+b+c=32$

$\therefore \quad 19+c=32$ and $a+20=32$

$\therefore \quad c=13$ and $a=12$

Hence, $12+b=19$

$\therefore \quad b=7$

**a** P(plays tennis)
$= \dfrac{12+7}{40}$
$= \frac{19}{40}$

**b** P(does not play netball)
$= \dfrac{12+8}{40}$
$= \frac{1}{2}$

**c** P(plays at least one)
$= \dfrac{12+7+13}{40}$
$= \frac{32}{40}$
$= \frac{4}{5}$

**d** P(plays one and only one)
$= \dfrac{12+13}{40}$
$= \frac{25}{40}$
$= \frac{5}{8}$

**e** P(plays netball, but not tennis) $= \frac{13}{40}$

**9**

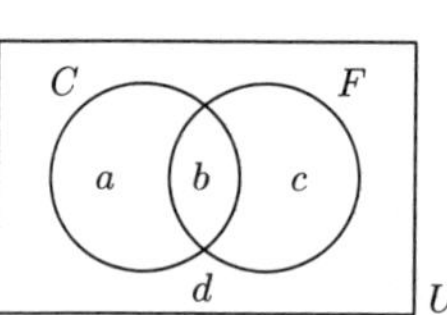


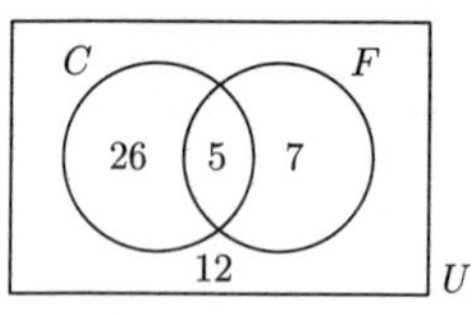


$C$ represents men who gave chocolates.
$F$ represents men who gave flowers.

$$\therefore \begin{cases} a+b+c+d=50 \\ a+b=31 \\ b+c=12 \\ b=5 \end{cases}$$

Thus $c=7$, $a=26$ and $26+5+7+d=50 \quad \therefore \quad d=12$

**a** P($C$ or $F$)
$= \dfrac{26+5+7}{50}$
$= \frac{38}{50}$ or $\frac{19}{25}$

**b** P($C$ but not $F$)
$= \frac{26}{50}$
$= \frac{13}{25}$

**c** P(neither $C$ nor $F$)
$= \frac{12}{50}$
$= \frac{6}{25}$

**10**

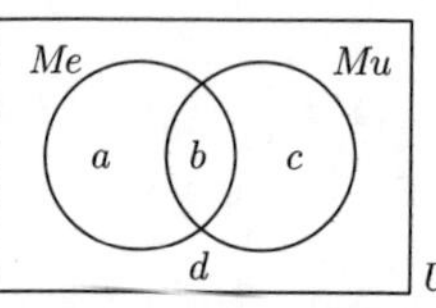


Me
Mu
12
12
2
4
U

$Me$ represents children who had measles.
$Mu$ represents children who had mumps.

$$\therefore \begin{cases} a+b+c+d=30 \\ a+b=24 \\ b=12 \\ a+b+c=26 \end{cases}$$

$\therefore \quad 26+d=30 \quad \therefore \quad d=4$
$24+c=26 \quad \therefore \quad c=2$
and $a+12=24 \quad \therefore \quad a=12$

**a** $\text{P}(Mu)$
$= \frac{14}{30}$
$= \frac{7}{15}$

**b** $\text{P}(Mu\text{, but not }Me)$
$= \frac{2}{30}$
$= \frac{1}{15}$

**c** $\text{P}(\text{neither }Mu\text{ nor }Me)$
$= \frac{4}{30}$
$= \frac{2}{15}$

**11** **a** $4 + 2 + 1 + a = 10$ {10 watched a movie}
$\therefore\ a = 3$

$4 + 2 + 1 + 3 + 6 + 12 + 9 + b = 40$ {40 individuals in total}
$\therefore\ 37 + b = 40$
$\therefore\ b = 3$

**b** **i** $\text{P(sport)} = \dfrac{6+2+1+3}{40} = \frac{12}{40} = \frac{3}{10}$

**ii** $\text{P(drama and sport)} = \dfrac{3+1}{40} = \frac{4}{40} = \frac{1}{10}$

**iii** $\text{P(movie but not sport)} = \dfrac{4+3}{40} = \frac{7}{40}$

**iv** $\text{P(drama but not movie)} = \dfrac{12+3}{40}$
$= \frac{15}{40} = \frac{3}{8}$

**v** $\text{P(drama or a movie)} = \dfrac{12+3+3+1+4+2}{40} = \frac{25}{40} = \frac{5}{8}$

## EXERCISE 22H.2

**1** **a**

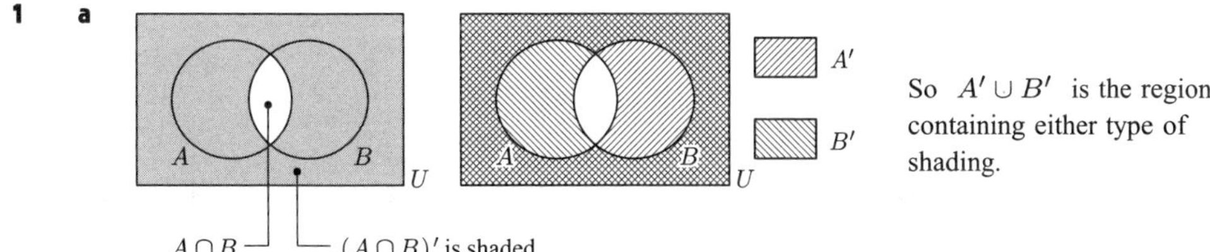


So $A' \cup B'$ is the region containing either type of shading.

Thus, as the regions are the same, $(A \cap B)' = A' \cup B'$ is verified.

**b**

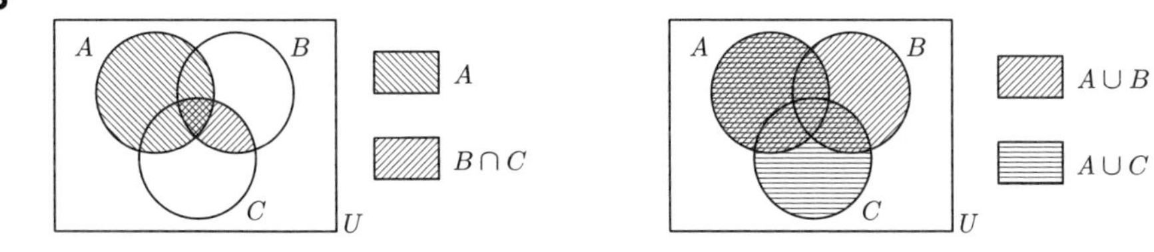


$A \cup (B \cap C)$ consists of the shaded region

$(A \cup B) \cap (A \cup C)$ consists of the 'double shaded' region.

As the two regions are identical
$A \cup (B \cap C) = (A \cup B) \cap (A \cup C)$ is verified.

**c**

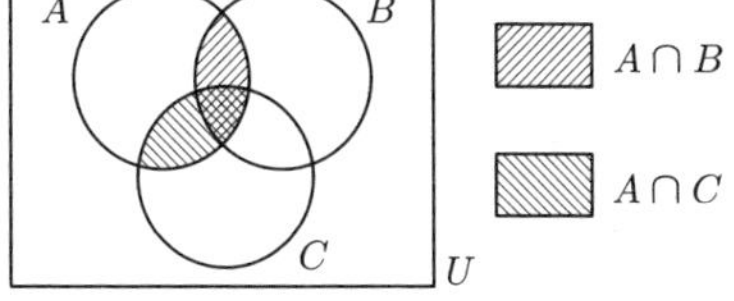


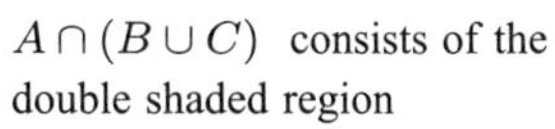

$A \cap (B \cup C)$ consists of the double shaded region

$(A \cap B) \cup (A \cap C)$ consists of the region shaded. (all forms ▧ and ▨ )

As the regions are identical, $A \cap (B \cup C) = (A \cap B) \cup (A \cap C)$ is verified.

**2** **a** $A = \{7, 14, 21, 28, 35, ...., 98\}$
$B = \{5, 10, 15, 20, 25, ...., 95\}$

**i** as $98 = 7 \times 14$, $n(A) = 14$ **ii** as $95 = 5 \times 19$, $n(B) = 19$

**iii** $A \cap B = \{35, 70\}$ $\therefore$ $n(A \cap B) = 2$

**iv** $A \cup B = \{5, 7, 10, 14, 15, 20, 21, 25, 28, 30, 35, 40, 42, 45, 49, 50, 55, 56, 60, 63, 65, 70, 75, 77, 80, 84, 85, 90, 91, 95, 98\}$ $\therefore$ $n(A \cup B) = 31$

**b** $n(A) + n(B) - n(A \cap B)$
$= 14 + 19 - 2$
$= 31$
$= n(A \cup B)$ ✓

**c** From the diagram, $n(A) + n(B) - n(A \cap B)$
$= (a + b) + (b + c) - b$
$= a + b + c$
$= n(A \cup B)$

**3** **a** **i** $\text{P}(B)$
$= \dfrac{n(B)}{n(U)}$
$= \dfrac{b + c}{a + b + c + d}$

**ii** $\text{P}(A \cap B)$
$= \dfrac{n(A \cap B)}{n(U)}$
$= \dfrac{b}{a + b + c + d}$

**iii** $\text{P}(A \cup B)$
$= \dfrac{n(A \cup B)}{n(U)}$
$= \dfrac{a + b + c}{a + b + c + d}$

**iv** $\text{P}(A) + \text{P}(B) - \text{P}(A \cap B) = \dfrac{a + b + b + c - b}{a + b + c + d}$
$= \dfrac{a + b + c}{a + b + c + d}$

**b** $\text{P}(A \cup B) = \text{P}(A) + \text{P}(B) - \text{P}(A \cap B)$ {using **iii** and **iv**}

## EXERCISE 22I

**1** $\text{P}(A \cup B) = \text{P}(A) + \text{P}(B) - \text{P}(A \cap B)$
$\therefore$ $0.9 = 0.4 + \text{P}(B) - 0.1$
$\therefore$ $\text{P}(B) = 0.6$

**2** $\text{P}(X \cup Y) = \text{P}(X) + \text{P}(Y) - \text{P}(X \cap Y)$
$\therefore$ $0.9 = 0.6 + 0.5 - \text{P}(X \cap Y)$
$\therefore$ $\text{P}(X \cap Y) = 0.2$

**3** $\text{P}(A \cup B) = \text{P}(A) + \text{P}(B)$ {$A$ and $B$ are mutually exclusive}
$\therefore$ $0.8 = \text{P}(A) + 0.45$
$\therefore$ $\text{P}(A) = 0.35$

**4** **a**

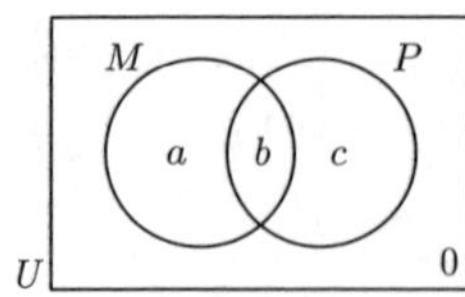


$a + b + c = 50$ $\therefore$ $a + 32 = 50$, $\therefore$ $a = 18$
$a + b = 40$ $\therefore$ $18 + b = 40$, $\therefore$ $b = 22$
$b + c = 32$ $\therefore$ $22 + c = 32$, $\therefore$ $c = 18$

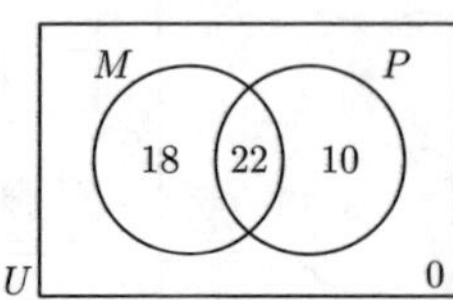


So 22 study both.

**b** **i** $\text{P}(M \text{ but not } P)$
$= \frac{18}{50}$
$= \frac{9}{25}$

**ii** $\text{P}(P \text{ given } M)$
$= \dfrac{22}{18 + 22}$
$= \frac{22}{40}$
$= \frac{11}{20}$

**5**

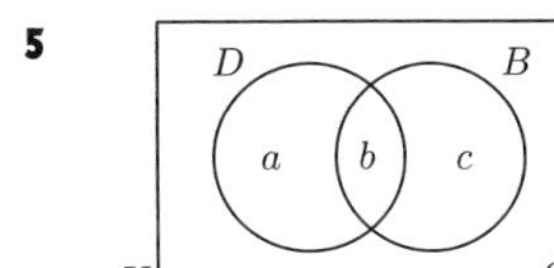


$a+b+c+d=40$ .... (1)

$a+b=23$ .... (2)

$b+c=18$ .... (3)

$a+b+c=26$ .... (4)

$\therefore\ d=14$ {using (1) and (4)}

$23+c=26$ and $a+18=26$

$\therefore\ c=3$ and $a=8$

Thus $b=18-c=15$

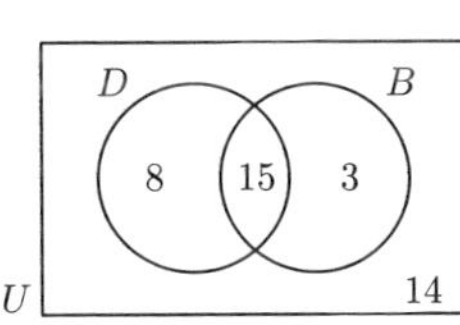


**a** P($D$ and $B$)

$=\frac{15}{40}$

$=\frac{3}{8}$

**b** P(neither $D$ nor $B$)

$=\frac{14}{40}$

$=\frac{7}{20}$

**c** P($D$, but not $B$)

$=\frac{8}{40}$

$=\frac{1}{5}$

**d** P($B$ given $D$)

$=\frac{15}{23}$

**6**

$a+b+c+d=50$

$a+b=23$

$b+c=22$

$b=5$

$\therefore\ c=17,\ a=18$

and $18+5+17+d=50$

$\therefore\ d=10$

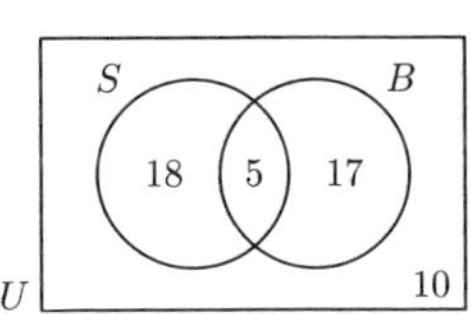


**a** P(not $B$)

$=\text{P}(B')$

$=\frac{28}{50}$

$=\frac{14}{25}$

**b** P($B$ or $S$)

$=\dfrac{18+5+17}{50}$

$=\frac{40}{50}$

$=\frac{4}{5}$

**c** P(neither $B$ nor $S$)

$=\frac{10}{50}$

$=\frac{1}{5}$

**d** P($B$, given $S$)

$=\dfrac{5}{18+5}$

$=\frac{5}{23}$

**e** P($S$, given $B'$)

$=\dfrac{18}{18+10}$

$=\frac{18}{28}$

$=\frac{9}{14}$

**7**

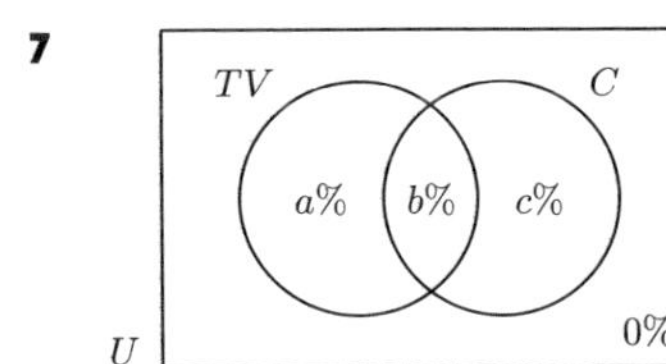


$$\begin{cases} a+b+c=100 \\ a+b=90 \\ b+c=60 \end{cases}$$

$\therefore\ c=10$ and $a=40$

$\therefore\ b=50$

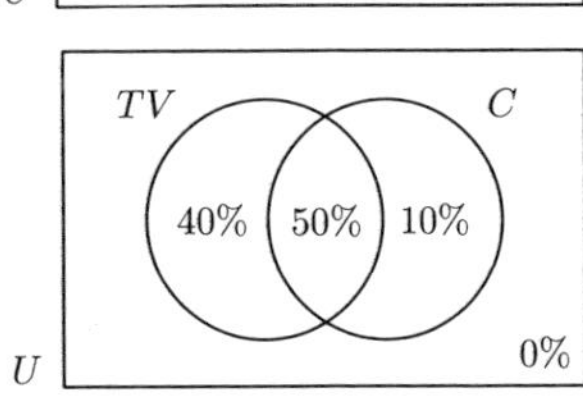


$$\text{P}(TV,\text{ given }C)=\frac{50}{50+10}=\frac{5}{6}$$

**8**

$a+b+c+d+e+f+g+h=100$

$a+b+d+e=20$

$b+c+e+f=16$

$d+e+f+g=14$

$b+e=8$

$d+e=5$

$e+f=4$

$e=2$

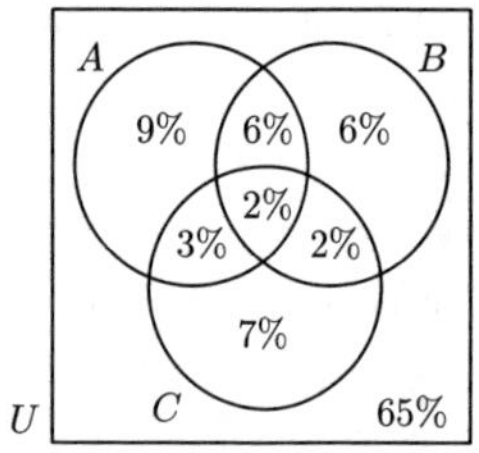


$\therefore\ e = 2,\quad f = 2,\quad d = 3,\quad b = 6,\quad \begin{cases} a+6+3+2 = 20 \\ 6+c+2+2 = 16 \\ 3+2+2+g = 14 \end{cases} \quad \therefore \begin{cases} a = 9 \\ c = 6 \\ g = 7 \end{cases}$

**a** $\text{P(none)}$
$= \frac{65}{100}$
$= \frac{13}{20}$

**b** $\text{P(at least one)}$
$= 1 - \text{P(none)}$
$= 1 - \frac{13}{20}$
$= \frac{7}{20}$

**c** $\text{P(exactly one)}$
$= \dfrac{9+6+7}{100}$
$= \frac{22}{100}$
$= \frac{11}{50}$

**d** $\text{P}(A \text{ or } B)$
$= \frac{9+6+6+3+2+2}{100}$
$= \frac{28}{100}$
$= \frac{7}{25}$

**e** $\text{P}(A, \text{ given at least one})$
$= \frac{9+6+2+3}{35}$
$= \frac{20}{35}$
$= \frac{4}{7}$

**f** $\text{P}(C, \text{ given } A \text{ or } B \text{ or both})$
$= \frac{3+2+2}{9+6+6+3+2+2}$
$= \frac{7}{28}$
$= \frac{1}{4}$

**9**

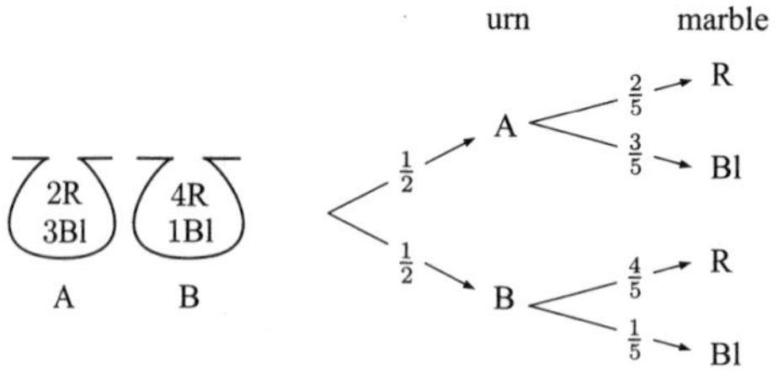


**a** $\text{P(R)} = \frac{1}{2} \times \frac{2}{5} + \frac{1}{2} \times \frac{4}{5}$
$= \frac{3}{5}$

**b** $\text{P(B | R)} = \dfrac{\text{P(B} \cap \text{R)}}{\text{P(R)}}$
$= \dfrac{\frac{1}{2} \times \frac{4}{5}}{\frac{3}{5}}$
$= \frac{2}{3}$

**10**

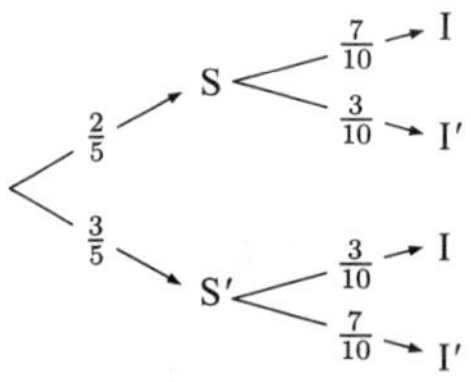


**a** $\text{P(I)} = \frac{2}{5} \times \frac{7}{10} + \frac{3}{5} \times \frac{3}{10}$
$= \frac{23}{50}$ (or 0.46)

**b** $\text{P(S | I)} = \dfrac{\text{P(S} \cap \text{I)}}{\text{P(I)}}$
$= \dfrac{\frac{2}{5} \times \frac{7}{10}}{\frac{23}{50}}$
$= \frac{14}{23}$

**11**

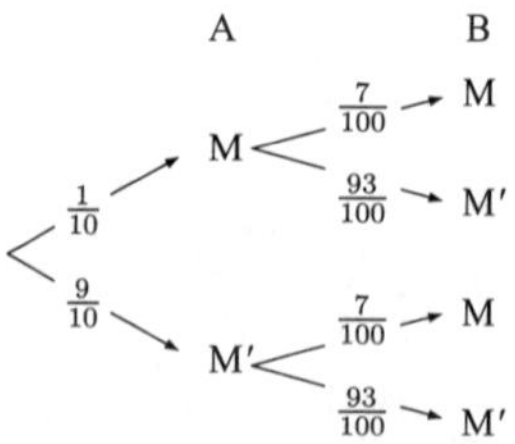


$\text{P(B | at least one malfunctions)}$
$= \dfrac{\text{P(B} \cap \text{at least one malfunctions)}}{\text{P(at least one malfunctions)}}$
$= \dfrac{\frac{1}{10} \times \frac{7}{100} + \frac{9}{10} \times \frac{7}{100}}{\frac{1}{10} \times \frac{7}{100} + \frac{1}{10} \times \frac{93}{100} + \frac{9}{10} \times \frac{7}{100}}$
$= \frac{7+63}{7+93+63}$
$= \frac{70}{163}$

**12** $\text{P}(B) = 0.5,\quad \text{P}(G) = 0.6,\quad \text{P}(G \mid B) = 0.9,$ where $B$ is "the boy eats his lunch" and $G$ is "the girl eats her lunch"

**a** $\text{P(both eat lunch)}$
$= \text{P}(B \cap G)$
$= \text{P}(G \mid B) \times \text{P}(B) \qquad \left\{ \text{as } \text{P}(G \mid B) = \dfrac{\text{P}(G \cap B)}{\text{P}(B)} \right\}$
$= 0.9 \times 0.5$
$= 0.45$

**b** $\text{P}(B \mid G)$
$= \dfrac{\text{P}(B \cap G)}{\text{P}(G)}$
$= \dfrac{0.45}{0.6}$
$= 0.75$

**c** $\quad \text{P(at least one eats lunch)}$
$= \text{P}(B \cup G)$
$= \text{P}(B) + \text{P}(G) - \text{P}(B \cap G)$
$= 0.5 + 0.6 - 0.45$
$= 0.65$

**13**

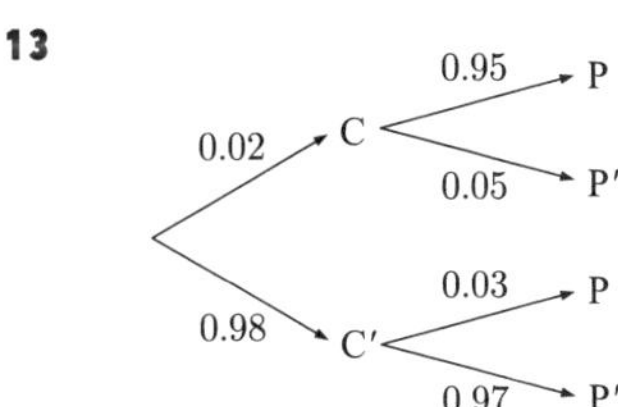


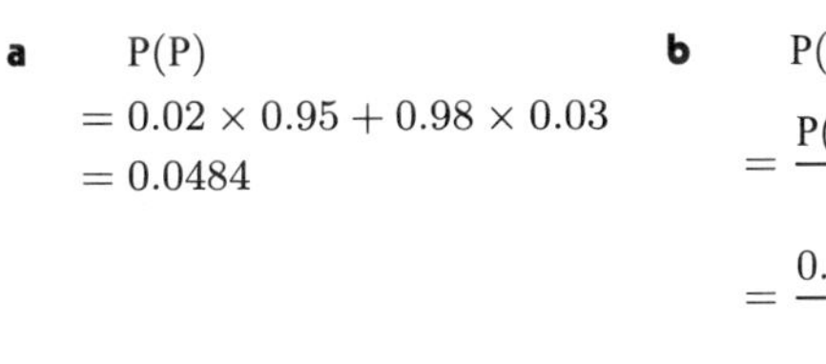

**a** $\quad \text{P(P)}$
$= 0.02 \times 0.95 + 0.98 \times 0.03$
$= 0.0484$

**b** $\quad \text{P(C} \mid \text{P)}$
$= \dfrac{\text{P(C} \cap \text{P)}}{\text{P(P)}}$
$= \dfrac{0.02 \times 0.95}{0.0484}$
$\approx 0.393$

**14** The coins are H, H T, T and H, T.

Any one of these 6 faces could be seen uppermost, $\therefore$ $\text{P(falls H)} = \frac{3}{6} = \frac{1}{2}$

Now $\text{P(HH coin} \mid \text{falls H)} = \dfrac{\text{P(HH coin} \cap \text{falls H)}}{\text{P(falls H)}}$
$= \dfrac{\text{P(HH)}}{\text{P(falls H)}}$
$= \dfrac{\frac{1}{3}}{\frac{1}{2}}$
$= \frac{2}{3}$

## EXERCISE 22J

**1** $\quad \text{P}(R \cap S)$
$= \text{P}(R) + \text{P}(S) - \text{P}(R \cup S)$
$= 0.4 + 0.5 - 0.7$
$= 0.2$

Also, $\quad \text{P}(R) \times \text{P}(S)$
$= 0.4 \times 0.5$
$= 0.2$

So, $\text{P}(R \cap S) = \text{P}(R) \times \text{P}(S)$ and hence $R$ and $S$ are independent events.

**2** **a** $\quad \text{P}(A \cap B)$
$= \text{P}(A) + \text{P}(B) - \text{P}(A \cup B)$
$= \frac{2}{5} + \frac{1}{3} - \frac{1}{2}$
$= \frac{7}{30}$

**b** $\quad \text{P}(B \mid A)$
$= \dfrac{\text{P}(B \cap A)}{\text{P}(A)}$
$= \dfrac{\frac{7}{30}}{\frac{2}{5}}$
$= \frac{7}{12}$

**c** $\quad \text{P}(A \mid B)$
$= \dfrac{\text{P}(A \cap B)}{\text{P}(B)}$
$= \dfrac{\frac{7}{30}}{\frac{1}{3}}$
$= \frac{7}{10}$

$A$ and $B$ are not independent as $\text{P}(A \mid B) \neq \text{P}(A)$.

**3** **a** As $X$ and $Y$ are independent
$\text{P}(X \cap Y) = \text{P}(X) \times \text{P}(Y)$
$= 0.5 \times 0.7$
$= 0.35$
$\therefore$ $\text{P(both } X \text{ and } Y) = 0.35$

**b** $\quad \text{P}(X \text{ or } Y)$
$= \text{P}(X \cup Y)$
$= \text{P}(X) + \text{P}(Y) - \text{P}(X \cap Y)$
$= 0.5 + 0.7 - 0.35$
$= 0.85$

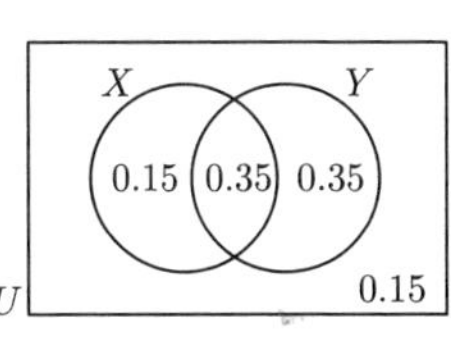


**c** $\quad \text{P(neither } X \text{ nor } Y)$
$= 0.15$

**d** $\quad \text{P}(X \text{ but not } Y)$
$= 0.15$

**e** $\text{P}(X \mid Y) = \dfrac{\text{P}(X \cap Y)}{\text{P}(Y)} = \dfrac{0.35}{0.70} = \frac{1}{2} \quad (= 0.5)$

**4** $\quad$ P(at least one solves it)
$= 1 - $ P(no-one solves it)
$= 1 - \text{P}(\text{A}' \text{ and } \text{B}' \text{ and } \text{C}')$
$= 1 - \frac{2}{5} \times \frac{1}{3} \times \frac{1}{2}$ {each student's ability to solve the problem is independent}
$= 1 - \frac{1}{15}$
$= \frac{14}{15}$

**5** **a** $\quad$ P(at least one 6)
$= 1 - $ P(no 6s)
$= 1 - \text{P}(6' \text{ and } 6' \text{ and } 6')$
$= 1 - \frac{5}{6} \times \frac{5}{6} \times \frac{5}{6}$
$= 1 - \frac{125}{216}$
$= \frac{91}{216}$

**b** $\quad$ P(at least one 6 in $n$ throws)
$= 1 - (\frac{5}{6})^n$
So we want $1 - (\frac{5}{6})^n > 0.99$
$\therefore \quad -(\frac{5}{6})^n > -0.01$
$\therefore \quad (\frac{5}{6})^n < 0.01$
$\therefore \quad n \log(\frac{5}{6}) < \log(0.01)$
$\therefore \quad n > \dfrac{\log(0.01)}{\log(\frac{5}{6})}$ $\quad$ {as $\log(\frac{5}{6}) < 0$}
$\therefore \quad n > 25.2585$
$\therefore \quad n = 26$

**6** $A$ and $B$ are independent, so $\text{P}(A \cap B) = \text{P}(A)\,\text{P}(B)$ .... (1)

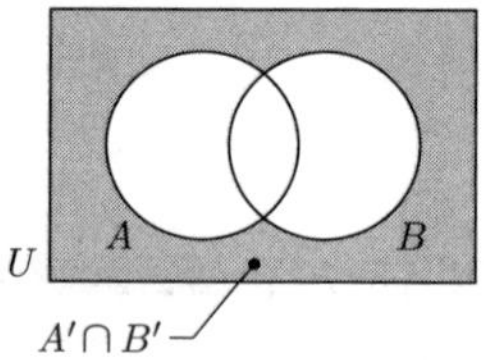

Now $\quad \text{P}(A' \cap B')$
$= 1 - \text{P}(A \cup B)$
$= 1 - [\text{P}(A) + \text{P}(B) - \text{P}(A \cap B)]$
$= 1 - \text{P}(A) - \text{P}(B) + \text{P}(A \cap B)$
$= 1 - \text{P}(A) - \text{P}(B) + \text{P}(A)\,\text{P}(B)$ $\quad$ {using (1)}
$= [1 - \text{P}(A)]\,[1 - \text{P}(B)]$
$= \text{P}(A')\,\text{P}(B')$



$\therefore \quad A'$ and $B'$ are also independent.

**7**

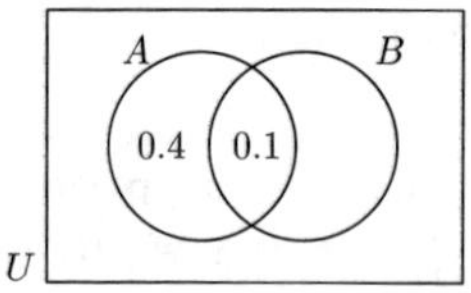

$\therefore \quad \text{P}(A) = 0.5$
and $\quad \text{P}(A \cap B) = \text{P}(A) \times \text{P}(B)$ $\quad$ {$A$ and $B$ are independent}
$\therefore \quad 0.1 = 0.5 \times \text{P}(B)$
$\therefore \quad \text{P}(B) = 0.2$

Now $\quad \text{P}(A \cup B') = \text{P}(A) + \text{P}(B') - \text{P}(A \cap B')$
$= 0.5 + 0.8 - 0.4$
$= 0.9$

**8** **a** **i** $\text{P}(C \mid D) = \dfrac{\text{P}(C \cap D)}{\text{P}(D)}$, so $\text{P}(C \cap D) = \text{P}(C \mid D)\,\text{P}(D)$

Similarly, $\quad \text{P}(C \cap D') = \text{P}(C \mid D')\,\text{P}(D')$

Now $\quad \text{P}(C \cap D) + \text{P}(C \cap D') = \text{P}(C)$
$\therefore \quad \text{P}(C \mid D)\,\text{P}(D) + \text{P}(C \mid D')\,\text{P}(D') = \text{P}(C)$
$\therefore \quad \frac{6}{13}\,\text{P}(D) + \frac{3}{7}\,[1 - \text{P}(D)] = \frac{9}{20}$
$\therefore \quad \frac{6}{13}\,\text{P}(D) + \frac{3}{7} - \frac{3}{7}\,\text{P}(D) = \frac{9}{20}$
$\therefore \quad \frac{3}{91}\,\text{P}(D) = \frac{3}{140}$
$\therefore \quad \text{P}(D) = \frac{91}{140}$ or $\frac{13}{20}$

**ii** $P(C \cap D) = P(C \mid D)\,P(D) = \frac{6}{13} \times \frac{13}{20} = \frac{3}{10}$

Now $P(C' \cup D') = 1 - P(C \cap D)$
$= 1 - \frac{3}{10} = \frac{7}{10}$

**b** $P(C \mid D) = \frac{6}{13}$ and $P(C) = \frac{9}{20}$

$\therefore$ $C$ and $D$ are not independent as $P(C \mid D) \neq P(C)$

*or*

$P(C \cap D) = \frac{3}{10}$ and $P(C)\,P(D) = \frac{9}{20} \times \frac{13}{20} = \frac{117}{400}$

$\therefore$ $C$ and $D$ are not independent as $P(C \cap D) \neq P(C)\,P(D)$

## REVIEW SET 22A

**1** ABCD, ABDC, ACBD, ACDB, ADBC, ADCB, BACD, BADC, BCAD, BCDA, BDAC, BDCA, CABD, CADB, CBAD, CBDA, CDAB, CDBA, DABC, DACB, DBAC, DBCA, DCAB, DCBA

**a** There are 24 possible orderings.

$\therefore$ P(A is next to C)
$= \frac{12}{24}$ {12 have A next to C}
$= \frac{1}{2}$

**b** P(exactly one person between A and C)
$= \frac{8}{24}$ {8 have one person between A and C}
$= \frac{1}{3}$

**2** **a** $P(A') = 1 - P(A)$
$= 1 - m$

**b** $m$ can be any value between 0 and 1 inclusive.
$\therefore$ $0 \leqslant m \leqslant 1$

**c** **i** $P(A \text{ exactly once})$
$= P(AA') + P(A'A)$
$= m(1-m) + (1-m)m$
$= 2m(1-m)$

**ii** $P(A \text{ at least once})$
$= 1 - P(A'A')$
$= 1 - (1-m)(1-m)$
$= 1 - (1 - 2m + m^2)$
$= 1 - 1 + 2m - m^2$
$= 2m - m^2$

**3**

coin
T • • ✓ •
H • × × ×
A B C D spinner

**a** Consonants are B, C and D
$\therefore$ P(H and a consonant)
$= \frac{3}{8}$ {those with a ×}

**b** P(T and C)
$= \frac{1}{8}$ {those with a ✓}

**c** P(T or vowel)
= P(T or A)
= P(T) + P(A) − P(T and A)
$= \frac{4}{8} + \frac{2}{8} - \frac{1}{8}$
$= \frac{5}{8}$

**4** $P(M) = \frac{3}{5}$, $P(W) = \frac{2}{3}$, where $M$ is the event "the man is alive in 25 years", and $W$ is the event "the woman is alive in 25 years".

**a** $P(M \text{ and } W)$
$= \frac{3}{5} \times \frac{2}{3}$
{assuming independence}
$= \frac{2}{5}$

**b** P(at least one)
$= P(M \text{ or } W)$
$= P(M) + P(W) - P(M \text{ and } W)$
$= \frac{3}{5} + \frac{2}{3} - \frac{2}{5}$
$= \frac{13}{15}$

**c** $P(M' \text{ and } W)$
$= (1 - \frac{3}{5}) \times \frac{2}{3}$
$= \frac{2}{5} \times \frac{2}{3}$
$= \frac{4}{15}$

**5** **a** $P(X \cap Y) = 0$ {$X$ and $Y$ are mutually exclusive events}

**b** $P(X \cup Y) = P(X) + P(Y) - P(X \cap Y)$
$\therefore$ $0.8 = P(X) + 0.35 - 0$
$\therefore$ $P(X) = 0.45$

**c** $P(X \text{ or } Y \text{ but not both}) = P(X \text{ or } Y)$ {$X$ and $Y$ mutually exclusive}
$= P(X \cup Y)$
$= 0.8$

**6** **a** Two events are independent if the occurrence of either event does not affect the probability that the other occurs. For $A$ and $B$ independent, $\text{P}(A) \times \text{P}(B) = \text{P}(A \text{ and } B)$.

**b** Two events $A$ and $B$ are mutually exclusive if they have no common outcomes.
$\therefore$ $\text{P}(A \text{ and } B) = 0$ and so $\text{P}(A \text{ or } B) = \text{P}(A) + \text{P}(B)$.

**7** **a** die 2, die 1

P(sum of 7 or 11)
$= \frac{8}{36}$
$= \frac{2}{9}$

**b** die 2, die 1

P(sum of at least 8)
$= \frac{15}{36}$
$= \frac{5}{12}$

**8** 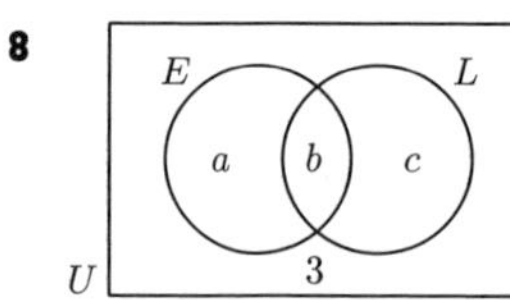


$a + b + c = 37$
$a + b = 22$
$b + c = 25$

$\therefore$ $22 + c = 37$ and $a + 25 = 37$
$\therefore$ $c = 15$ and $a = 12$
Hence, $b = 22 - a = 10$

**a** P($E$ and $L$)
$= \frac{10}{40}$
$= \frac{1}{4}$

**b** P(at least one)
$= \frac{12+10+15}{40}$
$= \frac{37}{40}$

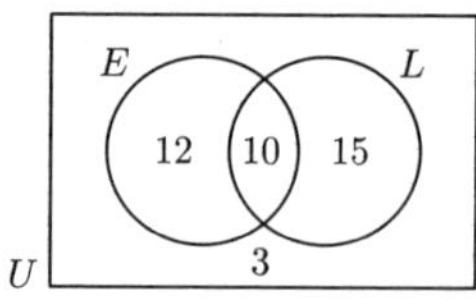


**c** $\text{P}(E \mid L) = \frac{10}{15+10} = \frac{10}{25} = \frac{2}{5}$

**9** Let $L_i$ be the event that the salesman leaves his sunglasses in store $i$.

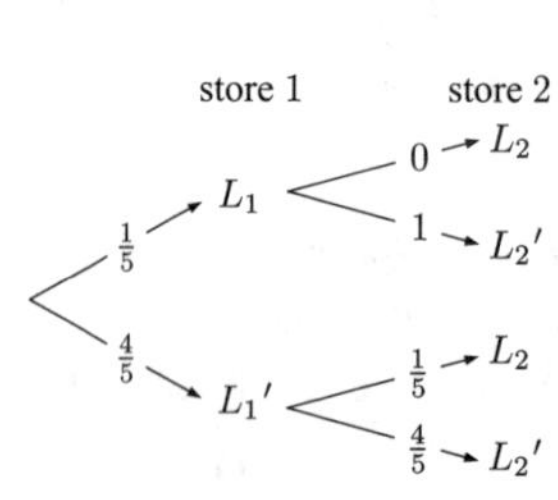


$$\text{P}(L_1 \mid L_1 \text{ or } L_2) = \frac{\text{P}(L_1 \cap (L_1 \text{ or } L_2))}{\text{P}(L_1 \text{ or } L_2)}$$
$$= \frac{\text{P}(L_1)}{\text{P}(L_1 L_2' \text{ or } L_1' L_2)}$$
$$= \frac{\frac{1}{5}}{\frac{1}{5} + \frac{4}{5} \times \frac{1}{5}}$$
$$= \frac{5}{9}$$

## REVIEW SET 22B

**1** 0.4 N (0.4 N, 0.6 R (0.4 N, 0.6 R)); 0.6 R (0.4 N (0.4 N, 0.6 R), 0.6 R)

P(Niklas wins)
$= (0.4)(0.4) + (0.4)(0.6)(0.4) + (0.6)(0.4)(0.4)$
$= 0.352$

**2** **a** P(win first 3 prizes)
$= \text{P(WWW)}$
$= \frac{4}{500} \times \frac{3}{499} \times \frac{2}{498}$
$\approx 0.000\,000\,193$

**b** P(win at least one of the 3 prizes)
$= 1 - \text{P(wins none of them)}$
$= 1 - \text{P(W}'\text{W}'\text{W}')$
$= 1 - \frac{496}{500} \times \frac{495}{499} \times \frac{494}{498}$
$\approx 0.0239$

**3**

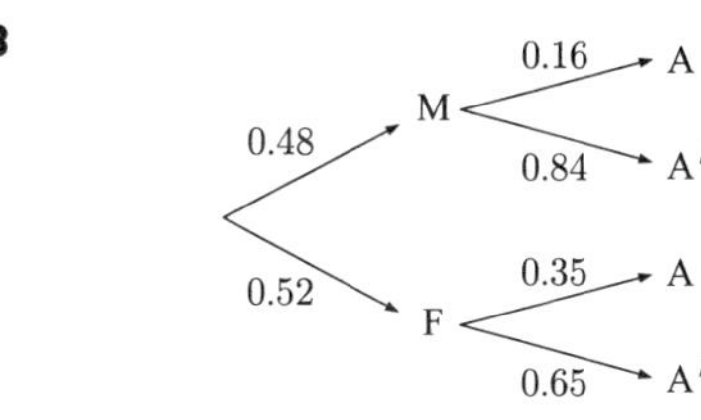


**a** $\text{P(A)} = \text{P(M} \cap \text{A or F} \cap \text{A)}$
$= 0.48 \times 0.16 + 0.52 \times 0.35$
$= 0.2588 \quad (\approx 0.259)$

**b** $\text{P(F | A)} = \dfrac{\text{P(F} \cap \text{A)}}{\text{P(A)}}$
$= \dfrac{0.52 \times 0.35}{0.2588}$
$\approx 0.703$

**4** **a**

R (0.25): W (0.36), W′ (0.64)
R′ (0.75): W (0.36), W′ (0.64)

**b** **i** P(R and W)
$= 0.25 \times 0.36$
$= 0.09$

**ii** P(R or W)
$= \text{P(R)} + \text{P(W)} - \text{P(R and W)}$
$= 0.25 + 0.36 - 0.09$
$= 0.52$

*or* $\text{P(R or W)} = 1 - \text{P(R}'\text{W}')$
$= 1 - 0.75 \times 0.64$
$= 0.52$

**c** We have assumed that the two events (rain and wind) are independent.

**5** $\text{P(A)} = 0.1, \quad \text{P(B)} = 0.2, \quad \text{P(C)} = 0.3$

$\therefore$ P(group solves it) = P(at least one solves it)
$= 1 - \text{P(no-one solves it)}$
$= 1 - \text{P(A}' \text{ and B}' \text{ and C}')$
$= 1 - (0.9 \times 0.8 \times 0.7)$
$= 0.496$

**6** **a** **i** $P(A') = 1 - P(A)$
$= 1 - 0.11$
$= 0.89$

**ii** $P(B) = 0.7$
$\therefore \dfrac{14}{n(U)} = 0.7$
$\therefore n(U) = \dfrac{14}{0.7}$
$= 20$

**b** **i** $P(A \cap B) = P(A) \times P(B)$
$= 0.11 \times 0.7$
$= 0.077$

**ii** $P(A \mid B) = P(A)$
$= 0.11$

**c** $P(A \cup B) = P(A) + P(B)$
$= 0.11 + 0.7$
$= 0.81$

**7** **a**

| | *Female* | *Male* | *Total* |
|---|---|---|---|
| *Smoker* | $60-40=20$ | 40 | 60 |
| *Non-smoker* | $90-20=70$ | $110-40=70$ | $200-60=140$ |
| *Total* | 90 | $200-90=110$ | 200 |

**b** **i** P(female non-smoker) $= \frac{70}{200}$ $(= 0.35)$

**ii** P(male given non-smoker) $= \frac{70}{140} = \frac{1}{2}$

**c** **i** P(two non-smoking females)
$= \frac{70}{200} \times \frac{69}{199}$
$\approx 0.121$

**ii** P(one is a smoker and the other is not)
$= \text{P}(\text{SS}') + \text{P}(\text{S}'\text{S})$
$= \frac{60}{200} \times \frac{140}{199} + \frac{140}{200} \times \frac{60}{199}$
$= \frac{42}{199} + \frac{42}{199}$
$= \frac{84}{199}$
$\approx 0.422$

**8** **a**

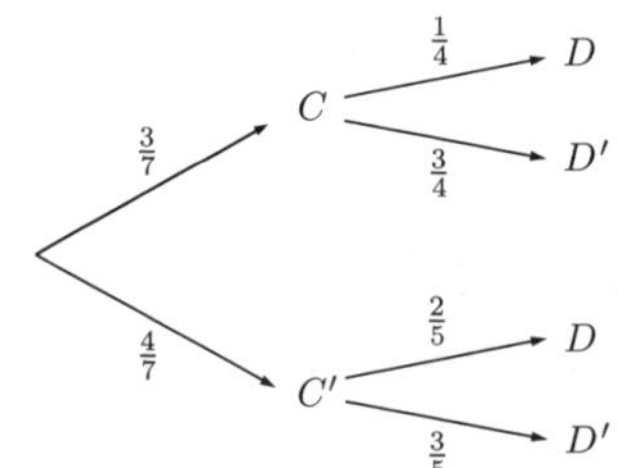


**b** **i** $\text{P}(CD) = \frac{3}{7} \times \frac{1}{4}$
$= \frac{3}{28}$

**ii** P(at least one pet) $= 1 - \text{P}(C'D')$
$= 1 - \frac{4}{7} \times \frac{3}{5}$
$= \frac{23}{35}$

## REVIEW SET 22C

**1** BBBB, BBBG, BBGB, BGBB, GBBB, BBGG, BGBG, BGGB, GBBG, GBGB, GGBB, BGGG, GBGG, GGBG, GGGB, GGGG

P(2B and 2G)
$= \frac{6}{16}$ ← 6 have 2B and 2G
$= \frac{3}{8}$

**2** **a** P(both blue)
$= \text{P(BB)}$
$= \frac{5}{12} \times \frac{4}{11}$
$= \frac{5}{33}$

**b** P(both same colour)
$= \text{P(BB or RR or YY)}$
$= \frac{5}{12} \times \frac{4}{11} + \frac{3}{12} \times \frac{2}{11} + \frac{4}{12} \times \frac{3}{11}$
$= \frac{19}{66}$

**c** P(at least one R)
$= 1 - \text{P(no reds)}$
$= 1 - \text{P}(\text{R}'\text{R}')$
$= 1 - \frac{9}{12} \times \frac{8}{11}$
$= 1 - \frac{6}{11}$
$= \frac{5}{11}$

**d** P(exactly one Y)
$= \text{P}(\text{YY}' \text{ or } \text{Y}'\text{Y})$
$= \frac{4}{12} \times \frac{8}{11} + \frac{8}{12} \times \frac{4}{11}$
$= \frac{16}{33}$

**3** **a**

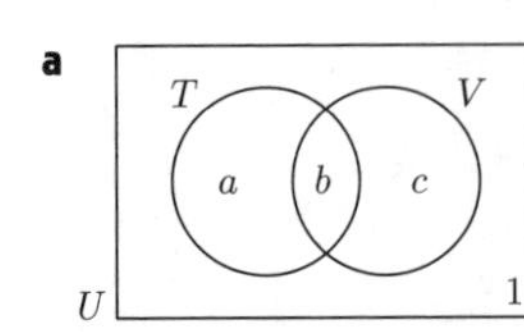


$a + b + c = 24$
$a + b = 13$
$b + c = 14$
Also $b = 13 - a$
$= 3$

$\therefore$ $13 + c = 24$ and $a + 14 = 24$
$\therefore$ $c = 11$ and $a = 10$

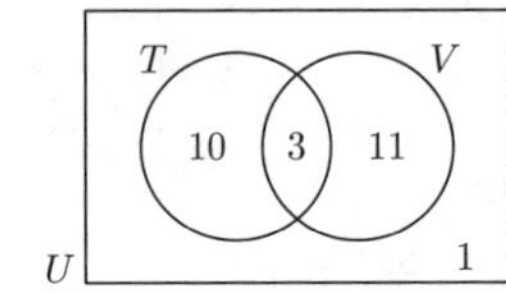

**i** P($T$ and $V$)
$= \frac{3}{25}$

**ii** P(at least one)
$= 1 - $ P(neither)
$= 1 - \frac{1}{25}$
$= \frac{24}{25}$

**iii** $P(V \mid T')$
$= \dfrac{11}{11+1}$
$= \frac{11}{12}$

**b** **i** $P(T'T'T')$
$= \frac{12}{25} \times \frac{11}{24} \times \frac{10}{23}$
$= \frac{11}{115}$

**ii** P(at least one plays tennis)
$= 1 - $ P(none play tennis)
$= 1 - \frac{11}{115}$
$= \frac{104}{115}$

**4** **a** There are now 3 red and 5 blue balls remaining.
$\therefore$ P(blue) $= \frac{5}{8}$

**b**

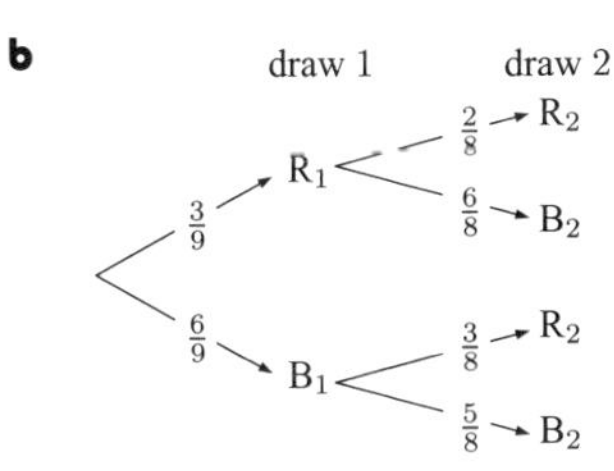


$P(R_1 \mid R_2) = \dfrac{P(R_1 \cap R_2)}{P(R_2)}$
$= \dfrac{\frac{3}{9} \times \frac{2}{8}}{\frac{3}{9} \times \frac{2}{8} + \frac{6}{9} \times \frac{3}{8}}$
$= \frac{1}{4}$

**5**

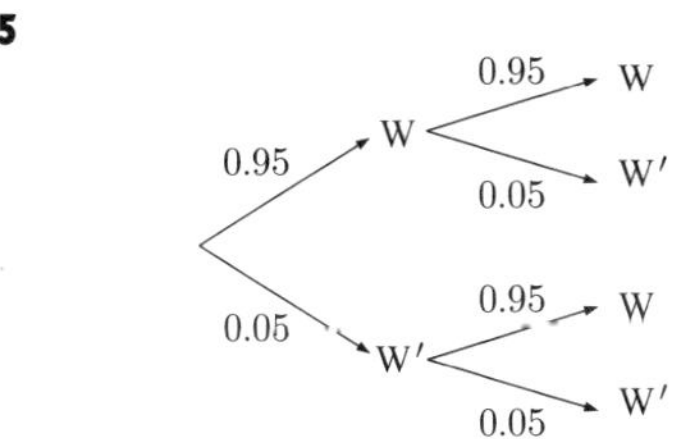


P(works on at least one day)
$= 0.95 \times 0.95 + 0.95 \times 0.05 + 0.05 \times 0.95$
$= 0.9975$

**6**

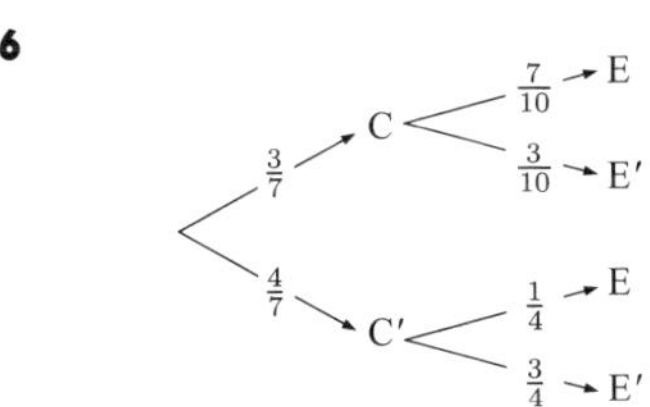


**a** $P(E) = \frac{3}{7} \times \frac{7}{10} + \frac{4}{7} \times \frac{1}{4}$
$= \frac{3}{10} + \frac{1}{7}$
$= \frac{31}{70}$

**b** $P(C \mid E) = \dfrac{P(C \text{ and } E)}{P(E)}$
$= \dfrac{\frac{3}{7} \times \frac{7}{10}}{\frac{31}{70}}$
$= \frac{21}{31}$

**7** **a**

| | *Men* | *Women* | *Total* |
|---|---|---|---|
| *Tea* | 15 | 24 | $15 + 24 = 39$ |
| *Coffee* | $50 - 15 = 35$ | $50 - 24 = 26$ | $35 + 26 = 61$ |
| *Total* | 50 | 50 | 100 |

**b**

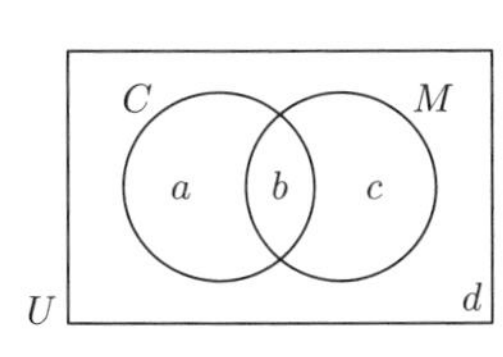


$b = 35$ {35 men prefer coffee}
$a + 35 = 61$ {61 people prefer coffee}
$\therefore\ a = 26$
$35 + c = 50$ {50 men were surveyed}
$\therefore\ c = 15$
$26 + 35 + 15 + d = 100$ {100 people were surveyed}
$\therefore\ d = 24$

$\therefore$ the Venn diagram is:

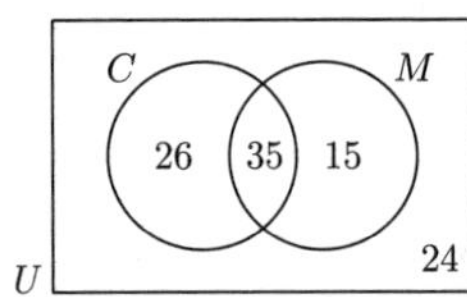


**c** **i** $P(C') = \dfrac{15+24}{100}$

$= \frac{39}{100}$

$= 0.39$

**ii** $P(M \mid C) = \dfrac{35}{26+35}$

$= \frac{35}{61}$

$\approx 0.574$

**8** **a**

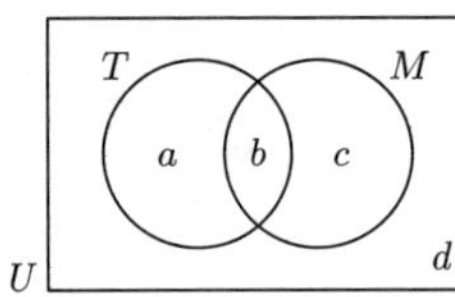


$$\begin{aligned} a+b &= 10 && \{n(T)=10\} \\ b+c &= 17 && \{n(M)=17\} \\ d &= 5 && \{n((T \cup M)')=5\} \\ \therefore\ a+b+c &= 30-5 && \{n(U)=30\} \\ \therefore\ a+17 &= 25 \\ \therefore\ a &= 8 \\ \text{and}\quad 8+b &= 10 \\ \therefore\ b &= 2 \\ \text{and}\quad 2+c &= 17 \\ \therefore\ c &= 15 \end{aligned}$$

T, M, 8, 2, 15, 5, U

**b** **i** $P(T \cap M) = \frac{2}{30}$

$= \frac{1}{15}$

**ii** $P((T \cap M) \mid M) = \frac{2}{17}$

**9** **b** Of 100 000 females born, 98 956 are still alive at age 15.
Of 100 000 males born, 98 555 are still alive at age 15.

$$\therefore\ \text{P(reaching the age of 15)} = \frac{98\,956 + 98\,555}{200\,000} = \frac{197\,511}{200\,000} = 0.987\,555 \approx 0.9876$$

**c** **i** There are 98 555 boys alive at age 15, and 53 942 still alive at 75.

$\therefore$ probability $= \dfrac{53\,942}{98\,555}$

$\approx 0.547$

**ii** There are 98 956 females alive at age 15, and 72 656 alive at age 75.

$\therefore$ P(15 year old girl does *not* reach 75)

$= 1 - \dfrac{72\,656}{98\,956}$

$= \dfrac{26\,300}{98\,956}$

$\approx 0.266$

**d** A 20 year old of either gender is expected to live for longer than 30 years, so it is unlikely the insurance company will have to pay out the policy.

# Chapter 23
## DISCRETE RANDOM VARIABLES

### EXERCISE 23A

**1** **a** The quantity of fat in a sausage is a continuous random variable.
**b** The mark out of 50 for a geography test is a discrete random variable.
**c** The weight of a seventeen year old student is a continuous random variable.
**d** The volume of water in a cup of coffee is a continuous random variable.
**e** The number of trout in a lake is a discrete random variable.
**f** The number of the hairs on a cat is a discrete random variable.
**g** The length of hairs on a horse is a continuous random variable.
**h** The height of a sky-scraper is a continuous random variable.

**2** **a** **i** The random variable $X$ is the height of water in the rain gauge.
**ii** $0 \leqslant X \leqslant 400$ mm **iii** The variable is a continuous random variable.
**b** **i** The random variable $X$ is the stopping distance.
**ii** $0 \leqslant X \leqslant 50$ m **iii** The variable is a continuous random variable.
**c** **i** The random variable $X$ is the number of times that the switch is turned on or off before it fails.
**ii** $X$ can be any integer $\geqslant 1$ **iii** The variable is a discrete random variable.

**3** **a** Since $X$ is the number of weighing devices that are accurate, $X = 0, 1, 2, 3$, or 4.
**b**

| | | YYNN | | |
|---|---|---|---|---|
| | | YNYN | | |
| | YYYN | YNNY | NNNY | |
| | YYNY | NYYN | NNYN | |
| | YNYY | NYNY | NYNN | |
| YYYY | NYYY | NNYY | YNNN | NNNN |
| $(X = 4)$ | $(X = 3)$ | $(X = 2)$ | $(X = 1)$ | $(X = 0)$ |

**c** **i** If two are accurate then $X = 2$.
**ii** If at least two are accurate then 2, 3, or 4 are accurate $\therefore$ $X = 2, 3$, or 4.

**4** **a** If 3 coins are tossed then the number of heads $X$ can be 0, 1, 2, or 3.
**b** Suppose H represents heads, T represents tails.

| | HHT | TTH | |
|---|---|---|---|
| | HTH | THT | |
| HHH | THH | HTT | TTT |
| $(X = 3)$ | $(X = 2)$ | $(X = 1)$ | $(X = 0)$ |

**c** $\text{P}(X = 0) = \frac{1}{8}$ $\text{P}(X = 1) = \frac{3}{8}$
$\text{P}(X = 2) = \frac{3}{8}$ $\text{P}(X = 3) = \frac{1}{8}$

**d**

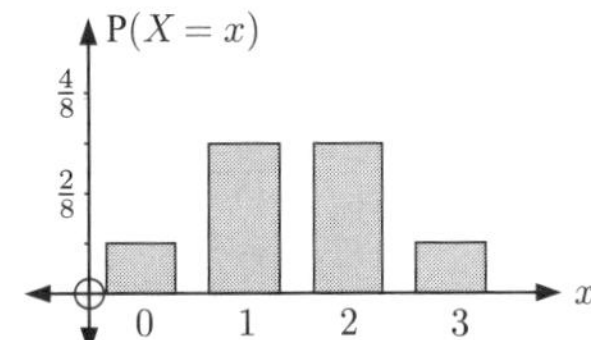


### EXERCISE 23B

**1** **a** $$\sum_{x=0}^{2} \text{P}(X = x) = 1$$
$\therefore$ $0.3 + k + 0.5 = 1$
$\therefore$ $k = 0.2$

**b** $$\sum_{x=0}^{3} \text{P}(X = x) = 1$$
$\therefore$ $k + 2k + 3k + k = 1$
$\therefore$ $7k = 1$
$\therefore$ $k = \frac{1}{7}$

**2** **a** $P(2) = 0.1088$ (from table)

**b** Since this is a probability distribution, $\sum P(x_i) = 1$

$\therefore \quad a + 0.3333 + 0.1088 + 0.0084 + 0.0007 + 0.0000 = 1$

$\therefore \quad a + 0.4512 = 1$

$\therefore \quad a = 0.5488$

This is the probability that Jason does not hit a home run in a game.

**c** $P(1) + P(2) + P(3) + P(4) + P(5) = 0.3333 + 0.1088 + 0.0084 + 0.0007 + 0.0000$
$= 0.4512$

This represents the probability that Jason hits at least one home run in a game.

**d**

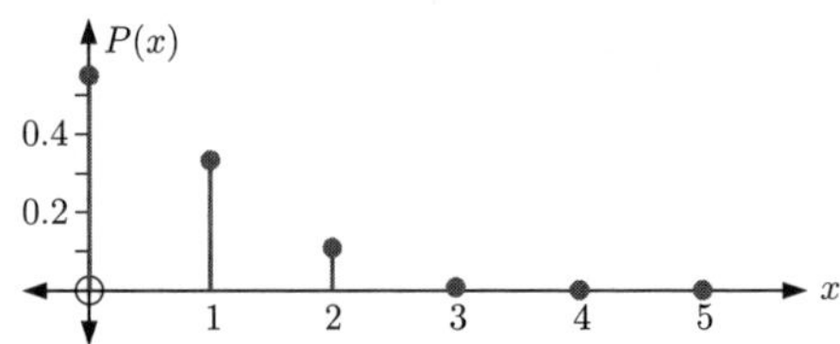


**3** **a** Sum of probabilities $\sum P(x_i) = 0.2 + 0.3 + 0.4 + 0.2 = 1.1$
Since this sum $\neq 1$, this is not a valid probability distribution.

**b** $P(5) = -0.2$, so not all of the probabilities lie in $0 \leqslant P(x_i) \leqslant 1$.
$\therefore$ this is not a valid probability distribution.

**4** **a** The random variable represents the number of hits that Sally has in each game.
$X = 0, 1, 2, 3, 4,$ or $5$.

**b** $0.07 + 0.14 + k + 0.46 + 0.08 + 0.02 = 1 \qquad \{\text{since } \sum_{x=0}^{5} \text{P}(X = x) = 1\}$

$\therefore \quad k + 0.77 = 1$

$\therefore \quad k = 0.23$

**c** **i** $\text{P}(X \geqslant 2)$
$= \text{P}(X = 2 \text{ or } X = 3 \text{ or } X = 4 \text{ or } X = 5)$
$= P(2) + P(3) + P(4) + P(5)$
$= 0.23 + 0.46 + 0.08 + 0.02$
$= 0.79$

**ii** $\text{P}(1 \leqslant X \leqslant 3)$
$= P(1) + P(2) + P(3)$
$= 0.14 + 0.23 + 0.46$
$= 0.83$

**5** **a** When rolling a die twice, the sample space is:

| roll 1 | | | | | | |
|---|---|---|---|---|---|---|
| 6 | (6, 1) | (6, 2) | (6, 3) | (6, 4) | (6, 5) | (6, 6) |
| 5 | (5, 1) | (5, 2) | (5, 3) | (5, 4) | (5, 5) | (5, 6) |
| 4 | (4, 1) | (4, 2) | (4, 3) | (4, 4) | (4, 5) | (4, 6) |
| 3 | (3, 1) | (3, 2) | (3, 3) | (3, 4) | (3, 5) | (3, 6) |
| 2 | (2, 1) | (2, 2) | (2, 3) | (2, 4) | (2, 5) | (2, 6) |
| 1 | (1, 1) | (1, 2) | (1, 3) | (1, 4) | (1, 5) | (1, 6) |
| | 1 | 2 | 3 | 4 | 5 | 6 |

roll 2

**b** $P(0) = 0 \qquad P(1) = 0$

$P(2) = \frac{1}{36} \qquad P(3) = \frac{2}{36}$

$P(4) = \frac{3}{36} \qquad P(5) = \frac{4}{36}$

$P(6) = \frac{5}{36} \qquad P(7) = \frac{6}{36}$

$P(8) = \frac{5}{36} \qquad P(9) = \frac{4}{36}$

$P(10) = \frac{3}{36} \qquad P(11) = \frac{2}{36}$

$P(12) = \frac{1}{36}$

**c**

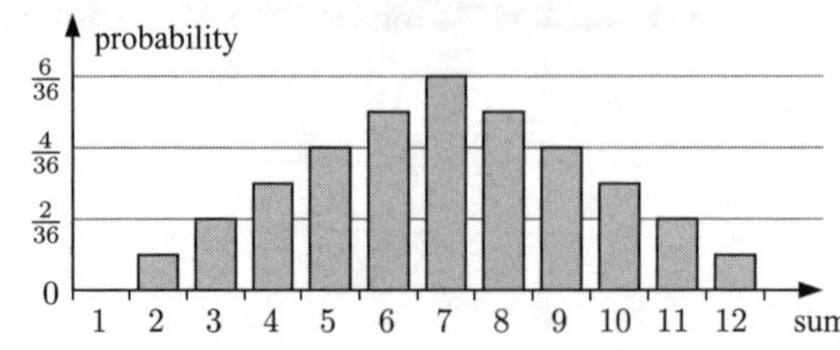

**6** **a** $P(x) = k(x+2), \quad x = 1, 2, 3$

$\therefore \quad P(1) = 3k, \quad P(2) = 4k, \quad P(3) = 5k$

Since this is a probability distribution, $3k + 4k + 5k = 1$

$\therefore \quad 12k = 1$

$\therefore \quad k = \frac{1}{12}$

**b** $P(x) = \dfrac{k}{x+1}, \quad x = 0, 1, 2, 3$

$\therefore \quad P(0) = k, \quad P(1) = \dfrac{k}{2},$

$P(2) = \dfrac{k}{3}, \quad P(3) = \dfrac{k}{4}$

Since $\sum P(x_i) = 1, \quad k + \dfrac{k}{2} + \dfrac{k}{3} + \dfrac{k}{4} = 1$

$\therefore \quad \dfrac{12k + 6k + 4k + 3k}{12} = 1$

$\therefore \quad \dfrac{25k}{12} = 1$

$\therefore \quad k = \frac{12}{25}$

**7** **a** $\mathrm{P}(X = x) = k\left(\frac{1}{3}\right)^x \left(\frac{2}{3}\right)^{4-x}, \qquad x = 0, 1, 2, 3, 4$

$\mathrm{P}(X = 0) = k\left(\frac{1}{3}\right)^0 \left(\frac{2}{3}\right)^4 = \dfrac{16k}{81} \approx 0.1975k$

$\mathrm{P}(X = 1) = k\left(\frac{1}{3}\right)^1 \left(\frac{2}{3}\right)^3 = \dfrac{8k}{81} \approx 0.0988k$

$\mathrm{P}(X = 2) = k\left(\frac{1}{3}\right)^2 \left(\frac{2}{3}\right)^2 = \dfrac{4k}{81} \approx 0.0494k$

$\mathrm{P}(X = 3) = k\left(\frac{1}{3}\right)^3 \left(\frac{2}{3}\right)^1 = \dfrac{2k}{81} \approx 0.0247k$

$\mathrm{P}(X = 4) = k\left(\frac{1}{3}\right)^4 \left(\frac{2}{3}\right)^0 = \dfrac{k}{81} \approx 0.0123k$

**b** Since $\sum \mathrm{P}(X = i) = 1,$

$\therefore \quad \dfrac{16k}{81} + \dfrac{8k}{81} + \dfrac{4k}{81} + \dfrac{2k}{81} + \dfrac{k}{81} = 1$

$\therefore \quad \dfrac{31k}{81} = 1$

$\therefore \quad k = \frac{81}{31}$

$\therefore \quad k \approx 2.61$

$\therefore \quad \mathrm{P}(X \geqslant 2) = P(2) + P(3) + P(4)$

$= \dfrac{4k}{81} + \dfrac{2k}{81} + \dfrac{k}{81}$

$= \dfrac{7k}{81} = \dfrac{7}{81} \times \dfrac{81}{31}$

$= \frac{7}{31} \quad (\approx 0.226)$

**8** **a** P(no faulty component)

$= \mathrm{P}(X = 0)$

$= \binom{10}{0}(0.04)^0(0.96)^{10-0}$

$= (0.96)^{10}$

$\approx 0.665$

**b** P(at least one faulty component)

$= \mathrm{P}(X \geqslant 1)$

$= 1 - \mathrm{P}(\text{none are faulty})$

$\approx 1 - (0.96)^{10}$

$\approx 0.335$

**9** **a**

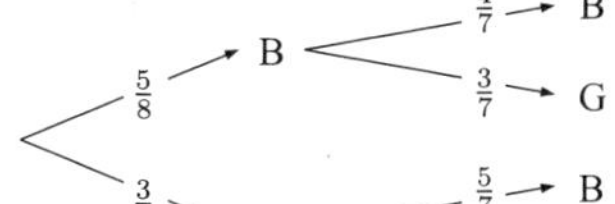


| Event | $X$ | Probability |
|---|---|---|
| BB | 2 | $\frac{5}{8} \times \frac{4}{7} = \frac{20}{56}$ |
| BG | 1 | $\frac{5}{8} \times \frac{3}{7} = \frac{15}{56}$ |
| GB | 1 | $\frac{3}{8} \times \frac{5}{7} = \frac{15}{56}$ |
| GG | 0 | $\frac{3}{8} \times \frac{2}{7} = \frac{6}{56}$ |

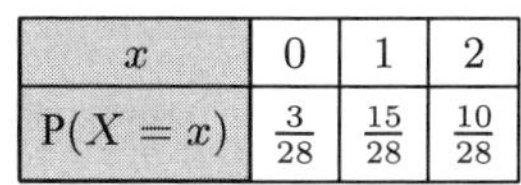

| $x$ | 0 | 1 | 2 |
|---|---|---|---|
| $\mathrm{P}(X = x)$ | $\frac{3}{28}$ | $\frac{15}{28}$ | $\frac{10}{28}$ |

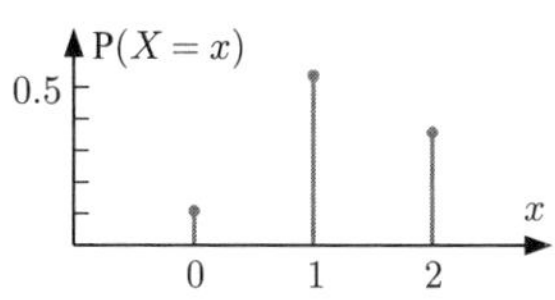

**b**

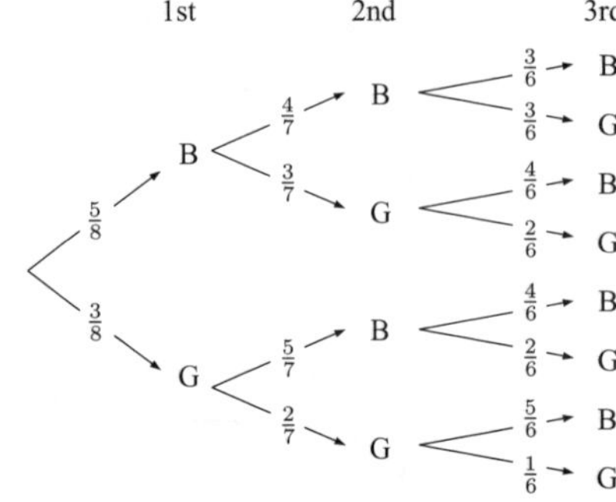


| Event | $X$ | Probability |
|---|---|---|
| BBB | 3 | $\frac{5}{8}\times\frac{4}{7}\times\frac{3}{6}=\frac{10}{56}$ |
| BBG | 2 | $\frac{5}{8}\times\frac{4}{7}\times\frac{3}{6}=\frac{10}{56}$ |
| BGB | 2 | $\frac{5}{8}\times\frac{3}{7}\times\frac{4}{6}=\frac{10}{56}$ |
| BGG | 1 | $\frac{5}{8}\times\frac{3}{7}\times\frac{2}{6}=\frac{5}{56}$ |
| GBB | 2 | $\frac{3}{8}\times\frac{5}{7}\times\frac{4}{6}=\frac{10}{56}$ |
| GBG | 1 | $\frac{3}{8}\times\frac{5}{7}\times\frac{2}{6}=\frac{5}{56}$ |
| GGB | 1 | $\frac{3}{8}\times\frac{2}{7}\times\frac{5}{6}=\frac{5}{56}$ |
| GGG | 0 | $\frac{3}{8}\times\frac{2}{7}\times\frac{1}{6}=\frac{1}{56}$ |

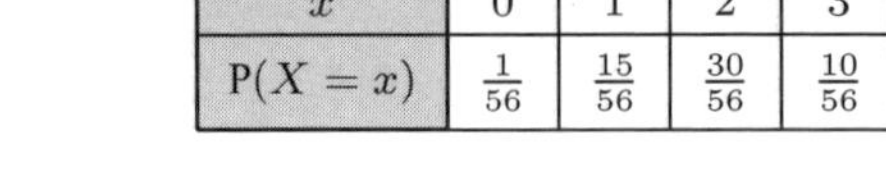

| $x$ | 0 | 1 | 2 | 3 |
|---|---|---|---|---|
| $P(X=x)$ | $\frac{1}{56}$ | $\frac{15}{56}$ | $\frac{30}{56}$ | $\frac{10}{56}$ |

**10 a**

| | | Die 2 | | | | | |
|---|---|---|---|---|---|---|---|
| | | 1 | 2 | 3 | 4 | 5 | 6 |
| | 1 | 2 | 3 | 4 | 5 | 6 | 7 |
| | 2 | 3 | 4 | 5 | 6 | 7 | 8 |
| Die 1 | 3 | 4 | 5 | 6 | 7 | 8 | 9 |
| | 4 | 5 | 6 | 7 | 8 | 9 | 10 |
| | 5 | 6 | 7 | 8 | 9 | 10 | 11 |
| | 6 | 7 | 8 | 9 | 10 | 11 | 12 |

36 possible results

**b** $P(D=7)=\frac{6}{36}=\frac{1}{6}$

**c**

| $d$ | 2 | 3 | 4 | 5 | 6 | 7 | 8 | 9 | 10 | 11 | 12 |
|---|---|---|---|---|---|---|---|---|---|---|---|
| $P(D=d)$ | $\frac{1}{36}$ | $\frac{2}{36}$ | $\frac{3}{36}$ | $\frac{4}{36}$ | $\frac{5}{36}$ | $\frac{6}{36}$ | $\frac{5}{36}$ | $\frac{4}{36}$ | $\frac{3}{36}$ | $\frac{2}{36}$ | $\frac{1}{36}$ |

**d**
$$P(D\geqslant 8 \mid D\geqslant 6)=\frac{P(D\geqslant 8 \cap D\geqslant 6)}{P(D\geqslant 6)}=\frac{P(D\geqslant 8)}{P(D\geqslant 6)}=\frac{15}{36}\div\frac{26}{36}=\frac{15}{26}$$

**11 a**

| | | Die 2 | | | | | |
|---|---|---|---|---|---|---|---|
| | | 1 | 2 | 3 | 4 | 5 | 6 |
| | 1 | 0 | 1 | 2 | 3 | 4 | 5 |
| | 2 | 1 | 0 | 1 | 2 | 3 | 4 |
| Die 1 | 3 | 2 | 1 | 0 | 1 | 2 | 3 |
| | 4 | 3 | 2 | 1 | 0 | 1 | 2 |
| | 5 | 4 | 3 | 2 | 1 | 0 | 1 |
| | 6 | 5 | 4 | 3 | 2 | 1 | 0 |

**b**

| $N$ | 0 | 1 | 2 | 3 | 4 | 5 |
|---|---|---|---|---|---|---|
| $P(N=n)$ | $\frac{6}{36}$ | $\frac{10}{36}$ | $\frac{8}{36}$ | $\frac{6}{36}$ | $\frac{4}{36}$ | $\frac{2}{36}$ |

**c** $P(N=3)=\frac{6}{36}=\frac{1}{6}$

**d**
$$P(N\geqslant 3 \mid N\geqslant 1)=\frac{P(N\geqslant 3\cap N\geqslant 1)}{P(N\geqslant 1)}=\frac{P(N\geqslant 3)}{P(N\geqslant 1)}=\frac{12}{36}\div\frac{30}{36}=\frac{2}{5}$$

## EXERCISE 23C

**1** P(rain) $=0.28$ $\quad\therefore$ we would expect rain on $0.28\times 365.25\approx 102$ days a year.

**2 a** $P(\text{HHH})=\frac{1}{2}\times\frac{1}{2}\times\frac{1}{2}=\frac{1}{8}$

**b** For 200 tosses, we expect $200\times\frac{1}{8}=25$ to be '3 heads'.

**3** P(double) $=$ P(1, 1 or 2, 2 or 3, 3 or 4, 4 or 5, 5 or 6, 6)
$=\frac{6}{36}$ {6 of the possible 36 outcomes}
$=\frac{1}{6}$

$\therefore$ when rolling the dice 180 times, we expect $180\times\frac{1}{6}=30$ doubles.

**4**

| *result* | *win* |
|---|---|
| H | \$2 |
| T | −\$1 |

For playing *once*,

we would expect to win $\frac{1}{2} \times \$2 + \frac{1}{2} \times (-\$1) = \$0.50$

$\therefore$ for 3 games we would expect to win \$1.50.

**5** Udo could expect to see snow falling on $\frac{3}{7} \times 5 \times 7 = 15$ days.

**6** The goalkeeper would expect to save $\frac{3}{10} \times 90 = 27$ goals.

**7** **a** $165 + 87 + 48 = 300$ **i** P(A) $\approx \frac{165}{300} = 0.55$ **ii** P(B) $\approx \frac{87}{300} = 0.29$ **iii** P(C) $\approx \frac{48}{300} = 0.16$

**b** **i** We expect $7500 \times 0.55 = 4125$ to vote for A.
**ii** We expect $7500 \times 0.29 = 2175$ to vote for B.
**iii** We expect $7500 \times 0.16 = 1200$ to vote for C.

**8** **a** **i** P(wins \$10) = P(rolls a 6) $= \frac{1}{6}$ **ii** P(wins \$4) = P(rolls 4 or 5) $= \frac{2}{6}$ (or $\frac{1}{3}$) **iii** P(wins \$1) = P(rolls 1, 2, or 3) $= \frac{3}{6}$ (or $\frac{1}{2}$)

**b** **i** Expectation $= \frac{2}{6} \times \$4 \approx \$1.33$ **ii** Expectation $= \frac{3}{6} \times \$1 = \$0.50$ **iii** Expectation $= \frac{1}{6} \times \$10 + \frac{2}{6} \times \$4 + \frac{3}{6} \times \$1 = \frac{1}{6}(\$21) = \$3.50$

**c** It costs \$4 to play and the expected return is \$3.50.
$\therefore$ you expect to lose \$0.50 per game. (50 cents)

**d** Over 100 games you expect to lose $100 \times \$0.50 = \$50$.

**9** **a** Expect to win $\frac{1}{6} \times €1 + \frac{1}{6} \times €2 + \frac{1}{6} \times €3 + \frac{1}{6} \times €4 + \frac{1}{6} \times €5 + \frac{1}{6} \times €6$
$= \frac{1}{6} \times €21 = €3.50$

**b** The expected gain is $€3.50 - €4 = -€0.50$
$\therefore$ the player should not play several games, as on each occasion he would expect to lose an average of €0.50.

**c** **i** The game is fair when the expected gain is 0.
$\therefore$ $3.50 - k = 0$, so $k = 3.50$.
**ii** $k > 3.50$

**10**

| *result* | *win* |
|---|---|
| HH | £10 |
| HT or TH | £3 |
| TT | −£5 |

**a** Expectation $= \frac{1}{4} \times £10 + \frac{2}{4} \times £3 + \frac{1}{4} \times (-£5) = £2.75$

**b** Expected win per game (payout) = £2.75
$\therefore$ the organiser would charge $£2.75 + £1.00 = £3.75$ to play each game.

**11**

| $x$ | 0 | 1 | 2 | 3 | 4 | 5 | $> 5$ |
|---|---|---|---|---|---|---|---|
| $P(X = x)$ | 0.54 | 0.26 | 0.15 | $k$ | 0.01 | 0.01 | 0.00 |

**a** $0.54 + 0.26 + 0.15 + k + 0.01 + 0.01 = 1$
$\therefore k + 0.97 = 1$
$\therefore k = 0.03$

**b** $\mu = \sum x_i p_i$
$= 0 \times 0.54 + 1 \times 0.26 + 2 \times 0.15 + 3 \times 0.03 + 4 \times 0.01 + 5 \times 0.01$
$= 0.26 + 0.30 + 0.09 + 0.04 + 0.05$
$= 0.74$ So, over a long period the mean number of deaths per dozen crayfish is 0.74.

**12** $P(X = x) = \dfrac{x^2 + x}{20}$ for $x = 1, 2, 3$

| $x$ | 1 | 2 | 3 |
|---|---|---|---|
| $P(X = x)$ | $\frac{2}{20} = 0.1$ | $\frac{6}{20} = 0.3$ | $\frac{12}{20} = 0.6$ |

$\mu = \sum x_i p_i$
$= 1 \times 0.1 + 2 \times 0.3 + 3 \times 0.6$
$= 2.5$

**13**

Die 2

| Die 1 | 1 | 2 | 3 | 4 | 5 | 6 |
|---|---|---|---|---|---|---|
| 1 | 1 | 2 | 3 | 4 | 5 | 6 |
| 2 | 2 | 2 | 3 | 4 | 5 | 6 |
| 3 | 3 | 3 | 3 | 4 | 5 | 6 |
| 4 | 4 | 4 | 4 | 4 | 5 | 6 |
| 5 | 5 | 5 | 5 | 5 | 5 | 6 |
| 6 | 6 | 6 | 6 | 6 | 6 | 6 |

**a**

| $m$ | 1 | 2 | 3 | 4 | 5 | 6 |
|---|---|---|---|---|---|---|
| $P(M = m)$ | $\frac{1}{36}$ | $\frac{3}{36}$ | $\frac{5}{36}$ | $\frac{7}{36}$ | $\frac{9}{36}$ | $\frac{11}{36}$ |

**b** $\mu = \sum m_i p_i$
$= 1\left(\frac{1}{36}\right) + 2\left(\frac{3}{36}\right) + 3\left(\frac{5}{36}\right) + \ldots\ldots + 6\left(\frac{11}{36}\right)$
$= \frac{1}{36} + \frac{6}{36} + \frac{15}{36} + \frac{28}{36} + \frac{45}{36} + \frac{66}{36}$
$= \frac{161}{36} \approx 4.47$

**14** **a**

Die 2

| Die 1 | 1 | 2 | 3 | 4 | 5 | 6 |
|---|---|---|---|---|---|---|
| 1 | 2 | 3 | 4 | 5 | 6 | 7 |
| 2 | 3 | 4 | 5 | 6 | 7 | 8 |
| 3 | 4 | 5 | 6 | 7 | 8 | 9 |
| 4 | 5 | 6 | 7 | 8 | 9 | 10 |
| 5 | 6 | 7 | 8 | 9 | 10 | 11 |
| 6 | 7 | 8 | 9 | 10 | 11 | 12 |

36 possible results

| $x$ | 2 | 3 | 4 | 5 | 6 | 7 | 8 | 9 | 10 | 11 | 12 |
|---|---|---|---|---|---|---|---|---|---|---|---|
| $P(x)$ | $\frac{1}{36}$ | $\frac{2}{36}$ | $\frac{3}{36}$ | $\frac{4}{36}$ | $\frac{5}{36}$ | $\frac{6}{36}$ | $\frac{5}{36}$ | $\frac{4}{36}$ | $\frac{3}{36}$ | $\frac{2}{36}$ | $\frac{1}{36}$ |

So, $P(X \leqslant 3) = \frac{1}{36} + \frac{2}{36} = \frac{1}{12}$
$P(4 \leqslant X \leqslant 6) = \frac{3}{36} + \frac{4}{36} + \frac{5}{36} = \frac{1}{3}$
$P(7 \leqslant X \leqslant 9) = \frac{6}{36} + \frac{5}{36} + \frac{4}{36} = \frac{5}{12}$
$P(X \geqslant 10) = \frac{3}{36} + \frac{2}{36} + \frac{1}{36} = \frac{1}{6}$

**b** The expected gain is
$\left(\frac{1}{12} + \frac{5}{12}\right)\left(-\frac{a}{3}\right) + \frac{1}{3}(7) + \frac{1}{6}(21) - a$
$= -\frac{a}{6} + \frac{7}{3} + \frac{21}{6} - a$
$= -\frac{7a}{6} + \frac{35}{6} = \frac{1}{6}(35 - 7a)$ dollars, as required.

**c** The game is fair when the expected gain is 0.
$\therefore\ \frac{1}{6}(35 - 7a) = 0$
$\therefore\ 35 - 7a = 0$
$\therefore\ a = 5$

**d** If $a = 4$, expected gain $= \frac{1}{6}(35 - 7(4))$
$= \frac{7}{6}$ dollars

So, the people playing would expect to win about \$1.17 per game, which means the organisers expect to lose \$1.17 per game.

**e** If $a = 6$, expected gain $= \frac{1}{6}(35 - 7(6))$
$= -\frac{7}{6}$ dollars

Expectation from 2406 games is $-\frac{7}{6} \times 2406$
$= -2807$

$\therefore$ the organisers would expect to gain \$2807.

## EXERCISE 23D.1

**1** **a** $(p + q)^4$
$= p^4 + 4p^3q + 6p^2q^2 + 4pq^3 + q^4$

**b** $P(\text{3 heads}) = 4p^3q$
$= 4\left(\frac{1}{2}\right)^3\left(\frac{1}{2}\right)$ {as $p = q = \frac{1}{2}$}
$= \frac{1}{4}$

**2** **a** $(p + q)^5 = p^5 + 5p^4q + 10p^3q^2 + 10p^2q^3 + 5pq^4 + q^5$

**b** **i** P(4H and 1T)
$= 5p^4q$
$= 5\left(\frac{1}{2}\right)^4\left(\frac{1}{2}\right)$
$= \frac{5}{32}$

**ii** P(2H and 3T)
$= 10p^2q^3$
$= 10\left(\frac{1}{2}\right)^2\left(\frac{1}{2}\right)^3$
$= \frac{10}{32}$
$= \frac{5}{16}$

**iii** P(HHHHT)
$= \left(\frac{1}{2}\right)^4 \times \frac{1}{2}$
$= \frac{1}{32}$

**3** **a** $\left(\frac{2}{3}+\frac{1}{3}\right)^4 = \left(\frac{2}{3}\right)^4 + 4\left(\frac{2}{3}\right)^3\left(\frac{1}{3}\right) + 6\left(\frac{2}{3}\right)^2\left(\frac{1}{3}\right)^2 + 4\left(\frac{2}{3}\right)\left(\frac{1}{3}\right)^3 + \left(\frac{1}{3}\right)^4$

**b** $P(S) = \frac{2}{3}$, $P(S') = \frac{1}{3}$ $\quad S'$ represents an almond centre

**i** P(all S) $= \left(\frac{2}{3}\right)^4 = \frac{16}{81}$

**ii** P(two of each) $= 6\left(\frac{2}{3}\right)^2\left(\frac{1}{3}\right)^2 = \frac{8}{27}$

**iii** P(at least 2 strawberry creams)
$= $ P(all S or 3S, 1S′ or 2S, 2S′)
$= \left(\frac{2}{3}\right)^4 + 4\left(\frac{2}{3}\right)^3\left(\frac{1}{3}\right) + 6\left(\frac{2}{3}\right)^2\left(\frac{1}{3}\right)^2$
$= \frac{16}{81} + \frac{32}{81} + \frac{24}{81}$
$= \frac{72}{81}$
$= \frac{8}{9}$

**4** **a** $\left(\frac{3}{4}+\frac{1}{4}\right)^5 = \left(\frac{3}{4}\right)^5 + 5\left(\frac{3}{4}\right)^4\left(\frac{1}{4}\right) + 10\left(\frac{3}{4}\right)^3\left(\frac{1}{4}\right)^2 + 10\left(\frac{3}{4}\right)^2\left(\frac{1}{4}\right)^3 + 5\left(\frac{3}{4}\right)\left(\frac{1}{4}\right)^4 + \left(\frac{1}{4}\right)^5$

**b** P('normal' kiwi) $= \frac{3}{4}$, P('flat back') $= \frac{1}{4}$

**i** P(2 'flat backs')
$= $ P(3F′, 2F)
$= 10 \times \left(\frac{3}{4}\right)^3\left(\frac{1}{4}\right)^2$
$= \frac{135}{512}$

**ii** P(at least 3 'flat backs')
$= $ P(2F′, 3F or 1F′, 4F or 5F)
$= 10\left(\frac{3}{4}\right)^2\left(\frac{1}{4}\right)^3 + 5\left(\frac{3}{4}\right)\left(\frac{1}{4}\right)^4 + \left(\frac{1}{4}\right)^5$
$= \frac{53}{512}$ on simplifying

**iii** P(at most 3 'normal' kiwis) $= 1 - $ P(4 or 5 normal kiwis)
$= 1 - $ P(4F′, 1F or 5F′)
$= 1 - \left(5\left(\frac{3}{4}\right)^4\left(\frac{1}{4}\right) + \left(\frac{3}{4}\right)^5\right)$
$= \frac{47}{128}$

**5** Let $X$ be the number of Huy's hits.

**a** Using the binomial expansion,
$P(X=2) = 6\left(\frac{4}{5}\right)^2\left(\frac{1}{5}\right)^2 \approx 0.154$

**b** $P(X \geqslant 2)$
$= 1 - P(X \leqslant 1)$
$\approx 1 - \left(4\left(\frac{4}{5}\right)\left(\frac{1}{5}\right)^3 + \left(\frac{1}{5}\right)^4\right)$
$\approx 0.973$

## EXERCISE 23D.2

**1** **a** The binomial distribution applies, as tossing a coin has two possible outcomes (a head or a tail) and each toss is independent of every other toss.

**b** The binomial distribution applies, as this is equivalent to tossing one coin 100 times.

**c** The binomial distribution applies as we can draw out a red or a blue marble with the same chances each time.

**d** The binomial distribution does not apply as the result of each draw is dependent upon the results of previous draws.

**e** The binomial distribution does not apply, assuming that ten bolts are drawn without replacement, as we do not have a repetition of independent trials.

**2** Let $X$ be the number of defective light bulbs.

**a** $P(X=2) \approx 0.0305$ {using technology} **b** $P(X \geqslant 1) \approx 0.265$

**3** If $X$ is the number of questions Raj answers correctly, then $X$ is binomial. There are $n = 10$ independent trials with probability $p = \frac{1}{5}$ of a corrrect answer for each.

P(Raj passes) $= P(X \geqslant 7)$
$\approx 0.000\,864$ {or about 9 in 10 000}

**4** $X$ is the random variable for the number working night-shift.

$\therefore$ $X = 0, 1, 2, 3, 4, 5, 6, 7$ and $X \sim \text{B}(7, 0.35)$.

**a** $\text{P}(X = 3)$
$= \binom{7}{3}(0.35)^3(0.65)^4$
$\approx 0.268$

**b** $\text{P}(X < 4)$
$= \text{P}(X \leqslant 3)$
$\approx 0.800$

**c** P(at least 4 work night-shift)
$= \text{P}(X \geqslant 4)$
$\approx 0.200$

**5** $X$ is the number of faulty items.

$\therefore$ $X = 0, 1, 2, 3, ...., 12$ and $X \sim \text{B}(12, 0.06)$.

**a** $\text{P}(X = 0) = \binom{12}{0}(0.06)^0(0.94)^{12}$
$\approx 0.476$

**b** P(at most one is faulty) $= \text{P}(X \leqslant 1)$
$\approx 0.840$

**c** P(at least two are faulty) $= \text{P}(X \geqslant 2)$
$\approx 0.160$

**d** P(less than four are faulty) $= \text{P}(X < 4)$
$= \text{P}(X \leqslant 3)$
$\approx 0.996$

**6** $X$ is the random variable for the number of apples with a blemish.

$\therefore$ $X = 0, 1, 2, 3, ...., 25$ and $X \sim \text{B}(25, 0.05)$.

**a** $\text{P}(X = 2)$
$= \binom{25}{2}(0.05)^2(0.95)^{23}$
$\approx 0.231$

**b** $\text{P}(X \geqslant 1)$
$\approx 0.723$

**c** $\text{E}(X) = np$
$= 25 \times 0.05$
$= 1.25$ apples

**7** $X$ is the random variable for the number of times in a week that the bus is on time.
Since it is late 2 in every 5 days, and on time 3 in every 5 days,
$X = 0, 1, 2, 3, 4, 5, 6$ or $7$ and $X \sim \text{B}(7, 0.6)$.

**a** $\text{P}(X = 7) = \binom{7}{7}(0.6)^7(0.4)^0$
$\approx 0.0280$

**b** P(on time only on Monday) $= 0.6 \times (0.4)^6$
$\approx 0.002\,46$

**c** $\text{P}(X = 6) = \binom{7}{6}(0.6)^6(0.4)$
$\approx 0.131$

**d** $\text{P}(X \geqslant 4) \approx 0.710$

**8** $X$ is the random variable for the number of students with the flu.

$\therefore$ $X = 0, 1, 2, 3, ...., 25$ and $X \sim \text{B}(25, 0.3)$.

**a** **i** $\text{P}(X \geqslant 2) \approx 0.998$

**ii** P(test cancelled) $= \text{P}(X \geqslant 6)$ {20% of 25 = 5}
$\approx 0.807$

**b** Expected absentees from 350 students $= 0.3 \times 350$
$= 105$ students

**9** $X$ is the random variable for the number of successful shots from the free throw line.

$\therefore$ $X = 0, 1, 2, 3, ...., 20$ and $X \sim \text{B}(20, 0.94)$.

**a** **i** $\text{P}(X = 20) = \binom{20}{20}(0.94)^{20}(0.06)^0$
$\approx 0.290$

**ii** $\text{P}(X \geqslant 18) \approx 0.885$

**b** $\text{E}(X) = np = 20 \times 0.94$
$= 18.8$ successful throws

**10** P(M wins a game against J) $= \frac{2}{3}$ $\therefore$ P(M wins) $= \frac{2}{3}$ P(J wins) $= \frac{1}{3}$

P(J wins a set 6 games to 4) = P(J wins 5 of the first 9 games **and** J wins the 10th game)

(J wins 5 of the first 9 games: this is binomial with $n = 9$ trials of probability $p = \frac{1}{3}$)

$\approx 0.1024 \times \frac{1}{3}$
$\approx 0.0341$

**11** If there are $n$ dice thrown, $\text{P(no sixes)} = \left(\frac{5}{6}\right)^n$

$\therefore \text{P(at least 1 six)} = 1 - \left(\frac{5}{6}\right)^n$

$\therefore$ need to find the smallest integer $n$ such that $1 - \left(\frac{5}{6}\right)^n > 0.5$

$$\therefore \left(\frac{5}{6}\right)^n < 0.5$$

$$\therefore n \log\left(\frac{5}{6}\right) < \log(0.5)$$

$$\therefore n > \frac{\log(0.5)}{\log\left(\frac{5}{6}\right)} \qquad \{\log\left(\tfrac{5}{6}\right) < 0\}$$

$$\therefore n > 3.80$$

$\therefore$ at least 4 dice are needed.

**12** If a fair coin is tossed 200 times, then $n = 200$ and $p = \frac{1}{2}$.

**a** $\text{P}(90 \leqslant X \leqslant 110)$
$\approx 0.863$

**b** $\text{P}(95 < X < 105)$
$= \text{P}(96 \leqslant X \leqslant 104)$
$\approx 0.475$

**13** $n = 38$, $p = 0.75$
$\text{P}(24 \leqslant X \leqslant 31) \approx 0.837$

## EXERCISE 23D.3

**1** **a** **i** $\mu = np = 6 \times 0.5 = 3$

$\sigma = \sqrt{np(1-p)} = \sqrt{6 \times 0.5 \times 0.5} \approx 1.22$

**ii**

| $x$ | 0 | 1 | 2 | 3 | 4 | 5 | 6 |
|---|---|---|---|---|---|---|---|
| $P(x)$ | 0.0156 | 0.0938 | 0.2344 | 0.3125 | 0.2344 | 0.0938 | 0.0156 |

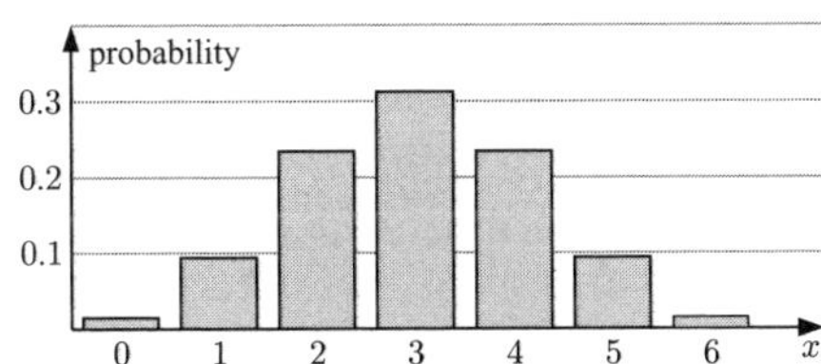


**iii** The distribution is bell-shaped.

**b** **i** $\mu = np = 6 \times 0.2 = 1.2$

$\sigma = \sqrt{np(1-p)} = \sqrt{6 \times 0.2 \times 0.8} \approx 0.980$

**ii**

| $x$ | 0 | 1 | 2 | 3 | 4 | 5 | 6 |
|---|---|---|---|---|---|---|---|
| $P(x)$ | 0.2621 | 0.3932 | 0.2458 | 0.0819 | 0.0154 | 0.0015 | 0.0001 |

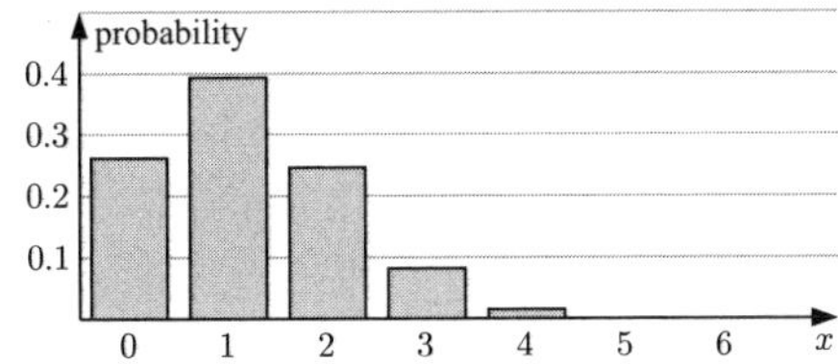


**iii** The distribution is positively skewed.

**c** **i** $\mu = np$
$= 6 \times 0.8$
$= 4.8$

$\sigma = \sqrt{np(1-p)}$
$= \sqrt{6 \times 0.8 \times 0.2}$
$\approx 0.980$

**ii**

| $x$ | 0 | 1 | 2 | 3 | 4 | 5 | 6 |
|---|---|---|---|---|---|---|---|
| $P(x)$ | 0.0001 | 0.0015 | 0.0154 | 0.0819 | 0.2458 | 0.3932 | 0.2621 |

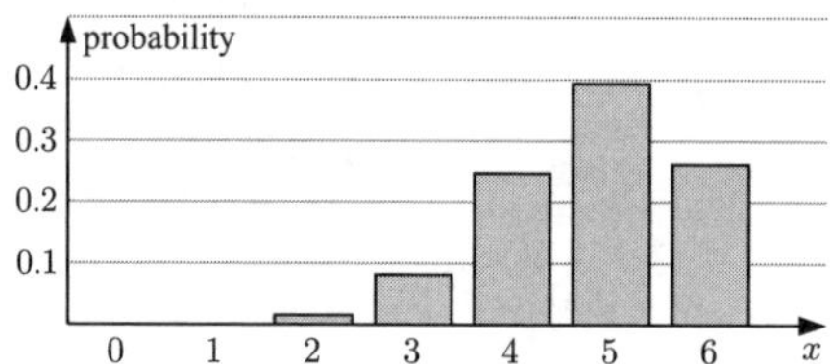


**iii** The distribution is negatively skewed, and is the exact reflection of the distribution in **b**.

**2** $n = 10$, $p = \frac{1}{2}$, mean $\mu = np$
$= 10 \times \frac{1}{2}$
$= 5$

and variance $\sigma^2 = np(1-p)$
$= 10 \times \frac{1}{2} \times \frac{1}{2}$
$= 2.5$

**3** **a** $n = 30$, $p = 0.04$
$\mu = np$
$= 30 \times 0.04$
$= 1.2$

$\sigma = \sqrt{np(1-p)}$
$= \sqrt{30 \times 0.04 \times 0.96}$
$\approx 1.07$

**b** $n = 30$, $p = 0.96$
$\mu = np$
$= 30 \times 0.96$
$= 28.8$

$\sigma = \sqrt{np(1-p)}$
$= \sqrt{30 \times 0.96 \times 0.04}$
$\approx 1.07$

**4** $n = 30$, $p = 0.13$

$\therefore$ mean $\mu = np$
$= 30 \times 0.13$
$= 3.9$

and standard deviation $\sigma = \sqrt{np(1-p)}$
$= \sqrt{30 \times 0.13 \times 0.87}$
$\approx 1.84$

## REVIEW SET 23A

**1** **a** $P(X = x) = \dfrac{a}{x^2 + 1}$ for $x = 0, 1, 2, 3$

| $x$ | 0 | 1 | 2 | 3 |
|---|---|---|---|---|
| $P(X = x)$ | $a$ | $\frac{a}{2}$ | $\frac{a}{5}$ | $\frac{a}{10}$ |

Now $a + \dfrac{a}{2} + \dfrac{a}{5} + \dfrac{a}{10} = 1$ $\{$as $\sum_{x=0}^{3} P(X = x) = 1\}$

$\therefore$ $10a + 5a + 2a + a = 10$

$\therefore$ $18a = 10$

$\therefore$ $a = \frac{5}{9}$

**b** $P(X \geqslant 1) = P(X = 1, 2, \text{ or } 3)$
$= P(X = 1) + P(X = 2) + P(X = 3)$
$= \frac{5}{18} + \frac{1}{9} + \frac{5}{90}$
$= \frac{4}{9}$

*or* $P(X \geqslant 1) = 1 - P(X < 1)$
$= 1 - P(X = 0)$
$= 1 - \frac{5}{9}$
$= \frac{4}{9}$

**2** Let $X$ be the number of defective toothbrushes.

$\therefore \quad X \sim \text{B}(120, 0.04)$

$$\mu = np = 120 \times 0.04 = 4.8 \text{ defectives}$$

**3**

| $x$ | 0 | 1 | 2 | 3 | 4 |
|---|---|---|---|---|---|
| $\text{P}(X = x)$ | 0.10 | 0.30 | 0.45 | 0.10 | $k$ |

**a** If this is a probability distribution then $\sum P(x_i) = 1$

$\therefore \quad 0.1 + 0.3 + 0.45 + 0.1 + k = 1$

$\therefore \quad 0.95 + k = 1$

$\therefore \quad k = 0.05$

**b** $\text{P}(X \geqslant 3)$

$= \text{P}(X = 3) + \text{P}(X = 4)$

$= 0.10 + 0.05$

$= 0.15$

**c** $\text{E}(X) = \sum x_i p_i$

$= 0(0.1) + 1(0.3) + 2(0.45) + 3(0.1) + 4(0.05)$

$= 0 + 0.3 + 0.9 + 0.3 + 0.2$

$= 1.7$

**4** **a** $\left(\frac{3}{5} + \frac{2}{5}\right)^4 = \underbrace{\left(\frac{3}{5}\right)^4}_{4\text{B}} + \underbrace{4\left(\frac{3}{5}\right)^3\left(\frac{2}{5}\right)}_{3\text{B}\ 1\text{B}'} + \underbrace{6\left(\frac{3}{5}\right)^2\left(\frac{2}{5}\right)^2}_{2\text{B}\ 2\text{B}'} + \underbrace{4\left(\frac{3}{5}\right)\left(\frac{2}{5}\right)^3}_{1\text{B}\ 3\text{B}'} + \underbrace{\left(\frac{2}{5}\right)^4}_{4\text{B}'}$

$\text{P(B)} = \frac{12}{20} = \frac{3}{5}$

$\therefore \quad \text{P(B}') = \frac{2}{5}$

**b** **i** P(2 Blue inks)

$= \text{P(2B and 2B}')$

$= 6\left(\frac{3}{5}\right)^2\left(\frac{2}{5}\right)^2$

$= \dfrac{6 \times 9 \times 4}{5^4}$

$= \frac{216}{625}$

**ii** P(at most 2 Blue inks)

$= \text{P(2B and 2B}' \text{ or 1B and 3B}' \text{ or 4B}')$

$= 6\left(\frac{3}{5}\right)^2\left(\frac{2}{5}\right)^2 + 4\left(\frac{3}{5}\right)\left(\frac{2}{5}\right)^3 + \left(\frac{2}{5}\right)^4$

$= \dfrac{6 \times 9 \times 4 + 4 \times 3 \times 8 + 16}{625}$

$= \frac{328}{625}$

**5** **a**

| Event | $X$ | Probability |
|---|---|---|
| GG | 2 | $\frac{3}{5} \times \frac{2}{4} = \frac{3}{10}$ |
| GY | 1 | $\frac{3}{5} \times \frac{2}{4} = \frac{3}{10}$ |
| YG | 1 | $\frac{2}{5} \times \frac{3}{4} = \frac{3}{10}$ |
| YY | 0 | $\frac{2}{5} \times \frac{1}{4} = \frac{1}{10}$ |

| | i | ii | iii |
|---|---|---|---|
| $x$ | 0 | 1 | 2 |
| $\text{P}(X = x)$ | $\frac{1}{10}$ | $\frac{3}{5}$ | $\frac{3}{10}$ |

**b** $\text{E}(X) = 0 \times \frac{1}{10} + 1 \times \frac{3}{5} + 2 \times \frac{3}{10} = \frac{6}{5} \quad (= 1\frac{1}{5})$

**6** **a**

| *Result* | *Pays* |
|---|---|
| 1 | £2 |
| 2 | £4 |
| 3 | £6 |
| 4 | £8 |
| 5 | £10 |
| 6 | £12 |

Expectation

$= \frac{1}{6} \times £2 + \frac{1}{6} \times £4 + \frac{1}{6} \times £6 + \frac{1}{6} \times £8 + \frac{1}{6} \times £10 + \frac{1}{6} \times £12$

$= \frac{1}{6} \times £42$

$= £7$

**b** Expected gain is $£7 - £8 = -£1$.

$\therefore$ advise Lakshmi against playing several games, as £1 is expected to be lost per game in the long run.

**7** **a** $n = 7$ and $r = \{0, 1, 2, 3, ...., 7\}$

$= x$

$\therefore \quad k = \binom{7}{x}$

**b** $n = 7, \quad p = \frac{1}{3}$

$\therefore \quad \mu = np = \frac{7}{3} \quad (\approx 2.33)$

and $\sigma^2 = np(1-p) = 7 \times \frac{1}{3} \times \frac{2}{3} = \frac{14}{9} \quad (\approx 1.56)$

**8** **a**

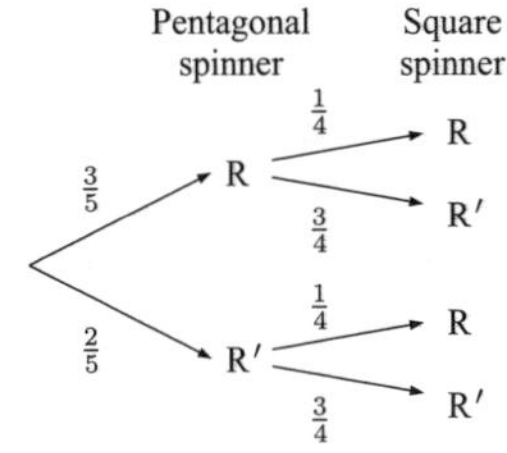


**b** $\text{P(exactly one red)} = \text{P}(RR') + \text{P}(R'R)$

$= \frac{3}{5} \times \frac{3}{4} + \frac{2}{5} \times \frac{1}{4}$

$= \frac{9}{20} + \frac{1}{10}$

$= \frac{11}{20}$

**c** **i** $n = 10, \quad p = \frac{11}{20}$

$\text{P}(X=1) = \binom{10}{1}\left(\frac{11}{20}\right)^1\left(\frac{9}{20}\right)^9$

$\text{P}(X=9) = \binom{10}{9}\left(\frac{11}{20}\right)^9\left(\frac{9}{20}\right)^1$

**ii** $\binom{10}{1} = \binom{10}{9} = 10$ so we need only consider the parts $\left(\frac{11}{20}\right)^1\left(\frac{9}{20}\right)^9$ and $\left(\frac{11}{20}\right)^9\left(\frac{9}{20}\right)^1$.

Now, $11 \times 9^9 < 11^9 \times 9$, and the denominators are the same in each case.

$\therefore$ it is more likely that exactly one red will occur 9 times.

## REVIEW SET 23B

**1** **a** $P(x) = k\left(\frac{3}{4}\right)^x\left(\frac{1}{4}\right)^{3-x}$ for $x = 0, 1, 2, 3$

$P(0) = k\left(\frac{3}{4}\right)^0\left(\frac{1}{4}\right)^3 = \dfrac{k}{64}$

$P(1) = k\left(\frac{3}{4}\right)^1\left(\frac{1}{4}\right)^2 = \dfrac{3k}{64}$

$P(2) = k\left(\frac{3}{4}\right)^2\left(\frac{1}{4}\right)^1 = \dfrac{9k}{64}$

$P(3) = k\left(\frac{3}{4}\right)^3\left(\frac{1}{4}\right)^0 = \dfrac{27k}{64}$

| $x$ | 0 | 1 | 2 | 3 |
|---|---|---|---|---|
| $P(x)$ | $\frac{k}{64}$ | $\frac{3k}{64}$ | $\frac{9k}{64}$ | $\frac{27k}{64}$ |

Now $\dfrac{k}{64} + \dfrac{3k}{64} + \dfrac{9k}{64} + \dfrac{27k}{64} = 1 \quad \{\text{as } \sum P(x_i) = 1\}$

$\therefore \quad \dfrac{40k}{64} = 1$

$\therefore \quad k = \frac{8}{5} \quad (= 1.6)$

**b** $\text{P}(X \geqslant 1) = 1 - \text{P}(X=0)$

$= 1 - \dfrac{k}{64}$

$= 1 - \dfrac{1.6}{64}$

$= 0.975$

**c** $\text{E}(X) = \sum x_i p_i$

$= 0 \times \frac{1.6}{64} + 1 \times \frac{3 \times 1.6}{64} + 2 \times \frac{9 \times 1.6}{64} + 3 \times \frac{27 \times 1.6}{64}$

$= 2.55$

**d** $n = 3, \quad p = \frac{3}{4} \quad \therefore \quad \sigma = \sqrt{np(1-p)} = \sqrt{3 \times \frac{3}{4} \times \frac{1}{4}}$

$= \frac{3}{4} \quad (= 0.75)$

**2** $X$ is the number of defectives. Then $X \sim \text{B}(10, 0.18)$. $X = 0, 1, 2, 3, ...., 10$.

**a** $\text{P}(X=1)$

$= \binom{10}{1}(0.18)^1(0.82)^9$

$\approx 0.302$

**b** $\text{P}(X=2)$

$= \binom{10}{2}(0.18)^2(0.82)^8$

$\approx 0.298$

**c** $\text{P}(X \geqslant 2)$

$\approx 0.561$

**3** Expected number of major knee surgeries $= np$

$= 487 \times 0.0132$

$\approx 6.43$

**4** If $X$ is the number of X-rays which show the fracture, then $X = 0, 1, 2, 3, 4$ and $X \sim \text{B}(4, 0.96)$.

**a** $\text{P}(X = 4)$
$= \binom{4}{4}(0.96)^4(0.04)^0$
$\approx 0.849$

**b** $\text{P}(X = 0)$
$= \binom{4}{0}(0.96)^0(0.04)^4$
$\approx 2.56 \times 10^{-6}$

**c** $\text{P}(X \geqslant 3)$
$\approx 0.991$

**d** $\text{P}(X = 1)$
$= \binom{4}{1}(0.96)^1(0.04)^3$
$\approx 0.000\,246$

**5** $X$ is the number of visitors who make a voluntary donation upon entry.
Then $X = 0, 1, 2, 3, ...., 175$ and $X \sim \text{B}(175, 0.24)$.

**a** $\text{E}(X) = np$
$= 175 \times 0.24$
$= 42$

**b** $\text{P}(X < 40) = \text{P}(X \leqslant 39)$
$\approx 0.334$

**6** $X$ is the number of players who turn up to a game.
Then $X = 0, 1, 2, 3, ...., 8$ and $X \sim \text{B}(8, 0.75)$.

**a** **i** $\text{P}(X = 8) = \binom{8}{8}(0.75)^8(0.25)^0$
$\approx 0.100$

**ii** $\text{P(team has to forfeit)} = \text{P}(X \leqslant 4)$
$\approx 0.114$

**b** Expected number of games forfeited in 30 $= np$
$\approx 30 \times 0.1138$ {from **a ii**}
$\approx 3.41$

**7** **a** If the mean is 30 then $np = 30$ .... (1)
If the variance is 22.5 then $np(1-p) = 22.5$ .... (2)
Substituting (1) into (2), we get $30(1-p) = 22.5$

$$\therefore \ 1 - p = \frac{22.5}{30}$$

$$\therefore \ p = 1 - \frac{22.5}{30}$$

$$\therefore \ p = 0.25$$

and so $n \times 0.25 = 30$

$$\therefore \ n = 120$$

So, $n = 120$ and $p = 0.25$ (or $\frac{1}{4}$).

**b** **i** $\text{P}(X = 25)$
$\approx 0.0501$

**ii** $\text{P}(X \geqslant 25)$
$\approx 0.878$

**iii** $\text{P}(15 \leqslant X \leqslant 25)$
$\approx 0.172$

## REVIEW SET 23C

**1** **a**

$$\sum \text{P}(X = x_i) = 1$$

$$\therefore \ \frac{k}{2 \times 1} + \frac{k}{2 \times 2} + \frac{k}{2 \times 3} = 1$$

$$\therefore \ \frac{k}{2} + \frac{k}{4} + \frac{k}{6} = 1$$

$$\therefore \ 6k + 3k + 2k = 12$$

$$\therefore \ 11k = 12$$

$$\therefore \ k = \tfrac{12}{11}$$

**b**

$$\sum P(x_i) = 1$$

$$\therefore \ \frac{k}{2} + 0.2 + k^2 + 0.3 = 1$$

$$\therefore \ 2k^2 + k + 1 = 2$$

$$\therefore \ 2k^2 + k - 1 = 0$$

$$\therefore \ (2k-1)(k+1) = 0$$

$$\therefore \ k = -1, \tfrac{1}{2}$$

If $k = -1$, then $P(0) = \frac{-1}{2} < 0$, so $P(x)$ would not be a valid probability distribution function.

$\therefore \ k = \frac{1}{2}$

**2** **a** $\text{P}(X = x) = \binom{4}{x}\left(\frac{1}{2}\right)^x\left(\frac{1}{2}\right)^{4-x}$

$\therefore\ \text{P}(X = 0) = \binom{4}{0}\left(\frac{1}{2}\right)^0\left(\frac{1}{2}\right)^4 = 0.0625$

$\text{P}(X = 1) = \binom{4}{1}\left(\frac{1}{2}\right)^1\left(\frac{1}{2}\right)^3 = 0.25$

$\text{P}(X = 2) = \binom{4}{2}\left(\frac{1}{2}\right)^2\left(\frac{1}{2}\right)^2 = 0.375$

$\text{P}(X = 3) = \binom{4}{3}\left(\frac{1}{2}\right)^3\left(\frac{1}{2}\right)^1 = 0.25$

$\text{P}(X = 4) = \binom{4}{4}\left(\frac{1}{2}\right)^4\left(\frac{1}{2}\right)^0 = 0.0625$

| $x$ | 0 | 1 | 2 | 3 | 4 |
|---|---|---|---|---|---|
| $\text{P}(X = x)$ | 0.0625 | 0.25 | 0.375 | 0.25 | 0.0625 |

**b** $\mu = \sum x_i \text{P}(X = x_i)$

$= 0 \times 0.0625 + 1 \times 0.25 + 2 \times 0.375 + 3 \times 0.25 + 4 \times 0.0625$

$= 2$

**c** $n = 4,\quad p = \frac{1}{2}\qquad \therefore\ \sigma = \sqrt{np(1-p)} = \sqrt{4 \times \frac{1}{2} \times \frac{1}{2}}$

$= \sqrt{1}$

$= 1$

**3** **a** $\left(\frac{4}{5}+\frac{1}{5}\right)^5 = \left(\frac{4}{5}\right)^5 + 5\left(\frac{4}{5}\right)^4\left(\frac{1}{5}\right)^1 + 10\left(\frac{4}{5}\right)^3\left(\frac{1}{5}\right)^2 + 10\left(\frac{4}{5}\right)^2\left(\frac{1}{5}\right)^3 + 5\left(\frac{4}{5}\right)^1\left(\frac{1}{5}\right)^4 + \left(\frac{1}{5}\right)^5$

**b** Let $X$ = the number of goals scored

**i** P(3 goals **then** 2 misses)

$= \text{P}(\text{GGGG}'\text{G}')$

$= (0.8)^3 \times (0.2)^2$

$\approx 0.0205$

**ii** P(3 goals and 2 misses)

$= \text{P}(X = 3)$

$= 10\left(\frac{4}{5}\right)^3\left(\frac{1}{5}\right)^2$

$\approx 0.205$

**4** $X \sim \text{B}(1200, 0.4)$

So, mean of $X = np$

$= 1200 \times 0.4$

$\therefore\ \mu = 480$

and standard deviation of $X = \sqrt{np(1-p)}$

$= \sqrt{1200 \times 0.4 \times 0.6}$

$\therefore\ \sigma \approx 17.0$

**5** $X$ is the number of trees that survive the first year.

$\therefore\ X = 0, 1, 2, 3, 4, 5$ and $X \sim \text{B}(5, 0.4)$

**a** $\text{P}(X = 1) = \binom{5}{1}(0.4)^1(0.6)^4$

$\approx 0.259$

**b** $\text{P}(X \leqslant 1) \approx 0.337$

**c** $\text{P}(X \geqslant 1) \approx 0.922$

**6** **a** **i** In the numbers 1 to 20, there are 10 even numbers.
However, '4' and '16' are square numbers, so 8 of the numbers in the bag win \$3.

$\therefore$ P(player wins \$3) $= \frac{8}{20} = \frac{2}{5}$

**ii** 1, 4, 9, and 16 are the only square numbers in the bag.
But '4' and '16' are even, so 2 of the numbers in the bag win \$6.

$\therefore$ P(player wins \$6) $= \frac{2}{20} = \frac{1}{10}$

**iii** 2 numbers are both even and square (4 and 16).

$\therefore$ P(player wins \$9) $= \frac{2}{20} = \frac{1}{10}$

**b** Expected winnings $= \frac{2}{5} \times \$3 + \frac{1}{10} \times \$6 + \frac{1}{10} \times \$9$

$= \$2.70$

$\therefore$ for the game to be fair, players should be charged \$2.70 per game.

**7** Suppose $X$ is the event that a 6 is rolled.

$\therefore\ n = 360,\quad p = \frac{1}{6}$

**a** $\text{P}(X < 50) = \text{P}(X \leqslant 49)$

$\approx 0.0660$

**b** $\text{P}(55 \leqslant X \leqslant 65) \approx 0.563$

# Chapter 24
## THE NORMAL DISTRIBUTION

### EXERCISE 24A

**1**

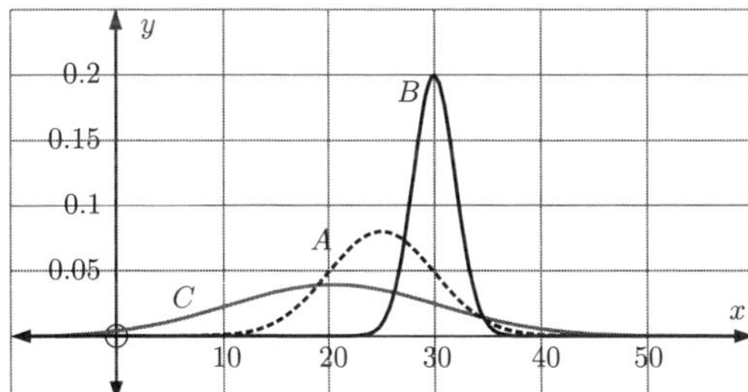


**2 a, b** The mean volume (or diameter) is likely to occur most often with variations around the mean occurring symmetrically as a result of random variations in the production process.

**3**

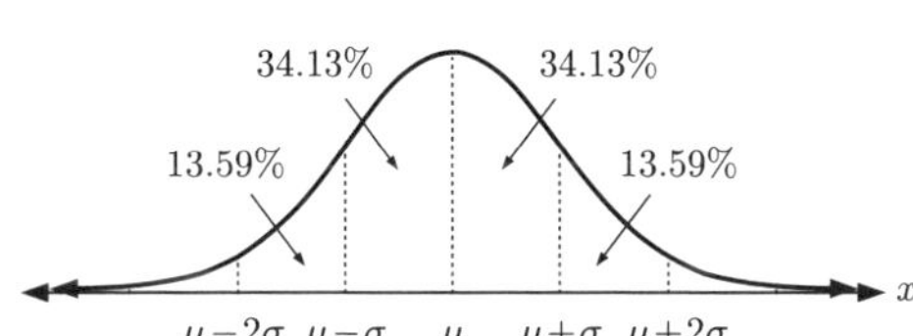


**a** P(value between $\mu - \sigma$ and $\mu + \sigma$)
$\approx 34.13\% + 34.13\%$
$\approx 0.683$

**b** P(value between $\mu$ and $\mu + 2\sigma$)
$\approx 34.13\% + 13.59\%$
$\approx 0.477$

**4 a**

We need the percentage greater than 189 cm.
This is $13.59\% + 2.15\% + 0.13\%$
$\approx 15.9\%$

**b**

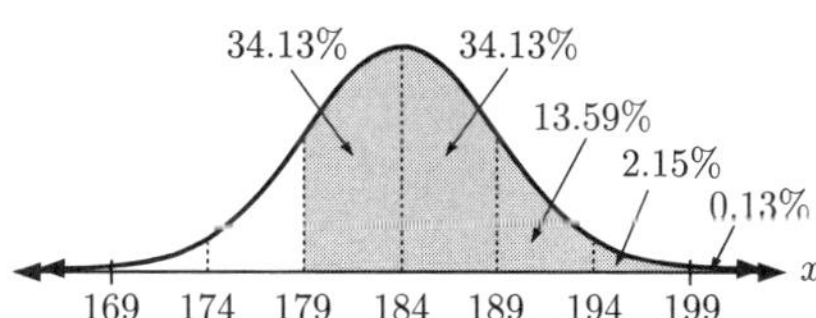


We need the percentage greater than 179 cm.
This is $34.13\% + 34.13\% + 13.59\%$
$+ 2.15\% + 0.13\%$
$\approx 84.1\%$

**c**

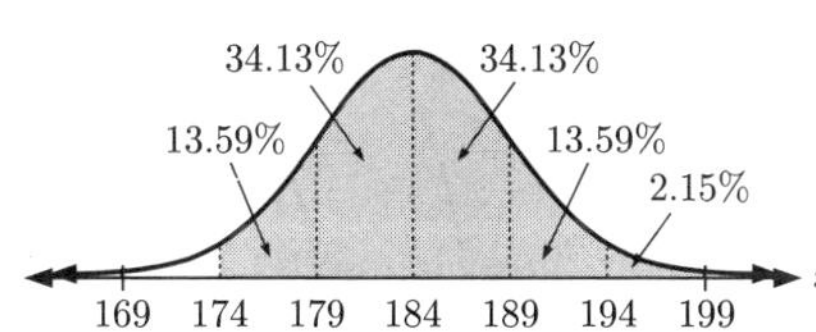


We need the percentage between 174 cm and 199 cm.
This is $13.59\% + 34.13\% + 34.13\%$
$+ 13.59\% + 2.15\%$
$\approx 97.6\%$

**d**

We need the percentage greater than 199 cm.
This is 0.13%.

**5**

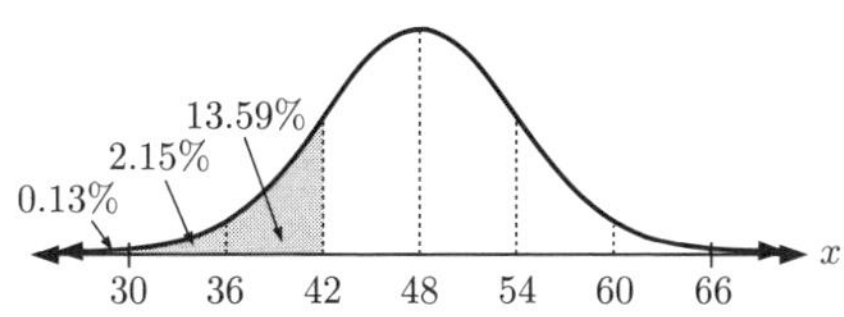


The chance of there being less than 42 mm of rain during August is
$0.13\% + 2.15\% + 13.59\% = 15.87\%$
and $15.87\%$ of $20 = 3.174$

So, over a 20 year period, there would be less than 42 mm of rain during August three times.

**6** **a**

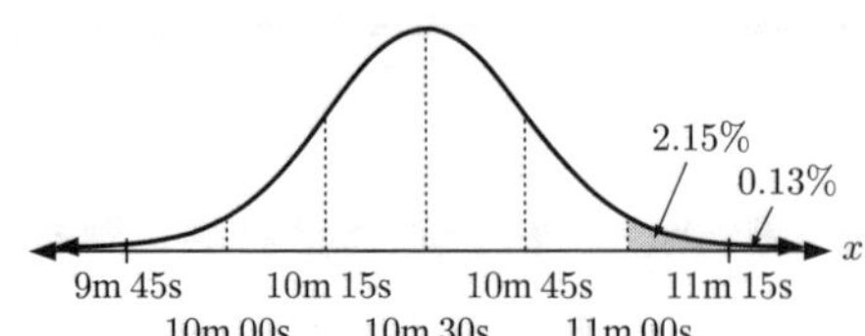


$2.15\% + 0.13\% = 2.28\%$ of competitors took over 11 minutes,
and $2.28\%$ of $200 = 4.56$
So, 5 competitors took longer than 11 minutes.

**b**

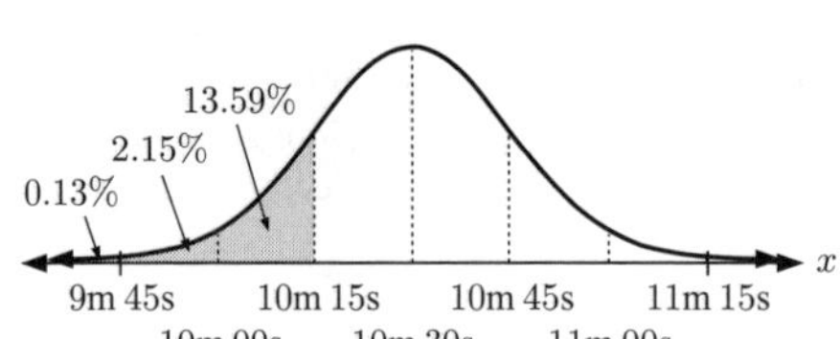


$0.13\% + 2.15\% + 13.59\% = 15.87\%$ of competitors took less than 10 minutes 15 seconds,
and $15.87\%$ of $200 = 31.74$
So, 32 competitors took less than 10 minutes 15 seconds.

**c**

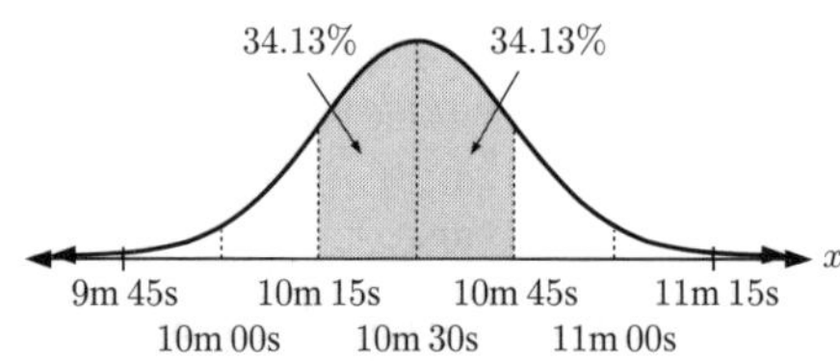


$34.13\% + 34.13\% = 68.26\%$ of competitors took between 10 minutes 15 seconds and 10 minutes 45 seconds,
and $68.26\%$ of $200 = 136.52$
So, 137 competitors took between 10 minutes 15 seconds and 10 minutes 45 seconds.

**7** **a**

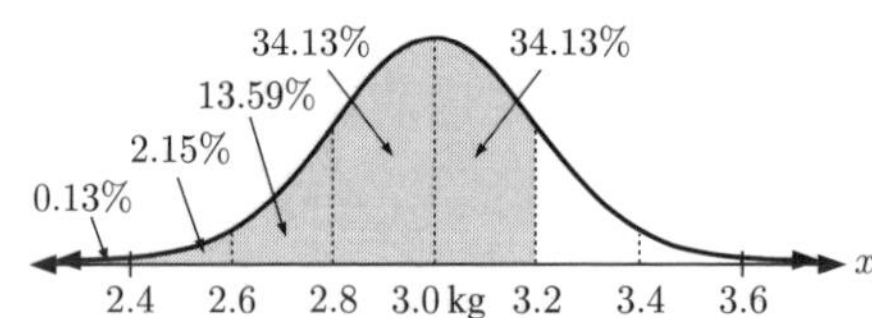


$0.13\% + 2.15\% + 13.59\% + 34.13\% + 34.13\% = 84.13\%$ of babies born weighed less than 3.2 kg, and
$84.13\%$ of $545 = 458.5085$
So, 459 babies born weighed less than 3.2 kg.

**b**

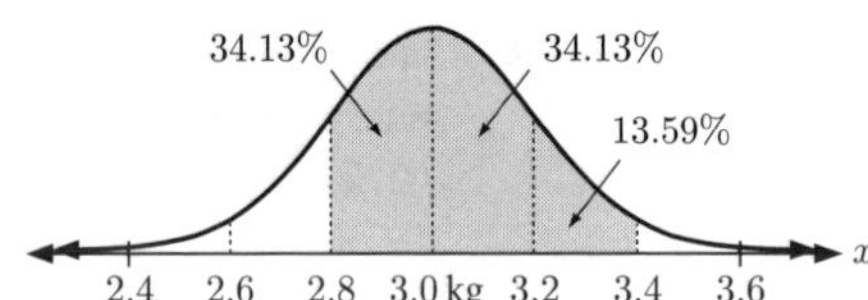


$34.13\% + 34.13\% + 13.59\% = 81.85\%$ of babies born weighed between 2.8 kg and 3.4 kg, and
$81.85\%$ of $545 = 446.0825$
So, 446 babies born weighed between 2.8 kg and 3.4 kg.

**8**

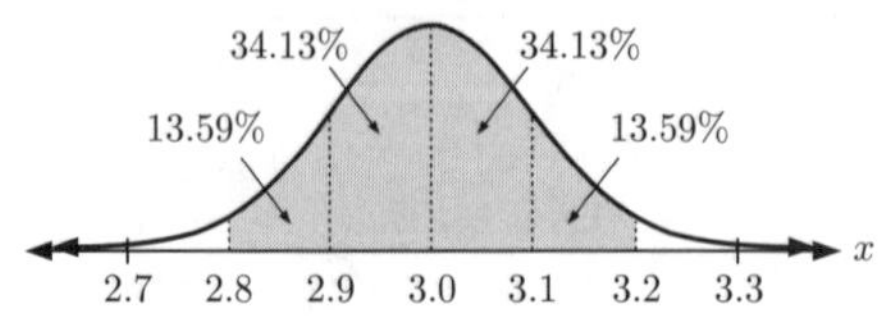


**a** P(value is within 2 standard deviations of the mean)
$= \text{P}(2.8 \leqslant X \leqslant 3.2)$
$\approx 13.59\% + 34.13\% + 34.13\% + 13.59\%$
$\approx 0.954$

**b** The value 1 standard deviation below the mean is $X = 3 - 0.1 = 2.9$

**9** **a**

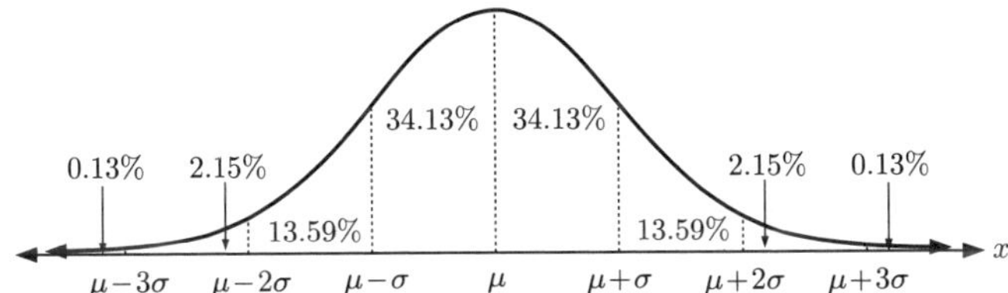


84% of the crop weigh more than 152 g $\therefore \quad \mu - \sigma = 152$
16% of the crop weigh more than 200 g $\therefore \quad \mu + \sigma = 200$ .... (1)
Adding: $2\mu = 352$, and so $\mu = 176$ g

Substituting $\mu = 176$ into (1) gives $\sigma = 200 - \mu = 24$ g.

**b** For $\mu = 176$ g and $\sigma = 24$ g, $152 \text{ g} = \mu - \sigma$, and $224 \text{ g} = \mu + 2\sigma$.
$\therefore$ between 152 g and 224 g, the percentage is $34.13\% + 34.13\% + 13.59\% \approx 81.9\%$

**10**

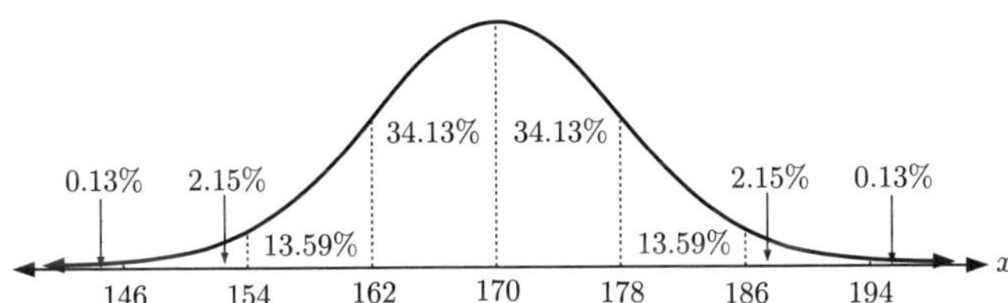


**a** **i** $P(162 < X < 170) \approx 34.1\%$

**ii** $P(170 < X < 186) \approx 34.13\% + 13.59\%$
$\approx 47.7\%$

**b** **i** $P(178 < X < 186)$
$\approx 13.59\%$
$\approx 0.136$

**ii** $P(X < 162)$
$\approx 1 - (0.5 + 0.3413)$
$\approx 0.159$

**iii** $P(X < 154)$
$\approx 0.0215 + 0.0013$
$\approx 0.0228$

**iv** $P(X > 162)$
$\approx 1 - 0.159$ {using **b ii**}
$\approx 0.841$

**c** 16% of students are taller than 178 cm $\{13.59\% + 2.15\% + 0.13\% \approx 16\%\}$
$\therefore \quad k = 178$

**11**

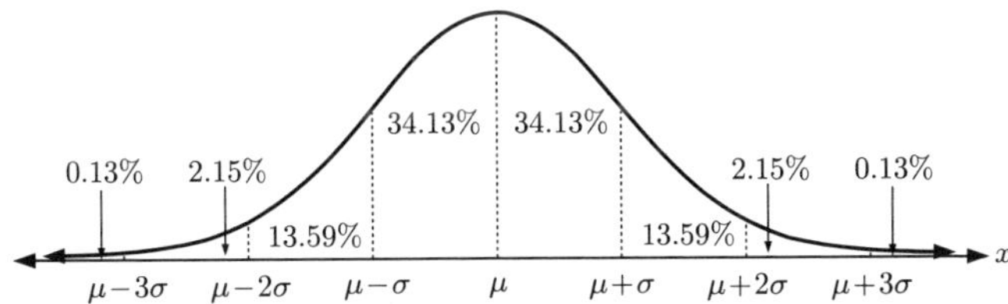


**a** 97.72% of 13 year old boys are taller than 131 cm $\therefore \quad \mu - 2\sigma = 131$
2.28% of 13 year old boys are taller than 179 cm $\therefore \quad \mu + 2\sigma = 179$ .... (1)
Adding: $2\mu = 310$, and so $\mu = 155$ cm

Substituting $\mu = 155$ cm into (1) gives $\sigma = \dfrac{179 - \mu}{2} = \dfrac{24}{2} = 12$ cm

**b** For $\mu = 155$ cm and $\sigma = 12$ cm, $143 \text{ cm} = \mu - \sigma$, and $191 \text{ cm} = \mu + 3\sigma$
$\therefore$ between 143 cm and 191 cm, the percentage is $34.13\% + 34.13\% + 13.59\% + 2.15\% \approx 84.0\%$
So the probability is 0.84.

**12** Without fertiliser

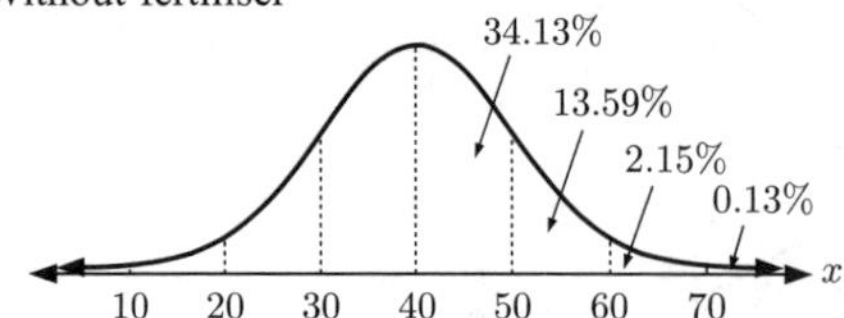


With fertiliser

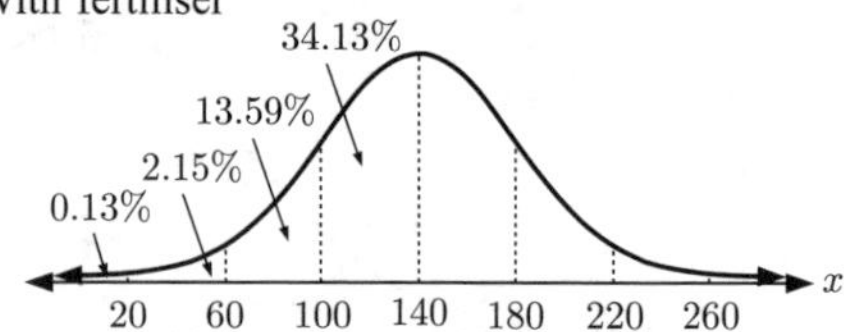


**a** P(without and $< 50$)
$\approx 50\% + 34.13\%$
$\approx 84.1\%$

**b** P(with and $< 60$)
$\approx 0.13\% + 2.15\%$
$\approx 2.28\%$

**c** **i** P(with and $20 \leqslant X \leqslant 60$)
$\approx 2.15\%$

**ii** P(without and $20 \leqslant X \leqslant 60$)
$\approx 2(34.13\% + 13.59\%)$
$\approx 95.4\%$

**d** **i** P(with and $X \geqslant 60$)
$\approx 13.59\% + 34.13\% + 50\%$
$\approx 97.7\%$

**ii** P(without and $X \geqslant 60$)
$\approx 2.15\% + 0.13\%$
$\approx 2.28\%$

**13**

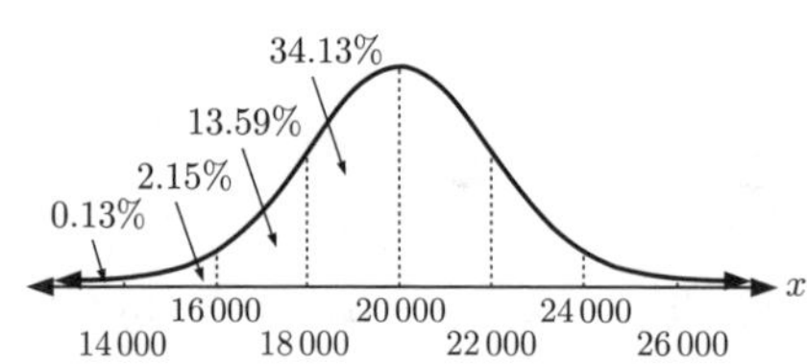


**a** $P(X < 18\,000)$
$\approx 1 - 0.5 - 0.3413$
$\approx 0.1587$
$\therefore$ we expect that less than 18 000 bottles are filled on $260 \times 0.1587 \approx 41$ days.

**b** $P(X > 16\,000)$
$\approx 0.1359 + 0.3413 + 0.5$
$\approx 0.9772$
$\therefore$ we expect that over 16 000 bottles are filled on $260 \times 0.9772 \approx 254$ days.

**c** $P(18\,000 \leqslant X \leqslant 24\,000)$
$\approx 0.3413 \times 2 + 0.1359$
$\approx 0.8185$
$\therefore$ we expect that between 18 000 and 24 000 bottles are filled on $260 \times 0.8185$ $\approx 213$ days.

## EXERCISE 24B

**1** **a**

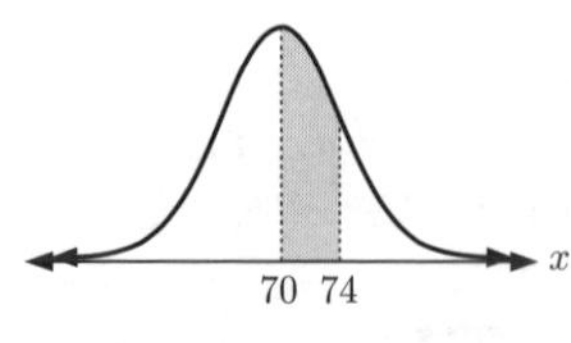


$P(70 \leqslant X \leqslant 74) \approx 0.341$

**b**

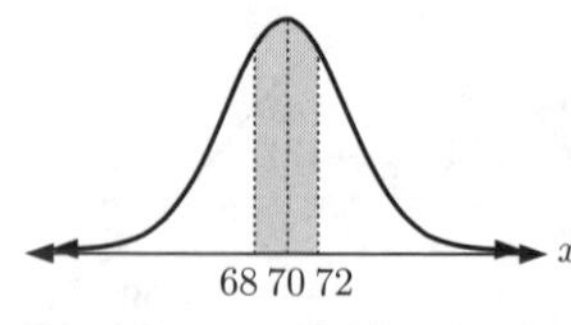


$P(68 \leqslant X \leqslant 72) \approx 0.383$

**c**

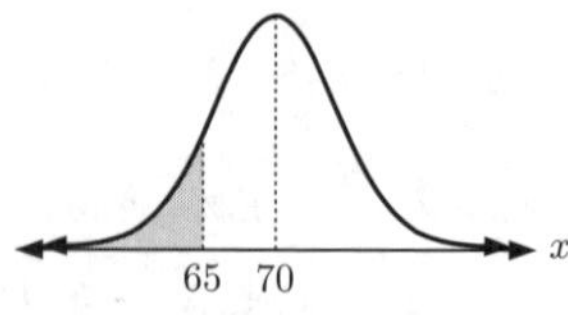


$P(X \leqslant 65) \approx 0.106$

**2** **a**

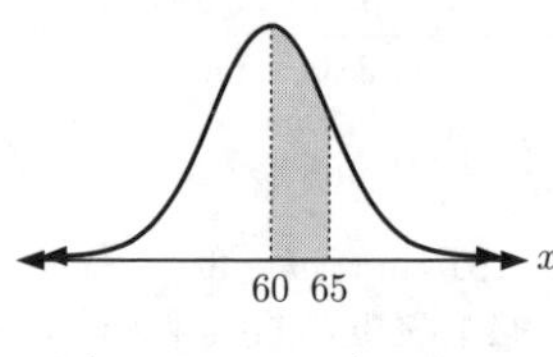


$P(60 \leqslant X \leqslant 65) \approx 0.341$

**b**

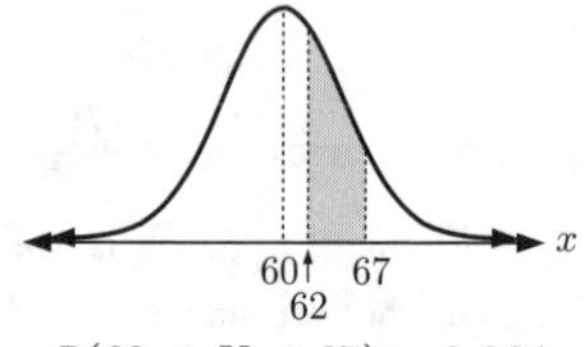


$P(62 \leqslant X \leqslant 67) \approx 0.264$

**c**

60 64

$P(X \geqslant 64) \approx 0.212$

**d**

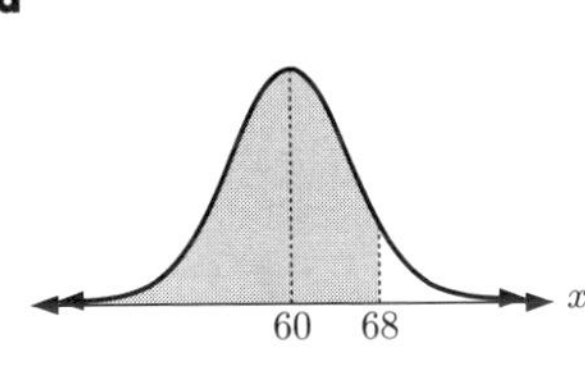


$P(X \leqslant 68) \approx 0.945$

**e**

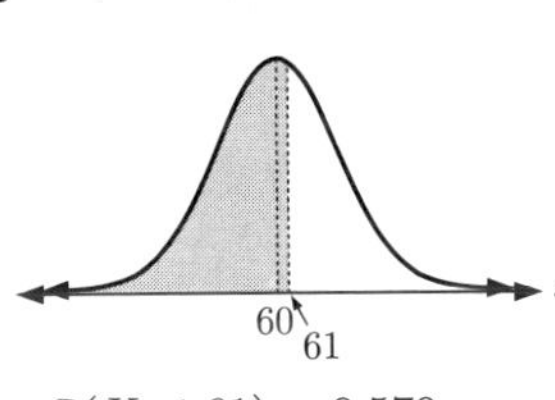


$P(X \leqslant 61) \approx 0.579$

**f**

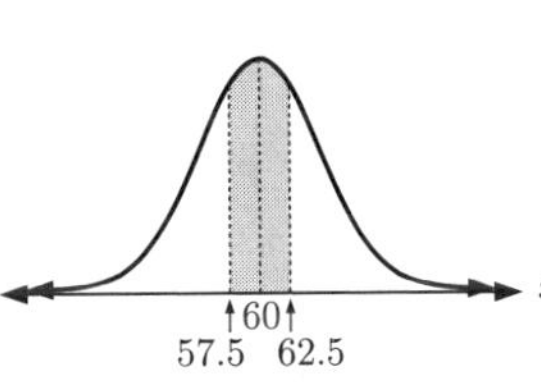


$P(57.5 \leqslant X \leqslant 62.5) \approx 0.383$

**3** If $X$ is the length of a bolt in cm, then $X$ is normally distributed with $\mu = 19.8$ and $\sigma = 0.3$.
$\therefore$ $P(19.7 < X < 20) \approx 0.378$

**4** If $X$ is the money collected in dollars, then $X$ is normally distributed with $\mu = 40$ and $\sigma = 6$.

**a** $P(30.00 < X < 50.00) \approx 0.904$
$\approx 90.4\%$

**b** $P(X \geqslant 50) \approx 0.0478$
$\approx 4.78\%$

**5** If $X$ is the length of an eel in cm, then $X$ is normally distributed with $\mu = 41$ and $\sigma = \sqrt{11}$.

**a** $P(X \geqslant 50) \approx 0.003\,33$

**b** $P(40 \leqslant X \leqslant 50) \approx 0.615$
$\approx 61.5\%$

**c** $P(X \geqslant 45) \approx 0.114$
So, we would expect $200 \times 0.114 \approx 23$ eels to be at least 45 cm long.

**6** If $X$ is the speed of a car in km h$^{-1}$ then $X$ is normally distributed with $\mu = 56.3$ and $\sigma = 7.4$.

**a** $P(60 < X < 75) \approx 0.303$ **b** $P(X \leqslant 70) \approx 0.968$ **c** $P(X \geqslant 60) \approx 0.309$

**7** If $X$ is the weight of an apple in grams, then $X$ is normally distributed with $\mu = 173$ and $\sigma = 34$.

**a**

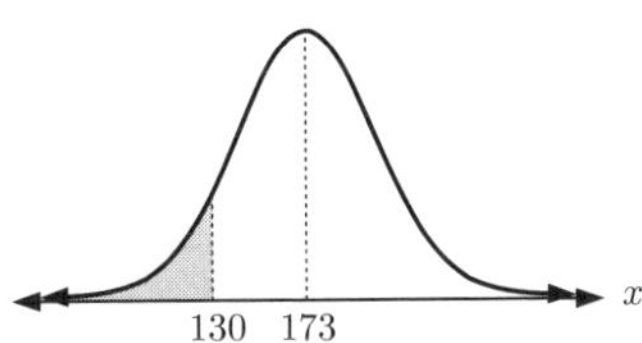


$P(X < 130) \approx 0.102\,988\,39$
$\approx 0.103$

So, 10.3% of the apples from this crop were too small to sell.

**b** The chance of one apple being too small to sell is $0.102\,988\,39$.
$\therefore$ the distribution is $B(100, 0.102\,988\,39)$
$\therefore$ $P(X \leqslant 10) \approx 0.544$
So, the probability that up to 10 apples were too small to sell is $0.544$.

**8**

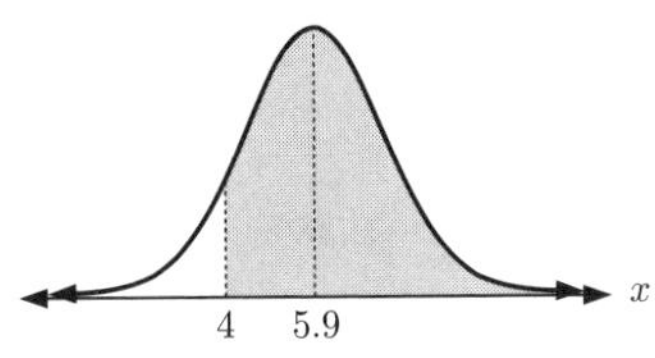


If $X$ is the drop in blood pressure (in units) then $X$ is normally distributed with $\mu = 5.9$ and $\sigma = 1.9$

**a** $P(X \geqslant 4) \approx 0.841\,344\,74$
So, 84.1% of people show a drop of more than 4 units.

**b** The chance of one person showing a drop of more than 4 units is $0.841\,344\,74$.
$\therefore$ the distribution is $B(8, 0.841\,344\,74)$
$\therefore$ $P(X > 5) = P(X \geqslant 6)$
$\approx 0.880$

So, the probability that more than 5 people show a drop of more than 4 units is $0.880$.

## EXERCISE 24C

**1** **a** For English, $z\text{-score} = \dfrac{48-40}{4.4} \approx 1.82$

For Mandarin, $z\text{-score} = \dfrac{81-60}{9} \approx 2.33$

For Geography, $z\text{-score} = \dfrac{84-55}{18} \approx 1.61$

For Biology, $z\text{-score} = \dfrac{68-50}{20} = 0.9$

For Maths, $z\text{-score} = \dfrac{84-50}{15} \approx 2.27$

**b** Mandarin, Maths, English, Geography, Biology

**2** **a** For Physics, $Z = \dfrac{73-78}{10.8} \approx -0.463$

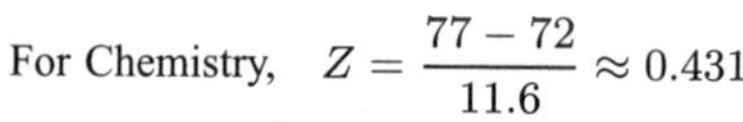

For Chemistry, $Z = \dfrac{77-72}{11.6} \approx 0.431$

For Maths, $Z = \dfrac{76-74}{10.1} \approx 0.198$

For German, $Z = \dfrac{91-86}{9.6} \approx 0.521$

For Biology, $Z = \dfrac{58-62}{5.2} \approx -0.769$

**b** German, Chemistry, Maths, Physics, Biology

**3** **a**

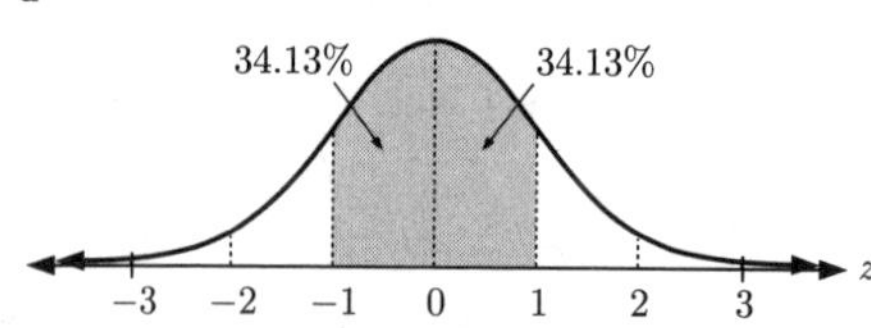


$\therefore$ $\text{P}(-1 < Z < 1) \approx 34.13\% + 34.13\%$
$\approx 0.683$

**b**

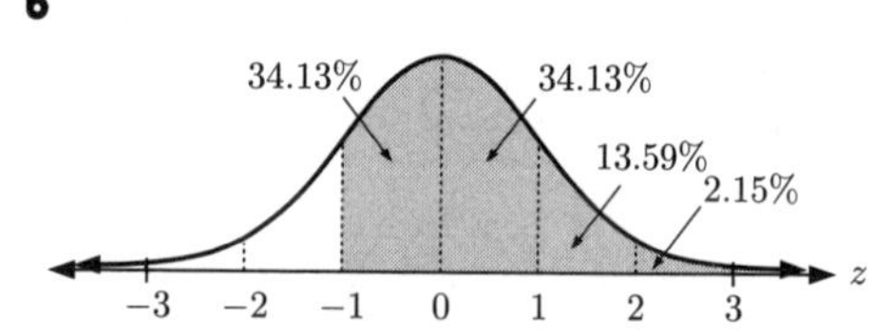


$\therefore$ $\text{P}(-1 \leqslant Z \leqslant 3)$
$\approx 34.13\% + 34.13\% + 13.59\% + 2.15\%$
$\approx 0.84$

**c**

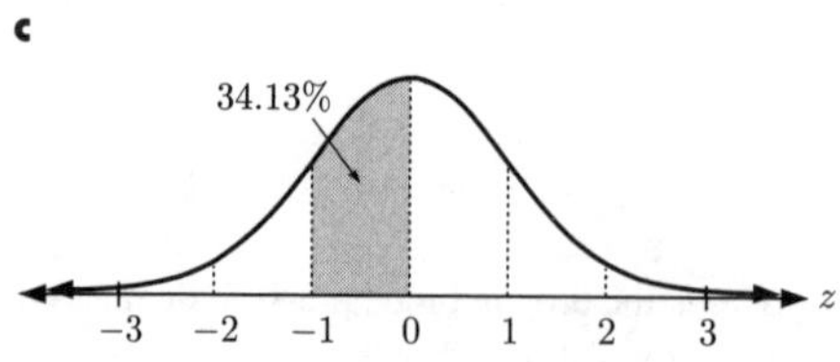


$\therefore$ $\text{P}(-1 < Z < 0) \approx 34.13\%$
$\approx 0.341$

**d**

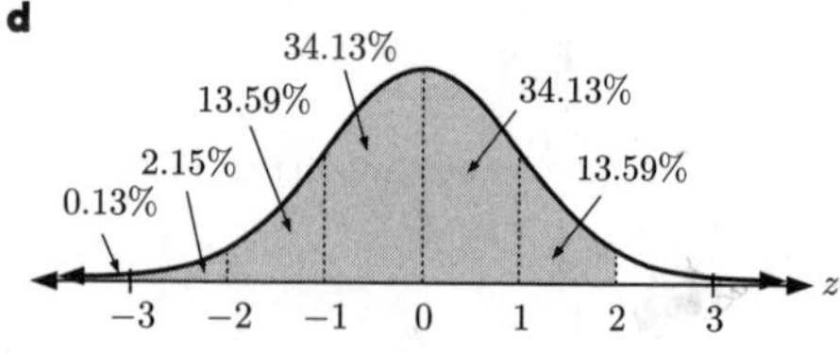


$\therefore$ $\text{P}(Z < 2)$
$\approx 0.13\% + 2.15\% + 13.59\% + 34.13\%$
$+ 34.13\% + 13.59\%$
$\approx 97.72\%$
$\approx 0.977$

**e**

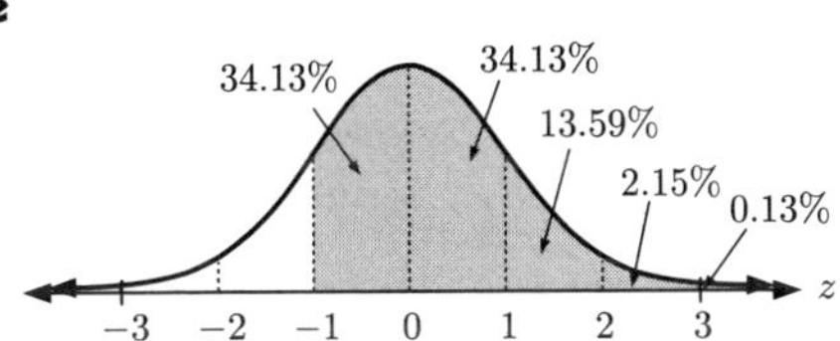


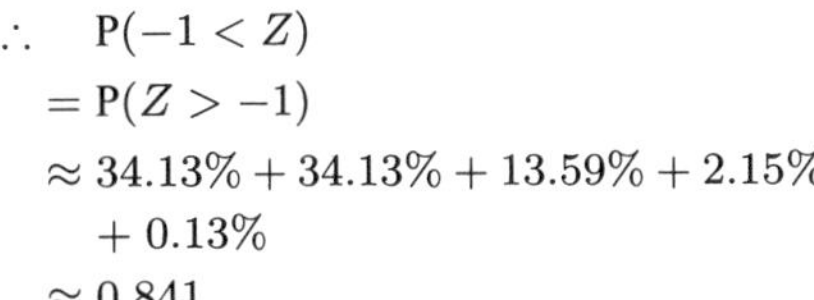

$\therefore \quad \text{P}(-1 < Z)$

$= \text{P}(Z > -1)$

$\approx 34.13\% + 34.13\% + 13.59\% + 2.15\%$
$+ 0.13\%$

$\approx 0.841$

**f**

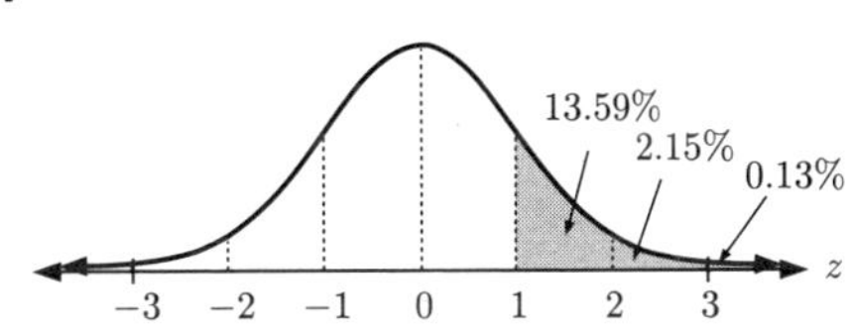


$\therefore \quad \text{P}(Z \geqslant 1)$

$\approx 13.59\% + 2.15\% + 0.13\%$

$\approx 0.159$

**4** **a** If $\text{P}(\mu - \sigma < X < \mu + 2\sigma) = \text{P}(a < Z < b)$

then $\dfrac{(\mu - \sigma) - \mu}{\sigma} = a$ and $\dfrac{(\mu + 2\sigma) - \mu}{\sigma} = b$

$\therefore \quad a = \dfrac{-\sigma}{\sigma} = -1$ $\qquad \therefore \quad b = \dfrac{2\sigma}{\sigma} = 2$

$\therefore \quad a = -1, \quad b = 2$

**b** If $\text{P}(\mu - 0.5\sigma < X < \mu) = \text{P}(a < Z < b)$

then $\dfrac{(\mu - 0.5\sigma) - \mu}{\sigma} = a$ and $\dfrac{\mu - \mu}{\sigma} = b$

$\therefore \quad a = \dfrac{-0.5\sigma}{\sigma} = -0.5$ $\qquad \therefore \quad b = 0$

$\therefore \quad a = -0.5, \quad b = 0$

**c** If $\text{P}(0 \leqslant Z \leqslant 3) = \text{P}(\mu - a\sigma \leqslant X \leqslant \mu + b\sigma)$

then $\dfrac{(\mu - a\sigma) - \mu}{\sigma} = 0$ and $\dfrac{(\mu + b\sigma) - \mu}{\sigma} = 3$

$\therefore \quad \mu - a\sigma - \mu = 0$ $\qquad \therefore \quad \mu + b\sigma - \mu = 3\sigma$

$\therefore \quad -a\sigma = 0$ $\qquad \therefore \quad b\sigma = 3\sigma$

$\therefore \quad a = 0$ $\qquad \therefore \quad b = 3$

$\therefore \quad a = 0, \quad b = 3$

**5** $Z \sim \text{N}(0, 1)$

**a**

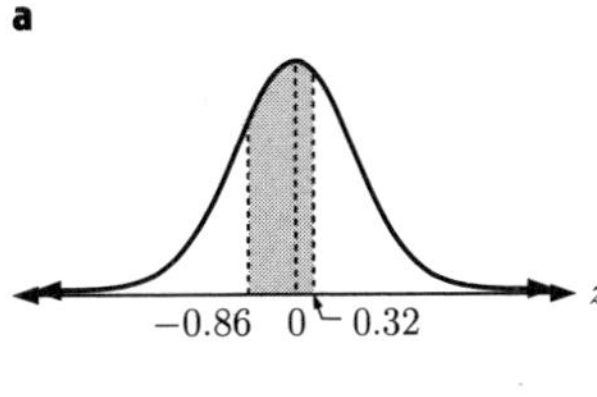


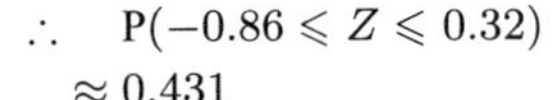

$\therefore \quad \text{P}(-0.86 \leqslant Z \leqslant 0.32)$

$\approx 0.431$

**b**

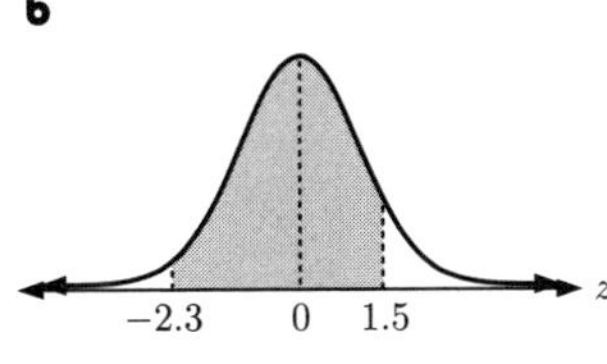


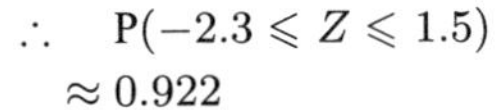

$\therefore \quad \text{P}(-2.3 \leqslant Z \leqslant 1.5)$

$\approx 0.922$

**c**

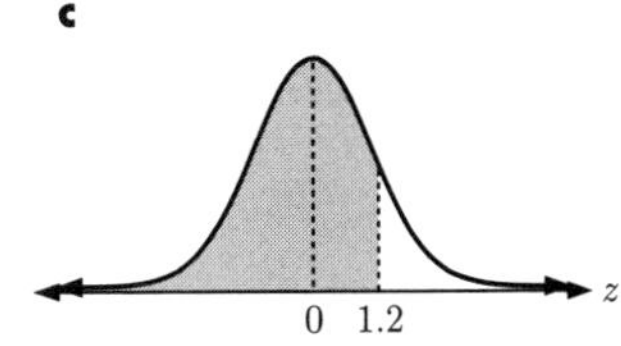


$\therefore \quad \text{P}(Z \leqslant 1.2)$

$\approx 0.885$

**d**

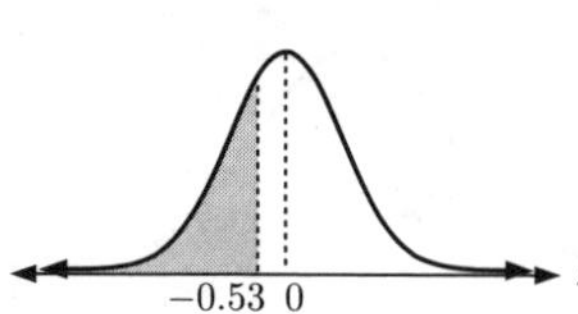


$\therefore \quad P(Z \leqslant -0.53)$
$\approx 0.298$

**e**

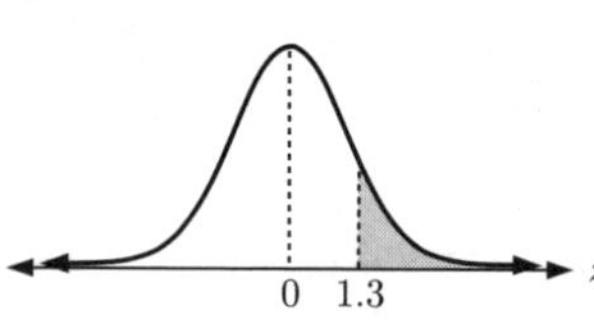


$\therefore \quad P(Z \geqslant 1.3)$
$\approx 0.0968$

**f**

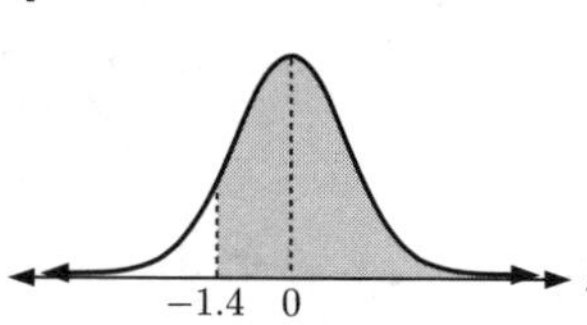


$\therefore \quad P(Z \geqslant -1.4)$
$\approx 0.919$

**g**

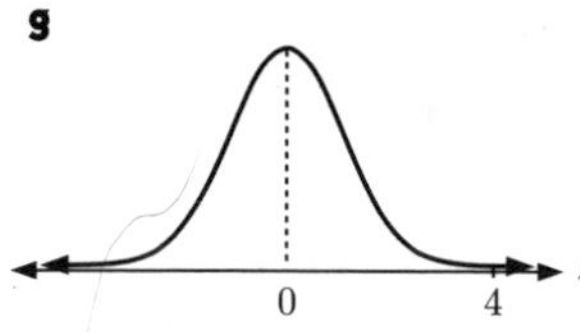


$P(Z > 4) \approx 0.000\,031\,7$
$(3.17 \times 10^{-5})$

**6** **a** **i**

$$z_1 = \frac{50.6 - 58.3}{8.96}$$
$$\therefore \quad z_1 \approx -0.859\,375$$
$$\approx -0.859$$
$$z_2 = \frac{68.9 - 58.3}{8.96}$$
$$\approx 1.183\,035\,714$$
$$\approx 1.18$$

**ii** $Z \sim N(0,\, 1)$

$$P(-0.859\,375 \leqslant Z \leqslant 1.183\,035\,714)$$
$$\approx 0.686\,535\,67$$
$$\approx 0.687$$

**b** $X \sim N(58.3,\, 8.96^2)$

$\therefore \quad P(50.6 \leqslant X \leqslant 68.9) \approx 0.687$ ✓

## EXERCISE 24D.1

**1** **a**

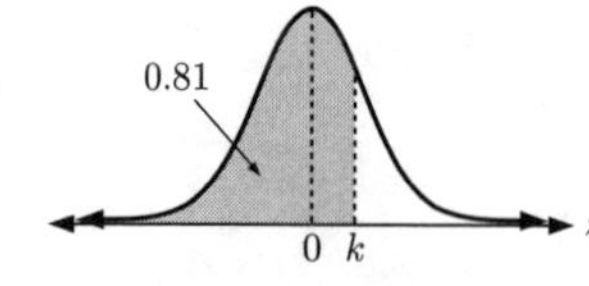


$P(Z \leqslant k) = 0.81$
$\therefore \quad k \approx 0.878$

**b**

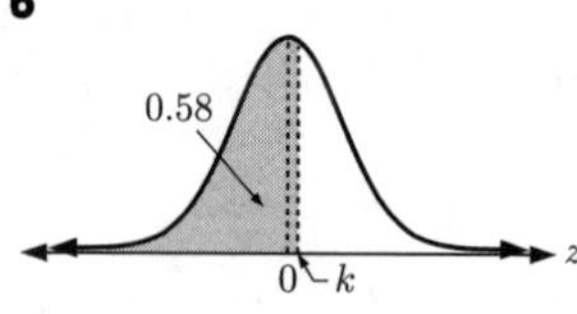


$P(Z \leqslant k) = 0.58$
$\therefore \quad k \approx 0.202$

**c**

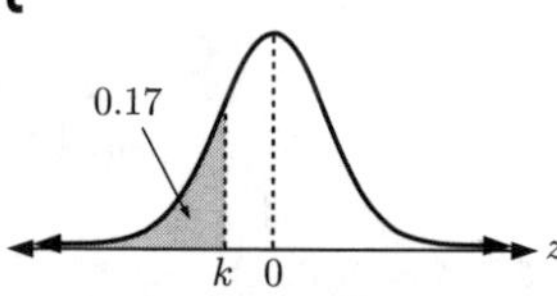


$P(Z \leqslant k) = 0.17$
$\therefore \quad k \approx -0.954$

**2** **a**

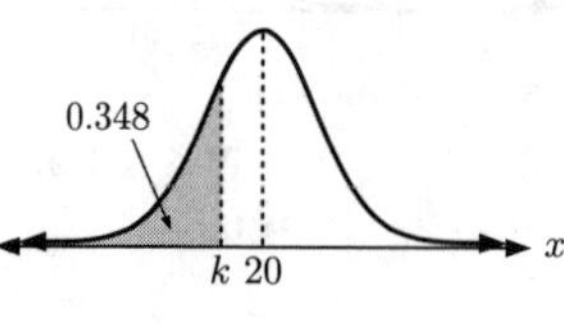


$P(X \leqslant k) = 0.348$
$\therefore \quad k \approx 18.8$

**b**

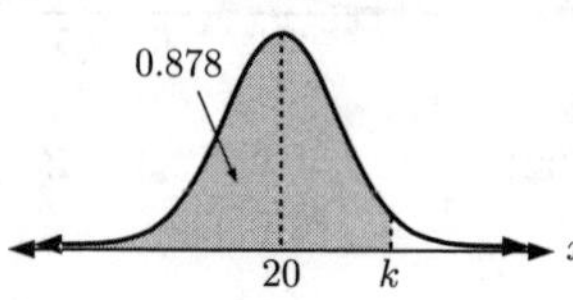


$P(X \leqslant k) = 0.878$
$\therefore \quad k \approx 23.5$

**c**

0.5
20
k
x

$P(X \leqslant k) = 0.5$
$\therefore \quad k = 20$

**3** **a**

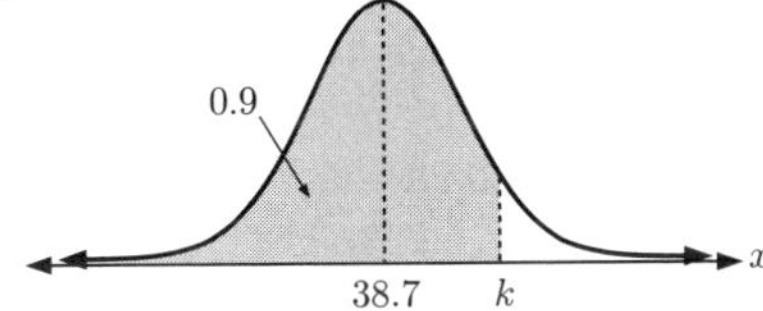


$\mathrm{P}(X \leqslant k) = 0.9$
$\therefore\ k \approx 49.2$

**b**

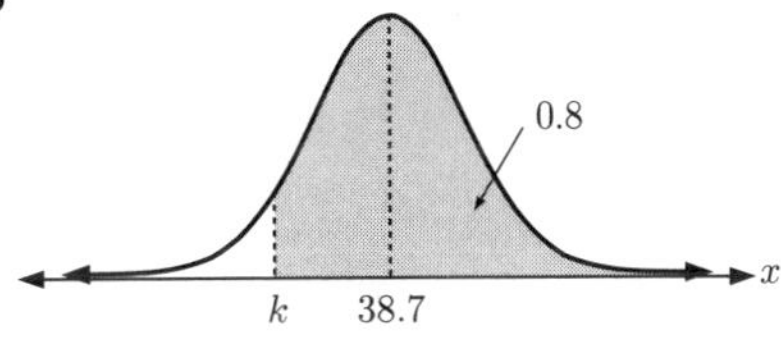


$\mathrm{P}(X \geqslant k) = 0.8$
$\therefore\ \mathrm{P}(X \leqslant k) = 1 - 0.8 = 0.2$
$\therefore\ k \approx 31.8$

**4** **a**

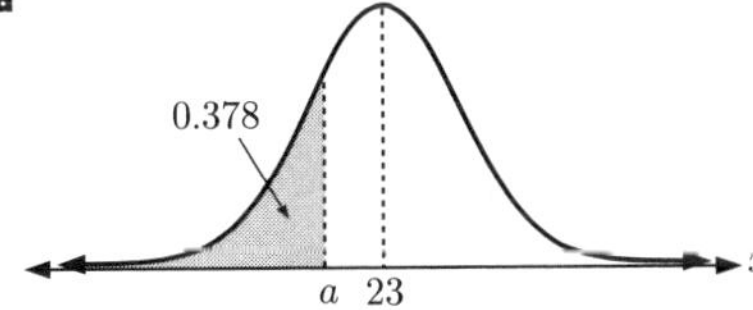


$\mathrm{P}(X < a) = 0.378$
$\therefore\ a \approx 21.4$

**b**

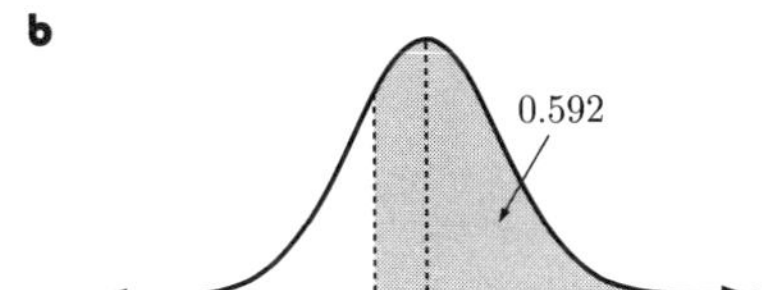


$\mathrm{P}(X \geqslant a) = 0.592$
$\therefore\ \mathrm{P}(X \leqslant a) = 1 - 0.592$
$= 0.408$
$\therefore\ a \approx 21.8$

**c**

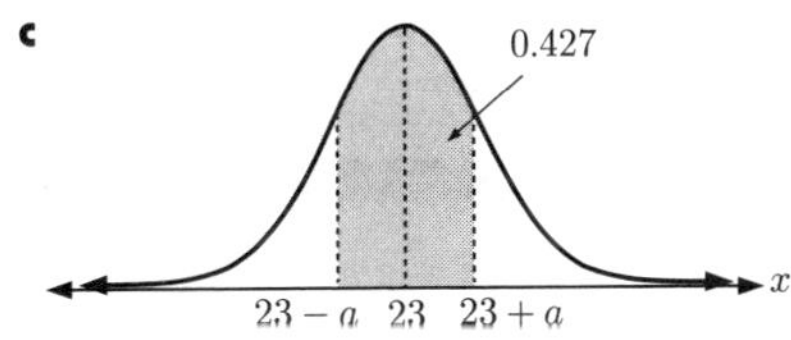


$\mathrm{P}(23 - a < X < 23 + a) = 0.427$
$\therefore\ 1 - 2 \times \mathrm{P}(X \leqslant 23 - a) = 0.427$
$\therefore\ -2 \times \mathrm{P}(X \leqslant 23 - a) = -0.573$
$\therefore\ \mathrm{P}(X \leqslant 23 - a) = 0.2865$
$\therefore\ 23 - a = 20.181\,806\,2$
$\therefore\ a \approx 23 - 20.181\,806\,2$
$\therefore\ a \approx 2.82$

**5** Let $X$ be the result of the Physics test, so $X \sim \mathrm{N}(46, 25^2)$.

We need to find $k$ such that $\mathrm{P}(X \geqslant k) = 0.07$
$\therefore\ 1 - \mathrm{P}(X < k) = 0.07$
$\therefore\ \mathrm{P}(X < k) = 0.93$
$\therefore\ k \approx 82.894$
$\therefore\ k \approx 82.9$

**6**

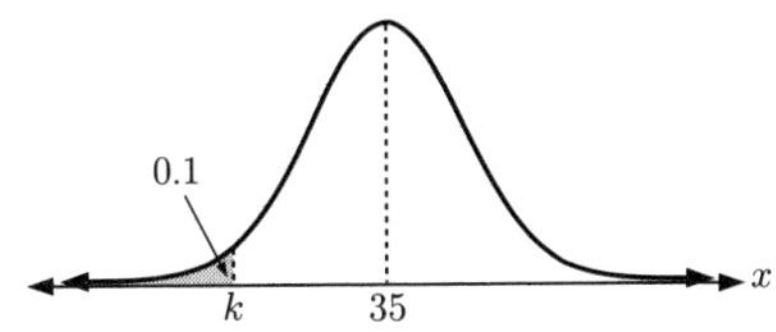


$X \sim \mathrm{N}(35, 8^2)$

We need to find $k$ such that

$\mathrm{P}(X \leqslant k) = 0.1$
$\therefore\ k \approx 24.747\,587\,5$

So, the length of the smallest fish to be harvested is 24.7 cm.

**7** $X \sim \text{N}(75, 0.1^2)$

We need to find $k$ such that

$\text{P}(X \geqslant k) = 0.01$

$\therefore \text{P}(X \leqslant k) = 0.99$

$\therefore k \approx 75.232\,634\,8$

So, the length of the smallest screw to be rejected is 75.2 mm.

**8** $Z$-score for algebra $= \dfrac{56 - 50.2}{15.8} \approx 0.3671$ $\qquad$ $Z$-score for geometry $= \dfrac{x - 58.7}{18.7}$

$\therefore$ we need to solve $\dfrac{x - 58.7}{18.7} = 0.3671$

$\therefore x - 58.7 \approx 6.86$

$\therefore x \approx 65.6$ $\qquad$ So, Pedro needs a result of 65.6%.

**9**

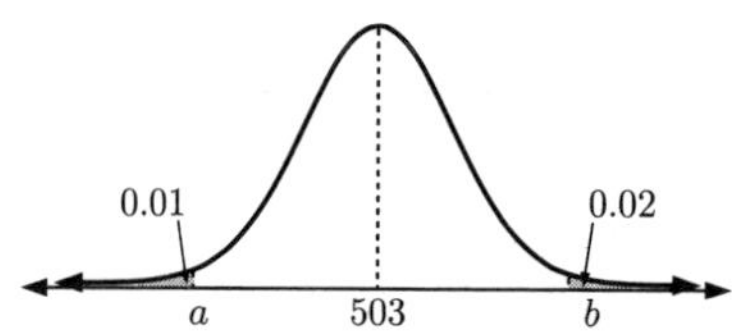


$X \sim \text{N}(503, 0.5^2)$

We need to find $a$ such that

$\text{P}(X \leqslant a) = 0.01$

$\therefore a \approx 502$

We also need to find $b$ such that $\text{P}(X \geqslant b) = 0.02$

$\therefore \text{P}(X \leqslant b) = 0.98$

$\therefore b \approx 504$

So, the range of volumes in the bottles that are kept is 502 mL to 504 mL.

## EXERCISE 24D.2

**1** Let the mean IQ of a student at school be $\mu$.

If $X$ is the IQ of a student at the school, then $X \sim \text{N}(\mu, 15^2)$.

Now, $\text{P}(X \geqslant 125) = 0.2$

$\therefore \text{P}\left(\dfrac{X - \mu}{15} \geqslant \dfrac{125 - \mu}{15}\right) = 0.2$

$\therefore \text{P}\left(Z \geqslant \dfrac{125 - \mu}{15}\right) = 0.2$

$\therefore \text{P}\left(Z < \dfrac{125 - \mu}{15}\right) = 0.8$

$\therefore \dfrac{125 - \mu}{15} \approx 0.8416$

$\therefore \mu \approx 112.4$

The mean IQ at the school is 112.

**2** Let the standard deviation of the distances jumped be $\sigma$ m.

If $X$ is the distance jumped by the athlete, then $X \sim \text{N}(5.2, \sigma^2)$.

Now, $\text{P}(X < 5) = 0.15$

$\therefore \text{P}\left(\dfrac{X - 5.2}{\sigma} < \dfrac{5 - 5.2}{\sigma}\right) = 0.15$

$\therefore \text{P}\left(Z < -\dfrac{0.2}{\sigma}\right) = 0.15$

$\therefore -\dfrac{0.2}{\sigma} \approx -1.036$

$\therefore \sigma \approx 0.193$

So, the standard deviation of the distances jumped is 0.193 m.

**3** Let the standard deviation of the weekly income be $\$\sigma$.

If $X$ denotes the weekly income of the bakery, then $X \sim \text{N}(6100,\ \sigma^2)$.

$$\text{Now,}\quad \text{P}(X \geqslant 6000) = 0.85$$

$$\therefore\ \text{P}\left(Z \geqslant \frac{6000-6100}{\sigma}\right) = 0.85$$

$$\therefore\ \text{P}\left(Z < \frac{6000-6100}{\sigma}\right) = 0.15$$

Using invNorm for $\text{N}(0,\ 1^2)$, $\dfrac{-100}{\sigma} \approx -1.036\,433\,4$

$$\therefore\ \sigma \approx \frac{-100}{-1.036\,433\,4}$$

$$\therefore\ \sigma \approx 96.5$$

So, the standard deviation is \$96.50.

**4** Let the mean arrival time be $\mu$ minutes after midday.

If $X$ denotes the arrival time of a bus, then $X \sim \text{N}(\mu,\ 5^2)$.

Now, $\text{P}(X \leqslant 235) = 0.1$ {3:55 pm $= 3 \times 60 + 55 = 235$ minutes after midday}

$$\therefore\ \text{P}\left(Z \leqslant \frac{235-\mu}{5}\right) = 0.1$$

Using invNorm for $\text{N}(0,\ 1^2)$, $\dfrac{235-\mu}{5} \approx -1.281\,551\,6$

$$\therefore\ 235-\mu \approx -6.407\,758$$

$$\therefore\ \mu \approx 235 + 6.407\,758$$

$$\therefore\ \mu \approx 241.407\,758 \text{ minutes after midday}$$

and $241.407\,758$ minutes $= 4$ h $1$ m $24$ s

So, the mean arrival time of buses at the depot is 4:01:24 pm.

**5** **a** $X \sim \text{N}(\mu,\ \sigma^2)$ where we have to find $\mu$ and $\sigma$.

We start by finding $z_1$ and $z_2$ which correspond to $x_1 = 30$ and $x_2 = 80$.

Now $\text{P}(X \leqslant 30) = 0.15$

$$\therefore\ \text{P}\left(\frac{X-\mu}{\sigma} \leqslant \frac{30-\mu}{\sigma}\right) = 0.15$$

$$\therefore\ \text{P}\left(Z \leqslant \frac{30-\mu}{\sigma}\right) = 0.15$$

$$\therefore\ z_1 = \frac{30-\mu}{\sigma} \approx -1.0364$$

$$\therefore\ 30-\mu \approx -1.0364\sigma \ \ldots.\ (1)$$

and $\text{P}(X \geqslant 80) = 0.1$

$$\therefore\ \text{P}(X < 80) = 0.9$$

$$\therefore\ \text{P}\left(\frac{X-\mu}{\sigma} < \frac{80-\mu}{\sigma}\right) = 0.9$$

$$\therefore\ \text{P}\left(Z < \frac{80-\mu}{\sigma}\right) = 0.9$$

$$\therefore\ z_2 = \frac{80-\mu}{\sigma} \approx 1.2816$$

$$\therefore\ 80-\mu \approx 1.2816\sigma \ \ldots.\ (2)$$

Solving (1) and (2) simultaneously, $\mu \approx 52.36 \approx 52.4$ and $\sigma \approx 21.57 \approx 21.6$.

**b** Let $X$ be the result of the mathematics exam.

$X$ is normally distributed with mean $\mu$ and standard deviation $\sigma$.

We know that $\text{P}(X \geqslant 80) = 0.1$ and $\text{P}(X \leqslant 30) = 0.15$.

So, from **a**, $\mu \approx 52.4$ and $\sigma \approx 21.6$.

If part marks can be given, $\text{P}(X > 50) \approx 0.544$

$$\approx 54.4\%$$

So, 54.4% of students scored more than 50.

**6** **a**

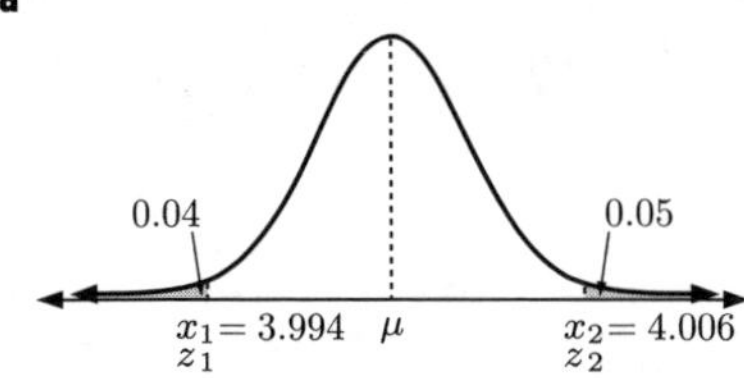


$X \sim \text{N}(\mu, \sigma^2)$ where we have to find $\mu$ and $\sigma$.
We find $z_1$ and $z_2$ which correspond to $x_1 = 3.994$ and $x_2 = 4.006$

Now $\text{P}(X \leqslant x_1) = 0.04$ and $\text{P}(X \geqslant x_2) = 0.05$

$\therefore\ \text{P}\left(Z \leqslant \dfrac{3.994 - \mu}{\sigma}\right) = 0.04$ $\qquad \therefore\ \text{P}\left(Z \leqslant \dfrac{4.006 - \mu}{\sigma}\right) = 0.95$

$\therefore\ \dfrac{3.994 - \mu}{\sigma} = -1.750\,686\,1$ $\qquad \therefore\ \dfrac{4.006 - \mu}{\sigma} = 1.644\,853\,63$

$\therefore\ 3.994 - \mu = -1.750\,686\,1\sigma$ .... (1) $\qquad \therefore\ 4.006 - \mu = 1.644\,853\,63\sigma$ .... (2)

Solving simultaneously, $\mu \approx 4.000\,187\,009$ and $\sigma \approx 0.003\,534\,047\,88$

$\therefore\ \mu \approx 4.00$ cm and $\sigma \approx 0.003\,53$ cm

**b** From **a**, $\mu \approx 4.000$ and $\sigma \approx 0.003\,534$

$\therefore\ X \sim \text{N}(4.000, 0.003\,534^2)$

$\therefore\ \text{P}(3.997 \leqslant X \leqslant 4.003) \approx 0.604$

So, the probability that a randomly chosen piston has diameter between 3.997 cm and 4.003 cm is 0.604.

**7** **a** $X \sim \text{N}(\mu, \sigma^2)$ where we have to find $\mu$ and $\sigma$.

We start by finding $z_1$ and $z_2$ which correspond to $x_1 = 1.94$ and $x_2 = 2.06$.

Now $\text{P}(X < 1.94) = 0.02$ and $\text{P}(X > 2.06) = 0.03$

$\therefore\ \text{P}\left(\dfrac{X - \mu}{\sigma} < \dfrac{1.94 - \mu}{\sigma}\right) = 0.02$ $\qquad \therefore\ \text{P}\left(\dfrac{X - \mu}{\sigma} > \dfrac{2.06 - \mu}{\sigma}\right) = 0.03$

$\therefore\ \text{P}\left(Z < \dfrac{1.94 - \mu}{\sigma}\right) = 0.02$ $\qquad \therefore\ \text{P}\left(Z > \dfrac{2.06 - \mu}{\sigma}\right) = 0.03$

$\therefore\ z_1 = \dfrac{1.94 - \mu}{\sigma} \approx -2.054$ $\qquad \therefore\ \text{P}\left(Z \leqslant \dfrac{2.06 - \mu}{\sigma}\right) = 0.97$

$\therefore\ 1.94 - \mu \approx -2.054\sigma$ .... (1) $\qquad \therefore\ z_2 = \dfrac{2.06 - \mu}{\sigma} \approx 1.881$

$2.06 - \mu \approx 1.881\sigma$ .... (2)

Solving (1) and (2) simultaneously, we get $\mu \approx 2.00$ cm and $\sigma \approx 0.0305$ cm.

**b** Let $Y$ be the number of tokens which will not operate the machine. This is a binomial situation with the probability $p = 0.02 + 0.03 = 0.05$ of failure to operate and $n = 20$. So, $Y \sim \text{B}(20, 0.05)$.

$\therefore$ P(at most one will not operate) $= \text{P}(Y \leqslant 1)$

$\approx 0.736$

## REVIEW SET 24A

**1** $X$ is the height of a 17 year old boy.

$X$ is normally distributed with $\mu = 179$ cm and $\sigma = 8$ cm.

**a**

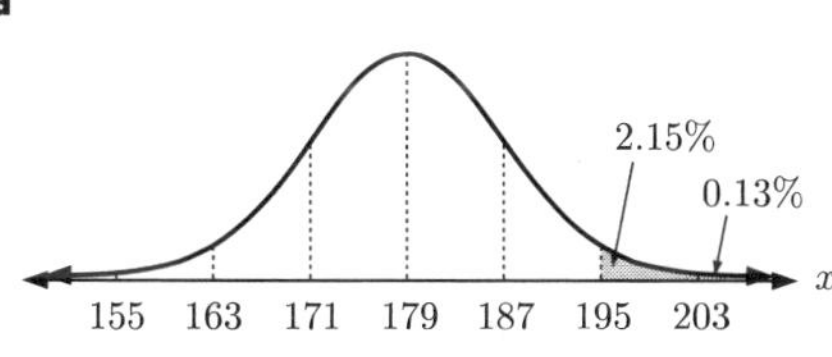


$P(X \geqslant 195) \approx 2.15\% + 0.13\%$
$\approx 2.28\%$

**b**

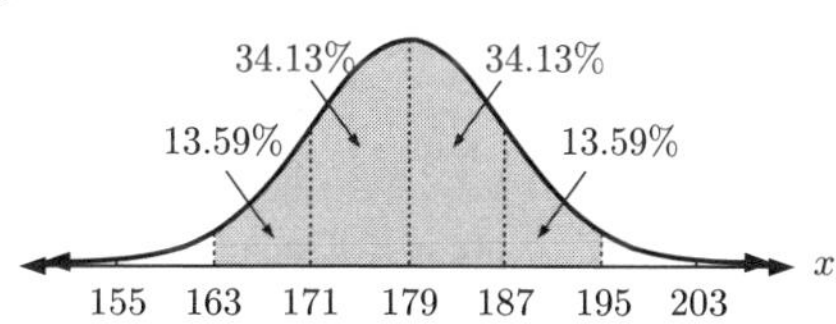


$P(163 \leqslant X \leqslant 195)$
$\approx 13.59\% + 34.13\% + 34.13\% + 13.59\%$
$\approx 95.44\%$
$\approx 95.4\%$

**c**

$P(171 \leqslant X \leqslant 187) \approx 34.13\% + 34.13\%$
$\approx 68.26\%$
$\approx 68.3\%$

**2** If $X$ is the contents of the container in mL, then $X \sim N(377, 4.2^2)$.

**a** **i** $P(X < 368.6)$
$\approx 2.15\% + 0.13\%$
$\approx 2.28\%$

**ii** $P(372.8 < X < 389.6)$
$\approx 2 \times 34.13\% + 13.59\% + 2.15\%$
$\approx 84.0\%$

**b** $P(377 < X < 381.2)$
$\approx 0.341$

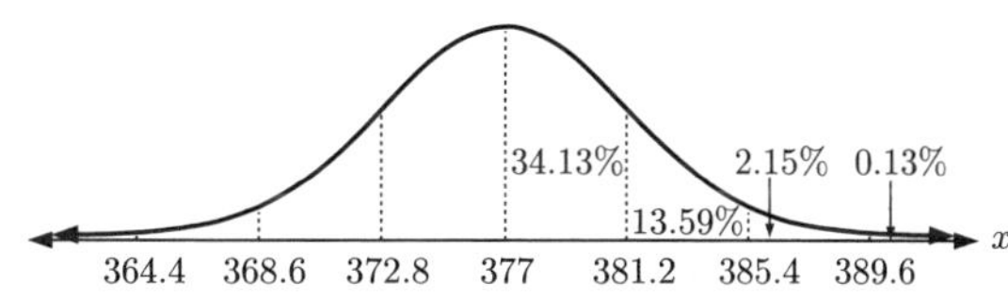


**3** If $X$ is the mass of a Coffin Bay Oyster, then $X \sim N(38.6, 6.3^2)$.

**a**

$$P(38.6 - a \leqslant X \leqslant 38.6 + a) = 0.6826$$

$$\therefore \quad P\left(\frac{38.6 - a - 38.6}{6.3} \leqslant \frac{X - 38.6}{6.3} \leqslant \frac{38.6 + a - 38.6}{6.3}\right) = 0.6826$$

$$\therefore \quad P\left(-\frac{a}{6.3} \leqslant Z \leqslant \frac{a}{6.3}\right) = 0.6826$$

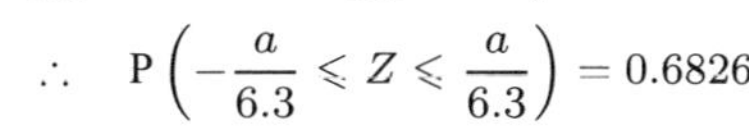

$$\therefore \text{ by symmetry, } \quad P\left(Z \leqslant -\frac{a}{6.3}\right) = \frac{1 - 0.6826}{2}$$

$$\therefore \quad P\left(Z \leqslant -\frac{a}{6.3}\right) = 0.1587 \quad .... (*)$$

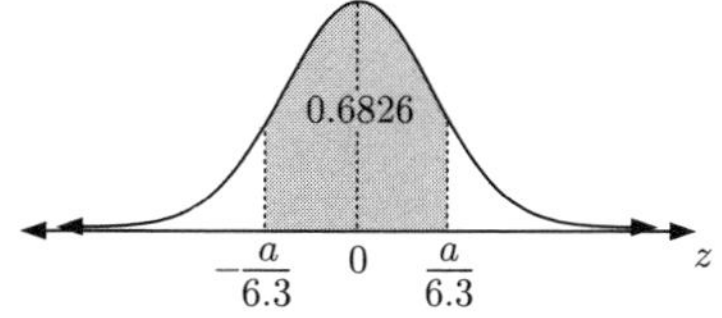


$$\therefore \quad -\frac{a}{6.3} \approx -1.00$$

$$\therefore \quad a \approx 6.30 \text{ g}$$

**b** $\quad \mathrm{P}(X \geqslant b) = 0.8413$

$\therefore \quad \mathrm{P}(X < b) = 0.1587$

$\therefore \quad \mathrm{P}\left(\dfrac{X - 38.6}{6.3} < \dfrac{b - 38.6}{6.3}\right) = 0.1587$

$\therefore \quad \mathrm{P}\left(Z < \dfrac{b - 38.6}{6.3}\right) = 0.1587$

Comparing with $(*)$, $\dfrac{b - 38.6}{6.3} = -\dfrac{a}{6.3}$

$\therefore \quad b - 38.6 \approx -6.30$

$\therefore \quad b \approx 32.3$ g

**4** **a** Harri's score is 2 standard deviations below the mean.

**b**

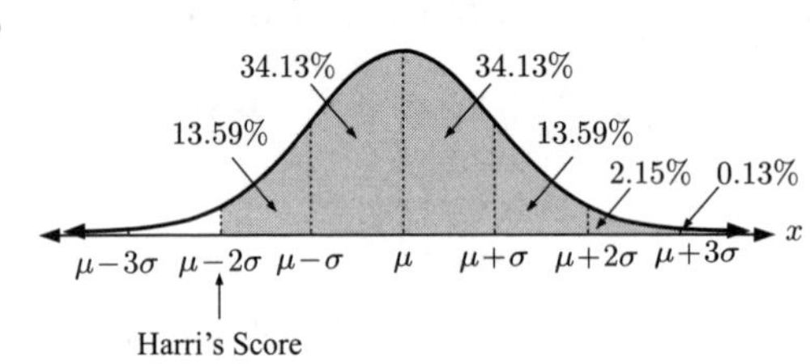


Proportion of students who scored better than Harri

$\approx 13.59\% + 34.13\% + 34.13\% + 13.59\% + 2.15\% + 0.13\%$

$\approx 97.72\%$

$\approx 97.7\%$

**c** $\mu = 151$ and $\mu - 2\sigma = 117$

$\therefore \quad 151 - 2\sigma = 117$

$\therefore \quad -2\sigma = -34$

$\therefore \quad \sigma = 17$

The standard deviation was 17.

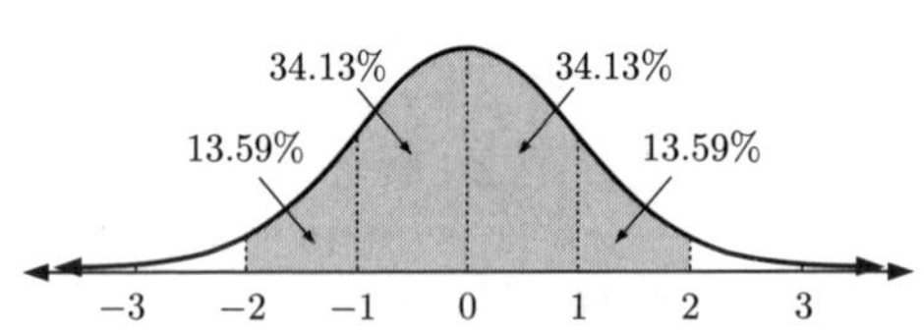


The shaded part of the diagram has an area of approximately 0.95.

$\therefore \quad \mathrm{P}(-2 \leqslant Z \leqslant 2) \approx 0.95$

$\therefore \quad k \approx 2$

**6** Jarrod's $z$-score is $\dfrac{41 - 35}{4} = 1.5$

$\therefore$ Paul needs $x$ such that $\dfrac{x - 25}{3} = 1.5$

$\therefore \quad x = 25 + 4.5 = 29.5$

Paul needs to throw a tennis ball 29.5 m to perform as well as Jarrod.

**7**

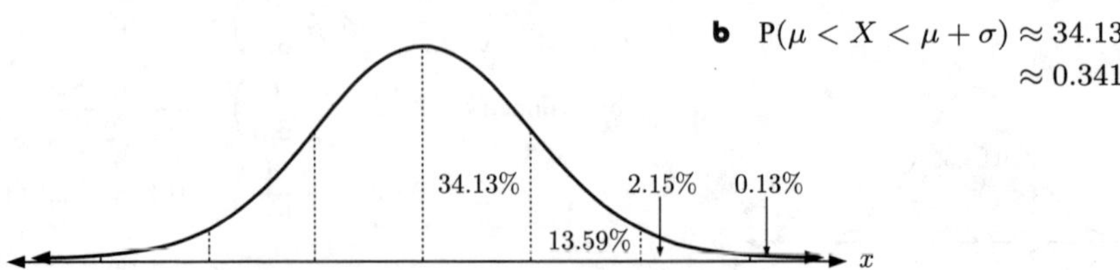


**a** $\mathrm{P}(\mu + \sigma < X < \mu + 2\sigma) \approx 13.59\%$

$\approx 0.136$

**b** $\mathrm{P}(\mu < X < \mu + \sigma) \approx 34.13\%$

$\approx 0.341$

**8** If $X$ is the number of bottles sold per day, then $X \sim \text{N}(2500,\ 300^2)$.

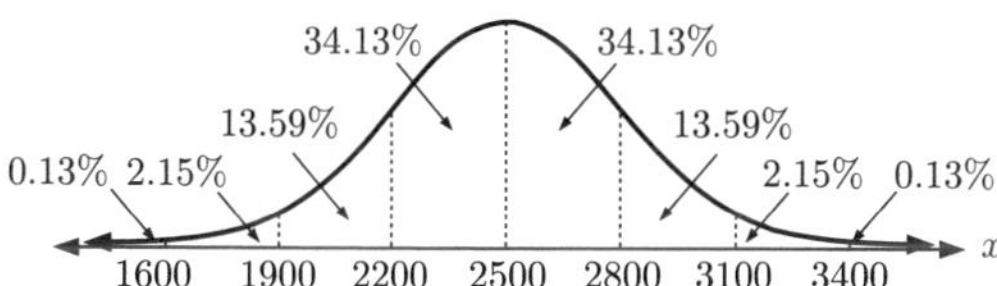


**a** $\text{P}(X < 1900) \approx 0.13\% + 2.15\%$
$\approx 2.28\%$

**b** $\text{P}(X > 2200)$
$\approx 2 \times 34.13\% + 13.59\% + 2.15\% + 0.13\%$
$\approx 84.13\%$
$\approx 84.1\%$

**c** $\text{P}(2200 \leqslant X \leqslant 3100) \approx 34.13\% + 34.13\% + 13.59\%$
$\approx 81.85\%$
$\approx 81.9\%$

## REVIEW SET 24B

**1** $X \sim \text{N}(150,\ 12^2)$

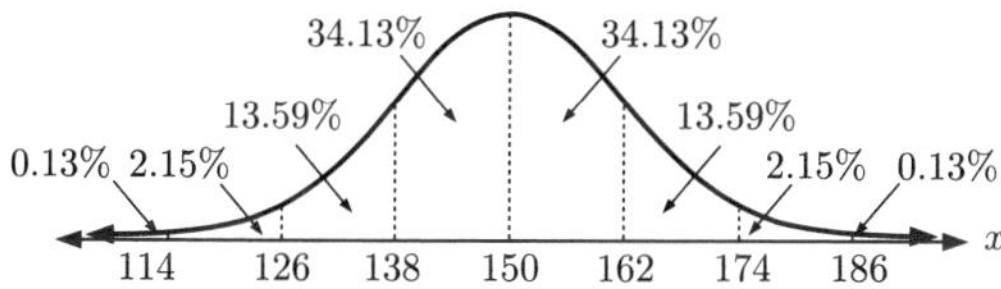


**a** $\text{P}(138 \leqslant X \leqslant 162)$
$\approx 34.13\% + 34.13\%$
$\approx 68.26\%$
$\approx 68.3\%$

**b** $\text{P}(126 \leqslant X \leqslant 174)$
$\approx 13.59\% + 34.13\% + 34.13\% + 13.59\%$
$\approx 95.44\%$
$\approx 95.4\%$

**c** $\text{P}(126 \leqslant X \leqslant 162)$
$\approx 13.59\% + 34.13\% + 34.13\%$
$\approx 81.85\%$
$\approx 81.9\%$

**d** $\text{P}(162 \leqslant X \leqslant 174)$
$\approx 13.59\%$
$\approx 13.6\%$

**2** If random variable $X$ is the arm length in cm then $X \sim \text{N}(64,\ 4^2)$.

**a** **i** $\text{P}(60 < X < 72)$
$\approx 2 \times 34.13\% + 13.59\%$
$\approx 81.9\%$

**ii** $\text{P}(X > 60)$
$\approx 50\% + 34.13\%$
$\approx 84.1\%$

**b** $\text{P}(56 < X < 64) \approx 0.3413 + 0.1359$
$\approx 0.477$

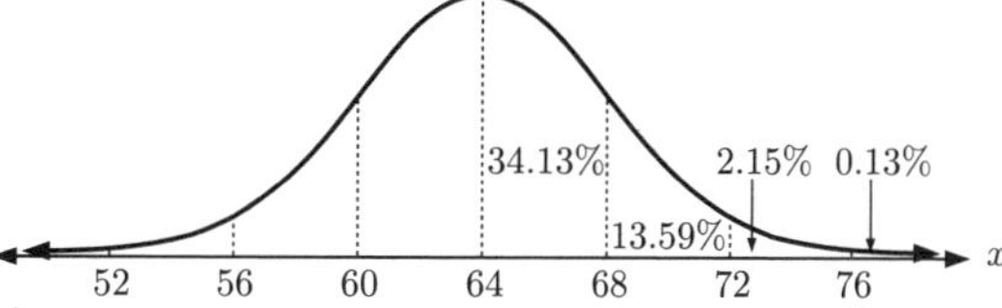


**c** $\text{P}(X > x) = 0.7$
$\therefore\ \text{P}(X \leqslant x) = 0.3$
$\therefore\ x \approx 61.9$ {using technology}

**3** If $X$ is the rod length in mm, then $X \sim \mathrm{N}(\mu, 3^2)$.

Now $\mathrm{P}(X < 25) = 0.02$

$\therefore \mathrm{P}\left(\frac{X-\mu}{3} < \frac{25-\mu}{3}\right) = 0.02$

$\therefore \mathrm{P}\left(Z < \frac{25-\mu}{3}\right) = 0.02$

$\therefore \frac{25-\mu}{3} \approx -2.0537$

$\therefore 25 - \mu \approx -6.161$

$\therefore \mu \approx 31.2$

$\therefore$ the mean rod length is 31.2 mm.

**4** **a** Since Area $A$ = Area $B$, 20 and 38 must be equal distances away from the mean $\mu$, because of the symmetry of the normal distribution.

$\therefore$ $\mu$ is halfway between 20 and 38, so $\mu = \frac{20+38}{2} = 29$.

Now $\mathrm{P}(X \leqslant 20) = 0.2$

$\therefore \mathrm{P}\left(\frac{X-29}{\sigma} \leqslant \frac{20-29}{\sigma}\right) = 0.2$

$\therefore \mathrm{P}\left(Z \leqslant -\frac{9}{\sigma}\right) = 0.2$

$\therefore -\frac{9}{\sigma} \approx -0.8416$

$\therefore \sigma \approx 10.69$

$\therefore \mu = 29, \quad \sigma \approx 10.7$

**b** Using the values obtained for $\mu$ and $\sigma$ in **a** and technology:

**i** $\mathrm{P}(X \leqslant 35) \approx 0.713$ **ii** $\mathrm{P}(23 \leqslant X \leqslant 30) \approx 0.250$

**5** $X \sim \mathrm{N}(503, 2^2)$

**a** $\mathrm{P}(X < 500)$

$\approx 0.066\,807\,2$

$\approx 0.0668$

So, approximately 6.68% of the bags are underweight.

**b** This is a binomial distribution where $X$ is the number of underweight bags,

$n = 20$ and $p = 0.066\,807\,2$

$\therefore \mathrm{P}(X \leqslant 2) \approx 0.854$ {using technology}

**6** If $X$ is the marks in the examination, then $X \sim \mathrm{N}(49, 15^2)$.

**a** $\mathrm{P}(X \geqslant 45) \approx 0.6051$

So, $2376 \times 0.6051 \approx 1438$ candidates passed the examination.

**b** Let $k$ be the minimum mark required for a '7'.

$\therefore \mathrm{P}(X \geqslant k) = 0.07$

$\therefore \mathrm{P}(X < k) = 1 - 0.07 = 0.93$

$\therefore k \approx 71.1$

So the minimum mark required to obtain a '7' is 71.1 marks.

**7** $X$ is the life of a battery in weeks.

$X$ is normally distributed with $\mu = 33.2$ and $\sigma = 2.8$.

**a** $\mathrm{P}(X \geqslant 35) \approx 0.260$

**b** We need to find $k$ such that $\mathrm{P}(X \leqslant k) = 0.08$

$\therefore k \approx 29.3$

So, the manufacturer can expect the batteries to last 29.3 weeks before 8% of them fail.

**8** **a** $\text{P}(X \leqslant 30) = 0.0832$ and $\text{P}(X \geqslant 90) = 0.101$

$\therefore\ \text{P}\left(\dfrac{X-\mu}{\sigma} \leqslant \dfrac{30-\mu}{\sigma}\right) = 0.0832$

$\therefore\ \text{P}\left(Z \leqslant \dfrac{30-\mu}{\sigma}\right) = 0.0832$

$\therefore\ \dfrac{30-\mu}{\sigma} \approx -1.383\,864$

$\therefore\ 30-\mu \approx -1.383\,864\sigma$ .... (1)

$\therefore\ \text{P}(X < 90) = 0.899$

$\therefore\ \text{P}\left(\dfrac{X-\mu}{\sigma} < \dfrac{90-\mu}{\sigma}\right) = 0.899$

$\therefore\ \text{P}\left(Z < \dfrac{90-\mu}{\sigma}\right) = 0.899$

$\therefore\ \dfrac{90-\mu}{\sigma} \approx 1.275\,874$

$\therefore\ 90-\mu \approx 1.275\,874\sigma$ .... (2)

Solving (1) and (2) simultaneously, we get $\mu \approx 61.218 \approx 61.2$ and $\sigma \approx 22.559 \approx 22.6$.

**b** $\text{P}(-7 \leqslant X-\mu \leqslant 7) \approx \text{P}(-7 \leqslant X - 61.218 \leqslant 7)$
$\approx \text{P}(54.218 \leqslant X \leqslant 68.218)$
$\approx 0.244$

**9** **a** The relative difficulty of each test is not known. We would need the mean mark and standard deviation for each test.

**b** Kerry's English $z$-score $= \dfrac{26-22}{4} = \dfrac{4}{4} = 1$

Kerry's Chemistry $z$-score $= \dfrac{82-75}{7} = \dfrac{7}{7} = 1$

Since the $z$-scores are the same, Kerry's performance relative to the rest of the class is the same in both tests.

## REVIEW SET 24C

**1** **a** The middle 68% of the distribution lies between 16.2 and 21.4, and the middle 68% of data lies between one standard deviation of the mean.

$\therefore\ \mu \approx \dfrac{16.2+21.4}{2}$

$\therefore\ \mu \approx \dfrac{37.6}{2}$

$\therefore\ \mu \approx 18.8$

and $\sigma \approx 18.8 - 16.2$

$\therefore\ \sigma \approx 2.6$

**b** The middle 95% of the data lies between 2 standard deviations of the mean.

$\mu - 2\sigma \approx 18.8 - 2 \times 2.6 \approx 13.6$ and $\mu + 2\sigma \approx 18.8 + 2\times 2.6 \approx 24.0$

$\therefore$ the middle 95% of the data lies between 13.6 and 24.0.

**2** Using technology:

**a** $\text{P}(X \geqslant 22) \approx 0.364$ **b** $\text{P}(18 \leqslant X \leqslant 22) \approx 0.356$ **c** $\text{P}(X \leqslant k) = 0.3$
$\therefore\ k \approx 18.2$

**3** $\text{P}(-0.524 < X-\mu < 0.524) = \text{P}\left(\dfrac{-0.524}{2} < \dfrac{X-\mu}{2} < \dfrac{0.524}{2}\right)$
$= \text{P}(-0.262 < Z < 0.262)$
$\approx 0.207$ {using technology}

**4** If $X$ is the length of a rod, then $X \sim \text{N}(\mu, 6^2)$.

$$\begin{aligned} \text{Now} \quad \text{P}(X \geqslant 89.52) &= 0.0563 \\ \therefore \quad \text{P}(X < 89.52) &= 1 - 0.0563 \\ \therefore \quad \text{P}\left(\frac{X-\mu}{6} < \frac{89.52-\mu}{6}\right) &= 0.9437 \\ \therefore \quad \text{P}\left(Z < \frac{89.52-\mu}{6}\right) &= 0.9437 \\ \therefore \quad \frac{89.52-\mu}{6} &\approx 1.5866 \\ \therefore \quad 89.52 - \mu &\approx 9.52 \\ \therefore \quad \mu &\approx 80.0 \end{aligned}$$

So, the mean is 80.0 cm.

**5**

$$\begin{aligned} \text{P}(X < 90) &\approx 0.975 \\ \therefore \quad \text{P}\left(\frac{X-50}{\sigma} < \frac{90-50}{\sigma}\right) &\approx 0.975 \\ \therefore \quad \text{P}\left(Z < \frac{40}{\sigma}\right) &\approx 0.975 \\ \therefore \quad \frac{40}{\sigma} &\approx 1.959\,96 \\ \therefore \quad \sigma &\approx 20.409 \end{aligned}$$

So, $X \sim \text{N}(50, 20.409^2)$

Now, the shaded area $= \text{P}(X \geqslant 80)$
$\approx 0.0708 \text{ units}^2$

**6** If $X$ is the weight of an apple, then $X \sim \text{N}(300, 50^2)$.

**a** $\text{P}(250 \leqslant X \leqslant 350)$
$\approx 0.682\,689\,49$
$\approx 68.3\%$

**b** This is a binomial distribution where $X$ is the number of apples that are fit for sale.
$n = 100$ and $p \approx 0.682\,689\,49$
$\text{P}(X \geqslant 75) = 1 - \text{P}(X \leqslant 74)$
$\approx 1 - 0.911\,645\,43$
$\approx 0.0884$

**7** If $X$ is the volume of drink in mL, then $X \sim \text{N}(376, \sigma^2)$.

$$\begin{aligned} \text{Now} \quad \text{P}(X < 375) &= 0.023 \\ \therefore \quad \text{P}\left(\frac{X-376}{\sigma} < \frac{375-376}{\sigma}\right) &= 0.023 \\ \therefore \quad \text{P}\left(Z < \frac{-1}{\sigma}\right) &= 0.023 \\ \therefore \quad -\frac{1}{\sigma} &\approx -1.995 \\ \therefore \quad \sigma &\approx 0.501 \end{aligned}$$

$\therefore$ the standard deviation is 0.501 mL.

**8** If $X$ is the height of an 18 year old boy, then $X \sim \text{N}(187, \sigma^2)$.

$$\begin{aligned} \text{Now} \quad \text{P}(X > 193) &= 0.15 \\ \therefore \quad \text{P}(X \leqslant 193) &= 0.85 \\ \therefore \quad \text{P}\left(\frac{X-187}{\sigma} \leqslant \frac{193-187}{\sigma}\right) &= 0.85 \\ \therefore \quad \text{P}\left(Z \leqslant \frac{6}{\sigma}\right) &= 0.85 \\ \therefore \quad \frac{6}{\sigma} &\approx 1.0364 \\ \therefore \quad \sigma &\approx 5.789 \end{aligned}$$

So, $\text{P}(X > 185) \approx 0.635$

$\therefore$ the probability that two 18 year old boys are taller than 185 cm $\approx 0.635^2$
$\approx 0.403$

# Chapter 25
## MISCELLANEOUS QUESTIONS

### EXERCISE 25A

**1** **a** $S_1 = u_1 = 2$ and $S_2 = u_1 + u_2 = 8$
$\therefore\ u_1 = 2$ and $u_2 = 6$
But $u_2 = u_1 r$
$\therefore\ 6 = 2r$
$\therefore\ r = 3$

**b** $u_{20} = u_1 r^{19}$
$= 2 \times 3^{19}$

**2** $\ln 2 + \ln 4 + \ln 8 + \ln 16 + ....$
$= \ln 2 + \ln 2^2 + \ln 2^3 + \ln 2^4 + ....$
$= \ln 2 + 2\ln 2 + 3\ln 2 + 4\ln 2 + ....$
which is arithmetic with $u_1 = \ln 2$ and $d = \ln 2$

Now $S_n = \dfrac{n}{2}(2u_1 + (n-1)d)$
$\therefore\ S_{40} = 20(2\ln 2 + 39\ln 2)$
$= 20 \times 41 \ln 2$
$= 820 \ln 2$

**3** $f(x) = be^x$ and $g(x) = \ln(bx)$

**a** $(f \circ g)(x) = f(g(x))$
$= f(\ln(bx))$
$= be^{\ln(bx)}$
$= b(bx)$
$= b^2 x$

**b** $(g \circ f)(x) = g(f(x))$
$= g(be^x)$
$= \ln(bbe^x)$
$= \ln(b^2 e^x)$
$= \ln b^2 + \ln e^x$
$= 2\ln b + x$

**c** $(f \circ g)(x^*) = (g \circ f)(x^*)$
$\therefore\ b^2 x^* = 2\ln b + x^*$
$\therefore\ (b^2 - 1)x^* = 2\ln b$
$\therefore\ x^* = \dfrac{2\ln b}{b^2 - 1}$

**4** 
**a** The vertex is $(b, 2)$.

**c**

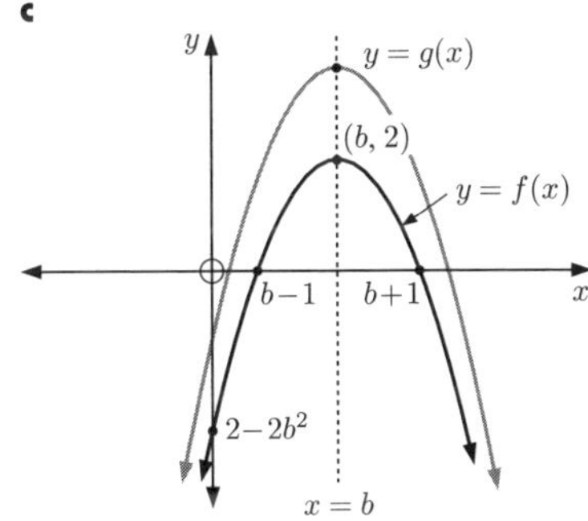


**b** $f(0) = -2b^2 + 2$
$\therefore$ the $y$-intercept is $2 - 2b^2$
$f(x)$ cuts the $x$-axis when $f(x) = 0$
$\therefore\ -2(x-b)^2 + 2 = 0$
$\therefore\ (x-b)^2 = 1$
$\therefore\ x - b = \pm 1$
$\therefore\ x = b \pm 1$
$\therefore$ the $x$-intercepts are $b - 1$ and $b + 1$

$g(x) = f(x) + b$
$= -2(x-b)^2 + 2 + b$
$= -2(x^2 - 2bx + b^2) + 2 + b$
$= -2x^2 + 4bx - 2b^2 + b + 2$
which has discriminant
$\Delta = (4b)^2 - 4(-2)(-2b^2 + b + 2)$
$= 16b^2 - 16b^2 + 8b + 16$
$= 8b + 16$

**i** $g$ has exactly one $x$-intercept if $8b + 16 = 0$
$\therefore\ b = -2$

**ii** $g$ has no $x$-intercepts if $8b + 16 < 0$
$\therefore\ 8b < -16$
$\therefore\ b < -2$

**iii** $g$ passes through the origin if $g(0) = 0$
$\therefore\ -2b^2 + b + 2 = 0$
$\therefore\ b = \dfrac{-1 \pm \sqrt{1 - 4(-2)(2)}}{-4}$
$\therefore\ b = \dfrac{1 \pm \sqrt{17}}{4}$

**5** **a** $(x-2)^3$
$= x^3 + 3x^2(-2) + 3x(-2)^2 + (-2)^3$
$= x^3 - 6x^2 + 12x - 8$

**b** $(3x^2-7)(x-2)^3$
$= (3x^2-7)(x^3 - 6x^2 + 12x - 8)$

$\therefore$ coefficient of $x^3$ is $3 \times 12 + (-7) \times 1$
$= 29$

**6** $f(x) = \sqrt{1-2x}$

**a** $f(0) = \sqrt{1} = 1$

**b** $f(-4) = \sqrt{1-2(-4)}$
$= \sqrt{9}$
$= 3$

**c** $f(x)$ is defined when $1 - 2x \geqslant 0$
$\therefore\ 1 \geqslant 2x$
$\therefore\ x \leqslant \frac{1}{2}$

So, the domain is $\{x \mid x \leqslant \frac{1}{2},\ x \in \mathbb{R}\}$

**d** As $f(x) = \sqrt{1-2x}$,
$f(x) \geqslant 0$ for all $x$ in the domain.
$\therefore$ the range is $\{y \mid y \geqslant 0,\ y \in \mathbb{R}\}$

*Check:*

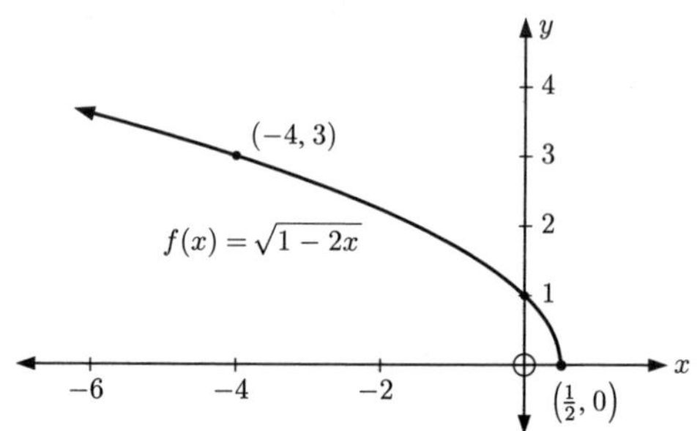


**7** **a** $\sin 160°$
$= \sin 20°$ $\{\sin\theta = \sin(180-\theta)\}$
$= a$

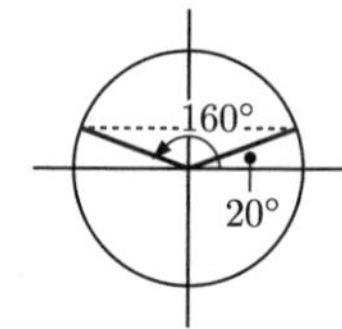


**b** $\tan(-50°)$
$= -\tan 50°$ $\{\tan(-\theta) = -\tan\theta\}$
$= -b$

**c** $\cos 70°$
$= \sin(90-70)°$ $\{\cos\theta = \sin(90-\theta)\}$
$= \sin 20°$
$= a$

**d** $\tan 20° = \dfrac{\sin 20°}{\cos 20°}$

Now $\cos^2\theta + \sin^2\theta = 1$, so $\cos\theta = \pm\sqrt{1-\sin^2\theta}$

But $\cos 20° > 0$, so in this case $\cos 20° = \sqrt{1-\sin^2 20°} = \sqrt{1-a^2}$

So, $\tan 20° = \dfrac{a}{\sqrt{1-a^2}}$

**8** **a** $(f \circ g)(x) = 1$
$\therefore\ f(g(x)) = 1$
$\therefore\ f(2x) = 1$
$\therefore\ \cos 2x = 1$
$\therefore\ 2x = 0 + k2\pi,\ k \in \mathbb{Z}$
$\therefore\ x = k\pi,\ k \in \mathbb{Z}$
$\therefore\ x = 0,\ \pi$, or $2\pi$

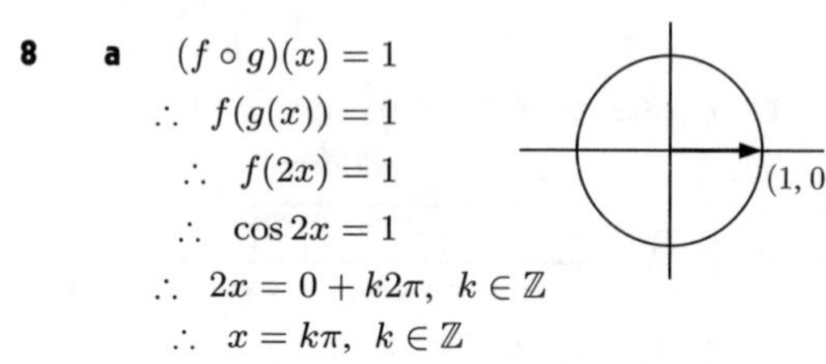


**b** $(g \circ f)(x) = 1$
$\therefore\ g(f(x)) = 1$
$\therefore\ g(\cos x) = 1$
$\therefore\ 2\cos x = 1$
$\therefore\ \cos x = \frac{1}{2}$
$\therefore\ x = \frac{\pi}{3}$ or $\frac{5\pi}{3}$

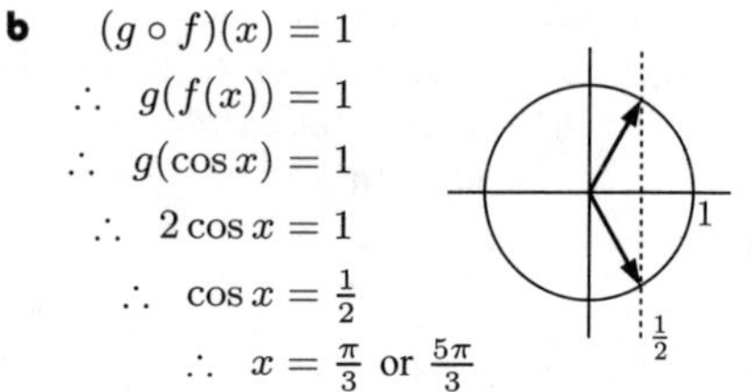

**9** Consider $f(x) = ax^2 + bx + c$.

$f(x)$ is concave up, so $a > 0$.

The $y$-intercept of $f(x)$ is positive, so $c > 0$.

The axis of symmetry is to the right of the $y$-axis, so $\frac{-b}{2a} > 0$

$\therefore\ b < 0 \quad \{a > 0\}$

$f(x)$ does not cut the $x$-axis, so $\Delta < 0$.

Consider $g(x) = dx^2 + ex + h$.

$g(x)$ is concave down, so $d < 0$.

$g(x)$ passes through the origin, so $h = 0$.

The axis of symmetry is to the right of the $y$-axis, so $\frac{-e}{2d} > 0$

$\therefore\ e > 0 \quad \{d < 0\}$

$g(x)$ cuts the $x$-axis twice, so $\Delta > 0$.

| *Constant* | $a$ | $b$ | $c$ | $d$ | $e$ | $h$ | $\Delta$ of $f(x)$ | $\Delta$ of $g(x)$ |
|---|---|---|---|---|---|---|---|---|
| *Sign* | $> 0$ | $< 0$ | $> 0$ | $< 0$ | $> 0$ | $= 0$ | $< 0$ | $> 0$ |

**10** **a**

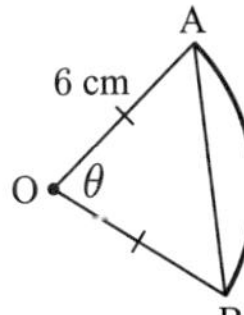


Perimeter of sector is $r + r + \theta r$

$\therefore\ 6 + 6 + 6\theta = 12 + 2\pi$

$\therefore\ 12 + 6\theta = 12 + 2\pi$

$\therefore\ 6\theta = 2\pi$

$\therefore\ \theta = \frac{\pi}{3}$

**b** Consider $\triangle$OAB:

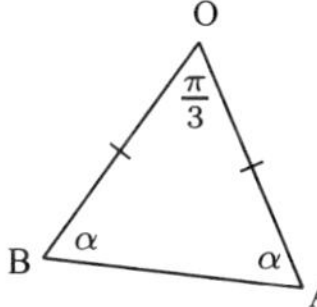


Now $\widehat{OBA} = \widehat{OAB} = \alpha$ {base angles of isosceles $\triangle$}

$\therefore\ \frac{\pi}{3} + \alpha + \alpha = \pi$ {angles of $\triangle$}

$\therefore\ 2\alpha = \frac{2\pi}{3}$

$\therefore\ \alpha = \frac{\pi}{3}$

$\therefore$ $\triangle$OAB is equilateral, since all angles are the same size.

$\therefore$ chord [AB] is 6 cm long.

**11** **a** $f(x) = 3$

$\therefore\ x^2 + 6 = 3$

$\therefore\ x^2 = -3$

which has no real solution.

$\therefore\ f(x) = 3$ cannot be solved.

**b** This tells us that 3 is not in the range of $f$.
In fact, the range of $f$ is $\{y \mid y \geqslant 6\}$.

**12** **a** $\mathbf{u} \perp \mathbf{v}$

$\therefore\ \mathbf{u} \bullet \mathbf{v} = 0$

$\therefore\ (1)(3) + (-2)(p) + (1)(-1) = 0$

$\therefore\ 3 - 2p - 1 = 0$

$\therefore\ 2p = 2$

$\therefore\ p = 1$

**b** $|\mathbf{u}|\,|\mathbf{v}| = \sqrt{1+4+1}\sqrt{9+p^2+1}$

$= \sqrt{6}\sqrt{p^2+10}$

$= \sqrt{6p^2+60}$

**c** $\mathbf{v} - \mathbf{u} = (3\mathbf{i} + p\mathbf{j} - \mathbf{k}) - (\mathbf{i} - 2\mathbf{j} + \mathbf{k})$

$= 2\mathbf{i} + (p+2)\mathbf{j} - 2\mathbf{k}$

For $\mathbf{v} - \mathbf{u}$ to be parallel to $\mathbf{u}$, there must exist some scalar $s$ such that

$s(\mathbf{i} - 2\mathbf{j} + \mathbf{k}) = 2\mathbf{i} + (p+2)\mathbf{j} - 2\mathbf{k}$

$\therefore\ s = 2,\ -2s = p + 2,$ and $s = -2$

There is no common solution to these equations, so no value of $p$ exists such that $\mathbf{v} - \mathbf{u}$ and $\mathbf{u}$ are parallel.

**13** **a** $\overrightarrow{BA} = \begin{pmatrix} 2-(-1) \\ 1-0 \\ 5-4 \end{pmatrix} = \begin{pmatrix} 3 \\ 1 \\ 1 \end{pmatrix}$ and $\overrightarrow{BC} = \begin{pmatrix} 0-(-1) \\ 1-0 \\ 1-4 \end{pmatrix} = \begin{pmatrix} 1 \\ 1 \\ -3 \end{pmatrix}$

**b** $|\overrightarrow{BA}| = \sqrt{9+1+1}$
$= \sqrt{11}$ units

$|\overrightarrow{BC}| = \sqrt{1+1+9}$
$= \sqrt{11}$ units

**c** ABCD is a parallelogram, so the opposite sides are equal in length, and $|\overrightarrow{BA}| = |\overrightarrow{BC}|$ from **b**.
$\therefore$ all the sides are equal in length, so ABCD is a rhombus.

**d** **i** $\cos(\widehat{CBA}) = \dfrac{\overrightarrow{BC} \bullet \overrightarrow{BA}}{|\overrightarrow{BC}||\overrightarrow{BA}|}$
$= \dfrac{3+1-3}{\sqrt{11}\sqrt{11}}$
$= \frac{1}{11}$

**ii** $\sin^2(\widehat{CBA}) + \cos^2(\widehat{CBA}) = 1$
$\therefore \sin^2(\widehat{CBA}) + \frac{1}{121} = 1$
$\therefore \sin(\widehat{CBA}) = \frac{\sqrt{120}}{11}$

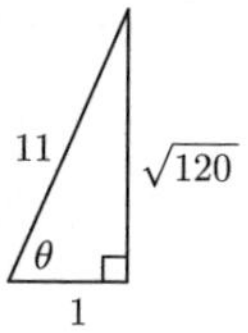


**iii** area $= 2 \times$ area of $\triangle$ABC
$= 2 \times \frac{1}{2} |\overrightarrow{BA}||\overrightarrow{BC}| \sin(\widehat{CBA})$
$= \sqrt{11}\sqrt{11} \times \frac{\sqrt{120}}{11}$
$= \sqrt{120}$
$= 2\sqrt{30}$ units$^2$

**14** **a** There are 13 data values, and $\dfrac{13+1}{2} = 7$
$\therefore$ the median $=$ 7th data value $= g$

**b** **i** range $= m - a$

**ii** $\underbrace{a\ b\ c\ d\ e\ f}_{\text{lower}}\ \underset{\uparrow\ \text{median}}{g}\ \underbrace{h\ i\ j\ k\ l\ m}_{\text{upper}}$

$Q_1 =$ median of lower half $= \dfrac{c+d}{2}$ $\qquad$ $Q_3 =$ median of upper half $= \dfrac{j+k}{2}$

$\therefore$ IQR $= Q_3 - Q_1 = \left(\dfrac{j+k}{2}\right) - \left(\dfrac{c+d}{2}\right)$

**15** The data set $\{a, b, c\}$ has mean 17.5 and standard deviation 3.2

$\therefore \dfrac{a+b+c}{3} = 17.5$ and $\sqrt{\dfrac{(a-17.5)^2+(b-17.5)^2+(c-17.5)^2}{3}} = 3.2$

**a** For the data set $\{2a, 2b, 2c\}$,

$\mu = \dfrac{2a+2b+2c}{3}$
$= 2\left(\dfrac{a+b+c}{3}\right)$
$= 2 \times 17.5$
$= 35$

$\sigma = \sqrt{\dfrac{(2a-35)^2+(2b-35)^2+(2c-35)^2}{3}}$
$= \sqrt{\dfrac{2^2(a-17.5)^2+2^2(b-17.5)^2+2^2(c-17.5)^2}{3}}$
$= 2\sqrt{\dfrac{(a-17.5)^2+(b-17.5)^2+(c-17.5)^2}{3}}$
$= 2 \times 3.2$
$= 6.4$

**b** For the data set $\{a+2, b+2, c+2\}$,

$$\mu = \frac{(a+2)+(b+2)+(c+2)}{3}$$
$$= \frac{a+b+c}{3} + \frac{6}{3}$$
$$= 17.5 + 2 = 19.5$$

$$\sigma = \sqrt{\frac{(a+2-19.5)^2+(b+2-19.5)^2+(c+2-19.5)^2}{3}}$$
$$= \sqrt{\frac{(a-17.5)^2+(b-17.5)^2+(c-17.5)^2}{3}}$$
$$= 3.2$$

**c** For the data set $\{3a+5, 3b+5, 3c+5\}$,

$$\mu = \frac{(3a+5)+(3b+5)+(3c+5)}{3}$$
$$= \frac{3(a+b+c)}{3} + \frac{15}{3}$$
$$= 3 \times 17.5 + 5 = 57.5$$

$$\sigma = \sqrt{\frac{(3a+5-57.5)^2+(3b+5-57.5)^2+(3c+5-57.5)^2}{3}}$$
$$= \sqrt{\frac{(3a-52.5)^2+(3b-52.5)^2+(3c-52.5)^2}{3}}$$
$$= \sqrt{\frac{3^2(a-17.5)^2+3^2(b-17.5)^2+3^2(c-17.5)^2}{3}}$$
$$= 3\sqrt{\frac{(a-17.5)^2+(b-17.5)^2+(c-17.5)^2}{3}}$$
$$= 3 \times 3.2 = 9.6$$

**16** Let the circle have radius $r$. The equal angles are $\frac{2\pi}{5}$ radians.

The area of each triangle $= \frac{1}{2} \times r \times r \times \sin(\frac{2\pi}{5})$
$= \frac{1}{2}r^2 \sin(\frac{2\pi}{5})$

$\therefore$ P(lands on shaded region | lands on board) $= \dfrac{\pi r^2 - \left(\frac{1}{2}r^2\sin(\frac{2\pi}{5})\right) \times 5}{\pi r^2}$

$$= \frac{\cancel{\pi r^2}^{\,1}}{\cancel{\pi r^2}_{\,1}} - \frac{5\cancel{r^2}\sin(\frac{2\pi}{5})}{2\pi \cancel{r^2}}$$
$$= 1 - \frac{5}{2\pi}\sin(\frac{2\pi}{5})$$

**17** $f(x)$ is a quadratic, so $f'(x)$ is linear and $f''(x)$ is constant. Let $f(x) = ax^2 + bx + c$.

**a** $f(x)$ is concave down, so $f'(x)$ is decreasing.
$f'(x) = 0$ when $f(x)$ is maximised
$f'(x) = 2ax + b$

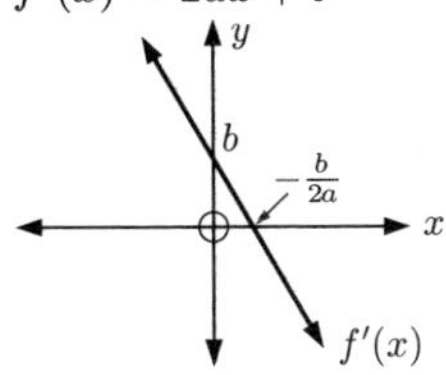


**b** $f'(x)$ is decreasing, so $f''(x)$ is negative.
$f''(x) = 2a$

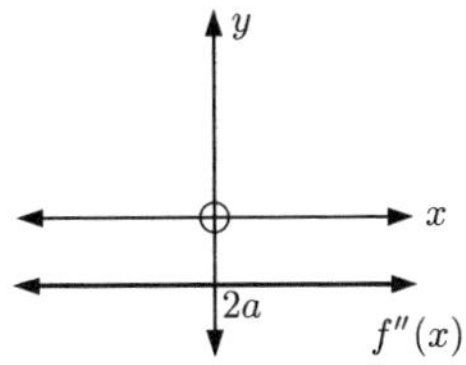

**18** $g(x) = 3 - 2\cos(2x)$

**a** $g'(x) = 0 - 2 \times -2\sin(2x)$
$= 4\sin(2x)$

**b**

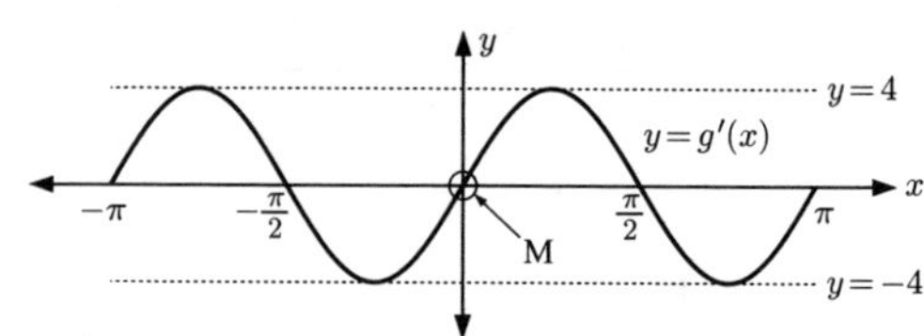


**c** $g'(x) = 0$ when $\sin(2x) = 0$
$\therefore\ x = -\pi, -\frac{\pi}{2}, 0, \frac{\pi}{2}, \pi$
$\therefore$ there are 5 solutions.

**d** If $g''(x) > 0$, $y = g'(x)$ is increasing
$\therefore$ M is at $(-\pi, 0)$, $(0, 0)$, or $(\pi, 0)$
(The point $(0, 0)$ is marked on the diagram above.)

**19** **a** $P(B) = 1 - P(B') = 0.57$
As $A$ and $B$ are mutually exclusive,
$A \cap B = \varnothing$ and so $P(A \cap B) = 0$.
$\therefore\ P(A \cup B) = P(A) + P(B)$
$= x + 0.57$

**b** We need to solve $x + 0.57 = 0.73$
$\therefore\ x = 0.16$

**20** $g'(0) = 0$ and $g''(0) = 0$
$\therefore$ there is a stationary inflection point at A(0, 2)
$g''(2) = 0$ and $g'(2) \neq 0$
$\therefore$ there is a non-stationary inflection point at B(2, 0)
$g'(4) = 0$ and $g''(4) > 0$
$\therefore$ there is a local minimum at C(4, −2)

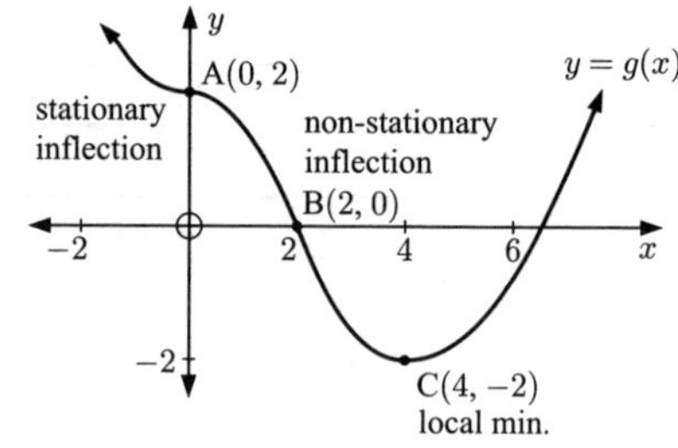


**21** $f(x) = xe^{1-2x}$

**a** $f'(x) = 1e^{1-2x} + xe^{1-2x}(-2)$
$= e^{1-2x}(1 - 2x)$

**b** $f'(x) = 0$ when $1 - 2x = 0$
$\therefore\ x = \frac{1}{2}$
and $f(\frac{1}{2}) = \frac{1}{2}e^{1-2(\frac{1}{2})} = \frac{1}{2}$
$\therefore$ the tangent is horizontal at $(\frac{1}{2}, \frac{1}{2})$

**c** **i** $f(x) > 0$ when $xe^{1-2x} > 0$
But $e^{1-2x}$ is always $> 0$
$\therefore\ f(x) > 0$ when $x > 0$

**ii** $f'(x) > 0$ when $1 - 2x > 0$
$\therefore\ 1 > 2x$
$\therefore\ x < \frac{1}{2}$

**22** $s(t) = 3 - 4e^{2t} + kt$ metres

**a** $v(t) = 0 - 4e^{2t}(2) + k$
$= k - 8e^{2t}\ \text{m s}^{-1}$

**b** $v(\ln 3) = 0$
$\therefore\ k - 8e^{2\ln 3} = 0$
$\therefore\ k - 8e^{\ln 3^2} = 0$
$\therefore\ k - 8 \times 9 = 0$
$\therefore\ k = 72$

**23** **a** $\log_3 27 = x$
$\therefore\ 3^x = 27$
$\therefore\ 3^x = 3^3$
$\therefore\ x = 3$

**b** $e^{5-2x} = 8$
$\therefore\ 5 - 2x = \ln 8$
$\therefore\ 2x = 5 - \ln 8$
$\therefore\ x = \dfrac{5 - \ln 8}{2}$

**c**
$$\ln(x^2-3)-\ln(2x)=0$$
$$\therefore\ \ln\left(\frac{x^2-3}{2x}\right)=0$$
$$\therefore\ \frac{x^2-3}{2x}=1$$
$$\therefore\ x^2-3=2x$$
$$\therefore\ x^2-2x-3=0$$
$$\therefore\ (x-3)(x+1)=0$$
$$\therefore\ x=3 \text{ or } -1$$
But $x^2-3$ and $2x$ must be $>0$
$$\therefore\ x=3 \text{ is the only solution}$$

**24** **a** $v(0)=1\ \text{m s}^{-1}$ and is the initial velocity.

**b**
$$v(t)=2, \quad \text{a constant for } 1\leqslant t\leqslant 3$$
$$\therefore\ v'(t)=0 \quad \text{on} \quad 1\leqslant t\leqslant 3$$
$$\therefore\ v'(2)=0$$
This is the acceleration at $t=2$.
The velocity is constant at this time.

**c**
$$\int_1^3 v(t)\,dt = \text{area under } y=v(t) \text{ from } t=1 \text{ to } t=3$$
$$=2\times 2$$
$$=4$$
This is the displacement in metres of the particle for $1\leqslant t\leqslant 3$.

**25** **a**
$$\int_{-1}^{2}(f(x)-6)\ dx$$
$$=\int_{-1}^{2}f(x)\,dx-\int_{-1}^{2}6\,dx$$
$$=10-[6x]_{-1}^{2}$$
$$=10-(12--6)$$
$$=10-18$$
$$=-8$$

**b**
$$\int_2^{-1}k\,f(x)\,dx=-5$$
$$\therefore\ \int_{-1}^{2}k\,f(x)\,dx=5$$
$$\therefore\ k\int_{-1}^{2}f(x)\,dx=5$$
$$\therefore\ k(10)=5$$
$$\therefore\ k=\tfrac{1}{2}$$

**26** **a**
$$(\sin\theta-\cos\theta)^2$$
$$=\sin^2\theta-2\sin\theta\cos\theta+\cos^2\theta$$
$$=\sin^2\theta+\cos^2\theta-\sin 2\theta$$
$$=1-\sin 2\theta$$

**b**
$$\int_0^{\frac{\pi}{4}}(\sin\theta-\cos\theta)^2\,d\theta$$
$$=\int_0^{\frac{\pi}{4}}(1-\sin 2\theta)\,d\theta$$
$$=\left[\theta+\tfrac{1}{2}\cos 2\theta\right]_0^{\frac{\pi}{4}}$$
$$=\left(\tfrac{\pi}{4}+\tfrac{1}{2}\underbrace{\cos\tfrac{\pi}{2}}_{0}\right)-\left(0+\tfrac{1}{2}\right)$$
$$=\tfrac{\pi}{4}-\tfrac{1}{2}$$

**27**
$$\mathrm{E}(X)=\sum p_i x_i$$
$$=-0.2-a+0+b+0.3$$
$$=b-a+0.1$$
But $\mathrm{E}(X)=0$
$\therefore\ a-b=0.1$ .... (1)

Also, $\sum p_i=1$
$\therefore\ 0.1+a+0.25+b+0.15=1$
$\therefore\ a+b=0.5$ .... (2)

Adding (1) and (2) gives $2a=0.6$
$\therefore\ a=0.3$ and so $b=0.2$

**28** **a** The common ratio

$$r = \frac{\frac{1}{e}}{e} = \frac{1}{e^2}$$

**b**
$$\begin{aligned} u_{101} &= u_1 r^{100} \\ &= e \times \left(\frac{1}{e^2}\right)^{100} \\ &= e \times e^{-200} \\ &= e^{-199} \end{aligned}$$

**c**
$$\begin{aligned} S &= \frac{u_1}{1-r} \\ &= \frac{e}{1-\frac{1}{e^2}} \\ &= \left[\frac{e}{1-\frac{1}{e^2}}\right]\frac{e^2}{e^2} \\ &= \frac{e^3}{e^2-1} \end{aligned}$$

**29** **a**
$$\begin{aligned} \binom{n}{r} &= \frac{n!}{r!(n-r)!} \\ \therefore \binom{6}{2} &= \frac{6!}{2!(6-2)!} \\ &= \frac{6 \times 5 \times \cancel{4!}}{2!\cancel{4!}} \\ &= \frac{30}{2} \\ &= 15 \end{aligned}$$

**b**
$$\begin{aligned} \binom{6}{4} &= \binom{6}{2} \\ \therefore \binom{6}{4} &= 15 \end{aligned}$$

**c**
$$\begin{aligned} &(x-2)^6 \\ &= \binom{6}{0}x^6 + \binom{6}{1}x^5(-2) + \binom{6}{2}x^4(-2)^2 + \binom{6}{3}x^3(-2)^3 + \binom{6}{4}x^2(-2)^4 + \binom{6}{5}x(-2)^5 + \binom{6}{6}(-2)^6 \\ &= 1 \times x^6 + 6 \times x^5 \times (-2) + 15 \times x^4 \times 4 + 20 \times x^3 \times (-8) + 15 \times x^2 \times 16 + 6 \times x \times (-32) \\ &\quad + 1 \times 64 \\ &= x^6 - 12x^5 + 60x^4 - 160x^3 + 240x^2 - 192x + 64 \end{aligned}$$

**30** **a, b**

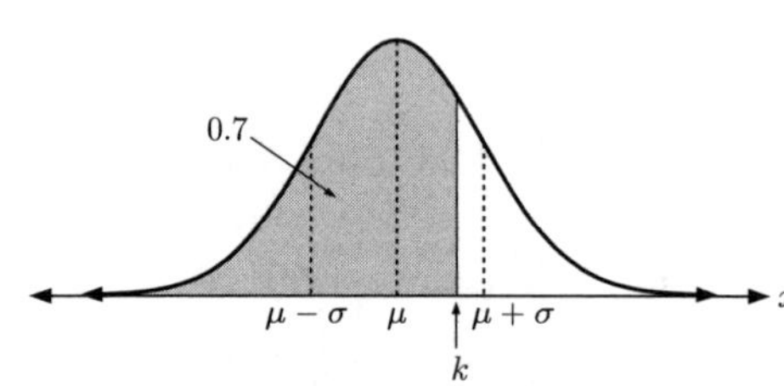


**c** **i** $P(X > k) = 1 - 0.7 = 0.3$

**ii** $P(\mu < X < k) = 0.7 - 0.5 = 0.2$

**iii**
$$\begin{aligned} &P(\mu - \sigma < X < k) \\ &= P(\mu - \sigma < X < \mu) + P(\mu < X < k) \\ &\approx 0.341 + 0.2 \approx 0.541 \end{aligned}$$

**d**
$$\begin{aligned} &P(k \leqslant X \leqslant t) \\ &= P(X \leqslant t) - P(X < k) \\ &= 0.8 - 0.7 \\ &= 0.1 \end{aligned}$$

**31** **a** $200 - 60 = 140$ out of 200 students passed

$\therefore$ 70% passed

**b** **i** $m = Q_1 \approx 27$

**ii** $n = Q_2 \approx 35$

**iii** $p = Q_3 \approx 42$

**iv** $q =$ maximum score $= 100$

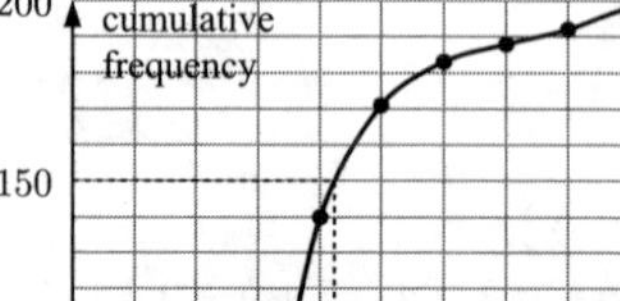

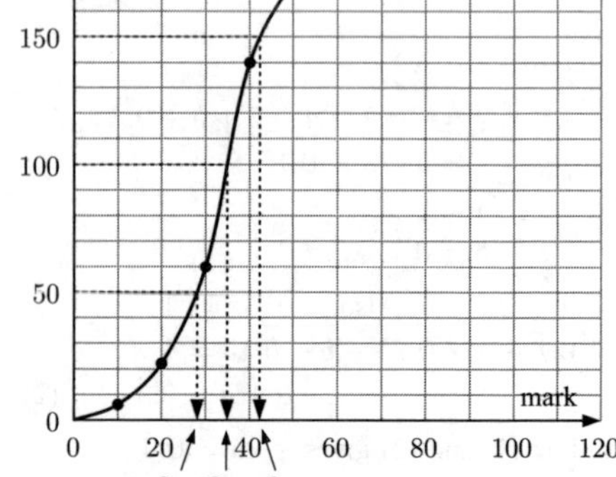

**32** $y = (\tan(\frac{\pi}{3}))x$ is $y = \sqrt{3}x$

**a** $(10, p)$ lies on $L$ if $p = \sqrt{3} \times 10$

$\therefore\ p = 10\sqrt{3}$

**b** The gradient of $L$ is $\sqrt{3}$

$\therefore$ the gradient of the perpendicular line is $-\frac{1}{\sqrt{3}}$

$\therefore$ its equation is $\dfrac{y - 10\sqrt{3}}{x - 10} = -\dfrac{1}{\sqrt{3}}$

$\therefore\ \sqrt{3}y - 30 = -x + 10$

$\therefore\ x + \sqrt{3}y = 40$

**33** **a** $a(t) = 1 - 3\cos(2t + \frac{\pi}{2})$ cm s$^{-2}$

$v(t) = \int \left[1 - 3\cos(2t + \frac{\pi}{2})\right]\, dt$

$= t - 3(\frac{1}{2})\sin(2t + \frac{\pi}{2}) + c$

But $v(0) = 5$ $\therefore\ -\frac{3}{2}\sin(\frac{\pi}{2}) + c = 5$

$\therefore\ -\frac{3}{2}(1) + c = 5$

$\therefore\ c = 6\frac{1}{2}$

$\therefore\ v(t) = t - \frac{3}{2}\sin(2t + \frac{\pi}{2}) + 6\frac{1}{2}$ cm s$^{-1}$

**b** $v(\frac{\pi}{4}) = \frac{\pi}{4} - \frac{3}{2}\sin\pi + 6\frac{1}{2}$

$= \frac{\pi}{4} + \frac{13}{2}$

$= \frac{\pi + 26}{4}$ cm s$^{-1}$

**34** **a** $f(x) = e^{3x-4} + 1$ has inverse

$x = e^{3y-4} + 1$

$\therefore\ e^{3y-4} = x - 1$

$\therefore\ 3y - 4 = \ln(x - 1)$

$\therefore\ 3y = \ln(x - 1) + 4$

$\therefore\ f^{-1}(x) = \dfrac{\ln(x - 1) + 4}{3}$

**b** $f^{-1}(8) - f^{-1}(3)$

$= \dfrac{\ln 7 + 4}{3} - \dfrac{\ln 2 + 4}{3}$

$= \dfrac{\ln 7 + 4 - \ln 2 - 4}{3}$

$= \frac{1}{3}\ln(\frac{7}{2})$

**35** **a** $\sin A = \frac{2}{5}$

Now $\cos^2 A + \sin^2 A = 1$

$\therefore\ \cos^2 A + \frac{4}{25} = 1$

$\therefore\ \cos^2 A = \frac{21}{25}$

$\therefore\ \cos A = \pm\frac{\sqrt{21}}{5}$

But $\frac{\pi}{2} \leqslant A \leqslant \pi$, so $\cos A$ is negative

$\therefore\ \cos A = -\frac{\sqrt{21}}{5}$

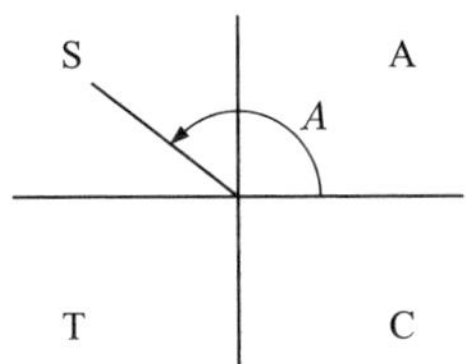


**b** $\tan A = \dfrac{\sin A}{\cos A} = \dfrac{\frac{2}{5}}{-\frac{\sqrt{21}}{5}}$

$= -\frac{2}{\sqrt{21}}$

**c** $\sin 2A = 2\sin A\cos A$

$= 2(\frac{2}{5})(-\frac{\sqrt{21}}{5})$

$= -\frac{4\sqrt{21}}{25}$

**36** **a** $u_1 = 160,\quad r = \frac{80\sqrt{2}}{160} = \frac{1}{\sqrt{2}}$

$u_{12} = u_1 r^{11} = 160 \times \left(\frac{1}{\sqrt{2}}\right)^{11}$

$= 160 \times \frac{1}{(\sqrt{2})^{11}}$

$= 160 \times \frac{1}{32\sqrt{2}}$

$= \frac{5}{\sqrt{2}} \times \frac{\sqrt{2}}{\sqrt{2}}$

$= \frac{5\sqrt{2}}{2} = \frac{5}{2}\sqrt{2}$

**b** **i** $S_n = \dfrac{u_1(1-r^n)}{1-r}$

$\therefore\ S_{10} = \dfrac{160\left(1-\left(\frac{1}{\sqrt{2}}\right)^{10}\right)}{1-\frac{1}{\sqrt{2}}}$

$= \dfrac{160\left(1-\frac{1}{(\sqrt{2})^{10}}\right)}{1-\frac{1}{\sqrt{2}}}$

$= \dfrac{160\left(1-\frac{1}{32}\right)}{1-\frac{1}{\sqrt{2}}}$

$= \dfrac{160\times\frac{31}{32}}{1-\frac{1}{\sqrt{2}}}$

$= \dfrac{5\times 31}{1-\frac{1}{\sqrt{2}}}\times\dfrac{\sqrt{2}}{\sqrt{2}}$

$= \dfrac{155\sqrt{2}}{\sqrt{2}-1}\times\dfrac{\sqrt{2}+1}{\sqrt{2}+1}$

$= \dfrac{155\sqrt{2}(\sqrt{2}+1)}{2-1}$

$= 310+155\sqrt{2}$

**ii** $S = \dfrac{u_1}{1-r} = \dfrac{160}{1-\frac{1}{\sqrt{2}}}\times\dfrac{\sqrt{2}}{\sqrt{2}}$

$= \dfrac{160\sqrt{2}}{\sqrt{2}-1}\times\dfrac{\sqrt{2}+1}{\sqrt{2}+1}$

$= \dfrac{320+160\sqrt{2}}{2-1}$

$= 320+160\sqrt{2}$

**37** **a** $\overrightarrow{PQ} = \begin{pmatrix} 1-3 \\ 3-1 \\ 4-(-2) \end{pmatrix} = \begin{pmatrix} -2 \\ 2 \\ 6 \end{pmatrix}$

**b** $|\overrightarrow{PQ}| = \sqrt{4+4+36}$

$= \sqrt{44}$ units

$\therefore$ speed $= \frac{\sqrt{44}}{2}$ units s$^{-1}$

$= \sqrt{11}$ units s$^{-1}$

**c** The line has equation $\mathbf{r} = \begin{pmatrix} 3 \\ 1 \\ -2 \end{pmatrix} + t\begin{pmatrix} -2 \\ 2 \\ 6 \end{pmatrix},\quad t \geqslant 0$

$\therefore\ x = 3-2t,\quad y = 1+2t,\quad z = -2+6t,\quad t \geqslant 0$

**38** **a**

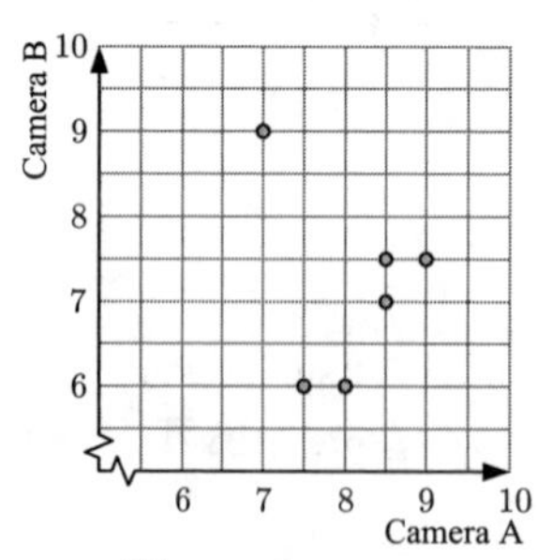


**b** The point $(7, 9)$ does not fit the general trend of the data.

$\therefore$ the data point (Camera A $= 7$, Camera B $= 9$) is an outlier.

**c** The online reviews for these cameras are moderately consistent.

**39** **a, c**

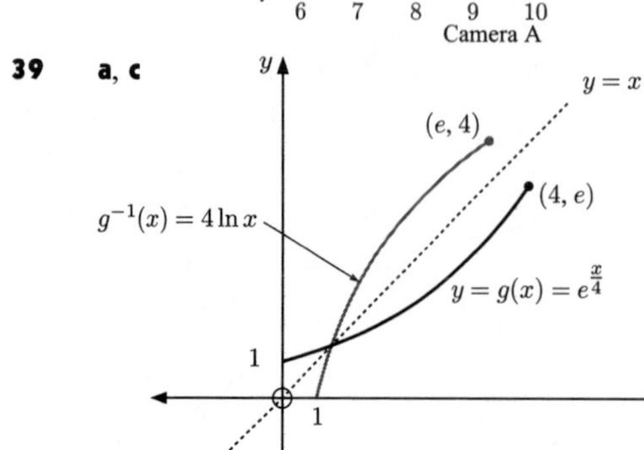


**b** The range of $g$ is $\{y \mid 1 \leqslant y \leqslant e\}$

**d** The domain of $g^{-1}$ is $\{x \mid 1 \leqslant x \leqslant e\}$

The range of $g^{-1}$ is $\{y \mid 0 \leqslant y \leqslant 4\}$

**e** The inverse of $y = e^{\frac{x}{4}}$ is

$x = e^{\frac{y}{4}}$

$\therefore\ \ln x = \dfrac{y}{4}$

$\therefore\ y = 4\ln x$

So, $g^{-1}(x) = 4\ln x$

**40** **a** By the cosine rule,

$$(\text{QS})^2 = 3^2 + 5^2 - 2 \times 3 \times 5 \times \cos\phi$$
$$= 9 + 25 - 30\cos\phi$$
$$= 34 - 30\cos\phi$$

$\therefore$ QS $= \sqrt{34 - 30\cos\phi}$ cm {QS > 0}

**b** **i** If $\phi = 60°$, QS $= \sqrt{34 - 30(\frac{1}{2})}$

$$= \sqrt{34 - 15}$$
$$= \sqrt{19} \text{ cm}$$

Now $\dfrac{\sin\theta}{7} = \dfrac{\sin 30°}{\sqrt{19}}$ {sine rule}

$\therefore$ $\sin\theta = \dfrac{7(\frac{1}{2})}{\sqrt{19}} = \dfrac{7}{2\sqrt{19}}$

**ii** Let RS $= x$ cm

$\therefore$ QS$^2 = 7^2 + x^2 - 2 \times 7 \times x \times \cos 30°$ {cosine rule}

$\therefore$ $19 = 49 + x^2 - 14x(\frac{\sqrt{3}}{2})$

$\therefore$ $x^2 - 7\sqrt{3}x + 30 = 0$

$\therefore$ $x = \dfrac{7\sqrt{3} \pm \sqrt{147 - 4(1)(30)}}{2}$

$\therefore$ $x = \dfrac{7\sqrt{3} \pm \sqrt{27}}{2}$

$\therefore$ $x = \dfrac{7\sqrt{3} \pm 3\sqrt{3}}{2}$

$\therefore$ $x = \frac{10\sqrt{3}}{2}$ or $\frac{4\sqrt{3}}{2}$

$\therefore$ $x = 5\sqrt{3}$ or $2\sqrt{3}$

$\therefore$ $x = 5\sqrt{3}$ {If $x = 2\sqrt{3}$, $\cos\theta = \dfrac{19 + 12 - 49}{2\sqrt{19} \times 2\sqrt{3}}$ which is $< 0$ $\therefore$ $\theta$ is obtuse}

So, the length of [RS] is $5\sqrt{3}$ cm.

**iii** Perimeter $= 5 + 3 + 7 + 5\sqrt{3}$

$$= 15 + 5\sqrt{3} \text{ cm}$$

Area $= \frac{1}{2}(3)(5)\sin 60° + \frac{1}{2}(7)(5\sqrt{3})\sin 30°$

$$= \tfrac{15}{2}(\tfrac{\sqrt{3}}{2}) + \tfrac{35}{4}\sqrt{3}$$
$$= \tfrac{50}{4}\sqrt{3}$$
$$= \tfrac{25}{2}\sqrt{3} \text{ cm}^2$$

**41** **a** $f(x) = -\frac{1}{4}x^2 + 3x + 4$

$\therefore$ $f'(x) = -\frac{1}{2}x + 3$

**b** **i** $f'(2) = -1 + 3 = 2$

$\therefore$ the normal has gradient $-\frac{1}{2}$, and passes through (2, 9).

$\therefore$ its equation is $\dfrac{y - 9}{x - 2} = -\dfrac{1}{2}$

$\therefore$ $2y - 18 = -x + 2$

$\therefore$ $x + 2y = 20$

**ii** The normal has equation $x + 2y = 20$

$$\therefore\ y = \frac{20 - x}{2}$$

$\therefore$ the normal meets $f(x)$ where $-\frac{1}{4}x^2 + 3x + 4 = \frac{20 - x}{2}$

$$\therefore\ -x^2 + 12x + 16 = 40 - 2x$$
$$\therefore\ x^2 - 14x + 24 = 0$$
$$\therefore\ (x-2)(x-12) = 0$$
$$\therefore\ x = 2 \text{ or } 12$$

when $x = 12$, $y = \frac{20-12}{2} = 4$

So, the normal meets $y = f(x)$ again at the point (12, 4).

**c** **i** area $= \int_2^6 \left(-\frac{1}{4}x^2 + 3x + 4\right) dx$

**ii** area $= \left[-\frac{1}{12}x^3 + \frac{3x^2}{2} + 4x\right]_2^6$

$= (-18 + 54 + 24) - (-\frac{2}{3} + 6 + 8)$

$= 46\frac{2}{3}$ units$^2$

**iii** volume $= \pi \int_2^6 \left(-\frac{1}{4}x^2 + 3x + 4\right)^2 dx$

**42** **a** **i** $r = \frac{-12}{4} = -3$

**ii** $u_{14} = u_1 r^{13}$

$= 4 \times (-3)^{13}$ or $-4 \times 3^{13}$

**b** **i** $r = \dfrac{x-2}{x} = \dfrac{2x-7}{x-2}$

$$\therefore\ (x-2)^2 = x(2x-7)$$
$$\therefore\ x^2 - 4x + 4 = 2x^2 - 7x$$
$$\therefore\ x^2 - 3x - 4 = 0$$
$$\therefore\ (x-4)(x+1) = 0$$
$$\therefore\ x = 4 \text{ or } -1$$

**ii** When $x = 4$, sequence is 4, 2, 1, .... with $r = \frac{1}{2}$

$\therefore$ series converges with $S = \dfrac{4}{1 - \frac{1}{2}} = 8$

When $x = -1$, sequence is $-1, -3, -9, ....$ with $r = 3$

$\therefore$ the series does not converge, and $S$ does not exist $\{S$ only exists if $-1 < r < 1\}$

**c** The sequence is arithmetic if $u_2 - u_1 = u_3 - u_2$

$$\therefore\ x - 2 - x = 2x - 7 - (x - 2)$$
$$\therefore\ -2 = 2x - 7 - x + 2$$
$$\therefore\ x - 5 = -2$$
$$\therefore\ x = 3$$

**i** When $x = 3$, the sequence is $3, 1, -1, ....$ which is arithmetic with $u_1 = 3$ and $d = -2$.

$\therefore\ u_{30} = u_1 + 29d$

$= 3 + 29(-2)$

$= -55$

**ii** $S_{50} = \frac{50}{2}\,[2(3) + 49(-2)]$

$= 25\,[6 - 98]$

$= 25 \times -92$

$= -2300$

**43** **a** $\overrightarrow{OA} = \begin{pmatrix} 3 \\ 2 \\ -1 \end{pmatrix}$, $\overrightarrow{OB} = \begin{pmatrix} 2 \\ -1 \\ -8 \end{pmatrix}$

$\overrightarrow{AB} = \overrightarrow{AO} + \overrightarrow{OB}$

$= \begin{pmatrix} -3 \\ -2 \\ 1 \end{pmatrix} + \begin{pmatrix} 2 \\ -1 \\ -8 \end{pmatrix}$

$= \begin{pmatrix} -1 \\ -3 \\ -7 \end{pmatrix}$

**b** $\overrightarrow{BA} = \begin{pmatrix} 1 \\ 3 \\ 7 \end{pmatrix}$

$\therefore\ |\overrightarrow{BA}| = \sqrt{1 + 9 + 49} = \sqrt{59}$

$\therefore$ the unit vector $\mathbf{u} = \frac{1}{\sqrt{59}} \begin{pmatrix} 1 \\ 3 \\ 7 \end{pmatrix}$.

**c** $\mathbf{u} \bullet \overrightarrow{OA}$

$= \frac{1}{\sqrt{59}} \begin{pmatrix} 1 \\ 3 \\ 7 \end{pmatrix} \bullet \begin{pmatrix} 3 \\ 2 \\ -1 \end{pmatrix}$

$= \frac{1}{\sqrt{59}}(3 + 6 - 7)$

$= \frac{2}{\sqrt{59}} \neq 0$

$\therefore$ **u** and $\overrightarrow{OA}$ are not perpendicular.

**d** $\overrightarrow{OC} = \begin{pmatrix} 1 \\ 1 \\ a \end{pmatrix}$

$\therefore \begin{pmatrix} 1 \\ 1 \\ a \end{pmatrix} \bullet \begin{pmatrix} a \\ -1 \\ 4 \end{pmatrix} = 0$

$\therefore a - 1 + 4a = 0$

$\therefore 5a = 1$

$\therefore a = \frac{1}{5}$

**e** M is $(\frac{5}{2}, \frac{1}{2}, -\frac{9}{2})$

$\therefore \overrightarrow{OM} = \frac{1}{2}(5\mathbf{i} + \mathbf{j} - 9\mathbf{k})$

**f** $\mathbf{r}_1 = \frac{1}{2}\begin{pmatrix} 5 \\ 1 \\ -9 \end{pmatrix} + t\begin{pmatrix} 3 \\ 2 \\ -1 \end{pmatrix}, \quad t \in \mathbb{R}$

**g** $\mathbf{r}_2 = \begin{pmatrix} m \\ 1 \\ -1 \end{pmatrix} + s\begin{pmatrix} 2 \\ -3 \\ 1 \end{pmatrix}$

**i** $L_1$ and $L_2$ are not parallel as $\begin{pmatrix} 3 \\ 2 \\ -1 \end{pmatrix} \neq k\begin{pmatrix} 2 \\ -3 \\ 1 \end{pmatrix}$ for any $k \in \mathbb{R}$.

**ii** If $L_1$ and $L_2$ intersect then $\frac{5}{2} + 3t = m + 2s$, $\frac{1}{2} + 2t = 1 - 3s$ and $-\frac{9}{2} - t = -1 + s$

$\therefore \begin{cases} 3s + 2t = \frac{1}{2} \\ s + t = -\frac{7}{2} \end{cases}$

$\therefore$

$3s + 2t = \frac{1}{2}$

$-2s - 2t = 7$

adding: $s = 7\frac{1}{2}$ and so $t = -11$

$\therefore \frac{5}{2} + (-33) = m + 15$

$\therefore -30\frac{1}{2} = m + 15$

$\therefore m = -45\frac{1}{2}$

**iii** Using $L_1$, when $t = -11$,

$\mathbf{r}_1 = \frac{1}{2}\begin{pmatrix} 5 \\ 1 \\ -9 \end{pmatrix} - 11\begin{pmatrix} 3 \\ 2 \\ -1 \end{pmatrix} = \begin{pmatrix} -30\frac{1}{2} \\ -21\frac{1}{2} \\ 6\frac{1}{2} \end{pmatrix}$

So, P is at $(-30\frac{1}{2}, -21\frac{1}{2}, 6\frac{1}{2})$.

**44** **a** The gradient of $y = \frac{1}{2}x - 1$ is $\frac{1}{2}$

$\therefore \tan\theta = \frac{1}{2}$

$\therefore \theta = \tan^{-1}(\frac{1}{2})$

**b** $(f \circ g)(x) = f(g(x))$

$= \frac{1}{2}(\sqrt{3}x) - 1$

$= \frac{\sqrt{3}}{2}x - 1$ which has gradient $\frac{\sqrt{3}}{2}$

$\therefore \tan\theta = \frac{\sqrt{3}}{2}$

$\therefore \theta = \tan^{-1}(\frac{\sqrt{3}}{2})$

**45** **a** **i** $y = -x^2 + 12x - 20$ has $a = -1 < 0$

$\therefore$ its shape is ∩, and so the quadratic has a maximum value.

**ii** $y$ is maximised when $x = \frac{-b}{2a} = \frac{-12}{-2} = 6$

**iii** When $x = 6$, $y = -6^2 + 12(6) - 20 = -36 + 72 - 20 = 16$

$\therefore$ the maximum value of $y = -x^2 + 12x - 20$ is 16.

**b** **i** The perimeter is 20.

$\therefore x + y + 8 = 20$

$\therefore y = 12 - x$

**ii** $y^2 = x^2 + 8^2 - 2 \times x \times 8 \times \cos\theta$

$\therefore y^2 = x^2 + 64 - 16x\cos\theta$

**iii** Since $y = 12 - x$, $(12-x)^2 = x^2 + 64 - 16x\cos\theta$

$\therefore\ 144 - 24x + x^2 = x^2 + 64 - 16x\cos\theta$

$\therefore\ 16x\cos\theta = 24x - 80$

$\therefore\ \cos\theta = \dfrac{24x-80}{16x} = \dfrac{3x-10}{2x}$

**iv** Area $A = \frac{1}{2} \times x \times 8 \times \sin\theta$

$= 4x\sin\theta$

$\therefore\ A^2 = 16x^2\sin^2\theta$

**v** $A^2 = 16x^2(1-\cos^2\theta)$

$= 16x^2\left[1 - \left(\dfrac{3x-10}{2x}\right)^2\right]$

$= 16x^2\left[1 - \dfrac{9x^2-60x+100}{4x^2}\right]$

$= 16x^2 - 4(9x^2-60x+100)$

$= 16x^2 - 36x^2 + 240x - 400$

$= -20x^2 + 240x - 400$

$= 20(-x^2+12x-20)$

**vi** $A$ is maximised when $A^2$ is maximised since $A > 0$.
From **a**, $-x^2+12x-20$ has a maximum value of 16 when $x = 6$.
When $x = 6$, $A^2 = 20(16)$

$= 320$

$\therefore\ A = \sqrt{320}$ $\{A > 0\}$

$= 8\sqrt{5}$

$\therefore$ the maximum area of the triangle is $8\sqrt{5}$ units$^2$.

**vii** When $x = 6$, $y = 12 - 6$

$= 6$

$\therefore$ the triangle is isosceles with AB = BC.

**46** **a** $f(x) = 4x - 3$ has inverse $x = 4y - 3$

$\therefore\ x + 3 = 4y$

$\therefore\ f^{-1}(x) = \dfrac{x+3}{4}$

$g(x) = x + 2$ has inverse $x = y + 2$

$\therefore\ y = x - 2$

$\therefore\ g^{-1}(x) = x - 2$

**b** $(f \circ g^{-1})(x) = f(g^{-1}(x))$

$= f(x-2)$

$= 4(x-2) - 3$

$= 4x - 11$

**c** $(f \circ g^{-1})(x) = f^{-1}(x)$

$\therefore\ 4x - 11 = \dfrac{x+3}{4}$

$\therefore\ 16x - 44 = x + 3$

$\therefore\ 15x = 47$

$\therefore\ x = \frac{47}{15}$

**d** **i, iv** $H(x) = \dfrac{4x-3}{x+2}$

$\therefore\ H(0) = \dfrac{4(0)-3}{0+2} = -\frac{3}{2}$

So the $y$-intercept is $-1\frac{1}{2}$.

When $H(x) = 0$, $4x - 3 = 0$

$\therefore\ x = \frac{3}{4}$

$\therefore$ the $x$-intercept is $\frac{3}{4}$.

$g(x) = 0$ when $x = -2$, so $x = -2$ is a vertical asymptote.

As $x \to -2^-$, $H(x) \to \infty$

As $x \to -2^+$, $H(x) \to -\infty$

$H(x) = \dfrac{4x-3}{x+2} = \dfrac{4-\frac{3}{x}}{1+\frac{2}{x}}$

As $x \to \infty$, $H(x) \to 4^-$

As $x \to -\infty$, $H(x) \to 4^+$

so $y = 4$ is a horizontal asymptote.

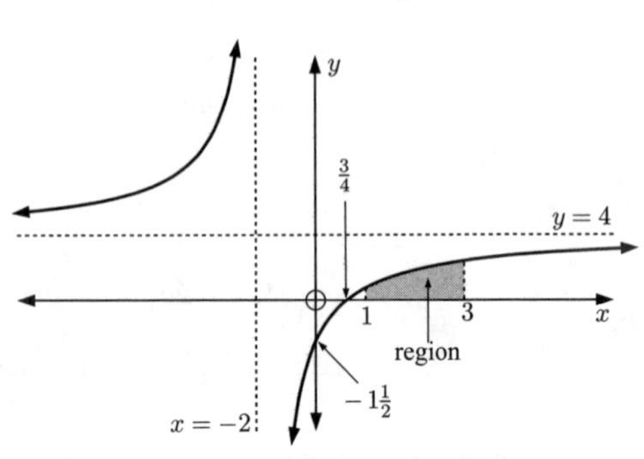

**ii**
$$\frac{4x-3}{x+2} = A + \frac{B}{x+2}$$
$$= \frac{A(x+2)+B}{x+2}$$
$\therefore \quad Ax + 2A + B = 4x - 3 \quad$ for all $x$
$\therefore \quad A = 4 \quad$ and $\quad 2A + B = -3$
$\therefore \quad 8 + B = -3$
$\therefore \quad B = -11$

**iii**
$$\int_{-1}^{2} H(x)\,dx = \int_{-1}^{2} \left(4 + \frac{-11}{x+2}\right)\,dx$$
$$= [4x - 11\ln(x+2)]_{-1}^{2} \qquad \{x+2>0 \text{ for } -1 \leqslant x \leqslant 2\}$$
$$= (8 - 11\ln 4) - (-4 - 11\ln 1)$$
$$= 8 - 11\ln 4 + 4$$
$$= 12 - 11\ln 4$$

**47** **a** Hannah is wrong as $0.1 + 0.3 + 0.3 + 0.2 + 0.2 = 1.1 \neq 1$

**b** $0.2 + a + 0.3 + b + 0.2 = 1$
$\therefore \quad a + b = 0.3, \quad 0 \leqslant a \leqslant 0.3, \quad 0 \leqslant b \leqslant 0.3$

**c** **i** $\text{P}(X=2) = \dfrac{2 \times 4}{50} = 0.16$

**ii** $\text{P}(X \neq 2) = 1 - \text{P}(X=2)$
$= 0.84$

**48** **a**

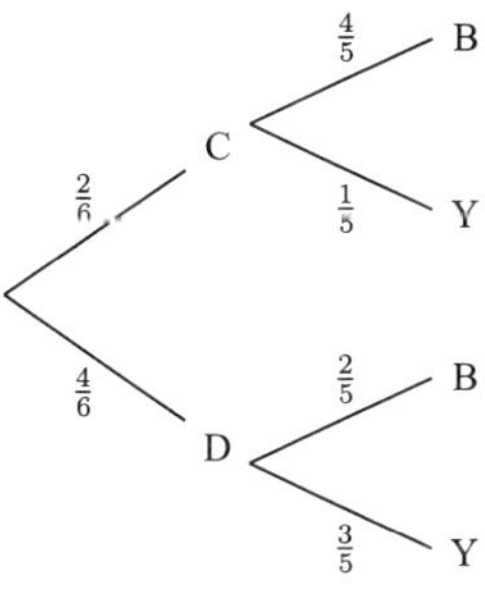


**b** P(a yellow drawn from D)
$= \text{P(D and Y)}$
$= \frac{4}{6} \times \frac{3}{5}$
$= \frac{2}{5}$

**c** P(a yellow drawn from either bag)
$= \text{P}(\text{C} \cap \text{Y} \text{ or } \text{D} \cap \text{Y})$
$= \frac{2}{6} \times \frac{1}{5} + \frac{4}{6} \times \frac{3}{5}$
$= \frac{14}{30}$
$= \frac{7}{15}$

**d** $\text{P}(\text{D} \mid \text{B})$
$$= \frac{\text{P}(\text{D} \cap \text{B})}{\text{P(B)}}$$
$$= \frac{\frac{4}{6} \times \frac{2}{5}}{\frac{2}{6} \times \frac{4}{5} + \frac{4}{6} \times \frac{2}{5}}$$
$$= \frac{8}{8+8}$$
$$= \frac{1}{2}$$

**e** $\text{P(B)} = \frac{2}{6} \times \frac{4}{5} + \frac{4}{6} \times \frac{2}{5} = \frac{16}{30}$
$\text{P(Y)} = \frac{2}{6} \times \frac{1}{5} + \frac{4}{6} \times \frac{3}{5} = \frac{14}{30}$
$\therefore$ expected return
$= \frac{16}{30} \times \$6 + \frac{14}{30} \times \$9$
$= \dfrac{\$96}{30} + \dfrac{\$126}{30}$
$= \$3.20 + \$4.20$
$= \$7.40$

**49** **a**

die 2
4
3
2
1
1 2 3 4
die 1

**b** The possible values of $X$ range from $1 + 1 = 2$ to $4 + 4 = 8$.
$\therefore$ the possible values of $X$ are 2, 3, 4, 5, 6, 7, and 8.

**c** **i** 3 of the 16 possible outcomes result in a sum of 4. {those shaded}

$\therefore$ $P(X = 4) = \frac{3}{16}$

**ii** 10 of the 16 possible outcomes result in a sum greater than 4.

$\therefore$ $P(X > 4) = \frac{10}{16} = \frac{5}{8}$

**d** From **c**, $P(X < 4) = 1 - \frac{3}{16} - \frac{10}{16} = \frac{3}{16}$

$$\begin{aligned}\text{Expectation} &= \tfrac{3}{16} \times 5 + \tfrac{5}{8} \times 1 - \tfrac{3}{16} \times d \\ &= \frac{25}{16} - \frac{3d}{16} \\ &= \left(\frac{25 - 3d}{16}\right) \text{ euros}\end{aligned}$$

The expectation = €0 when $25 - 3d = 0$

$\therefore$ $d = 8\frac{1}{3}$ $(\approx 8.33)$

**50** $a(t) = 3t - \sin t \text{ cm s}^{-2}$

**a** $a(0) = 3(0) - \sin 0 = 0 \text{ cm s}^{-2}$

$a(\frac{\pi}{2}) = 3(\frac{\pi}{2}) - \sin(\frac{\pi}{2}) = (\frac{3\pi}{2} - 1) \text{ cm s}^{-2}$

**b** $v(t) = \int a(t)\, dt$

$\therefore$ $v(t) = \dfrac{3t^2}{2} + \cos t + c \text{ cm s}^{-1}$

But $v(0) = 3$

$\therefore$ $0 + \cos 0 + c = 3$

$\therefore$ $c = 2$

Thus $v(t) = \frac{3}{2}t^2 + \cos t + 2 \text{ cm s}^{-1}$

**c** $$\begin{aligned}&\int_0^{\frac{\pi}{2}} v(t)\, dt \\ &= \int_0^{\frac{\pi}{2}} \left(\frac{3t^2}{2} + \cos t + 2\right) dt \\ &= \left[\frac{t^3}{2} + \sin t + 2t\right]_0^{\frac{\pi}{2}} \\ &= \left(\frac{\pi^3}{16} + 1 + \pi\right) - (0) \\ &= \left(\frac{\pi^3}{16} + \pi + 1\right) \text{ cm}\end{aligned}$$

As $\pi \approx 3.14$, $\dfrac{\pi^3}{16} + \pi + 1 > 0$

$\therefore$ $\int_0^{\frac{\pi}{2}} v(t)\, dt > 0$

**d** This integral represents the displacement of the particle in the first $\frac{\pi}{2}$ seconds of motion.

**51** $f(t) = a \sin b(t - c) + d$

**a** $a = \text{amplitude} = \frac{17-3}{2} = 7$

$\text{period} = \dfrac{2\pi}{b} = 13 - -3 = 16$

$\therefore$ $b = \frac{\pi}{8}$

$c$ = the $x$-coordinate of the point halfway between the first minimum and the following maximum

$= \dfrac{-3 + 5}{2} = 1$

The equation of the principal axis is $y = \dfrac{3 + 17}{2} = 10$ $\therefore$ $d = 10$

**b** **i** Under a translation of $\binom{2}{-3}$, $A(5, 17) \mapsto A_1(7, 14)$

Under a vertical stretch with scale factor 2, $A_1(7, 14) \mapsto A'(7, 28)$

$\therefore$ $A(5, 17) \mapsto A'(7, 28)$

**ii** Under a translation of $\binom{2}{-3}$, $y = 7 \sin \frac{\pi}{8}(x - 1) + 10$ becomes

$y = 7 \sin \frac{\pi}{8}(x - 1 - 2) + 10 - 3$

$\therefore$ $y = 7 \sin \frac{\pi}{8}(x - 3) + 7$

and then under a vertical stretch of factor 2 it becomes $y = 2\left[7 \sin \frac{\pi}{8}(x - 3) + 7\right]$

$\therefore$ $y = g(x) = 14 \sin \frac{\pi}{8}(x - 3) + 14$

**iii** To map $g$ back to $f$, we perform the inverse transformations in the reverse order.

So, the transformation is a vertical stretch of factor $\frac{1}{2}$ followed by a translation of $\binom{-2}{3}$.

**52** **a** $4^x - 2^x - 20$
$= (2^x)^2 - 2^x - 20$
$= (2^x + 4)(2^x - 5)$

**b** $2^x(2^x - 1) = 20$
$\therefore\ 2^{2x} - 2^x - 20 = 0$
$\therefore\ (2^x + 4)(2^x - 5) = 0$
$\therefore\ 2^x = -4$ or $5$

But $2^x$ is never negative
$\therefore\ 2^x = 5$
$\therefore\ \log 2^x = \log 5$
$\therefore\ x \log 2 = \log 5$
$\therefore\ x = \dfrac{\log 5}{\log 2}$
or $\log_2 5$

**c** **i** If $p = \log_5 2$ then
$p = \dfrac{\log 2}{\log 5}$
$\therefore\ x = \dfrac{1}{p}$

**ii** $8^x = 5^{1-x}$
$\therefore\ 2^{3x} = 5^{1-x}$
$\therefore\ 3x \log 2 = (1-x)\log 5$
$\therefore\ \dfrac{1-x}{3x} = \dfrac{\log 2}{\log 5} = p$
$\therefore\ 1 - x = 3px$
$\therefore\ x(3p+1) = 1$
$\therefore\ x = \dfrac{1}{3p+1}$

**53** **a** $f(x) = a\cos 2x + b\sin^2 x$
$= a\cos 2x + b[\sin x]^2$
$\therefore\ f'(x) = a(-2\sin 2x) + 2b[\sin x]^1 \cos x$
$= -2a\sin 2x + b\sin 2x$
$= (b - 2a)\sin 2x$

**b** $b < 2a \quad \therefore\ b - 2a < 0$
$\therefore$ the max value of $f'(x)$ is $2a - b$
when $\sin 2x = -1$
$\therefore\ 2x = \frac{3\pi}{2} + k2\pi,\ k \in \mathbb{Z}$
$\therefore\ x = \frac{3\pi}{4} + k\pi$

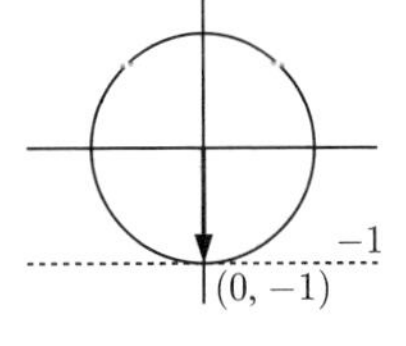


So, the maximum value of $f'(x)$ is $2a - b$ when $x = \frac{3\pi}{4}$ or $\frac{7\pi}{4}$ $\{0 \leqslant x \leqslant 2\pi\}$

**c** $f'(x) = 0$ when $\sin 2x = 0$
$\therefore\ 2x = 0 + k\pi, \quad k \in \mathbb{Z}$
$\therefore\ x = \frac{k\pi}{2}$
$\therefore\ x = 0, \frac{\pi}{2}, \pi, \frac{3\pi}{2}, 2\pi$

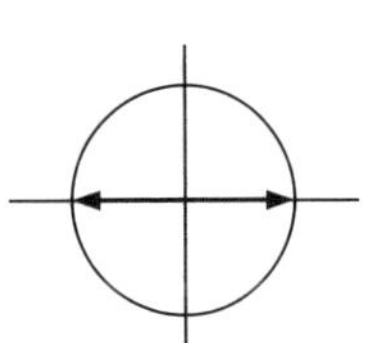

Sign diagram of $f'(x)$:

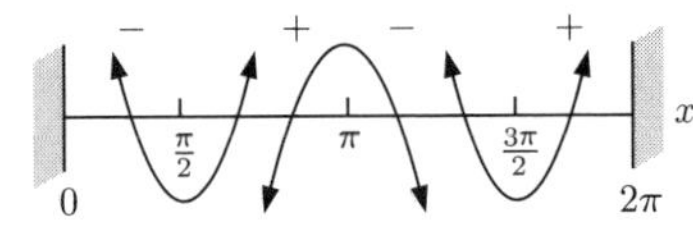


$f(0) = a\cos 0 + b\sin^2 0 = a$
$f(\frac{\pi}{2}) = a\cos\pi + b\sin^2(\frac{\pi}{2}) = b - a$
$f(\pi) = a\cos 2\pi + b\sin^2\pi = a$
$f(\frac{3\pi}{2}) = a\cos 3\pi + b\sin^2(\frac{3\pi}{2}) = b - a$
$f(2\pi) = a\cos 4\pi + b\sin^2(2\pi) = a$

So, the maximum turning points are $(0, a)$, $(\pi, a)$, and $(2\pi, a)$, and the minimum turning points are $(\frac{\pi}{2}, b-a)$ and $(\frac{3\pi}{2}, b-a)$.

**54** **a** $[C(x)]^2 - [S(x)]^2 = \left(\frac{e^x + e^{-x}}{2}\right)^2 - \left(\frac{e^x - e^{-x}}{2}\right)^2$

$= \left(\frac{e^x + e^{-x}}{2} + \frac{e^x - e^{-x}}{2}\right)\left(\frac{e^x + e^{-x}}{2} - \frac{e^x - e^{-x}}{2}\right)$

$= \left(\frac{2e^x}{2}\right)\left(\frac{2e^{-x}}{2}\right)$

$= e^0$

$= 1$

**b** $\frac{d}{dx}[S(x)] = \frac{1}{2}\left(e^x - e^{-x}(-1)\right)$

$= \frac{e^x + e^{-x}}{2}$

$= C(x)$

**c** $\frac{d}{dx}[C(x)] = \frac{1}{2}\left(e^x + e^{-x}(-1)\right)$

$= \frac{e^x - e^{-x}}{2}$

$= S(x)$

**d** $T(x) = \frac{S(x)}{C(x)}$

$\therefore \frac{d}{dx}(T(x)) = \frac{S'(x)C(x) - S(x)C'(x)}{[C(x)]^2}$

$= \frac{[C(x)]^2 - [S(x)]^2}{[C(x)]^2}$ {using **b**, **c**}

$= \frac{1}{[C(x)]^2}$ {using **a**}

**55** $P(t) = \frac{60\,000}{1 + 2e^{-\frac{t}{4}}}, \quad t \geqslant 0$

$= 60\,000\left(1 + 2e^{-\frac{t}{4}}\right)^{-1}$

**a** $P(0) = \frac{60\,000}{1 + 2e^0}$

$= \frac{60\,000}{3}$

$= 20\,000$

**b** $P'(t) = -60\,000\left(1 + 2e^{-\frac{t}{4}}\right)^{-2} \times 2e^{-\frac{t}{4}}\left(-\frac{1}{4}\right)$

$= \frac{30\,000e^{-\frac{t}{4}}}{\left(1 + 2e^{-\frac{t}{4}}\right)^2}$

**c** Since $e^{-\frac{t}{4}} > 0$ for all $t$ and $\left(1 + 2e^{-\frac{t}{4}}\right)^2 > 0$ for all $t$, $P'(t) > 0$ for all $t \geqslant 0$.

This means $P(t)$ is increasing for all $t \geqslant 0$.

**d** $P''(t) = \frac{30\,000e^{-\frac{t}{4}}\left(-\frac{1}{4}\right)\left(1 + 2e^{-\frac{t}{4}}\right)^2 - 30\,000e^{-\frac{t}{4}} \times 2\left(1 + 2e^{-\frac{t}{4}}\right)^1 2e^{-\frac{t}{4}}\left(-\frac{1}{4}\right)}{\left(1 + 2e^{-\frac{t}{4}}\right)^4}$

$= \frac{-7500e^{-\frac{t}{4}}\left(1 + 2e^{-\frac{t}{4}}\right) + 30\,000e^{-\frac{t}{2}}}{\left(1 + 2e^{-\frac{t}{4}}\right)^3}$

$= \frac{-7500e^{-\frac{t}{4}} - 15\,000e^{-\frac{t}{2}} + 30\,000e^{-\frac{t}{2}}}{\left(1 + 2e^{-\frac{t}{4}}\right)^3}$

$= \frac{15\,000e^{-\frac{t}{2}} - 7500e^{-\frac{t}{4}}}{\left(1 + 2e^{-\frac{t}{4}}\right)^3}$

$= \frac{7500e^{-\frac{t}{4}}\left(2e^{-\frac{t}{4}} - 1\right)}{\left(1 + 2e^{-\frac{t}{4}}\right)^3}$

**e** The growth is given by $P'(t)$ which is maximised when $P''(t)=0$

$\therefore\ 2e^{-\frac{t}{4}}=1$

$\therefore\ e^{-\frac{t}{4}}=\frac{1}{2}$

$\therefore\ -\dfrac{t}{4}=\ln(0.5)$

$\therefore\ t=-4\ln(0.5)$

$\therefore\ t=4\ln 2$

At this time, $P'(t)=\dfrac{30\,000(\frac{1}{2})}{\left(1+2(\frac{1}{2})\right)^2}=\dfrac{15\,000}{4}$

$\therefore$ the maximum growth rate is 3750 per year at $t=4\ln 2$ years.

**f** As $t\to\infty$, $e^{-\frac{t}{4}}\to 0$

$\therefore\ P(t)\to\dfrac{60\,000}{1+2(0)}$

$\therefore\ P(t)\to 60\,000^-$

**g**

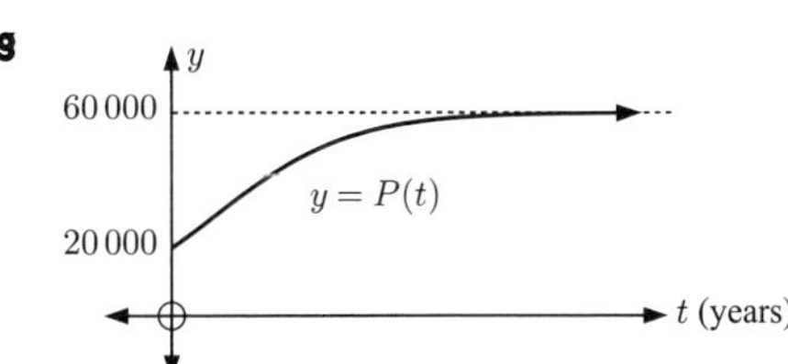


**56** **a** $\int x^2e^{1-x^3}\,dx$

$=\displaystyle\int -\tfrac{1}{3}\frac{du}{dx}e^u\,dx$

$=-\tfrac{1}{3}\displaystyle\int e^u\frac{du}{dx}\,dx$

$=-\tfrac{1}{3}\int e^u\,du$

$=-\tfrac{1}{3}e^u+c$

$=-\tfrac{1}{3}e^{1-x^3}+c$

$u=1-x^3$

$\therefore\ \dfrac{du}{dx}=-3x^2$

$\therefore\ x^2=-\tfrac{1}{3}\dfrac{du}{dx}$

**b** $\int_0^1 x^2e^{1-x^3}\,dx$

$=\left[-\tfrac{1}{3}e^{1-x^3}\right]_0^1$

$=-\tfrac{1}{3}e^{1-1^3}-\left(-\tfrac{1}{3}e^1\right)$

$=-\tfrac{1}{3}e^0+\tfrac{1}{3}e$

$=\dfrac{e}{3}-\dfrac{1}{3}$

$=\dfrac{e-1}{3}$

**57** $f(x)=\cos^3x=[\cos x]^3$

**a** As $-1\leqslant\cos x\leqslant 1$ then $-1\leqslant\cos^3x\leqslant 1$

$\therefore$ the range is $\{y\mid -1\leqslant y\leqslant 1\}$

**b** $8\cos^3x=1$

$\therefore\ \cos^3x=\frac{1}{8}$

$\therefore\ \cos x=\frac{1}{2}$

$\therefore\ x=\frac{\pi}{3},\ \frac{5\pi}{3}$

So, there are 2 solutions.

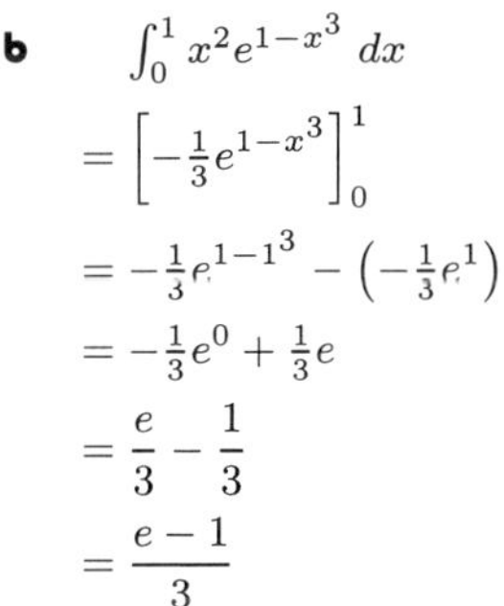


**c** $f'(x)=3[\cos x]^2(-\sin x)$

$=-3\sin x\cos^2x$

**d** Volume $=\pi\int_0^{\frac{\pi}{2}}\left[\sqrt{3}\cos x\sqrt{\sin x}\,\right]^2dx$

$=\pi\int_0^{\frac{\pi}{2}}(3\cos^2x\sin x)\,dx$

$=-\pi\int_0^{\frac{\pi}{2}}(-3\sin x\cos^2x)\,dx$

$=-\pi\left[(\cos x)^3\right]_0^{\frac{\pi}{2}}$

$=-\pi(0^3-1^3)$

$=\pi$ units$^3$

**58**

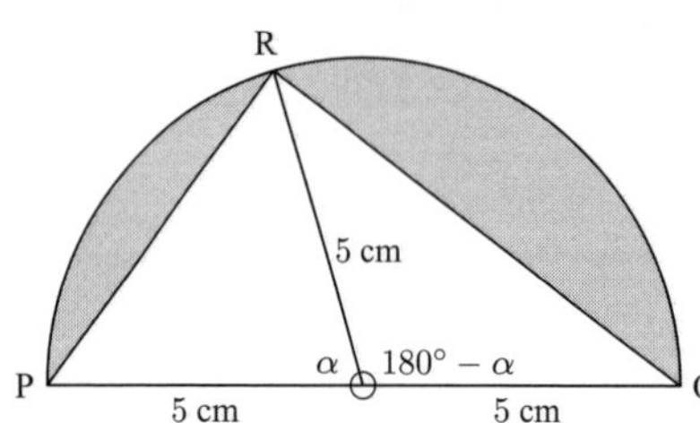


**a** Area of $\triangle$PQR

$= \frac{1}{2}(5^2)\sin\alpha + \frac{1}{2}(5^2)\sin(180° - \alpha)$

$= \frac{25}{2}(\sin\alpha + \sin\alpha)$

$= 25\sin\alpha \text{ cm}^2$

**b** $A = \frac{1}{2}\pi(5^2) - 25\sin\alpha$

$= (\frac{25\pi}{2} - 25\sin\alpha) \text{ cm}^2$

**c** Now $0 \leqslant \sin\alpha \leqslant 1$ $\{$as $0 \leqslant \alpha \leqslant \pi\}$

$\therefore$ the area is a minimum when $\sin\alpha = 1$

$\therefore$ $A_{\min} = \frac{25\pi}{2} - 25(1)$

$= 25(\frac{\pi}{2} - 1)$ cm$^2$ when $\alpha = \frac{\pi}{2}$

The area is a maximum when $\sin\alpha = 0$.

$\therefore$ $A_{\max} = \frac{25\pi}{2}$ cm$^2$ when $\alpha = 0$ or $\pi$.

**59** **a** The vertex is $\left(\frac{1+7}{2}, 18\right)$ which is $(4, 18)$.

Thus **i** $h = 4$ **ii** $k = 18$

$\therefore$ $f(x) = a(x-4)^2 + 18$

**iii** Now $f(1) = 0$

$\therefore$ $a(-3)^2 + 18 = 0$

$\therefore$ $9a = -18$

$\therefore$ $a = -2$

**b** $f(x) = -2(x-4)^2 + 18$

$= -2(x^2 - 8x + 16) + 18$

$= -2x^2 + 16x - 14$

$\therefore$ the shaded area $= \int_1^3 (-2x^2 + 16x - 14)\, dx$

$= \left[-\frac{2}{3}x^3 + 8x^2 - 14x\right]_1^3$

$= \left(-\frac{2}{3}(27) + 8(9) - 42\right) - \left(-\frac{2}{3} + 8 - 14\right)$

$= 12 + 6\frac{2}{3} = 18\frac{2}{3}$ units$^2$

**60** **a** **i** $2^{1-2x} = 0.5 = \frac{1}{2}$

$\therefore$ $2^{1-2x} = 2^{-1}$

$\therefore$ $1 - 2x = -1$

$\therefore$ $x = 1$

**ii** $\log_x 7 = 5$

$\therefore$ $x^5 = 7$

$\therefore$ $x = \sqrt[5]{7}$

**b** $25^x - 6(5^x) + 5 = 0$

$\therefore$ $(5^x)^2 - 6(5^x) + 5 = 0$

$\therefore$ $(5^x - 1)(5^x - 5) = 0$

$\therefore$ $5^x = 1$ or $5$

$\therefore$ $x = 0$ or $1$

**c** $2^x = 3^{1-x}$

$\therefore$ $2^x = \dfrac{3^1}{3^x}$

$\therefore$ $6^x = 3$

$\therefore$ $x = \log_6 3$

**61** **a** **i** $\dfrac{1 - \cos 2\theta}{\sin 2\theta} = \sqrt{3}$, $0 < \theta < \frac{\pi}{2}$

$\therefore$ $\dfrac{1 - (1 - 2\sin^2\theta)}{2\sin\theta\cos\theta} = \sqrt{3}$

$\therefore$ $\dfrac{\cancel{2}\sin^{\cancel{2}}\theta}{\cancel{2}\cancel{\sin\theta}\cos\theta} = \sqrt{3}$

$\therefore$ $\dfrac{\sin\theta}{\cos\theta} = \sqrt{3}$ $\{\sin\theta \neq 0$ on $0 < \theta < \frac{\pi}{2}\}$

$\therefore$ $\tan\theta = \sqrt{3}$

**ii** As $0 < \theta < \frac{\pi}{2}$, $\theta = \frac{\pi}{3}$.

**b** If $\cos 2x = 2\cos x$ then $2\cos^2 x - 1 = 2\cos x$

$$\therefore \quad 2\cos^2 x - 2\cos x - 1 = 0$$

$$\therefore \quad \cos x = \frac{2 \pm \sqrt{4 - 4(2)(-1)}}{4} = \frac{2 \pm \sqrt{12}}{4} = \frac{2 \pm 2\sqrt{3}}{4} = \frac{1 \pm \sqrt{3}}{2}$$

$$\therefore \quad \cos x = \frac{1 - \sqrt{3}}{2} \quad \{-1 \leqslant \cos x \leqslant 1\}$$

**62** $S_n = n^3 + 2n - 1$

Now $u_n = S_n - S_{n-1}, \quad n > 1$

$$= n^3 + 2n - 1 - \left[(n-1)^3 + 2(n-1) - 1\right]$$
$$= n^3 + 2n - 1 - \left[n^3 - 3n^2 + 3n - 1 + 2n - 2 - 1\right]$$
$$= \cancel{n^3} + \cancel{2n} - \cancel{1} - \cancel{n^3} + 3n^2 - 3n + \cancel{1} - \cancel{2n} + 3$$
$$= 3n^2 - 3n + 3, \quad n > 1$$

and $u_1 = S_1 = 2$

$$\therefore \quad u_1 = 2, \ u_n = 3n^2 - 3n + 3, \ n > 1$$

**63** $\sin^2 x + \sin x - 2 = 0, \quad -2\pi \leqslant x \leqslant 2\pi$

$\therefore \quad (\sin x + 2)(\sin x - 1) = 0$

$\therefore \quad \sin x = -2$ or $1$

$\therefore \quad \sin x = 1 \qquad \{$as $-1 \leqslant \sin x \leqslant 1\}$

$\therefore \quad x = \frac{\pi}{2} + k2\pi, \quad k$ an integer

$\therefore \quad x = -\frac{3\pi}{2}$ or $\frac{\pi}{2}$

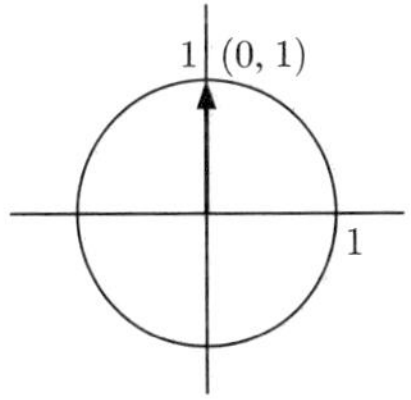


**64** $f(x) = \ln x$ has inverse $f^{-1}(x) = e^x$

$g(x) = 3 + x$ has inverse given by $x = 3 + y$

$\therefore \quad y = x - 3$

so $g^{-1}(x) = x - 3$

**a** $f^{-1}(2) \times g^{-1}(2)$
$= e^2 \times -1$
$= -e^2$

**b** $(f \circ g)(x) = f(g(x)) = f(3 + x)$
$= \ln(3 + x)$

$\therefore$ the inverse of $(f \circ g)(x)$ is $x = \ln(3 + y)$

$\therefore \quad 3 + y = e^x$

$\therefore \quad y = e^x - 3$

So, $(f \circ g)^{-1}(x) = e^x - 3$

and $(f \circ g)^{-1}(2) = e^2 - 3$

**65**

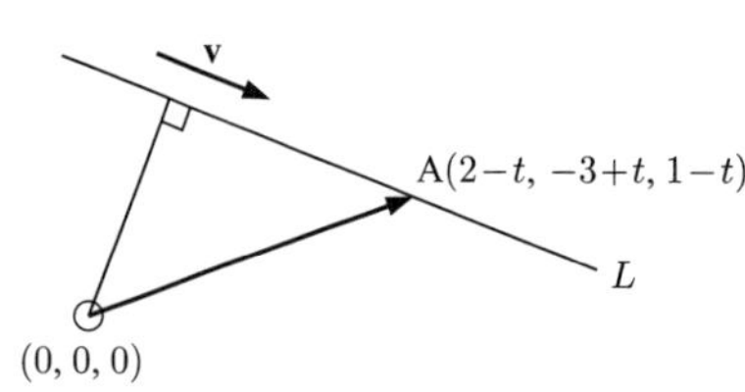


$\overrightarrow{OA} = \begin{pmatrix} 2-t \\ t-3 \\ 1-t \end{pmatrix}$ and $\mathbf{v} = \begin{pmatrix} -1 \\ 1 \\ -1 \end{pmatrix}$

The shortest distance occurs when $\overrightarrow{OA} \bullet \mathbf{v} = 0$

$\therefore \; -(2-t) + t - 3 + (-1)(1-t) = 0$

$\therefore \; t - 2 + t - 3 - 1 + t = 0$

$\therefore \; 3t = 6$

$\therefore \; t = 2$

So, the point on $L$ that is nearest the origin is $(2-2, -3+2, 1-2)$, which is $(0, -1, -1)$.

**66** $f'(x) > 0$ and $f''(x) < 0$ for all $x$

$\therefore$ $f(x)$ is increasing and concave downwards for all $x$.

**a** $f(2) = 1$ and $f'(2) = 2$

$\therefore$ $(2, 1)$ lies on the curve and the tangent at this point has gradient 2

$\therefore$ the equation of the tangent is $y = 2x + c$

and $1 = 2(2) + c$, so $c = -3$

$\therefore$ the tangent has equation $y = 2x - 3$.

**b**

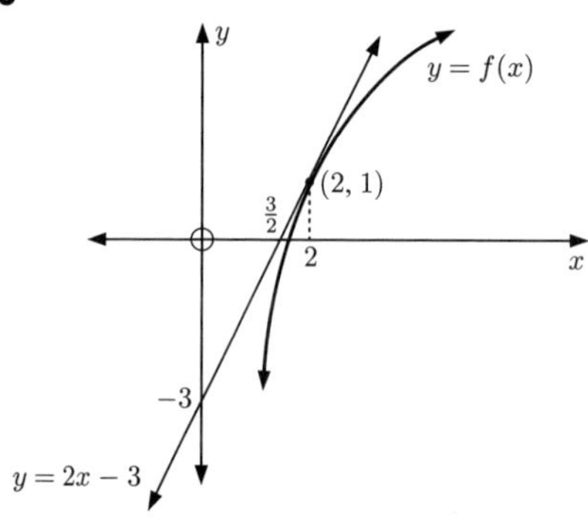


**c** As $f(x)$ is increasing it has *at most one* zero. But $f(x)$ is also concave downwards for all $x$, so it always lies below the tangent shown. So, for $x < \frac{3}{2}$, the tangent's $y$-values are negative and so $f(x)$ is also negative. Thus $f(x)$ has *exactly one* zero.

**d** From the graph, the $x$-intercept of $y = f(x)$ lies inside $\frac{3}{2} < x < 2$.

**67** **a**

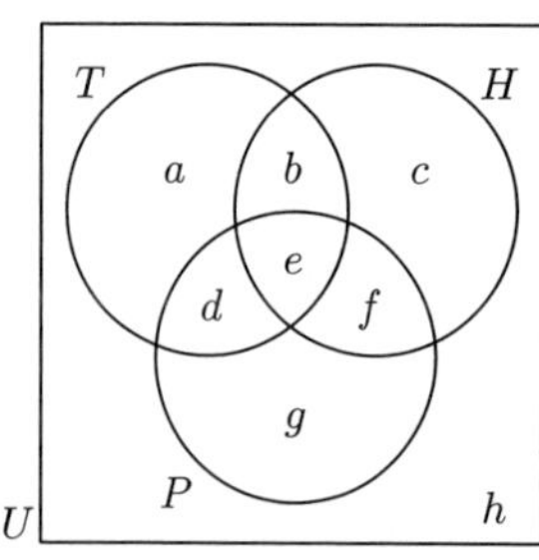


$a + b + c + d + e + f + g + h = 30$

$a + b + d + e = 13$

$b + c + e + f = 12$

$d + e + f + g = 13$

$e = 2$

$b + e = 5$

$e + f = 4$

$d + e = 3$

So, $b = 3$, $f = 2$, and $d = 1$

$\therefore \; \begin{cases} a + 3 + 1 + 2 = 13 \\ 3 + c + 2 + 2 = 12 \\ 1 + 2 + 2 + g = 13 \end{cases} \quad \therefore \; \begin{cases} a = 7 \\ c = 5 \\ g = 8 \end{cases}$

$\therefore \; 7 + 3 + 5 + 1 + 2 + 2 + 8 + h = 30$

$\therefore \; h = 2$

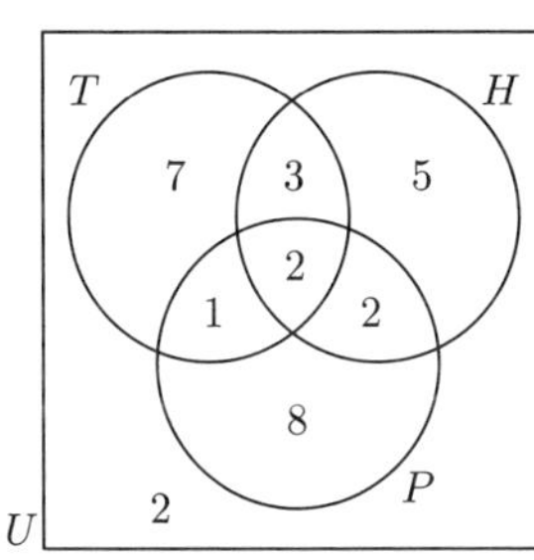


**b** **i** $P(T \cap H) = \dfrac{3+2}{30} = \frac{5}{30} = \frac{1}{6}$

**ii** $P(P) = \frac{13}{30}$

**iii** $P(H \cap P') = \dfrac{3+5}{30} = \frac{8}{30} = \frac{4}{15}$

**iv** $P(T \cup P) = \dfrac{7+3+1+2+2+8}{30} = \frac{23}{30}$

**v** $P(T \cap H' \cap P) = \frac{1}{30}$

**68** After labelling the triangles,

$\triangle$ABC and $\triangle$ABD are isosceles {equal base angles}

and $\widehat{BDC} = 2\alpha$ {exterior angle of $\triangle$ABD}

$\therefore$ $\triangle$BDC is isosceles {equal base angles}

Thus AD = BD = BC = $x$, say.

Now $\triangle$ADE and $\triangle$BDE are congruent {AAcorS}

$\therefore$ AE = EB

If we let AE = EB = 1 unit, then AC = 2 and DC = $2 - x$.

Now $\triangle$ABC and $\triangle$BCD are similar {equiangular}

$$\therefore \quad \frac{AB}{BC} = \frac{BC}{CD} \quad \text{and so} \quad \frac{2}{x} = \frac{x}{2-x}$$

$$\therefore \quad x^2 = 4 - 2x$$

$$\therefore \quad x^2 + 2x - 4 = 0$$

$$\therefore \quad x = \frac{-2 \pm \sqrt{4 - 4(-4)}}{2}$$

$$\therefore \quad x = \frac{-2 \pm 2\sqrt{5}}{2}$$

$$\therefore \quad x = -1 \pm \sqrt{5}$$

$\therefore \quad x = \sqrt{5} - 1$ {as $x$ must be $> 0$}

Now $5\alpha = 180°$ {angle sum of a triangle}

$\therefore \quad \alpha = 36°$

$$\text{so in } \triangle\text{BED}, \quad \cos 36° = \frac{1}{\sqrt{5}-1} = \frac{1}{\sqrt{5}-1}\left(\frac{\sqrt{5}+1}{\sqrt{5}+1}\right) = \frac{1+\sqrt{5}}{4}$$

**69**

$$\text{Area } = \int_a^{a+2} x^2\,dx = \frac{31}{6}$$

$$\therefore \quad \left[\frac{x^3}{3}\right]_a^{a+2} = \frac{31}{6}$$

$$\therefore \quad \frac{(a+2)^3}{3} - \frac{a^3}{3} = \frac{31}{6}$$

$$\therefore \quad \frac{\cancel{a^3} + 6a^2 + 12a + 8 - \cancel{a^3}}{3} = \frac{31}{6}$$

$$\therefore \quad 12a^2 + 24a + 16 = 31$$

$$\therefore \quad 12a^2 + 24a - 15 = 0$$

$$\therefore \quad 4a^2 + 8a - 5 = 0$$

$$\therefore \quad (2a-1)(2a+5) = 0$$

$$\therefore \quad a = \tfrac{1}{2} \text{ or } -\tfrac{5}{2}$$

But $a > 0$, so $a = \frac{1}{2}$

**70** If $X$ and $Y$ are independent events then $\mathrm{P}(X \cap Y) = \mathrm{P}(X)\,\mathrm{P}(Y)$

Thus $\mathrm{P}((A \cap B) \cap (A \cup B)) = \mathrm{P}(A \cap B)\,\mathrm{P}(A \cup B)$

$\therefore\ \mathrm{P}(A \cap B) = \mathrm{P}(A \cap B)\,\mathrm{P}(A \cup B)$ {since $A \cap B \subseteq A \cup B$}

$\therefore\ \mathrm{P}(A \cap B) = 0$ or $\mathrm{P}(A \cup B) = 1$

**71** $\sin\theta\cos\theta = \frac{1}{4}$

$\therefore\ \frac{1}{2}\sin 2\theta = \frac{1}{4}$

$\therefore\ \sin 2\theta = \frac{1}{2}$

$\therefore\ 2\theta = \frac{\pi}{6} + k2\pi$ or $\frac{5\pi}{6} + k2\pi,\ \ k \in \mathbb{Z}$

$\therefore\ 2\theta = \frac{\pi}{6}, -\frac{11\pi}{6}, \frac{5\pi}{6}, -\frac{7\pi}{6}$ {as $-2\pi \leqslant 2\theta \leqslant 2\pi$}

$\therefore\ \theta = -\frac{11\pi}{12}, -\frac{7\pi}{12}, \frac{\pi}{12}, \frac{5\pi}{12}$

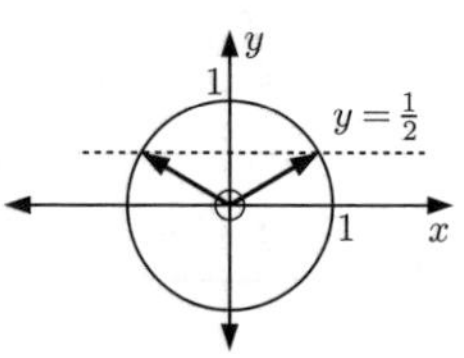


**72** **a** $f(x) = \ln(x(x-2))$ is defined when $x(x-2) > 0$

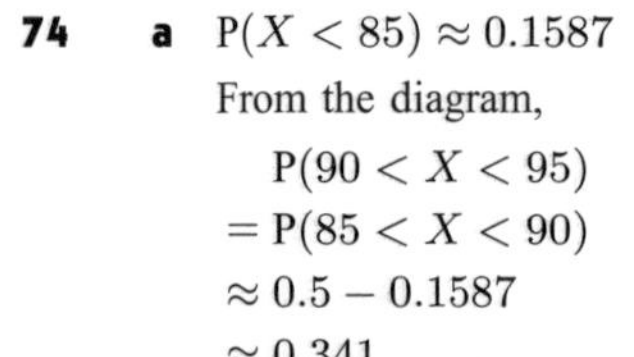


$\therefore\ x < 0$ or $x > 2$

So the domain is $\{x \mid x < 0 \text{ or } x > 2\}$

**b** $f(x) = \ln x + \ln(x-2)$ {log law}

$\therefore\ f'(x) = \dfrac{1}{x} + \dfrac{1}{x-2}$

**c** $f'(3) = \frac{1}{3} + 1 = \frac{4}{3}$ at $(3, \ln 3)$

$\therefore$ the tangent has equation

$\dfrac{y - \ln 3}{x - 3} = \dfrac{4}{3}$

$\therefore\ 4x - 12 = 3y - 3\ln 3$

$\therefore\ 4x - 3y = 12 - 3\ln 3$

**73** **a**

| 3G 4B | 4G 3B |
|---|---|
| (1) | (2) |

P(both same colour)

$= \mathrm{P(GG\ or\ BB)}$

$= \frac{3}{7} \times \frac{4}{7} + \frac{4}{7} \times \frac{3}{7}$

$= \frac{24}{49}$

**b** P(G from (2) | both different)

$= \dfrac{\text{P(G from (2) } \cap \text{ both different)}}{\text{P(both different)}}$

$= \dfrac{\text{P(G from (2) \textbf{and} B from (1))}}{1 - \frac{24}{49}}$

$= \dfrac{\frac{4}{7} \times \frac{4}{7}}{\frac{25}{49}}$

$= \frac{16}{25}$

**74** **a** $\mathrm{P}(X < 85) \approx 0.1587$

From the diagram,

$\mathrm{P}(90 < X < 95)$

$= \mathrm{P}(85 < X < 90)$

$\approx 0.5 - 0.1587$

$\approx 0.341$

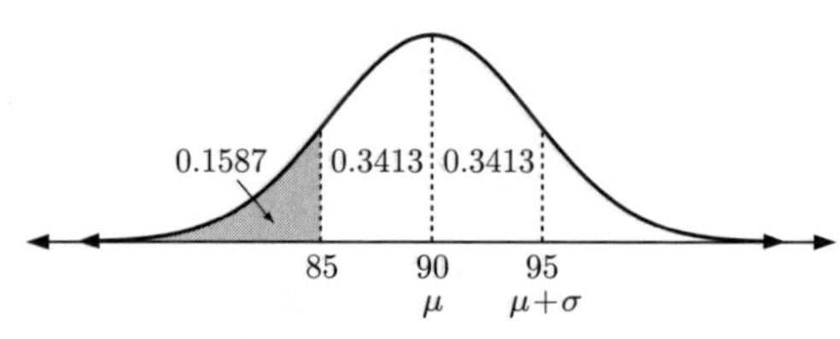


**b** As roughly 34.1% of scores lie between $\mu$ and $\mu + \sigma$ for the normal distribution, then $\sigma \approx 5$.

**75** $\mathrm{P}(X = x) = a\left(\frac{2}{5}\right)^x$ where $x = 0, 1, 2, 3, ....$

$\therefore\ a\left(\frac{2}{5}\right)^0 + a\left(\frac{2}{5}\right)^1 + a\left(\frac{2}{5}\right)^2 + .... = 1$ {$\sum \mathrm{P}(x) = 1$}

$\therefore\ a\left(1 + \frac{2}{5} + \left(\frac{2}{5}\right)^2 + ....\right) = 1$

$\therefore\ a\left(\dfrac{1}{1 - \frac{2}{5}}\right) = 1$ {infinite geometric series, $u_1 = 1,\ r = \frac{2}{5}$}

$\therefore\ \dfrac{a}{\frac{3}{5}} = 1$

$\therefore\ a = \frac{3}{5}$

**76** $x = \log_3 y^2$ $\quad \therefore \quad y^2 = 3^x = (81^{\frac{1}{4}})^x = 81^{\frac{x}{4}}$

$\therefore \quad y = (81^{\frac{x}{4}})^{\frac{1}{2}} = 81^{\frac{x}{8}}$

$\therefore \quad 81 = y^{\frac{8}{x}}$ and so $\log_y 81 = \dfrac{8}{x}$

**77**  **a, d**

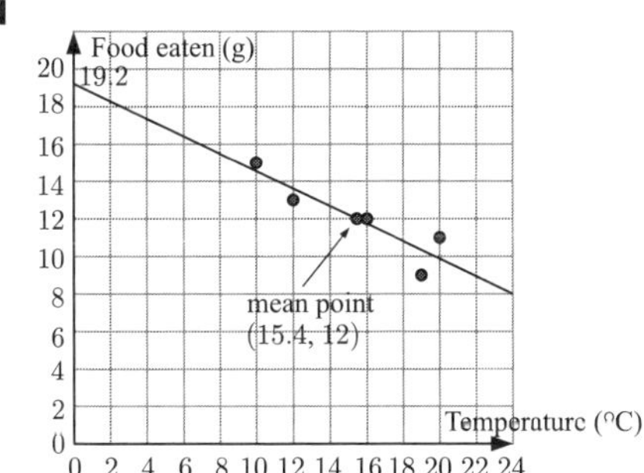


**b** There is a strong, negative, linear correlation between *temperature* and *food eaten*.

**c** $\overline{x} = \dfrac{\sum x}{n}$

$= \dfrac{10+16+12+19+20}{5}$

$= \dfrac{77}{5}$

$= 15.4$

$\overline{y} = \dfrac{\sum y}{n}$

$= \dfrac{15+12+13+9+11}{5}$

$= \dfrac{60}{5}$

$= 12$

$\therefore$ the mean point is $(15.4, 12)$.

**e** The $y$-intercept is $\approx 19.2$, so the line of best fit has the form $y = mx + 19.2$.

Now, the point $(15.4, 12)$ lies on the line, so $12 = 15.4m + 19.2$

$\therefore \quad 15.4m = -7.2$

$\therefore \quad m = \dfrac{-7.2}{15.4} \approx -0.47$

$\therefore$ the equation of the line of best fit is $F \approx -0.47t + 19.2$

**f** When $t = 5$, $F \approx -0.47 \times 5 + 19.2$

$\approx 16.9$

$\therefore$ Pug would eat about 16.9 g of food on a 5°C day.

**g** This prediction may be unreliable as it is an extrapolation.

**78** **a** $y = f(x-2) + 1$ is obtained from $y = f(x)$ by a translation of $\binom{2}{1}$.

$\therefore \quad \text{A}(-2, 3) \mapsto \text{A}'(0, 4)$.

**b** $y = 2f(x-2)$ is obtained from $y = f(x)$ by a translation of $\binom{2}{0}$ followed by a vertical stretch of factor 2. $\quad \therefore \quad \text{A}(-2, 3) \mapsto \text{A}'(0, 3) \mapsto \text{A}''(0, 6)$.

**c** $y = f(2x) - 3$ is obtained from $y = f(x)$ by a translation of $\binom{0}{-3}$ followed by a horizontal stretch of factor $\frac{1}{2}$. $\quad \therefore \quad \text{A}(-2, 3) \mapsto \text{A}'(-2, 0) \mapsto \text{A}''(-1, 0)$.

**d** Consider $y = f^{-1}(x)$. For an inverse function, the point is reflected in the line $y = x$.

$\therefore \quad \text{A}(-2, 3) \mapsto \text{A}'(3, -2)$.

**79**

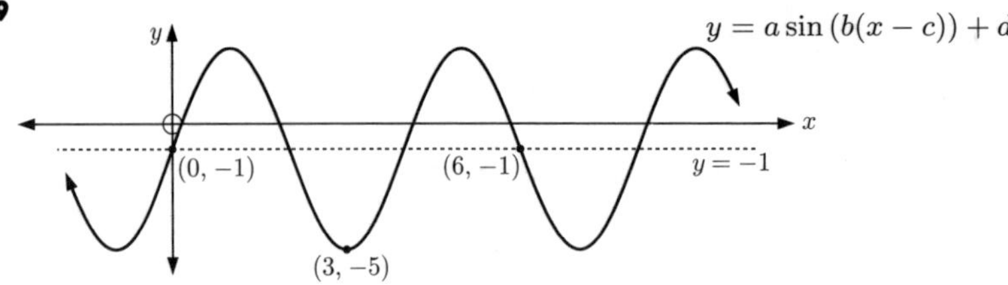


The amplitude $= a = 4$. The period $= 4 = \dfrac{2\pi}{b}$ $\therefore\ b = \frac{\pi}{2}$

The basic sine curve has been translated through $\begin{pmatrix} 0 \\ -1 \end{pmatrix}$. $\therefore\ c = 0,\ d = -1$

Thus $y = 4\sin\left(\frac{\pi}{2}x\right) - 1$

*Check*: $y(3) = 4\sin\left(\frac{3\pi}{2}\right) - 1 = 4(-1) - 1 = -5$ ✓

$y(6) = 4\sin(3\pi) - 1 = 4(0) - 1 = -1$ ✓

**80** **a** $A$ and $B$ are mutually exclusive if $A \cap B = \varnothing$. In this case $\text{P}(A \cap B) = 0$, so $x = 0$.

**b** If $A$ and $B$ are independent, then

$$\begin{aligned} \text{P}(A \cap B) &= \text{P}(A)\,\text{P}(B) \\ \therefore\ x &= (0.3 + x)(0.2 + x) \\ \therefore\ x &= 0.06 + 0.5x + x^2 \\ \therefore\ x^2 - 0.5x + 0.06 &= 0 \\ \therefore\ (x - 0.2)(x - 0.3) &= 0 \\ \therefore\ x &= 0.2 \text{ or } 0.3 \end{aligned}$$

**81** **a**

$$\begin{aligned} &9^{\log_3 11} \\ &= \left(3^2\right)^{\log_3 11} \\ &= \left(3^{\log_3 11}\right)^2 \\ &= 11^2 \qquad \{x = a^{\log_a x}\} \\ &= 121 \end{aligned}$$

**b**

$$\begin{aligned} &\log_m n \times \log_n m^2 \\ &= \frac{\log_n n}{\log_n m} \times \log_n m^2 \qquad \{\text{change of base rule}\} \\ &= \frac{1}{\cancel{\log_n m}} \times 2\cancel{\log_n m} \qquad \{\log A^n = n \log A\} \\ &= 1 \times 2 \\ &= 2 \end{aligned}$$

**82** **Cumulative frequency curve of watermelon weight data**

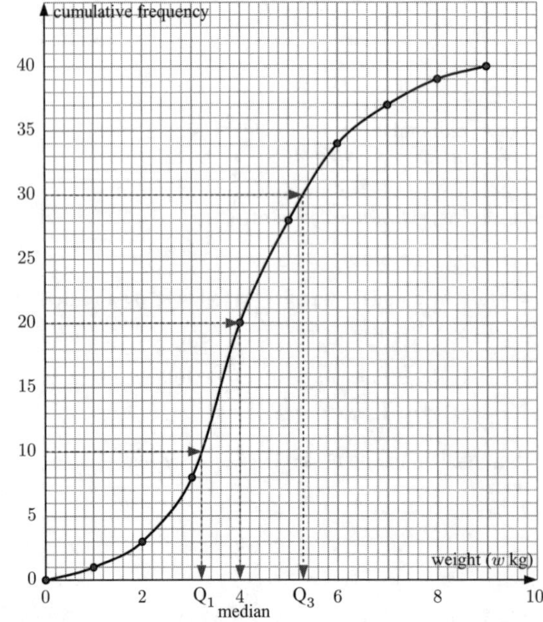


**a** **i** median $\approx 4$ kg

**ii** $\text{IQR} = Q_3 - Q_1$
$\approx 5.3 - 3.2$
$\approx 2.1$ kg

**b**

| *Weight* $w$ (grams) | *Frequency* |
|---|---|
| $0 \leqslant w < 1$ | $1 - 0 = 1$ |
| $1 \leqslant w < 2$ | $3 - 1 = 2$ |
| $2 \leqslant w < 3$ | $8 - 3 = 5$ |
| $3 \leqslant w < 4$ | $20 - 8 = 12$ |
| $4 \leqslant w < 5$ | $28 - 20 = 8$ |
| $5 \leqslant w < 6$ | $34 - 28 = 6$ |
| $6 \leqslant w < 7$ | $37 - 34 = 3$ |
| $7 \leqslant w < 8$ | $39 - 37 = 2$ |
| $8 \leqslant w < 9$ | $40 - 39 = 1$ |

**c**

| *Weight w* (grams) | *Frequency* | *Midpoint* ($x$) | $fx$ |
|---|---|---|---|
| $0 \leqslant w < 1$ | $1 - 0 = 1$ | 0.5 | 0.5 |
| $1 \leqslant w < 2$ | $3 - 1 = 2$ | 1.5 | 3 |
| $2 \leqslant w < 3$ | $8 - 3 = 5$ | 2.5 | 12.5 |
| $3 \leqslant w < 4$ | $20 - 8 = 12$ | 3.5 | 42 |
| $4 \leqslant w < 5$ | $28 - 20 = 8$ | 4.5 | 36 |
| $5 \leqslant w < 6$ | $34 - 28 = 6$ | 5.5 | 33 |
| $6 \leqslant w < 7$ | $37 - 34 = 3$ | 6.5 | 19.5 |
| $7 \leqslant w < 8$ | $39 - 37 = 2$ | 7.5 | 15 |
| $8 \leqslant w < 9$ | $40 - 39 = 1$ | 8.5 | 8.5 |
| *Total* | 40 | | 170 |

$$\therefore \quad \overline{x} = \frac{\sum fx}{\sum f} = \frac{170}{40} = 4.25$$

So, the mean weight of the watermelons is about 4.25 kg.

**83** **a** $(f \circ g)(x) = f(g(x))$

$$= f\left(\frac{x+1}{x-2}\right) = 2\left(\frac{x+1}{x-2}\right) + 1 = \frac{2x+2+x-2}{x-2} = \frac{3x}{x-2}$$

**b** $y = \dfrac{x+1}{x-2}$ has inverse $x = \dfrac{y+1}{y-2}$

$\therefore \quad xy - 2x = y + 1$

$\therefore \quad y(x-1) = 2x + 1$

$\therefore \quad y = \dfrac{2x+1}{x-1}$

$\therefore \quad g^{-1}(x) = \dfrac{2x+1}{x-1}$

**84** **a** If $A$ and $B$ are mutually exclusive then $\text{P}(A \cup B) = \text{P}(A) + \text{P}(B) = \frac{1}{3} + \frac{2}{7} = \frac{13}{21}$

**b** $\text{P}(A \cup B) = \text{P}(A) + \text{P}(B) - \text{P}(A \cap B)$

$= \text{P}(A) + \text{P}(B) - \text{P}(A)\,\text{P}(B)$ $\{A$ and $B$ independent$\}$

$= \frac{1}{3} + \frac{2}{7} - \frac{1}{3} \times \frac{2}{7} = \frac{11}{21}$

**85** $\log_a(x+2) = \log_a x + 2$

$\therefore \quad \log_a(x+2) - \log_a x = 2$

$\therefore \quad \log_a\left(\dfrac{x+2}{x}\right) = 2$

$\therefore \quad \dfrac{x+2}{x} = a^2$

$\therefore \quad 1 + \dfrac{2}{x} = a^2$

$\therefore \quad \dfrac{2}{x} = a^2 - 1$

$\therefore$ since $a > 1$, $x = \dfrac{2}{a^2 - 1}$

**86** **a** $(a-b)^5 = a^5 - 5a^4b + 10a^3b^2 - 10a^2b^3 + 5ab^4 - b^5$

**b** This expression is the binomial expansion of $(0.4 + 0.6)^5 = 1^5$
$= 1$

**c** $\left(2x + \dfrac{1}{x}\right)^5$

$$= (2x)^5 + 5(2x)^4\left(\frac{1}{x}\right) + 10(2x)^3\left(\frac{1}{x}\right)^2 + 10(2x)^2\left(\frac{1}{x}\right)^3 + 5(2x)\left(\frac{1}{x}\right)^4 + \left(\frac{1}{x}\right)^5$$

$$= 32x^5 + 80x^3 + 80x + \frac{40}{x} + \frac{10}{x^3} + \frac{1}{x^5}$$

**87** **a** $\left(x+\frac{1}{x}\right)^2 = a^2$

$\therefore\ x^2 + 2 + \frac{1}{x^2} = a^2$

$\therefore\ x^2 + \frac{1}{x^2} = a^2 - 2$

**b** $\left(x+\frac{1}{x}\right)^3 = a^3$

$\therefore\ x^3 + 3x^2\left(\frac{1}{x}\right) + 3x\left(\frac{1}{x}\right)^2 + \left(\frac{1}{x}\right)^3 = a^3$

$\therefore\ x^3 + 3x + \frac{3}{x} + \frac{1}{x^3} = a^3$

$\therefore\ x^3 + \frac{1}{x^3} + 3\left(x + \frac{1}{x}\right) = a^3$

$\therefore\ x^3 + \frac{1}{x^3} = a^3 - 3a$

**88** **a** **i** The ellipse cuts the $x$-axis when $y = 0$

$\therefore\ \frac{x^2}{16} = 1$

$\therefore\ x^2 = 16$

$\therefore\ x = \pm 4$

So, A is $(4,\ 0)$ and B is $(-4,\ 0)$.

**ii** The ellipse cuts the $y$-axis when $x = 0$

$\therefore\ \frac{y^2}{4} = 1$

$\therefore\ y^2 = 4$

$\therefore\ y = \pm 2$

So, C is $(0,\ 2)$ and D is $(0,\ -2)$.

**b** Since $\frac{y^2}{4} = 1 - \frac{x^2}{16}$

then $y^2 = 4 - \frac{x^2}{4}$

$\therefore\ y = \pm\sqrt{4 - \frac{x^2}{4}}$

But $y > 0$, so $y = \sqrt{4 - \frac{x^2}{4}}$

**c** $\int_0^4 \sqrt{4 - \frac{x^2}{4}}\, dx$ is the area of one quarter of the ellipse

$\therefore$ area of the ellipse $= 4\int_0^4 \sqrt{4 - \frac{x^2}{4}}\, dx$

**d** Volume $= \pi\int_{-4}^{4} y^2\, dx$

$= 2\pi\int_0^4 y^2\, dx$

$= 2\pi\int_0^4 (4 - \frac{1}{4}x^2)\, dx$

$= 2\pi\left[4x - \frac{1}{4}\frac{x^3}{3}\right]_0^4$

$= 2\pi\left\{\left(16 - \frac{16}{3}\right) - (0)\right\}$

$= 2\pi \times \frac{32}{3}$

$= \frac{64\pi}{3}$ units$^3$

**89** **a**

| $x$ | $0$ | $\frac{\pi}{4}$ | $\frac{\pi}{2}$ | $\frac{3\pi}{4}$ | $\pi$ | $\frac{5\pi}{4}$ | $\frac{3\pi}{2}$ | $\frac{7\pi}{4}$ | $2\pi$ |
|---|---|---|---|---|---|---|---|---|---|
| $f(x)$ | $0$ | $\frac{1}{2}$ | $1$ | $\frac{1}{2}$ | $0$ | $\frac{1}{2}$ | $1$ | $\frac{1}{2}$ | $0$ |

**b**

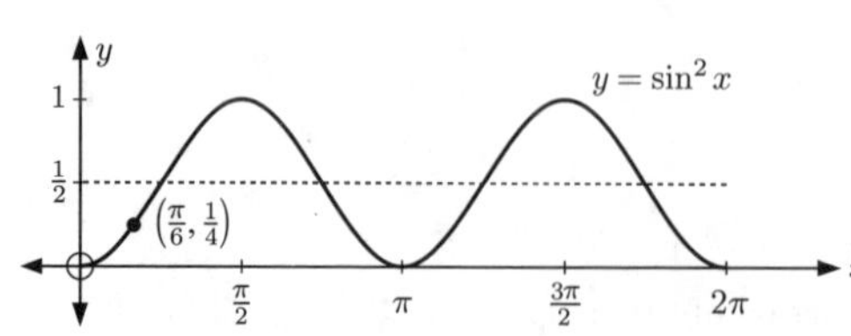


**c** When $x = \frac{\pi}{6}$, $\sin x = \frac{1}{2}$

$\therefore\ y = \left(\frac{1}{2}\right)^2 = \frac{1}{4}$ ✓

**d** The range is $\{y \mid 0 \leqslant y \leqslant 1\}$

**e** Area $= \int_0^{\pi} \sin^2 x\, dx$

$= \int_0^{\pi} \left(\frac{1}{2} - \frac{1}{2}\cos 2x\right)\, dx$

$= \left[\frac{1}{2}x - \frac{1}{2}(\frac{1}{2})\sin 2x\right]_0^{\pi}$

$= \left(\frac{\pi}{2} - \frac{1}{4}(0)\right) - (0 - 0)$

$= \frac{\pi}{2}$ units$^2$

**f** $f'(x) = 2(\sin x)^1 \cos x$

$= \sin 2x$

$\therefore\ f'(\frac{\pi}{4}) = \sin\frac{\pi}{2} = 1$

$\therefore$ the slope of the tangent is $\frac{1}{1}$ at $(\frac{\pi}{4},\ \frac{1}{2})$

$\therefore$ the equation of the tangent is $x - y = \frac{\pi}{4} - \frac{1}{2}$

**90** $f(x) = x + x^{-1}$

**a** $f'(x) = 1 - x^{-2}$, and $f'(x) = 0$

when $1 - \dfrac{1}{x^2} = 0$

$\therefore\ 1 = \dfrac{1}{x^2}$

$\therefore\ x^2 = 1$

$\therefore\ x = 1 \quad \{\text{as } x > 0\}$

**b** Sign diagram of $f'(x)$:

0   −   1   +   $x$

$\therefore$ $f$ has a minimum at $x = 1$.

$f(1) = 1 + \frac{1}{1} = 2$

$\therefore$ A is at $(1, 2)$

**c** Since the minimum value of $f(x) = x + \dfrac{1}{x}$ for $x > 0$ is 2, the sum of a positive number and its reciprocal is at least 2.

**d** **i** Since the minimum value of $f(x) = x + \dfrac{1}{x}$ for $x > 0$ is 2, $x + \dfrac{1}{x} = 1$ has no solutions.

**ii** The line $y = 2$ touches $y = x + \dfrac{1}{x}$ at A.

So, $x + \dfrac{1}{x} = 2$ has one positive solution.

**iii** The line $y = 3$ cuts $y = x + \dfrac{1}{x}$ in two places. So, $x + \dfrac{1}{x} = 3$ has two positive solutions.

**91** **a** $\mathbf{r} = \begin{pmatrix} 2 \\ 0 \\ -3 \end{pmatrix} + t \begin{pmatrix} 1 \\ -1 \\ 2 \end{pmatrix}, \quad t \in \mathbb{R}$    **b** $x = 2 + t, \quad y = -t, \quad z = -3 + 2t, \quad t \in \mathbb{R}$

**c** $(2 + t, -t, -3 + 2t)$ represents any point on the line.

**d** $\overrightarrow{\text{BP}} = \begin{pmatrix} 2 + t - -1 \\ -t - 3 \\ -3 + 2t - 5 \end{pmatrix} = \begin{pmatrix} t + 3 \\ -t - 3 \\ 2t - 8 \end{pmatrix}$

**e** $\overrightarrow{\text{BP}} \bullet (\mathbf{i} - \mathbf{j} + 2\mathbf{k}) = (t + 3)1 + (-t - 3)(-1) + (2t - 8)2$
$= t + 3 + t + 3 + 4t - 16$
$= 6t - 10$

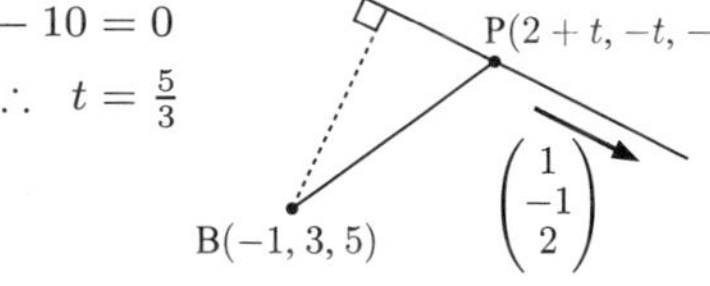


**f** [BP] is perpendicular to the original line when $6t - 10 = 0$

$\therefore\ t = \frac{5}{3}$

**g** When $t = \frac{5}{3}$, P is at $(2 + \frac{5}{3}, -\frac{5}{3}, -3 + \frac{10}{3})$

$\therefore\ (\frac{11}{3}, -\frac{5}{3}, \frac{1}{3})$ is closest to B.

**92** **a** **i** $a$ is the minimum value of $X$.   **ii** $b$ is $Q_1$, the lower quartile.

**iii** $c$ is $Q_2$, the median.   **iv** $d$ is $Q_3$, the upper quartile.

**v** $e$ is the maximum value of $X$.

**b** **i** maximum value − minimum value $= e - a$ is the range.

**ii** $Q_3 - Q_1 = d - b$ is the interquartile range or IQR.

**c** **i** 75% of the scores are less than $d$, and 25% of the scores are less than $b$.

$\therefore\ P(b < X \leqslant d) = 0.75 - 0.25 = 0.5$

**ii** 75% of the scores are greater than $b$.

$\therefore\ P(X > b) = 0.75$

**d**

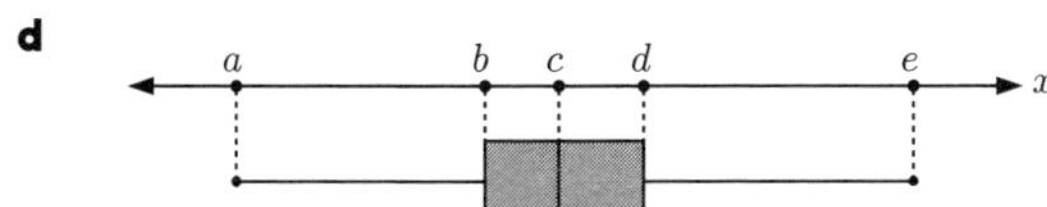

## EXERCISE 25B

**1** $\sum\limits_{k=1}^{n}(2k-31)=0$

$\therefore\ (-29)+(-27)+(-25)+....+(2n-31)=0$

The LHS is arithmetic with $u_1=-29,\ d=2$ and "$n$" $=n$.

$\therefore\ \frac{n}{2}(-58+(n-1)2)=0$

$\therefore\ \frac{n}{2}(2n-60)=0$

$\therefore\ n(n-30)=0$

$\therefore\ n=30$ as $n\neq 0$

**2** $f(x)=5\ln(x-4)+2$

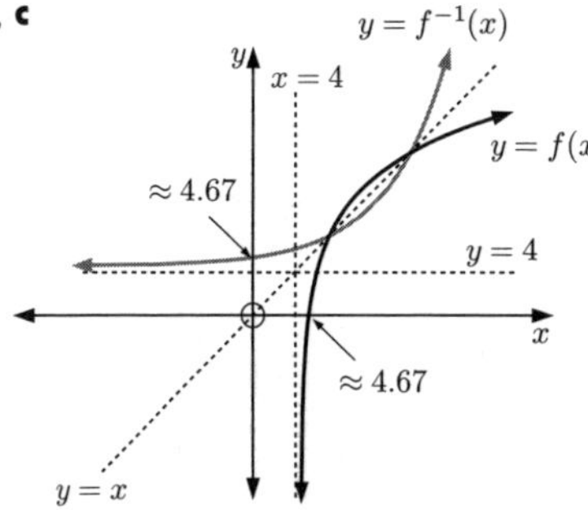


**b** $f(x)=1$ when $x\approx 4.82$ {technology}

**d** $f'(x)=5\left(\frac{1}{x-4}\right)+0$

$\therefore\ f'(5)=5$

$\therefore$ the normal has gradient $-\frac{1}{5}$ when $x=5$.

$f(5)=5\ln(1)+2=2$

$\therefore$ at $(5,\,2)$ the normal is $\frac{y-2}{x-5}=-\frac{1}{5}$

$\therefore\ 5y-10=-x+5$

$\therefore\ x+5y=15$

**3** The general term is $T_{r+1}=\binom{9}{r}x^{9-r}\left(-\frac{1}{5x^2}\right)^r$

$=\binom{9}{r}x^{9-r}\left(-\frac{1}{5}\right)^r x^{-2r}$

$=\binom{9}{r}\left(-\frac{1}{5}\right)^r x^{9-3r}$

If we let $9-3r=0$ then $r=3$

$\therefore\ T_4=\binom{9}{3}\left(-\frac{1}{5}\right)^3x^0$

$=84\times-\frac{1}{125}$

$=-\frac{84}{125}$

So, the constant term is $-\frac{84}{125}$.

**4** **a** When $t=0$, $V=7500\times 2^0$

$=7500$

So, the initial value of the cash investment is \$7500.

**b** **i** When $t=5$, $V=7500\times 2^{0.09\times 5}$

$\approx 10\,245$

The value of the investment after 5 years is \$10 245.

**ii** When $t=15$, $V=7500\times 2^{0.09\times 15}$

$\approx 19\,118$

The value of the investment after 15 years is \$19 118.

**c** Percentage increase $=\frac{10\,245-7500}{7500}\times 100\%$

$=36.6\%$

**d** If the investment doubles in value it will be worth \$15 000.

So, we solve $7500\times 2^{0.09t}=15\,000$ for $t$

$\therefore\ 2^{0.09t}=2$

$\therefore\ 0.09t=1$

$\therefore\ t\approx 11.1$ years

**5** $-900, -750, -600, -450, ....$ is arithmetic with $u_1 = -900$, $d = 150$

**a** $u_{20} = u_1 + 19d$
$= -900 + 19(150)$
$= 1950$

**b** $S_{20} = \frac{20}{2}[2(-900) + 19(150)]$
$= 10[-1800 + 19 \times 150]$
$= 10\,500$

**6** **a** $(f \circ g)(x) = -8$
$\therefore f(g(x)) = -8$
$\therefore f(1 - 5x^2) = -8$
$\therefore 2(1 - 5x^2) = -8$
$\therefore 2 - 10x^2 = -8$
$\therefore 10x^2 = 10$
$\therefore x = \pm 1$

**b** $(g \circ f)(x) = -8$
$\therefore g(f(x)) = -8$
$\therefore g(2x) = -8$
$\therefore 1 - 5(2x)^2 = -8$
$\therefore 1 - 20x^2 = -8$
$\therefore 20x^2 = 9$
$\therefore x^2 = \frac{9}{20}$
$\therefore x = \pm\sqrt{\frac{9}{20}}$
$= \pm\frac{3}{2\sqrt{5}}$

**c** $f(x) = 2x$ and $g(x) = 1 - 5x^2$
$f'(x) = g'(x)$ implies that
$2 = -10x$
$\therefore x = -\frac{1}{5}$

**d** $f(x)$ is $y = 2x$
$\therefore f^{-1}(x)$ is $x = 2y$
$\therefore y = \frac{x}{2}$
$\therefore f^{-1}(x) = \frac{x}{2}$
Thus $\frac{x}{2} = 1 - 5x^2$
$\therefore 5x^2 + \frac{x}{2} - 1 = 0$
$\therefore 10x^2 + x - 2 = 0$
$\therefore (2x + 1)(5x - 2) = 0$
$\therefore x = -\frac{1}{2}$ or $\frac{2}{5}$

**7** **a** $P = 50\pi$
$\therefore 2r + r(\frac{7\pi}{9}) = 50\pi$
$\therefore r(2 + \frac{7\pi}{9}) = 50\pi$
$\therefore r = \frac{50\pi}{2 + \frac{7\pi}{9}} \approx 35.3507$
$\therefore r \approx 35.4$ cm

**b** area $= \frac{1}{2}\theta r^2$
$\approx \frac{1}{2}(\frac{7\pi}{9})(35.3507)^2$
$\approx 1526.77$
$\approx 1530$ cm$^2$

**c**

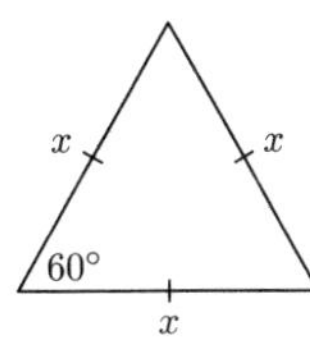


area of triangle $= \frac{1}{2} \times x \times x \times \sin 60°$
$\therefore \frac{1}{2}x^2(\frac{\sqrt{3}}{2}) \approx 1526.77$
$\therefore \frac{\sqrt{3}}{4}x^2 \approx 1526.77$
$\therefore x^2 \approx 3525.9$
$\therefore x \approx 59.4 \quad \{x > 0\}$
$\therefore$ the sides are 59.4 cm long.

**8** **a**

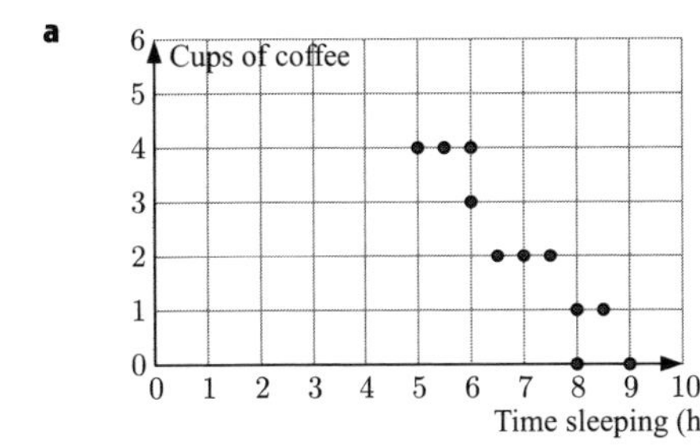


**b** Using technology, $r \approx -0.937$

**c** There is a strong, negative correlation between *Time sleeping* and *Cups of coffee*.

**9** **a**

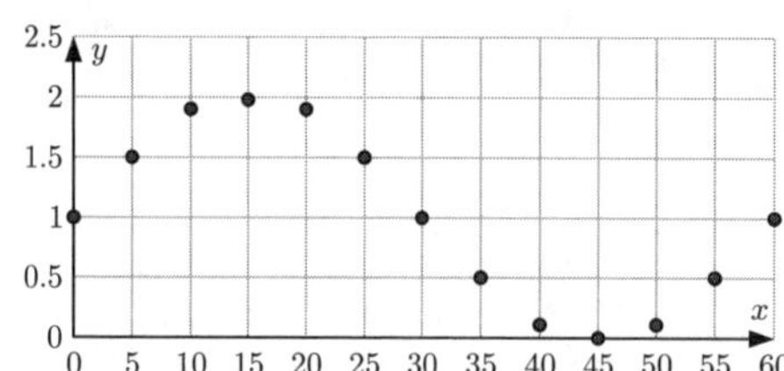


**b** **i** The equation of the principal axis is

$$y = \frac{0+2}{2} = 1$$

$\therefore\ y = 1$

**ii** The maximum value is 2.

**iii** Assuming that the pattern continues, the period is 60.

**iv** The amplitude is

$$\frac{2-0}{2} = 1.$$

**c** From **b**, we obtain $a = 1$, $b = \frac{2\pi}{60} = \frac{\pi}{30}$, and $c = 1$.

$\therefore$ the equation that models the data is $y = \sin\left(\frac{\pi}{30}x\right) + 1$.

**10** $f(x) = m\cos n(x-p) + r$

**a** The amplitude $m = \dfrac{12-4}{2} = 4$

$$\text{period} = 2(5-1) = 8$$

$$\therefore\ \frac{2\pi}{n} = 8$$

$$\therefore\ n = \tfrac{\pi}{4}$$

maximum value $= m + r = 12$

$\therefore\ 4 + r = 12$

$\therefore\ r = 8$

The maximum value occurs when $x = 1$.

$\therefore$ when $x = 1$, $\cos n(x-p) = 1$

$\therefore\ \frac{\pi}{4}(1-p) = 0$

$\therefore\ p = 1$

Thus, $m = 4$, $n = \frac{\pi}{4}$, $p = 1$, $r = 8$.

**b** $f(x) = 4\cos\frac{\pi}{4}(x-1) + 8$

$f(6) = 4\cos(\frac{\pi}{4} \times 5) + 8$

$= 4\cos(\frac{5\pi}{4}) + 8$

$= 8 - 2\sqrt{2} \approx 5.17$

**c** $f(x) = 10$

$\therefore\ 4\cos\frac{\pi}{4}(x-1) + 8 = 10$

$\therefore\ \cos\frac{\pi}{4}(x-1) = \frac{1}{2}$

$\therefore\ \frac{\pi}{4}(x-1) = \pm\frac{\pi}{3} + k2\pi$

$\therefore\ x - 1 = \pm\frac{4}{3} + 8k$

$\therefore\ x = 2\frac{1}{3}$ is the smallest positive $x$

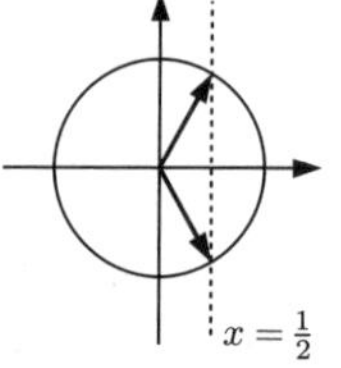


**11** **a** $L_1$ and $L_2$ have direction vectors $\begin{pmatrix} 4 \\ -1 \\ 3 \end{pmatrix}$ and $\begin{pmatrix} 6 \\ -5 \\ 15 \end{pmatrix}$ respectively.

Let $\theta$ be the acute angle between $L_1$ and $L_2$.

$$\therefore\ \cos\theta = \frac{|24+5+45|}{\sqrt{16+1+9}\sqrt{36+25+225}}$$

$$= \frac{74}{\sqrt{26}\sqrt{286}}$$

$$\therefore\ \theta = \cos^{-1}\left(\frac{74}{\sqrt{7436}}\right) \approx 30.9^\circ$$

**b** When $t = -2$, $\mathbf{r}_1 = \begin{pmatrix} -2 \\ 1 \\ 4 \end{pmatrix} - 2\begin{pmatrix} 4 \\ -1 \\ 3 \end{pmatrix} = \begin{pmatrix} -10 \\ 3 \\ -2 \end{pmatrix}$

$\therefore$ $(-10, 3, -2)$ lies on $L_1$.

**c** We require $\begin{pmatrix} -10 \\ 3 \\ -2 \end{pmatrix} = \begin{pmatrix} -5+6s \\ -5s \\ 7+15s \end{pmatrix}$ for some $s \in \mathbb{R}$

$\therefore\ 3 = -5s$ and so $s = -\frac{3}{5}$

But $-5+6s = -5+6(-\frac{3}{5})$
$= -8\frac{3}{5}$
$\neq -10$

$\therefore$ P does not lie on $L_2$.

**d** If $L_1$ and $L_2$ meet then $\begin{cases} -2+4t = -5+6s & \text{.... (1)} \\ 1-t = -5s & \text{.... (2)} \\ 4+3t = 7+15s & \text{.... (3)} \end{cases}$

From (1) and (2), $\begin{cases} 6s-4t = 3 \\ t = 1+5s \end{cases}$

$\therefore\ 6s - 4(1+5s) = 3$
$\therefore\ 6s - 4 - 20s = 3$
$\therefore\ -14s = 7$
$\therefore\ s = -\frac{1}{2}$
and $t = 1+5(-\frac{1}{2}) = -\frac{3}{2}$

Check in (3): LHS $= 4+3t = 4 - \frac{9}{2} = -\frac{1}{2}$
RHS $= 7+15s = 7 - \frac{15}{2} = -\frac{1}{2}$ ✓

$\therefore\ L_1$ and $L_2$ meet where $s = -\frac{1}{2},\ t = -\frac{3}{2}$

$\therefore$ they meet at $(-8, 2\frac{1}{2}, -\frac{1}{2})$.

**e** $\begin{pmatrix} 4 \\ -1 \\ 3 \end{pmatrix} \bullet \begin{pmatrix} a \\ 2 \\ 8 \end{pmatrix} = 0$

$\therefore\ 4a - 2 + 24 = 0$
$\therefore\ 4a = -22$
$\therefore\ a = -5\frac{1}{2}$

**12** **a** The $x$-axis intercepts are $-1$ and $\beta$.
$f(0) = (1)(-\beta) = -\beta$
$\therefore$ the $y$-intercept is $-\beta$

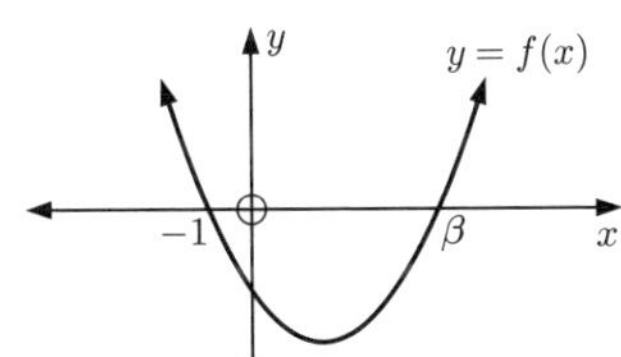


**b** $g(x) = -f(x-1)$ is obtained by translating $f(x)$ 1 unit to the right, then reflecting the result in the $x$-axis.

**c** The $x$-intercepts of $g(x)$ are 1 unit to the right of the $x$-intercepts of $f(x)$.
$\therefore$ the $x$-intercepts of $g(x)$ are 0 and $\beta+1$.
The $y$-intercept is 0.

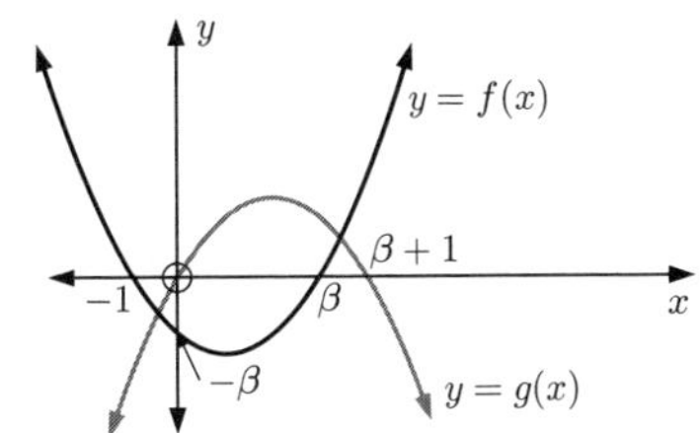

**13** **a**

| *Time* (min) | $f$ | *Cumulative freq.* |
|---|---|---|
| $0 < t \leqslant 10$ | 15 | 15 |
| $10 < t \leqslant 20$ | 10 | 25 |
| $20 < t \leqslant 30$ | 25 | 50 |
| $30 < t \leqslant 40$ | 50 | 100 |
| $40 < t \leqslant 50$ | 30 | 130 |
| $50 < t \leqslant 60$ | 20 | 150 |

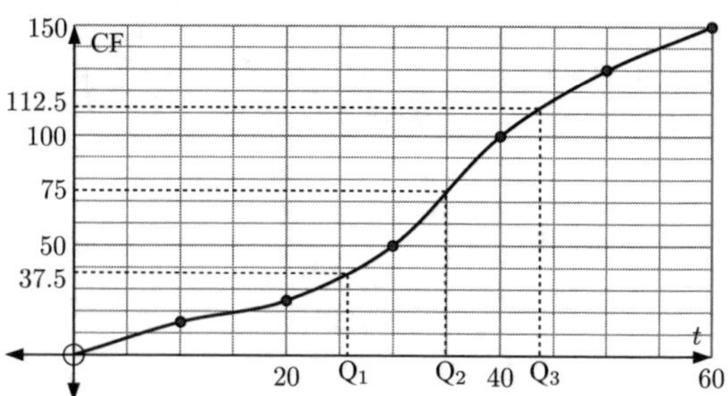


**b** **i** median $\approx 35$ min $(= Q_2)$

**ii** $\text{IQR} = Q_3 - Q_1$
$\approx 44 - 26.5$
$\approx 17.5$ min

**iii** $\approx 50\%$ of the data is less than 35 min
$\therefore$ the probability $\approx 0.5$

**14** Total number of seats $= 16 + 18 + 20 + 22 + ....$ $(n = 30)$
which is arithmetic with $u_1 = 16$, $d = 2$, and $n = 30$

$\therefore$ $S_{30} = \frac{30}{2}(2 \times 16 + 29(2))$
$= 15(32 + 58)$
$= 15 \times 90$
$= 1350$

and $u_{30} = u_1 + 29d$
$= 16 + 29(2)$
$= 74$

$\therefore$ P(seated in last row) $= \frac{74}{1350} = \frac{37}{675} \approx 0.0548$

**15** If $X$ is the number of successful shots, then $X \sim \text{B}(5, 0.86)$.

**a** $\text{P}(X = 5) = (0.86)^5 \approx 0.470$

**b** $\text{P}(X = 3) = \binom{5}{3}(0.86)^3(0.14)^2$ as there are $\binom{5}{3} = 10$ different ways of scoring 3 from 5.

These are:

SSSMM SMSSM
SSMSM MSSSM
SSMMS MSSMS
SMMSS MSMSS
SMSMS MMSSS

{S = score, M = miss}

Max used only one of these. So, he is incorrect.

**16** $(1 + 3x)^7$ has 8 terms in its expansion

$= 1 + \binom{7}{1}3x + \binom{7}{2}(3x)^2 + \binom{7}{3}(3x)^3 + \binom{7}{4}(3x)^4 + \binom{7}{5}(3x)^5 + \binom{7}{6}(3x)^6 + \binom{7}{7}(3x)^7$

$= 1 + 21x + 189x^2 + 945x^3 + 2835x^4 + 5103x^5 + 5103x^6 + 2187x^7$

So, the coefficients of the last 4 terms are greater than 1000.

$\therefore$ the probability is $\frac{4}{8} = \frac{1}{2}$.

**17** **a** Carl's $z$-score for the 100 m $= \dfrac{9.99 - 10.20}{0.113}$
$\approx -1.86$

His $z$-score for the 200 m $= \dfrac{17.30 - 18.50}{0.706}$
$\approx -1.70$

**b** His $z$-score is further from the mean 0 for the 100 m, indicating that his performance is better in that event.

**18** $f(x) = \sin(x^3), \quad 0 \leqslant x \leqslant \frac{\pi}{2}$

**a** $f(x)$ cuts the $x$-axis when

$\sin(x^3) = 0$

$\therefore \ x^3 = 0 + k\pi, \quad k \in \mathbb{Z}$

$\therefore \ x = 0, \ \sqrt[3]{\pi}, \ \sqrt[3]{2\pi}, \ ....$

$\therefore \ x = 0 \ \text{ or } \ x \approx 1.46 \quad \{0 \leqslant x \leqslant \frac{\pi}{2}\}$

So, the $x$-intercepts are $0$, $\approx 1.46$.

**b**

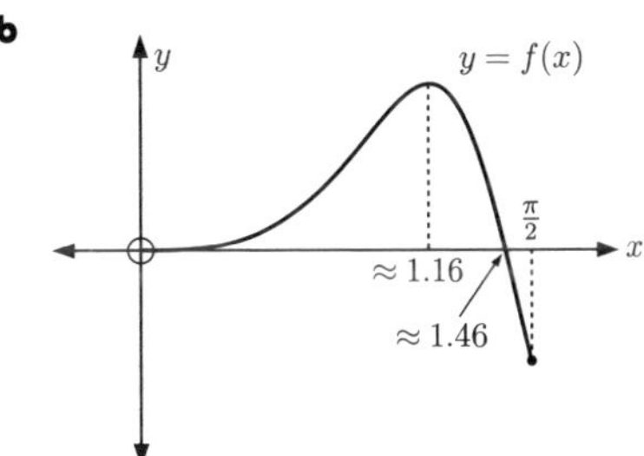


**c** $f'(x) = \cos(x^3) \times 3x^2$

$\therefore \ f'(\frac{\pi}{4}) = \cos\left(\frac{\pi^3}{64}\right) \times 3\left(\frac{\pi^2}{16}\right)$

$\approx 1.64$

So, at $\left(\frac{\pi}{4}, f(\frac{\pi}{4})\right)$ the tangent has gradient $\approx 1.64$

and since $f(\frac{\pi}{4}) = \sin\left(\frac{\pi^3}{64}\right) \approx 0.466$ the tangent has equation

$$\frac{y - 0.466}{x - \frac{\pi}{4}} \approx 1.64$$

$\therefore \ y - 0.466 \approx 1.64x - 1.286$

$\therefore \ y \approx 1.64x - 0.820$

**d** $f''(x) = -\sin(x^3)3x^2(3x^2) + \cos(x^3)6x$

$= 6x\cos(x^3) - 9x^4\sin(x^3)$

$= 3x\left[2\cos(x^3) - 3x^3\sin(x^3)\right]$

So, $f''(x) = 0$ when $x = 0$ or when $2\cos(x^3) = 3x^3\sin(x^3)$

$\therefore \ x \approx 0.903 \quad$ {technology}

From the graph in **b**, $x \approx 0.903$ is the only solution where $f(x) > 0$ and $f'(x) > 0$.

$\therefore$ P is $(0.903, 0.671)$

**19**

| | Under 35 | Over 35 | *Total* |
|---|---|---|---|
| Successful | 951 | 257 | 1208 |
| Unsuccessful | 174 | 415 | 589 |
| *Total* | 1125 | 672 | 1797 |

**a** **i** $\text{P(successfully treated)} = \dfrac{1208}{1797}$

$\approx 0.672$

**ii** $\text{P(over 35} \mid \text{unsuccessful)} = \dfrac{415}{589}$

$\approx 0.705$

**b** This is not strictly a binomial situation as the probability of selecting a successfully treated person changes with each selection.

However, as the population is very large, the binomial model provides a *very good approximation.*

So, Harry's method is valid to obtain a very good approximation.

**Note:** If $X \sim \text{B}(10, 0.672)$, $\text{P}(X = 8) \approx 0.202$

**20** $f(x) = 3e^{1-4x}$

**a** $f'(x) = 3e^{1-4x}(-4)$

$= -12e^{1-4x}$

**b** $\int f(x)\,dx$

$= \int 3e^{1-4x}\,dx$

$= 3(\frac{1}{-4})e^{1-4x} + c$

$= -\frac{3}{4}e^{1-4x} + c$

**c** $\int_0^2 f(x)\,dx$

$= \left[-\frac{3}{4}e^{1-4x}\right]_0^2$

$= \left(-\frac{3}{4}e^{-7}\right) - \left(-\frac{3}{4}e^{1}\right)$

$= \frac{3}{4}(e - e^{-7})$

$\approx 2.04$

**21** **a** $y = (\ln x)\sin x$ cuts the $x$-axis when $y = 0$

$\therefore \ln x = 0$ or $\sin x = 0$

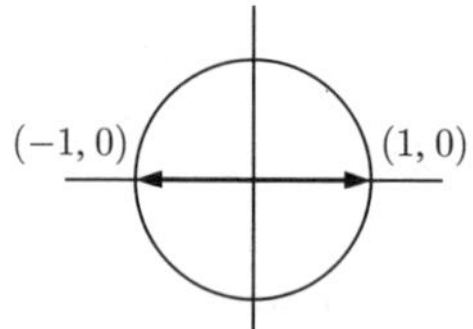


$\therefore x = 1$ or $x = k\pi,\ k \in \mathbb{Z}^+$ {since $x > 0$}

$\therefore$ A is $(1, 0)$ and B is $(\pi, 0)$

**b** C is $\approx (2.128, 0.641)$ {using technology}

**c** $\dfrac{dy}{dx} = \left(\dfrac{1}{x}\right)\sin x + (\ln x)\cos x$

From the graph, the point we are seeking is the point where $\dfrac{dy}{dx}$ is a maximum, between $x = 1$ and $x \approx 2.128$.

This is when $x \approx 1.101$ {technology}

$\therefore$ the point is $(1.101, 0.086)$

**d** A non-stationary inflection.

**22** **a**

$$r = 1 + \frac{0.045}{12} = 1.003\,75$$
$$u_1 = 2000$$
$$\therefore u_{n+1} = u_1 \times r^n = 2000 \times (1.003\,75)^n$$

The investment doubles when $u_{n+1} = 2000(1.003\,75)^n = 4000$

$$\therefore (1.003\,75)^n = 2$$
$$\therefore n\log 1.003\,75 = \log 2$$
$$\therefore n = \frac{\log 2}{\log 1.003\,75}$$
$$\therefore n \approx 185.19 \text{ months}$$

So, it will take 185 months for the investment value to double.

**b** The investment value will quadruple when $(1.003\,75)^n = 4$

$$\therefore n\log(1.003\,75) = \log 4$$
$$\therefore n \approx 370.37$$

So, it will take 370 months for the investment to quadruple in value.

**23** **a** $(x^2+2)^5$ has

$$T_{r+1} = \binom{5}{r}\left(x^2\right)^{5-r} 2^r$$
$$\therefore T_4 = \binom{5}{3}\left(x^2\right)^2 2^3 = 80x^4$$
$$T_5 = \binom{5}{4}\left(x^2\right)^1 2^4 = 80x^2$$
$$T_6 = 2^5 = 32$$

**b**

$$\int \left(x^2+2\right)^5 dx$$
$$= \int \left(x^{10} + 10x^8 + 40x^6 + 80x^4 + 80x^2 + 32\right) dx$$
$$= \tfrac{1}{11}x^{11} + \tfrac{10}{9}x^9 + \tfrac{40}{7}x^7 + 16x^5 + \tfrac{80}{3}x^3 + 32x + c$$

**24** **a** $f(x) = a(x-1)^2 + 4$

But $f(0) = 0$

$\therefore a(-1)^2 + 4 = 0$

$\therefore a = -4$

So, $f(x) = -4(x-1)^2 + 4$

**b** The curves meet where $(x-1)^2 = -4(x-1)^2 + 4$

$$\therefore 5(x-1)^2 = 4$$
$$\therefore x - 1 = \pm\sqrt{\tfrac{4}{5}}$$
$$\therefore x = 1 \pm \sqrt{\tfrac{4}{5}}$$
$$\therefore x \approx 0.106,\ 1.89$$

**c** **i** $A \approx \int_{0.106}^{1.89} [f(x) - g(x)]\, dx$

**ii** $A \approx \int_{0.106}^{1.89} \left[-4(x-1)^2 + 4 - (x-1)^2\right] dx$

$\approx 4.77$ units$^2$

**25** **a** $\text{P}(A \cup B) = \text{P}(A) + \text{P}(B) - \text{P}(A \cap B)$

But $A$ and $B$ are independent $\therefore$ $\text{P}(A \cap B) = \text{P}(A)\,\text{P}(B)$

$$\therefore\ 0.68 = \frac{\text{P}(B)}{3} + \text{P}(B) - \frac{\text{P}(B)}{3}\text{P}(B)$$

$$\therefore\ 2.04 = \text{P}(B) + 3\,\text{P}(B) - [\text{P}(B)]^2$$

$$\therefore\ [\text{P}(B)]^2 - 4\,\text{P}(B) + 2.04 = 0$$

**b** The solutions are $\text{P}(B) = 3.4$ or $0.6$ {technology}

But $0 \leqslant \text{P}(B) \leqslant 1$

$\therefore$ $\text{P}(B) = 0.6$ and $\text{P}(A) = 0.2$

**26** **a** $\overline{x} = \dfrac{63 + 76 + 99 + .... + 83}{12} = 70.5$ kg

**b** Total weight of the students $= 12 \times 70.5 = 846$ kg.

Let $x$ be the weight of the student who left $\therefore$ $\dfrac{846 - x}{11} = 70$

$$\therefore\ 846 - x = 770$$

$$\therefore\ x = 76$$

$\therefore$ the student weighed 76 kg.

**c** **i** $s \approx 15.1$ kg {technology}

**ii** The heaviest student is 99 kg.

The $z$-score $\approx \dfrac{99 - 70}{15.1}$

$\approx 1.92$

$\therefore$ the heaviest student is about 1.92 standard deviations above the mean.

**27** If $X$ is the number of correct answers, then $X \sim \text{B}(30, \frac{1}{5})$.

**a** $\text{P}(X = 10) \approx 0.0355$ {technology}

**b** P(no more than 10 correct) $= \text{P}(X \leqslant 10)$

$\approx 0.974$ {technology}

**28** **a** If $D$ is the number of defective batteries, then $D \sim \text{B}(20, 0.03)$.

**i** $\text{P}(D = 0) \approx 0.544$

**ii** P(at least one is defective) $= \text{P}(D \geqslant 1)$

$\approx 0.456$

**b** $X \sim \text{B}(n, 0.03)$

**i**

$$\text{P}(X = r) = \binom{n}{r}(0.03)^r(0.97)^{n-r}$$

$$\therefore\ \text{P}(X = 0) = \binom{n}{0}(0.03)^0(0.97)^n$$

$$= (0.97)^n$$

**ii**

$$\text{P}(X \geqslant 1) \geqslant 0.3$$

$$\therefore\ 1 - \text{P}(X = 0) \geqslant 0.3$$

$$\therefore\ 0.7 \geqslant \text{P}(X = 0)$$

$$\therefore\ (0.97)^n \leqslant 0.7$$

$$\therefore\ n\log(0.97) \leqslant \log(0.7)$$

$$\therefore\ n \geqslant \frac{\log(0.7)}{\log(0.97)}$$

$\{\log(0.97) < 0\}$

$$\therefore\ n \geqslant 11.709$$

$\therefore$ the smallest $n$ is 12.

**29** **a** **i**

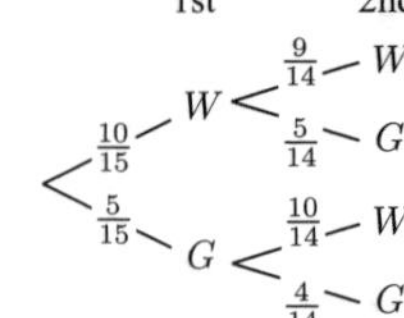


**ii** P(same colour)

$= \text{P}(WW \text{ or } GG)$

$= \left(\frac{10}{15}\right)\left(\frac{9}{14}\right) + \left(\frac{5}{15}\right)\left(\frac{4}{14}\right)$

$= \frac{110}{210}$

$= \frac{11}{21}$

**b**

1st 2nd

$\frac{5}{5+n}$ W: $\frac{4}{4+n}$ W, $\frac{n}{4+n}$ G

$\frac{n}{5+n}$ G: $\frac{5}{4+n}$ W, $\frac{n-1}{4+n}$ G

If $\text{P}(WW) = \frac{2}{11}$, then

$\left(\frac{5}{5+n}\right)\left(\frac{4}{4+n}\right) = \frac{2}{11}$

$\therefore\ (n+4)(n+5) = 110$

$\therefore\ n^2 + 9n - 90 = 0$

$\therefore\ (n-6)(n+15) = 0$

$\therefore\ n = 6 \text{ or } -15$

As $n > 0$, the only solution is $n = 6$.

**30** If $X$ is the mass of a sea lion, then $X \sim \text{N}(\mu, \sigma^2)$.

We start by finding $z_1$ and $z_2$ which correspond to $x_1 = 500$ and $x_2 = 900$.

$\text{P}(X < 500) = 0.15$

$\therefore\ \text{P}\left(Z < \frac{500-\mu}{\sigma}\right) = 0.15$

$\therefore\ z_1 = \frac{500-\mu}{\sigma} \approx -1.0364$

$\therefore\ 500 - \mu \approx -1.0364\sigma$ .... (1)

Also $\text{P}(X > 900) = 0.1$

$\therefore\ \text{P}(X \leqslant 900) = 0.9$

$\therefore\ \text{P}\left(\frac{X-\mu}{\sigma} \leqslant \frac{900-\mu}{\sigma}\right) = 0.9$

$\therefore\ \text{P}\left(Z \leqslant \frac{900-\mu}{\sigma}\right) = 0.9$

$\therefore\ z_2 = \frac{900-\mu}{\sigma} \approx 1.2816$

$\therefore\ 900 - \mu \approx 1.2816\sigma$ .... (2)

Solving (1) and (2) simultaneously, we obtain $\mu \approx 679$ kg and $\sigma \approx 173$ kg.

**31** **a** When $d = 0$,

$H = 4 \times e^{0.3442 \times 0}$

$= 4$

So, 4 horses were initially infected.

**b** We need to solve

$4 \times e^{0.3442t} = 200$ for $t$

$\therefore\ e^{0.3442t} = 50$

$\therefore\ 0.3442t = \ln 50$

$\therefore\ t = \frac{\ln 50}{0.3442}$

$\approx 11.4$ days

**32** **a** $\text{P}(A') = 0.2 \quad \therefore\ \text{P}(A) = 0.8$

$\therefore\ x + y = 0.8$ .... (1)

$\text{P}(A \cup B) = x + y + z = 0.9$

$\therefore\ 0.8 + z = 0.9$ {using (1)}

$\therefore\ z = 0.1$

**b** $\text{P}(A \mid B) = \frac{y}{y+z} = 0.5$

$\therefore\ y = \frac{1}{2}(y+z)$

$\therefore\ 2y = y + z$

$\therefore\ y = z = 0.1$

**c** From (1), $x + 0.1 = 0.8$

$\therefore\ x = 0.7$

**33** **a** $y = \frac{1}{x}$ meets $y = x + 2$ where $x + 2 = \frac{1}{x}$

$$\therefore \; x^2 + 2x = 1$$
$$\therefore \; x^2 + 2x - 1 = 0$$
$$\therefore \; x = \frac{-2 \pm \sqrt{4 - 4(1)(-1)}}{2}$$
$$\therefore \; x = \frac{-2 \pm \sqrt{8}}{2}$$
$$\therefore \; x = \frac{-2 \pm 2\sqrt{2}}{2}$$
$$\therefore \; x = -1 \pm \sqrt{2}$$
$$\therefore \; m = -1, \quad n = 2$$

**b** **i** $y = \frac{1}{x}$ under a translation of $\binom{-2}{0}$ becomes $y = \frac{1}{x+2}$ which under a reflection in the $x$-axis becomes $y = -\frac{1}{x+2}$

$$\therefore \; g(x) = -\frac{1}{x+2}$$

**ii** $g(x)$ is undefined when $x + 2 = 0$

$\therefore \; x = -2$ is a vertical asymptote.

As $x \to \infty$, $g(x) \to 0^-$

As $x \to -\infty$, $g(x) \to 0^+$

$\therefore \; y = 0$ is a horizontal asymptote.

**iii** $g(0) = -\frac{1}{2}$

$\therefore$ the $y$-intercept is $-\frac{1}{2}$.

**iv**

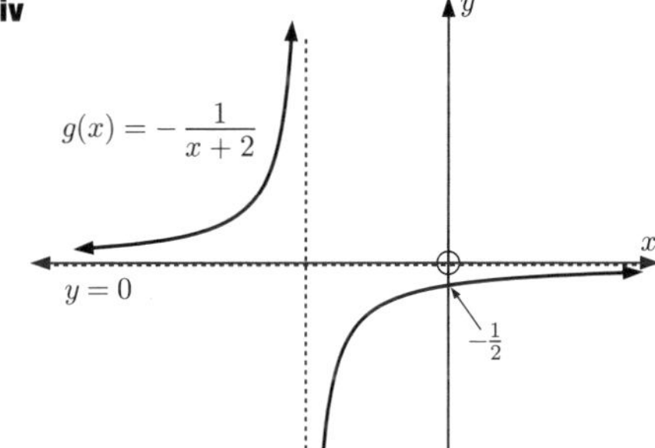


**34** **a** **i** When $t = 0$, $\mathbf{r}_1 = \binom{-1}{3}$

So, the initial position of the first object is $(-1, 3)$.

**ii** When $t = 10$, $\mathbf{r}_1 = \binom{-1}{3} + \binom{40}{20}$

$$= \binom{39}{23}$$

So, the first object is at $(39, 23)$ after 10 seconds.

**b** distance $= \sqrt{(39 - -1)^2 + (23 - 3)^2}$

$= \sqrt{40^2 + 20^2}$

$\approx 44.7$ m

**c** When $t = 1$, $\mathbf{r}_2 = \binom{0}{4} + \binom{9}{4}$

$= \binom{9}{8}$

$\therefore$ the second object passes through $(9, 8)$.

**d** We consider $\binom{9}{8} = \binom{-1}{3} + t\binom{4}{2}$

$\therefore \; 9 = -1 + 4t$ and

$8 = 3 + 2t$

$\therefore \; t = 2.5$ is the common solution

So, both objects pass through $(9, 8)$, but they do not collide since they pass through $(9, 8)$ at different times.

**35**

$$\text{area of } A = \text{area of } B$$

$$\therefore\ -\int_{-2}^{k} f(x)\,dx = \int_1^3 f(x)\,dx$$

$$\therefore\ \int_{k}^{-2} f(x)\,dx = \int_1^3 f(x)\,dx$$

$$\therefore\ \left[-\frac{x^4}{4}+\frac{2x^3}{3}+\frac{5x^2}{2}-6x\right]_k^{-2} = 5\tfrac{1}{3} \quad \{\text{technology}\}$$

$$\therefore\ (-4-\tfrac{16}{3}+10+12)-\left(-\frac{k^4}{4}+\frac{2k^3}{3}+\frac{5k^2}{2}-6k\right)=5\tfrac{1}{3}$$

$$\therefore\ 12\tfrac{2}{3}+\frac{k^4}{4}-\frac{2k^3}{3}-\frac{5k^2}{2}+6k=5\tfrac{1}{3}$$

$$\therefore\ \frac{k^4}{4}-\frac{2k^3}{3}-\frac{5k^2}{2}+6k+7\tfrac{1}{3}=0$$

$$\therefore\ 3k^4-8k^3-30k^2+72k+88=0$$

$$\therefore\ k \approx -0.969 \quad \{\text{technology}\}$$

**36** $y = e^{-x^2}$

**a** When $x = 0$, $y = e^0 = 1$ and
when $x = 2$, $y = e^{-4}$
$\therefore$ A is $(0, 1)$ and B is $\left(2, \dfrac{1}{e^4}\right)$.

**b** Area $= \int_0^2 e^{-x^2}\,dx$
$\approx 0.882$ units$^2$

**37** **a** As the results are independent,
$P(H \cap H) = P(H) \times P(H)$
$\therefore\ [P(H)]^2 = 0.64$
$\therefore\ P(H) = 0.8 \quad \{P(H) > 0\}$

**b** Let $X$ = the number of heads obtained.
$X \sim B(10, 0.8)$

**i** $P(X = 6) \approx 0.0881 \quad \{\text{technology}\}$

**ii** $P(X \geqslant 6) \approx 0.967$

**38** If $X$ is height of a maize plant, then $X \sim N(\mu, 6.8^2)$.

**a**

$$P(X < 45) = 0.75$$

$$\therefore\ P\left(\frac{X-\mu}{6.8} < \frac{45-\mu}{6.8}\right) = 0.75$$

$$\therefore\ P\left(Z < \frac{45-\mu}{6.8}\right) = 0.75$$

$$\therefore\ \frac{45-\mu}{6.8} \approx 0.6745$$

$$\therefore\ 45-\mu \approx 4.59$$

$$\therefore\ \mu \approx 40.4$$

**b** $X \sim N(40.41, 6.8^2)$
$\therefore\ P(X < 25) \approx 0.0117$

**c**

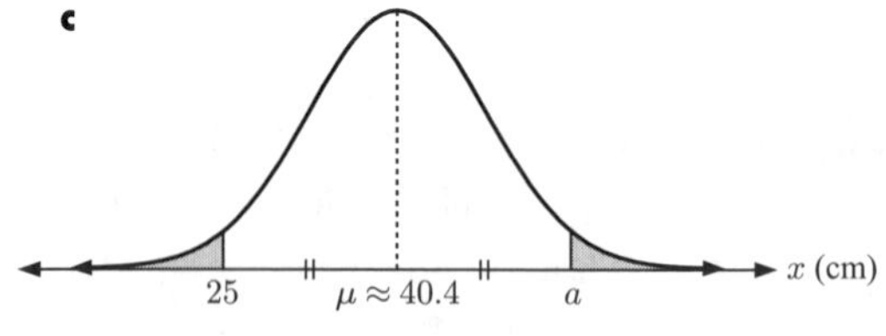


If $P(X < 25) = P(X > a)$ then

$$\mu - 25 = a - \mu$$

$$\therefore\ a = 2\mu - 25$$

$$\therefore\ a \approx 55.8$$

**39** **a**

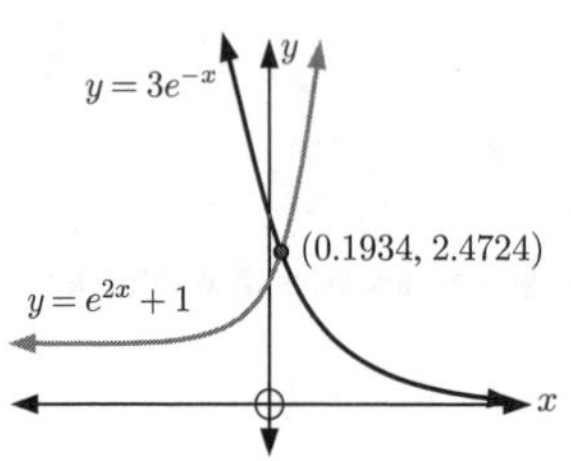


**b** Using technology, $x \approx 0.1934$

**40** $f(x) = x\sin(2x), \quad 0 < x < 3$

**a**

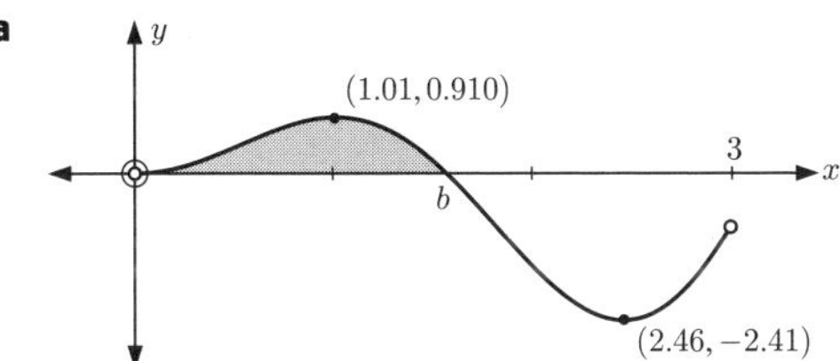


**b** Using technology, the maximum value of $f(x)$ is 0.910, and the minimum value is $-2.41$.

$\therefore$ the range is $\{y \mid -2.41 \leqslant y \leqslant 0.910\}$

**c** $f(x) = 0$ when $\sin(2x) = 0 \quad \{0 < x < 3\}$

$\therefore \; 2x = 0 + k\pi, \quad k \in \mathbb{Z}$

$\therefore \; x = \dfrac{k\pi}{2}$

$\therefore \; b = \frac{\pi}{2}$

**d** $A = \int_0^{\frac{\pi}{2}} x\sin(2x)\,dx$

$\approx 0.785 \quad \{\text{technology}\}$

**41** $f(x) = e^{-3x}\sin x, \quad -\frac{1}{2} \leqslant x \leqslant 3$

**a** $f'(x) = e^{-3x}(-3)\sin x + e^{-3x}\cos x$

$= e^{-3x}(\cos x - 3\sin x)$

**b** $f'(\frac{\pi}{2}) = e^{-\frac{3\pi}{2}}\left(\cos\frac{\pi}{2} - 3\sin\frac{\pi}{2}\right)$

$= e^{-\frac{3\pi}{2}}(0 - 3(1))$

$= -\dfrac{3}{e^{\frac{3\pi}{2}}}$

$\therefore$ the tangent has equation

$y - e^{-\frac{3\pi}{2}} = -\dfrac{3}{e^{\frac{3\pi}{2}}}\left(x - \frac{\pi}{2}\right)$

$\therefore \; e^{\frac{3\pi}{2}}y - e^{-\frac{3\pi}{2}}\left(e^{\frac{3\pi}{2}}\right) = -3x - 3(-\frac{\pi}{2})$

$\therefore \; e^{\frac{3\pi}{2}}y - 1 = -3x + \frac{3\pi}{2}$

$\therefore \; 3x + e^{\frac{3\pi}{2}}y = 1 + \frac{3\pi}{2}$

**c**

area $= \int_0^1 e^{-3x}\sin x\,dx$

$\approx 0.0847$ units$^2$

**42** $f(x) = \dfrac{\sin x}{\cos x} + \dfrac{\cos x}{\sin x}, \quad 0 \leqslant x \leqslant \frac{\pi}{2}$

**a** $f(x) = \dfrac{\sin^2 x + \cos^2 x}{\cos x \sin x}$

$= \dfrac{1}{\sin x \cos x} \times \left(\dfrac{2}{2}\right)$

$= \dfrac{2}{\sin 2x}$

**b** $\sin 2x = 0$

$\therefore \; 2x = 0 + k\pi, \quad k \in \mathbb{Z}$

$\therefore \; x = k\frac{\pi}{2}, \quad k \in \mathbb{Z}$

$\therefore \; x = 0$ or $\frac{\pi}{2}$

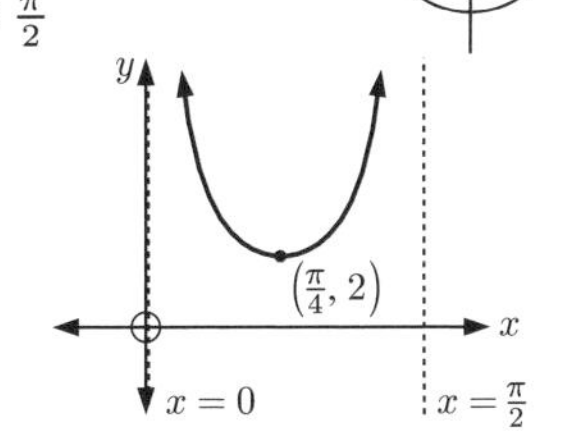


$\therefore \; x = 0$ and $x = \frac{\pi}{2}$ are vertical asymptotes of $f(x)$

**c** The least value of $f(x)$ is $\frac{2}{1} = 2$
when $\sin 2x = 1$

$\therefore\ 2x = \frac{\pi}{2}$ $\{0 \leqslant 2x \leqslant \pi\}$

$\therefore\ x = \frac{\pi}{4}$

**d** $\sin a = \frac{1}{3}$ so $a = \sin^{-1}(\frac{1}{3})$

$$\therefore\ f(2a) = \frac{2}{\sin(4a)} = \frac{2}{\sin\left(4\sin^{-1}(\frac{1}{3})\right)} \approx 2.046$$

**43** If $F$ is the weight of a female and $M$ is the weight of a male,
then $F \sim \text{N}(78.6, 5.03^2)$ and $M \sim \text{N}(91.3, 6.29^2)$.

**a** **i** $\text{P}(M < 80) \approx 0.0362$ **ii** $\text{P}(F < 80) \approx 0.610$ **iii** $\text{P}(70 < F < 80) \approx 0.566$

**b** $\text{P}(F < k) = 0.2$

$\therefore\ k \approx 74.4$

**c**

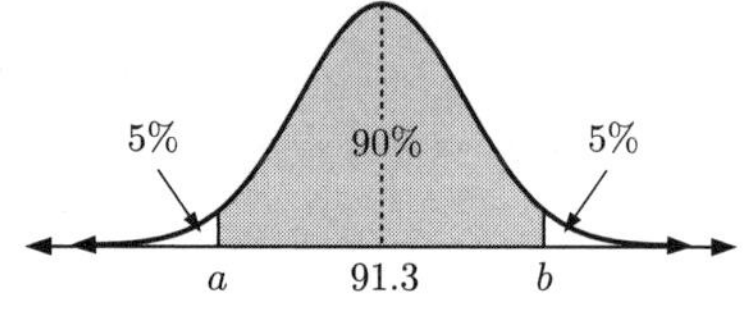


$\text{P}(M < a) = 0.05$

$\therefore\ a \approx 81.0$

and $\dfrac{a+b}{2} = 91.3$

$\therefore\ 81.0 + b \approx 182.6$

$\therefore\ b \approx 102$

**d**

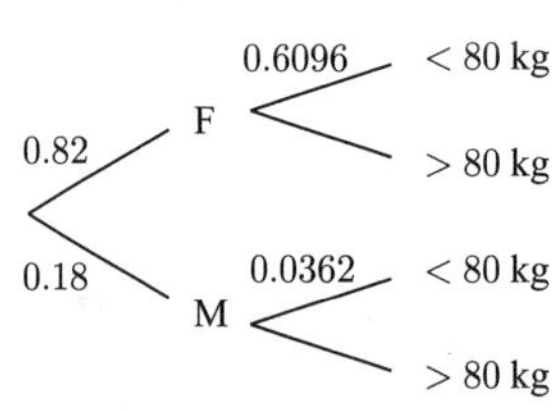


$\text{P}(< 80 \text{ kg})$
$\approx 0.82 \times 0.6096 + 0.18 \times 0.0362$
$\approx 0.506$

**44** **a** $\overrightarrow{\text{BA}} = \begin{pmatrix} 3-1 \\ 2--1 \\ -1-4 \end{pmatrix} = \begin{pmatrix} 2 \\ 3 \\ -5 \end{pmatrix}$ $\therefore\ |\overrightarrow{\text{BA}}| = \sqrt{4+9+25} = \sqrt{38}$ units

**b**

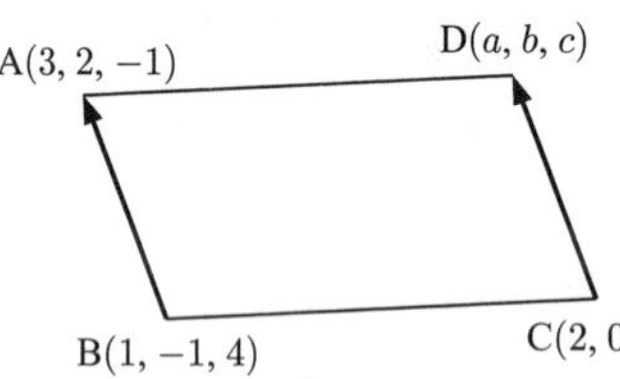


$\overrightarrow{\text{CD}} = \overrightarrow{\text{BA}}$

$\therefore\ \begin{pmatrix} a-2 \\ b-0 \\ c-7 \end{pmatrix} = \begin{pmatrix} 2 \\ 3 \\ -5 \end{pmatrix}$

$\therefore\ a = 4,\ b = 3,\ c = 2$

$\therefore$ D is at $(4, 3, 2)$.

**c** Let F be $(p, q, r)$.

$\therefore\ \overrightarrow{\text{BF}} = \begin{pmatrix} p-1 \\ q+1 \\ r-4 \end{pmatrix} = \begin{pmatrix} 6 \\ -3 \\ -6 \end{pmatrix}$

$\therefore\ p = 7,\ q = -4,\ r = -2$

So, F is at $(7, -4, -2)$.

**d** $\overrightarrow{\text{BA}} \bullet \overrightarrow{\text{BF}} = \begin{pmatrix} 2 \\ 3 \\ -5 \end{pmatrix} \bullet \begin{pmatrix} 6 \\ -3 \\ -6 \end{pmatrix} = 12 - 9 + 30 = 33$

$$\cos(\widehat{\text{ABF}}) = \frac{\overrightarrow{\text{BA}} \bullet \overrightarrow{\text{BF}}}{|\overrightarrow{\text{BA}}||\overrightarrow{\text{BF}}|} = \frac{33}{\sqrt{38}\sqrt{36+9+36}} = \frac{33}{\sqrt{38} \times 9} = \frac{11}{3\sqrt{38}}$$

**e** Area ABFE $= 2 \times$ area $\triangle$ABF

$$= 2 \times \tfrac{1}{2}\, |\overrightarrow{BA}||\overrightarrow{BF}| \sin(A\widehat{B}F)$$
$$= \sqrt{38} \times 9 \times \sqrt{1 - \cos^2(A\widehat{B}F)}$$
$$= \sqrt{38} \times 9 \times \sqrt{1 - \frac{121}{9 \times 38}}$$
$$= \sqrt{38} \times 9 \times \sqrt{\frac{9 \times 38 - 121}{9 \times 38}}$$
$$= \sqrt{38} \times 9 \times \frac{\sqrt{221}}{3\sqrt{38}}$$
$$= 3\sqrt{221} \text{ units}^2$$

**45** **a**
$$\sum p_i = 1$$
$$\therefore\ \tfrac{1}{10} + 2t + \tfrac{3}{20} + 2t^2 + \frac{t}{2} = 1$$
$$\therefore\ 2t^2 + \tfrac{5}{2}t - \tfrac{3}{4} = 0$$
$$\therefore\ 8t^2 + 10t - 3 = 0$$
$$\therefore\ (4t-1)(2t+3) = 0$$
$$\therefore\ t = \tfrac{1}{4} \text{ or } -\tfrac{3}{2}$$
But $t \geqslant 0 \quad \therefore\ t = \tfrac{1}{4}$

**b** 
$$\mathrm{E}(Y) = \sum y_i p_i$$
$$= 1(\tfrac{1}{10}) + 2(\tfrac{1}{2}) + 3(\tfrac{3}{20}) + 4(\tfrac{1}{8}) + 5(\tfrac{1}{8})$$
$$= 2.675$$

**c** 2.675 is the mean of the $Y$ distribution.

**46** $f(x) = 5x + e^{1-x^2} - 2, \quad -1 \leqslant x \leqslant 2$

**a** $f(0) = 0 + e^1 - 2$

$\therefore$ the $y$-intercept is $e - 2 \approx 0.718$

**b**

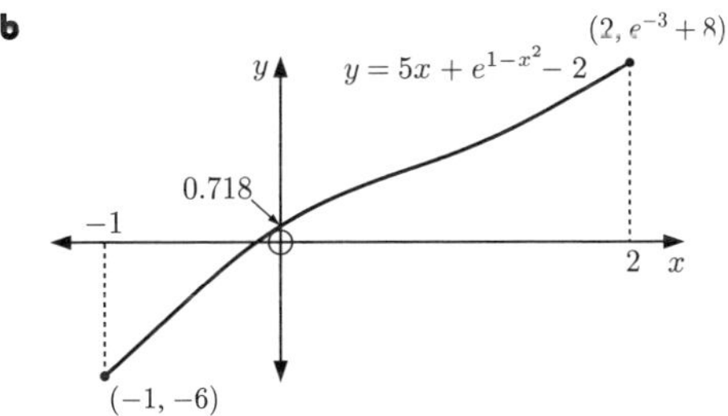


**c** $x$-intercept $\approx -0.134$ {technology}

**d** $f'(x) = 5 + e^{1-x^2}(-2x)$

$\therefore\ f'(1) = 5 + e^0(-2) = 3$

So the tangent has gradient 3 when $x = 1$.

**47** If $X$ is the length of a zucchini, then $X \sim \mathrm{N}(24.3,\ 6.83^2)$.

**a** $\mathrm{P}(X < a) = 0.15$ and $\mathrm{P}(X > b) = 0.2$

$\therefore\ a \approx 17.2$ $\quad\quad \therefore\ \mathrm{P}(X \leqslant b) = 0.8$

$\therefore\ b \approx 30.0$

**b** Using technology:

**i** $\mathrm{P}(17.2 < X < 30.0)$
$\approx 0.649$
(Compare this with
$1 - 0.15 - 0.2 = 0.65$.)

**ii** $\mathrm{P}(20 < X < 26)$
$\approx 0.334$

**iii** $\mathrm{P}(X < 24.3)$
$= 0.5$

**48** **a**

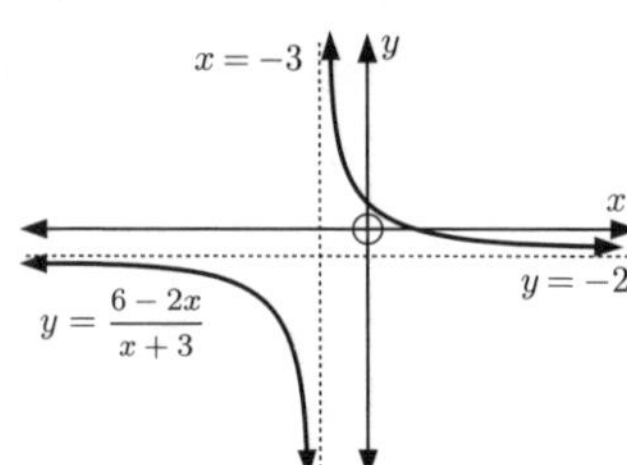


**b** As $x \to -3^-$, $y \to -\infty$
As $x \to -3^+$, $y \to \infty$
$\therefore$ $x = -3$ is a vertical asymptote.
As $x \to \infty$, $y \to -2^+$
As $x \to -\infty$, $y \to -2^-$
$\therefore$ $y = -2$ is a horizontal asymptote.

**c** $\lim\limits_{x \to -\infty} \dfrac{6-2x}{x+3} = -2$, $\lim\limits_{x \to \infty} \dfrac{6-2x}{x+3} = -2$

**49** **a** $3 + s = -2$ $\therefore$ $s = -5$

$$\therefore \mathbf{r}_1 = \begin{pmatrix} 2 \\ -1 \\ 3 \end{pmatrix} + (-5)\begin{pmatrix} 1 \\ -1 \\ 1 \end{pmatrix} = \begin{pmatrix} -3 \\ 4 \\ -2 \end{pmatrix}$$

So, A is at $(-3, 4, -2)$.

**b** They meet if $2 + s = 6 + 2t$ .... (1)
$-1 - s = -3$ .... (2)
$3 + s = 4 - t$ .... (3)
$\therefore$ $s = 2$ and in (3), $5 = 4 - t$
$\therefore$ $t = -1$
So in (1), $2 + s = 4$ and $6 + 2t = 4$ ✓
$\therefore$ $L_1$ and $L_2$ meet at $(4, -3, 5)$.

**c** The lines have direction vectors
$\begin{pmatrix} 1 \\ -1 \\ 1 \end{pmatrix}$ and $\begin{pmatrix} 2 \\ 0 \\ -1 \end{pmatrix}$.

$$\therefore \cos\theta = \frac{|2 + 0 - 1|}{\sqrt{1+1+1}\sqrt{4+0+1}} = \frac{1}{\sqrt{3}\sqrt{5}}$$

$$\therefore \theta = \cos^{-1}\left(\tfrac{1}{\sqrt{15}}\right) \approx 75.0^\circ$$

**50** $f(x) = e^x(x^2 - 3x + 2)$

**a** $y = f(x)$ cuts the $x$-axis when $y = 0$
$\therefore$ $x^2 - 3x + 2 = 0$ {as $e^x > 0$}
$\therefore$ $(x-1)(x-2) = 0$
$\therefore$ $x = 1$ or $2$
So, A is at $(1, 0)$ and B is at $(2, 0)$.
$y = f(x)$ cuts the $y$-axis when $x = 0$
$\therefore$ $y = e^0(2) = 2$
$\therefore$ C is at $(0, 2)$.

**b** As $x \to -\infty$, $f(x) \to 0^+$
$\therefore$ $y = 0$ is a horizontal asymptote.

**c** $f'(x) = e^x(x^2 - 3x + 2) + e^x(2x - 3)$
$= e^x(x^2 - 3x + 2 + 2x - 3)$
$= e^x(x^2 - x - 1)$

**d** $f'(x) = 0$ when $x^2 - x - 1 = 0$
$\therefore$ $x \approx -0.618$ or $1.62$

$\therefore$ the local maximum has $x$-coordinate $\approx -0.618$ and the local minimum has $x$-coordinate $\approx 1.62$.

**e** $f'(1) = e(1 - 1 - 1) = -e$
$\therefore$ the normal at A$(1, 0)$ has gradient $\dfrac{1}{e}$,
and the equation of the normal is
$\dfrac{y-0}{x-1} = \dfrac{1}{e}$
$\therefore$ $ey = x - 1$
$\therefore$ $x - ey = 1$

**f** $x - ey = 1$ meets $y = e^x(x^2 - 3x + 2)$
where $\dfrac{x-1}{e} = e^x(x^2 - 3x + 2)$
$\therefore$ $x - 1 = e^{x+1}(x-1)(x-2)$
$\therefore$ $(x-1)\left[1 - e^{x+1}(x-2)\right] = 0$
$\therefore$ $x = 1$ or $x \approx 2.0475$ {technology}
$\therefore$ the $x$-coordinate of D $\approx 2.05$

**g** Area $\approx \int_1^{2.0475} \left[\frac{x-1}{e} - e^x(x^2 - 3x + 2)\right] dx$
$\approx 0.959$ units$^2$

**51** **a** $f(x) = xe^{-x}$

$f(0) = 0 \times e^0$

$= 0$

$\therefore$ the $y$-intercept is 0.

**b** $f'(x) = e^{-x} + xe^{-x} \times (-1)$

$= e^{-x} - xe^{-x}$

which is 0 when $e^{-x} - xe^{-x} = 0$

$\therefore\ e^{-x}(1-x) = 0$

$\therefore\ 1 - x = 0 \quad \{e^{-x} > 0 \text{ for all } x\}$

$\therefore\ x = 1$

and $f(1) = 1 \times e^{-1} = \dfrac{1}{e}$

$\therefore$ A is at $\left(1, \frac{1}{e}\right)$.

**c** $f'(x) = e^{-x}(1-x)$

$\therefore\ f''(x) = -e^{-x}(1-x) + e^{-x} \times (-1)$

$= -e^{-x} + xe^{-x} - e^{-x}$

$= xe^{-x} - 2e^{-x}$

which is 0 when $xe^{-x} - 2e^{-x} = 0$

$\therefore\ e^{-x}(x-2) = 0$

$\therefore\ x - 2 = 0 \quad \{e^{-x} > 0 \text{ for all } x\}$

$\therefore\ x = 2$

and $f(2) = 2 \times e^{-2} = \dfrac{2}{e^2}$

$\therefore$ B is at $\left(2, \frac{2}{e^2}\right)$.

**d** Area $= \int_1^2 xe^{-x}\,dx$

$\approx 0.330$ units$^2$

**52** **a**

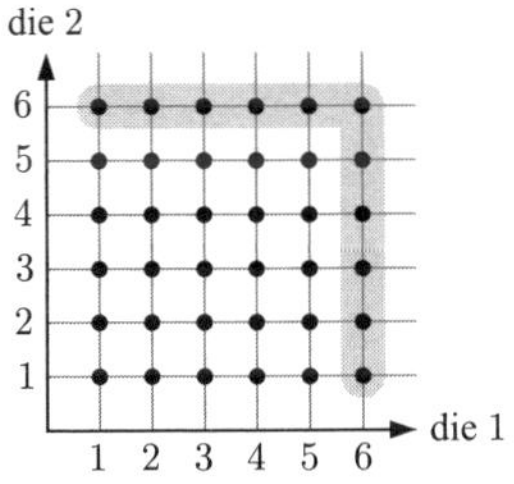


**b** $\text{P}(E) = \frac{11}{36}$ {enclosed points}

**c** If $X$ is the number of times $E$ occurs, then $X \sim \text{B}(10, \frac{11}{36})$.

**i** $\text{P}(X = 2) \approx 0.227$

**ii** P(at most 3 times) $= \text{P}(X \leqslant 3)$

$\approx 0.635$

**53** **a** $\overrightarrow{\text{AB}} = \begin{pmatrix} -2-2 \\ 4--1 \\ -1-3 \end{pmatrix} = \begin{pmatrix} -4 \\ 5 \\ -4 \end{pmatrix}$

$\therefore$ $L_1$ has equation

$\mathbf{r}_1 = \begin{pmatrix} 2 \\ -1 \\ 3 \end{pmatrix} + t\begin{pmatrix} -4 \\ 5 \\ -4 \end{pmatrix} = \begin{pmatrix} 2-4t \\ -1+5t \\ 3-4t \end{pmatrix}$

$\therefore\ \mathbf{r}_1 = (2-4t)\mathbf{i} + (5t-1)\mathbf{j} + (3-4t)\mathbf{k}$

**b** $\begin{pmatrix} 4 \\ a \\ b \end{pmatrix} = \begin{pmatrix} 2-4t \\ -1+5t \\ 3-4t \end{pmatrix}$

$\therefore\ 2 - 4t = 4$

$\therefore\ -4t = 2$

$\therefore\ t = -\frac{1}{2}$

So, $a = -1 + 5(-\frac{1}{2}) = -3\frac{1}{2}$

and $b = 3 - 4(-\frac{1}{2}) = 5$.

**c** They meet if $2 - 4t = 3s - 5$ .... (1)

$5t - 1 = s - 16$ .... (2)

$3 - 4t = 16 - s$ .... (3)

From (2) and (3), $5t - 1 = -(3 - 4t)$

$\therefore\ 5t - 1 = -3 + 4t$

$\therefore\ t = -2$

and $5(-2) - 1 = s - 16$

$\therefore\ -11 = s - 16$

$\therefore\ s = 5$

In (1), LHS $= 2 - 4t = 2 + 8 = 10$

RHS $= 3s - 5 = 15 - 5 = 10$ ✓

$\therefore$ all 3 equations are satisfied by $t = -2,\ s = 5$

So, the lines meet at

$(2 - 4(-2), 5(-2) - 1, 3 - 4(-2))$

$\therefore$ D is at $(10, -11, 11)$.

**54** **a** $DB^2 = 7.6^2 + 8.1^2 - 2 \times 7.6 \times 8.1 \times \cos 30°$

$\therefore\ DB \approx 4.09$ m

$BC^2 = 16^2 + 8.1^2 - 2 \times 16 \times 8.1 \times \cos 30°$ $\{AC = 7.6 + 8.4 = 16\}$

$\therefore\ BC \approx 9.86$ m

**b** Let $A\widehat{B}E = \theta$

In $\triangle ABD$, $\dfrac{\sin\theta}{7.6} \approx \dfrac{\sin 30°}{4.092}$

$\therefore\ \sin\theta \approx \dfrac{3.8}{4.092}$

$\therefore\ \theta \approx 68.2°$

$\therefore\ A\widehat{B}E \approx 68.2°$

Let $D\widehat{B}C = \phi$

$\therefore\ \dfrac{\sin\phi}{8.4} \approx \dfrac{\sin(30° + 68.22°)}{9.856}$ {exterior angle of a $\triangle$}

$\therefore\ \sin\phi \approx \dfrac{8.4 \times \sin(98.22°)}{9.856}$

$\therefore\ \sin\phi \approx 0.8435$

$\therefore\ \phi \approx 57.5°$

$\therefore\ D\widehat{B}C \approx 57.5°$

**c** Area BCD

$= \frac{1}{2} \times BC \times DB \times \sin(D\widehat{B}C)$

$\approx \frac{1}{2} \times 9.86 \times 4.09 \times \sin(57.5°)$

$\approx 17.0\ m^2$

**d** In $\triangle AME$,

$\cos 68.2° \approx \dfrac{4.05}{x}$

$\therefore\ x \approx \dfrac{4.05}{\cos(68.2°)}$

$\therefore\ x \approx 10.9$

$\therefore\ AE \approx 10.9$ m

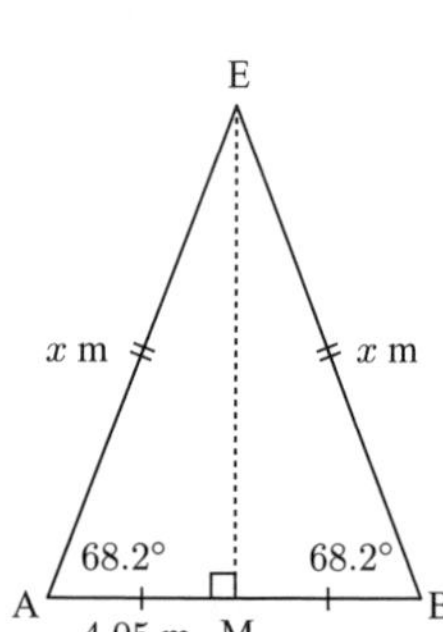


**55** **a** As $x \to \infty$, $\dfrac{3}{(x-1)(x+b)} \to 0$

$\therefore\ y \to a$

But $y = -1$ is the H.A.

$\therefore\ a = -1$

The vertical asymptotes are $x = 1$, $x = -2$

$\therefore\ b = 2$

**b** $f(x) = -1 + \dfrac{3}{(x-1)(x+2)}$

$\therefore\ f(0) = -1 + \frac{3}{-2} = -2\frac{1}{2}$

$\therefore$ the $y$-intercept is $-2\frac{1}{2}$.

**c** The function cuts the $x$-axis when $y = 0$.

$\therefore\ \dfrac{3}{(x-1)(x+2)} = 1$

$\therefore\ (x-1)(x+2) = 3$

$\therefore\ x^2 + x - 2 = 3$

$\therefore\ x^2 + x - 5 = 0$

$\therefore\ x = \dfrac{-1 \pm \sqrt{1 - 4(1)(-5)}}{2}$

$\therefore\ x = \dfrac{-1 \pm \sqrt{21}}{2}$

$\therefore$ $x$-intercepts are $\frac{-1-\sqrt{21}}{2}$ and $\frac{-1+\sqrt{21}}{2}$.

**d** $f(x) = -1 + 3(x^2 + x - 2)^{-1}$

$\therefore\ f'(x) = 0 - 3(x^2 + x - 2)^{-2}(2x + 1)$

$= \dfrac{-3(2x+1)}{(x^2 + x - 2)^2}$

$\therefore\ f'(x) = 0$ when $-3(2x + 1) = 0$

$\therefore\ x = -\frac{1}{2}$

and $f(-\frac{1}{2}) = -1 + \dfrac{3}{(-\frac{3}{2})(\frac{3}{2})}$

$= -1 + \dfrac{3}{-\frac{9}{4}}$

$= -1 - \frac{4}{3}$

$= -2\frac{1}{3}$

$\therefore$ D is at $(-\frac{1}{2}, -2\frac{1}{3})$.

**e** **i** $A = -\displaystyle\int_{\frac{\sqrt{21}-1}{2}}^{k} \left(-1 + \frac{3}{x^2 + x - 2}\right) dx$

**ii** $A = \displaystyle\int_{3}^{\frac{\sqrt{21}-1}{2}} \left(-1 + \frac{3}{x^2 + x - 2}\right) dx$

$\approx 0.558\ \text{units}^2$

**56** **a** **i** $P(G \cup T) = P(G) + P(T) - P(G \cap T)$

$\therefore\ 1 - 0.42 = 0.3 + 0.4 - P(G \cap T)$

$\therefore\ 0.58 = 0.7 - P(G \cap T)$

$\therefore\ P(G \cap T) = 0.7 - 0.58$

$= 0.12$

The probability of a randomly selected customer buying both a goldfish and a tortoise is $0.12$.

**ii**

$a + 0.12 = 0.3$

$\therefore\ a = 0.18$

$b + 0.12 = 0.4$

$\therefore\ b = 0.28$

So, the completed Venn diagram is:

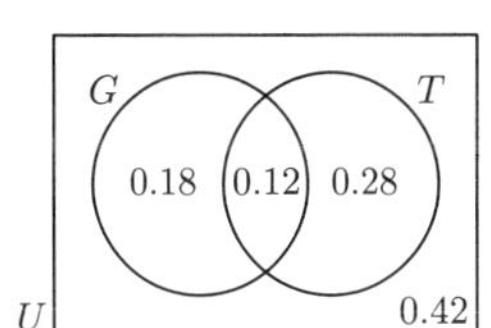


**iii** $P(G) \times P(T) = 0.3 \times 0.4$

$= 0.12$

$= P(G \cap T)$

$\therefore$ $G$ and $T$ are independent events.

**b** **i** Since $P(T) = 2\,P(G)$,

$b + c = 2(a + b)$

$\therefore\ b + c = 2a + 2b$

$\therefore\ c = 2a + b$

**ii** Since $G$ and $T$ are independent,

$P(G \cap T) = P(G)\,P(T)$

$\therefore\ b = (a+b)(b+c)$

$\therefore\ b = (a+b)(b+2a+b)$ {from **i**}

$\therefore\ b = (a+b)(2a+2b)$

$\therefore\ b = 2(a+b)^2$

$\therefore\ (a+b)^2 = \dfrac{b}{2}$

$\therefore\ a + b = \sqrt{\dfrac{b}{2}}$ {as $a + b > 0$}

$\therefore\ a = \sqrt{\dfrac{b}{2}} - b$

**iii** As $a + b + c = 0.58$,

$a + b + 2a + b = 0.58$

$\therefore\ 3a + 2b = 0.58$

$\therefore\ 3\left(\sqrt{\dfrac{b}{2}} - b\right) + 2b = 0.58$

$\therefore\ 3\sqrt{\dfrac{b}{2}} - b = 0.58$

So, $b \approx 0.104$ $\{0 < b < 1\}$

and $a \approx 0.124$

**iv** $P(G) = a + b \approx 0.228$

**57** **a** When $t = 0$, $W_0 = 1800 \times (0.95)^0$

$= 1800$ g

$\therefore$ the initial quantity of DDT was 1800 g.

**b** **i** When $t = 7.5$, $W_{7.5} = 1800 \times (0.95)^{7.5}$

$\approx 1225$ g

**ii** When $t = 40$, $W_{40} = 1800 \times (0.95)^{40}$

$\approx 231$ g

**c**

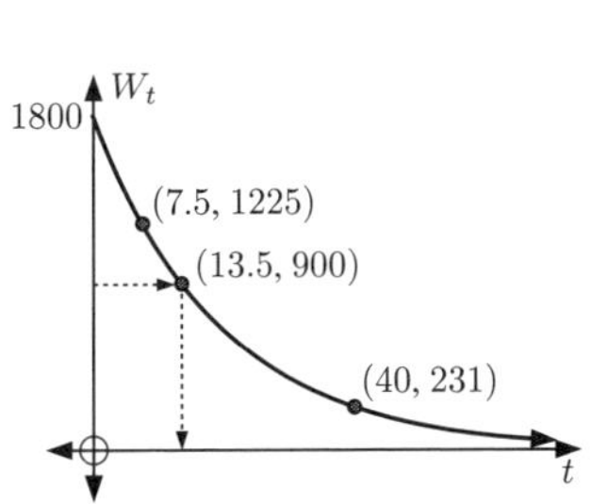


**d** From the graph, when $W_t = 900$ g,

$t \approx 13.5$

$\therefore$ the half-life of DDT is approximately 13.5 years.

**e** We need to solve $1800 \times (0.95)^t = 100$ for $t$

Using technology, $t \approx 56.3$ years.

**58** **a** Let the smaller semi-circle have radius $r$ cm.
In $\triangle$PTR, we have:

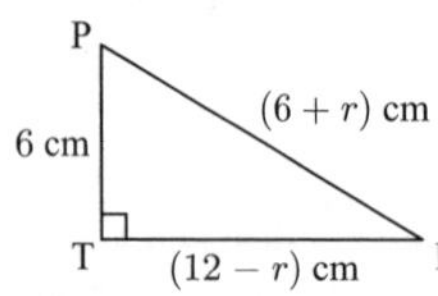


Thus $(6+r)^2 = 6^2 + (12-r)^2$
$\therefore\ 36 + 12r + r^2 = 36 + 144 - 24r + r^2$
$\therefore\ 36r = 144$
$\therefore\ r = 4$

**b** **i** $\cos(\widehat{TPR}) = \frac{6}{10} = 0.6$
$\therefore\ \widehat{TPR} \approx 53.1° \approx 0.927^c$

**ii** $\widehat{PRT} \approx 90° - 53.1°$
$\approx 36.9° \approx 0.644^c$

**c** **i** Area of $A$ = area $\triangle$PTR − (area sector PQT + area sector RQS)
$\approx \frac{1}{2}(8 \times 6) - \left(\frac{1}{2}(0.927)(6^2) + \frac{1}{2}(0.644)(4^2)\right)$
$\approx 24 - 16.69 - 5.15$
$\approx 2.16$ cm$^2$

**ii** Area of $B$ = area of quarter circle − area of semi-circles − area of $A$
$\approx \frac{1}{4}\pi(12^2) - \frac{1}{2}\pi(6^2) - \frac{1}{2}\pi(4^2) - 2.16$
$\approx 36\pi - 18\pi - 8\pi - 2.16$
$\approx 10\pi - 2.16$
$\approx 29.3$ cm$^2$

**59** **a** $\log_2(x^2 - 2x + 1) = 1 + \log_2(x-1)$
$\therefore\ \log_2(x-1)^2 - \log_2(x-1) = 1$
$\therefore\ 2\log_2(x-1) - \log_2(x-1) = 1$
$\therefore\ \log_2(x-1) = 1$
$\therefore\ x - 1 = 2^1$
$\therefore\ x = 3$

**b** $3^{2x+1} = 5(3^x) + 2$
$\therefore\ 3(3^x)^2 - 5(3^x) - 2 = 0$
$\therefore\ 3m^2 - 5m - 2 = 0 \quad \{m = 3^x\}$
$\therefore\ (3m+1)(m-2) = 0$
$\therefore\ m = -\frac{1}{3}$ or $2$
$\therefore\ 3^x = -\frac{1}{3}$ or $3^x = 2$
The first equation is impossible as $3^x > 0$ for all $x$.
$\therefore\ 3^x = 2$
$\therefore\ x = \dfrac{\ln 2}{\ln 3}$ or $\log_3 2$

**60** **a**

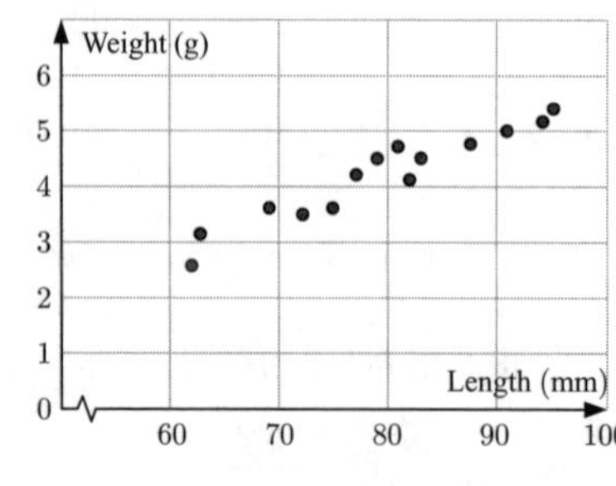


**b** Using technology, $r \approx 0.962$.

**c** There is very strong, positive, linear correlation between *Length* and *Weight*.

**d** Using technology, the equation of the least squares regression line is $y \approx 0.0729x - 1.57$.

**e** **i** When $x = 110$ mm,
$y \approx 0.0729 \times 110 - 1.57$
$\approx 6.45$ g

**ii** When $x = 70$ mm,
$y \approx 0.0729 \times 70 - 1.57$
$\approx 3.53$ g

**f** The prediction in **e ii** is more likely to be reliable, as it is an interpolation.

**61** **a**

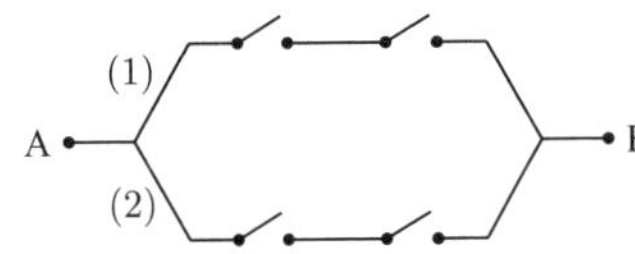


P(current flows)

$= \text{P}((1) \text{ closed} \cup (2) \text{ closed})$

$= \text{P}((1) \text{ closed}) + \text{P}((2) \text{ closed})$

$\quad - \text{P}((1) \textbf{ and } (2) \text{ closed})$

$= p^2 + p^2 - p^4$

$= 2p^2 - p^4$

**b** We need to solve $2p^2 - p^4 \geqslant \frac{1}{2}$

We graph $y = 2x^2 - x^4$ for $0 \leqslant x \leqslant 1$.

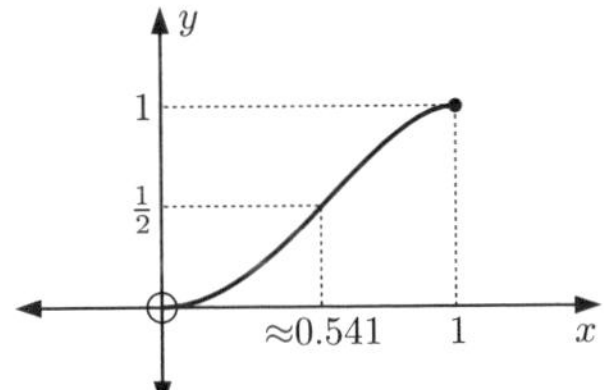


So, for $2p^2 - p^4 \geqslant \frac{1}{2}$, $p \geqslant 0.541$

$\therefore$ the least value of $p$ is $\approx 0.541$

**62** **a**

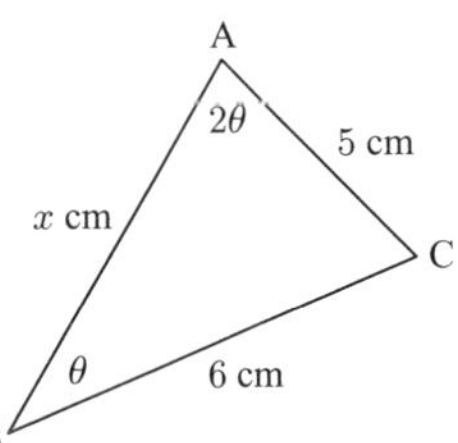


Let the angle at B be $\theta$ and at A be $2\theta$

By the sine rule: $\dfrac{\sin 2\theta}{6} = \dfrac{\sin\theta}{5}$

$\therefore \dfrac{2\sin\theta\cos\theta}{\sin\theta} = \dfrac{6}{5}$

$\therefore \cos\theta = \frac{3}{5}$ {as $\sin\theta \neq 0$}

**b** Let AB $= x$ cm.

$\dfrac{5}{\sin\theta} = \dfrac{x}{\sin(180 - 3\theta)}$

$\therefore x = \dfrac{5}{\sin(\cos^{-1}(\frac{3}{5}))} \times \sin(180 - 3 \times \cos^{-1}(\frac{3}{5}))$

$\therefore x = 2.2$

$\therefore$ the length of [AB] is 2.2 cm.

**c** Using the cosine rule,

$5^2 = x^2 + 6^2 - 2x(6)\cos\theta$

$\therefore 25 = x^2 + 36 - 12x\left(\frac{3}{5}\right)$

$\therefore x^2 - \frac{36}{5}x + 11 = 0$

$\therefore 5x^2 - 36x + 55 = 0$

$\therefore (x - 5)(5x - 11) = 0$

$\therefore x = 5$ or $\frac{11}{5}$

$\therefore$ AB $= 5$ cm or $2.2$ cm

So, using the cosine rule would give two answers.

**63**

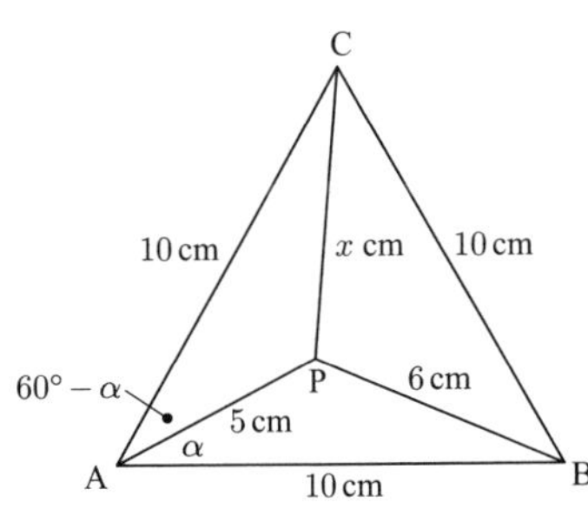


Using the cosine rule in $\triangle$ABP,

$\cos\alpha = \dfrac{5^2 + 10^2 - 6^2}{2(5)(10)} = \frac{89}{100} = 0.89$

$\therefore \alpha \approx 27.127°$

$\therefore 60° - \alpha \approx 32.873°$

So, in $\triangle$APC, $x^2 = 10^2 + 5^2 - 2(10)(5)\cos 32.873°$

$\therefore x^2 \approx 41.0127$

$\therefore x \approx 6.40$

$\therefore$ P is about 6.40 cm from C.

**64** Let $F$ be the event of a faulty chip.

$\therefore$ P$(F) = 0.03$ and P$(F') = 0.97$

If $X$ is the number which are faulty, then $X \sim$ B(500, 0.03).

So, P$(5 \leqslant X \leqslant 10) \approx 0.114$ {1% is 5, 2% is 10}

**65** $X \sim \text{N}(\mu,\ 2.83^2)$

$$\therefore \quad \text{P}(-4 < X - \mu < 4) = \text{P}\left(\frac{-4}{2.83} < \frac{X-\mu}{2.83} < \frac{4}{2.83}\right)$$
$$\approx \text{P}(-1.4134 < Z < 1.4134)$$
$$\approx 0.842$$

**66** **a**

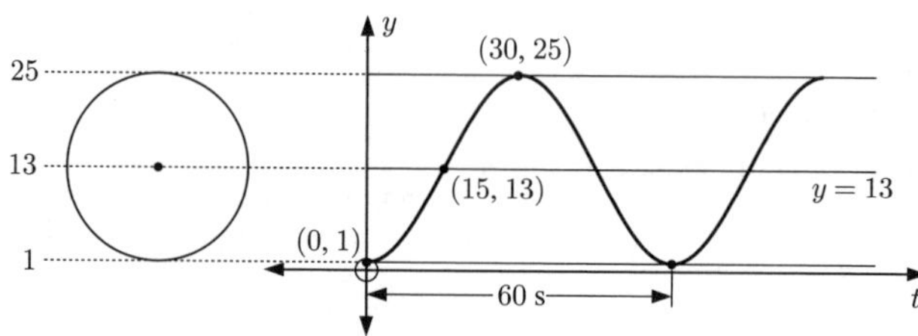


We model the Ferris wheel using $h(t) = a + b\sin(c(t-d))$.

The amplitude $= b = 12$. The period $= \dfrac{2\pi}{c} = 60 \quad \therefore \quad c = \frac{\pi}{30}$

$d =$ the $t$-coordinate of the point halfway between the first minimum and the following maximum

$= \dfrac{0+30}{2} = 15$

The equation of the principal axis is $y = \dfrac{1+25}{2} = 13$

$\therefore \quad a = 13$

Thus $a = 13, \quad b = 12, \quad c = \frac{\pi}{30}, \quad d = 15.$

So, $h(t) = 12\sin\left(\frac{\pi}{30}(t-15)\right) + 13$

*Check*: $h(0) = 12\sin\left(\frac{-\pi}{2}\right) + 13 = 12(-1) + 13 = 1$ ✓

$h(30) = 12\sin\left(\frac{\pi}{2}\right) + 13 = 12(1) + 13 = 25$ ✓

**b** When $t = 91$, $h(91) = 12\sin\left(\dfrac{\pi \times 76}{30}\right) + 13 \approx 24.9$ m

**67** **a** The period is $\dfrac{\pi}{a} = \dfrac{\pi}{2}$

$\therefore \quad a = 2$

The $y$-intercept is $-\sqrt{3}$.

$\therefore \quad b = -\sqrt{3}$

$\therefore \quad a = 2, \quad b = -\sqrt{3}$

**b** $x$-intercepts occur when

$\tan 2x - \sqrt{3} = 0$

$\therefore \quad \tan 2x = \sqrt{3}$

Now, if $-\frac{3\pi}{4} \leqslant x \leqslant \frac{3\pi}{4}$,

then $-\frac{3\pi}{2} \leqslant 2x \leqslant \frac{3\pi}{2}$

$\therefore \quad 2x = -\frac{2\pi}{3}, \frac{\pi}{3}, \frac{4\pi}{3}$

$\therefore \quad x = -\frac{\pi}{3}, \frac{\pi}{6}, \frac{2\pi}{3}$

$\therefore \quad x \approx -1.05, 0.524, 2.09$

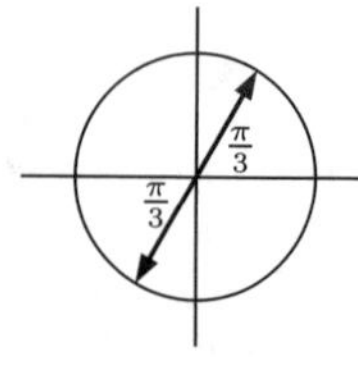


**68** $f(x) = e^{\sin^2 x}, \quad 0 \leqslant x \leqslant \pi$

**a** $f'(x) = e^{\sin^2 x} \times 2\sin x\cos x$

$= e^{\sin^2 x}\sin 2x$

which is 0 when $\sin 2x = 0$

$\therefore \quad 2x = 0 + k\pi, \quad k \in \mathbb{Z}$

$\therefore \quad x = 0 + \dfrac{k\pi}{2}$

$\therefore \quad x = 0, \ \frac{\pi}{2}, \ \pi$

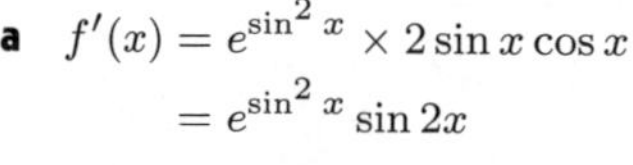

Sign diagram:

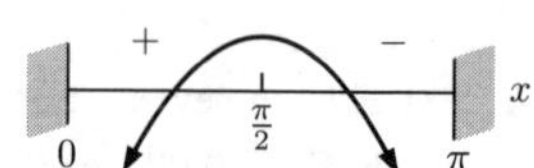


$\therefore \quad f(x)$ has maximum value when $x = \frac{\pi}{2}$, and this maximum value is $e$.

**b** $f''(x) = e^{\sin^2 x} \times \sin 2x \sin 2x + e^{\sin^2 x} 2\cos 2x$
$= e^{\sin^2 x}(\sin^2 2x + 2\cos 2x)$

**c** $f''(x) = 0$ when $\sin^2 2x + 2\cos 2x = 0$ on $0 \leqslant x \leqslant \pi$
$\therefore \sin^2 2x = -2\cos 2x$
Using technology, $x \approx 0.999$ or $2.14$

So, $f(0.999) \approx 2.03$ and $f(2.14) \approx 2.03$
$\therefore$ the points of inflection are $(0.999, 2.03)$ and $(2.14, 2.03)$.

**69** $\dfrac{u_1}{1-r} = 49$ and $u_1 r = 10$

$\therefore \dfrac{10}{r} = 49(1-r)$
$\therefore 10 = 49r - 49r^2$
$\therefore 49r^2 - 49r + 10 = 0$
$\therefore (7r-2)(7r-5) = 0$
$\therefore r = \frac{2}{7}$ or $\frac{5}{7}$

When $r = \frac{2}{7}$, $u_1 = 35$, $u_2 = 10$, $u_3 = \frac{20}{7} = 2\frac{6}{7}$
When $r = \frac{5}{7}$, $u_1 = 14$, $u_2 = 10$, $u_3 = \frac{50}{7} = 7\frac{1}{7}$
Thus $S_3 = 35 + 10 + 2\frac{6}{7} = 47\frac{6}{7}$
or $S_3 = 14 + 10 + 7\frac{1}{7} = 31\frac{1}{7}$

**70** $f(x) = xe^{1-2x^2}$

**a** $f'(x) = 1e^{1-2x^2} + xe^{1-2x^2}(-4x)$
$= e^{1-2x^2}(1-4x^2)$
$f''(x) = -4xe^{1-2x^2}(1-4x^2) + e^{1-2x^2}(-8x)$
$= e^{1-2x^2}(-4x + 16x^3 - 8x)$
$= e^{1-2x^2}(16x^3 - 12x)$

**b** $f'(x) = e^{1-2x^2}(1+2x)(1-2x)$

$\therefore$ there is a local minimum at $(-\frac{1}{2}, -\frac{\sqrt{e}}{2})$
and a local maximum at $(\frac{1}{2}, \frac{\sqrt{e}}{2})$.

**c** $f''(x) = 0$ when $16x^3 - 12x = 0$
$\therefore 4x(4x^2 - 3) = 0$
$\therefore x = 0$ or $\pm\frac{\sqrt{3}}{2}$

**d** As $x \to \infty$, $f(x) \to 0^+$
As $x \to -\infty$, $f(x) \to 0^-$

**e**

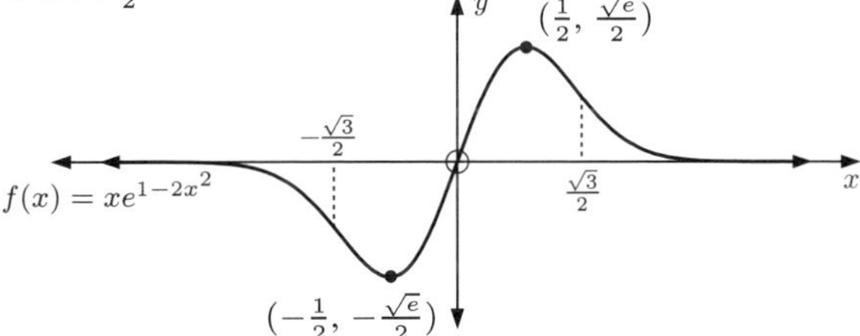


**71**

| *Score* | *Midpoint* ($x$) | $f$ |
|---|---|---|
| $9 \leqslant X < 11$ | 10 | 2 |
| $11 \leqslant X < 13$ | 12 | 7 |
| $13 \leqslant X < 15$ | 14 | 6 |
| $15 \leqslant X < 17$ | 16 | 21 |
| $17 \leqslant X < 19$ | 18 | 17 |
| $19 \leqslant X < 21$ | 20 | 5 |
| | *Total* | 58 |

**a** mean $\approx \dfrac{10 \times 2 + 12 \times 7 + .... + 20 \times 5}{58}$
$\approx 16.0$

**b** standard deviation $\approx 2.48$ {technology}

**72** **a** $|\mathbf{a}| = \sqrt{\cos^2\theta + \sin^2\theta + \cos^2\theta}$
$= \sqrt{1+\cos^2\theta}$

**b** As $-1 \leqslant \cos\theta \leqslant 1$ for all $\theta$,
then $0 \leqslant \cos^2\theta \leqslant 1$
$\therefore\ 1 \leqslant 1+\cos^2\theta \leqslant 2$
$\therefore\ 1 \leqslant \sqrt{1+\cos^2\theta} \leqslant \sqrt{2}$
$\therefore\ 1 \leqslant |\mathbf{a}| \leqslant \sqrt{2}$

**c** $\mathbf{a} \bullet \mathbf{b} = 0 \quad \{\mathbf{a} \perp \mathbf{b}\}$
$\therefore\ \cos\theta\sin\theta - \sin^2\theta + \cos^2\theta = 0$
$\therefore\ \frac{1}{2}\sin 2\theta + \cos^2\theta - \sin^2\theta = 0$
$\therefore\ \frac{1}{2}\sin 2\theta + \cos 2\theta = 0$
$\therefore\ \frac{1}{2}\sin 2\theta = -\cos 2\theta$
$\therefore\ \tan 2\theta = -2$
$\therefore\ 2\theta \approx -1.1071 + k\pi, \quad k \in \mathbb{Z}$
$\therefore\ \theta \approx -0.5536 + \dfrac{k\pi}{2}$
$\therefore\ \theta \approx 1.02,\ 2.59,\ 4.16, \text{ or } 5.73$

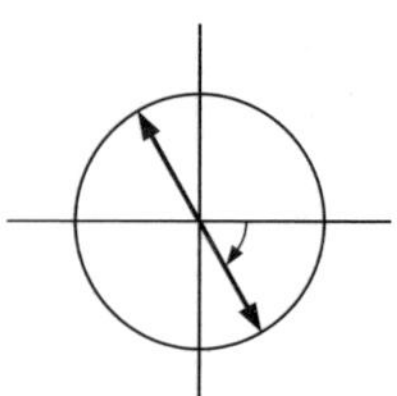

**73** **a** $3 + 2\sin x = 0$
$\therefore\ \sin x = -\frac{3}{2}$ which is impossible as $-1 \leqslant \sin x \leqslant 1$
$\therefore$ no solutions exist

**b** $3\cos\left(\frac{x}{2}\right) + 1 = 0$
$\therefore\ \cos\left(\frac{x}{2}\right) = -\frac{1}{3}$
$\therefore\ \dfrac{x}{2} \approx \pi \pm 1.231 + k2\pi, \quad k \in \mathbb{Z}$
$\therefore\ x \approx 2\pi \pm 2.462 + k4\pi$
$\therefore\ x \approx 3.82 \qquad \{0 \leqslant x \leqslant 2\pi\}$

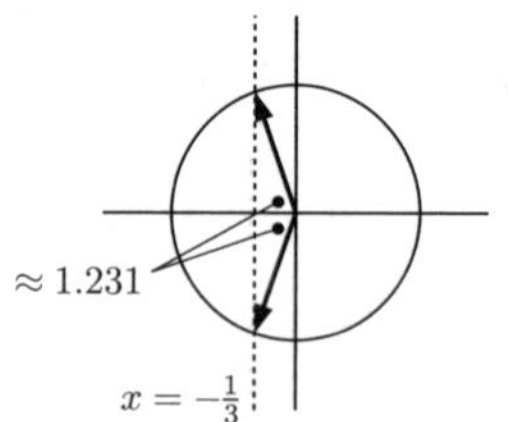


**74** **a** As $\sum_{x=0}^{4} \mathrm{P}(X = x) = 1$,
$\dfrac{1+k}{50} + \dfrac{4+2k}{50} + \dfrac{9+3k}{50} + \dfrac{16+4k}{50} = 1$
$\therefore\ \dfrac{30+10k}{50} = 1$
$\therefore\ 30 + 10k = 50$
$\therefore\ k = 2$

**b**

| $x$ | 0 | 1 | 2 | 3 | 4 |
|---|---|---|---|---|---|
| $f(x)$ | 0 | $\frac{3}{50}$ | $\frac{8}{50}$ | $\frac{15}{50}$ | $\frac{24}{50}$ |

$\mu = \sum x_i f_i$
$= 0 + \frac{3}{50} + \frac{16}{50} + \frac{45}{50} + \frac{96}{50}$
$= \frac{160}{50}$
$= 3.2$

**c** $\mathrm{P}(X \geqslant 2) = \frac{8}{50} + \frac{15}{50} + \frac{24}{50}$
$= \frac{47}{50}$

**75** $f(x) = \sin x\cos(2x), \quad 0 \leqslant x \leqslant \pi$

**a** $f'(x) = \cos x\cos(2x) + \sin x\,(-2\sin(2x))$
$= \cos x(2\cos^2 x - 1) - 2\sin x(2\sin x\cos x)$
$= 2\cos^3 x - \cos x - 4\sin^2 x\cos x$
$= 2\cos^3 x - \cos x - 4\cos x(1 - \cos^2 x)$
$= 2\cos^3 x - \cos x - 4\cos x + 4\cos^3 x$
$= 6\cos^3 x - 5\cos x$

**b** $f'(x) = 0$ if $6\cos^3 x - 5\cos x = 0$

$\therefore\ \cos x(6\cos^2 x - 5) = 0$

$\therefore\ \cos x = 0$ or $\cos^2 x = \frac{5}{6}$

$\therefore\ \cos x = 0$ or $\pm\sqrt{\frac{5}{6}}$

**c** $\cos x = 0$ or $\pm 0.912\,87$

$\therefore\ x = \frac{\pi}{2}$ or $x \approx 0.421,\ 2.72$

Sign diagram of $f'(x)$:

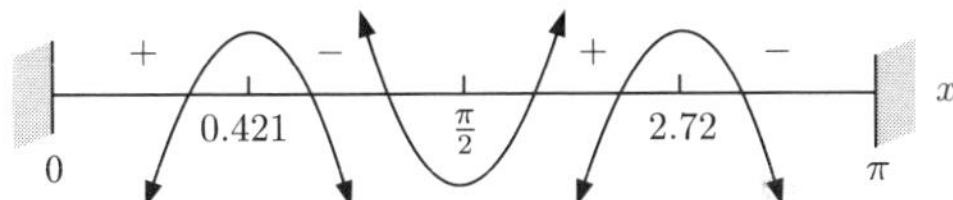


$\therefore$ there is a local maximum at $(0.421, 0.272)$ and $(2.72, 0.272)$ and a local minimum at $(\frac{\pi}{2}, -1)$.

**d**

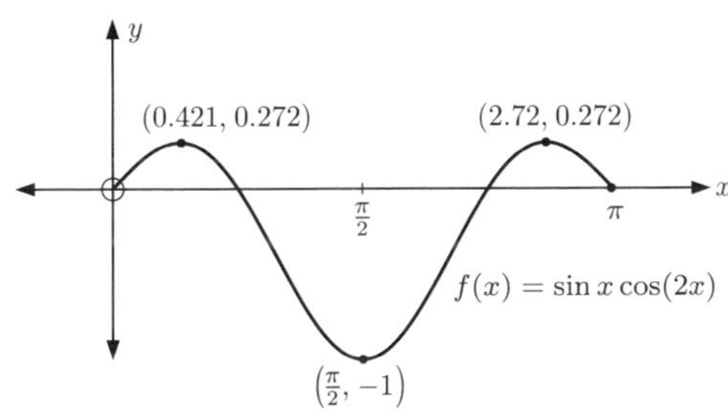


**76** **a** **i**

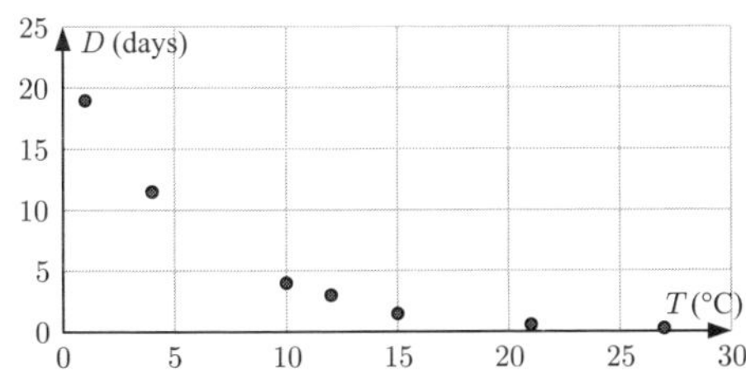


**ii**

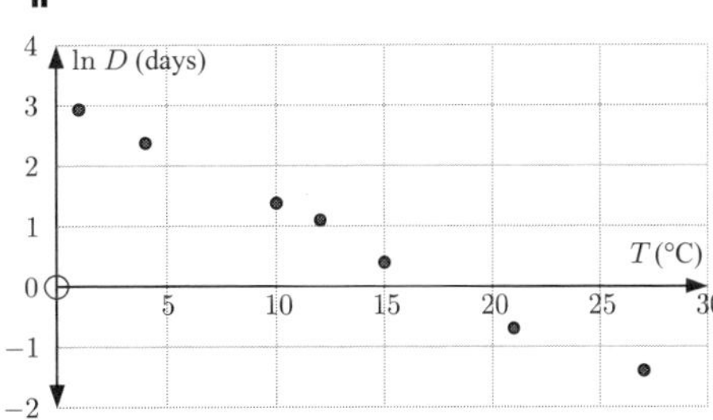


**iii**

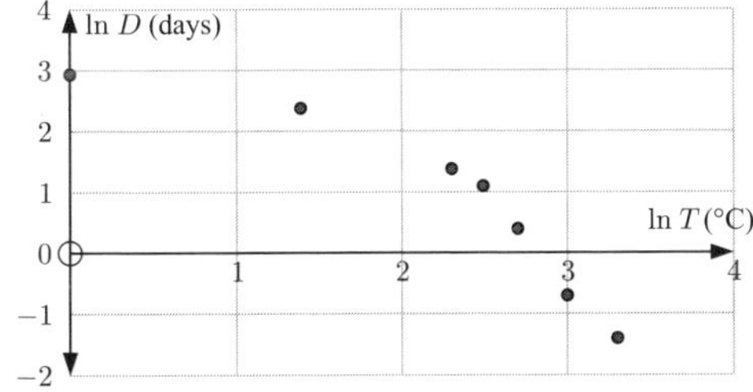


**b** The graph of $\ln D$ against $T$ (part **a ii**) illustrates a linear relationship.
The equation of the least squares regression line in this case is $\ln D \approx -0.172T + 3.10$.

**c** $\ln D \approx -0.172T + 3.10$

$\therefore\ D \approx e^{-0.172T+3.10}$

$\therefore\ D \approx e^{3.10} \times e^{-0.172T}$

$\therefore\ D \approx 22.2 \times (0.842)^T$

**d** When $T = 7$, $D \approx 22.2 \times (0.842)^7$

$\therefore\ D \approx 6.66$ days